DEMOCRACY IN DANGER! Authoritarians advance!

Kendall Hunt
publishing company

Illustrations on the cover and cartoons throughout book created by Klaus Shmidheiser.

PAK ISBN: 978-1-7924-7302-9
Text alone ISBN: 978-1-7924-7303-6

Published in the United States of America

GO TO THE WEBSITE...
OR PUTIN WILL HAVE TO
PAY YOU A VISIT.
PLAID AVENGER
LOG ON TO
WWW.PLAIDAVENGER.COM
FOR UP TO DATE INFORMATION AROUND THE WORLD

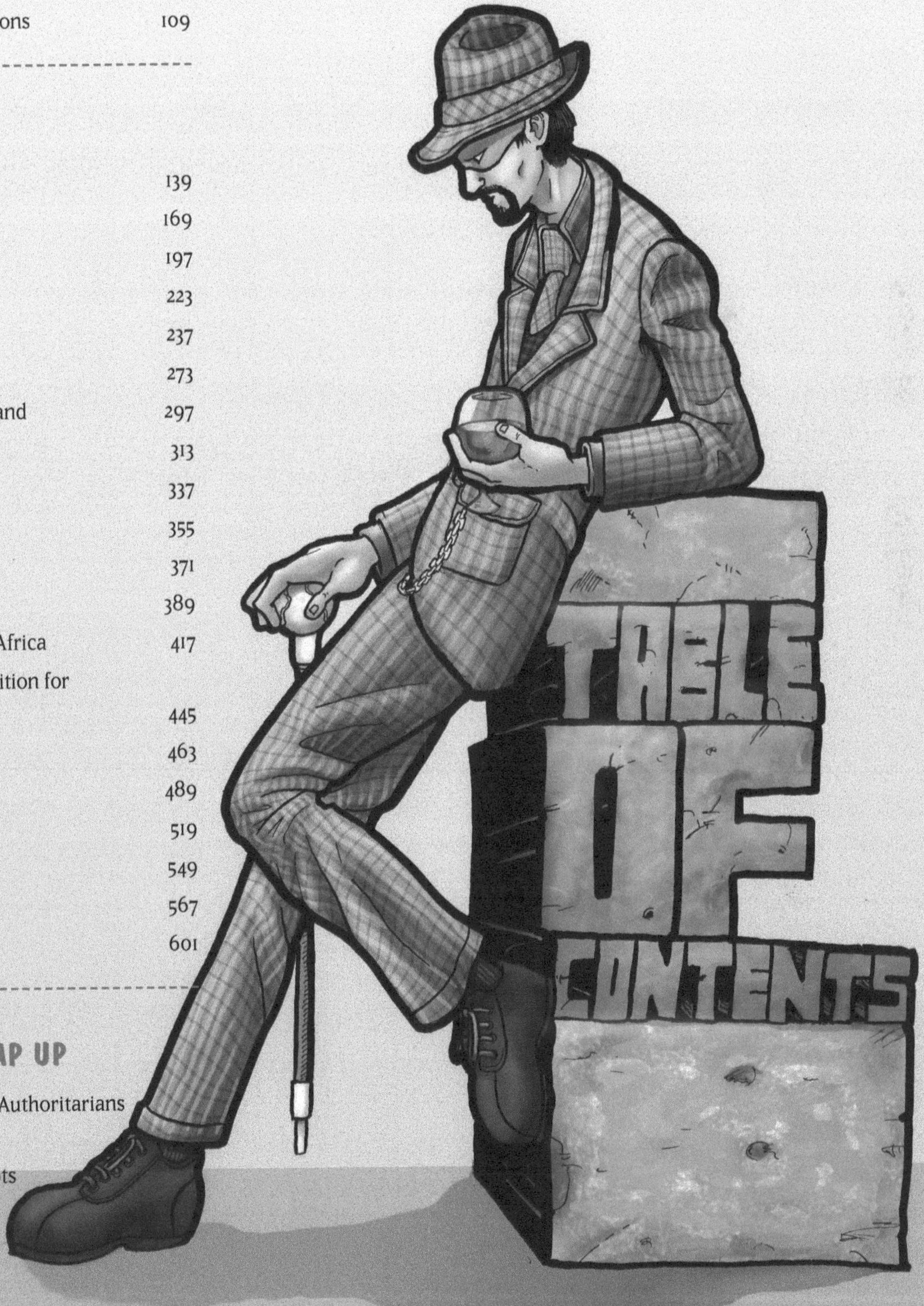

PART ONE
UNDERSTANDING THE PLAID PLANET

PART TWO
THE REGIONS

PART THREE
THE GLOBAL WRAP UP

Acknowledgments

Contributing Writers: Ashleigh Breske, Steven Rich, Flash Clark, Josette Torres, Amber Zoe Smith, Alexander Reniers, Nicholas Reinholtz, Jason Hushour, and Julia Sherry

Chief Editor: Noelle Henneman

Editors: Josette Torres, Chris Drake, & Lauren Beecher

Image Acquisitions: Katie Pritchard

Graphics Creation: Katie Pritchard, David Fox, & Andrew Shears

Film Critic: Steven Rich

Interactive Maps Creator: Andrew Shears

Personal Assistant to the Plaid Avenger: Katie Pritchard

And a special thanks to the most awesome plaid artist on the planet:

Cartoonist: Klaus Shmidheiser

PART ONE
UNDERSTANDING THE PLAID PLANET

CHAPTER OUTLINE

1. Age of Globalization
2. A Wealth of -ations
3. What Is Geography?
4. What Is a Region?
5. A Matter of Scale
6. So Who Is the Plaid Avenger?

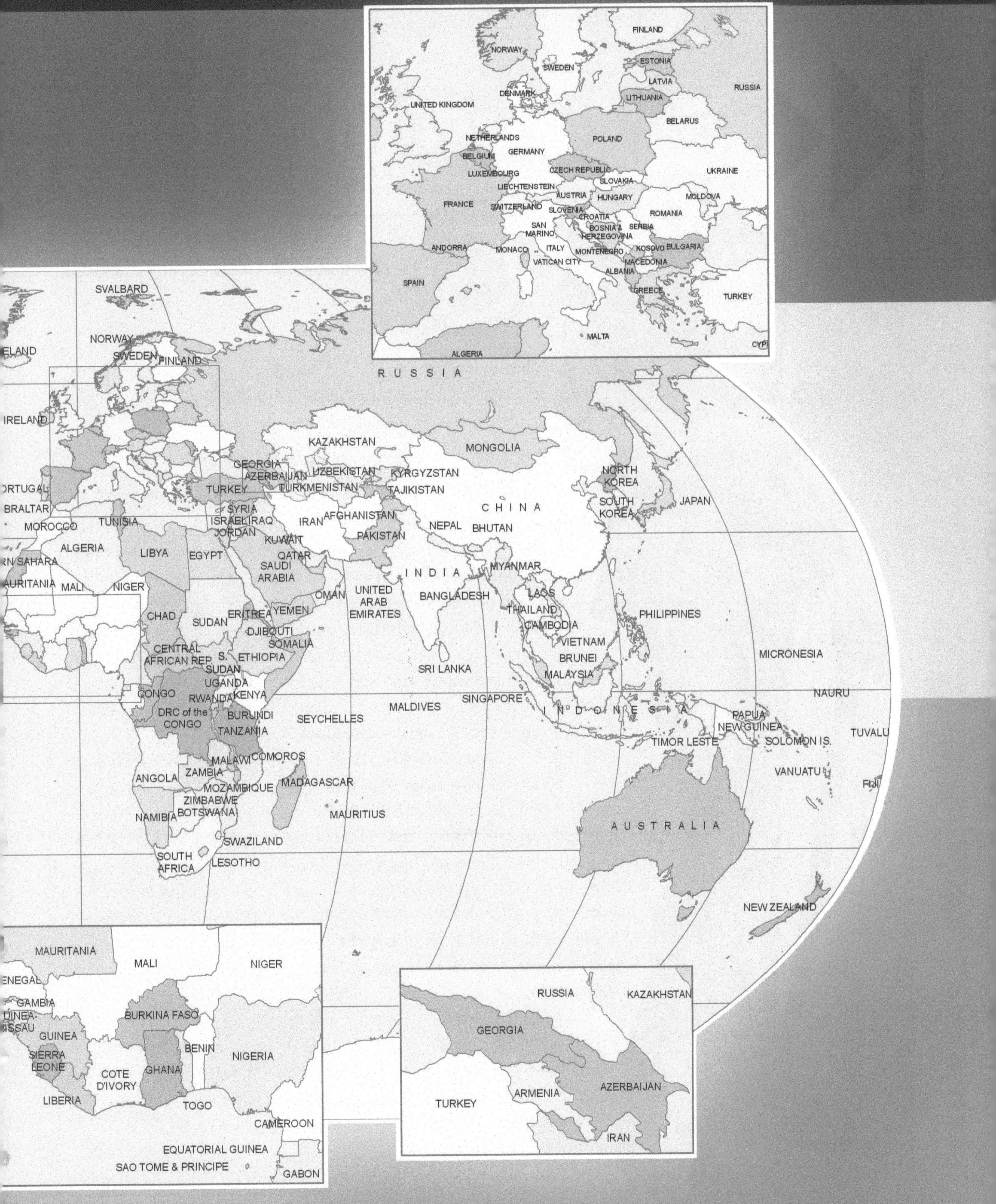
FINLAND
NORWAY
SWEDEN
ESTONIA
LATVIA
RUSSIA
DENMARK
LITHUANIA
UNITED KINGDOM
BELARUS
NETHERLANDS
POLAND
BELGIUM
GERMANY
LUXEMBOURG
CZECH REPUBLIC
UKRAINE
SLOVAKIA
LIECHTENSTEIN
AUSTRIA
HUNGARY
MOLDOVA
FRANCE
SWITZERLAND
SLOVENIA
ROMANIA
CROATIA
SAN MARINO
BOSNIA & HERZEGOVINA
SERBIA
ANDORRA
MONACO
ITALY
MONTENEGRO
KOSOVO
BULGARIA
VATICAN CITY
MACEDONIA
ALBANIA
SPAIN
GREECE
TURKEY
MALTA
ALGERIA
SVALBARD
NORWAY
SWEDEN
FINLAND
ICELAND
R U S S I A
IRELAND
KAZAKHSTAN
MONGOLIA
GEORGIA
AZERBAIJAN
UZBEKISTAN
KYRGYZSTAN
NORTH KOREA
PORTUGAL
TURKEY
TURKMENISTAN
TAJIKISTAN
GIBRALTAR
SYRIA
C H I N A
SOUTH KOREA
JAPAN
MOROCCO
TUNISIA
ISRAEL
IRAQ
IRAN
AFGHANISTAN
JORDAN
NEPAL
BHUTAN
KUWAIT
PAKISTAN
ALGERIA
LIBYA
EGYPT
QATAR
WESTERN SAHARA
SAUDI ARABIA
I N D I A
MYANMAR
MAURITANIA
MALI
NIGER
OMAN
UNITED ARAB EMIRATES
BANGLADESH
LAOS
CHAD
SUDAN
ERITREA
YEMEN
THAILAND
PHILIPPINES
DJIBOUTI
CAMBODIA
SOMALIA
VIETNAM
CENTRAL AFRICAN REP.
S. SUDAN
ETHIOPIA
BRUNEI
MICRONESIA
SRI LANKA
MALAYSIA
UGANDA
CONGO
RWANDA
KENYA
SINGAPORE
NAURU
MALDIVES
DRC of the CONGO
BURUNDI
SEYCHELLES
I N D O N E S I A
PAPUA NEW GUINEA
TANZANIA
TUVALU
TIMOR LESTE
SOLOMON IS.
COMOROS
MALAWI
ZAMBIA
ANGOLA
VANUATU
MADAGASCAR
MOZAMBIQUE
FIJI
ZIMBABWE
NAMIBIA
BOTSWANA
MAURITIUS
A U S T R A L I A
SWAZILAND
SOUTH AFRICA
LESOTHO
NEW ZEALAND
MAURITANIA
MALI
NIGER
SENEGAL
GAMBIA
GUINEA-BISSAU
BURKINA FASO
GUINEA
BENIN
SIERRA LEONE
NIGERIA
COTE D'IVORY
GHANA
LIBERIA
TOGO
CAMEROON
EQUATORIAL GUINEA
SAO TOME & PRINCIPE
GABON
RUSSIA
KAZAKHSTAN
GEORGIA
AZERBAIJAN
ARMENIA
TURKEY
IRAN

1 A Plaid World Intro

What is globalization?

How do the interconnections of the global economy and world politics affect my life?

Why are some places rich while others are poor?

What does the future of the planet look like?

All great questions, fellow world watchers—questions that deserve great answers. Thoughtful, intuitive and well-researched answers contained in a well-ordered and glibly constructed textbook.

This is not that book.

Welcome to my world. The world of the Plaid Avenger, where seeking knowledge of our planet is imperative, where blissful ignorance is not accepted, and where truth and justice can only be achieved by those willing to learn . . . willing to work . . . willing to fight. It is not a place most are ready to enter yet. But you are here. Good. Read on.

In *The Plaid Avenger's World*, we will strip off the shallow window dressing in which you have been trained to see the world. We will lay it bare to see what is really happening around the planet. We do this in order to gain enough insight about the current state of the world to truly understand the how and why and where things are happening right now. In this world, no single government or press dictates our views; no single political party shapes our opinions; no single religion or ethnicity tints our not-so-rose-colored glasses. We will see the world in plaid: a mystical weaving of facts, figures, cultures and viewpoints from every corner of the planet, culminating into the fabric that is today.

Many, if not most, in our society would say, "Who cares? Why bother?" Oh, you silly simpletons. Here's why: because the world is changing your life whether you like it or not, no matter who you are or where you live. It's a new age. A global age. And you should be in it to win it. Read on . . .

1

AGE OF GLOBALIZATION

Globalization: what is it and what does it mean to me? Economically? Technologically? Politically? Socially? Culturally? Morally? One of these should resonate with you . . . hopefully more than one.

We constantly hear about how the world is getting smaller—is it? Pure poppycock! The world is the same size it has always been . . . *but* it is becoming more connected and more interdependent than ever before. Goods and services and information are exchanged in our local economies from every nook and cranny of our cram-packed planet. For the first time ever, we can travel to any part of the world virtually overnight. Corporations move capital and jobs from one country to the next in a matter of days. News of international significance is reported seconds after it happens. We can communicate in real time with any part of the globe. The world is now one system . . . mostly.

We are the first generation of humans who enjoy foreign travel as a casual part of life, who communicate by direct-dialing to any country on all continents, who receive instant news of world happenings, who expect to work overseas or work for a company that deals overseas. Let me reiterate that—we *expect* you to interact with the entire world; to not do so would be an exception. This is a really important concept, especially to you—the first generation that is living in the postindustrial, highly interconnected age.

Many, if not all of you, will work for multinational companies whose business is all over the world. Many, if not all of you, will work and live outside the United States at some point in your careers. Businesses and jobs are internationalizing as we speak—almost all jobs, not just the fancy ones. You are the people who are going to be running the world. You are the decision makers—when all is said and done, I'm just a single pseudo-superhero out thwarting international intrigue. But *you* will be the ones building the bridges, and electing leaders, and stabilizing governments, controlling monetary exchange rates; you may even be setting up all sorts of private, national, or even international businesses/programs/projects that will shape the world and its population.

Make no mistake about it, the AIDS rate in Africa *does* affect you, the increasing coal consumption in China *does* affect you, an earthquake in Japan and the price of cocaine in Colombia *does* affect you. (allocation of your tax dollars, your jobs, the price you pay for goods, global pollution which affects your health, etc.) Globalization is pretty much a one-way street. We are not going back to medieval times, no matter what isolationists say, do, or think. Ignore the rest of the world at your own peril—you won't be hurting anyone but yourself. How did this globalization happen anyway . . . and how has it shaped our planet?

A WEALTH OF -ATIONS

Of all the organisms hanging out on this blue marble we call Earth, us human-types have been the biggest modifiers, movers, shakers, benders and breakers of our fair planet. We grow food, we congregate in cities, we move around, we plan, we build machines, we communicate, we educate, we procreate . . . and do it faster and more thoroughly year after year after year at a larger and larger scale that has inevitably incorporated the entire world. The idea that globalization is merely a modern phenomena is blasted balderdash; all of human history can be seen as a relentless drive to expand our tribe to every nook and cranny of the planet, and to increase the non-stop interactivity between peoples, for better or for worse. Think about it. Civilization, migration, urbanization, industrialization, modernization, communication: all have served to organize, spread, and interconnect us. What a wealth of "-ations" that all feed into one grand scheme of globalization!

Take a bite of this apple, and let's get this globalization ball rolling!

You probably have learned about a lot of these "-ation" terms in other places, but I want you to think about them again for a few minutes in the context of this globalization concept. Without getting into too much tedious detail, I tell you that humans have been doing this globalizing gig since the birth of the species (or since a naked dude and a naked dudette were plopped down in a garden, if you prefer). **Migration** of humans started when the first modern *Homo sapiens* started trucking their tribal units to new turf outside of Africa over 70,000 years ago. Humans subsequently spread to all continents and all major islands, and this process of movement and interconnection is still going strong today.

Humans used to migrate in order take advantage of new unpopulated lands, untapped resources, or to follow game; in today's world, they migrate for jobs, for security, for a better life in a richer place. Wait a minute—is there a difference between those migration motivations? Maybe not. Did you really think that "illegal" Mexican migration to the US or "unwanted" African migration to Europe was some new thing? Ha! We've been on that route forever. But let's be civilized about this, which brings me to . . .

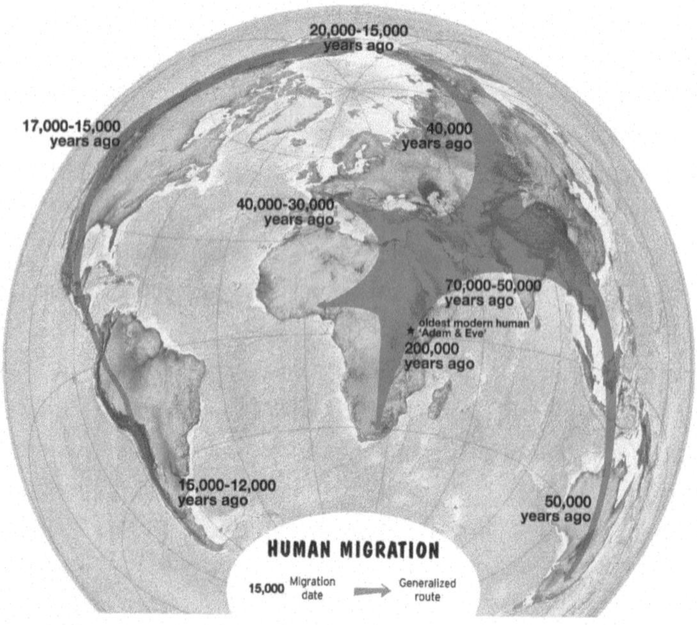

The Migratory Circle of Life is nearly complete.

Civilization, a process that wildly impacted the human experience. See, we used to be the ever-expanding, ever-migrating, hunter-gatherer types living in small unconnected bands scattered across the planet, but around 12,000 years ago, some folks stumbled upon agriculture, and that was a game changer. Humans did a radical lifestyle shift from hunters to farmers. From that point forward, we hairless-ape types turned into peeps who were developing cooler and cooler tools, growing more and more food, congregating together in bigger and bigger pools. We call that human civilization, fools!

Specifically neolithically, we refer to this slow cumulative process that occurred independently in many different locations between 10,000BC and 3,000 BC as **The Neolithic Revolution**. The nifty Neolithic! Neo for "new" and -lithic for "stone." The New Stone Age! And those humans were totally new-rocking it out!

Early iPad prototype: the iPetroglyph.

See, all that extra food surplus as a result of growing plants in a predictable cycle led to folks settling down in a permanent place. Then some smart peeps who did not have to waste their whole lives just finding food began to pursue their passions . . . and thus they invented the **domestication** of animals, the wheel, pottery, tin-smithing, the Bronze Age, written languages, architecture, engineering, legal structures, religious institutions . . . you starting to get the picture here? Oh yeah, 'cuz they would have been painting pictures as well.

That's what the extra food is really important: it increasingly allowed people to do their own thing . . . to specialize their skills and inventions. This in turn created even better tools/technologies, specialization of work, and increasingly complex societies and cultures which encouraged trade and cooperation. And not just trade within a single society, but between different ones in different places.

Aha! That produced intentional interconnections between populated areas in an effort to trade not just goods and services, but ideas and technologies and peoples themselves! Globalization game on! They call this stuff "out-sourcing" and "economic activity" and '"technology transfer" in today's world . . . but it's all the same stuff as way back when. From the Flintstones to Futurama: Same shizzle, different millennium.

This civilization gig also transforms the geographic organization of human life: we go from small bands of folks out in the wild, to hanging out with each other in bigger in bigger numbers in things called villages, and then towns, and then cities. The first major civilizations that popped up 6,000 years ago in China, Indus Valley, Mesopotamia, and Egypt began a perpetual process of population growth and **urbanization** that continues unabated to this day. Translation: humans have increasingly chosen to live in big population clusters in built-up areas that we call cities as opposed to hanging out in the sparsely-populated rural areas we refer to as "the sticks." Humankind reached a milestone in 2009 when, for the first time in human history, over 50% of the world's population lived in an urban area. That urban-dwelling percentage is growing bigger, faster than ever, still to this day.

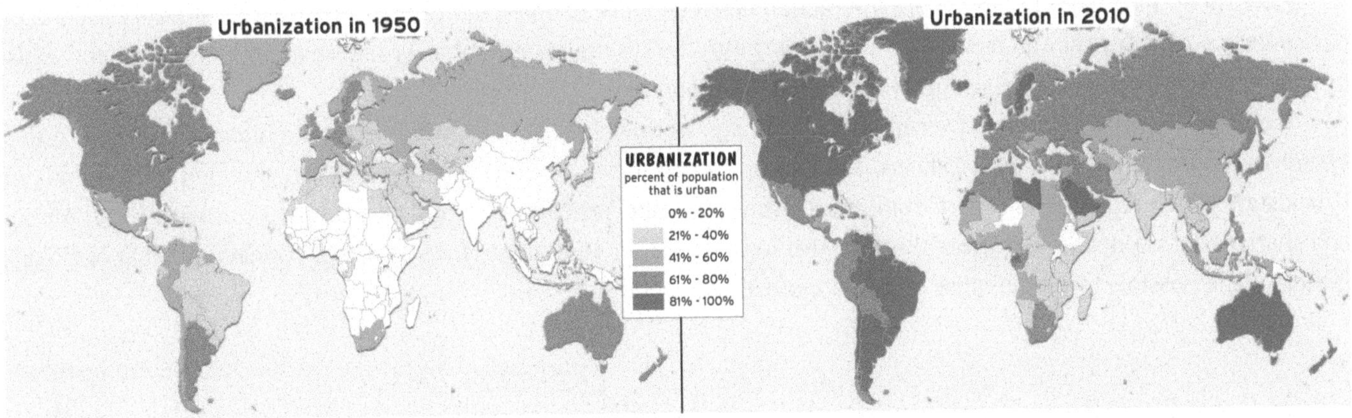

This urbanization has intensified human interaction exponentially over time; interaction which has accelerated the creation and transfer of ideas and technologies. New ideas, new inventions, new stuff! Which prompted **industrialization,** of course! Did you ever stop and wonder how it is possible for so many people to now cram themselves into cities, thus not being able to grow their own food nor collect their own water nor hunt their own bison? Because we invented machines to do the heavy lifting that used to require all our labor, and we discovered energy sources to fuel those machines. Bottom line: one dude on a tractor can grow enough food to feed a million people in the city; ten kids in a sweat shop can make clothes for that million, and oil-powered big machines can build skyscrapers and sewer systems and water lines for that million people. What the heck is left for those city folks to do?

Much like at the beginnings of the Neolithic Age, those folks have more time to think. To interact. To create. To be inventors, doctors, artists, engineers, scientists, industrialists, priests, rock stars, and whatever else we come up with. That's how we got electricity, the polio vaccine, the Hoover Dam, mobile phones, and computers. Unfortunately, with so much free time and such diverse and bizarre human motivations, we also have nuclear weapons, mustard gas, lawyers, and Sham-Wow. Such is the yin and yang of life. But I digress. This urbanization/industrialization combo has served to link up the globe like never before: more humans compacted into bigger, concentrated urban areas has caused even more interaction within the city, but also increasing interaction between cities worldwide. Cities are both the engines and the nodes of globalizing forces; the conduit for transfers of ideas, money, technologies, and power, everywhere, all the time.

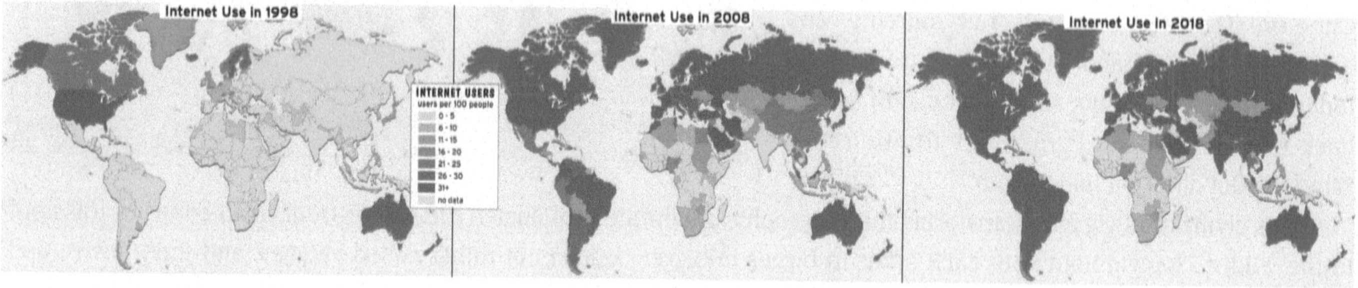

This "connectedness" is now happening at light speed, thanks to advances in transportation and **communication**. Yep, our last -ation is really the power-booster icing on this globalizing cake. From smoke signals to the Pony Express to trans-oceanic telegraph cables, we humans have spent a lot of time and effort figuring out how to better transmit information to each other across the planet. Let me be the first to tell you, my brothers and sisters: we have almost reached the end-game on this one. With mobile phones, the Internet, and satellite TV/radio, we can pretty much transmit any type of information almost instantaneously to any part of the globe. It's totally radical if you think about it.

More ideas, more information, more interaction between more people than *ever* before. Twittering every second of our collective experience! And more folks are jumping into the system faster than ever before as well. Just look at the figure above to see the speed at which the world is hooking up on-line!

We have essentially set ourselves up as a single huge worldwide computer, wherein all the humans attached to this systems are interacting to solve problems, sell goods, provide news, and everything else! You can chat with friends in Uzbekistan, watch a live news report from Brazil, vacation at the Vatican, have business transactions with South Africa, and share research ideas via Skype with associates in Switzerland. That, my friends, is global connection. That is globalization, made possible by all the other -ations of note in this section.

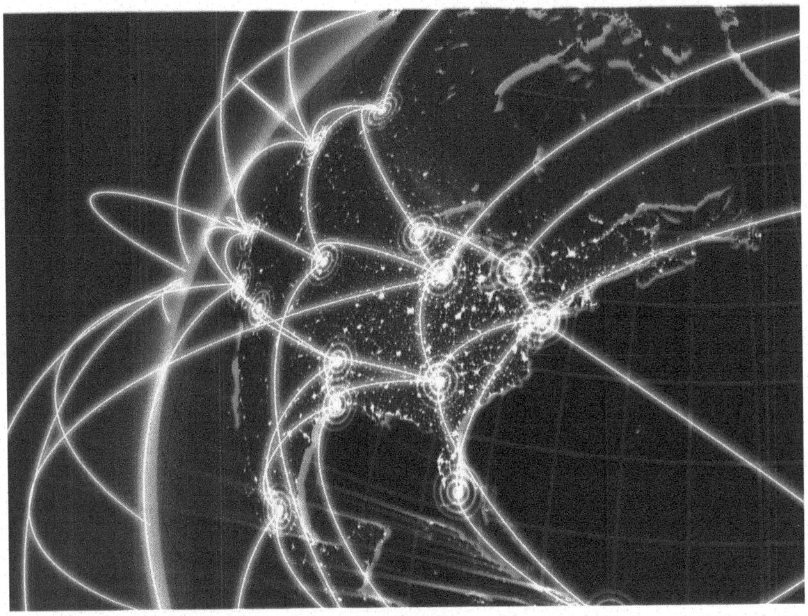

What does it all add up to? A more fully *connected* world in every sense of the word, in which the actions of any one person affects the lives of everyone. All of the aforementioned -ations have gone from local, isolated or internal concepts to completely planetary themes in scope and practice. In this increasingly populated and interconnected world, it is becoming clear that all "local" problems are actually "global" problems, from the movement of narcotics, to an outbreak of an infectious disease, to environmental pollution, to the existence of nuclear weapons, and so on.

And there are global problems a'plenty on our poor little planet! However, the solutions lie within this connected framework as well. See, this globalization is neither bad nor good. It just is, and will continue to be. Will globalization result in a homogenization of culture? In a modernization of all societies? In a pacification of the planet? In a disintegration of the concept of the sovereign state? Oh, my oh my, what a delicious stew of other -ation possibilities that I will leave to you to debate . . . and we will return to the tensions of the local versus the global in the final chapters of this book, to see how globalization is playing out in the 21st century plaid world.

For now, I just wanted to spill some of that crazy globalization knowledge on your skullcap to see the sparks fly.

Knowledge is power, or at least empowerment. The more you know about the globalizing world in which we live, the more power you have. It's good to have at least a minimal geographical understanding of our planet. What's that? You don't know what geography is?

An all-inclusive neural network now . . . from a nebulous Neolithic nucleus!

WHAT IS GEOGRAPHY?

Geography is one of those words, and subsequently one of those fields of study, which has become so generic that it seems to have lost its own definition in the modern world. The term is so truly holistic in meaning that many other social sciences, as well as a lot of the physical ones, are actually sub-branches of it, as opposed to geography's current designation as a sub-branch of one of them. What am I talking about? Consider for a moment the origin of the word and the discipline; **geography** has its roots in the ancient world, roughly translated as "describing the Earth," and every culture and society with a written record has done just that—described both physically and culturally the environment around them as they understood it.

Be it Greek philosophers calculating the size of the known world in the 2nd century BCE, Chinese diplomats considering trade ties with Southeast Asia in the 11th century, military strategists planning the Boer War in Africa, or American scientists assessing the impact of the loss of Brazilian rainforest on world climate in the 21st century—all are geographers in the sense that they are studying the physical and/or cultural components of their environments to gain understanding and make decisions—just as all of us do every day in our own lives. How do I get from here to there? Should I buy an American car to support the American economy, even though their fuel efficiency is worse than imported cars? What is the foreign policy of the political party I support? Should I donate money to alleviate hunger in Ethiopia? Is this neighborhood I'm in a high-crime area? Paper or plastic? All questions require us to consider economic, social, political, and environmental knowledge and repercussions of that knowledge on the world around us.

We are unique! In culture, climate, and so much more!

I am intentionally pointing out here that the world around us—and every individual, every place on the planet no matter where you are—has both physical traits and cultural traits that make it unique. Every place has a certain climate, particular landforms, and some kind of soils, vegetation and animal life. These are its **physical** traits, much like every human has some natural hair color, skin color, a certain height and weight, and particular physical abilities—maybe to run fast or jump high. At the same time, every place has languages being spoken, religious practices, economic activities, political organizations, and human infrastructure like roads and buildings. These are the **cultural** traits of the place, just like a human has cultural traits like certain religious beliefs, spoken language, a job, a learned skill like archery, and a favorite flavored ice cream. Just like every human in the world, every place in the world is unique in its own right, kind of like snowflakes. Defining any place in the world, or any region of the world, involves looking at both of these aspects, as well as their interaction with each other.

Every place on the planet is unique in that even when many of these factors are identical—say between two small towns in the Midwestern USA located only five miles apart—there will still be tangible differences. Each town has a different history. The weather may be pretty much the same, but one will get more rainfall than the other. The people may all be of the same religion, but there will be different churches that do things just slightly different. The economies may both be based on corn, but there will be different business names, and different storefronts.

There will be at least one Chinese restaurant in both towns, but they will definitely have different tasting General Tso's chicken. Like human identical twins, no matter how much is the same, there are always distinguishable differences upon closer examination. To understand our world, we will look at the physical and cultural traits of regions of the planet, how these traits converge to form a distinct region, and perhaps more importantly for our assignment, how these regions interact with each other.

So what the heck is a region?

WHAT IS A REGION?

The world is just too darn big and filled with a heck of a lot of things going on and way too many facts and figures and images and names and places for us to know and comprehend everything all of the time. You can't possibly even know all the facts and histories and physical variables of your own home town, much less your county, your state, or your planet. There's just too much, and the story is added to and updated daily. But don't give up hope! Nil desperandum, my dear friends! Across the desert lies the promised land! The human mind has a coping mechanism for this overflow of knowledge, which of course has gotten much worse with the advent of global communications and the 24-hour news cycle: We assess importance. We filter. We generalize. We are going to do the same geographically for the planet. By synthesizing and systematizing vast amounts of information from parts of the world and making pertinent generalizations, we create a unit of area called a **region.**

Regions are areas usually defined by one or more distinctive characteristics or features, which can be physical features or cultural features, or more often than not, a combination of both. We could identify a strictly physical region such as a pine forest, the Sahara Desert or the Mississippi drainage basin. Conversely, we could form up an area that we would identify culturally like the Bible Belt, a Wal-Mart store service area, or even Switzerland (defined by human-created political borders). However, since we have already pointed out that every part of the planet has both types of traits, most regions are identified as a combination of physical and cultural characteristics, such as the regions we refer to as the Midwestern US, tropical Africa, or Eastern Europe. These names typically make one think of both physical and cultural traits simultaneously, and may actually be meaningless in a context of just one or the other. This last type of world regional delineation is what we will mainly utilize in our journey.

That is a good jumping off point for what we will consider a region for our global guidebook. A region has three components:

1. A region has to have some area. Otherwise, it's just a point. And what's the point of that?
2. A region has to have some boundaries—although these boundaries are typically fuzzy, or imperfectly defined. Where does the Middle East stop and Africa start? What US states are in "The South"?
3. A region has to have some homogeneous trait or traits that set it apart from surrounding areas. This is the most important component to consider!

The user (that is you!) defines what trait is homogeneous. You can define any place on earth as being in an infinite number of regions, depending on what trait you pick. Your current exact position could be described as being in a distinct political region like Charleston, California, or Canada. Or perhaps you're in a distinct physical region like the Everglades or the Appalachians or the Badlands. Simultaneously you may be in a distinct socially defined region like the Bible Belt, the Rust Belt, or the "The Beltway"—what region do you think you are in right now? Play this exciting "name your region" game with all your friends, and you will be the toast of the town. Or perhaps a big geek. But I digress.

Here is a quick breakdown of the world regions we will be examining:

FIGURE 1.1 WORLD REGIONS AS DEFINED BY THE PLAID AVENGER

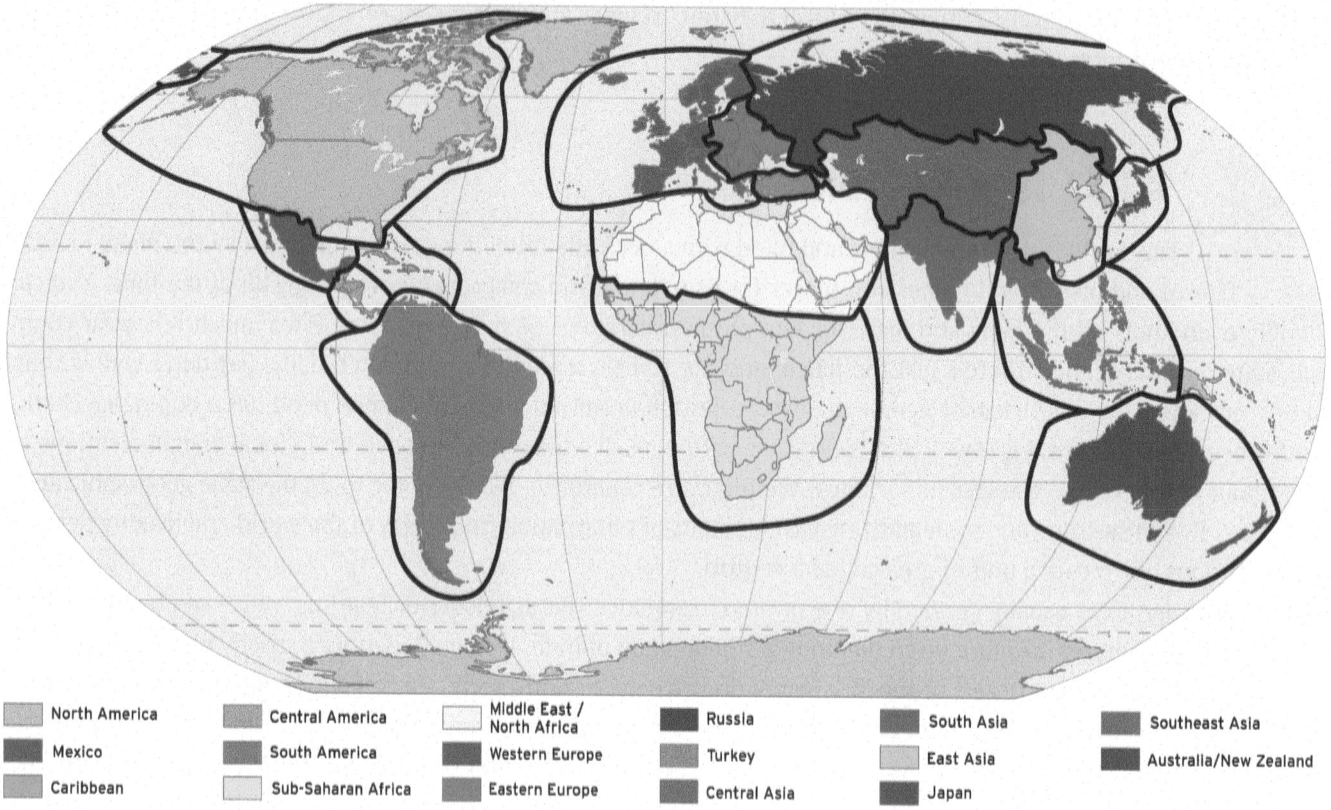

Why these areas? Why these borders? Why these regions? Because they are defined by homogeneous traits as picked by me. These are the Plaid Avenger's world regions. The bulk of this book will be explaining these regions and their homogeneous traits. Your world regional map may be different than mine depending upon what traits you want to focus on. Don't like my regions? Then go write your own bloody book!

A MATTER OF SCALE

When we are being geographers, or defining regions, or even just trying to get from point A to point B, we must always keep the **scale** of our endeavor in mind. How much area are we talking about? Are we examining someone's backyard in Australia, or the entire Australian Outback? That is a big shift in scale! Does this description of the environment in my hometown scale apply to a larger scaled area like my state or country? It rarely does. Thus, we must always be wary about how far we can push our analysis. Changing scales typically calls for reassessment of the area being considered.

The other reason to pay attention to scale is because it plays an important component of our definition of regions. Since we have already expressed that regions have some sort of homogeneous factor that defines them, we must consider at what scale does this homogeneity apply—because the scale itself defines it. Let me give you an example.

A couple presidential elections ago the USA, by majority, elected Barrack Obama—a Democrat—to the presidency. Since more than half voted Democrat, we could say that the US, at the country scale, is a Democratic region, based on that singular homogeneous trait. However, if we looked at the state of Texas, it voted predominately Republican—so at a smaller scale, you are looking at a Republican region. The city of Austin, a smaller region within Texas, is a hip liberal art-sy town that voted predominately Democrat, thus, in that smaller region, you are back to a Democrat-defined area. Maybe most of the people in a certain city block in Austin, a smaller region still, voted Republican, so at the block scale you are in Republican territory again. Thus, defining regions based on voting preferences *demands* that you state the scale of focus. Most importantly, the larger the region you define, the more exceptions to your homogeneous trait you will find within your region.

This is what generalizations are all about—we are going to discuss and define our regions with *generally homogeneous* traits within the region, knowing full well our generalization won't apply to everyone and every place. For example: by any definition, the Middle East region would be identified as an area dominated by Islam. Oops, except for a radical and extremely important exception—that of the Jewish state of Israel. For our travels, we will be pointing out and elaborating on those homogeneous traits which define the entire region, but will also include those glaring exceptions to the rule when they are of particular significance to today's headlines. The other main goal of this guide is to describe each of the world regions' interactions with each other, and their role in the world at large. This is a tall order to be sure, but a goal worthy of pursuit by the mightiest of global superheroes, the Plaid Avenger.

SO WHO IS THE PLAID AVENGER?

Yes, it seems that everything is growing into a singular world system. We speak of a world economy in which goods and services and businesses move all over the planet; they even have a club for everyone: the WTO. We have a great global transportation network that can transport us faster than ever to any point on the planet. We have the United Nations: a world legislative body that sets standards and rules for conduct on the planet (and I've heard rumors that they are also supposedly peace-keeping enforcers of these rules . . . although I won't swear to this). Thanks to mass media and global telecommunications, we can even speak of movement towards a more homogeneous world culture—where in the world can you *not* talk on your cell phone while you watch *Game of Thrones* and sip on a Coca-Cola?

But wait, there seems to be something missing. Hmmm. . . . Global leadership . . . check. Global legislature . . . check. Global economy . . . check. Global judicial system . . . Global judicial system . . . Global justice??? Bueller . . . Bueller . . . Bueller? Where is it? I knew something was missing! No justice to be found!

Just as the world continues to become more interconnected, and every event across the globe becomes more pertinent to our daily lives, we also gain more knowledge about inequalities and unfairness around the planet. This comes at a time here at the dawn of the 21st century when conflict proliferates around the globe, multinational corporations grow unchecked and unhindered by law, diseases have the capacity to truly create an unprecedented planetary epidemic, and trade in guns, drugs, and people continue unabated. Yes, even the trade in people . . . you know, slavery. Global inequality may be reaching a new zenith; that is, the gulf between the rich and the poor widens as every day passes, and those poor folks are growing in numbers. You heard of the "Occupy Wall Street" movement? You may live to see an "Occupy the Globe" movement as well!

The Plaid Avenger is a product of this age. Somewhere at a major university on the eastern seaboard of the US, a meek but smartly dressed college professor by day, he toils in an effort to educate the youth of America about the wider world, and their role in it. By night, he roams the planet fighting organized international crime, abusive multinational corporations, and corrupt governments, wherever they may be. The Plaid Avenger: international equalizer and educator. A fighter for truth, global justice, and the international way, he also possesses an unstoppable urge to bring plaid back into fashion.

That brings us to your first assignment: Your first mission to become globally literate—that is, smart—and know the locations of the states of the world. Most Americans call them "countries," but you should start calling them "states" now and just get over it. Look back to Figure 1.1, and get to work. The reason? While I'm not an advocate of memorizing every town, district, and province on the map, we do need to have a working vocabulary to discuss things intelligently. Not that you need to be able to draw a map from memory, but you will be amazed how much more intelligent you appear when you are in a discussion about a news event and you know with authority that Senegal is in western Africa, not in Central Asia. Trust me, your date will dig it. So dig this . . .

Analyzed an atlas or wall map and now know all the states of the world? Got them all down? Then look at the following figures. A straight-up matching game with some sassy style. As the Plaid Avenger must often work undercover around the planet, he has an endless array of outfits to help him blend into the local environs he is investigating. He also

A
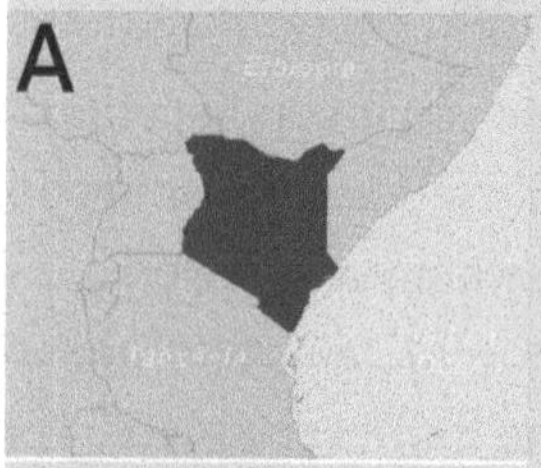

B

C
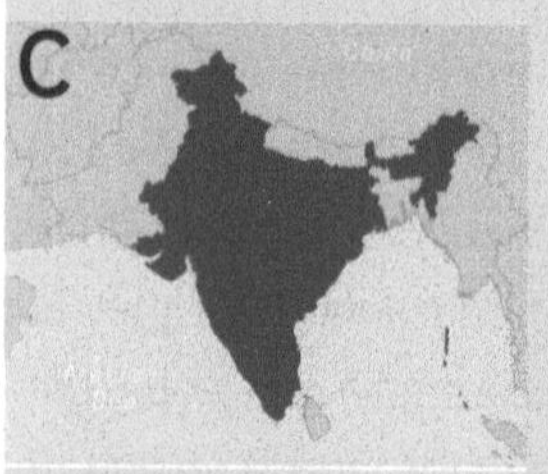

D

E

F
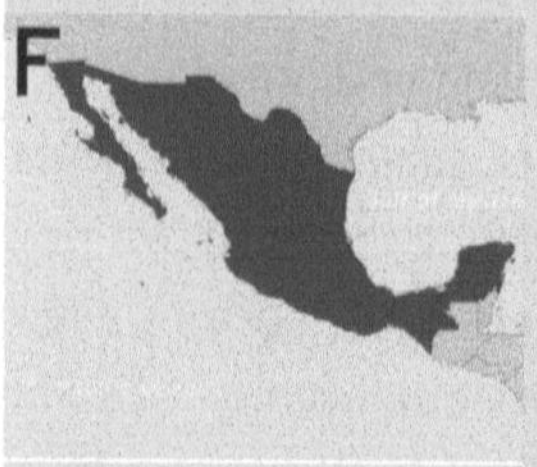

1

2

3

frequently has to foil dastardly plots at famous local landscapes. Study the costumes, maps and photos below. Your mission: match the appropriate highlighted country map to the outfit the Plaid Avenger would be wearing and the appropriate landscape he would be lurking in. Good luck, and see you in Chapter 2.

6

5

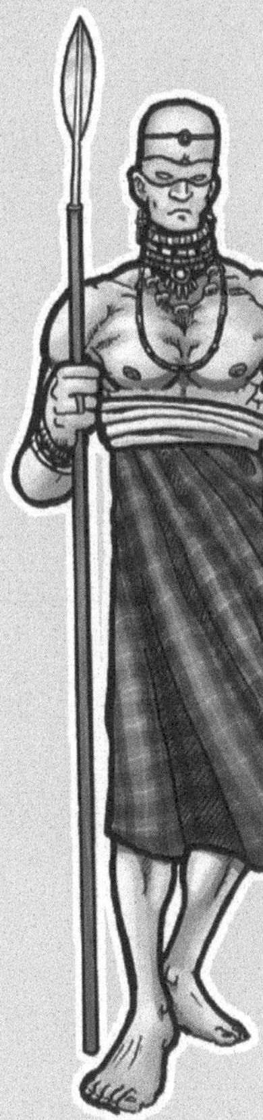
4

a

b

c

d

e

f

CHAPTER OUTLINE

1. Intro to People aka World Sex Ed 101
2. Regional Differences in Population
    - 2.1. How Population Is Changing
3. How/Why/Where Population Is Changing: The Demographic Transition
    - 3.1. Demographic Transition Stage One
    - 3.2. Demographic Transition Stage Two
    - 3.3. Demographic Transition Stage Three
    - 3.4. Demographic Transition Stage Four
4. Population Pyramids
    - 4.1. The Classic Pyramid
    - 4.2. The Column Shape
    - 4.3. Inverted Pyramid . . . You Mean It's Upside Down?
    - 4.4. It's Hard to Stop Having Babies
5. A Prickly Problem of Skewed Sexes . . .
6. Peeps Won't Stay Put!
7. Some Final Thoughts on People . . .

World Population Dynamics

2

BEFORE we get to the regions, we will focus on several topics that are better approached at the global scale; they involve traits that all regions possess equally (like people or religions), or that all regions participate in as a singular global unit (like the world economy or international organizations). Since it is *people* that create, define, and operate all the cultural aspects of our planet, let's start with them. What's up with the peeps?

INTRO TO PEOPLE AKA WORLD SEX ED 101

People, people, people, all over the world. Old ones, young ones, rich ones, poor ones, African ones, American ones, Asian ones, and even plaid ones. Some places got lots of people, while others have just a few. Are there too many people? Perhaps not enough in some places? Some states have growing populations, while other states actually have shrinking populations. What's that all about? Although there are great differences around the world in numbers of people as well as growth rates of populations, it is best to approach the subject by looking at it systematically. That is, we can look at how human population dynamics operate as a whole, because the rules are essentially the same anywhere you go on the planet. Let's get to know how it works, and then you can apply your knowledge to understand what's happening in any state or region of interest. Game on!

For starters, you should know that we just crossed over the 7 billion mark in terms of how many peeps are currently alive on the planet. This has not always been the case. In fact, numbers this huge for human population are actually quite recent. Consider Figure 2.1.

FIGURE 2.1 WORLD POPULATION GROWTH OVER HISTORY

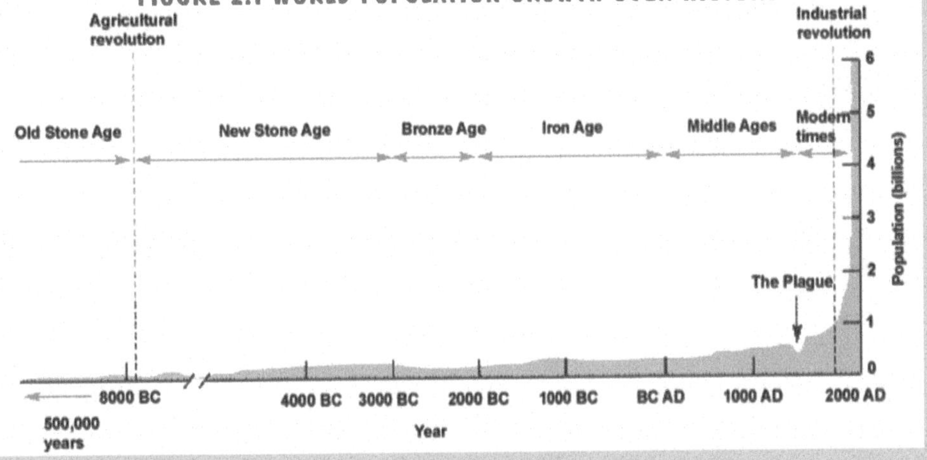

As you can see, for most of humankind's existence, population totals have been relatively small. It took approximately one million years, from the beginning of time until about the year 1800, for the first billion humans to appear on the earth at the same time. From then, it's roughly

- 1927 that the second billion showed up
- 1960 saw the crossing over to 3 billion
- 1974 picked up number 4 billion
- 1987 flipped the pop-odometer to 5 billion
- 1999 the predawn of the 21st century we reached 6 billion
- And Happy Halloween! On that costumed holiday in 2011, 7 billion peeps were alive for the tricks and treats!

What you may be detecting here is a radically accelerated population increase in the last fifty years. The time it takes to add another billion people gets shorter every cycle, with the 8 billion mark already rapidly approaching. Why is this happening? Population growth is exponential, not mathematical. Adding ten more fertile females to the population pool doesn't equate to adding just ten more babies, but more like one hundred babies. Each woman has the potential to spawn many more offspring, who in turn can produce many more offspring themselves down the road. Get it?

As a result of this exponential growth in the human population, many folks believe that the planet is already overpopulated. Is that true? I can't give you an answer for that, because it is a relative question. Where are we talking about? Siberia? It's certainly not overpopulated. Calcutta, India? Yeah, maybe they have maxed out on the peeps. Maybe. Many assume that Africa as a whole is overpopulated, but given the size of the place and its current population totals, it is actually quite sparsely populated, particularly if you compare it to a place like Western Europe. And if Western Europe is overpopulated, why do many of its governments encourage their citizens to have more babies? Hmmmm . . . things get complicated fast. Plaid Avenger rule of thumb: a place seems to be considered truly overpopulated only when not enough resources exist to supply the people who live there. Thus, the 86 million people in DR Congo may all agree that their country is overpopulated, but the 82 million people in Germany would probably not consider themselves so, even given that DR Congo is roughly six and a half times the size of its schnitzel-eating friend.

WHAT THE WHAT? WHAT THE H? WHAT THE WHAT? WHAT THE WHAT? WTH?

CHART 1: TOP 20 POPULOUS STATES

COUNTRIES RANKED BY POPULATION: 2020

Rank	Country	Population
1	China	1,420,000,000
2	India	1,370,000,000
3	United States	330,000,000
4	Indonesia	270,000,000
5	Brazil	215,000,000
6	Pakistan	204,575,000
7	Nigeria	201,000,000
8	Bangladesh	169,000,000
9	Russia	144,000,000
10	Mexico	132,500,000
11	Japan	126,800,000
12	Ethiopia	111,100,000
13	Philippines	108,050,000
14	Egypt	101,100,000
15	Vietnam	97,500,000
16	DR Congo	86,750,000
17	Turkey	83,000,000
18	Iran	82,900,000
19	Germany	82,500,000
20	Thailand	69,500,000

But enough for now on the theme of over- or underpopulation. Let's look at where people are in the first place.

REGIONAL DIFFERENCES IN POPULATION

Where are people, and where are they not? In some parts of the world, harsh climates and terrains are too formidable for large numbers of humans to hang out in. The cold Arctic areas in Canada and Russia, the great desert and steppe regions in North Africa and Central Asia, and the high Andes and Himalayan ranges serve to keep population numbers low. Humans tend to proliferate in well-watered areas and along coastlines. Generally speaking, human settlements favor the mid-latitudes; there are far more people in Eurasia and North America than in tropical areas. Draw on your own experiences to figure out this trend. Would you like to live in a tropical rainforest, a desert, or a mountain top? Why or why not?

FIGURE 2.2 POPULATION DENSITY, MAJOR CENTERS

People have adapted to living in just about every extreme environment on our planet—just not in great numbers. What is the deal with the great numbers? It has a lot to do with history, culture and current population dynamics, which we'll get to in just a second. First, a few points about the map in Figure 2.2. As you can see, I've circled just six big population clusters on the planet for our discussion, but of course there are others . . . that you should discuss amongst yourselves. I can't do it all!

For starters, over half the world's population are found in Eurasia, particularly a people-packed arc from Japan to Eastern China, through Southeast Asia to South Asia. Currently, China is the most populated state with over a billion people. But watch out! India has a billion people as well, and, more importantly, is growing at a faster rate than China, which means that it will certainly take over the top slot within a decade or so. And don't overlook India's neighborhood—Pakistan and Bangladesh are both members of the top ten most populous states in the world as well.

The monstrous Asian population centers are, in large part, a product of history. People in these areas have been getting busy—in more ways than one—for thousands of years since the innovation of agriculture and the birth of civilization as we know it. They have existed as stable civilizations for long periods of time. They also happen to mostly be in physically conducive environments—well-watered, mid-latitude lowlands. This also helps to explain why the cradle of Western Civilization—Mesopotamia, which is just as old—did not form huge populations over time. Its physical environment is more arid and unable to support large numbers of people.

However, due to great leaps in technology during the Industrial Revolution, including lots of technologies that helped keep more humans alive longer, Europe's population boomed wicked fast in the last several hundred years. And now they are another significant center for people-packing on the planet. By contrast,the Americas have historically had low population numbers, pre-contact. When Europeans did arrive in the New World, they brought diseases that wiped out a vast number of people, thus leading to even lower numbers on our side of the sphere. But the Western Hemisphere has now been going and growing for long enough to have significant population concentration in eastern USA, along with the mega-cities area of eastern South America centered in southern Brazil. History, technology, and the physical world have a lot to do with where most people are today.

Two more global population points of interest: Africa, largely believed to be overpopulated, is a gi-normous place with half the population totals of South Asia or East Asia, but note the large concentration of peeps in West Africa, centered around Africa's most populous state of Nigeria. My favorite population fact deals with Russia and the US, the two Cold War adversaries. Look again at the map and realize this: these two regions account for less than 10 percent of the world's population, but have effectively shaped the political and economic fate of the other 90 percent during the last 70 years. Interesting, isn't it? Okay, maybe not to you, but it certainly was during the Cold War. Yes, those rascally Ruskies used to be a world power, but now Russia is a state currently in population decline. Which brings us to our next point: where are populations growing, where are they shrinking, and why?

HOW POPULATION IS CHANGING

To complicate matters further, not just total population but also **population growth** rates are unevenly distributed around the world. **Population growth rate** refers to how fast or slow a group's population total is expanding . . . or shrinking, when referring to our Cossack friends. When you see a number like 3.5 for a population growth rate, it indicates that, by this time next year, the population total for that country will be 3.5% bigger than it is right now. A negative number is the inverse: population growth of −0.5 means that population total for next year will be 0.5% smaller than it is presently. Check out Figure 2.3 below for some insights . . .

FIGURE 2.3 WORLD POPULATION GROWTH RATES

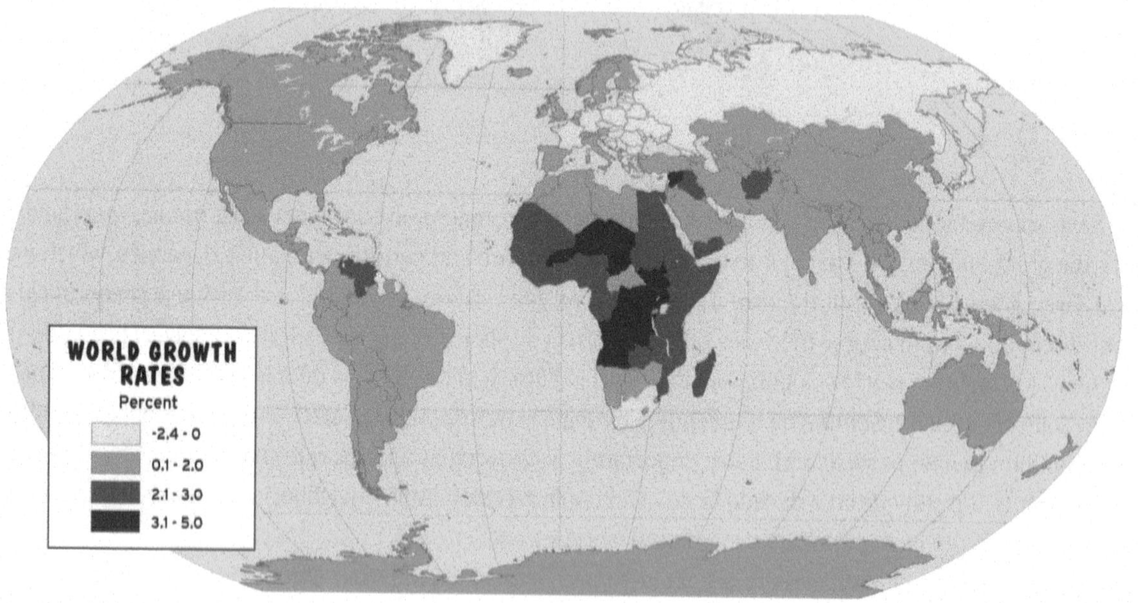

Source: CIA World Factbook 2021

Look! Ethiopia and Yemen's population total is expanding very quickly, which means those countries have a high population growth rate. Both the US and China's total population gets bigger every year, but not by much percentage-wise; the countries have a moderate to slow growth rate. Russia and Japan's population totals are getting smaller every year—a negative growth rate. You may have also identified a major trend: the highest growth rates typically occur in states that we consider underdeveloped, a.k.a. the poor ones. Say again? You mean the poorest areas of the planet are where more people are being added faster than ever? That is correct. The developed, or richer, states all seem to have low growth rates. So the places that could afford to provide for more kids, have less kids? Yep, that's true as well. Why is this so? Good question. Answer: the Demographic Transition.

HOW/WHY/WHERE POPULATION IS CHANGING: THE DEMOGRAPHIC TRANSITION

The Demographic Transition is a lovely little model that goes a long way in explaining lots of things about human population change around the world today. Be forewarned: it is just a theory, but man oh man, it makes a whole lot of sense when applied to just about anywhere, anytime. It helps us understand why population is booming in the poorer parts of the world while it shrinks in richer areas, and even why women in Laos may have ten kids, when women reading this book here in America may not want to have kids at all.

The model is based on the experience of the currently, fully developed states, which underwent a population surge on a smaller scale beginning about 1700 in Europe, and then later stabilized. Other states—typically ex-colonies of Western European powers such as the United States or Australia, or states in close proximity to Western European powers, such as Russia—followed suit in the last several hundred years. Every other place on the globe can be seen as somewhere in the transitional process that these states have already gone through.

FIGURE 2.4 BASE MAP, DEMOGRAPHIC TRANSITION

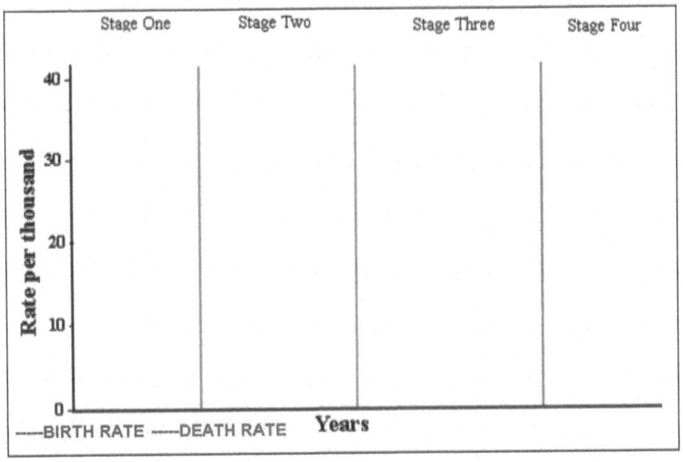

Generally speaking, this "transition cycle" begins with high birth rates and high death rates, passes into a high birth rate/lower death rate period for a variety of reasons, and ends with low birth rates and low death rates. The total population at the beginning and end of this cycle is stable or has a very low, perhaps even negative, growth rate. However, it is the massive increases in population during the middle phases that makes the model so compelling, and explains so much about what is happening in today's world. But I'm getting ahead of the story. Let's take it one step at a time.

For the rest of this discussion, **birth rate** refers to how many children are born every year per 1000 people. To give you some context, the birth rate in the US right now is about 12/1000 every year. **Death rate**, of course, refers to how many people kicked the bucket that year per 1000 people. The US death rate is currently around 8/1000. Now, on to the transition.

DEMOGRAPHIC TRANSITION STAGE ONE

This whole concept hinges on the idea that all societies want to go *from* premodern, hunter-gatherer, stick-collecting goobers *to* postmodern, latte-sipping, Vespa-driving goobers. It does seem to make sense. Given the option anywhere in the world, I think most folks would choose the latte; that is, most societies are striving to become industrialized, richer, and all-around better off. You don't have to buy my theory, because quite frankly, I'm not selling it. There are those that argue that we would all be a lot better off living in grass huts somewhere weaving baskets from giraffe hair, because that would be true sustainable development in harmony with Mother Earth. Good luck with that one. Give anyone a chance not to live on a dirt floor, and I'll bet they pack their bags, set the giraffes free, and head out to a better life for themselves and their kids. But I digress. Where were we? Ah! Stage One . . .

STAGE ONE of societal development finds us making baskets from giraffe hair. We typically associate this stage with premodern times; most folks are hunter-gatherers, living solely by collecting food naturally occurring out in, um . . . nature. This is pretty much the way things were for a great number of humans for a good long time in human history. Small groups of folks on the move, searching for food, waiting around for civilization to pop up. Of note for our model is that both the birth rates and the death rates are extremely high and erratic. Essentially you have a situation where lots of kids are born per 1000 people, and lots of people die per 1000 people, with some years being really good, and others being really bad. Why would that be?

Some Stage 1 hold-outs: Kalahari Bushmen in Namibia.

Why the high death rates? Because this lifestyle is hard, and it sucks! It's easy to die from just about anything: lack of health care, food shortages, poor food containment so things go bad fast, lack of regular clean water, lack of sewage disposal or worse yet, your sewage disposal plan involves your drinking water source, animals that want to eat you, animals that just want to kill you, animals just having fun with

I liked your Granny too. She was delicious.

you, diseases of all sorts, simple infections you could contract from a hangnail . . . ew, this sounds like no fun at all. **Infant mortality**, the number of children that don't make it to their first birthday per 1000, is also high because of lack of immunizations and/or inadequate diet. **Life expectancy**, or the average age to which the population is expected to live, is low. Old people get sick more easily, can't pull their own weight out picking berries and therefore don't eat as much, and in general, are slow enough to get caught by the animals that want to eat them. Poor Granny. I really liked her. Now she is cheetah food.

Why are the birth rates high? The same reasons that they are high in parts of today's underdeveloped world. For starters, it's a mindset; if you expect half your kids to die before they reach adulthood, then you should have twice as many. Makes sense. Also, children are often seen as a resource in these societies: maybe Junior starts picking berries at an early age to help out the fam. More kids = more labor = more food = good. Plus, no health care = no health education = no sex education = no contraceptive usage. Well, I guess people could always just abstain from having sex . . . yeah, right.

But why are the birth and death rates so erratic? For the simple fact that some years are good, while others years are bad. A drought or a plague would cause births to drop and deaths to rise. A very good, wet season with lots of food available would cause the spikes to move in the opposite direction.

Because both birth and death rates are extremely high, they offset each other, equating to a total population that is low and a population growth rate that is slow or stable. Looking back at Figure 2.1, you can see that for most of human history, population growth rate has been very slow, or stable, right on up to the Industrial Revolution. Before we leave this rather boring phase, just a quick note: there really are no more societies like this left anywhere on the planet. You have to dig deep into the Amazon rainforest, the remote savannas of tropical Africa, or into the highlands of Papua New Guinea to find folks still living this lifestyle. Even then, they will be very small numbers of people in isolated pockets. No state economy on the planet today would be classified in Stage One.

FIGURE 2.5 DEMOGRAPHIC TRANSITION STAGE ONE

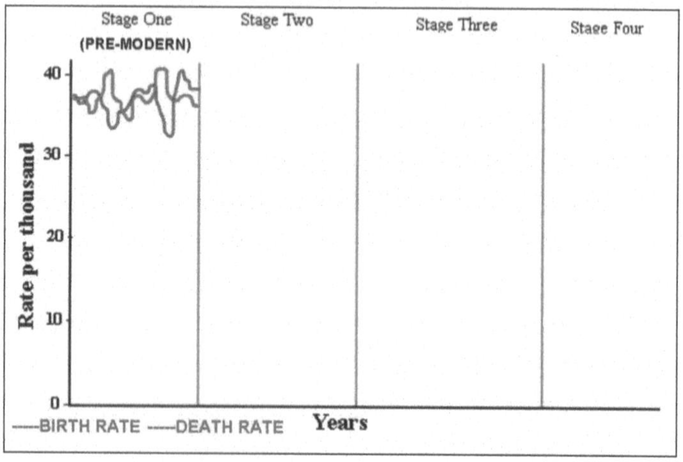

DEMOGRAPHIC TRANSITION STAGE TWO

STAGE TWO is the trickiest phase to consider, because we are going to pack a veritable smorgasbord of different human experiences under this single banner. They occur over long periods of time, but have this one big Stage Two result: the death rate declines while the birth rate remains high and stabilizes. Fewer people die; same amount of people being born. What happens to the total population numbers in such a situation?

Before we answer that, let's define the stage a little more. I said it is fairly all-inclusive of the human development spectrum, and I meant it. Stage Two includes the transition of humans from hunter-gatherers to agriculturalist to factory workers. Yes, this is a long stretch of time that also entails the formation of what we call technical innovation, civilization, and urbanization. Essentially, dudes figure out that staying put in one place and growing food is more productive than moving around and hunting all the time—an agricultural revolution. With increases in

FIGURE 2.6 DEMOGRAPHIC TRANSITION STAGE TWO

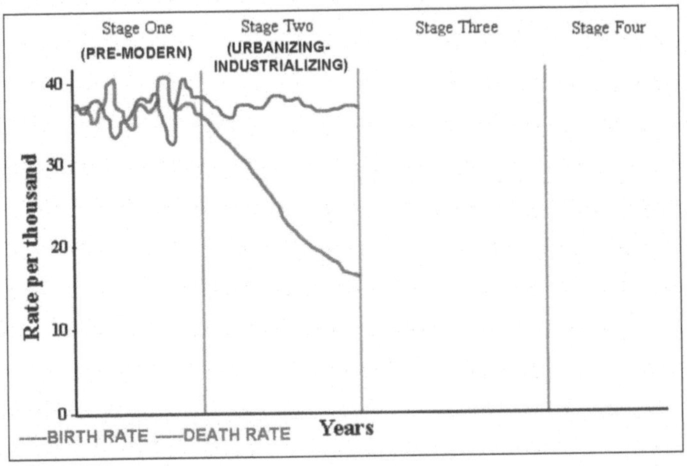

technology, so much extra food is produced that not everybody has to be farmers. Some become blacksmiths, butchers, priests, traders, artists, and inventors—that's specialization. This leads to the formation of villages, towns and eventually cities where larger and larger groups of dudes hang out and exchange products and services with each other—that's urbanization. Further technological advancements lead to the creation of machines to automate our work, which leads to the creation of machines that make stuff, and even make other machines—that's industrialization. Eventually taken to its logical extreme: robots that make robots that kill all the humans. Yikes! The Matrix! This is worse than the Granny-eating cheetah! But that is for a future textbook, so let's stick with Stage Two for now. . . .

What's all this got to do with people having kids? During this part of the societal transition, death rates absolutely plummet. Why? Remember how I told you that everything sucked in Stage One? In Stage Two, everything gets way better. Increases in food production and increases in food storage technology allow for a steady, stable food supply. There are no more "bad" food years or, at the very least, not as frequent as they were in Stage One.

Technological advances in water resources, sewage treatment, and health care, all based on growing scientific knowledge, serve to keep more people alive for longer periods of time. Child mortality drops and life expectancy rises. We have developed the shotgun to ensure that Granny is not eaten by the cheetah. More important, however, is the fact that more kids stay alive than ever before; young children become an increasingly bigger percentage of the total population. Education across the board, at all levels, also significantly increases quality of life and survivability.

During this vast sweep of progress, death rates drop and birth rates remain solidly high and even stabilize high. What is going on here? Basically, people are caught in a mental time warp and old habits die hard. If Mom had ten kids, and her mom had ten kids, and her mom had ten kids, it is highly likely that you would be of the mindset to have ten kids as well. It's just what people do. The society as a whole, and certainly not the individuals within it, does not understand that it is going through a transition. The mentality that producing large numbers of kids is good because half of them are going to die and the rest will help gather berries remains the same, even though conditions have changed. At the end of a Stage Two society, half the kids are now NOT going to die; maybe only one or two of them will die, maybe none at all. Granny had ten kids and only five survived. Mom had ten kids and seven survived. I had ten kids and . . . holy crap in the kiddie pool, all these little brats are still here!

This mind warp is called **cultural lag**. Conditions have changed, but the culture is lagging behind. Folks with Stage One mentality are thinking "Wow, this is just a good year," without realizing that it is a good year after a good year after a good year, but they are still having kids like it's a boom-and-bust cycle. Result: **population explosion!** More kids beget more kids, and people just don't die like they did in Stage One. Mo' peeps having mo' peeps having mo' peeps.

Stage 2 Shenanigans: A Heaping Helping of Historical Human Activities

FIGURE 2.7 CULTURAL LAG

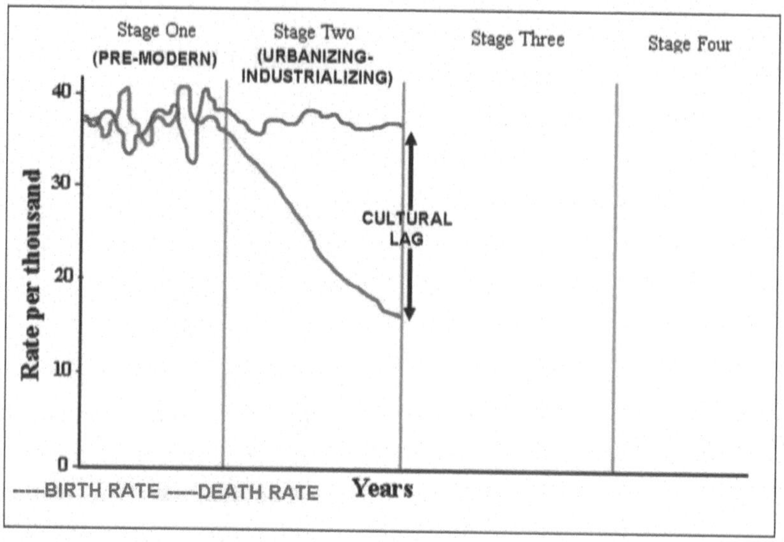

Stage Two main points: death rates decrease, while birth rates remain the same, total population explodes, and the structure of the total population becomes more youthful. This last point is the most important for the rest of this story, as we have more potential baby-makers entering the scene. Many states in the world may fall into the later fringe of Stage Two, mostly in Africa, Central America and parts of Southeast Asia. These states are still heavily reliant on agricultural production as a main economic activity. They also may have their death rates drop more from foreign external aid, importation of life-saving technologies, and humanitarian relief than from true upward evolution of the society. An important note: all of these things will help the death rate drop immediately, which means the period of cultural lag will likely be longer, significantly increasing total population overall. Angola and Guatemala are examples of states in Stage Two.

DEMOGRAPHIC TRANSITION STAGE THREE

As we move to **STAGE THREE**, we approach more familiar ground. Stage Two ends with the beginning of the industrialization of the society. Stage Three takes us the rest of the way through it, ending with what we refer to as the "mature industrial phase." Most states on the planet are currently located here, with varying degrees of development that can be quite radically different, depending upon whether they are early Stage Three or late Stage Three.

At this time, the Stage Three state continues down the modern industrial path. This is marked particularly by a shift in what the majority of folks do for a living. It is at this point that the scale gets tipped; more people are working in the processing, manufacturing and service sectors than are working as farmers. This development is significant because this typically means that agricultural technology has superseded the need for vast amounts of human labor in the fields.

So long farming! We are manufacturing minions now! As we also shift to services. . . .

This equates to less farmers, which equates to more people leaving the countryside and heading to the big city to get a job—thus increasing urbanization.

Even today, agriculture is still the number one job on the planet for humanity as a whole. However, states in which agriculture is not the predominant occupation are further down the development road. That is, the richer a state is = fewer number of farmer peeps; the poorer it is = greater the number of farmer peeps. Food for thought—pun intended.

We're just getting warmed up with the impacts of this stage (agriculture-to-industry, and rural-to-urban shifts). What happens to the ideas about having lots of kids when this shift occurs? Plenty! Life in the big city is more costly. Having more kids to feed costs more money. Plus, you've got to clothe them and house them and buy school books for them and throw birthday parties for them and eventually buy them cars. Wow! This is starting to suck as bad as Stage One. But it gets worse! The value of kids has changed as well. Junior used to be an asset picking berries and plowing the fields. Now Junior is an added cost who only picks his nose and plows the family car into the side of the garage. The cultural lag is over thanks to you, Junior! No more kids for us!

In addition to these changes, several other things of note are on the rise. Health care technology and accessibility continues to increase, and especially with regards to increasing knowledge about birth control and contraceptive use in general. Increasing education across the board helps more and more people, but I want you to think more specifically of the education of women and its impact on the whole equation: more educated women = more women entering the workforce = more contraceptive use = more women delaying family formation = fewer kids.

Educating the women of any society decreases fertility rate instantaneously. Okay, maybe instantaneously is a bit strong, but it's a HUGE factor in affecting the **fertility rate** in today's world. Wait, what's that new term?

Fertility rate is simply how many kids on average one woman will have in her lifetime in a particular state. The fertility rate of the US is about 2.1. In Italy, the fertility rate is about 1.3. But in Mali, the fertility rate is about 7.5. The average woman in Mali will have seven kids and one half, say from the waist down. In the US, a woman will have two kids and

FIGURE 2.8 WORLD FERTILITY RATES

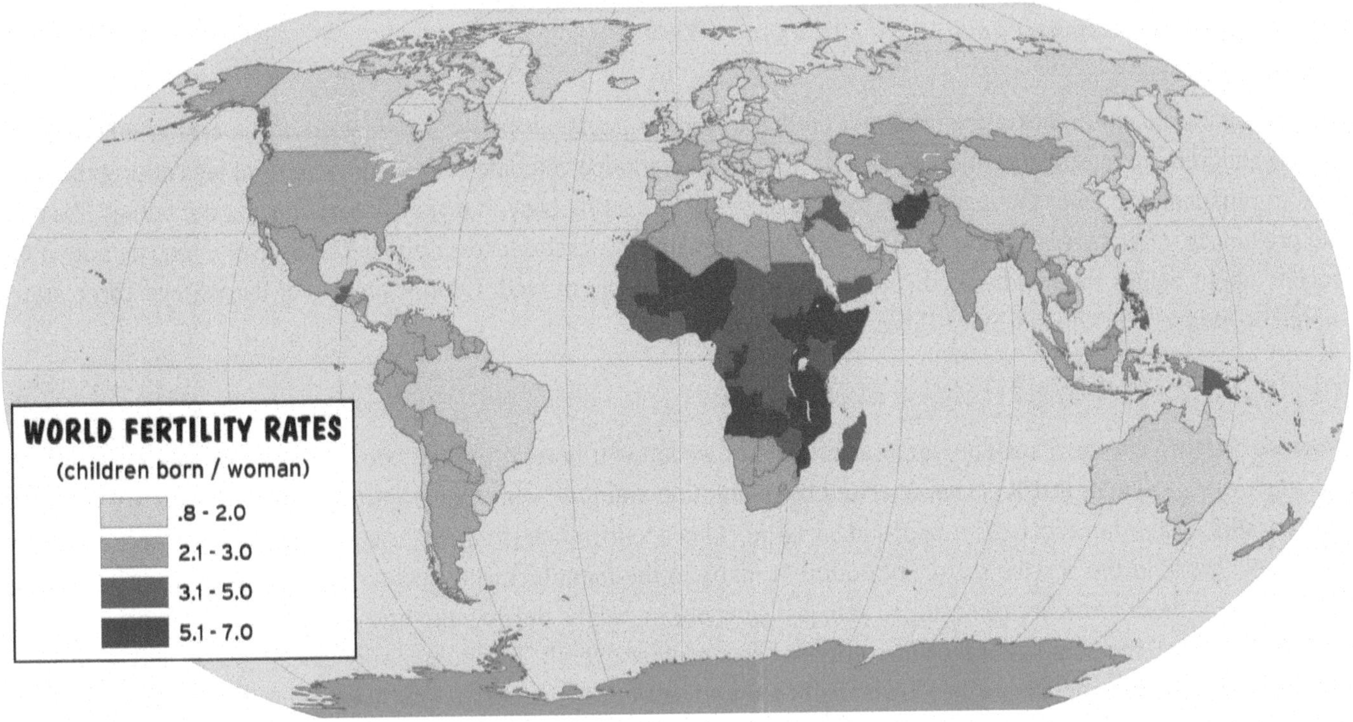

Source: CIA World Factbook 2014

a leg. For our international readers, that is what we call a joke. Actually, the US figure is a quite important fertility rate; that number also happens to be something called the **replacement level**. A fertility rate of 2.1 is exactly what it takes to "replace" the current population. See, two people have to get together to create kids (at least still for now, thankfully) and if two people produce 2.1 kids, then when the parents die, there are still two humans they've created to take their place. Mom and Dad have "replaced" themselves in the population. Sweet! We'll return to this idea in a little bit.

Back to Stage Three. The result of increases in health care, technology, childcare and education serve to knock the death rates down even further. However, there is only so much that modern medicine can do, and eventually that line flattens out. Child mortality decreases and life expectancy increases a bit more, but we all have to die sooner or later, so we'll say goodbye to Granny once again.

FIGURE 2.9 DEMOGRAPHIC TRANSITION STAGE THREE

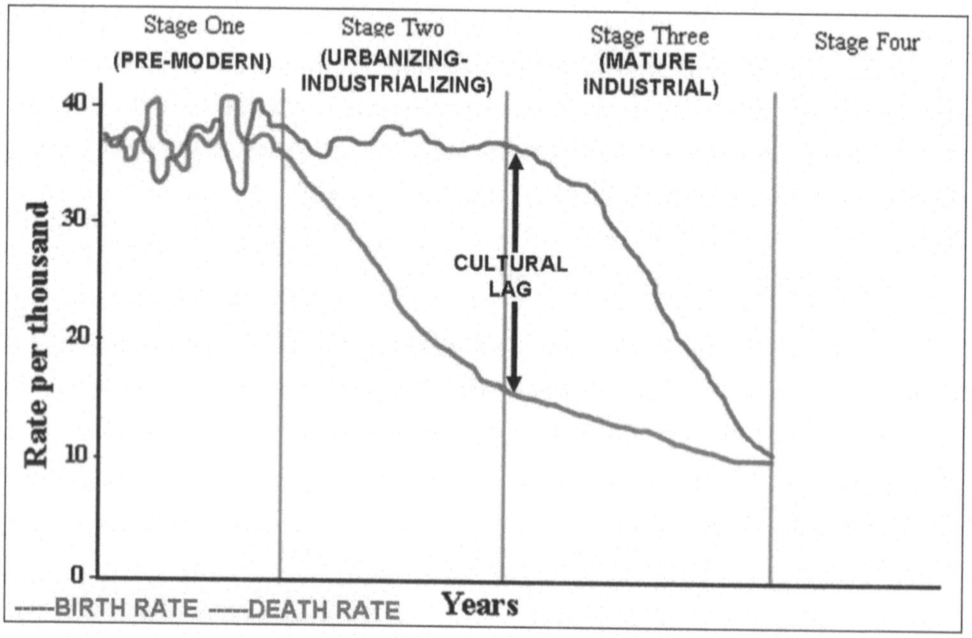

Of significant note: Increasing urbanization due to the employment shift, re-evaluation of the cost and benefits of having children, increased education of women (and all that entails) combine to overcome cultural lag causing the birth rate to plummet and meet the death rate. People may still be getting busy, but they are not having the babies like they did previously. After several generations, the fertility rates sink down closer to 2.1, the percentage of young people in the society about equals the middle aged cohort, and the older people as well. Good examples of early Stage Three states would be India and Brazil; late Stage Three, China and Chile.

DEMOGRAPHIC TRANSITION STAGE FOUR

Now we are through the transition, approaching life as we know it here in the fully developed, post-industrial, mostly Western world. **STAGE FOUR** is characterized by population stability, where death rates and birth rates are both low and parallel each other nicely. This post-industrial world is one in which yet another economic shift has occurred; now most folks work in the service sector, not quite as many in the manufacturing sectors, and virtually no one works in agriculture. If these terms are confusing to you, just hang on; we will be delving into economics in the next chapter.

In this stage, the population age structure has become older overall. Technology and education may still be increasing, but only in minute detail can they lower death rates anymore. The population of a Stage Four state is overwhelmingly urban and educated; they use some form of birth control and spend way too much on coffee.

The US, France, and Japan are all great examples of Stage Four states. As a result of expected higher standard of living and higher costs of living, many people will plan to have one or two children at most. Some people will decide not to have any children at all.

The Service Sector Circle of Life. You will likely live here.

This leads us to a possible expansion of the Demographic Transition to a Stage Five, in which birth rates actually dip below death rates. The effect? Net population loss—the state's population shrinks every year, and unless supplemented by **immigration**, the state would eventually disappear. Immigration is when people not born in the state move into it. **Emigration** is just the opposite perspective: when people leave your state, they are emigrating out of it. These concepts are increasingly critical in today's world, because many Stage Four and Five states rely on immigrants to fill jobs, pay taxes and boost slumping population growth rates. The US is a prime example whose population is increasing partly due to immigration. This is actually more pronounced in Europe where states may be fading away because the locals just aren't having many kids. If any at all.

Have another cup . . . and get back to Stage 4 work!

Of course, the disappearing state thing has never happened before, but there are several states that are currently in this Stage Five category. Russia, Sweden, Italy, and Japan are all losing people. Italy and Japan are both in dire straits, but for different reasons; both have declining fertility rates, but in Italy it's because no one wants to have kids at all, while Japan refuses to allow any immigration into its pristine palace. We shall see how that works out in the long run. Serves them right for creating Pokemon.

FIGURE 2.10 DEMOGRAPHIC TRANSITION STAGE FOUR

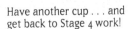

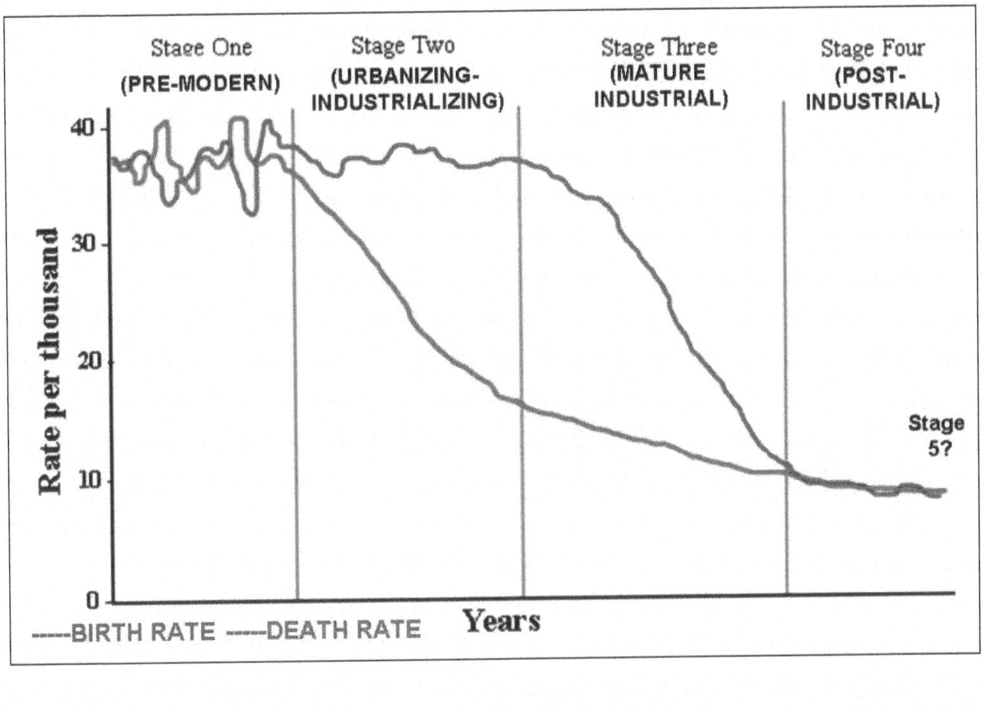

POPULATION PYRAMIDS

A final tool for our consideration of population dynamics on the planet is called a population pyramid. Population pyramids are constructed to show the breakdown of the population for gender and age groups. The x-axis across the bottom displays either a percentage of total population or actual population numbers, and the y-axis shows age cohorts, typically in 5-year increments. The two sides of the pyramid are divided up with males on one side, females on the other. Kind of like your junior high prom dance floor. In the Figure 2.11, you can see that in Nepal in 2019, there are about 2 million males aged 0–4 years old, about 1.5 million females aged 10–14 years old, and about 0.5 million males aged 50–54.

FIGURE 2.11 A POPULATION PYRAMID

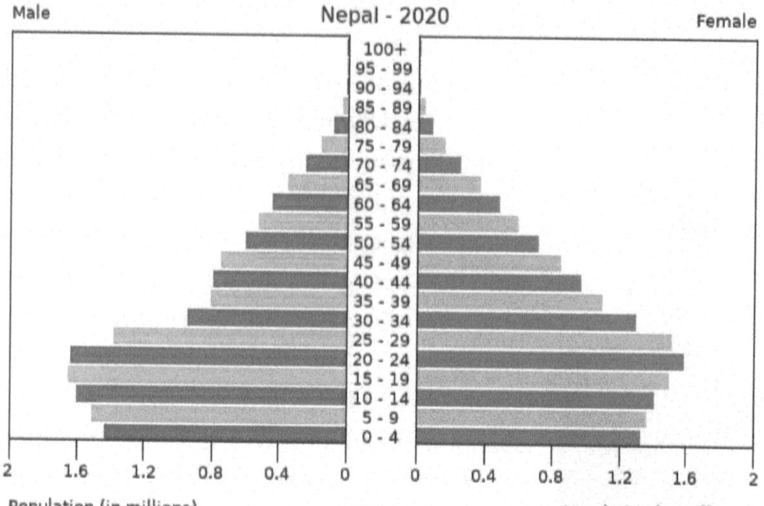

Source: U.S Census Bureau, International Data Base

Like I give a crap about how many Nepalese women are over 80 years old! What do these pyramids really tell us? They tell us all sorts of things about a state: current population, current economic conditions, and quite a bit about the standard of living as well. But that's not all! These pyramids tell us a whole lot about what the future will hold and you don't even need a crystal ball. Consider the three basic shapes that these pyramids can take: a **classic pyramid**, a **column** shape, or an **inverted pyramid**.

THE CLASSIC PYRAMID

The classic pyramid shape is, um . . . shaped like a perfect pyramid. The wide bars at the bottom taper up gradually to smaller and smaller bars as age increases, displaying that there are more people in the younger brackets and the size of the group diminishes steadily with age. Broad-based pyramids (Ethiopia in Figure 2.12 below) are characteristic of populations with high birth rates and lower life expectancies. Why? You already know the answer! Look back at the Stage Two and Stage Three descriptions of the Demographic Transition.

FIGURE 2.12 THREE TYPES OF PYRAMIDS: CLASSIC, COLUMN, AND INVERTED

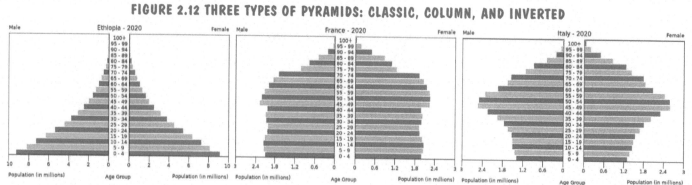

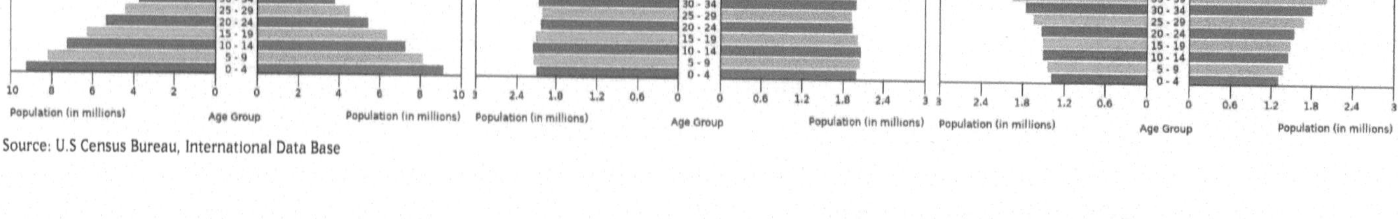

Source: U.S Census Bureau, International Data Base

Industrialization and agricultural innovations, as well as increases in technologies across the board, result in advances in food supply and public health overall. This results in recently reduced infant and childhood mortality rates, and slightly increased life expectancy. However, they are definitely in the cultural lag mode, as fertility rates continue to be high. This in turn results in high population growth rates and in many cases, outright population explosion. Each year, the bottom bracket of 0–5 year olds gets wider than the previous year. Of particular significance: the numbers of kids under the age of 15 is larger than the number of folks in the 15–35 brackets.

Why is this significant? Because the under-15 are dependents in the society; that is, they absorb resources. Typically, the 15–35 year olds are the biggest providers in the society, both in economic means as money-makers for the state (gross domestic product, or GDP) and as family sustainers of young people and the elderly. That is, they provide most of the resources. Perhaps now you are starting to see why countries with this type of pyramid are often poorer or less-developed; the dependents often vastly outnumber the providers. More often than not, Stage Two and early Stage Three societies exhibit the pyramid shape demographic. The best examples of this type are any African country.

There is ever so much more we can say about a society with this type of pyramid. It is probably heavily reliant on agriculture or other primary industries for much of its economic earnings. Most people definitely grow some food, either as an occupation, for sustenance, or both. The society probably has lower literacy rates and little to no social safety nets like welfare, and also lacks adequate infrastructure, like good roads or sewer systems. Also, as pointed out earlier, many of the gains in health care, food availability, and life-sustaining technologies may be attributed to foreign aid and humanitarian aid programs. This adds a distinct 21st century dimension to the transition model; no one really knows what impact this will have in the long run for less-developed countries.

THE COLUMN SHAPE

All columns are not created equal, and there can be radical variations on this shape, but the main thing to identify this type of population pyramid is the more overall fuller figure. No, I'm not talking about a waist size here—that's a different book altogether. The real distinguishing mark here is that the size of the 15–50 year cohorts are *roughly* the same size as the under-15 cohorts, giving the overall shape more of a cylindrical look. The older age brackets also grow slightly, and upwards, as more people stay alive longer, adding to the elongated shape.

When this is the case, we realize that the fertility rate must be much lower than in the true pyramid shape, usually hovering between 2.0–3.0. This means that those child-bearing folks (typically between 15–50 years old) are having just two or three kids each, roughly replacing themselves or maybe even just a bit more. Remember that term **replacement** level? Well, the closer a society gets to that 2.1 fertility rate, which constitutes the replacement level, the more perfect the column shape will become. Some examples of this are the US, Lebanon, and Norway.

The operative words in these states is "stability." The total population growth rate is low, or perhaps even zero, with the state adding just a few percentage points of population every year. What else can be said about these countries? They are fully developed industrialized or post-industrialized societies. The highs: GDP per capita, standards of living, health care quality and access, available social programs, life expectancy, education levels, technology levels, urbanization, use of birth control, service sector jobs, and SUV ownership. The lows: fertility rate, infant mortality, illiteracy, farmers, miners, and people who die from infectious diseases.

INVERTED PYRAMID . . . YOU MEAN IT'S UPSIDE DOWN?

Indeed. This is an easy one to describe, because its essential ingredient is that the younger cohorts are smaller than the older ones; child-bearing peoples are creating less people than themselves; and total population is actually shrinking! Fertility rates are under 2.1 and thus the replacement level is not being reached; left unchecked, the population would shrink into nonexistence.

This has never really happened before, and is not likely to ever come to its full conclusion, either. To counter this effect, immigration is increased or some other government policy is put into place to encourage fertility rates to rise.

Why does this scenario happen? Perhaps due to several different reasons. The first and foremost explanation is due to the same processes at work which serve to end cultural lag in developing societies. Namely, the higher cost of living in highly urbanized areas, combined with kids becoming a drain on resources, changes peoples' attitudes towards family size. Smaller is better in industrialized societies. They only have one or two kids, and invest heavily in them, as opposed to having lots of kids. In some places here in the 21st century, this has gone to an extreme; many people totally opt out of the family thing altogether; they want no children in order to maintain their own high standard of living. End result: fewer or no kids, and thus replacement level for the state is not reached. The best examples of this today are Japan, Italy, and Sweden.

Another reason for declining populations in the world may be due to economic circumstances. Some countries, while considered part of the developed world, have gone through radical changes and economic collapse due to the end of the Soviet Empire. Russia, Belarus, and Ukraine are all in this inverted pyramid category due to lack of resources, jobs, health care, and many other services that the government used to supply. This is particularly evident in some of the measures of well-being for Russia, whose current life expectancy more approximates a poor, underdeveloped African country than a former world superpower. Hard times can also cause fertility rates, and a state's population, to decline as life becomes too difficult and large families too expensive to maintain.

IT'S HARD TO STOP HAVING BABIES

Many of you will look at these definitions and numbers and come up with some puzzling questions. You may wonder how China is the most populous state on the planet, at 1.4 billion people, yet their fertility rate is only 1.73. Or perhaps you may discover that projecting ahead in India, they will continue to grow their population rapidly for some time to come, and indeed will surpass China as the most populous state very soon, but their fertility rate is only 2.73. That's not very much more than replacement level! How is this possible?

Simple. **Population Momentum** is the answer you seek. Consider India for the next 50 years:

What these three images from 2000, 2025, and 2050 in India are showing you is the Demographic Transition model in action. In 2000, India was early- to mid-Stage Three, as you can just detect the perfect pyramid starting to

FIGURE 2.13 INDIA: THE SNOWBALL GROWS

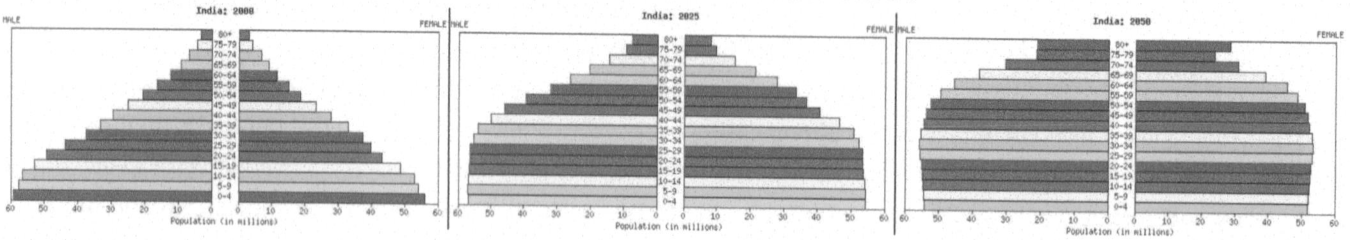

round away the edges of the bottom age cohorts in an obvious sign that cultural lag is wearing off and fertility rates are starting to sink from what were highs of four to six kids average per fertile female. We see that, in 2025, this has taken full effect; India is certainly in the column shape as its late Stage Three or early Stage Four industrialization has knocked down fertility rates to a precise 2.1, the replacement level. It's a perfect column. How can population totals continue to expand after this?

Population momentum is like a snowball rolling down a steep hill. Even after fertility rate stabilization, the population pyramid will have to "fill out its figure," as you can see by projecting further ahead to 2050. Everyone may be only having 2.1 kids, but a much bigger number of them are doing it, as represented by those bottom age cohorts back in 2000, where the stabilization actually starts. By the time the snowball reaches the bottom of the hill, it is massive—as will India be by the time its population growth rate total flatlines.

A PRICKLY PROBLEM OF SKEWED SEXES . . .

I would be remiss in my duties to fully educate and illuminate on population issues without shining some light upon some more disturbing manipulations of the Demographic Transition which are spelling social and sexual disaster for some states! Of what do I speak? The un-natural selection of the sexes, no less!

Say what? Well, in the natural way of things, boy babies slightly outnumber girl babies at birth, usually on the order of 105 boys to 100 girls. Why is this so? Hmmm . . . maybe it's because boys are crazier, riskier, and the ones that usually go to war with each other and therefore die/kill off each other in greater numbers than girls do. Thus, you need a few more of them around so that after some of them die off, your society will still have enough men to match up with women for the procreation gig. It's like that for most animals, and our species ain't no exception. Human females also have slightly longer life spans on average than their male counterparts, all over the world, thus the ladies naturally outnumber the lads simply because they outlast them.

"Yo, where are all da ladies at?"

But dig this: in some countries around the world there is a significant cultural bias that favors males over females, especially in Asia, but the trend is spreading. Male babies are prized over females because males carry the family name, are a symbol of virulence, get higher paying jobs, are stronger physically for work on the farm, and are seen as the potential extended family providers when the parents get old. In India, it is the traditional role of the bride's family to 'pay' the groom a **dowry** when the couple gets married, so having a daughter actually 'costs' the family all the way up to adulthood. A family stands to gain a lot of money in India if they have a bunch of sons, and no daughters.

Basically, in these societies men are seen as the continuation of the family lineage and the major breadwinners, but that belief alone is not the problem. It becomes a problem when parents start to skew the sexual ratio on purpose to get more males into the mix. With the advent of sonograms/Ultrasound in the 1980's, coupled with easy access to safer abortion options, there has been a real "gendercide" underway, where female fetuses have been aborted in greater and greater numbers in an effort to get more male offspring . . . and it is starting to radically affect the demographics of the states, in increasingly negative ways. . . .

Particularly in China and India, males outnumber females at alarming rates. Right now in China, there are 35 million more boys than girls. In India, there are only 915 daughters born to every 1000 sons, resulting in 20 to 30 million more men than women. These ratios are seriously skewed from the natural norm. And while Asian cultures seem to be the epicenter of this syndrome, even they are not alone anymore: the practice has spread to Europe, where skewed sex ratios now can be found in Albania, Armenia, Azerbaijan, Serbia, Bosnia and Georgia.

So there are way too many penises in some countries. So what?

Ummm . . . can you say "life without a wife," and more critically, can you think about the potential problems that will cause? Not enough women for all the men means that an increasing number of males in these societies will not be able to find a wife and start families. That will affect fertility rates for sure (and wedding cake sales), but more importantly it leaves a gigantic pool of unmarried, unattached, uncontrolled, undersexed, and testosterone-fueled males in your country. Whoa! That ain't nothing but trouble!

Indeed! The affects are becoming obvious: there is unrest among young men unable to find partners. Unmarried men in China unable to find partners are nicknamed "bare branches" . . . what do you think that refers to? Reports are increasing from China and India of antisocial behavior and violence, threatening societal stability and security, as unattached males are more likely to be involved in harassment, gang activity, gambling, and even violent crime. But business opportunities have sprang up to serve the needs of these men as well: prostitution is on the rise, as is the trade in buying foreign wives. A quiet mass 'migration' may be in effect wherein Thai, Vietnamese, and Burmese women are being brought into China as mail-order brides. (Market price for Burmese bride is currently between $1,000 and $2,400.)

"I don't care if we are out! I want a woman, now!"

So consider the repercussions of this skewed battle of the sexes in these societies, and how it affects not just the demographics, but also the cultures, economies, government policies, social stability, and even immigration policies in these places. All of these states have already outlawed abortions based solely on preferential sex of the fetus, but things like Indian tradition, the aftermath of China's 'One Child Policy,' and current economic situations in the Caucuses trump the law and social convention . . . but with grave repercussions imminent. Penis propensity problems will proliferate! And speaking of problems with peeps . . .

PEEPS WON'T STAY PUT!

That's right! Unlike rocks and trees, some of these pesky peoples keep moving around the planet, from one place to another! Now, when you just move from one side of your town to the other side, or from one city in your country to another city, that is just called moving. But when you move yourself and/or your family to a whole different continent or country with intent to stay permanently, that is called **migration**. And humans have actually been doing that migration thing for millennia!

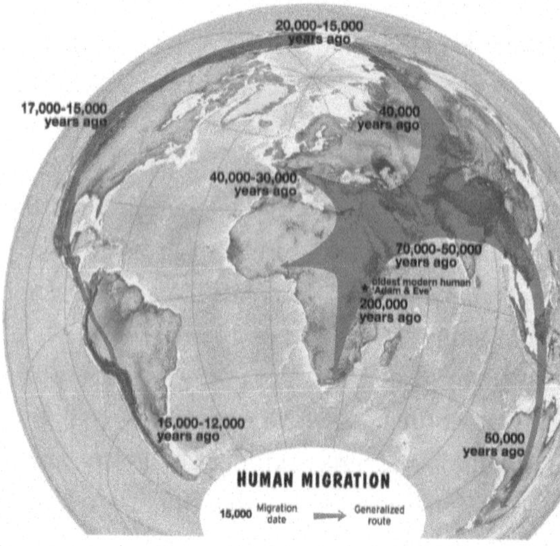

The O.G. Migration Situation

If you accept the whole evolution of our species idea, then you already know that pre-humans and humans *migrated* out of Africa long ago to eventually occupy every nook and cranny of the globe. Heck, even if you don't believe in the evolution of our species, you know that in the last 2,000 years, the Huns migrated across the Roman Empire and sacked it, the Vikings migrated across northern Europe and looted it, and the Mongols migrated across Eurasia and conquered it. Just in the last 500 years, millions of Europeans *migrated* to North America, South America, and Australia. Millions of Chinese have *migrated* to Southeast Asia and Central Asia and even southeastern Russia. And in much more recent times, Africans and Syrians have *migrated* to Europe; Indians and Pakistanis have *migrated* to Bahrain and the UAE; Mexicans and Cubans have *migrated* to the USA . . . actually, peeps from just about everywhere on the planet have migrated to the USA, thus its characterization as a "nation of immigrants." Wait, *immigrants*? Is that the same as *migrant*?

Ah! That brings us to some basic definitions to know! **Immigration** is the international movement of people into a destination country of which they are not natives or where they do not possess citizenship in order to settle and live there permanently. When people cross national borders during their migration, they are called **immigrants** from the perspective of the country which they enter . . . on the flip side of that, from the perspective of the country which they leave, they are called **emigrants**. Easy way to remember: in the country they are going **I**nto they are an **I**mmigrant. From the country they are **E**xiting, they are an **E**migrant. If a guy named Ibrahim from Iran moves to India, he is an immigrant to India. If a girl named Eddel from Ethiopia moves to Eritrea, she is an emigrant from Ethiopia. lol I could do this all day long! If a guy named Ralph runs away from Russia for fear of being rubbed out, then he is a **refugee**. Wait, what? Hang on Ralph, we will get to you in a minute.

So why would people uproot themselves from the place they were born, and live, and have citizenship in, to migrate to a place foreign to them . . . and in most cases to a place where they do not have a job, a house to live in, or any resources at all to speak of? And they often have to risk life and limb to do it! Well, it's probably pretty obvious to you. Migrants are motivated to leave their countries for a variety of reasons, including a lack of local access to resources, to find a better job (or a job, period), to better their standard of living, to reunite with family that is already in that country, to retire to a more awesome place, to move to a climate/environment that is more awesome, to escape from prejudice (religious, ethnic, or otherwise) and/or persecution, to escape a conflict, to escape a natural disaster, or simply the wish to change one's quality of life . . . presumably for the better.

And that brings us to a few more important distinctions. If people are migrating specifically for the purposes of seeking employment and an improvement in quality of life and access to resources, we refer to them as **economic migrants**. If people are migrating specifically to escape from being persecuted, imprisoned, or killed, we refer to them as a **refugee**. Oh, and if people are migrating to retire to a nicer place, or a nicer climate, or because it's a cheaper cost of living, we refer to them as **filthy rich**. Ok, they don't necessarily have to be filthy, but they are definitely rich.

Back to the point: An economic migrant is distinct from someone who is a refugee fleeing persecution, from a social and legal standpoint. And it is a distinction that is increasingly important in today's world. See, a **refugee** is specifically someone who flees their home country, because if they stay . . . they might be killed! And if they are returned . . . they might be killed! And death sucks! People are given official refugee status because they are in an active war zone (like in Afghanistan and Syria right now); or in a place where the government is possibly conducting genocide against them (like in Burma/Myanmar right now); or in a place where the government will kill/imprison them for having dissident political activity (for decades the USA accepted all

Syrian refugees: Some come by land...

...some come by sea; all yearn to be free.

Cuban people as refugees, assuming they were all fleeing from political persecution); or in a place where the government may persecute or kill them for being the wrong religion/ethnicity/sexuality (e.g. ISIS kills everyone that does not follow their interpretation of Islam; in Uganda being convicted of being a homosexual carries a life in prison sentence). Once you officially gain the status of a refugee, the UN and most all the countries of the world have agreed to help you out and almost all countries around the world take in an allocation of refugees every year. According to the United Nations High Commissioner for Refugees, at the end of 2019, the total number of refugees of all statuses reached 86.5 million . . . that's one out of every 88 people on the planet. Yikes. Depending on when you are reading these words, it is very likely even higher now.

HOWEVER not all countries are equally keen on taking on all refugees, and it has become a hot-button political issue in the USA and many European countries, especially Germany. There is increasing fear (whether it is well founded or not) that accepting refugees from Islamic countries may result in "infiltrating terrorist" types that will take advantage of the world's goodwill to perpetrate evil. There is also an increasing fear that allowing in too many "foreigners" who do not share your country's culture and values system will have a diluting effect on the host country over time . . . will Germany be just as "German" if they let in 10 million Turks or Syrians or Somalians? Of even greater concern to many in the "rich" world is the effects of allowing too many "poor" economic migrants into their countries. "They're taking our jobs!" is the oft heard mantra from those opposed to economic migration, along with the idea (again, whether justified or not) that immigration causes increased crime, social disorder, lower wages for locals, and a loss of local culture as "multiculturalism" prevails. But can you blame peeps for wanting to go to greener pastures to seek out a better life for themselves and their children?

I think not, and quite frankly migration has been the way of the world for far longer than its critics want to recognize. Peeps have forever moved to places with increased opportunities and more resources . . . the need for larger and untapped hunter-gatherer territory was the primary drive for humans to leave Africa in the first place. Phoenicians and Greeks creating colonies across the Mediterranean Sea; Vikings pushing out of Scandinavia to England and Iceland; powerful dynasties in China and Russia pushing territorial expansion and migration of their peoples into those areas to cement their sovereignty; Europeans colonizing the Americas and Africa; and even America's westward expansion across a continent all speak to peoples migrating elsewhere to acquire more space or resources for themselves.

The benefits of grabbing new territory and resources are obvious, but that was back in the day and does not apply so much in the modern world. But benefits of migration still do exist, even for the host countries. Migrants do much of the "dirty work"—that is, they supply the labor for jobs that most locals don't want to do. In the UAE and Qatar, the grunt work oil industry jobs are done by millions of Pakistanis and Indians who have migrated there. In fact, there are more economic migrants in Qatar than there are Qatari citizens! In Europe, an increasing amount of janitorial work, factory work, and restaurant jobs are filled by African migrants. The US agricultural sector relies so heavily on Mexican immigrant labor that it would effectively be shut down if migrants all left. On the flip side, migrants often send money back home, a term called **remittances**, which helps distribute wealth and development globally in a much more tangible way than big international projects or loans.

To wrap this up, you should also be aware that migration has really only become a global "issue" in modern times, the result of several colliding factors: 1) The ideas of the nation-state and state sovereignty (explained in full in Chapter 3), which have only been around for a few hundred years but are now the international standard. They are the reason we now have hardened state borders, and the concept of legal citizenship . . . and if you ain't from here, you ain't a citizen and you can't come across our borders! 2) We now have over 7.7 billion people on the planet as you have learned in this very chapter, and that is more peeps than ever before that want and need more space and resources . . . but those things are finite and have mostly been claimed/owned! 3) Wealth disparity has increased dramatically, globally. That means there are fewer and fewer peeps that have a huge share of the world's wealth, while millions—nay, billions—of folks are living in poverty, or something not far from it. This concentration of wealth in different regions is a primary reason people move . . . to go from a poor area to one of the rich areas. And finally 4) with more borders than ever before, and more people than ever before, and more wealth disparity than ever before, we are now having more conflict than ever before . . . which of course is the other primary reason people migrate: to get the hell out of a war zone!

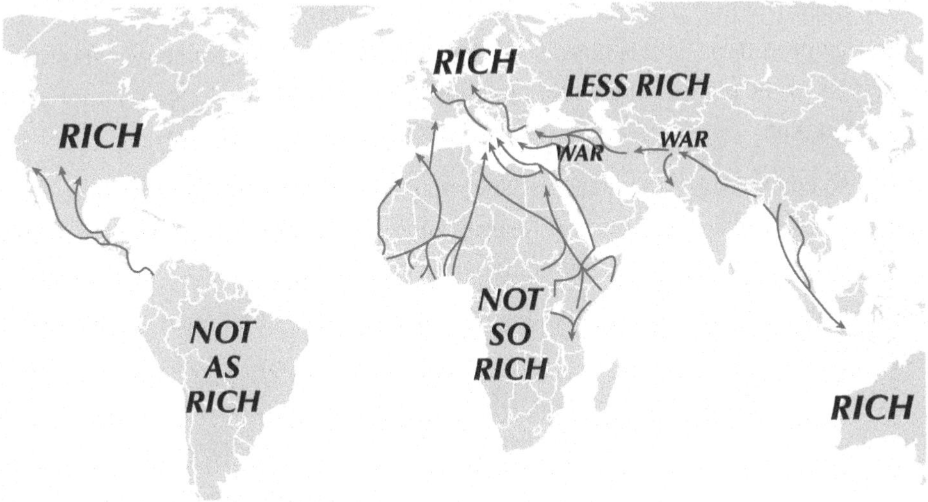

Major Modern Migration Arteries

From poor places to rich places, from war zones to peace zones, peoples remain on the move . . . much to the chagrin of many, and for better *and* for worse. Since this is a book on world regions, let's finish this journey with a few really big generalizations about modern migration. The obvious economic migration routes: People move from poorer states to richer states; lots of folks from Subsaharan Africa and the Middle East have been migrating to Europe; lots of folks from Central America and Mexico have been on the move to the US and Canada; lots of Asians have tried to get to Australia. As a side note, people all over the world have been migrating from rural areas to urban areas.

On the war front, ever since the Arab Spring uprisings across the Middle East in 2011, multiple events have led to mass migrations of folks to Turkey and all parts of Europe. Specifically, Libya devolved into a low grade civil war and Syria devolved into a gigantic civil war that has hastened the exodus of hundreds of thousands. The War in Iraq and the rise of ISIS have also contributed to the Middle East mayhem that fuels emigration from the region. The decade-old conflict in Afghanistan has pushed a plethora of peeps into Pakistan and surrounding countries. And the decades-long state of chaos in Somalia has kept it as one of the hot spots to escape from, with most of those refugees heading next door to Ethiopia. Here in 2021, Turkey is currently the host to the greatest number of refugees in the world, followed by Pakistan, Lebanon, Jordan, and Ethiopia. Those are just the highlights (or low points) of the tens of millions of peeps on the move, because it would take a whole book just to mention all the current conflicts that compel, and economic incentives that entice, the migration marathon that our species has been running. And we are fast running out of book space for this chapter already, so . . .

SOME FINAL THOUGHTS ON PEOPLE . . .

They disgust me! No, I'm joking. I love all the plaid peoples of the planet. What can we say about the future of our plaid population? It shouldn't be too hard to figure out some general trends. Most countries will continue down the development path, and thus we should be able to track what's happening in societies and within population dynamics through the lens of our Demographic Transition model. Everybody in the world wants a good job, a good standard of living, access to health care, and a better life for their kids. It's just the human way.

Having said that, there is nothing to lead me to believe that poorer states won't continue to try and modernize and industrialize like the rich ones have. Is this possible? That's a whole other ballgame we won't get into yet. But what will this mean for the peoples of the planet? The world population total will continue to go up for some time to come, but we can be more specific than that:

- The least-developed countries are the places where population will truly grow the fastest, and the most. Regions like sub-Saharan Africa, the Middle East, Central America, and parts of Asia will continue to pile on the humans, even though they are the regions perhaps most ill-equipped to handle more.
- Regions further down the development path, like India, Mexico, and South America, will continue to grow, albeit at a slower pace as fertility rates stabilize.
- Fully developed regions, like Western Europe, US/Canada, China, and Australia are stabilized populations with no growth, but with the potential to grow bigger mostly through allowing immigration from other regions.
- Regions that are currently shrinking in populations, like Russia, Eastern Europe, some parts of Western Europe, and Japan, will almost certainly not disappear—but most certainly will encourage folks to have more kids and allow more immigration as well. A disturbing thought: By 2020, more adult diapers will be sold in Japan than baby diapers.
- However, we should never underestimate the power of culture to undo all our tidy demographic transition assessments! Some states in the Middle East and South America are maintaining their high birth rates even despite increased development and climbing costs of living. This is due primarily to powerful religious influences (Islam and Catholicism) which prize large family size. Also, there are folks bucking the trend in places like China, where the state-sponsored One-Child Policy predominated for decades, because the wealthy and growing middle class can now simply afford to have more kids, and are doing so. Quite the opposite of anything the model would have predicted. These wacky humans can be so impulsive when it comes to procreation!
- **2021 UPDATE!!!** In 2016, China relaxed its controversial "One-Child Policy"—one of the world's strictest family planning regulations—to a "two-child policy" due to widespread concerns over an aging workforce and economic stagnation. It didn't move the needle at all. So in 2021, China made it a "three-child policy" in the hopes of encouraging those that want even more babies to go for it.

This is the current state of the plaid world. This is the future state of the plaid world. How many kids do you want? How many can the plaid world support? All of the factors explained in this chapter combine to create the world population of today. Check out the regional totals in Figure 2.14 to see how these things are reflected in today's world, and know these figures well. . . .

FIGURE 2.14 WORLD REGIONAL MAP WITH POPULATION TOTALS

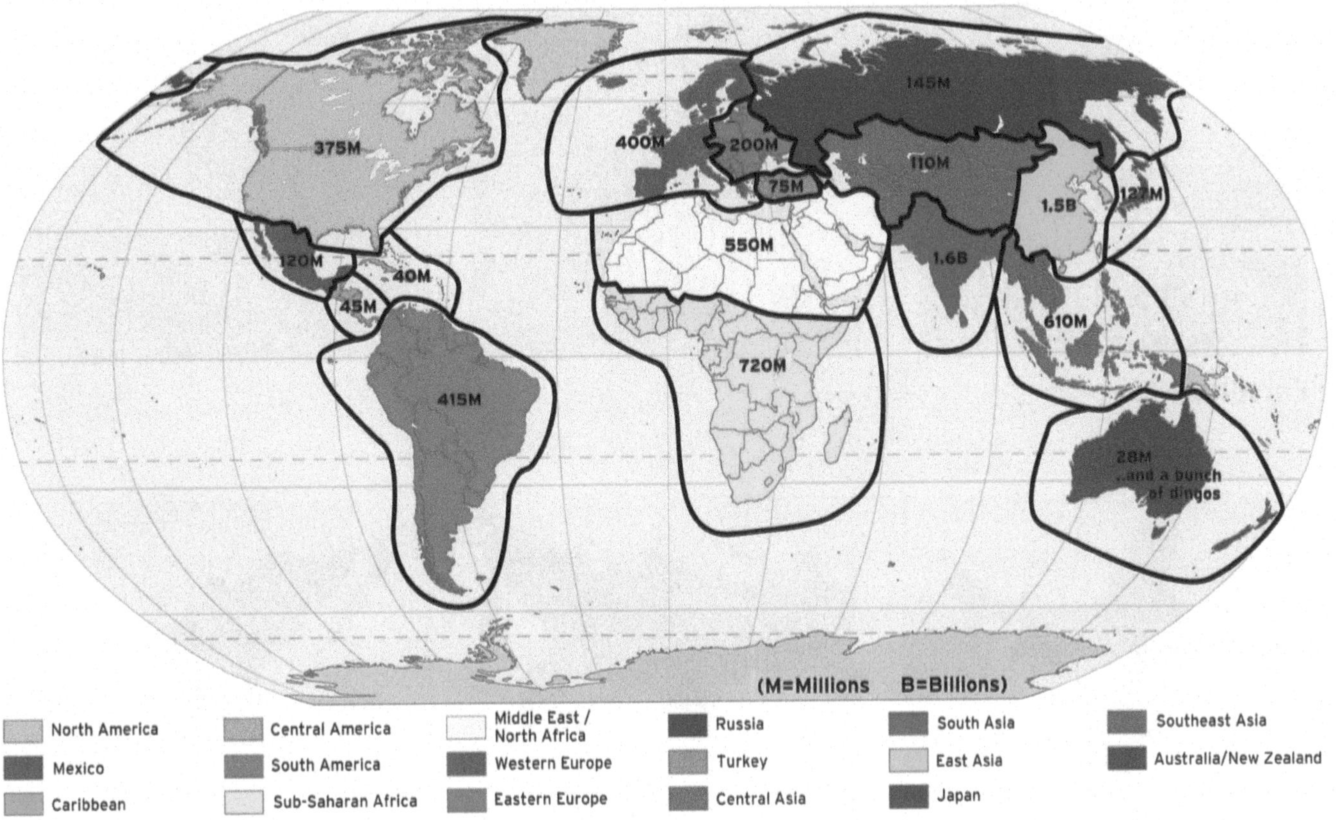

Now let's see what you've learned about population dynamics, population pyramids, and fertility rates. Help the Plaid Avenger find his way home with his family. Match the Plaid Avenger fertility rate to the appropriate population pyramid and then to the country to which they belong. Be sure to notice the partial children!

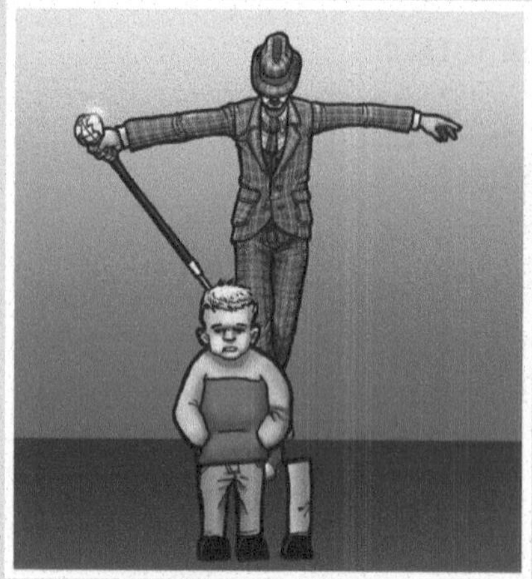

Fertility Rate

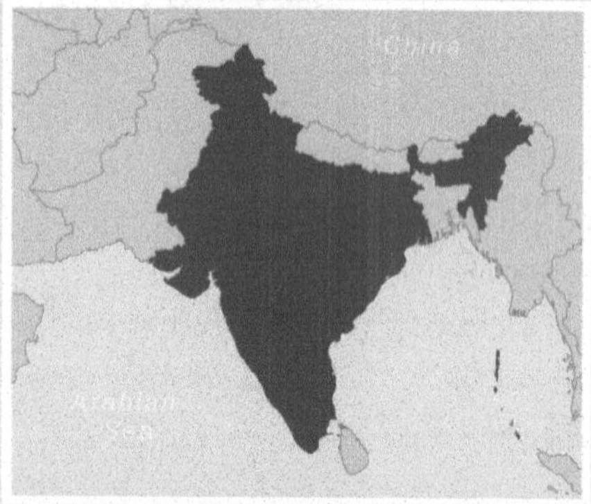

Country of Origin

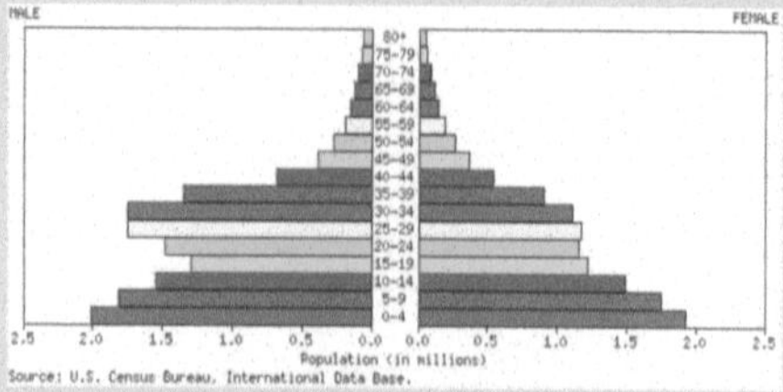

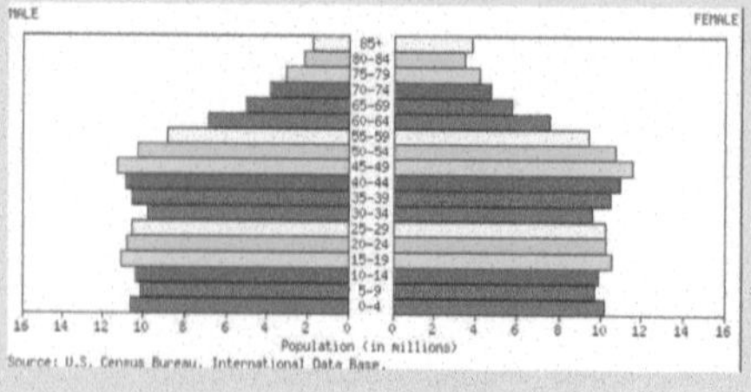

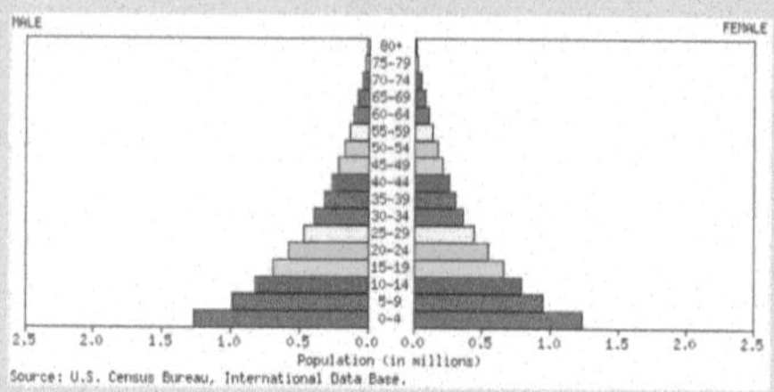

Population Pyramid

Source: U.S Census Bureau, International Data Base

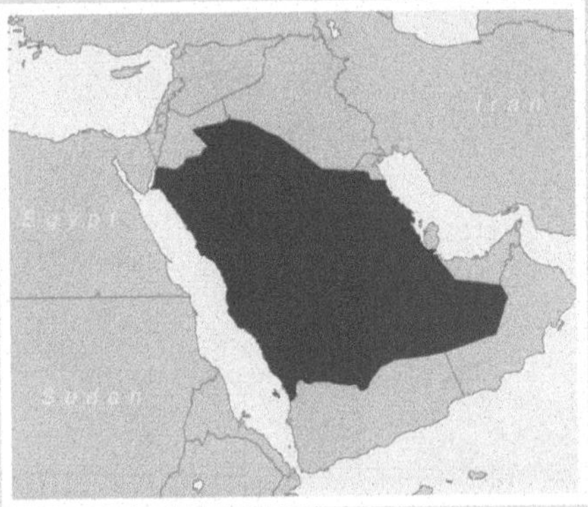

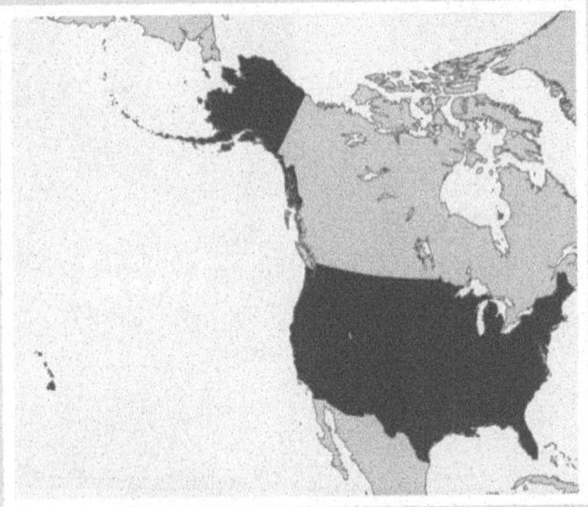

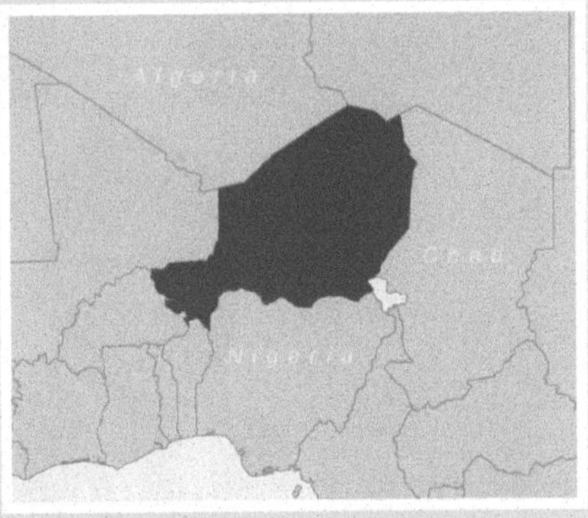

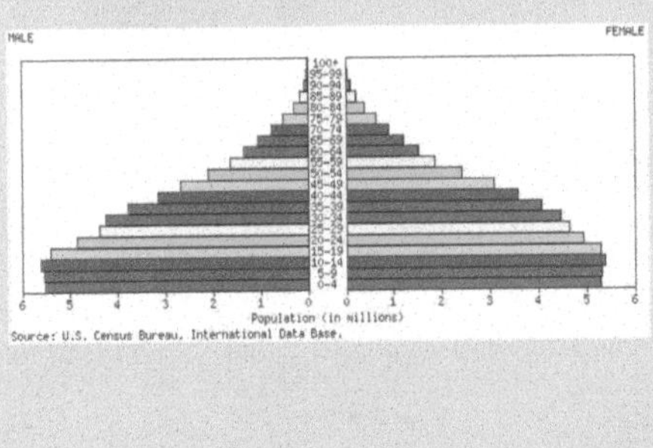

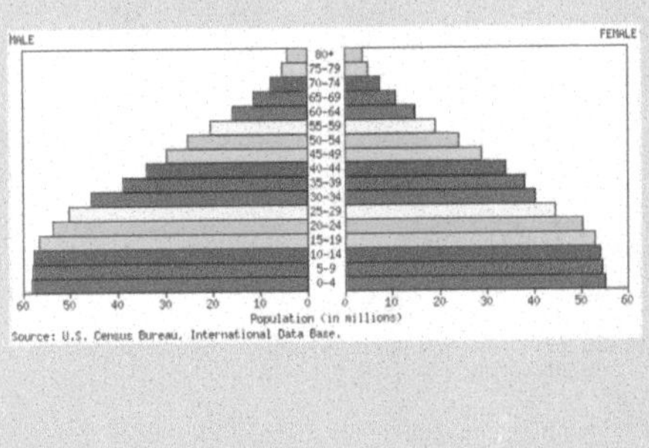

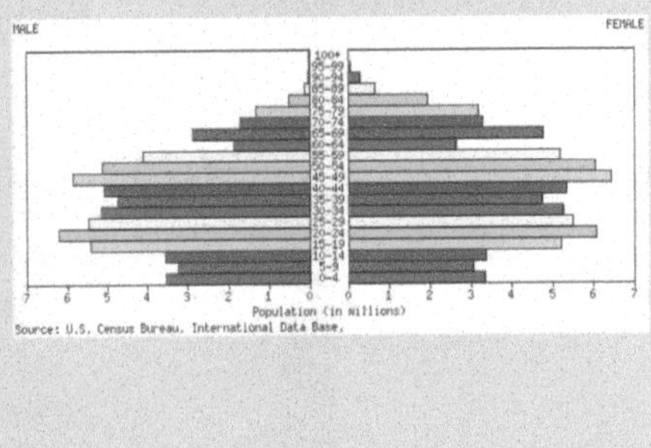

Source: U.S Census Bureau, International Data Base

CHAPTER OUTLINE

How many of these planetary principles can you put a name to?

The State of States

3

WE'VE talked about peoples in the world, how fast those peoples are pro-creating, and how those peoples' societies evolve—but, where are they doing all this? Mostly they are doing it in some place we call a **state**. This textbook is all about the regions of the world, and indeed we're going to tackle 18 or 19 of those regions here in a little bit. That's first and foremost how the Plaid Avenger wants you to understand what's going on around the planet. But regions are user-defined, many edges around regions can be fuzzy, and not all of us would agree on how to form them nor draw those regional lines. On the other hand, everybody is located in a legally and physically defined entity called a **state**. And the borders of a state are not up for debate. What the heck is a state anyway? What's that all about? Where did it come from? Why is it important for understanding the world? Let's state some state facts here, and understand the state of states . . .

IDEA OF A STATE

The idea of a defined territory over which a political entity would rule is a fairly new concept. If you think back, for most of human history there have been kings, or rulers, or leaders of *peoples*, not of geographically defined political *spaces*. Think back to, let's say, the French. I know most of us try not to think about the French, but bear with me for a moment. Throughout most of history, the king or leader has been referred to as the *King of the Francs*, not the *King of France*. Only around the 16th and 17th centuries did the innovation of legally defining a space on Earth evolve—the idea that we're going to define a place on the planet in which those Francs cohabitate and call it "France." In the 18th century, the concept spread to the European colonies and ex-colonies, and then on to the rest of the planet.

FIGURE 3.1 SOVEREIGN STATES BY AGE

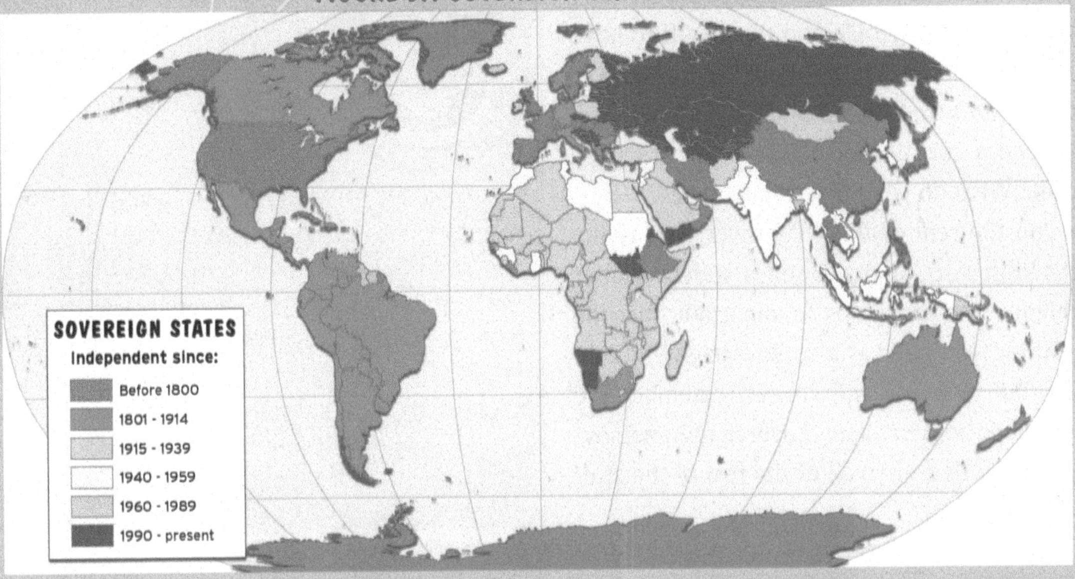

There are some truly old states in the world. China and Persia (Iran), in particular, are ethnically distinct nations of folks who marked the borders of their empires for a good long time. While we can recognize this fact, the idea that these borders are recognized by all other states is the truly European addition to the idea; it is also why most European states are counted among the world's "oldest," even though they may be relatively new in the bigger picture of human history. One has to look no further than the United States to see how new this concept is; the US is in the "oldest" category of states despite the fact it is truly one of the youngest countries on the planet! It has only been around for 200 years total, but is counted one of the oldest continuous sovereign states!

A 19th century wave of "new" states cropped up as European colonies started declaring independence: like in the countries of Latin America, South Africa, and Australia. Some others popped up as a result of disintegrating empires in Europe: e.g. when the Austro-Hungarian Empire dissolved in 1918, Austria, Hungary, Czechoslovakia, and Yugoslavia were born. Another wave cropped up as the Ottoman Empire, which controlled a large chunk of the Middle East, dissolved in the early 20th century, which later "birthed" the modern states of Turkey, Saudi Arabia, Iraq, and many others. Sovereignty didn't really get to places like Africa and India and Asia, large chunks of which were European colonial holdings, until they too became independent countries—that is, states—in the second half of the 20th century. If we look around the planet, most states in the world are fairly new.

A state, in its basic definition, is a politically and legally and territorially defined geographic area that is recognized by other sovereign states across the planet. Most sovereign states on the planet are fairly new, having been reshaped or declared independent in the last fifty years or so. Maybe we should start defining these terms that the Plaid Avenger is throwing out here. What is a state? What is a nation? What is a nation-state? Before we even get to those, we have to tackle the critical issue of **sovereignty**, which is one of . . .

THE ESSENTIAL INGREDIENTS

We hear these terms all the time, but what do they mean? Well, we kind of already hinted at it. A state has something to do with some defined boundaries. That's easy. A **state** is a legally defined and recognized political area on the planet. Legally defined lines are drawn (like a deed to a property), and all the neighbors agree to its parameters—at least for the most part. A few of those lines are still disputed.

But it's not just about a legal description, is it? When we think about the term **"the state,"** and if you ask people what was exactly needed to have a state, you would get a wide variety of answers. Many would say you need a government to head up the state, or a military to protect the state, or a fence around the state. Others would say you need an economy, or money, or an official currency, or jobs to have a state. Some might point out that you obviously need some area and some people and some stuff to have a state. Maybe you need roads and buildings and sewer systems and hospitals and McDonalds to have a state. Perhaps you need a constitution, or laws, or courts to have a real state. What do you think a state needs?

I'll give you the real-deal brief answer: you really only need two things to have yourself a state. Number One is sovereignty. The concept is so important, that the Number Two thing you need is also sovereignty. Yeah, I'm a goofball, but now you will always remember the two things you need to be a state: Sovereignty and sovereignty. If you have that, then all of the rest of the stuff doesn't even matter.

> WHAT THE H? WHAT THE WHAT? WHAT THE WHAT? WHAT THE WHAT? WTH?
>
> **What state am I in, when I'm in the United States?**
>
> By the way, the term ***state*** is what we use synonymously with *country* here in the United States of America. It gets confusing. Here in the US, we say we're in the state of California or New York, but we're not really. We're in the state of the USA. We say *countries* because it makes it easier for us. Country = state, state = country; just use the term **state** when you're dealing with international politics. Everybody understands it better. What state of mind are you in right now as you are contemplating which state of the United States you are in, which of course is a state in its own right, right now. Right?

ALL YOU NEED IS LOVE . . . NO, I MEANT SOVEREIGNTY

So what is this **sovereignty** stuff all about? Well, the concrete definition of sovereignty is really up for grabs nowadays. For the last few hundred years, it has been a concept that everybody has agreed on and understood. However, in the last couple of decades, there's been some serious debate about what it truly means. Times have changed, and it appears the idea of sovereignty is also changing. This redefinition is causing considerable consternation in the world, among big states, small states, and all states in-between—but I'm getting ahead of myself.

First, we want to find out what it *truly* means for an entity to have sovereignty, and thus be called a state. The Plaid Avenger's interpretation is this: when an entity (a government, king, dictator, etc.) has a defined territory on which it exercises TOTAL internal and external control . . . then they gots the sovereignty. No other state can have any power over your territory or the people in it. How can a government demonstrate that it holds real ultimate power over its territory and the peoples in that territory? Kill them. Straight up. If the state can legally get away with that, then it has sovereignty, and it is a sovereign state in the world.

Wait . . .What?

Let me say that again.

In the end game, does the political entity in charge have the ultimate power to do the ultimate deed within its borders? And what is the ultimate deed? The ultimate, beat all, true test of power is the taking of life from its citizens. When you think about it, everything else is quite trivial, isn't it? The right to tax them, to enact laws over them, to jail them, to make inferior products at crappy jobs, or sing sappy national anthems . . . that's all pretty tame stuff compared to death. If you can legally kill people within your state—that is, the leaders and government can kill their own citizens legally and nobody else in the world can do anything about it—then you're in something we call a state that has sovereignty.

People want to say that sovereignty means a system, a government, some leadership, some people who vote, some laws . . . well, you'll have all of that too. But if you want the real test of "Is this place a state?" then find out if they can kill their own citizens. Now that sounds extreme, and . . . it really is! Are we a sovereign state in the US? Can the US pass the ultimate sovereignty test? Absolutely. They kill people all the time; it's called capital punishment. The death sentence.

So the government kills people. They put criminals to death. But that's too easy. Let's take it a step further: How about innocent people? What would happen if the government of this country, or any country, just decided to kill 10,000 of its own citizens? Let's say the FBI or the CIA bungled some information—like that's never happened before—and decided the entire state of New Jersey had to be eliminated by presidential directive, and then nuked it. Millions of lives taken by the state. (On the plus side: no more 'Jersey Shore.') Now, again, that's extreme. Nothing like that's going to happen . . . yet. But let's say it happened. What would France's response be? What would the United Kingdom do? What would China do? What would Angola do?

Save for giving the offender a great big collective hairy eyeball, the answer is . . . they would do nothing *(especially if it were France)*. Even though it's an extreme event, and it may be of horrific consequence, the US is a sovereign state. They can legally kill their own citizens, as can all sovereign states. That is an agreed upon principle, and again, it's the extreme. If you can kill someone in your state, then you can pretty much do anything else: tax them, tell them what to do, make some laws, all that stuff. If you can do the extreme, you can do all the lesser stuff, and that's the first big principle rule of sovereignty. The *Number One Rule of Sovereignty* is: Does the state have the ultimate right to do the ultimate deed to its citizens, without the interference of other states? If the answer is "yes," then you're in a sovereign state.

How do you get that status? What if I, The Plaid Avenger, created a Little Plaid Nation over here, and I said, "Well I'm giving myself the authority to kill all people who enter the royal domains of my backyard, which I have christened Plaidtopia." Obviously people would say, "No, you can't do that, man, you are crazy!" Even if Plaidtopia had a constitution, some laws, an economy, some area, a fence, a military, a police force, and a harem of women bodyguards who were trained to kill interlopers on my behalf, it still wouldn't be a sovereign state. The reason that *I* could not do it is because other political entities, namely the other sovereign states, wouldn't agree with it. That's the *Number Two Rule of Sovereignty*.

The *Number Two Rule of Sovereignty* is that all the other states in the sovereign state club have to recognize your sovereignty. You are in the "country club," so to speak, but without the golf benefits. Indeed, the United Nations is kind of the country club, and if all the other sovereign states say, "Yes, you kill your own citizens, and we respect your right to kill your own citizens," then you possess sovereignty. If you are in a legally defined political territory, recognized by other sovereign states, and no other state has power over your territory—as ultimately tested by killing your own citizens—you are a sovereign state. Everything else is inconsequential.

BUT SOVEREIGNTY MAY BE REDEFINED

Why would I suggest that this concept is increasingly up for grabs? One of the reasons that today's world is so complicated is because the nature of the world has changed. For centuries everyone has said, "Yeah, you can kill your own citizens. As long as you're not attacking another sovereign state, no one's got a problem with it." That's been the agreed upon, core, central component to sovereignty. Part of the base principle here is **reciprocity**: we will not intervene in your state affairs when you mess with your own citizens . . . and in return, we expect you to respect our ability to mess with our citizens as well. That's the way it all goes. Even today we say, "Hey, the US doesn't like what China's doing to its citizens; we think their human rights are atrocious." But the US is not going to invade. No one's going to do anything about it, really. The US or the UN might protest and say, "Hey! China! Or Russia! You suck for the way you treat people in your state but . . . here . . . be the host of the Olympics anyway." At most, an official condemnation issued from a head of state or perhaps even a trade embargo would be the extent of any disapproval. No active military intervention is going to happen, because that's what it is to be a sovereign state. We respect their sovereignty so that they will respect ours. At least, that is the way it was . . .

Former President Bill Clinton: sovereignty side-stepper.

But it is morphing fast in the 21st century . . . mostly due to the international shenanigans of the US and NATO! How so?

Well, back in March of 1999, during the Clinton administration, the United States led NATO on a bombing campaign of Yugoslavia. Why did the United States and NATO start bombing Yugoslavia, a sovereign state?

Because Yugoslavia invaded the US? No.

Threatened US citizens? Nah.

Because Yugoslavia invaded another fellow NATO country? Nope. Threatened or invaded any other sovereign state? Ummm . . . no.

Because Clinton wanted to defer attention from his White House sex scandal? Mmmm . . . good one, but no.

Does that attack seem important to you? Do you remember hearing anything about it? I doubt it. Let me be the first to educate you on how big a deal it was to sovereign states. Here's an excerpt from the British *Financial Times:* "The enormity of NATO launching its first attack against a sovereign state is not to be underestimated. Unlike Iraq, Belgrade has not invaded another country. Nor is the situation akin to Bosnia, where the legitimate government invited outside intervention. Nor, finally, has the United Nations Security Council specifically authorized NATO to bomb."

Slobodan sez, "I can kill them if I want . . ."

The thing that happened was that this crazy guy, **Slobodan Milosevic**, started killing his own citizens. The who or why or how is not important right this second. What is important is that "killing your own citizens" was one of the baseline agreements of sovereignty. States respect the authority of every other sovereign state to kill its own citizens. You may remember that from seven paragraphs ago. It's the cornerstone. In this circumstance, it was possibly an ethnic cleansing situation, where Slobodan—then leader of Yugoslavia, currently worm food pushing up daisies—may have been persecuting and encouraging his national army to outright kill members of a specific ethnic group. In hindsight, there is not much debate about it. Slobodan was actively pursuing genocide/ethnic cleansing . . . but only against his own citizens. But he was the leader of that sovereign state. Should be no problem, but . . .

Apparently, it got to be too much to bear. There was a large outcry about the situation in Europe and even in the United States. Perhaps it was the memory of the Jewish Holocaust that occurred in Europe, or perhaps the instability in this very region that had previously launched World War I, or even as some have pointed out, it was simply because there were white people being slaughtered, and other white people didn't like that idea. (And I include that statement to point out the horrific irony that the US/NATO/UN has stood on the sidelines plenty of times while genocides occurred in Africa and Asia.) For whatever reasons, President Bill Clinton decided to enact a bombing campaign, coordinated with NATO, to stop the genocide. Long story short—the United States bombed Yugoslavia into submission for what it was doing to its own citizens.

Why am I telling this story? Because this changed the whole idea of a sovereign state! As we suggested, this whole "kill your citizens" thing used to be the standard. What a state did to its own people was not up for debate, but apparently this was the straw that broke the camel's back—and it's not an isolated incident. Since the 1999 bombing campaign on Serbia, the US and/or NATO have played an active role in Afghanistan, Bosnia, and Iraq under the umbrella of liberating the locals. Meanwhile, serious genocides have occurred in Sudan, Burma, and the Congo, for which many in the international community are calling for invasive action to remedy. Those are on the heels of the now infamous Rwandan genocide of 1994, which no state, international organization, or otherwise, stepped in to stop.

This trend has continued and intensified here into the 21st century: in 2014 the UN-sponsored a US/UK/France/NATO-enacted invasion of Libya! This is a super-gigantic big deal when it comes to the definition of sovereignty, and my friends, I believe it is the last nail in the coffin of the concept in the modern era. Why do I make such bold assertions? Well, why was Libya bombarded by NATO? Because a political protest rose up internally against the government. That's it! There is nothing else to claim here! No genocide, no invasion of another country, no terrorism . . . not even a threat of any of those things! While I have no love of former Libyan dictator Muammar Qaddafi, he had done absolutely nothing to break the old rules of soveriegnty . . . which is why I am suggesting that those old rules don't mean diddly-squat anymore.

When Libyans took to the streets to protest against their government, Muammar made the fatal mistake of announcing out loud he would crush the uprising, and apparently even this audible "threat" to civilians has now become the new standard to incite international intervention into a sovereign state! How bizarre! For the first time (maybe ever) you had an entire collective global movement to affect an internal political situation in a state. New stuff. Interesting stuff. But that is a tricky can of worms that has been opened . . . aren't there other states with equally crappy, repressive leaders that harm their citizens? Ummm . . . North Korea? Sudan? Zimbabwe? Will they be next on the list of countries to be invaded on the premise of pathetic leadership? Time will tell. Since 2011, that international debate is currently hottest regarding what to do about the situation in Syria; will 'Team West' pull another intervention, or will classic sovereignty be upheld with the help of Russia and China? Still too close to call as of this writing . . . and the Syrian Civil War rages on. . . .

All of the listed above are actions, or inactions, based upon unacceptable sovereign state behavior in dealing with their own citizens. Again, it used to be the hallmark of what defined a sovereign state; now it's totally up for grabs. How much misbehavior will the international community allow before intervening even when that misbehavior is confined inside a single state?

This is causing a lot of confusion on the planet. If the golden rule for what defines a sovereign state is up for question, then what exactly is a sovereign state? Sorry, my friends, the Plaid Avenger doesn't have an answer for you right now, because nobody does. The most recent unraveling of the concept came when Russia annexed (aka "took") the Crimean Peninsula from sovereign state Ukraine in 2014 under the auspices of "the Crimeans are mostly ethically Russian peoples, and therefore it was really our property anyway." Wow. That is bringing back in a super-old-school idea that sovereignty is about leading a group of people, not just a geographic area. Leave it to Vladimir Putin to take us back in

Want to hear more about the most recent sovereignty shizzle? Watch this:

3.3.1 2016 Sovereignty Shizzle Updates (28:26) at https://vimeo.com/180646003

time. We may be fast returning to a "might makes right" world in which the only true sovereign states are those that can defend their territory physically from all other sovereign states.

With the current movement of UN troops around the planet, and unilateral actions by the United States and/or NATO and/or Russia in today's world, we don't really know. You can see this in today's news as a major point of contention particularly with the Chinese, and a lot of times with the Russians. Why those two? Just because they hate the United States? No, it's got nothing to do with that at all. These two countries, particularly China, are really big into the classic sovereignty thing because they want to protect their own self-interest and international image. Remember, forever it's just been: 'don't invade other countries, respect everyone else's sovereignty, and you can get away with whatever you want in your own country.' Well, maybe. The Chinese and the Russians think, "We are not really keen with the US involvement in other countries because it is violating the sacred rights of sovereignty to which we've all agreed." And the Chinese in particular are very anxious to avoid any outsiders from intervening into what they consider their own sovereign territories of Tibet and Taiwan . . . more on that later. . . .

Sparked by the US/NATO bombing campaign in Yugoslavia (then Afghanistan, then Iraq, then Libya), sovereignty has also been called into question by human rights groups and lots of folks around the planet who say, "Okay, you're right. We're in the 21st century, we cannot allow acts of genocide to occur, even if it is in a sovereign state." While Yugoslavia was the test run for this, in terms of a trial flight on intervening into a sovereign state to protect citizens, it will probably be used more. Indeed, the former Bush administration said, "One of the reasons we went to Iraq was to protect the Kurds and the Shi'ites there from what their own government was doing to them." There are folks around the world who are calling for more action from the UN and the world to stop genocide in other places. Right now, people are saying, "Hey! What's the deal? You guys bomb Serbia to protect *those* guys, and you're currently in Iraq protecting *those* people from their government, so how come you didn't interject in Rwanda? Why aren't you in Sudan or Burma or Palestine right now?" This is a big debate on the world forum right now. It's another reason why, you'll see, that people hesitate big time in the UN to use the word "**genocide**." Why is this? Read the inset box below to understand more fully . . .

What's the Deal with Genocide?

Genocide is when a government, or a group of people, kills or tries to completely exterminate another specific group of people, usually based on ethnicity or religion. The worst example of this in modern history occurred during World War II, when the Germans tried to exterminate all of the Jews out of German territory, actually out of all of Europe. This was such a horrific act that, after World War II, the world agreed that we would never allow this to happen again. All the states in the world passed a United Nations law making genocide an international crime. The ***Convention on the Prevention and Punishment of the Crime of Genocide*** says this: "Any of the following acts committed with intent to destroy, in whole or in part, a national, ethnic, racial or religious group, as such: Killing members of the group; Causing serious bodily or mental harm to members of the group; Deliberately inflicting on the group conditions of life calculated to bring about its physical destruction in whole or in part; Imposing measures intended to prevent births within the group; and forcibly transferring children of the group to another group."

But the old rules of sovereignty were still in play. Things floated along pretty well, until the appearance of possible genocide in Serbia/Yugoslavia, in Rwanda, and in Sudan. Now wait a minute. Didn't we all agree we were not going to let this happen again? Shouldn't we do something?

That is one of the calls on the world stage right now. The reason you did not hear the Rwandan situation referred to as genocide is because the use of that particular word demands a call for action. In other words, as long as everybody in the UN is just saying, "Yeah, well, they've got a civil war," or "They have some sort of internal conflict," or perhaps "They may have isolated acts of ethnic cleansing," (which was said about Rwanda) then there is no need to act. Sovereignty takes precedence. They say these things very intentionally because everyone is scared to whisper, "It's genocide." If everyone in the UN agrees that genocide is occurring, then they have to act. It's their law. "World law" if there is such a thing . . . and that trumps sovereignty.

That's the deal with genocide and why people gingerly dance around the term, particularly at the UN. You also now know how genocide is calling into question the whole sovereignty theme and how the modern definition of sovereignty is being called into question in places where citizens are being persecuted politically even when it is obviously not a genocide. But let's move on to less lethal topics. . . .

NATION, STATE, OR NATION-STATE?

Now we've looked at what constitutes a state. Perhaps now you know more about what constitutes a state than you ever wanted to know. Yeah, me too. But it's my job. However, what about these other terms? What is a *nation?* Is it different from a *state?* If so, then what in the wide wide world of sports is a *nation-state?* It's easy. Let's break it down.

A **state,** you now know, is a legally defined political unit that possesses **sovereignty.** Fair enough. I've beaten you over the head with that for several pages now.

A **nation,** straight up, has nothing to do with a state. You think it does, but it doesn't. A nation is about the people. Remember the intro paragraph of this chapter? It used to be the standard unit of global identity: a leader led a specific group of people. The Romans, or the Huns, or the Aztecs, or the Zulus. It's all about the peoples. A nation has much more to do with the people than any specific area. In fact, some peoples don't even have an area—but I get ahead of myself. A nation refers to a group of people who share a common culture, and who may even want to have their own government, and may want to rule themselves. Common culture could be just about anything, but often it is a combination of a lot of things: a common ethnicity, religion, historical background, diet, shared beliefs and customs, traditional attire, sports and games . . . the list goes on.

The number of nations in the world is undefined. There could be thousands, nay, millions. It depends on what group of people and on what scale you're looking. Perhaps in every town there's a small nation of people who think that they're different; that they share a common culture, and perhaps would like to form their own country. Again, it's an undeterminable number, but it is all about the people. Let's think about some true examples on the planet that we can all agree are distinct groups of people who share a common culture. This is actually quite easy.

Distinct nations are all over the place . . . but are they in nation-states?

What common cultures are there around the world that are tied to people? How about the French? How about the Germans? Wow, this is easy; let's stay in Europe. How about the Italians? Think of all these guys. They all have distinct common cultures from each other; just think about cuisine or their languages: Germans eat sauerkraut, while the French dine on snails. BTW: ew. The Czechs, the Poles, the Russians are all distinct cultures and thus distinct peoples, and thus distinct nations. But it can get even more specific. We could look at the UK as a UK culture. Maybe, but there's definitely a Scottish culture within the UK, and an English culture, too, but now we're getting stuck in the Old World. Is there a Saudi nation? Yes, I believe so. Is there an Argentinean one? That one gets a little fuzzier, but it's possible. Japanese? Yes, definitely. The Japanese are a nation, a distinct group of people and distinct common culture. The Chinese? Yes. The Vietnamese? Yep. Thai, Laotian, Kurds, the Turks, the Cherokee in the US, the Maasai tribesmen in Kenya? Sure. We could go all over the planet and see pockets of common culture, and an idea of a group of people who share something in common who, may or may not, want to rule themselves, under their own government. That's a nation.

Let's get to the final definition then: what is a nation-state? Simple. Let's just add a **nation** to a **state**, and presto, you got a nation-state. A **nation-state** is a group of people who share a common culture, who want to have their own government and want to rule themselves—a nation—*and do so* in a defined and recognized political area—a state—which has sovereignty to boot. That is a nation-state. Let's give an example of a nation-state. We look around the world, and see a lot of the ones we've already talked about. Germany: perfect nation-state. This group of people with a common culture in a distinct area that we call Germany. The French: same deal. The Japanese: even better. The Koreans are the best example yet; a 100 percent ethnically homogeneous group of folks with a distinct Korean culture and cuisine. That's it: that's a nation-state.

FIGURE 3.2 TYPICAL NATIONS WITH STATES

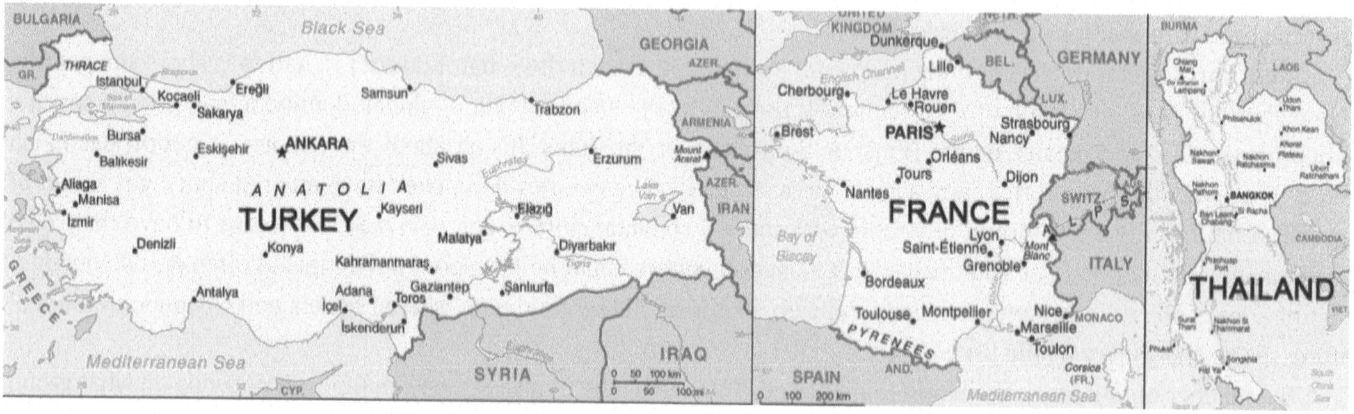

Turks in Turkey, Thais in Thailand. Yep, that's a nation-state!

The list can go on and on, all over the world. Most of the states of the world are close to what we call a nation-state. Of greater significance than listing all the nation-states: let's look at those places in the world that are not nation-states.

Which brings up the question: can we have a **nation without a state**? Indeed, we certainly can, and this is one of the causes for conflict on the planet today. There are lots of groups of people with a shared common culture, and a desire to rule themselves in a legally defined territory, but they ain't got their own state. A prime example is the Palestinians in a place called Palestine, which is not yet a state—it might be soon—but who are still largely under the control of Israel.

Let's get more complicated. Places like Chechnya—yeah, that's a good one! A very distinct culture in a place that's called Chechnya, but it's not sovereign. They'd like to be sovereign; they would love to not be part of Russia, because they are not ethnically or culturally Russian. Or how about the Kurds: a nation of people scattered across 4 different states. They've been petitioning to become their own sovereign nation-state for decades, but no one wants to give up territory to let that happen.

FIGURE 3.3: SOME EXAMPLES OF NATIONS WITHOUT STATES

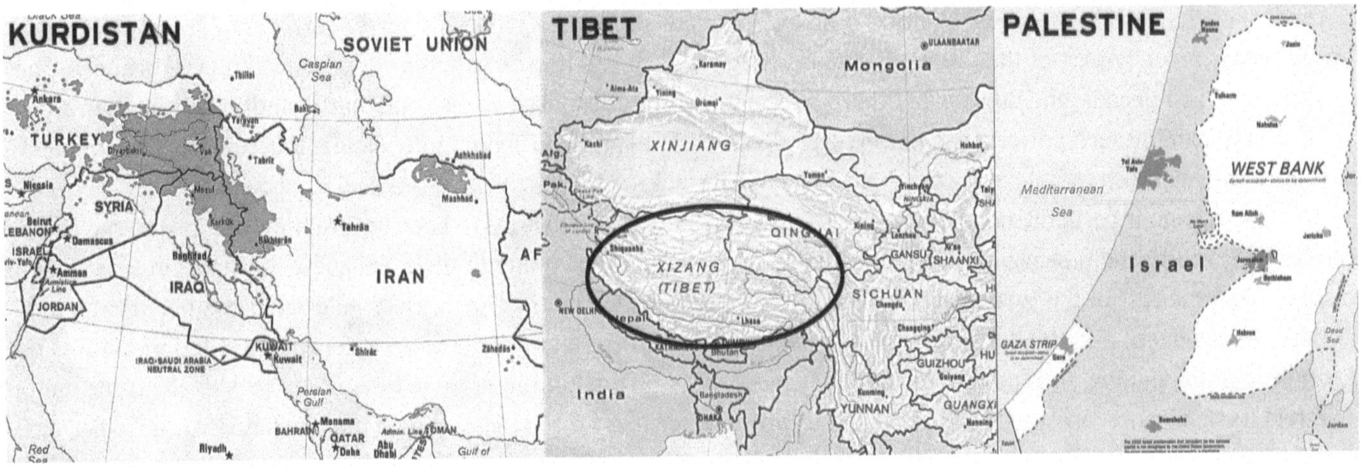

Nations sans statehood

What about Tibet? You've heard of them. The leader of the Tibetans, the Dalai Lama, is the spiritual leader of the Tibetan Buddhist on the planet. Their spiritual and traditional homeland has been Tibet. Most of the people there share this Tibetan and Buddhist culture, but the area called Tibet is now part of China. Thus, they are a nation, but not a nation-state. It's one that's likely to *never* become a nation-state. That's a little Plaid insider tip for you. Sorry, Lama. China will not let Tibet go.

Lama has no state.

These are all very good examples of nations without states. Some of these are areas quite active in the world in the terms of conflict, with those people trying to become stand-alone nation-states, and full-fledged sovereign states at that.

SO, HOW MANY STATES ARE THERE?

There are an unidentifiable number of nations. Groups of people that are self-defined, and at multiple scales, are all over the place. Every country will have small pockets of folks that consider themselves culturally different. Like the Basque in Spain, the Scots in the UK, or the French in Canada—maybe we should just say the French everywhere. It's impossible to determine the number of nations. Now nation-states, that's a little bit easier, but even that definition gets a little muddled if you think about it too much. Is the US a nation-state? What's our common culture? Besides a shared loathing of Shia LaBeouf, do we share any unequivocal commonality?

However, the number of states we can identify with some confidence. Exactly how many states are there here in 2019? The magic number we're using right now is 194. We're talking about sovereign states in the "country" club. They all recognize each other, and they recognize each other's sovereignty.

Let's get back to the numbers. There are only 193 members of the United Nations. Why 194? Because the Vatican City, which is a sovereign state—and though all the other 193 recognize it as a sovereign state—is not a member of the UN. That's why there's a little confusion here. The Pope and his buddies have said, "Nah, that's okay, the UN is cool, but we're kind of a religious people. We're going to keep to ourselves."

Pope sez, "Nope" to UN.

So there are 194 fully recognized states in the world . . . although there are about 10 other geographic entities that have either unilaterally proclaimed themselves as sovereign, or are legitimately working toward sovereign status. Places like Transnistria, Palestine, and Kosovo are in this nebulous non-state, maybe trying to be a state status.

In particular, Taiwan for decades has tried to get international recognition as an independent state, and at one point had perhaps half the world's states recognizing it as independent of China; China, of course, considers Taiwan as a 'rogue' entity that is actually part of their state. But times they are a-changing, and as China's economic and political clout has grown in the last decade, the number of other states recognizing Taiwanese sovereignty has dwindled to less than a couple dozen (including such powerhouse players as Palau and Haiti) and you should fully expect that number to continue to shrink, in the mad rush for all countries on the planet to kiss up to China.

While Taiwan is certainly not going to get a sovereign state spot, there have been several new editions to the club in the last few years. The first tag-along is East Timor. It was formerly a colony of the Portuguese—a situation that made it distinct from the rest of the Indonesian territory that was controlled by the Dutch. Over a decade ago, they had an independence movement and the Portuguese finally pulled out, and the Timorese declared independence. Indonesia was not keen on that happening, so it turned into a bloody mess. We won't get into it here, but East Timor has finally come out on top, had its application accepted at the UN, and is fully sovereign since 2002. Based on a diplomatic treaty, a former territory of Yugoslavia named Montenegro became fully independent and sovereign in 2006. The most recent addition is "South Sudan," which voted for succession from Sudan back in 2010, and the Sudanese government peacefully agreed to the split too—well, agreed to it after decades of brutal civil war in the region. It became fully official at the UN in 2011: South Sudan is officially a state as well, despite its wildly uncreative choice of state name. Seriously? South Sudan? That's the best you could come up with? Dudes, you guys should have had a state-naming write-in competition or something.

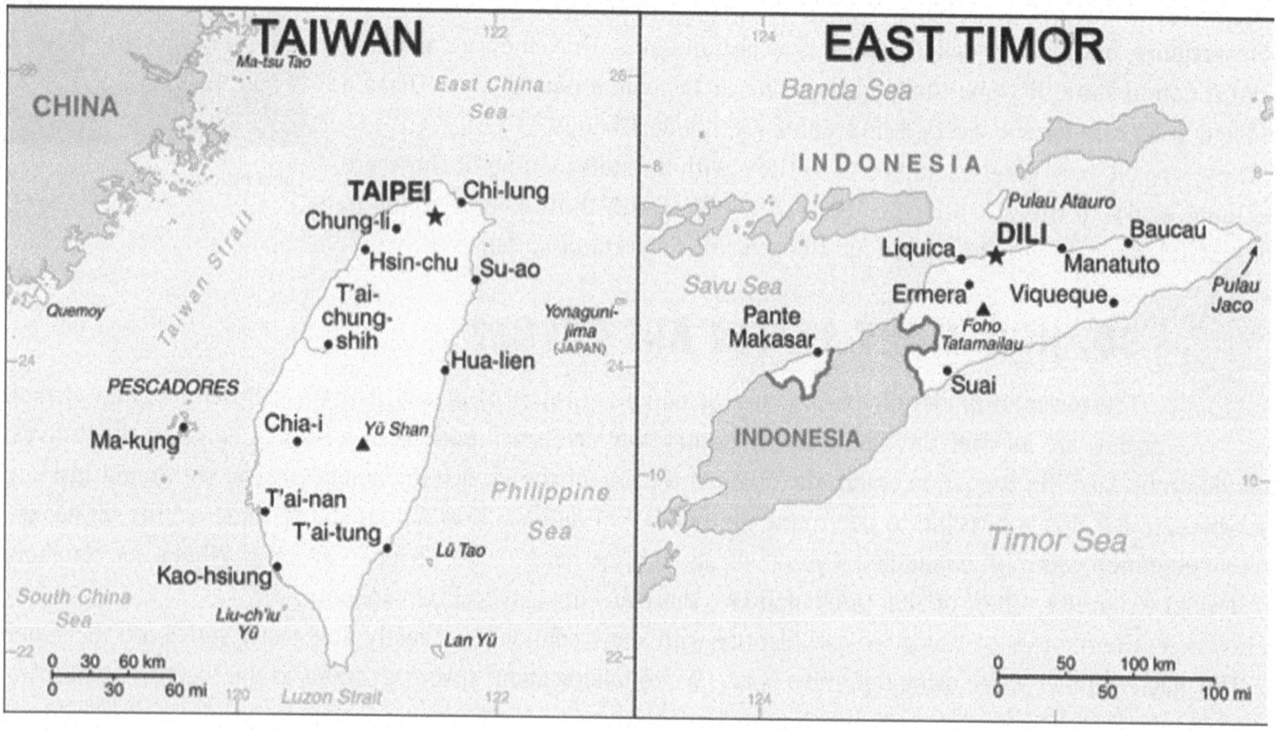

Future statehood? Taiwan out, Timor in!

Number 195 will perhaps be Palestine. This is a big "IF," since it's the Middle East, and you can't predict its future from one second to the next, much less year to year. For the last decade, there has been a general movement toward true Palestinian independence from the state of Israel. Indeed, given the track record of the last two years, it does seem, with a lot of trouble and turbulence, that they are heading toward that goal. We might be looking at two years out, we might be looking at ten years out, but certainly at some point, it will be a independent state. Maybe. Here in 2021, the tide has turned against it quite a bit, as current Israeli leadership is much more in a mood to absorb parts of Palestine rather than recognize its sovereignty, but as I said, things can change fast on this front depending on which way the political winds are blowing. More on this later.

Speaking of Israel and Palestine, it is also important to note that the recognition of a state is not always a given, depending upon to whom you are asking the question. For instance, while Israel is certainly fully accepted as a state by most governments in the world and the UN, many Arab and Islamic countries continue to shun it; in particular, Iran outright refuses to recognize it at all, as does Saudi Arabia and most other Arab neighbors. But that is not an isolated sovereignly-denying incident. North Korea refuses to recognize South Korea, and vice versa. Turkey refuses to recognize Cyprus. In what can only be called the most un-exciting recognition refusal of all time, Liechtenstein and the Czech Republic refute each other's existence because of some ancient WWII-era decree. How hilarious.

Kosovo causing consternation.

The most recent sovereignty showdown erupted in early 2008, when Kosovo declared independence from Serbia, and about 40 countries, including the US and most European Union states, immediately recognized it. This would perhaps be no big deal, like the Taiwan issue is not anymore, but for one small fact: this showdown is pitting major powers against each other. What do I mean? The US, UK, and France all recognize Kosovo, while China and Russia are firmly opposed. Hey, that's a major rift at the UN Permanent Security Council, a topic for later discussion!

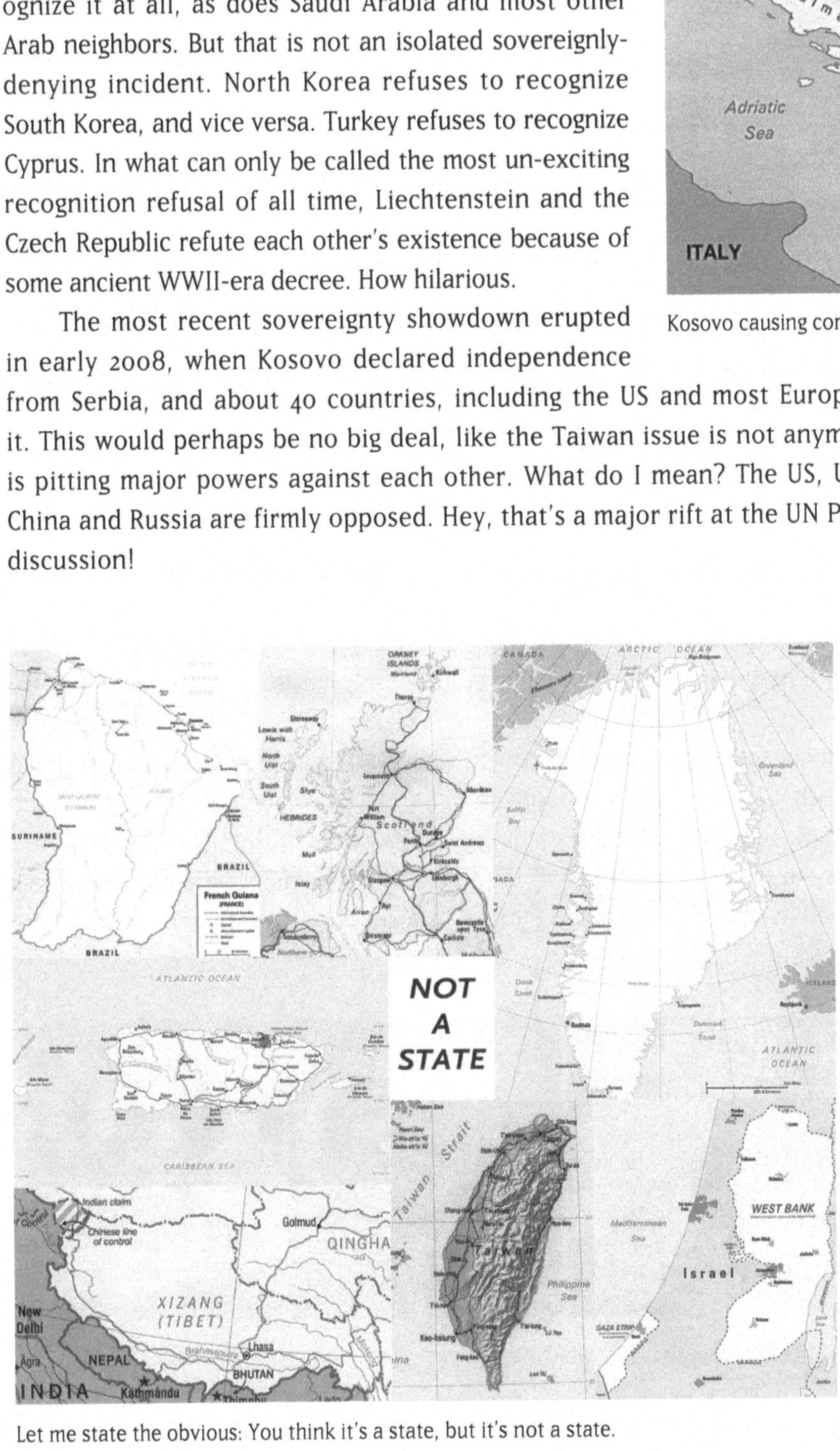

Let me state the obvious: You think it's a state, but it's not a state.

One more thing: let me point out a few places that are not states, but are often mistaken for states. Some of them are still colonies of other countries. For example, French Guiana, a fairly big place in South America, is a French colony; it's not a state. Greenland is still a Danish colony. There are some extenuating circumstances and also some strange ones, like Western Sahara, which Morocco is claiming and attempting to absorb legally and politically. It's in some nowhere land status in the middle. Bermuda is an overseas territory of the UK, not a sovereign state. Puerto Rico is not a state, not really an independent territory, and not really a colonial holding, either. It's just kind of associated with the United States. That's a bit of a muddled situation there as well.

On top of that, you have a lot of nations around the world *without* a state, which are sometimes mistaken for states. Places like Tibet in China we have referenced already, or like the Scots in Scotland. Scotland is a political subdivision of the UK, not an independent sovereign state. It's a nation of Scots, a group of people who definitely think they're culturally distinct. You know, Scottish people, we'd all agree to that—the guys that wear kilts, play bagpipes, and eat haggis. What the h is that? Haggis consists of a pudding made of onions, oatmeal and sheep's entrails, then stuffed into a sheep's stomach and roasted. Ummm . . . and then they eat it? Ugh, that is definitely a distinct culture—and let's make sure they stay that way. Give them the state status, as long as they keep the haggis!

The idea of an independent Scotland has been just narrowly defeated in a couple of referendums in the last few years, but may be back on the ballot soon! The same can be said for a part of Spain called Catalonia! They maybe want to become their own state too! Stay tuned!

WHO CONTROLS THE STATE?

We talked about what was a state, what is a state, who is not a state, it's all about the state . . . but who controls the state? Well, that's what the rest of this chapter is all about. World governments and world political systems: the peeps in charge of these sovereign entities. Before we get to that, we have to throw out a few terms that we hear all the time, but are wildly confusing to most. And many of these messy and muddled terms refer not to just politics, but sometimes economics and religions and social issues as well. They are terms like the **Left** versus the **Right, Liberal** versus **Conservative, Communism** versus **Capitalism.** We'll get to some of these terms in later chapters, but let's start with the **Left** versus the **Right** for now. Throughout this book, we're going to refer to about 'leftist' governments or 'right-wing' movements, so let's go ahead and flush these things out politically right this second.

What does it mean to be on the **Left** or the **Right** in the political spectrum? Maybe you've heard reference to "lefty commies," or "right-wing fascist" or "leftist guerrillas." So what are these terms referencing? Do I have to be ambidexterous to understand this stuff? Dig this from the get-go: we're talking about the *world* political stage here, not the politics within independent sovereign states. It's easy to get the terms mixed up; you might hear reference to "the liberal Left" or the "conservative Right" in America, but that internal domsetic name-calling stuff doesn't mean the same thing that we are talking about on the world stage.

Check out the world stage, the world political spectrum, from left to right represented across the bottom of these pages. Maybe you're starting to get a sense of what it means. Here's what I want you to know about it: On the extreme **left** would be government styles like communism or anarchy. These are ruling systems where there is (supposedly) little to no central government control. Another way to put it, as you head to the left: less government power, more power in the hands of the people. This is best displayed by anarchy, which we think of as a chaotic state, but indeed there are

Extreme Left Anarchy . . .

Communism . . .

Direct Democracy . . .

Representative Democracy . . .

One-Party State . . .

Power Increasingly Distributed to the Peoples

anarchists—political anarchists, that is—and they believe that there should be no government at all. Every single person within every single state in every single society should have the right to do anything they want. That's what true anarchy really means. It's never existed anywhere, and probably never will, but the theory is that the people themselves have *all* of the power. Never mind the bollocks, the government, or the rulers of the state, who have no control over individual people. That's an extreme. And it's an extreme on the left.

What's on the other side of the political spectrum, on the other extreme? That's on the total extreme **right**, which is just the reverse. All of the power is in government hands, or a single person's hands. Control of the country is in the hands of the few, the people in charge. Most of the population has zero or limited political voice in an extreme right system. To think of it in perspective—and extremes are always good to clarify the perspectives—is that the left-leaning power pyramid is upside down in anarchy. The people are on the top; they get everything, and can do anything they want. On the extreme right are fascists like Hitler, who actually said something along the lines of: "*I* am the state. An individual person. Me. I am the state. Everyone here in this state of Germany is here at my disposal at my will. You work for *me*." On the left, is just the reverse; the state works for *you*. The state is an apparatus that exists solely for the people in a left-leaning system. In an extreme right-wing system, you all exist for the state. That's a huge difference to consider. That's what the world spectrum is all about. All of the political systems in the world fall somewhere between these two extremes, because—and here's a little insider tip—these extremes don't exist anywhere. In fact, some of these systems hardly exist anywhere at all anymore. Let's go through them really quickly.

ANARCHY

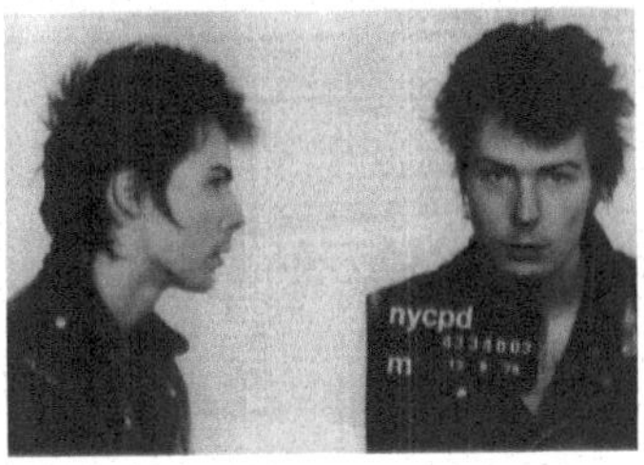

He did it his way.

We'll start on the far left. **Anarchy**, as I already suggested, is essentially no government at all. Maybe the Libertarian party in the United States is close to ideological anarchy, but only in that they want absolute minimal government in terms of taxes, services, and laws. Yeah, Ron Paul partying with Sid Vicious—now there's a good time! No superstructure, no hierarchy of folks that are in control of anything. In an anarchy state, there would be really no government, no taxes, no national defense, no one taking care of roads, or any of that kind of stuff. You're starting to get the flavor of this; high schoolers with punk rock jackets may think that anarchy's really cool, but it doesn't sound too good when you really begin to define what's going on in this type of society. It's pretty much every man for himself. We already know in human society that this deteriorates fairly rapidly into people beating the living daylights out of each other for a cigarette. In the political spectrum, when we're just thinking about this abstract idea politically, it comes down to this: "It's the individual's right to do anything he or she wants." Long story short, anarchy never happened anywhere, and anarchy will never happen anywhere. Not as a legitimate state system. Are there any anarchy states anywhere in the world today?

Theocracy . . .

Monarchy . . .

Military Government . . .

Dictatorship . . .

. . . Extreme Right Fascism

Power Increasingly Concentrated into Fewer Hands

There is perhaps one, but it's not intentional, and that would be Somalia. It is a state that has fallen into chaos—and we typically consider anarchy and chaos synonymous—and indeed, that is the case for Somalia. But they aren't really an anarchy; Somalians didn't intentionally become anarchic; they've just deteriorated into chaos. In the sense of an intentionally instituted system of anarchy, the whole concept humorously contradicts itself, much like Ron Paul running for president. (To be fair, American Libertarianism emphasizes freedom, individual liberty, voluntary association and generally advocate a minimizing of government power . . . not total abolition of it. I think.) But besides the few disintegrating states, there aren't any true intentional anarchic states in the world.

COMMUNISM

Communism is another hilarious system because it doesn't exist anywhere either. What a barrel of laughs. The Plaid Avenger wonders if it actually ever did exist anywhere. Why are we talking about it at all if it doesn't exist? Because the ideology is powerful, and it has greatly shaped history, especially the last century. Communism as an idea is still a potent force that shapes judgments of decision makers around the planet. There are several confusing factors about Communism, and before I get to those confusing points, let's just say what it is.

Famous dead communist dudes.

Communism was a concept that had been around forever, but some folks in 19th century Europe, namely Karl Marx and Friedrich Engels, really put it into its modern interpretation. These guys were living in the environment of the Industrial Revolution. They were witnesses to flagrant excesses of capitalism, ruled by monarchs on the throne who controlled all political power, and rich factory owners who controlled the economy. Marx and his crew looked around and said, "This really sucks. We have no power over our lives. We need more power to the people." So they wrote their manifestos and all that jazz, and long story short, came up with this: In a Communist political system there is a government, but the government is solely there (remember the inverted pyramid) to provide everything that the citizens need. To facilitate this provision of economic and political needs, the government would set up small **soviets**: that is, small groups of folks in cities of which every citizen is a member. Everybody is attached to a specific soviet, kind of like voting districts here in the US, but much much smaller. These soviets would debate on every issue that affects the people. In this manner, the government listens to the people directly and distributes resources or makes decisions based on the citizens' needs and requests.

How can a government afford to do that? In this society, there is no private property; the state controls it all. In other words, the government is kind of "by the people, for the people"—we think about that in democracy, but it's even beyond that in theocratical communism. It exists solely to distribute everything as per the needs of the people in the society, and that means economic as well as political power. That's one of the reasons it's extremely confusing. Dig the confusion:

- Confusing point #1: Communism is both a political and economic term, because it combines politics and economy into one system in which all ownership of property is communal and all resources are controlled by the government; the entire economy and the political system is run by a singular government. It's the only kind of system that works that way. All of the other political systems we'll talk about are just politics, and the economy goes its own way and does its own thing. In true Communism, it's all one big happy family where the government, Big Brother, controls everything.
- Confusing point #2: When we think about the manifestations of Communism in the world, we typically think of dictatorship. The big one, of course, is during the Russian commie experience: the USSR under Josef Stalin. Or even

Chairman Mao. Communism in China turns into Maoist China, a singular leader that ends up being a dictator. Even when it's not a well-recognized singular leader, these systems tend to perpetuate a small group of people in charge of everything . . . which seems like the opposite of a government "of the people."

Indeed, that's exactly what has always happened. During the Russian experience, Vladimir Lenin came to power in 1917 with his commie posse. They were attempting to implement a system that had never been tried before. To become politically Communist meant to be politically and economically Communist. Lenin said, "Well, okay, this has never been done before. We want to get to this utopian society where everybody's equal, and all decisions are made communally, and all property is held communally, and all resources are distributed communally—that's what we're shooting for. But we're not there yet. In fact, we're very far from it. We need to set up a small group of people in order to get from point A to point B. In other words, to get from where we are now to our perfect commie society, we basically have to become an oligarchy/dictatorship . . . just for a while." I'm paraphrasing here. That is not a quote from Lenin. He is not as good-looking as me, although his goatee is stylin'.

Main point: this is supposed to be a temporary thing, a transition, a transitory government in which a few people hold all the power, supposedly just for a brief period of time until everything is 'fixed.' Here's the problem, folks: Of all of the societies that have tried to go down the Communist path, not one single one has actually finished the transition. In other words, they start the transition, and they want to get to this pure utopian society, but it always ends up as a power grab by few people, or typically one person, who we then call a dictator. Some good examples are in the Soviet system, where Josef Stalin became a singular dictator. In the Cuban system, Fidel Castro, who's still there hanging out, is a singular man in charge of everything. In China, Chairman Mao, and then his successors, were singular people. It never fully develops into equality for all. Plaid Avenger Tip: Communism will never, has never, and, did I mention, will never work the way it ideally should. Which brings us to. . . .

- Confusing point #3: There are no Communist countries on the planet right now. They don't exist. Even if we look at places like today's China, or even today's Cuba, these are something closer to what we would call one-party systems, because *purely* political Communist states can't exist—remember, the system is a political and economic combination. It's like peanut butter and jelly or Batman and Robin: they got to happen together. Today, everybody pretty much now has some form of capitalist economies. In other words, the economies of the supposedly "Communist" countries are not communist. Everybody's given up on that (with the possible exception of North Korea, which is a place so bizarre as to not even be claimed by ardent communists). Since Communism in its pure form is both politics and economy, it does not exist. We've got to re-classify everything that *was* categorized as commie in some other category.

Those are the three confusing points about Communism. Let's move onto something more sound, and that is . . .

No libation without representation!

DEMOCRACY

Now we're getting into familiar territory; this is the stuff that we all get and understand, because we live in it. **Democracy**: *ocracy* = "ruled by" and *demo* = "the people." We are still in this inverted pyramid in which the government is here mostly for the people, as a service for the people, and not the other way around. We are not here to serve the US government, it is here to serve us. It protects us; it does things that are in our interest, not its own interest. That may be up for debate on some issues, but that is the system as it is in the abstract form. **Direct**

democracy doesn't really exist anywhere; we don't even have it in the United States. Direct democracy says, "We think everybody's so equal that every single issue and every single person we vote for and every law we enact must be voted upon by every single person involved in democracy, that is, everyone in the state." You can probably figure out why there's not too many of these in the world—because it's impossible. In countries that have millions of people, you can't have millions of people trying to come to a common consensus on what's going on, even if it were something as simple as a dog-leash law. Direct democracy, that is, democracy that involves every single person, really only comes close in one place, and these guys are strange on several fronts: Switzerland. I mean, even their cheese is full of holes, so you can't expect too much from the rest of the culture.

The Swiss have the closest manifestation of direct democracy on the planet. All citizens have to spend a large component of their time dealing with all the laws that have to be passed. Sounds like that **soviet** stuff described above, doesn't it? There is a much larger percentage of people who serve in some sort of elected office, like almost all of them. They don't just have a single President, they have a presidential council of like 7 members. I won't say it's a chaotic system because it's not, but it's definitely a time-demanding system that most of the planet is not willing to ascribe. That's why most of the planet is something else: a multiparty state/**representative democracy**.

This is exactly the same as a direct democracy, in terms of the government working for the people. But, instead of every single one of us voting on every single thing, we just vote for a few people and say, "You guys go decide. We will elect you into positions of power, knowing that you are here for our good. If you tick us off, or you do something wrong, we will remove you from your positions by voting for somebody else." That's the way it works, but the main term here is "representative." That is, one person is going to represent our county, or district, or state, and we're giving him or her the power to decide what's best for us. Does that make sense? It should—it's not that complicated.

On a side note: Multiparty states and representative democracy are all the rage on the planet right now. It is the government option of choice for most, and one that is growing. In the last thirty years, since the demise of the Soviet Union, this is the one that all the cool kids do. It is the most accepted, and the most perpetuated by the UN, and by big states like the United States and various European countries. This one is heading up in the numbers column. Most of the planet is in this category already.

We should also reference another type of democracy that is prevalent throughout the world, but befuddles us Americans because the system still has kings and queens: that would be **constitutional monarchy**. It really is a democracy, but for reasons of tradition (and pompousness), the state has maintained the presence of their royal family as figureheads, most of which have no real power at all . . . but man, they look so cool at fancy frock balls and palace ceremonies! These systems typically have a constitution, a parliament, and a Prime Minister who actually holds the real power, but they will still 'consult' the monarch on vital issues. Mostly all pomp and circumstance. Whatevs. Places still under this Cinderella spell include the UK, Australia, Canada, Belgium, Sweden, Norway, Denmark, Netherlands, Spain; non-European examples include Bhutan, Bahrain, Kuwait, Malaysia, Thailand, and even Japan, which still has it's antiquated Emperor!

God save the Queen! Fake power must be maintained!

I want to point out another thing under this representative democracy description: How could direct democracy and representative democracy be right next to communism on the political spectrum? This is confusing, and we've already pointed out why the term communism is so confusing. If we are not leftists here in the US, then why are the two systems beside each other on the left side of the spectrum? People in the US don't consider themselves to be leftists, but indeed, if you look at the whole spectrum, you are. That is, you're for power to the people, and that's more liberal, to the left. Liberate the people, freedom for the people, everything's for the people. In that sense, communism, on paper, is right beside democracy. There's just slightly different takes on how to go about getting power in the hands of the people. It's confusing, because communism has never worked. In theory, it's all about people, just like democracy. Now it gets a little simpler, as we're going to start heading to the right side of the spectrum with one-party states.

ONE-PARTY STATES

One-party states are exactly what they imply: there is only one political party that holds all the cards. Something that looks like elections happen from time to time. Sometimes there are things that look like choices on the ballot, but there's only one real political party that always seems to win. There's only one political party that is in charge. Some of these are in a transition path from their failed attempts at communism, like China and several other Central Asian states that used to be part of the USSR. Some of them are bordering on military led governments or full-on dictatorship, like former President Hosni Mubarak of Egypt, who took power in a military coup, then ruled the state for 30 years by holding "elections" which only had his name on the ballot, and the names of all the pre-selected politicians as well, all from the same political party. And that is fairly common in one-party states. That is, they do have parliaments, and they do have groups of people come together who make laws, but it's pretty much only one group of people or one political party with only one viewpoint of how to run things. In other words, you can go vote at the polls, but there's only one group of folks for which you can vote.

The One-Party Party Boys!!! The only ones on the ballot.

And now there are other states joining this one-party party due to the overwhelming dominance of a single political party in the state, even though they started as a multi-party democracy system. I am referring to two super states in particular: Russia and Turkey. Ever since their rise to prominence in the last decades, the leaders of these states have been extremely popular and have used that popularity to entrench their political party's power and their personal position for perpetuity! Namely, President Vladimir Putin of Russia and President Recep Tayyip Erdoğan of Turkey both started as popular politicians within a multi-party system, but since gaining the top slots of power both have assumed control of the press, crushed all political opposition, and even manipulated their countries' constitutions in order to give themselves more power . . . all while maintaining extremely high popularity ratings in their states! The people love them, and thus their political parties have dominated the scene at the expense of all other opposition parties. They didn't start out as military coups, nor communist experiments, but the situation that these two strongmen have created now firmly places their government style in the one-party state category. With no one else to really vote for, how can they be labeled as anything else, even when they hold "elections"?

One-party states have the illusion of choice; some of them might have some choices for some of the legislature and parliamentary sections, but it's really one group of folks that kind of have most of the power. They do some things for the people—they are not necessarily completely top-down hard-core dictatorships, but they are not true democracies. Not all people are created equal, and not all opinions of running the state are heard. It perhaps would only take a nudge to get them closer to true democracy, though. As opposed to theocracy. . .

THEOCRACY

Now we're getting to more easily identifiable, right-sided systems. Here's where the pyramid now has tipped. Indeed, there are some individual people who rule the state, and they do it for their interest in what they consider the right way to do things with little to no input from the people. This will get worse as we progress right-ward.

We are holy, and we are in charge.

Theocracy is fairly straightforward. It's an *ocracy,* "ruled by," *theo,* "religion." Some sort of religion controls the state. Our best examples would be Iran and Vatican City. These are states which are not **secular**. There is no division between the church and the state; in fact, it is one singular unit. The church *is* the state. The person at the top—which in other places we would call the king, or the president, or the head cheese—is a religious person in a theocracy. Like the Pope: he is the head of the Catholic Church, and also the head of Vatican City. In Iran, you have the Ayatollah: a supreme religious leader. He is also the leader of the state; they are one in the same.

Now the Pope has been around for a while, heck, I don't know, like since biblical times I think. (Actually, his title is "Bishop of Rome," established in the first century A.D., and since 1929 the position is also the head of the sovereign state of Vatican City.) But the Iranian theocracy is something quite new on the world stage, and represents a whole new take on how states should be managed. Indeed, most consider the Iranian Revolution of 1979 the most recent truly revolutionary political idea of human history. It's a big deal. The idea that a modern state can be recast and re-organized with religion as the central theme has not been tried lately, if ever. In the past we have had kings or queens or governments that are inherently tied to religion, or that even share power with religion. But this version of theocracy is totally unique in that ALL other power structures are subservient to the religion. It is a new idea, and one that has not totally figured itself out yet, even in the Iranian example, but the concept is finding fertile ground in many Islamic countries around the globe.

As you might expect within these states, religious law or biblical/religious text law is the state's law. This is of particular significance for Muslim states, as there are lots of folks around the planet in Muslim countries who would like to see their countries become theocracies. In doing so, they want to enact religious law as their state law. The Muslim manifestation of this is something called **sharia.** Sharia law is Koranic law: the laws and punishments described in the Islamic holy text, the Koran. The big problem with this, and why I'm pointing this out in this section, is because this is becoming a world issue. There are very few states in the world that have just Muslim people in them, or just *any* single ethnic group, or just *any* single religious group. To adopt the religious law as state law means to enforce that religious law on people who possibly are not of that religion. In their defense, many proponents of sharia would argue that western democracies embrace biblical law as a major component of their legal systems, and that is a valid point. Thou shall not kill, and all those other 10 Commandments is the basis for western law, so what's the difference? We'll look at this in more detail when we look at the Middle East and Sub-Saharan Africa, as this is a particular problem in Nigeria and the **Sahel**. But we need to finish off our political systems. Let's go further right to . . .

MONARCHY

Now these are getting extremely easy. **Monarchies** are the royal families, who, by the blessing of God, are smarter and better than all the peasants within the country that they lead. To be honest, this has been the most utilized system of human rule for the better part of human history. The emperors of China, the Shahs of Iran, the Maharajahs of India, the Sultans of the Ottoman Empire: all family lineage positions of leadership. And don't forgot those wacky Euro-duders! This used to be all the rage in Europe; everybody loved their monarchs. In fact, those chip-eating, goober British still love their monarchs so much that they still

Royal goobers: A dying breed. Sultan of Brunei and King of Saudi Arabia

have them there, hanging out, even though they have no real power (i.e. constitutional monarchies). If you want to get into a good fight in Australia, just make fun of the Queen of England. For whatever reason, there is a strong attachment to this idea, an idea that the Plaid Avenger not only finds preposterous, but quite frankly, revolting. The concept that, at birth, someone is better than me and has the power by the blessing of God to rule me, I find repugnant.

We don't have to describe the system any further; it's simple. The original royal family was touched by the divine hand of The Creator to have all the power, which is then reinforced by papal decree, then they begat some sons and daughters, creating the kings and queens who are far superior to us mortals. They get to rule until they beget more sons, who beget more sons; and pretty soon, they're all begetting each other, and that's why half of them are insane anyway. What is the power structure of the state? Why, it exists for the frolicking and pomp and circumstance of the crown-wearers, of course! By your leave, your majesty? Yeah, I'll leave you something of great majesty—my plaid boot in your hind-quarters! In states ascribing to monarchy, the royal family has all the political power. Back off, peasants. Just do what you are told.

There is one important point to make about this government type: it was repudiation and disgust with monarchy that led to the creation of most of those systems to the left of it on our spectrum. The concept of "divine" rights reserved for this elite group of people, particularly the rights to rule the rest of us, became so infuriating to so many on the planet that it became the catalyst for revolutionary thinking—thinking that created a lot of the other government types that we just talked about. Democracy (as we know it today), communism, Marxism, socialism, and probably a handful of other –isms on the left side came about because people were so ticked off and tired of the monarchy system which dominated Europe and most of the world at the time.

Can anyone think of any examples of this? Oh, how about the United States of America? Yeah, they had a revolution and kicked those tea-sipping suckers out. Later on, the French took their fight one bloody step further than the Americans did by outright killing their royals during the French Revolution. Vladimir Lenin and his commie crew assassinated the royal family as part of the Communist Revolution. Wrap your head around this idea in particular: many of the alternate systems of government around the world today are the rebellious progeny of former monarchies. Before we go any further, let's point out that democracy is an ancient concept; Socrates and the other philosophizing Greeks came up with this stuff long ago. However, democracy in its modern form, in its modern manifestation on the planet, was a result of aversion to monarchy, as was communism. Monarchy is top-down, power-concentrated concept all the way around, but there are some other political systems that leave even a worse taste in the Plaid Avenger's mouth, like military governments.

MILITARY GOVERNMENTS

My guns are bigger than the politicians!

Military governments are exactly what they say they are: governments run by people with guns. Every country on the planet has a military, or something that looks extremely similar to a military. Maybe called a self-defense force, or an emergency reserve, or a ground self-defense force, or a national guard—places like Japan and Costa Rica don't have an "official" military. Whatever name it goes by, the military in most states is subservient to the central government. That is, it's there for national defense, maybe even to internally put down revolts, but it works for, or at the request of, the government. Typically, the head of the state is also the ultimate commander of the military, but is not active militarily. Typically.

However, in states run by a military government, the military has actually superseded the political leaders. Often a small group of military elites takes over effective, real control of the government, directly or indirectly, by military coup. The best examples of this on the planet today are places like Egypt, Sudan, and Fiji. In those states, people who are from a military background, or are still active in the military, have wiped out the government for whatever reasons—maybe the government was corrupt, maybe as a move for national security—and have taken control. Does this mean they control everything and everyone at gunpoint? Not necessarily, but the military is the true power in any country.

I know this is confusing. People in the United States might say, "No, not here," but just ask yourself this—if the entire military of the United States wanted to take over the country, could they? Well, obviously, but we have laws and stuff in place here that all of us, including the military, respect. Their role is to serve, and to uphold the law. In other parts of the world, that's gotten a little fuzzy. Oftentimes when a state is taken over by the military, the government may continue to run as a functional entity, and we would say, "Hey, they've got a senate or legislature; they have courts, so it's still a legit semi-democracy place" but the real power is from a small group of people, and if it's a singular person, we typically refer to them as a military dictator. If it's a group of military people, we refer to them as a **junta**. You might think that this style of government is just a historic relic, but not so! There have been a myriad of military men in charge in very modern history: Idi Amin, Joseph Mobutu, Saddam Hussein, Than Shwe, just to name a few, were all military madmen that grabbed the top reigns of power in their respective countries and made a lasting impact . . . but unfortunately a negative one. The last few years have seen a turnover of military governments as well: Hosni Mubarak rose thru the military before taking over Egypt and ruling it for 30 years, mostly under martial law, until he was deposed in the Arab Spring of 2011. That same Arab uprising saw the demise and death of one Muammar "Colonel" Qaddafi of Libya, who pulled the same stunt as Mubarak but lasted 42 years as the military head of state. The best current examples of this in 2019 are in Egypt (again) where the old military dictator Mubarak has since been replaced (after another military coup) by a new military man, namely one Abdel Fattah el-Sisi. Meet the new boss, same as the old boss. And in the most bizarre and current military coup on the planet, after years of political gridlock and chaos, Thailand was taken over in 2014 by Prayut Chan-o-cha . . . formerly of, yep you guessed it, the military. He initially led a military junta, as the Commander in Chief of the Royal Thai Army, but now his title is Prime Minister Prayut. But make no bones about it: he is former military, having taken power as part of a military coup, and I believe that Thailand is still operating under martial law! They may still have a government in Thailand, but this one military dude has all the real power. That is the textbook example of a military government bordering on dictatorship . . . well, at least in this textbook anyway. And speaking of bordering dictatorships . . .

PM Prayut previously professional pugalist

DICTATORSHIP

Dictatorships are typically, but not always, started as military governments, where the military has taken over for whatever reason. A central strong figure rises from the ranks, and, sometimes through cults of personality, sometimes through brute force, become the sole leader of the country. A true dictatorship exists when all decision making occurs at one individual's whim.

He has the final say on all laws, all actions, all everything. The state apparatus functions only through him. For example, Hitler said, "I am the state," and he meant it! He decided if they were going to invade another country, or what the tax system would look like, or what rights individuals had. It's usually implemented by force of the military, but that's not the defining feature—it's the singular person

Saddam and Mugabe: They held all the cards!

holding all the cards that sets it apart. It can be military, or it could be religious—these lines get fuzzy. We can say the Ayatollah kind of looks like a religious dictator, and perhaps that's kind of true, but even he has a group of religious folks who work for him that have to come to a consensus on certain things. That does not exist in a true dictatorship, where a single guy has all the power.

The best example of this in the modern world was Saddam Hussein, currently residing somewhere in Hell. He was a single dude, insulated by a cult of personality, who built up and protected his power through use of force from 1979 all the way up to 2003. There are other even more recent examples, like the former president of Zimbabwe, Robert Mugabe. Mugabe—and again, all dictators are not active military, although he started as a military guy—was a rebel leader who helped fight off the colonial powers, helped liberate his country into independence, and originally was a hero. He was elected, assumed elected power, and then over the years and decades, he consolidated that power and just couldn't seem to get enough of it. From 1980 up to 2017, he increasingly held all the cards until he was physically forced from office at the age 94!!! Even more recent than that: Omar al-Bashir, the military dictator of Sudan from 1989 to 2019, just got overthrown in a military coup. . . the same way he had taken power himself decades earlier! Live by the sword, die by the sword, I suppose.

I am un-sure what my leadership style even is!

One man, with all the power; that's a dictatorship. All these guys refused to give up or share power, as is often the case in these scenarios, particularly in Africa. Power seems to ultimately corrupt, and the longer you hold on to it, the greater the hunger for it, like "the one true ring" that Bilbo Baggins found. But I digress. Those filthy hobbits always get me off track!

The point here is that dictators can come from a variety of backgrounds, often military, but not necessarily. The best living example may be Kim Jong-Un, supreme leader of North Korea, a place so screwed up it's actually difficult to attach a label to it. Historically, we might have the tendency to look at Fidel Castro or Joseph Stalin or Mao Zedong as a dictator. This leads to some questions about what it means to be a dictator vs. a fascist.

FASCISM

I am the State!

As has already been reflected in the extreme left, anarchy doesn't exist anywhere. **Fascism** doesn't really exist in today's world either. What separates fascism from dictatorship? A very thin line, I suppose. What the Plaid Avenger calls the manifestation of true fascism is that it *is* kind of a true dictatorship, but it's a dictatorship that's not necessarily perpetuated by the use of force. In other words, it's typically a cult of personality. It is a dictator—it is one person with all the cards—but one who has been put there because people like him so much. He's put there by popular mandate. And true fascists invariably want to spread their vision of a perfect society (according to themselves) out to the rest of the world . . . by conquest. So while your typical dictator in the modern world is satisfied with conquering and controlling his own state, the fascist feels the need to take over everyone else's state, too.

This is where your best examples are Adolf Hitler and Benito Mussolini, perhaps the only real examples in modern history. While we consider most of their dictatorial actions pretty bad in hindsight, at the time, they were very popular within their countries. They were rulers of the military, but they did not need the military to maintain control, nor to gain power originally. Hitler certainly was the leader of the military, and certainly using that military knocked the living crap out of Europe. But, in his rise to power, he did not hold the Germans under the sway of the gun and say, "I want you to agree to all these policies." Most people just said, "We think you're great, do whatever you want, this is awesome!" Then Hitler got the smack-down, committed suicide in a ditch, and that was pretty much that.

Let's face it: fascism sucks on every level. On the one hand, you have a generally charismatic leader (who is a few fruit loops short of a complete breakfast) who rises to power by mentally mesmerizing a bunch of needy folks who are only looking for a solution to their miserable lives. On the other hand, you have a large population of people who more or less blindly allow their government to get away with ultimately controlling their lives while destroying the lives of others. And that story always seems to have a happy ending. Or not. Ever.

Famous fascists biting off more than they could chew.

ANOTHER PLAID AVENGER INSIDE TIP: Neither Hitler nor Stalin drank alcohol. Both were teetotalers in societies well-known for their drinking habits. Think about it.

SO WHO IS WHAT, AND WHERE?

Check out the map in Figure 3.4 of current political systems around the globe. Can we detect any patterns? Any problems? Regional variation? As you can see, we have some dominant political systems with some variations on theme from region to region all around our governed globe. . . .

FIGURE 3.4 POLITICAL SYSTEMS OF THE WORLD

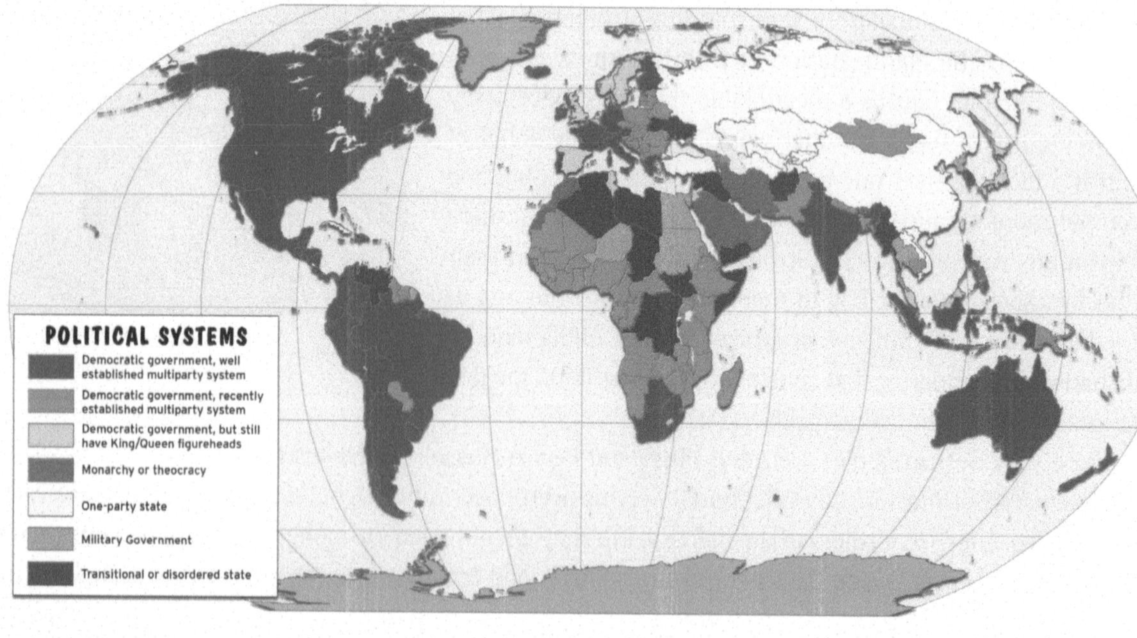

Look at the Americas. All of North America is staunchly democratic with a well-established government. Europe is also staunchly democratic with well-established governments—even the ones that still have kings and queens hanging around. As previously referenced, call them **constitutional monarchies**, but the kings and queens have no real power, they are simply figureheads; old habits die hard. Virtually all of Latin America is also staunchly, long-term democratic. India is the largest democratic nation on the planet, and places like Japan and the UK are considered staunch democracies despite the continues presence of their figurehead monarchs. Don't forget about Turkey, easily the most die-hard democracy on the planet that happens to be totally composed of followers of Islam. . . it really was a true, full-on democracy until very recently as Recep Erdoğan has obsessively consolidated power unto himself and his political party to turn Turkey into a one-party state. All is not lost though; signs indicate that Turkey may yet revert back to democracy once he departs.

New or newly established democracies on the planet are found in Eastern Europe and Russia. Since the fall of the Soviet Union, they've embraced democracy, but they are still what we call in transition, only about ten to twenty years old. **BIG 2019 UPDATE:** And those transitions don't always work! I'm officially putting Russia back into the One-Party State category after some serious anti-democratic shenanigans here as of late, but more on that in the Russia chapter. Most of Africa, both in the Middle Eastern sections and sub-Saharan Africa, is sort of in the democracy column, but they are kind of new as well . . . of course with the 2011–12 "Arab Spring" that occurred across many states of the Middle East, we may see true democracy blooming, but it is still too early to call on how it will turn out. To summarize sub-Saharan Africa, there's a little bit of everything going on down there. There are some chaotic states, some military dictatorships, some new democracies, and a couple of old democracies; it's quite the patchwork quilt. The Middle East is characterized by a patchwork quilt of a different flavor: they have most of the theocracies and true monarchies on the planet. Iran is a theocracy, but Saudi Arabia, Yemen, and Morocco have some true old-school kings and queens still in charge. Which brings up another **BIG 2019 UPDATE:** Turkey, long a bastion of secular democracy in the region, has, under the leadership of Recep Erdoğan, fallen firmly back into a one-party state status!!! Oh no! Democracy takes another hit!

We get to Central Asia to find a real crapshoot. Some states have put up the new democracy façade, but it's too new and too close to call which way they are going. One of the themes we'll talk about in Central Asia is that they're kind of leaning back toward one-party states. Folks are getting in charge there and hemming up power and staying in control. When we get around to places like Southeast Asia; they've got a little bit of everything. Some places are staunchly democratic, like the Philippines. Some places are in rapid transition, like in Burma where a military dictatorship suddenly broke out into a democracy! And then there's every place in between: some military takeover in Thailand, and even some hold-out commies in Vietnam running a one-party state. **BIG 2021 UPDATE:** Made that call too early on Burma! In 2021 it devolved once more into a chaotic/military state after the generals staged another coup. History keeps repeating itself in this poor place, much to the chagrin of the citizens of the country which are essentially all held hostage by the military, which more closely resembles a mafia. More on this issue in Chapter 25: The Cover Story.

In East Asia, we have China, which is definitely a one-party state. I know that the Communist China label is confusing to a lot of people because politically, they have a party called the Communist Party, and that is the party that is in charge, but they are not an economically communist state. Therefore, it's better classified as a one-party state. North Korea is definitely a distinct crazy wacko one-party/dictator state, perhaps better described as a one-psycho state, and South Korea is just the opposite, a staunchly democratic place. That's just a quick summary of what's going on politically in the world. We'll go into more detail about this in each of our regions.

A FINAL WORD ON THE STATE OF STATES

A few general points to consider across the planet here in the dawn of the 21st century:

The number of states is actually growing. That's the first thing to consider. Why is that? The bigger entities are breaking down into smaller ones. Smaller entities perhaps better fit the people within (e.g., the demise of the Soviet Union, the breakdown of Yugoslavia). The best example of all is the division of Czechoslovakia into a Czech state and a Slovak state.

The other main trend to consider is the demise of communism in the global sphere, and the movement away from dictatorships and one-party states to democracies. That is, there is a growing trend of democratization across the planet, something the leaders of the United States are very happy about and promote, of course. You can see this particularly in places like Eastern Europe—the Ukrainian and the Georgian Revolutions, respectively referred to as the Orange and Rose Revolutions. What about places like Afghanistan and Iraq? Well, I suppose they are heading towards a democracy, even if it has to be implemented at gunpoint. This whole 'Arab Revolution' thing that was occurring across the Middle East may result in a whole slew of new democracies . . . or at least something different from the autocratic, rightist ruling regimes of the past. Still too early to call on that one, as some places like Egypt are already sliding back to military rule, and others like Syria are in full-on civil war. **BIG 2021 UPDATE:** Democracy worldwide is actually currently taking a hit, as the re-rise of autocratic, anti-democratic strongmen are surging in power in multiple places across the globe. Democracy is not in full-fledged retreat, but is being challenged significantly in the current era. Much more about this in Chapter 25: The Cover Story.

A minor trend of which you should also be aware: the Iranian experiment with theocracy is both new to the world stage, and being exported abroad. Other states with large Islamic populations may be gravitating towards this system of government. I should say the *people* in those states are interested in some form of theocracy, but certainly not the current governments of those states who like the power structure that has them in the driver's seat right now, thank you very much. One need only look to an extreme manifestation of theocracy, the Taliban in Afghanistan, to see that this idea is not just an Iranian thing. There are also numerous political parties in virtually all Muslim states, from Nigeria to Egypt to Saudi Arabia to Pakistan to Indonesia, which are pushing for a ruling system along the lines of the Iranian model.

So democracy is being challenged, autocracy is on the rise, and lots of places are in transition, awaiting to see which system will dominate in the future or the mid-21st century, but we need to know a little more. The Plaid Avenger has put together a little high school yearbook of faces for you to identify, memorize, and be able to recognize if you pass them on the street. These are good people to know—not because they are necessarily good people, but because knowing them will make you a savvy global citizen. This yearbook is certainly not exhaustive, but contains particular faces that will be movers and shakers, newsmakers, and deal-breakers in the year to come. Know them well, my plaid friends, and join the ranks of those in the know, globally speaking. I've provided the names and places. You figure out who is who, and to which type of government they belong. I've provided some superlatives as clues to help you on your way. Good luck!

You simply must know your heads of state to be globally hip.

THE PLAID AVENGER WORLD YEARBOOK

Names
Kim Jong-Un
Joko Widodo
Salman bin Abdulaziz Al Saud
Angela Merkel
Xi Jinping
Recep Erdoğan
Andrés Manuel López Obrador
Boris Johnson
Joe Biden
Ebrahim Raisi
Michel Temer
Nicolás Maduro
Emmanuel Macron
Cyril Ramaphosa
Pope Francis
Hassan Rouhani
Scott Morrison
Vladimir Putin
Moon Jae-in
The Dalai Lama
Muhammadu Buhari
Yoshihide Suga
Abdel el-Sisi
Benjamin Netanyahu

Places
Iran
Australia
Russia
South Korea
Mexico
Nigeria
Israel
Egypt
North Korea
Tibet
Saudi Arabia
Germany
China
Turkey
Indonesia
United Kingdom
United States
India
Brazil
Venezuela
Japan
France
Vatican City
South Africa

Government Types
Established Democracy
Young Democracy
One-Party State
Theocracy
Monarchy
Dictator
Anarchy
Communism
Fascism
State of Insanity

Most Likely to Fahrvergnügen

Chancellor ____________
State: ____________
Type of Gov't: ____________

Singularly Sovereign Sino-CEO

President ____________
State: ____________
Type of Gov't: ____________

Indispensable Islamic Executive of Immense Island Empire

President ____________
State: ____________
Type of Gov't: ____________

Fish n' Chipiest

Prime Minister ____________
State: ____________
Type of Gov't: ____________

Overzealous Ottoman Oligarch

Prime Minister ____________
State: ____________
Type of Gov't: ____________

The Most Popular Politician on the Planet

President ____________
State: ____________
Type of Gov't: ____________

Most Likely to "Curry" Favor with Investors

Prime Minister ____________
State: ____________
Type of Gov't: ____________

Cleanest Linens

King ____________
State: ____________
Type of Gov't: ____________

Most Likely to Lambada

President ____________
State: ____________
Type of Gov't: ____________

Socialist Sledgehammer

President ____________
State: ____________
Type of Gov't: ____________

Most Likely to Run Screaming from a Large Lizard Creature Who was Awakened from Its Underwater Slumber

Prime Minister ____________
State: ____________
Type of Gov't: ____________

Most Likely to Surrender

President ____________
State: ____________
Type of Gov't: ____________

Most Likely to be Mistaken for the Kid from "Up"

Great Successor ____________
State: ____________
Type of Gov't: ____________

Most Mod American Amigo

President ____________
State: ____________
Type of Gov't: ____________

Most Likely to Forgive

President ____________
State: ____________
Type of Gov't: ____________

Most Likely to Eat Vegemite, Crikey!

Prime Minister ____________
State: ____________
Type of Gov't: ____________

Hardcore Cleric Future Ayatollah?

President ____________
State: ____________
Type of Gov't: ____________

Most Likely to Judo-Chop a Bear

Prime Minister ____________
State: ____________
Type of Gov't: ____________

Awesome Anti-Apartheid Activist & Billionaire Businessman

President ____________
State: ____________
Type of Gov't: ____________

Most Likely to Tae Kwon Do

President ____________
State: ____________
Type of Gov't: ____________

Most Reincarnated

President ____________
State: ____________
Type of Gov't: ____________

Most Kosher

Prime Minister ____________
State: ____________
Type of Gov't: ____________

Most Likely to Lead Africa's Biggest Economy

President ____________
State: ____________
Type of Gov't: ____________

Leading the Return of the Pharaohs

President ____________
State: ____________
Type of Gov't: ____________

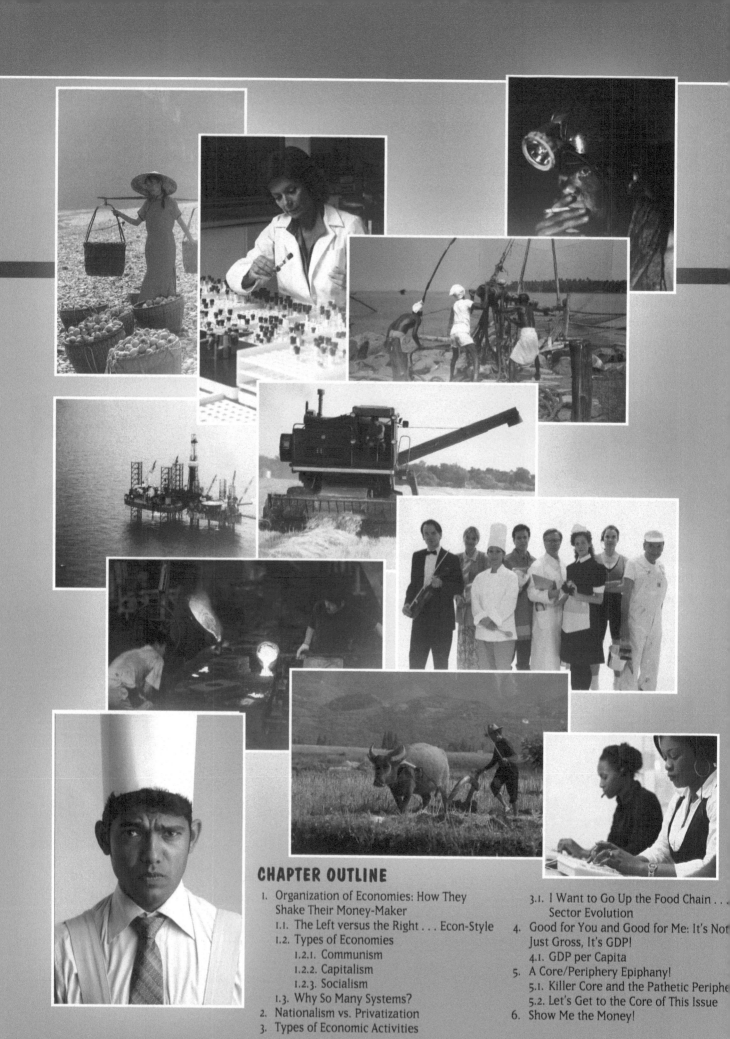

CHAPTER OUTLINE

The Plaid World Economy

4

OK we talked about people on the planet, where they are, how many there are, and how they procreate. We then looked where those people live, the state, and how they are ruled within that state. Now let's look at what people do. People like to eat. People like to buy clothes. People like to buy services. People like to buy stuff for the kids. People like to buy stuff for themselves. For all this stuff people want to buy on the planet, they have to have money to do it. All this buying and selling of stuff worldwide makes the economic world go round. So how do people make money in the world? Of course, we don't want to look at each individual person; we want to look at the entire society. How do the peoples in each individual state make money? And how does that relate to the global economy? Let's talk dollars and sense about world economic systems, shall we?

ORGANIZATION OF ECONOMIES: HOW THEY SHAKE THEIR MONEY-MAKER

First, we have to look at the overall organization of the economy within each state, and subsequently each region. We are going to revisit a theme we talked about last chapter: the left versus the right. There's a left and a right in political systems and there's left and right in terms of social idealism; there's also a left and right spectrum when it comes to economies. What does it mean in the economic context when we refer to the left versus right?

THE LEFT VERSUS THE RIGHT . . . ECON-STYLE

Often, you see terms such as communism versus capitalism in the economic spectrum and indeed that's a good place to start. Think about the Cold War. Not only was it the "free" democracies versus the "evil" communists in terms of politics, but was also defined as "free market" capitalism versus communist "command economy."

When we talked about the left versus the right politically, we defined things like communism as a system that was all about the individual, wherein the state was functioning in a capacity to serve the individual: the state exists only to serve the people. The opposite, right side of that politic spectrum was exemplified by a person or small group of people at the top that had all the power and the state (the people, that is) worked for them. We can kind of look at the economic spectrum in a similar fashion by discussing how many/how few peeps have the economic power, and how much the government plays a role in manipulating the economy.

TYPES OF ECONOMIES

COMMUNISM

We'll start on the extreme left, just like the last chapter, which makes sense since communism is at once a political system and an economic system all rolled into one. In the economic sense, communism refers to a society where everything in the state is owned, possessed, controlled and operated by the government. What? I thought you just said that 'the left' was all about the people having all the power? I know it is confusing, but dig this: the government in a true communist society simply controls all these things in the name of the people. The commie interpretation is that all the people in the state *are* the state, and thus are the true owners of everything. The government is simply composed of a small group of people that are passing all the stuff out to you, as you need it, because it's your stuff. Bottom line: all stuff is public stuff; no private ownership of anything allowed.

What stuff are we talking about here? All the goods, all of the services, all of the resources, all the land, and all the factories/means of production. Everything. With resources, it's pretty obvious: all the oil, all the wheat, all the coal, all the gold, all the forest, all the land—which is an important one. Means of production include all the factories and processing plants, all the plants that make cars, and tanks, and refrigerators, and light bulbs. Everything is owned and dolled out by the state. This includes all services too . . . the doctors and lawyers and building contractors . . . all those services would also be organized by the state and handed out to the peoples as they are needed. So in the big picture, the farmer grows food that the state redistributes to factory workers and doctors, and the factories make tractors that are given to the farmers to grow that food, and the farmers and factory workers get free health care from the doctors when they need it. Everybody in the society working together as a team, each getting what they need while helping fulfill the needs of others. And the state is in charge of all these transactions.

Again, the state is doing this *for* the people; all the people truly own all the stuff. People, people, people. Why do I keep reiterating that the people own all this stuff in a communist system? Because all the people sharing all the stuff don't freakin' work—never has, never will. It's just an abstract idea on how to create a "fair" economy. As we already referenced in the last chapter, communism has never manifested itself truly in this way—in fact, it has never even come close. Every attempt at it has gone awry. A small group of folks have gained power in order to make the transition to the commie utopia, but it has never worked, because money and power ultimately (and sometimes quickly) corrupt the small group of people in charge.

In addition, an entire economy—of even the smallest state—has way too many variables to try to control through a central administration, and winds up being more complicated than any small group can handle. The Russian description of this experience was **command economy**. You can see, just from the words themselves, that the government attempted to control and thus command all aspects of everything. A tall order for even the smartest and least corrupt.

Corruption and complication have combined to make pure communism pretty much an abject failure everywhere it has been attempted. True, pure economic communism is, on paper, possession of everything by everyone. In the real world, it never gets remotely close to that.

Confusing commie points:

- Communism is both a political and economic term because it combines politics and economics into one system in which ownership of all property is communal.
- You may think, and correctly so, that Russian communism's political and economic power was very concentrated in the hands of few. This was supposed to be a temporary position while they "fixed" the economy to a true communist utopia. As you now know, they never got it right.

- Even in today's world you hear references to "Communist China," "Communist Cuba," or even "Communist Vietnam." Make no mistakes about it, my plaid friends: these countries' economies are much more capitalist in nature. Their political leadership may still be called "communist," but their economies are def not.

CAPITALISM

Let's head right to the other extreme. On the other side of the spectrum is something of which we are much more familiar: **capitalism**. In capitalist society, all the resources, all the stuff, all the oil, all the land, all the fish, gold, and all the means of production, all the factories, all the car plants, all the doctor's offices are owned by private individuals. **Private ownership** is the key in a purely capitalist society.

Where did this system come from? Capitalism is an amazingly popular and resilient system, mostly because it happens quite naturally. Humans naturally gravitate towards marking off territory and claiming stuff. There is also an inherent understanding that the harder you work and the more risks you take, the more money you should accrue, or at least deserve to accrue. Everybody takes care of themselves, and in doing so, the economy takes care of itself. What this equates to over time is ownership by the few—which reinforces why these systems are on opposite sides of the spectrum. In a communist society, everybody owns everything; in a capitalist one, just a percentage of the population owns things.

Hard-core capitalists would argue for free trade, for free markets, and for minimal government intervention. "Let the market forces run their course" is a mantra of the capitalist. However, in a purely capitalist system in which the government plays no role, it would pretty much be every man for himself. Dog eat dog world. Only the strong, smart, and/or rich survive. Wealth would continue to be further and further concentrated into fewer and fewer hands. That would eventually equate to large segments of the society as destitute, impoverished, uneducated, and perhaps even unemployed. Is that a society that anyone wants? Probably not. Thus . . .

Just like communism, where does true capitalism exist? Nowhere. We live in a predominately capitalist-inspired world and capitalism is a popular system on the planet right now. But where does a *purely* capitalist or a *pure* communist state exist? Not anywhere, because virtually every place on the planet is in the middle. What is in the middle? Socialism. That's next.

Confusing capitalist points:

- We think that all the rich countries, and even many of the poor ones, are pure capitalist societies. Pure capitalism is an extreme in which the central government of the state plays absolutely no part in the economy. This is not true anywhere on the planet, not even in the US which is resoundingly the biggest proponent of capitalist expression on the globe. Even "heavily" capitalist countries like the US or Germany provide some services and benefits to their peeps, and regulate businesses at least a little.
- Even hard-core capitalist countries end up having anti-monopoly laws for big businesses. Why? Because if businesses and capitalist entities are extremely successful, they will keep getting bigger and bigger and absorbing other businesses until their power becomes unchecked, and uncheckable. Isn't that what pure capitalism is all about? Indeed. But no state wants to become second fiddle to a corporation. And "too big to fail" is becoming too dangerous to be be sustained.
- We like to think of the capitalist system as the "best" or most equal or even the most catering towards individual rights because everyone has an equal opportunity to own the stuff. However, having the same opportunity does not make for equal, or even fair, distribution of stuff. In the end, capitalism will concentrate wealth into fewer hands—many would say this is appropriately based on how smart you are, level of skill, your hard work, etc. Others say that concentration of wealth perpetuates stagnation in the masses, as idiot-savant kids inherit all the wealth from their rich mommies and daddies (sorry country-clubbers!) and opportunities to access that wealth by the masses becomes diminished. What do you think?

SOCIALISM

Now I know some of you ardent conservatives are saying to yourselves right now, "Blasphemy! There is no such thing as socialism in the US. We're not socialists here in this country; that's a bad word! Socialism is just communism in disguise!" But I'm here to tell you that when we're looking at things economically on the planet, virtually everybody is somewhere in the socialist sphere. Having said that, there is a broad range of what it means to be socialist and it goes all the way from close to the extreme left to close to the extreme right. In other words, that's where everybody really is: in a shade of grey socialism.

And there are many, many shades, maybe even 50 shades of grey? They all have the root of the word in their interpretation: *social*. In other words, they look out for, or have some role in, the societies over which they rule; societies made up of people.

What is economic socialism? You might have figured this out already: it's somewhere between the extremes. That is, the government (i.e., the state), owns *some* stuff and controls *some* stuff, maybe some lands, industries, maybe some of the resources. At the same time, private interests or private individuals own *some* stuff as well: some oil, some factories, some lands, and indeed, that's pretty much where the world is at this moment in time. Let's delve into this in a little more detail.

Why would the state want to control any of this economic stuff? Short answer: to provide services for its peoples. Even the most ardent capitalist would agree that the state should provide things like an army to defend the country, and maybe road systems or postal systems to facilitate economic growth. Others in a society think that the state should do a whole lot more, like provide social security, welfare systems, and unemployment benefits, maybe even health care. Different states provide different amounts of these things to their peeps.

In either case, how is the state going to pay for these things? Another short answer: by way of resource/industry control or collecting taxes, or some combination of both. Some states are very heavy-handed in this approach, perhaps controlling the most lucrative industry in its entirety . . . like in the state operated oil industries of Saudi Arabia or Venezuela. Other states like the US make some money from resources (mostly by selling the rights to drill to private companies) and some money from taxation. Other states, like Sweden, tax the living daylights out of everybody to make money. Different states have different approaches to making money. Look at the left-to-right systems spread at the bottom of this page to see what I'm talking about.

Confusing socialist points:

- Socialism comes in many packages; you may see references to **democratic socialism**, or **social democracy**, or mixed systems which are prevalent in Western Europe and increasingly prevalent in Latin America.
- Socialism can refer to heavy state ownership and control of some industries—the South American model. Socialism can also refer to states that generate revenue by heavy taxation in order to provide lots of services to citizens—the European model.
- The term is often misused by the conservative right in the US who confuse regulation of certain aspects of the economy with government ownership of business—you will see this when any laws are passed which limit big businesses in any way; those on the right will say the US is drifting into socialism. You may also have heard "socialism" used a lot when when the Obama administration was passing its health care legislation in 2010.
- Even though the US and many other states control/own aspects of their economy, we still refer to them as "capitalist" simply because they are more toward the right side of this spectrum. Conversely, places like Venezuela are openly referred to as "socialist" because they are much closer to the left side of the spectrum, as they control many more aspects of the economy. Well, that and the fact that they call themselves socialists.

Let's look at this spectrum from left to right again in terms of some active interpretations of this socialist sphere because really, the entire planet is somewhere in here—albeit in varying degrees.

If we think of all-the-way-left commies like Lenin, we see the attempted manifestation of true communism. He wanted the state to control everything in order to provide all stuff and all services to all the people all the time. Then, he died. Didn't work out so well in the long-term, or the short-term, or really any term in between.

Extreme Left, Full State Control . . .

. . . increasing government control of the economy

Castro is the modern-day Leninist attempt at communism. In Cuba's case, the state does control quite a bit; all of the resources in Cuba are owned supposedly 'by the people' and are administered and controlled by the government 'for the people'. But it's not working out so well, and there is a large underground free-market/black-market economy at work to supplement the services and stuff that is supposed to be supplied by the government, but is not—because they're broke.

King Salman and Nicolas Madura are slightly more to the right. Neither embraces communism, and indeed there are a lot of private businesses and private ownership in their countries. However, in both cases, the state controls the extremely lucrative oil industries and uses the profits generated from oil to provide all sorts of services to their people. Saudi Arabia happens to be a little bit better at it right this second, and everything from health care to education is provided to its citizens. Maduro's Venezuela has been trying to duplicate that model, but starting to stumble badly in the process. However, the 'socialist' cause is widely supported by the poor people because of the propaganda, which promises more stuff to the masses.

Xi Jinping's China comes next, even though it is supposedly a "communist" state that one would assume should be on the extreme left side of this spectrum. Not so. China initiated capitalist reforms for the last three decades and privatization ensued, but the government does still play a heavy hand in many industries and policies, although they continue to move quickly further right on the spectrum. Next we see Vladimir "the man" Putin from Russia. What? Russia? They were the center of the commie sphere! How can they be 'right of center' now? Because they fully embraced the capitalist way after the collapse of the USSR, and they did it with such a vengeance that it became a "wild west" of capitalism. Most things went into private ownership, and the government became so broke that it could not provide all those services that it once did. As a result, Russia has since re-taken control of its lucrative oil industry once more, leaning back towards the left side of the spectrum. More on that later. This is the classic example of how the world as a whole is firmly capitalist in outlook, even if the individual states exhibit varying characteristics of socialism.

We can go a little bit further to the right and we get perhaps a lot of European states like Sweden, France, and Norway that—while they don't control a lot of resources outright—heavily tax luxury items like cars in order to raise revenue to supply stuff to their citizens. In Sweden, (I'm picking on Sweden in particular because everybody likes those boxy Swedish cars, the Volvos) almost no one owns a Volvo because the government taxes the heck out of them. Why do they do this? It's considered a luxury good like alcohol or mink furs. They heavily tax those items to raise revenue, because they supply some serious services to their citizens. When a woman has a baby in Sweden, she gets two years off—PAID—maternity leave. Two years? Paid? Wow. In the US, women get like two hours off after having a child. Then it's "Okay, it's out. Get cleaned up, and get back to work! And by the way, those sheets you messed up, yeah, we're gonna have to get five hundred bucks from ya for those. Thanks bunches."

Now we're on familiar ground. The US is certainly much closer to the purely capitalist camp, but through limited resource control and taxation does provide a lot of services to the people. Quite a bit, actually, but not as much as many European states or even Canada, which has free health care. None of the government activities of the developed countries would be considered resource re-allocation, as maybe we would define activities in Venezuela or Cuba, but a lot of stuff is provided to the citizens nonetheless. Ever heard of welfare or retirement benefits, or public shcools?

It is always up for debate in the US between Republicans and Democrats as to how many of these services the government should supply. In fact, it's one of the main dividing lines between these two parties. Republicans are considered as

. . . decreasing government control of economy . . . to the Right, Full Private Control

favoring big business because they support free market, free trade, and privatization of perhaps all government services. Democrats are typically associated with wanting the government to provide these services, and maybe even more. This issue is evident in the Republican effort to repeal "Obama-care," which is a heavy state regulation and restructuring of health care in the US . . . but not a "socialist" government nationalization of the entire health care industry, as described by its opponents. This comes on the heels of President Obama partially nationalizing some banks and even auto manufacturers as the economic meltdown unfolded in 2008. Hmmm . . . tricky business for the undisputed champion of free market capitalism.

Adam Smith rounds out our line-up on the extreme capitalist right, and he is here intentionally as the symbol of pure capitalism—he is the dude that wrote *Wealth of Nations* in 1776, the handbook on how and why to keep governments out of the economy altogether. Because as I've already pointed out, pure capitalism doesn't exist anywhere. And Mr. Smith is long since dead, so he doesn't exist anywhere either. No state on the planet does nothing for its people—because that's actually a really good way to get thrown out of office, chucked off the throne, and/or guillotined.

WHY SO MANY SYSTEMS?

Speaking of getting guillotined, (were we?), where did these systems come from? Just like the concentration of political power in the hands of a few was the undoing of monarchy in Europe, there was the same type of impetus in the economic systems. This largely arose because of the Industrial Revolution. The Industrial Revolution, as it occurred in the 17th, 18th, and 19th century European experience, enabled a vast concentration of wealth in the hands of the few industrialists—the factory owners—and simultaneously pathetic circumstances for the peasants, the peons—the factory workers.

Industrialization: Rise of the Machines. And concentrations of wealth.

During this societal evolution, there were not many laws; in fact, they were nonexistent for things like child labor or workplace safety or minimum wages. The industrialists were all for the 195 hour work week—hey, that sounds good. How about we only pay people a dime a day? That sounds good too! Five-year-olds working in the factory? Why not? Got to give the little people something to do!

What you had was unregulated growth and crappy working conditions that really tested the limits of what societies would tolerate. Needless to say, this really pissed off lots of people and got them thinking about possible alternative systems to unregulated capitalism. Dudes like Marx and Engels, and then Lenin and others, thought, "Hey, this is unfair. These few factory owners have concentrated wealth and power. And the workers are powerless." This sentiment was not unlike the active aversion felt by peoples ruled under monarchies at the time as well. Result? Marxism, communism, socialism, and perhaps a few more –isms I don't even know about, were created to offer other options. These new economic systems evolved to offer alternatives to the unchecked power of capitalism—just like alternative government types evolved to counter unchecked power of monarchy at the time. And now, strangely enough, we have come full-circle. Whereas many of these systems originally evolved to check concentration of power and wealth in state systems, the states are now looked to by many to help counter wealth and power which has accumulated somewhere else . . . where would that be? Answer: the corporation!

Corporations are a significant thing to consider, if for no other reason than they've kind of replaced governments as the true holders of real economic power on the planet. You have to understand this; governments were seen as holding all the real power 200 years ago, which fired up enough people who started revolutions to unseat or redistribute that power (e.g., **American Revolution**, **French Revolution**, **Russian Revolution**, etc.). In today's world, multinational corporations are seen as the real powers on the planet and thus we've had a kind of move from protesting against a government or saying, "Hey, this government is unfair, we should change it or do something about it," to "Hey, this

What Is the Deal with . . . Multinational Corporations?

An executive with Dow Chemical recently said, "I have long dreamed of buying an island owned by no nation and of establishing the World Headquarters of the Dow Company on the truly neutral ground of such an island, beholden to no nation or society." Such is the story of multinational corporations. For most of history, the world economy has been controlled by nation-states. They made the rules, set taxes, and imposed regulations that corporations had to follow. This is changing.

In today's world of globalization, multinationals have become the primary actors in the world economy. Free trade agreements are reducing the barriers for companies to operate in other countries, and the global economy is becoming more interdependent. Many countries want multinational corporations to do business within their borders because they boost the local economy, bring jobs, and pay taxes. Therefore, countries, and even regions of the world, compete with each other to attract multinational corporations by offering tax breaks, lax environmental or labor standards, and improved infrastructure. Multinational corporations may have their headquarters in one country, but they do business in multiple countries of their choosing. In this way, the multinationals now make the rules.

Multinational corporations, which include Exxon, Microsoft, Pepsi, and Nintendo, have become very powerful; some of them have higher revenues than most sovereign states. However, they also have their critics. Nationalists and patriots are suspicious of them because corporate focus is solely on making money (corporations would never do that!) and taking advantage of the host country in the process. Profits over people. Antiglobalization protesters claim that the great power the multinationals wield is forcing countries to bend over backwards to please them by doing things like looking the other way when they ruin the environment or take advantage of poor people. There have been arguments that multinationals ruin local culture (so you won't be able to tell Cairo from Tokyo in a few years) and that they destroy local businesses because locals just can't compete with the big multinationals.

Fun Fact: Some consider the Dutch East India Company, founded in 1602, to be the first multinational company.

multinational corporation has all the power; they've got more money than our government does and they work outside the laws of our government because they're operating in ten or twelve different countries."

When you see protests against multinational corporations, or outcry against entities like the WTO, you are seeing a reaction to the idea that true power is not held solely by governments anymore. These protesters feel that the new oppressing force on the planet is not any particular state or government, but these powerful economic entities. In their opinion, corporate power is the real threat to human rights and human pursuit of happiness. Now it appears that these multinational corporations are bigger, badder and hold more money than most states on the planet.

The result? People are turning to back state governments to counter the unchecked power of multinational corporations. Confusingly enough, the liberal movement started as anti–big government, and now in most countries is credited with being very pro–big government. How did that happen? Part of that answer lies in how the world has changed in the last few hundred years; 200 years ago, governments were the primary sources of all power, with economic entities like corporations running a very distant second. Is that how it is in today's world? Not hardly. The multinational and even just the national companies in our world have become the major powers, with governments running a close second. In this atmosphere, the liberal attitude has shifted from "liberating" people from an oppressive government to attempting to "liberate" them from the economic powers of today's world. They attempt to do this by using the powers of government to counter those of big business.

NATIONALISM VS. PRIVATIZATION

We need to further define a few terms that we've been tossing around here while describing all of these economic systems. You will see these terms frequently, and they often cause consternation for folks on either side of the economic spectrum. These forces in action have often been the cause of public unrest, strained international relations, and have even been the impetus for invading countries or assassinating leaders. What's the deal?

Nationalization is the state acquisition and operation of businesses previously owned and operated by private individuals or corporations. It is usually done in the name of social and economic equality, often as part of a communist or socialist doctrine. Nationalization of foreign owned property, like the Suez Canal by Egypt in 1958 or the copper mining industry by Chile in 1971, typically attempts to end foreign control of an industry or asset and poses complex problems for international law.

In other words, the state assumes control of an industry, kicks out owners/operators, and starts taking the profits to the state bank account. This seriously ticks off corporations, who in turn usually get their home government to kick up a fuss, take the case to an international arbitrator like the WTO, embargo the nationalizing state, or maybe even invade the country outright to get their stuff back. Sound preposterous? It's happened plenty in the last hundred years. France and Great Britain invaded Egypt to try to regain the Suez Canal, and the US has destabilized or overthrown whole governments in Chile, Guatemala and Cuba to satisfy pressures by corporations who fell victim to nationalization.

As you might have guessed already, **privatization** is just the opposite. Selling of businesses to private ownership after they have been the property of the state is the process of privatizing. Since the collapse of the USSR, Russia and all of its former areas of control have been in a mad scramble to privatize industries. But they are not alone. India and China are fertilized fields for mass privatizations, and the trend is still occurring in the developed world. Western European countries and Japan, who just privatized their postal service, continue to push more government operations to the private business sphere. In the US, the term has often been broadly applied to the practice of outsourcing the management of public schools, prisons, airports, sanitation services, and a variety of other government-owned institutions to private companies. Sometimes this contracting of services does not entail the outright sale of the industry to private hands.

Why the pervasive theme of privatization in the world? The popular argument is that government-run business is like a monopoly—and lack of competition makes the business not-so-productive and noncompetitive and wasteful, and maybe even ripe for corruption. That is certainly a valid point. The other reality is that bigger and bigger corporations have much more leverage than ever before in the past to put pressure on states to privatize. Who funds lobbying groups to persuade congresspeople to privatize? Who has money to fund big studies that show that government operations are wasteful? Who would have the money to buy the senator a Ferrari for his teenage daughter's graduation present? And then who would be wealthy enough to bid on the sale of government resources and services that were being auctioned off? Um, maybe massively rich multinational corporations? You can call the Plaid Avenger jaded, you can even call him a leftist, but he is merely a realist. It's the way the world works.

Privatization is the growing theme in the 21st century, but nationalization is far from extinct. Hardcore capitalists and businessmen and multinational corporations across the planet still shudder when they hear the word. Ten years ago, I would have said that nationalization was completely dead and that you wouldn't see this happening much anymore. However, it is making a slight resurgence, particularly in the left-leaning South American region—I should say left-leaning Latin America because it rolls off the tongue much more glibly. You'll also see some of this happening perhaps in Africa in the very near future, but I digress.

Especially after the global economic meltdown which started in 2008, governments around the globe went scrambling to nationalize key industries, even in the hardcore capitalist countries. Forget socialist Venezuela; one need look no further than the US, which partially or fully nationalized a bunch of banks and even some auto manufacturers in an effort to stabilize the crashing economy. Russia and China are following suit, quickly leaning back towards more state control in all things economic in order to prevent the financial chaos which has ensued in the West. Adam Smith is rolling in his grave. Even in the capitalist strongholds, when recession or depression hits, nationalization becomes vogue again.

That brings up a confounding question: why would states which advocate free-market capitalism ever stoop to the depths of nationalization? Several reasons crop up: (1) In times of national emergency or war, states often nationalize critical industries like steel or energy in order to ensure national security (so they don't have disruptions in bomb and tank production during war time). (2) It can also occur if a state deems the social benefits far outweigh the economic costs, such as running a postal or healthcare system . . . one could argue that having a healthy workforce in your country makes the whole economy work much better, so spending on free health care for everyone brings in a much bigger societal reward. (3) Sometimes it is done for purely economic reasons, i.e. government takeover of a business or industry about to fail, which would equate to lost jobs and revenue for the country.

Put all those reasons together into one big pot and stir, and you have the answer to why US President Barack Obama and many other European leaders got back in the nationalization game in the last decade. Partial or full government takeover of financial institutions and car companies was done under the banner of "saving the economy/saving the country" from further collapse and preserving social benefits like jobs, while also alluding to a sense of maintaining national security which would inevitably be threatened if critical components of the economy like the whole banking system were to crash. Interesting stuff.

TYPES OF ECONOMIC ACTIVITIES

No matter what the type of economy, or how far to the left or right in the socialist world they may be, or even if things are being nationalized or privatized, people on the ground are always doing something to make this thing called an economy go forward. What is it those peeps are doing?

I'm not talking about economic structure at the state scale. I'm talking about what real people do on the ground. In real life. What the heck do you do? You chop down trees for a living or do you make toothpicks? What do you do to earn your paycheck? What businesses provide the most jobs in your hometown? What kind of stuff do those businesses make? Now we're on familiar ground here, and this chapter gets exceptionally easy because we can really classify all economic activity that happens in the world into four distinct types of activity.

Why are we doing this? Just so we can call out people and see what level they're on? To point and laugh? Hahahaha—your mama sells seashells by the seashore in the service sector. No! We want to identify these things because it's important to understand what different states are doing and how many people are dedicated to certain activities within the state. This in turn tells us a lot about how the economy operates, why some places are richer than others, and how this is currently changing in different regions around the globe. And it is oh so simple. They all go in order, and it is even numerical. Let us proceed with speed: on to the Primary!

Primary economic activities are as simple as it gets. Literally. This level includes anything and everything that involves natural resource extraction from the Earth. Just taking stuff from Momma Earth and then selling it is a primary activity. The list is short:

- Agriculture production: Cucumbers, cocoa, squash, cows, pigs—I don't care what it is; it's a primary economic activity if it's growing stuff we eat.
- Timbering/logging: Chop down a tree and take it. That's extraction of a raw commodity.
- Mining: Oil, coal, gold, diamonds—mining of everything. That's all simple extraction. Just pull it out of the ground and give it to me.
- Fishing: Yep, just taking a fish out of the water. That's a primary economic activity, even if you use dynamite to do it.

Pretty straightforward, pretty simple stuff. Resource extraction is the key. And that is primary. Now when you hear things like gold or oil, you think, "Ooooooweee . . . that is worth a lot of money!" Here is what the Plaid Avenger wants you to know: No it really ain't. Main point: primary activities produce low-value raw commodities.

Virtually all primary activities aren't worth squat in the big picture. How can I say that? Oil is worth a lot of money, isn't it? Well, it's critical for everything; it's used for everything . . . but it's worth less per gallon than homogenized milk, and that isn't worth much either, is it? Speaking of milk, what are all agricultural products worth per pound? Per boatload? How about wood? Quick question: would you rather have a business that grew oranges, or made orange juice concentrate? Why? Because one makes more money than the other due to processing of the product, which leads us to . . .

Secondary economic activities involve the processing of everything you got from the primary economic activities. All secondary activity consists of somehow manipulating, altering, refining, making better, doing something to the stuff you got from the primary activities—that is, the raw materials are modified in some way. Here's the main point: modified in such a way so it is worth more money.

You are adding value, and that is something that will continue to happen as you go up this chain of economic activities. As I've suggested earlier, products of primary activity are not worth much. You can cut down a tree, and a tree lying down on the forest floor is worth something, but not a lot. How about if we take that tree, rip off the bark, cut it into planks and then further cut them into boards. That's worth a lot more! Contractors and construction people will pay for boards, not for a tree.

Processing something of a big quantity to a smaller quantity is simple processing, but it still adds value. Then there is also refining. If we take oil out of the ground, it doesn't go straight in your gas tank. It has to be processed. It has to be cleaned up. Taking the oil out of the ground is our primary activity, processing it at a refinery is a secondary activity that adds a lot more value. Further manipulate it into gasoline and it's worth quite a bit more.

This can include a vast array of activities, which you don't often think about. Simple things like roasting coffee beans. Are roasted coffee beans worth more than raw coffee beans? Indeed they are. How about mud? Mud? Mud is just mud—but form it into squares, let it dry out, and now you have bricks. You can't sell mud but you can sell bricks. You've processed it; you've added value. Processing of any raw material counts—including taking a fish and chopping out its guts. Is a gutless fish worth more than a fish with guts? Ask anybody who wants to cook a fish. Most importantly: all "basic manufacturing" items fall into this category: from textiles to furniture to cars to paper to toys. If it is modified or made, it's a manufactured item, and manufactured/factory goods are all secondary activity stuff.

Now let's get up the chain another notch: let's get to tertiary level activities. **Tertiary level activities** are also known as service sector activities, so stage three includes all services on the planet. Perhaps we're going to take some of the raw commodities, or even some finished commodities, and maybe even some manufactured commodities, and do something with them, maybe move them around, further manipulate them, or sell them—provide a service of some sort.

The service sector is what most of us in the developed world are used to and what most of us have done for a living; you really have to go out of your way to find agriculturalists or lumberjacks or people who work on assembly lines in rich countries. But a service sector job? Yeah, we've all done those.

Every hamburger that gets fried, every surgery that gets performed, every used car that gets sold, every fire that gets put out—someone has provided a service. Even simple transportation is a service. Truck drivers are an excellent example of this: taking commodities from one place and simply moving to another place. I'm sorry, what have we done here? We've added value to the commodity! A bunch of lumber sitting in the middle of Saskatchewan, Canada doesn't do a thing for anyone in the United States—but if you put it on trucks and move it to a Home Depot in Des Moines, now you've added value to it. They can use it; it's more accessible; it's worth more. Value has been added.

That's just one division of the service sector. We tend to think of everything from construction to police to doctors to teachers to army people, navy people, air force people, marine people as all providing a service. Perhaps the biggest one you think of is sales. Workers at McDonald's, at the mall, at Walmart, cashiers, wait staff; all these people are providing a service to you. They are moving around stuff from the first two activity levels and getting it to customers and/or increasing the value in some way, shape, or form.

Again, this process entails a ton of different activities, but its main feature is its tendency to increase the overall value of stuff from stages one and two in some discernible manner. Reconsider our hamburger: a cow that we killed in primary activity terms is ground into hamburger during secondary activity—worth more than the cow to be sure, but even raw hamburger is not super valuable. Now take it to a restaurant and provide some services in the form of a cook and then maybe a maître d' that seats you at the table and maybe a waiter or waitress that brings you that hamburger. Now you've provided a bunch of services and you will pay a lot more for that hamburger in a restaurant than you will for the raw hamburger at the store, which is even more than you'll pay for the dead cow somewhere out in the field.

You see, as we progress up this chain we keep providing more, adding more value to commodities. Providing services adds value to commodities. Here's another important thing to consider: how much more do people get paid in the service sector than, say, farmers or lumberjacks or even people who work in a mill? Well, sometimes we have comparative salaries between occupations, but oftentimes as we progress up this chain, people get paid more to do, in real terms, what is a little less. Every level adds value to the commodities themselves, as well as to the salaries of the people who are doing it.

So what's left? **Quaternary sector** is kind of a new one. Some analysts and economists don't even recognize this one yet, due mostly to the fact that these types of jobs have not been on the planet for very long, and don't constitute a big percentage of the labor force. But they are on the rise, and do tell us some important things about the societies in which they occur. Quaternary sector provides something new. Not a natural resource, nor a processed commodity, and not really a service either. In essence, the quaternary sector deals with the creation and manipulation of intangibles—and the main intangible that I am thinking of here is information.

In the technologically advanced countries, an increasing number of people aren't dealing with hamburgers or trees or oil or services so much anymore as they are with the idea of information for information's sake. Computer programming, researchers, lab technicians, even astronauts are all in the realm of data creation, but don't necessarily "provide" a service to anyone. The results of their work may lead to a commodity that is mass produced, but their work in and of itself is neither a commodity nor a service.

To a certain extent, you can even say stock market people are in the quaternary sector because they are not really producing anything. Perhaps they are providing a service, but

it's more likely that they are manipulating information and moving abstract things around to create or add value to them (i.e., stocks).

It's a fuzzy one to be sure, but the reason I like to point it out is because we can make some easy assumptions about any country that has folks working in this sector. How would you describe a state that has a lot of astronauts and archaeologists and research scientists? Hmmmm . . . I don't know much about space travel, but a state that has astronauts must be technologically advanced, rich, and stable. No two ways around it. How much do those folks get paid? Bangin' bank, that's what. What are those products worth? Genetic code, computer programs, cures for diseases . . . are you kidding me? That's real money!

If you can critically examine what economic activities that different countries and regions focus on, you can go a long way to understanding why rich countries are rich, why poor countries are poor, and how the global economy is working. This insight, or should I say sight beyond sight, gives you the inside scoop on the real mechanics of money flow in the world. What? You don't have the sight beyond sight yet? Better call in Lion-O and proceed directly to. . . .

I WANT TO GO UP THE FOOD CHAIN . . . SECTOR EVOLUTION

We can look at any state on the planet, just in terms of their economic activities, to get a fuller picture of what's going on, how rich they really are, and indeed, what their future holds. It's not only about how much money they've got and how much money they are going to make but also what's going on in their society, how developed or developing they are, and lots of other things that we can typically tie to other demographic and cultural factors. And it's easy. Check it:

FIGURE 4.1 ECONOMIC SECTOR RATIOS

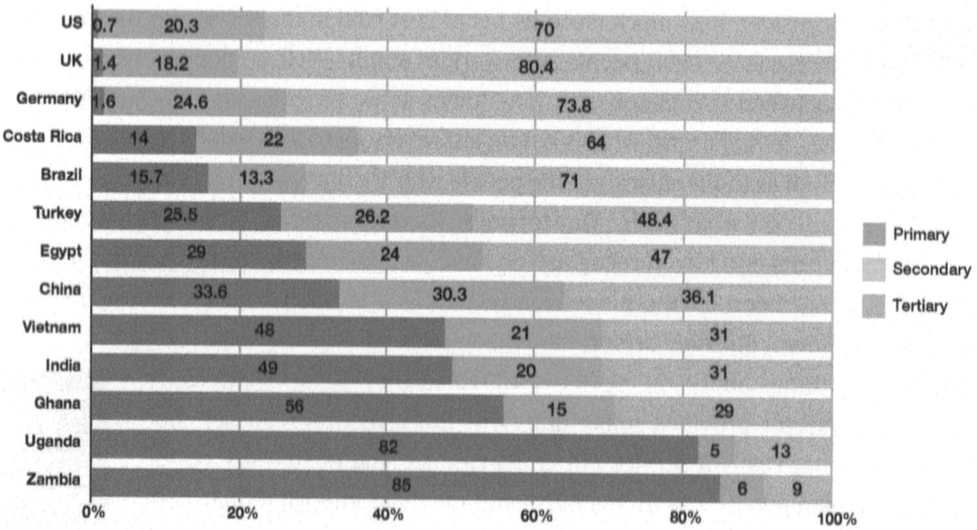

As you can see in the above graphic, the percentage of people within each state who work in each of these levels of economic activity tells us volumes about what's happening. We can define some obvious patterns between developed and developing countries, as well as explain anomalies to these patterns. First, the trends:

The fully developed countries, that is, the richer countries in the world, have a much greater percentage of their workforce in the tertiary/service sectors than in primary and secondary sectors. Check out Luxembourg, the UK, and Germany on the top of the chart. In fact, primary industries in the developed world account for almost nothing, but that should make sense to you now. Advanced technology and machinery in these countries equates to small percentages of folks needed to grow all the food and mine all the coal.

In the US, just under two percent of the population are farmers, yet those farmers supply the other 98 percent with food, and export tons of food on top of that. They are all the way through the demographic transition; one big tractor in the US operated by one guy can do the work of 1,000 small farmers in Africa—heck, more like the work of 10,000.

Secondary/manufacturing sector jobs are continuing to decrease in the rich, developed world due mostly to the cost of labor. Developed countries have labor unions and health benefits and minimum wage laws; all of which serve to increase the cost of labor. Developing countries typically have none of these things, thus, labor is cheaper. Manufacturing jobs of virtually all things—cars, toasters, toys, luggage, computers—have been migrating from developed countries to developing ones to take advantage of this cheaper labor. What does that leave for everyone to do in the rich states? Service sector employment of some specialized, skilled jobs (doctors, lawyers, firemen) and a lot of un-specialized jobs (clerks, janitors, salespeople). Good money for most. Of course, some of the folks in these societies have the know-how and opportunity to create and manipulate information, leading the world in quaternary level activities as well. In developed regions, there is:

- hardly anyone in the primary sectors,
- typically less than a quarter in the secondary processing sector,
- a majority of the workforce in the service sector,
- and a small but significant number of people in the high-end quaternary sector.

Just the reverse is happening in the developing countries. To point out these major differences, let's start with some of the countries that are the least developed, and subsequently are the poorest parts of the planet. Check out the bottom brackets filled by the African countries in Figure 4.1. They are overwhelmingly primary industry-based, very limited secondary and service sector employment; quaternary is out of the question. What can we say about a place that is almost exclusively agriculture- or fishing- or mining-based? What are those products really worth? As pointed out earlier, not much. In the least-developed regions:

- most folks are in primary—and most of them are farmers,
- less than a quarter are in manufacturing—sometimes much less than that,
- even fewer are in the service sector—with virtually none in quaternary.

Most states on the planet are somewhere in between. The trends do seem to indicate that the more developed a place becomes, the more tertiary and secondary will be gained at the expense of the primary sector. More service, more manufacturing, fewer farmers. End game: more money. We can see this reflected in the previous chart. Costa Rica and Brazil are diversified across sectors more in tune with developed countries, and are doing pretty well. We wouldn't think of those countries as impoverished or undeveloped. India and Vietnam are hustling and transforming, but they are still overwhelmingly agricultural-based societies, and as such, are not making the big bank money yet—not per capita anyway. China and Turkey are stable, prosperous societies that are perhaps the most balanced across the board in economic activities. Plaid Avenger inside tip: this balance and diversity will probably serve both these countries well in the future, and you may see other countries try to emulate this approach as opposed to stacking up solely in the tertiary category. Just a hypothesis. But I'm usually right. China has actually stated that they want to keep this diversity in order to maintain their competitive edge in manufacturing.

Confucius say: Show me the money!

Confucius say: Get astronaut, but keep farmer. Get scientists, but keep miners. Keep low wage to keep manufacturing sector; make China strong.

He could be onto something.

Now to look at an anomaly. What's the deal with Egypt up there? Looks pretty balanced, perhaps even more developed than Turkey or China. Are they really approaching full development as their economic activity breakdown suggests? Probably not. We have to look at the real world in this case to decipher the truth; Egypt has a low percentage of folks in primary sector, not because it's an evolved economy, but because they don't have a lot of agricultural land to farm! It's a desert! Except along the Nile, of course. They have some oil to mine, but not very much, and it just doesn't create many jobs anyway. So, low primary activities. Why high service sector, then? One of the moneymakers for Egypt is tourism. You know: King Tut and the Pyramids. How much money are those tourism jobs worth? Eh, not too much. We can make sense out of these economic activities on a case by case basis when some of these countries buck our trends.

But the trends do make sense. They do seem to support the concepts outlined in the Demographic Transition Model from Chapter 2. Most states are working toward increasing manufacturing and service sectors, and lessening dependence on the primary sector, because that makes them more money. Currently, most of the value added to primary commodities happens in the developed world—to primary products shipped to them from the developing world. You should be able to see how this puts the poorer countries at a "permanent disadvantage"—they export low value stuff and have to import high value stuff. That's why the more successful countries in the 21st century are trying to catch up, not by producing even more cheap stuff, but by changing percentages in their economic activities sectors. China and India are changing fast and will be making big bucks. Saudi Arabia and Equatorial Guinea are making big bucks on primary sector stuff (oil) but not changing or diversifying internally. Who is going to win in the long term?

GOOD FOR YOU AND GOOD FOR ME: IT'S NOT JUST GROSS, IT'S GDP!

How does what all these folks do for a living in all these different states in the world affect real wealth—that is, real money and real dollars. That something is called **GDP**. GDP is **Gross Domestic Product**. You may have also heard of GNP, or Gross National Product. What's the difference between the two? Are you ready for this? Not much; they are two terms which mean pretty much the same thing. GDP has been growing in popularity and GNP has been falling out of use. Okay, so what do they mean?

GDP is simply all the goods and all the services within any state that are created and sold in that year. What do I mean by all the goods and all the services? Well, everything. Every single thing that is bought and sold. Every transaction in terms of a final sale. Every hamburger that's sold, every car that's sold, every service that's provided, every employee salary at McDonald's, all of it. All of it added together for the year is the GDP.

You do have to put this in context of the *final sale of the commodity* and what that means. For GDP, we only count the final sale and not all the transactions that led up to that final sale. Example: a car. A car gets sold on a lot for $20,000. That $20,000 car transaction is the final sale because it is going to a consumer, who will then "consume" it. That $20,000 goes into the total for GDP for that year. Now the reason we say only the final sale counts is because you can't sell things twice or three times or five times in the case of commodity chains anywhere in the world. The car has tires on it when the guy sold it off the lot. Well, those tires were made somewhere, and then sold to a distributor, who then sold them to a retailer, and then the retailer maybe sold them to a car manufacturer and that's three separate transactions, but you wouldn't count those other transactions because then you would be counting the same tires three times within one year. Only the last sale counts.

However, the car salesman's salary, the distributor's salary, the assembly-line worker's salary—all these services are a part of GDP for the year. The final sale and all worker salaries go into this equation. It doesn't matter if twenty people were involved in the movement, creation and distribution of that car; all their salaries go into GDP.

Now we know what GDP is, but what can we say about patterns of GDP on the planet? Keep your world regions in mind when analyzing the GDP map that follows. . . .

If we look at the GDP planetary totals, we see that the heavyweights who make the most money on the planet are represented by the developed world, with the United States economy weighing in at number one in the world (for now). Second in line: China has the number two economy on the planet. The third is now Japan, with the European powerhouse of Germany holding steady with the fourth biggest GDP. They all have GDP that total in the trillion-dollar range. That is a million million, or is it a billion million? I can't remember; it's a one with a butt-load of zeros behind it. Over a trillion dollars per year.

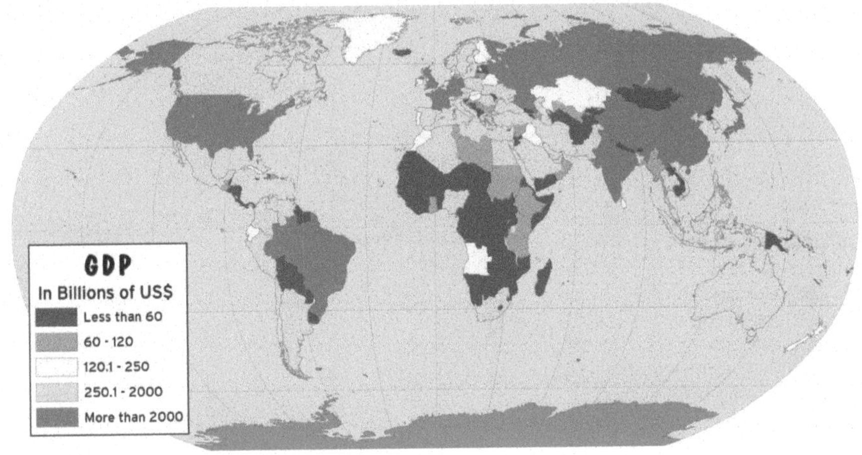

But this is where the lines are drawn and we have to stop making comparisons about what we consider to be the developed and developing world based on just GDP. Mainly because China really makes things interesting: China surpassed France and the UK in total GDP in 2008, then bumped the Germans in 2009, and then Japan in 2010 . . . surpassed 3 huge economies in just three years! So is China fully developed? Many would say not. I personally argue the affirmative, but we will get to this in the next chapter. They sure are making the dolla' dolla' bills y'all!

We see that most of the Western European states are still in the high GDP category, and most of Africa and Asia are in the low GDP category, with Latin America, South Asia, and East Asia falling somewhere in between.

I do want to point out a couple things as we look at these totals around the planet. Maybe you are thinking, "Wow, the rich countries—they are really, really rich. The poor countries, they really stink. They don't make squat. I guess everybody's sitting around there not making any money. They're not providing any goods or services. What losers!" Reality check: we have to throw in a disclaimer here. When examining GDP totals, in relations to a state's actual wealth, we have to consider what is *not* included in GDP. Check out the little green box, yo!

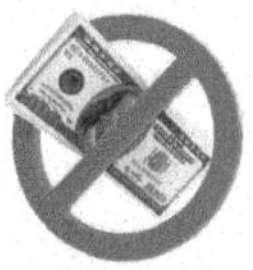

What is Not Included in GDP?

We said all the goods and all the transactions of all the goods and all the services are included in GDP–that's everything, isn't it? What would this not include?

First, think of some things that don't have monetary value in an exchange sense, but are actually hard work. This list would include housework, child rearing, any labor done in the home, food grown for home consumption, food gathering, food preparation, firewood gathering, the list goes on. Why would I point this out? Because that's what most of the people on the planet are doing. Every single day. Day in, day out. None of these activities are included in GDP.

Second: anything that the government doesn't know about. Any transaction that occurs between folks under the table. This can refer to anything as innocent as a flea market or a yard sale to items or services traded, bartered or sold between people that is untaxed and therefore not included in the tally. But this also includes anything illegal–stuff the government prohibits and therefore wouldn't be able to tax or know about, including all illicit narcotic production and trade, all moonshining, gunrunning, and human trafficking. All of these activities, which do generate large amounts of wealth, are not included in GDP.

I'd like to point out one example in particular. If we look at Mexico on the map above, we see that they are in the second tier of GDP ranking. Were you to include the illegal drug economy, they would instantaneously be in the top ten nations on the planet for GDP. Don't undervalue the stuff that's not being counted.

I bring this up because when we look at the poorest places on the planet, they're actually working pretty hard—but a lot of the stuff they do doesn't count in the official statistics. GDP is a Western-derived and Western-measured definition of wealth, and simply does not always adequately reflect reality. What do I mean by this? Well, when we look at Africa, it looks like nothing's going on there but indeed, let's face the facts; a whole heck of a lot of work goes on there to keep that place going; it's just not measured.

In summary, anything that's done at home, for home, anything that happens between friends, any type of barter transactions, and any illegal, illicit activities are not counted as part of GDP. That would radically change the map of what's going on here. There's one more thing to consider: the comparison between GDP and GDP per capita.

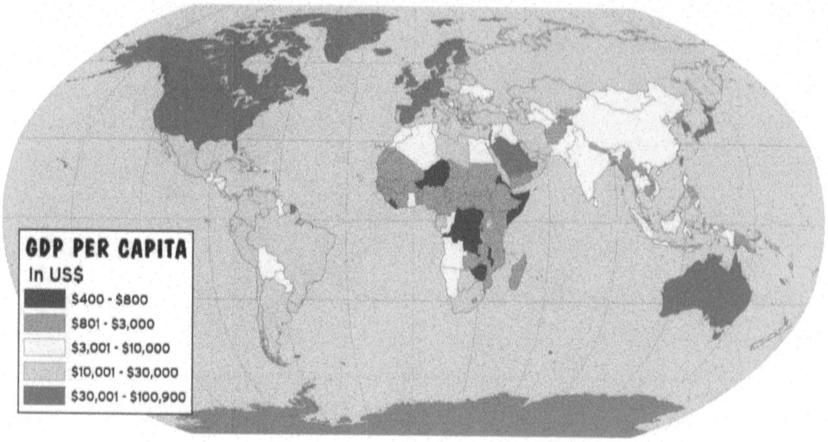

GDP PER CAPITA

Total GDP we covered earlier. GDP per capita is simply the total GDP divided by the population total for the state. In other words, it tries to approximate how much wealth there is per person in the society; this is perhaps a much better indicator of how the average Joe is doing in the world. Looking at these two maps side by side offers a contrasting picture of what's happening on the planet in terms of levels of wealth and economic activity. As to be expected, the most developed countries maintain themselves in the highest brackets, and the most undeveloped countries maintain themselves in the lowest brackets of both maps, but look what happens to some interesting places in between.

If we look at total activity, that is, the total GDP map on the previous page, we see total wealth as a singular number that a society or state produces. All these numbers look really big—and wow, would you look at that? China and India are right up there with the developed countries! So is Saudi Arabia! Sweet! They're making bank! They're not doing too bad, right? Brazil, Argentina, they're doing pretty well. Everyone in Saudi Arabia and India and Brazil are rich, aren't they? Eh. Um. Well. Maybe not. How do I know everyone there is not rich?

Well, let's check out a more realistic interpretation in the GDP per capita map. What you see is a bit of a shakedown. India and China in particular have gone from the top classes of total GDP, to the lower classes of GDP per capita, a significant shift when broken down to the individual level. Why does this happen? Well, because they have big populations, and they're producing lots and lots of valuable stuff, *but* broken down to an individual level, it ain't really that much per person. GDP per capita is a much better gauge of what's happening in the society in terms of how well the place is really doing. A trillion dollars doesn't really make as much impact if everybody in the society is only getting a buck each. It's even worse if just a handful of people in the society have most of that trillion bucks, leaving even way less to be divided up amongst the penniless masses.

This high GDP/low **GDP per capita** scenario possibly could be a very equal society, but unfortunately it never happens that way, at least not on this planet. The reality is a concentration of wealth in certain businesses and certain sectors and certain hands, with most people not really cutting much of a piece of the pie. This comparison becomes a much better measure of wealth disparity, and as such, is a much better indicator of how this society or state is really doing.

Sometimes these measures can be skewed radically. In other words, even if a state has a huge number in terms of GDP, and perhaps even when a state has a big number in the GDP per capita category, we still may not be getting the real picture. I'll pick on a few states in particular here. Places like Equatorial Guinea, Saudi Arabia, even Oman; what's happening in these places? High total GDP, high GDP per capita. Are they really so rich? Are they fully developed like the other rich, developed countries?

I think you already know the answers to these questions. Something doesn't add up. Equatorial Guinea, a rich place? Look how much money per person! On the books, they make $50,000 GDP per capita. That makes them one of the world leaders in that category . . . oh, except for the fact that most folks are starving to death under a thieving dictator. Indeed, the majority possess squat and a handful of people own virtually everything. **Wealth disparity** is massive, but the official economic indicator figures remain high because of a double whammy; the GDP total is absolutely massive, and the total population is low, which equates to an exceptionally high GDP per capita . . . on paper.

That's when you have to take the economic sector breakdown of labor in conjunction with GDP and GDP per capita to get a real sense of what's happening in these states. The real deal that's happening in these examples is that these societies have high numbers based on one single thing: the export of oil. They export tons and tons and tons of oil which means they make tons and tons and tons of money, a big number. Even if you divide that by the population, it is still a big number, but all of the wealth is, in reality, in the hands of very, very few people.

Equatorial Guinea is a particularly nasty case where none of these numbers mean squat because most of the people there are impoverished beyond belief, while a single corrupt ruler and his cronies get 100% of the money from the sale of the state's resources. Saudi Arabia, Oman, UAE, and many other OPEC countries fall into similar situations, though not as bad as the Equatorial Guinea extreme. You have to consider other things—looking at the economic sector breakdown of a place in conjunction with its GDP per capita gives us a good sense of what's going on, as well as a good sense of development and standards of living.

A CORE/PERIPHERY EPIPHANY!

Let's add up all of the stuff we have learned so far in this chapter about economic activities and apply it to the singular global economy of which we are all now a part. What do I mean by that? Well, whether you want to accept it or not, the whole human world has been ever-so-slowly moving towards increased global connectivity, with economic activity being the most pervasive and successful system thus far.

Booting up the global economy!

The creation and movement and buying and selling of stuff now happens across the entire globe as a single system. Oil from Kuwait is processed into gasoline in Burma and then fills the tanks of Toyotas in Tokyo. Auto engines made with German parts are constructed in Romania and then shipped to Brazil to be installed in VW Jettas which are sold in Chile. Argentinean leather and Indonesian rubber is sent to southern China to be made into Nike shoes which end up on the shelves of Walmarts in Wisconsin. And that is just the easily identified tangible stuff; there are also billions of dollars worth of services, stocks, investments, and currency exchanges which happen every single day among all the countries of the world. Which brings us to an interesting concept, appropriately named **world-systems theory**, which tries to make sense of the winners and losers of such a global economic scenario.

The development of the current world economy is the result of intensification of world trade, interconnections, and industrialization at a much higher level than in the past. For the last 500 years, the Europeans, and then the Americans (and maybe the Japanese, et al.), have dominated and shaped this global system, and many would argue to their ultimate advantage. Hey, wait a minute! Europeans, Americans, and Japanese . . . that's the rich Team West! You knows this! Through past colonial/imperial activity, coupled with modern industrialization, the Team West core is certainly a huge beneficiary of this world economic situation. *Core* is a great term to use, because it is at the heart of world-systems theory.

KILLER CORE AND THE PATHETIC PERIPHERY

This dude named Immanuel Wallerstein formulated this functional flowchart back in the 1980s, and it continues to evolve even today. He characterizes the world-system as a set of mechanisms which redistributes wealth from the economic **periphery** to the **core.** Another word for periphery would be "edge" or "fringe." In other words, the core gets stuff from the periphery, either resources or money via economic exchanges; in this scenario, the core always gains and the periphery primarily loses, particularly in the long run. Typically, the core country buys the raw resource for cheap, let's say it's steel, from a periphery country. The core makes that steel into cars, and then they sell those cars back to the periphery country for a much higher price. Core wins, core wins, core wins!

In Wallerstein's terminology, the core is the developed, advanced, industrialized, rich part of the world, while the periphery is the not-so-developed, raw materials-exporting, poorer part of the world. The free-market and financial systems are the means by which the core exploits the periphery. Maybe some of you think that this is the natural order of things, and that the word "exploit" is too harsh. That's cool. I'm not trying to make a political point here, I just want you to understand how the economy works, and this world-systems approach does do a good job explaining the current flows of wealth in the world, as well as why rich places get richer while poor places mostly get poorer. Pointing out the winners and losers makes the game easier to follow.

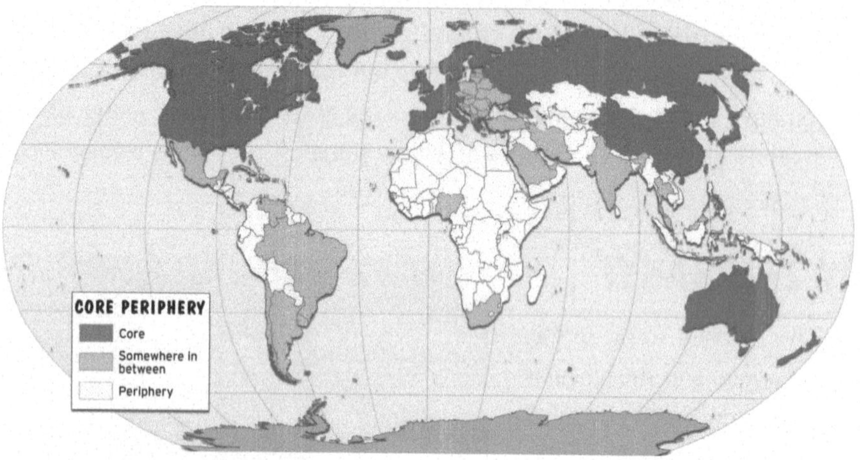

Check out the map to the left, and then let's add some more descriptors to these elements to enlighten even further.

Core states have the maximum level of industrialization and technology, which of course makes them centers for innovation and creation . . . and manipulation of capital; you know, money and credit. Whether you dig this theory or think it's leftist propaganda, there is no disputing that for all of the modern era, the core "Team West" countries have controlled global economic institutions like the IMF, World Bank, Inter-American Bank, the WTO and a host of others. Heck, "Team West" freakin' created all those institutions to begin with! The core runs the show, although that is now starting to change a bit (see G-20 in Chapter 6). The core has the smarts and the industrial capacity to make raw resources into much more valuable commodities. The core also has the cash to reinvest in its own infrastructure and businesses continually in order to keep itself in the top slot. The core is all about keeping free-trade free and opening up new markets to the global system . . . because, of course, all these things benefit them tremendously!

The periphery is the total opposite. They are poorer countries that may lack industrial capacity, financial means, and/or technological savvy to process their resources into finished goods. But they do have natural resources and tons of workers ready to slave away for pennies a day. They end up selling raw resources/cheap labor, which are cheap, and then importing finished goods, which are expensive. You don't need to be a math major to figure out how this story ends. Also, most of the industrial capacity and resource extraction that happens in periphery countries is usually owned fully or partially by multinational corporations from the core, which means that the core countries sometimes benefit economically even from the raw resources too. Ouch! Double-whammy! Most of Sub-Saharan Africa would be the best examples of periphery.

Because they are on the losing side of this equation, periphery countries are often locked into a cycle of never having enough money to invest in their own education systems or businesses in order to break out of this gerbil-wheel. Epic economic fail! When the situation gets bad enough, or when a strong leader comes to power in a periphery country is when you sometimes have things like **nationalization** occur. Remember that stuff from ten pages ago? A charismatic leader like a Fidel Castro (Cuba) or Hugo Chavez (Venezuela) or Mohammad Mosaddegh (Iran) can rise to power under the banner of "kick out the foreign companies and take our country's stuff back." Yep, that's nationalism, and it still may have a role to play in peripheral economies that are ticked off enough to strike back if the game board becomes too stacked against them.

Of course, not every single state in the world falls nicely into one of these two categories. There are those countries that benefit more from their resources, like most of the Middle Eastern states which fully control their oil resources, and there are some places that have effectively nudged their way closer to the core via crafty economic policies and strategic political alignment. Which reminds me to get back to point:

LET'S GET TO THE CORE OF THIS ISSUE

Can a state get to the core if they are seemingly stuck in the periphery? Oh yes, my friends! Things are changing very fast here in the 21st century! While Team West is currently the richest and certainly has an outstanding competitive edge in today's world, this was not always the case. In times past, both India and especially China would have been considered as the economic core of the global system, positions that they are working very hard at reclaiming in the modern era.

China has been strategically controlling its own resources, as well as acquiring cheap resources from other periphery states. They have used these resources along with their massive pool of cheap labor to become the manufacturing "workshop of the world," producing every imaginable industrial and commercial product. They are now core winners. India has taken a slightly different path: investing heavily in service and quaternary sector jobs like computer programming is what is currently pulling them out of the periphery realm.

They are not alone. Some other states have taken alternative paths in an effort to break out of the "poor/loser" category. South Africa, Turkey, UAE, and Brazil are a few examples of up-and-coming regional economic powers whose paths we will look at closer next chapter. Speaking of which, let's get to that. Someone . . .

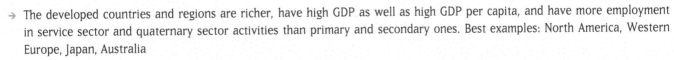

The periphery are the base floor of the wealth of the core!

SHOW ME THE MONEY!

In summary, what people do around the planet and how the economy of the state is structured, in large part, determines how well or how poorly the state is doing—and will be doing in the near future.

We have to look at several more factors to determine the true level of development of a place. Numbers aren't always enough, but we can look around the planet and see some obvious trends. Some of the ones we've already pointed out:

- The developed countries and regions are richer, have high GDP as well as high GDP per capita, and have more employment in service sector and quaternary sector activities than primary and secondary ones. Best examples: North America, Western Europe, Japan, Australia
- The developing countries and regions typically have lower GDP as well as lower GDP per capita. However, in some circumstances, even when GDP total is high, the GDP per capita will still come out on the lower end of the spectrum
- The developing countries and regions typically have more people employed in the primary and secondary activities than tertiary and quaternary. Examples: Sub-Saharan Africa, the Middle East, Central Asia, Central America, the Caribbean
- The states and regions that are developing the fastest are changing this equation more successfully than other states who are still heavily reliant on primary activities. Rapidly innovating regions include South America, East Asia, Turkey
- Some states and regions are currently lagging in GDP not due to economic activity structure, but because of turbulence attributed to shifting economic systems. They are slowly rebounding. See Russia and Eastern Europe

There's a diverse mix of what's going on economically in the world. Know these factors and features and what people are doing for the green, so as to stay keen about what's happening in today's world and which way the states are heading. Who has the money? Who will have the money? Who is developed? Who is developing?

Developed or developing? Hey, that's our next topic . . .

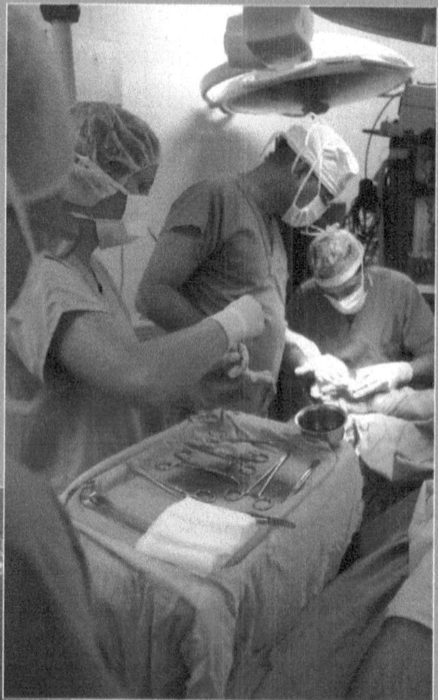

CHAPTER OUTLINE

Developed or Developing?

5

STINKING RICH places. Dirt POOR places. Some states are really well-off, and a whole lot more are struggling day to day to make ends meet. Why are some countries/regions rich, and others poor? As this manual has pointed out, it is not simply a matter of size of the country or the resources it contains. Reality, as always, is much more complicated than that. Before we can even identify traits of these developmental differences around the globe, we should first sort out some basic definitions and terms. Have you ever wondered. . . .

IF THERE IS A "THIRD WORLD," WHERE ARE WORLDS ONE AND TWO?

In order to understand these crazy catch-phrases, let's start with a close-up of some other common conundrums. These words and descriptors are often tossed around to describe the state of states—that is, how well off they are, or inversely, how close to being in the global gutter they are.

A **developed country** is one that we would consider first and foremost . . . rich! Most texts, news articles and smart people somehow forget to speak the plain truth and just call a spade a spade. They will describe lots of different variables that account for how a state is rich, without ever outright just saying it. The developed countries are the rich ones! Admit it! Have you heard of a fully developed country that's flat broke? I suppose it's possible, but I've not partied in one yet.

Developed also carries the connotation that these countries are matured, at the end of the cycle, and therefore stable. I think that's a very good way to put it. But what cycle are we talking about? We'll get to that in a bit. You might also hear developed countries called *industrialized countries, post-industrial countries, more economically developed countries* (MEDC), *fully developed countries, Advanced Industrial Economies (AIE)*, or even the *First World*.

A **developing country** on the other hand, is perhaps not so rich. However, it may not necessarily be totally poor either, which could be indicated by such measures as GDP and GDP per capita. Most of the planet would be in this category (both numbers of states as well as numbers of people) and, as such, there is tremendous range in the spectrum of "developing-ness." Some states are close to being fully developed. Others are desperately poor and stagnant or not really "developing" at all.

Perhaps developing countries are best described as all the regions that are obviously not fully developed. Is that a nebulous enough description? Yes, but it kind of works. The operative word here is *developing*, which means that these countries are in transition, are changing, are developing into something newer, better, and richer. No state is actively trying to un-develop are they?

Synonyms for developing countries include *less-developed countries* (LDCs), *least economically developed countries* (LEDCs), *underdeveloped nations, undeveloped nations, emerging national economies, newly industrialized countries (NICs)*, and the most popular term that everyone recognizes: *the Third World* . . . although that last term has largely fallen out of favor, and is considered derogatory to some.

The **Third World**. What a hilarious term. Everyone recognizes and uses it, but no one knows what it really means. And it really doesn't mean anything anymore. Here's the real deal: During the **Cold War**, there were two opposing camps: Team 1 was the capitalist democracies led by the US, aka "The West." Team 2 was the communist countries led by the USSR, aka "The Soviet Bloc." They became known, respectively, as the **First World** and the **Second World**. All the other countries were encouraged to join one of these teams, and many did—or at least would allow themselves to be associated with Team 1 or Team 2. In hindsight, it was comparable to siding up with Thing 1 or Thing 2 from "The Cat in the Hat." But I digress.

However, there was a group of holdouts who refused to side up. They wanted nothing to do with the Cold War nonsense, and identified themselves as non-aligned. India, Egypt, Ethiopia, and Yugoslavia led this **Non-Aligned Movement (NAM)**. (Created in 1961, NAM is actually still around, has 120 member states, 17 observer states, and 10 observer organizations as of 2019.) I point out those particular 4 NAM leader states by name because their status then, as now, has a lot to do with how the term "the Third World" evolved. Many other similar states joined this movement, and as you might imagine, their poor, less-developed economies became an identifying mark . . . a mark that still sticks today. There is no more Cold War, and there are no more First and Second Worlds, but somehow the Third World is still out there batting. Can someone please tell them the game is over?

But seriously, why was it that these countries were so much poorer than the First and Second Worlds at the time? And why have some of them not changed much since the Cold War era, which is why the Third World terminology has stuck? Ah! I'm so glad you asked! To understand today's world of development, we must travel back in time just for a bit and pick up a theory we introduced at the end of the last chapter. . . .

A PITHY HISTORY OF THE PERIPHERY

Think back to that Immanuel Wallerstein stuff I threw at you in chapter 4. Remember the **world-systems** approach I alluded to? It was that set of mechanisms which concentrates wealth from the economic **periphery** to the **core**. Time to dust it off and take a second look through a historical lens, so that we can more clearly see why places like India and Vietnam and Brazil came to be on the peripheral edges of the world economy prior to the Cold War as well as how the US and Europe were at the core.

In doing so, we can also elaborate on why these descriptions have persisted into the modern era, along with some examples of major changes which are currently underfoot. Ha! Watson, the game is afoot! And Sherlock Holmes' Britain is a great place to start the investigation. . . .

Deucedly developmental deductions, Watson!

COLONIAL PERIOD

In terms of world history, the 1500s to the early years of the 1900s are a total European heyday. Prior to this age, neither European states individually, nor the region as a whole made a significant impact on the rest of planet earth. They certainly were not anywhere near the 'core' of a global economy, nor a center of political power, nor a center of technical innovation. You would have had to go to India or China or the Middle East to get those things. Why? Because Europe was a backwater; the US was as of yet "undiscovered"; and Japan was a samurai-infested, isolated state.

But what a difference a few hundred years makes! As the 1500s and 1600s unfolded, European traders, scientists, businessmen, sailors, and military men all increasingly adopted and refined technologies from other societies. Namely, Arab sailing technologies from the Middle East and gunpowder from the Chinese. As a major case in point, the Chinese had invented explosive powder, and employed it for . . . ready for this? . . . fireworks! The Europeans borrow this explosive powder stuff from China and they make . . . guns! I think you may be starting to see how the core/periphery tide was starting to turn already . . .

Chinese firework make big bang for Europeans!

The Europeans take their newly modified and improved technologies and build grand navies and militaries which inevitably make them masters of the seas, both militarily and for trade. And with that mastery came colonization and imperial takeover . . . you know, the Europeans setting sail and taking over the rest of the planet. North America, South America, Australia, and eventually Sub-Saharan Africa, South Asia, even parts of China et al, come under the direct control and exploitation of the Europeans. Raw resource wealth, wealth from any existing manufacturing, along with wealth created by trade itself, now flowed from the colonized areas back to Europe. Europe is quickly becoming the core.

Tools of the "trade"

Then starting in the 1700s, the Europeans are the first to undergo this whole process we now call industrialization. Making machines to do stuff that humans and animals used to do . . . and doing it better, faster, cheaper! Virtually all aspects of agriculture, manufacturing, mining, and transportation are made way more productive and profitable. And this also means that Europe is now the center of innovation and the technology leader . . . which of course makes them richer still. Europe sustains this revolution by fully mastering trade relationships around the globe: absorbing the raw resources of the planet, taking them back home to be processed (thus creating jobs and wealth at home) and then pushing their manufactured goods to their colonies.

Example: the UK forces India to stop producing cloth by insisting that they instead sell the UK all their cotton. The UK takes the cheap cotton back to the textile mills in London, produces it into cloth, and then sells the finished goods back to India, at a higher price of course. Buy low, sell high. Europe is now the core and India has become part of the periphery. It is important to note that China undergoes a similar devolution at this time, having their core status severely eroded by European, Russian and even Japanese trade dominance during the 1800s . . . culminating in the catastrophic **1839 Opium Wars** in which the UK totally dismantled Chinese authority altogether in their bid to forcibly sell opium under the banner of "free trade" to China. Wow. Put that in your pipe and smoke it, China. And they did.

Please keep in mind, in this scenario Europe's strong naval presence is perhaps more significant in its successful take-over of global trade systems than it is for its takeover of physical territory itself. Because, inevitably, they will lose those colonies, but the trade dominance remains for some time after that. Here is an equation to consider:

Dominating world trade creates wealth at home = more taxes collected by government = stronger government = government invests more back into technology, infrastructure, military and businesses = stronger military and businesses = more power abroad to colonize and control trade = more wealth = go back and repeat equation. Repeat for centuries.

Not bad for a few hundred years work by folks who previously were groveling in the mud waiting for King Arthur to find the Grail. But changes are coming. . . .

Got Grail?

INTO THE MODERN ERA

Of course, the good old days don't last for the Europeans forever. Independence movements worldwide were kicked off by the American Revolution in 1776, but they didn't happen all at once. It took from 1800 right on up to the 1960s for those European imperialists to lose control of all of their overseas possessions, with many modern African states the last to shake off the colonial hangover. These movements coincide with the abolition of slavery and serfdom worldwide, and thus the end of free labor, which also put a crimp in the European wealth-generating machine.

In addition, some former colonies and others learned the lessons of European history very well . . . and then duplicated them! The United States in particular underwent its own industrial revolution and rise to global power, and thus entered the core arena itself by the 1900s. Russia and Japan modernized as well, and quickly became core players too, albeit on rockier paths that we will investigate further in later chapters. Australia and Canada followed suit. Point is, the core expanded to include others now in competition with the European masters, and the periphery became even weaker as these new global power players had more economic strength than ever imaginable in the past.

But hang on! Do not interpret this as the end of European economic dominance! Because, as I suggested above, the trade patterns had been well-established for hundreds of years, and they proved to be a much more tenacious beast than simple ownership/control of colonies. Colonial labor may not have been free anymore, but let's face it: former-colony labor remained extremely cheap, as it does into today's world. Basic manufacturing has shifted from the core to the periphery in order to take advantage of this cheap labor, but the profits still go to the core. Same as it ever was.

In addition, while the UK or France or Spain may have officially left the colony of India or Algeria or Chile, they usually left behind multinational corporations under their sponsorship which still controlled and/or profited from the raw resource of the former colony. This is a situation that is still common today. Almost all of the biggest and richest multinational companies of the world originate in Europe or the US (China is now catching up fast), and operate with huge competitive advantage in the "poorer" parts of the planet . . . you know, the periphery! Local yokel companies in Africa or Asia can't possibly compete with the likes of Wal-Mart or Exxon. Core wins!

End result: core countries still import cheap raw resources from the periphery and produce high-value commodities which they then export to the periphery. Core wins! Quite frankly, the equation has not changed much, and neither have the goods traded. The periphery exports cotton, tea, bananas, oil . . . and imports finished goods. The core imports raw stuff and basic manufactures, and exports guns, machines, finished goods, etc. Same as it ever was.

Let's be brutally honest here: the rich, core countries have absolutely no vested interest in changing the system. Why would they? No government or business is going to actively try to change the rules of a game that ensures their own victory. So is economic history finished? Is the system now set forever? Ha! Nothing stays the same forever, and even in the most structured system. . . .

ESCAPE IS POSSIBLE!

As you have now seen in this historical overview, these core/periphery labels are not stagnant statuses my friends! Oh no! China and India in particular used to be core, then were subverted to periphery during the rise of the Europeans, but now they are breaking out again. And speaking of the Europeans: for sure, they are still rich parts of the core . . . but they are starting to slip a bit, and no one here in the 21st century would label Western Europe as a center of innovation or technology anymore. Many analysts are now speculating that perhaps even the top-slotted USA is taking it on the chin from the rising Asian titans. Times are a-changing. However, it is difficult to be dethroned from the core once you are there, so there is no point to focus too much on the top dogs . . . they will maintain their monopoly for some time to come. More to the point, let's look at those entities that are climbing out of the periphery to join the kings of the core. . . .

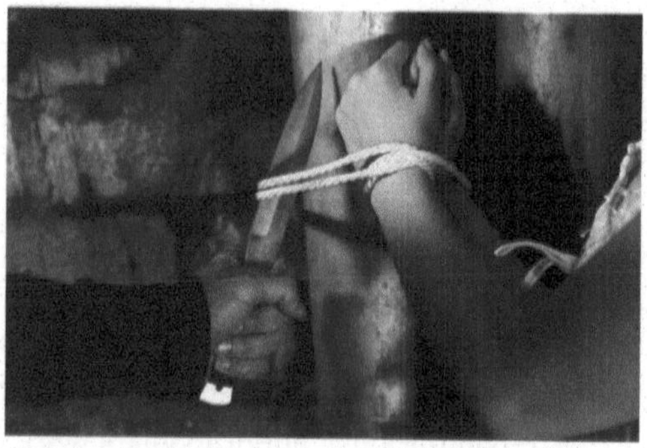

Let's get the heck out of the periphery!

ESCAPEES OF NOTE

There are a variety of countries in the world right now that don't fit nicely into the definite core or definite periphery categories. Some states do have significant control over their own resources, or have competitive local industries that do compete on the international stage, or do make more high-valued finished goods for export. Of note, Brazil, Argentina,

Chile, Turkey, South Africa, Mexico, South Korea, Indonesia, Bahrain, and the UAE could all be considered much further down the development path, and I even refer to them as power-players within their respective regions.

A specific cluster of escapees is referred to as the BRIC countries (Brazil, Russia, Indian, and China), a term initially coined in 2001 by an economist that would later become BRICS when South Africa was added in 2010. These countries were considered emerging national economies given their significant influence within their regions. The trajectory for India and China was more positive than for the others since the 2008 financial crisis, but this grouping just goes to show how the rest of the world is starting to recognize the strategic shift underway. The shift I am referring to is the rise of non-Western economies into the core.

With no reservation though, China and India are the future core members to consider. How did they climb out of the core/periphery cycle? Well, both of these countries are unique in that they were previous core members, countries with huge populations, and countries with significant periods of technological innovation and know-how. It's not like they have had to start from scratch. The current re-rise of these Asian giants has more to do with throwing off the yoke of colonial powers and reasserting political, and then economic independence.

India chucked out the Brits in 1947; the Chinese started their modern era after the defeat of their Japanese oppressors in WW2, and their bloody civil war, which ended in 1949. Both states have used their massive pool of cheap labor to increase agricultural production, create a manufacturing sector, and invest in their infrastructure. As pointed out in previous chapters, China has really focused heavily on its manufacturing sector to make a myriad of products for cheaper than anyone else on the globe, and they have done this exceptionally well, to the point of now easily beating the core countries in the competition. Ever see *Made in China* on the tag of any products you own? Ha! You probably can't find anything you currently possess that was NOT made in China!

We take this back now.

With all the massive inflow of wealth, China has invested heavily in its infrastructure, technology and industries to make themselves even more competitive . . . hey! Just like the core countries do! Yep. China is now in the core. While India is not quite as far along, it has a vast (and growing) reserve of labor coupled with high projected growth and expansion plans. Instead of focusing on manufacturing, India has been specializing in service sector and high-end stuff like computer programming. Ever call your Dell computer 1-800 help line? Who did you talk to? Yeah, you know it! Someone in India! Their famous technical support centers are even featured in *Slumdog Millionaire*! Oh yeah! Citing Indian specialization reminds me to include this next section. . . .

ESCAPE ROUTES

I have now alluded to different strategies which India and China have taken in order to become more developed, and less peripheral. Let's go ahead and flesh these things out totally so you can better assess the strategies that other countries have taken (or will be taking in the future) to attempt to better their situation. There are some major paths that have been followed by poorer peripheral states in order to bust out of the cycle:

1. **Nationalize** a particular resource in the country, focus on efficiency and technology for that one industry, become way more competitive globally for that one industry, and thus reap a bigger percentage of economic benefits. We've talked about this concept already. Middle Eastern countries, and almost all other OPEC members, did the nationalization thing with their oil. It sometimes results in alienation by the core countries, and less competitive and poorly run local industries. But no one would argue that Saudi Arabia, UAE, Qatar, or even Russia is worse off for nationalizing their oil. It has reaped large profits in the modern era, and gained them some measure of control in the global economy. Other states have done it with other commodities as well, even into the modern era.

2. **Import substitution** has been experimented with ever since the **Great Depression** crunched the whole global system. This is a policy in which the government works with private industry to set up production of a variety of higher-valued manufactured goods like cars, washing machines or textiles. The point is not to export them, but sell them to their own citizens, thus decreasing the reliance on importing expensive finished goods from other countries. These local goods are given an advantage by the government putting high taxes or **tariffs** on similar goods being imported from other countries, thus making the local goods cheaper and a better deal. More of a short-term fix than long-term solution, but has had mixed successes.
3. **Specialization** on a few commodities which, with government help, can become efficient enough to compete on the export market globally. China focused on making a literal boatload of manufactured commodities at cheap prices, and has done very well. But most other countries usually focus on just a handful of things: India is focusing on service sector jobs and computer programming, while Brazil focuses on an extremely competitive steel industry, as well as military computer applications.

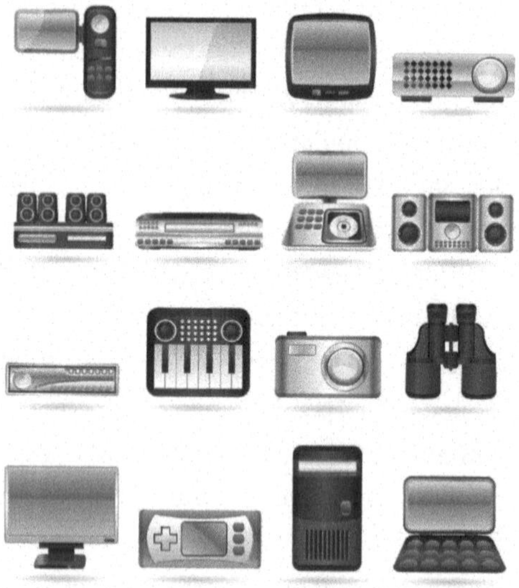

Consumer electronics: Asian Tiger escape route.

Historical example: Beginning in the 1970s and 1980s, South Korea, Singapore, and Taiwan were all considered newly industrialized countries (NICs). These countries have several things in common: rapid industrial/economic growth, political and social reforms, and trade reforms. Add Japan to that threesome, and you have an Asian core that has made trillions focusing on consumer electronics and automobiles. Those country names are just synonymous with high quality products! When a country does something the best, everyone knows it, and everyone wants it . . . especially when they do it the best for the cheapest . . . then that state is making the money! And thus heading out of the periphery, which many of those **Asian Tigers** have done!

Make no bones about it; in all of the examples above, former peripheral states challenged and out-competed core states in these industries, which is why they can't be counted among the totally undeveloped, not-as-developed, or underdeveloped states in the world today. Egads! I have totally lost sight of the development of this chapter! We were supposed to be talking about the developed versus the developing world! Let's get back to topic. . . .

DIFFERING DEGREES OF DEVELOPMENT

What are some common measures of the fully developed state versus the fully undeveloped one? The richness versus poorness? Core versus periphery? People in a developed state typically enjoy a high standard of living—i.e., life is good. Food in the fridge, healthy children, bills paid, access to health care, good housing, good environment, and there's even some leisure time to party. Yeah, that sounds about right. And developing? Those folks are probably not eating as much as they should, maybe not getting access to health care, maybe not having clean water or electricity, and their leisure time is otherwise known as "too sick to work."

But we can do even better than that. Let's look systematically at some categories of human life to better compare and contrast the standards of living on our planet. Following is a brief outline of some of the contributing factors of current levels of development in the wider world. This list is not exhaustive, as we have not included all the historic factors that have influenced current state of affairs, or physical resources that these regions possess. However, I want you to think about the bigger developed/developing picture here—if we can understand the mechanics of what is going on in these places, we can better understand the present and better predict the future.

Nukes are high-tech!

TECHNOLOGICALLY SPEAKING . . .

The fully developed state has the highest levels of technology across the board. I'm not just talking about computers here either. Agriculturally, they can grow more food with way fewer people; one big tractor can do the work of hundreds of humans. Industrially, they can produce more and better stuff with machines in factories that typically displace workers as well. Machines do the work; people run the machines. They are also the creators of information; they have the best military technology like cyber warfare and drones, and they do extreme stuff which requires extreme cash, like space exploration.

The newest and coolest stuff is created in the fully developed world. Infrastructure like roads, bridges, buildings, and communications are at the highest quality and highest safety levels. Emergency and medical services provide lightning response and insane possibilities of keeping people alive. You can reattach a severed arm? No way! The frontiers of science, medicine, and technology are pushed forward here. The developed world produces all the newest information, is responsible for 99% of new patents, new trademarks, and new copyrights. Meanwhile, *Grand Theft Auto IV* offers mind boggling graphics to thousands of glassy-eyed users.

The developing state does not have all of these technological advances, and some states may have little to none of these things. Agriculture and industry are still labor-intensive, and typically not as productive. Infrastructure is not as good; roads and communication systems may be lacking. When an earthquake or flood occurs, sub-standard structures crumble. Buildings, sewer systems, and electric grids may be not as safe or efficient as possible; blackouts and back-ups occur regularly. They may possibly be at minimum standards, which means they don't function as well, or last as long. Life-saving technologies and services operate at less than favorable standards. There is no 9-1-1 to call. You will lose the arm.

INDUSTRIALLY SPEAKING . . .

The developed state uses more machine labor than human labor. As such, they have higher **labor productivity**, which means one human can do a crap-load of work. One guy on a tractor can plow a thousand acres. One guy in a semi can move twenty tons of cargo 500 miles in a day. Because of this, they get a heck of a lot more work done per person, but the developed world is also much higher in energy consumption per person; just think of all the energy you use in one day for electricity in your home, at the office, driving your car around, and in making your skinny, half-caff, sugar-free, mocha lattes. In addition, most of the energy produced in the developed world is based on fossil fuels. As a result, there is a high fossil fuel dependency in the developed world: Iraq will have to be invaded . . . hahahahaha I'm joking! (oops, bad example)

The fully undeveloped world can vary greatly in the amount of machine versus human labor. In places where development is low, there is a greater percentage of human labor involved, thereby reducing the rate of productivity in total work done. How many humans does it take to plow that same thousand acres? And how much longer does it take them? Just using elbow grease, the energy consumption per person is significantly lower. Even the type of fuel used changes in the developing world; fossil fuels are too costly, so less efficient fuel sources like coal, wood, and even dried dung are used more often. Dung? Dang.

Dung: What a load of . . . inefficient fuel.

There is a plus side, though: a lower dependency on fossil fuels, and dried feces is cheap, as it should be. However, those fuels are not very efficient in outputting energy and are overly efficient in outputting pollution. But do realize this little known fact: many oil-exporting countries actually don't use a lot of oil themselves . . . it is too expensive for them! They make too much money exporting the stuff to actually want to burn it! That's why places like Iran are investing in nuclear energy, even though they are a major exporter of oil. How bizarre!

Looking broadly at the economies of each of these societies, we have to point out that the fully developed world has not only undergone a full industrial revolution in their society . . . but most of them have now long since left the industrial economy behind. They took full advantage of the technological advances and profits from their industrial/manufacturing era, and are now onto the service and quaternary sector activities. Totally undeveloped states have yet to undergo this revolution, or perhaps only done it in patches . . . mostly due in part to foreign multinational corporations setting up industrial capacity for specific manufacturing purposes. In that scenario, the true benefits of an industrial revolution are never realized. They just get the crappy jobs and the pollution.

ECONOMICALLY SPEAKING . . .

Speaking of those crappy jobs, how do we differentiate the developed and developing states economically? Well, the easy descriptors include what people do for a living, which is the stuff we covered in the previous chapter. It never hurts to summarize again, though.

In the developed state, higher technology and industrial capacity equate to increased labor productivity, which in turn equates to higher salaries. The economy is more diversified. There are more desirable types of labor, more choices of occupations, of typically safer types of labor. You can get a job just using your brain. What a novel idea. More of the economy is focused in the service sector, with some in the quaternary sector as well. Primary and secondary jobs are available, but are not the primary earners of GDP for the state.

More importantly, GDP per capita is high. **Corporate earnings** are higher. The developed state produces value added products which are more expensive. Processed goods and high-skilled services are apparent: computers, cars, lawyers, investment bankers. Exported high-value goods equate to a **positive trade balance** for the state; it sells more stuff than it buys every year. It has a surplus in their economic bank. All of these factors lead to increased investments that are funneled back into the economy and infrastructure, perpetuating the positive cycle.

In the developing state, lower labor productivity usually leads to limited options for everyone economically. Most of the economy is focused in the primary and secondary sectors. Lots of agriculture and basic manufacturing exists, which produces goods of less value. Harder labor, and laxer environmental and safety laws mean more unhealthy/dangerous working conditions. It should also be noted that many developing countries have a very large chunk of their GDP based on a single commodity, like oil or diamonds or copper. This is an extremely risky situation for the state. What happens when the price of oil on the world market plummets . . . sometimes overnight? Yep, that's right: that state loses money, instantly. Makes it hard to plan for the long term. Very volatile economics for the developing state.

Lower salaries, lower corporate earnings, and a special twist—many of the companies operating in the developing states are foreign multinationals, so most profit exits the state and goes back to the multinationals' country of origin. Low value goods are exported, high value goods are imported, resulting in an overall **negative trade balance**. Put it all together and it spells out a lower investment in the infrastructure and economy, less personal wealth, usually less GDP, and almost always less GDP per capita—although we have seen some exceptions to this rule as well.

DEMOGRAPHICALLY SPEAKING . . .

Just to reinforce some concepts from Chapter 2, we should also highlight some obvious differences in the demographic makeup of a state based on its level of development. Fully developed states have fully undergone the demographic transition, and both their fertility rates and their population growth rates are relatively low. Fertility rates usually hover around the 2.1 replacement level, and the overall population total is either stable or perhaps even shrinking, as is the case of Japan, Russia, and many Western European states. The column or inverted pyramid shape of the population pyramid indicates that there are typically a smaller number of folks under the age of 15 (dependents) as there are working age cohorts 15 to 55 (the suppliers of labor and wealth).

By contrast, less developed states are somewhere in late Stage 2 or Stage 3 of the demographic transition, and their classic pyramid shape suggests that they are currently in a cultural lag/population explosion scenario. The later in Stage 3 they are, the more rounded the pyramid will appear, especially in the younger age cohorts. Fertility rates remain well above the 2.1 replacement level, usually in the 2.5 to 4 range . . . and the higher the number, the more

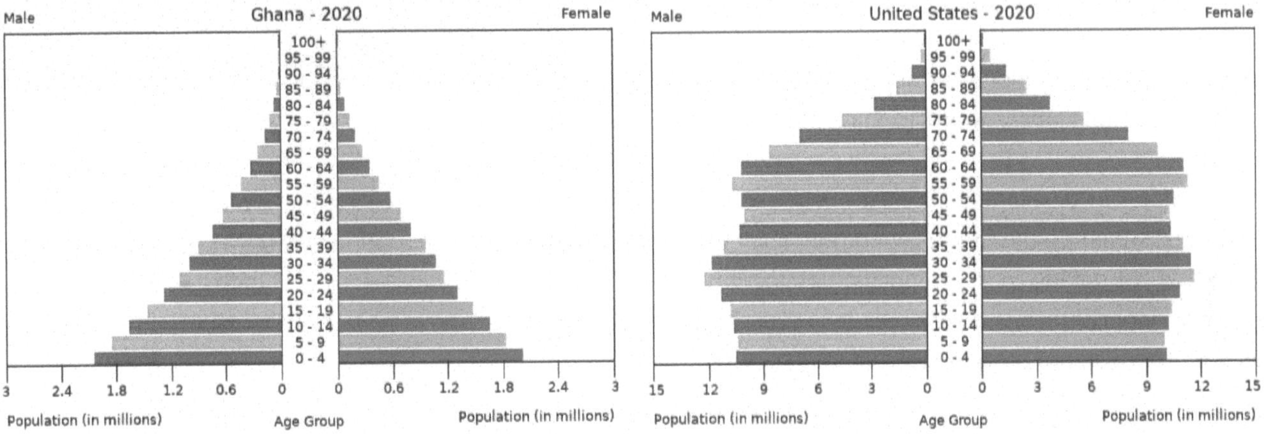

Developing state pyramid . . . develops to . . . a fully developed column state.

explosive the overall population growth. Of particular note, the classic pyramid shape indicates that there are way many more folks under the age of 15 (dependents) than in the working class/provider brackets.

What does this mean in real life, for real families? In the developed world, people have just one or two kids, invest a lot of time and money into keeping those kids alive, healthy, educated, and happy . . . and then eventually pass all their remaining wealth to the kid once mommy and daddy kick the bucket. This type of high investment per kid equates to a huge competitive advantage for children of the developed world. How can they fail? Well, I'm sure some of your fellow students are figuring out new creative ways to fail despite their advantage. We call them slackers.

BUT . . . slackers aside . . . After the economic crash of 2008 and the slow ongoing recovery, it is getting harder for students (like you all) to get ahead financially since you're all taking on massive amounts of debt, becoming more entwined in the "gig economy" that you may or may not get a good job in . . . which in turn will likely decrease the number of kids you have, if you have any at all! How will this affect the fully developed world moving forward now that future generations may not be doing as well as the previous generations, for the first time in a century or two?

And in the undeveloped world? Large family size means less investment per kid, less resources per kid, and much less wealth to pass on after death. And they don't have that much to pass around to begin with. As total population continues to swell year after year, there are less and less resources to divide up among an ever increasing number of folks. For even an exceptionally smart or gifted child, the climb out of poverty is fraught with major challenges.

SOCIALLY SPEAKING . . .

In the developed state people make decent money. Those folks have higher salaries, which equates to more disposable income, which equates to higher consumption rates of everything and higher rates of saving. It also means people can invest in their own future, as well as provide better health care and education to the next generation. That is crucial. Literacy rates are high. Developed countries heavily invest in education—each generation gets smarter, richer, better, and more cognizant of the Plaid Avenger's World.

The developed world also has higher mobility and social safety nets, like welfare, which increase **risk-taking** and increase opportunities. You can risk quitting your job to start a new business or go back to school without your family starving to death. This increases your opportunities for employment and income-making potential. It also makes the country a center of innovation and wealth, as these risk-taking entrepreneurs become successful. Think about it: Could Bill Gates the kid have become Bill Gates the trillionaire had he been born in Uzbekistan or Ethiopia? Doubtful. Infrastructure and education provide a tremendous amount of opportunities that allow the citizenry of the developing world to make their dreams come true.

We are not so fully developed. Daily life is more challenging.

By contrast in developing countries, less expendable income means less food on the table, less overall consumption, less investment in the future generation, less education, and generally lower literacy rates. Lack of social safety nets leads to less risk-taking, fewer opportunities in employment, and in life. Availability and quality of infrastructure and educational opportunities are generally much lower, thus locking in the cycle of poverty from generation to generation. Focus is on survival, not upward social mobility.

HOW IS YOUR HEALTH?

In the developed state, people should be pretty healthy overall. Increased education, increased access to health care, and increased public hygiene equate to decreased infant mortality, increased longevity, and decreased susceptibility to epidemic outbreaks. You will live longer in the developed world, but what will it eventually get you?

Due to changes in work types, increased consumption, and wider variety of dietary choices, a distinct health shift occurs as a state becomes fully developed. People become more sedentary as they use their brains instead of their brawn to work. They have stress from using their brains too much. People over-consume in caloric intake. In diet, they move from whole grains and fruits and vegetables to processed foods with a radical increase in sugars, fats, and meat. Result? Around 75 percent of people in a developed world die from diseases of the circulatory system (heart disease, stroke) and cancers.

Poorer people eat more grains and veggies. The richer you get, the more the average (increasingly obese) citizen's diet sucks!

poor → rich

In a bizarre irony of the "rich" world, access to fresh fruits and vegetables has become a marker of wealth. There are a lot of **food deserts** in poorer areas of wealthy countries that also add to obesity and poor health rates. Fast food is now so much cheaper than organics and healthy options, that a increasing "wealth gap" can almost literally be measured by the waist-lines of different groups within a single state.

In the developing world people are not as overall healthy, but for different reasons. Less education, less access to health care, poor public sanitation and hygiene, and even little or no access to immunizations equate to high infant mortality and lower life expectancy. Epidemics can be catastrophic. Diets consist of basic grains and fruits and vegetables, which is really good, but their caloric intake may not be sufficient for optimal health. Result? 50 to 60 percent of the people will die of an infectious or parasitic disease, or in childbirth. The infectious diseases are easily preventable and curable. Access to health care and existing medical techniques/technologies becomes a true life and death situation for billions across the underdeveloped globe.

PLAID AVENGER SURE-FIRE WEIGHT LOSS PROGRAM: Move to a developing country. You will certainly eat less than you do now, and likely eat stuff that is actually way more healthier for you when you do eat. And hell-tons of physical labor tends to keep the pounds off too.

HOW BIG IS YOUR MIDDLE?

No, I'm not talking about your waistline . . . we just talked about that in the previous section . . . I'm referring to the size of the middle class. I want you to think critically about this concept, because it ends up being a core component of how most of us now classify the levels of development around the planet . . . especially for those states that do not easily fall into either the full-on developed nor the totally un-developed categories.

In fully developed states, there is much less wealth disparity. That is, there is a roughly equal distribution of wealth across the society. Now, every society and every state has some wealth concentration in the hands of a few rich folks, i.e., the upper class, or aristocracy. And every society and state has a lower, poorer class of folks with not much wealth at all. That is true the world over. However, the percentage of the haves to the have-nots differs wildly from state to state, and I'm suggesting here that fully developed states have a lower percentage of wealth concentration in the hands of the few, with greater distribution of that wealth across the whole society. That's where the middle comes into play.

Poor peeps exist even in rich states, as do the few ultra-rich in poor states. But in the fully developed states, the middle class constitutes the biggest group overall. You know, it's the folks that have a good job, a nice house, a decent car, can go on vacations, and can send their kids to college, and can eventually do this awesome thing called "retire." The ability for a majority of folks in the state to achieve adequate levels of wealth is an extremely powerful stabilizing factor, and one that may be the the newest, best indicator of how developed a state is here in the 21st century. Because as you have seen in previous chapters, a state like Equatorial Guinea has a huge GDP, and even a big GDP per capita figure. BUT, the wealth is totally concentrated in the hands of the state dictator and his family . . . while the masses are starving. Therefore, there is no middle class at all. Speaking of which. . . .

In the developing world, there are much greater differences between the haves and have-nots . . . in some societies it may be best described as the few have-all versus the masses of have-nothing. **Wealth disparity** is huge. Wealth is overwhelmingly concentrated in the hands of a few, who then, of course, have the power and influence to make sure it stays that way! The impoverished masses have no real competitive avenues to accumulate wealth, and for them "vacation" is another word for unemployed, and "retirement" occurs only at death. Kind of like killing a *Replicant* in the movie ***Blade Runner***. This makes for an unstable economic and political situation in the state as a whole, which brings us to. . . .

GOVERNMENTALLY SPEAKING . . .

In the developed states, democracy is the undisputed champion. Almost all of the fully developed, rich states on the planet are democracies. The political/economic structure may have varying degrees of socialist policy, but all are staunchly democratic. These systems are more dynamic, more open to change, and they prove it by rotating out leadership in a timely and regulated manner. In addition to their dynamic demeanor, developed states are politically stable, leading to stable economies which actively participate in global trade and investment. This stability is one of the primary reasons that international investment is typically higher in developed countries: because your money and your investment are safe in a stable environment. It should also be noted that since the mid-20th century, democracies don't generally fight other democracies. The can/do join forces to fight non-democratic states but use political/economic tactics to deal with each other instead of military action. More stability, more money.

Power to the few = revolution overdue.

In the developing world, most other types of governments can be found: military dictatorship, theocracy, monarchy, one-party states, and chaotic states abound. It should be noted that many developing states are also democracies, but usually not well-established with long track records. Why is this important to note? Aside from the newer democracies, these systems are more static, or closed to change; some exhibit no possibilities for change at all. The more closed a system is, the more opportunity arises for corruption. This creates an unstable situation in the long term because underlying tensions and forces usually manifest themselves in violent upheaval, civil war, or outright implosion of the state. Check into the effects of civil strife, civil war, and right-wing crack-downs across the entire Middle East as your best most modern example of this.

Unstable governments can often lead to unstable economies, and unstable economies are not where international investors put their money. These states are often insular, protecting their industries or economic interest above all, which discourages global trade and investment. As alluded to in the middle class section previously, impoverished masses themselves are always a potential flashpoint for total chaos within a state: when conditions get horrific enough, and people have nothing left to lose, they sometimes band together and topple governments. Check out the Haitian or Russian Revolutions for details.

MILITARILY SPEAKING . . .

There's one last category that most people don't really look at to assess development status, but the Plaid Avenger does lots of things that other people don't. I think it's a keen insight into what's happening in the society . . .

In the developed world, military technology is at the maximum. Nuclear capabilities are either possessed or easily acquired. Like in industry and agriculture, most manual labor is done with machines. Soldiers operate big machines and the "death-wielding labor productivity" is very high. Developed states prefer to use their military technology as opposed to large numbers of humans on the ground. Human casualties and injuries are minimized at all costs. A major final point: while total expenditures on the military may be high, in a developed state this expenditure is a fairly low percentage of their GDP total. Check out the box to the right to see what I'm sayin'. (And holy hand-grenades! Be sure to note how many Middle Eastern states are on this astounding arming agenda! We will follow up on that issue in chapter 18.)

In the developing state, military technology can be advanced, but typically is not. They usually buy hand-me-down weapons and last year's camouflage fashion from the developed states, and to be sure, the fully developed states never sell them the top-of-the-line gear. Some states, like India, Pakistan, and China, have nuclear capability, but they are pretty much alone in the developing world in this respect. Lots of other states may want it, but will have to try and buy it, since they lack the infrastructure to develop it themselves.

In developing countries, manpower is still the primary component of the military, and most real action will be dudes on foot, firing guns. China has the largest free-standing army on the planet right now. The last major point: while total expenditures on military expenses may be lower in the developing world, it's a fairly high percentage of GDP and/or the government's budget. Yep. You got it. The less money a state has, the bigger % it usually spends on guns.

"We got a man down! But military expenditures are way up!"

WHAT THE H? WHAT THE WHAT? WHAT THE WHAT? WHAT THE WHAT?

MILITARY EXPENDITURES AS A PERCENT OF TOTAL GDP

(THIS IS NOT AN EXHAUSTIVE LIST)

Country	% of GDP
North Korea	33.0
Saudi Arabia	8.8
Oman	8.2
Algeria	5.3
Kuwait	5.1
Jordan	4.8
Israel	4.3
Ukraine	4.1
Pakistan	4.0
Russia	3.9
USA	3.2
South Korea	2.6
Turkey	2.5
India	2.4
France	2.3
China	2.0
Australia	1.9
UK	1.8
Brazil	1.5
Canada	1.3
Germany	1.2
Japan	0.9
Switzerland	0.7

Source: Trends in World Military Expenditure, 2021; SIPRI, Stockholm International Peace Research Institute

THE BOTTOM LINE: WHO IS DEVELOPED? WHO IS DEVELOPING? WHO IS SOMEWHERE IN BETWEEN?

Let's get this chapter wrapped up already! Given the terminology, the historical precedent, and the specific categories of comparison, can we now officially point out exactly who the developed and developing countries are? Well, why not? Let's give it a go! Following are a few different systems for classifying these levels of development, and we might as well start is out with the best one . . . namely, mine!

PLAID AVENGER DEVELOPMENTAL RULE OF THUMB

The rich countries are easy to see, so let's not dally on those developed dudes; the rest of the planet is much more interesting to consider. You won't find this theory in a textbook anywhere, besides this one, because it's a Plaid Avenger original. Lots of different states have lots of the developed world features, but still don't seem to be in the fully developed category. Like who?

Pakistan has advanced weaponry and nuclear bombs, but is not fully developed economically or demographically. India has nuclear bombs and is a well-established democracy, but is not fully developed. Saudi Arabia has a huge GDP and free health care, but is not fully developed. China is getting rich as Midas, has high technology, has nuclear bombs, a stable population growth rate, a diversified economy, and even put a man in space . . . maybe they are getting really close, but even they don't seem to be in the fully developed column yet. Okay, maybe they are. Its a close call at this point. So what makes the Plaid Avenger difference?

For me, it's all about how the average dude or dudette is doing. How are most people doing in the society? How much health care or good jobs or education do the *majority* of folks have access to? What is the real GDP per capita of the bulk of the workers? It kind of gets back to that theme I've hit you with about the middle class. There will always be a small percentage of excessively rich folks, and there will always be a lot of poor folks. But how are folks doing in the *middle*? Any country can have any amount of any of the factors listed previously, but if it's not accessible to the majority of folks, then I classify the state as still developing. The fully developed states are always marked by the majority of people doing well and having access to all of those great things.

Make sense? That's why places like India may be booming economically right now, may be focused on quaternary computer jobs, have a space program, be a democracy, have a strong military and a huge GDP . . . but most people in India still live on a buck or two a day. And "most" of a billion people is a lot of peeps! India may have lots and lots of trappings of the developed world, but how can you call them fully developed when half a billion people don't have access to health care, or lack a proper diet? My answer: I don't.

HDI: A UN TAKE ON HOW FOLKS ARE DOING

I think the good folks at the UN agree with the Plaid Avenger on this one, too. Here is their take on things:

The UN Human Development Index (HDI) is some sort of complex mathematical formula that takes into account poverty, literacy rates, education, life expectancy, fertility rates, and a host of other factors. It has become a standard

FIGURE 5.1 HDI

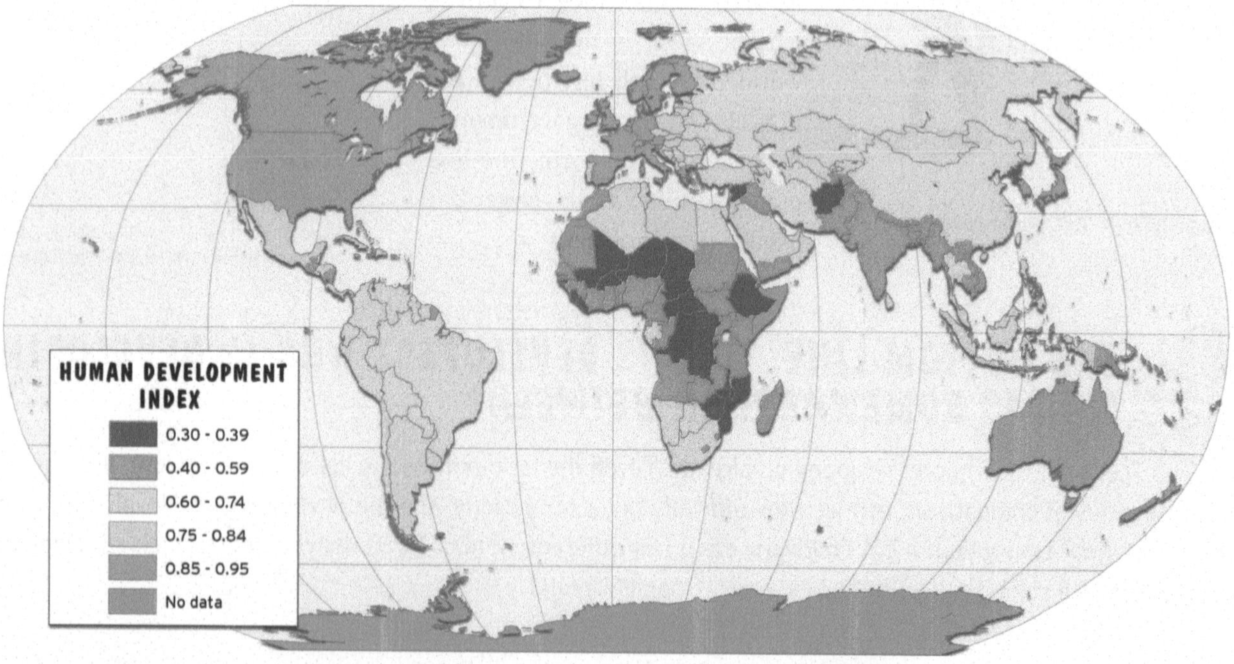

means of measuring overall human well-being. The index was developed in 1990 by a Pakistani economist, which gives it more credibility in the Plaid Avenger's eyes, and is used by the United Nations Development Program in its annual Human Development Report.

As you can see from the map, the closer you are to a perfect score of 1, the better off the humans are in that state. Take what you like from the map, but I just wanted to point out a few trends. Obviously, the developed, richer countries stand out near the top of the spectrum in North America, Europe, Australia and Japan. South America is not in bad shape by this measure, especially the southern states on the continent. Eastern Europe and Russia have slipped; they probably used to be in higher brackets, but factors such as life expectancy in some of these places has plummeted, as has their HDI. China, Central Asia, and some parts of the Middle East and Southeast Asia are also in that bracket, as their fortunes rise while the former states of the USSR fall. As we approach the bottom end of the HDI numbers, we see all of South Asia (esp. India) and big parts of Africa in orange. And finally, the countries in red include swathes of Africa, and failed states the world over.

AVENGER ASSESSMENT OF STATES' STATUSES

So tell us, Plaid Avenger: who is developed? Who is developing? Who is disorganized? Who is darn near deceased? Name the names! Alright then, since you have been so patient and asked so nicely, let's wrap this bad boy up. Here's a final word from the UN about development statuses:

> There is no established convention for the designation of "developed" and "developing" countries or areas in the United Nations system. In common practice, Japan in Asia, Canada and the United States in northern America, Australia and New Zealand in Oceania and Europe are considered "developed" regions or areas. In international trade statistics, the Southern African Customs Union is also treated as developed region and Israel as a developed country; countries emerging from the former Yugoslavia are treated as developing countries; and countries of eastern Europe and the former USSR countries in Europe are not included under either developed or developing regions.

Well, that's nice, I guess. And totally safe. But I'm all about the real deal. Many textbooks and news sources and academics simply lack the spine to actually point things out for what they are. But you know that I will throw down, and I'll tell you the shades of developmental plaid that I view the world in. Consider the following map, as created by yours truly:

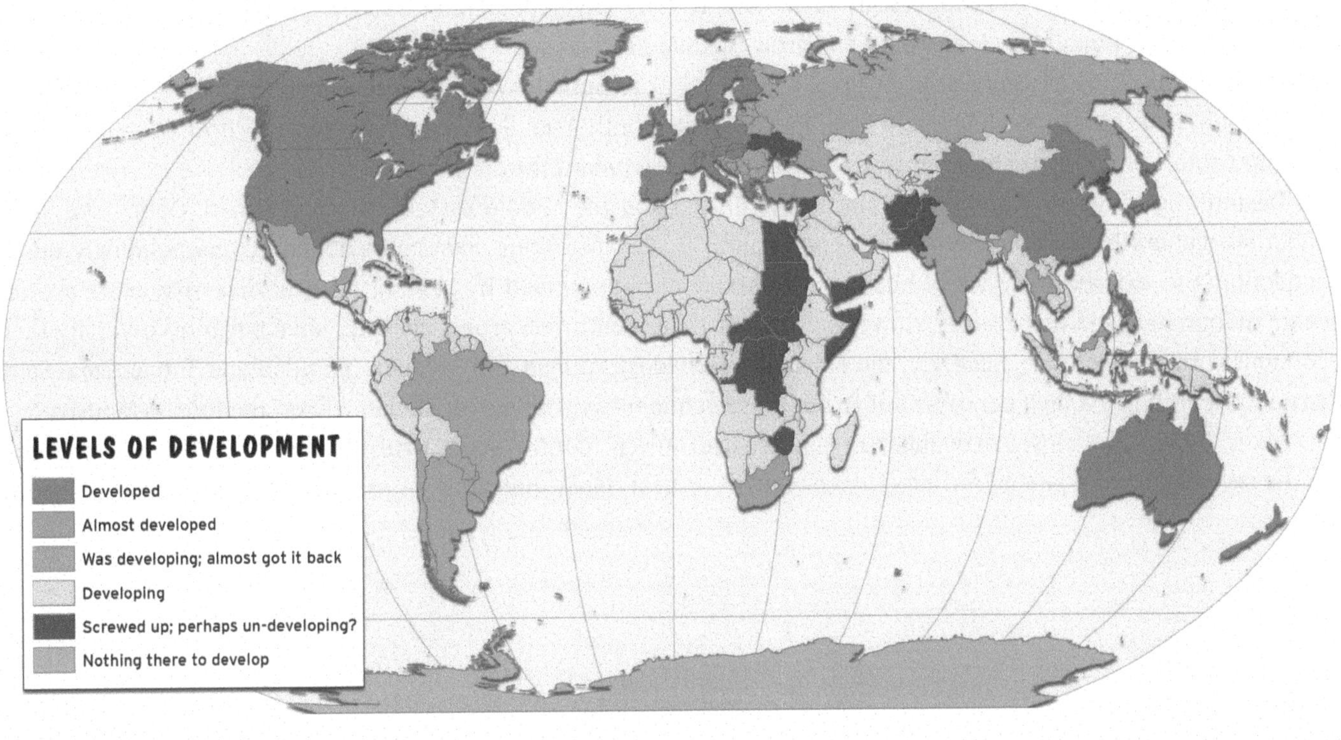

Here's the Plaid interpretation:

Full-on Developed: The easy category. We've beaten these guys down enough this chapter. North American region (US and Canada), Western Europe region, Australia-New Zealand region, Israel, and Japan are the pedal to the metal, full-on fully developed regions/countries. A few others in the club: South Korea, Taiwan, Singapore. Fully developed. No doubts. And one from Eastern Europe to watch: Poland, which is rocking right now!

You know what? I'm going to go ahead and go out on a limb. Let's pull the tooth and add China to the fully developed category. I know, I know, I know! Most other analysts and books are not ready to do this yet. That's why I'm ahead of my time. China is booming; and Macau, Hong Kong, and Taiwan are already considered fully developed, and are being reintegrated back into China proper. The Chinese have also pulled something on the order of 400 million people above the poverty line since 1980. Nothing suggests that these trends won't continue. Many economists have now slated China to be the world's largest economy by some time in 2020 . . . that's like now! High levels of technology, they are planning to put a human on the moon and have active exploratory missions to Mars, huge GDP, internal infrastructure investments, growing middle class. . . . Smells like fully developed to me!

Almost developed: This can be a bit tricky, but you can handle it given all you have learned in this chapter. Many South American countries are getting close, like Brazil, Argentina, and Chile. Mexico wavers back and forth—sometimes closer to full development, sometimes falling back to developing status. Check it year to year to see how it's doing. South Africa, Egypt, UAE and Turkey are beating down the door of full development as well. They have most of the features, but their GDP per capita is still fairly low and their middle class still thin in comparison to other developed places. India is booming, but with a much longer way to go, and a much bigger gap to close. The Philippines, Thailand, and Malaysia are our best Asian examples that are rounding out those countries getting close to full developed status.

Was developed, almost got it back: Eastern Europe and Russia are the only players in this category. And a distinct category it is. We don't really consider any place becoming "re-undeveloped," particularly when those places are nuclear powers and former world superpowers. However, the economic and political meltdown caused by the dissolution of the USSR in 1991 has left many of these countries with GDP numbers and life expectancy numbers more similar to impoverished African nations than European states. But they have the know-how, they have the technology . . . and like the six-million-dollar-man . . . they can rebuild it. The Plaid Avenger is assuming that they will: go for it, former commies! Russia could easily jump back to fully developed status soon, especially if they can encourage their declining population to get busy and grow the middle class some more.

Developing: This category still includes a whole lot of countries with a whole lot of people. Comprising big areas of sub-Saharan Africa, the Middle East, Central Asia, South Asia, Southeast Asia and South America, there can be tremendous variability on how far down the path of development they are. Or aren't.

Despite their high per capita GDP, places like Brunei and the Middle Eastern states of Oman, Qatar, and Saudi Arabia are generally not considered developed countries because their economies depend overwhelmingly on oil production and export. This lack of economic diversity is compounded by lack of political diversity; most are old school monarchies and theocracies. However, some of these countries, especially UAE, have begun to diversify their economies and democratize. Similarly, the Bahamas, Barbados, Antigua and Barbuda, Trinidad and Tobago, and Saint Kitts and Nevis enjoy a high per capita GDP, but these economies depend overwhelmingly on the tourist industry.

Many other countries—particularly in sub-Saharan Africa, Central Asia, Central America, the Caribbean, and Southeast Asia—have a patchy record of development at best. Most indicators of progress discussed in this chapter would be on the lower end of the development spectrum. Many have a long way to go, and are not currently traveling that fast on the footpath.

Screwed up; perhaps un-developing? This category is comprised of countries with long-term civil wars, large-scale breakdown of rule of law, or leaders who are totally insane. All could be described as nondevelopment-oriented economies or just outright chaotic systems. You may also see some of them referred to as "failed states." Examples: Yemen, Somalia, Democratic Republic of Congo, and North Korea. With an ongoing civil war, huge numbers of external/internal refugees, and a dictator/leader who refuses to leave office, Syria is currently the best (or is it the worst?) of this category. I hate to have to do this, but my friends in Pakistan are currently on a serious rocky road which may end catastrophically for them, so they get bumped to this category. Pakistan is not so much focused on developing so much as simple survival right now! Good luck, guys!

And add to this list here in 2021: Burma, which has once more been taken over by it's military in yet another coup... but "the military" of that place is more like an outright criminal mafia than a protective force for the nation, so development is spiraling down the sewer once more at the hands of those crooks. And the country of Mali in Africa is suffering through its second military coup just this year, with neighboring Chad also in coup status!

Just to give you another opinion of the state of troubled states: The Fragile States Index is a report researched and released jointly by the Fund for Peace and Foreign Policy magazine. Released annually since 2005, the report assesses the vulnerability of sovereign states throughout the world.

The report uses indicators across four broad categories (Cohesion, Economic, Political, Social) to determine if a state is vulnerable to conflict or collapse. Within that framework are 12 different indicators used to determine the vulnerability of the states (e.g. human rights, public services, demographic pressures, refugees and internally displaced persons, and security).

For 2021, the top ten most troubled spots according to The Fragile States Index are:

- Yemen
- Somalia
- South Sudan
- Syria
- Democratic Republic of Congo
- Central African Republic
- Chad
- Sudan
- Afghanistan
- Zimbabwe

... Burma would have likely made the list, but it was formed before the military coup took place. Let's hope that all of these places (and the peoples within them) make it through 2021 without fully disintegrating into chaos, war, or worse.

WORLD AT NIGHT: ANOTHER FUN PLAID AVENGER MEASURE OF HOW THE WORLD IS DOING

There's a final image to consider. It can show you a lot about the levels of development on the planet. Plus, it's just so awesomely cool.

FIGURE 5.2 EARTH AT NIGHT

From the Visible Earth: A catalogue of NASA images and animations of our home planet:
This image of Earth's city lights was created with data from the Defense Meteorological Satellite Program (DMSP) Operational Linescan System (OLS). Originally designed to view clouds by moonlight, the OLS is also used to map the locations of permanent lights on the Earth's surface. The brightest areas of the Earth are the most urbanized, but not necessarily the most populated. (Compare western Europe with China and India.) Cities tend to grow along coastlines and transportation networks. Even without the underlying map, the outlines of many continents would still be visible. The United States interstate

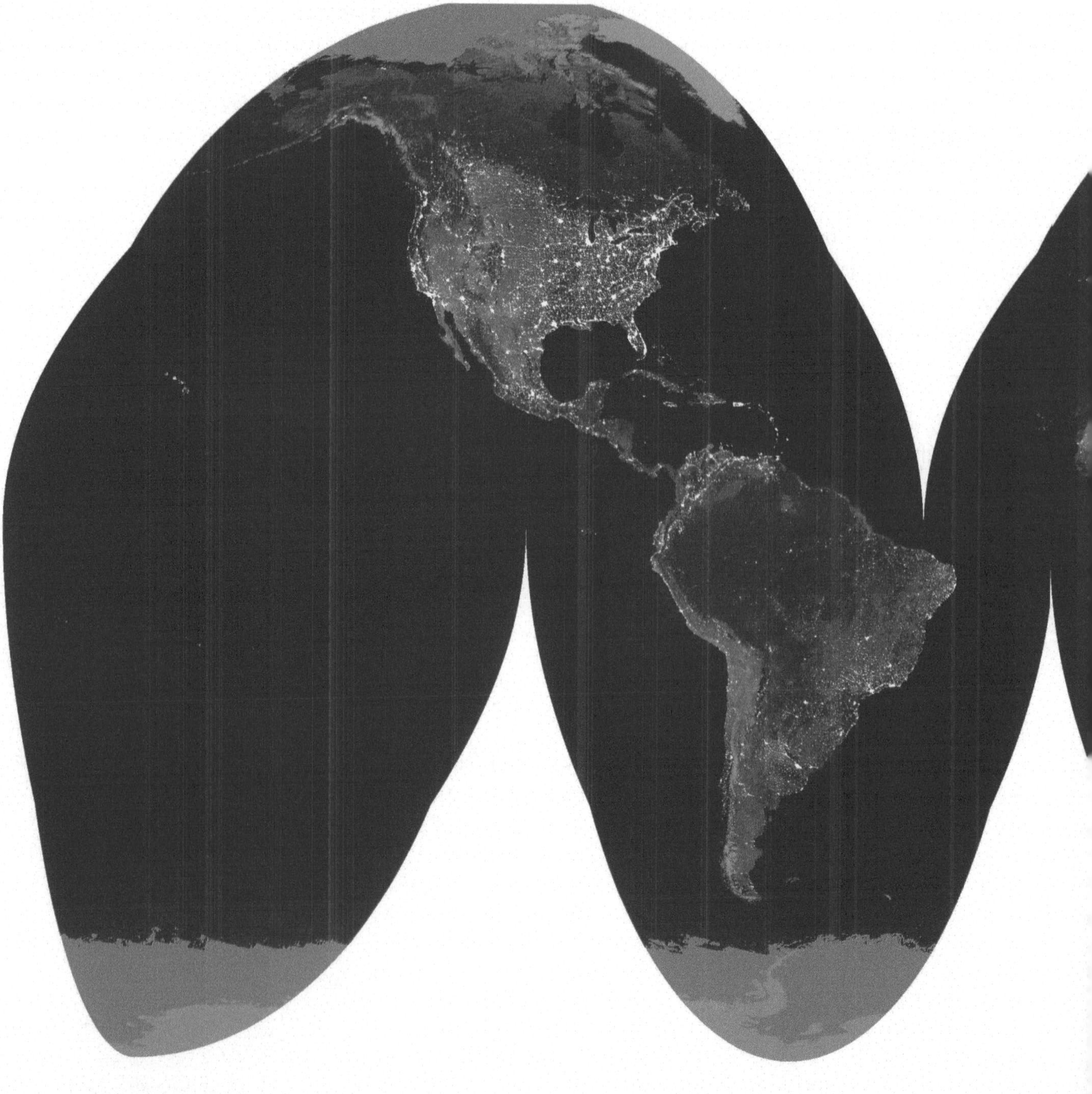

highway system appears as a lattice connecting the brighter dots of city centers. In Russia, the Trans-Siberian railroad is a thin line stretching from Moscow through the center of Asia to Vladivostok. The Nile River, from the Aswan Dam to the Mediterranean Sea, is another bright thread through an otherwise dark region. Even more than 100 years after the invention of the electric light, some regions remain thinly populated and unlit. Antarctica is entirely dark. The interior jungles of Africa and South America are mostly dark, but lights are beginning to appear there. Deserts in Africa, Arabia, Australia, Mongolia, and the United States are poorly lit as well (except along the coast), along with the boreal forests of Canada and Russia, and the great mountains of the Himalayas.

Credit: Data courtesy Marc Imhoff of NASA GSFC and Christopher Elvidge of NOAA NGDC.

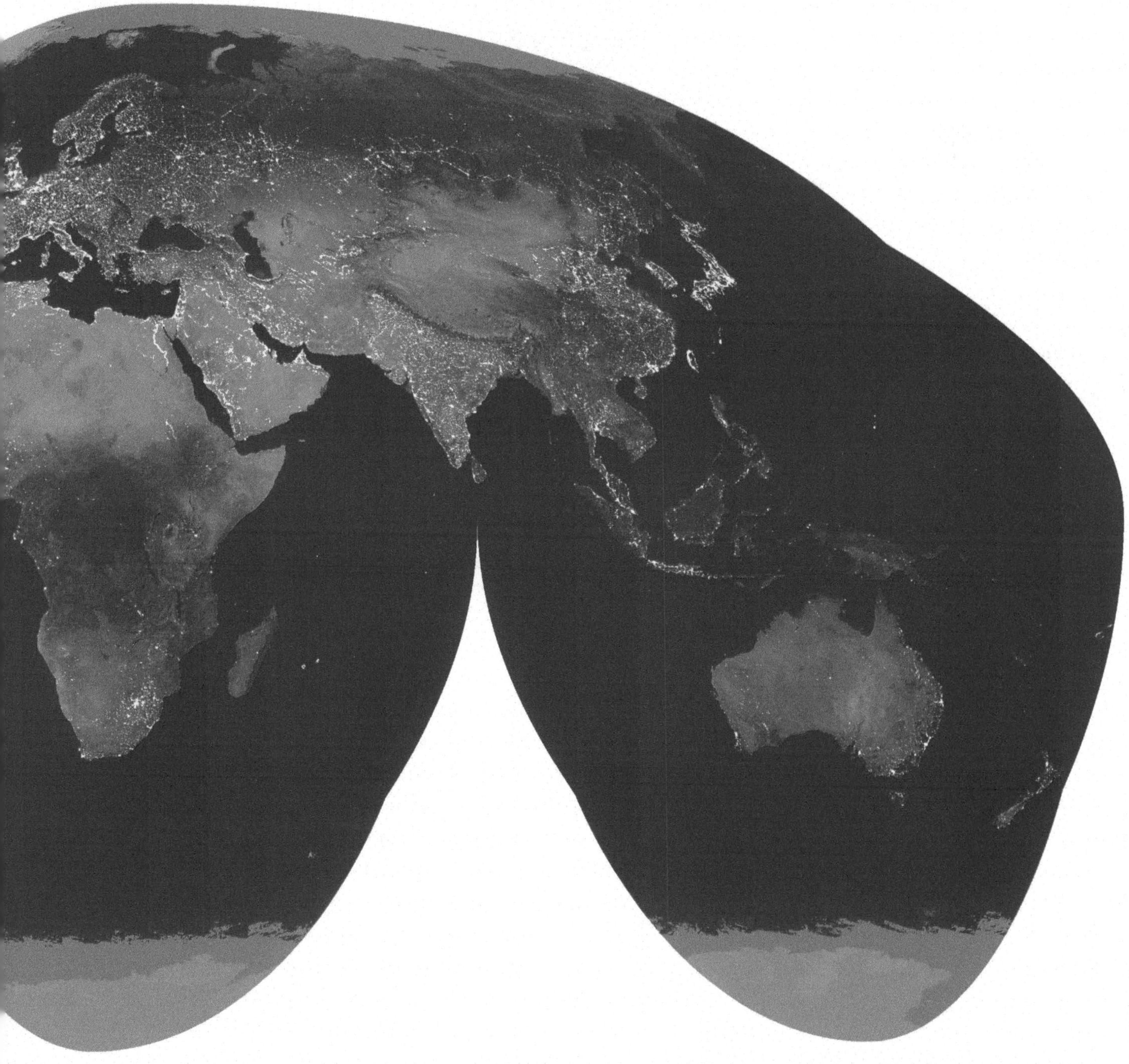

I'M DEVELOPMENTALLY SPENT . . .

There it is. That's the Plaid round-up. Hope you now understand the true differences in what is going on in the world in terms of levels of development, and that you have a better handle on how the world will be changing in the coming decades. I know I do. But dig this reference grid whenever you need a handy guide to assessing a developmental status of a state. It will make you think!

QUICK REFERENCE ROUND-UP

	Developed	Developing
Technology	• Highest levels of technology • Machines work faster and better • Extensively developed infrastructure • Effective emergency response	• Little to no technology • Without machines, people work harder and accomplish less • Unsafe or inefficient infrastructure • Poor emergency response
Industry	• More machine than human labor • High productivity • High energy consumption • Dependent on expensive, efficient fossil fuels	• More human than machine labor • Low productivity • Low energy consumption • Dependent on cheap, inefficient alternative fuels
Economics	• High salaries, high GDP per capita • Diverse economy • Large service sector, some quaternary • Value-added goods and services • Positive trade balance for state	• Low salaries, low GDP per capita • Limited economic options • Large primary and secondary sectors, no quaternary • Low value primary goods • Negative trade balance for state
Society	• Majority of population in middle class • Widespread investment in health care and education; high consumption • Can afford to take risks • Column or inverted population pyramid; fertility rate around 2.1 • Each generation gets better	• Majority of people in lower classes; middle class might exist • Little money to invest in health care, education, or consumption • Can't afford to take risks • True pyramid-shape population; high fertility rates (variable) • Generational evolvement is stagnant, or getting worse

	Developed	Developing
Health	• Decreased infant mortality, increased longevity • Access to effective health care • Sedentary work with mental stress • More processed foods • Most people die from circulatory disease or cancer	• Increased infant mortality and lower life expectancy • Susceptible to epidemics • Physical work with physical stress • Healthy food but insufficient calories • Most people die from preventable and curable disease
Government	• Solid democracy • Politically stable • Open to change; partially protected against corruption • Full participation in global trade and investment	• All forms of government found • Politically unstable • Closed to change; very susceptible to corruption • Insular economies; absent from global trade and investment
Cultural Cues	• Lots of overweight people • They eat animals • People go to specific areas named "gyms" in order to stay lean • People have extra time to "recreate" "vacation", and "chill" at malls, parks,beaches, and foreign countries • Some people are so bored that they take drugs for fun • People do this cool thing called "retire" after they have worked for most of their lives	• Mostly lean or skinny people • Animals eat them • People bust butt trying to stay alive 24/7 to stay lean. It's called life • People generally "hang out" only with their families at home or in the village. Most never leave their local area, much less their state • Some people grow drugs for the bored rich people in order to earn money so they can eat • People do this thing called "die" after they work their entire life
Military	• Maximum military technology • Actual or possible nuclear power • Weapons preferred to standing army • Casualties minimized • High military spending is low percentage of GDP	• Typically limited military technology • Nuclear power impossible (with some exceptions) • Standing army is main military power • Higher casualties • Low military spending is high percentage of GDP
Plaid Rule	• Most people are doing pretty well	• No matter what the stats say, most people aren't doing that well
HDI	• Close to score of one	• Close to score of zero
Night View	• Lit up like a Christmas tree!	• Lights out

CHAPTER OUTLINE

International Organizations

6

AND now we get to the trendiest world trend of globalization. This chapter consists of brief explanations of some entities that fall outside, or rather across, state boundaries—global players in a global age. We call them **supranationalist** organizations. Above and beyond the national level, these organizations play an increasingly important role in what is happening across our planet. But who are they? Where did they come from? How are we supposed to know this stuff? I don't know, friends. If the Plaid Avenger doesn't tell you these things, who will?

Supranationalist organizations are groups of states working together to achieve a common, or outlined, objective. This is another fairly new concept in human history, as states or nations have spent most of their time doing the opposite: beating the daylights out of each other or undercutting each other at every available opportunity. So why do (some) countries now work together? The Plaid Avenger sees order in the universe; we can classify global cooperation teams into three main groups: economic, defensive, and cultural.

I'll introduce you to the more important and happening entities here, but by no means is this list exhaustive. This chapter will also serve as a functioning reference for you as you progress through the rest of the book; come back often to refresh your memory when you see these acronyms appear in the regional chapters.

ECONOMIC ENTITIES—SHOW ME THE MONEY

Money. Who doesn't want it? Not any of the states of the world, that's for sure. A great way to make more money, if you are a country, is to make some trade deals with other countries. I'll buy all my bananas from you if you buy all of your wheat from me—sound good? On top of that, I won't put an import tax on your bananas, but if any other countries try to sell bananas here, I'll tax the heck out of them. Deal? This is the essence of **trade blocks** which are, as you might have guessed, a dandy vehicle for increasing trade between two countries . . . or perhaps among a whole bunch of countries, depending upon how many new kids are in your bloc.

Many economists believe that **free trade** between countries increases competition, which decreases prices for consumers, which in turn increase consumption of products . . . which ultimately benefits producers and consumers alike! Get governments out of the way, and let the market rule! For this reason, both the United States and the European Union are trying their hardest to promote "trade blocs" or "free trade zones" with neighboring countries, so that they can improve economic performance and increase sales. Even countries in Latin America, Africa and Asia have caught the economic bloc bug.

However, there is a tug-of-war going on. Independent sovereign states naturally want to protect their own economies, so they are reluctant to sign up for free trade when they think that their local industries may lose the trade game. For example, if Chinese shoe companies make cheaper and better shoes than French shoe companies, France is not going to want free trade in shoes with China. Everybody in France might want to buy the less expensive Chinese shoes, so the French shoe companies would go out of business. Historically, the cheaper imported products are hit with a **tariff**, an import tax, which subsequently makes the price of the product more expensive, and thus the local products can compete better.

Many times, poor countries accuse rich countries of trying to take advantage of them by using free trade agreements. These poor countries argue that free trade isn't equally beneficial for both sides; that it's mostly beneficial for the fully developed, industrialized country because their companies are already more competitive. Furthermore, they accuse developed countries of cheating, and they point to agricultural **subsidies** in these rich countries as an example. Farmers in Europe and America produce crap tons of food using lots of big equipment and fertilizers, thus their costs are high, and subsequently the food they make costs more. Farmers in poor countries don't use that expensive stuff and have cheaper labor, therefore their food should cost much less, giving them a competitive advantage in the world market. However, the rich farmers still "win" on the international market because Europe and America give their farmers huge subsidies to offset the higher costs of production they face. Sometimes Uncle Sam gives American farmers money just to be farmers, and the farmers can turn around and sell their food for cheaper prices and still make money. You dig? And if you dig a lot, maybe you should become a farmer.

Poor countries argue that the only reason that rich countries became rich in the first place is by protecting their domestic industries by using things like tariffs and other forms of **protectionism**. Also, it can be argued that fully developed mega-rich companies from mega-rich countries are so technologically superior that they have a competitive advantage that can never be overcome . . . meaning that the less developed states will always be stuck buying finished goods and selling primary level commodities, thus always losing money. On the other hand, free trade usually does mean more trade, so the less developed countries do stand to sell much more oil or lima beans or flip-flops or beef lips. Poorer countries are torn as to whether or not it is in their best interest to join these trade blocks with the fully developed states.

Perhaps it's on these grounds that we are seeing many new trade blocks springing up that are comprised solely of states in "developing status," with no "rich kids" invited to the party. It certainly is the reason for the foot-dragging with the FTAA, but once again, I have gotten ahead of the story.

Check out these economic entities that you will be hearing a heck of a lot more about, as they will play an increasingly larger role in the way the global economy operates:

NAFTA

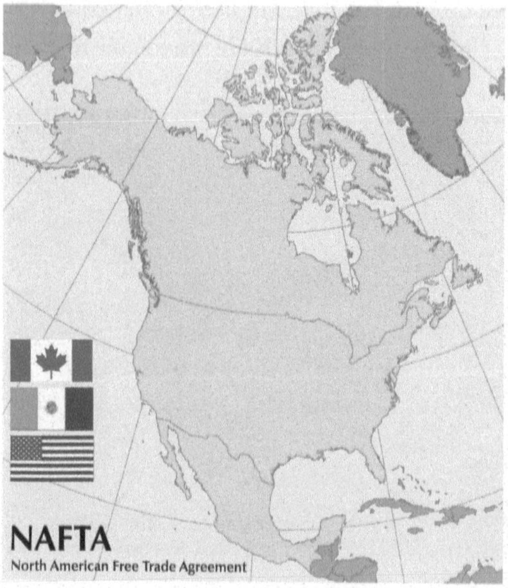

3 Members: United States, Canada, and Mexico

Summary: NAFTA, which stands for North American Free Trade Agreement, is between the United States, Mexico, and Canada, enacted in 1994. This agreement is meant to gradually eliminate all duties and tariffs on all goods and services between these three countries. However, the three nations are resisting lowering specific barriers that would hurt specific components of their economies. For example, the United States and Canada have been bickering because the United States imposes a duty on Canadian lumber that goes to the United States. The Canadians are accusing the Americans of not sticking to the treaty and are considering imposing duties on American goods to retaliate.

NAFTA has been very controversial in other ways too. Generally, multinational corporations support it because lower tariffs mean higher profits for them. Labor unions in the United States and Canada have opposed it because they believe that jobs will go from the United States and Canada to

Mexico because of lower wages there. They were all correct: manufacturing jobs, particularly from the auto industry, migrated rapidly south of the border where wages were significantly lower, but this did make the costs of products cheaper to the consumer as well. American jobs were lost, but Americans pay lower prices for goods. Also, farmers in Mexico opposed it because agricultural subsidies in the United States forced them to lower the prices on their goods. But make no bones about it: NAFTA has been an incredible plus to the 3 countries, as trade has exploded; more goods and more services now flow between the North American titans than ever before, and it looks set to expand into the future.

Trump Bump Alert!!! After US President Donald Trump took office in 2017, he worked to replace NAFTA with a new agreement. In September 2018, the United States, Mexico, and Canada reached an agreement to replace NAFTA with the United States-Mexico-Canada Agreement (USMCA). NAFTA will remain in force, pending the ratification of the USMCA . . . so as of this writing in 2021, I will still refer to it as NAFTA. Honestly, besides some minor tweaks, no one is really quite sure what the differences are yet.

DR-CAFTA

7 Members: United States, Costa Rica, Dominican Republic, El Salvador, Guatemala, Honduras, Nicaragua. Currently, the US Administration is pushing hard to get Colombia and Panama into this club as well.

Summary: DR-CAFTA stands for Dominican Republic—Central America Free Trade Agreement and is an international treaty to increase free trade. It was ratified by the Senate of the United States in 2005. Like NAFTA, its goal is to privatize public services, eliminate barriers to investment, protect intellectual property rights, and eliminate tariffs between the participating nations. Many people see DR-CAFTA as a stepping stone to the larger, more ambitious, FTAA (Free Trade Agreement of the Americas).

The controversy regarding DR-CAFTA is very much like the controversy regarding NAFTA. Many people are concerned about America losing jobs to poorer countries where the minimum wage is lower and environmental laws are more lax. Also, some people are concerned that regional trade blocs like DR-CAFTA undermine the project of creating a worldwide free trade zone using organizations like the WTO.

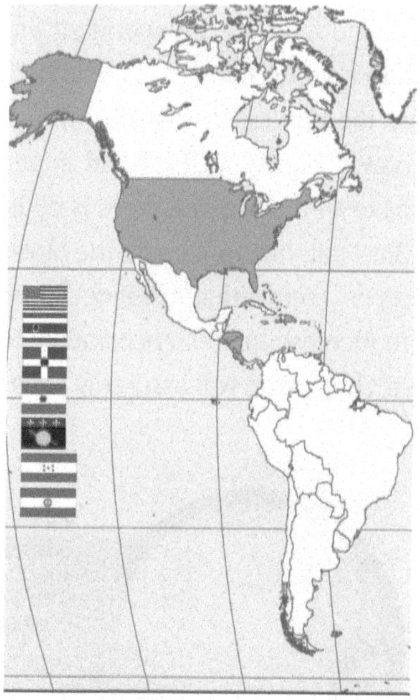
CAFTA members

Fun Plaid Fact: Many Washington insiders see DR-CAFTA as a way of reducing the influence of China in Central America.

2021 Trump Bump/Biden Time UPDATE! After re-working NAFTA, the Trump administration set their sights on renegotiating DR-CAFTA, with a specific focus on kicking out Nicaragua, the Dominican Republic, and El Salvador, while keeping the rest of the agreement in place. As of this writing in 2021, no major moves have yet been made by the former Trump team, nor the current Biden administration. Things are stagnant all the way around.

FTAA (PROPOSED ONLY!)

35 Members: *PROPOSED* All of the nations in North and South America, except Cuba. 'Cause the US hates commies. Dirty pinko commies.

Summary: The FTAA, which stands for Free Trade Area of the Americas, is a proposed agreement to end trade barriers between all of the countries in North and South America. It hasn't been ratified yet, because there are some issues that need to be worked out by the participating countries. The developed (rich) countries, such as the United States, want more free trade and increased intellectual property rights. The developing (poorer) countries, especially powerhouse Brazil, want an end to US/Canadian agricultural subsidies and more free trade in agricultural goods.

The key issue here for poor countries is agricultural subsidies. Farmers in the United States, and in rich countries generally, produce agricultural goods at a higher price than poor countries do. However, to keep their goods cheap, and thus competitive on the world market, the government of the United States pays their farmers subsidies. These subsidies make developing countries very angry because they believe subsidies give American farmers an unfair advantage. For this reason, some Latin American leaders have stalled the agreement. Former Venezuelan President Hugo Chavez called the agreement "a tool of imperialism" and proposed an alternative agreement called the Bolivarian Alternative for the Americas . . . more on that in a minute.

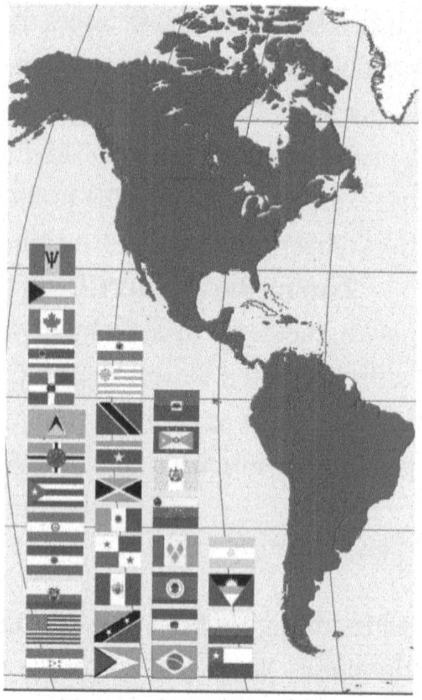
Proposed FTAA members

For rich countries, the issue is intellectual property rights, which is best exemplified by copyright laws. Less developed countries sometimes oppose these rights because they believe that if they are enacted, they will stifle scientific research in Latin America and widen the gap between the rich and poor countries in the Americas.

It should be noted, as of this writing in 2021, that this tentative agreement is still stalled by more independent and "leftist" Latin American states that don't want to join a group that the US would likely dominate . . . especially Brazil, which sees itself as the natural true leader of an economically united Latin America. In addition, the now inwardly-focused and increasingly anti-immigrant "MAGA" attitudes in the USA have all but assured there is no immediate future for the FTAA . . . if any future at all. But I still like to include the block so you understand how things change on the global scene even within just a decade or two. Who knows? Maybe the concept makes a resurgence in the future if the USA wants to increase its influence and build a stronger, more unified "American Block" to better compete with the growing influence of China, who will assuredly keep building bigger economic blocks of its own across Asia.

FUN PLAID FACT: The only country that would not be included in the FTAA is Cuba, because the United States has an economic embargo that prohibits all trade with the communist regime. For this reason, Cuba more fully supports the Bolivarian alternative for the Americas, which is MERCOSUR . . . we will get to that soon.

EU

27.5 Members: Austria, Belgium, Bulgaria, Croatia, Cyprus, Czech Republic, Denmark, Estonia, Finland, France, Germany, Greece, Hungary, Ireland, Italy, Latvia, Lithuania, Luxembourg, Malta, Netherlands, Poland, Portugal, Romania, Slovakia, Slovenia, Spain, Sweden, United Kingdom. But why would I say that there is 27.5 members instead of 28? Because the UK is leaving soon, but can't quite figure out how to do it properly. So they have one foot in the EU, and one foot out.

Summary: For years, European philosophers and political observers have recognized that the best way to ensure peace on the European continent while also increasing trade is to politically and economically integrate the nations. After the destruction and loss of life caused by World War II, European nations finally began taking small steps toward interdependence. They started by integrating their coal and steel industries in the ECSC (European Coal and Steel Community). What a long way they have come since then! Currently, the European Union, which has 28 member states, has a common market, a common European currency (the euro), a European Commission, a European Parliament, and a European Court of Justice. The nations of the EU have also negotiated treaties to have common agricultural, fishing, and security policies. More so than any other free trade agreement, the European Union covers way more areas other than just trade.

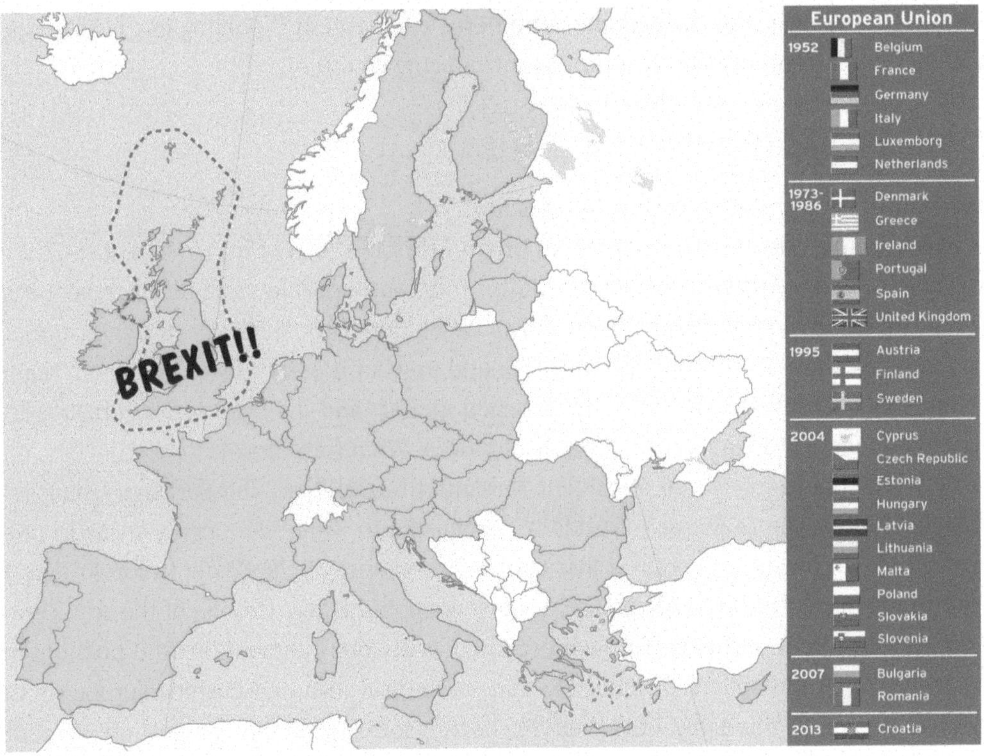

Consequently, the EU is the most evolved supranationalist organization the world has ever seen—perhaps a "United States of Europe." Free movement of people across international borders of the member states makes it unique in the trade block category. Of greater importance is an evolving EU armed force, a single environmental policy, and increasingly, a single foreign policy voice. That is a very big deal!

However, there has been some resistance to integration within European countries. Some countries, like Norway and Switzerland, have refused to join, and others, like the United Kingdom, have refused to fully adopt the euro: the Brits do use the euro, but maintain their traditional pound as well. Also, in 2005, a constitution for the European Union was rejected by French and Dutch voters, putting the future of European integration into question. Many Europeans simply don't care about the European Union and others see it as a secretive, undemocratic organization that is taking away power from their home countries. Some of the richer countries in Europe are afraid that adding nations with weaker economies will take away money from them just to give it to less productive economies.

2021 BREAKING BREXIT UPDATE!!! In a June 2016 referendum, 51.9% of British citizens voted to withdraw from the EU! Thus, BREXIT = BRitain EXITing the trade block! The separation process is super complex, causing political and economic changes for the UK and other countries, and they are still trying to figure out how to proceed since it has never been done before. Is this the beginning of the end of the block, or will they rally back to get on track? All these vexing complexities will be covered in the Europe chapter, but for now just be aware that the times (as always) are a'changin'! After almost 5 years of negotiations, this exodus officially occurred on Jan. 1, 2021, and now the real changes will start to take affect. Will this move hurt, hinder, or help the UK into the future? We shall be finding out fast in the years to come.

But make no bones about it, the EU unification and expansion has made Europe a global power once again. Individually, these European countries are rich places, but none could compete on their own with the likes of the US or even China. However, as a unit, the EU has the largest GDP on the planet. After a serious growth spurt in the last decade, it appears that the expansionism may be played out. Future candidates include Bosnia, Macedonia, Albania, and possibly Turkey . . . but Ukraine and Georgia are now seen as too hot to handle for the EU given Russian resurgence of influence in these areas. Russia itself is usually invited to big EU talks as kind of an associate member already, although the idea of Russian ascension into the EU will never happen. Not everybody likes it, but the EU makes Europe a player in the world economy and world political terms. Divided, they are not much. United, Europe still has a big voice. And maybe they better stay unified a bit longer and stronger, because fierce competition is coming in to challenge their block from the east! Let's get to that beast from the east right now by introducing you to . . .

EURASIAN ECONOMIC UNION (EEU OR EAEU)

5 Members: Russia, Belarus, Kazakhstan, Armenia, Kyrgyzstan

Summary: A Soviet Union, take 2? USSR 2.0? Soviet Union Reboot? Project Putin? The anti-EU? Whatever you want to call it, Russia's Vladimir Putin wants you to respect it! Respect his authority! Respect his backyard! Respect Russia! And this is Russia's way of regaining its power and influence over areas that it considers strategically and economically their home turf . . . by building an alternative to the EU, to NATO, and to what it perceives as European/Team West expansion into it's historic 'hood. That alternative was officially launched in 2015, and they call it the Eurasian Economic Union! A more "Asian" alternative to the "European" block! A battle of blocks! What fun! Or not . . .

While originally an idea posed by Kazak President Nursultan Nazarbayev, the Eurasian Union is without a doubt the fanatical fantasy football league of Russian President Vladimir Putin, who now openly seeks to reassert a resurgent Russia's influence and outright power into areas it lost during the collapse of the Soviet Union in 1991. Putin for his part has actually referred to the demise of the USSR as "the greatest geopolitical catastrophe of the 20th century," . . . and it is a catastrophe he now seems intent on fixing. Through economic coercion, intimidation, and possibly force if necessary, Russia looks to be regaining its old sphere of influence here in Central Asia and Eastern Europe. Is this an attempt to rebuild the Soviet Union proper? Is this a new Cold War? Eh. I'm not sure it is all that, at least not yet, but here is the deal:

In 2011, the presidents of Russia, Belarus, and Kazakhstan signed an agreement to create the Eurasian Economic Union—a custom union (like a free trade zone) with partial economic integration between the states, with the intent of becoming something very close to the European Union (you know, the EU) in the future with full-on economic, political, and military cooperation. Just like the EU, it is supposed to safeguard regional economical interests and facilitate trade and business within member countries. As of 2015, the Union has an integrated single market of 176 million people and a gross domestic product of over $4 trillion (US$). It also supposedly has free movement of goods,

capital, services, and people and provides for common transport, agriculture, and energy policies, with provisions for a single currency and greater integration in the future.

Putin's grand gamble here is that his union (and yes, it is referred to as his personal grand strategy at this point) will become a successful alternative to the European Union, and that the states of Central Asia, Eastern Europe, and even the Caucuses Region will want to sign up for it. I suppose it is being marketed as a great way to boost economic growth, without having to bother with all that pesky democracy and human rights stuff. And it is definitely being pushed as an alternative to joining Team West. And it definitely is Russia trying to regain "influence" in areas it previously held as its own. The Ukraine was to be a jewel in the crown of all of this, thus Putin's interest in keeping that country in the Russian orbit, and the Eurasian Union specifically. That whole taking over Crimean Peninsula stuff? Yeah, that was the kick-off of the game, not the finish of it.

This is a now fervently fired-up Russian response to the last two decades of the EU and NATO creeping closer and closer to its borders. The Bear wants its backyard back! Will it work? Who knows yet, but everyone in the 'hood is worried, as Russian pressure to get on board this Eurasian party train is growing. The EU, the US, and even China are watching these developments warily. States that have lots of ethnic Russian people in them (like Kazakhstan) are starting to fear a Russian take-over of parts of their territory, much like what happened to Ukraine with Crimea. But the other Central Asian states are now also stuck in this battle for control. Because Kyrgyzstan and Tajikistan are highly dependent on Russian foreign aid and economic trade . . . they will likely join. The other "–stans" seem to be hedging their bets, as they like their sovereignty more than they like Russia. Armenia and Kyrgyzstan just officially joined the ranks in 2015, and the fate of Ukraine as a whole is (as of this writing) still unsettled . . . a full Russian takeover of that state will see it bolster the Eurasian Union's ranks into a fully legit power block. But will Putin want to go that far to see his 21st century Ruskie dream realized? Stay tuned . . .

MERCOSUR

5 Members: Brazil, Argentina, Uruguay, Paraguay, Venezuela

Associate Members: Bolivia, Chile, Colombia, Ecuador, Peru, & Surinam.

Summary: MERCOSUR is a free trade agreement between several South American countries that was created in 1991 by the Treaty of Asuncion. Like other free trade agreements, its purpose is to promote free trade and the fluid movement of goods and currency between the member countries. Many people see MERCOSUR as a counterweight to other global economic powers such as the European Union and the United States.

MERCOSUR's combined GDP is only 1/12 of the United States', standing at 1 trillion dollars. But watch out! Brazil is becoming a serious global player with a booming economy, and actually the entire group is prospering. They are further integrating on economic and political decision, and even talking about free movement of workers and a standard labor law throughout the block. Good stuff! However, keep an eye on this one as leftist events unfold in Latin America. It could become a viable force. Already, MERCOSUR has now added all members of the Andean Community (another economic bloc) as "observer states" with the grander goal of uniting the entire continent in super-block called the Union of South American Nations (**UNASUR**)!!! Wow!

FUN PLAID FACT: I have been watching to see how serious this MERCOSUR/UNASAR concept is, for years, and it was really starting to take shape as a continent-wide super-bloc that was going to supercede MERCOSUR itself. But here in 2019 the whole thing has suddenly collapsed due to shenanigans happening in Venezuela. Let's pick up that story in the Latin America chapter. Sí?

ASEAN

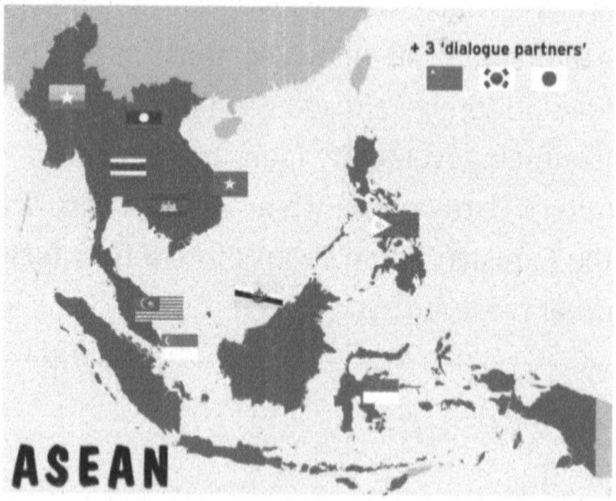

10 Members: Brunei, Cambodia, Indonesia, Laos, Malaysia, Burma, Philippines, Singapore, Thailand, Vietnam

Summary: ASEAN, which stands for **Association of Southeast Asian Nations**, is a free trade bloc of Southeast Asian countries. Like the European Union, ASEAN may possibly evolve into more than just a free-trade zone; it aims for political, cultural, and economic integration. It was formed in 1967 as a show of solidarity against expansion of Communist Vietnam and insurgency within their own borders. During that time, many countries around Vietnam were turning communist, and capitalist governments were extremely worried that communism might infect them as well. However, even Vietnam has joined ASEAN since then.

ASEAN is significant because of the heterogenous nature of its constituent countries. ASEAN countries are culturally diverse, including Muslims, Buddhists, and other religions. Governments in ASEAN range from democracy to autocracy. The economies of ASEAN countries are also very diverse, but they mainly focus on electronics, oil, and wood.

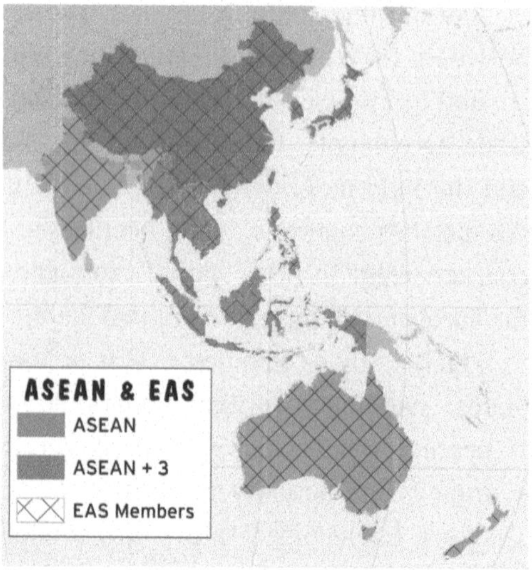

These guys are increasingly modeling themselves after the EU experiment, too. Even though they remain much more nationalistic than the European countries, the ASEAN group has much bigger goals than to simply be trade block. A common electric grid across the member countries has been proposed; an "open-sky" arrangement is soon taking effect (free movement of all aircraft among member states); and common environmental policies are being adopted across the region. The most interesting ASEAN prospect is now focused on, of all things, human rights! They actually wrote and adopted a legitimate constitution/charter which mentions the idea of protection of equal rights for all. This from a club which, at the time, had the dictatorship of Burma as a member! Interesting Asian times ahead for the ASEAN.

In addition, during annual ASEAN meetings, the three 'Dialogue Partners' (China, Japan, and South Korea) meet with ASEAN leaders . . . and when this happens, it is referred to as **ASEAN+3**. Wow! Those 3 economic titans as semi-associate members of ASEAN? That is a serious economic situation!

And I can do you one better: now ASEAN has become the core of the annual **East Asia Summit** which is the ASEAN+3 plus India, Australia, and New Zealand! The summit discusses issues including trade, energy, security and regional community building. Is this the early formation of an Asian EU? Who knows! But it's a fascinating development!

PLAID ALERT! Holy trade tips! Keep your eyes on AEAN in the future. This could turn into the largest, richest, baddest block on the planet in your lifetime. Booming economies in most member states with more growth in the future. Bigger deal than that: China, South Korea, Japan and India as "associates" members to create the ASEAN+3? Are you kidding me? That's like over half of the entire planet's population under one economic umbrella. Watch out! They are going to be hot!

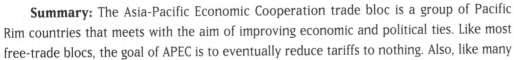

APEC

21 Members: Australia, Brunei, Canada, Chile, China, Indonesia, Japan, Korea, Malaysia, Mexico, New Zealand, Papua New Guinea, Peru, Philippines, Russia, Singapore, Taiwan, Thailand, United States, Vietnam—pretty much all the guys with Pacific coastline.

APEC
VIET NAM 2006

Summary: The Asia-Pacific Economic Cooperation trade bloc is a group of Pacific Rim countries that meets with the aim of improving economic and political ties. Like most free-trade blocs, the goal of APEC is to eventually reduce tariffs to nothing. Also, like many free trade agreements, agricultural subsidies have become a point of controversy. The leaders of all APEC countries meet annually in a summit called "APEC Economic Leaders' Meeting" which meets in a different location every year. The first of these meetings was in 1993 and was organized by US president Bill Clinton.

The countries in APEC are responsible for the production of about 80 percent of the world's computer and high-tech components. The countries in many of the Pacific Rim are also significant because the population in many of these countries is increasing dramatically. This trade bloc could possibly become a huge force in the global economy in the near future. Or it could totally be replaced by a proposed Pacific pact being pushed hard by the USA named the TPP. The **Trans-Pacific Partnership**: a very similar economic union that makes even deeper ties between the states, while simultaneously excluding China from the group. Currently only proposed, but being put together at such a frantic pace that it will likely be included in the next edition of this book . . . in which case APEC gets deleted! **2021 Trump Bump Alert!!!** The US administration of Donald Trump almost immediately killed US participation in TPP upon taking office in 2017. The rest of the TPP players are still in the game, but minus Uncle Sam now. It remains to be seen if the Biden Administration will make moves to rejoin it, but likely might just counter the influence of a growing and aggressive China within the Pacific realm.

OECD

OECD

35 Members: Australia, Austria, Belgium, Canada, Chile, Czech Republic, Denmark, Estonia, Finland, France, Germany, Greece, Hungary, Iceland, Ireland, Israel, Italy, Japan, Latvia, Luxembourg, Mexico, Netherlands, New Zealand, Norway, Poland, Portugal, Slovakia, Slovenia, South Korea, Spain, Sweden, Switzerland, Turkey, UK, & USA

Summary: Kind of a "Team West" roster. The Organisation for Economic Co-operation and Development (OECD) is an international organization of countries that accept the principles of democracy and free markets. After World War II, when Europe was in ruins, the United States gave European countries aid in the form of **the Marshall Plan** to rebuild the continent and repair the economy, while also ensuring that European countries remain democracies. The Organisation for

European Economic Co-operation (OEEC) was formed in 1948 to help administer the Marshall Plan. In 1961, membership was extended to non-European countries and renamed the OECD. Because it contains most of the richer, more developed states around the world which are all democracies to boot, the OECD is really kind of the core of what I refer to as "Team West" throughout this text.

Like many trade agreements, the purpose of the OECD is to promote free trade, economic development, and coordinate policies. The OECD also does a lot of research on trade, environment, agriculture, technology, taxation, and other areas. Since the OECD publishes its research, it has become one of the world's best sources for information and statistics about the world.

FUN PLAID FACT: While you will still see many references to OECD in news/literature, it has never really been a really pro-active group like the other ones discussed so far, and is of decreasing significance altogether here in the 21st century as regional (specifically Asian) trade blocks have proliferated and dominated.

DEFENSE

Why should countries get together defensively? If they are all on the same team, then they won't fight—right? Well, that's the emphasis of the UN. But perhaps more pertinent are regional defense blocks that have cropped up between countries throughout history. Their thinking is more along the lines of: "I'll help you if you get attacked by an outsider, if you help me if I'm attacked by an outsider." If this sounds like trivial schoolyard thinking, don't laugh; the basis for World War I was a whole host of such pacts between European countries—once one country was attacked, virtually every other country was immediately pulled in as a consequence of defense agreements. Here are the big three that are pertinent in today's world—even though one of them is now gone—plus a fascinating newcomer with tremendous future potential growth. But first, the easy ones . . .

THE UNITED NATIONS

UN Secretary General António Guterres

193 Members: All the sovereign states in the world except Vatican City. There are currently 193 of them. Even the Swiss finally joined a few years ago.

Summary: The United Nations, or UN, was founded in 1945 as a successor to the League of Nations. Like the League, the goal of the UN is to maintain global peace. Unlike the League, no major world wars have happened on the UN's watch. This is not to say that the United Nations has achieved global peace. In fact, UN "peacekeepers" have been on hand to witness some of the most egregious violations of human rights in recent history.

The UN is made up of several bodies, the most important of which is the Security Council (see Security Council section). The second most important body in the UN is the General Assembly where each of the 193 member nations has a representative and a vote. The General Assembly has produced gems such as the *Universal Declaration of Human Rights* and the lesser known *International Convention on the Protection of the Rights of All Migrant Workers and Members of Their Families*. The

General Assembly is clearly the home of utopian thinkers, but not of any real international power. This leaves the major world powers like the US and China free to ignore everything that the General Assembly says, without even having to waste the time vetoing it.

The UN also includes hundreds of sub-agencies that you've heard of before, such as the World Health Organization (WHO) and UNICEF. The WHO is in charge of coordinating efforts in international public health. UNICEF (The United Nation's Children Fund) provides health, educational, and structural assistance to children in developing nations. Both agencies are supported by member nations and private donors. UNICEF also receives millions of pennies collected each year by children on Halloween. Just a handful of the hundreds of other UN agency acronyms you may have heard of include the FAO, IAEA, UNESCO, IMF, WMO, and the WTO.

Critics often charge the UN with being ineffective. This is largely true, but the United Nations was never really intended to be a global government. The best way to view the UN is a forum in which nations can communicate and work together. The UN is ill-equipped to punish any strong member for violations. If a member is especially naughty, a strongly worded UN resolution might recommend voluntary diplomatic or economic sanctions. Perhaps after World War III, the United Nations will be once again renamed and given stronger international authority. If there is anything left of us.

THE REAL POWER AT THE UN: THE UN PERMANENT SECURITY COUNCIL

5 Members: US, UK, Russia, China, and France and 10 other rotating positions.

The Security Council is composed of five permanent members (the United States, the United Kingdom, Russia, China, and France) and ten other elected members serving rotating two year terms. The Security Council is charged with responding to threats to peace and acts of aggression. Basically, for anything to get done, the Security Council has to do it. But things rarely get done because each of the five permanent members has the power to veto and prevent any resolution that they do not like. A single veto from any one of the permanent members kills the resolution on the spot. This group of rag-tag veto-wielding pranksters is currently the ultimate source in interpreting international law. Most of the Cold War saw little to no consensus on anything, as Team US/UK faced off against Team Russia/China. Whatever one team tried to push, the other team generally would veto. The Frenchies vetoed according to mood and lighting of the room. Even today, votes tend to fall along these same alliance lines.

The other ten rotating members of the Council do not have veto power, but are often used as a coalition building tool to get things done. E.g.: During the build-up to the 2003 invasion of Iraq, the US worked very hard to get as many members of the Council as possible to back the resolution for war in Iraq, knowing full well that China and Russia would veto it. This was a strategic move to show broad support for the war, even though the US accepted up front that the resolution would not be passed.

There is currently speculation that new members may be added to the UN Permanent Security Council. The prime candidates are Germany and Japan. The United States supports their candidacies; maybe because they have over 270,000 military personnel (including dependents of military) in Germany and Japan combined, and they are staunch US allies. There is also talk of including Brazil or India, or even more remotely, an "Islamic member" or an "African member." But seriously, what incentive does the Security Council have to dilute their powers? Remember, all five would have to agree to let a new member in, so while the United States would certainly support the incorporation of Japan, China would be more likely to tell Japan to go commit **Seppuku**, veto-style. However, the four most likely members (Japan, Germany, Brazil, and India) have released a joint statement saying that they will all support the others' entry bids. The best argument for enlargement is that Japan and Germany are the second and third largest contributors to the UN general fund, and thus deserve more power. Regardless, don't count on the Security Council getting any bigger unless serious global strife starts going down, which it will, sooner or later.

NATO

28 Members: Albania, Bulgaria, Estonia, Latvia, Lithuania, Romania, Slovakia, Slovenia, Croatia, the United States of America, France, the United Kingdom, Iceland, Spain, Portugal, Germany, Italy, Belgium, Luxembourg, Poland, the Czech Republic, Hungary, Greece, Turkey, Norway, the Netherlands, Denmark, and Canada. Newest members inducted in 2017: Montenegro.

NATO, which stands for the North Atlantic Treaty Organization, is a military alliance between 26 European countries, Canada, and America. It was originally created in 1949 to serve as a discouragement to a possible attack from the Soviet Union (which never occurred). The most important part of NATO is Article V of the NATO Treaty, which states, "*The Parties agree that an armed attack against one or more of them in Europe or North America shall be considered an attack against them all. . . .*" This is called a **mutual defense clause** and basically means that the United States must treat an attack on Latvia the same as it would treat an attack on Tennessee.

Although NATO is a multilateral organization, the United States is clearly the captain of the ship. As a rule, US troops are never under the command of a foreign general. NEVER. Because of this, NATO troops (mainly American) are ALWAYS under American command. The United States also uses NATO countries to base its own troops and station nuclear weapons. Many historians blame the United States for provoking the Cuban Missile Crisis, saying that the Russians only wanted to put nukes in Cuba because the United States had at that time stationed nukes in Turkey (a NATO member).

NATO Secretary General Jens Stoltenberg. Don't make him angry. You wouldn't like NATO when he's angry.

Since the Cold War, NATO has been looking for a new role in the world. Many of the former Soviet republics have since been admitted to NATO—which, by the way, really ticks off Russia. NATO expansion was promoted as an expansion of democracy and freedom into Eastern Europe. More likely, it was to make sure Russia would never be able to regain the territory. NATO has also been increasingly active in international police work, although there is no real justification for this in the NATO charter. NATO forces were heavily involved in the Bosnia conflict in 1994 and the Yugoslavia conflict in 1999, although in reality these were just American troops under a multinational flag. After September 11th attacks on the US, NATO has also become involved in the anti-terrorism game, even invoking Article V for the first time with regard to Afghanistan. Remember, the war in Afghanistan is a NATO mission, not a US mission. But let's be honest here; the US does most of the heavy lifting, as usual.

And the NATO role continues to become broader and more bullish lately: NATO was the central organizing entity in the 2011 invasion of Libya. Say what? What the heck did Libya do to any NATO country? Answer: Nothing, which goes to show how the entity is rapidly redefining itself here in the 21st century. It appears that NATO is fast becoming the military muscle for the objectives of 'Team West,' wether those objectives are defense, economic, or purely political. Interesting stuff, eh? And infuriating stuff to those not aligned with the NATO countries.

FUN PLAID FACT: The only country in NATO without a military force is Iceland. The Icelandic Defense Force is an American military contingent stationed permanently on the island.

NATO members ☐ ☐ NATO members since 2004

Belgium (1)	Germany (8)	Luxembourg (15)	Slovakia (21)
Bulgaria (2)	Greece (9)	Netherlands (16)	Slovenia (22)
Canada (3)	Hungary (10)	Norway (17)	Spain (23)
Czech Republic (4)	Iceland (11)	Poland (18)	Turkey (24)
Denmark (5)	Italy (12)	Portugal (19)	United Kingdom (25)
Estonia (6)	Latvia (13)	Romania (20)	United States (26)
France (7)	Lithuania (14)		

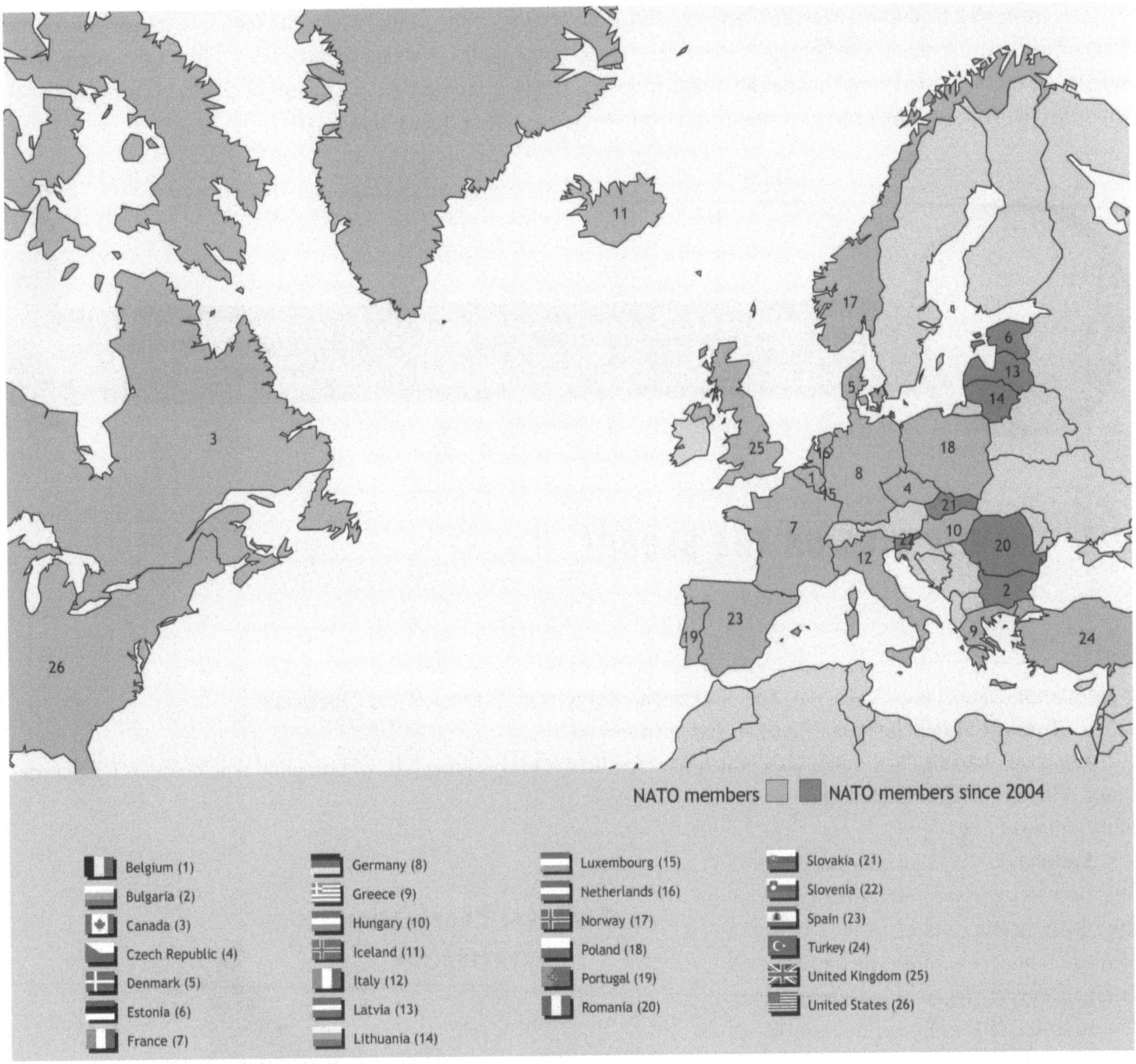

WARSAW PACT—DEFUNCT!

PAST Members: Soviet Union (club president), Albania (until 1968), Bulgaria, Czechoslovakia, East Germany (1956–1990), Hungary, Poland, and Romania.

The **Warsaw Pact**, or if you prefer the more Orwellian Soviet name—the "Treaty of Friendship, Co-operation and Mutual Assistance," was the alliance formed by the Soviet Union to counter the perceived threat of NATO. The Warsaw Pact was established in 1955 (six years after NATO) and lasted officially until 1991 (two years post-Berlin Wall). Much like the Bizarro-Superman to the United State's real Superman, the Warsaw Pact never had the teeth or the organizational strength of NATO. Perhaps this is because many of the members actually hated the dominance of the Soviet Union. Two countries, Hungary (1956) and Czechoslovakia (1961), tried to assert political independence and were subsequently crushed by Soviet military forces in exercises that would make **Tiananmen Square** uprising look like an after school special.

The main idea behind the Warsaw Pact was mutual protection. If the United States attempted to invade any of the Warsaw members, it would guarantee a Soviet response. In this way, the Warsaw countries acted like a tripwire against the expansion of Western-style capitalism and democracy, firmly establishing the location of the Iron Curtain. Shortly after the Cold War, most Warsaw Pact countries either ceased to exist or defected to NATO.

FUN PLAID FACT: The Soviet Union despised American acronyms. Instead of taking the first letter from each word, the Soviets preferred taking the entire first sound. For example, Communist International was "Comintern."

NEW KIDS ON THE BLOC!!!

SCO

8 Members: China, Russia, India, Pakistan, Kazakhstan, Kyrgyzstan, Tajikistan, and Uzbekistan

 4 Observer States: Belarus, Mongolia, Afghanistan, and Iran

Goodbye, Warsaw Pact! Hello, SCO! Watch this new BLOC! It is evolving fast, with gigantic repercussions for international relations and world balance of power in the future of Asia!

Summary: The Shanghai Cooperation Organization grouping was originally created by 5 member states in 1996 with the signing of the *Treaty on Deepening Military Trust in Border Regions* . . . however with the addition of Uzbekistan in 2001, which brought them up to 6 members, they can't use that wicked cool nickname of "Shanghai 5" anymore. Too bad. But they are are the hottest defense block to keep an eye on, as they are the newest on the scene; so new, in fact, that they have not exactly quite figured out what they are yet. Part military, part economic, part cultural . . . but 100% important to know. Let's just focus on military aspects for now.

Created with that *Deepening Military Trust* issue, they have also agreed to a *Treaty on Reduction of Military Forces*

SHANGHAI COOPERATION ORGANISATION

■ Members ■ Observers ■ Observer applicants ■ Dialogue Partners

in Border Regions in 1997 and in July 2001, Russia and China, the organization's two leading nations, signed the *Treaty of Good-Neighborliness and Friendly Cooperation* . . . wow, could they get any more sickening sweet with the descriptors? What an Eurasian love-fest. The SCO is primarily centred on its member nations' security-related concerns, often citing its main threats/focuses as: terrorism, separatism, and extremism. You can easily read into this as a great vehicle for all these governments to help each other crack down on any internal political dissidents as well.

They work together thwarting terrorism and stuff, but they are also quickly absorbing other avenues of cooperation, like in domestic security, crime, and drug trafficking. Over the past few years, the organization's activities have expanded to include increased military cooperation, intelligence sharing, and counterterrorism. Of significant note: there have been a number of SCO joint military exercises, and while it is very early to think that the SCO is a serious strategic military power, please keep in mind that the group is still very young and has a long way to go and to grow. We may just be seeing the beginning of their military prowess. Dig this: both Russia and China are nuke powers, Russia has a ton o' of weaponry, and China has the largest standing army on Earth. Oh, and the club is getting bigger: in 2017 both India and Pakistan officially acceded into the group as full members. SO make that four nuclear powers in the club! So this organization could become a serious defensive force, if they put their minds to it.

Is this thing sizing up to be a counterbalance to US power or as an overt anti-NATO? Perhaps. But one thing is for sure: its eight full members account for 60% of the land mass of Eurasia and nearly half of the entire world's population. The SCO has now initiated over two dozen large-scale projects related to transportation, energy and telecommunications and held regular meetings of security, military, defense, foreign affairs, economic, cultural, banking and other officials from its member states. No multinational organization with such far-ranging and comprehensive mutual interests and activities has ever existed on this scale before. The SCO is widely regarded as the "alliance of the East," due to its growing centrality in Asia-Pacific, and has been the primary security pillar of the region. A combo EU/NATO of Eurasia for this century? Could be! So you gotsta' know the SCO! They are fast becoming a playa'!

CULTURAL ORGANIZATIONS

Some supranationalist organizations form out of a desire to maintain a cultural coherence with like countries, or to promote certain aspects of their culture among their member states. In other words, monetary gain is not the driving force behind the organization, although economics usually sneaks in there as well. Here are three very different such organizations to compare and contrast.

ARAB LEAGUE

22 Members: Egypt, Iraq, Jordan, Lebanon, Saudi Arabia, Syria, Yemen, Libya, Sudan, Morocco, Tunisia, Kuwait, Algeria, United Arab Emirates, Bahrain, Qatar, Oman, Mauritania, Somalia, Palestine, Djibouti, Comoros. Syria currently SUSPENDED!

The Arab League is an organization designed to strengthen ties among Arab member states, coordinate their policies, and promote their common interests. The league is involved in various political, economic, cultural, and social programs, including literacy campaigns and programs dealing with labor issues. The common bond between the countries in the Arab League is that they speak a common language, Arabic, and they practice a common religion, Islam. The Charter of the Arab League also forbids member states from resorting to force against each other. In many ways, the Arab League can be seen as a regional UN. It was formed in 1945.

Let's get this Arab party started!

However, the Arab League is better known for their lack of coherence and in-fighting more so than any unifying activities they have had so far to date. In fact, Libyan leader Muammar Qaddafi threatened to withdraw from the League in 2002, because of "Arab incapacity" in resolving the crises between the United States and Iraq and the Israeli-Palestinian conflict. If the Arab League ever gets its act together, it could be a powerful force in the world. Right now, it is not.

Radical Arab League Update for 2012: Holy Middle Eastern Madness! The League finally agreed to something, and that something significantly impacted world events! I am referring to the international invasion of Libya in the early months of 2011, a move that was supported by the Arab League! They voted against one of their own too . . . all the other Arab leaders so despised Muammar Qaddafi that they supported outside intervention to help have him deposed! The Arab League's support of this measure was crucial because US/UK/French/NATO/UN intervention would have likely not happened at all if the League would have voted against it. No western power would have wanted to be seen as a foreign, anti-Islamic, imperialist force invading Libya . . . but once all the other Arab states backed it, then game on! And now the Arab League has united once more: they group is standing with a single opposition voice against fellow Arab Bashar al-Assad, President of Syria currently cracking down brutally on his own people. Saudi Arabia in particular is making plans to arm the Syrian rebels, and the rest of the League will likely support that action! A new era for the Arab group may be dawning. **2021 UPDATE:** Never mind. Not much came of their united efforts to help Libya, nor end the Syrian Civil War. Both countries are currently still in chaos. Oh well, got that one wrong back in 2012. That's why I have to keep updating this darned book, but I wanted you to actually see my thought process as it evolved over the last decade.

FUN PLAID FACT: Egypt was suspended from the Arab League from 1979 to 1989 for signing a peace treaty with Israel. Libya has sporadically quit and re-joined the group multiple times depending on how Muammar Qaddafi was feeling at the time. (See more on that in the AU section on next page.) Syria has currently been suspended as well, due in part to the government mass killing of Arabs, aka Syrian citizens.

OAS

35 Members: Argentina, Bolivia, Brazil, Chile, Colombia, Costa Rica, Cuba, Dominican Republic, Ecuador, El Salvador, Guatemala, Haiti, Honduras, Mexico, Nicaragua, Panama, Paraguay, Peru, United States, Uruguay, Venezuela, Barbados, Trinidad and Tobago, Jamaica, Grenada, Suriname, Dominica, Saint Lucia, Antigua and Barbuda, Saint Vincent and the Grenadines, Bahamas, Saint Kitts and Nevis, Canada, Belize, Guyana

Summary: The OAS, which stands for Organization of American States, is an international organization headquartered in Washington, DC. According to Article 1 of its Charter, the goal of the member nations in creating the OAS was "to achieve an order of peace and justice, to promote their solidarity, to strengthen their collaboration, and to defend their sovereignty, their territorial integrity, and their independence." Other goals include economic growth, democracy, security, the eradication of poverty, and a means to resolve disputes. Historically, the first meeting to promote solidarity and cooperation was held in 1889 and was called the First International Conference of American States. Since then, the OAS has grown, through a number of small steps, to become the organization it is today. Like the Arab League, this organization is a kind of regional UN.

Unlike free-trade blocs, the OAS encompasses many areas other than just trade. For example, it oversees elections in all of its member countries. However, it has also been criticized as a means for America to control the countries in Latin America. For example, when America wanted Cuba kicked out of the OAS, the organization quickly did so. However, many dictatorships that America has supported have remained within the OAS.

But this "US as imperialist" attitude is changing fast in the OAS. US President Barack Obama personally addressed the group in April 2009, pledging to bring more unity to the hemisphere and even hinting at thawing relations with the Cubans. At that meeting, Obama even shook hands with legendary US-hating (now-deceased) Venezuelan President Hugo Chavez! The OAS may actually be gaining ground as a legit forum for problem-solving in the Americas. As of this writing, Cuba is being reconsidered for membership, and it appears that things are moving forward rapidly. During the 2016-2020 Trump administration, the US retracted from this group as well, specifically hardening their attitudes towards Cuba being a part of anything. As previous comments allude to, the Biden administration has not yet gotten around to addressing their attitudes towards this grouping either, but likely will soon.

AU: AFRICAN UNION

55 Members: Algeria, Angola, Benin, Botswana, Burkina Faso, Burundi, Cameroon, Cape Verde, Central African Republic, Chad, Democratic Republic of the Congo (Kinshasa), Djibouti, Egypt, Equatorial Guinea, Eritrea, Ethiopia, Gabon, Gambia, Ghana, Guinea Bissau, Guinea, Ivory Coast, Kenya, Lesotho, Liberia, Libya, Madagascar, Malawi, Mali, Mauritania, Mauritius, Morocco, Mozambique, Namibia, Niger, Nigeria, Republic of the Congo (Brazzaville), Reunion, Rwanda, Senegal, Seychelles, Sierra Leone, Sao Tome & Principe, Somalia, South Africa, South Sudan, Sudan, Swaziland, Tanzania, Togo, Tunisia, Uganda, Western Sahara*, Zambia, Zanzibar, Zimbabwe. Every sovereign state on the entire continent!

Suspended Members: Mauritania

Summary: The Organization of African Unity was established in 1963 at Addis Ababa, Ethiopia, by 37 independent African nations to promote unity and development; defend the sovereignty and territorial integrity of members; eradicate all forms of colonialism and promote international cooperation. This organization changed its name to the African Union in 2002. Institutionally, the AU is very much like the EU with a parliament, a commission, a court of justice, and a chairmanship which rotates between the member countries. The AU is also beginning to deploy peacekeepers, and it has sent over 2500 soldiers to the Darfur region of the Sudan. Every country in Africa is a member of the AU except for Morocco, which withdrew in 1985. Also, Mauritania was suspended in 2005 after a coup d'etat occurred and a military government took power. The new government has promised to hold elections within two years, but many observers are doubtful that it will.

African military problems to be increasingly solved by African troops. How novel!

There are many problems facing Africa such as civil war, disease, undemocratic regimes, poverty, and the demographic destabilization caused by the AIDS epidemic . . . these issues are more than even a rich, well-established regional block could handle, and the AU has thus far proved incapable of positively impacting any of these problems successfully. However, its most successful component to date is their military wing. Major powers from around the globe have hailed the common

AU military as the most awesome thing the continent could do, and have supported it wholeheartedly. Why? Because the AU military can thus be tasked with solving African conflicts, without the outside powers becoming directly involved. After the US debacle in Somalia (go watch *Blackhawk Down*) no one wants to send actual troops to stop African conflicts, even in the face of titanic humanitarian disasters like the Rwandan genocide. And maybe now they don't have to! Because the AU has become the choice d'jour for outside support to deal with these issues. While the AU may not get outside support for anything else, look for the military to be bolstered with funding and training from the US, the EU and maybe even China and Russia, in lieu of actual participation in conflicts on the ground.

"Up yours, Arab League. I'm African now!"

FUN FACT: The idea of an African Union separate from the OAU came from Muammar Qaddafi, who wanted to see a "United States of Africa." He was sick of developments in the Arab world and publicly gave up on being an Arab. In 2011 he publicly gave up on living, when he was assassinated during the Libyan uprising.

HUGE AU 2019 UPDATE: What, what, what? How did my tingling plaid spidey-sense totally miss this one? The AU just had a major economic breakthrough, one which is running counter to the current anti-bloc trend happening in other parts of the world, and it affects the entire continent of Africa! What is this surprise move of which I speak? Why, **AfCFTA** of course!!! Which I just heard about the day it was officially ratified. Dig this:

The African Continental Free Trade Agreement (AfCFTA) is a trade agreement between African Union member states, with the goal of creating a single market followed by free movement of people, and then a single-currency union. Wow! That sounds like the EU of Africa! The AfCFTA plan was only signed onto in Kigali, Rwanda on March 21, 2018, and as of May 2019, 52 of the 55 African Union states have signed the agreement, with only Benin, Eritrea, and Nigeria so far still not participating fully in the negotiations. (This is actually quite important since Nigeria has the continent's largest economy. That's the equivalent of saying Germany is not joining the EU. But Nigeria may just be holding out for better concessions, and word is they will eventually jump in.)

Ratification by 22 countries was required for the agreement to actually enter into force and for the free trade area to become effective. Kenya and Ghana were the first countries to officially be in after ratification through their parliaments in May 2018. With ratification by Sierra Leone and the Saharawi Republic on April 29, 2019, the threshold of 22 ratifying states for the free trade area to formally exist was reached. Several more states have ratified since then, so as of May 30, 2019 this thing is now totally real!!! It's so new, I have no idea what else to teach you about it yet! We will revisit AfCFTA's potential in more detail in the Sub-Saharan Africa chapter.

INTERNATIONAL ODDBALLS

In addition to these, other entities have been formed at the international level for specific functions. Many of these organizations are frequently in the news, so I feel a brief introduction to them is merited. Haven't you ever wondered who the heck is the G7, the G8, the IMF, the World Bank, or the WTO, and what is an NGO?

G-7 GROUP OF SEVEN

Members: Canada, France, Germany, Italy, Japan, UK, and the United States

The **Group of Seven** (or G-7) is the "country club" of international relations. It's a place where the richest industrialized nations go to talk about being rich, form strategies for staying rich, and—like any other country club—figure out how to keep everyone else poor. The leaders from G-7 countries meet each year for a summit. The summits are often widely protested for reasons such as global warming, poverty in Africa, unfair trade policies, unfair medical patent laws, and basically, for a general sense of arrogance.

Originally, the group was called the G-6, which was the G-7 minus Canada. They formed out of the **Library Group** because rich countries were enraged about the **1973 oil crisis**. Then Canada joined the group in 1976, making it the G-7. Even though this group has since evolved into the G-8, the G-7 group still meets annually to discuss financial issues.

G-8 GROUP OF EIGHT

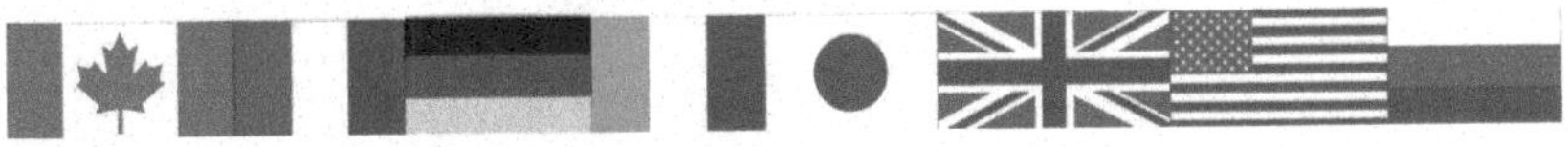

Members: G-7 plus Russia

Same as above, with Russia in attendance. Russia is still a nuclear power, still the largest territorial state on the planet, and perhaps most significantly a major energy resource provider to the world—especially to the G-7 countries. They have to invite the big boy to the party every now and again, or the big boy might feel neglected and go play with China. The G-7 doesn't want that. They need their energy! In 1991, Russia's entry into the group turned it from G-7 to the G-8. Who knows how much longer the G-7 or G-8 will even last given the rapidly changing face of our planet. What do I mean? Well, in our day and age, what point is there to having a meeting of the "biggest" world economies that doesn't include China? Or India? Or Brazil? It's getting to be kind of outdated and pathetically sad for all the old white-boy countries to get together to chat about world affairs without these other up-and-coming-non-white states having input, so we are probably witnessing the end of days of both the G-7 and G-8, although old traditions die hard and they will survive for a bit longer. And the G-8 as a whole maybe be down-graded back to the G-7, as Russia continues to lean more back towards Asian ties over Western ones . . . as evidenced by their more staunch support for the SCO and BRICS (more on that in a minute.)

G8 UPDATE: Russia was officially suspended from the group in 2014 after its annexation of the Crimean Peninsula as well as its meddling in Ukraine in general. The group will vote again in 2017 to decide on restoring Russia's status . . . but quite frankly I can't see Russia really giving a Crimean crap one way or the other about the 8. And a follow-up 2019 update on that update: Before the 2017 vote even took place, Russia declared themselves out of the G-8 permanently. So I guess it is officially defunct . . . for now.

G-20 GROUP OF TWENTY

PLAID ALERT!!!

Great googly global groups! We have a new international organization titan in our midst, my friends! The G-20 (formally, the Group of Twenty Finance Ministers and Central Bank Governors).

Members: Technically, it is comprised of the finance ministers and central bank governors of the G7, 12 other key countries, and the European Union Presidency (if not a G7 member). Also at the meetings are the heads of the European Central Bank, the IMF, and the World Bank.

The literal translation: the G-8 rich states + Argentina, Australia, Brazil, China, India, Indonesia, Mexico, Korea, Saudi Arabia, South Africa, Turkey + all those aforementioned bank heads. Collectively, the G-20 economies comprise 85% of global GDP, 80% of world trade and two-thirds of the world population. Whoa! That's all the money!

Summary: The G-20 is a forum for cooperation and consultation on matters pertaining to the international financial system which was created in 1999. It studies, reviews, and promotes discussion among key industrial and emerging market countries of policy issues pertaining to the promotion of international financial stability, and seeks to address issues that go beyond the responsibilities of any one organization. The key here is to "promote international financial stability," but I'm thinking they are quickly turning into a whole hell of a lot more. Why would I think that?

Because all the prime ministers and presidents are now showing up for these meetings! A Washington, D.C. summit was held in November 2008 and a London summit in April 2009, which were attended by all the heads of the respective states. Barack was there baby, chillin' with other heads of states! And these folks are truly transforming the power structure of the planet: the G-8 guys realize that they can't fix nothing without the cooperation of titans like China and India. In London, the G-20 adopted a collective approach to solving the global recession, but also are reorganizing the IMF and World Bank to incorporate other non-G8 voices, which have dominated global financial institutions for years. This is exciting stuff!

I truly believe this G-20 unit has now fully superseded the G-7, the G-8, and any other G's as the premier problem-solving institute on earth. The UN is too big and bureaucratic, and the G-7 too white and too exclusive, but the 20 seems to be a great balance of peoples and powers from across the planet. I would speculate that this much more representative and simultaneously streamlined group will be where all sorts of international treaties and agreements will have their foundation stones set; from economic crises to global warming issues to nuclear proliferation problems, look for the talking points to be brought up at these G-20 pow-wows.

This star is new, but shining already. And its future is bright indeed!

FUN PLAID FACT: There actually has been another G-20 gang comprised of an array of developing nations from Africa, Latin America and Asia. Their primary goal has been fighting for trade rights, particularly in the agricultural sector. You probably won't hear much from that group anymore, but just be aware in case you do . . . they are the poor G-20, not this huge rich G-20 described above.

BRICS

Members: Brazil, China, India, Russia, & South Africa

The Avenger has his plaid panties in a bunch about this unofficial "group" because it represents a serious global shift that has started in the last several years: a move to more representative economic and political power on the world stage. You can possibly look at the BRIC as an alternative center of power to the "Team West" rich countries that have run the global show for a century. It is a new, dynamic, and growing coalition which you will hear a lot more about in the future . . . . just as you will increasingly hear less and less about the G-7 or G-8. So what is this BRIC house all about?

An acronym for the states of Brazil, Russia, India and China combined, BRIC was first coined in 2003 by a financial analyst at Goldman Sachs who speculated that by 2050, these four economies would be wealthier than most of the current major economic powers put together. For sure, China and India will become the world's biggest suppliers of manufactured goods and services, with Russia and Brazil becoming some of the biggest suppliers of natural resources. All four are developing and industrializing rapidly and making bank. Because of their emerging statuses, their increasingly educated populations, and their lower labor costs, the BRIC has also become a magnet for foreign investment, outsourcing, and development of R&D centers.

But this thing has already grown well beyond a simple classification of future rich countries, and this is the deal I really want you to know. It is consistently pointed out that these four countries have not formed a structured political alliance like the EU, NAFTA or NATO . . . . . yet! I'm telling you friends, these guys are thinking about it! They started having face-to-face summits in 2009, and are increasingly agreeing on a whole lot of economic AND foreign policy issues which they then announce to the world under a common voice. Craziness!

What am I spouting about? At the 2011 meeting of the G-20, the BRIC countries demanded, and received, more voting power in global financial institutions like the IMF and World Bank (which they should, since they are increasingly funding them). And what is certainly the boldest and most telling strategic political initiative they have taken to date, all four BRIC countries have jointly declared that they do not support harsh sanctions against Iran and that they support Iran's right to

BRICS 2019

develop nuclear power. What's that got to do with economic policy? Nothing! But it certainly does fly in the face of Team West's opinion of Iran, and that is why I think its important that you know that the BRIC is fast evolving into something much more than a simple economic group . . . . they are becoming a global entity on par with Team West. It's not official, it's not on paper, and it's not heavily coordinated yet, but for sure the BRIC is going to be a significant entity of the 21st century. Get hip to the BRIC!

BODACIOUS BRIC ADDITION: Exciting expansion has been enacted! In 2012 the BRIC became the BRICS my friends! Say what? A new 'S' in the bric-house? That's right, the S is for South Africa which just joined the ranks of this increasing important global group. Now the gang includes the second biggest economy of Africa, to join the biggest economy of South America, along with the 3 biggest economies of Asia.

At their last meeting, the members worked on a slew of trade issues, promised to invest heavily in each other's economies, and also made a public statement that the use of force "should be avoided" in Libya. What's that got to do with economics? Nada, which is why its important to note. This group, while still young, is already making unified political statements about world events . . . world events on which they share common cause, a common cause that is distinctly NOT the Team West view. Hmmmm . . . do I see the formation of a bi-polar world forming? Let's keep a close eye on this entity to find out.

WTO

World Trade Organization
Organisation Mondiale du Commerce
Organización Mundial del Comercio

Dudes In Club—164 total members

Dudes Observing the Club—Iran, Iraq, Sudan, Vatican City . . . and 16 others

Dudes NOT In the Club—Palestine, Somalia, North Korea . . . and 12 others

The World Trade Organization (WTO) is an international, multilateral organization that makes the rules for the global trading system and resolves disputes between its member states. The stated mission of the WTO is to increase trade by promoting lower trade barriers and providing a platform for the negotiation of trade. In principle, each member of the WTO is a privileged trading partner with every other member. This means that if one member gives another a special deal, he's got to give it to everyone else—like in elementary school if you were caught with candy.

Things that make the WTO sigh in delight are "open markets," "tariff reductions," and "long walks on the beach at sunset." The WTO is basically the application of capitalism on a global scale. The key idea is that competition creates efficiency and growth. *Any* country should be able to sell *any*thing it can, *any*where it wants, at *any* price that *any*one will pay. The WTO can boast some successes in growing international trade, but these have also been accompanied by increased wealth disparity between rich and poor nations AND between the rich and poor within nations.

Formally established in 1995, the WTO is structured around about 30 different trade agreements, which have the status of international legal texts. Member countries must ratify all WTO agreements to join the club. Many of the agreements are highly criticized including the Agreement on Agriculture, which reduces tariffs hurting small farms in developing countries. One of the most famous anti-globalization protests occurred around the 1999 WTO meeting in Seattle that galvanized various groups like pro-labor, pro-environment, anti-fur, etc. under a single anti-WTO banner.

FUN PLAID FACT: The Kingdom of Tonga became the 150th member in 2005. The oldest animal ever recorded, a tortoise named Tu'i Malila, died in Tonga in 1965. Okay, this has nothing to do with the WTO, but turtles are cool.

IMF

Members: Everyone. Seriously. Okay, you got me! Everyone except North Korea, Cuba, Liechtenstein, Andorra, Monaco, Tuvalu and Nauru.

The primary responsibility of the International Monetary Fund (IMF) is to monitor the global financial system. The IMF works at stabilizing currency exchange rates. Doing this, they provide security to overseas investors and help promote international trade. The IMF's policies are also aimed at reducing the phenomenon of "boom and bust," where economies grow rapidly, then stagnate, then grow rapidly, then stagnate, then grow rapidly, et cetera. The main tool of the IMF is "financial assistance" (aka loans), which they provide to countries with "balance of payment problems" (aka big time debt). As a condition of the loans, the IMF mandates "structural adjustment programs." These programs are designed to turn a cash profit, allowing the borrowing country to repay its debt to the IMF. Here is a short glossary of "structural adjustment" terms and how they are interpreted by the locals:

IMF sez . . .	Locals sez it means . . .
Austerity	cutting social programs
User Fees	charging for stuff like education and health care and water
Resource Extraction	selling stuff out of the ground that rich countries want
Privatization	selling state owned stuff to rich companies (usually foreigner-owned)
Deregulation	removing domestic control over stuff
Trade liberalization	allowing foreigners to open sweatshops and exporting stuff made in sweatshops

Much like the WTO and World Bank, the IMF is often congratulated for growth in the global trade and production and simultaneously scorned for increasing the poverty gaps within countries and between countries.

THE WORLD BANK

Similar to the IMF, the mission of the World Bank (actually the World Bank Group) is to encourage and safeguard international investment. All the while, the World Bank attempts to help reduce poverty and spawn economic development. The World Bank works primarily with "developing countries" helping them develop in a Westerly fashion. Also, like the IMF, the World Bank loans are contingent on adopting "structural adjustments." While the IMF deals primarily with currency stabilization, the World Bank is primarily like a real bank, loaning countries money for very specific development projects like a hydropower plant or a disease-eradication program.

Unlike the IMF, which is headquartered in Switzerland, the World Bank Group is headquartered in Washington, D.C. and the US plays a very heavy hand in both its leadership and loaning activities. At least for now . . .

It should be noted that the World Bank is breaking new ground on the redefinition of the sovereignty issue that we covered in chapter 3. Due to government corruption in many states, particularly Africa, the World Bank is now putting further stipulations and oversight on all the loans it makes to countries. They are going in and making sure that the money they give a country to build an AIDS clinic doesn't end up getting used to buy weaponry for the state. The Plaid Avenger is proud as punch at such a bold move, but many states are hopping mad that this type of intense oversight violates their sovereignty—I'm with the World Bank on this one. Suck it up, sovereign states. If you want the dollars, prove it's going to be used responsibly.

Watch for this issue to gain world attention soon. As of this writing, it's causing consternation in Chad as we speak . . .

NGOS

NGOs are Nongovernmental Organizations. These are basically every private organization that is not directly affiliated with a government, like Amnesty International or Greenpeace. We are talking about a growing number of highly influential international groups that are playing an ever important role in transnational politics. These groups transcend borders and unite common interests. NGOs are set up to represent a diverse array of special interests (environmental protection, human rights, et cetera). Here are a few examples of influential NGOs:

- Human Rights Watch—With a budget of ~$20 million a year, this NGO aims to document violations of international humanitarian law by sponsoring fact-finding research. HRW recently waged a successful campaign against the use of land-mines, which the U.S. government opposed.
- Freedom House—This NGO supports research for democracy promotion. Also, each year FH ranks countries on a scale from "Free" to "Not Free." Luckily, as of 2014, the United States is still "Free."
- Greenpeace—As the name suggests, this NGO hopes to achieve greenness using peaceful means. Greenpeace is also the only NGO to own a ship *(The Rainbow Warrior)* that the French government intentionally sunk. Funny story, actually; google it and find out. The greatest French military victory since the Napoleonic era.
- Amnesty International—This NGO is committed to protecting the human rights enshrined in the UN Universal Declaration of Human Rights.
- International Red Cross (and Red Crescent)—The sole function of the Red Cross is to protect the life and dignity of victims in armed conflict. The Red Cross is independent, neutral, and all that other crap. They help anyone and everyone.

We will discuss some of these NGOs in more detail in later chapters. Party on.

THE NUKE GROUP

7 Members: US, Russia, UK, France, China, India, Pakistan

3 Questionable Members: Israel, North Korea, Iran

Last, but certainly not least, is a most important gang of states on planet earth with this homogenous trait: they got nuclear bombs! Or perhaps maybe they do. Or perhaps maybe they are trying to get them. While not an "official" entity in the spirit of the other groups discussed in this chapter, I would be remiss without sticking this information in somewhere . . . . and what better place than a chapter on clubs, as this particular club has enough firepower to blow up all of the other clubs on the planet? The club with the biggest clubs! But it's much more important than that if you want to understand how the world works and how geopolitical power is actually wielded in real life in our day and age. What do I mean?

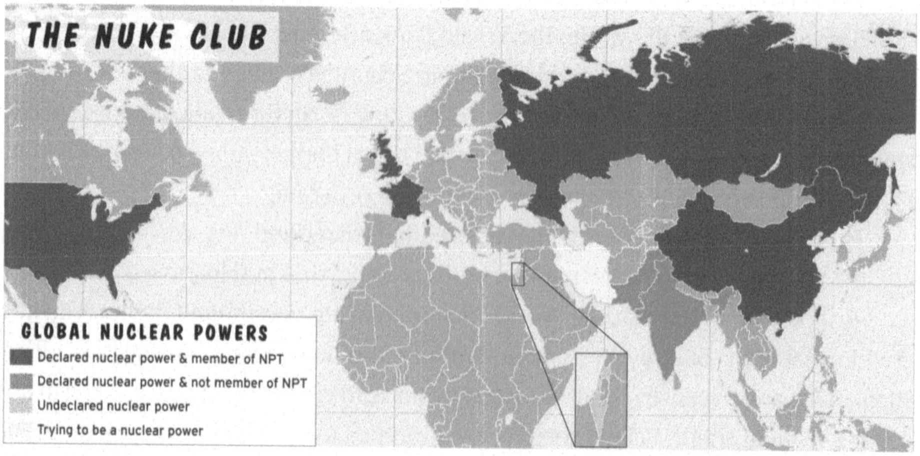

Well, who is in the club? Let's start with the original declared nuclear powers: US, Russia, UK, France, and China. Hey! Wait a minute! That's the exact same group that as the UN Permanent Security Council! How true, my quick-witted friends, and that is no coincidence! These 5 countries were the first to develop, test, and therefore prove that they possessed nuclear weapons. The fact that they admitted this to the world means that they have "declared" their status. Kind of like coming out of the nuclear closet. Because of their "first-ness" and openness, these nuke powers have the veto powers at the UN, as per their permanent status on the security council.

Soon after the development of nuclear weapons, most everyone agreed it would be a horrible idea for all countries on earth to have access to these weapons, so a movement to limit them emerged, culminating in the 1970 **Nuclear Non-Proliferation Treaty** (or **NPT**). This treaty has been signed by 189 countries, including the aforementioned big 5 declared nuke powers . . . only four countries are not signed up for it (more on that in a sec). The treaty has three basic pillars it strives to deal with:

1. non-proliferation, or non-spread, of nuclear weapons and/or nuclear weapon information;
2. disarmament, or getting rid of existing nuclear weapons; and
3. the right to peacefully use nuclear technology, which means all signers of the treaty are allowed to have nuclear energy, but not nuclear bombs.

As you can see, this treaty is mostly to ensure that nobody is making nuke weapons, but nuke power technology for energy production is allowed by the NPT, which makes enforcement of it tricky and a hot potato in current events. But before we get to that, what about the other possible nuke powers not yet named? That would be India, Pakistan, North Korea, and Israel. Ah! Yes! Those would be the exact four countries that have not signed the NPT (or dropped out of it in the case of North Korea). What a coincidence! Not! Here is the deal:

India developed and tested its nuclear weapons in 1974, and not to be outdone, Pakistan followed suit in the late 1990's. Why would these countries need nuclear weapons? Because they hate each other and have already fought three wars so far, with more to come. Once India became nuclear, Pakistan could not rest until it got the bomb as well. That's the way rivalries work. The tension between these two countries is so great that neither will give up their right to possess, and even create more, nuclear weapons, which is why both countries have refused to sign the NPT. But they have declared and proven that they have bombs, so they make the list of nuke powers, even though they cannot legally pursue nuclear energy industries since all that stuff is also regulated by the NPT.

Who is left? Israel, of course! With no reservations, everyone on the planet knows that Israel is also a nuclear power, but they have never (openly) tested, proven, or declared it. The US and many others never want them to declare it either, for fear it would spark a regional **arms race**. Israel has probably been a nuclear power since the 1960's, and it wants to have that nuclear edge mostly in order to ensure its survival, were it to ever be attacked by surrounding states again (see chapter 20). Because of this "secret" status, Israel has refused to sign the NPT, because to do so would mean it would have to open up for inspections and declare their stockpile . . . which, again, might cause a regional arms race.

Finally, there are those who might be trying to get into this nuclear club, but aren't quite there yet. North Korea originally signed the NPT but has subsequently quit, mostly because having the elusive illusion of possibly having nuclear material is the only bargaining chip they have left to play to get international attention as their state nears collapse. The koo-koo North Koreans exploded something underground a few years back and claimed it was a nuclear device, but no one is really quite sure what the heck they actually have. Including themselves. And I'm sure you have heard plenty about Iran recently too . . . they actually are a signatory of the NPT, which gives them the right to develop nuclear energy. Which they are, while claiming peaceful intent. But some of Iran's semi-crazy leaders have already made reference to themselves as a "nuclear power," which of course they are technically not. Yet. Maybe. Wowsers, this is confusing!

Why am I throwing you all this info on the Nuke Group? Well, unless you have been hiding out in an underground nuclear bunker since the 1950's, you know that the issue of nuclear energy/nuclear weapons has become an extremely hot topic in today's world. Like radioactive hot. Like 1.21 gigawatts hot. Crazy North Korea may be developing something which it then may use to blow up a neighbor. And there is great fear of terrorist organizations obtaining nuclear material and doing something nasty with it. But most of all, the world is currently at arms with itself over what to do with Iran: half the world thinks Iran is just developing nuclear power, the other half think Iran is trying to get nuclear weapons . . . it is possibly the most divisive issue of current times, and may result in war. That's why the UN permanent Security Council is debating these issues non-stop, why Team West is at odds with the BRICS, why former US President Barack Obama held the first-ever "Nuclear Security Summit" in Washington DC in 2010, and why former US President Donald Trump ramped up intense pressure on Iran, thus setting the stage for a full-on war. Yikes! More on this later!

No state currently with nuclear weapons (and the lion's share of states that don't) wants to see more states go nuclear; these things are dangerous and deadly and could spell the end of humanity! Dudes! I've seen *The Road Warrior* and *The Matrix!* It's scary! At the same time, there are those states that want nuclear weapons. With no question, nukes are the absolute best deterrent that would prevent your country from being attacked. Who wouldn't want a nuke? I mean, let's be honest here: the main reason that the Cold War never turned into a hot war was because both sides had nuclear weapons; therefore, neither side could attack without suffering the same nuclear annihilation itself. Would the US have dropped atomic weapons on Japan in World War II if Japan was equally armed? Would the US administration be openly talking about invading/bombing Iran if Iran was already a nuclear power? Having a nuclear weapon is a game-changer. No wonder some countries may still want one.

This brief section was only to alert you to the nuclear status of our planet, not to fully explain and engage all the complicated topics surrounding these weapons of ultimate destruction. Hopefully, you now at least have a handle on the hotness and how it plays out in current events and the world regions. Regions? Oh yeah! That's what this book is about! Let's get to the regions soon, shall we? But allow me just one more paragraph or two of ranting to close on this chapter first . . .

SLIDING BACK FROM THE SUPRANATIONAL

I would be remiss in my teaching duties if I did not at the very least reference some significant—and very recent—reversal of fortune for the concept of supranationalism that this entire chapter has focused on. I try to enlighten about very large and long-term global themes, so I'm not really trying to expound about the domestic politics of a single leader in a single country at any given time . . . unless that particular leader and that particular country somehow have huge global impact on overall trends. Of course, I am referring here to former US President Donald Trump and his administration. Now, again, I'm not really interested in his domestic agenda for his country, nor his personal life, nor his political beliefs, nor how much you personally like him or hate him. I only want you all to consider his opinion and policy on supranational organizations, and the impact his opinion has had, even though he is no longer in office. What is that particular opinion?

To be concise: he HATED them! And I don't think I am being overly simplistic in that assessment. He made it quite clear through his words and actions that he much more favored a return to the old-school world in which sovereign states fend for themselves, be it economically, militarily, or even culturally. As you have seen in the references to him peppered throughout the chapter in the updates, President Trump immediately threw out the TPP, re-worked NAFTA, and pulled the US back from CAFTA and the FTAA. On top of that, he was vocally no big fan of the WTO, even less of a fan of the UN, berated and perhaps even intimidated members of NATO, and encouraged the UK on its BREXIT break-up.

The Donald sez: "International Organizations . . . BAD! They make me . . . SAD!"

The important point here is not that one single leader has an anti-supranationalist opinion, it's that the most powerful country in the world had these opinions . . . and it's the same country that actually helped build all these suprantionalist organizations in the first place! This is a major reversal of decades-long policy! Added to that, President Trump's opinions on the concept are also shared by a lot of other world leaders, who are now empowered to follow his lead; think of champions of BREXIT, anti-EU political parties of Europe, and even the current disintegration of the once-promising integration movement in South America.

But not everyone on the planet is following the anti-globalization populist parties and policies that President Trump espoused. China still leads the charge in building bigger and bigger blocs; ASEAN, APEC, and the SCO are all growing; the AU just created AfCFTA; and the TPP is still moving forward despite US withdrawal. So, the big picture here is that the "Western countries" may be pulling back from supranationalism (for now), but it is still alive and well and expanding in Asia, Africa, and the Pacific.

This shift of "the west" versus "the rest" is a major theme covered in this 11th edition of *The Plaid Avenger's World;* a new fight between **globalism** and **nationalism** has broken out in the last few years, and has very identifiable regional patterns and impacts of note. Added to that is the retraction or diminishing of democratic principles and norms, which is also happening in various places and is manifesting itself in very real ways as existing democracies are literally being demolished across the planet. It is so important a theme that it is the subject of Chapter 25: The Cover Story, and we will elaborate on it in much more detail then. You just have to navigate through 19 more chapters to get to it! So let's get to those regional chapters now!

PART TWO
THE REGIONS

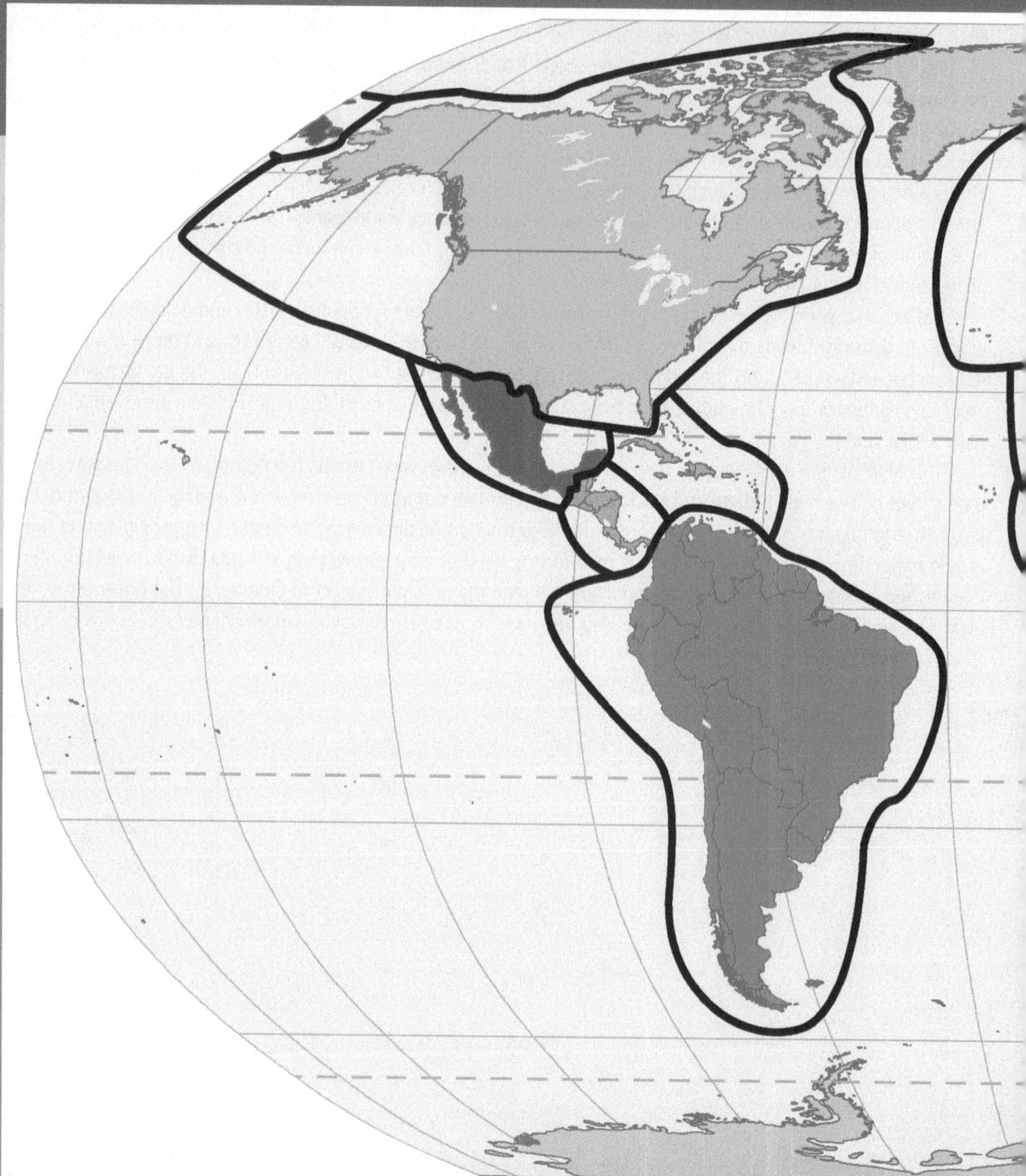

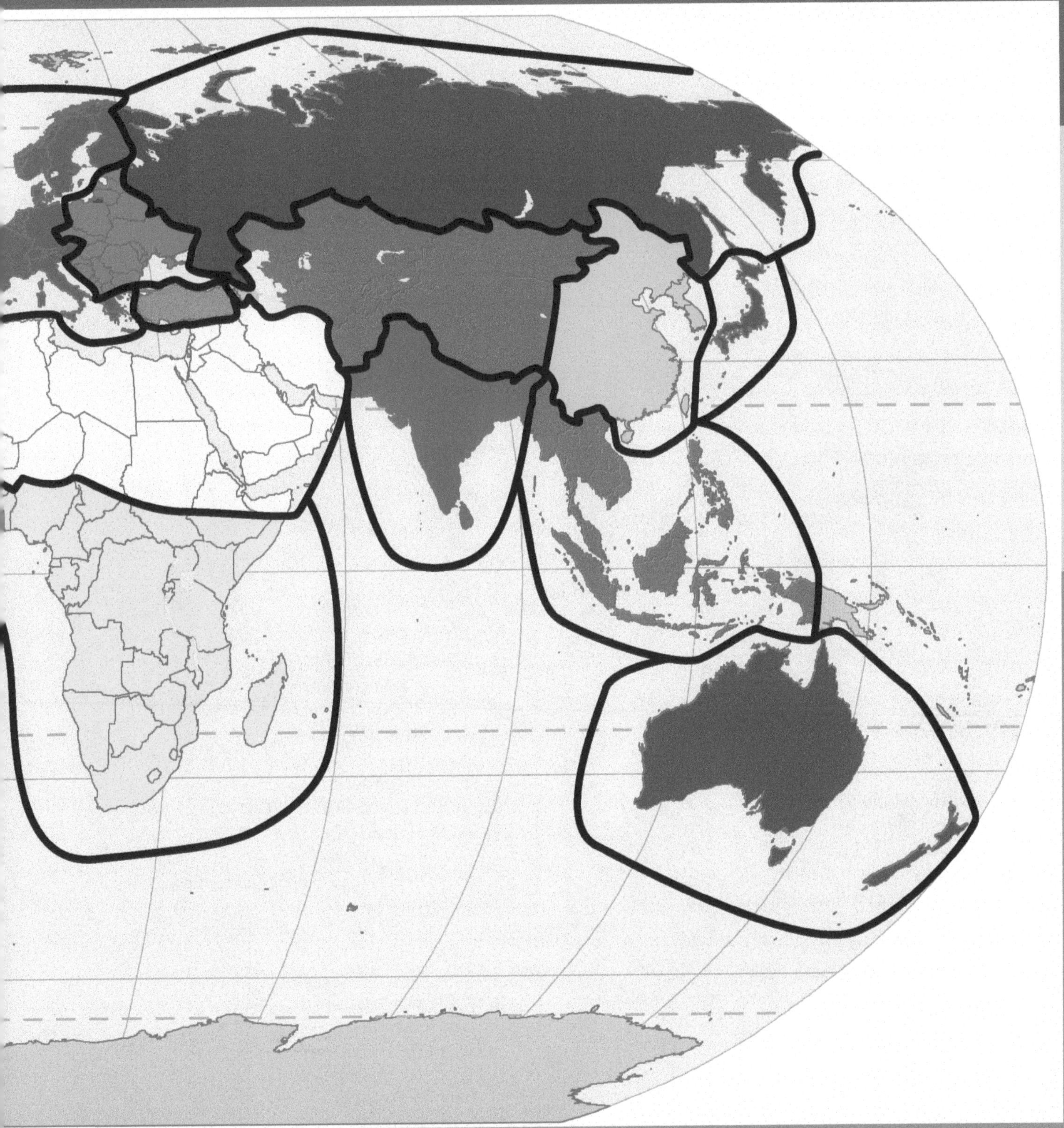

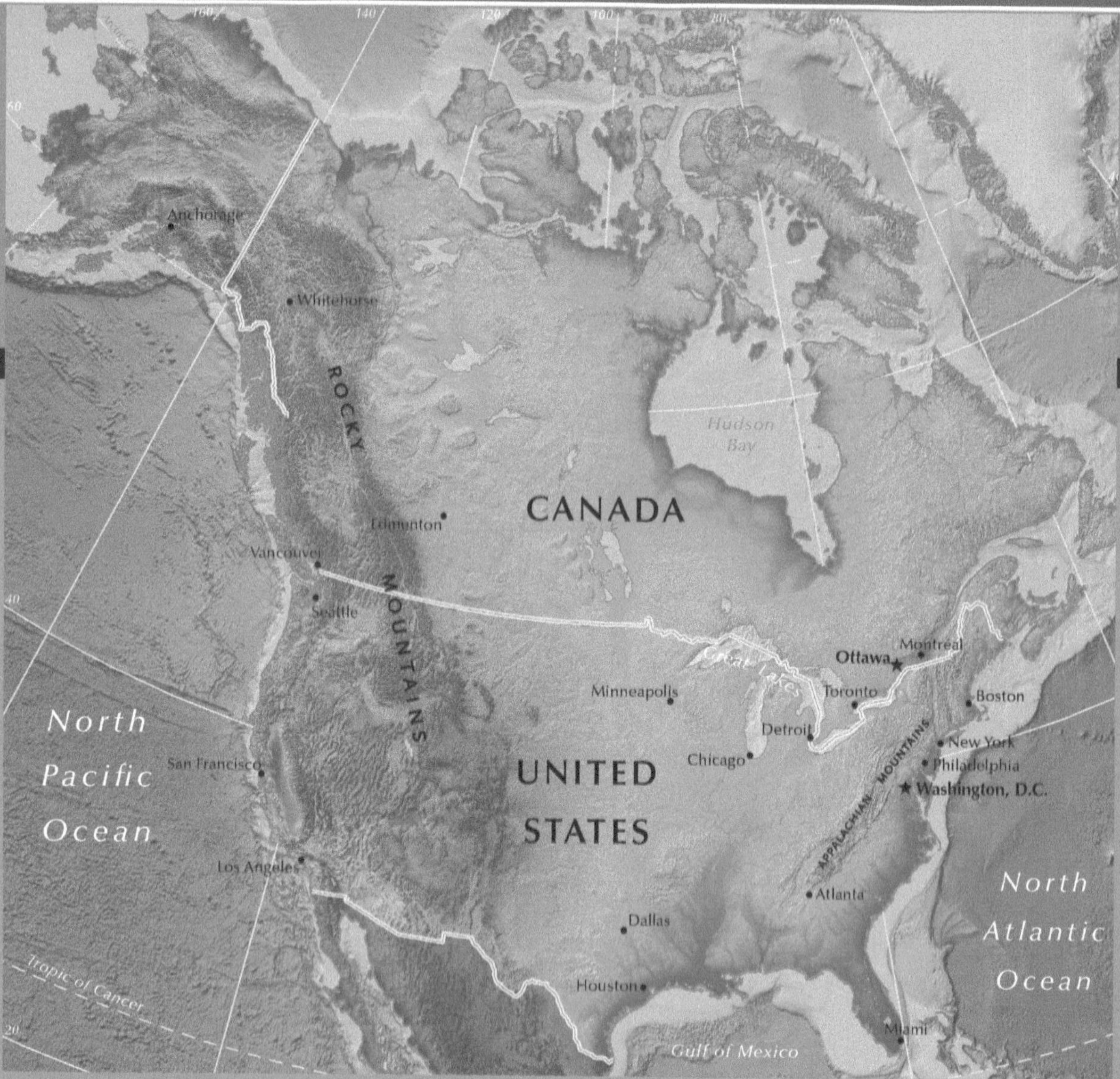

CHAPTER OUTLINE

Anglo-America

7

NOW that we've discussed all the preliminary stuff, we can get to the real meat and potatoes of this text: the regions. And I start with the biggest meat and potato eaters: the Anglo-American region.

This region consists of just two huge countries: the United States and their redheaded stepchild to the north, Canada. But wowsers, what an incredible impact on the rest of the world this region makes. Of course, I'm referring mostly to the powerhouse of the planet, the United States of America. For most of the following discussion, I'll be talking about the United States, but always keep in mind that their chilly Canadian cohorts are generally on board for all things Anglo-American, even US foreign policy. So is their population; 90% of Canadians live within 100 miles of the US/Canadian border. Heck, I'm not even sure why it's a separate country at all sometimes. I think the only real difference between these states is styles of bacon. But I digress . . .

Canadian Prime Minister Justin Trudeau: "Don't forget about us, eh!"

WHY IS THIS A REGION?

Why is this a region at all? Think back to the 3 things that constitute a region. A region needs some space, of which North America has plenty. A region must also have some borders, which may be fuzzy, but this region's borders are actually very clear cut; it's everything north of the US-Mexican border. Also necessary for a region is homogeneity. It must have some traits that are roughly the same throughout the region, across the entire area.

Done and done: this is one of the most homogenous regions on the planet based on a cornucopia of common characteristics. It has a high standard of living; it is one of the wealthiest regions in the world. If we head south of the border, we see that Mexico is distinctly different economically, which is one reason why it is not included in this region.

Another factor is language. There is, for the most part, a single language that dominates. "Anglo" actually roughly translates to "English": maybe you have heard of Anglo-Saxons, the Anglican Church, Anglophones, or perhaps Anglia . . . which was the Latin language name the Romans gave to England itself. Although there are many people who speak Spanish in the US—and of course there are also those wacky French Canadians—primarily both countries are English-speaking. Canada and the US also share historical and cultural backgrounds. Both countries were British colonies at one time. This cultural baggage, which includes religion and the aforementioned trait of language, also solidifies the homogeneity. The fact that the US and Canada are the oldest and most stable democracies is also important. Human rights, the belief in individual freedoms, economic activities, and similar governmental structures also make this a solid region. And a "great" region at that! Wait . . . great?

AWESOME ANGLO ASSETS: PHYSICAL GEOGRAPHY

Yes! It is "great" in many ways, not the least of which is it's fantastically great physical geography! From the amber waves of grain, to the purple mountain majesty, above the fruitful plains . . . wait, what the heck is a fruited plain? Whatever that may be, Anglo-America as a region is one powerhouse of fabulous physical geography of diversity and richness, which has played into their success story from the start. Let's do a brief overview of the physical nature of the region from sea to shining sea and all the way up to the Canadian Great White North, eh! Anglo-America the beautiful and fruited, and here's why:

BODACIOUS SPACIOUS

The North American continent makes up about 17% of the world's land surface, with our Anglo-American region of Canada and the US dominating most of that space. This place is the host with the most—nearly every climate type can be found on this single continent, and virtually every type of natural resource that humans want to extract from the planet. And why is that? Well, frankly, because it is a huge place! From the western Aleutian Islands off Alaska to Newfoundland is about 5,000 miles and from Miami, Florida to Point Zenith (the farthest point north still on the North American mainland) in far northern Nunavut, Canada is another 4,500 miles. Tons of stuff in tons of space, especially when you consider that there really aren't that many people here: 325 million Americans + 35 million Canadians = 360 million peeps total across a gigantic continent. (Sub-Saharan Africa and Southeast Asia have twice that amount; both India and China each have four times more peeps.)

Just as important is the relative location of the region: Anglo-America is a continent crosser! Just two countries that span an entire continent, and they both have access to two coastlines. Why so coastal cool? Oceans are superhighways of trade, which Anglo-America has taken great advantage of over the centuries, as well as extension of their naval military power. Think about other major world powers: China . . . only one coast, Europe . . . only one coast, Russia . . . no coasts at all! lol poor Ruskies. But Anglo-America can go east or west, can connect to the Atlantic world AND the Pacific world, can shift either way economically and militarily! Simultaneously, those two great oceans serve as great buffer zones from all other regions; it's tough to get to the place either as a migrant or an invading power. The Anglo-American region can easily defend itself, as well as choose when and where it wants to interact with the rest of the world. Quite a unique situation. More on this later.

WATER WORLD

Speaking of those coasts, this region has got the most! Okay, maybe that is an exaggeration, but Anglo-America certainly has a tremendous strategic advantage in its over 135,000 miles of coastlines. When you think of coasts, you likely are reminded of beautiful beaches, of which this region has many. Much more important is having great access to the oceans via great natural port facilities, and Anglo-America is awash with such cities: L.A., Newark, Seattle, Norfolk, Charleston, Houston, Miami, Baltimore, Boston, St. John's, Montreal, Vancouver . . . these are not just great cities, they all started as great ports! The movement of goods in and out of a country is vitally dependent upon natural harbors and ports, and the more you have, the more you are connected domestically and internationally, and typically the more trade you do, the more money your economy makes. But what about all the places not located on the coasts?

Well, in this region, there is water, water, everywhere, and in various ways! They have huge amounts of "water property" in terms of areas that are territorially owned by them: think of the Northwest Passage, Hudson Bay, Great Bear Lake, Great Slave Lake, the Great Lakes, Chesapeake Bay, and large territorial waters off their coasts, including parts of the Gulf of Mexico. Combined, Canada and the USA would make the largest nation on Earth with the largest amount of fresh water supply. Fresh water is of course a necessity for life itself, but also is critical for large-scale food production as well as transportation. Generally speaking, the more water you have, the more food you can make and the more interconnected you are. Having said that: Anglo-America has one of the longest rivers in the world—the Mississippi—as well as a system with one of the greatest water capacities on the planet—the Great Lakes–St. Lawrence system. I didn't even mention Chicago as a great port city in this list . . . because that seems weird, right? I mean Chicago is in the center of the continent after all!

https://commons.wikimedia.org/wiki/File:Mississippiriver-new-01.png

But think about it: Chicago exists because of the great waterways that so interconnect this region that one can move great amounts of cargo just about anywhere via water. This is a region of immense rivers and water systems. The Great Lakes region is not only a huge fresh water reserve, but it connects even the deepest interior parts of the continent to the Atlantic Ocean and the world. And the vast Missouri-Mississippi-Ohio river system connects almost the entire vast central area of America and allows for navigation and trade out to the Gulf of Mexico. A trivia fact to support this watery importance: the Mississippi River carries 60% of U.S grain shipments, 22% of oil and gas shipments, and 20% of coal from the interior of the continent. That's just one river system. Add to that the Great Lakes. Then add to that long rivers (from the Rocky Mountains) such as the Colorado, Columbia, Fraser, and Yukon that flow west to the Pacific. A connected continent, indeed!

TERRIFIC TERRAIN

And the connection of the continent is also not hampered much because of the terrific terrain of the place. Anglo-America is somewhat unusual among the continents in having large, geologically stable interior lowlands that are almost completely enclosed by younger mountain chains . . . none of which are high or formidable enough to become barriers to movement, like the Andes and Himalayan mountains are in South America and Asia. From the eroded and ancient Appalachians in the east to the Rocky Mountain highs in the center to the volcanically active and growing Coastal Ranges in the west, the north to south trending mountain systems of the region have shaped the landscape and the peoples on it. Providing challenges but not insurmountable obstacles; providing a wealth of minable resources and arable land; providing regional identity but not separateness: the terrain is a crucial component of regional success.

The continent can be divided into four great terrain areas: 1) the Great Plains/Interior Lowlands stretching from the Gulf of Mexico to the Canadian Arctic; 2) the geologically young, mountainous west, including the Rocky Mountains and the Great Basin, along with the geologically distinct Coastal Ranges that run from Mexico thru California and Canada up to Alaska; 3) the raised but relatively flat plateau of the Canadian Shield in the northeast; and 4) the varied eastern region, which includes the Appalachian Mountains, the coastal plain along the Atlantic seaboard, and the Florida peninsula extending along the Gulf of Mexico.

http://www.nationsonline.org/oneworld/map/north_america_reliefmap.htm

Coastal plains and huge areas of interior lowland made for ease of movement, (coupled with the aforementioned water systems which did the same), and seemingly limitless land resources . . . The Great Plains especially became a vast production area of food crops. The Appalachians are a minor, older chain that originally provided a nice little backyard barrier for the 13 Original Colonies while they grew up and became an independent state, but were small enough for the growing new country to overrun and expand into the vast continental interior. The Rockies presented technological challenge that was overcome by railroads, and later interstate highway systems. And the major mountain chains run down both coasts, providing an almost perfect natural defense from any invading armies. Not that that matters much in today's world of intercontinental ballistic missiles, but it did back in the day! But on to biomes . . .

BANGIN' BIOMES!

The size and location of Anglo-America makes for a diverse climate, which gives us even more diversity in plants, animals, and landscapes. Latitude shapes it heavily, and this region got a lot of latitude! The region is less than 10 degrees of latitude away from both the North Pole and the Equator, resulting in a huge range of temperatures north to south. Places like Central America and Mexico are hot and semi-tropical year round; the "middle ground" of the US and southern Canada is subtropical and temperate, with enough precipitation and temperature variation to grow all kinds of vegetation and crops. The far northern parts of the region in northern Canada and Alaska are subarctic, arctic, and even permanent ice cap in places. You can find just about every climate and biome type on earth in this one region! Okay, it's cheating a little because we are including Hawaii, which has some full on tropical rainforest type stuff there, but truly outside of the tropics, this region boasts the most when it comes to biomes. Anglo got it all: everything from the tropical swamps of Florida to the temperate rainforest of Seattle to the deserts of New Mexico to the arctic tundra of Labrador . . . and every four seasons-style climate in between! Hardwood forests, mixed forests, rainforests, grasslands, prairies, aquatic biomes, tundra—you name it, they got it. Such great diversity in a single region, and makes for a great resource base . . .

ftp://newftp.epa.gov/EPADataCommons/ORD/Ecoregions/cec_na/NA_LEVEL_I.pdf

RESOURCES & RISKS

To finish off this rant on the friggin' fabulous physical world of Anglo-America, consider how stable and rich it is physically. Of course there are risks, but the resources far exceed them. But even the risks are minimal when compared to the wider world. How so? Well, what destructive forces do Anglo-Americans fear the most? The region has crazy tornado action in the Great Plains; it gets hit by hurricanes in the southeast, Nor'easters in the northeast, blizzards in the midwest and deep north; flooding and fires in California; earthquakes and volcanoes along the western coast margins. But notice how we have to actually name a specific sub-region when we talk about any of these individual risks; none of them occur region-wide. It is a rare event indeed that affects more than the people of a single state, much less everyone in the entire country or region. Why is that?

Because of the region's size and diversity and stability. Key words to remember. While the occasional earthquake can be severely destructive in California, most of the continent is geologically stable as it sits in the middle of the North American plate, if you know that plate tectonics stuff. The real tectonic action is at the edges (and that is where California is), thus the earthquakes associated with the San Andreas fault and active volcanic activity (from Oregon/Washington to Alaska) are confined to the western margins of the region where the North American plate rubs up against the Pacific plate. The other risks to the region are weather related, and therefore sporadic and erratic (i.e. hurricanes, blizzards) and affect limited areas. Given the region's wealth, high levels of infrastructure, high levels of rapid mobility, high levels of information and preparedness, and high levels of technology, the death and destruction these risks present are minimized greatly.

Thus, none of those challenges has kept the region's inhabitants at bay for very long, because the riches so outweigh the risks! And now to the spoils of riches! Since Anglo-America has nearly every climate zone on the entire planet, the wide ranges in weather allow them to grow many different kinds of crops. In tropical areas like Florida, farmers harvest a variety of tropical produce such as citrus and bananas. In the mid latitudes, the more humid climates can support fruits and vegetables, tobacco and cotton. More northerly latitudes such as the central US and southern Canada support vast fields of grain: corn, wheat, rice, and soybeans as well. These humid continental landscapes are some of the most productive breadbaskets of the world. And across the vast grassland prairies of both countries, cattle has become king. That's on top of being some of the biggest pig and chicken producers in the world as well.

Can you dig the wealth of Anglo-America? You know it! The region has a wealth of just about all mineral and metal and minable resources. From nickel to iron ore to silver to copper to coal, Anglo-action is on it! Extraction resources like mining and drilling lead Anglo-America's economy and have made them a major player in the world energy market since there was a world energy market! Coal is mined in several key areas of the US and Canada—especially the Appalachians (historically) and currently in Wyoming and the Canadian provinces of British Columbia and Alberta. Coal mining still contributes about 5 billion dollars to Canada's economy and over 225 billion to the US economy. Fossil fuels lead the US energy market, and most Americans probably remain oblivious to the fact that the USA is still one of the biggest producers of oil in the world, with operations from Alaska to California to Texas that have been in production for decades or longer. The Athabasca tar sands in Canada (Alberta) are the world's largest reservoir of heavy crude oil, although it is dirtier oil and thus has to be processed more to make it usable. The "next big thing" in fossil fuel production is natural gas via a process called "**fracking**." As fossil fuels go, natural gas is the cleanest and most efficient burning option and gas basins are currently being drilled in southern Canada, the Appalachians and Eastern US, Oklahoma, and the Dakotas (where they're also drilling for oil, too). So much energy wealth, although of course all that carbon-based energy may be the death of us all soon via global climate change. Can't we all just go back to living in the trees? Ah!

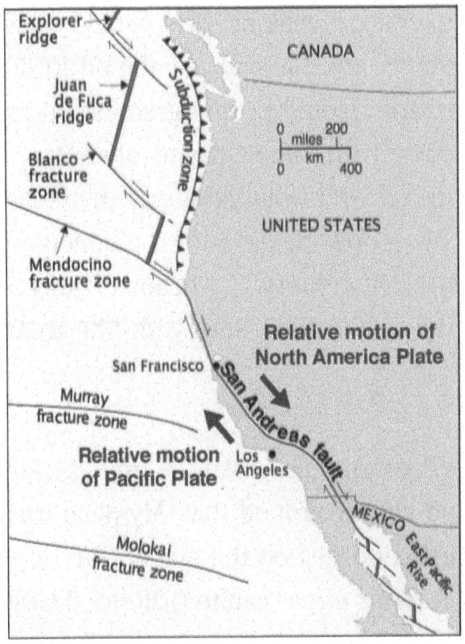

Let's not forget the trees! Forestry is a major player in Anglo-American economies, too. In the Pacific Northwest, the boreal forest of Canada, and the gulf/southeastern US states, huge amounts of lumber are exported worldwide. Parts of Washington state and the Pacific NW get so much rain, they are considered a "temperate rain forest." Some of the world's biggest paper/pulp mills are located in these areas because trees grow so well in the abundant rainfall and moderated temperatures. As for the numbers, the US is the world's largest producer and consumer of lumber, and exports more wood than anyone else. Canada is closing the gap; because their government has subsidized the industry so heavily, companies are more attracted to the Canadian side of timbering than the US at the moment. You never knew how good you had it for wood, did you?

SUCCINCT PHYSIQUE SUMMARY

Enough of the grand Anglo-geography tour! Sorry for all the driveling detail on the physique of this fab region, but it is important to understand just how stacked the deck is in favor of the Anglo-American region. It contains almost every desirable element that all other countries across the globe would kill for: huge amount of land, natural buffers of the oceans, great coasts and harbors, great river systems, great resource base for water, great resource base for food, great resource base for metals/minerals/forests, diversity of biomes, diversity of landscapes, diversity of flora and fauna, diversity of crops, generally geologic stability, minimal large scale natural disaster threats . . . the list goes on and on! Just remember the key words: stability, diversity, and size. The region is stable, has just about everything, and a lot of it! So keep in mind the awesome physical geography of Anglo-America as you go about understanding why it is the greatest region on earth!!!! Oh right, we should get back to that now . . .

THE GREATEST REGION ON EARTH!

Canadian lights not too far from the border.

Wow, that's a bold claim that the Plaid Avenger is making. Why would I suggest that the United States and Canada, combined into our Anglo-American region, would be the greatest on the planet? That does come with some caveats, of course. The greatest at many things, and simultaneously the not-so-greatest at others. Two opposing sides of a uniquely powerful coin here in the 21st century . . . heads or tails . . . call it in that air! Heads it is: We'll start with "the greatest."

+THE RICHEST

Why would I say that the United States and Canada is the greatest region on Earth? For starters: they're the richest. There's no getting around that. My plaid friends, I think you all accept this, as do most, that the United States is one of the richest countries on the planet. We forget their Canadian cohorts to the north, though. The Canucks are also pretty well off in terms of per capita GDP, total GDP are a top 20 world economy, top 10 on the Human Development Index (HDI), and of course both countries are at the core of the G-7 and G-20. Why, and in what respect, are they together two of the richest countries on the planet? Let's break this down.

PHYSICAL RESOURCES

One: physical resources. These two countries are the second and fourth largest countries on the planet in terms of size. Lots of land in which to move around and lots of space to grow. Space/land is a valuable commodity in and of itself for future potential growth of a country, and both these big boys have it. The US and Canada contain tons of natural resources too . . . and I mean everything across the board. These countries produce tremendous amounts of food, be it grain or grapes or chickens or corn or anything else. They have tremendous amounts of water resources. H_2O is an oft overlooked resource when people think about a place's richness, but hey, you've got to face it, friends—if you don't have water, you have nothing. The US and Canada have lots of food and lots of water . . . but also lots of the other natural resources. North America is virtually saturated with almost every type of resource possible: coal, oil, tobacco, lumber, copper, steel, gold, fertile croplands, and almost anything else. You use it, you want it—Anglo-America has got it.

PHYSICAL DISTANCE

This Anglo-American region also has a strategic advantage in that it is far from other places. What's so special about that? Why would that make them rich? If for no other reason, you have to consider the fact that while the US and Canada have been involved in virtually every major world conflict of the last century, the fights have never occurred on their own soil. To my recollection, Pearl Harbor and the September 11, 2001 attacks are the only times that anyone for a century has even made it close to North American shores for a strike. While those were horrific events to be sure, they were pinpricks in comparison to the larger wars of which they were a component, for the most part creating death and destruction in far-off foreign lands. Having two great oceans separating this region from the rest of the planet is a big security and strategic plus, indeed.

Specific example: After World War II, Europe was destroyed. China was decimated. Parts of Russia were blasted. Japan was completely leveled. The United States and Canada? Sitting pretty and virtually untouched. Besides Pearl Harbor, nothing happened on the soil of this region during those hugely destructive wartime eras.

Also, consider terrorism in the modern world: for those with malicious intent, it is very difficult to get to Anglo-America from other parts of the world. This is one of the reasons why the 9/11 attacks in the USA were a really big deal. People just couldn't believe it. Folks around the rest of the world are used to bombings, terrorist attacks, uprisings, but it's very unique in the Anglo-American region—it just doesn't happen. Terrorist attacks occur in Europe, Asia, the Middle East, and across Africa all the time, but not in Anglo-America. After 9/11, people in the United States were freaking out big time, because that type of thing just seems really extraordinary to Americans. . . .

Why do they happen in other places and not here? Distance! People that may be upset with the United States and Canada are very far away. Anglo-America is not an easy place to get to if you're a disgruntled Afghan. You can hate the US all you want, but you're not going to get there to do anything about it unless you are well-funded and organized, which rules out most of the people on the planet.

The US and Canada have chosen to get involved in world affairs when it suits them. Very few other places have this convenience. The result? Untouched for virtually all global wars and confrontations of the last 200 years, and even now in the 21st century, they have a wide safety buffer from the rest of the world's problems. Their distance is a key feature.

EVOLUTION OF AN ECONOMIC POWERHOUSE

The Anglo-American region has experienced 200 years of continuous growth. That's not something that any other region can brag about. Look through the rest of this blasted book. There is no other place on the planet that has essentially gotten richer decade after decade, century after century, like the US and Canada have for 200 years straight.

The formation of the modern states of the USA and Canada, of course, goes back to the colonial era when the British, the Spaniards, and the French rolled across the Atlantic in boats and staked claims to the continent. From this nice little nest egg of an area, they proceeded to invest heavily into infrastructure and developing industries. It was a beautiful little place to get a start-up country going—a nice little fixer-upper.

The physical geography played a role in this too, namely the Appalachian Mountains. European settlers were all over the few hundred miles of land from the shores of the Atlantic to the mountains: that was the 13 original colonies. This became a perfect little incubator for the embryonic United States. It insulated the colonies away from Amerindians, who were on one side. The protection that the mountains provided allowed the colonists to gain control of the region and build up resources and population which, in turn, set up the right conditions for westward expansion.

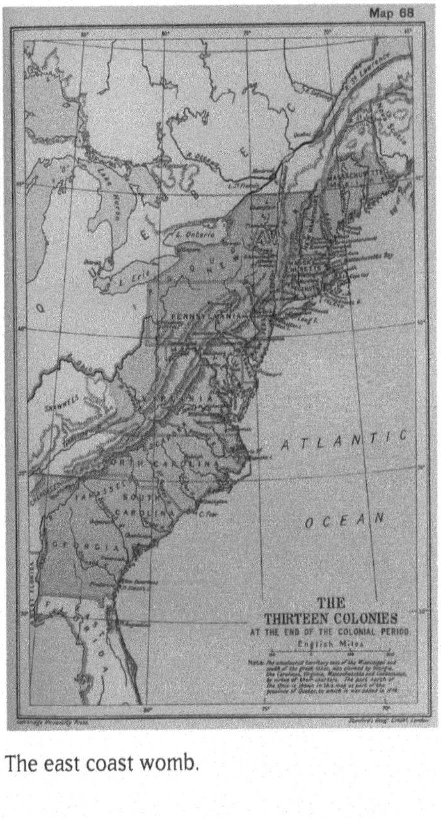

The east coast womb.

This is where distance begins to tie in to the US history. As the colonies began to develop their own ideas about government, freedom, and self sufficiency, they realized that they didn't need the British overseeing them anymore. Being an ocean away from the colonial power afforded them the 'space' necessary to evolve and enact their ideas about independence.

During a little thing called the **American Revolution**, though the British had superior firepower and numbers, the distance was still too great to effectively fight a prolonged war. Distance helped win the war, and also kept foreign powers at bay while the fledgling country strengthened itself in the early years.

No other country in the history of the world has had such a large amount of growth continuously sustained for 200 years than the United States. This place has gone steadily upwards since its inception. Here's how:

From that point forward 200 years ago, the United States grew westward, as did Canada, adding more resources, more water, more fields to grow more food, more coal, more oil, more gold, expanding all the way to the west coast of the continent. The US and Canada had essentially 100 years of continuous physical and economic growth; gaining more stuff, more resources, more land. It was **manifest destiny** at its best, and mo' money, mo' money, mo' money. Running out of land or opportunity in Virginia? Then move to Tennessee, then Indiana, then Oregon! And did the economic expansion stop when physical expansion maxed out? No way, dude! It just took a new direction. In the 1860s to early 1900s, the American Industrial Revolution began to kick off. Luckily, the Americans had witnessed a similar revolution in Britain almost a hundred years earlier, when the tea-sippers were the top dog.

"You all have little girly economies."

By the way, a Plaid Avenger note: if you broke off California from the rest of the continental US and made it its own sovereign state, it would have the 8th largest economy on the entire planet. Also, former Gov. Arnold Schwarzenegger himself would be the 18th largest landmass.

Ahhh . . . the good ol' days of the manufacturing era. Trump tried and Biden now hopes to restore it.

The Americans took all the best elements from the European industrial revolution, and minimized most of the worst. While Europe was very destructive during theirs, the Americans were able to avoid a lot of the pitfalls, and go straight for the goods. The US shifted from an *agricultural* based economy to an *industrial* one. After the land and resource grabs, they exploded industrially.

In the 1960s and 1970s, there was another change. When other countries also began to industrialize, they could produce things cheaper than the US, due primarily

US States Renamed for Countries with Similar GDPs, 2014

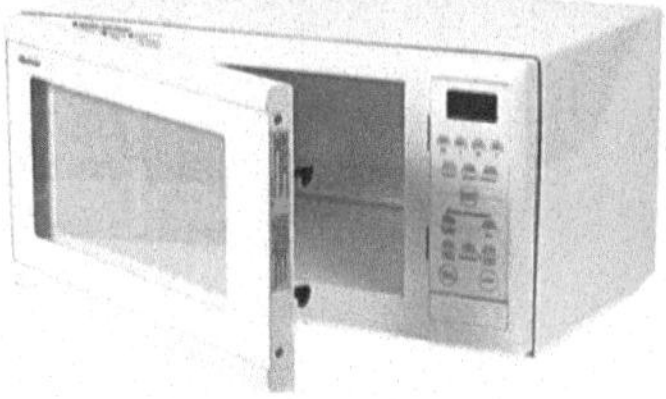

to their lower labor costs. Things like automobile, textile, and steel production shifted to developing countries. Was this the end of American economic expansion? Are you kidding me? The US once again shifted gears and changed to a *service sector economy* with an emphasis on *technology*. You know: computers and softwares and stuff. Nowadays, the next wave of transformation has the US as a leader of the *information age* . . . in which the economic focus is on manipulation and creation of computer programs and new medical breakthroughs and patenting DNA codes. Instead of making cars and textiles, the US now produces information as one of its key resources. This region is truly the first to undergo this technological transformation.

Long story short: over 200 years of almost continuous growth in various ways, either increasing land or resources, then increasing to an industrial capacity, and then increasing into the service sector, and now transforming the world as the leader of the technological and information age. Of course, you have the Great Depression in there, we don't want to forget that little tidbit; that was a decade blemish on a basically 200 year perfect record of economic growth. Even now, here at the outset of the 21st century, the Anglo-American region has the biggest total GDP on the planet, with no close second. (Okay, China is closing the gap, and fast.) If money talks, then this region has a lot to say.

BIRTHPLACE OF IDEAS; TECHNOLOGY TITAN

Why else is the Anglo-American region the greatest and richest region on Earth? It has do with what we just talked about: this is also a birthplace of ideas. The Anglo-American region is the place that came up with things like the telephone, the computer, and the Internet (thanks Al Gore!). This region has the freedom to do lots of things and grow industrially and technologically, and it comes up with all kinds of new crazy great stuff. This region has the highest levels of technology because it is always on the forefront of innovation. Japan makes cool stuff cooler, and China is growing fast, and India is becoming a software hub, but Anglo-America is still the top dog when it comes to the forefronts of science, engineering, and computer innovation. However, the competition is getting tougher. You students reading this book better get on your A-game and keep the innovations flowing . . . or they are going to flow right over to Asia!

Since World War II, the United States has been not only an industrial leader but on the forefront of science and technology in all capacities: the Internet, the computer, the telephone, the microwave, nuclear bombs. Put a man on the moon? You betcha! This region did it. That is one of the things that makes the US really rich. People come from around the world to invest here, go to school here, start businesses here, and they do all these things because this is the place to do it. Also, it has the highest levels of technology across the board in fields such as science, math, computers, military and nuclear capabilities, right on up to the upcoming space race. The US, because it is a technology leader, is still the only country that has made it to the moon. It will be joined soon by China, but that's for another chapter.

BIG REGION; SMALL POP

Anglo-America actually has a fairly small population given its size; this equates to a very high GDP per capita. Both of these countries are fully industrialized and developed, huge in terms of area, heavily urbanized, with a low population density overall. This may be argued by folks in the United States right this second who have problems with immigration: "We've got too many people here! They're takin' our jobs! Wah wah wah!" Whatevs.

The original—and equally unsuccessful—anti-immigration faction.

There's around 330 million people in the United States, while Canada has only about 38 million; and both countries continue to grow population-wise, but pretty slowly and in a stable fashion . . . another big plus! About 361 million people total live in two of the biggest countries on Earth. There's a lot of space in this place and not really that many people in it, relatively speaking. There's tremendous room for growth. Even if growth does not occur, you're looking at tons of resources and a huge economy, both in the US and Canada, that's divided up amongst not that many folks on the global stage. If you're looking out at China, they've got almost a billion and a half people. India is working on 1.3 billion people; Sub-Saharan Africa has three-quarters of a billion, and even richy-rich Europe has over half a billion residents. The North American region? Ha! A fraction of the big boys, especially given the amount of space.

In conclusion: lots of space, lots of room, and lots of stuff per person makes the Anglo-American region fairly well off. The United States has the biggest GDP total on the planet. Canada is in the top twenty as well; both countries have some of the highest GDP's per capitas, and that's not likely to change anytime soon. Yes, China and India are growing fast, but neither will ever have the amount of richness per person that the United States and Canada enjoy right this second, if ever.

+THE STRONGEST

We just talked about why Anglo-America is the richest. Now let's talk about why it's the strongest.

Team America on patrol.

MILITARY POWERHOUSE

The Anglo-Americans are also def the best and the strongest when it comes to military might. You can talk about whether you like the US or not, you can talk about whether you agree with its policies or not, but that's of no consequence to the Plaid Avenger. I'm just trying to explain why this is one of the richest regions on the globe, or the greatest region on the globe. Militarily speaking, you cannot dispute that the US has the strongest single military power on the planet. This is not by size, by the way. We all know size matters, but in this case, the US does not have the largest standing military, meaning troops on the ground. China holds the top slot with 2.5 million people in their military. The United States and Canada don't need to have the largest amount of people in the military because—technologically speaking, of course—we all know they are the best.

The Anglo-American region has the most weapons per capita on the planet. We're not just talking about handguns. We're talking about everything from shotguns, right on up to intercontinental ballistic missiles, tanks, and warships. The United States has the most aircraft carriers on the planet and this is a very real projection of power. Militarily speaking, no country can compete with the United States, not even Russia, their old Cold War adversary. In terms of military spending, no one else is even in the US ballpark. Yearly military expenditures for the US now top 3/4 trillion dollars. Wow. That's twice what China spends, and almost half the world total. There is no contest when it comes to who's the strongest or most technologically

WOPR asks: Shall we play a game?

advanced military on the planet: it's the Anglo-American region, particularly the United States.

PROJECTION OF POWER ABROAD

However, it's not just numbers and budgets. It's also about the ability for the Americans to project this power abroad. This is what makes them the strongest. Aircraft carriers are a great item to consider. With an aircraft carrier, you can push your power to all points on the globe, and the US projection of power to any point on the planet exceeds anything ever even previously imagined by former empires. Sure, the British had their fleet that did a pretty darn good job, and the Spaniards had their global Armada before that, but that was centuries ago. The United States has the technological capacity to annihilate so many more humans than the British did that no comparison can really be made. With a carrier, you can essentially "move" your country anywhere in the world on the water to set up an island to launch a strike. That is some serious global power. Fun fact: the US has 10 "super carriers"—more than the rest of the planet combined!

Country	Military Expenditures 2020 (US$)
United States	$778,000,000,000
China	$253,000,000,000
India	$73,000,000,000
Russia	$61,500,000,000
United Kingdom	$60,000,000,000
Saudi Arabia	$57,700,000,000
France	$52,800,000,000
Germany	$52,500,000,000
Japan	$49,200,000,000
South Korea	$45,800,000,000
Italy	$29,100,000,000
Brazil	$25,100,000,000
World Total	$1,435,200,000,000

http://www.sipri.org/

The Anglo-American region is the absolute best at moving troops, people, guns, and missiles anywhere on the planet. That is real ultimate power. One need look no further than current conflicts in which the American region is involved in: Afghanistan and Iraq, which are on the other side of the planet from their home turf.

If you start to look at American military stationed abroad, it's mind boggling. The US has active troops, dudes and dudettes with guns, in about fifty countries. That's astounding if you think about it. That you would have military people stationed all over the planet, even in countries where there is not an active conflict, is quite amazing. This region is certainly the most powerful when you consider the current active conflicts and military deployments, and consider that the United States is a natural leader of NATO (re-read chapter 6 if you forgot about NATO), and thus inherently involved in all NATO missions as well. The Plaid Avenger argues that NATO is easily the most successful defensive organization of all time. I stand by that claim, and the United States is at the core of it.

The aircraft carrier: Carrying war straight to your door!

This region also possesses the most intercontinental ballistic and nuclear missiles. As I suggested earlier, the US is the only country that has landed men on the moon. I won't say the US dominates space, but it has the strongest presence in space . . . for now. There is actually a new 'space race' on right now, which may see Asian powers surpass the US domination of space. Someday. Also, factor in the

Maverick! Goose! Take to the skies!

fact that the US is a big proponent of some sort of anti-missile defense shield, which will certainly, if it ever comes to fruition, make the United States even more powerful. Go learn about the Ronald Reagan era when he wanted to put lasers in space to shoot down other countries' missiles. Only a great power could even consider such a plan and then to put resources toward actually making it happen. I keep harping on space because that is the next frontier; the US is promising to build a moon base and perhaps a missile defense shield in space, and lots of other things in space as well. That just takes tremendous resources, and you've got to be great to pull off those kind of things.

By the way: I keep talking about the United States, but Canada is right beside the US in every single conflict in which it has been involved. All of them. Therefore, we can logically speak of foreign policy/military might as a singular Anglo-American venture.

MULTINATIONAL CORPORATIONS

Now let's shift to Anglo-American corporations in the world, both US and Canadian. These are some of the biggest, richest entities on the planet. I'm speaking of multinational corporations which operate globally. Every multinational corporation needs to have a home base somewhere, for many, this base is Anglo-America. Of the top 2000 largest corporations on the planet right now, over one-third call Anglo-America home (BTW: that is down from over half a decade ago). Exxon, Wal-Mart, General Motors: these are some of the richest entities on the planet, and with that wealth comes power. These folks have a lot of power to do the things they want to do in other parts of the world . . . and at home.

Consider an entity like Exxon. Sure, it's just a company, not a government, and it doesn't have political power like the US. Well . . . yes and no. Exxon makes like a quadrillion dollars a year; if it wants to look for oil in Namibia, it is going to get what it wants, regardless of Namibian opinion. Let's face it, the company makes a hundred or perhaps a thousand times more money than the GDP of Namibia; with that money comes tremendous weight to throw around. Whether through open economic pressure or illegal back-room bribery of officials, huge corporations can get their way.

Some of these Anglo-American companies are the biggest and the richest in the world, and many are getting richer and more powerful by the day. Back in 2013, Exxon captured a world record by earning the highest profits in a single economic quarter ever—thank $100 a barrel for oil for that one! Folks are starting to openly debate whether or not multinational corporations are stronger than governments. The answer: of course! Uber-rich corporations are stronger and more powerful than small countries and weak governments. There's no doubt about that, and it's a situation that's likely to increase in the future. To

2019 FORBES RANKING OF AMERICAN-BASED MULTINATIONAL COMPANIES

World Rank (Overall Value)	Company	2019 Sales (USD)
#2	JPMorgan Chase	$132.9 billion
#5	Bank of America	$111.9 billion
#6	Apple	$261.7 billion
#10	Wells Fargo	$101.5 billion
#11	ExxonMobil	$279.2 billion
#12	AT&T	$170.8 billion
#14	Citigroup	$100 billion
#16	Microsoft	$118.2 billion
#17	Alphabet	$137 billion
#19	Chevron	$158.7 billion
#20	Verizon	$130.9 billion
#26	Berkshire Hathaway	$247.8 billion
#28	Amazon	$232.9 billion
#29	Walmart	$514.4 billion

From: http://www.forbes.com/global2000/list/ *value calculated May 2019 restate: 1/3 of the top 2000 global companies are US or Canadian in origin; almost half of the 30 biggest companies in the world are from the US alone; and half of the top 50 biggest are from this region . . . yes, Canada has some big boy companies on this list, too! (Very important note: 7 of the top 30 world's largest public companies are now from China, including 5 of the top 10. China had zero companies in the top ten a decade ago. Now that is some major growth, and a major change in global wealth distribution.) But let's shift to another topic, and the final subsection of why this region is the greatest.

+THE FREE-EST

OLD MAN DEMOCRACY

Both the United States and Canada are two of the oldest democracies on the planet. This is an entire region that prides itself in a somewhat egalitarian light; everybody is equal regardless of race, sex, creed, religion, or anything else, and everyone has an equal voice. That is one of the cornerstones of this entire region. They are the free-est. There are other countries around the world that are democracies, but the Anglo-American ones have been around the longest and have been the most successful for the longest stretch. In fact, these countries, mostly the US—sorry Canada, I have to poke a little fun at you—were so determined about democracy that they had a little revolution and kicked the Europeans out. They said: "Hey, Europeans, that's cool, we're glad you came here and set up shop for us, sent us some immigrants, set up some industries, but you guys can all go home now because you're not being fair enough to us."

Maybe you've heard of it: the American Revolution. 1776, yeah, baby! American-style full-on democracy was born in the US! Meanwhile, Canada invented hockey. Hahahahaha! Why am I making fun of the Canadians? They kissed some British monarchy butt for quite a bit longer, like another hundred or so years, before they threw off the yoke of the Commonwealth. But even during their Queen-kiss-up years, Canada was pretty much into the democracy thing too, kind a variant on the constitutional monarchy deal. To restate, Anglo-America is the home of two of the oldest, strongest, longest lasting democracies on the planet.

In fact, these guys have really set the trend for everyone else. You can look at Europe and say that they are also rich, developed democracies with a long track record. Yeah, that's true, dudes. But starting when? Was it before the American Revolution or after? That's right—it was after. The American experience was kind of the impetus for the French Revolution. The French Revolution then begat other revolutions across Europe, which eventually led to all of them becoming something closer to democracy. If you look at it as a chain of events, the Americans started things up. George Washington, Thomas Jefferson and my main man Ben Franklin totally rocked the hizzle with their democracy dizzle. Nice job, guys!

Got tea?

FOCUS ON INDIVIDUAL LIBERTIES

A focus on individual liberties is also a cornerstone of this region from its inception. This is something quite distinct from democracy itself. Democracy is a ruling system in which the people have a say in who's ruling them. That's cool. But on top of that, civil rights and individual liberties are extremely important to this region. Maybe this is confusing to you, but I just want to point this out: there are other democracies in the world, pseudo-democracies or real democracies, where they don't have this focus on individual liberties as much.

Countries like Malaysia or Uganda or Tunisia are totally democracies. They vote, they elect people, but there is nothing in their constitution that says that they believe every single person is exactly equal, and that the state is there to protect their individual rights. Look at Iran, which has a semi-democracy. People vote; in fact, their voter turn-out shames most US election participation. They do elect many of their leaders, but they definitely don't have the focus on individual liberties. The government still controls ultimately what people can read or say, and even the way they can dress. Perhaps making comparisons to Iran or Malaysia is too extreme, because they aren't as developed and rich like the US. Okay, how about Singapore or Russia? Both countries are rich and developed, but done so at the expense of individual liberties. They are tightly controlled democracies where too much individual expression may be detrimental to your business.

"We come for individual freedoms—and native babes."

Only in the Anglo-American region do they say first and foremost, "We have a democracy, but it's based on individuals having liberties." By the way, if you think about how the United States came to be, it's because people were fleeing from places that weren't giving them true liberty. Those Pilgrim peeps who went to Plymouth Rock were fleeing religious persecution back in Europe. They wanted more freedoms to individually do their own thing here in North America. That's how the thirteen original colonies started, and quite frankly, it's still going on today. People flock from around the world to head to the Anglo-American region because they know when they get there, they are free to live the way they want. This is not the case across the entire planet, even in established democracies.

I'm not suggesting all these liberties have been there since the beginning. Freedom of thought, religion, and speech were there early on. More recently, civil rights and liberties regarding race and sex were fought for and won by activists and concerned citizens. And the US not only sets the world standard for freedoms, but continues to be on the front lines of battles over civil liberties and rights: look at the fight for gay rights or the rights of an unborn child for examples of how passionate this region gets about individual liberty.

And part of the reason Anglo-Americans feel this way is that, quite frankly, they don't give a crap. Wait, what?

PERSONALLY THEY DON'T GIVE A CRAP

This region doesn't give a shake about you as a single individual. That sounds a little backwards. What do I mean exactly? We can make fun of Anglo-America for being a region of imperial power, of gun-loving nuts, and of isolationism, (lol ok that applies to the USA only, not Canada). But if you get there—when you are actually in the hizzle—it's really freakin' cool. Anglo-Americans just generally don't give a crap who you are or where you are from on a personal basis. They are laid back. Do what you like! It's cool. Just don't ruin any of my stuff. This is a phenomenally great component of Anglo-American society.

If you look around the planet at the history of human society around the world, we have a pretty consistent record of kicking the living daylights out of each other. Why do we do this? It has a lot to do with intolerance of differences between people. Sometimes we kill each other over resources or land or whatever. However, conflict mostly arises between people of different religions, ethnicities, races, or nationalities—and sometimes it escalates to some pretty nasty business. Think of the atrocities of Nazi Germany or the Rwandan genocide or the implosion of Yugoslavia. The list goes on and on. People killing people who were not like themselves.

This is a common theme across the planet, except in Anglo-America. By no means are all Americans sitting around eating s'mores and singing "Kumbayah" together. Of course, there are people here who don't like other people, but this is at nearly the absolute minimum in the history of the world. Americans don't care where you're from, what color you are, how smart or stupid you are, what god you pray to, or even what you do with your leisure time. Quite frankly, American tolerance toward their fellow citizens may be the best example the rest of the world should follow.

In real life in America, not caring equates to everyone being pretty much equal. Some people call it a meritocracy, meaning you only advance by working hard and doing good stuff—by earning your way. Your position in society is not, or isn't supposed to be, based on your religion or skin color; Anglo-America doesn't really have organized ethnic or religious clashes. Many people from around the world emigrate to this region to escape those very issues.

Let me elaborate on why I'm saying this is a positive thing. In most other parts of the planet—and this is today's world, not historically—people don't have the luxury of not giving a crap, meaning that there is serious ethnic, religious, and racial strife all around the world. If you are in the minority, practice a religion that's not fashionable, or are from the wrong

Britney Spears: a symbol of a tolerant America. We even give K-Fed a pass!

ethnicity or tribe, you may not have equal rights at all in your country. Even if the country pretends you do, you may be discriminated against on a daily basis. You might be thrown out of the country or you could be caught up in ethnic genocide or civil war or religious persecution, which happens all the time across the planet.

That doesn't happen in Anglo-America because individuals are doing their own thing. Maybe it's because the countries' very foundations were as immigrant nations, but whatever the reason, they don't give a crap about the guy sitting next to them on the bus. They just don't, and that's a very powerful thing. I can't stress it enough: it's one of the reasons why folks from around the world say, "Wow. If I could just make it to North America, then I truly am free. Nobody cares that I'm from a Hmong ethnicity. Nobody cares that I'm a Zoroastrian. Nobody cares if I tattoo my face, dye my hair blue, and blast Justin Bieber on my boombox. I can do what I want to when I get to those shores." By and large, that is true and that is one of its real benefits. It is one of the free-est places on the planet. Okay, maybe they should pass a law against listening to Justin Bieber, but for now it is still protected behavior.

PLATFORM FOR SUCCESS

What this sets up, essentially, is that the Anglo-American region is a kind of platform for success. People know that once they make it to Anglo-America, all other factors are equal. You work hard, you pay your dues, you put your nose to the grindstone, then you can be successful. No, the streets aren't paved with gold in Anglo-America, as the old saying goes. They are paved with opportunity, because immigrants don't have to worry about things that they'd have to worry about in their home countries. The Anglo-American region is a real platform for success. Which reminds me to introduce you to a concept called **brain drain** (see inset box).

For example, after World War II, the Russians took half of Germany's rocket scientists and America got the other half: the foundations of both the US and Russian space programs. Brilliant inventors arrive and they use the US as a platform for success. There are other great platforms, but the US has been a very important one for the last 100 years.

This has a lot to do with immigration. People want to come to Anglo-America to express themselves. It has become a self-perpetuating cycle; the US is so rich and doing so well that people all over the globe want to come here with their ideas. People want to start businesses with the highest probability for success—and that is in the US. Success breeds success, and more people want to come. American success is a strong magnet in the world. Rich and poor alike have a strong attraction to the Anglo-American region for all of its opportunities. It's an extremely competitive place, but it does seem to work for the vast majority of its inhabitants. Poverty numbers are very low, compared to other regions; opportunities exist for even the most destitute.

What's the Deal with Brain Drain?

Any person given a free plane ticket from their lesser developed country to any other destination in the world will most likely choose the Anglo-American region.

The type of people that actually make it to the region are smart, educated people with options. Also, wealthy people who want to become wealthier figure out ways to come to the North American region.

This is actually a problem, because the elites of other societies are leaving their native regions to emigrate to wealthy regions in order to become successful. There are all sorts of political, religious, and economic reasons to flee native lands and set up shop in Anglo-America. When the political problems arise, people leave. When people are afraid of losing their assets or afraid of religious persecution, they will go to a place that accepts them and allows them to continue growing. For a hundred years, the destination of choice has been the United States.

Brain Drain has been a fantastically successful concept in the development of Anglo-America. Whenever bad things happens around the globe, the US and Canada stand to gain.

Political crackdown in Iran? Economic meltdown in Thailand? Religious persecution in Sudan? Anglo-America gains in refugee and immigrant populations, usually of skilled, hard-working, and/or asset-laden peoples. Conversely, the country that loses the peeps is impacted quite negatively, thus the "drain" part of the term.

2019 UPDATE: Wealth disparity is on the rise, even in the mighty Anglo-American region, and it is a major issue that is getting attention from leaders and protestors alike; decades of wage stagnation and decreasing opportunities for the working middle class were a main reason for the rise of the populist President Donald Trump. He was expected to remedy this, but how successful he was at accomplishing it is now a matter of opinion and conjecture. The lost Covid year certainly didn't help. Trump at least gave the perception of fighting the good fight for coal-miners and auto-workers who have seen declining opportunity for decades, but it's too early to tell if any concrete advances have been made to combat wealth disparity even within these groups . . . and now President Biden hopes to take up this issue with his grand infrastructure and green energy industries plan to spend trillions on rebuilding the country while simultaneously creating tons of blue collar jobs again. We'll see.

If you live in an impoverished society, one with no welfare program or unemployment benefits, you cannot take risks. New things come about by taking risks, and environments like that do not enable people to take risks. America will take care of you if you or your idea fails. You can afford to work hard, save money, start your own business. No wonder people think the streets are paved with gold—golden opportunities, am I right?

Another great thing about this region is its **hypermobility**. The size of the region allowed for the interstate system, developed by US President Dwight Eisenhower, whereas other countries that are smaller rely more heavily on public transportation. People can go anywhere and do anything, anytime they want in Anglo-America, via a variety of transportation options. They even have four-wheel drive vehicles to take them places where roads are nonexistent.

Mobility allows people to take advantage of opportunity and expands individual choice. Any job, any opportunity, any time, any place. Not all folks in other countries have this convenience. There is an exactly proportional relationship between mobility and choice. The more mobile you are, the more choices you have; as a result, the US is more successful.

We Like Ike! What's the Deal with the Interstate System?

The formal name of the United States interstate system is the "Dwight D. Eisenhower National System of Interstate and Defense Highways." The US interstate system was created by the Federal-Aid Highway Act of 1956, which, as the name suggests, was championed by Eisenhower. It was built for both civilian and military purposes. Some of you might be thinking, "Yeah! One in every five miles of road must be straight so that military aircraft can land and take off on them!" Well, maybe. The true military aspects of the interstate system are primarily to facilitate troop movement and to allow for the evacuation of major cities in the event of nuclear war.

The civilian aspects of the Interstate Highway System have helped shape American culture in more ways than we can possibly imagine. Most US interstates pass through the center of cities, which allows people to live outside the city and commute in for work, which has also played a huge role in the creation of suburbs and urban sprawl. The highway system also gives the federal government power over state governments. The US government can withhold interstate highway funds, which are huge amounts of money, from uncooperative state governments. The US government used this tactic to increase the national drinking age to 21 and to lower the blood alcohol level for intoxication to 0.08%. According to the Constitution, both of these issues should be decided by states. However, if they choose to disobey, they don't get their highway money.

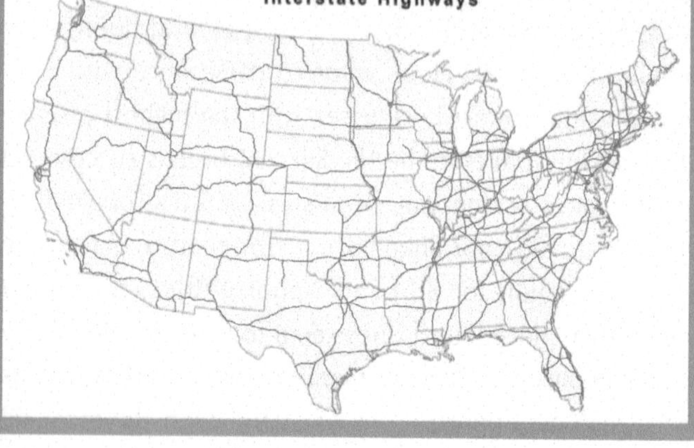

Also, the system allows high speed transportation of consumer products. This makes the prices of everything, from bananas to concrete, cheaper. The result: the US is an automobile culture heavily reliant on fossil fuels for the movement of everything and for virtually all aspects of their lives. Americans are quite unique in this respect. It's also why they are the best stock car racers in the world. Who else has NASCAR? Who is more worried that the price of oil is reaching $100 a barrel?

Living in the Anglo-American region furnishes infinite choices and provides opportunity to take risks; this is unprecedented anywhere else on the planet. If you have a great idea and you could have any choice of where to cast the dice to really make it work for you, it's a no-brainer for you to come to Anglo-America. It just makes sense. The odds are stacked in your favor. Not true in Europe for immigrants, not true in Asia or Africa as a whole, (Chinese, Japanese, and Russian business cultures are notoriously difficult to penetrate as a foreigner)—but def true in Anglo-America.

CORE OF TEAM WEST

To wrap up this free-est section, all of these factors put together are what make the Anglo-American region the true core of what I will refer to as Team West. This is a concept we'll come back to later in the book. Anglo-America, along with Europe and Australia, form a common block of ideology that sets them apart from other cultures in the world. You know what? Let's throw Japan in there as well. Even though they're way out in the east, they more closely follow the western tradition.

What are the identifiable traits of this team? As exemplified by Anglo-America, Team West has a basis in the western civilization tradition that values democracy in their governments, a focus on individual rights in their societies, and are strident free-market capitalists when it comes to their economies. This team consists of regions and states that are fully developed, largely rich, technologically superior, and focused on service sector and information age types of employment. Other attributes include rampant materialism, an obsession with pop culture, and unproportional power and leverage over the rest of the globe—but those are more negative attributes.

How appropriate, because it's time to get negative on the team leader right about now!

THE WORST PART OF BEING THE GREATEST

We're going to shift gears now and expose the underbelly of the greatest region. I've told you about how awesome the Anglo-American region is and why it's the greatest and the leader in terms of richness and power and freeness. Now let's take a look at why it's perhaps not so great. What's the worst part of being the greatest region? It's an easy schemata, because we're just going to follow the same outline we did before. We'll start with number one.

-THE RICHEST

Wait a minute! I thought I just said that being rich was great? Everybody thinks being rich is great, but the bad side of being the richest region on the planet is that US and Canadian policies, of course, favor their own peeps and companies.

UNSPORTSMANLIKE CORPORATE CONDUCT

In other words, they take care of their own. Why wouldn't they? Every country has its own interest as the number one priority . . . that is completely natural. US foreign policy is no exception: be it intervention into another country or manipulating the World Trade Organization or some sort of other pressures on other countries in order to help their own multinational corporations succeed. It happens. I'm not making fun of anyone, but I'm trying to get you to understand why other folks around the world may have discontent about American power. All states have self-interest, but being the richest most powerful region means you can achieve those self-interests better than anyone else can! The cards are stacked in your favor!

The United States and Canada have some of the biggest, richest multinational corporations in the world. Simultaneously, the US and Canada are two of the richest countries in the world. Therefore, it's almost a set of unprecedented power players that have really never occurred before—that the richest country can help out the richest multinational corporation. This gives those corporations quite a bit of leverage and power in the world. Power to do what? Well, most free-marketers would say to do what business does: go out and acquire resources, produce products, and sell those products. That's true, but they also have a lot of power to screw with other people or governments, if they so choose. They have the power to ignore other countries' environmental regulations or to bend the rules to favor themselves as frequently as possible. There is no way that a local company in a less-developed country could possibly compete with a Wal-Mart or an Exxon; these big boys can pretty much take what they want, when they want it.

This does present a problem. It is a situation where there is a fairly lopsided trade balance around the world where multinational corporations have the power to take a lot more from poor countries and give a lot less back. It's just simply the way it is. It's not good. It's not bad. In my book, that's just the way it is. US and Canadian policy favors their corporations, which are already rich. This helps them perhaps take advantage of other countries. Also, the Anglo-American's global policy typically ignores the rest of the globe to maintain its own richness. Let me put in a disclaimer: every country does that. China certainly is going to try to favor Chinese policy. The French look out for France. But the Anglo-Americans are so powerful and so rich that, here in the dawn of the 21st century, their attitudes toward the globe are now looked upon with some animosity by others on the planet. And that is because they have . . .

SELECTIVE SELF-SERVING SELF-RIGHTEOUSNESS

Let me give you some examples, and there are many to choose from. The idea of free trade may be the most blaring example of Anglo-America not practicing what it preaches. The US and Canada are fervent supporters of this concept, which essentially equates to trade based on the unrestricted international exchange of goods. This means no tariffs (taxes) should be levied on other countries' goods that are coming into your country, and no helping out industries within your country to make their goods more competitive than imported goods. Anglo-America regularly touts, praises, and leads the charge for free trade to the rest of the planet. In particular, the US wants a free trade area for the entire Western Hemisphere and it pushes hard to beat down the countries of Latin America to support the idea. Also, the US is the unrivaled leader in bringing up lawsuits against other countries at the WTO for infractions of the world trade rules. (**2021 Update:** Again, the former Trump administration actually hated those groups, so things are on hold for now until the Biden crew figure out what direction they want to go in as a hemispheric leader.)

Your taxes already paid for it, so eat it, just eat it!

But get this—the US and Canada regularly cheat on free trade all the time! Both countries support agricultural subsidies for farmers, which essentially equates to paying big agribusinesses money every year in order to make them more profitable. Corporations can then charge less money for the food they produce, making it cheaper than food imports from Africa or Asia. Both countries also regularly slap tariffs on imported steel or textiles in order to kiss the butts of political lobby groups and voting constituencies at home, especially in election years.

Another example of this selective support on issues: US/Canadian policy on global warming. For around two decades, most of the other countries on the planet have been saying, "Hey! This warming is a global thing, part of a global commons, and we need to look after the globe by all working together to reduce carbon dioxide emissions." The US and Canada have basically been very frank by saying, "No, we're not doing a darn thing. We refuse to do anything because it's going to hurt our companies. It's going to hurt our richness. We don't want to change our consumption patterns. We don't want to change how much money we have, and therefore, we don't give a carbon crap about reducing CO_2 emissions. The rest of y'all can go do whatever you want. We're going to keep doing our own thing." As always, I am paraphrasing policy here, but that does pretty much cover American attitudes.

It's also important to point out that the Anglo-American region is the highest user of world resources per capita. People in the US and Canada use more oil, more food, more plastic, more everything per person than any other folks on the planet. Whatever! It's a free market economy and people can do whatever they want, right? But other folks on the planet, and there are quite a few of them, kind of look at this and say, "Dudes. You guys are really selfish. You use the most, but you don't want to help out with any global problems." Being the richest has come at some cost, mostly at a cost to the reputation of the region, which is increasingly not seen as a real world leader on world issues.

-THE STRONGEST

What's wrong with being the strongest? Somebody's got to be the strongest, right? When it comes to Anglo-America, and I'll mostly pick on the United States right here, there is a perceived or real threat of US imperialism by many folks across the planet. That is, as I suggested earlier, the US is the undisputed, strongest country on the planet; because of this, people think, "They scare me! They are unstoppable!" The US is so powerful, it appears that it could do anything it wanted. You know what? Quite frankly, that's true. If the US wants to bomb Uzbekistan tomorrow, it has the capacity to do it, and current events have shown that sometimes this does occur . . . e.g. Iraq, Afghanistan, Libya . . .

It is a very real consideration that people around the world say, "The US is so powerful, they cannot be checked by any other entity." Truth. Who would stop them? Russia? China? The UN? Give me a break! The UN stopping something? . . . hahahaha, that's a good one.

We need look no further than the most recent Iraq war. Most folks around the world said, "Hey! We think it's a bad idea." Most countries of the UN said, "No, we're not going along with that. We're not doing it." The US said, "Well, we're doing it anyway." Again, I'm not making fun of them, I'm not trashing the Iraq war or the Bush administration. I'm just saying, think about the perceived light of imperialism that most folks around the world now view the United States bathed in. It does cause more than a bit of consternation and concern for both enemies and friends of North America alike.

DOESN'T PLAY WELL WITH OTHERS

The most recent Iraq war is worthy of further consideration for a moment more. This story gets at a larger point . . .

There is a particular disdain of the United Nations by the US. Let me put it more bluntly: the US as a whole totally hates the UN. Why? Because the US is so powerful and resourceful on its own, that the only thing the UN does for it is to get in the way of US objectives. Does that make sense? The US has a certain opinion about things in the world, and it wants to act on its opinion to fulfill its own self-interests. Many in America would argue that indeed this is exactly the job of their government: to look out after their own self-interests, the rest of the world be damned. Since the US *can* do anything it wants, it *should* do anything it wants to benefit itself. Following this train of thought, what exactly is the point of asking the UN for permission, advice, or for any thing else? Again: We witnessed this attitude to an extreme during President Trump's tenure.

You almost can't blame it either. The US did help save Western Europe's butt from Hitler without asking permission from anybody. They saved the world from communist takeover as well during the Cold War. Other folks don't like this unchecked influence, even if it is supposed to be for good. This is something that people in the US just don't get. They don't get it because they see themselves as good: "We are the good guys! We help people! We helped the French and Brits in World War II! We saved the Koreans and Kuwaitis and Kurds! We have always helped to defeat bad guys around the planet!"

Maybe, that is true. Maybe. The US seems to be fighting the good fight, but that's not always the perception around the world, particularly in this day and age. Back in World War II, everybody understood that the Nazis were the bad guys, and recognized the US as fighting the good fight. In the current War on Terror, however, the enemies and objectives are not so well-defined. This unchecked ability of the United States to make judgment calls on all these issues bothers many people.

US Nuclear Hit List still in play.

Another example involves nukes. Even though the Cold War is now over, and the US is busy fighting terrorism and not communism, the United States still maintains the **nuclear hit list option**. What is that? During the Cold War, all of the countries that had nuclear capabilities had lists of who they would blow up first. Number one on the US list was Russia, then maybe China because they were pinko commies too, then maybe France just for good measure, etc. All countries with nuclear capabilities had these lists during the Cold War.

Recently, most countries have said, "Now that the Cold War is over, we'll get rid of the hit list and start to disarm our nuclear warheads. We're not going to point them at anybody." All the countries got together, particularly Russia, and said, "Hey, let's all do this! It'll be a good thing!" The United States said, in essence, "You guys are so cute! You should do that! But not us! We are going to hang on to our nuclear hit list, just in case."

At a time when many states want to ratchet down the nuclear issue, the US stands alone in its opinion. That has increasingly been the case for the last two decades. The US maintains its hit list, is not a full player in UN procedures, and openly refuses to sign treaties that the vast majority of countries in the world support. Things like the Law of the Sea Convention, which defines territorial rights of the ocean, or the Kyoto Protocol to reduce greenhouse gases, even legislation banning the production of land-mines. In all of these circumstances and many more, the US stands virtually alone in its non-participation. Oh, and if there is any other country to not play ball, it would be the Canadians. Especially on the global warming issue.

Is the Plaid Avenger trying to indoctrinate you into thinking the US is bad, or too powerful, or too pompous? Not at all. I'm trying to educate you on perhaps why the US has a bit of a bad rap on the world stage right now. Sometimes standing alone is a sign of strength; sometimes it is a sign of selfish stubbornness. You have to decide for yourself, as will the rest of the world.

POWERFUL POWER PROJECTIONS

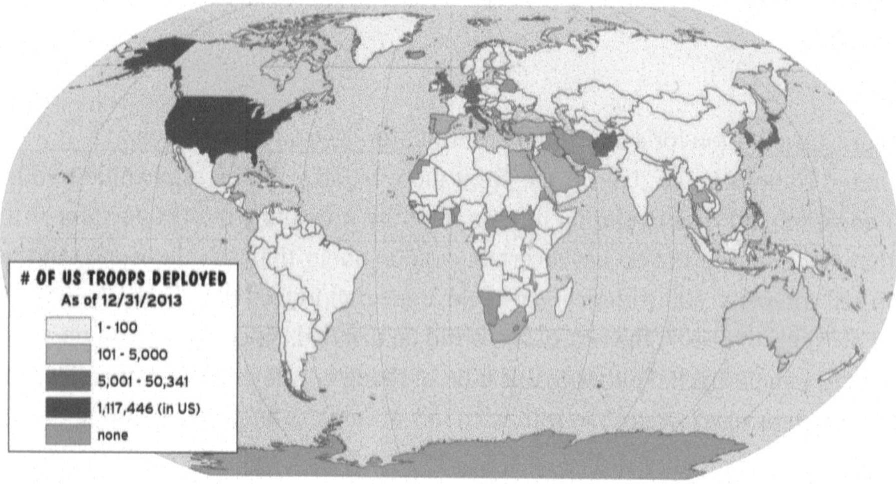

The overwhelming strength of Anglo-America has also brought a sense of helplessness and outright fear to some parts of the globe. Specifically, the US has a basically open, identifiable, and (some folks would say) scary projection of real power as evidenced by the active wars in Iraq and Afghanistan. These active wars are possible, and proceed rapidly, because of the active US military presence around the globe. Whether for good or bad, there are Anglo-Americans with guns all over the planet.

More often than not, you see US soldiers more than you see any other country's soldiers. You can see UN soldiers, NATO soldiers, and EU soldiers sometimes, but those are groups, associations of countries, not single countries like the US. US soldiers are the only ones from a specific country that you can find all over the planet. This military presence is perceived as unchecked strategic military power. In a very real sense, you won't find Chinese soldiers in Iraq, or Afghanistan, or in the United States. Can you even conceive of having foreign soldiers on US soil? The thought is comical. Other countries accept that there are US troops around the world, on foreign soil in Europe, Asia, South America, just about everywhere. It is a reality elsewhere, but inconceivable in Anglo-America.

There are no other countries that have that type of real power evidenced around the entire planet, not just the active wars, but in deployment of troops. Factor in the largest nuclear arsenal on the planet, the most sophisticated and largest air force, and more aircraft carriers than the rest of the world combined, and you have the basis for other countries soiling themselves at the thought of US intervention. Sound fanciful? Ask Saddam Hussein of Iraq or Muammar Qaddafi of Libya how they feel about it. Well, Saddam and Muammar are very much dead, and so is Osama bin Laden . . . all thanks to US power projection. And don't even get me started on the drones! Now the US is the undisputed leader in unmanned death vehicles in the world . . . a powerful tool of death projection across the planet that is rising in popularity, and polarization!

"Special delivery for . . . wait, who are we killing today?"

CULTURAL IMPERIALISM

Finally, under the strongest category is something that you probably have never thought of before critically: **cultural imperialism**. With no doubt, Anglo-America is the most successful, if you can call it that, at this thing called cultural imperialism. Whether the US and Canada are true imperial powers is debatable. However, it isn't really up for debate that the culture of the US is the most successful thing being spread around the planet. What am I talking about here?

We just discussed power-driven imperialism, including an actual military presence, so what is this cultural imperialism thing? Imperialism usually means that people step in with guns and take over saying, "We are an imperial power; your stuff is now ours." Think about Darth Vader and the imperial storm troopers. They are taking over and ruling by force.

However, cultural imperialism is much more subtle. There are no guns, there is no iron-fisted dark lord, and there are no storm troopers. Cultural imperialism simply means that people begin adopting your culture. Think about pop culture. This is the one that most often springs to mind. People around the world from France to South Korea to Australia to South Africa like watching crappy American movies, listening to crappy American pop songs, and watching ridiculously crappy TV shows like *American Idol*. Think about the title: it's *American* Idol and it's watched and *idolized* by people all over the planet! How crazy is that?

Cultural imperialism is much more subtle, because nobody has a gun to anybody's head telling them to accept it. For some strange reason, people just like it. There have been millions of movies made in America that have been exported to every other country on the planet, many of which you'll never see because they suck so bad that nobody in this country will watch them—but people in other countries will, just because they're American-made. People are attracted to that successful vibrant American pop culture. You can go to Tibet and watch *Titanic*, you can go to Russia and listen to Run DMC (sorry, the Plaid Avenger is old school), or you can go to Burma and see an episode of *Baywatch*.

A lot of countries in today's world are passing laws to try to limit the amount of American pop culture that enters their countries. France is a good example, because every few years they attempt to ban the importation of American-made films. It never lasts long, but they try. Middle Eastern countries try to keep all that American pop music and porn out. Even China tries to tightly control Internet action so that none of those American democratic-human-rights-and-freedoms ideas can seep into their culture, which brings us to another point to consider. Pop culture is the main component of cultural imperialism that you know of, but here's one that is much more intriguing and important. . . .

Cultural imperialism from an *ideological* standpoint is much more important to acknowledge, especially considering how world events are going down right this second. I'm talking about the ideologies of capitalism and democracy being spread. As you know from current events, unless you have been living under a rock, the United States has an active military presence in Iraq and Afghanistan and perhaps will occupy some other places in the very near future. They don't necessarily import Britney Spears at gunpoint, but what is happening is that they are heavily influencing the societies to become like the US in terms of adopting democracy, individual liberties, and capitalist culture.

The capitalist part is not that difficult, since every country makes money in some way or another on capitalism. However, some people in the world tend to have a problem with the democracy part. These are ideas Uncle Sam pushes across the planet quite openly. The US says, "Hey! We're going into Iraq because we want a democracy here. We're going to help them have a democracy. We're going to manipulate or push or try our best to get other countries to embrace capitalism because it benefits us and we think it benefits the world. We're going to try to get everyone to do it." Indeed, the US and Canada in most of their endeavors say, "Yes, and we believe that all people are equal and that we think individual rights and civil liberties should be the norm across the planet."

This is not something up for dispute. You can see this every day when the US government says, "Hey! We think China has horrific human rights abuses and we're going to complain about it at the UN." The US may complain about the Chinese

IT'S UP TO YOU
PROTECT THE NATION'S HONOR
— ENLIST NOW —
ASSOCIATED MOTION PICTURE ADVERTISERS

You're not gonna stand there and let Lady Liberty get abused, are you? Of course not!

treatment of the Tibetans and say that their human rights stink. That is essentially saying, "We believe that our system is right and yours is wrong," and that's something to consider.

Let the Plaid Avenger go on record as a freedom fighter myself by stating that I personally uphold the idea of democracy in the world and think human rights are the most basic necessity for every human being on Earth. I'm a bit biased since I'm from this region. I have this ideal myself. However, not everyone in the world does, and I'm willing to recognize that. I may not like it, but I at least recognize that another point of view exists.

The US planting seeds of democracy in the Middle East is of great consequence in today's world. It's the reason that US tax dollars are going to support troops who will continue to prop up governments and establish democracies in places that have no history or background in such cultural phenomena. Lots of people in the Middle East and other places in the world have no experience with democracy and don't necessarily want it. They do, however, feel as if the world power, i.e. the United States, may be pushing them toward it.

NAVY!
Uncle Sam is calling YOU
ENLIST in the Navy!
Recruiting Station
"I WANT YOU IN THE NAVY
and I WANT YOU NOW"

Now you know where the Village People got the idea for "In the Navy."

I don't think it's an exaggeration to describe it that way, and I think that's a fair assessment. It's up to you as American and world citizens to decide if that's kosher. However, I just want you to understand that democracy is a major part of your culture that you accept as right and good and proper. Other people in other places may not share that sentiment. Food for thought.

Look no further than the 2011 Arab Revolutions across the Middle East where grassroots movements pushed for a change to western-style democracy too . . . and used US technologies of the Internet and Facebook to facilitate the transition! Fascinating! Think those Arab leaders being toppled from power are thanking the US for its cultural influence in the region? Me thinks not!

-THE FREE-EST

What could possibly be bad about being the free-est? The first thing I want to point out is the freedom to be clueless. Anglo-Americans as a whole (and we do have to pick on the Americans more than the Canadians in this respect) are generally speaking completely clueless about what's going on in the world.

FREE TO BE CLUELESS

Most Americans have no idea where Tibet even is on a map, much less what human rights abuses are happening there. We pointed out that peeps in this region are physically far away from other regions; therefore, they're not as heavily impacted by other regions, even in this global interconnected world. North Americans focus on their own thing. I won't call them isolationists, but they have a tendency to focus on themselves at the cost of simply just not caring about the rest of the world.

To sum this up: internationally, they don't give a crap.

It seems bizarre for the Plaid Avenger to say because this entire chapter so far has been spent explaining how the US is a central key player, deeply involved in the world's political and economic systems. But by and large, American citizens

are utterly clueless about other countries. Many around the world would say that American politicians, whose job it is to care about foreign affairs, are about as clueless as those they represent. Well, I guess in that respect they are truly representin'. Representin' ignorance that is.

As the Plaid Avenger has traveled around the planet and talked to lots of people everywhere, a common theme reveals itself: many people perceive America as an unchecked power, and around the world people feel that America doesn't care about them. Perhaps they are too small or not enough like the United States, or they have no economic resources that the United States cares about, and so their perception is that the United States doesn't care about them. Perhaps they're right.

In this interconnected and globalized world where everyone knows everything all the time, citizens of the United States, by and large, know the least about what's going on in the world. If that's a radical statement to you, then stop surfing for shoes on Zappos and open your eyes.

If people in Sudan can get to the Internet, they know everything that's happening in the world and they pay attention to what's happening in global movements of ideas and people. Of course, they especially know the details of the relationship between the US and Sudan, and what the UN Security Council is doing about Sudan. If you ask anyone in the United States, they won't have an utter clue. Sudan? Security Council? Who? Huh? Is *Celebrity Apprentice* on yet?

That bothers a lot of people around the world. They say, "Wow, this is the richest group of people who have the most influence on the entire planet, and they don't give a crap because they don't even know what's going on. How could they not care enough to know, for instance, that this genocide is going on?" Isolation is a dual edged sword. We say the US is great because it has been isolated, but at the same time, that allows its people to be lazy. The general US population does not know what's going on, and that really burns people's bottoms around the planet simply because citizens of the United States have so much power individually. They could easily change some things if they wanted, but they don't because either they don't know and/or they don't care.

EXPORTER OF FREENESS . . . BY FORCE

As I suggested just above, this region is also a huge exporter of democracy and capitalism and individual civil liberties and rights or at least the concepts of them. That sounds like a pretty good deal, and the Plaid Avenger is all about individual liberties and justice, so I'm all for it. Let's get this game on! Mostly, the Anglo-American region exports these themes in its literature, ideas, technology, and pop culture. However, we can't overlook the fact that sometimes these good ideas—democracy, capitalism, and individual rights—are also exported by force if necessary. That can be a little problematic.

Sure, it's awesome that you want people to have a democracy, but perhaps not so awesome that you go there and kill them in order to make them into one. Current events are playing out right now to determine how successful forced democracy is going to work out in the Middle East, and it doesn't look too good so far for the forces of freedom. In terms of freedom, can it be forced upon a society? Can it be given, or can it only be earned? We shall see, unfortunately, at the expense of many lives.

Anglo-America is the free-est, with lofty and admirable goals for others, but being blinded by ambition has led to a big fat negative for Anglo-America. Which leads us to the next theme . . .

HUBRIS AND HYPOCRISY

For some of you reading this book, don't bother writing letters saying that the Plaid Avenger is some blatant tree-hugging bunny kissing liberal wannabe commie who's just making fun of the US. I'm not, okay? I'm trying to get you to understand what's going on here!

It is problematic around the globe right this second that the United States and Canada say, "Hey! We're all about democracy. We're all about individuals. We're egalitarian. We think democracy is great." They do. I truly believe that. However, when you have things like the Abu Ghraib prison scandal in Iraq, as well as increasing civilian death counts in Afghanistan, drone strikes in Pakistan and Yemen, and then you have people being held in Guantanamo Bay without individual rights forever, you start to get a bad rap. The world, not the Plaid Avenger, is looking upon this and saying, "Wait a minute. You guys are talking this big game about equality and democracy, but you're not even doing it yourselves." Given current events, in the modern context in the 21st century, there's a bit of a problem with being the free-est when you're not living up to it.

FREENESS ENVY

Well, maybe not in the Middle East . . .

Finally, there might be something going on that I like to call "freeness envy." Oh, man, am I good or what? Freeness Envy . . . ha! And that is something that some other cultures don't necessarily dig. What am I talking about here? Well, the Anglo-Americans are all about the freeness, man. Equality, democracy, freedom of speech, religion, choice, peace, love, dirty hippies, whatever. It's all about equal rights. Everybody's equal. Everybody's cool. Yeah, man, pass the greenery Snoop Dogg . . . . oops, I digress . . .

However, don't assume that the rest of the world holds these ideals as highly as the Anglo-Americans do. Let me give you some examples. The Saudi Arabian government and most of the Saudi Arabian population is not necessarily in favor of equal rights for women. The idea that women are completely equal in all aspects doesn't really play there. Let's go with something else like gay rights. You know, in Anglo-America, everybody's equal, but that doesn't play in Africa. In Sub-Saharan Africa, and parts of the Caribbean, gays are persecuted heavily; most folks in those societies would say, "No. We are not about those people being as free as we are." You can go around the world and look at different societies—and this is the people I'm talking about here, not just the governments—and lots of folks would say, "No, we're not about that type of freedom." Russia and China are not really into freedom of the press. Countries in the Caribbean aren't really into gay rights. The Vatican isn't really into freedom of choice. The Middle East may not be really into sexual equality or freedom to draw cartoons.

Freeness envy may be happening around the globe. The Plaid Avenger is all about individual liberties, so I am a big proponent of it. But when I go into other cultures, I don't assume that everybody around the entire planet wants democracy. Well, I think they should have it, but that doesn't mean they all want it. I think that all women should be equal, because in the Plaid Avenger's world, all women are equal in my eyes, but I don't assume that all cultures necessarily accept that, either. You have to keep this in context.

Freeness envy, huh? Great term.

Go get 'em Cap!

UNCLE SAM WRAP-UP

Let's wrap this thing up. We've looked at the good side of the greatness of Anglo-America. We've looked at the negative side of the greatness of Anglo-America. As awesome as Canada is, we have been mostly focused on the much more potent United States of America, and Uncle Sam is really a bipolar fellow here lately. Sammy, my man, has a real Jekyll and Hyde complex going down.

Dude, you are scaring me.

You know Captain America . . . . He's out there fighting the good fight, standing up for the impoverished, the people who don't have democracy or with no rights. Yes, the Cap'n does quite a bit; even though he's taken a bad rap in today's world, the United States and Canada continue to be the biggest foreign aid donors, the biggest funders of the UN, and also do a heck of a lot for the planet. Even though they're sometimes a bit heavy handed about it, they think they're doing it for good reasons. Who's to dispute that? No other country in the world is stepping up to be the leader for truth, justice, and the global way. You think China or Russia or even the EU is going to lead the way to global stability? Not likely.

There is of course, the perceived American super-villan that comes out every now and again. Sometimes this region does bad things abroad in the world, even when supposedly done for a good cause. However, there are other times that this region is quite openly selfish in terms of looking after its own interests. Since it's the most powerful region, it can do it in the most aggressive manner across the globe. Again, no judgement call here, that's just the way it is. The US and Canada are going to look out for their own corporate and economic interests. If that means telling people to kiss their gas on the global warming issues, then that's what they're going to do. And that's typically, what they've done.

What next, Sammy?

The bigger issue to think about: we are in the midst of big changes in the world power structure that will most greatly affect the Anglo-American region more than any other. Why? Because the United States has been used to being the sole superpower for twenty or thirty years now. Ever since the Soviet Union went away, and the US won the Cold War, the Anglo-American region has been uncontested for power on the planet. It has the most political, military, and economic power, as I pointed out earlier.

But times are a-changing: China and India are on the rise, the EU is in deep ideological trouble, Russia is back and bolder than ever, and other international players like Iran, Turkey, and even the Philippines are all making regional impacts on the balance of power and re-forming global alliances as well. The planet is morphing into a multi-polar world, a place with more than one superpower. This isn't up for debate, nor a future prediction—this is happening now. The world of the future is a place where the US is not the only primary actor making global decisions. The question then becomes, how is Uncle Sammy going to deal with this new world, this up-and-coming, changing world right now in the 21st century?

Will Sammy go lightly into this world? Will he see the changes that are coming and work with other big powers and other big countries to get the job done and still push his own agenda? Or is it going to be more of a rougher road, where the United States has a tough time giving up its top slot and is perhaps bound and determined to have its way at all costs? This could result in potential conflict, war, stonewalling economically or in other ways, or becoming isolationist and just putting up a big wall and looking inward to try to focus on itself only.

That is the big question. But this is a region of greatness and the US/Canadian team is not going to go away anytime soon, meaning it's going to maintain its powerful position as a major player in planetary affairs for some time to come. But will it be a bigger force for good or more of an internally-looking isolationist state? Only time will tell.

As the insightful Uncle Ben once told Peter Parker, "With great power comes great responsibility." Good advice for Spiderman, and probably for North America as well.

THE TRUMP SPEED BUMP?

Wow! What a difference one election can make! The 2016 presidential election in the USA witnessed the rise of America's foremost populist to the most powerful position on the planet! President Trump held the top slot from 2017 until 2021 when Joe Biden took the helm, and while he may be gone, he is certainly a million miles away from being forgotten. Still dominantly in charge of the openly populist wing of the Republican Party (which may constitute 30-40% of voting Americans), his opinions on all things from domestic policy, foreign policy, and even the existence of the US democracy itself will continue to influence the possible future paths of the country. Below are impacts I predicted back in 2017, with update notes to allow you to see my train of thought . . . which was exactly on track!

1) In 2017: Trump touts an anti-globalism agenda that could radically alter the current global economic order in which trade blocks and trade pacts operate . . . possibly causing trade wars with other entities like China, or diminishing trade ties with neighbors like Mexico and Canada. **2020 Update:** Yep. All of that happened already, or is currently still happening!
2) In 2017: Trump would be renegotiating these trade pacts in order to bring manufacturing jobs back to the USA—a tall order to be sure, but one that would radically affect the US economy and wealth disparity that has become a major issue in the region. **2020 Update:** Yep. That's been happening, too. Unemployment is at historic lows in the USA, but it is less understood if that is a result of Trump policies and if the targeted coal, auto, and general manufacturing sectors are the ones actually benefiting from job creation—and then of course the lost Covid year completely tanked everything.
3) In 2017: Trump may embrace authoritarian states like Russia, thus US/Russian relations could be re-made in their entirety and in a positive direction, which could re-shape NATO, Eastern Europe, and the Middle East in big ways. **2020 Update:** Ummmm . . . also: yep. With the exception of the US/Russia relationship, which is still on the frosty side.
4) In 2017: Trump's foreign policy philosophy could most succinctly be summarized as: Might makes right. You are likely to see much more saber-rattling and more aggressive actions of the US military around the world in order to achieve whatever specific objectives Trump's administration wants in certain parts of the planet. **2019 Update:** As of June 2019, the US is ramping up pressure on Iran, and just sent an additional 1,000 troops to the Persian Gulf to demonstrate force and possibly to prepare for war!

Those are just a few big deal themes which would affect the Anglo-American region, and the world, in very big ways. A "New World Order" may be at hand! Economically and politically, we very well could be at a major turning point historically, as the rise of the right and populist candidates are being brought to power all over the world, with President Trump who was the most powerful force of them all, as the former leader of the most powerful (and at the time populist-led) state on the planet. Will Trump be a bump-up for the USA, or a speed-bump that derails liberal democracy the world over? **2021 UPDATE:** With the return to "normalcy" under current US President Joe Biden, this move to the autocratic right and remaking of the World Order is on hold . . . but will it be on hold in perpetuity, or just until the 2024 election? We shall see soon enough!

ANGLO-AMERICAN RUNDOWN & RESOURCES

View additional Plaid Avenger resources for this region at http://plaid.at/na

BIG PLUSES

- Biggest economy in the world
- Most powerful military in the world
- Leader of technological innovation in most fields
- Leader of democracy and human rights promotion worldwide
- One of the most solid infrastructures in the world
- Great educational system, decent social safety nets
- High standards of living, high degree of personal freedoms
- Multi-cultural, multi-ethnic, multi-religious region where everybody pretty much gets along
- Politically stable, solid rule of law

BIG PROBLEMS

- Overzealous focus on materialism has made region the biggest consumer of everything on the planet and a giant polluter
- Overzealous focus on materialism has made region's citizens most obese on planet and increasingly disillusioned
- Economic, political, and foreign policies typically focus solely on short-term solutions
- Two-party system has become entrenched, unresponsive, corrupt and grid-locked; citizen apathy is reaching all-time high
- Has a tendency towards over-doing it on the global stage, or alternatively, intense isolationism
- Chronic dependence on fossil fuels and illegal narcotics imported from other regions; you decide which one of those is worse

DEAD DUDES OF NOTE:

George Washington: Father of the US, first US President, dollar-bill guy, refused to be appointed King in favor of establishing the democracy, warned against two-party system.

John A Macdonald: Father of Canada, first Prime Minister, Canadian ten-dollar bill guy, helped forge the 2nd largest nation on the planet, thus ensuring it wouldn't be absorbed by the US.

Ronald Reagan: Commie-hater extraordinaire; claimed by some to have single-handedly won the entire Cold War. A demi-god to the conservative right.

PLAID CINEMA SELECTION:

Atanarjuat aka *Fast Runner* (2001) The first feature film to be written directed and featuring only actors using the language Inuktitut, used by the Canadian Inuit population. Excellent look at some of the indigenous people of Canada from their perspective rather than that of outsiders. And assuredly, it is the only film you can see a naked man running across icebergs and insane Arctic landscapes for over two hours.

Bon Cop, Bad Cop (2006) Archetypal buddy cop film which pairs two policemen from Quebec and Ontario who are nothing alike. Film shows great contrast of real cultural differences between these two provinces as well as examples of both Francophone and Anglophone humor. Second highest grossing Canadian film behind *Porky's*.

The Corporation (2003) Full-on documentary only for those serious about understanding how the business world works. Be forewarned: it does not paint a pretty pro-corporate picture, but is a worthwhile watch even for the future board members among you. While corporations are not an exclusive American phenomenon, this film pieces together their rise in the modern world, which has been extremely US-influenced and defined.

Roger & Me (1989) You've heard all the hype about the ultra-liberal Michael Moore, but so few have actually ever checked out his first, and easily best, docu-drama. Like him or hate him, this film is an excellent look at a by-gone era of American manufacturing culture, and will help you understand why the US car market crashed and burned, and also how the "Rust-Belt" formed in the now abounded post-industrial northeastern US. It also really helps explain the rise of President Trump. Seriously, check it out to see why "Rust Belt" voters went his way in the 2016 election!

Smoke Signals (1998) Atypical buddy/road-trip movie that features great insight into current Native American reservation culture, with fantastic diverse biome shots of western and northwestern US.

Super Size Me (2004) Documentary/comedy on eating patterns of Americans, and the real-life consequences when the nastiness is pushed too far. Features an attempt at an all-fast food diet!

LIVE LEADERS YOU SHOULD KNOW:

Joe Biden: Current US President—yeah, the head honcho of the world's most powerful economy and military. Has grand plans for huge investments in infrastructure and green energy development. While a centrist, hails from the liberal/Democratic camp.

Donald Trump: Former US president that maintains the premier leadership role within the right-wing/conservative populace and will continue to shape cultural and populist influences for years to come.

Justin Trudeau: Canadian Prime Minister—huge US ally and head honcho of that other country just north of the US.

MAP GALLERY

View Map Gallery & additional Plaid Avenger resources for this region at http://plaid.at/na

CHAPTER OUTLINE

1. World Maker, Shaker, and Breaker
2. Who Is in the Region . . . and Why?
3. Physically: Water, Water Everywhere...
   - 3.1. A Continent It Is Not
   - 3.2. Water World
   - 3.3. Conducive Climate
   - 3.4. Some Plain Terrain
   - 3.5. Resource & Risks
4. Plaid Avenger's History of Western Europe
   - 4.1. Cashing in on Colonies
   - 4.2. Colonial Cultural Consequences
   - 4.3. The Beginning of the End: The 20th Century
5. The EU
   - 5.1. The Rise of the EU
   - 5.2. What's So Cool about the EU?
   - 5.3. Power Struggles & Power Structure of Europe
   - 5.4. Today's EU
6. Euro-Problems: Immigration and Culture Friction
   - 6.1. Culture Friction
   - 6.2. Immigration
7. New Faces, New Directions
8. Summary—Western Europe
9. Map Gallery

Western Europe

8

WORLD MAKER, SHAKER, AND BREAKER

What the heck's that supposed to mean? Well just this: while the United States and the Anglo-American region in general may enjoy prominent status in today's world, it's Western Europe that really has shaped the current cultural, social, economic, and political world in which we live today. They've been at it for about 500 years longer than the US, and have had a tremendous impact worldwide in determining what's happening right now.

Let's face it—English is an international language not because of America, but because of the British, and their influence. Soccer is the world sport, not American football, for the same reasons. When we look at today's world, Western Europe may not be the world leader anymore, but it certainly was the maker and shaker. What about the breaker part then?

Well as you know, Western Europeans spent a lotta time in the last thousand years, and particularly the last century, beating the living crap out of each other. WWI, WWII . . . need I say more?! But today's world is very much different for Western Europe, as it is united as never before—making, and shaking, without breaking, in the 21st century. Let's get to it!

WHO IS IN THE REGION . . . AND WHY?

Everybody knows and understands that Western Europe is a region. It is the western side of the Eurasian landmass. But being a Plaid Avenger world region is not solely based on just the physical geography, so what is it that makes this a homogenous unit? We need to back up before we ask that question and ask who, what, and where is this region? What states? All the western European states? Which ones are those? All those cool ones that we take vacations to visit. Places like Norway, Sweden, Iceland, UK, Ireland, Germany, Belgium, Luxembourg, Switzerland, Austria, France, Spain, Portugal, Italy, Greece. Oh, wait, and Greece . . . Greece? But it's on the eastern side of Europe!

Look at the map here on the opposing page. All of those that are connected somehow are all Western Europe, but Greece is disconnected; it's over on the other side all by itself. Indeed, I think all of us would say, "Yes of course. Greece is definitely part of Western Europe." That gets at the heart of what it is to be Western Europe, of what it is to be in this region, and the homogeneous traits therein.

So what characteristics do these countries share that categorize them as a region? What could possibly be the same about all of these countries? Because, really, isn't it easier to point out the differences between France and Germany, or between the UK and Italy? Yeah . . . that's what makes the regional definition kind of tricky here.

What we have is a set of extremely distinct nation states. In fact, all of these countries are exceptionally different from one another. When we look back at the US and Canada, we said they share a common culture and language and background. When we get over to Western Europe, we have Spain; with Spanish people who speak their own language and have their own culture, Spanish everything. Go next door and there's France, with French people who speak French; in Germany, there are German dudes and they speak German. All of these things bring to mind extremely distinct cultures. The Germans eat sausages; the French drink wine. The Germans attack, while the French surrender. Backgrounds and histories are all very different. On top of all that, Western Europe is a collection of states that vary in size: big states like France and Germany, and micro states like Luxembourg and Andorra.

However, there are a whole lot of homogeneous things about every state in this region, even though they have radical differences in backgrounds, histories, cultures, food, and drink. But there is a bunch of stuff that is quite the same. Most of the culture here in the North American region is based on cultures from Western Europe, not Eastern Europe or Russia, but predominantly Western European backgrounds. What are the similarities?

"Western civilization" is one. These are the places that created the big ideas on which our lives are based. I'm talking about the big things here: philosophies, legal systems, economic systems, medical practices, writing systems, and particularly religions. Christianity may have been born in the Middle East, but Western Europe became the cradle of Christianity and all its subsects. Europe expanded and grew the religion into a global movement.

As it did for so many other ideas. This is why Greece is thrown into the fray; it's the birthplace of all the Western philosophies. People like Socrates, Plato, and all those dudes from Greece that affected history in a major way. Ideas of democracy and western medicine even have their origins in Western Europe. All of the "-isms" that you learned in high school: political and economic systems like socialism, Marxism, communism, and capitalism were created or evolved radically in Western European countries. These countries have shaped the way we think about life and how we want to live it.

There are great cultural differences between these separate and distinct nation-states that make up the birthplace of western civilization, but they are *united in their ideologies*. Another uniting factor in this region is that it is a rich place. The quality of life is similar to North America, even exceeding that quality of life in many places. It is a very interconnected place and is very urban like the United States too. There are high levels of technology, high levels of consumption, and high GDP overall. Compared to the rest of the world, this region is rich! Considering the combination of quality of life and its ideological background as the birthplace of Western civilization, Western Europe is a distinct and homogenous region in and of itself. While many Eastern European states are increasingly 'catching up' with their western counterparts, and I soon may just refer to a single European region, I think here in 2019 there are still enough significant differences to talk about them as two distinct regions.

Jesus, Socrates, Marx, Descartes: bearded dudes of note in Western Civilization.

PHYSICALLY: WATER, WATER EVERYWHERE . . .

When it comes to Western Europe and its profound impact on the planet for the last several hundred years, we have to consider the physical geography for at least a page or two. Water has played a significant hand in creating a super-mobile society, but the climate, terrain, and resources of this region have also contributed to Europe's particular evolution and overall success.

A CONTINENT IT IS NOT

For starters, "Europe" is a group of countries on the western end of the Eurasian continent . . . but a stand-alone continent it is not. For those of you that heard Europe is a separate continent, that's poppycock! It's not a separate large landmass, nor is it a distinct continent in terms of geology, biomes, or hydrography. It's just the western peninsula of the great Eurasian landmass. Physically, there is no way to justify "Europe" as a distinct land mass or continent, but culturally perhaps we can . . . and perhaps that's what Europeans have been doing for centuries in order to set themselves apart from other civilizations and peoples. Fair enough, I suppose. Especially if you consider your civilization as "superior" to the other ones you are setting yourselves apart from!

Now I said that Western Europe is a peninsula at the western end of Eurasia, but it may be better described as a "peninsula of peninsulas." You know that peninsula thing: any piece of land surrounded by water on three sides. In the simplest sense, all of "Europe" is a peninsula sticking out from the western edge of the Eurasian supercontinent, bordered on three sides by the Arctic Ocean to the north, the Atlantic Ocean to the west, and the Mediterranean, Black, and Caspian Seas to the south. Then you look a little closer and you see that the place is composed almost entirely of peninsulas itself! Spain and Portugal are the Iberian Peninsula; Sweden and Norway are the Scandinavian Peninsula; Denmark is the Jutland Peninsula; Greece is the tip of the much bigger Balkan Peninsula; Italy itself is the Italian Peninsula; and Turkey mostly is a peninsula formed by the Black and Mediterranean Seas. What a plethora of peninsulas!

WATER WORLD

So . . . so what? Well, combine all those peninsulas with a bunch of islands in the region (Great Britain, Ireland, Iceland, Malta, etc.) and you have major mondo access to water. Europe has a higher ratio of coast to landmass than any other subcontinent or region in the world! Its maritime borders abound, as do great port facilities, which means this is a place physically perfect for fishing, shipping, interregional trade, and even global trade. And boy oh boy, have the Europeans found success with the whole trade and shipping stuff! Over the centuries, partly inspired by their super sea access, European powers built vast fleets of commercial ships and strong armadas and navies to protect them, eventually expanding their economic and

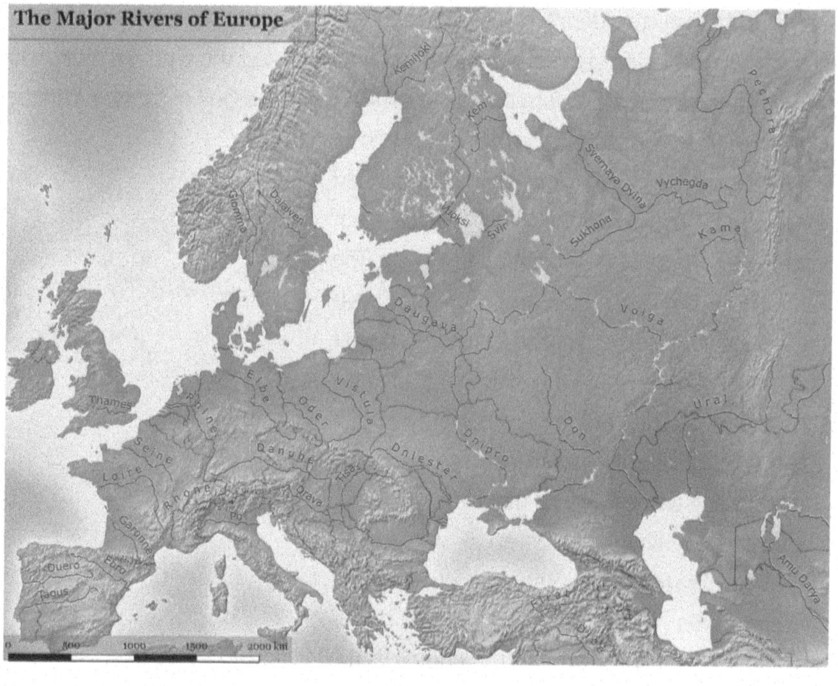

military power across the globe. But I am getting ahead of myself! Back to the physical geography, because we ain't even done with the water part yet . . .

See, Europe also possesses another fantastic feature based on the H_2O: freshwater fabulousness. The plains are cut by many important rivers like the Loire, Rhine, and Vistula in the west; the Northern Dvina and Daugava flowing northwards in Eastern Europe and Russia; and the Volga, the Don and the Dnieper flowing southwards of the European Russia. And let's not forget the bodacious blue Danube! The Danube is Europe's second-longest river, after the Volga River, and also the longest river in the European Union region. The Danube was once a long-standing frontier of the Roman Empire, and today flows through ten countries, more than any other river in the world. Originating in Germany, the Danube flows southeast for 2,860 km (1,780 mi), passing through or touching the borders of Austria, Slovakia, Hungary, Croatia, Serbia, Romania, Bulgaria, Moldova, and Ukraine, (and passing through four state capital cities), before emptying into the Black Sea. Its drainage basin extends into nine more countries.

Combine all those busy, and extensive, coastlines with Europe's wildly interconnected waterways and you have a place that is very fluid, indeed! lol pun intended! Mobility, mobility, mobility. Mobility is everything in a region where so many nations live in such close proximity, and Western Europe always was always well connected naturally via its waters. Connections create networks. Networks facilitate and promote exchange of ideas, of goods, of services, of science, of medicine, and of technologies. Highly mobile Europeans have been interchanging and exchanging ideas for centuries . . . any wonder that the "Scientific Revolution" occurred here? Now the flip side is that all that interconnection and mobility can also bring warring parties together on a frequent basis, but more on that later.

For now, consider water as Europe's greatest liquid asset: a catalyst for trade and commerce and military power. It's this closeness and association with water that helped propel Europe into the limelight 500 years ago. Virtually all of the countries in this region were once major maritime powers, or at least maritime explorers. This makes Western Europeans some serious seagoing folk. Think about those explorer-dude types from Italy, Spain, Portugal, and Britain, even the Vikings of Scandinavia, going out and poking into the rest of the world. Now, peeps did this in other parts of the world, too, but the Western Europeans took it to the next level by taking over the entire globe with a passion and speed unrivaled in human history. And they did it via naval power.

CONDUCIVE CLIMATE

Another physical consideration about Western Europe is that while it's very far north—most of it is north of the US/Canadian border—it has a more moderate climate because of something called the **Gulf Stream**, which morphs into the **North Atlantic Drift.** This is a warm water current which forms in the tropical areas in the Gulf of Mexico, Caribbean Sea, and Atlantic Ocean. It actually sweeps up the Eastern Seaboard of the United States and flows very far north, next to Ireland and the United Kingdom, all the way to the top of the Scandinavian Peninsula around Sweden and Norway. The Gulf Stream is nicknamed "Europe's central heating," because it makes Europe's climate warmer and wetter than it would otherwise be given how far north it is. The Gulf Stream not only carries warm water to Europe's coast but also warms up the prevailing westerly winds that blow across the continent from the Atlantic Ocean.

With water all around acting as a climate modifier, what we have is a place that is warmer in the winter and a little cooler in the summer than other places at equal latitudes.

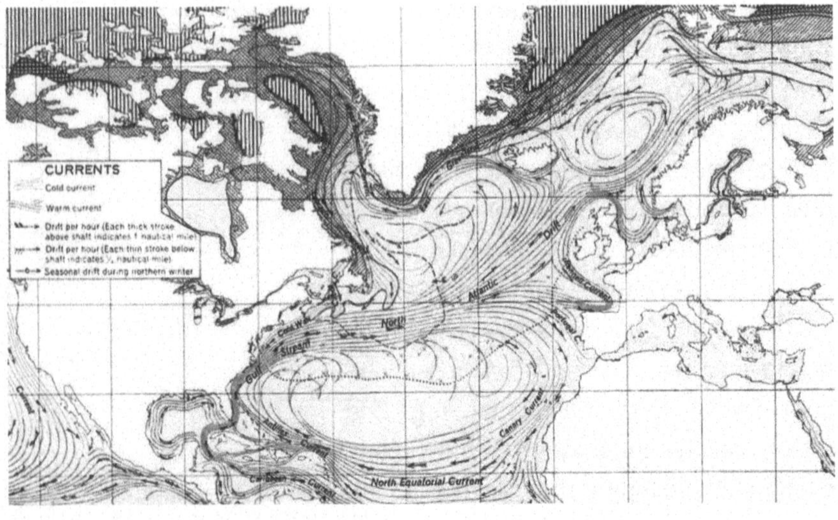

Gulf Stream/North Atlantic Drift: moderates even the far north areas of Europe.

It is a very "user-friendly climate." As we move into the continent towards Eastern Europe and Russia, we are going to see the places at the same latitudes get quite a bit colder. What's so great about that? As we saw in America, a well-watered, moderate climate is conducive to human settlement and population growth. While a few thousand years ago nobody was hanging out in Western Europe, in the last thousand it has become increasingly populated, and has become one of the cores of population on the planet because, physically speaking, the situation is right. The agricultural heartland of the western European plain profits the most from this, with fairly warm growing conditions most of the year, and plenty of rain to water crops. Southern Europe's weather also benefits from the surrounding oceans, with sufficient rainfall to nourish some amount of crops, but generally dry, mild, and pleasant weather most of the year . . . maybe you have heard of a whole climate type/biome that was named for the area: Mediterranean. Scandinavia isn't so lucky, with frigid weather most of the year along with occasionally ferocious winter storms, but even there the climate is just moderate enough (thanks to the Gulf Stream) to allow some crops to grow, and has allowed human populations to thrive for centuries, which is true of all of Europe.

SOME PLAIN TERRAIN

While water is crucial to Europe's overall picture, and their modified climate is an absolutely killer app as well, we should also point out a major terrain factor that has shaped the pseudo-continent, and that is the **Great European Plain.** The European Plain or Great European Plain is an expansive lowland area plain in Europe and is a major feature of the place. It is the largest mountain-free landform in Europe, although a number of highlands are identified within it. The plain stretches from the Pyrenees mountains and the French coast of the Bay of Biscay in the west, and gets broader and broader heading east all the way through the northern edges of Western and Eastern Europe and deep into Russia, up to the Ural Mountains. Its shores are washed to the west and northwest by waters of the Atlantic basin; to the northeast, the Arctic basin; and to the southeast, the Mediterranean basin.

To the south of the Middle European Plain stretch the central uplands and plateaus of Europe elevating to the peaks of the classic Alps and the Carpathian Mountains. To the northwest across the English Channel lie the British Isles, and some other upland areas known as the Scottish Highlands. (Geology fun fact: the Scottish Highlands are part of the same orographic system that created the Appalachians in the USA. Same mountain range, different continents!) Of course there are other areas of mountains, likely some famous ones you had to memorize in school, mostly in southern Europe: the Pyrenees separating Spain and France; the Apennines going through the spine of Italy; and virtually all of the Balkan peninsula is defined by its rocky, rugged mountainous terrain . . . the Albanian Alps, the Diniric Alps, and the Balkan Mountains, just to name a few, cover Greece, the states of the former Yugoslavia, Albania, and go all the way over to Romania.

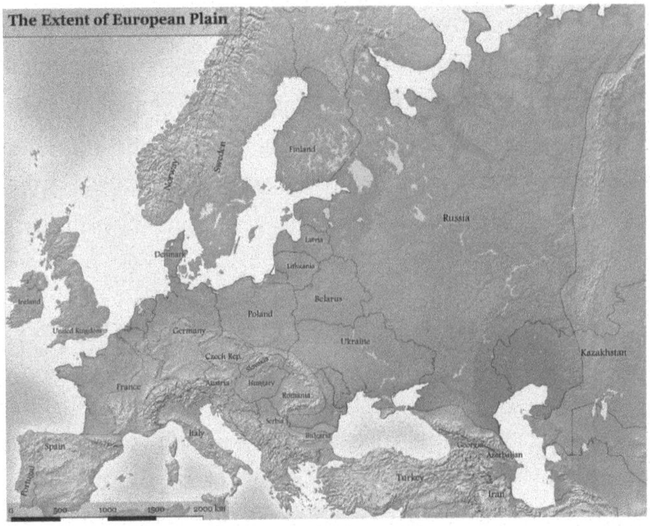

But let's stick with the main plain: Most of the Great European Plain consists of rolling lowlands below 150 meters (500 feet) in elevation. It is host to almost all the biggest and most navigable rivers I mentioned before, including the Rhine, Weser, Elbe, Oder, and Vistula. The European Plain was once largely covered by forest (temperate broadleaf and mixed forest biome) and mad amounts of wildlife, before human settlement and the resulting deforestation that occurred. The climate on the plain supports a wide variety of seasonal crops; it is a variable breadbasket of food resources, the most agriculturally productive region of Europe. Taken together, these physical features allowed for early communication, travel, and eventually massive agricultural and industrial development, and to this day it remains the most densely populated region of Europe. How's that for some plain talk?

RESOURCE & RISKS

Speaking of agricultural productivity in the plain, let's wrap up our Euro tour on physical geography with the resources and risks to the region. This flat, fertile land, coupled with a near-ideal climate, formed the perfect spot for abundant agricultural production—and with it, organized societies able to go beyond just feeding themselves. There's a reason Europe was the place that spawned the most recent radical Agricultural Revolution! Even outside the huge grain producing areas of the European plain (that also raise cattle, pigs, and make cheese!), there are ample agricultural production areas down around the Mediterranean coastlines, where citrus, olive, and wine dominate. And fish is always on the menu down there, too . . . ah, the Mediterranean! A biome, a diet, and a lifestyle, all in one word! As the Italians say: Que bella!!!

Back to the fishes: Beyond crops, Europe is also particularly rich in fishing grounds, a staple of their development in both northern (North Sea, Baltic, Atlantic) and Mediterranean waters. Staying on the North Sea, oil and natural gas have become a minor resource for extraction for the UK and Norway in particular. And Norway doesn't even use the oil it produces! They export it all, because they have so many other great alternative energy options, specifically hydroelectric . . . dang, somehow we came back to water again! And while the entire European plain used to be forested, not so much now; but timber and specialty wood products are still a major export from many countries, especially Scandinavia. Why would I point them out in particular? Ever heard of IKEA in Sweden? I think they might export a few specialty wood products . . . or a megaton of them. Additionally, coal and iron ore used to be a major resource for the region, especially in a belt from southern England to northern Italy, and those raw resources fueled the Industrial Revolution that made Europe great. But they largely played out of most of those materials now, and are not major exporters of many raw goods. That's okay though, because they are now super rich and provide other things to the world now.

On to the risks: Western Europe is a pretty benign place in terms of natural hazards, leaving the populace free to focus on other things . . . like conquering and exploiting most of the known world. There have been occasional disruptions to that routine, however, especially in the south. A major historical threat, especially in Italy, has been isolated volcanic activity, with infrequent but devastating eruptions from legendary mountains like Vesuvius and Etna. Thunderstorms and tornadoes can be sometimes found sweeping through Western Europe's central plains, though of not nearly the intensity or frequency of those found in America. Snowstorms have always been a threat to contend with in the north, as well as in Europe's mountain chains, and visitors to the latter have always had to contend with the risk of landslides and avalanches. However, as of late, the biggest threat to Europeans has come from a recent series of devastating, nearly continent wide heat waves that have killed thousands over the years. Droughts and floods have also had some level of impact recently, especially in the south.

However, given their mild and well-watered climate, their geologic stability and lack of seismic disasters (outside of Italy), and given their levels of technology and development, Western Europe is an incredibly stable, "risk-free" zone. At least when referring to the natural world. The humans, on the other hand, have made this place an epic zone of death and destruction for centuries, culminating in the bloodiest decades thus far in human history! What? They got all those physical geography advantages, they ended up taking over the world, and they still screwed it up? Yep. Let's turn to that next . . .

PLAID AVENGER'S HISTORY OF WESTERN EUROPE

We don't need to waste too much time talking about European history, as you have probably already had a thousand classes on it in during your scholastic career, but the Plaid Avenger can summarize the last couple thousand years for you very quickly. The big pluses: the **Renaissance** (they innovate heavily in art, literature, education, and political diplomacy), the **Age of Discovery**/Age of Colonization (they learn to explore and then take over the world and grab all its stuff), the **Second Agricultural Revolution** (they learn to produce a lot more food), the **Age of Enlightenment** (they wildly expand scientific knowledge and philosophy), and finally the **Industrial Revolution** (they invent machines that make a lot more stuff, better/cheaper/faster). Yeah: all of that happened in Western Europe. And, yeah: put it all together and it made them one rich region.

It should be noted for the record that the transformation of this region into the dominant world leader is really only a product of the last 500 years. Yes, I know the Greek and Roman Empires were really hip in their day a couple of millennia ago . . . but they were also very local . . . and then they crashed and burned, leaving Europe a provincial little playground until the "restoration" of the Renaissance. The region was NOT a global leader, and it was NOT a global force, prior to 1500AD. Other civilizations around the globe made Western Europe look like a backwater of buffoons. And it was.

But then those Europeans got savvy after their **Dark Ages**. They learned about sailing and navigation and shipbuilding from Muslim traders. They learned about the global trade network from Asian merchants. They learned about gunpowder from the Chinese. The Europeans took all this information and applied it in new and exciting ways which subsequently made them masters of the seas, masters of the global economy, and especially masters of military. China made fireworks with their gunpowder. Europeans made guns with it. You can figure out the rest from there.

If these guys were taking over the world by storm, then what were the negatives? In a word: competition. All these different ethnic groups, religious groups, and different nationalistic groups have spent a whole lot of time in the last couple centuries beating the living hell out of each other. You have probably heard of war after war after war that you had to memorize in high school; all of them occurred on European soil. The last couple of biggies were the World Wars of the 20th century, but there were hundreds of other confrontations before those.

Virtually all of those contests were based on competition between these European states for more power, more money and more influence on both the European and the world stage. The states of this region, as a whole, have spent a good deal of their time figuring out how to out-maneuver, out-flank, and generally beat up their nation-state neighbors. Why is that? How did it get to be this way? And how has it radically changed in the last 70 years? These are the important questions that you need to consider. The whole concept of a nation-state was formed in Western Europe. How did this concept form here if they have spent all this time fighting? Hmmm . . . perhaps it was precisely because all these groups were fighting with each other that the nation-state concept evolved! See . . .

There are groups of people identifying themselves as distinct cultures, like the French, the Italians, the Swiss. Those terms mean something to us. It means something to them as well. The idea that you would have a distinct group of people and that you would outline a political area in which you reside (you know, nation-states) is a Western European phenomenon, just like all the "-isms" and philosophies we talked about earlier. Unique, competitive cultures sitting side-by-side in the little chessboard of Europe. Yep. That's how conflict started.

I should also reiterate that there are people who begat people who begat people to make Western Europe one of the most populous regions on the planet, especially in the last 500 years. People are all over Europe. The coastlines and river systems are packed to the point that Europe as a whole is about half the size of the United States with about twice the population. (See figure to the right: area and latitude are held constant for comparison.) We talked about the North American region having tons of space that fostered the evolution of the automobile, interstate system, suburbs, and individual transport systems; Europe is the reverse. You don't have all that space. You have double the population in half the space. We can see the physical manifestation of this with public transportation. People use trains and buses; not everybody owns vehicles there. It's quite a different society just from the standpoint of land and how little of it there is.

Land and latitude comparison: US and Europe.

CASHING IN ON COLONIES

I said something about the European Age of Discovery/Colonization at the outset of this section, and it now needs to be revisited. You already know from a previous Plaid chapter that the North American region is the number one region in terms of military, economy, wealth, and all that stuff. But that only came about in the last 100 years. Who used to hold those top slots? Western Europe did, and they had their day in the sun for a good 500 years. From Columbus right on down until about World War I, Western European countries had the tiger by the tail and controlled most of the world.

What does the Plaid Avenger mean by "controlled most of the world?" You mean like cultural imperialism from the United States? No. I mean *direct imperial control*. Because of the physical geography we talked about, Western Europe was full of maritime nations, and because of the certain situation at that time in their history, they started going out and exploring the world. Their technology increased and surpassed the technology available in other places on the planet. Thus, their exploration and exploitation potential increased.

These guys, at the exact right time in history, went out and bumped into the New World. They took over all the New World and colonized it. They said, "Hey, look! North America! That's ours! South America! We'll take that too! Divvy that up between Spain and Portugal! North America? We'll divvy that up between the French and the British!" That's not all. Throughout the last 500 years, a tiny handful of states from Western Europe controlled most of the rest of the planet.

As you can see from the graphic to the right, eight nations—UK, Germany, France, Spain, Portugal, Italy, Netherlands, and Belgium—controlled all the countries in pink. This is a massive territorial holding. Again, the Plaid Avenger is trying to get you to understand how it is that Western Europe has some of the richest countries in the world right now. Look at this map again. For 500 years, these guys siphoned the material wealth from the world.

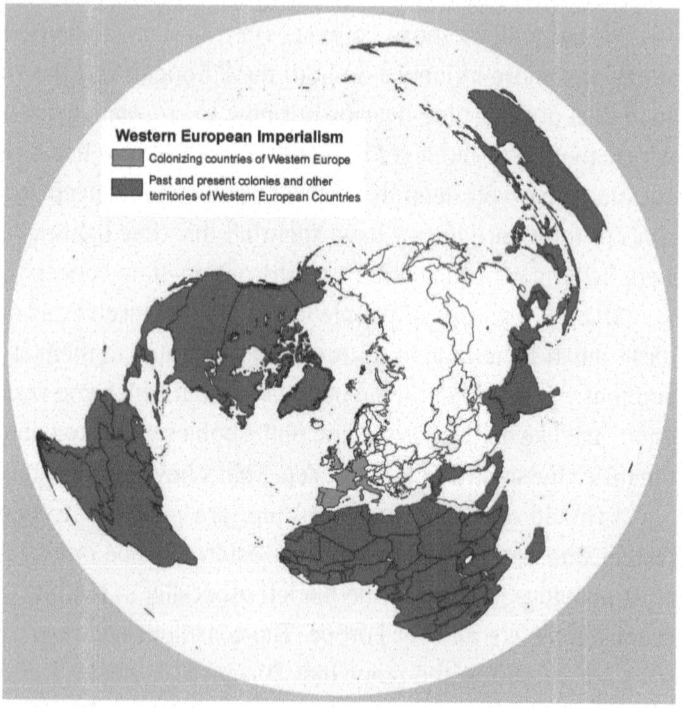

Colonial masters of disaster!

Let's think about a place like the Netherlands. There isn't anything there, my friends. It's flat. They grow tulips! How are they, of all countries, one of the richest in the world? You have to understand this much about history: it's because they controlled massive territorial holdings, many of which were exclusively economically exploited colonies. The Dutch just siphoned off all the resources.

Consider the Spanish Empire, which controlled all of South America, Mexico, Central America, and the Caribbean. They brought back gold and silver bouillon . . . the booty! It was all there in the New World, my friends. What did the Spaniards do with it? Did they dig it up in Mexico and redistribute to the natives? Hell no! They took everything of value and brought it back to Western Europe. All of it. Meanwhile, the British were busy bringing back all the tobacco they could smoke up, as much cotton as they could weave into fabric, and as much rum as they could pull out of the Caribbean and disperse to their sailors.

All of this economic exploitation hugely benefitted Western European countries. This is not the case with Eastern Europe. You do not see Eastern European or Russian or Middle Eastern or Chinese colonies in the world. Taking over the planet and exploiting it on a grand scale was a uniquely Western European phenomena, a fact that has been largely overlooked in textbooks, for whatever reason. All of the world's wealth went floating back to a small group of countries. That may sound pretty negative. Maybe it is. I think if I were somewhere in sub-Saharan Africa and had been a colonial holding for the last 500 years, which left me and my people poor while those European countries are rich as Midas, I might be a little bitter. Fortunately, the Plaid Avenger grew up in the United States, where we kicked those Brits out before they exploited us for too long. The United States is in a much different situation than many other ex-colonies in the world today precisely because of that fact. I do not believe that it is much of a coincidence that most of the poorest parts of the planet in today's world were the ones that suffered under the colonial legacy the longest . . . namely African ones.

COLONIAL CULTURAL CONSEQUENCES

But let's not beat up our European brothers anymore right now. Let's think about some of their other impacts on the world. What are the other things to consider when we look at the Western European region that had such a tremendous influence on the planet? What's the number 1 sport in India? Cricket. Brazil? Soccer. What language do they speak in Pakistan? What language do they speak in Chile? What religion do they follow in Mexico? What are the religious practices in the Philippines? What types of government do all the countries of the world follow? What are the major philosophies that most people in Africa ascribe to? These questions all have answers of Western European origin.

Virtually all the types of government across the globe, many modes of thought, major religions, and major languages were all spread throughout the world via the conquests of Western Europe. People in Brazil speak Portuguese because it used to be a colony of Portugal. A lot of people in India speak English because the British used to colonize those dudes. All sorts of little things, including dietary habits, holidays, and particularly the world's sports that people love for unknown reasons, have their origins in Western Europe.

And it's not just customs and habits, it's the people too: Why are significant portions of the populations of North America, South America, and Australia white? In addition to the movement of ideas and culture, you also have the mass movement of peoples. You have English and French and German and Italian and Irish and Scottish enclaves all over the planet. As a result of this movement, nearly all of European thought and culture is accepted as part of today's world culture all over the globe. You need to know the good and the bad. Why is Western Europe rich? It has something to do with them exploiting the rest of the world. Why can you communicate using English all over the world? Same reason.

If you are a pilot that flies between countries, you must speak English. It's an international standard. Why? It's not because America is so great, but because Western Europe, Britain in particular, has had so much influence on the planet over the last 500 years. America was just the heir to the throne of global power and influence that was created by the period of Western European colonial imperialism.

This colonial period is really the most important in European history, probably in world history. It led to the amalgamation of wealth that essentially made Western Europe the rulers of the world. We can debate whether the United States is an imperial power, but there's no debate about Western Europe. They were a hegemonic imperial power that controlled most of the world and amassed incredible wealth from it. They don't have that position anymore, of course. The US has that top slot. What are some of the reasons for this? Well, the Europeans lost their colonial holdings and ran out of land. They could not continually expand as the United States has done over the last 200 years. They didn't have any more resources because they have been using those resources for generations upon generations. Europe is like a geriatric version of the US. It ran out of its room to grow and is a young buck no longer. But they accelerated their decline greatly by their shenanigans of the last century . . .

THE BEGINNING OF THE END: THE 20th CENTURY

And that's enough history to set the stage. Western Europe is not number one today, although it is still very rich. It still has a lot of influence in the world, though what we are seeing in the 21st century is that European power is actually on the decline. How did it go from top banana to just another in the bunch? Europeans may not like to hear the Plaid Avenger say that, but it's true. They are slipping. They have a very unique solution on how to stop that slip, by the way, and we'll discuss that in the last part of this chapter. But back to the slip . . . wait a minute. Banana . . . slip. Am I good or what? But I digress . . .

As we already pointed out, they were on top of the world for quite a while, but the 20th century was unkind to not only Western Europe, but Europe as a whole. Just as America was exploding onto the world scene, Europe decided to beat the living daylights out of itself for 50 years solid—and did a darn good job of it. These nation-states with their independent cultures and separate identities smashed themselves not once, but twice, in World Wars I and II. The Cold War settled in after that, and that didn't help matters at all. The same 100 years that Europe was destroying itself, it was losing its colonial holdings. They had already lost America long ago, but in the 19th and 20th centuries, they lost all the rest. South America fought wars for independence, a lot of Africa petitioned and fought for sovereignty, and Canada and Australia became mostly independent. And the few colonies that they retained to the bitter end were giving the Europeans nothing but a headache too: India for the British, French Indochina for the French, the Middle East for everybody. These places broke out into independence wars, civil disobedience, and colonial chaos in general. So little by little, as we approach the modern era, the Europeans lost these territories that, for a long time, gave them wealth accumulation and serious competitive edge.

Massive WWII destruction in Europe.

As the Plaid Avenger suggested already, America has continuously expanded economically because of its incredible amount of resources and territory. You have that with the Western European powers as well, but this reversed in the last 100 years. They lost all their overseas territory. They lost the unfettered access to the resources of those places as well. This reversal of fortunes happened at the same time they battled each other in wars, destroying their entire infrastructure. Make no bones about it: World War II was the most destructive thing that could have happened to any region as a whole. There was total devastation across the entire continent. Factories, roads, and bridges were entirely wiped out. Again, this plays back into why America became #1 post-WW2, because none of that type of destruction happened in America (excepting Pearl Harbor, of course).

So the first half of the 20th century nearly destroyed this once-globally-dominant region. In just that 50 year time span, Western Europe lost nearly all its colonies abroad (in wars as varied as the Spanish-American War, the Indian Independence Wars, the First Indochina War, etc.), thus losing the massive inflows of resources, wealth, and world dominance that those colonies afforded them. The nationalistic-fueled competition between European countries spawned costly conflicts across the world even a century before (Seven Years' War, Boer Wars, US Independence War) which finally culminated in massive destruction on the home soils (Napoleonic Wars, WWI, WWII).

After their near total self-annihilation of that last war, some very wise thinkers from many countries realized that they could not continue down this uber-nationalistic competitive path anymore. They needed to change and try something different, otherwise the next war would indeed end it for all of them. Plus, the rest of the world was changing, and European countries could no longer compete effectively as individual states. So from the ashes of WWII, those wise thinkers invented a new path for their region, one that involved their nation-states working together as a singular whole . . . and strangely enough we can trace the origins of multinationalism in Europe to a single geographic point at a single place in time . . .

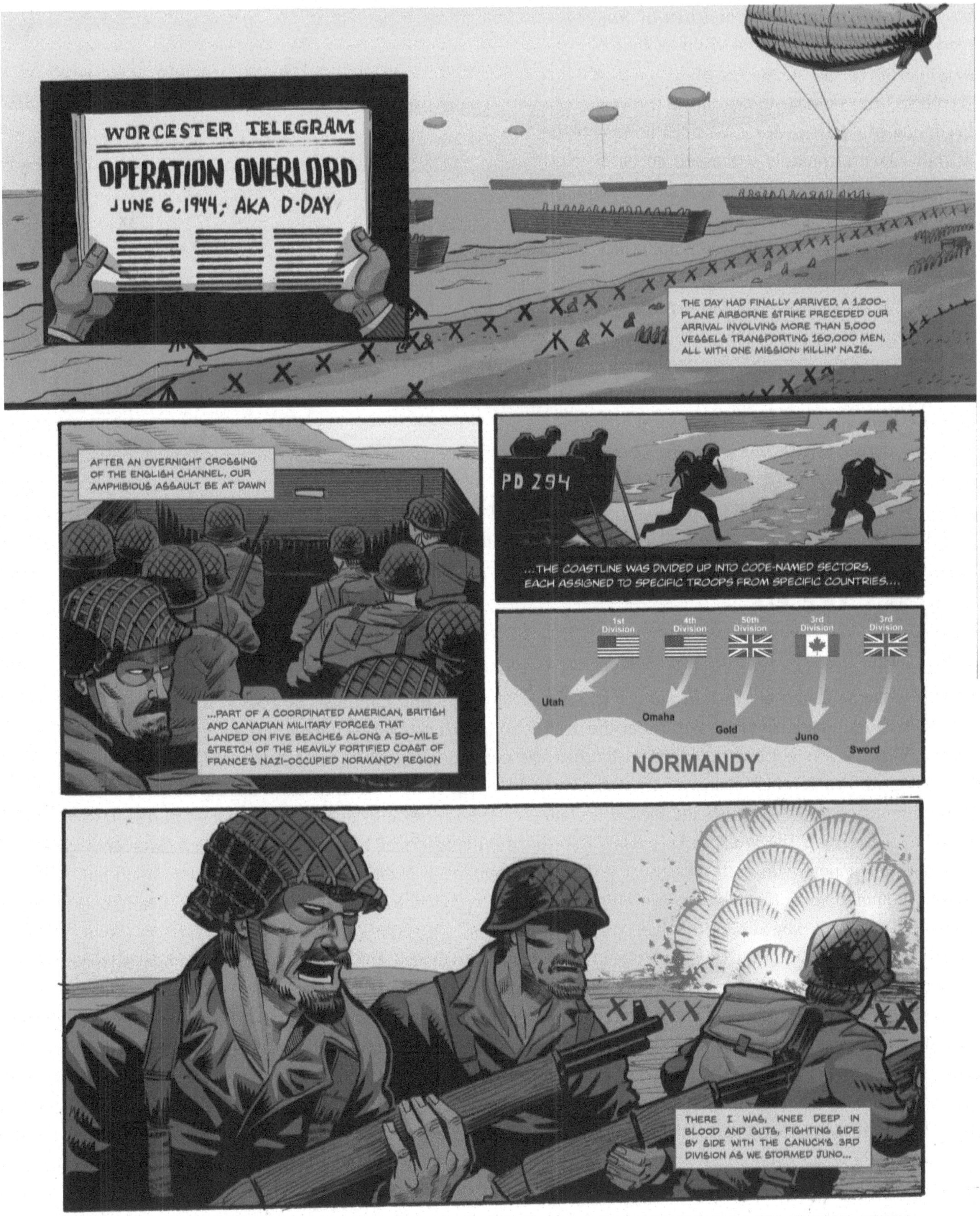

WORCESTER TELEGRAM
OPERATION OVERLORD
JUNE 6, 1944; AKA D-DAY
THE DAY HAD FINALLY ARRIVED. A 1,200-PLANE AIRBORNE STRIKE PRECEDED OUR ARRIVAL INVOLVING MORE THAN 5,000 VESSELS TRANSPORTING 160,000 MEN, ALL WITH ONE MISSION: KILLIN' NAZIS.
AFTER AN OVERNIGHT CROSSING OF THE ENGLISH CHANNEL, OUR AMPHIBIOUS ASSAULT BE AT DAWN
...PART OF A COORDINATED AMERICAN, BRITISH AND CANADIAN MILITARY FORCES THAT LANDED ON FIVE BEACHES ALONG A 50-MILE STRETCH OF THE HEAVILY FORTIFIED COAST OF FRANCE'S NAZI-OCCUPIED NORMANDY REGION
PD 294
...THE COASTLINE WAS DIVIDED UP INTO CODE-NAMED SECTORS, EACH ASSIGNED TO SPECIFIC TROOPS FROM SPECIFIC COUNTRIES....
1st Division
4th Division
50th Division
3rd Division
3rd Division
Utah
Omaha
Gold
Juno
Sword
NORMANDY
THERE I WAS, KNEE DEEP IN BLOOD AND GUTS, FIGHTING SIDE BY SIDE WITH THE CANUCK'S 3RD DIVISION AS WE STORMED JUNO...

THOUSANDS OF MEN FELL THAT VERY DAY, A PART OF AN INTERNATIONALLY COORDINATED, MULTI-NATIONAL OPERATION ON A NEVER-BEFORE-WITNESSED SCALE. THE BATTLE OF NORMANDY AND THE LIBERATION OF EUROPE WAS UNDERWAY, BUT LITTLE DID ANY OF US KNOW THE LONG TERM IMPACTS OF THIS HISTORIC EVENT. THIS WAS THE BEGINNING OF THE END OF NAZI GERMANY. THIS SMALL STRETCH OF BEACH WAS SIMULTANEOUSLY THE ORIGIN POINT OF TWO OF THE WORLD'S MOST POWERFUL FUTURE ENTITIES: NATO AND THE EU....

Indeed! Storming the beaches of Normandy by the Allied forces in WWII on June 6, 1944 in order to liberate the rest of Europe from the Nazis, can now be looked at as the very beginning of a new path for Western Europe! A path which ultimately witnessed all European states working together to 1) defeat fascist Italy and Nazi Germany, then 2) ensure that regional conflict at that scale would never happen again. The US, UK, and Canadian forces met up with French, Belgian, and Dutch resistance forces to eventually win that war, but then they did something extraordinary: they worked collectively to rebuild the place (including the defeated Italy and Germany), and formed both a military partnership binding them all together as well as an economic partnership which bound them all together even tighter. NATO is the military part, and you learned all about that in Chapter 6. The European Union was the economic part, and it proved to be even more successful than its original planners had envisioned . . .

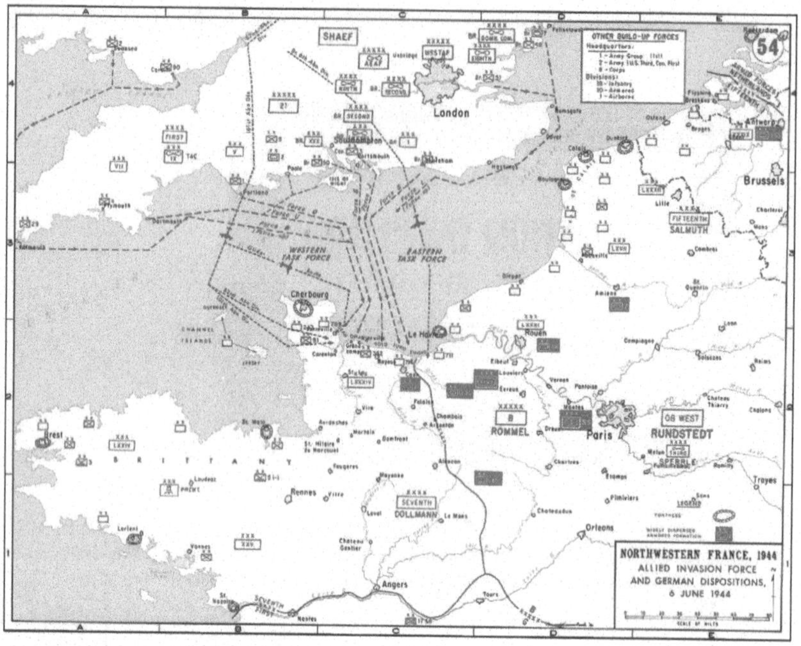

THE EU

THE RISE OF THE EU

The EU is the salvation of Western Europe as a powerful region and the only real reason it is still a global player, period. What do I mean by that? Well, after the destruction of World War II, some smart dudes—incredibly, they were French—said, "We've got to stop killing each other. We need to economically integrate ourselves so that we are not competing with each other anymore." They thought that part of the problem was that the Germans hated the French and the French hated the British and the British hated everybody because they were in competition with each other. And they were right. The Western Europeans were once the masters of the universe, but those days were gone. To continue to compete with each other, just within the continent, was disastrous, as they already found out. So they said, "We should work together, we can all get a little bit richer if we all try and work together better. Let's start rebuilding from WWII by pooling our coal and steel industries."

That's how the EU started. Everybody needed coal and steel to rebuild stuff, and all of Western Europe had to be rebuilt. The smart guys who thought this up knew it would evolve into something else. They said, "We'll do this as a start, and it will tie us to each other so much that we won't be able to afford to attack each other. If we make this into a common industry, Germany won't be able to attack France, because it would be hurting its own economic needs." You can't beat up the people you're actively trading with because you will lose their trade. It was an interdependence that they formed in the coal and steel industries, but they wanted to make it something else. They wanted it to reach farther.

This is where we have the real rise of supranationalism in Europe following WWII. The EU grew and it said, "Well, let's not stop here; let's continue to make things more and more solid as every day passes so we can't compete anymore. We are so far behind now; we are so wiped out. America is totally whipping our butt, and we lost our colonies. How can we effectively compete on the world stage?" The answer was to expand the coal and steel thing into a free trade zone thing—a free trade block.

What does this have to do with world trade? Can Belgium compete with the United States' economy? No. Can Italy? No. How about Ireland? Not a chance—they are all too busy drinking Guinness. How about Germany, the number four economy on the planet, can they compete? Nope. Bidding on a contract or trying to make some sort of international tie or trade tie with another country, the United States will win every time. So these countries got together and said, "Wait a minute! What if we all act under a common umbrella? Can all of our countries together as a unit effectively compete against the United States?" Yes, and they can compete with lots of other trade blocks around the world as well.

The EU evolved into a **trade block**, and, as you read about in chapter 6, trade blocks are all about decreasing tariffs and taxes between states to encourage trade. The French said, "Hey, Germany, we won't tax your sausages if you don't tax our wine!" They all agreed not to tax each other's stuff and therefore, people were able to afford to buy more of it. People bought more, people sold more. Mo' money. Once that got going and they got back on their feet during the last 50 years, the EU expanded into something that the world has never seen. That's why I am going to spend the rest of this chapter just talking about what's so cool about the EU. **2021 UPDATE:** So in 2016 the UK voted to leave the EU, and after years of wrangling it finally took full effect in 2021. This may incite other countries to exit the EU as well. But maybe not. More on this later.

What's the Deal with . . . Coal and Steel?

In 1951, six European countries–France, Belgium, West Germany, Italy, the Netherlands, and Luxembourg–signed the Treaty of Paris establishing the European Coal and Steel Community (ECSC). The ECSC created a common market, free of trade barriers, for steel and coal. These two commodities were especially important, because both were key resources for industrialization and rebuilding after WWII. Trading coal and steel freely was the first move to quell European power rivalries by establishing this economic unity and interdependence. The ECSC is often credited with creating a community between former World War II enemies. In fact, the European Union can directly trace it roots to the ECSC. The ECSC member countries soon after established the European Economic Community, which was renamed the European Community; this became the first, and most important, "pillar" of the European Union in 1992.

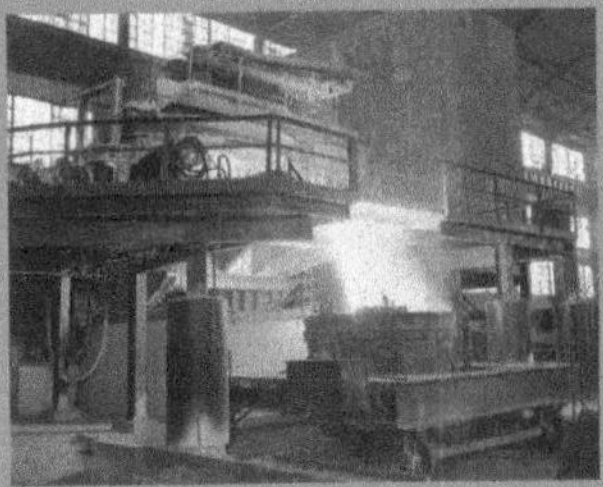

WHAT'S SO COOL ABOUT THE EU?

Why is it of such great significance? Thanks to the EU, Europe is turning from a bunch of nation-states that competed with each other for a thousand years into the **United States of Europe**. What does that mean? The independent sovereign nation-states of Europe are becoming much more like sub-states of the United States than independent countries, in terms of free movement of peoples and elimination of borders. Borders are still there, but they don't really mean anything. If you walked in a straight line across Europe thirty years ago, you would get your passport checked twenty different times according to how many countries you crossed.

That policy is largely disappearing. In fact, the number one thing for you to remember about what's different from the EU and all other trade blocks, or even supranationalist organizations around the planet, is the free movement of peoples. Nowhere else on the planet do you have a coalition of countries in which the people in your country can move anywhere within all the countries of the club without having their passport checked. This free movement of people is a big, big deal. A 2012 note: This free movement of people is being seriously re-evaluated right now, especially for the most recent EU inductees, but it is still a critical and revolutionary idea within a trade block. **2021 UPDATE:** Yeah . . . so that whole free movement of people has seriously been put on hold since BREXIT in tandem with the whole year of Covid crisis which saw borders actually fully closed. Future = uncertain.

Compare this to NAFTA, the free trade union between the US, Canada, and Mexico. Do we have a free movement of people? No! You can go back and forth between Canada and the US and they will check your passport, but you can't just do anything you want. Can you move back and forth from Mexico to the United States? Oh, no way! The US has a huge wall that it is, in fact, trying to make bigger between the countries. The unique factor about the EU is free movement of people. Because of this fact, they have to have a common law system as well. They are building a common judicial system, and a common legislative system. Thus, the EU is becoming a United States of Europe.

1952 ECSC: European Coal & Steel Community

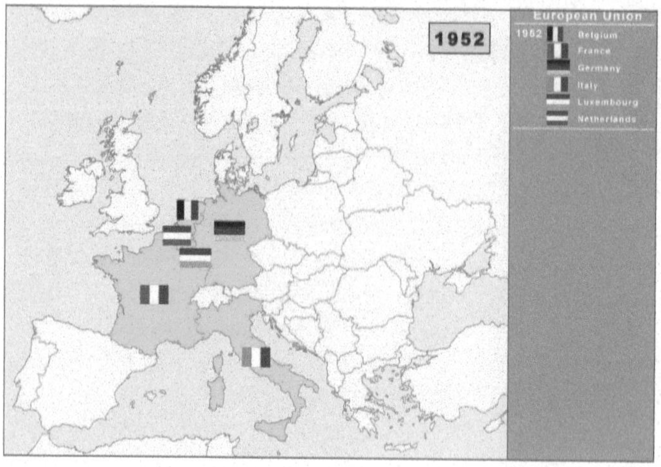

1957 EEC: European Economic Community, Expansion Phases

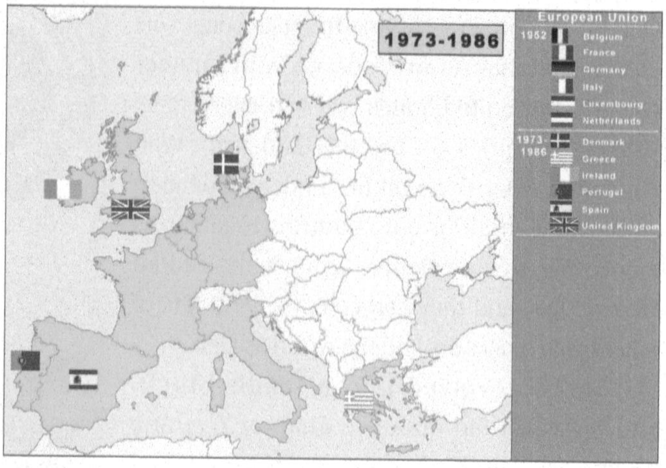

Before I go any further, does everybody in Europe think this is the most awesome idea ever, with everybody doing shots of Jägermeister while extolling the great virtues of their fellow Europeans? Oh, no. Political leaders think a United States of Europe would be totally sweet, particularly leaders whose countries participated in the wars. Businesses and a lot of other people like the idea too. However, not everybody digs it, and the reason for the animosity lies in that cultural imperialism thing that we talked about in the last chapter.

What does that have to do with the EU? People in Italy are Italians and they have their Italian heritage. A lot of the French are fiercely defensive of French culture. Those Brits will defend the Queen and teatime with their lives. Because the EU is comprised of nation-states with independent cultures and identities, there are lots of folks within them that see the EU as a threat to the survival of their culture. Sometimes you'll see a McDonalds restaurant being set on fire in protest of the EU. You can interpret any sort of anti-conglomeration or anti-homogenization attack on a multinational organization as a protest against the EU. The anti-EU opposition has been gaining ground as of late, too. They will tell you, "This EU thing sucks! I don't want to be in the United States of Europe! I want to be Italian!" . . . or Croatian, or whatever nationality they are.

The EU is definitely continuing despite the small amount of dissent (**2021 UPDATE:** that original small amount of dissent from a decade ago has become a huge amount—more on this later), but where's it going? Well, it's got a common defense system, which is a huge stride. Each country, for the most part, has its own little army, but the EU also has a common defense army. You can hear about the EU forces being sent into places like Yugoslavia, or EU forces being requested in Iraq. The idea of a unified army for these countries is a massive change in a region that has spent a thousand years beating itself up. They now have a common defense force, which pretty much eliminates the possibility of any future conflict. You can't fight wars with each other if you have the same army.

POWER STRUGGLES & POWER STRUCTURE OF EUROPE

The EU was growing for decades because most member countries think that for their organization to compete better on the world stage, their economic numbers need to be higher. Those members think the EU should become bigger. The bigger the club, the more world power it has, and the more power they have as individual countries. That is the common perception in the EU, so they expanded into Eastern Europe after the collapse of the Soviet Union. This means places like Poland, the Czech Republic, and Latvia are now in the EU. Bulgaria and Romania and Croatia were the most recent additions, and the rest of the Balkan Peninsula is trying to follow suit. This block is expanding eastward, unifying Europe more than ever before.

1992 EU: European Union, Post-USSR Expansion Phases

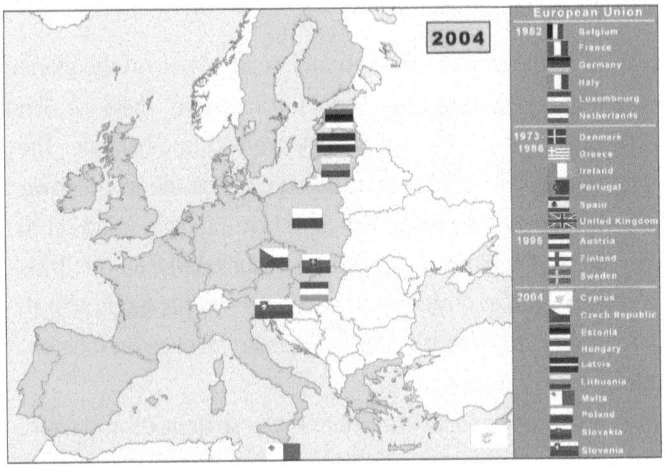

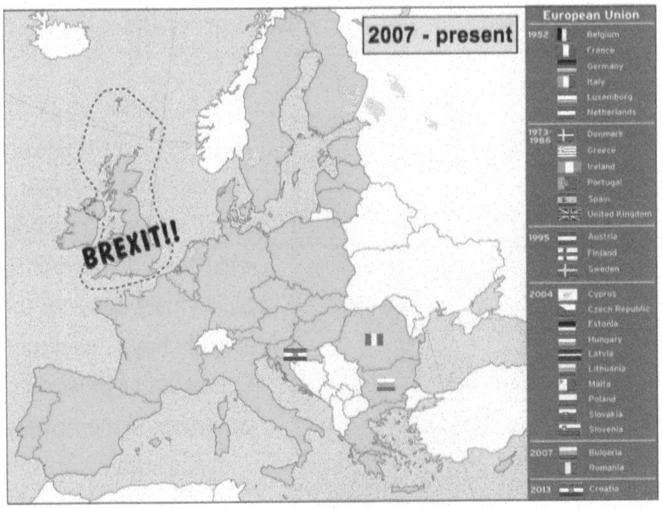

Here in the Plaid Avenger's book, we have been talking about Western Europe and Eastern Europe in separate chapters, but in another five or ten years there may be no difference; it may just be Europe. We will get to the remaining differences for the next chapter, but for now, we will leave it at that.

This has set up some internal power struggles within the EU. There is something the Plaid Avenger calls **Europe's Continental Divide of Power**. These internal struggles for power within the EU have a lot to do with how the system is set up. It's a lot like the United States in that there is a legislature, a judiciary system, and something similar to a House of Representatives in which each nation-state has a say according to the size of its population. The EU also has something like a Senate, in which each nation-state has a single representative. We can obviously tell who has the power by looking at populations. Who's got the power? Big surprise: France and Germany. They're the big power players in population and economics within Europe, so they have much more say in what the EU does. A lot of smaller states within the EU say, "Hey wait a minute, we are not getting as much of a voice! This is unfair! We don't like this! I'm Liechtenstein! Hear me roar!"

While the French and the Germans have battled each other throughout history, they are actually a team now. And what a team! It's the **Franco-German Alliance**, if you want to have some really cool terminology to throw around at dinner parties. These guys are also on the continent, so I call them **the continentals**. Germany is the number four economy on the planet, and France is number six, possibly seven at this point. This makes them a huge power core of the EU. These two states, while they have historically agreed on nothing, pretty much agree on everything nowadays. They are big pro-EU states quite frankly because they are power players in it. A lot of the EU's smaller states complain that if France and Germany don't like a proposal or idea, then it's not going to happen. There is definitely some truth in that. If France and Germany are against an idea, it can get stymied easily. This causes a true power divide between big states and small states.

The EU's number two power core is the UK (the #5 biggest world economy). Perhaps you never thought of the UK this way, but it is kind of distinct from the rest of Europe. "Say what? Plaid Avenger, I thought you just said the UK was part of Western Europe?" It is physically; it's there off the coast of the Eurasian continent, and all of its history is European history. However, the British have, for quite some time, been slightly separate. Snobbish even. Tea-sippers, with the pinky in the air. Remember the old British blue-blooded aristocracy—the same ones who still think the Queen is the bomb. A bit of snobbery remains, and the British have seen themselves as distinct from the rest of Europe proper throughout history. The Brits actually never fully adopted the official EU currency of the Euro, opting instead to keep the British Pound as their monetary unit. How is that for keeping an arm's distance away from the continent? **2021 UPDATE:** Wow. I wrote the paragraph above a decade ago, and called that one early. Need proof that the UK considers itself distinct from Europe? They left the EU in 2021. Mic drop.

Side by side again indeed! Trump-approved slogan!

As I said before, the British were the predecessors to American power on the planet. They were on the world throne for a very long time, but not anymore. Here's a Plaid Avenger interpretation: the attitude of the British is that they are still number one. They would still very much like to be the singular world power. If you can't be the world power, what should you do? Hang out with the world power! If you're not the biggest kid on the block, hang out with the biggest kid on the block and make him your buddy. So the British actually have much more in common with, and are much more eager to hang out with the United States in world politics than they are with the Franco-German alliance on their own continent, in their own trade block.

Big states and small states argue over constitutional power structures and representation. The older EU states argue with the newest EU states over immigration policies. And the Franco-German axis may argue with the British over economic and foreign policy. But the times are a'changin' on this front as well, which brings us to . . .

TODAY'S EU

Western Europe's going through some fairly radical changes here in the dawn of the 21st century. The first that's already been alluded to, and that's the EU. Let's elaborate a bit more on the EU in today's world, because it really has become the face of Western Europe and Europe as a whole, led by Western Europe, though. And it could now be in some trouble here in 2019 and beyond . . . but let's start with the positive repercussions of their experiment in Union.

The EU, if taken as a singular entity, is the largest economy on the planet. It has the largest GDP. Yes, indeed, a Gross Domestic Product bigger than that of even the United States. Now again, that's putting all the member countries of the EU together into a single block, but that's of course, what they are. And although it perhaps started as trade, perhaps morphed a little bit into politics, it is something entirely new on the world landscape, and it's becoming more powerful every day.

What's so new? Well, the EU has been working hard to frame a constitution for itself, an endeavor which has not met with ultimate success—so far. But it's quite important to consider, if for nothing else, that when it does get passed eventually, and it will, it's going to dawn an entire new age for Europe. It really will lay the foundations of Europe being the United States of Europe, a different beast altogether than anything that has come before in continental Europe. And they're doing a whole lot of things to cement this idea in place and take it to the next level.

A constitution, and also a unified judicial system, which is becoming stronger every day. They have unified, singular environmental policies. They've already unified their monetary system. The borders between the original EU states are fairly solvent, meaning you can move around within them—free movement of people. We've already talked about that. With the introduction of lots of new states in Eastern Europe, the free movement of people is changing a bit, but we'll get to that in just a second.

The thing I want to finish on, relating to the EU and its future in the world, is that it also has a common defense. This appears to be growing rapidly here very recently, as well. They want to have a defense structure—essentially an army—an EU army, that behaves and acts independently of the United States and even of NATO, of which most EU countries are already a member.

To finally finish, the big point: foreign policy needs to be considered as well. Increasingly, and this is bizarre, the EU is more and more speaking with a single voice when it comes to foreign policy. That's important because here lately, the EU foreign policy line does not agree with, or contradicts, the US foreign policy line. This is, again, quite important to understand what's going on in the world. Meaning that the EU, a singular entity, puts out political statements about its attitude about things that are happening on the planet. It's not always completely unified, but the fact that they are doing it is what's important. Maybe a few examples are warranted here.

As President Robert Mugabe of Zimbabwe was destroying his own economy and harassing his own peoples, the EU stepped forward with a statement and said, "We do not condone this behavior. We are going to have sanctions against Zimbabwe." That "we" being the entire EU. We're seeing that increasingly the EU is acting as a singular entity to make statements about, say, Russian aggression in Eastern Europe, Chinese human rights, or other issues that are occurring around the planet. This, combined with this singular EU national defense, are the most important points that are being developed rapidly here at the dawn of the 21st century, and are what's making the EU a real political, as well as economic, force in the world.

Why would they do this? Just like with the economic stuff, no singular European nation can compete with the US or increasingly can even compete with China, or Japan, or many other places in the world on their own. So they got together, and as a singular economic unit, they can compete quite well as one of the biggest players in their own right.

Now the same thing has happened with politics. They realize that, yes, while France has a big voice in the world, by itself it can't do as much, and certainly Belgium can't, and there's no way Poland can. But as a singular unit, politically their voice is much, much stronger. But they still don't agree on everything . . .

EURO-PROBLEMS: IMMIGRATION AND CULTURE FRICTION

Western Europe also has some big problems ahead of it. The main BIG issue that it's currently confronted with is: people. What? Why are people a problem? People are a double-edged sword in Western Europe because the populations of most Western European countries are stagnant, which means they're not growing, and some are actually declining—they're getting smaller. (Remember that demographic transition stuff from way back in chapter 2?) Therefore, to keep their economies going, moving forward and growing, you have to have people. You have to have people filling jobs—working on farms, working in restaurants, washing dishes, putting down the beds at the hotels. Western Europe, with its declining local populations, has had to rely on immigration from other regions in order to fill those jobs. It has to happen. That's a fact. But how is it happening?

We talked about distance being one of the United States' great physical attributes; it is far away from most other places in the world, and only borders one other distinctly different region. On the other side of the spectrum, Europe is very close to every place else. Western Europe can almost be considered the epicenter of every place nearby, in that it is infused with people from Eastern Europe, Russia, the Middle East, Africa, and pretty much anybody else that can very easily get to Europe. Europe, like the US, is extremely rich. The Middle East, and Africa, and parts of Eastern Europe could be put in the poorer column. Maybe even the extremely poor column. This results in a great movement, a great migration wave, of people that are hitting Europe, and causing tremendous problems. Europe has generally welcomed them, but it is reaching a kind of critical breaking point now.

The French in particular have always had a sort of an open-door immigration policy with all of the citizens of its former colonies, particularly in Africa. "You were French citizens when you were a colony, so we are going to consider you French citizens if you want to come here." Well, that's all fine and dandy, except that the French state is now going broke! A more lefty, socialist state which provides a lot of government benefits is starting to feel the economic strain as its population continues to swell with Arab and African immigrants. In addition, collisions of culture are starting to ferment.

CULTURE FRICTION

Cultural collision—what does that mean? That sounds a bit strong. Perhaps cultural friction is better. A lot of people are getting quite worried across Europe right now, that they're losing their European-ness. That perhaps the Belgians are losing their Belgian-ness, that the French are losing their French-ness. Why? Because the larger and larger numbers of immigrants from other places are bringing their cultures and languages and religions into the European homeland, and in more deeply integrated ways than ever in the past. Let's consider this for just a moment more.

A big debacle in worldwide news in September of 2005 was the **Danish cartoon of the prophet Mohammad** that originated in Western Europe. Why was that such a big deal? Because so many immigrants are of Middle Eastern Muslim and African Muslim background, both of which constitute growing percentages of the population in Western Europe. But cartoons are only the tip of the iceberg when contemplating the current European friction with Muslim culture.

Why would I pick on those folks in particular? Well, what these African and mostly-Arab Middle Easterners that are of Muslim religion have brought to Europe is a whole new kind of layer of different culture into the European experience. There are mosques being built all over Europe. There are folks who dress differently being incorporated into Europe. There are folks with different ideas about, say, individual liberties and rights that are moving into Europe. Liberties like the value of freedom of speech to draw a cartoon of a religious figure. Ah! Get it now?

Result? An anti-immigration backlash has started to creep into European society. There has been a rise of more ultra-conservative rightist political factions across Europe—in places like Austria, Germany, and even France in particular. Those political parties are gaining popularity by saying, "We're losing our Austrian-ness, we're losing our German-ness. And we wanna kick out all the foreigners or at least limit them coming into our state to prevent this."

Legislation has been passed in Switzerland and in Austria banning the construction of new mosques. In France, laws banning Muslim headscarves have been passed, and the French and Dutch are debating about banning the burqa altogether. In 2011, France actually did officially ban the burqa, and other states are following suit. Speaking of the Dutch, a politician from the Netherlands now has 24-hour a day protection for life since he released an anti-Islamic film in 2008 mostly as an ultra-rightist political stunt, a stunt which resulted in Muslim countries around the world boycotting Dutch goods. Starting to see the problem here?

Burqa high fashion not hip in Holland.

Long story short, Western Europe is torn right now. Its governments (being part of Team West, the core of Western Civilization) say, "Hey, we protect individual rights to do crazy stuff like offending Muslims by drawing a cartoon of the prophet Mohammed." However, they also have larger and larger Muslim constituencies, which is causing real conflict and clash in different parts of Europe who find offense in some of these things.

This cultural friction is not limited to the Muslims either. Let me give you another example, and we'll even stick with our Dutch wooden-shoe-wearing-tulip-growing friends. The Netherlands is also having problems with Catholic Christians from Poland! What? What is that all about? Well the Netherlands is . . . I won't say it's an atheist state, but it's not really a fervently religious group of folks. In fact, most of Western Europe is fairly blasé when it comes to attending church or being very religious, but the Poles sure are! Staunch Catholics, that bunch! As lots of Polish folks have migrated to western Europe seeking better economic conditions, their numbers are starting to tip the balance on societal views of everything from drug use to abortion rights.

That's a good topic to pick on. Catholic Poles are staunchly anti-abortion. Well, the Netherlands is known as being staunchly whatever-you-wanna-do! "Hey, we have a Red Light District—go have a hooker! It's all legal. Smoke some hemp! We don't care! You wanna have an abortion? Have three." Well, wait a minute. Now you have a bigger constituency of Catholic Poles in the Netherlands saying, "No, we don't agree with that." The Dutch government has to consider these things. So the Dutch may be losing their Dutch-ness to Catholic Poles now, as well as Muslims from North Africa. At least, that is the perception of some.

IMMIGRATION

One last point on the whole immigration and cultural friction issue that you should be aware of is that immigration is not equal across all of Western Europe, and that means immigration problems are not equal across all of Western Europe.

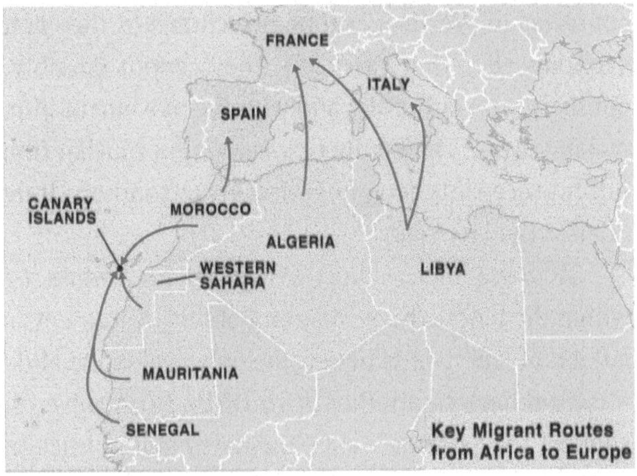

I just wanted to point out this spatial feature of who's the most concerned about immigration within the greater EU. As might be evident, the closer a state is to, say, North Africa or the Middle East, the more concerned it is with immigration, because that's the most used and easiest traveled avenue to get into Europe.

Meaning, if you want to migrate from North Africa to Western Europe, you go through the countries that are closest to you, namely Spain, Italy, and Southern France. As fate would have it, the countries that are most concerned with immigration, most worried about it, raise up the most fuss about it at the EU, are Spain, Italy, and France. Surprise! The further that European countries are from other world regions, the less likely they are to be complaining about any immigration problems, because the less likely they are to have any. Again, let's start with the evident. Norway, Sweden, Finland—these places are far away. They're also a lot colder. Who wants to go from North Africa to Finland? And Iceland? Don't even get me started! Like those guys would ever have an immigration problem. Even Leif Ericson bailed out of there.

You should be aware that there's also internal tribulations when the EU gets together and meets. Because states like France and Spain say, "Hey, we have a real immigration problem and we need to pass more stringent, strict laws to control this." Other places say, "Well, no, we don't really think it's that big an issue." It should be noted that the EU lacks a singular immigration policy for the block, which makes all of this a bigger debacle, as each state still does its own thing when it comes to this issue. But look for the EU to form a unified policy soon, and I'm willing to wager it's going to make getting into Europe a lot more difficult for potential immigrants.

Titanic 2017 Immigration Update: Holy Arab exodus, I was dead on the mark talking about the impending immigration crisis in Europe, which fully boiled over due to the Arab Revolution across the Middle East! Since 2011, boatload after boatload of Tunisian, Egyptians, Syrians, and Libyans made their way, or tried to make their way, to Italian and French shores in an effort to escape the bloodshed and economic chaos of their home countries (especially Syria, still in a bloody civil war since 2011 that has killed over 500,000 and seen five million flee the country. That was five MILLION) . . . and this sudden influx has caused open rifts within the EU.

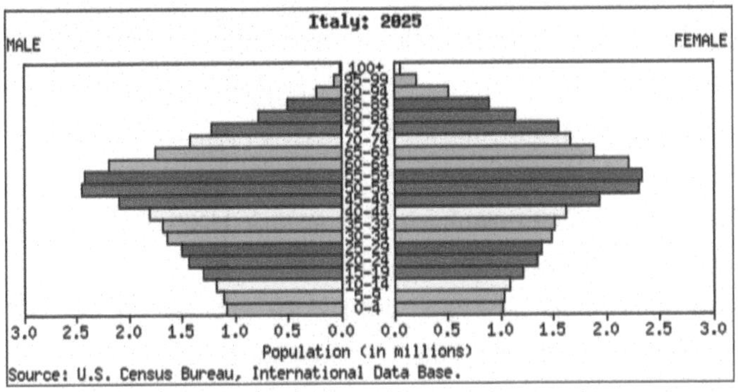

Bye-bye Italians! Hello immigrants!

Namely, Italy has been inundated with refugees, which they have started shipping to other parts of the continent, and thousands of immigrants have now died trying to make the trip across the Med in crappy boats, making it an embarrassing humanitarian disaster on the global stage. Watch this story continue to unfold this year, and EU immigration policy to be radically impacted.

Of course, the complicating factor here is that Europe is dependent on at least some immigration. European population dynamics, as we discussed in chapter 2, indicate that population is declining in Western Europe. In Italy, the population is declining because people aren't having enough babies. (Check Italian pop pyramid above.)

Now lots of folks are saying, "Isn't that good? Less people means more stuff for the people there, right?" Not necessarily . . .who's going to work at the factories that prop up the country's GDP? Who's going to make money for the economy? Who's going to work in the service sector? Population is completely stable or declining in all of these countries in Western Europe, which means they actually need immigrants to come in and keep things moving forward economically. That's the rub. They cannot possibly stop immigration, and at the same time, it's causing tremendous conflicts both culturally and with the movement of jobs and people. This is Western Europe's biggest problem. Think I'm making it up? The conflict over the anti-Muslim film in the Netherlands, the rise of skinhead movements in Germany, full-on race riots targeting Africans in southern Italy, and even anti-Semitic crime in France have all been major headlines in the last years.

Actually, this has been simmering for a while . . . Back in 2011, the EU moved to reverse decades of unfettered travel within the block when a majority of EU governments agreed the need to reinstate national passport controls amid fears of a flood of immigrants fleeing the upheaval in the Middle East. It did not help.

And back to another **2019 UPDATE:** A wave of populism broke out across Europe, resulting in a huge upsurge of support for extreme right candidates and politicians who are calling for a reversal of immigration openness and perhaps an exodus out of the EU altogether! And the immigration issue was the #1 concern fueling this movement! Called it! This is a extremely significant political event! Europe takes a step back from a border-less world, and a big step to the conservative right! Wow! The economic recession that has been pummeling the continent, combined with the catastrophic Covid crisis of 2020, will likely cause increased protectionism, more conservative ideology, and yet even tighter immigration policies in the near future. Yikes!

NEW FACES, NEW DIRECTIONS

To sum up and finish up our tour, we just need to talk about the power structure of what's happening in Western Europe right now as new folks have arrived. No, not the immigrants, but the fresh faces on the political scene. Leadership has rolled over all across Europe in the last few years, but let's just focus on the folks that make the biggest impact worldwide, and for the US in particular. Before we go any further, please, please, please remember this political fact about Western Europe: as a region, it is currently undergoing a significant shift to center-right all across the continent, especially in leadership. There are many reasons for this shift, such as economic recession, immigration issues, and cultural frictions; reasons which typically make people get all conservative and traditional, and vote for like-minded leaders. It could also be that all politics is cyclical, and its just time for the other team to have a go at it. Europe is definitely moving to right of center politically, and possibly socially.

Sammy needs you, baby!

There has also been a resurgence in extreme-right groups like skinheads in Germany, anti-immigration hate-groups in France, and borderline-fascist groups in Austria; many of these groups have been gaining support by playing up the immigration and cultural friction fears discussed in the previous sections. But those are just the extremes. But yes, they are still growing here in 2019.

Whatever the logic, more conservative, center-right (and sometimes far-right) politicians are getting elected almost everywhere in the region. What I want you to know about for this section are the main folks that are in power in that all-important **EU-3**. In UK, France, and Germany, leadership is truly, and literally, up for grabs right now, with the conservative right appearing to be winning out. Maybe. Let's look at these playas and how this shift is playing out.

First, the way it was: As we have previously discussed, the US and the UK are tight allies, and the UK has typically, in the last century at least, gone along with all things the US desires, more so than they have with the folks in their own neighborhood, namely the rest of continental Europe. This staunchly close US-UK situation has been referred to as the

Special Relationship. You could see this highlighted at its best back in 2003 as the US was ramping up for its invasion of Iraq to take out Saddam Hussein, under the (since discredited) idea that they had weapons of mass destruction (WMDs) and had to be stopped. The US went to the EU and said, "Hey, we're all buddies, we're all in NATO, hey EU countries, help us out. Support us in this second Iraq War." The British, of course, said, "Absolutely! We're lap dogs of the US foreign policy, we'll do anything you want. Yes sir, yes ma'am, sure thing Uncle Sam."

However, the Franco-German alliance said, "No, absolutely not. We're not on board with that Iraq policy at all." Indeed, most of the rest of Western Europe went along with the Franco-German alliance on that. France threatened to veto anything the US put forward at the UN on the matter. That's very typical of the way things have been for the last 10 to 20 years, maybe even more like seventy years since the end of WWII. Yes, the Europeans in general are US allies and friends, but when it comes to global issues and foreign policy, the US has only been able to rely on the UK year in and year out. And it could generally count on the French to oppose it year in and year out. This situation is becoming much more tricky here in 2019, as the "Trump Effect" has rattled even the most solid relationships, and the EU situation in general has thrown leadership across the continent into disarray, and that leadership has been flipping over fast as well. Faster than a textbook can keep up with sometimes! But I shall do my best to point out some important particulars . . .

Cameron called for vote on BREXIT. Then quit.

May muffed implementing BREXIT. Then quit.

Boris "the British Trump" Johnson championed BREXIT, now in charge.

UK first: The Brits just put conservative David Cameron into the Prime Minister position in May 2010, the first center-right (Tory party) guy to hold the post for nearly 14 years! And then he got re-elected by a resounding margin again in 2015! And then he pulled the biggest bonehead move ever and was forced to quit! Wait . . . what? Yeah, so Cameron was the dude who decided to let the people of the UK vote in a referendum on whether or not to stay in the EU. He thought it was a no-brainer that would pass easy . . . him, and the rest of the world actually. He said that if the British population voted to quit the EU, then he himself would quit as their leader. And then that is exactly what happened on June 23, 2016. So, good to his word, Cameron quit and Theresa May became the new Tory leader of the UK. She was then put in charge of the BREXIT debacle being implemented.

And if you followed any UK news of the last year, you may already know that May utterly tanked at the task. After multiple failed legislative attempts to establish an orderly BREXIT, she finally called it quits in May 2019. So May quit in May. It appears likely that the next to take the reigns of this mess will be one Boris Johnson, who is basically the "Donald Trump of the UK." No, I don't mean he is a billionaire with a reality TV show. I mean he is a flamboyant, outspoken, anti-globalist, anti-immigration politician that politically and socially shares the same mindset as the current US president. And apparently they share the same barber as well. Hahahaha . . . seriously, what's up with his hair? And, yes, I am referring to both of them!

Boris was actually one of the major campaigners to get BREXIT passed, so he will ironically be in charge of making it happen now. He will take power in the summer of 2019, so it remains to be seen what he will actually get done. However, given his anti-EU and anti-immigration tendencies, it is not hard to predict that the UK will likely be isolating or at least downplaying itself more from Europe, and potentially from other international organizations as well. Given that he and Trump have the same brain on many things, it is also likely that US/UK relations could get tighter than ever before. Assuredly a separate US/UK trade pact will be in the works; possibly even more foreign policy collaboration as well. But enough speculation on Boris and the Brits . . .

Let's shift to the continent: France and Germany—the biggest states, populations, and economies in the EU. Together they form a significant power center, and to restate, one that is typically not always in agreement with all things British, or American. What of our Franco-German friends? Well, there is significant strain and imminent turnover happening here, too!

President Macron, focus of fury.

The Frenchie first: the debonair and dashing President Emmanuel Macron of France was the youngest man ever elected to this high office back in 2017. He is ardently pro-Europe and pro-EU and has stated that he "has proudly embraced an unpopular European Union." He also does not consider immigration to be a major issue, and downplays the fears surrounding it, as well as downplaying nationalist sentiment in the country. Macron is a centrist or possibly center-right conservative politically, financially, and socially. (Do keep in mind that when it comes to cultural issues, like gay rights or workers' rights, all of France is pretty liberal.)

When he came to power, much hope was placed on his conservative fiscal policies "fixing" the ailing French economy, and that has been the focus of his policy-making. That meant tightening the belt of the socially progressive French budget, increasing retirement age, pulling back on some benefits provided by the state, cutting the amount of civil servants, weakening labor unions, cutting corporate tax rates, etc., all in a bid to shrink public debt and increase productivity/job creation. How's it going so far?

Bad. Macron's coolly calculated plans seem to have infuriated the middle to lower classes, and his popularity was already rapidly declining before he decided to raise taxes on petrol (gas) by January 2019. This projected increase sparked the longest and most destructive protest in the history of modern France. Starting in November 2018, and still happening up to this writing in summer 2019, the **Yellow Vest** movement has involved demonstrations and the blocking of roads and fuel depots, some of which developed into major riots. Named for the yellow safety vests worn by many workers, the movement is a populist, grassroots, political movement for "economic justice" that sees Macron's moves as anti-worker, pro-urban elite, and increasing the cost of living for the common citizen. It has galvanized into an anti-Macron, and simultaneously anti-EU, platform that has supporters across the political and economic spectrum. However, it is really most beneficial to the **National Rally** which has gained much ground and many supporters since this mess all started. Who?

Safety first for the protest!

The far right, anti-EU, anti-immigrant political party formerly named the National Front has re-branded itself as **National Rally**, and just recently (June 2019) became the most voted for party in the European Parliament elections to the EU. The party leader, Marine Le Pen, has led the group in France for over a decade, a period that has seen the far-right go from fringe obscurity to now-power-player-party that could possibly soon control the French government!!! Dang! Given her anti-EU, anti-globalist, anti-immigrant stance, she could be best described as the "female Donald Trump of France," and the antithesis of Macron.

Far-right Le Pen FTW

Much like in the United States, there's a split between the two Frances: between Mr. Macron's France (cosmopolitan, pro-global, pro-diplomacy, pro-trade, pro-urban elite) which is in charge and the peripheral France (rural, depressed, deindustrialized areas, now also with anti-EU and anti-migrant attitudes) which considers itself a victim of globalization. The face-off between nationalists and globalists is now in place in France . . . hey, this really is starting to look a lot like the USA, right? Maybe Le Pen will make France great again? Well, that may have to wait until the next edition of this book, but you should know for now who Le Pen is, and that Macron may not be around for much longer. Which is a shame for the EU, since he is a big fan of it, and was actually trying to position France as the most powerful leader of the block in the future, maybe even superseding Germany. Ah! Ze Germans! That leads us to the last of the EU leaders to elucidate about . . .

She's Merkel-icious! But possibly in trouble in this new age of Euro-populism!

The Teutonic titan, Angela Merkel! The most powerful woman in the world, a title she has held for over a decade! Conservative, center-right, German Chancellor Angela Merkel has had much success in the last decade in power swinging the Germans back to the conservative side of the spectrum. Her predecessor Chancellor Gerhard Schroeder was much more liberal and much more anti-US. Germany's first woman leader has strengthened its economy, pulled in the reigns of immigration, and become a serious force in world affairs in her own right. When she talks, people listen. Merkel is more of a supporter of US policy, although not enough to get involved in any more 'wars on terror', as most Germans on the street were vehemently opposed to both the US involvement in Iraq and in Afghanistan. But do remember this, my plaid friends: increasingly Germany has been invited to sit in on the workings of the UN Permanent Security Council. Even though it does not have an official vote, the German voice is becoming of greater significance on the world political stage by affecting UN policy decisions. Quite frankly, Germany is holding Europe and the EU together right now, almost single-handedly. They have the biggest and strongest economy, the most economic clout, and the most control over the European banking system . . . thus, Chancellor Merkel has unprecedented power to determine economic and fiscal policy for the entire EU, and she is now looked to as the leader and possible savior of not just Greece or the Spanish economy, but of the whole EU experiment as well! Germany holds the purse strings of Europe, and is demanding austerity policies as the solution to Europe's woes. But Chancellor Merkel was also the leader that decided on an open door policy when it came to the Syrian refugee crisis, a crisis that has seen over 3 million immigrants move into Germany in just the last couple years! Intense! It has been straining the German budget and resources, but more importantly it is straining the average Germans' patience and acceptance. Her once super high approval rating has been sliding as of late, and she will face her first true challenge at the polls when she goes up for re-election in 2017. The far right parties have been gaining ground in Germany, too, and Merkel is no longer untouchable even within her own party! Challenge phase is on! Can the last bastion of liberal democracy on the planet hold her position in the face of the populist onslaught? We shall see, and right soon! **2019 UPDATE:** Yep, Merkel is having right-wing headaches, too! The rise of anti-immigrant and anti-globalization forces in Germany caused major setbacks and election losses to her political party (the CDU or Christian Democratic Union of Germany) across the country. She announced in October 2018 that she will definitely be stepping down as chancellor in 2021. No idea as of yet who will fill that power vacuum, neither personally or politically. Too early to call.

So, big changes going on. New faces, new directions. These three leaders taken collectively carry a lot of weight in Europe, and in fact their countries together are often referred to as the ***EU-3***. While Hollande is now leading a charge back to the liberal left in France (and currently failing at it), the vast majority of European states are still staunchly in the center-right column economically and socially, and that is a trend that is still growing . . . thus look for more conservative measures to pass in each of their countries, but also in the EU as a whole. I have only hit the big three; there are also center-right and even some extreme right leaders in Austria, Finland,and Switzerland. Not to belabor the point, but limiting immigration and "cultural preservation" legislation, like banning burqas, will almost certainly be percolating through the EU in the very near future. You heard it here first.

SUMMARY—WESTERN EUROPE

Increasingly, Western Europe and Eastern Europe are simply becoming Europe. However, we can still tease out enough stuff in today's world to make some finishing remarks about this Western Europe region. It is one of the richest regions in the world. It is one of the most politically powerful regions in the world. Along with the United States, it is the foundation stone of Team West: one of the most important economic, political, and even cultural forces on the planet. While the US certainly leads Team West in the world today, Western Europe is where Team West started, thus the term Western Europe . . . Team West: go figure!

It does have some new faces, some new attitudes, but the main big theme in today's world is the EU becoming somewhat of a United States of Europe. What that means for the future in terms of not just a Constitution, but defense and foreign policy, are important points to consider. Following that with these new faces in today's world, we're seeing a Western Europe that is being increasingly supportive of the United States, but still quite independent in its own right.

Keep us in the game, coach!

However, I must now be brutally honest with you, my plaid friends . . . this region has passed its prime, and its not likely to return to the glory days anytime soon. Yes, Western Europe is rich. Yes, Western Europe is a core component of the powerful and potent Team West. Yes, Western Europe is probably one of the most awesome places to party on the planet. BUT, Western Europe is no longer a center of innovation and creation; not for technology, not for business, and not even for demographics! They can't even crank out enough kids to keep up with replacement level! Which is, of course, why immigration is both a necessity and a problem for their region.

Western Europe is great, but mired in bureaucracy and tradition that simply does not facilitate it as a revolutionary center of activity anymore. Given its imperialistic phase and its centuries of conflict, this may not necessarily be a bad thing. Either way, these Europeans will continue to focus on EU evolvement to keep themselves in the global game as a serious player, knowing full well that its better to still be on the sidelines as the seasoned second-stringer than out of the ballpark altogether. At the core of the EU, the IMF, NATO, World Bank, UN, and Team West, this region has still got game, even if it gets winded easily.

2021 Final Update Note: While the 2008 economic recession that rattled Europe is now winding down, the economies of the continent have not fully recovered and growth is slow to stagnant in some places. Add to that the inflow of hundreds of thousands of economic migrants and war refugees in the last decade. Add to that the whole lost year of 2020 due to the catastrophic Covid crisis, which sees Europe suddenly way behind other developed countries in terms of vaccination programs . . . increasingly being blamed on bureaucratic bungling and ridiculous red tape tied to EU regulations. Then add to that a persistent and growing fear amongst the masses of many states that they are "losing" their culture, or "losing" their identity. Stir all those ingredients together, and bring to a boil. This ain't no shepherd's pie being made; this is big trouble for the EU as a whole. This battle between pro-EU and anti-EU attitudes (in combination with the crappy economic situation) played itself out in the 2016 BREXIT vote for the UK to leave the EU. Radical far-right political parties in virtually every country made significant gains in the last year . . . meaning they won seats in the EU legislature. These far-right fringe folks are calling for immigration blocks, a return of powers from the EU to the sovereign states, and perhaps even pulling out of the EU altogether! Keep in mind, that the majority of seats held in the EU are still quite sane people, but the big gains of this fringe have caused every single leader in every single EU country to re-evaluate their policies to address this growing chorus of anti-EU peeps. It is kind of a big deal, because what we are talking about here is a full-on debate about the EU project itself, and a fight over the concept of the nation-state . . . remember, that is an idea that these Western European countries invented! And now some are fighting to return to the old-school definition of it!

The fate of the Eurozone and the entire EU experiment are completely unpredictable at this point, which is why I won't predict anything. But follow close, and the next edition of this text will likely have to throw out this entire chapter and start anew. Let's hope the Europeans and the American economy survive the radical and imminent transition intact! Whatever happens, Europe will not be regaining any of its former glory anytime soon . . . and a more cynical forecast would say that the region is about to enter a new "Dark Ages" Let's hope that's not the case!

Dark times ahead for Europe: Is it knight-knight time for the EU?

WESTERN EUROPE RUNDOWN & RESOURCES

View additional Plaid Avenger resources for this region at http://plaid.at/we

BIG PLUSES

- Biggest economy in the world
- The EU is the most highly organized and successful political organization of all time
- Interstate armed conflict is virtually impossible; very stable place
- The EU as a singular entity is the biggest economy on the planet
- Western Europe has some of the highest standards of living on the planet
- Multi-ethnic, diverse societies that generally get along with each other
- Fantastic regional rail transportation network which interconnects the region and runs extremely efficiently (except for Italy lol)

BIG PROBLEMS

- Rough EU roads ahead as the richer economies (Germany, France) have to continue to take care of the poorer economies (Greece, Portugal)
- Move to conservative center-right will certainly exacerbate anti-immigration friction that is starting to rear its head across the region
- Aging populations, over-extension of state benefits, and high costs of living will make economic growth extremely challenging in the future
- Western Europe will continue to lose its once overwhelmingly-predominate position as a major center of technological innovation, finance, and investment . . . mostly to Asia
- Cultural friction will continue to grow as folks feel they are losing their cultural identity to a monolithic EU and/or foreign immigration
- BREXIT vote combined with the rise of populism and right-wing political groups could very well see the unraveling of the European Union project

DEAD DUDES OF NOTE:

Winston Churchill: The British Prime Minister guy that held the UK and Allied powers together during WWII. Great statesman, orator, historian, artist, and drinker by all accounts. Coined the phrase Iron Curtain and fostered the tight Special Relationship with the US

Adolf Hitler: Great public speaker and snappy dresser that tried to take over the world and kill everyone that wasn't white and German. His idiocy transformed Europe and became the impetus for forming the EU, the UN, and the UN convention against genocide.

Charles de Gaulle: National hero and dominant political leader of France during and after WWII, he was one of the few Frenchies who didn't surrender. Charismatic and proudly nationalistic, he pursued an independent path from both Europe and the US, which is why France was not a major payer in NATO, or great friends with the UK or US for decades.

Margaret Thatcher: Conservative Prime Minister of UK from 1975–90, was nicknamed the "Iron Lady" for her aggressive stance towards commies. She was the "Ronald Reagan of the UK" in that the two were identical in political and economic ideology.

PLAID CINEMA SELECTION:

(Look, I know there are ten million awesome classic films from Europe that are probably way better than this list. I'm just giving you a handful of more modern picks that will intrigue the younger generations a bit more. Frankly, I usually can't stand anything labeled as a "classic" of European cinema.)

GoodBye, Lenin! (2003) Fantastic drama-comedy that takes place in Germany as the Iron Curtain falls, and the soviet economy/culture of East Berlin is replaced by western materialism/consumerism. A crowd pleaser.

Trainspotting (1996) Scottish cult heroine film, made Ewan McGregor famous pre-Obi Wan Kenobi. Great look at blue collar life in UK and drug culture, and more relevant than ever for Americans, as the US deals with a very similar situation called the opioid crisis, along with rising heroin use. "Toilet scene" is classic insanity you won't forget.

Pan's Labyrinth (2006) Bizarre whacked-out insane fantasy film set in Franco's fascist Spain in the 1940's. Gives some insight into Spain's fascist period, and will cause hours of sleeplessness.

Gomorra (2008) A contemporary mob drama set in Naples that exposes Italy's criminal underbelly by telling five stories of individuals who think they can make their own compact with Camorra, the area's Mafia. Oh, and the Camorra actually exist in real life.

Amélie (2001) Wildly popular, clever, unconventional, light-hearted comedy from France. Nothing too serious, but a pleasant look at modern French life, and will make you not hate the French so much.

Run Lola Run (1998) Awesome action/adventure/time warp film that perhaps has no great geographic utility, but is a great high energy MTV-style flick that I use to introduce hesitant students to foreign film. They love it, despite the subtitles.

Volver (2006) Director Pedro Almodóvar is in love with the entire country of Spain, including its women, and it shows in this comedy of errors involving murder and family secrets. If you don't find the gorgeous Spanish scenery compelling—and why not?—then the gorgeous live-action Spanish scenery that is Penélope Cruz will keep you entertained.

Ocean's Twelve (2004) Okay, it's just a guilty pleasure, but I love watching this heist film which jumps across Western Europe. No great learning material here, although it does feature Interpol and a variety of excellent landscapes across the continent. And most people despise the "trick ending," which makes me like it even more.

LIVE LEADERS YOU SHOULD KNOW:

Emmanuel Macron: President of France, centrist to center-right, pro-EU, pro-globalist leader of the world's 6th largest economy. Focusing on France's economy, but now widely despised for policies.

Marine Le Pen: the "female Donald Trump of France." Leads the far-right, anti-EU, anti-globalist, anti-immigrant political party named National Rally, which is gaining much ground lately and is a serious political power now.

Angela Merkel: Center-right Chancellor of Germany, 4th largest economy, EU power player, routinely voted the most powerful woman on the planet. Anti-immigration mood in Germany has led to her stepping down in 2021.

Theresa May: Previous, center-right Prime Minister of UK, 5th largest economy. Couldn't pass BREXIT legislation, so quit the job in 2019.

Boris "the British Trump" Johnson: UK Prime Minister, championed BREXIT, center-right, anti-EU (but not as anti-globalist), anti-immigration. Huge ally of the USA, and likely eats fish n' chips, but too soon to tell.

MAP GALLERY

View Map Gallery & additional Plaid Avenger resources for this region at http://plaid.at/we

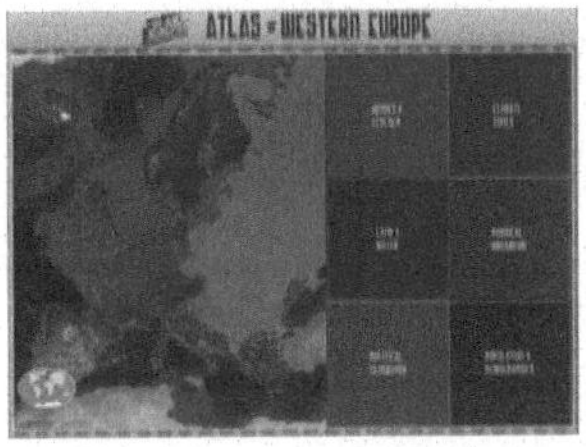

Tallinn
ESTONIA
LATIVIA
Riga
Baltic Sea
LITHUANIA
Vilnius
Minsk
BELARUS
Homyel
POLAND
Warsaw
Poznan
Wroclaw
Krakow
Kharkiv
Kiev
UKRAINE
Dnipropetrovs'k
L'viv
Prague
CZECH REPUBLIC
SLOVAKIA
CARPATHIAN MTS
MOLDOVA
Chisinau
Odesa
Bratislava
Budapest
HUNGARY
ROMANIA
SLOVENIA
Ljubljana
Zagreb
Constanta
Bucharest
Black Sea
CROATIA
BOSNIA AND HERZEGOVINA
Belgrade
SERBIA
Sarajevo
BULGARIA
Sofia
Adriatic Sea
MONTNEGRO
KOSOVO
Sarajevo
Skopje
MACEDONIA
Tirana
ALBANIA

CHAPTER OUTLINE

Eastern Europe

9

THE BOOKENDED REGION

We just wrapped up Western Europe. We now know of its extremely rich and multifaceted world influence. And we are aware of how it destroyed itself in the 20th century. Let's move on to Eastern Europe. It has a slightly different physical geography, a slightly different cultural geography, and it's a place that has been radically different from Western Europe in the recent past. But things are changing, and they may not be that much different from the West anymore . . .

Eastern Europe is the *bookended* region, as the Plaid Avenger refers to it, and bookended in a variety of ways. The bookend reference means there is something in the middle being held up, propped up, or pinched between two bigger sides. In this respect, physically, Eastern Europe is between the giant Russia to its east, and those rich and powerful Western European states to its west. Ideologically, it's also in the pinch in the middle: Eastern Europe has been a buffer zone between these two same giants—Russian culture to the East, Western European ideas to its west. Because of this, it has essentially been the battleground between these two teams, and sometimes even a fuse to the powderkeg of big conflicts. Where did the major transgressions of the 20th century start? World War I, World War II, the Cold War? Yep . . . all actually got launched in the Eastern Europe territory.

This region had been marked for a very long time by **devolution**, or shattering apart; big political entities breaking down into smaller units. This region is also distinct because it has been a **buffer zone** and a battleground of ideologies throughout the 20th century. It was a buffer between Russia and Western Europe, and as such, was a battleground for the Cold War. Commies in Russia thought one way, dude-ocracies in Western Europe thought another way, and everyone in Eastern Europe was caught in the middle. The book getting squished by the bookends.

On top of that, we're going to see that there are a lot of historical influences that have permeated Eastern Europe proper, making it extremely diverse in every aspect you can imagine: linguistically, ideologically, religiously, ethnically, culturally . . . It's all here. This diversity is one of the reasons Eastern European states have historically been pulled in lots of different directions and sometimes shattered apart. That's why it's a region, and just one of the many reasons why it's distinct from surrounding regions. Russia is Russia, Western Europe is Western Europe, and Eastern Europe is something in-between, for now.

But that was then and this is now. If you want a key word, a single word that keeps coming up in the discussion of Eastern Europe, the Plaid Avenger has one: **transition**. This is a region in transition. Of course, every place on the planet is changing and moving around, but this is an entire region in which just about all facets of life are on the shift: with an obvious place they came from, and with an obvious place they're going to. All the states in this part of the world are going through this process of transition. What sort of transition, Plaid man? A transition from what to what? We'll talk about that in more detail as we go along.

THE PHYSICAL MIDDLE GROUND

When we think of Eastern Europe, we think of that region that's sandwiched between two much bigger and more powerful other world regions. But if we just look at it physically, just physically for a minute, it's also kind of a buffer zone in a very natural sense, in a variety of different ways.

CLIMATE

The temperature and precipitation patterns of Eastern Europe are more **continental** in character than areas to its west. Being further and further away from the Atlantic Ocean, and particularly its most moderating Gulf Stream effect, makes these states more prone to increased temperature disparity: that is, it can get much colder in the winters and perhaps even more hotter in the summers. As we progress further into the continent and away from large bodies of water, this continentality effect becomes even more pronounced. We'll see this played out to the extreme when we get into Russia, deep in the continent's interior, where it's way colder still. Nothing is quite like Russia, where the only thing that's not frozen is the vodka. Even the Plaid Avenger's beard froze up in that place multiple times on secret subversive missions. When we think of the inverse, we think about the coastlines of Western Europe, which has a much more moderated climate. It's not as extreme in the west: it rarely freezes, even in places as far north as Great Britain or parts of Norway.

Eastern Europe is in between, in that it's not quite as bad, not quite as hard core cold as Russia but not quite as moderated as the West either. As you progress from Western Europe inward, into the continent, you get more continentality—meaning less moderation, more temperature extremes during the year, and certainly cooler, if not downright colder, in the wintertime. Climatically, they're in a zone of transition, as well, from the West to Russia. But don't let me mislead you: this place does have some nice climates, and in fact some areas are exceptionally rich in natural resources and soils, too. In fact, Eastern Europe actually produces a heck of a lot of food: Ukraine has classically been one of the breadbaskets of all of Europe, and it continues to be. There is a lot going on here physically, but it's just not quite as moderated as its Western European counterpart.

LAND SITUATION

Another big physical feature to consider with Eastern Europe is that big sections of these countries are landlocked. When we think of Western Europe, we think of maritime powers—all of them have sea access, all of them have coastlines, and, therefore, they all have big navies and armadas, and they're all big traders and were all big colonizers of other parts of the world. Not so much so for the Eastern Europeans, if at all. Yes, there is some coastline along the Black Sea for some countries like Bulgaria and Romania. Yes, some of them have Mediterranean Sea access like Albania and Croatia. And yes, some have Baltic Sea frontage like Poland and Estonia. But none of these states in what we're considering Eastern Europe were ever big maritime powers, and they're still not. Unlike their Western European counterparts, none of these states were big colonial forces in the world. They didn't have the advantages we talked about in Western Europe of sapping off the resources and riches of the planet by controlling vast areas outside of Europe proper. Those Eastern European empires and states were never major maritime powers, nor colonizers, nor absorbers of wealth from other points abroad. And that sets them quite apart from their Western European counterparts, even up to this day.

CULTURE CLUB SANDWICHED

Order up! Who wanted the Turkey, Polish sausage with Russian dressing on rye?

The last big physical thing to consider is that they were physically in the middle of some very big powers, that have very different cultures. What do I mean by this? Well, Eastern Europeans have their own culture, obviously: Polish folk have their Polish culture, Ukrainians have their Ukrainian culture, etc, etc. But all these folks in all the countries in this region are sandwiched between entities that have *bigger* histories, *bigger* cultures, and are *bigger* powers. By bigger, don't misinterpret, I'm not talking about "bigger is better." I'm just suggesting here that they're *bigger* in terms of having more of an impact within their regions, but also beyond their regions as well. A main one to consider is Mother Russia: a very distinct culture, language, religion, and ideology that by its very size and strength has dominated areas close to its Moscow core. And what's close to the Moscow core? That would be all the Eastern European countries.

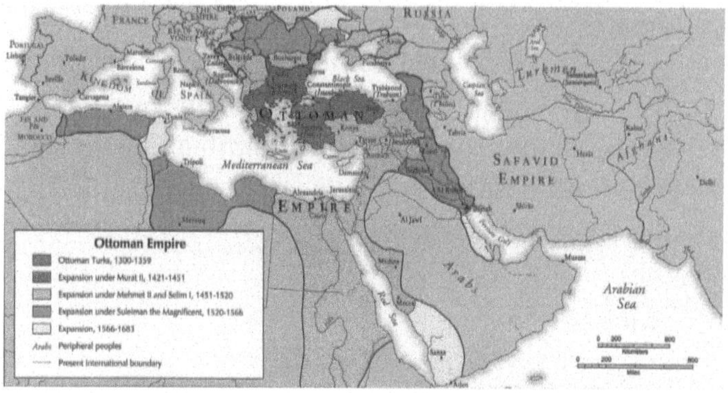

But Russia is not the only big power heavily influencing this regions. Team West and its cultures influenced the region, as has North Africa. You also have the former Ottoman Empire down south. That empire is long since gone and left the sole state of Turkey in its place, and so there's a Turkish culture that abuts the region as well. An important part of that Turkish culture is the fact that they are Muslim, and the Ottomans brought Islam into the Balkan region all the way to Austria starting way back in the 14th century, and there are pockets of Islamic culture still today in Albania and parts of former Yugoslavia. That's important to note, by the way, because perhaps unlike any other world region or other part of the world, you have a variety of world religions that come together in one place, albeit, not always on friendly terms. You do have Islam which did penetrate into Christian Europe via the Ottomans. But even before you get to that, you have to think about the divisions of Christianity.

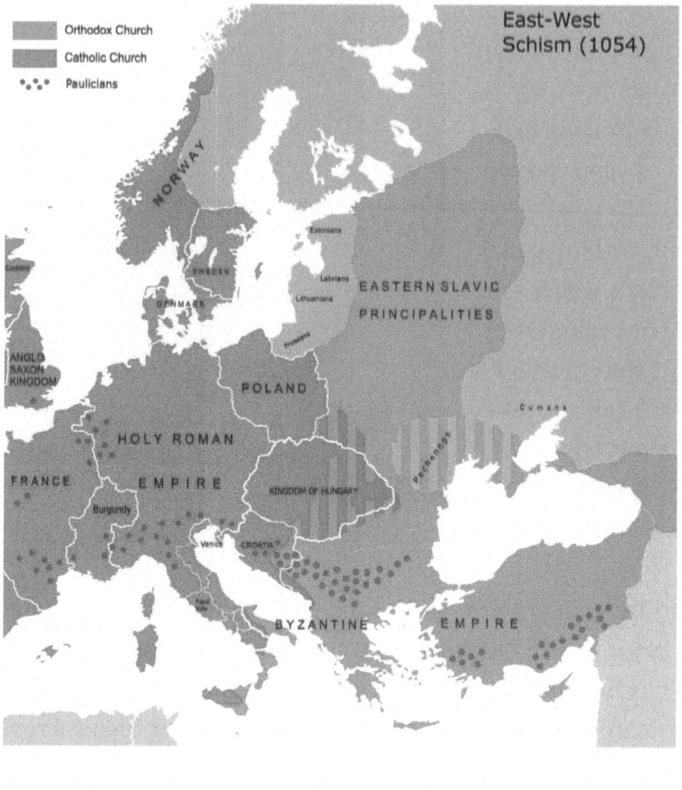

All of Europe, in general, was the cradle of Christianity, but the religion had a fight with itself that culminated in the Great Schism of 1054: that theological and political split between the Christian East and West which created the two sects we now refer to as Eastern Orthodox and Roman Catholic. Later in the 16th century, the Christian West would once more have an internal struggle within itself named the Protestant Reformation (and subsequent bloody Counter-Reformation) which resulted in a further division creating the Protestant sect. Why am I talking about all these religious schisms and shizzle? Because Eastern Europe is home to all of them, and indeed many of the division lines fell exactly across this region, as did the battles between the groups for dominance of their particular brand.

A quick summary here: Eastern Europe in the middle of major world cultures—Russian versus Western versus Arab/Islamic. Also, it's the middle ground of major world religions—Christianity, Judaism, Islam, and all the divisions of Christianity itself, as well. These divisions all coming together—different ethnicities, different people, different religions, different cultures, different histories—has served to play out within Eastern Europe, more often than not, as a battlefield between competing peoples, powers, empires, ideologies, and religions. This leads us to the concept of Eastern Europe as, not just a battleground, but a **shatter belt** . . . a zone of breaking down, breaking apart bigger entities to smaller and smaller ones. Perhaps I'm getting ahead of myself. Before we get to this idea of a shatter belt and a breaking down, a.k.a. **devolution** (another great word), let's back up the history boat for just a second and take these things one at a time as they proceeded throughout history.

IDEOLOGICAL MIDDLE GROUND

As pointed out above, Eastern Europe is in the middle of lots of different cultures, different religions, and different ethnicities who have battled it out across the plains of Eastern Europe over the centuries. I'm not going to bore you with all the details; I'll just set the stage so you understand the modern era, how some of these outside influences came to cohabitate within this region over time, and why perhaps it's still causing some conflicts today.

HISTORICALLY

TEAM WEST VS. RUSSIA VS. THE OTTOMANS

Historically speaking, and I'm only going back a few hundred years, Eastern Europe was kind of a battleground of big empires. We have some of the Team Western players like the Austrian empire, the Prussian empire (which evolved into Germany), the Russians, and the Ottoman Empire. Four big entities which virtually controlled all of what we consider now Eastern Europe. Just so you know the Russians and the Ottomans fought it out for long periods of time vying for control of these territories. The Russians and the Austrians did the same thing, as did the Prussians and the Russians.

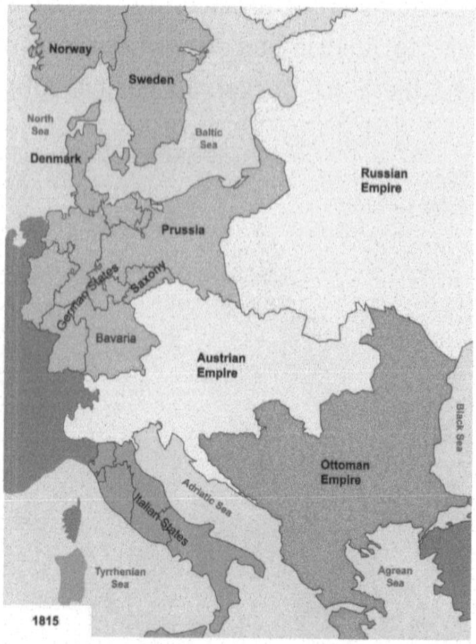

1815: Four main players.

These empires have radically different cultures as well, in things like language, ethnicity, economies, politics, and religions. Differences which sow the seeds of conflict for generations. Consider only the religious differences for a moment more. The Germans and Austrians were mostly Protestants and Catholics; the Russians were staunchly Christian Orthodox; the Ottomans were Islamic. They are all vying for cultural and political influence in this region. Are you starting to get a sense of the confluence of conflicting ideas in this area? Empires trying to expand, fighting each other, in this fringe zone we're going to call Eastern Europe, as we get a little further along.

Now that's the background set-up. We need to get into the modern era, and in the modern era there are some major devolutions that occur. Evolution has the connotation of growth, of building into something bigger and more complex. **Devolution** is just the opposite: a term that simply means breaking down—devolving into smaller, simpler pieces. Devolution in a political sense involves a big empire or a big country breaking down into smaller countries; that's a real big theme for Eastern Europe historically and perhaps even into today's world.

WORLD WAR I

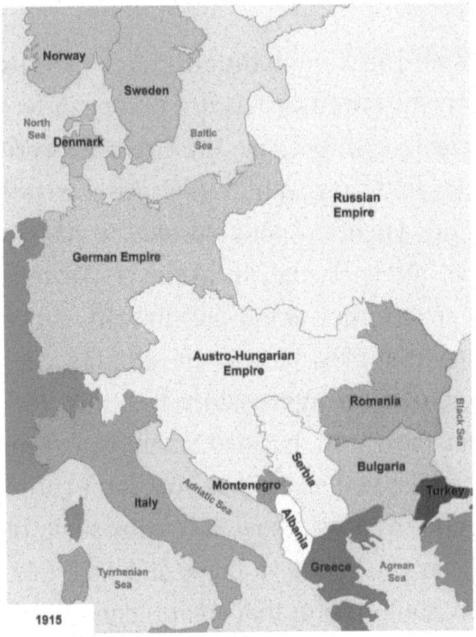

1915: Goodbye, Ottoman Empire!

The starting phase of devolution in the modern era was with the Ottoman Empire. The Ottoman Empire, which was a Turkish-Islamic empire, introduced Islam into southeastern Europe. Places like Bosnia, Albania, and Kosovo are still predominately Muslim today as a result of the cultural influences that the Ottomans brought in. You have to remember, the Ottomans were actually knocking on the door of Austria for a very long time in a bid to expand their empire into Europe, but they were weakened considerably and were on the brink of collapse by the time the 20th century rolled around. Let's fast forward: this declining empire most unwisely allied themselves with the Germans during the lead up to World War I (you can read more about that in the chapter on Turkey). But the Germans lost that war, which means their Turkish allies were also losers, which resulted in the carving up of the Ottoman territory. By 1915, there was no such thing as the Ottoman Empire. It's subdivided, as you see in the map into a variety of new states already—places like Romania, Yugoslavia, and Bulgaria.

1919: Goodbye Austro-Hungarians!

The Austria-Hungarian Empire, which kind of started World War I—or at least the assassination of their leader Franz Ferdinand triggered it—they lost the war as well, and thus their Empire became the next shatter zone. Ottomans shattered first, then the Austria-Hungarian Empire shattered next, which created several more countries—individual entities which declared independence.

We also have to point out that Russia was unwittingly pulled into World War I, and it got the crap kicked out of it by the Germans. World War I was just so bizarre! All of the major, and minor, powers in Europe had a complex web of military alliances with each other that—once the war started—obligated everyone to jump into the fray to protect their buddies. Pretty much everyone declared war on somebody within days of the Austro-Hungarian assassination spark. Thus, the Russians got sucked into a land war to help their Serb allies, just at a time when dissent and revolution were internally brewing back at home. The results of which were that Russia, under the command of the inept Tsar Nicholas, sucked so bad on the battlefield that they lost big chunks of their western fringe to the Germans . . . and of course this did not sit well with already-aggravated Ruskies back home. And that was when our main Commie friend, Vladimir Lenin came to power in Russia in 1917, he essentially said, "Hey, we want out! We're out of World War I. We surrender. Germany can have all the territory that we have ceded thus far. They can have it. Take it! We're done!" This was actually a very popular policy back home in Russia, which was on its way to becoming the USSR. We will pick up the importance of this WW1 Russian territorial loss in a couple of paragraphs from now . . .

Even after the Russian withdrawal, Germany didn't really win the war either, and therefore, they lost that recently-gained Russian territory, as well as some of their own. The result of which was a bunch of new countries that popped up around 1920—places like Estonia, Latvia, Lithuania, Poland, even parts of Ukraine. They all declared independence. They were in this middle zone, again, this Eastern European middle zone between these major battling powers. A whole bunch of new countries popped up then, but they ain't gonna last . . .

1945: Goodbye, German Empire!

WORLD WAR II

Because, of course, Russia then became the Soviet Union under Lenin's tutelage and, as the USSR grew in power, it wanted its Eastern European territories back. That brings us up to World War II, because the Germans were still over there hopping mad under Hitler and they wanted their lost territories and prestige back as well. Under "der Fuehrer," they decided to have a re-conquest of Eastern Europe and, what the heck, they'll take over Western Europe too, just for good measure. To ensure they could pull this off, Hitler got together with Stalin—what a fun party that must have been—and together these two signed what is called the **Pact of Non-Aggression**. Basically, Hitler said, "Uhh, okay, I'm gonna take over the world, but hey Stalin, you seem cool enough to me. Yeah, so we won't attack you as long as you don't get in the way of us wiping out Western Europe. And we know you want some of these Eastern European territories back, so you can have those and we'll take everything else." Stalin, in all his wisdom, said, "Word; that sounds cool to me. Game on!"

However, about halfway through the war, Hitler reneged and decided to attack the Soviet Union, pulling them into the war. Why would Hitler have done that? Oh . . . that's right: because he was an insane megalomaniac. You know the end of the story: Since Hitler attacked the USSR, the Soviet Union sided up with the US and Western Europe to beat the Nazis. During this Nazi smack-down, the Soviet Union swept in from the East, the US and allies swept in from the West, and the Nazi smack-down finished up with everyone high-five-ing right square in the middle of Eastern Europe.

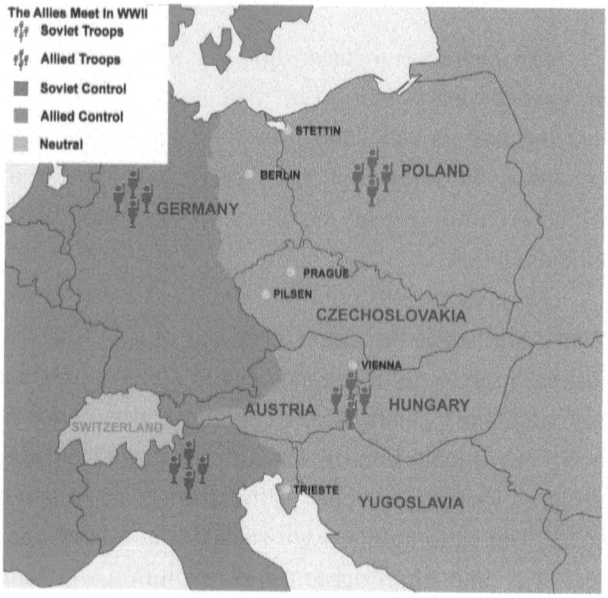

As Borat would say: "Aaah-High-Five" in Europe.

Amazingly, if you look at the line where the Allies met, it's right in the middle of what we call Europe today. The troops met and high-fived each other when they finished killing all the Germans, or at least the completely crazy ones who continued to fight, then that was it. World War II was over. Peace! Peace out!

Well, kind of. It was great, for all of five minutes, before it became a face-off, again, but with new players . . .

THE COLD WAR

Hooray! That successful military partnership between Team West and the Soviet Union crushed the Nazi Empire, won the war, liberated Europe . . . and then promptly disintegrated almost overnight. All the goodwill between the former allies evaporated and mistrust set in, setting the stage for the next 60 years of ideological struggle. You've got the Soviets controlling Eastern Europe, where they swept in and cleaned out the Nazis, and the West who swept in from the West and cleaned out the rest of the Nazis now controlling Western Europe. Although they were playing for the same side during World War II, where they met became the new face-off of major powers. This is more familiar ground we're getting into—the modern era. You, of course, know that this is the Commie Russian realm squaring off against Team

West: the freedom-loving democratic capitalists in Western Europe. But how did this go down? How did this happen since they were all allies during the War?

SOVIETS TAKE OVER

Well, as I suggested, we had the Soviets, who were buddies of Team West at the time and helped wipe out the German threat. When it was over the Soviets said, "Well, you know what, you guys over in Western Europe are kind of crazy. We keep getting invaded by you. This Hitler guy was just the most recent one; the German Kaiser Wilhelm before that, and even Napoleon did it before him! You Western Europeans are nuts!"

A PLAID AVENGER TIP ON WHAT NOT TO DO: Never invade Russia. No one has ever successfully invaded Russia. Everyone who invades Russia loses. Always. Period. Napoleon. Yep. Hitler. Yep. They sucked at it. If megalomaniacs can't pull it off, who can?

To continue the Soviet thought process: "Hey, you people over in Western Europe keep invading us, so we're going to stay here in Eastern Europe, for a couple of different reasons." One: they wanted to reclaim a lot of these territories they lost in World War I. That is Estonia, Latvia, Lithuania, parts of Poland and Ukraine were pulled back into Mother Russia—now the Soviet Union. Two: Russia very openly said, "We're going to stay in Eastern Europe to create a buffer zone so that there are no further intrusions from the West." Now this essentially was a Soviet land grab guised as the creation of a buffer zone, but it was done in such a way that it wasn't as imperialistic-looking as it truly was.

Here's how they did it: just as the Americans, French and British hung around Western Europe to help with reconstruction, the Soviets claimed they were following the same playbook by chilling out in Eastern Europe to mop up and clean up. The Soviet official line was, "Okay, we're here helping out. We'll stay in Poland. We'll stay in Estonia. We're going to stay here in Hungary. We're going to help rebuild just like you guys are going to do in the West. And ya know what? We'll even help them run elections. This'll be great!" Lo and behold, every Soviet-overseen election that occurred in Eastern Europe resulted in a landslide victory for Communist candidates. What a surprise! Go figure. The Communist party just went through the roof in popularity!

Now was any of this real? I don't know, perhaps a little. I wasn't there personally. I was busy deep undercover protecting the Champagne stocks in the underground vaults of France. In the east, the Russians really were liberators of many of these places during the war, so perhaps there was some sympathy, some empathy, and some popularity of communism at the time, but certainly not as much as would have swept a tide of elections across all of these countries. Basically what I'm saying is, this was kind of a farce. Soviets held up these elections and said, "Well, they all elected Soviet leaders, they all elected Communist leaders, so they're just under our umbrella now. We're just here helping out. Isn't that great?" Maybe not so much.

People in the Western European realm, and the US as well, were exceptionally unhappy about these developments. Maybe you younger generations today might think, "Why didn't they just go in and get the Commies then?" You have to remember that they were the allies during World War II; they were on our team, and we could not have won the war without them. And World War II had just finished. Nobody wanted to fight a new war. Nobody wanted to then declare war on the USSR and have to do all that fighting all over again.

So this Soviet 'takeover' of Eastern Europe happened as Team West was busy rebuilding the west—rebuilding and restructuring: laying the foundations for the EU (to rebuild Western Europe), and NATO (to help protect it). It happened when the eastern side was being rebuilt and occupied by the Soviet Union. Where these two teams met during the war, of course, became the division between these two forces . . . and is where something called the Iron Curtain then fell. See the map and inset box below . . .

What's the Deal with . . . the "Iron Curtain"?

In 1946, British Prime Minister Winston Churchill delivered his "Sinews of Peace" address in Fulton, Missouri. The most famous excerpt:

> From Stettin in the Baltic to Trieste in the Adriatic an "iron curtain" has descended across the Continent. Behind that line lie all the capitals of the ancient states of Central and Eastern Europe. Warsaw, Berlin, Prague, Vienna, Budapest, Belgrade, Bucharest and Sofia; all these famous cities and the populations around them lie in what I must call the Soviet sphere, and all are subject, in one form or another, not only to Soviet influence but to a very high and in some cases increasing measure of control from Moscow.

Iron Curtain. But which side was Iron Maiden on?

This speech introduced the term "iron curtain" to refer to the border between democratic and communist (soviet controlled) states. Because of this—and because most Americans have no clue what a "sinew" is—Churchill's address is commonly called the "Iron Curtain Speech."

The "Iron Curtain Speech" was received extremely well by President Harry Truman, but much of the American public was skeptical. Throughout the speech Churchill's tone was very aggressive towards the Soviet Union. Many Americans and Europeans felt that this was unnecessary and that peaceful coexistence could be achieved. The United States had recently considered the USSR an ally. In fact, Stalin, whom the American press had dubbed "Uncle Joe" to boost his popularity during WWII, was probably the most pissed about the speech, feeling he was betrayed by his allies. The "Iron Curtain Speech," besides coining an important Cold War term, set the tone for the next 50 years of US-Russian relations—which is ironic, considering it was given by a British dude. (Note: On the map, Austria appears to be divided, but that was only temporary. The east side of the state was administered by the Soviets only from 1945 to 1955, but then via diplomatic resolution, the Commies left and Austria was united again on the "Team West" side of the curtain after promising to be militarily neutral.)

When all this action was going down, all of those countries in Eastern Europe fell into one of several different categories of which I want you to be familiar. First, some of this territory of Eastern Europe was simply absorbed back into the Soviet Union and became part of other entities and simply just disappeared as independent states. Parts of Poland were worked back into the Lithuania SSR and the Ukrainian SSR. So **absorption** was one option. But what is an SSR?

Ah! That's the second option: some of these briefly independent states became **republics** of the USSR. See, USSR stands for the Union of Soviet Socialist *Republics*. Kind of like states of the United States of America, like Wisconsin is a political component of the USA. Some of these entities in Eastern Europe actually went from being an independent

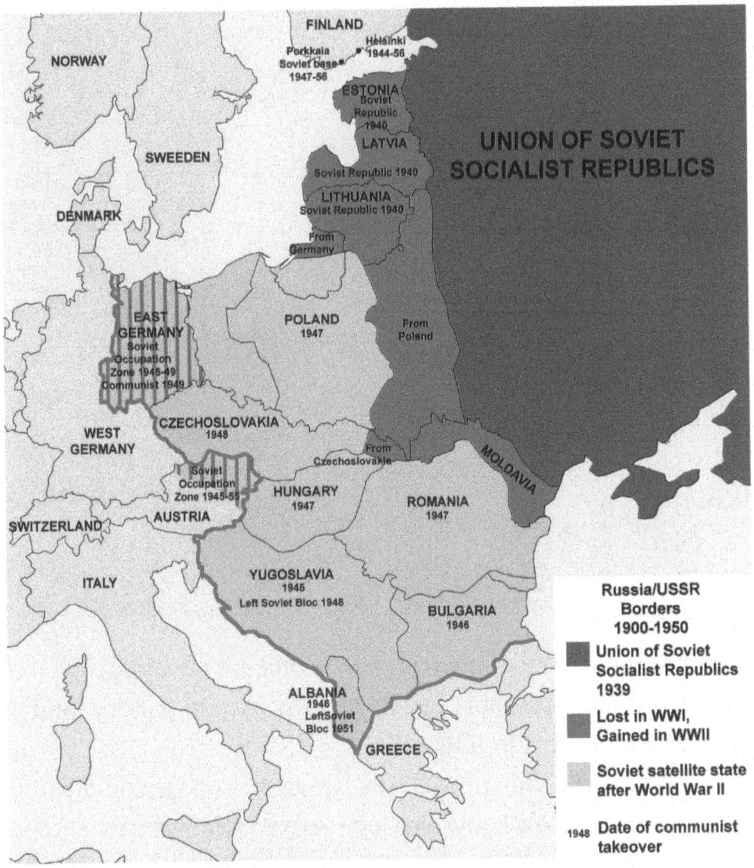

Soviet territory regained, and soviet satellites acquired after WW II.

sovereign state, to becoming a republic of the USSR. I'm thinking specifically of Moldova, Estonia, Latvia, and Lithuania. They ceased to be sovereign states, and became a part of the USSR.

The third and final possibility, and perhaps the most important option, was that some of these sovereign states actually retained their sovereign state status but became **Soviet satellites**. Mainly, I'm thinking of Poland, Romania, Bulgaria, Hungary, Czechoslovakia, and East Germany. These were countries. They were sovereign states. They had a seat at the UN. However, they didn't really have control over their own countries, and everybody knew it. Everybody knew the Soviets were pulling the strings of the Polish government, of the Romanian government, of the Czech government. *Everybody* knew it. They did have seats at the UN, they were supposedly sovereign states, but the Soviet Union truly controlled them, because the USSR had their patsy commie officials running the show in the governments of these places. Does that make sense?

That's the scenario as it was for about fifty years during the Cold War. The Soviets either controlled directly or indirectly (due to their massive influence) every place that is now called Eastern Europe. Team West was on the other side of the curtain: the democracies, the free market capitalist economies that were supported by the US. Those are your two teams during the Cold War. Eastern Europe was close enough to see all the economic growth and political freedoms stuff going on in the democratic countries, all while they were getting paid visits by the KGB to make sure they still loved being Commies. Now do you see how Eastern Europe is a battleground in the middle of this most recent ideological game? Stuck dead in the middle of the commie vs. capitalism showdown!

So long sovereignty, hello SSR!

MARSHALL REJECTED

The other term I want you to think about as well, or at least understand and know, is that the Soviet satellite states did not accept the **Marshall Plan:** a US-sponsored program of aid, loans, and material support to rebuild Europe. Just after WW2, the United States, in its infinite wisdom, said, "You guys are screwed over there! Your whole region has been leveled!" Some smart folks in the United States government said, "We should help them out." In particular, US Secretary of

State George Marshall said, "Look, we gotta help these dudes out. We need to get them back on their feet, because if we don't, their economies will get worse, collapse and then turn into complete chaos, and if we don't help Western Europe out, the Soviets—who we are already worried about—are right next door. The Soviets will sweep all the way through if we don't do something!" Probably a fair statement. That Marshall was a smart cookie.

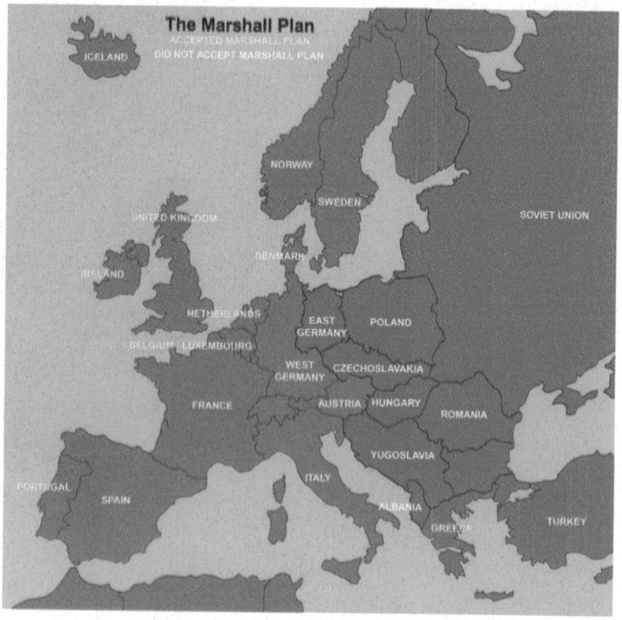

More on the M-Plan: this aid package was offered to all of Europe. The US went to every single one of these countries and said, "Here, we'll give you a ton of money to help you out with industrial capacity and infrastructure to get you back on your feet economically." It was even offered to the Soviet Union. The countries in yellow text on the map accepted the Marshall Plan and took the aid, the sweet cash. Lo and behold, these are the richest countries in Europe right now. The countries labeled in white refused the Marshall Plan, and those are the not as rich countries of Europe right now. While we certainly can't pin everything in the modern era on a singular aid package from the US, it certainly had its impacts back in the day. And it reinforced the division between east and west, as the teams seemed to be now fully set.

If you look at this map of who accepted the Marshall Plan and who didn't, you'll see our same Iron Curtain division between Eastern and Western Europe again demonstrated perfectly. We know that one reason why Eastern Europe is poorer is because the Soviet Union eventually lost the Cold War, mostly because their economy sucked so bad. The second reason is they were stymied out of the gate because they were influenced by Russia to decline the Marshall Plan.

We've already talked about what a Soviet satellite was, and the Soviet influence in Eastern Europe. Make no bones about it; Poland would have probably loved the Marshall Plan. The citizens of the Czech Republic would have loved the Marshall Plan, but they were under Soviet influence, and they were talked into declining it. Maybe there was stricter coercion than just talk going on. Who knows? That's why Eastern Europe is, generally speaking, not quite as developed and rich as Western Europe even today, but they are trying to catch up quickly.

NATO VS. THE WARSAW PACT

Back to the story. Once those Soviet Union and Soviet puppet states did not accept the Marshall Plan, the writing is really on the wall at this point. The lines are drawn. The Iron Curtain has fallen, the rift is exposed, and this is most evident by this nice Cold War relic map shown below of the formation of NATO versus the Warsaw Pact. **NATO** was formed by the US and all of those Western European states who were looking over at the USSR and Eastern Europe saying, "Oh, we feel threatened by them. They might invade us. They might come and kill us. Therefore, we're going to form up NATO, which says if any of you Soviet-type people attack any Western European state, the US is going to come to their aid—in fact, all the NATO countries will come to their aid." That's why NATO was formed, on one side of that Iron Curtain.

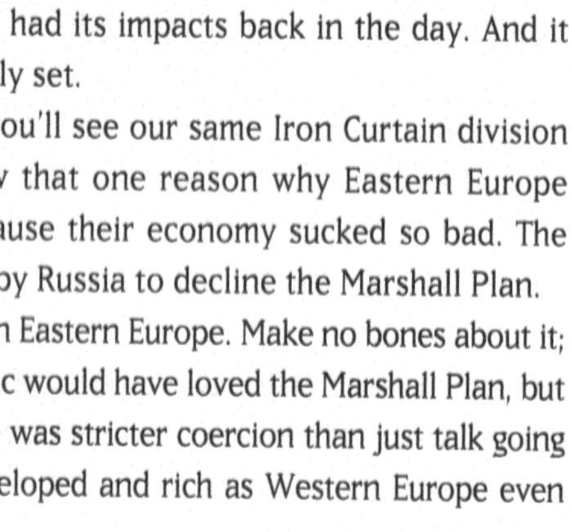

This is a sweet Cold War relic!

The USSR, not to be outdone, said, "Oh yeah? You guys have your own NATO club? Fine! We'll start our own club. We're calling it the Warsaw Pact." The **Warsaw Pact** was essentially an anti-NATO device based on the same premise that an attack on one constituted an attack on all. Of course, everyone knows it's bunk, since the participating countries were simply puppets of the USSR anyway. But the Soviets said, "Oh, no, no, no. Everybody's free around here. So we asked Poland if they wanted to join, and they said yes. We asked Romania, and they said yes too." The goofball Soviets even went out of their way to name it the Warsaw Pact to try and demonstrate that it was the Poles who had thought it up. Yeah, right.

What you had was an entrenchment of ideas—NATO on one side, Warsaw Pact on the other. Free western democracies on one side, the Communist Commies on the other. Why am I talking about this in detail? Because that's what Eastern Europe has been through for the last seventy years. Now you know the end of the story, though. It won't last for too much longer as we approach the modern era . . .

THE THAWING

What's happening today? What we've had since around 1988 is that the Soviet Union began to **devolve**. There's that word again! The Soviet Union fully collapsed in 1990. They finally figured out, "We can't do it anymore. We give up." During this entire Cold War, the Soviets kept insisting that the Eastern European states were voluntarily part of their sphere of influence, and specifically told the states themselves, "If you guys don't want to be part of the Soviet Union, just have a vote and we'll let you out." Basically, that was a bluff. Occasionally, the peoples of some of these countries would take them up on it, and have street demonstrations and stuff like that. Every time that happened, the Soviet Union sent in the tanks and shut it all down. Everyone knew the Soviets were lying. They'd say, "No, really, you guys are sovereign states. Really!" Wink wink, nudge nudge. Not really.

As 1989 approached, when the Soviet Union started to collapse, this notion crept back up again. You may have heard of this **Mikhail Gorbachev** guy. He was floating around saying, "The Soviet Union's in trouble for lots of different reasons. However, if you guys over in Poland want self-determination, maybe we'll *think* about letting you guys go again. If you really want self-determination, go ahead and have yourselves a vote." This caused an explosion. Once a little bit of the leash was let out, everybody ran. What you had was an instantaneous devolution, where Eastern Europe pulled away as rapidly as possible from the Soviet Union. You see here in this independence dates map is that the USSR, a single political entity, had turned into fifteen different countries virtually overnight.

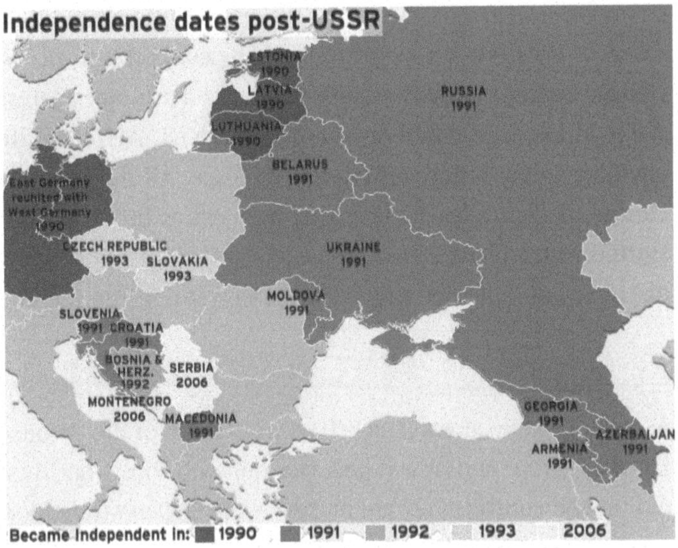

Free at last!

Czechoslovakia, Poland, Hungary and lots of other places said, "Yeah! The Soviet Union is going to let us out! Good, we vote to be out, we're out! We're truly sovereign again, we're free!" Places like Estonia, Latvia, and Lithuania said, "We want out too!" Romania, Bulgaria, Ukraine, "We're out!" Everyone wanted out. But it didn't stop there! Even after all these states are out, places like Czechoslovakia said, "Uhh, we're already out, and we want to devolve further into the Czech Republic and Slovakia for lots of different reasons." (See **What's the Deal with . . . the Velvet Revolution?** box on page 210.) There is another, separate wave of devolution in what was Yugoslavia, which all occurs at roughly the same time, but for different reasons, but we will get to that later.

FLIGHT OF FANCY

This flight from Soviet influence didn't end with declarations of independence. Oh no! That was just the beginning of their run for the border! Let's take a closer look at three states in particular that set the trend for what was to happen next: Poland, Czechoslovakia, and Hungary. These three states crystallize all of the stuff we've been talking about. During the

Hotspots of Uprising

Soviet occupation and the era of Soviet influence, every time the Soviets bluffed about possible independence, people in these three countries took to the streets to challenge it. As you can see from the map, the biggest street protests and riots in the last 45 years of the Cold War era took place in Poland, the Czech Republic, and Hungary. These countries were never entirely happy with being under Soviet influence. In fact, after World War II, they initially were like, "Yay! We're free! We can be just like Western Europe now!" It's hard to be free when the Soviets set up a puppet government for you.

Now, guess which three countries pushed to get out immediately after the devolution process began in the early 1990's? Guess which three countries wanted in NATO immediately? Guess which three countries petitioned to be in the EU immediately? That's right, kids! It's the same three countries: Poland, Czech Republic, and Hungary. As soon as it was possible, they immediately voted themselves out of Soviet influence. They were the first out of the blocks. But that wasn't enough for them. The peoples within these countries had been so unhappy being under the Soviet yoke for over 50 years that they wanted to ensure that it didn't happen again. All three countries immediately petitioned to join NATO.

We talked about NATO in chapter 6, so you know about the impacts of being a NATO member. These three Eastern European states wanted to become NATO members immediately to ensure no further Russian influence. I say again, *ensuring no further Russian influence.* These guys did not want to be anywhere near the Soviets; they wanted to be a part of Western Europe as fast as they could run to it! You know that NATO Article 5 says, "Hey, anybody that's in our club gets attacked, then we take that as an act of war against all of us." Poland, Czech Republic, and Hungary said, "Sweet! We're in the NATO club! Russia can't touch us. We're like the UK, France . . . we might as well be Canada! Russia can't do anything to us!" That was exactly the case. All these countries also immediately petitioned to get into the EU. Eastern Europe, being under Soviet sway, was just as broke as the Soviets. They had not done well during the Cold War. They started with a blank slate in the early '90s—broke, ideologically bankrupt, embracing the west, embracing NATO, wanting out of Soviet influence, and particularly wanting into the EU as quickly as possible for the sake of their economies.

RE-ALIGNMENT WHILE RUSSIA IS WEAK

As you probably already picked up from some of the details in this chapter, we are in an era of transition for this Eastern Europe region. And now it becomes quite obvious what they are transitioning from and to—from that Soviet era of occupation and control, they are mostly realigning and transitioning to become adopted into the capitalist democracies of the west. This happened at a brisk pace after 1991 when the Soviet Union officially voted itself out of existence, and the entity broke up from one huge power into fifteen sovereign states, as already pointed out.

We also already pointed out, many of the states, Czech Republic, Hungary, and Poland, were quick to embrace the West; Estonia, Latvia, and Lithuania were right on their heels, by the way. Because, as you now have learned in this chapter, they used to be independent sovereign states in between World War I and the end of World War II, but then were reabsorbed by the Soviets. Those guys were always chafing under Soviet rule, so Estonia, Latvia, and Lithuania were the second round of states that jumped ship into the arms of the West. Many others have since followed. All of this realignment and transitioning of these countries of Eastern Europe was occurring right after the Soviet crash . . . quite frankly, when Russia (and Russia became a new country, as well, at this time) was excessively weak. I mean, they just lost the Cold War, and they mostly lost the Cold War because they were freaking broke. You'll read more about that in the next chapter. Their economy was in shambles, their power structure was shattered, and their government was in chaos. It was precisely in this period of Russian weakness that so many of these Eastern European countries just *ran* to the west.

EU creep into Russia's former sphere of influence. Those Euro-creeps!

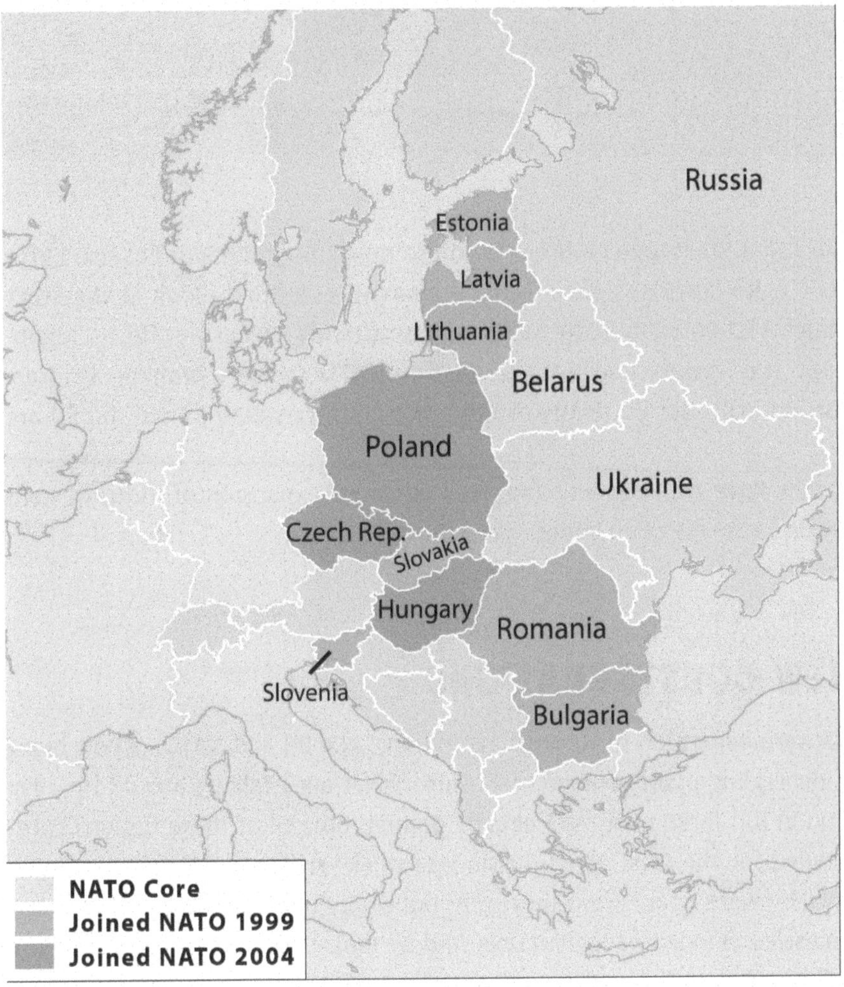

This transition, of course, happened in a couple of distinct ways by some distinct entities we've talked about many, many, many times already in this great text. Number one was the EU—the European Union. Most, if not all, of these countries *immediately* applied for EU entry status. Many of them were soon granted it. As you can see by the maps on this page, there has been a progressive wave of Eastern European countries entering the EU since the very beginning of their "freedom" from Soviet domination.

As already suggested, Hungary, Poland, and the Czech Republic were the first three that jumped in. They were followed by a whole host of countries that jumped in by 2004, and the fun's not over yet. Romania and Bulgaria entered in 2007, and there are still multiple potential candidates. Most of the countries that constituted the prior Yugoslavia have asked for entry. It's quite important to note that Ukraine, a huge country and a former part of Russia historically, has applied for EU membership, an issue which has divided the country and caused it to fall into civil war. We'll come back to Ukraine in a bit.

Perhaps the most telling tale to tell about the push for the EU in Eastern Europe is Czechoslovakia, a country that used to be one, that's now two—another shattering within Eastern Europe—simply to get into the European Union. That sounds bizarre—you've got to bust up your country just to join a supranationalist organization? Indeed, it is a bit bizarre, and you should know why, when, and where that happened. It was called the **Velvet Revolution**.

What's the Deal with . . . the Velvet Revolution?

What's up with Czechoslovakia? Why did it split into two, the Czech Republic and Slovakia? Were there two different ethnic groups that wanted their own self-determination and their own countries? Not really. While there are ethnic Slovaks and ethnic Czechs, neither group had much of an idea that a partition was even occurring. Most couldn't have cared less. There was no animosity between these ethnic groups. So what's the deal? Why did they do it if there wasn't a problem? Why did they split?

They split because the western side of Czechoslovakia, where the Czech Republic is now, was much more like the West. It bordered Germany; it had the industrialized sector, and was richer, with a higher GDP. Slovakia, the eastern side of the state, was much more Russian/Soviet influenced. It was poorer and more agricultural. Folks in the Czech Republic, as the Soviet Union was collapsing, said, "Hey, we want to get into the EU. Help us out, Plaid Avenger! What should we do?"

I said, "Well, let's look at this strategically. To get into the EU, your economy has to be decent and it has to be stable. As a whole, your economy is neither. But if you were to lop off the poor suckers on the eastern side of your state, your GDP per capita would go up! All your averages would go way up!" Indeed, that's exactly what happened. People in what is now the Czech Republic said, "Hey, we'd be better off on our own without Slovakia dragging us down!" Czechoslovakia underwent a transition called **The Velvet Revolution** in 1993. This revolution was like velvet, nice and smooth. There were no shots fired. From the stroke of a pen, one country became two.

Where did it go?

That brings us to the other avenue for the Team West love embrace by these newly independent Eastern European countries in the 1990s, and that was NATO. No surprises here. In fact, you can essentially look at the story told by those maps of EU entry and see that the NATO entry tells the same tale. As already suggested, Czech Republic, Hungary, and Poland were quick to escape the Soviets and jump under the NATO security blanket. Estonia, Latvia, and Lithuania right on their heels—just like with the EU. Many of the other countries who joined the EU are also now NATO members.

This has been aggravating Russia a bit as they have watched the continued eastward expansion of those western clubs to their Russian borders. And that, my friends, is worthy of an entire new section. . . .

RE-INVIGORATED RUSSIA CAUSING NEW CONSTERNATIONS

What's all this about? Well, this eastward expansion of western institutions, like EU and NATO, is now being slowed, stalled, or fully stopped here in 2019, depending upon your point of view. What am I talking about? Hey, I'm talking about Russia, man! The Bear is back, and in full force now! As I already pointed out, all of these Eastern European countries jumped the Soviet ship and headed for the west when Russia was weak—right after the Soviet Union crashed, when Russia was down and out, when they were poor, when they were politically bankrupt. But that time has passed. They're back and they're flexing their muscles in more ways than one. And by that, I am thinking specifically of Vladimir Putin. Have you seen this dude topless? Watch out, the man is a menace! But seriously . . . okay, seriously, he is ripped . . . Russia as a great power is back on the world stage, and is no longer allowing encroachment into their arena of influence. We can look at four distinct regional themes of the last decade that underscore this Russian resurgence. A fearsome foursome of Russian re-encroachment back into Eastern Europe, starting with . . .

ISSUE 1: UKRAINE YEARNING: THE ORANGE REVOLUTION

Like the Velvet in Czechoslovakia, the Orange Revolution in Ukraine was defined by the east/west differences of folks within the state, although unlike the Velvet, it did not end with the permanent division of the state into two new states. Yet. While the Velvet Revolution occurred over two decades ago, the Orange one just happened in 2004–2005. Russia has exerted influence and/or control over Ukraine off and on throughout most of its history. Russian influence is very strong in areas physically closer to it, like eastern Ukraine with whom it shares a border. Back in 2004, Ukraine had a big move towards true democracy. The scenario included two dudes. The first was Victor Yushchenko: the pro-democracy, pro-EU, pro-NATO candidate. A real poster child for the west. The EU supported him, and the United States thought he was awesome. They loved this dude!

His opponent was Viktor Yanukovych: an old-schooler, very conservative, more Russian influenced, and in fact, he's the candidate that Vladimir Putin from Russia came over and campaigned for. To try to help lock up the election before it even was held, former elements of the KGB even unsuccessfully poisoned Yushchenko! Talk about a street fight!

How hilarious is this? I can almost hear the boxing announcer: "In the red, white and blue trunks, fighting for Team West, hailing from western Ukraine, it's Victor Yush! And his opponent, in the red trunks, representing the red Russians, from eastern Ukraine, its Victor Yanu!" Hahahaha dudes, could you make this stuff up? Victor vs. Victor—I wonder who will be the victor?

Why's the Plaid Avenger bringing up this election in Ukraine of all places? Because it exemplifies what's still happening in Eastern Europe, which is this battle between the West and Russia; it's a battle that's not so much for outright control as it is for *influence* within this region. Russia still has serious economic and security interests in this area, and still likes to think of it as within Russia's sphere of influence. When this election went down, the pro-Russian candidate won, and everyone in the world said it was a fraudulent election. There were massive street protests, and this ultimately turned into the 2004/05 **Orange Revolution**. So much heat got put on the Ukrainian government, they threw out the election results, re-ran it, and then the pro-western Yushchenko won. We can see this as part of an ongoing battle for influence in Ukraine, which is representative of the broader battle for influence in the region as whole.

The Russian candidate was downed because the pro-western candidate won—downed, but not out. In the Post-Orange Revolution Ukraine, there are still a lot of pro-Russian people. There's a pro-Russian political party, and in March of 2006, the same party gained enough seats to win back control of the Congress. The Plaid Avenger knows

Bush

Yushchenko

Yanukovych

Putin

Bush supported Victor Y. while Victor Y. was backed by Putin. Y ask Y?

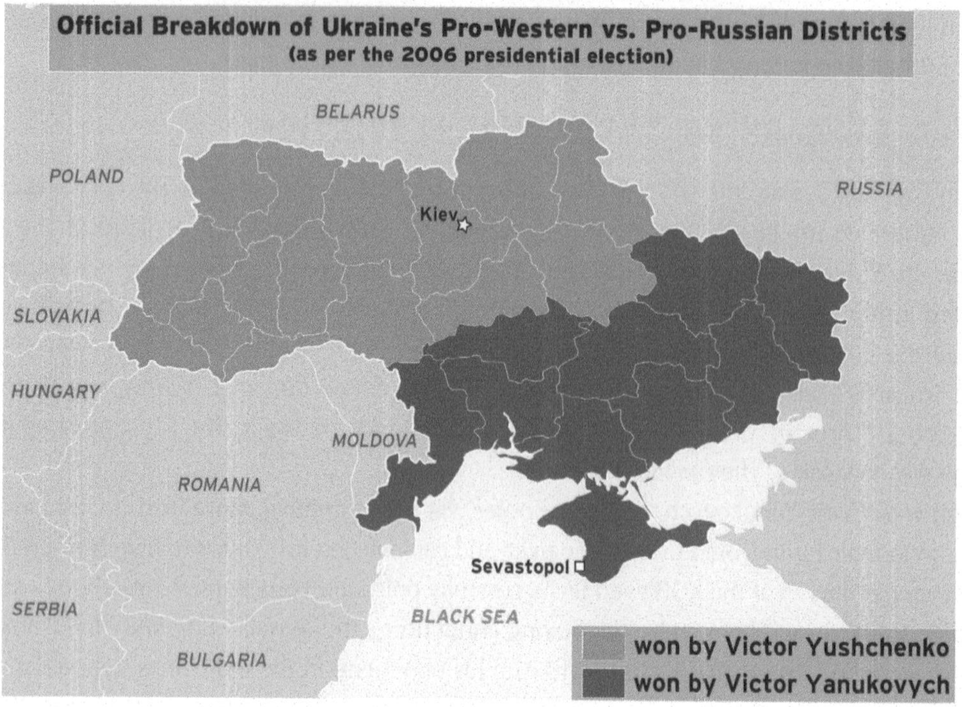

that those in the US think democracy's great, that the good guys won, and that's that. Not so! This is an ongoing battle for control. It's not over! Most other countries have gone the way of the west, but it's definitely not over yet for the Ukrainians.

And the country is about evenly split between the western pro-West and the eastern pro-Russia, as you can clearly see from the results of the 2006 election cycle in the map above. Could the divisions in this country be any geographically clearer?

Because of the pro-Russian political party gaining the majority of seats in their Congress, they got to choose the Prime Minister position . . . which promptly went to Yanukovych! (The President is voted in by direct election, Prime Minister appointed by majority of Congress.) So after the 2006 election, you had this bizarre situation where President Yushchenko was pro-West, and Prime Minister Yanukovych was pro-East! Talk about a country with a split personality!

This internal division EXACTLY symbolized the fight for influence in Eastern Europe between Team West and Russia. Yushchenko continued to push for EU and NATO entry; Yanukovych actually announced that no way in hell his country would join NATO. The US sent aid to Yuschchenko and his allies; the Russians sent aid to Yanukovych and his party. The Russians and pro-west Ukrainians even had several economic battles over Russian supplies of oil and natural gas to the country.

Second time's the charm for now-President Yanukovych

Meanwhile, the entire country's economy has tanked and they are more reliant on outside aid than ever to keep themselves afloat, which means this pitched battle for influence has gotten even more important. Are you Yanuk-ed out yet? I hope not because the story gets crazier as we play it forward to the present, as the truth is always stranger than fiction. Dig this . . .

The Ukrainians went back to the presidential polls in 2010, and take a wild stab who won. Victor Yanukovych, of course! Seriously? Yep, seriously, and apparently legitimately this time. So is the Orange Revolution dead? Umm . . .

kind of, I guess. I mean, Ukraine is still a democracy, and they did have a free and fair election, but unfortunately for those revolutionaries, the same pro-Russian guy they had the revolution to throw out, won the next election and got back in!

The first order of business for Yanukovych as president was to solidly state there would be no NATO in Ukraine's future, which coincided with Russia giving a huge price break to Ukraine on oil/gas contracts and promising them a bunch of foreign aid as well. Isn't it funny how coincidences like that happen? Yanukovych also immediately extended leases on critical Russian naval bases located on the Crimean Peninsula, a move which infuriated pro-Western politicians who retaliated by throwing eggs at the Speaker of the House when the bill was passed. No. Really. Google it.

In particular, there is a serious fight brewing over the future of Ukraine's southernmost province otherwise known as the Crimean Peninsula, but we will save that story for world hotspots in chapter 25. By all means, flip to it now if you can't wait to find out more! **2012 UPDATE:** tensions within Ukrainian continue to simmer, as European leaders are currently boycotting the Euro 2012 football finals held in Ukraine, in protest against Yanukovych's jailing of former Prime Minister Yulia Tymoshenko. Also, a massive fist-fight broke out in their Parliment over a law put forward by Yanukovych to recognize Russian an official language of their state. I told you this fight was far from over!

Ummmmmmm . . . BOOM! Called It!

2015 Insane on the Brain, Ukrainian Unraveling Update! So glad I pointed out the Ukrainian rift for all these years to all the plaid students who used this text! Because that prediction was dead on the mark . . . literally. Unless you have been comatose since your bobsledding accident in the 2014 Winter Olympics, you have at least heard that Ukraine melted down politically once more. Massive street protests culminated in the overthrow of President Viktor Yanukovych, which prompted the Crimean Peninsula to declare independence and ask for absorption into Russia! Which then happened! The Ruskies have had tens of thousands of troops massed on the Russian/Ukrainian border for months, and now there are similar "join Russia" protests across eastern Ukraine that have turned bloody and violent. Will Russia invade and convert other parts of the state into Russian territory? Who knows? It is all such convoluted chaos right now, that anything could happen! Team West and NATO are furious, the Ruskies are indignant and flexing their muscles further, and some are predicting that this is the start of a new Cold War!

Ukraine aflame!!!

I think that may be a bit of an overstatement, but this is certainly a titanic regional game-changing event. Here in late 2019, civil war in Eastern Ukraine is still open, active, and hot! Russian troops are still massed at the border, who knows how many guns and covert operatives Russia has already sent into the mess, and there is talk of the US and NATO training and arming the Ukrainian government. This civil war is far from over, and will shape the region for some time to come. And it is Russia that is doing the shaping. . .

Want to learn all about the Ukrainian situation in more detail? Visit: http://plaid.at/russia to read **The 2015 Cover Story: Ukraine Unraveling, Resurgent Russia**.

Russia keeps laying the pipe.

ISSUE 2: RUSSIAN PETRO POWER

By this, I mean that Russia possesses vast amounts of natural gas and oil, and they are increasingly using this commodity as a political pressure point, or an outright political weapon of choice! For now, know this: Russia provides something on the order of about one-third of all European demands for energy. That's a lot, dudes. By Europe, I mean Eastern and Western Europe depend on Russia for about one-third of its energy needs. What's that got to do with current events? What's that got to do with a reinvigorated Russia? Just this: they have petro power now. To reassert their strength and influence in Eastern Europe, they are playing the fuel card. Meaning, when those Eastern European states get a little too uppity, when they do things Russia doesn't like—for example, if the Ukraine says they want to join NATO—Russia says, "Ho, ho, ho . . . hold on there comrades. You get your oil from us. You make us angry—and the price of your heating fuel might double overnight. You make us more angry—and you get no oil at all. You dig that? And it gets awfully cold there in the winter too, don't it? How do you like them apples?"

Does this sound like fiction? It shouldn't, my friends; it's already been occurring. Russia has arbitrarily raised prices at various points in the last decade to essentially punish states that are doing stuff that they don't want them to do. They did it to Ukraine and Belarus in 1994 and again in 2006 when they doubled the prices to their Belarusian brothers. They completely dried up shipments to Latvia in 2004–06 to pressure the Latvians into a port deal. Most recently, the Russians shut down a section of pipeline that supplied a Lithuanian refinery under the guise that the infrastructure was just too old and needed to be replaced. I'm sure that was totally unrelated to the fact that Lithuania was threatening to veto an EU-Russian economic pact. Total coincidence, I'm sure. Oh, and if you beleive that, then you will also believe that Russia tripling the price of oil to the pro-West government of Ukraine in 2014, and demanding full payment of all outstanding bills, is also a huge coincidence. But of course you don't believe that, because you now know how the rascally Ruskies operate!

Even though the price of oil has dropped drastically since its historic highs of $150/barrel back in 2008, Russia has made trillions over the last decade, with more rubles to come . . . but it's not just about that. Yes, they get rich from it, but more importantly, it has given them political power due to the economic leverage that controlling that commodity brings. Russia is like the crack dealer to Eastern Europe in particular: they've got the stuff that those Europeans need. If you anger the dealer, you might not get your energy fix. That's not a situation the Eastern Europeans are happy about. The Russian influence is back. That's not all that Russia is doing to flex its muscles in this region.

ISSUE 3: NYET! TO NUKE SHIELD

Another thing that's extremely topical—and we haven't seen the end of this story yet, either—is the proposed missile defense shield system that is being pushed hard by the United States to be implemented specifically in the Eastern European countries that are NATO members. Specifically right this second, we are talking about the Czech Republic and Poland. This is specifically incensing the Russians.

Maybe I should back up—what is this *missile defense shield* nonsense about? This is a concept the United States has been toying with for 60 odd years. It has never really come to fruition. But the idea is, you set up a bunch of missiles and a radar system which can detect if any missiles are coming into Europe. Then, you launch your missile to blow up the incoming missile before they land. Again, it's never worked; out of perhaps ten million tests maybe they've done it once successfully, but that's not the point of this rant right this second.

It's the fact that these new NATO members in Eastern Europe have said, "Uhh, okay, yeah, we'll do it since we still support the US." At a NATO summit which occurred in February of 2008, all the NATO countries—all of them—agreed to go forward with the US plan for the missile defense shield. And just to kind of wrap this up, Russia is seriously *aggravated* about it. They have been pounding their fists on the table, saying, "Nyet, this is not going down. This is seriously threatening us and we are going to use economic leverage and political leverage to intimidate the Eastern European countries that are going along with this."

Early version of European Missile Defense Shield.

What do they mean by that? Well, they've already made the fuel threat. When the Ukraine said they wanted to join NATO, Russia said, "Yeah, that's fine. You join NATO and we will re-target nuclear missiles at you." This is something that Russia has also promised to do to any of the states that jump into the missile defense shield program. This is serious stuff, man. Again, this story is still ongoing. The United States says, "Hey, look Russia. Come on, dudes. The Cold War's over. We're not enemies anymore. This is mostly a defense system against . . . Iran, ooh, Iran. Yeah. Iran's going to send a missile somewhere, so this is about that. Or other terrorists might bomb our NATO allies."

The Russians are still riled up about that game.

But Russia says (and you've got to empathize with them a little bit here), "Uh, yeah right. Come on. NATO's been encroaching to Russian borders for the last twenty years. Now you're putting up this big missile defense system. It's an obvious attempt to neutralize or marginalize Russian power in this region and in the world."

The Plaid Avenger's take on this mess is that the United States is probably telling the truth, but Russia sees it in a very different light . . . and, again, I always try to empathize. Not sympathize, not agree with, but empathize. The Russians do have a long history of being screwed with by folks over in Western Europe, from Napoleon, to Hitler, right on up through the 1980 US hockey team's miracle victory over the Soviets at the Winter Olympics. So in no way, shape, or form, or any time soon, are they going to be okay with an encroaching military technology that is set up on their borders. And they're really ticked off about it. Vlad "the man" Putin has been saying, "No, we're going to do everything we can in our power to make sure that does not happen, including veiled threats about missiles and open threats about energy issues." As always, sucks to be in the middle of the mess, but Eastern Europe seems to have a historical niche for it. **2012 UPDATE!!!** At the big NATO summit in 2012, the NATO and the US officially announced that the first phase of the shield network in now up and operating! Two more phases in the future will add more radar stations and missile interceptor launch sites! Russia no like! Yikes! Are we entering a new Cold War phase for Eastern Europe? Let's watch and find out! **2015 UPDATE!!!** Given the new tensions between Russia and the US over Ukraine, look for NATO to increase exercises and troops deployment to Eastern Europe, and look for the US to invest billions more into the missile defense system . . . even if the system don't work, it will infuriate the Russians, so money well spent either way. **2016 UPDATE:** US goes live with the long-awaited missile defense systems in Romania. In response, the Russians have radically increased war-game exercises and even moved ballistic missiles to Kaliningrad, which is a tiny Russian province sandwiched between Poland and Lithuania along the Baltic Coast. **2021 UPDATE:** During the 2017-20 reign of US President Trump, and his attempts to become best-ies with Vladimir Putin, NATO was widely dissed by the USA, and talk of this missile shield has halted entirely. No shield for now, but watch for this issue to pop back up under future presidents.

ISSUE 4: YUGO-SLOB-IAN MESS: KOSOVO CHAOS

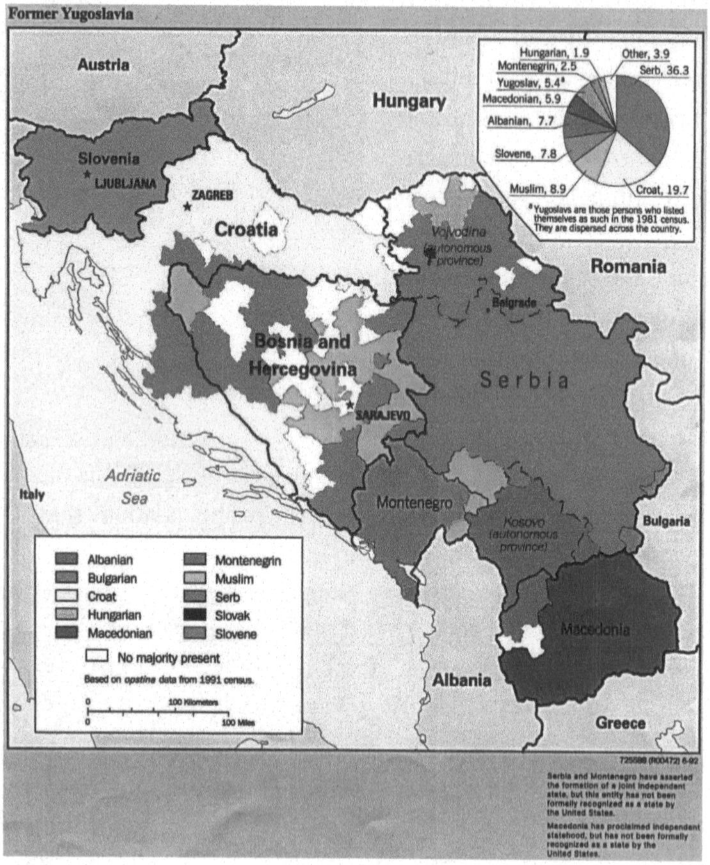

The fourth and final theme about what's happening in Eastern Europe with this reinvigorated Russia, comes the latest chapter in the disintegration of what was Yugoslavia. FYI: "**Balkanization**" is a term for the shattering apart of a place because of precisely what happened here in the Balkan Peninsula. I've intentionally kept the mess out of the chapter up until this point because it's a debacle that muddies the Eastern European waters a little too much to tackle it up front. But now that you understand the rest of the region, I am confident that you can now figure out this confounding coalition of convoluted crap with the greatest of ease. Well, we at least have to give it the old college try, since events here are once more crystallizing the fight between the West and the East. Let's do it.

The Plaid Avenger said earlier that Yugoslavia was different than the Soviet Union. Let's revisit them so you'll understand what's happening in today's former-Yugoslavia. First off, they were never part of the Soviet umbrella. They were commies, with a commie-run government after WW2, but not officially in the union. It was a communist state under the tutelage of **Marshall Joseph Tito**. "Marshall" is really just a nicer sounding title than "oppressive dictator extraordinaire." The Plaid Avenger suggested earlier, as we were talking about the historical background of this region, that perhaps no other singular country in Eastern Europe has more different ethnicities and religions and cultures as did the former Yugoslavia. Emphasis on *did*.

They had Slovenes there, Croats, Bosnians, Herzegovenians, Serbians, Montenegrins; I don't even know what half these terms mean! On top of that, you have Christian Serbs, Muslim Serbs, also Christian Bosnians, a group that encompasses Orthodox and Catholic Bosnians, Muslim Bosnians . . . It's confusing! It was a multi-ethnic, multi-religious, multi-media mess.

Why has there only been serious conflict and devolution there in the past thirty years? It's all because of Joseph Tito; not the Tito of the Jackson family, the other Tito, the dictator. Tito was a strong-armed dictator, like Stalin and lots of other folks in the communist realm. FYI: Tito was the only reason Yugoslavia ever worked. He convinced all of these ethnic groups that they were a new nation called "Yugoslavs" (all the slavs) and his charisma basically kept them together. When ethnic or religious tension reared its ugly head in what was Yugoslavia, he sent in the army and crushed any uprising or disturbance with the iron boot. Through force, Tito held together this state full of all these different ethnicities and religions. Tito was the only reason it didn't devolve years ago. However, he died in 1980. Since Yugoslavia was a military state, it was able to perpetuate itself for another decade, but eventually imploded from the stress of diversity.

Tito double vision: Josip or Jackson?

Look at the time frame and do the math. Tito died in 1980; the state held on for about ten years. That brings us to about 1990, the same time that all the other parts of Eastern Europe were devolving, splitting, splintering, fracturing, shattering into separate countries because of the implosion of the USSR. It all hit the fan in Yugoslavia as well. Without the strong-armed dictator Tito there to hold the different groups together by force, the whole country collapsed on ethnic lines, followed by wars fraught with ethnic discrimination, vicious bloodshed, and human rights violations. Long story short, Eastern Europe in the last 20 years saw the USSR devolve from one entity to fifteen sovereign states, Czechoslovakia devolve from one country to two, and Yugoslavia devolve from one country to five in the 1990's. In 2006, Montenegro became #6. And the fat lady has not sung yet, my friends, which brings us up to current events. . . .

What's the Deal with Slobodan?

You should know this dead dictator dude. He was arrested for war crimes, and died in The Hague before he even got through his trial. What did Slobodan do that was so important that the Plaid Avenger has to tell you about it? We know from Chapter 3 that the ultimate test of sovereignty is that you can kill citizens in your own country with no international repercussions. The government can do anything it wants to in its country if they are a sovereign state; it's their right.

So what's the deal with Slobodan? He's another strong-armed guy like his predecessor pal Tito. Except when he came to power, his state was disintegrating. He helped perpetuate the conflict that led to the breakdown of Yugoslavia. During that period, he essentially allowed genocide to occur. He was an Orthodox Serb, which was the majority ethnic and religious group. He allowed the Serbian people to start picking on and beating the crap out of the ethnic Albanian-Muslim minorities in the Kosovo region, which were perhaps petitioning for independence. He not only allowed violence; he promoted it.

This started to spiral out of control in the early 1990s to the point that people in the outside world—particularly Billy-boy Clinton from the United States and lots of folks from the UN—began debating the limits of sovereignty. The rest you know from our talk on sovereignty back in chapter 3. After the US/NATO intervention, Slobodan was eventually arrested for war crimes and died while on trial in 2006. Oddly enough, when his body was brought back to Yugoslavia, lots of people were cheering him. He's considered a war criminal by most of the world, yet lots of Serbs have him held up as a hero. That's typically a bad sign. We'll talk more about that later.

"I'm dead, and I'm still pissed off."

As you can see, the Yugoslavians have had kind of their own tale that's a bit separate from the USSR, but of course related, and that brings us full circle to today's world. And the shattering may not be over yet! Slippery #7 may be on its way, but this time it won't be without a fight. Kosovo may be the next to go. You by now know the story of Slobodan, genocide in Kosovo, and the US/NATO invasion of Serbia to stop it. The end of the tale goes like this: In February of 2008, Kosovo declared independence. Immediately, a bunch of Western European states and the US recognized their independence, which, of course, *infuriated* both Serbia and Russia. Russia is still not very happy about this. NOTE: Not all Western European states recognized Kosovo, and therefore, the EU as a singular entity did not recognize it either. The big boys like the UK and France did, but many other minor European powers did not, for reasons we shall get to in a minute.

Kosovo still causing consternation

We have this situation that, again, crystallizes this new power struggle within Eastern Europe that's related to the old power struggle. Russian influence versus Western influence. The Russians, their historically Serbia (that is, what's left of Yugoslavia), and other entities around the world said, "You can't recognize Kosovo. You guys were promising all along you wouldn't." Because, of course, the United States and NATO intervened in Kosovo, kicked out Slobodan, and protected the Kosovars. But they always claimed that they were merely preventing genocide, and not carving out a new state. Russia argued, "Hey, all along when you guys did that intervention thing, you said that you weren't setting up a new state, that you weren't going to let them have independence, that you were just going to stop the civil war, and basically you're liars."

Russia is not alone on this issue either. China, as well as many European countries, have refused to support Kosovar independence because they think it sets a bad precedent. If there's one small, little group of people that are ticked off at the government of a country and you're going to allow them to declare independence, then there's going to be a whole lot more shattering going on, and not just in Eastern Europe. You're setting a precedent for other folks to do it around the world. Russia don't want the Chechans declaring independence; China don't want Tibetans declaring independence either. It's interesting to note that places even in Europe like Spain did not support Kosovo independence because they said, "Hey look, we have a small group of radicals who want an independent Basque country, so no way. There's no way we're going to support an independent Kosovo, because then the Basque people will say, 'We're going to be independent too.'" You see how complicated this issue has become. And while the Basque peeps have stopped petitioning for independence from Spain, now another area named Catalonia has been pushing to split off from the state. . . Spain can't catch a break!

Back to the point: this is all about Russian influence here. The Russians are supporting their old Serb allies, saying, "No, this is nonsense. We ain't recognizing Kosovo and nobody else should either." And the West is saying, "Oh, but, you know, they want to be a democracy, and peace-loving and all that stuff. We are going to support 'em.'" Yep. Eastern Europe right back in the middle again. I guess old habits die hard.

The point of this last section is that Russia is back. They are re-flexing their muscles. They are reasserting influence, increasingly by using outright force, over their old Eastern European sphere of influence, and the West is still pushing in as well. **2021 UPDATE**—Ok, I guess we can now report that the Russians are actually also reasserting real control, since they officially re-absorbed the Crimean Peninsula of Ukraine back into the Russian motherland in 2015! And perhaps they are going to "re-absorb" more parts of Ukraine, as the state is still in the middle of a nasty, low-grade civil war. Maybe the Ruskies will take back Moldova too! Who knows? Honestly, not even the Russians know quite yet what they are going to do, and all of us will just have to watch as events unfold. But Eastern European states are once again back in the middle ground battle-zone between competing powers . . . that much we can say for sure. Same as it ever was.

CONCLUSION: THE TRANSITION IS ALMOST FINISHED!?

2021 UPDATE: Nope, it is not. By 2005-2010, it seemed that the Eastern Europe transition was nearing an endgame, one which saw virtually all of its states back firmly in the Western Europe orbit. But then Russia reasserted itself back into the scene. Many of these states had joined the western block institutions, with perhaps just a few more to go. Many countries are in the EU. Even more are in NATO. So many states are officially (kind of) part of Team West now. Some may be leaning back towards Mother Russia. Some. And now Mother Russia is leaning back into them, big time.

So Eastern Europe has historically been squeezed in a vice and perhaps it may be ready to explode again. This region does have a strange history of being under the influence of these greater powers to its east and its west, which is why I started the chapter by calling it "bookended." Somehow, Eastern Europe is the scene that sparks all the trouble between the sides. Seriously, think about it. Go back to World War I: It was started in what is now Yugoslavia, when Archduke Franz Ferdinand (it's not just a band) was assassinated in Sarajevo by a young Bosnian Serb from the

Serbian radical group named the **Black Hand**. World War II started when Hitler invaded Poland. Then the Cold War battle lines were drawn in Eastern Europe, with most of the missiles either deployed on it, or pointed at it. The current Ukrainian Civil War is still pitting the pro-West and pro-Russia teams against each other, and as usual the blood is spilling on Eastern European soils.

Why list all these transgressions again? To stress a not-so-obvious point: all major conflicts of the last century were sparked in this bookended region. Hmmm . . . that certainly is food for thought, isn't it? Major players seem to get sucked into confrontation over this swath of land between Russia and Western Europe. Sucked in, and then life really sucks for all involved.

The disintegration of both Ukraine and Yugoslavia also started here in Eastern Europe—go figure, because they are located there—and it's not done yet. That's why I bring up this Kosovo situation, which is once again pitting an east versus west, a Russia versus Team West. Both the Ukrainian and Kosovo situations are promising to polarize these sides unlike really anything else that's going on here at the dawn of the 21st century. I know what you're thinking: "Hahahahaha what a load of bull, like a major war is going to start over some insignificant little hole in the wall like Kosovo, or Crimea, or **Transnistria**?" Yeah, you're probably right. My bad. How stupid am I being? I mean, the odds of that happening are about as likely as a global war that killed millions being started over the assassination of an unknown Austrian Archduke while motor-cading through Sarajevo. Ummm . . . oops. Bad example.

The question is, when will the conflicts conclude? How many more subdivisions can occur? How much more shattering before it's over? Quite frankly, the answer is perhaps never, because it's outside forces which seem to promote, antagonize, and continue to manipulate this battleground buffer zone that we call Eastern Europe. As alluded to earlier, Ukraine is now the country to watch as the next chapter of Eastern European devolution, and it promises to be the most exciting, action-packed episode yet! And never, ever discount the possibility of Bosnia or Serbia blowing back up again; tensions continue to simmer under the lid of the former Yugoslavia. We keep talking about Russia being involved as one of the players in this battleground of ideologies in Eastern Europe, so perhaps we should now turn our attention to the Bear. Oh Vlad . . . are you out there . . .

Eastern Europe today . . . at least for now.

EASTERN EUROPE RUNDOWN & RESOURCES

View additional Plaid Avenger resources for this region at http://plaid.at/ee

BIG PLUSES

- Due to lower wages and overhead, Eastern Europe is attracting lots of foreign investment and jobs from Western Europe
- Eastern Europe produces a lot of agricultural products and is becoming a manufacturing hub for Western Europe
- Incorporation into the EU promises to help raise standards of living and the economy for most Eastern European states to eventually reach Western European standards
- Incorporation into NATO promises to ensure political and military stability for its Eastern European members

BIG PROBLEMS

- Many states of the region have a long way to go to achieve western-style standards of living
- Possibility of ethnic/religious/nationalist conflict rearing its head again is highly likely, particularly in the Balkans
- Non-NATO states in the region (Ukraine, Belarus) and even some NATO members (Poland, the Baltic states) will continue to be torn between their western allies and having to suck up to Russia
- Many Eastern European states are heavily reliant on energy supplies from Russia
- Ukraine. New Cold War warming up? 'Nuf said for now.

DEAD DUDES OF NOTE:

Archduke Franz Ferdinand: Leader of Austria-Hungarian Empire who was assassinated by a radical Serbian nationalist in 1914; the Empire then declared war on Serbia, which triggered all of Europe to declare war against each other, thus World War I.

Marshall Josip Tito: The communist strong-armed dictator that held together Yugoslavia for decades, despite the wildly diverse ethnic/religious mix of the state. And when he died, it all came unraveled. He was Serbian and an ally of the Russians, but never a part of the USSR.

Slobodan Milošević: Former Serb President of Yugoslavia, this is the dude whose encouragement of Serbs to persecute Muslim Albanians in Kosovo resulted in the US/NATO bombing of his country into submission. He died in jail during his war crimes trial at the Hague. Still is looked upon as a nationalist hero by some Serbs.

PLAID CINEMA SELECTION:

Everything is Illuminated (2005) A freaky Jewish-American man (played by hobbit Frodo Baggins aka Elijah Wood) road trips to find the woman who saved his grandfather from Nazis during WWII in a Ukrainian village. Bizarre and intriguing, and great landscapes shot on location in Ukraine and Czech Republic.

No Man's Land (2001) Set in war-torn Bosnia-Herzogovina in 1993, a Serb and a Croat are caught in a catch-22 situation in a trench between opposing armies' with a guy laying on a live land mine. Also features: ineffective UN, physical scenery of Slovenia, and realistic rendition of how confusing this war is to outsiders and insiders alike.

Before the Rain (1994) Time twisting tale in 3 parts that shows the internal political turbulence of Macedonia as it pulled away from war-torn Yugoslavia and became independent. Shot on location in Macedonia, builds on the futility of the cyclical nature of violence endemic in this region, pitting families against each other, and their neighbors. After you watch it, retrace the timeline, and also figure out who actually killed who.

4 Months, 3 Weeks and 2 Days (2007) Set in 1987, on of the last years of the communist/dictatorship Ceausescu era, in Bucharest, Romania, this film documents the struggles of two adolescent girls, as one goes through an illegal abortion. At times, mildly disturbing, the power of cinema is evident throughout the entire film.

I Served the King of England (2006) Filmmakers from both the Czech Republic and Slovakia joined to create this film depicting Czechoslovakia during the 1940s. Nice historical sweep of the area from pre-war to German occupation to soviet occupation , and an excellent look at a variety of landscapes in Eastern Europe and Germany.

Katyn (2007) War torn Poland during WWII and the events leading up to and eventually culminating in the Katyn Massacre, where the soviets murdered 15,000 Polish POW officers and citizens. Includes actual German and Soviet newsreels from the time. This is as topical as it gets: did you hear about the Polish President dying in a plane crash in April 2010? He was heading to a Katyn memorial service at the time. Freaky!

Force 10 from Navarone (1978) Awesomely cheesy, action WWII film set in Yugoslavia and other parts of Eastern Europe. Shows little-known Nazi action in the Balkan Peninsula and actually filmed in parts of Yugoslavia and Montenegro. Stars a young Harrison Ford, as he shot it at the same time Star Wars was being made. Sorry, it's a plaid guilty pleasure.

LIVE LEADERS YOU SHOULD KNOW:

Viktor Yanukovych: His fraudulent election was the cause of Orange Revolution and, amazingly enough, was also the cause of the second uprising/revolution in Ukraine in 2014. His pro-Russian, anti-Western leanings brought him to power, and were the reason for him being thrown from power. Talk about a double-edged sword.

Viktor Yushchenko: Previous pro-West president of Ukraine that surfed to power on the Orange Revolution wave. Tried unsuccessfully to incorporate his country into EU and NATO, but political stagnation and inefficiency got him tossed out of office. Oh, and he's the dude that got poisoned by ex-KGB during his run at the presidency.

Yulia Tymoshenko: International political hottie extraordinaire and previous Ukrainian Prime Minister. Has her own pro-Yulia party, and is adept at playing both sides of the pro-West and pro-Russia sides of the game to her advantage. You will see her rise again in Ukrainian politics, but from the sidelines, not as a major player.

Volodymyr Zelensky: Outsider, pro-Western, populist President of Ukraine since May 2019. His previous occupation was as a comedian/actor that starred in a TV show in which he played the President of Ukraine. Yes, seriously.

MAP GALLERY

View Map Gallery & additional Plaid Avenger resources for this region at http://plaid.at/ee

CHAPTER OUTLINE

1. Middle East Mayhem, Especially Syria. . .
2. A North/South Divide
3. Rise of the Right
4. BREXIT
5. Reverting to Type
6. Tea Time: Summary

Euro-Addendum: Europe Teeters, in Tatters?

8/9A

HOLY ROMAN EMPIRE! Europe in crisis! Economic recession and stagnation for years! Millions of immigrants streaming into the strained resource base of Europe! Terrorist attacks terrify the population! The rise of populist right-wing political parties in country after country! And Great Britain just BREXIT'ed out of the EU! Is this the beginning of the end for Europe? Is the European Union crumbling? Is Euro-chaos and Euro-calamity going to crush the continent?

Well . . . maybe.

Or maybe it is a European evolution that most of us just didn't expect or predict, which makes everything seem so scary. Humans in general are terrified of unpredictability and change, which honestly are about the only things you can always count on to be present in all situations, throughout all human endeavors. Unpredictability and change. Both are currently consuming Europe, making its future very hard to figure out at this current juncture. Which is precisely why I wanted to add an addendum to this year's edition of the textbook: to alert you in more detail to what is happening in this crucial world region, why it is happening now, and perhaps suggest some possibilities of its future . . . while hopefully also allaying some of the panic and fear that is gripping the globe.

For full disclosure, I, the Plaid Avenger, for years now have been debating merging Chapters 8 and 9 of this very book into a singular chapter named "Europe." You may have seen reference to that idea in previous pages. My reasoning (for decades) has been that as the European Union has grown and thrived, it had been pulling the 28 separate sovereign states of Europe into a more singular, cohesive whole. Outside of explaining the Cold War history stuff, it was getting harder and harder over the years to justify telling the stories of Western Europe and Eastern Europe separately, when both stories seemed inevitably to be ending in a harmonious, single, shared chapter of union. The European Union to be exact, making east and west differences increasingly inconsequential since they were all in the same club, playing by the same rules, building a shared culture of values and governance.

But then shizzle started hitting the fan. Voices of discontent about the EU started growing. Those voices became markedly louder as the 2008 global economic recession kept dragging on in Europe . . . a place that was most crippled by that very recession. FYI: the global recession was caused by really horrific practices of major investment banks and financial institutions in the USA, which European banks and institutions then readily followed . . . right off a cliff. Anglo-America has started to slowly rally back from this recession, but most of Europe just can't seem to get back on their feet to get back to economic growth rates of the past, much less bigger and better growth rates for the future.

These economic doldrums were definitely part of the reason that anti-EU voices become louder and more pronounced. Remember, a lot of folks across Europe have never been entirely enthralled with the idea from the start. Many Italians want Italy to just focus on Italy. Many Brits want their government to just help the British. Many French people think their leaders should just focus on "making France great again," (lol it worked for Donny!), etc., etc., so on and so forth for all the countries of the continent. When economic times get bad, voices like those point out the obvious: why spend their tax dollars on helping ALL of the collective EU whole, when they would be better off using their tax dollars to help just their own country. This is a return to the past: back to the concept of the nation-state as the premier unit to organize political and economic life. A backing away from globalization and the interconnection it brings across nation-state boundaries . . . a reversal away from supra-nationalism, of which the EU was the shining star example. This is a current battle between the forces of globalization versus nationalistic tendencies that is actually playing out in other regions of the world, too.

However, economics alone cannot account for the sudden spiral of fortune the EU, and Europe as a whole, has suffered in the last couple years. There have always been ardent nationalists, and there always will be. There have always been anti-globalists, and there always will be. There have always been EU-naysayers, and there always will be. So what else has conspired to shake the very core of the European construct? AH! So glad you asked! Here is a short list of situations that have currently conspired against the continent and are exacerbating the radical revision of Europe itself . . .

MIDDLE EAST MAYHEM, ESPECIALLY SYRIA . . .

First off, the situation in an entirely different region has spilled over to affect Europe in profound ways. The region in question of course is the Middle East, and the proliferation of problems there has been greater than usual since 2011, especially in Syria . . . and that is really saying something since this region has been prone to conflict for decades already, having the greatest number of open and active conflicts of all the regions on the planet. I won't even complicate the matter by covering all of the bloody Middle Eastern conflict history—we will just pick up with the very recent events in 2011 and play it forward. Let's start with the Spring . . . of an Arab variety. . .

The Arab Spring was a revolutionary wave of demonstrations and protests (both non-violent and violent), riots, and civil wars in the Arab world that began on December 18, 2010 in Tunisia and spread throughout the countries of the Arab League . . . mostly fueled by regular Arab peeps who were tired of decades of economic stagnation and the right-wing governments of their countries that had never allowed for any political participation. The Arab peeps on the streets had just had enough, and once the powder keg had been lit, it exploded in country after country across the Arab world. Some governments were overthrown (Tunisia), some leaders were jailed/killed (Egypt, Libya), and some places fell into a state of chaos and civil war that they have not yet gotten out of yet (Yemen, Syria) . . . which brings us back to Syria. . .

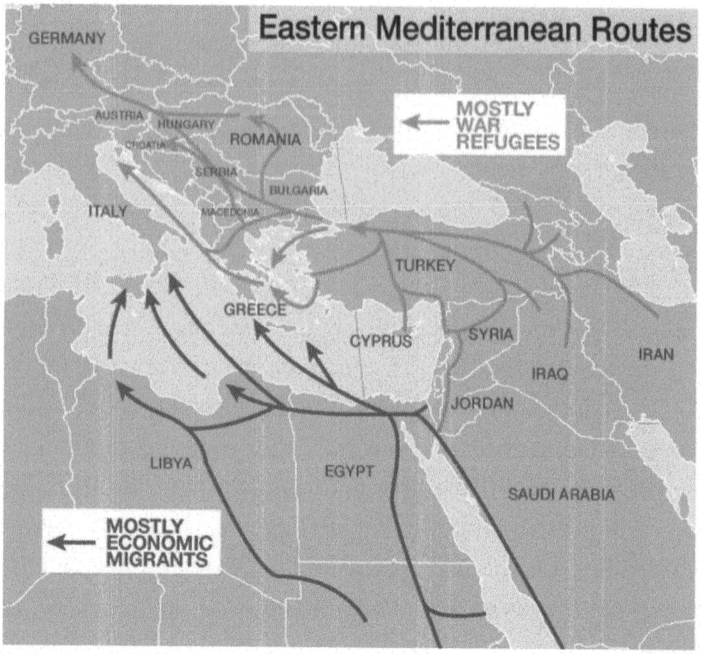

The Syrian Civil War is an ongoing armed conflict taking place in Syria since March 2011. The unrest began in the early spring of 2011 within the context of Arab Spring protests, with nationwide protests against President Bashar al-Assad's government, whose forces responded with violent crackdowns. Violence escalated and the country descended from prominent protests into an armed rebellion into a full-on civil war as rebel brigades were formed to battle government forces for control of cities, towns, and the countryside. Then Iran and Russia started helping the Syrian government. Then the US and its Arab allies started helping rebels. Then a whack-nut-job group of yahoos named ISIS took advantage of the chaos to form a new terrorist organization hell-bent on overthrowing the entire Middle East and establishing an Islamic caliphate in its place. Then ISIS took over territories in Syria and Iraq and unleashed hell on earth on the civilian populations and anyone else who got in their way. Then everyone started bombing everyone all over the place. Repeat daily.

Yep. That about sums up Syria from 2011 right on up to 2019, with the only change being that ISIS was officially beaten by all other parties involved and is not a major player in the chaos anymore. For now. But the Syrian Civil War is still the biggest human tragedy on the planet, with the death toll now surpassing 250,000 people killed (according to the UN) with many agencies believing it is closer to half a million. 13.5 million people are in urgent need of humanitarian assistance. 6–7 million people have fled the fighting as refugees. What. Stop. Refugees? Okay now we can bring it back to Europe . . .

Because that is precisely where millions of refugees from Syria have been fleeing to. And other economic and political refugees from Africa and the Middle East have been moving to Europe for decades before this, so this isn't even a new trend. And it was already causing social division within many states in the EU, and was considered one of the biggest problems for the EU countries to come to grips with. But they never did come up with a coherent policy for the entire block to follow. And the numbers of refugees arriving in Europe in the past were never as huge, nor rapid, as they have been since this Syrian catastrophe exploded. I believe the official numbers are now around 3–4 million peeps moved into Europe from the Middle East in the last five years, mostly to Germany. Why? Because Germany is the richest economy with the most opportunities. And Chancellor Angela Merkel of Germany said it was the right thing to do from a humanitarian perspective. And it is. But it was 3–4 MILLION peeps.

So mondo peeps moving in to Europe has strained the resources of many states, and caused fruitless bickering between EU states about how to respond to the asylum issue and the resource allocation to help states deal with it. The lack of a strong, singular EU policy on the matter encouraged states to develop their own strategies: some countries just opened the floodgates to allow refugees to pass through at will (Greece, Italy), some countries opened the doors totally to absorb them (Germany), some states imposed limits on numbers that could enter daily (Austria), some states started building border fences to deny them totally (Hungary). So different states in the EU have differing opinions on different issues. . .

Aha! That brings us to a long-standing riff within the EU that has been internally dividing the block long before the current crisis. But the current crisis has been exacerbating the situation, and badly . . . and that situation is further exacerbating. . .

A NORTH/SOUTH DIVIDE

As suggested earlier, for years now I have been thinking about when to pull the trigger on re-writing those Europe chapters to minimize the whole idea of the west/east divide, and turn it into two chapters named Northern Europe and Southern Europe, because that is increasingly where the European divide lies in the modern world. Say what?

Yeah, as I referenced above, for decades it made sense to do the whole east/west thing because I had to teach you all about the Cold War and its consequences, the main one being the division of Europe. But since the fall of communism in 1991, the continent has been coming together under this unifying construct called the European Union. So it has increasingly been just one big happy Euro-family; a family so unified and so equal and so great that everyone else wanted to join the club, so the family has been growing to include those Eastern European states. Right? Well . . . not so much.

Indeed the European Union has been fantastic at decreasing regional rivalries, preventing war in Europe, guaranteeing democratic stability, improving living standards, and boosting trade and economic activity of all its member states. That's always been part of the appeal of the EU experiment. But one big happy family where all members are equals? Absolutely not, never was, and increasingly more "not" as we come to current events. While the EU project has brought 28 separate, sovereign nation-states together into a singular union, it has not necessarily made all states equal nor made everyone happy.

I believe I can explain this to you best with a simple comparison to America. Think about the USA: all the states are in a political union together, but does Mississippi have the same health care access as Maryland? Do citizens of Connecticut get the same education as those in Arkansas? Does Wyoming have as much of a voice in Congress as Texas? Is Alabama as rich as California? Hmmmm . . . that would be a gigantic NO. The USA is one political and economic union, but great disparity exists between the parts of that union. In fact, the disparity of economy and of culture became so great in America that at one point it almost ripped itself in two . . . you know, the American Civil War thing.

And we have the same-same going on in Europe. No, they are not bordering on a civil war yet, but there is great disparity between states of the E-union on a variety of topics, and what we are witnessing right now is a "reckoning" of sorts about how the EU and its member states are going to deal with all that disparity. Some want to strengthen the central governance to help solve problems, some want to weaken the central governance and re-strengthen the individual states' powers to help solve problems, some want to secede from the union altogether (that is, leave). Dang! That is really starting to sound like the pre-American Civil War scenario again, right? Am I good or what?

Now I don't want to go into all the major issues that were already dividing the EU before the recent crisis, save one: economics. I opened this addendum talking about the 2008 financial crisis which turned into a now eight year-long recession. Europe got hit worst, but not all of the EU got hit the same way. First off: the northern European countries are the richest countries of the continent; I am referring specifically here to Germany, the UK, Scandinavia (Finland, Sweden, Norway), BENELUX states (**BE**lgium, **NE**therlands, **LUX**embourg), and maybe even France (France is one of the richest states, but they have so many problems it makes them more complicated). The biggest, richest, most developed, biggest exporting economies are in the north; highest levels of development, access to social benefits like health care, highest standards of living, etc., etc., etc. The south on the other hand is where the more problematic, poorer states reside, namely the **PIGS: P**ortugal, **I**taly, **G**reece, **S**pain. BTW, this is a relative comparison within Europe itself: as poor as Greece is right now, their GDP per capita is still way higher than most African and Asian countries. Like way higher. But back to the European comparison . . .

When the financial crisis hit Europe, the rich countries could absorb the shock; the poor ones could not. Just stick with the best examples within the EU: super-rich #4 world economy Germany versus super-destroyed wildly bankrupt Greece which is so far in debt it can't ever possibly pay back what they owe . . . and who do you think they "owe?" That's right: the rich countries' banks (led by Germany) are the ones that have lent money to the poorer countries for decades. As the recession has dragged on, the northern countries (led by Germany) have demanded that the poorer countries (exemplified by Greece) tighten their belts by introducing "austerity measures" which force the state to balance their budgets and cut government spending on things like health care and education, thus saving enough money to pay back their debts. Greece counter-argues that the austerity measures have undercut their entire society and have also stymied economic growth . . . thus, they are in a vicious circle where they need to spend government money to encourage economic growth to get out of the hole they are in, but the austerity measures prevent them from spending money, so their economy shrinks more, making them less able to pay back debts. Repeat cycle for eight years.

And that is why there has been a growing rift between "the North" and "the South" in Europe for a decade, or more. Wow, we are back to US Civil War comparisons once again! It's not intentional! I swear! With no reservations though, the animosity between the rich north and the broke south has been reaching a boiling point, as most Greeks now openly despise Germany as being their fascist-banker overlords, and most Germans see the Greeks as a lazy, slacking society that mismanaged its finances so bad that is has to be regulated like a small child. Yeah, it's gotten that ugly. Check the Greek protest images to the right to see what I mean.

Why did it have to go down like this? Because as I've suggested before, all states of Europe are not equal, but just as important: the EU did not do enough to put their financial house in order as a single economy. In the USA, if Alabama were to default on loans and collapse, the federal government would step in to save them, using tax dollars collected from all 50 states. The EU does not have such a financial structure. The EU does not have that kind of top-down, strong, financial safety net. The EU's refusal to develop a real, foundational, common policy to help new/poorer members reduce their economic drawbacks compared with the richest countries is largely responsible for structural discrepancies that work against the process of European integration . . . and in fact are contributing to the current tearing apart of the union itself.

And while there are other struggling countries in the EU (Romania, Bulgaria, Ireland), I intentionally just named the ones in the south (the PIGS: Portugal, Italy, Greece, Spain), for a couple reasons: 1) Their economies have floundered the worst and they are in the most amount of debt and have had to adopt the austerity measures just to keep getting more loans from the EU/northern countries, and taken all together, many of their citizens have started to despise the EU as a whole, and 2) These exact countries are the ones that have had to deal the most with the flood of millions of refugees, and feel as if the north is not helping enough . . . creating yet another division between North and South. Dig it:

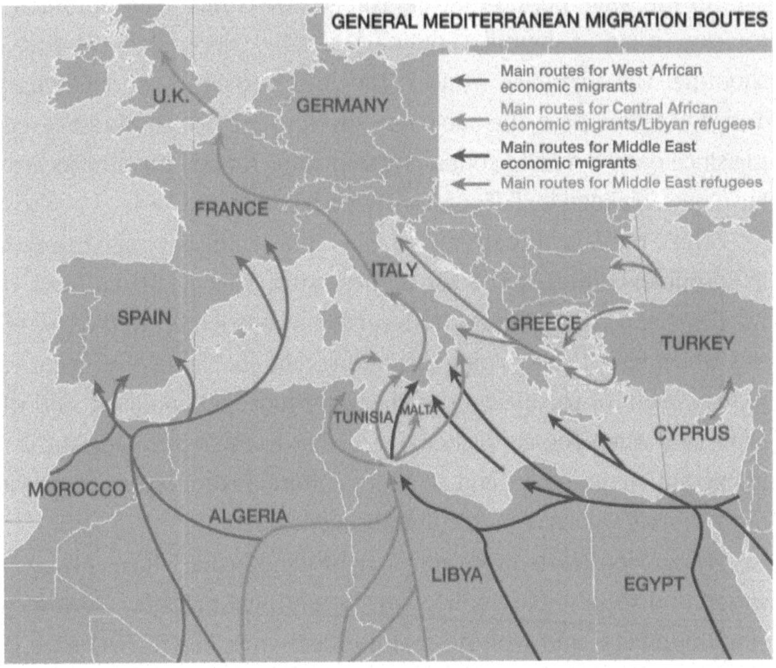

As referenced previously, economic migration to the EU from Africa/the Middle East has been happening for decades. And mostly over the Mediterranean Sea. Southern Europe is just a stone's throw away from those other regions, and honestly the Med has been a super-highway of travel, economies, and interactions for most of history. Remember the old Roman Empire? Yeah, that was a Mediterranean Empire that held territory on all coasts of that Sea, and it united all the peeps around that Sea, and humans have gone back and forth over that Sea for centuries. Oh, but that was then, and this is now, so the rich EU no longer wants those poor Africans and Middle Easterners to use those transit routes. But let's not delve on the moral or ethnic issue of limiting migrations, nor even give details about the thousands and thousands of peeps that have died in the last decade trying to make that trip north across the Med. No, let's just focus on how the issue is dividing Europe:

See, EU migration policy was already a hotly debated issue within the block, and the southern states have long been arguing that they need more resources to deal with what is in effect an EU problem. The southern states have to use their navies and coast guards to help these migrants not die; the southern states have to house, clothe, and feed these migrants when they get there; and the southern states have to process their applications for asylum and eventually deal with them legally. All that takes resources and time. And of course many migrants never go through the system: they just show up on the shores and blend into society, thus exacerbating social issues of cultural conflict, job competition with locals, homelessness, squatter camps, and possibly increases in crime. Many in the rich north have fought against allocation of huge amounts of resources to the south, considering it a local problem. So many governments in the south started just saying: "Oh, really? A local problem? Okay then, we are opening our borders and letting these peeps go wherever they want . . . because we all know they want to go north to the rich countries!" See the north/south rift opening here?

Okay then: now add in the 3–4 million Syrians trying to get to Europe since 2014. Yeah. Shizzle just got real. An existing fire that was dividing Europe got a million gallons of gasoline poured on it. Yikes. Do you get why I am increasingly seeing Europe as a north v. south as opposed to east v. west?

So an economic crisis, followed by a migrant crisis, in a Union that already had a small group of haters that *never* liked the EU, with a growing number of discontented people who were *starting* to not like the policies (or lack of good policies) of the EU. Mix that all together and add in continued flow of refugees and you have the recipe for . . .

RISE OF THE RIGHT

Because these refugees are not just from another place, but another culture, another language, and another religion. That in turn has inflamed the cultural friction I alluded to in the Western Europe chapter: locals feel like they are "losing" their country, and their culture, in the onslaught of all the "outsiders" moving in. And those outsiders are "takin' our jobs!" Add to that the idea that some of these immigrants may be crazy ISIS terrorists who are sneaking into Europe disguised as refugees with the intent on carrying out terrorist atrocities. Oh yeah . . . back to ISIS.

In the same period that the refugee crisis has been unfolding, Europe has been plagued with dozens of terrorist attacks, (some of them exceptionally deadly), particularly in 2015–2016, mostly in UK and France. The Charlie Hebdo shooting. The 2015 Paris Attacks. The 2016 Brussels Airport Bombing. The 2016 Nice Cargo Truck Attack. **The Wave of Terror in Europe** is the name now given to the period of increased terrorist activity on the European continent, erupting since early 2014 and continuing into 2019. Though the attacks are unrelated to each other, virtually all the major ones have been claimed by ISIS, or have been ISIS-inspired.

Thus, the Islamic extremist/immigration/cultural conflict issue has not just taken center stage, it is the entire stage in European politics now. Some of these attacks have killed dozens, or hundreds, made international headlines, and paralyzed societies. These issues have certainly contributed to the rise of radical right political parties all across Europe, parties which have been gaining ground big time for the last few years already. And these issues have demanded an overall reassessment of Europe's super-liberal attitudes and policies. And what's the opposite of left liberalism? Conservativism, especially when you go further and further to the right. Radical right parties are typically anti-EU, pro-nationalism, anti-immigrant, xenophobic, and all about cultural preservation . . . preserving their culture, of course, which is the opposite of the liberal values of multiculturalism and openness.

Now, Neo-Nazis and radical right-wing groups have always existed in Europe, but have typically been super small, and totally marginalized in the political process. In other words, haters have always been out there, but in small numbers, and with no real political voice. BUT THAT VOICE IS GROWING NOW. And an increasing number of people who are not particularly even radically right nor fascist have become alarmed at the numbers of refugees coming in, the speed at which they are coming in, and the surge in recent terrorist attacks. So more and more Europeans (many of whom maybe like the EU concept) have become more conservative about the direction that the block, and the direction their individual countries, are heading. The conservative right has been rising slowly for a decade, and now growing at breakneck speed in the last couple years. The smaller and much more radical right fringe is growing just as fast.

More and more of those "right" minded folks believe that the entire EU experiment should be called into question, and they want their governments to focus only on their country specifically; again, this is a return to the old concept of the nation-state as the highest and most powerful form of political organization. "Our nation first!" "Pro-nationalism!" "Our country first!" Hmmm . . . do these slogans ring any bells? It should. It is very similar in sentiment to "Make America Great Again!" While not technically Americans, many in Europe agree. Because more and more "right" minded peeps have also decided that part of the problem is the EU itself; that it has become a bloated bureaucracy of career politicians that have lost touch with the regular people they are supposed to be representing. Fair enough assessment I suppose. That sentiment is exactly the same reason that **populism** is currently exploding across the planet.

For now, I just wanted you to be aware of the real gains the conservative and ultra-conservative right have made in Europe in just the last few years. Radical right wing parties have gained ground in local and state elections in Finland, France, UK, Austria, Hungary, Poland, Greece, Netherlands, Italy . . . heck, really all over. Of note:

- Andrzej Duda is the current President of Poland and leader of the right-wing Law and Justice party (PiS). Openly expresses distaste for liberal policies, especially if pushed by the EU, and has a hard line on migration and a firm defense of national sovereignty. Poland is quickly slipping into authoritarianism under his rule.
- Viktor Orban is Hungary's current President, and openly right-wing in political views. Was bitterly opposed to Syrian refugees seeking to enter Europe. He declared migrants to be "poison" and fretted that the refugee exodus would threaten the continent's "Christian roots." Gave the order to build fences around Austria to cease the refugee flow.
- Marine Le Pen is the leader of France's National Rally (formerly called the National Front), a far-right group that calls for pulling France out of the EU, out of NATO, and halting immigration totally. She has been quoted saying that marine patrols should "push migrants who want to come to Europe back into international waters." Her party captured 17% of the votes in the 2012 presidential election, and then 33% of the vote in the 2017 presidential election. In 2019, the party actually won the highest percentage of votes for France's election of representatives to the European Parliament. So, yeah, France's anti-EU party has the largest number of France's seats at the EU now. #irony. Of additional importance: that 2019 election was the first time France's major conservative party and its major liberal party completely lost the game, with neither one of them finishing in the top 5 of vote-getters.
- Geert Wilders is an overtly anti-Islam Dutch politician, leader of the pro-nationalist Party for Freedom, who has campaigned to have the Koran banned in Holland. And that is one of the friendlier policies he spouses. Wow.
- Frauke Petry is leader of the Alternative for Germany (AfD) party, openly anti-EU, anti-immigrant, and has compared multiculturalism to a "compost heap" and said border police should "use firearms if necessary" when dealing with refugees.
- Norbert Hofer just narrowly lost the presidential election of Austria in 2017. Narrowly. His campaign slogan was "Putting Austria First." That ring any bells? He leads the Austria Freedom Party, and has in the past recommended that the public arm themselves with guns as a logical reaction to the refugee crisis. Although he lost, his party made huge gains in elections, thus they are here to stay.

Finally: the Prime Minister of Italy Matteo Renzi stepped down in December 2016 after voters punished him badly by defeating a referendum he staked his reputation on: one that would have seen a streamlining reform to the Italian government giving it more power to get stuff done. Think about that: the voters rejected even a moderate reform that would have empowered the government to do more within the EU. This is a conservative rejection of the entire system at this point. And that isn't even the biggest display of anti-EU behavior. For that we got to head back north to Great Britain, to check out their not so great decision . . . or was it great? That's for the Brits to decide, because they had a. . . .

BREXIT

In June of 2016, the citizens of Great Britain were given a referendum vote to decide if their country would stay in the EU or exit it. Oops. Much to the shock of the EU and Team West and the world, they voted to bail. The BREXIT is on. BTW: Brexit is a portmanteau of "Britain" and "exit." lolol how many times have you heard the term portmanteau? I just couldn't resist!

And apparently neither could the 51.9% of Brits who voted to bail on the EU block. The UK government intended to invoke Article 50 of the Treaty on European Union, the formal procedure for withdrawing, by the end of March 2017. This, within the treaty terms, would have put the UK on a course to leave the EU by March 2019. But unless you have been in a self-induced medical coma for the last two years, you have probably heard that all attempts to make BREXIT a reality have utterly failed. Failed so miserably in fact, that the acting Prime Minister Theresa May (whose sole job was to figure out how to make Article 50 happen), just quit the job and is stepping down in June 2019. Apparently 50% of the population of the UK wanted BREXIT, but 0% of them can agree to any policies to enact it. But one way or another, Article 50 is coming, and the Brits will leave (or be booted) from the EU soon.

To restate, the idea of being part of the EU was never universally beloved by all Brits, but for sure never the majority of them since the start. The UK joined the European Economic Community (EEC), a predecessor of the EU, in 1973, and confirmed its membership in a 1975 referendum by 67% of the votes. Historical opinion polls from 1973–2015 tended to reveal majorities in favor of remaining in the EU, but again, it was never even close to 100%. In the 1970s and 1980s, withdrawal from the EU was advocated mainly by some Labour Party and trade union figures who were all about protecting jobs from foreign competition (the same thing Donald Trump is now advocating in the US by hating on NAFTA and TPP). From the 1990s onward, withdrawal from the EU was supported mainly by some Conservatives and by the newly founded UK Independence Party (UKIP). So what changed from 1973 to now?

Just all the stuff I have now mentioned in this European Addendum. **Euroscepticism**, (criticism of and strong opposition to the European Union), has been on the rise for all those reasons, and more. The failure of the EU block to manage any of the recent crises, (Greece debt debacle, recession, immigration, Ukraine), has not only not instilled much confidence in the block, but brought outright disdain of the EU as a bureaucratic backwater. Of course the economic downturn and immigration issues for the whole of the EU played a part in the BREXIT outcome, but much analysis suggests that the main concerns were domestic. That is, 52% of those British voters were highly troubled by the high levels of immigration to their country, and were concerned that British funds that go to the EU would be much better spent at home to create jobs and infrastructure. It remains to be seen if any of that will change, or if the financial figures are even accurate, but in either case what won't remain is the UK itself.

The immediate consequences? The British pound fell, stock markets the world over suffered losses, businesses in Britain are having to assess whether to stay or leave the country since it will no longer be a part of a single market. Things seem to be stabilizing now, but the big debate is all about what happens next when it's actually time for the UK to leave the party . . . physically walk out the door of the EU. Of course the UK wants to make the transition as smooth as possible by maintaining all of the beneficial trade ties, but getting rid of all the pesky EU rules they have to follow, especially the open borders thing. Will the EU allow for that, or will it punish the UK in order to set the precedent for other countries intending on leaving, too? That is the real question here, but it is all speculation at this point, so let's leave it at that.

The major take-away point here is that the UK has opted not just to leave the EU, but to return to that concept of nation-state sovereignty. The way it used to be in Europe for centuries. And that is a path that many other EU states may want to follow as well. Do the benefits of being an EU member outweigh the costs of losing total, ultimate, sole control of your economic/social/immigration policies? For Great Britain, the answer was "no." For Germany and many others, the answer is still "yes." For some, it's a "maybe." That's where we are right this second. A lot of soul-searching and self-evaluation is happening in country after country across Europe. Some may be encouraged to also try their hand at going it alone again, as "the people" continue to raise their voice of opposition to globalization and supranationalism. This wave of populism that has struck the entire world stage is actually happening the hottest and heaviest in Europe, and with pretty heavy consequences for the future of the EU itself, and Europe's place in world affairs.

So maybe this is the beginning of the end of the EU? And of a peaceful and stable and rich Europe? Eh. Maybe. Or maybe not. I never like to end on such a nefarious note. I will point out the positive: All is not lost! I think we could just be witnessing a historic. . .

REVERTING TO TYPE

If you pan up to the beginning of this chapter, you will see I started these shenanigans by exclaiming: **Holy Roman Empire**! Hahahaha was that just a random Charlemagne shout out? Not at all. There is always a bit of method to my madness, and in this case I want everyone to take a really deep breath about the BREXIT bombshell as a possible EU disintegration starter, and chill for a second to realize this: dynamics of the EU have been evolving for a millennium, not even a century, and change is sometimes good! A millennium? What is that? It's 1,000 years, my time-challenged friends, and I believe the EU concept started even longer ago than that! And we may be heading back . . . to the future!!!

Okay, Doc Brown aside, what am I getting at here? Just this: there has always been a slow-motion process to unite the kingdoms/countries of the continent into a singular whole for centuries. Let's get holy about the whole! The first time a "united Europe" concept came into play was back in the year 800 AD with something that would eventually be called the Holy Roman Empire . . . an agglomeration of territories which has been repeatedly pointed out as being neither holy, nor Roman, nor an empire. Now I won't bore you with all the historical details, but on December 25, 800 (wow! what an easy year to remember! Christmas Day, 800!), Pope Leo III crowned the Frankish king Charles the Great, (otherwise known as **Charlemagne**), as Emperor, reviving the title of "Roman Empire" in Western Europe, more than three centuries after the fall of the O.G. Roman Empire.

Why was Charles so "Great"? Well, Chuck is just one of those mythical figures in history that wins at everything he tries. He fought back against the Moors in Spain, thus stopping Islamic expansion into Europe. He took the Frankish throne in 768 and became King of Italy in 774. He fought campaigns in every corner of Europe and mostly won. He ended up as a protector of the Papacy in Rome. His rule spurred the Carolingian Renaissance, a period of energetic cultural and intellectual activity within the Western Church. All Holy Roman Emperors up to the last Emperor Francis II (1792–1835), as well as both the French and German monarchies, considered their kingdoms to be descendants of Charlemagne's empire.

But here is what I want you to know for sure: Charlemagne has been called the "Father of Europe" as he united most of Western Europe for the first time since antiquity . . . and even back then it was mostly centered around Mediterranean Europe. This dude was from the north, and united a huge area from France to northern Italy to Germany, and a bunch in between; he really laid the foundations for modern France, Germany, and the BENELUX states, and this marked a significant shift for Europe as a whole: from henceforth, the real power of Europe would be the north, not the south.

Think about it. From the very roots of "Western Civilization" and European history proper, the story was about the south. Ancient Egypt. Phoenicians. Ancient Greece. Alexander the Great. Rome. Caesar. The Roman Empire. Constantinople. The Byzantine Empire. These are all Mediterranean-centered civilizations. That is where all the action was. 800 AD is a crucial break point because from then on, most of the economic and political action was going to come from northern Europe. The French. The German Empire. Swedish kingdoms. The Dutch traders. The Hanseatic League. The Reformation. Napoleon. Hitler. Churchill. The UK. The Royal Navy. Getting the drift of my word association here? The north is where the action shifted to, a fact made clear by the current state of the EU countries down south, and the rich countries up north that I have alluded to previously.

And while I was suggesting the north European countries are the wealthiest and best off, let me assure you that it was not just the countries of the south that mark this internal wealth disparity in the continent. This is a real core versus periphery situation! Remember that stuff from the Economics chapter? The hierarchical relationship at global level that I described to you with a rich **Core** consisting of the US, the EU, Japan, and Australia (now joined by China) and a **Periphery** consisting of the "less-developed" or "developing" countries is also happening within Europe itself. The *EU Core* consists of the most

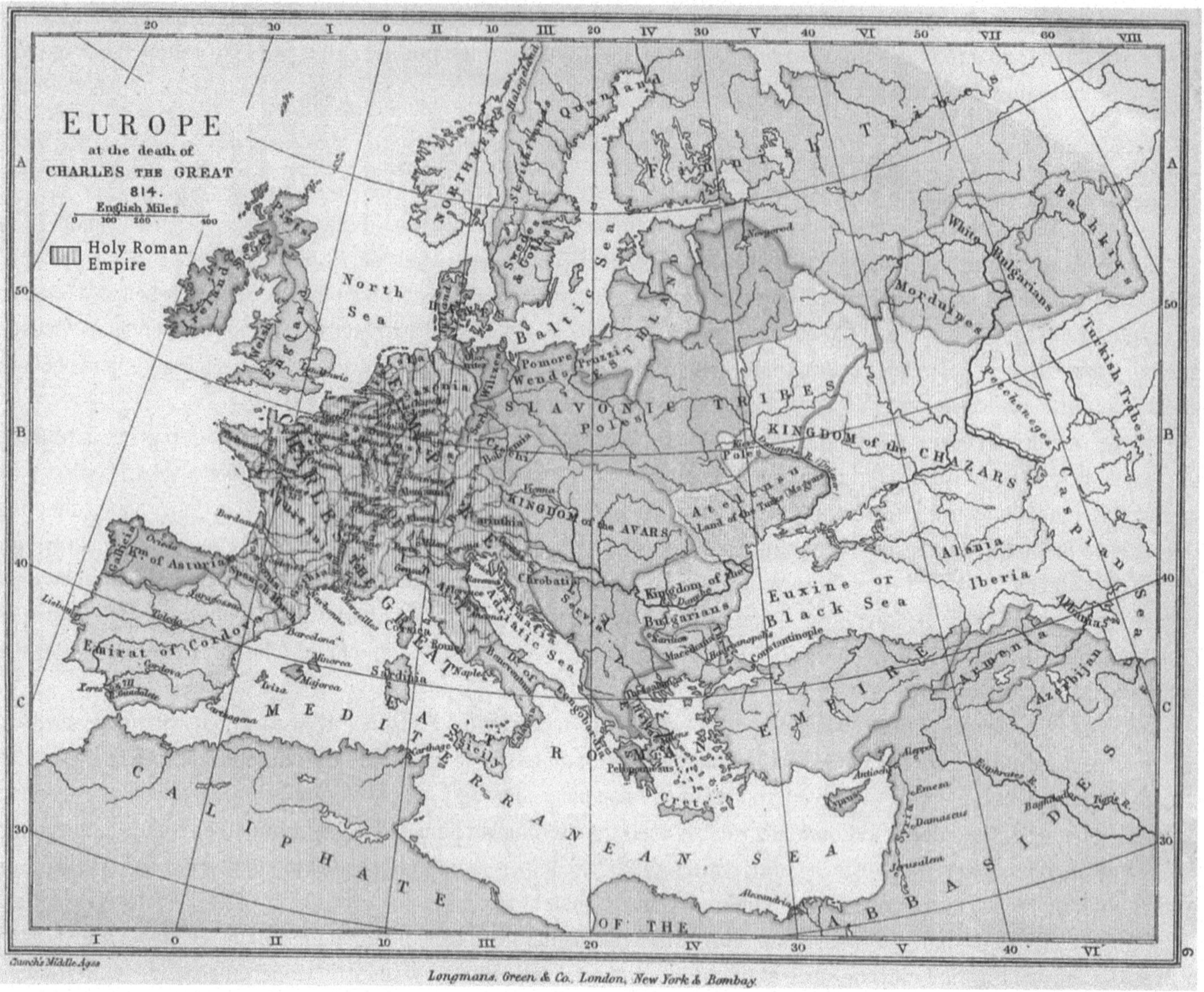

The OLD old school original EU Carolingian core.

powerful countries like Germany, France, the UK (for now), Italy (mostly the north), and the BENELUX. The *EU Periphery* is subjected to decisions made by this hegemonic Core and mainly consists of countries lying to the south and east of the EU: Portugal, Spain, southern Italy, Greece, Romania, Bulgaria, the Balkan states, and also including Ireland to the West. Through their brute economic and financial force, the Core gets what it wants, at the expense of the Periphery. The EU Core does this by dominating the EU political structure as well as controlling the most powerful financial institutions in Europe.

What's one got to do with the other? lolol Here is where it gets interesting to me: check out the side-by-side maps of Europe above and on the next page and tell me what you see.

Isn't that crazy? The original northern-focused unification of Europe under Charlemagne ends up being almost exactly the first grouping of countries to come back together under the original iteration of what would become the European Union! The core of 800 AD remains the core of 1950 AD . . . and beyond. While it did not exist very long in the same shape or by the same name, the Holy Roman Empire lasted in one form or another until 1806 when Napoleon devised the Confederation of the Rhine and talked 16 German sub-states into leaving the Holy Roman Empire and joining the Confederation . . . leading eventually to the formation of a unified and sovereign state of Germany. Later, that same

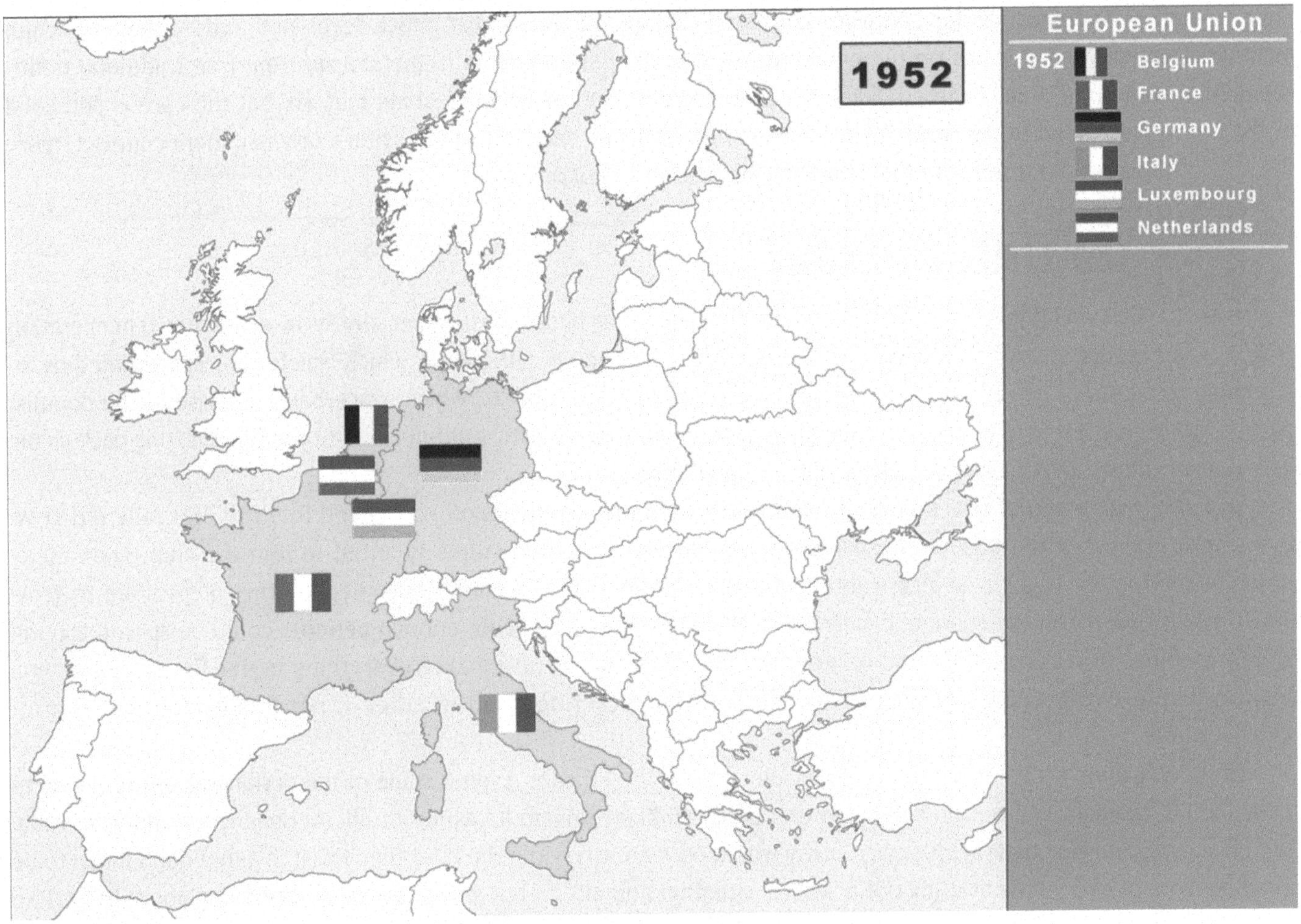

More than a millennium later, the same core players put together the modern EU.

Germany tried to unify Europe once more . . . by brutal force. Remember those Nazi dudes? Yeah, not so cool. Thanks for getting the ball rolling on that, Napoleon.

The point is: Europe has been on again off again trying to unite all these separate warring sovereign states together for a mighty long time. Sometimes it works, sometimes it fails. Sometimes it fails spectacularly (again, check out WW2). The European Union is just the latest iteration of this experiment, one that traces itself back to Charlemagne. And just like in Charlemagne's day, the core of Europe centers around those rich northern states. That's how the EU started: the rich northern states binding themselves together economically to coordinate resources, help rebuild themselves, and work together politically to ensure stability and limit conflict. In 1951, six European countries—France, Belgium, West Germany, Italy, the Netherlands, and Luxembourg—signed the Treaty of Paris establishing the European Coal and Steel Community (ECSC). The UK was not in it. Neither was Greece, nor Portugal, nor Romania. All those countries (even the UK) were in the periphery in 800 AD. One could argue that in many ways they still are . . . the UK is now by choice!

The economic recession, the anti-EU sentiment, the "invasion" of immigrants, the North/South divide, all of which has expressed itself in the rise of populist fury across Europe . . . all these issues are really related, and can be best understood by recognizing the core/periphery situation in which the current European Union evolved, and is trying to figure itself out in. The core of Europe is still astoundingly strong, even after being pummeled with challenge after challenge. The core has for centuries consisted of some of the richest, most developed, most interconnected, technologically advanced states on the entire planet. The core still benefits tremendously from free trade and political stability that the EU has fostered. Some states in the periphery may come, some may go, but the core is still solid to the . . . well . . . solid to the core! The EU may just be reverting back to form . . . to it's original, more compact, more homogeneous territory that Charlemagne himself would recognize.

TEA TIME: SUMMARY

To sum up this European Addendum: Mayhem in the Middle East, especially Syria, exacerbated immigration to Europe . . . which exacerbated the north/south divide in Europe . . . which was further exacerbated by an economic recession . . . which exacerbated the core/periphery disparity . . . which exacerbated the rise of the populist right . . . which exacerbated BREXIT . . . which may just be bringing us historical full circle to the EU reverting back to the Holy Roman Empire, part 2 I suppose. Hahahaha . . . get all that?

So I say, grab a cup of tea, take a sip, and relax. Perhaps a contraction was called for. The EU really did grow at a frantic rate once the USSR collapsed and most Eastern European states sprinted to join the club. Then many states of the former Yugoslavia also wanted to jump into the party. It made sense at the time to continue to grow the EU ranks: the more, the economic merrier! And while times were good, enough benefits could be spread around to help stabilize other economies and other governments, which in turn makes everything better for everyone, in a positive loop cycle. But now times are tough, and thus the bickering about whether or not to abandon the EU project as a whole.

Eh. I don't think it's going anywhere, especially not for the EU core. Even if some of the peripheral states decide to leave the EU, the core will remain solid, and that core will likely remain a bastion of liberal democracy and free trade. And once everyone gets their act together, many will likely want to re-join the EU if they left it. Again: I don't mean to be so relaxed about everything as if it's not a serious situation and all . . . but why is everyone panicking about BREXIT so much? It's not the end of times! It's not like they will be banned for life! If times change and the UK doesn't do smashingly well all by itself, it can certainly re-apply to be re-admitted! It's just an economic/political club for goodness sake! It's not a life and death scenario here! Geez!

And if it makes other Euro-skeptics think it is going to be easy to abandon the club, also consider that the UK is a unique EU member in this respect. As I suggested, the UK was usually on the European periphery anyway, so their departure can be seen as a "reverting to form" as well. And they demonstrated this for decades by not even fully adopting the official EU monetary unit of the Euro. They accepted Euros, but kept their own money, the British Pound. Which made the idea of returning to full sovereignty and leaving the EU even easier . . . both from an ideological standpoint (they had their own money, so they really were never fully dedicated to being "in") and economically (they already have control over their monetary policy, so fiscally being back in charge of their own finances and economy is no big deal at all). The same IS NOT true for most of the other EU countries, on one or both of those points. SO don't expect a mass exodus from the EU, despite the dire warnings from naysayers, populists, and pundits. Well, I suppose I am a plaid pundit myself, so feel free to ignore my words as well . . . but I just don't buy all the negative hype on the imminent collapse of the EU or Europe. Here's why:

Most Europeans—excluding young people in the peripheral Mediterranean countries—still enjoy the safest, fairest, and most comfortable daily life on earth. According to the WHO, most countries where people can expect to live to age 82 or longer are European. On the UN's human development index, Estonia, Slovakia, and even Greece still outrank Qatar, despite its wealth. Most "emerging economies" lag Europe by decades. As "poor" as Greece is, and despite all the financial chaos there, Greek income per capita is double Brazil's, more than three times China's and 15 times India's, according to the World Bank. EU is the most equitable region for the gender gap in pay. It has the least corrupt governance in the world, with some of the greatest amount of transparency. The EU still contains almost half of the world's top 25 economies. Standards of living, mobility, interconnection, personal freedoms, and health care opportunities are all exceptionally high.

Add it all up and you get a Euro-dream: dozens of sovereign states living together in harmony and freedom, with interconnections and free-flow of ideas, money, technology, and peoples across borders, and with the world's highest quality of life if not highest incomes. Why is it all the migrants and refugees want to go there? Why is it that most people even from the "rich' world want to go there to vacation? With all of this going on, why is it we think this place is "collapsing?" Well, I for one do not believe that, but I do believe it is evolving. The EU may have hard times ahead; the EU may even contract into a smaller group; but the EU will still stand as a very successful attempt by humans to at least try to form a more perfect political union. And striving for perfection is never easy, is always costly, and is ultimately never fully successful. But that don't mean you shouldn't strive for it!

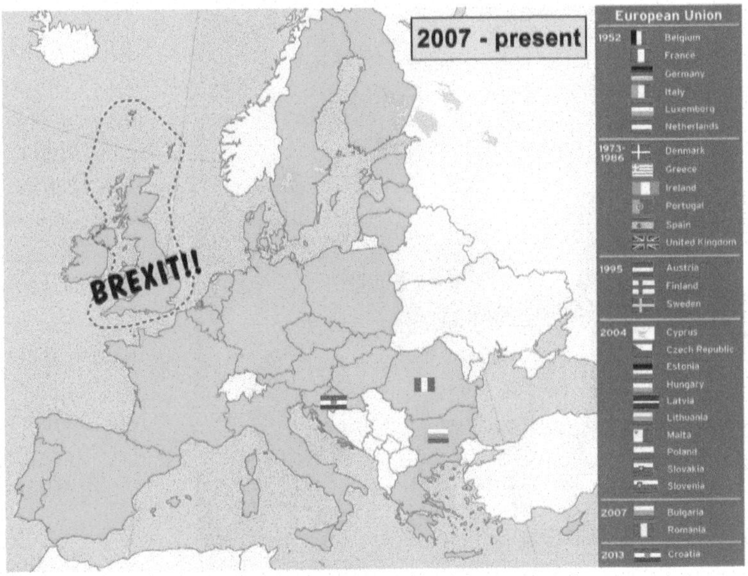

More exits after BrEXIT? Meh. Doubtful.

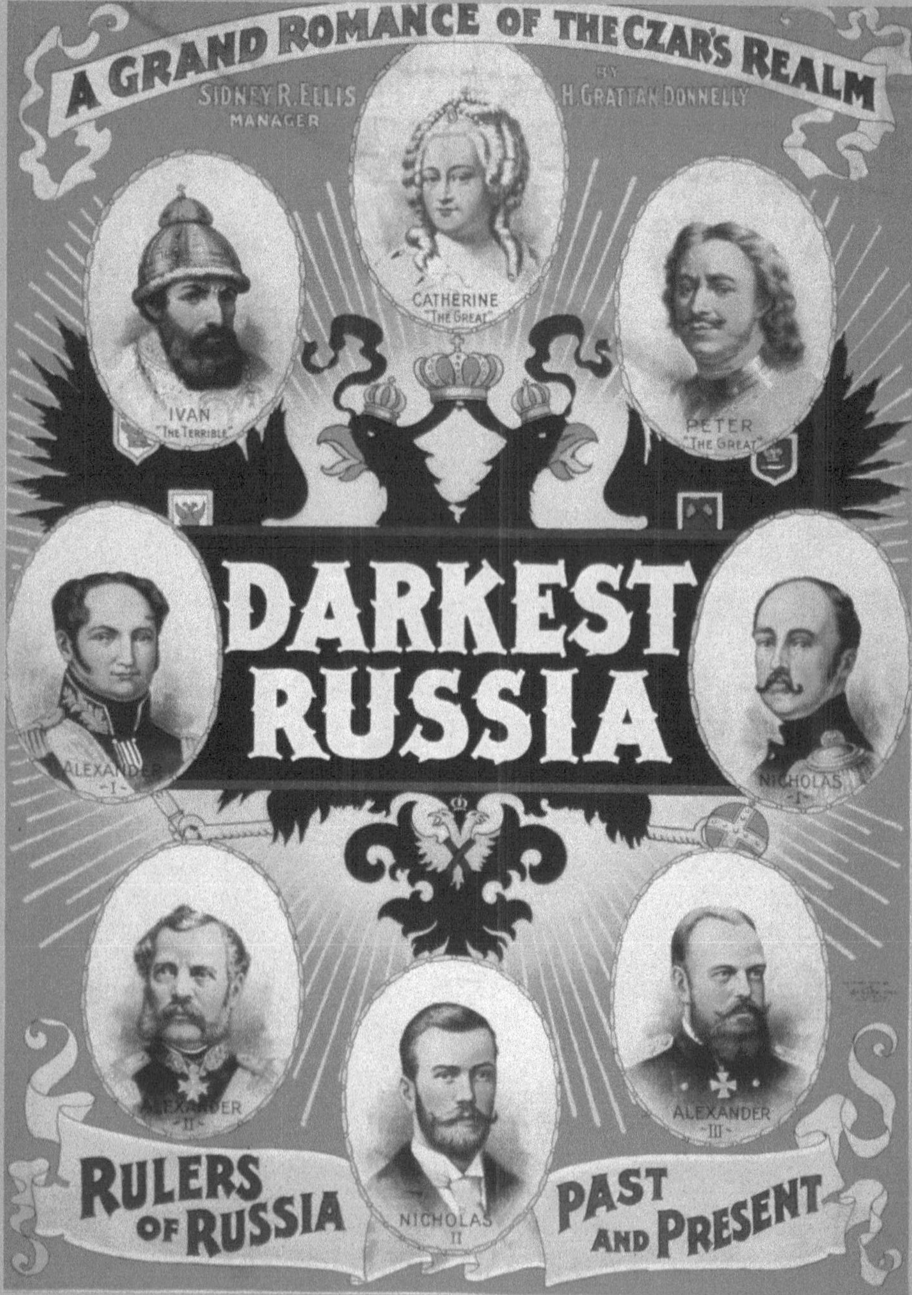

CHAPTER OUTLINE

1. The Bear Is Back
2. Hair of the Bear
3. From Cub to Grizzly
   - 3.1. Historic Growth
   - 3.2. Serf's Up, Dudes!
4. The Bear Takes a Beating
   - 4.1. Godzilla Attacks
   - 4.2. World War I
   - 4.3. That's Revolting! Commies Take Command
5. The Bear Battles Back
   - 5.1. Stylin' with Stalin
   - 5.2. The Bear Enters World War II
   - 5.3. Is It Getting Chilly in Here? The Cold War
   - 5.4. Commie Econ 101
6. Catastrophic Commie Implosion
   - 6.1. Ruble Rupture
   - 6.2. Afghan Body Slam
   - 6.3. The Great Communicator Says: "I Hate Commies"
   - 6.4. Holy Glasnost, This Is Driving Me Mad! Pass the Salt, Perestroika
   - 6.5. So Long to the Soviet: The End
7. Radical Russian Transition
   - 7.1. Economy Takes a Dump
   - 7.2. Oligarchs Over-Do It
   - 7.3. Purgin Population
   - 7.4. Environment Endangered
   - 7.5. Stolen Sphere of Influence: The Realignment
8. Resurgent Russia: The Bad Boys Are Back!
   - 8.1. Prez to Prime; Prime to Prez: Putin Power!
   - 8.2. Petro Power!
   - 8.3. Global Warming? No Sweat!
   - 8.4. Geopolitical Jump Start
   - 8.5. Don't Like Putin? Crimea River!
9. Re-Established World Role
   - 9.1. "Team West" Alternative Offered?
   - 9.2. Trump Speed Bump: Biden Back at It
   - 9.3. Ghost of Stalin Returns
10. Map Gallery

Russia

10

THE BEAR IS BACK

Russia is the largest country on the planet. They were the Cold War adversary of the US; you know, the bad guys. We refer to Russia as the Bear, and this wild woolly bear seems to change focus and direction every semester that passes, just as a wild bear hunts in the woods. I'm scared! Not really, but it is a region that is in transition, much like Eastern Europe; in this regard, it is changing every day. However, unlike Eastern Europe which has a very distinct direction of change, Russia is what the Plaid Avenger calls a fence-straddler—that is, it is playing the field in terms of with whom it does business, with whom it has a political alliance, and with whom it will throw in its future.

Time to bear all!

Russia has been a distinct culture, a world power player, the core of the Soviet Union, and a major shaper of world history and events of the 20th century. It has experienced tremendous political and economic upheavals and can almost be considered behind the times here in the 21st century . . . but hang on, because the Putin-inspired takeover of the Crimean Peninsula in 2014 proves that Russia is back on course for a possible huge power grab! How did this region go from a global power to a globally broken basket-case? More importantly for our understanding of today's world, how is it successfully getting back into the game? All of these things and more, we will explore as we tell the story of the Bear.

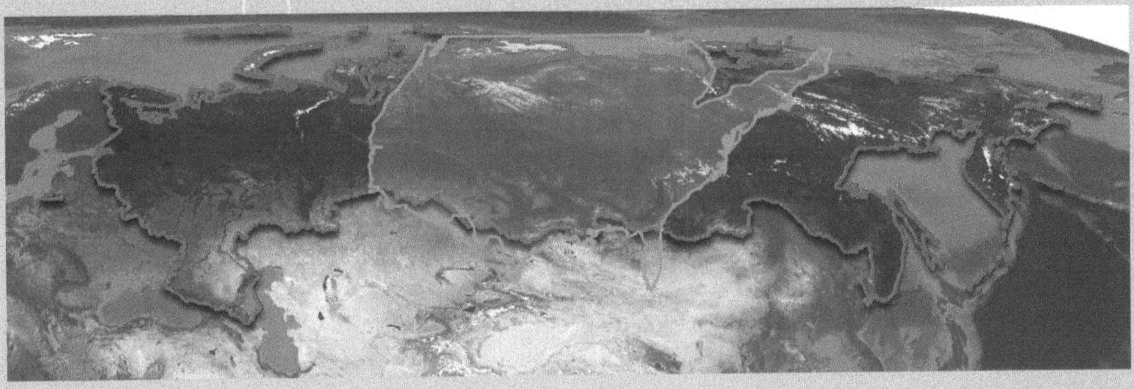

One big bear. US/Russia size comparison.

HAIR OF THE BEAR

We always start our regional tour with a breakdown of what's happening in the physical world. When we think of Russia, several words typically come to mind: big, cold, and vodka. There's some truth in this. Russia is a cold place, a big place, and those Russians do love their vodka. But kids, just say "No" to the vodka . . . unless you are in Russia, of course, and then you say, "Nyet" . . . although apparently few of them do.

Russia is the largest state on the planet, twice as big as Canada, the second largest. Does size matter? Do I even need to ask? Of course it does! Size matters in most aspects of life when you are a state. This huge country crosses eleven time zones, spanning almost half of the globe. When the sun is setting on one side of Russia, it is rising on the other side! Just think of all the problems associated with infrastructure and communications in a state this big. Think of all the problems associated with just trying to keep all of your people within your country aware of what's going on at any given second, much less having to move things around in it, like military equipment.

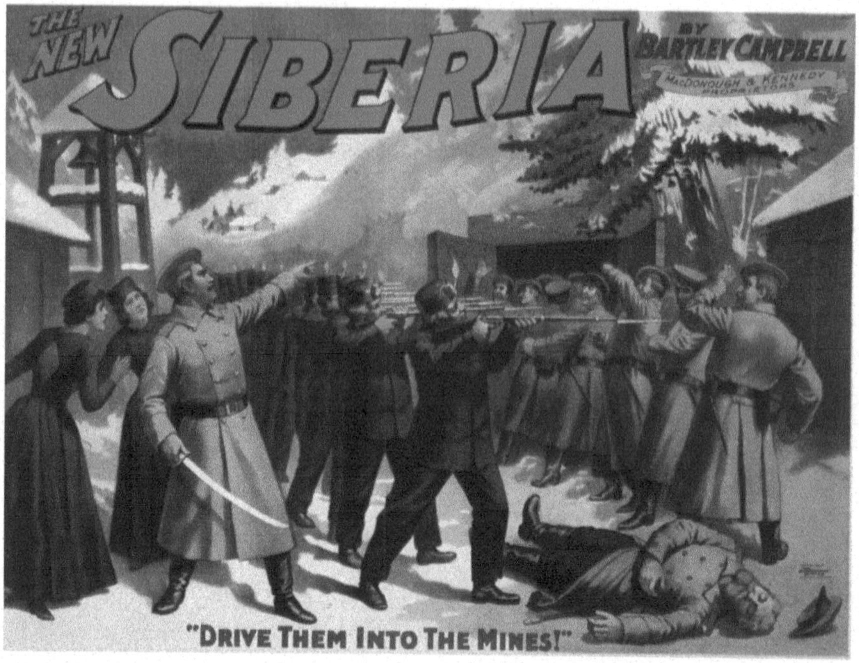

Siberia! The Spring Break destination of the Soviets!

Don't even try to think about having to defend its vast borders. Your head will explode. Russia has a top slot in terms of how many different countries it borders, which I believe is now fourteen in all. But it's not all bad. This massive territory contains loads of stuff—what stuff? All kinds of stuff: oil and gold, coal and trees, water and uranium and lots and lots of land. What do lots of people do with lots of land? They typically can grow lots of food, but that's not always the case in Russia because of its second physical feature—it's cold.

Maybe that's why the bear has such a thick coat. Virtually all of Russia lies north of the latitude of the US-Canadian border. This high latitude equates to the cooler climates found on our planet; indeed, some of Russia lies north of the Arctic Circle, which means most of its northern coast remains frozen year round. To compound matters, Russia lies at the heart of the Eurasian continent; because of this fact, it has perhaps the greatest extremes of **continentality** expressed by its extreme temperature ranges, both season to season and also day to day. There are some parts of Russia in which the temperature within a 24-hour period may change 100 degrees: that is, it could be 60 degrees Fahrenheit at 3:00 in the afternoon and 12 hours later it might be −40 degrees Fahrenheit. That's extreme. Maybe this tough climate has something to do with Russian attitudes and their dour outlook on life in general. A tough place with a tough history which brings us to our last physical descriptor: vodka.

Russia is the greatest consumer of vodka on the planet. Both in terms of total quantity and per capita consumption, no one can touch the Russians when it comes to drinking vodka; alcoholism rates are high, and this mind-numbing stuff continues to impact society and health in a variety of sobering ways. Perhaps it's the bleak and challenging climate that makes the Russian life so hard and vodka consumption a necessity to deal with the challenge. Perhaps this is why vodka is such an integral part of Russian history. Did someone say Russian history?

FROM CUB TO GRIZZLY

How did the Russian state get to be the largest country on the planet? How did it get so much stuff? Where did these guys come from? What does it mean to be Russian? The "original Russians" were actually Swedish Vikings who moved into the area around modern day Moscow to Kiev, probably around 1000 to 1200 C.E. Thus, Russians look European and share many of their cultural traits.

From the land of Volvo, we come to conquer Kiev.

HISTORIC GROWTH

Looking at the maps on the next page, we can see that Russia, in 1300, was more like a kingdom than an empire; indeed, it was not a huge territory. It was a small enclave of folks that called themselves ethnically Russian but were politically subservient to the real power brokers of that era: the Mongols. **The Golden Horde** was a political subunit of the Mongol Empire that controlled parts of Russia and Eastern Europe at the time. It was not until 1480, under the leadership of Ivan the 3rd, a.k.a. **Ivan the Great**, that Russia stood up to the Mongols, threw off the yoke of Mongol oppression, and began its growth into the land juggernaut we know today.

Ivan united his people, kicked some Mongol butt, and expanded the empire. His son, Ivan the 4th, a.k.a. **Ivan the Terrible**, continued this trend by feeding the Bear and expanding the empire further. Ivan the Terrible earned his name by having an incredibly nasty temper, eventually murdering his son and heir apparent with an iron rod. Nice guy. Good family man.

Soon after the Ivans created this empire called Russia, an imperial family line was put into place that lasted uninterrupted up until the 20th century. They were the Romanovs, to whom you've probably heard references before. I'm not going to go into great detail on all of them, but just to put a few into perspective, let's list some of the more famous names. **Peter the Great** popped up around 1700 and was known for many things, notably creating and building the town of St. Petersburg. Hmmm . . . I wonder where he got the name?

Why is this significant? Because it shows what Peter was all about, and that was embracing Europe. Even back then, Russia lagged significantly behind its European counterparts in terms of industrialization, technology, military, and economy. Peter was all about catching the Bear up. St. Petersburg was often called "a window to the West" because the city embraced Western technology and architecture and there was a real sense that it was opening Russia to interaction with the rest of Europe. Here's another example of how Russia is tied historically to the West: the term "Czar" in Russia was taken from the Roman Empire's "Caesar." In addition, Moscow was often referred to as "the third Rome" throughout history, as it viewed itself as the protector of true Christianity, in its distinctive Eastern Orthodox tradition. (Rome #1 was Rome; after the Roman Empire collapsed, then Rome #2 was Constantinople; after the Byzantine Empire crashed, then Rome #3 was Moscow . . . or so they want the world to think. BTW: Current President Putin still pushes this theme, big

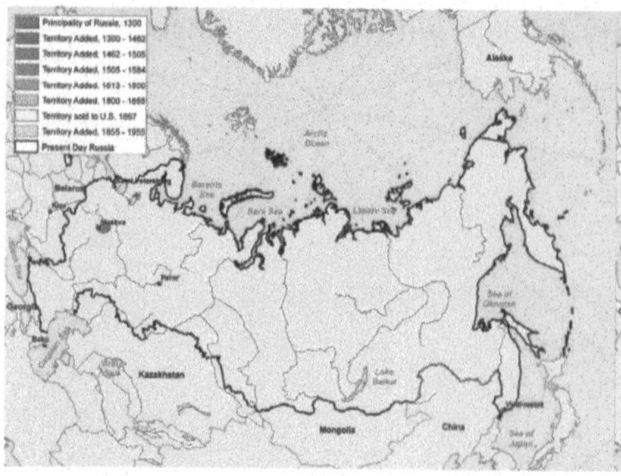
Russia in 1300

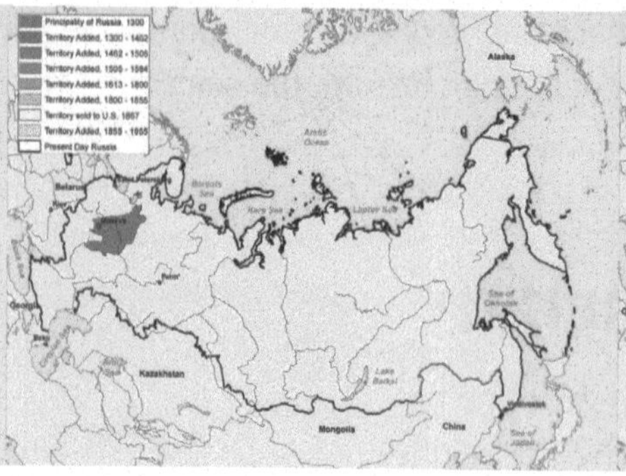
Russia in 1462

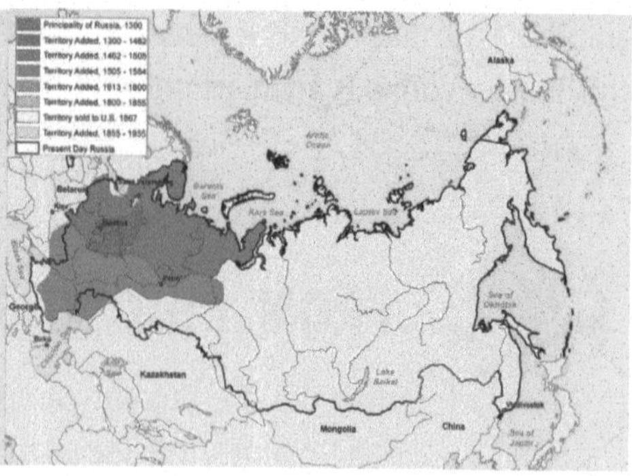
Russia in 1584

time). Russia has embraced and associated itself with the West in many ways over the years, although in most cases, it lagged behind.

Peter expanded the empire in all directions. This brings us to an interesting component of the Russian Empire's growth: Why expand? Many historians have speculated that there has been an inherent drive to expand the state in an effort to gain coastline. All major European nations with world dominance during this era had a lot of coastline and were maritime powers. Because of Russia's position on the continent, it is conceivable that most of its growth occurred in part to reach the coast in order to become a world power as well. This is speculative to be sure, but it does make some common sense.

Russian ambassador dealing with Napoleon.

One resource that's certainly driven Russian continental expansion was **fur**. Given the climate of this part of the world and the way people dressed, it certainly is the case that the trading of fur—that is, stripping the exterior of live animals and making coats out of it—was the main economic engine behind the continued growth of the land empire. Cold climate = animals with lots of fur = animals we want to kill for their fur = fur coats worth lots of money = more land = more furry animals to continue driving the economy. Speaking of continued land growth, we come to another familiar name in history, **Catherine the Great**. She radically expanded the empire across the continent, was a familiar face in European politics, and helped propel the Russian state into serious power player status on the continent during the 1760s and 1770s. Catherine continued to modernize in the Western European style, and her rule re-vitalized Russia, which grew stronger than ever and became recognized as one of the great powers of the continent for the first time ever.

Russia in 1800

Russia in 1867

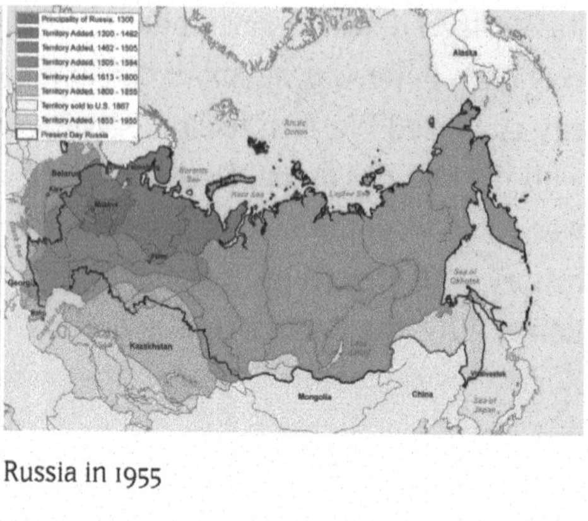
Russia in 1955

Ivan the Great: Mangled the Mongolians.

Ivan the Terrible: Skewered his son. Expanded the Empire.

Peter the Great: Wanted to be Western.

Catherine the Great: Made Russia into a European power playa'.

The last face to consider during our old school Russian tour is that of **Napoleon**. Wait a minute! He's not Russian! True, but that Frenchie did invade Russia in 1812 in one of the many great blunders in history. Napoleon was doing pretty well conquering Europe until he invaded Russia. Here's a little Plaid Avenger tip from me to you: don't invade Russia. Don't ever invade Russia. No one has ever successfully invaded Russia. As Napoleon—and later, Adolf Hitler—found out, it's easy to get *into* Russia, but it's impossible to get *out*. As the French forces advanced to Moscow and eventually took over the city, they did so in the middle of a horrifically cold Russian winter, and because of their dwindling supplies, were forced to retreat. They were attacked continuously on their way out. Bad call, Napoleon. You silly short dead dude! Russian Trix are for kids!

I bring this up to reiterate a common theme in Russian history: Russia always seems to come out stronger at the end of extreme turmoil. After Napoleon's invasion, it comes out sitting pretty as the world's largest territorial empire. We will see this theme again after World War II.

SERF'S UP, DUDES!

I've painted a pretty picture with lots of faces and names that you've heard before, but don't let me suggest that life and times in Russian history were good. Mostly they really sucked, for the lower classes in particular. During the imperial reign I've described thus far, life was fine for the royal court, but life as a commoner was brutal. We have a sort of a feudal system in Russia for most of its history. Serfdom in this country really turns into something more equivalent to full-on slavery. What we're talking about here is a typical feudal structure where there is a lord who owns the land and under him, the mass of people in the country are workers tied to that land. As such, they have no rights to hold landed notes, no human rights, and are basically slaves of the landholder. In combination with the fantastic climate of Russia, one can see how living here for most folks was not a fun time. Chronic food shortages and mass starvations were common occurrences as particularly brutal winters, bad crop harvests, and repressive taxation systems kept the peasants in a perpetual state of misery.

Why talk about the plight of Russian peasantry? Because the dissatisfaction among a vast majority is soon going to culminate in a revolution; a revolution the likes of which the world has never seen. It wasn't like the aristocracy didn't see it coming, either.

In an attempt to placate the masses, Tsar Alexander II passed the sweeping Emancipation Reform of 1861, which was supposed to end this miserable peasant situation. Say what? Emancipation didn't occur until 1861? That's late in the game! Consider: that was the same year that the US started their Civil War in order to emancipate slaves! Unfortunately for Russia, this proclamation on paper didn't amount to much in real life, and the crappy peasant existence continued. By the turn of the century, freed slaves in the US had more rights than supposedly freed peasants in Russia, not to mention that the vast majority of Europe as a whole had long since abolished both slavery and feudalism.

As we progressed toward the year 1900, many Russian folks were still essentially slaves of the land. This was a time when the rest of Europe was industrializing and getting more connected. Europe had political revolutions, which created states that emphasized equality for individuals, even the lowest classes. People in Russia knew what was going on

Dimitri! Where's the party, dude? Looks like the fun never stopped for the Russian peasantry!

in the rest of Europe. They heard about the French Revolution and Britain's change to a constitutional monarchy. If you look at a political map of Europe's changes over time, you'll see that things started in the West, progressed eastward, and these changes took a long time to make their way to Russia. Indeed, economic and political revolution reached there last, but when it did come, it came big time.

As changes were enacted in Europe, as common folk in Europe gained more and more rights and perhaps even more and more wealth, Russia was stagnant. Russia was falling behind yet again. Dissatisfaction was also on the rise, and things in the 20th century made it quite a bumpy ride for the Russian Bear, particularly for its royal line.

THE BEAR TAKES A BEATING

The 20th century, taken as a whole, was not kind to Russia. It started the century on a downward slide. As Russia fell farther and farther behind the times and as popular dissent increased, the imperial government did the worst thing possible: it lost a war. One of the things that kept peasants in line was the concept of a strong central government, even though it may not be popular or even good. Citizens wanted a government to protect them from foreign powers, invasion or destruction from invaders. Even though the monarchy was not popular, at least they maintained a strong military presence to guard their citizens and territory. When they failed to do that, full failure of the state was not far behind, and Japan provided the first kink in the armor of the **Romanov** line.

GODZILLA ATTACKS

Nicholas II rides to the front lines of WWI: "The Monk made me do it!"

In 1905, the Japanese declared war on Russia. The Japanese, having risen in power for the previous fifty years up to that point, began encroaching on Russian lands in the east. This came to a head when the Japanese took over part of mainland Asia claimed by Russia. **Tsar Nicholas II** deemed it necessary to go to war to reclaim this land and to put the Japanese in their place. Easier said than done. The Russians sent their entire fleet from Europe to take on the Japanese in the Sea of Japan. After this extensive voyage, the Russian fleet arrived to be beaten down by the Japanese fleet in a matter of minutes. It was the most one-sided naval battle in history. The entire Russian fleet was obliterated. Back home in Moscow, popular dissent turned into popular hatred after this stunning defeat. The **Russo-Japanese War** was costly and unpopular and served to really get people thinking about replacing the monarchy with a whole new system altogether. But the fun had only just begun!

Back home in Russia, the peasants and workers alike had about enough. Later in 1905, a general strike was enacted empire-wide to protest the slow pace of reform and general discontent with the aristocracy. The system was broken, and the people knew it. This protest was met with an iron fist by the government, and things turned nasty, fast. The 1905 Russian Revolution is best remembered for the legendary Bloody Sunday massacre; cross-toting, hymn singing protesters were slaughtered by government troops as they marched to the Tsar's Winter Palace. Brutal! The Romanovs had restored order, but not for much longer.

WORLD WAR I

The Russo-Japanese War may have been damaging and unpopular, but at least it was far away. However, the next phase of fun happened closer to home, when Archduke Franz Ferdinand of Austria was assassinated in what later became Yugoslavia. This was the spark that ignited World War I and everybody declared war on each other. Russia had diplomatic ties with Serbia, who was pulled into the war immediately, so the Ruskies had to jump in as well. As we pointed out, this came on the heels of a loss to the Japanese on the other side of the Russian empire. From the onset, World War I was extremely unpopular because people at home were saying, "Hey, you already lost to the Japanese! Life here is sucking! Things are going down hill fast and now we are in another war? Nobody even knows what we are in it for! Serbia? Serbia who?"

Archduke Franz Ferdinand

My death started WWI. Sorry!

Tsar Nicholas II

My death made the Revolution irreversible. Sorry!

Vladimir Lenin

My death launched Stalin into power. Sorry!

The Mad Monk, Rasputin

I'm still not dead. Someone get me the hell out of this coffin!

World War I was fought on Russia's front doorstep; for three long years, there was a catastrophic death toll on the Russian side. The Germans made huge gains into Russian territory and snapped up large parts of the front. Meanwhile, back at the bear cave, popular dissent was growing wildly, fueled in part by the shenanigans of yet another famous Russian: **Rasputin**. "The Mad Monk" as he was referred to by many Russians, was a shaman/conman who worked himself and his magic into the inner circles of aristocratic power, including being a direct advisor to the royal family themselves. In point of fact, Tsar Nicholas II was strongly advised by Rasputin to personally lead the charge on the front lines of WWI, even though everybody knew good old Nick was not up for the challenge. Chaos then ensued both on the war front, but more importantly, back home as well. (Check out the inset box on Rasputin, to the right.)

I don't need to get too much more into World War I history after this, because the Russians didn't have a lot to do with it. About halfway through the war, internal dissent within Russia reached an all time high. The Bear reeked of revolt.

What's the Deal with Rasputin?

The biography of Grigori Rasputin, a.k.a. the Mad Monk, is often fortified with folklore, which is fine by the Plaid Avenger because the folklore that surrounds Rasputin is hilarious. What is clear is that Rasputin practiced some sort of Christian-like religion and gained favor with Tsar Nicholas II by helping to medically treat his son through prayer. The Russian Orthodox Church didn't like Rasputin—mainly because churches never like competing ideology, but also because Rasputin loved sinning. Much of this sinning involved prostitutes.

Big City Russian Pimpin'.

During World War I, Rasputin advised Tsar Nicholas II to seize command of the Russian military. This turned out poorly for two reasons: (1) Tsar Nicholas II wasn't a good army commander and (2) while Tsar was away, Rasputin gained considerable control of the Russian government and helped screw up the Russian economy. Needless to say, a lot of people were not happy with him. In December of 1916, a group set out to assassinate the Mad Monk. First, they attempted to poison him. The would-be assassins loaded two bottles of wine with poison, which Rasputin drank in their entirety. The assassins waited and waited, but Rasputin continued to display lively behavior. In a panic, they shot him point-blank in the back. When they came back later to deal with the body, Rasputin jumped up, briefly strangled his attempted killer, and then took off running. The group of assassins gave chase, shot Rasputin AGAIN—this time in the head with a large caliber bullet, and then beat him with both blunt and sharp objects. Finally, they wrapped his body in a carpet and threw it off a bridge into the Neva River, but, the river was frozen over, so Rasputin's body just smashed into the layer of ice. The assassins climbed down to the frozen river and broke the ice under Rasputin so his body would sink into the water. They eventually succeeded and Rasputin disappeared into the freezing depths. Three days later, authorities found Rasputin's body and performed an autopsy. The autopsy revealed that Rasputin had died from drowning and, in fact, his arms were frozen in a position that suggested he died while trying to claw his way through the ice and out of the river.

Me-oww! That cat was a freaky-freak. And his lives ran out. Maybe . . .

THAT'S REVOLTING! COMMIES TAKE COMMAND

The revolt occurred in February of 1917 and Tsar Nicholas II abdicated, meaning he quit before he got fired. When the Tsar, who had been out fighting in the battlefield during World War I, returned to Russia, he discovered his people were in open revolt. He abdicated, and put his brother on the throne. Faster than you can say, "I hereby resign from the throne," his brother abdicated as well. A temporary government was set up, but in terms of who is really going to take power, nobody could tell.

"Let's redistribute some wealth, people!"

On October 25, 1917, this culminated in an event that people all over the world know, **the Bolshevik or Communist revolution**, led by our good friend, **Vladimir Lenin**. Communism was not something that everybody in Russia were just jumping up and down about . . . it is likely that most people had never even heard of it. Karl Marx and Friedrich Engels wrote about the concept in Germany and Austria; they weren't Russians. Lenin was a revolutionary even in his youth, and came in contact with a lot of these ideals during his college years. His older brother, who was executed because of his socialist activism, also influenced him. Lenin became a lawyer, but was eventually exiled to Siberia because of his radical revolutionary activities.

In Siberia, he wrote and published a lot of socialist literature, and became a prominent figure in the revolutionary movement. He eventually fled to Finland, and later Switzerland. While there, Lenin made appearances and speeches to other socialist groups in Europe. When the open revolt in Russia started to happen, Vladimir and other revolutionaries boarded a train and headed back into Russia after a long absence from their homeland. Once Lenin returned, he became a prominent figure in the Bolshevik party and leader of this "Soviet Revolution."

One of his ideas that gained popular support was not necessarily that communism was the greatest thing ever, but instead, "We need to get the heck out of World War I!" He received popular support for his Soviet ideas by default, because everyone pretty much agreed with his other political stances that monarchy, peasantry, and Russian life in general sucked, and change was necessary. The power structure was completely unfair. What also gained popularity was the promise to get Russia out of the war in exchange for the support of his cause. Long story short, the Communist revolution became an internal power struggle, and the Bolsheviks came out on top. "**Bolshevik**" means "majority," but in reality, the movement was anything but a majority. The Bolsheviks had many political rivals in the struggle to control Russia. The commies were actually a very small group of people who enacted this revolution and took over the whole nation.

True to their word, as soon as the Bolsheviks took over the government at the end of 1917, they immediately bailed out of World War I. They agreed to let the Germans have any invaded territory and signed a peace treaty. As you can see from this map, this resulted in huge territorial losses. The Russians lost territories that later became Estonia, Latvia, Lithuania, Moldova, and large parts of Poland. I am telling you this specifically because it is going to become Russian territory again after World War II. The Ruskies never truly intended to give away those lands permanently. In their minds, it was always a temporary fix.

Now as I just suggested, not everybody in Russia immediately embraced this communism gig. In fact, there were many holdouts. As you might expect, there were conservative people in the aristocracy and the military who thought that the monarchy, the old established way, should return and things should revert back to the way they were.

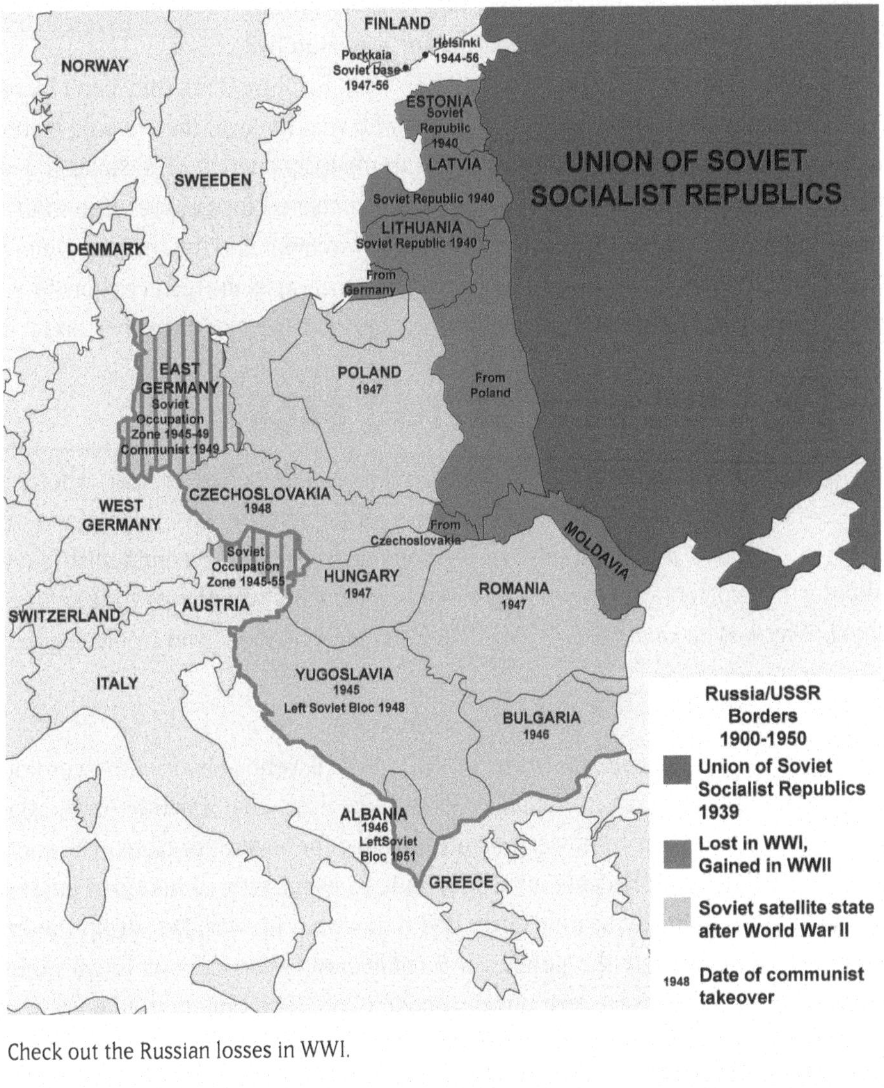

Check out the Russian losses in WWI.

As soon as the Bolsheviks took power, one of the first things Lenin did was put the entire royal family under house arrest. Shortly thereafter, a civil war broke out; from 1918 to about 1920, there was open fighting throughout the country. The **Russian Civil War** was fought between the new party in charge, the Communists, and the old conservative holdouts and much of the military. The parties who fought the Russian Civil War referred to it as "the Reds versus the Whites." Obviously, the Reds were the communists and the Whites, the loyalists. The Red Russians versus the White Russians!

Again, our story here is that the 20th century was not very kind to Russia. It had a disastrous World War I, and then an internal revolution which turned into a civil war. The civil war is eventually won by the Reds. BTW: once the civil war broke out, Lenin ordered the assassination of the entire royal family. Why did Lenin do this? I thought he was a nice commie. Lenin believed that in order to succeed, Communist Russia had to sever all ties with the past: "You people are fighting for a monarchy? You want to restore the imperial royal line? Well, I will fix that! They can't be restored to the throne if they're all dead!"

What's the Deal with Anastasia? Anastasia was born the Grand Duchess of Russia in 1901. When the Bolsheviks took power in Russia, Anastasia's entire family was executed because of their links to the Romanov dynasty. However, legend is that Anastasia and her brother Alexei might have survived the execution. This legend is supported by the fact that Anastasia and Alexei were not buried in the family grave. Several women came forward in the 20th Century claiming to be Anastasia, but guess what: she's dead. Regardless of the intrigue and the crappy animated kid's movie, The Plaid Avenger is certain that Anastasia died of lead poisoning with the rest of her family. Lead poisoning induced by bullets. Lots of bullets.

Got Dead? Yep.

This made the statement that no matter who won the war, there would be no return to monarchy because there were no more monarchs. The Romanov line ended with the assassination of Nicholas II, his wife, and five children including Anastasia; rumors suggesting Anastasia escaped execution circulated for decades (see inset box on right). The civil war was finished by 1920, and the Communists had power over the government and the country. Unfortunately, fate was still not smiling on Mother Russia, as the country saw back to back famines in 1921–23 as the result of horrific winter weather—like the weather is even good in the good years? However, by the mid-1920's, the worst seemed to be over and it was time to get that commie transition on!

THE BEAR BATTLES BACK

The next event of great significance happened in 1924, when Lenin died. For vehement anti-communists, that was a day of celebration; for Russians, it was a day of mourning, because Lenin was the founder of the party, a rallying post, and a popular leader of the communist movement in Russia. He had the grand designs about where society was going, with an ideal of equality where everyone shared the wealth, and the whole utopian society thing. Why was his death titanic? Well, because after he died, who would succeed him? The answer was **Joseph Stalin**.

STYLIN' WITH STALIN

From 1924 to 1953, Joseph Stalin, one of the biggest psychos history has ever known, was in power of what was now called the Union of Soviet Socialist Republics (USSR), the post–civil war title for Russia.

Stalin's ascension to power was problematic for several reasons. For starters: he was crazy. After he gained power, he was psychotic and became more psychotic as every year of his reign passed. He consolidated all power to the center under himself, basically as a dictator, and that's what he was. He set up the infamous secret police force, the **KGB**, who sent out spies amongst the people, and conducted assassinations of all political rivals, their families, their cats, their dogs, and their neighbors. Anyone who spoke ill of Stalin ended up in a grave with a tombstone over his head. No, I take that back. They ended up in a mass grave with dirt over their heads. He was a nasty guy in a tough place in an extremely gut-kicking century for Russian history. But our story is not over yet, kids—the fun is still not done!

Before I go any further, I should say there is always a bright side to every individual. Smokin' Joe Stalin's silver lining was that part of the commie plan was to consolidate agricultural lands and speed industrialization. The Reds wanted to catch the country up with the rest of the modern world economically, technologically, and militarily. As much as I hate to admit it, that is exactly what occurred during the Stalin era.

Essentially Stalin oversaw a "crash course" of modernization, industrialization, and collectivization. These hugely ambitious goals could not have been achieved just by planning, though; the Russian people themselves have to be given a large share of the credit to have pulled this off. Perhaps Stalin's biggest contribution to these gains was not just organization and planning, but he somehow instilled a die-hard sense of uber-patriotic nationalism in the people that made for explosive societal gains. Specifically, the Russian people were convinced to work harder, work longer, make more out of less, consume less, and totally work hard for the country, not for themselves. And they did!

Of course, Stalin used all sorts of coercion and open force to achieve these aims when necessary. And again, he was nuts. But if we are going to credit Stalin with anything other than intense insanity, we can credit him with the Soviet infrastructure moving forward at breakneck speed, and they indeed caught up in a matter of a few decades. They created the atomic bomb shortly after the US did, and even beat the US into space with the launch of the **Sputnik** satellite. Fifty years earlier, they were all beet farmers. That is some fast progress!

THE BEAR ENTERS WORLD WAR II

By World War II, the USSR was starting to be a major world force. Before the war began in 1939, Joseph Stalin and Adolf Hitler met and signed a nonaggression pact. The meeting went down like this, brought to you Plaid Avenger style: Germany said, "We are going to take over the world, but here's the deal: we won't attack you as long as you don't attack us." Smokin' Joe Stalin, in all of his wisdom, said,

Joseph Stalin: Scariest Psycho of the 20th Century
Unfortunately, the award for "Scariest Psycho" is highly competitive. There's Hitler, Pol Pot, Mussolini, Idi Amin, Pinochet, Bob Barker, and many more. But the Plaid Avenger is fairly certain that Joe Stalin is the nastiest of them all. Stalin was one of the Bolshevik leaders who brought communism to Russia in 1918. When Vladimir Lenin, who was the original leader of the Bolsheviks, unexpectedly died in 1924, Stalin slowly assumed power. In 1936, Stalin initiated **The Great Purge** which lasted for two years. During this time, "dissenters" were shipped off to Gulag labor camps and often executed. Most "dissenters" were actually just normal folks who had been wrongly accused by other normal folks who were being tortured at a Gulag labor camp. As you can see, this was a self-perpetuating cycle. If everyone had to sell out eight friends to stop getting tortured, and each friend had to sell out eight friends, pretty soon you would have a ton of political "dissenters." It's like an incredibly violent email pyramid scam. Anyway, The Great Purge resulted in the death of perhaps 10 to 20 million Russians.

While The Great Purge was probably Stalin's sentinel work of psycho-sis, it was far from his last. Stalin organized a giant farm collectivization which left millions homeless and millions more hungry. He also continued purging dissenters until he died in 1953. Historians estimate over 45 million Russians died directly from Stalin's actions (this figure does not include the estimated 20 million Russians that died during WWII). Norman Bates got nothing on Stalin.

Russian soldiers line up for their daily vodka ration—of three liters.

"Okay, go ahead and take over Western Europe and the rest of the world. That sounds good to us. They were always kind of bothersome anyway. Also, we won't attack you as long as you don't attack us. Deal!" The war officially kicked off in 1941, when Germany attacked everybody. Hitler later made the fantastically idiotic mistake, after taking over most of Europe, of reneging on the nonaggression pact. In other words, he attacked the USSR anyway.

As discussed before about Napoleon, one of the greatest historical blunders for anybody is to invade Russia. No one's ever done it successfully, and Hitler was one in a long line of idiots who tried. After he attacked the USSR, the USSR was then obliged to defend itself and was pulled into World War II. This is of great importance: the USSR was the real winner of the war in Europe. Once the Germans declared war on the United States, who joined the rest of the Allies, it became the Germans versus the world. The Germans were really in a pinch as they were attacked on all flanks, the Allies on one side and the Russians on the other. While I am certainly not going to devalue the role of the Allies in the west, Russia really took out a lot of Germans and caused them to divert resources and men to the Russian front, which is the reason why World War II was won by the good guys.

Had the Russo-German nonaggression pact been upheld by Germany, it would be hard to tell what the map of Europe would look like today. It certainly wouldn't look like it does now. And the Russian death toll was astronomical. There were almost 9 million Russian military deaths, and at least 20 million civilian casualties—29 million people total! Take 29 and multiply it by a thousand, then take that and multiply it by another thousand! The Plaid Avenger's not trying to insult your intelligence. I just want you to take a second and think about how many folks died in this thing. Just on the Russian side!

This was some of the most brutal fighting on the entire continent. The Russian front was bloody, nasty, and deadly, and when all was said and done, the Russians beat the crap out of Germany. It was part of the main reason the Germans had to surrender in the end. I really want to stress the Russian losses. 29 million killed. We think about just the US's role in World War II (which was great by the way), but the US suffered maybe half a million to Russia's nearly 30 million deaths! As a result of WWII, Russia experienced an epic population loss and made a major impact on the European theater of war. The war ended in 1945, the end of which set up the scenario for the next fifty years.

IS IT GETTING CHILLY IN HERE? THE COLD WAR

At the conclusion of the war, the Allies came in from the west and the Russians were mopping up in the east and they met each other in the middle of Europe. They all did high fives and then they split Germany in half. Hitler was dead in the bunker, and there was a line where the Allies met after the war. The Allies were the United States, Britain, France, and the USSR. That's right! Smokin' Joe Stalin was on our team! Where the two forces met, as you see on this map, is halfway through Europe. That line quickly became known as **The Iron Curtain**.

What does this mean? Well, the Russians occupied Eastern Europe; after the war was over and everyone finished celebrating, the Soviets said, "Maybe we will just stay here for a little while. We'll make sure everything has settled down. We'll help rebuild this side of the continent." The United States, under the **Marshall Plan**, was building up Western Europe. Both parties occupied Germany to make sure it wouldn't start up trouble again—thus we had an East Germany and a West Germany, which reflected events happening in Europe as a whole.

The metallic drapes were drawn.

In their effort to do this, the Soviets were pretty sinister. Under Soviet tutelage, "elections" were held all over, in places like Poland, Czechoslovakia, Hungary, Yugoslavia, and parts of Austria. We all know the outcome: lo and behold, the Communist party won resounding victories in every single place they held a vote! Everyone in the West knew this was a farce, but nobody wanted to start another war by calling out the Russians publicly about this scam. Remember, they were allies at the time, and they helped defeat the Germans together. The United States just wanted to get the heck out of there. They were not about to start a new war with Russia. The Western governments hesitantly went along with these newly "converted" communist states.

This set up a scenario where puppet governments replaced the previous governments. There's a term for this you should already know called **soviet satellites**. Some examples were sovereign states like Poland, Czech Republic, and Hungary that actually had seats at the UN—but everybody knew that the Soviets had the real control. These were puppet governments, to give the illusion of being expressive governments of their people. The situation simply evolved into Soviet occupation and control at this point. One of the reasons Western Europe and the United States allowed this to happen is that the Soviets very adeptly pointed out that they were staying in these Eastern European nations to make them a buffer zone because countries in Western Europe historically kept attacking them. That was true (e.g., Napoleon's France, Hitler's Germany).

These Eastern European states behind the curtain, that ended up being Soviet satellites, were actually parts of the territory that the Soviets had lost at the end of WWI, so the Soviets regained that loss. I like to point this out because it plays into what is happening in today's world. Whole countries that were sovereign states for a while between World War I and World War II (like Estonia, Latvia, Lithuania, Belarus, and the Ukraine), were totally reabsorbed by the Soviet Empire. These places declared independence after World War I, and then suddenly found themselves no longer sovereign states but soviet republics or sub-states of the Soviet Union. That's why the map of Eastern Europe changed so rapidly. Places came and went as sovereign states were reabsorbed.

COMMIE ECON 101

The first half of the 20th century was fairly repressive, brutal, chaotic, and completely violent for Russia, but things stabilized for them internally during the Cold War. We have already set the stage in Europe for the East versus West ideological battle for control of Europe, which is going to turn into a global phenomenon that has shaped virtually every aspect of every other region that we talk about in this book. What's the deal with the USSR as full-fledged commies after World War II?

As I have suggested previously, the Soviet Union experienced massive growth during Stalin's regime. He died shortly after World War II, but a lot of the industrialization which was radically successful under him continued on. Many of the programs, while oppressive and brutal, made the state a stronger world power. Indeed, it was the rise of the USSR which created the bipolar world that defined the Cold War. Stalin did a lot to get them up in the big leagues, and then he checked out. He brought about massive industrialization, particularly in the weapons sector, which continued after his death.

What happened under the Soviet experiment post–World War II? What did they do? What did it look like there? What did they make? What did they produce? How would that be part of their eventual undoing? During the Soviet era one of the primary goals, even under Lenin, was to catch up industrially. They did this by centralizing everything. That's what communism is known for. The state runs everything, all aspects of politics—of course that's easy, it's what all governments do—but also all aspects of the economy—and that's typically not what the governments do. That is the communist way: full political and economical control of the entire country.

Tanks for the memories, commies!

This is called **central economic planning**, and it entails the government's approval of where every 7-Eleven is located, where every ounce of grain is grown, where railroad tracks are laid down, where coal will be mined, where whole cities will be built to support something like an automotive industry. Every single aspect of life that we take for granted in a capitalist system was controlled in the Soviet sphere. This was done very early in the Soviet Empire to accomplish specific goals, one of which was called **forced collectivization**. Russia had been a peasant society, land based, rural, farmers digging around in the dirt, for hundreds of years. It couldn't do that in the 20th century if it was going to catch up and become a world power; it needed to get people off the land and into the cities. Why the cities? To work in the big factories that were industrializing and creating stuff.

Goal number one was forced collectivization. Pool all the land together, because it now belonged to the state: "You used to have rights to it and a deed on it, Dimitri, and maybe you used to grow cabbage and beets on it, but those days are gone now, comrade! You've got two options: go to a city or get in a collective, government regulated, growing commune." The state controlled all the land and agricultural production, because the state had one huge tractor to do the work of 1,000 men, and it had pesticides and fertilizers, which made huge monoculture crops. This new system produced tons more food. Whereas 1,000 dudes had to farm a piece of land before, now just ten can get the job done. 990 dudes headed to the city to accomplish the next goal: **rapid industrialization**.

To achieve rapid industrialization, the Soviet government needed to get everybody in and around the city into big factories. What were the factories going to produce? All of the things the West has that the Soviets needed to catch up on. Things like tanks, petrochemicals, big guns, rifles, maybe some big fur coats every now and again because it's cold there, aircraft carriers, missiles, bulldozers, heavy machinery, etc. The question you need to ask yourself at the same time is: what were they not making? The big difference between the communist expression and the capitalist expression in the 20th century comes down to this question. The answer is that the Soviets didn't make the items that normal people buy. Instead, they made huge stuff that only countries used. Like I told you earlier, the Soviet system stressed more work and *less consumption* by the individual, so that the state can become stronger.

You have to remember, the Soviet way was not focused on individual citizens or individual rights. The Soviet government didn't care about personal expression or fashion statements or if you liked building model airplanes. For them, the Soviet Union was an awesome state and an awesome idea and its citizens should be ardent nationalists and do everything for their state. By and large, people were pretty cool with that; at least most of them were, because to not be cool with that meant the KGB would run you down in the middle of the night and shoot you. People accepted it. Lots of them even liked it.

What wasn't produced in the Soviet economy? Consumer goods like cool clothes, sunglasses, lawn chairs, microwave ovens, refrigerators, independent-use vehicles. Anything that you'd go to a mall to buy today did not exist in the Soviet world. Knick-knacks did not serve the greater goal of the government, which was to catch up with the capitalists. That's just the way it was.

As much as the Plaid Avenger likes to make fun of knick-knacks and other crap you don't need, it played a role in why the Soviet Union unraveled—but that's for the next section. They did indeed catch up and became a world power under this system. One of the other things the Soviets did with the big things like tractors and bulldozers and weapons was to export them abroad. These are items that the Soviet Union is still known for. Russia is, to this day, a huge manufacturer and exporter of arms. Example: the Russian made AK-47 is easily the most popular and most widely distributed rifle on the planet.

The AK-47; Russian export extraordinaire. Seriously. Still used the world over.

Russia also exported a lot of this stuff at cut rates to other countries around the world. Why would you export a bunch of missiles to Cuba? That's right, for the Cuban Missile Crisis! You also want to send around machinery and petrochemicals and similar items that you produce to make money and recoup costs, but more importantly, to promote your influence in the world. That is what the Cold War was all about: coaxing other countries to join your side by providing them with things they really need, building bridges, selling them weapons and chemicals, and maybe even lending money. Anything to get people on your side. The United States and the Soviet Union both did this all over the place.

Only in a few places does this battle of ideology turn hot and people start shooting at each other. Two places of note where this occurred were in Korea, resulting in the Korean War and in Vietnam, resulting in the Vietnam War. We might even go as far to say as this also happened in Central America for all the Central American wars, during which both these two giants funneled in weapons so that the locals could fight it out between their opposing ideologies. The commies versus the capitalists in all these places was really what the wars were all about, and the two giants funded their respective sides.

CATASTROPHIC COMMIE IMPLOSION

How did the Cold War end for the Soviets? Why did I suggest that knick-knacks on the shelf may have been part of the winning strategy of the United States, when all is said and done?

RUBLE RUPTURE

The Soviet Union, from the 1960s through the 1980s, simply overextended itself. The United States and Western Europe's economies were based on producing a little bit of everything: tanks, aircraft, and petro-chemicals, but also consumer goods like cars, washing machines, toys, and Slap-Chops.

Hurry! Buy now! These things are flying off the shelves!

When you have stuff that individual citizens can buy and sell, you don't have to solely rely on big ticket items to fuel the economy. How many bulldozers can you sell in one season? I mean, The US system ended up championing in the end, because people buy consumer goods all the time. People like to buy refrigerators and new cars and blow dryers and pet rocks; but the US also made and sold bulldozers and missiles too. Long story short; US/Team West got richer, Soviets stagnated.

Russia's economy stank because it was based on flawed principles and was focused solely on items that eventually made it unsustainable. But there were other reasons these guys were going broke. Namely, the USSR liked to give lots of money and support to their commie allies worldwide. During this expansion of Soviet influence across the planet, they really started to overextend themselves. They were giving stuff to Angola and the Congo in Africa, buying sugar from Cuba that they didn't even need, just so Cuba would be their ally, and lending money to places in Southeast Asia. By the 1980s, they were going broke while their economy was also stumbling.

Another ruble-draining effort was being stupid enough to get involved in a war in Afghanistan. Geez! Who would be dumb enough to get bogged down in an unwinnable war in Afghanistan? Oh, ummm, sorry US and NATO, I forgot about current events for a second. You guys should have really learned more from the Soviet experience, which was an . . .

AFGHAN BODY SLAM

Let's go into more detail on this mess, because it does play into current events more than most want to admit. Ever since the USSR came into existence, it crept further and further into control of Central Asia. eventually absorbing all of those -stan countries and making them into Soviet republics. All except one: Afghanistan. The USSR stayed very cozy with the leadership of the country, though, and gave them lots of foreign aid to ensure that they stayed under the commie sphere of influence.

However, there were other forces within Afghanistan that didn't like the Soviets and, for that matter, their own leaders. A group of Muslim fighters we'll call the **mujahideen** decided to wage a war to overthrow the Afghan government. The government feared a greater coup was at hand, so it invited the USSR to invade them to help out. Sounds good so far, huh? Yeah, right. This was a huge freakin' mistake for the Soviets.

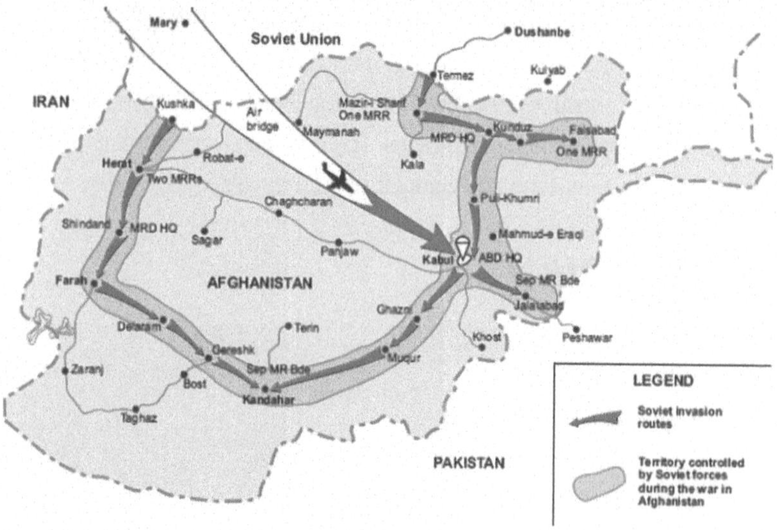

In 1979, the Soviets invaded Afghanistan, and started a decade-long war which will serve to demoralize and humiliate the USSR. Remember, this was during the Cold War, so the US CIA—under orders from the government—funneled tons of weapons and training to the mujahideen to keep the Soviets pinned down. It worked like a charm. In fact, it worked too well! After the mujahideen eventually repelled the Soviets, they then proceeded to have a civil war, and the winners of that civil war became the Taliban. Crap. Any of this starting to ring a bell yet? The Taliban are buddies with and shelter al-Qaeda, and after 9/11, the US declared war on both groups. Now the US finds itself fighting against forces it helped arm and train back in the 1970s. Oops.

But that is a tale for another time; back to our soviet story. The Soviet/Afghan campaign of 1979–89 was a devastating war for Russia, comparable to the Vietnam War for America. The active conflict in Afghanistan started to siphon away resources at a rapid rate. At the end of ten years of fighting, many Russian lives were lost, many Russian rubles were spent, and all for nothing; they were forced to walk away from the whole mess with their tail between their legs. Check out an awesome flick which shows how this went down: *Charlie Wilson's War* starring Tom Hanks. As a bonus, it has male rear nudity—of Hanks, himself! Actually, ew.

THE GREAT COMMUNICATOR SAYS: "I HATE COMMIES"

Once the 1980s rolled around, a new component is introduced: the Reagan Factor. Ronald Reagan was the most hilarious US president ever, in the Plaid Avenger's opinion, and was certainly one of the reasons for the demise of the USSR. He is often credited as being the entire reason, which is preposterous, but his administration did a lot to accelerate the processes in play regarding Russia's economic decline. The drive for military buildup during the Cold War was all fine and dandy when the Soviet Union was expanding and growing industrially and economically, but this started to take its toll on them in the stagnant 1970s and 1980s. When the Americans developed a new type of missile, the Russians would be forced to develop a similar one in order to stay competitive. Then the Americans built a new type of tank, so the Russians built one as well. It was a "keeping up with the Joneses" sort of situation. You have to be as cool as your neighbors by having the same stuff, the newest and coolest gadgets. That had been happening since the 1950s. **The Space Race** was a classic example of that competitiveness. Then in 1980, Ronald Reagan was elected president of the United States, and hilarity ensued and accelerated.

Reagan hated commies. If you learn one thing about Reagan, it's that Reagan hated commies. He hated commies when he was a Hollywood actor and ratted out a bunch of other actors as communists during the McCarthy hearings. He hated commies when he was the governor of California, and he really hated commies when he became the President of the United States. He hated them so much that he was willing to send money and arms to anybody on the planet under the guise that they were fighting commies. In conclusion: Ronald Reagan hated commies. Have I made that clear yet? When he came in power in 1980, he had to deal with domestic issues because the US economy sucked. One of the primary things that he and his administration did very early on was accelerate military spending. This was a very clever move because it did a few things at once: It helped the American economy by creating jobs building more bombs, and also, Reagan knew that the Soviets were trying to keep up militarily, so he was making them spend more money as well. Money they didn't have.

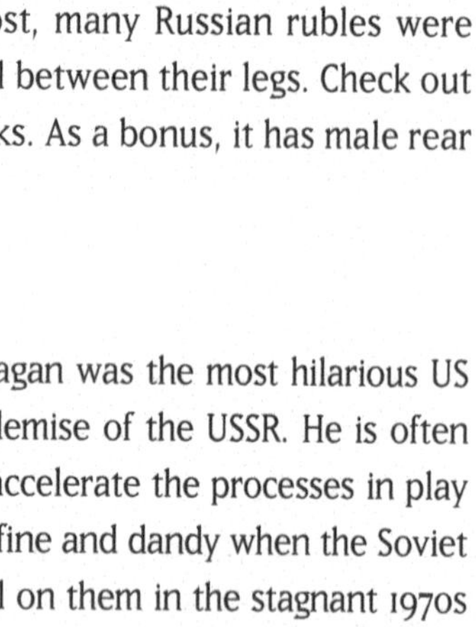
"I hate commies."

20TH CENTURY COMMIE COUNTDOWN

Lenin: 20th century premier commie and snappy dresser.

Stalin: 20th century premier psycho with a serious 'stache.

Krushchev: Placed missiles in Cuba. Nice job, laughing boy.

Brezhnev: Patented the Unibrow, kin to Herman Munster.

Andropov: He "dropped ovv" after only 16 months in office.

Chernenko: Not to be outdone, he croaked after only 13 months.

Gorbachev: "Oh crap, this place is falling apart."

This over-spending craze happened during the entire decade of the 1980s, just as internal dissent became more prevalent within the Soviet Union. Rioting broke out in places like Czechoslovakia, Poland, and Hungary, and as the economy continued to worsen in the Soviet Union, people began to starve. The US media showed scenes of people in long bread lines in the Soviet Union. These guys weren't just broke; they were going hungry at the same time they were spending millions, if not billions, on weaponry to keep up with the US. It was a crafty move that paid off huge dividends for the US. When the final Soviet leader came to power, Communist Party Secretary General **Mikhail Gorbachev**, he inherited a state that was broken and riddled with holes. Gorbachev knew his administration was screwed. The comrades continued to lie to themselves and used the KGB to scare everyone so they wouldn't utter it out loud, but the people at the top knew that the party was over. The Communist Party, that is.

HOLY GLASNOST, THIS IS DRIVING ME MAD! PASS THE SALT, PERESTROIKA

What Gorbachev did was enact three things. Two of them are words you should know. One is **glasnost**, which means "openness." The second is **perestroika**, which means "restructuring." The third thing he did was try and limit military funding. Glasnost meant Gorbachev was tired of the secrecy of the Soviet government. He didn't want to lie to people any more, and he wanted to stop the KGB from terrorizing people. He wanted citizens to be more open with the government so it could improve. Places in Eastern Europe such as Poland, who wanted their own votes and no longer wanted to be a part of the Soviet Union, were let out. Under perestroika, Gorbachev knew they had to restructure the economy and reel in military spending. They also needed to cut off subsidies to countries they could no longer afford to prop up. This all leads to number three: not being able to spend any more money on the military. Or at least not as much.

This third enactment led to the **SALT**, first in the early 1970s and then again late in the decade. The **Strategic Arms Limitation Talks** were about curbing weapons production—not stopping weapons production altogether, not getting rid of weapons—just slowing down the speed at which they were being made. What you have to understand about the Cold War is that, if humans survive for another 1000 years, it will be looked on with hilarity and humor; the whole reason that we haven't killed each other yet is because there are so many bombs on the planet between the Soviets and the USA. There was a principle that basically sustained our life on earth, known as **MAD**. **MAD** stood for **Mutually Assured Destruction**, which essentially meant if anybody lobs a bomb, then the other side will throw a bomb, in which case both sides will throw all their bombs and everyone will die. Who is going to do that? The insane logic of MAD is what stopped us from attacking each other, but it made us continue to make more bombs. Both sides already had enough bombs to annihilate the other side, but they just kept making more. The US would say, "We have 1000 bombs that can each destroy five countries at once!" Then the Soviet Union would build 1001. That's where you get into these astronomical numbers of bombs, a bomb for every darn person on the planet. Why? I can't really answer that, because it makes no sense. But I digress. I get a little agitated thinking about MAD; how fitting.

Back to our story: this equated to Gorbachev realizing that they had to stop doing this. During the SALT talks, they agreed, "Let's not build this type or class of bomb anymore, or build half as many as you were going to and we will build half as many. Also, how about we eliminate this type of submarine, and you guys eliminate that type of submarine." When these talks started, there was some capitulation, but for the really big thing, Gorbachev would say, "Let's not build these big intercontinental ballistic missiles," but Reagan (because he hated commies so much) built more of those missiles because he knew the Soviets were going broke. Indeed, not only did he *not* agree to weapons elimination, but accelerated production and built more. In the same year, Reagan appeared in front of the US Congress during the State of the Union address and said that the USSR was an "evil empire" and that the USA must do everything in their power to combat evil. He was really throwing all the cards down on the table—what a huge bluff in the global game of poker.

Here is the bluff that won the game: Somebody made mention to Reagan of how cool it would be if the US put missiles in space; thus when the Soviets shot nuclear weapons at anyone, the US could shoot their bombs out of the sky with these space missiles. This missile system evolved further to become a space laser to shoot out bombs and was called **SDI**, the Strategic Defense Initiative, given the nickname **Star Wars**.

When this idea came out it was merely that: an idea. Then some people started throwing around some funds for this idea, which was most likely sketched on a cocktail napkin by a science fiction writer. Somehow Reagan got his hands on this cocktail napkin and started to circulate it around. Anybody heard about this Star Wars thing? It sounds pretty good. When news of this reached Gorbachev in the Soviet Union, the comrades said, "Oh, crap! They are going to have an orbiting missile defense system! In space! We are going to have to spend millions and billions of research dollars to figure out how to get one of these too and we are already broke. We can't do it!"

SO LONG TO THE SOVIET: THE END

Gorbachev went to Reagan and said, "We will stop making these bombs, guns, and tanks, just don't do this Star Wars thing!" Reagan, in all of his wisdom, said, "Sorry, we can't; it is already built. We are too far along and can't stop now." This was a complete line of bull. There was nothing in space. Nothing was even tested. This huge bluff paid off because Gorbachev decided to throw up his hands and said, "Well, that's it then; we can't keep up. If we can't have a missile defense, that means you guys win. You could nuke us, so we quit. MAD is over and so are we!" This was when the USSR really fell down, around 1988 to 1989.

"We're not MAD anymore; we're broke!"

Several things collide here. The Soviet economy was in trouble, they had overextended themselves with their propaganda around the world and with Soviet aid in other countries. They had a war in Afghanistan with active fighting for 10 years, from 1979 to 1989, which proved to be disastrous. It took a huge toll in not only millions of dollars of equipment, but also the death of 25,000 Russian soldiers. When all was said and done, they had not done anything there for 10 years. They failed to control Afghanistan, so they had to walk out of it. Glasnost and Perestroika are starting to be interpreted literally by Eastern Europeans and the Soviet people themselves, who were now calling for restructuring and/or independence. In 1989, the Soviet Union had to throw in the towel: "Poland, you want out? Go ahead. Estonia, Latvia, Lithuania, you want out too? I guess you guys are out as well. Central Asian republics, we're done with you too." It all disintegrated in very short order, and all those countries declared independence.

Fifteen new sovereign states emerged from the ashes of the USSR. The strings of the Soviet satellites' puppet governments were cut. The map was redrawn in one fell swoop. In 1990, they adopted economic and democratic reforms into what is left of the USSR, which became the United Federation of Russia. In 1991, Boris Yeltzin was elected president of Russia. He recognized the independence of all the countries that had declared it. In that same year, the newly formed Congress voted for the official and permanent dissolution of the USSR. It was no more. From 1991 forward, it was called Russia.

RADICAL RUSSIAN TRANSITION

By the 1990s, the commie threat was gone and the USSR was dismembered, so everything in Russia should be all champagne and caviar, right? Unfortunately, everything still sucked for Russia. It has gotten a lot better in the last few years, but we are going to carry it forward from 1991, when all this becomes official in today's world. That first decade of post-Soviet independence for the Ruskies sucked as bad as the decade before it. The pain of a radical transition had to be suffered through as they approached the 21st century. What kind of transition, and what kind of pain?

ECONOMY TAKES A DUMP

As we pointed out during the final decades of the Soviet Union, their economy was in bad shape, and changing their name doesn't change the facts. Now it's the "Russian" economy that sucked. Their industries were still in Cold War mindset and only produced big stuff like tanks and missiles that didn't have much use in today's world. You could still sell such items, but not nearly enough to rebuild an economy. They had tremendous problems with shifting their entire industrial sector to more capitalist-centered stuff, normal consumer goods. Capitalism was still a new concept for them, and they were still breaking in their capitalist cowboy boots. The transition from commie to capitalist was full of pitfalls and road bumps and full-on collisions, which made for an extremely rocky economic road to recovery.

The Bear gets skinned.

On top of that, Russia lost its international status. They used to be "the other world power" but in the 1990s nobody would give them the time of day, much less a bank loan. This has changed in the last few years, but right after 1991, no one was willing to help Russia. Russia used to be a somebody, but now it was a nobody. It's also important to note that the USSR went from a singular sovereign entity to fifteen separate sovereign states, of which Russia was just one. It lost tremendous amounts of territory. Those other fourteen countries that weren't Russia were plots of land it no longer owned and there was the agriculture produced on that land, and mineral and energy resources under it, and the people who live on it were a workforce. Those are the basic foundations of economies. This territorial/resource loss was a tremendous blow to the Russian economy, an economy that was already hurting.

On top of this, you have to think about the Soviet era, during which the Soviets used a lot of resources from these countries. For example, most of the food production for the Soviet Union came from the Ukraine and Kazakhstan, which became separate countries. A lot of oil and natural gas came from Central Asia, and not only was Russia no longer profiting from these resources, it now had to buy those resources with cash money . . . and they used to own the stuff! It was a double-edged sword with both edges toward Russia. They were getting sliced and diced. Finally, most of the nuclear and weapons stockpiles were in what was Ukraine and Belarus, and there were a lot of expensive complications involved in destroying and/or relocating those weapons.

The second problem with the economy was the influence of organized crime: the Russian Mafia. Crime is everywhere, so organized crime is everywhere as well . . . but it is really strong in Russia. In fact, if you have ever seen adventure movies or international crime dramas, there is always a Russian mafia somewhere. They have always had a presence, even back in the days of communism when they created the Russian black market, which moved untaxed alcohol, weapons, and hardcore drugs. Indeed, they are still doing a tremendous job trafficking drugs and weapons internationally. They're also trafficking people. Yes, I did say people. A lot of the Russian mafia deals with the movement of women and children for sexual exploitation purposes. It's a fairly nasty business, but I guess that's what mafias do.

During the confusion and transformation of the Russian economy in the 1990's, these illegal criminal elements had a complete field day. The Soviet system was fairly corrupt internally to begin with, and that was perpetuated under free market capitalism where there are not a lot of rules. In addition, the government was broke and couldn't pay federal employees, such as the military and police. In such circumstances, corruption becomes endemic and easy to do at all levels. People do what they have to do to stay alive. The influence of organized crime is a major function of what is happening, even in today's Russia. Vladimir Putin himself came to power on a platform of being tough on crime, and in fact, nobody would be elected who is not.

This powerful criminal element in Russia not only stymies local business, but has negatively influenced international investment as well. But the open and outright blue-collar criminals were not alone in the 1990s. A new dimension of white-collar crime arose to give the underworld a run for their money (ha! Pun intended!): **the Oligarchs**.

OLIGARCHS OVER-DO IT

The **oligarchs** were white-collar criminals who, during the transition period of communism to democracy and capitalism in the 1990s, worked for the government or had insider ties with the government and bought whole industries and businesses under the table before they would go to auction. In communism, everything is owned by the state—all the oil, land, timber, energy production, everything. In capitalism, individuals own everything and the state owns virtually nothing. When a government goes from communism to capitalism, it starts selling everything. Ideally, this process would allow any individual

Top Five Russian Oligarchs

Who?	What did he steal from the people?	How much? (USD)	Where is he now?
Roman Abramovich	Oil (Sibneft)	$14.7 Billion	In England, owns big-time soccer team Chelsea FC. Also owns four giant yachts and is building a fifth.
Vladimir Lisin	Aluminum and Steel	$7 Billion	Most likely sleeping in a pile of hundred dollar bills.
Viktor Vekselberg	Oil (Tyumen)	$6.1 Billion	Buying Faberge Eggs. Seriously, he digs them.
Oleg Deripaska	Aluminum (RUSAL)	$5.8 Billion	Hanging out with good friend Vladimir Putin.
Mikhail Fridman	Conglomeration of Valuable Things (Alfa Group)	$5.8 Billion	Lost much of his wealth in Russian tort suit—for stealing from the people, probably not enough evidence for criminal case.

the opportunity to buy former state industries at fair market prices. However, in Russia, this did not happen; most purchases were made by government insiders at rock bottom prices.

Here's an example: Yukos was Russia's oil company, a government owned oil company that was going to go privatize. If you knew some government officials when it was going up for sale and could get the price, and you knew people who worked at a bank, here is what you would do. You find out the day before the sale occurs and go to your friend at the bank and say, "Lend me half a billion dollars," with the promise that you will personally give him 100 million bucks as a bonus after repaying the loan. He smiles and you take the check for 500 million bucks from the bank over to the natural resources department to your buddy who is overseeing the sale of the oil company. You hand him the check for 500 million; he stamps it "sold" and hands you the deed to the oil company. You and your conspirators walk out together the next day as owners of this oil company, and sell it for 2 or 3 billion dollars, its true value, and then you go back and settle up your loan for 500 million with your banking friend and you split the rest. Congratulations! You have joined the ranks of the oligarchs! This was a really seedy big business tactic in the wild west of capitalism where there are no rules. These guys became ultra-millionaires and billionaires overnight.

What is the big deal with this? What's the problem? The Russian people got screwed, that's what. You have state owned resources whose sale should have contributed to the economic growth of an entire country, which instead turned into a billion dollars in a Swiss bank account under one man's name overnight. It was a legal transaction, apparently, but also a massive loss to the government.

In his rise to power, Vlad "the Man" Putin brought some of these crooks to justice—including Yukos head Mikhail Khodorkovsky, formerly the second richest man in Russia—although his motivation for doing so is questionable. It was likely a move designed to consolidate power more than a move to achieve justice.

Another platform in Vladimir Putin's campaign was, "I'm going after those Oligarchs! Going after crime and those chumps that robbed the Russian people!" That is one of the reasons he's such a popular leader. He is seen as a strong anticrime, pro-government force, and he's backed it up because the owner of Yukos is now sitting in jail. As a result, Putin helped Russia stabilize itself a little more. But maybe I am "Put-in" the cart before the horse in this story. . . .

PURGIN POPULATION

Another issue that is extremely problematic for Russia is the population itself. What's happening to the people in Russia? Russia's population is one of the classic examples, if not the classic example, of a population in decline: negative fertility rates and dropping birth rates. The population total is declining, with only about 150 million people in Russia; that number keeps getting smaller every year, which doesn't do much for Russia's economy. You need working people to keep things moving. A lot of things are going on that can account for Russia's demographic decline. It is hard to pinpoint a single factor, but nevertheless, the population is shrinking. Shrinkage is just the start of problems.

If we looked at some of the social indicators for what's happening in Russia, you would swear that you were looking at an African nation. The life expectancy for males is around 47 years, a little more than half that of the United States. You might guess this life expectancy for sub-Saharan Africa or Central Asia, but you typically would not think of Russia. Several things may be contributing to Russia's decline. One is *lack of health care*. Because of the collapse of the economy and the corruption associated with it, health care has become a huge issue. Access to good health care is almost nonexistent, but again, this may be changing. People aren't as healthy

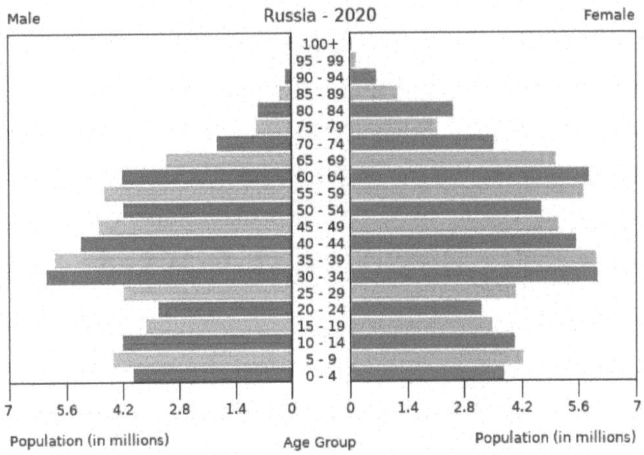

Numbers are going down.

as they should be and don't have access to certain operations or preventive care. Drinking an average of 30 liters of vodka a year probably isn't helping matters either. Other unhealthy attributes include . . .

ENVIRONMENT ENDANGERED

In its rapid industrialization, forced collectivization, and striving to catch up with the West to be a global power phase, the Soviet system completely ignored the environment. There was no limit to the amount of toxic spilling and dumping, or atomic weapons testing. Anything the Soviets could do to get ahead was done with the thought that once they won, they'd go back and clean everything up. Oops, they lost, and it's all still sitting there. Parts of the Kara Sea glow green sometimes from all of the accumulated toxic waste dumped into the river systems. Also, be sure to check out the Aral Sea as soon as you can, because it is disappearing due to misuse and pollution during the Soviet era. There are no less than fifty sites in Russia with pollution of catastrophic proportions. Don't forget to include all of the old nuclear weapons silos that are sitting around, deteriorating. There is impending environmental disaster on a large scale here, just based on past pollutions and degrading weapons. Russia has dealt with this huge problem during the last twenty years, with some international help from the United States and several others.

STOLEN SPHERE OF INFLUENCE: THE REALIGNMENT

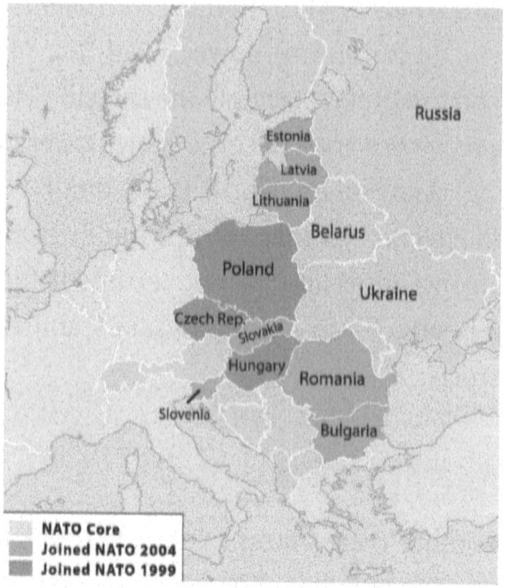

NATO creeps ever closer to the Ruskies.

The last thing you should be aware of in Russia right now is that a lot of its ex-territories and ex-states in its sphere of influence have now disappeared under radical realignment—much to the chagrin of Russia, because it still sees itself as a world player. What am I talking about? Eastern Europe is now desperately trying to become Western Europe by realigning with the West. Russia is losing its sphere of influence all over the place, though it's desperately trying to hold on. It is still actively courting Central Asia politically because there is a lot of oil and natural gas there. They are also courting China and want to keep good strategic ties there. It is really losing influence over Eastern Europe, and what's happening now only rubs it in their face.

That NATO expansion we talked about in previous chapters is still troubling the Russians. Something quite radical, that would have been unthinkable a decade ago, has happened: places like Poland, Hungary, and the Czech Republic joined NATO very quickly after they declared independence in 1991, mostly to ensure no further Russian influence. The next wave of folks was 2004 when Estonia, Latvia, Lithuania, Romania, and Bulgaria, places that were once firmly in Soviet control, joined NATO. Russia now finds that its former territories, which it sometimes harshly subjugated, are sovereign political entities that want nothing to do with it. They are NATO members and could theoretically have NATO weaponry pointed at Russia. This is a big loss of status and influence for Russia. Think about this in an abstract context. The Baltic States go from being owned by Russia, to ten years later being sovereign states with weapons pointed at Russia. That's a fairly big change. But the scales are starting to tip back to the Russian side, because. . . .

RESURGENT RUSSIA: THE BAD BOYS ARE BACK!

Watch out! Things continue to change fairly rapidly for Russia, and it is the understatement of the century to say that things are looking up. Yes, I've talked about how crappy its economy has been. Yes, I've told you all about its people problems, its environmental problems, its crime problems, its international status problems. But what a difference one man can make, and the Russian man of the century is, with no doubt, Vladimir Putin. Putin, along with petroleum, has brought the Russians from the brink of the abyss back to world power status in less

"I can snap your neck with my bare hand. Enjoy your dinner."

than a decade. There are a few other factors at play here as well that are serving to bring the Russian region back to a starring role on the world stage. Let's wrap up this chapter looking at this incredible reversal of Russian fortune.

PREZ TO PRIME; PRIME TO PREZ: PUTIN POWER!

This guy is unstoppable! So rarely in life can you truly credit a single individual with changing the course of history but, for better or for worse, that is an apt description for our main man Vlad. It's no wonder that Time magazine voted him Man of the Year for 2007—he really has made that big of a difference in his country, and in the world. Vladimir Putin is also easily the most bodacious beast of a leader on the planet: as a former KGB agent and judo blackbelt, he could handily whip the butt of any elected official on this side of the Milky Way. And his people love him! Putin held 60 to 80% approval ratings his entire first eight years in office, and stepped down as President with closer to 90% love from the Russians. Stalin would have killed for that kind of popular support—oh, wait a minute, he did kill to get that kind of support. But I digress. Why all the love?

Putin is only the second president Russia has had. He was the hand-picked successor to Boris Yeltsin, a fairly popular leader in his own right. In his eight year run from 2000–2008, Putin oversaw the stabilization of the economic transition process that had thrown Russia into chaos for its first decade. He was tough on the oligarchs, and tough on crime, which helped stimulate international investment. He also took a fairly pro-active government role in helping Russian businesses, especially those businesses dealing with petroleum and natural gas. More on that in a minute. GDP grew six-fold, and propelled Russia from the 22nd largest economy up to the number ten slot. The economy grew 6–8% on average every year he was in office. Investment grew, industry grew, agricultural output grew, construction grew, salaries grew; pretty much every economic indicator you could look at has gone up during his tenure.

President? Prime Minister? It's all Putin to me!

Unlike most world leaders, Putin was very savvy with this cash flow too. He paid off all Russian debts. Imagine that: Russian National Debt = 0. Dang. That's legit. Putin also stashed billions, if not trillions, in a Russian "rainy day fund" that is set aside for any future hard times. How refreshing! Most leaders usually stash that extra cash in their personal Swiss bank accounts. But not Putin. No wonder Russians love him!

Back off, NATO! Putin is protectin' his 'hood!

However, the love is not just based on economics. Vladimir Putin re-instilled a great sense of national pride in his citizens. After the demise of the USSR, Russia was a second rate power with even less prestige, at the mercy of international business and banking and stronger world powers like the US, the EU, and NATO. Russia pretty much just had to go with the flow, even when events were inherently against their own national interest, like the growth of NATO, for example. Putin brought them back politically to a position of strength. He played hardball in international affairs, and re-invigorated a Russian nationalism that had long laid dormant.

Of course, this has come at a cost: Putin achieved a lot of this by consolidating power around his position, controlling a lot of state industries, cracking down on freedom of the press and free speech, and manipulating power structures of the government. In fact, he stepped down from the presidency in May 2008, and stepped immediately into the Prime Minister position the same day, maintaining a lot of command and control of the system simultaneously. That crafty Russian fox! He is increasingly reviled and even hated by Team West because people see him as leaning back towards totalitarianism, but also because he increasingly clashes with western foreign policy on the international stage.

But never fear, Team West, because Russia is a democracy, and one that has presidential term limits, so Putin won't be in power much longer, right? Wrong: After a little constitutional "correction" in 2014 which changed the term length of office from 4 to 6 years, Vladimir Putin was once again elected President in 2012 and assumed the top office. In 2018 he easily won election again, and is currently in his fourth (and final?) term until 2024 . . . unless of course they change the constitution again to let him run indefinitely, or if they just throw the whole constitution out entirely and start a new system. Which they might.

My main man Medvedev.

Medvedev and Putin: the Russian Dynamic Duo

But hold the phone! I keep talking about Putin, and have completely dissed the other guy who held the top slot in the "in-between Putin" years. Let's remedy that right now! Dmitry Medvedev was the hand-picked successor to Putin, and was President from 2008 to 2012 when he basically stepped down by refusing to even run for a second term, allowing Putin to walk back into the position. He pretty much followed and even strengthened virtually all of Putin's policies, and was therefore considered by many in the west as a mere Putin puppet. There could be some truth to that. However, he did have the popular support of the Russian population, and why on earth would he stray too far off of the successful path that Putin had paved? Answer: he didn't. What he did do in office was to start a massive military makeover which seeks to revamp and remodel the Russian armed services into a state-of-the-art ultra-modern military that will certainly be causing consternation for the US and Team West in the future. When Medvedev stepped down, he became the Prime Minister (a position he holds to this day), so this dynamic duo will be holding the top spots for some time to come. Like maybe forever.

In conclusion, Putin power is not to be underestimated. At least six more years as President, and after that he can hold the Prime Minister position indefinitely either way, so he will be large and in charge for a long Russian while. The Russians love him for making them strong, and making them rich and making them proud again. Let's look at some specifics of how he pulled this off and the implications for the future.

PETRO POWER!

Putin is awesome. On that point we are clear. However, we do have to at least partially, if not fully, credit his great success in turning the Bear around on this one single commodity: petroleum. Dudes! Russia has made total bank on oil and natural gas in the last decade. To understand the importance of oil for Russia, one need only consider this: when the USSR crashed in 1991, the price of a barrel of oil was about $10. In 2008, the price of a barrel of oil approached $150. The price may have dropped a bit since, but oil is still the preferred energy source of choice for the entire planet. My friends, you should know this: Russia has a ton of oil! Look at the map! You do the math!

Increasingly, Russia has also reeled in all control of the petro industries to the state. They **privatized** a lot of those industries back in the 1990s, but they have been busy **nationalizing** a lot of them back since 2000. Remember those terms from Chapter 4? Russia sure does. It now has controlling interest in virtually all the oil and natural gas businesses in the country, which even more of the profits swing back directly to the state.

To reiterate a few points from this chapter and the last: oil = power for Russia. Economic and political and international power, that is. Not only have they paid off their foreign debt with oil money, they have re-invested that oil money back into their economy, and also set up their rainy day fund with oil dollars. Since the Russians supply one third of European energy demand, this gives them all sorts of economic and political leverage over their Eurasian neighbors. Mess with Putin, and you might not have heating oil next winter. But the Europeans aren't the only ones who need oil. Look eastward and you see Japan and China both vying for petro resources. Russia is really sitting pretty right now. Energy master of the continent!

Russian Oil and Natural Gas at a Glance

Oil	2006	Natural Gas	2006
Oil reserves	80 billion barrels	Gas reserves	48 trillion cubic meters
Oil reserves, as percentage of world	7 percent	Gas reserves, as percentage of world	26 percent
Saudi Arabian reserves	264 billion barrels	Iranian reserves	28 trillion cubic meters
US reserves	30 billion barrels	US reserves	6 trillion cubic meters
Oil production	10 million barrels per day	Gas production	612 billion cubic meters
Oil production, as percentage of world	12 percent	Gas production, as percentage of world	21 percent
US oil production	7 million barrels per day	US gas production	524 billion cubic meters
Oil exports	7 million barrels per day	Gas exports	263 billion cubic meters
Oil exporter, rank	2	Gas exporter, rank	1
Oil exports, to US	370,000 barrels per day	Gas exports, to Europe	151 billion cubic meters

Sometimes having a lot of gas is a very good thing: Russia may become the center of a natural gas OPEC-like cartel!

GLOBAL WARMING? NO SWEAT!

I can't get out of this section without referencing the global warming situation. While the rest of the planet may be wringing their hands in a collective tizzy about rising temperatures, melting polar ice caps, and rising sea levels, let me assure you that Russia is not sweating the situation at all. Ha! Not sweating the warming—am I good or what?

Seriously though, think of the strategic benefits that are being bestowed upon Russia as the thermometer continues to creep upward every year. Russia has forever been a land empire due to its lack of accessible and navigable coastlines. Additionally, almost all of its river systems empty into the ice-locked Arctic Ocean, making them essentially useless for transportation and exporting commodities. But just wait! As the permanent ice cap covering the North Pole disappears, Russia will become more open to the world like never before in recorded human history!

You may not be paying attention to any of this, but the Russians sure are. In preparations for its ice-free northern coastlines of the future, Russia is staking a claim to the entire Arctic sea east of the North Pole. Seriously, I'm not making this up. The race for the Arctic has begun! Russia stands to be the biggest beneficiary as well. Not only will it open up their territory for greater sea and land access, and greater export power, but it is widely believed that there are massive oil reserves in the Arctic basin, reserves which will become accessible once the ice is gone. Russia is all about the global warming! Let's get this party started! No, actually let's end it with. . . .

"Comrades, the North Pole is ours now. Bye-bye, Santa!"

GEOPOLITICAL JUMP START

You got your Putin, and you got your petroleum, and you add to that a huge dose of nationalistic pride. Bake for twenty minutes, and you end up with an extremely resurgent and resilient Russian cake on the world political table. These guys are playa's once again. Their huge economy, their huge oil reserves, and their huge potential for the future is fast becoming the envy of the world. They can afford to play the fence when suitors come calling. What do I mean by that?

Russia has been invited to become a strategic partner state of the EU— not a full-fledged member, mind you, but a partner. Europe realizes that the Russian economy and Russian energy is such an integral part of their own lives, that to not have Russia at the table for major decisions would be folly. Russia has a real impact on some EU policy. The same can be said of NATO. While Russia is not a member, the group realizes that almost all major decision-making should involve the Russians since they play such an increasingly important role in Eurasian affairs. Russia is typically invited to major NATO summits now. Why doesn't Russia just join the EU and NATO fully?

Because the Russians are forging economic and strategic political ties to their east as well, mostly with China. They are at a historical east/west pivot point right now; they are straddling the fence about which relationships to make or break. I think they're being extremely savvy and are going to avoid taking sides in anything. Could Russia join the EU? It's already a strategic partner of the EU. I bet it will be invited to join as a full-fledged member, but I also bet it will decline.

Why? It is also strategic partners with China, the other huge economy on the other side of the continent. Russia will not be stymied economically by joining a club that can possibly limit their economic options. The Bear is also still a major player in Central Asia, an area with tons of natural gas and oil that has to be moved out of Central Asia to the rest of the world that wants it. That oil and natural gas usually moves out through Russian territory, supplying most of the fuel for Eastern and Western Europe, in addition to increasingly supplying fuel for Japan, China and India.

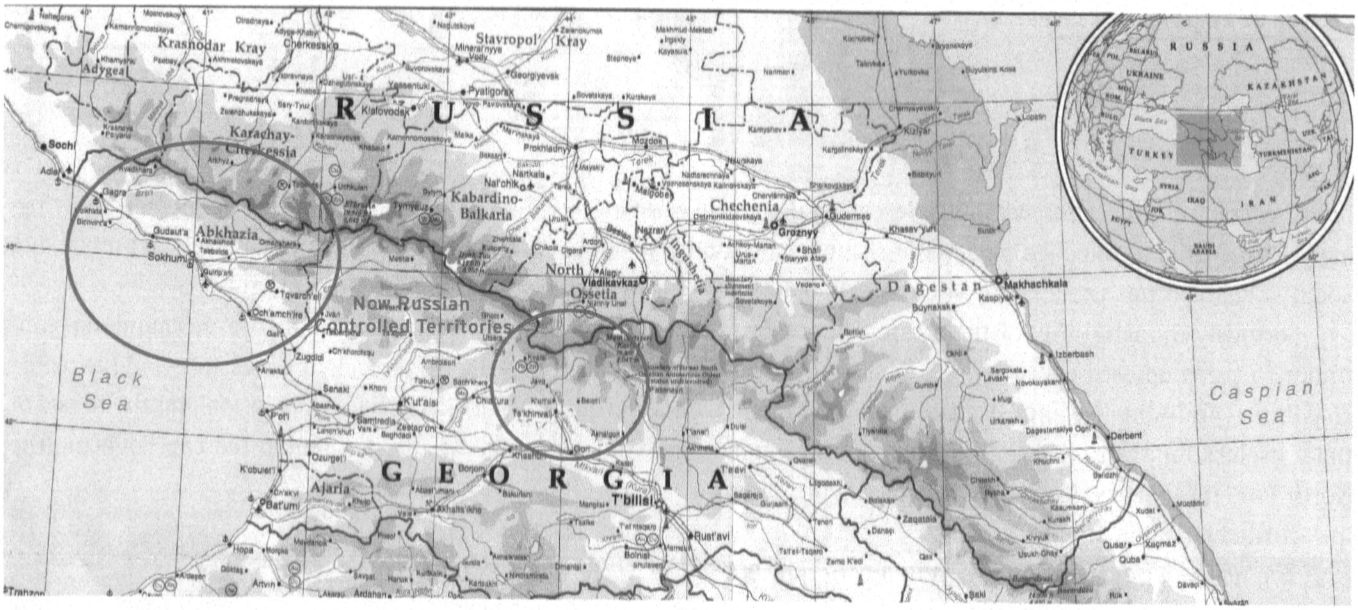

Abkhazia and South Ossetia circled in red: Russian "liberated" and occupied areas that only Russia recognizes as "sovereign."

Beyond the geopolitical ramifications of energy and economics, Russia is back in full action militarily speaking! Big time! Were you aware that in August 2008, Russia invaded the Georgian regions of South Ossetia and Abkhazia? It currently still occupies these areas, and has gone as far as to recognize their declared independence, which is infuriating the US and Team West as a whole.

You will have to turn to chapter 25 to get all the details of this conflict and its future ramifications, but I just want you to know this for now: Russia is back! The Russian–Georgian War did not change the balance of power in this region, it simply was an announcement that the shift in the balance of power had already occurred!

Russia is now openly and powerfully reasserting its influence in its own backyard, and that is a big deal. Russia now feels that it has regained a position of strength to not only stand up to what it sees as threatening Team West expansion to its borders, but to stop that advance with force if necessary. With guns! This open, bold, and even proud use of force by the Russians has changed the whole attitude of the US, NATO, and even the EU when it comes to political maneuvers near the Russian border. The bear has lashed out!

At the risk of redundant repetition, what supranationalist organizations are the Russians all about if they don't really want to play with the EU and NATO? Well, remember the BRIC? They dig that one, and it should be noted that BRIC is specifically not a "Western" institution. Of even greater significance are the Eurasian Union (EEU) and SCO. Flip back to chapter 6 to get the full run-down on these groups, but just let me tease you for now by telling you this much: the EEU is a Putin-led attempt at creating a Russian-centered EU and the SCO is a cross between an Asian version of NATO and a Russian version of OPEC, and Russia has every intention of being the leader of this grand experiment. Stalin would be proud.

DON'T LIKE PUTIN? CRIMEA RIVER!

Holy wowsers! Huge things happened in 2015 that demonstrate this resurgent Russia! Unless you have been hiding under Vladimir Putin's pectoral muscles, you have heard about Russia invading, occupying, and then completely absorbing the Crimean Peninsula, in what was formerly part of Ukraine! Go back to the Eastern Europe chapter to get all the gritty details, but know for now that in the chaos of the second Ukrainian civil meltdown of the last decade, a major political move was made by Putin that has repercussions for the region and the world. How so?

See, there are a significant number of ethnically Russian peeps that live outside of Russia proper since the collapse of the USSR. . . and in this case, a very high concentration of them on the Crimean Peninsula. In addition, Russia has an extremely important naval base in Crimea and used to outright own the whole peninsula back in the day, too. And I want you to think about this whole mess from the Russian perspective for just a minute: what they were seeing (and still are) is a totally dysfunctional Ukrainian state, in borderline chaos, that is overthrowing democratically elected leaders, while trying to 'side up' with Team West—the very group that was the cause for Soviet Union collapse. If you were Russia in this situation, what would you think? In addition, NATO is perceived by most Crimeans (and Russians in general) as encroaching upon Russia's borders. This weighed heavily upon Moscow's decision to take measures to secure her Black Sea port in Crimea. And she secured it by taking it.

Here's how: In the wake of the collapse of the Yanukovych government and the resultant 2014 Ukrainian revolution, a secession crisis began on Ukraine's Crimean Peninsula. . . which is 70% ethnically Russian. Large protests erupted there in February calling for independence from Ukraine, and absorption into Russia proper! This is unheard of in the modern age! Well open your ears and hear it now, cuz it just happened. Unmarked, armed Russian soldiers began being moved into Crimea on February 28. The next day exiled Ukrainian President Viktor Yanukovych (still hiding somewhere in Russia to this day) requested that Russia use military forces "to establish legitimacy, peace, law and order, stability and defending the people of Ukraine." Later that same day, Vladimir Putin requested and received authorization from the Russian Parliament to deploy Russian troops to Ukraine and took control of the Crimean Peninsula by the next day.

On March 6, the Crimean Parliament voted to "enter into the Russian Federation with the rights of a subject of the Russian Federation" and later held a referendum asking the people of these regions whether they wanted to join Russia as a federal subject, or if they wanted to restore the 1992 Crimean constitution and Crimea's status as a part of Ukraine. The "join Russia" option passed with an overwhelming majority. Crimea and Sevastopol formally declared independence as the Republic of Crimea and requested that they be admitted as constituents of the Russian Federation. On March 18, 2014, Russia and Crimea signed a treaty of accession of the Republic of Crimea and Sevastopol in the Russian Federation.

So shouldn't all of that stuff been in the Eastern Europe chapter, and not the Russian one? Nyet! Here is the most important part: Later that day, President Putin gave an emotional speech in the Grand Kremlin Palace announcing that Crimea had been peacefully returned to Russia, reversing a "historic injustice" of the Soviet Union. He also peppered his speech with references to the restoration of Russia after a period of humiliation following the Soviet collapse. . . and blamed most of that nasty era on the domination of the "global superpower and its allies," which of course means the US, the EU, and NATO. In addition, Putin at this point started referring to the duty of the Russian government to protect Russian people, everywhere. . . not just Russian citizens. Do you understand the difference? "Russian people" is not the same as "Russian citizens," and implies that Putin's Russia is giving themselves the authority to protect and possibly unite this proud tribe of people, wherever they are. This speech was given to thunderous applause, standing ovations, and tear-filled eyes of people chanting "Russia! Russia!"

Yes, loss of life and societal chaos are terrible. And, yes, I guess it's a big deal whether Crimea is part of Russia or part of Ukraine. And, yes, the future of Ukraine is important. But. . . none of those things has a big a global impact and will affect the future events on our fair planet as much as the part where Vladimir Putin, wildly popular President of Russia, suggested that he had the duty to "protect Russian people." I mentioned above that this turn of phrase is quite important. Prior to this, Putin, and every other world leader, would say that their duty is to protect the *citizens* of their sovereign state. . . not the members of their particular tribe. Because that is what he is saying: that he, Vladimir Putin, as the President of a sovereign state of Russia not only has the duty of protecting Russians in his sovereign state, but all Russians, the entire tribe called Russian, no matter what other sovereign state they happen to be in!

Dudes! Dudettes! That is the old school definition of sovereignty! Remember that from back in Chapter 3? For most of human history there have been kings, or rulers, or leaders of *peoples*, not of geographically defined political spaces. Think back to, let's say, the French. Throughout most of history, the king or leader has been referred to as the King of the "Francs," not the King of "France." But times changed, and we all accepted this concept of sovereignty attached to states, that is, defined geographic spaces. . . but not so much to peoples anymore. The leader of every sovereign state has complete control over all the peoples in their defined geographic area, their state. . . but no one outside that state! Those outsiders are controlled by the leader of whatever state they are in! Well, not for Russians, according to Vladimir Putin. All the Russians, everywhere, are his state's responsibility, and it is their duty to protect them. . . wow, he is so old school that he rides a dinosaur to work.

And why would he do that? For starters, to justify Russian intervention in Ukraine, and justify them re-absorbing Crimea. . . after all, it was the Crimeans who asked for it, and it just so happens they are ethnically Russian, so to not help them would be to turn his back on fellow Russians—at least in Putin's interpretation. But it is also about something bigger: legitimizing Russia's dream of expanding its influence back out into the world after it lost that influence abroad when the USSR collapsed. A Putin quote from that same Putin speech I referenced before, and he is referencing the end of the USSR: "Millions of Russians went to bed in one country and woke up abroad. Overnight, they were minorities in the former Soviet republics, and the Russian people became one of the biggest — if not the biggest — divided nations in the world." And he wants to protect his Russian brood, wherever they reside. And check out the map below to see where that would be. Think Kazakhstan or Latvia may be a little worried by Putin's declaration and actions in Crimea? I would be. Think they will be going out of their way to kiss the butts of their ethnically Russian citizens? I would be.

This Putin Peninsula Pinch is a huge deal. A re-defining of sovereignty in the modern world deal. A re-defining of Russia's place in the world deal. And a possible start to an expansionist era of Russia that sees them take back territories that contain ethnically Russian peoples. These peeps are partying like it's 1799, because no one has seen such old-school style empire building for almost a century. And they want to roll back time in order to become a world power again, in order to have a. . .

RE-ESTABLISHED WORLD ROLE

Without getting into all the terrible details (and there are many), I have to also alert you to the fact that Russia has once again started to establish itself as a global power player that is affecting events outside of their own country, and even outside of their near abroad. No, Russia does not yet have full global reach as it did as the USSR at the height of the Cold War, but it has recently reasserted itself as a power that can radically affect events in other regions, and in serious deadly ways. . . specifically in Syria.

In brief: The Syrian Civil War has been raging since 2011 between multiple opposition (anti-government) groupings and the Assad government. Since 2014, a significant part of Syria's territory was claimed by that insane clown posse ISIS group as well. The US and its allies have been trying to help rebels overthrow the Syrian government, as well as crush ISIS, by conducting air strikes and funding rebel groups. But Russia is actually friends with the Bashar al-Assad led Syrian government, and has thus entered the war on their behalf, much to the chagrin of the US and others. . .

The Russian military intervention in the Syrian Civil War began in September 2015 after an official request by the Syrian government for military help against rebel and jihadist groups. The intervention initially consisted of air strikes fired by Russian aircraft against militant groups opposed to the Syrian government. Prior to the intervention, Russian involvement in the Syrian Civil War had mainly consisted of supplying and arming the Syrian Army. Apart from fighting terrorist organizations such as ISIS, Russia's goals included helping the Syrian government retake territory from various anti-government groups. . . maybe you have heard of the bloody campaign against rebels in Aleppo?. . . which is definitely part of a broader geopolitical objective to push back US influence. And it has been disastrous for fighters and civilians alike: in September 2016, the pro-rebel Syrian Observatory for Human Rights stated that Russian airstrikes have killed around 3,800 civilians, about a quarter of them children, along with 2,746 ISIS fighters, and 2,814 rebels.

I point this out not as a well-rounded assessment of the Syrian Civil War, but only to point out that Russia is again quite confident in pushing its influence out into other regions, and specifically to be a foil to American power and American foreign policy. This speaks to its military power, its willingness to use force, its willingness to go against global opinion, its willingness to be an alternative to US world leadership, and its willingness to even be seen as "the bad guys" again. . . which is an indicator of extreme confidence that may just be getting warmed up. Russia absolutely wants to be viewed as a world power, not just a regional one, although it is going to continue to build up its clout by attempting to re-unite the surrounding regions under its authority. . . and as a. . .

"TEAM WEST" ALTERNATIVE OFFERED?

According to the current Russia's National Foreign Policy Concept (basically, an annual 'white paper' or report on the objectives of Russian foreign policy), a major priority of Moscow's diplomacy for the next decade is the strengthening of regional integration in the post-Soviet territories, and the grand formation of this Eurasian Union among the countries of the region. Is this a Soviet Union, take 2? USSR 2.0? Soviet Union Reboot? Eh. Doubtful. But it is 100% being offered as a Team West alternative that Putin wants to build into something akin to the EU, or perhaps more. And with the show of force by Russia in Ukraine and Georgia, it looks to be a grouping that many countries will feel compelled to join just to stay on Russia's good side! So, what is it again?

The Eurasian Economic Union (or EEU), is without a doubt the wet dream of Russian President Vladimir Putin, who now openly seeks to reassert a resurgent Russia's influence and outright power into areas it lost during the collapse of the Soviet Union in 1991. Putin for his part has actually referred to the demise of the USSR as "the greatest geopolitical catastrophe of the 20th century,". . . and it is a catastrophe he now seems intent on fixing. Through economic coercion, intimidation, and possibly force if necessary, Russia looks to be regaining its old sphere of influence in Central Asia and Eastern Europe.

Is this an attempt to rebuild the Soviet Union proper? Is this a new Cold War? Eh. I'm not sure it is all that, at least not yet, but here is the deal as of 2019:

The formations of it are entirely logical and benign, at least on paper: In November 2011, the presidents of Russia, Belarus, and Kazakhstan signed an agreement to create the Eurasian Economic Union: a customs union (like a free trade zone) with partial economic integration between the states, with the intent of becoming something very close to the European Union (you know, the EU) in the future with full-on economic, political, and military cooperation. Became fully functional in 2015. Just like the EU, it is supposed to safeguard regional economic interests and facilitate trade and business within member countries. Since Russia is obviously the powerhouse player in all of this, Putin could not be happier. But he isn't satisfied just as yet. . .

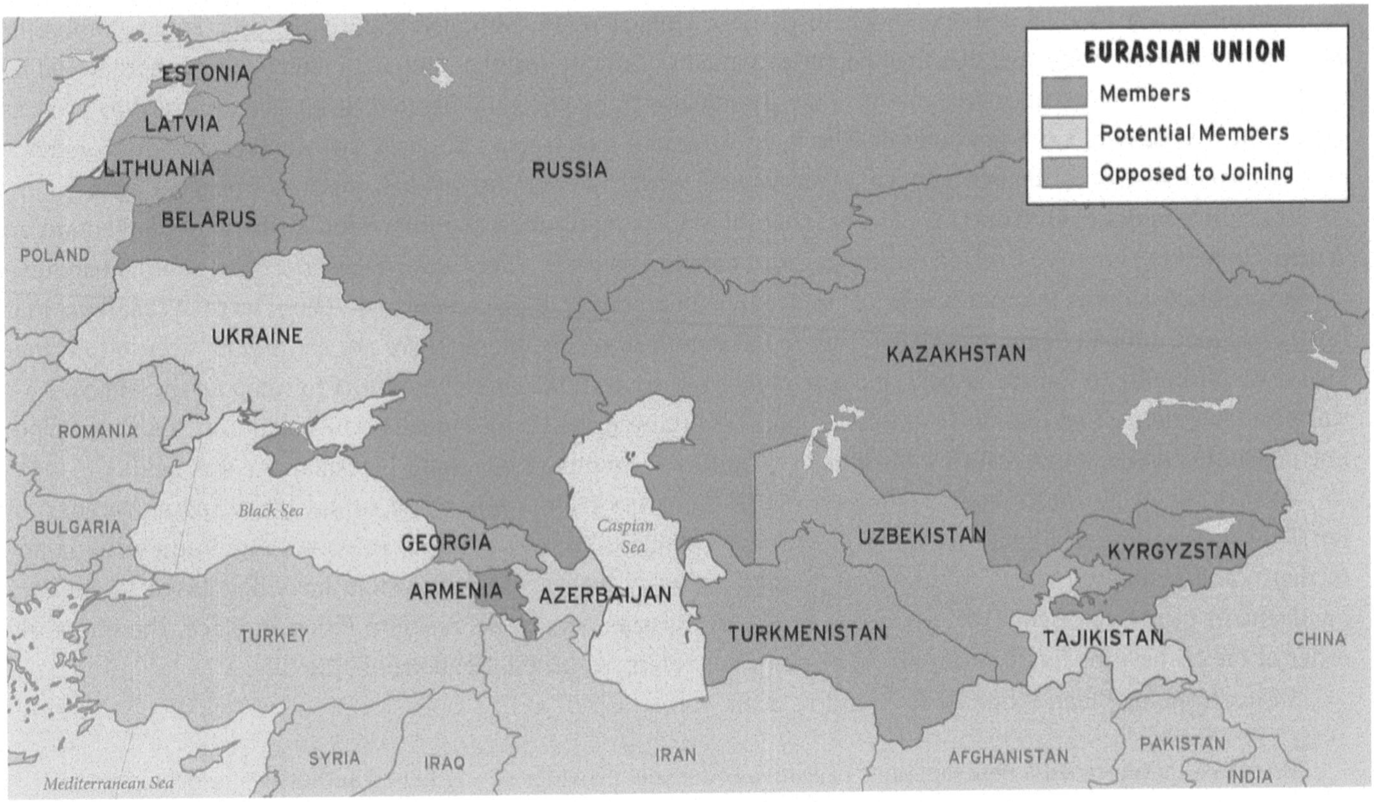

Putin's grand gamble here is that his union, (and yes, it is referred to as his personal grand strategy at this point), will become a successful alternative to the European Union, and that states of Central Asia, Eastern Europe, and even the Caucuses Region will want to sign up for it. Given the 2016 BREXIT vote of Great Britain to leave the EU, along with the rising tide of anti-EU populist sentiment across many other European countries, this Russian EEU has suddenly a LOT more traction and power than it did a mere five years ago. With all of the chaos unfolding in the EU and the rise of the Trump anti-globalist agenda, Russia is now posed to expand greatly back into its periphery, and this EEU thing looks like it may actually work.

I suppose it is being marketed as a great way to boost economic growth, without having to bother with all that pesky democracy and human rights stuff. And it is definitely being pushed as an alternative to joining Team West. And it definitely is Russia trying to regain "influence" in areas it previously held as its own. Ukraine was to be a jewel in the crown of all of this, thus Putin's interest in keeping that country in the Russian orbit, and the Eurasian Union specifically. That whole taking over Crimean Peninsula stuff? Yeah, that was the kick-off of the game, not the finish of it. And what game would that be? Ah! Back to the Great Game of course! This is a now fervently fired-up Russian response to the last two decades of the EU and NATO creeping closer and closer to its borders. The Bear wants its backyard back! Game on! How far will Putin go to see his 21st century Ruskie dream realized? We will soon find out, and it may even get accelerated by. . .

TRUMP SPEED BUMP: BIDEN BACK AT IT

2021 UPDATE: During the 2016 US election cycle, and the subsequent 4 years of US President Donald Trump's administration, there was a radical shift in the US attitude towards Putin's Russia... and many, including my plaid self, speculated that a major attitude change in relations between the two former Cold War adversaries was becoming a reality. The budding "bromance" between Vladimir Putin and Donald Trump should be no great surprise to you, since it was covered ad nauseam during the election and first couple of years of Trump's reign. Both men publicly declared an admiration for each other, both men suggested that their countries should work more together, and both men believed there would be a huge "thaw" in relations once Trump took office. Trump also threatened and berated NATO as a historical by-product whose time has passed, much to Putin's delight. Putin suggested that Trump's business savvy and conservative/authoritarian overtones are to be admired and that he was looking forward to working with him, much to Trump's delight.

BUT... it was discovered and investigated that Russia has intervened and interfered in the 2016 US election on Trump's behalf (a charge reasserted during the 2020 election as well). There were accusations that Russia had murdered foreign dissidents in other countries, typically with an old KGB poisoning trick. There were accusations that Russia had paid Taliban fighters in Afghanistan to kill American troops there. There are continuous accusations that Russia is continuing to foment dissent and arm radicals in eastern Ukraine in an effort to ramp up the almost-civil war occurring there. And then in 2021, Russian hackers attacked the US oil industry which shut down a major pipeline on the eastern seaboard causing gas shortages, until a $5 million ransomware blackmail fee was paid.

As a result of all these things and more, the US Congress collectively acted to put increased embargoes on Russia as well as severely limit the actions of then US President Trump to "warm" relations with Putin's Russia any further... and actually actively voted against Trump on many measures which would have decreased US military involvement in critical areas that are in place to deter Russia. With new President Biden in place, the old world order of the US treating Russia as an outright enemy has totally returned. Need proof? Dig this:

Direct from the Biden White House, April 15, 2021:

> President Biden signed a new sanctions executive order that provides strengthened authorities to demonstrate the Administration's resolve in responding to and deterring the full scope of Russia's harmful foreign activities. This E.O. sends a signal that the United States will impose costs in a strategic and economically impactful manner on Russia if it continues or escalates its destabilizing international actions. This includes, in particular,

efforts to undermine the conduct of free and fair democratic elections and democratic institutions in the United States and its allies and partners; engage in and facilitate malicious cyber activities against the United States and its allies and partners; foster and use transnational corruption to influence foreign governments; pursue extraterritorial activities targeting dissidents or journalists; undermine security in countries and regions important to United States national security; and violate well-established principles of international law, including respect for the territorial integrity of states.

Wow. Talk about a policy shift that is a blast from the past! US/Russian relations may now be at an all time low, including during the height of the Cold War!!! Russia is now seen by the USA and most other democracies around the world as merely a spoiler state, whose main role seems to be disrupting elections, sponsoring economic and political cyber-attacks, killing dissidents within and outside of their territory, aggressively agitating conflicts to increase their own territory... and in general being an all-around disturber of the peace and pain in the ass. Cold War Part 2: The Re-Chilling will be in a theater near you soon. But hopefully it's not a theater of war.

Still, the solid majority of Russians think might makes right and are proud of Putin for his assertive actions, and don't seem to mind at all the continued rightward march of their state. Stalin himself would be stoked! Much more about this worldwide shift to the autocratic right, and Russia's role in this phenomenon, in Chapter 25: The Cover Story.

GHOST OF STALIN RETURNS

Which is a good jumping off point. I guess any mention of Joseph Stalin is a good point to get the heck out of Dodge. I want to leave you with this: the memory of Smokin' Joe Stalin is actually being softened in today's Russia, and that is a very telling sign of what's happening in the society. Some folks in Russia are looking back with pride at what most of us consider one of the worst dictators in history. Why? This resurgent world role and restored sense of Russian greatness has not been felt since the Stalin era, and, quite frankly, the Russians are quite proud to be Russian again, even though they are paying a hefty price in terms of losing political freedoms and individual human rights. Given Putin's consolidation of political power, repression of individual human rights, increased militancy abroad, and rabid support of most of the public, Russia has firmly fallen back into the ONE-PARTY STATE category of political rule. But you add in trillions of petro dollars, increasing empire due to global warming, and a renewed sense of global political power, and you got yourself a region that will be shaping world events of the future. Russia is riding high! There is a restored Russian pride, happiness, nationalism, and even tenacity to a level that has not been seen in decades. They feel as if they are a world power again, for reals! Growl! The Bear is back!

Yeah. That should do it. Problem solved.

RUSSIAN RUNDOWN & RESOURCES

View additional Plaid Avenger resources for this region at http://plaid.at/ru

BIG PLUSES

- Biggest state by land area, one that is extremely rich in virtually all natural resources
- Specifically, a huge oil and natural gas producer: energy resources critical to world economy
- Virtually no national debt
- Regained sense of national pride and political world leadership
- Has veto power at the UN Permanent Security Council
- Has as many nuclear bombs as the US, if not more, as well as a serious military, second only to the US.
- Is a founding member of some of the most promising future international coalitions like BRIC and SCO
- Is still considered quite important to the EU and NATO too, enough to get invited to the meetings.
- Global warming? No problem for Russia! Will increase usable land area and resource base

BIG PROBLEMS

- Gi-normous size problematic for infrastructure, defense, and cultural cohesiveness.
- Economy and social structures, like health care, still not fully recovered from Soviet era crash
- Male longevity rates are lower than most African countries; alcoholism rates among the highest in the world
- Population overall is shrinking.
- Organized crime, rampant corruption, and tax evasion have seriously stymied international investment and hinder internal development
- Economy still heavily reliant on exporting natural resources, mostly energy.
- Has some of the worst environmental disaster areas on the planet
- Anti-immigrant, anti-foreigner, and flat out racism are significant issues in Russia
- Has moved significantly to the political right: is back to a one-party state status

DEAD DUDES OF NOTE:

Tsar Nicholas II: Last of the royal Romanov line, and last monarch of Russia, who was assassinated along with his whole family in 1918. His piss poor rule transformed Russia into an economic and military disaster which helped fuel the revolution which left the commies in charge.

Rasputin: The un-kill-able "Mad Monk" who recommended that Tsar Nicholas get Russia involved in World War I. He didn't instigate the royal family's demise, but he sure as hell accelerated it. May still be alive somewhere in hell.

Vladimir Lenin: "Father of the USSR" Passionate organizer, orator, and propagandist who led the Russian Revolution of 1917, in which he implemented the first ever communist experiment at the state scale. Then he died.

Joseph Stalin: The semi-psychotic who took over control of the USSR after Lenin's death. Formed the KGB, purged out all political adversaries, and may have been responsible for the deaths of 30 million Russians. But, he also oversaw the industrialization and expansion of the USSR into a world power, and helped kill the Nazis in WWII. So it's a mixed bag.

Nikita Khrushchev: Leader of the USSR at what was perhaps the height of the Cold War. Was the bonehead that decided to put Soviet missiles in Cuba pointed at Florida, which brought the world to the brink of nuclear war aka the Cuban Missile Crisis of 1962.

PLAID CINEMA SELECTION:

Prisoner of the Mountains (1996) Set in the Caucus Mountains with incredible scenery, this quiet "war film" is set during the First Chechen War . . . while the war itself was still happening! Themes to watch for: cultural differences in the region and the piss-poor state of the Russian military in the early 90's. Still topical, as tensions with Chechnya are still part of current events.

Tale in the Darkness (2009) A Russian independent film, which strives to explore the life of a female police officer in Far East Russia. This drama reaches a sort of "Men are from Mars, women are from Venus" conclusion.

Burnt by the Sun (1994) Set in USSR circa 1936, this film shows the effects of Stalin's insane political purges of his own peoples, even those loyal to him. A Soviet war hero on summer retreat at his dacha with his family becomes the target of KGB investigation and elimination.

Siberiade (1979) A film about the small village of Yelan hidden in the backwoods of Siberia. This epic, four part film shows the evolution of the small village until at the end of the film, it is no longer remote, as oil has been found in the village. Show realistic evolution of Russian culture as well as providing an early reflection of the processes of globalization.

Newsmakers (2009) A crime drama set in Moscow whose plot involves a savvy female detective setting up a real-life reality TV show as a ruse to flush out a criminal gang that got the better of the cops in a live shoot-out.

Oligarkh aka *Tycoon* (2002) Not a great film, but does chronicle the conversion of soviet command economy to full-on unrestrained capitalism that occurred as the USSR imploded. Greed, corruption, and cronyism via a personal storyline.

Brat aka *Brother* (1997) Hardcore bleak and gritty realistic gangster film in St. Petersburg that epitomizes the way it works at the local level in Russia. No big money or flash here, just desperate people who do desperate things. Spawned a whole culture of glorifying young gangsters in films, often called "keelers," much like gangsta' rap did in the US.

Nochnoy dozor aka *Night Watch* (2004) Okay this one is just for fun, and to throw a monkey-wrench into the film mix. This is a Russian vampire/wizard/good vs. evil/horror film that is now one of the highest grossing films in their history. It's *The Matrix* of Moscow.

LIVE LEADERS YOU SHOULD KNOW:

Mikhail Gorbachev: Oversaw the collapse of the USSR, implemented the glasnost and perestroika policies, and helped bring the Cold War to a peaceful end. Is a hero to the West and won the Nobel Peace Prize, but looked upon almost as a traitor in Russia.

Vladimir Putin: Current President of Russia since 2012, having previously served as such from 2000–2008, got Russia's mo-jo back, got them rich, and got them powerful again on the world stage. Look for him to be in Russian politics indefinitely.

Dmitry Medvedev: Current Prime Minister of Russia, the Putin right-hand man. Many think he is merely a Putin puppet, but has so far shown himself to be an adept and strong leader in his own right who is continuing the successful Putin policy legacy.

Sergey Lavrov: Current Russian Foreign Minister, which is the equivalent of the US Secretary of State, the head diplomat of the country. This dude is wick-edly smart, savvy, and has been behind the scenes making sure Russia regains its power status on the world stage.

MAP GALLERY

View Map Gallery & additional Plaid Avenger resources for this region at http://plaid.at/ru

THE BATTLE FOR TOKYO IS ON

CHAPTER OUTLINE

Japan

11

KONNICHIWA!

Welcome to the islands! And a most unique set of islands they are—so unique, in fact, that we will classify Japan not just as an independent state, but as an entirely unique region of the world as well. One of the classic mistakes that all textbooks make, and that a lot of people in the West make, is overgeneralizing Asia. They look at this place and say, "Oh well, you know, it's just Asian people there and six of one, half dozen of the other. Japanese, Chinese, Koreans—whatever." This couldn't be further from the truth when we are talking about Japan and mainland Asia. Japan has an extremely distinct culture from all the countries around it, as do the Chinese, the Koreans, and the Vietnamese, but unlike the others, Japan has followed a more western trajectory in the last century that has set them quite apart from even their closest neighbors. They are the most "western" of the eastern cultures. Huh? Hang on. We'll get to that in a bit.

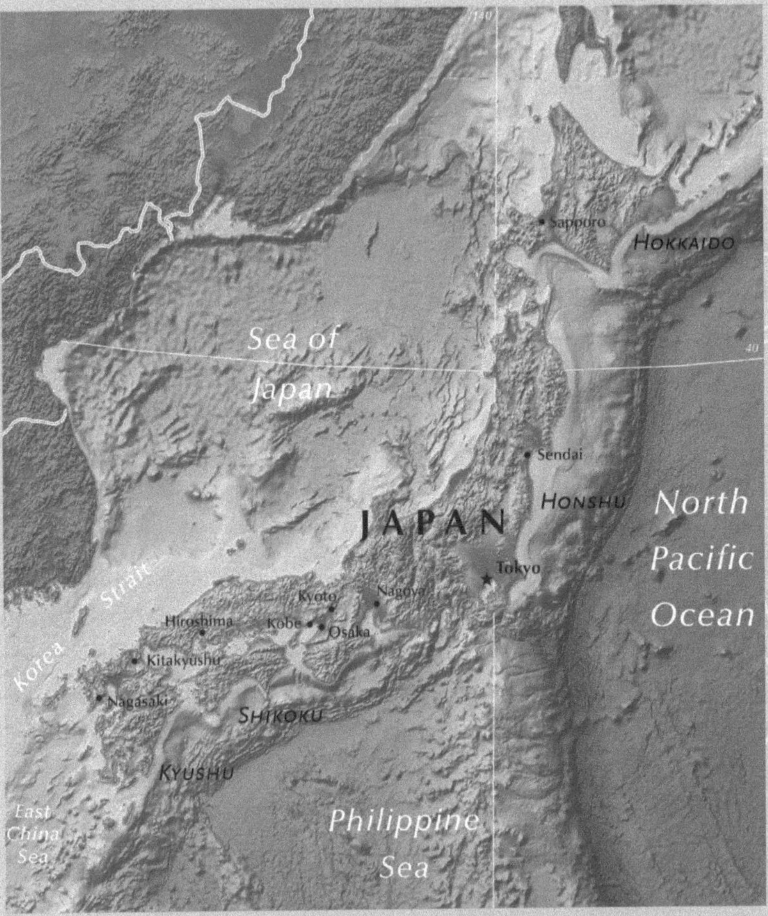

Just know this for now, we are coming up on the Clash of the Titans of the 21st Century: The Clash of the Asian Titans. The titans are, of course, Japan and China, two states with huge economies, two states shaping global events, and two states with a long history of animosity between them.

One of the most critical relationships that will be affecting 21st century events is how these two Asian giants will deal with each other as China rises to full superpower status. But I am getting ahead of the story, as usual. Let's get some background on Japan itself, and then look at its evolution into the modern era, and finally finish with that all important relationship with China. Ready? Then Godzilla game on!

PHYSICAL

Ahhh! Most honorable Mt. Fuji!

Japan is a group of four main islands: Hokkaido, Honshu, Shikoku, and Kyushu, from north to south. Japan was created by three or four different tectonic plates coming together in this area. Because of this, Japan is volcanic in origin. Another feature that goes along with this is its mountainous terrain. There are very few flatlands, and this equates to very little natural agricultural land. Due to plate tectonics, Japan is also earthquake prone—always has been, always will be. It is a cyclical thing and it will not go away anytime soon. It is not a question of *if* Japan is going to be hit by another major earthquake; it's only a matter of *when*. **2011 Update:** ummm . . . yep. It just happened again. Big time.

"Honey, I think this section of the interstate is closed."

Why should you care about this? Well, Tokyo's stock exchange, and Japan as a whole, holds tremendous amounts of foreign investment, foreign currency, and bank reserves. When the day comes that a major earthquake shuts them down, depending on the scale of the disaster itself, it could plunge the entire world into recession almost instantaneously. If it is a tremendously bad earthquake, it could plunge the world into a depression in the long term. It's another instance of globalization at its worst, in that a big hit here from the physical world could equate to disaster for the entire planet. However the Plaid Avenger doesn't want to scare anybody; it will happen, they will rebuild, life goes on.

The other physical thing to consider with Japan is climate. Climatologically speaking, this is an easy one for the Plaid Avenger to describe to you; just superimpose Japan, keeping latitude constant, on to the eastern seaboard of the United States (see image below). As you can see, latitude and size being held constant in this image, the climate of Japan is identical to the eastern seaboard of the United States. Hokkaido is around the upstate New York area and they get a lot of snow up there. Southern Honshu, as well as Shikoku and Kyushu, reach down into the southern United States and indeed, there are actually smaller islands that go down to what would be tropical Florida. Every place else is similar to some place else in between. For instance, Tokyo is roughly the latitude of Washington, D.C. These two regions, though on opposite sides of the planet, have very similar climates.

Japan/US: size and latitude comparison.

One disclaimer here is that Japan is an island nation. As an island, it is surrounded by water, and water is a major modifier of climate. As such, it has a similar climate to the east coast of the United States, though somewhat moderated due to the influence of the surrounding water.

Lastly, physically speaking, Japan is lacking in natural resources. I just suggested that Japan is mostly mountainous, which means it does not have a lot of agricultural resources. The Japanese have found ways throughout the centuries of getting around that by creating terraced rice paddies and fishing like champs. There is still no way it can supply itself with its own food needs at this point due to its lack of agricultural resources, and lack of arable land. More importantly, it lacks just about everything else. Say what?

Japan is a real conundrum in that it is the third largest economy on the planet that, at the same time, lacks virtually everything that most countries with powerhouse economies have. Japan has to import most of its metals, ores, and virtually all of its energy resources. Japan is entirely dependent on the rest of the planet for many things. That has something to do with its history.

SAMURAI TO JACK

Bring on the bamboo.

What about natural resources, Plaid Avenger? Not having any has something to do with Japan in the modern world. I'm going to go through the early history of Japan in fairly short order to point out one thing in particular: In terms of unique Asian cultures and unique ethnic groups, Japan is really a laggard. It's the latecomer. It's the last one to show up on the world scene.

Chinese civilization has been around for 5,000 years. The Koreans have been around for 3,000–4,000 years. Lots of people from Central Asia and Southeast Asia, like the Thais, the Vietnamese, and other ethnic groups with distinct cultures, have been around for a while too. If you search hard, you will find that Japan's culture cannot really be called a distinct culture until about 1400–1500 years ago, and even at that point, it wasn't anything closing in on what we consider a nation-state.

Japan started as a small group of people that migrated from mainland China and through the Korean Peninsula over the course of the previous thousand years. They did not become a unique entity themselves until comparatively late in the Common Era. The Plaid Avenger points this out not because I'm that concerned about you learning the history of every place in the world, but because this history plays into the modern world. In what way?

Virtually all of Japanese society is borrowed from other societies. All of the writing systems, religious systems, and philosophical systems all have roots over in mainland Asia. However, if you read Japanese literature or government propaganda over the last 100 years, it's all reversed. The Japanese are fiercely proud of their independence and their distinct culture, so given that they have been a powerhouse over the past fifty years, they have rewritten history to make it look like as if they were there first and everybody else borrowed stuff from them. This couldn't be further from the truth.

Move over, UK! Japan has inbred snobs as well!

What we have is this group of folks that started up in 500–700 CE, and it was only at this point that they formed into something vaguely resembling a state. They started to have a head of state, a prince. They started a royal/imperial line, which actually stays in place to the current day. For those of you saying, "What? They still have a royal family there?" Yes, there is still an Emperor and there is still an Empress. They have the longest standing monarchy/royal family in world history, and it continues. You can trace it back to 600–700 CE, same family, same dudes. The faces change, but the DNA remains

Shoguns? Sho' nuff!

the same. It is very much like Great Britain and its constitutional monarchy—the idea of royals hanging out. They don't have any real power now, but they are still there.

But let's get back to history. We can fast forward very quickly because Japan turned into, for the next thousand years, something very similar to a feudal state in medieval Europe. That is, there was the Emperor on top, there were Dukes underneath him, there were Lords underneath the Dukes, and eventually, we get to the peasants that were essentially tied to the lands controlled by the Lords. There are different names and faces in Japan. They are Asians, not Europeans. They have sweet Samurai outfits and swords, instead of knights, and long shafts and jousts, but was all essentially the same system playing out on opposite sides of Eurasia.

The peasants lived on the lands and were essentially owned by the Lords, who paid tribute to the Dukes, who had one master, the Emperor. The easy reference is to what we call **Shoguns**, essentially militaristic governors which were the Dukes and the Lords. Just like in Europe, they competed for lands and ownership of things to gain political influence. They still recognized the Emperor as the main dude, but they all fought for places right below him on the totem pole.

Why was this so great? Nothing was really great about it, unless you're the Emperor. He allowed this feudal system to continue because, as all of his subordinates are fighting each other, no one was ever actually threatening him. That's the point. The kings of Europe did the same exact thing. The point of feudalism was to keep people beating the crap out of each other so they can't mess with you. This became standard operating procedure in Japan fairly early on, and it continued for most of Japanese history, which we have suggested, began pretty late in the Asian game.

We can tie some dates to this age of the Shogun, this really entrenched take on the feudal system. From around 1000 to 1580 CE, they had multiple competing shogunates all vying for power and influence. After 1580, one group rose to the top. They were called the **Tokugawa Shogunate**. The Tokugawa Shogunate unified the whole country under their single command from about 1600 to 1800. A single Duke or Lord from this line of shoguns became so powerful that they actually stopped all the infighting. Japan had a period of stability and peace where everybody was under the same blanket because of those Tokugawa guys. Here's the big thing: they unified the whole country, united all the smaller armies battling each other into a singular army, but most importantly, they plunged the whole country into global isolation.

What do you mean, Plaid Avenger? Well, these guys were tough. You've seen the movies! They were so awesome that they repelled a Mongol invasion in 1274—no one else pulled that off! They had their own little world going on and they were not interested at all in the outside world. They cut ties with everybody—not that they ever had a lot of contact. Historically speaking, they had always been a subservient state, or **tribute state** to the Chinese, though the Chinese never took them over physically. Japan knew that China was the big power, and therefore, kept itself on the fringe. The Japanese have escaped takeover by all other entities historically, mostly because they were a fringe state of no great strategic value, and had no great economic resources that other powers would want.

What was happening in the rest of the world from 1600 to 1800? Europe was colonizing the world, was taking over other big parts of the planet and seeing the beginnings of its industrialization period. You had a foreign influence making its way over to the Far East, even into Japan. With industrialization occurring, these European guys had better weapons, better ships, and better navigation. By the 1800s, Europe virtually controlled three-quarters of the planet and finally arrived in Japan. Europeans started making in-roads to establish trade with Japan.

During the Tokugawa Shogunate in 1600–1800, missionaries started showing up at Japan's doorstep. The Shogunate said, "What are you peddling? What are you trying to convert us to? Really? Christianity? That sounds great. And now we'll chop your head off! Get out of our country! We are not interested in your beliefs. We are isolating ourselves! We are the Shoguns, we are the Samurais! We don't need your stuff, so go away! If you don't go away when we tell you to, then we'll kill you!" That was the way it went for a good long time, but it was only going to be perpetuated for so long.

Commodore Perry, official Japan-opener.

This European colonization story is told again and again when we go through the rest of the regions on the planet. As Europe's technology and military hardware became more advanced, neither Asia nor Japan could effectively counter the foreigners forever. Through its self-isolation, Japan got behind the times technologically. It came to the realization that it couldn't compete with the rest of the outside world. The Japanese were still fighting with swords, fighting dudes with guns. Their isolated world came to a quick halt. The Plaid Avenger is not big on dates, but this is a good one to note: 1853, which is not long ago. In 1853, a dude named Commodore Matthew Perry showed up on the scene. Commodore Perry wanted to open up trade so he pulled his ironclad battleship into Tokyo Harbor and came ashore. He said, "Let's meet with the local head-honchos, these Samurai guys with their swords." Perry said, "Hey, we want to establish trade."

For the previous 250 years, the Shogunate answered that with, "Get back on your ship and get out of here or we'll chop your head off!" Perry said, "Hold on, fellows, hold on. Take a look out in the harbor. Do you see that? That's an ironclad vessel. See those things poking out of it? Those are cannons. We can blow up you and all of your people right now and you are completely defenseless." Remember, these Samurai fought mostly with swords; now they were stacked up against modern military hardware. The Samurai figured out pretty quickly that this was going to be a one-sided affair. In one fell swoop, in 1853, Commodore Matthew Perry forcibly opened up trade with Japan. They kind of said, "Yeah, those guns probably do better than our swords. We probably are not going to be able to survive this."

What the . . . ? Commodore Perry, stop this Sumo madness!

It is important to note which country the Commodore hailed from: the United States of America. It's actually the United States that forcibly opened up trade in Japan, and luckily (for Japan) not a typical European colonizing power. It is historically significant that it was *not* the land-grabbing, colonizing Europeans who busted Japan open to the world, but the Americans who went over there to establish a trade relationship. This was the beginning of one of the most amazing occurrences in modern history, as far as the Plaid Avenger is concerned.

When a country is confronted with an outside world in which they are severely behind, they can take one of two roads. All other regions faced this exact same choice as the onslaught of European colonization of the world occurred. Road One: "We can fight it out! Let's try to kick them out! Let's have protests and shoot guns at them and do whatever we can do to kick out the European powers!" Most countries went that way. Road Two, which only the Japanese seem to have found: "Let's embrace the West! Let's do it their way! And let's do it better!" This precipitates a fascinating time in *world* history, not just Japanese history. This is what makes the difference between Japan and China in today's world. This is the crossroads. The Plaid Avenger is telling you, it's rare that you can say, "This is a definitive turning point. Everything is different after this!" The Plaid Avenger is telling you: *this is it.* And the "it" is named Meiji.

MEIJI MANIA!

Meiji Makeover: Rickshaw to Railways.

In 1868, the Japanese decided to undergo something now referred to as **The Meiji Restoration**, named for the Emperor at the time, the Head Honcho Meiji as he came to power. He was known as being a fairly savvy guy, but it was probably the collaborative effort of his counselors and the people around him that, once they were forced to embrace the outside world, understood how far ahead the Europeans and the Americans were ahead of them. They made this radical decision, saying, "Wipe the slate clean. Let's redo everything, EVERYTHING, in light of what we know about the outside world." They embarked on this fantastic re-invention of their society, a "restoration" if you will. . . .

They sent what would be the equivalent of college students, learned men, as well as government workers, all over the planet. They sent thousands of them to Europe, to America, to the Middle East, everywhere there were centers of learning, centers of technology, centers of industry. Their mission was to find out every single thing they could. They not only learned stuff at universities and through businesses, but they also brought stuff back. They brought back military hardware, guns, tanks, whole steam engines, railroad equipment. They shipped it back to Japan, and teams of smart dudes ripped the stuff apart to figure out each individual component. They started the individual industries to make the pieces, the wheels, steam engines, and everything else on their own, from the ground up. This was a massive innovation, rejuvenation, a catching-up period, for technology in particular, but also for government and military.

The Japanese did this fantastically well. Think about what Japan does today. It's still like this, isn't it? They don't really make anything unique—they make everything better! They started this tradition way back during the Meiji Restoration. They are reverse engineers and efficiency experts. They did this in very short order, in around ten years. In two or three decades, they caught up from a lifetime of technological stagnation. It's insane! It's crazy! It's the Meiji Restoration!

At the same time, they modified a lot of things internally. You can see great movies about this, such as *The Last Samurai* with Tom Cruise. Tommy's Scientology aside, it's a pretty decent movie. The film shows the extremely turbulent times in Japan when everything changed. Not just the technology stuff we talked about, that's easy enough to understand. The social restructuring was the big thing. They actually got rid of the Samurai. They reinvented their society in the likeness of Western Europe and America. They banned the Samurai, which is what *The Last Samurai* is all about. They wanted to rid themselves of the old ways. They said, "No, we're going to rejuvenate! We are doing everything to catch our military up, which means we're getting rid of the old Shogun ways. We're going to consolidate into a bureaucratic government, just like all these other places. We're going to restructure our banking system. We're going to redo EVERYTHING!" And they did it, in the image of the West.

This is why Japan is a very Western looking nation, as opposed to its neighbors. The Meiji Restoration was the critical turning point in history. Japan became Western-like. Every place else in Asia did not. Japan currently has the number three economy in the world. Every place else in Asia isn't even close, with the exception of China (number two), who has stepped up in the last few decades. But watch out! South Korea will soon break into the top ten as well. This region has three very important 21st century power-player economies. That is a fact worth remembering.

TERRIBLE LIZARD

Back to the story. The Meiji Restoration was the complete, inside-out do-over of Japanese society on all levels, and it was extremely successful. Like I already suggested, they caught up within a few decades. Japan went from the 1850s, where the Samurais were in charge and fighting dudes with swords, to 1895 when, having been a subservient vassal state of China for their entire history, they attacked China. Why? Because once the beast awakened, it was hungry! The beast must be fed! What does it want? Resources! The Japanese adopted another very Western concept in order to get these resources: imperialism!

GODZILLA IS BORN

In 1895, during the Sino-Japanese War, the Japanese went into parts of modern-day Korea and Manchuria and started conquering. They attacked China, its colossal powerhouse neighbor, *and they won*. In 1905, ten years later, they attacked Russia, the largest state on the planet, *and they won*.

Minute 1: battle begun. Minute 5: no Russians alive.

What were they doing this for? They were taking parts of Russian territory on the east coast of Asia. The Russians sent their entire naval fleet around to sink the Japanese fleet because they said, "Hey, we're Russia, we've been around for like 500 years. We're the biggest state in the world! We're European; we know what's going on." They sent their entire fleet around the entire continent, and it was entirely sunk by the Japanese in five minutes. Boom. To put this in perspective: within 40 years, Japan went from having NO navy to having a navy that wipes out Russia without batting an eye—the biggest one-sided naval defeat in all of history! **The Battle of Tsushima** was an embarrassment for the Russians, and established the arrival of Japan in the modern world. Japan goes on to advance further into Korea, and in 1910 it took over all of modern-day Korea. In 1917, it annexed more of Russia, while the Russians were busy doing their Communist Revolution thingy.

We can stop for a second and ask, "What were they doing all this for? Why were they taking all this land, inciting fights with their neighbors?" This gets back to what I was talking about in the physical section: this place had no resources. To become a world power, you need resources. To build railroad cars and tanks and airplanes and industries, you need resources. To build an army, you need resources. What Japan did, as it started to come into power, was acquire resources. Take them from other places. Take over lands. Japan becomes really the only Asian colonizing power in modern history. They were the only Asian imperial power in the Western European sense of the word (i.e., by forcibly taking over other parts of the world). Once the beast is awakened, he must feed. An insatiable hunger develops which inevitably leads to . . .

GODZILLA ATTACKS!

With Japan's acquisition of all this territory and all these resources in mainland Asia, you might say, "That's enough. You've got enough now, Japan." But once Godzilla is awake, he's not going back to sleep very easily! The Japanese continued to meet

with success after success. In 1914, before World War I was over, Japan strategically joined the Allies against Germany, though they played no real part in the war. As a result of their maneuver, they gained control of German territories in China and German-controlled islands in the Pacific. Without doing anything, they gained more territory at the end of World War I. We already said they started to pull out more pieces of Russia during Russia's civil war. After Russia got decimated in World War I, the Japanese took over some more of their territory in the interim period.

In 1931, the Japanese took over all of what is now Manchuria, the entire Northeastern part of China, and during the rest of the 1930s, they established more footholds in China and virtually every other Southeast Asian nation. In each new place, they set up shops and military camps.

In this period between World War I and World War II, all the Europeans were worried about themselves. No one was paying much attention to Japan, except the United States. The United States was very worried about Japanese movement and the growth of their Empire. Even before the WWII was officially launched, there was animosity between these two Pacific powerhouses, Japan and the United States.

The United States, being an anti-imperial and anti-colonial force, was looking over the Pacific and saying, "That doesn't look good. How could this end well? This isn't good for us or anybody!" Indeed, they were right. A lot of folks said, "Whoa, we're really surprised that Japan attacked Pearl Harbor." Nothing could be further from the truth. These countries may not have declared war at the time, but they weren't on friendly terms either. In fact, the United States already had an oil embargo against Japan. That's one of the reasons Japan said it attacked the United States. Let's get to this attack.

In 1941, the Empire continued to expand, taking over mainland China and places in Southeast Asia such as Vietnam, Cambodia, Laos, Thailand, Burma, Indonesia. Japan continued to expand its Pacific Island holdings as well. In 1941, it became so big that it decided to attack the United States over in Hawaii. Why didn't it attack mainland United States? That's anybody's guess. In hindsight, it may have been a better idea, but who knows why any of this occurred? What if the Japanese had not attacked the United States? Maybe the United States wouldn't have done anything to them and there would still be a huge Japanese Empire in what is China today. It's hard to tell.

The Plaid Avenger's take on Pearl Harbor is that it was a warning signal. Japan was setting up a fence around its territory. Japan wasn't trying to take Hawaii over; it was establishing where the fence was. The Japanese said, "Okay, America, you're a powerhouse; but we're also a powerhouse now. We can bomb you in Hawaii, so don't come on our side of the fence. This stuff over here, this is ours. This is our imperial holding over here, so just stay on your side." Japan apparently didn't reckon that would be a failure. We know the ultimate end to this. Maybe we've talked about World War II too much already in this book. But in the official war years, 1940–1945, Japan invaded every other nation in Southeast Asia and the Pacific: a full-on Asian Empire led by Japan was the plan.

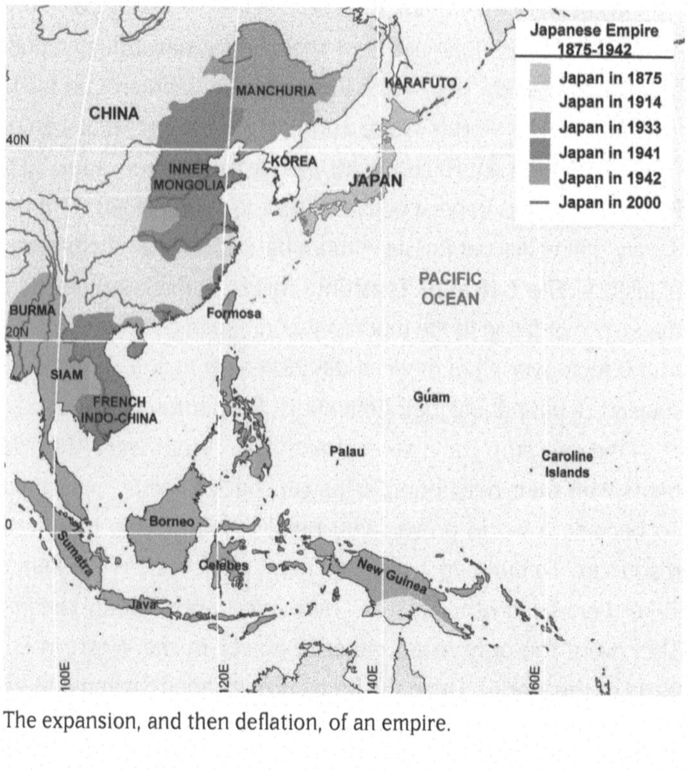

The expansion, and then deflation, of an empire.

Here's a little known fact that the Plaid Avenger wants you to understand and know: Japan also had designs to invade India and Australia. When I said they had designs, I don't mean some sketches on a cocktail napkin. We're talking about war plans. They had their little Risk board marked with the movements of soldiers, and their troops were in place for this to become reality. It was only a matter of time. They already had small landing parties in Australia, who were camping and scoping things out on reconnaissance.

The other concrete item that still exists in the world today as evidence of Japan's dream to take over all of Asia is the Burma Railway, nicknamed "Death Railway" because 16,000 Allied POWs (mostly British, Australian, Dutch, and American) and over 100,000 Asian laborers died making it. It is the subject of a movie, *The Bridge on the River Kwai*. The laborers were forced to build this railroad system as a supply line for the planned Japanese invasion of India. It ran from Thailand, through Burma, right up to the Indian border.

ANIMOSITY IN ASIA

In these ambitious Japanese plans were a few more details we should flush out in order to understand today's world a little better. How was it that such a small state with a small army ended up rolling over all of Asia in the first place?

The Asian Liberators?

Before all the war crimes and atrocities, Japanese forces were often welcomed to intervene in internal affairs of many of these places. The Japanese Army in the 1930s didn't have to fight tooth and nail to take over many of these countries. Why? By and large, the Japanese came in under something called **The Greater East Asia Co-Prosperity Sphere.** The Japanese used this concept to say, "Hey, we're not here to take over; we're here to free you from your Western European colonizers! These guys are jerks! They're white men! Look at us! We look like you, man; we're Asians! We are here to liberate you!" Japan was often welcomed, particularly in French Indochina, where Vietnam, Laos and Cambodia are today.

Japan went into these places and threw off the yoke of European colonialism. However, they simply replaced the colonizers, without any real liberation effort. It was all a big sham. This made a lot of folks within Asia pretty miffed because the Japanese ended up being, during the war years, very brutal. The Plaid Avenger can tell you, war is brutal, and war sucks for everybody involved. However, the Japanese in WWII had a particularly nasty style of warfare and the war crimes and atrocities that were committed against a whole lot of Asian people by the Japanese has not been, and will not be, forgotten. Like ever.

A horrific example is the **Rape of Nanking** in China from 1937 to 1938. The Japanese took over a city with millions of people, and since they didn't really feel like having a prison camp, they just killed everybody. They took their bayonets and chopped people's heads off. They lined people up in a row so one bullet would go through more than one of them, so as to save on ammunition. For several weeks, the Japanese slaughtered tens of thousands, if not hundreds of thousands, of civilians. This is something the Chinese will never forget, and actually they haven't: the date of invasion is marked every year in China as a day of remembrance. And that was not the only atrocity that Asians still remember . . . see the Japanese War Atrocities (inset box that follows).

Let's not forget about our friends in the middle, the Koreans. Many still hate the Japanese with a passion as well. Japan took over and extracted all of the resources out of Korean mines. How did they do that? They enslaved the Koreans! The whole country essentially became a work camp for Japan. Everyone in Korea was also forced to change their names to Japanese names. Every single person in the country! This was an incredibly difficult time for the Koreans. Trust me, they have not forgotten about it.

The occupation was also pretty nasty for the Chinese and the Koreans, in that some of the worst documented use of biological warfare took place. The Japanese said, "Let's poison a water supply in a town of five million and see how long

What's the Deal with Japanese War Atrocities?

Bataan Death March

Marching to nowhere . . .

Approximately 75,000 Filipino and US soldiers, commanded by Major General Edward P. King, Jr., formally surrendered to the Japanese under General Masaharu Homma, on April 9, 1942, which forced Japan to accept emaciated captives outnumbering its army. Captives were forced to make a weeklong journey, beginning the next day, about 160 kilometers to the north, to Camp O'Donnell, a prison camp. Most of the distance was done marching, a smaller distance was a ride packed into railroad cars. Prisoners of war were beaten randomly, and then were denied food and water for several days. The Japanese tortured them to death. Those who fell behind were executed through various means: shot, beheaded, or bayoneted. Over 10,000 of the 75,000 POWs died. Check out the old-school films *Bataan* (1943) and *Back to Bataan* (1945) with John Wayne.

Bridge on the River Kwai

Great freaking movie, you have to check it out. Based loosely on these real life events: In 1943, the Japanese used POWs to build a railroad across Burma (this is against the Geneva Convention on treatment of prisoners). This was part of a project to link existing Thai and Burmese railway lines to create a route from Bangkok, Thailand to Rangoon, Burma (now Myanmar) to support the Japanese occupation of Burma. The railway was going to be a critical connection and supply line for the planned invasion of India. About 100,000 conscripted Asian laborers and 16,000 prisoners of war died on the whole project. And the guy who played Obi Wan Kenobi in *Star Wars* is in the movie. That helps.

The Rape of Nanking

The Nanking Massacre, commonly known as **The Rape of Nanking,** refers to the most infamous of the war crimes committed by the Japanese military during World War II—acts carried out by Japanese troops in and around Nanjing (then known in English as Nanking), China, after it fell to the Imperial Japanese Army on December 13, 1937. The duration of the massacre is not clearly defined, although the period of carnage lasted well into the next six weeks, until early February 1938.

During the occupation of Nanjing, the Japanese army committed numerous atrocities, such as rape, looting, arson, and the execution of prisoners of war and civilians. Although the executions began under the pretext of eliminating Chinese soldiers disguised as civilians, a large number of innocent men were wrongfully identified as enemy combatants and killed. A large number of women and children were also killed as rape and murder became more widespread. Final toll: 10,000 to 80,000 raped; 15,000 to 120,000 slaughtered.

Want to know why there is animosity between the Chinese and Japanese to this day? Look no further.

it takes to kill everybody!" Here's one of the Plaid Avenger's personal anti-favorites: "Let's get a bunch of plague-infested rats, put them in a bomber, and drop them over the countryside to see how the disease progresses through the country!" They also did some other "little" things like mustard-gassing whole towns. This was pretty nasty business.

Now you know why some Asian countries view Japan with extreme suspicion, apprehension, and/or outright hatred. The Chinese government and some other governments still want more **war retributions:** monetary payments from the Japanese for the atrocities they committed during World War II.

The Plaid Avenger had to elaborate on this animosity that exists to this day between Japan and many other Asian countries, particularly China and Korea, because it continues to play a part of politics in today's world. Every year when the Japanese educational systems releases their history textbooks for the K-12 schools, Korean and Chinese people protest. Why would Koreans care about what school kids in Tokyo read? Just this: the Japanese textbooks continue to gloss completely over the war years, thus marginalizing all the bad stuff that was perpetrated on the other Asians groups. War crimes, biological warfare, and Japanese sex slave operations in Korea are all conveniently missing from the history books in Japan, and that really enrages some Asians.

This animosity also bubbles to the surface whenever the Japanese prime minister visits a war shrine located in Tokyo: **The Yasukuni Shrine**. Why do you need to know this? It will continue to be a point of friction between these two countries who will be the two big power players of the 21st century. It's good to know why they hate each other and this is a great example. This war memorial graveyard/shrine commemorates all of the Japanese soldiers who died fighting for, or in service of, the Emperor, no matter what the circumstances.

You have to know about these bad things in order to get the attitudes that these countries and their peoples have about each other. From the Japanese standpoint, you need to think about how big this goal of dominating Asia proper really was. From the Asian standpoint, think about how far the Japanese demonstrated that they were willing to go to achieve it. This isn't a trivial thing. This gets into why Japan is viewed with suspicion, animosity and sometimes hatred by parts of Asia to this day. The Plaid Avenger started off this chapter by saying China vs. Japan, Japan vs. China for a reason. These are places that do not like each other at all. And it could get worse.

This is an actual (insanely racist) US goverment wartime propaganda poster. I don't think America was happy about Pearl Harbor. Do you?

I'm sure you think that all this stuff is ancient history and that it really doesn't affect today's world at all. There hasn't been an official heads of state meeting between China and Japan for almost a decade because of the Yasukuni Shrine business. Trade deals have been blocked between the countries when tensions rise. Lawsuits between the countries on war reparations are still active. Now they may go to war over a group of rocks in the Pacific named the **Senkaku** or Diaoyutai Islands. Is that real enough for you? Let's go ahead and get past the war years and bring it on home into the modern era.

What's the Deal with the Yasukuni Shrine?

Yasukuni Shrine (literally "peaceful nation shrine") is a controversial Shinto shrine, located in Tokyo, that is dedicated to the spirits of soldiers who died fighting on behalf of the Japanese emperor.

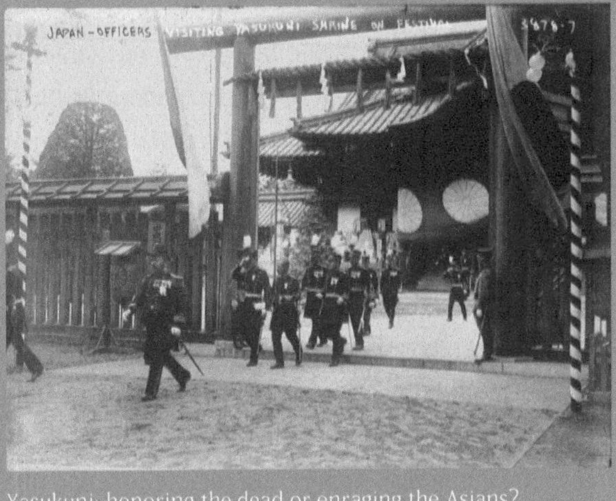

Yasukuni: honoring the dead or enraging the Asians?

So what's the problem? Every country honors its war dead, don't they? The Yasukuni Shrine also honors a total of 1,068 convicted Japanese war criminals, including 14 executed Class A war criminals, a fact that has engendered protests in a number of neighboring countries who believe their presence indicates a failure on the part of Japan to fully atone for its military past. After a prolonged period of leaders not visiting the shrine, in 2014 former Prime Minister Shinzo Abe pulled the trigger (no pun intended) and officially visited Yasukuni. Thus, his administration made their attitude crytal clear.

It's the equivalent of the Germans visiting a war shrine to pay respects to Hitler and the other Nazi war dead. How do you think Europeans would feel about that? Sound extreme? Yep. I call 'em like I see 'em.

GODZILLA IS DEAD

Nagasaki go boom.

The aftermath.

After the Germans were defeated in Europe, Uncle Sam then turned all his attention to finishing the job in the Pacific theater of war. And it was brutal. Island by island, enclave by enclave, the Japanese were beaten back by the Americans, but at a horrific cost of human lives on both sides. The US was pressing Japan for total, unconditional surrender, something that was not going to come at an easy price.

To finish up, we know in August 1945—Hiroshima and Nagasaki go boom. The other part of World War II history, which is mostly unknown yet crucial, is the fact that the US had been firebombing the entire country for months leading up to the actual dropping of the atomic bomb on Hiroshima from the Enola Gay. After the second bomb was dropped on Nagasaki, the Emperor gave the unconditional surrender, and for the first and only time in its entire history, Japan was not only defeated, but occupied.

Japan, much like Europe after World War II, was totaled. The two cities that the atomic bombs dropped on were gone, but the rest of Japan was leveled as well. There was nothing; it was utter destruction. This is important to note before we go on. Before we get to the rebirth, the rejuvenation of Godzilla!

AFTER THE FIRE

MacArthur: Japanese breaker, taker, and re-maker.

Post-World War 2 was another fascinating time because—just like the Meiji Restoration—the Japanese remade themselves again. Inconceivable! Their country was smashed, destroyed, and occupied by a foreign power. The United States, led by MacArthur, was in effective control of the entire government and they set about redoing it again, in a more democratic fashion. MacArthur and his team helped them write a new constitution, fashioned them a new society, and set about helping them rebuild so that Japan could become a strong, stable US ally in the area.

The Japanese were again faced with the same two paths. "What should we do? Should we take Option 1: resist the foreign invaders or bide our time and wait for them to leave, should we have some underground resistance movement? Or Option 2: should we embrace them, do everything they say and do it better?" Of course, they went with Option 2 again. There was another major restoration, but it had no catchy name like "Meiji." It was just a do-over of everything this time, completely under United States guidance. They went from a monarchy to a constitutional monarchy, which is basically a fair equivalent of a democracy in today's world. In fact, it was exactly like American democracy except they paid tribute to the royal line by keeping their Emperor, just like the British kept their Queen. There are quite a few other British-Japanese similarities of which you should be aware . . . (see box on the next page).

One side note of note: The Japanese themselves added a clause to their new constitution which banned them from establishing a military. Apparently they felt so bad about their wartime activities that they wanted to completely eliminate any possibility that they would repeat their mistake in the future. McArthur and crew were adamantly *against* this clause, but Japan insisted. Why would McArthur not want this? Well, the old General was pretty savvy: he was already looking over at China as a potential Cold War adversary, and he wanted Japan to be the counter-balance to that growing commie threat. However, instead of getting a militaristic Japan under US direction, he got a pacificist Japan that needed US protection. This sets up the situation which still exists today: the US is the primary protector of Japan should any outside force attack it.

United Kingdom and Japan: Siamese Twins?

Why are the two countries similar? **Number One:** Island nation, **Number Two:** Off the coast of Eurasia, flanking opposite sides of the continent. **Number Three:** Stand-offish from their neighborhood international organizations: Great Britain is distant from the EU, Japan is distant from ASEAN and other Asian groups. **Number Four:** Lapdogs for US foreign policy; they do pretty much whatever the US tells them to do. The UK does it because they love the US; Japan because they owe the US and are still defended by the US. No one will attack them because the United States is their military. **Number Five:** Both countries are constitutional monarchies which retain their regal, royal figureheads despite their utter uselessness, while both having prime ministers running the real show.

Queen Elizabeth II of England

Emperor Hirohito

UK Prime Minister Johnson

Japanese Prime Minister Yoshihide Suga

USA's loyal pets—oops, I mean allies.

The Japanese embraced capitalism, democracy, re-industrialization and rebuilding of their society with a passion under the tutelage of the US. Uncle Sam gave them billions of Yankee dollars to achieve this goal, much like an Asian version of the **Marshall Plan**. However, one big difference was that the US stuck around to oversee the rebuilding and restructuring of Japan . . . particularly because they had assumed a "big brother" role since Japan had officially banned itself from possessing a military of its own. This is one of the main reasons that these two countries are still tight when it comes to foreign policy issues.

One final thing about the military issue: think of the big bonus and boon to Japanese society due to not having to invest in a military. Most of the countries in the world have to provide for self-defense. They have to spend at least *some* money on it. Not so in Japan. Although they do, investing billions into their Self-Defense Force. . . .

ECONOMY NOW

This is one of the reasons, among many others, why Japan is so rich today. They've had "big brother" United States there. They had the luxury of investing every ounce of capital straight into the infrastructure to make themselves better and better and better. Let's make a list of why Japan has the number three economy in the world today:

- **Number One:** They have not *had* to invest in any military expenditures for seventy years now (although they increasingly do).
- **Number Two:** They have enjoyed US influence and economic ties with the number one economy in the world for the last seventy years.
- **Number Three:** Just like BASF, the Japanese don't make anything—they make everything better. They are a value-adding society. While they never invented the automobile, the computer, the video game, television, or cell phones, they make all those things better. They continue to add value in the high spectrum of things.
- **Number Four:** They are almost completely technology and service sector oriented, and they are very good at it. They're so good at it that they run **positive trade balances** with every other country in the world with whom they trade. They make more money on the sale of the stuff they make than they spend on the stuff they buy. They are always in the plus. I will come back to the list.

What else is going on with Japan? The role of Japan in Asia is very similar to the United States' role in our part of the world. While there is some animosity between Japan and some other Asian countries because of the past, it is still looked to as an economic leader. The US is the number one economic GDP powerhouse, Japan is number three, and both these countries are looked to as models to be mimicked. Just like when the US economy takes a turn for the worse, places like South America and Caribbean follow suit, when Japan's economy takes a dive, all the surrounding economies in Korea, Taiwan, and Indonesia do too, because there is so much trade from one country to the next. In this respect, Japan is to Asia what America is to the Western Hemisphere. Although at this point, China (with the number 2 economy on the planet) is quickly eclipsing Japan in economic importance in this 'hood: a trend that will likely continue the rest of our lives.

We can draw a few more parallels to the United States. Japan is an extremely rich society, so it is industrialized and fully urbanized, which means labor is really expensive, just like in the United States. You cannot get cheap labor in Japan. Like the USA, Japan has minimum wage laws and generally high labor costs. So what? What is the parallel here, Plaid Avenger? This high cost of labor has caused jobs and industries to move out of Japan, just as they have in the United States.

Just like all the car manufacturing companies that were once in the United States are all in Mexico or other nearby places with cheaper labor, Japan's service sector jobs have been moving to places like China, Korea, and Indonesia. Corporations make more money if they manufacture things where labor is cheap. While Japanese corporations might still be the biggest moneymakers on Japanese cars, most of them aren't making them in Japan anymore; they are made in all points south: China, Indonesia, Vietnam, and even India.

POPULATION NOW

What's up with the Japanese peeps? For starters, the total population of Japan is around 126 million. That is actually quite a few people on a handful of islands, that is mostly covered with mountains and forests. Japan is overwhelmingly urbanized, post-industrialized, and sanitized. They have some of the highest living standards, levels of technology, and GDP per capitas on the planet. Sounds good so far! However, the population of Japan is shrinking. This is a double whammy. People are increasingly coming up short on the 2.1 child **replacement level**, which you remember from Chapter 2. Every year, the population of Japan, while still stable, gets slightly smaller. What does this have to do with anything?

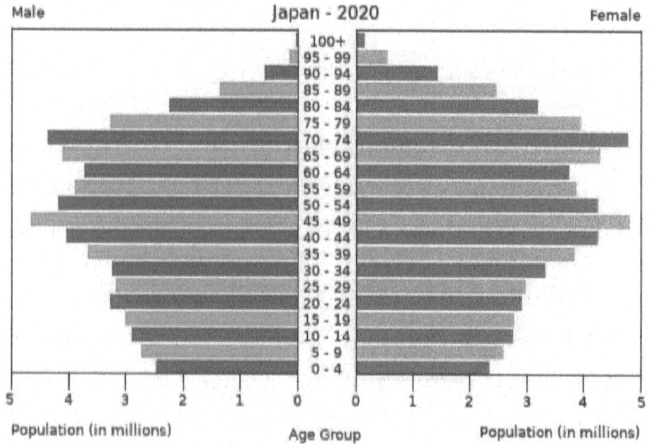

Like the Russians, it's going down.

You need workers in the factories to make stuff, to make money, so that the economy will benefit. If you don't have workers, employers have to go somewhere else. And boy, have they ever! They can't even *maintain* the technology and service sectors there at this point in time because the population is getting smaller. But wait, it gets compounded even further!

"Yes Plaid Avenger, I need your help to change Japanese demographic trends."

As I suggested earlier, the Japanese are extremely, fiercely, culturally independent and ethnically distinct. They're *Japanese*, not Chinese and not Korean. Everybody else in the neighborhood would agree with that statement. By the way, this is not a racist slant by The Plaid Avenger. Both Japanese and Korean societies are almost 100% purely Japanese or Korean. This is even more true in Korea, the most ethnically pure place on the Earth. It's easy to take the census in Korea because there is only one box to check: the Korean one. In Japan,99 percent of the population is Japanese. They actively discourage immigration there. When you visit there, you will have a great time, but if you move there permanently don't expect many people to be very friendly, particularly if you are a few shades away from Asian. This may sound blunt, but racist or not, the Plaid Avenger is telling you how it is around the world.

Kabuki theater actor, or psychotic killer?

This is not a place embracing a lot of people from around the planet. That might be all fine and dandy with them, but as I suggested, their work force is going down, birth rates are falling, thus they have a shrinking population. In the United States, maybe people are not having huge families either, but immigrants come from other parts of the world to fill those work spots and keep the economy growing. Japan is coming to a critical juncture at which they need to figure out what to do. These negative demographic trends cannot go on indefinitely.Although, due to strong work ethic and demand, many Japanese are continuing to work into their 70s, 80s, and beyond! Hey Granny Fu, out of the rocking chair, and back to work!

Not only is the declining population affecting the workforce, but it also has impacts on the consumption side of the economy. Many car manufacturers and other retail businesses are re-focusing their sales to accommodate consumers in other parts of Asia. What's the point of having a Honda dealership in Tokyo, where everybody already has a car and the population, and therefore, consumer base—is shrinking? Better to invest in India or China or Vietnam where consumers are plentiful, getting richer, and the potential consumer base is growing. This is affecting development of new products line as well; Honda and Toyota are introducing new lines in India that cost less than $2500 per vehicle! Sweet!

The Japanese government is really struggling with this issue right now. They need workers, they need a tax base to draw from, and they need their economy to stay vibrant so they can maintain their great standards of living. Unfortunately, people don't want to have big families, and they also don't really want to change their extremely restrictive immigration policy. Something's gotta give. Don't be surprised to hear about the government either promoting people to get busy in the bedroom, or "re-educating" their population to be more accepting of foreigners.

If it looks like a military, and smells like a military. . . .

MILITARY NOW

Another thing to consider in modern Japan is the fate of the **Japanese Self-Defense Forces**. What's this all about? As I suggested to you, after World War II, under MacArthur, the Japanese signed away their desire and ability to have an active military. Like the Germans, the Japanese conceded that they screwed up bad and that they were going to from that point forward be peace-loving people, and to ensure that they could never go to war again, they banned themselves from having a pro-active military. So they do not have a military that has the option of pre-emptively striking an adversary first, but they do have a bunch of dudes that can protect the country from an attack. Their current national Self-Defense Forces are exactly that: the equivalent of the US National Guard that serves as an emergency 911 for the whole country in times of national disaster and national emergency—or, theoretically, to repel any attack on the nation from foreign powers.

Anchors away! But don't leave the harbor . . .

It is important to note that it is implicit in their charter that they do not possess any powers to pro-actively attack anybody. They have absolutely no offensive role, no offensive tactics, no offense, period. No troops abroad, no missions abroad, no action abroad. Only self-defense of the state. It's the "home team" in every sense of the phrase. You do have people wearing helmets and camouflage outfits, sometimes with guns, on ships and in tanks and airplanes and helicopters. It sure looks like a military, but their only function is helping out when natural disasters like earthquakes strike.

The reason that the role of the Self-Defense Forces was so specifically defined was because right after World War II, everyone in the neighborhood was scared that the Japanese might re-militarize and start its imperial rampage all over again. Apparently, Japan was scared of this as well, and that's why they neutered themselves. But everyone was willing to accept the Self-Defense Forces. That's no problem. End of story, right? Well . . .

This force has been evolving rapidly into something that looks more and more like a military lately. Important to note is that this only came to the attention of China and the world a decade ago. How'd they find out? During the 2003 US invasion of Iraq, Japan volunteered its "troops" to go and support the US in Iraq. This was done primarily as a show of support for their US buddies; to be a part of the "Coalition of the Willing." In reality: all they were doing was refueling and resupplying US naval vessels out in the Indian Ocean, but it's the thought that counts. The United States may say, "Oh, that's a good thing; they're our ally and they owe it to us." But this is taken with great apprehension by other Asian nations, particularly China, who has reinforced the notion that if Japan says it doesn't have an army, it shouldn't. They say, "How are the Self-Defense Forces defending Japan in Iraq?" The rest of Asia has also concluded that change is afoot. The fact that Japan has sent its Self-Defense Forces abroad challenges the definition of self-defense. You send armies abroad, not self-defense forces.

With no uncertainty, the Plaid Avenger wants you to know this for sure: While many Asian nations may be apprehensive about the Japanese Self-Defense Forces, there is a country on this planet that wants Japan to throw the anti-war clause out of its constitution altogether and get busy re-arming. I wonder who that could be? The United States is practically *demanding* that they remove it. Why would the United States want this? The United States would prefer Japan be a countering force to the rise of China in Asia. It's a checkmate. In this respect, Japan's role is very much like Great Britain.

Sounds strange, but that's what it is. That's why the Japanese Self-Defense Force status are a big issue. The Plaid Avenger knows that anytime anyone even hints at this issue, China gets enraged that Japan is morphing its self-defense forces into an army, almost as enraged as China gets when any Japanese prime minister visits that war shrine we talked about earlier.

One more point on the Japanese changing their constitution for their military: There are several states in the world that the UN is considering allowing on the UN Permanent Security Council. Japan is one of the candidates to be on this council and the US is its biggest sponsor. The US is all in favor of Japan being on the UN Permanent Security Council. I wonder why? Maybe because Japan will do everything the US says. Meanwhile China is flipping out because it is vehemently opposed to this.

Here's the funny thing about it: The US has told Japan, more or less, that it will not support its candidacy until Japan gets a pro-active military. If it gets one, it would be no surprise if it developed nuclear capacity within days. Japan already knows how to produce nuclear power. It is feasible that if it changes to have an offensive capability military, it would be the next declared nuclear bomb power in the world. That is exactly what the US wants: to have a nuclear station off the coast of Eurasia to be on our team. This is a check and checkmate situation for the US. It is using Japan to keep checks and balances against the rise of Chinese power.

Along with the Chinese checkmate, the US would very much like Japan to reassert itself militarily for much more practical reasons too:

1. The US military has become over-committed, over-stretched and over-used in other parts of the world, and they simply can't keep tens of thousands of troops hanging out in Japan and South Korea indefinitely. Redeployment of troops is already underway, mostly to places in the southern Pacific where countering Chinese influence is becoming a main mission.
2. If Japan gets its army back, then they could actually help out the US in all of its current active conflicts. I'm sure the United States would welcome the change from being their protector to being a joint partner in military activities. The US would warmly welcome any permanent addition to their "Coalition of the Willing" in the War on Terror.
3. If Japan re-arms, especially in the nuclear capacity, then the US could let them deal with the North Korean freaks more effectively. Japan was always the most probable target of a North Korean attack anyway. And a nuclear armed Japan would def be a nice card to play against increasing Chinese power in this 'hood.

Back in 1945, Japan decided that they weren't going to have a military anymore, so they couldn't get into any more trouble. Unfortunately for Japan, trouble has found them. The status and activities of their Self-Defense Forces has them in a pinch between their pro-military, US ally and their anti-military Asian neighbors. Anchors away, my samurai friends!

In July 2014, the Japanese government approved a reinterpretation which gave more powers to its Self-Defense Forces, allowing them to defend other allies in case of war being declared upon them, despite concerns and disapproval from mainland China and South Korea . . . of course, the USA loved the move. **2021 UPDATE:** Former President Donald Trump expressed repeatedly that he thought Japan should be more independent, pay more for US military protection, and possibly even get their own nuclear weapons. Wow. It is still too early to call, but the effects of the Trump era may force Japan to re-militarize themselves to pre-WW2 levels, a development which will certainly upset the delicate balance of power in the region. Or worse.

FOREIGN POLICY NOW

All that Self-Defense Forces talk is intimately related to Japanese diplomacy in the modern era. As I have suggested earlier and often in this chapter, relations between Japan and its neighbors have been tense, to say the least, since the war years. At times, downright hostile might be a better description. Every year when the school books are released and every time a high-ranking official hits up the Yasukuni Shrine, the Chinese and Koreans get severely steamed and relations get further strained. Japanese troops sent abroad to support the US is just more fuel for the fire.

As of late, a few of the previous Prime Ministers were pushing for a more positive Japanese relationship within their Asian 'hood . . . and had taken a much more conciliatory stance towards their Chinese cohorts in particular. Their

Koizumi to Suga: Make Japan strong!

thinking has been that Japan cannot afford to anger the economic giant that China is becoming: both exports and imports have been increasing dramatically between these two super-states, and no one wants to jeopardize that. In fact, China has now superseded the USA as Japan's biggest trade partner . . . which of course makes perfect sense given the proximity of the countries, as well as the sheer gi-normous size of the Chinese market. (These statements apply to South Korea as well, since it will soon be the tenth largest economy on the planet.) With Japan still carrying around that bad rap due to their nasty imperial period, China increasingly has the upper hand as the future economic and political leader of the wider Asian region. To stay in the game, Japan simply cannot risk entirely isolating itself from its neighbors in this crucial wider Asian-Pacific region . . .

But. . . .

From 2012 to 2020, Japan had a new sheriff in town, and he was ready to rumble! That main man of Japan was one Shinzo Abe, and a more proud and unapologetic and hawkish figure has not held the post since the beloved Junichiro Koizumi (Prime Minister 2001–2006) was in charge. Koizumi was a wildly popular, center-right conservative, a regular Yasukuni shrine visitor, and the dude that sent Japanese defense force ships to help out the US in Iraq. Yeah, that guy. And Shinzo went even further down that "rah-rah Japan is awesome" pro-nationalist path than even his prestigious predecessor. How so?

Well, for starters, the awesome Abe was a political legacy: his father, grandfather, and even father-in-law were all prominent politicians . . . and all of them fairly adamant about making Japan strong again, even in some cases (his grandpa) wanting to scrap the constitutional clause on having no offensive military, decades ago. And we all know how family influences our attitudes and outlook on life, thus . . .

Shinzo can best be described as a right-wing nationalist, and one who decided that Japan must take a much more aggressive, pro-active role in the region in order to stay on top. I'm paraphrasing a bit here, but Shinzo was essentially of the opinion that "Japan's era of apologizing was over!" Meaning: he was not going to keep feeling guilty about the WW2 atrocities, was not issuing any more apologies, was not going to talk about it in textbooks, was not going suck up to China, and wanted to work towards changing the constitution to allow for increased military options! Boo-yah! Back up Asia, Japan may be coming back!

And that is going to cause huge tensions. During his 8-year tenure, there was a blow-up over the typical textbook release. There was a fracas over Shinzo supposedly modifying an official state apology to wartime sex slaves in Asia. Then he created a flap when he visited the Yasukuni Shrine. Anti-Japanese sentiment and protest abounded in China and Korea . . . mostly over the biggest issue between these states: a handful of rocks in the ocean!

Believe it or not, these old rivalries have become quite active again as Japan is now embroiled in two differential territorial disputes—one with Korea, and a much bigger one with China—over ownership of some dinky little islands. The rocks in question with Korea are named Takeshima Isles; the MUCH bigger possible flash point to war with China is over the rocks named Senkaku or Diaoyu Islands . . . this issue is important enough to describe in greater details in the "Global Hot Spots" chapter in part 3 of this text. Fast forward to it now if you want to know more asap!

Luckily for the pro-military faction within Japan, the totally whacked-out North Koreans under the leadership of the ever-tubby Kim Jong-Un is also changing the equation. The increased nuclear activity and missile launching and saber-rattling by the North Koreans in the last few years is starting to severely swing public support to change that constitutional clause before it's too late! Those asinine antics of the kooky North Koreans may be what totally tips the scales and causes Japan to officially re-arm with a pro-active military. The Japanese government has been all for the change; the US government supports this as well, and the former Trump administration accelerated movement in this direction. And now, maybe the Japanese public too. Look for this to be hotly debated in the coming year . . . or until the North Koreans attack.

2021 UPDATE: But honestly, not much of one! Shinzo Abe unexpectedly retired from office citing health issues in late 2020, and a dude name Yoshihide Suga became the new Prime Minister. Why is it not much of an update? Because Suga has been a lifelong ally and political operative under Abe, and actually served as a Chief Cabinet Secretary (pretty much the second-in-command) under Abe from 2012-2020. He is basically an Abe doppelgänger that will be completely following the Abe game plan already in place. In his first address of the Diet, Suga stated that his premiership will focus first on responding to the COVID-19 pandemic, then on protecting employment and preserving economic conditions in light of the pandemic, and he will do that by continuing the Abenomics suite of economic policies. He also stated that his premiership will focus on continuing the policies and goals of the Abe administration, including the Abenomics suite of economics policies and even the revision of Article 9 of the Japanese Constitution. Meet the new boss, same as the old boss.

SAYONARA!

With no doubts we can say that Japan will have considerable impact on world affairs for some time to come. As an economic leader in the world and in its neighborhood, it has set the example that other Asian economies are rapidly living up to. That role as an economic powerhouse will hold for a while longer, and it has company coming to dinner! Japan is the number three economy. China is the number two economy and eventually will be number one, South Korea is number eleven soon to be number ten. Wow. Three top ten economies all within spitting distance of each other. This 'hood is hot!

Japan is also staunchly in the Team West camp as well, and one need look no further than the spill-over in commercial and popular culture to see how intimately tied they are to the west. Ninjas, karate, **Atari**, judo, manga, sushi bars, sake shots, Panasonic, Nintendo, Sony, Honda, Toyota, Kawasaki, Speed Racer, Samurai Jack, Tamagotchi, Totoro, Akira, the Matrix, Kill Bill, 4chan, Pocky, J-Pop, and the godforsaken Pokemon—there was a blasphemous Pikachu blimp floating in the Macy's Thanksgiving Day parade right beside Snoopy, for pete's sake. Japan not only adopted western institutions and technologies, but has the unique ability to morph the Eastern and Western traditions together to come out with interesting products that appeal to consumers in both cultures. They do it like no other.

As they try to improve relations in Asia, and continue their strong relations with the US, the EU, and other rich entities on the planet, Japan will certainly have a unique perspective on the events of the 21st century. A shaper of the global economy, global politics, and most importantly, global culture, our Pokemon playing friends have a critical role to play.

Now, only if we can get them to un-invent karaoke. . . .

2019 SAMURAI TSUNAMI UPDATE

Let's start with the bad news: Earthquake! Tsunami! Nuclear Disaster! Unfortunately for Japan, how about all three . . . at once! The **2011 Tōhoku Earthquake** (referred to in Japan as the **Great East Japan Earthquake**) was the most powerful earthquake ever recorded in Japan (magnitude 9.0-9.1), and the fourth most powerful earthquake in the world since modern record-keeping began in 1900. The quake and subsequent tidal surge tsunami shocked our Japanese friends and their economy into near submission. Over 15,000 casualties, economic cost of over 235 billion dollars, and horrific long-term impacts including level 7 meltdowns at three reactors in the Fukushima Daiichi Nuclear Power Plant complex. Some areas in its vicinity will be toxic for decades.

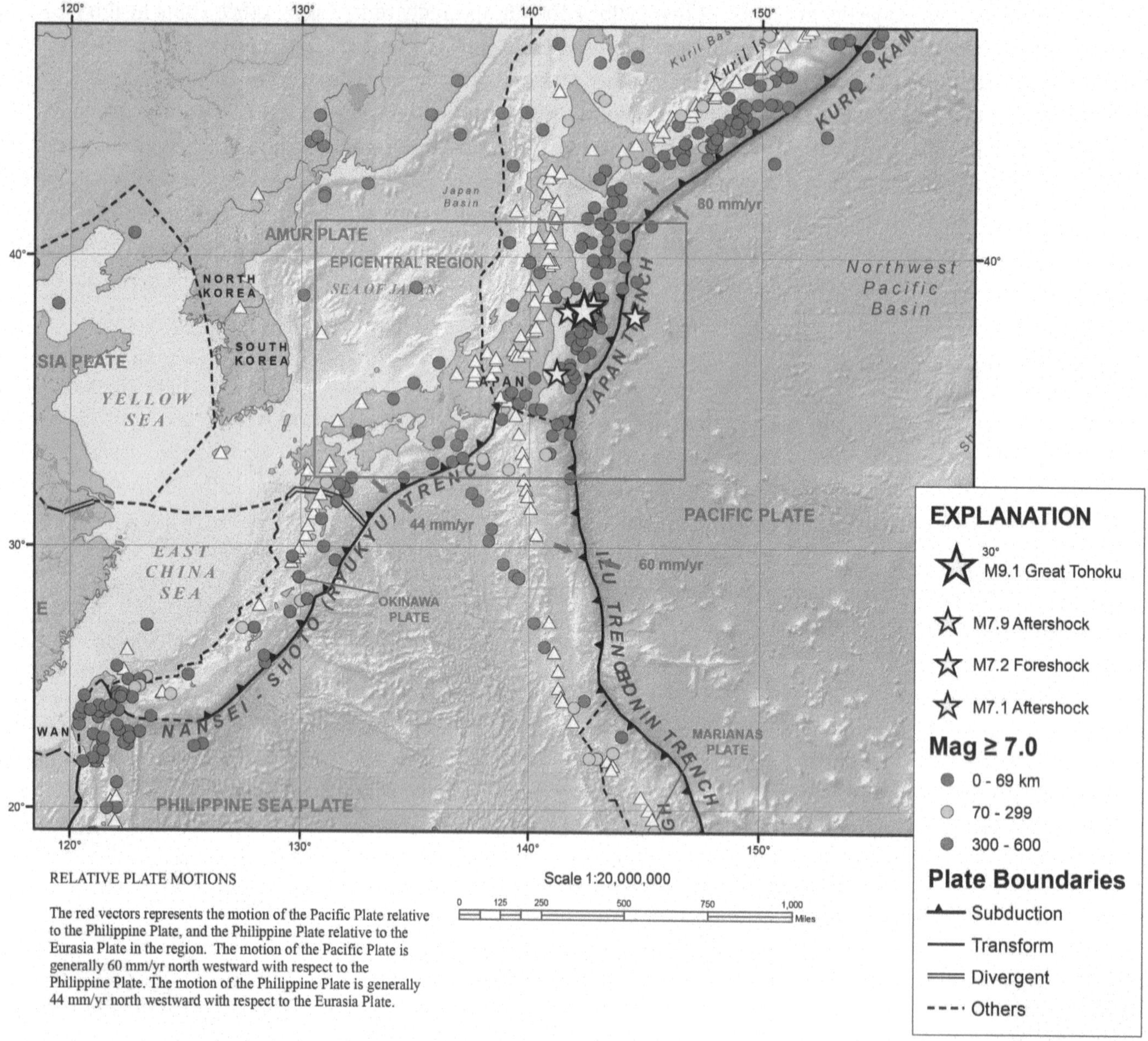

RELATIVE PLATE MOTIONS

Scale 1:20,000,000

The red vectors represents the motion of the Pacific Plate relative to the Philippine Plate, and the Philippine Plate relative to the Eurasia Plate in the region. The motion of the Pacific Plate is generally 60 mm/yr north westward with respect to the Philippine Plate. The motion of the Philippine Plate is generally 44 mm/yr north westward with respect to the Eurasia Plate.

EXPLANATION

⭐ 30°
M9.1 Great Tohoku

⭐ M7.9 Aftershock

⭐ M7.2 Foreshock

⭐ M7.1 Aftershock

Mag ≥ 7.0

● 0 - 69 km

○ 70 - 299

● 300 - 600

Plate Boundaries

▲—▲ Subduction

—— Transform

═══ Divergent

- - - - Others

→https://earthquake.usgs.gov/archive/product/poster/20110311/us/1552332847859/poster.pdf

Their economy took a huge hit from the disaster, and was stagnant for 2 decades before that, but then Shinzo Abe unleashed an economic stimulus program that has them back in the positive growth column (nicknamed "Abenomics"). In 2012, Japan voted to phase out all nuclear power, but Shinzo and crew mostly undid that move as well, since they understood the repercussions of Japan having no home-grown energy in an era that may see oil prices continue to go higher and higher.

Another challenge we have referred to earlier: the Japanese are disappearing. In some 2020 updated demographic forecasts, Japan's population of 126 million could shrink to a scant 85 million by 2100; that is troubling news to an economy already facing contraction. While still super rich, with reasonable growth and a super low unemployment figure, disposable income levels have stagnated and wealth/income disparity is on the rise. It may also be losing its once-dominant role as tech leader of Asia, both as a producer and innovator of high tech stuff. But enough of the grim Godzilla forecasting, what is still looking good in Japan?

Japan is rightly famous for its accomplishments in the fields of science and technology. It remains a major economic, science, and technology hub of the world, has a world-class education system, and invests 3.3% of its GDP in research and development. And it still has the mojo to pull off global superstar events: Japan hosted the G20 summit on June 28-29, 2019 and then attempted to follow that up by welcoming hundreds of thousands of sports fans and athletes to the Tokyo Olympics in 2020, but as you are well aware of at this point, the lost Covid crisis year not only postponed those Olympic Games, but here in 2021 even though they occurred, the stands were virtually empty. . . adding up a total loss of up to $20 billion investment the government made in infrastructure in combination with lost tourist revenue. The country has made special efforts to cultivate strong economic ties with the US, India, and others; just inked a free trade agreement with the EU; and has signed the agreement for the Trans-Pacific Partnership (TPP) with Canada, Japan, Singapore, and other Pacific economies (despite the USA bailing out of it). The land of the samurai is well-equipped to maintain its position as a prosperous, stable, and innovative global player.

So Japan is on the upswing from the disaster, and spirits are high under the Shinzo self-pride program. But, in the end, I want to go ahead and offer this Plaid Avenger assessment: Japan will rebound, but it will not in our lifetime regain its predominant role as the biggest economy and/or most important Asian political power with world impact. China is now richer, and helps Japan out in a crisis . . . that is a significant reversal of fortune to consider. And there are many other Asian titans rising fast too, many of which will soon catch or surpass Japan in international importance and material wealth. This is no diss of Japan (I love the country! And they will continue to be awesome!) but a pragmatic assessment of their future position of one of many Asian powers, not as a leader of Asian powers. Dig?

21st century: Is the last stand of the Japanese economy upon us?
Samurai have some experience with this. . . .

JAPAN RUNDOWN & RESOURCES

View additional Plaid Avenger resources for this region at http://plaid.at/ja

BIG PLUSES

- Ninjas
- Samurai
- 3rd largest economy in the world
- Center of technological innovation in elec-tronics, auto industry, robotics, etc.
- Some of the highest standards of living on the planet; highest longevity rates
- One of the most solid infrastructures in the world
- Environmentally awesome; 70% of the country is essentially a forested national park
- Politically stable, solid rule of law
- Huge US ally, and would certainly have the entire US military as back-up if they ever get into trouble

BIG PROBLEMS

- Earthquakes, Tsunamis, & sproadically, Godzilla
- Population shrinking, and immigration not really accepted
- Very high cost of living and high wages causing industry/job movement to Asia
- Heavily dependent on natural resources, fossil fuels and energy in general imported from other regions.
- Has been economically/financially stagnant for decades; little to no economic growth
- Has been politically gridlocked and governmentally ineffective for several years
- Friction still exists between Japan and Asia; particularly China and Korea. This has stymied closer economic/political integration with its nearest neighbors. (Both of which are huge economies)
- Would likely be the #1 target of a missile if/when North Korea goes completely nuts

DEAD DUDES OF NOTE:

Commodore Matthew Perry: Commodore of the United States Navy that forcibly opened Japanese ports/trade to the West with the Treaty of Kanagawa in 1954, ending centuries of self-imposed isolation under the Shogun.

Emperor Meiji: The main man in charge of Japan during its revolutionary overhaul to Western ways, which then set the stage for its imperial rise to world power. Thus, his name is attributed to the entire time period of his rule, and the 'Restoration' itself.

Hideki Tōjō: General in the Japanese Imperial Army and Prime Minister of Japan during most of WWII, Tojo is held responsible for the bombing of Pearl Harbor which pulled the US into the war. Tojo was later tried for war crimes and executed.

General Douglas MacArthur: General of the US Army, played a major role in the Pacific Thater of WWII, officially accepted Japan's full surrender at the end of the war, and oversaw the US occupation of Japan from 1945 to 51. As the effective ruler of Japan, he implemented radical economic, political, and social changes to the country. That's why they are kind of like the US!

PLAID CINEMA SELECTION:

Note: Japanese cinema tradition is as rich, diverse and numerous as American and European industries, so there is no real hope of selecting just a handful to represent. Instead, here is the plaid amalgam of picks from across the board.

Caterpillar (2010) A critique of the right-wing militarist nationalism that guided Japan during the Sino-Japanese War and into WWII. This bizarre story has a limbless war hero returning from the front lines of Japan's invasion of China, and depicts the ultra-nationalistic and militaristic attitudes of the country at the time.

Godzilla (1954) Documentary on the effects of giant lizards and their impact on a major city. Includes an indictment of use of weapons of mass destruction by the United States during World War II. 2004 US re-release is considered by some to be a cinematic masterpiece.

Lost in Translation (2003) A movie star with a sense of emptiness, and a neglected newlywed meet up as strangers in Tokyo, Japan and form an unlikely bond. Great quirky look at modern Japanese culture through the eyes of foreigners. And Bill Murray—total awesomeness.

Seven Samurai (1954) A film about a village of farmers who hire seven masterless samurais to help protect their farms from bandits. Considered the greatest film of Akira Kurosawa, arguably the greatest film director in Japanese history, and one of the best in the world. Influenced Western cinema for several decades after its release.

Shall We Dance? (1996) Another fantastic insight into the prim and properness of Japanese societal norms, this time in Toyko, and all about ballroom dancing. I know, sounds boring, right? But it will make you laugh out loud in parts, and is great at portraying the distinctly Japanese take on family life, work life, and even household life. Yes, this is the original that the idiots in Hollywood remade with Richard Gere, which does not work at all outside the Japanese cultural context.

Twilight Samurai (2002) Set only a few years before the Meiji Restoration, the film follows the life of a low-ranking samurai who is employed as a bureaucrat. Paints an accurate portrait of the life of a subordinate samurai at the end of the Tokugawa Shogunate. Set in same time period as Tom Cruise's *The Last Samurai*, and a million times better.

Udon (2006) A simple story of noodles. But man oh man, it's a great subtle comedy that shows so much more. Fantastic insight into today's Japanese culture. Shot on location all around Kagawa Prefecture on Shikoku. Long, but a student crowd pleaser, and will make you hungry.

LIVE LEADERS YOU SHOULD KNOW:

Junichiro Koizumi: Extremely popular former Prime Minister of Japan from 2001-06. Was "the Ronald Reagan of Japan": conservative, pro-business, pro-US, and had a penchant for wild hair and Elvis. Even visited Graceland. Sent Japanese Self-Defense Forces abroad for the first time in its history to help the US in Iraq.

Shinzo Abe: Current Prime Minister of Japan since 2012 and an ardent nationalist who is likely to increase tensions with neighboring Asia . . . especially China. Wants Japan to be get past the 'era of apologizing,' has re-juiced their economy, and enjoys widespread support. For now.

MAP GALLERY

View Map Gallery & additional Plaid Avenger resources for this region at http://plaid.at/ja

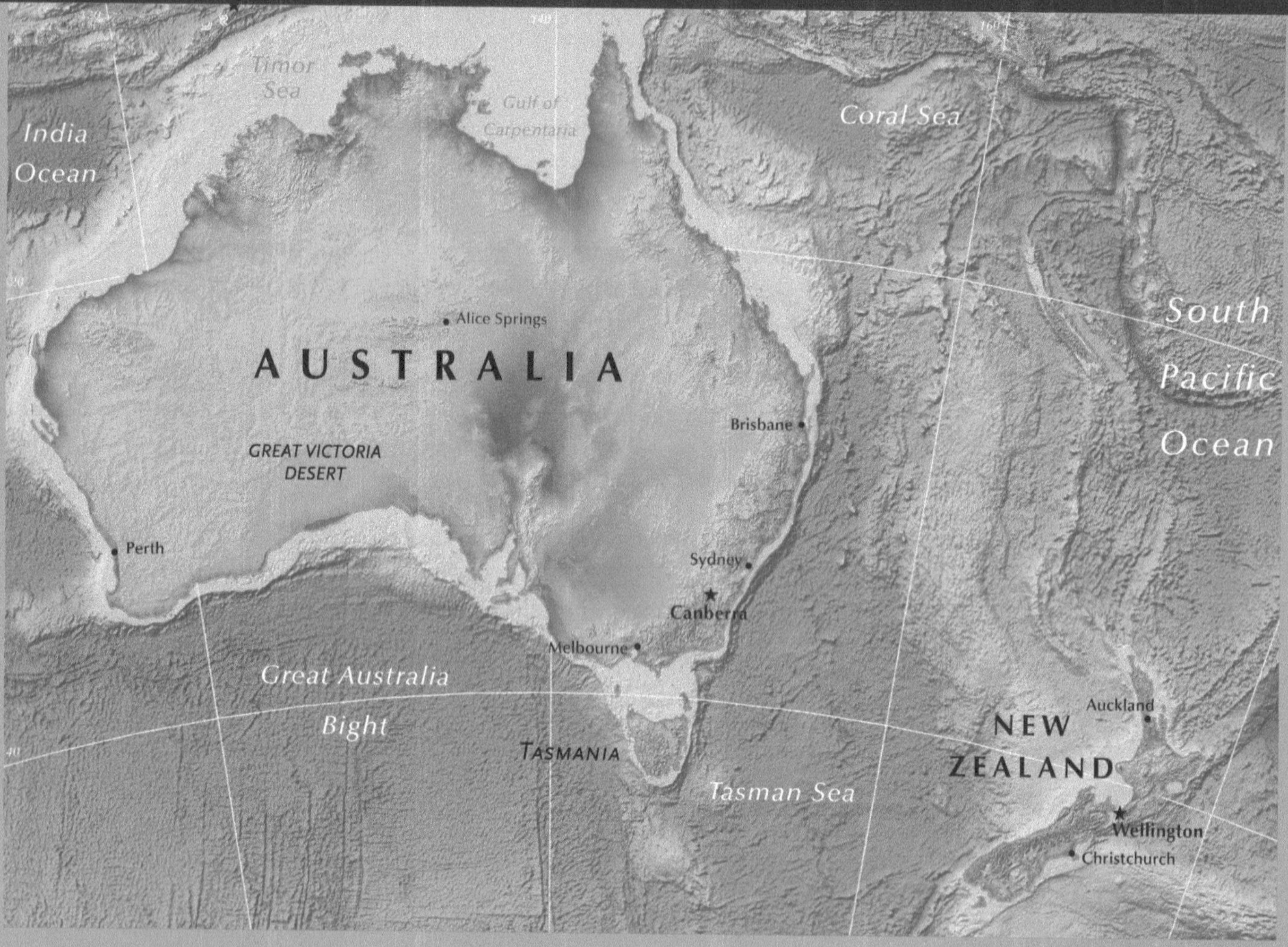

CHAPTER OUTLINE

Australia and New Zealand

12

G'DAY, MATES!

We're heading south from Japan to our friends in the Australia and New Zealand region. When the Plaid Avenger is traveling around the world and can't make it back home to America, I just stop off in Australia, or as I like to call it, the **Mini-America of the southern hemisphere**. If you ever want to fire up people from Australia, just pass that along. It infuriates them for some reason. They tell the Plaid Avenger that they're distinctly Australian and unique in their own way. Yeah, right! You dudes and dudettes are exactly the same as us: albeit an European and American 21st-century cultural-amalgamation-sensation south of the Equator!

The Plaid Avenger is always straight shooting, so I am not going to lie to you; this region is not one of tremendous significance on the world stage. But like Japan, it does play its part to effect current events and regional activities. Its primary role right now is as a platform of European culture and US foreign policy in the southern hemisphere. Even that is changing quickly of late as the mates from "down under" are shifting their political and economic focus to be more in tune with their Asian 'hood. But I am getting ahead of myself. Let's first focus our attention on the physical traits of the region.

LET'S GO FOR A WALKABOUT

Physically, Australia and New Zealand are two island countries, even though Australia is classified as a continent. If you look at it, it's really just a big island. They are two islands of extremes. In Australia, the vast majority of the interior is desert and steppe. It's very similar to the American Midwest. There's a bunch of cattle ranching

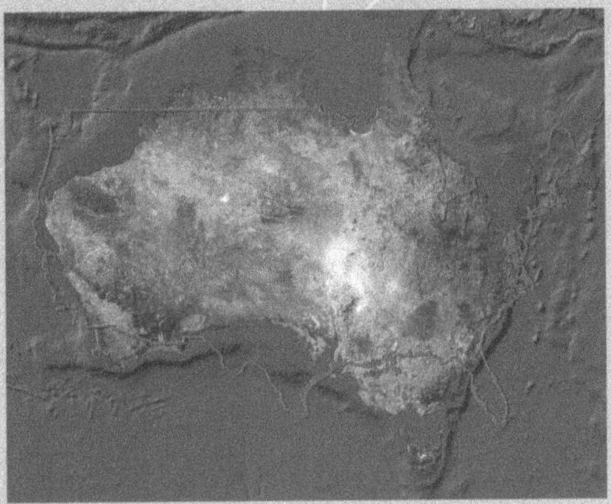

Area comparison.

Lights out in the middle!

Deserted desert interior in Australia

going on; it's the outback, after all, but there aren't too many people there. In fact, a massive percentage of the interior of Australia is so dry, it has promoted a primarily coastal settlement pattern. Everybody's hanging out on the coast and the vast majority of Australia's population is located on these coastal margins, particularly the southeastern coastal margin.

As you can see from the lights at night, virtually everybody is on the east coast of Australia. Before we go any further, let me begin pointing out the parallels between Australia and America. Here is one outlined for you on the map: population primarily in coastal areas, with a high concentration on the eastern seaboard. In New Zealand, we have the same pattern, but for different reasons. New Zealand, much like Japan, is a mountainous country. In fact, Japan and New Zealand are almost identical. They are both island nations, made of several different islands, and both are on the borders of major continental plates; both are volcanic in origin, so the interior of the country is very mountainous. The terrain makes it a no-brainer for most people to live on the coast, especially those little hobbits which seem to be over-running the country here lately. Gandalf, help us!

New Zealand: mountainous center spine

Climatically speaking, these two countries are opposites. New Zealand is cooler and wetter, with a climate more like the Pacific Northwest section of the US, or even like the UK. Australia has some Mediterranean style climates on the southern coastlines, and even a little tropic savanna-type areas in the north, but the interior is largely steppe and desert; a dry area with limited rainfall, few trees, and scrub and short grass vegetation. What are those climates good for? Cattle and sheep production. There is a lot of grazing action going on in Australia. Big exports of wool, lamb, and beef.

One last physical note: what Australia lacks climatically, it more than makes up for in physical resources. They have tons of coal, iron, copper, opal, zinc, and uranium. Put that together with the agricultural commodities, and you have a serious export base to work with. Think about that for a second; what types of countries mainly export primary products? Hold that thought, we'll be back to it soon enough.

AMERICA, JUNIOR

Why does the Plaid Avenger refer to this region as the mini-America of the south? Where do many people from America who like to travel abroad want to go? Australia! Why? Because it's just like home! People want to see the same stuff, eat the same stuff, and speak the same language as in the US, because they are more comfortable in that setting. It's far enough away that you think you're actually being adventurous! There's actually something to the idea of Australia just being a transplanted southern piece of America or Europe, but just "down under." Down under the equator, that is.

Australia is quite unique in the southern hemisphere of this planet for a couple of reasons. For one, it's the only region in the southern hemisphere that is habitated primarily by white, English-speaking, European descent-type people. It's also the only region south of the equator that is totally rich and developed. It has standards of living and material wealth on par with that of Western Europe, Japan, and the US, and no other region south of the equator can say that, yet. The standard of living thing is another reason why Australia is the Mini-America of the south. But oh, there's so much more.

BACKGROUND

Why is Australia so white? Because it was colonized by the pasty British white peoples, of course! Both the US and Australia were originally British **convict colonies.** In the UK back in the day (in the 1600's and 1700's), one of the subsidiary purposes of Britain's colonies was to get rid of British criminals. Britain was perhaps slightly overpopulated, and laws were very strict.

The historical anecdote that explains this fairly well is the establishment of the **baker's dozen.** The baker's dozen is 13 instead of 12. Why? There was actually a law that stated if a baker shorted you a donut, he could be put in jail. So the baker's dozen was born. Throw in an extra one, because who the hell wants to go to jail over a donut?

This was a pretty legally strict society where lots of people got tossed into jail for sometimes menial offenses. Among the options given to convicts were the gallows or a boat ride to a colony. Guess which one people chose? You know there is a 'criminal' element to America's background, and the same goes for Australia. Here's the funny part: the British used to ship their convicts to their American colony, so when did they start shipping them to Australia? Right around the 1780s. Why is that? In 1776, America had a revolution, and we kicked those British tea-sipping, prisoner-exporters out, eliminating the US as their felon-absorbing resource. The British, who were floating around the world, figured out that they had claimed a huge chunk of land down in the southern hemisphere. Look at that, a place to send convicts! Australia was established in 1788 as a convict colony, in part because the British couldn't send them to America anymore.

SOCIETAL STRUCTURE

How else is Australia like America? Let me count the ways. If we look at virtually any facet of life in this part of the world, we can look at America and see that it's pretty much the same. Let's start listing them off, shall we? Australia is a highly **urbanized** society. There are roughly identical urbanization rates between Australia and the US. This point is of significance because everybody has the opposite impression about Australia. Everybody thinks they're all about the outback, the Crocodile Hunter, and Crocodile Dundee. There are all these wild and crazy guys down there! However, if you ask any Australian how much time they spend in the outback, "You've got to be kidding me, mate," is a likely response.

Peeps are largely coastal urbanites.

Why is that? Quite simply, the outback sucks. It's crazy that anybody would want to go to the outback. It has all kinds of dangerous creatures in it, spiders that will jump onto your face, the top ten poisonous snakes in the world, and so on. Nobody wants to go out in "the bush." It is a classic mythology of Australia. It sounds cool, makes them seem all brave and manly, but nobody hangs out in the interior of Australia. Most people live in cities and suburbs, which you are all probably familiar with. They are all content to stay out of the outback, and have the nice little picket fence around their manicured suburban lawns.

And this relates to the standard of living. This is a very rich place. Per capita, it is on par with the US and Western Europe and any other rich place. Standards of living, levels of technology, material wealth: this region is staunchly in the 'fully developed' category. People live there just like they live in the US, and have all of the same creature comforts. This reiterates why nobody is out in the wild, because rich people go and see the wild when they're on vacation, and then they get the heck out of there because the wild sucks.

One of the reasons why they have a high standard of living: there are not a whole lot of people living in Australia. Australia and New Zealand populations are extremely low, only about 28 million folks across this entire region . . . 25 mil in Aussie-land, and less than 5 mil in NZ. New York City, L.A. and Tokyo each have bigger populations than the entire continent of Australia! Look at the population pyramid! As stable as stable gets. Only through immigration will the population of the region increase, much like the US. But there's more!

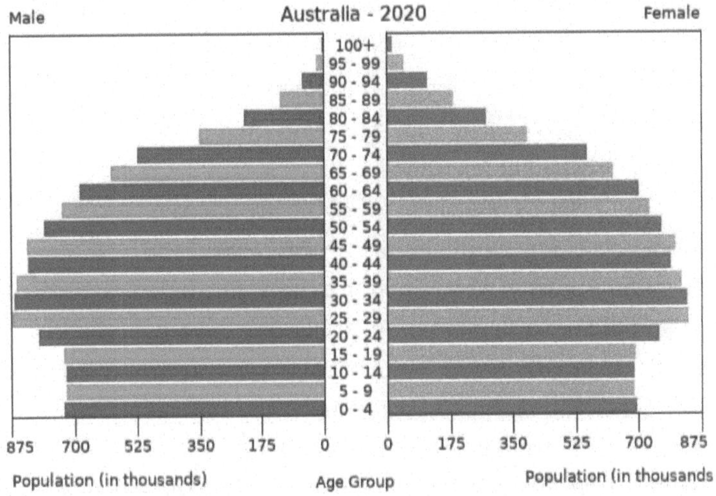

A stable yet small population.

Australia also has similar dietary habits, which ties back in with standards of living. They are big meat consumers, just like in the US. The only big difference is that the Americans like to slay the cows and eat hamburgers, but the Australians are all about lamb. They love to silence the lambs. And then eat them . . .

Lambs! How cute.

And how delicious!

They are also big beer and wine drinkers, like the US. In most places that were colonized by Western Europeans, the beer and wine tradition still exists. Australia is one of the booming wine regions now, and despite their remoteness and smaller sizes, the Aussies are the 4th biggest wine exporters on the planet, and New Zealand is the 7th! That's a lot of juice from down under!

Perhaps the Plaid Avenger is getting too shallow, talking about standards of living and beer and meat consumption. Let's get back on point. What other parallels are there?

This may just seem like a funny side note, but this may play a part in why the two nations are so similar today. Both countries have a criminal background, a criminal profile; remember, they started as convict colonies. How has that played into today's society? This is a stretch, but stay with me. What do people like to do in both of these societies? We've already pointed out the beer drinking thing: both are beer drinking societies, and both like to party. What else do they like to do while consuming alcoholic beverages? I don't know—watch brutal sports? Yes! The Americans have their football, Australians have Australian-rules football and rugby. The main difference between the football styles? The Aussies don't have any rules, and don't wear any pads! They just don woolen jerseys and beat the crap out of each other, while everyone who isn't playing drinks beer and watches.

On a more serious note, each country has a rugged, individualistic archetypal character. The Americans have the Marlboro man, a lone, rugged warrior out on the plains, wrangling cows and looking out for coyotes. That's part of the common symbology, the perception of what it is to be an American. He embodies individualistic pride. They've got the same stuff down in Australia, except they wrangle crocs and look out for dingoes. Do these types of characters represent the majority of either place? No, but these characters are popular symbols in this society. There really aren't these archetypal figures anywhere else in the world. It may be partially due to the fact that both places have large expanses of land with low population density. They have frontiers, and there has to be the rugged individual to explore those frontiers.

Here's another similarity between Australia and the US: the settling convicts exterminated nearly the entirety of their respective native populations. There is a history of repression of native peoples in both places. In the US, it was the Native Americans, the Amerindians, that were displaced and/or annihilated. In Australia they had the Aborigines; in New Zealand the Māori. All of those peoples suffered immensely in the onslaught of British colonization in exactly the same ways. And it was particularly bad in Australia.

American Cowboy.

Australian Cowboy.

See the difference?

In Tasmania, an island off of the south coast, there were some indigenous people called the Tasmanians who were openly hunted during a period of the settlement of Australia. The government essentially said, "Hey, settlers, want some free land? Go down to Tasmania! If the natives give you any trouble, don't feel bad about killing them." It became a sport to go out and hunt the indigenous Tasmanians. This was done so efficiently that they killed all of them. Every. Last. One. That's the criminal background manifesting itself in one of the nastier ways in history.

Another historical consideration of why both societies evolved out this frontiersman mythology has to do with how both countries grew. Both started on the eastern seaboard of a large uncharted land mass, and proceeded to expand and conquer the continents in a continuous westward expansion. Interesting, isn't it? Australia even underwent a western "Gold Rush" at almost the exact same time as the famous Californian rush of the same name, but with one slight difference: theirs was more profitable!

Brutal sports, slaughter of indigenous populations, mostly white, cowboy image, and even the English language all play into the cultural baggage of the US and Australia. This is why the Plaid Avenger refers to it as the Mini-America of the south.

Put all these historical, physical, and cultural factors together and you are starting to get a sense of why the Plaid Avenger draws so many parallels between the US and Australians region. But you don't have to take my word for it! There is proof a'plenty that these two states have a tight relationship: let's just take a look at their foreign policies for a minute.

AWESOME ALLY

When it comes to Team West foreign policy, one could not find a better friend than those Aussie mates. They have been there year in and year out for their western allies, for every single international conflict and conflagration. We can be even more specific: first, Australia has always supported every endeavor of the old British motherland, and then later came on board for all things American. This bunch really sticks together.

The 'Roo Brigade will always be ready to serve freedom!

I'm not talking about supporting them on paper, or in a UN vote, or economically. I'm talking about Australians participating side by side with their English-speaking brothers-in-arms in World War I, World War II, the Korean War, the Vietnam War, the Cold War, and the War on Terror, which has spawned the two newest Aussi-supported wars: the campaign in Afghanistan and the war in Iraq. Every conflict, every time, they are there for the West. Check out one of Mel Gibson's first starring roles in a movie named *Gallipoli* (1981) about Aussie troops fighting the Turks way back in WWI.

Being a small country with a small population, the numbers of Australian forces that have participated in these conflicts is not a huge number, but it's the thought that counts, especially when the thought is a bunch of dudes with submachine guns. But I digress. Point is, that they have regularly been looked to as a vote of support for western ideology worldwide; most currently, counting themselves as one of the "Coalition of the Willing" in the lead up to the Iraq War. They are the only folks south of the equator you can say that about.

So historically strong are the Australian/American ties that the Plaid Avenger counts them among the three most reliable, solid platforms of US foreign policy in the world. The UK, Japan, and Australia can be counted on to support US endeavors, almost with no questions asked. In the past, this essentially made these countries the American lapdogs in terms of foreign policy, but times do change.

While I stand by the description of Australia as a loyal ally of the US, modern times have forced them to be more pragmatic. Sometimes, being a loyal ally has to be pulled back due to popular opinion. Most Australians were opposed to the US-led invasion of Iraq, and the government at that time did take a popularity hit for its staunch participation. In addition, the Aussies and the Kiwis are having to adapt to a world which sees the rise of Asian powers, as is the rest of the world. But while this region remains entrenched as part of Team West, they have to start being much more sensitive to the feelings of their Asian trading partners . . . which is increasingly putting them between a US rock and a Chinese hard place. More on that later.

ANZUS on the ready, sir!

But one need look no further than ANZUS to prove the tight relationship between these Pacific partners. ANZUS—The Australia, New Zealand, United States Security Treaty—is a non-binding military alliance or collective security agreement signed in 1951. It is, in spirit, similar to NATO Article 5, meaning an attack on any is considered an attack on all. (It was followed later by the Southeast Asian Treaty Organization in 1954 and included members of ANZUS and Britain, France, and other Asian states. Acronym: SATO. . . Ah! even rhythms with NATO!) It primarily focuses on the security in the Pacific, although the treaty in reality is related to conflicts worldwide.

Unlike NATO, the countries do not have a fully integrated defense structure, nor a dedicated active force waiting on the sidelines ready to go. However, this three-way defense pact is a commitment to defend and coordinate defense policy in the Pacific, and includes joint military exercises, standardizing equipment, and joint special forces training. In addition, the Aussies are host to several joint defense facilities, mainly ground stations for spy satellites and signals intelligence espionage for Southeast and East Asia. Oh China! The Aussies are watching you! Well, maybe its the Americans in Aussieland that are watching you!

As an interesting side note, New Zealand was "uninvited" from ANZUS from 1984 to 2012, because back in the 80's they declared their country a "nuclear-free zone" and thus refused to let any US navy ships that were nuclear-powered or nuclear-armed come into their territorial waters. Of course that ticked off Uncle Sam mightily, and got the kiwis kicked out of the club. But the world has changed, and the US needs their Pacific allies more than ever, so in 2012 Barack Obama officially ended the ostracization, and the ANZUS crew is stronger than ever . . . and what timing! Why would I say that?

Because also in 2012, the US and Australia announced the FIRST EVER permanent US military base on Aussia soil. It was originally proposed as a small coastal station way up north in Darwin (Northern Territory) that would be home to a few dozen Marines . . . and within 6 months it was expanded to house closer to 2500 US military service personnel. At this writing, the agreement for the Darwin base is to have US troops in Australia until 2040 and a contingent of 1,500 US marines arrived in 2018. I also have insider tips from some Marine friends of mine that there are covert bases already in NZ as well. Can you say "countering Chinese influence in the Pacific"? I knew you could! Interesting times my friends!

A WORD ON THEIR WEIRD WAY TO WEALTH

Australian number one export. Talk about a return on investment! Ha! Boomerang? Return? Get it?

This is where things get a bit trickier, so the Plaid Avenger is going to have to sort a few things out for you. We've made it expressly clear that the US is the powerhouse economy of the planet, but China, in the number two slot, is catching up fast. Japan is at the number three biggest, and Germany's at number four. Where's Australia? It's in 13th place, which is super respectable for their small pop and overall size! But in terms of total GDP, how rich is this place? It's doing okay. However, in levels of material wealth per person, they are totally awesome. How could this be? How could they have a standard of living identical to ours if they are not a huge exporter of manufactured goods or technology or other rich stuff? It is a strange circumstance, and perhaps a unique circumstance to Australia, that when we think of features that make it the most unique, it's not an exceptionally wealthy place as a total.

What do they produce? What do they really do? What do you buy from there? Do they make computers? Cars? Linens? DVD players? Software? Video games? I don't think so. What have you ever bought that said, "Made in Australia"? Here's the quick answer: didgeridoos and boomerangs. That's not much of an export economy to put them among some of the wealthiest countries in the world. Well, that's not what has made them wealthy.

Australia is perhaps the only country in the fully developed world that has gotten rich from the export of *primary products.* And it is still a big part of what they do. Remember that primary products are things extracted from the earth like coal, oil, minerals, diamonds, uranium, wool, cattle, and meat. What is all this stuff worth? When it comes right down to it, not worth much. They are a bunch of unprocessed materials. How can the Aussies be rich if they are doing stuff that poor countries do? Well, for starters, there are only 25 million peeps in the entire country, and it's the size of the US. They export so many goods because of the resource-rich land, and the wealth generated from that is spread between so few that they are rich overall. A pretty big GDP based on natural resources, divided by a small number of peeps, makes for a really high GDP per capita number.

But don't let the Plaid Avenger lead you too far astray. If you look at the numbers, over 60% of the workforce is in the service sector, but who are they serving? Pretty much just themselves. Seriously, no innuendo implied. They work in malls and mailrooms and breweries and bakeries and accounting and telemarketing centers. They produce computers and cars, but many of the manufacturing facilities are owned and operated by multinational companies (my favorite example: the Subaru Outback—a car named for Australia, made by a Japanese company), and the products are only built to be consumed in Australia. In other words, they make Apple computers there, because people there want Apple computers. They are not exported out of Australia. There are no cars or refrigerators exported from Australia. That's because they are too busy making serious bank on exports of primary commodities. Their internal economy has enough juice to keep them going and growing, but it is a matter of low population and lots of primary resource wealth that makes for such high GDP per capita. They are the only place on the planet that I know of that has gotten away with that. Bottom line: they may have 65% of the people in the service sector, but they make 65% of their total GDP on exports of primary products.

You'll often see information about sheep and lamb being a huge export to places like the Middle East, since they consume more lamb than they do things like beef and chicken. You'll also see information about the coal and oil that Australia has being exported to places like China and Japan, both of which are hungry for those energy resources. Australia also signed a uranium deal with China on April 4, 2006, to export raw uranium to be processed and used. No other "rich" country on the planet is exporting raw commodities at this scale. Everybody else processes things, because that's where the money is. It's fascinating that their primary source of income is from raw products. They are definitely an economic conundrum of the rich/fully developed world.

Throw him on the barbie, I dare you.

PACIFIC SHIFT

What is in Australia's future? To answer that we might ask ourselves what's changing, because while we've been saying that this is the mini-America of the south, things are definitely on the move.

ECONOMICALLY

As China and other Asian states increases in world power and wealth and are requiring more resources, Australia is reorienting its economic focus to nearby countries in Asia—instead of trading primarily with other white English-speaking countries like the US and Great Britain. In the past, these two countries were the primary trading partners with Australia. Seriously, what a pain it

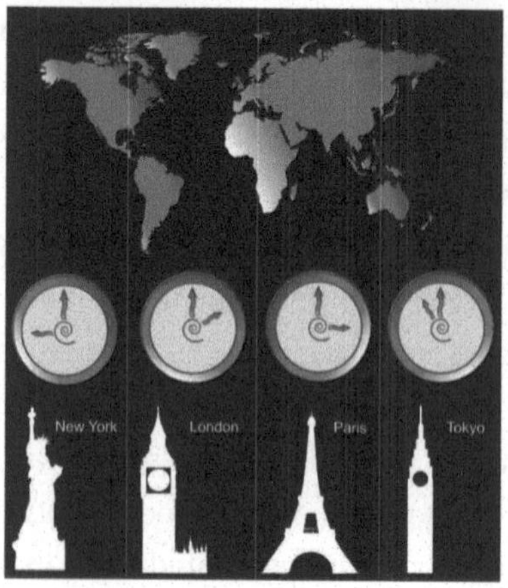

Aussie's really on Tokyo time

is to ship exports that far. It's a long commute for meetings and it's difficult to maintain trading relationships with their white, English-speaking allies. They're on the other side of the planet! Nothing, other than fine Australian wine can be shipped all the way across the world anymore and still turn a profit.

Also, consider what impact longitude has on trade relationships. If you are at the same longitude, you are in the same time zone. Now, Australia is doing business in real time, during their business day, instead of having to deal with the massive time difference between it and other trading partners like the US and Europe. It's a tremendous complication to be awake when your trading partners are asleep.

They have reoriented their trade relationships to more local areas are neighboring regions like southeast Asia, China and Japan. "Why continue trading with the white people that look like us, when we can make lot more money trading with the Asian people that don't look like us?" Money talks. This is the way it is. Plus, the booming economies and large populations of Asian countries make for a much bigger demand on Aussie resources. Team West economies have been stagnant or slow growing, while China and ASEAN countries have been on fire with economic growth. And they are closer too! So there is much more money to be made right now supplying Asian demand, not propping up old trade ties on the other side of the planet. Trade between Australia and ASEAN countries grew by 10% just from 2016-2018. There's been a promotion of trade, migration, education exchanges, and tourism according to the government's own description of the relationship.

In conclusion: Aussie economic future lies in Asia. Over the next few years, look for the portion of Australia exports of raw goods to Asian countries to increase, and dramatically. Raw material exports are not the only thing on the menu, my friends. Many Australian businesses, like so many others across the planet, are moving manufacturing facilities as well as research and development operations to Asian locations to take advantage of both cheap labor and the large local talent pool. I suppose the future of the continent may also be as a giant national park serving tourists from around the world serving shrimp on the barbie after you scuba on the Great Barrier Reef. And there is something to that: tourism is a big money maker, particularly since Australia and New Zealand are beautifully scenic areas, with the added bonus of being fully developed and politically stable . . . which means people love to come visit to live well, be safe, and buy great stuff.

Those same reasons are why many multinational companies (even Asian ones) are locating their corporate headquarters in Australia/NZ; headquarters that run their global operations in Asia. Say what? Heck yeah: corporate heads want to do business in Asia to make their companies rich....but they want to live in an awesomely nice and clean and stable place like Australia while they do it! High end sectors like finance, banking, and research are taking hold in Oz as well, since a critical mass of highly educated, internationally-connected peeps has evolved there.

Even more so than America, Australia's future is tied closely with Asia. Asia is a booming region on the planet, a place where a lot of things are going on economically, and Australia is right there to help them along and make some profit at the same time. As we will see in the Mexico chapter, it is good to be next to a gigantic economic engine, and Australia's proximity to China, who will be the largest consumer of raw goods on the planet in the coming century, will enable Australia to turn huge profits. Former Prime Minister John Howard (in office 1996–2007) was quoted in the past about Australia refocusing their efforts and economy to better compliment China's rising power. He has called Australia "an anchor of stability" in this region. He also said, "When we think about the world, we inevitably think of a world where China will play a much larger role."

POLITICALLY

Let's stay with Former Prime Minister John Howard for just a minute more, so I can explain to you how Australia's role in the vicinity is changing politically as well. To do this, I must elaborate on the most hilarious "story of the sheriff" which ruffled many feathers. It goes a little something like this:

About twenty years ago, former US President George W. Bush *complimented* Former Australian Prime Minister Howard during a press release by referring to Australia's status as the "sheriff of Asia." What the heck was that supposed to mean? Most Asian countries interpreted this as "the white dudes in the area are in charge," and that the long arm of the US law was being stretched through Australia to keep order in the area. This seriously ticked off all of Australia's neighbors. Malaysia, which is a awesome state in its own right, nearly declared war over it. To paraphrase the Malaysian prime minister, "Think that all you want, but if you ever set foot on our soil we will declare war on you instantaneously, no matter what the US says about it."

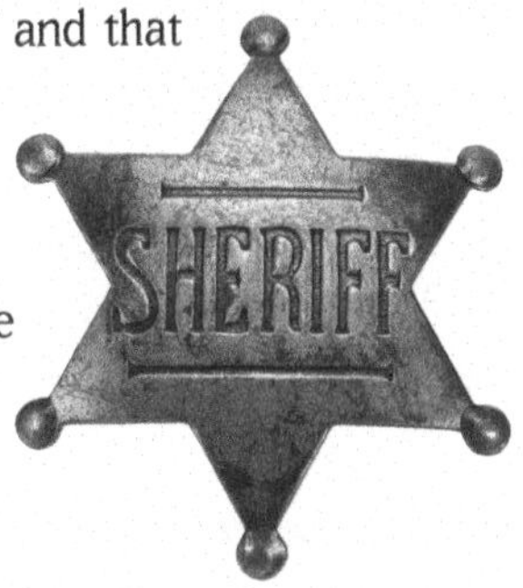

Whoa there USA! Leave us out of it!

Even John Howard was a little miffed. He was probably thinking, "Thanks for the compliment, Mr. Bush, but try not to ever say anything about me in public again." This was a problem especially since Australia is reorienting itself toward Asia. These are the guys with whom they are trying to buddy up! The last thing they need is negativity about their foreign policy ties to the US. They still want to be allied with the US, but they don't want to be throwing it in people's faces.

The Plaid Avenger suggested in an earlier chapter that Europeans are slightly apprehensive to support US foreign policy, as they have an increased risk of terrorist threats due to their proximity to the Middle East and Central Asia. Australia is in the same boat. They're closer to other regions that may be slightly hostile to the US foreign policy. Australia is right there beside Southeast Asia, parts of which are known hotspots for extremism and terrorism. There are some extreme fundamentalist Muslim groups in this area; even my Muslim friends would concur. Terrorist cells are known to exist in lots of places in Southeast Asia (e.g., Indonesia, the Philippines, or Malaysia), and Aussies have been targeted by terrorist attacks in the past (look up 2002 Bali Bombing for an example).

This is why John Howard was quick to offer a modification of Bush's comment, because there is a real threat nearby. The US is far enough away to be safe from imminent threat, but it's not hard for terrorists to make it from Southeast Asia to Australia. Because Australia is seen as the face of American foreign policy in the area, it is very surprising that Australia has not seen any terrorist activity on their soil already. However, there have been several foiled terrorist plots in recent history and this situation is not likely to let up anytime soon. The Aussies remain ever vigilant.

Bush was quoted as saying that he didn't think of Australia as a deputy sheriff, "but as a full sheriff." In fact, Australia does supply soldiers in most UN activities in the countries surrounding them. In many cases, they are actually doing patrols and peacekeeping directly for the UN in places like the Solomon Islands, Indonesia, and other hotspots where the UN needs peacekeepers. The idea of them as sheriff does have some weight, and is going to cause them some problems in the future. Many countries in the 'hood remain perpetually pissed that the white man is on their soil at all, even as peacekeepers.

To end this rant, know this: this is changing fast. Another Former Prime Minister Kevin Rudd (who served after Howard from 2007–2010, and again in 2013) was a bit more liberal and conciliatory than his conservative predecessor, and he made moves to soften the Australian image in this part of the world. He decided to pull Aussie troops from Iraq, and was much more engaged with local Asian countries' leaders to head-off any friction that might arise due to Australian presence, as UN representatives or otherwise. Oh yeah, and he spoke fluent Chinese, and not just to order from a take-out menu either. Given Australia's changing focus, I assume it may be a common attribute for future leaders to be just as fluent. . . .

THE PEEPS

First off: there hardly ain't none of them! Say what? Yep. There are only about 25 million Aussies, and 5 million Kiwis (plus or minus a few Hobbits.) Altogether this region has less peeps than New York or Tokyo or Rio, but spread out over an area the size of the continental US! And it's not like the equation is going to change much in the near future either, as both countries have full-on stabilized, low pop growth rates. As in the US and Europe, the only way their pop gets any bigger is via immigration from other places, and (just like in US and Europe) this inflow of people is changing their ethnic and demographic outlook here in the 21st century . . . but how?

Because Australia is doing a lot of business with China, Japan, Indonesia, ASEAN, etc, they will become more like those countries culturally, as interaction with and immigration from Asian countries increases. This is changing very rapidly just in the modern era, because quite frankly it was not allowed to happen any earlier. How so, Plaid Avenger?

At the expense of aggravating my Aussie friends, I will have to tell you that a racist streak has run through the society since its inception—a bad habit that they probably inherited from the Brits. Perhaps racist is too harsh, but certainly they had a superiority complex when it came to other peoples in their neighborhood. The Australian treatment of the **Aboriginals** was bad from the get go, mostly treating them as third rate citizens, at best. There was also an official state policy from 1870 to 1970 which made it perfectly legal for the state to take Aboriginal children from their parents to re-educate and 'civilize' them in white families. These folks are now referred to as the **Stolen Generation**.

Aborigines: Not held in high regard.

I guess that is better than what the Aussies did to the Tasmanians, which was to kill them all in an open hunting season. On top of that, Australia had a white-only immigration policy in place for most of its history. What? Yeah, I'm afraid it's true. If you were of European or American descent, then you had an open door to the country; but that same door was slammed shut for Asians or Africans of any stripe. Pretty nasty business.

What is this madness? No wonder Rudd apologized.

To be fair to our friends "down under," they have made great strides in the last few decades to make amends. Much has been done to alleviate the impoverished plight of many Aboriginal communities, and even more has been done to overcome the negative attitude and stereotypes of the group. In fact, Australia picked an Aboriginal woman to carry the flag into the introductory ceremonies for the 2000 Olympics. That brings me back to what former Prime Minister Kevin Rudd did his first month in office: in February 2008, he made an official government apology to the Aborigines for all past reprehensible deeds of the Australian government, specifically citing the Stolen Generation. This was big news, and perhaps a critical turning point for the society. **2019 UPDATE:** Unfortunately, like many other rich parts of the world, anti-immigration sentiment and outright racism are once again rearing their ugly heads in this region. There is a rise of white supremacy and associated violence/terrorism, as most recently evidenced by the horrific 2019 Christchurch mosque shootings, which were perpetrated by an Aussie.

Back to the story: Since 1976, when their historically strict immigration laws became more relaxed, the country's Asian population has increased. Surprise, surprise. More people that are close by, from regions adjacent to Australia, are coming there because it is a rich place where there is more opportunity to succeed than there may be in their home countries. Again, it's America, Jr.! All of the poorer people nearby want to get there so they can set up shop. Australia may not be full-on encouraging it, but at least they are allowing it. They are changing direction, changing focus, and becoming more Asian. When you visit Sydney now, you can bump into a very vibrant and growing Chinatown for the first time in its history. These changes are occurring nationwide. As you may remember from the international organizations chapter, ASEAN is now a rallying entity for inter-regional action in this part of the world, and Australia is a part of the ASEAN +3 dialogue. They are also a new member of the EAS: that East Asia Summit that is a open trade/politics/planning talk shop that everyone in the Asian 'hood is on . . . and now so are Australia and New Zealand, despite their lack of Asian-ness.

Who else have we talked about that is becoming more Asian due to increasing immigration from Asian states? That's right: America, Sr.!

NO WORRIES

This is a region that plays a dual role as a US foreign policy anchor and as an economic player with China. They have a lot of natural resources and stable population growth, that is being bolstered by Asian immigration too. There is still the terrorist threat, however; we'll just have to see how they balance their security against their economic interest.

Australia would like to be a power broker between China and the US. This is promising; they have far fewer complications to worry about when dealing with China, as opposed to Japan, which has loads of historical emotional baggage. Japan may be a little more abrasive than Australia when dealing with Asian relations in general, and this gets the Aussies ahead.

Australia is going to have a leg up dealing with China and all the other Asian economic giants, because of proximity as well as the fact that they don't carry a lot of cultural baggage. What do I mean by that? They haven't dissed. They haven't invaded anyone. They never colonized anyone. They are pretty antiseptic all the way around. That is a big plus in today's world, particularly for any state wanting to get in on the action in Asia, which is, of course, everyone. And it sure don't hurt that they have an absolute ton of resources and services that those growing Asian economies will be snapping up indefinitely. Dudes! The koalas and the kiwis are sitting pretty down there right now! Down under is the place to be! Down under Asia, that is.

Uncle Sam's little brother down south is still a staunch ally of the US . . . and the awesome ANZUS pact makes them secure, all while US troop presence is increasing across the region. Luckily, they remain just distant enough to dodge a lot of the negative press associated with such a role. Of course, the Aussies are distancing themselves slightly from US foreign policy of the last decade . . . but dang, who isn't? Australia is a pretty chill place all the way around: politically, economically, socially. That's why we like to go vacation there, because Australia is a laid back sort of place. Throw a shrimp on the barbie, crack open a Foster's, and enjoy the prosperous future, America Junior.

A SINISTER SHOW AND TELL

Here's a great game that all the kids love to play! It's called "Quickest Kill." Known for it's exotic and often lethal wildlife, the Australia/New Zealand region is hot to a whole slew of the world's deadliest animals. Of the following creatures listed below, which one do you think could take out the kid sitting next to you in geography class the fastest? And which one do you think is responsible for the most deaths? Have a round table debate in your class, and you get bonus points for letting one or more of these creatures loose while defending your choice. It's the Aussie version of 'Hunger Games'! Good luck!

Salt Water Croc

Eastern Brown Snake

Stonefish

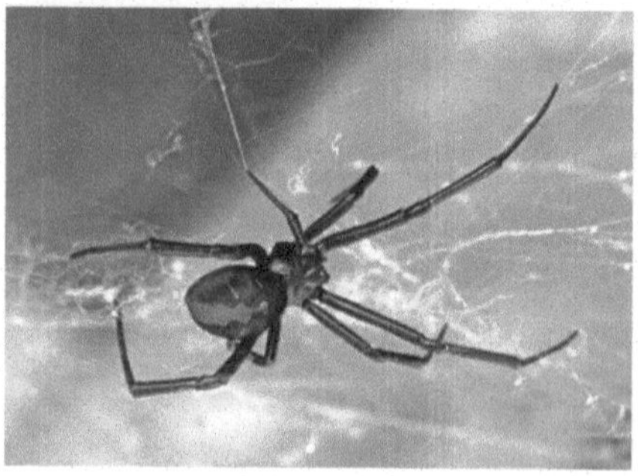
Red-backed Spider

Death Adder

Great White Shark

Funnel-Web Spider

Inland Taipan

Blue-ringed Octopus

Tiger Snake

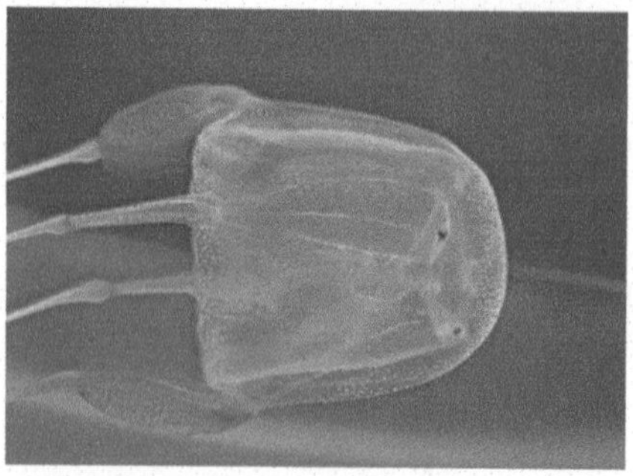
Box Jellyfish

Gollum

AUSTRALIA RUNDOWN & RESOURCES

View additional Plaid Avenger resources for this region at http://plaid.at/aus

BIG PLUSES

- Koala bears and kangaroos and hobbits. Who doesn't love them? Ok, besides Sauron.
- Small population that is about evenly balanced with resource/industrial base
- Location, location, location! In the neighborhood of the hottest, fastest growing Asian economies, to which Australia sells tons of goods
- 13th largest economy in Australia; very high standards of living across region
- Resource/agricultural product rich
- Stable democratic governments
- No real enemies on the planet; no international conflicts with anyone

BIG PROBLEMS

- Crocodiles, the Sydney funnel-web spider, 9 of the top 10 most deadliest snakes in the world. Crikey!
- National legacy of racism: still has friction with its Aboriginal groups, and immigration is becoming a hot button issue much like it is in US
- Economy not very diversified; heavily dependent on natural resource exports
- By mere association with Team West, combined with its proximity to SE Asia, Australia is a high-risk target for international terrorism

LIVE LEADERS YOU SHOULD KNOW:

John Howard: Popular former Prime Minister of Australia from 1996–2007. Kind of the "Ronald Reagan of Australia": Center-right conservative, pro-business, pro-US, and presided over a period of strong economic growth and prosperity. Strongly supported US in Iraq and Afghanistan. Mostly lost in 2007 simply because voters were bored.

Kevin Rudd: Prime Minister of Australia from 2007–10, and the polar opposite of Howard. Center-left liberal who was more focused on establishing stronger ties with Asia, and possibly not following US foreign policy so adamantly. Speaks Mandarin Chinese fluently, supports Kyoto Protocol, made official apology to Aboriginals, and pulled troops out of Iraq.

Scott Morrison: Current prime minister of Australia. Center-right (or further), social conservative, deeply religious conservative. Anti-immigration (including a "Stop the Boats" policy), anti-gay rights, and likely anti-environmental issues. I wouldn't call him radical right, but he is the furthest right political leader the country has had for a long while.

Jacinda Ardern: Prime Minister of New Zealand, youngest serving woman head of state; liberal and progressive agenda on domestic and international policies; active advocate of same-sex marriage, abortion rights, self-determination of MĐori, and feminism. In the news recently after the 2019 mass shooting in Christchurch and her successful move to ban most semi-automatic and military-style weapons.

PLAID CINEMA SELECTION:

Australia (2008) Australian cowboy version of *Gone With the Wind*. At nearly three hours in length, this epic is based on real life events that occurred in Northern Australia during World War II. Shows, at length, much of the northern territory of Australia during wartime, including several aspects of Aboriginal life, and the national legacy of racism to boot.

"Crocodile" Dundee (1986) Probably the most well known Australian film of all time. Has two versions, an Australian version and an international version. The Australian version features more regional slang and extended scenes. Despite its comedic aspect, the film shows a lot of great examples of the outback, including both scenery and wildlife.

Matariki (2010) Complex film tells the plot through five interweaving stories, all in the days leading up to Matariki, the Māori New Year. The film shows a wonderful view of the landscape in South Auckland, New Zealand. Funded completely by the New Zealand Film Commission.

The Piano (1993) Dramatic film about a mute pianist and her daughter living during the mid-19th century in New Zealand. This wonderful film covers many aspects of language, as three languages are spoken in the film if you include sign language. Great scenery as the purples and greens of New Zealand capture the eye and don't let go.

Romper Stomper (1992) Drama about a group of bored Neo-Nazi skinheads clashing with Vietnamese immigrants in a Melbourne, Australia suburb. Inspired by actual events, this film stars a young Russell Crowe in his pre-Gladiator days. The violent closing scenes take place within the natural beauty of Victoria's Twelve Apostles rock outcroppings (of which there are not actually twelve).

Walkabout (1971) Total hardcore art-house piece with little dialogue and loads of symbolism, this film was extremely controversial when released due to its overt sexual themes and nude scenes with a minor. But it is just awesome at showing the physical beauty and bizarreness of Australia, Aboriginal culture, and the disconnect between the "modern" human life and the "uncivilized" human experience. Not for those who need explanation and closure in films.

Rabbit-Proof Fence (2002) Closely based on real events, a real fence, real people, and real national racist policies of the time. The story of three girls who are forcibly taken from their home and put in a camp because it was believed that the Aboriginals were a danger to themselves and should be bred out of existence. Shows struggles of Aboriginals in western Australia during the 1930s. And it inspired the artist, Plaid Klaus, to create this single-page historical piece on the opposing page.

MAP GALLERY

View Map Gallery & additional Plaid Avenger resources for this region at http://plaid.at/aus

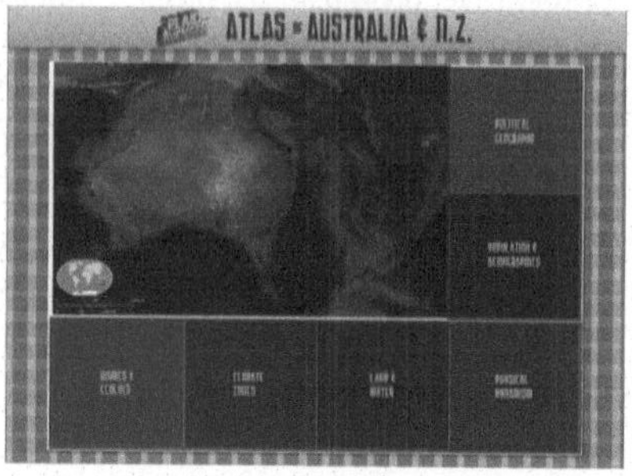

BARWON PARK
AUSTRALIA, 1859
ARE YOU SURE ABOUT THIS?
THE INTRODUCTION OF A FEW RABBITS COULD DO LITTLE HARM...
IT DID. THIS SPARKED THE CONSTRUCTION OF A SERIES OF FENCES TOTALING TWO THOUSAND MILES FROM 1901 TO 1907 KNOWN AS THE RABBIT PROOF FENCE.
BUT THE ANIMALS WERE PERSISTENT. IN 1932, MIGRATING EMUS FOUND THEIR WAY THROUGH THE FENCE. HUMANS DECLARED WAR ON THE BIRDS ONLY TO SUFFER DEFEAT.
MEANWHILE, SOME ABORIGINAL AUSTRALIANS WERE BEING FORCED TO ASSIMILATE AND MANY WERE SEPARATED FROM THEIR FAMILIES.
SAY...HAVE YOU EVER THOUGHT ABOUT *PLAID* UNIFORMS?
THE RABBIT PROOF FENCE WOULD BE USED AS A COMPASS AND GUIDE SOME BACK TO THEIR HOME.
THE RABBIT PROOF FENCE, NOW KNOWN AS THE STATE BARRIER FENCE OF WESTERN AUSTRALIA, HAS SERVED MANY PURPOSES THROUGHOUT HISTORY.

CHAPTER OUTLINE

Latin America

13

A SINGLE REGION?

Latin America is a region with a question mark after it. The Plaid Avenger has never really considered it a region in the past, because as you know, creating a "region" involves identifying some sort of homogeneous singular trait that you can apply to the geographic space in question. I've said for years this place is too big. You can't apply singular homogeneous traits across all of Latin America because it encompasses everything from the south of the US-Mexican border all the way to Tierra del Fuego, the tip of South America, which is almost in the Antarctic. It's just too darned big, and too darned complex! A single same-ness that applies to everything from the Rio Grande to Rio de Janeiro, or the Atacama to the Aztecs to the Andes? Not hardly. Except maybe the term 'Latin America' itself . . . and BTW, where did that come from? A:

What's the Deal with . . . the Latins in Mexico?

What happened was this freaky-freak French king Napoleon III (not Napoleon the short, dead dude we all know, but one of his later kin) sent over this dude named Maximilian (actually, an Austrian) to sit on the Mexican throne in a feeble effort to reestablish France's presence in the New World in 1862. The French were actually invited to do this by an elite rightist core of Mexican aristocracy who wanted to revive the Mexican monarchy and simultaneously dismantle the leftist movements occurring in what was independent, sovereign Mexico at the time. So this French dude Maximilian was in charge, and when he was out scouting around he said, "Hmmm. . . . We're French, we're not even Spanish. What rallying point can we possibly use to get the locals to think that we should be in charge here?"

What our brilliant French brethren came up with was that the Spanish, Portuguese, and the French languages are all linguistically linked. They are all part of the Romance, or Latin-derived, languages. In a fit of what must have been desperation at the time, Napoleon III told ol' Maxi to say, "Look, I may be just the French guy here in Mexico, but we are all brothers under the Latin language, so we're all family here in ***Latin America.***" Again, a totally bogus, politically made-up term, but somehow it stuck. The Mexicans deposed Maximilian's derriere fairly quickly, but the term stayed—in fact, whipping the French is what the Cinco de Mayo holiday is all about; it's not Mexican Independence Day! That is a different holiday altogether! The date is observed to commemorate the Mexican army's unlikely victory over French forces at the Battle of Puebla on May 5, 1862. That is a "cinco" worth celebrating!

After thinking about the French fool Maximilian's antics for a while, the Plaid Avenger has realized that, indeed, there are a lot of traits that can be recognized across the whole of Latin America. We're going to look at some of those traits in this introduction pre-chapter to the more unique subregions of Latin America.

WHAT IS LATIN AMERICA?

So . . . this "Latin America" is a term that everybody recognizes, everybody gets, you know where it is, but it's based on a word that is completely and utterly meaningless: Latin. Do all of these people speak Latin south of our border? Non certe! Latin is a dead language. Nobody speaks that Caesar-ized stuff anymore. In fact, there are really only two predominate languages spoken in the region: Spanish and Portuguese. That, and a smattering of small states that speak English or French in the Caribbean.

Are there other attributes defining this term Latin America: are they all of Latin or Greek ancestry? Obviously not. As with many things on the planet, we can blame the French. In that pretty little piece of propaganda, the French called for a Latin brotherhood based on all their languages (French, Portuguese, Spanish) being of Latin roots. Pretty shaky foundation for a regional identity, but there it is. But we can identify some real regional qualities that stick.

Geographically, the term means everything south of the US/Mexican border. That's pretty common knowledge across the planet. Mexico, which we teased out for obvious reasons earlier, is not part of the US/Canadian region. There are too many differences, including its Latin-based language, levels of development, culture, and lots of other issues which distinguish Mexico from the United States. So even though it is a NAFTA member (and southwestern US is increasingly becoming part of a "greater Tex-Mex" region), for now Mexico still shares more commonalities with states south of that border. . . . Before we get to those similarities, let's identify frequently referenced regions in this part of the planet.

LATIN LOCATIONS

There are a few terms that the Plaid Avenger wants you to know before we get into these subsequent chapters. These definitions will help us understand a lot of terminology used on the global stage. We already talked about Latin America encompassing every single country south of the US/Mexican border, in the entire western hemisphere. Now let's break down some more specific regions within Latin America.

SOUTH AMERICA

The first one is easy enough. South America is both a continent and a region. The South American continent is that other major chunk of land in the western hemisphere. It starts with Colombia and then heads east to Venezuela, Guyana, Suriname, French Guyana, and then turns down south to Brazil. South of Colombia is Ecuador, Peru, Chile, Bolivia, Paraguay, Uruguay, and Argentina. It is important to note that most all of the countries of South America are fairly good sized; Brazil itself is a monster country, being the fifth largest in the world. This makes South American states quite territorially, physically, and economically distinct from the slew of much smaller micro-states to their immediate north. Food for thought, and I'm not talking about salsa either.

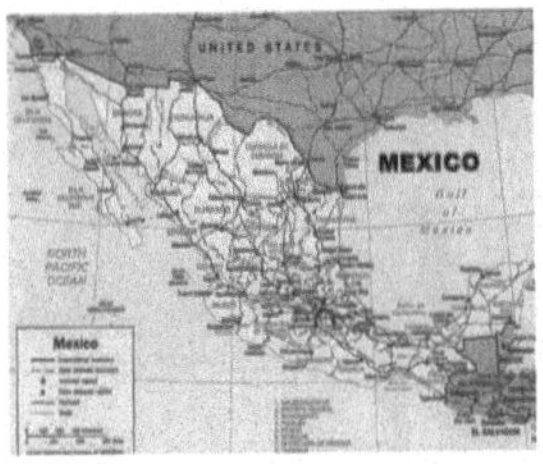

+

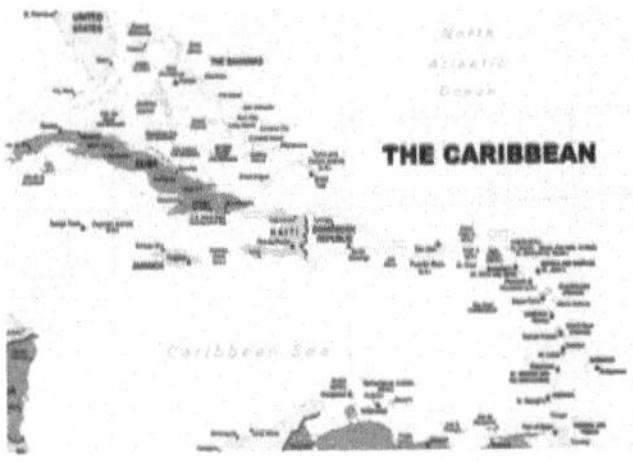

+

=

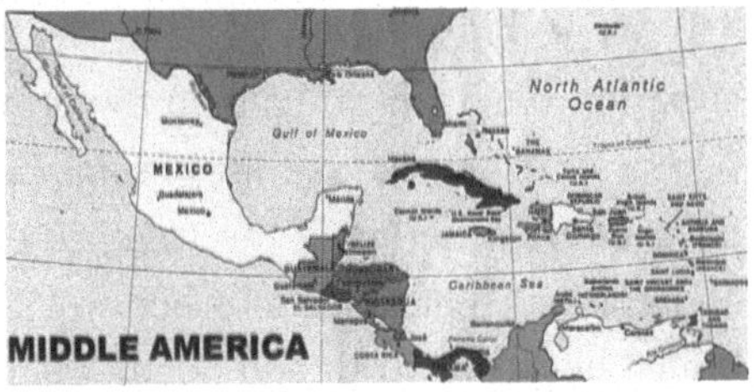

MIDDLE AMERICA

The term **Middle America** is everything between the southern border of the United States and the South American continent. It's a fairly generic definition that usually only gets play in high school and college level textbooks anymore, but I suppose it doesn't hurt to know the reference when you see it pop up. For this magnificent treatise of learning, I will further break this area down to more manageable and meaningful subregions, described below. Study the Middle America map equation above to be cartographically clever. And here are the components of that equation:

MEXICO

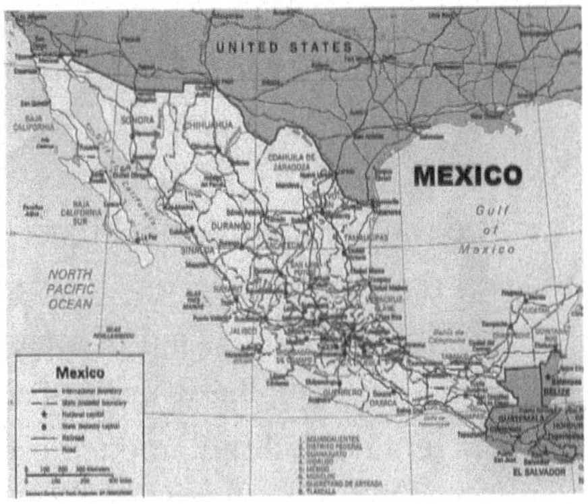

The first subdivision is Mexico itself, solo. Mexico is a country, a state in its own right, but it's also radically different from all the states around it, including the US, and it's also quite different from the Central American and Caribbean states. Mexico is its own subregion that we're going to define in the next chapter. In respect to size, resources, economy, population and level of development, Mexico stands apart from all its neighbors in Middle America: it is a giant of population, economy, and resources when compared to the rest of the Middle America 'hood. And its shared border with the USA make it quite distinct from all the rest of the Latin American states as well . . . as does it's NAFTA membership. Combine that huge economic interaction with the US with the huge cultural interaction with the US, and you have a sub-region that really straddles two different worlds.

THE CARIBBEAN

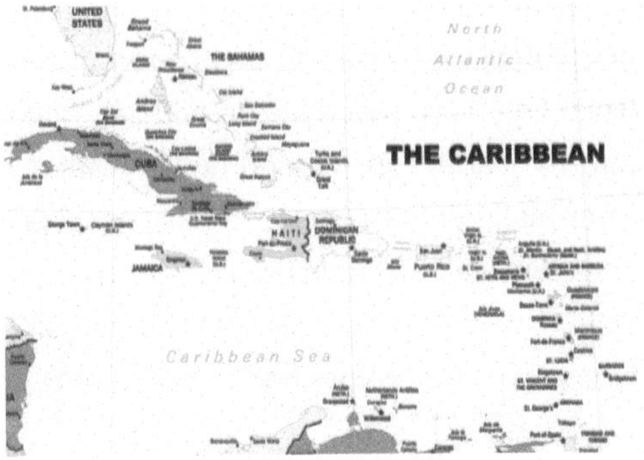

The Caribbean is a group of island states comprised of the Greater Antilles and the Lesser Antilles. The Caribbean, when we think about it, calls to mind a distinct culture in terms of . . . everything. Cuisine, language, stuff they drink, how they party—it's all different. Caribbean means something to us, and it means something that's different than Mexico. We all understand that, right? We go to a Caribbean restaurant or a Mexican restaurant, but we never really see the two put together. (Note to self: that might be a good idea. Instead of Tex-Mex cuisine, how about Car-Mex cuisine—or is that a lip balm?) The island nations in the Caribbean Sea, south of the United States, are a distinct subdivision.

CENTRAL AMERICA

The last sub-region of Middle America is **Central America**. This is the one that causes the most confusion for younger students of the world. It is everything between Mexico and South America, noninclusive. Mexico and South America are the bookends. There are seven distinct countries in what is widely accepted as Central America: Belize, Guatemala, El Salvador, Honduras, Nicaragua, Costa Rica, and Panama. You may hear reference to Central American civil wars or Central American gangs; this group of countries is the origin for that descriptor. Yeah, good times.

We've got a bunch of Americas here, and the current Central America situation has little in common with Mexico, the Caribbean, or even South America, besides the sharing of some borders. It has a lot of unique challenges, particularly in terms of the violence endemic to the region that we will talk about in much more detail when we get to the Central America chapter. It's a bridge between North and South America, but an easily passable bridge it is not. Because of its peculiar social, economic, and political challenges, it stands apart from its neighbors, and even from the rest of the Americas.

South America plus the three regions of Middle America—Mexico, Central America, and the Caribbean—that's Latin America all the way around. Now that we know where it is and where the term itself comes from, let's talk a little bit more about what Latin America *is*. What homogeneous traits can we use to define this vast region?

WHAT IS LATIN ABOUT LATIN AMERICA?

COMMON CULTURE

First, we can point out some distinct elements so we don't have to keep repeating ourselves in subsequent chapters. Number one is culture. There is kind of a common culture south of the US/Mexican border, be it in the Caribbean, Central America, Chile, Argentina, or Brazil; there are some things that do remain the same. One of the kind-of-the-same things that we already pointed out is language. While Latin America is a bizarre term in and of itself, Latin Americans do primarily speak Spanish, and the only other really big language is Portuguese.

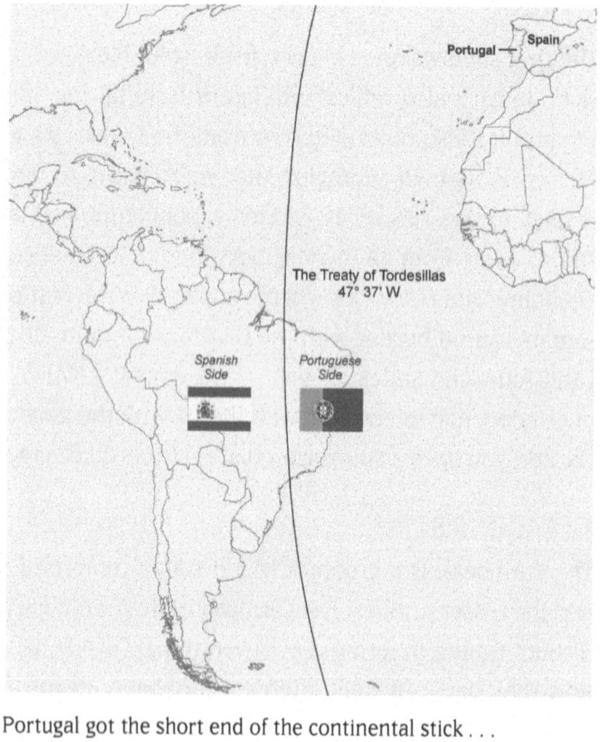

Portugal got the short end of the continental stick . . .

How did this come about? Easy enough. It came about because of colonial endeavors in this region that date as far back as 1492 when Columbus sailed the ocean blue, came over here and bumped into the Caribbean . . . and he was sailing under the Spanish flag. After his "great discovery," the big naval powers at the time, Spain and Portugal, started floating over here in fairly short order as well. (Great Britain didn't come along until later. It wasn't as big of a naval power as the Spanish at the time, so it was a latecomer. It therefore had to head north for colonial expansion which is why it gets the leftovers—North America.)

Pope responsible for Tordesillas . . . and Catholicism in Latin America.

Let's get back to our story here. Why do we have these two main languages, Spanish and Portuguese? Columbus came over in 1492. Two years later, a fairly important event occurred which I want you to understand and know about: **The Treaty of Tordesillas**. The Treaty of Tordesillas occurred in 1494 when these European countries were bumping into the New World, colonizing, taking over, and staking claim to pieces of it. Of course, there was friction between the countries for hundreds of years at home anyway; now they were just taking the fight abroad. Much of the friction was between Portugal and Spain, two of the main colonizing powers in this part of the world. Who was going to own what? Are we going to fight over it? The scramble for stuff was fast turning into a full-scale fracas. Both these countries, being predominantly Catholic—which is another common cultural tie we'll see more about in just a second—would listen to the papa, the **Pope**, the main man in the Vatican. And therefore to alleviate conflict, the Pope said, "Hey guys, come on now, we're all civilized colonial imperial masters here. Let's all get on the same page, get around the same table here, let's work together! We don't need to fight! We're going to settle this fair and square. Papa is going to just draw a line on the map and everything on one side we'll give to the Spanish and everything on the other side we'll give to the Portuguese. Now you guys be good!" The line happened to be 45 degrees western longitude. Check out the map on the previous page. And the caption.

As you can see from the map, the Portuguese look like they got shafted on this deal. They only got the tip of what is now known as Brazil. Why is this? Did the Pope just not like the Portuguese? Well, maybe a little. But you also have to keep in mind, this was during early exploration in 1494, only a few years after Columbus got there. They didn't even know what was there yet. They were just bumping around the coastline; with no comprehension about the actual size of the continents . . . neither North nor South America, quite frankly. What they thought was a good deal was based on known circumstances at the time, which really wasn't much. As we all now know, this is a pretty big place. South America is the fourth largest continent on the planet. As exploration continued and the true scope of the land's magnitude unfolded, the Portuguese did pick up what is now modern-day Brazil. Its boundaries naturally fell back to the Andes—a nice, easily defined natural border. Anyway, the Treaty of Tordesillas is the cornerstone of why the Spaniards ended up controlling so much of the New World—basically, all the rest of it outside Brazil.

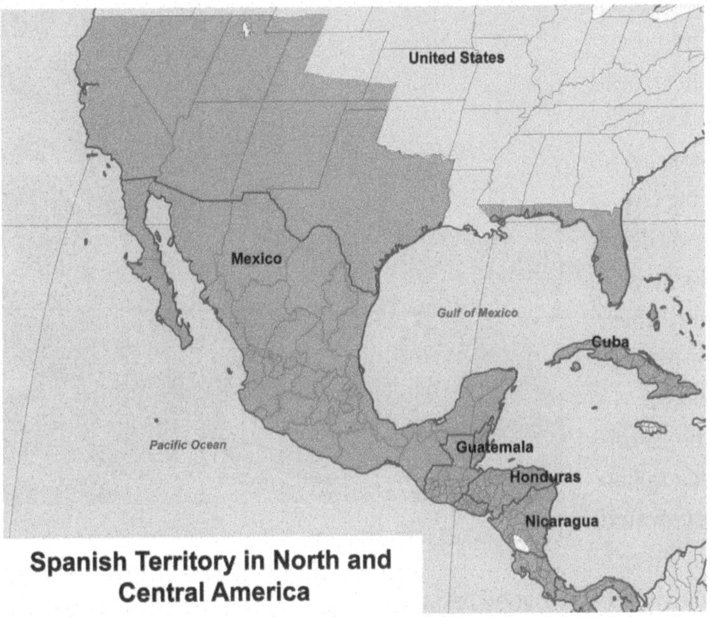

Spanish Territory in North and Central America

And it all started with Mexico. We think of today's Mexico as that place "south of the border," but Mexico in its days as part of the Spanish Empire included all of western North America, as well. It was everything that's now California up to Washington State, to Utah, to New Mexico, to Texas. Spanish territory also included Mexico, all of Central America, most of the Caribbean, Florida, and all the way down the western seaboard of South America. Check it out on the map to the left. That is why today these territories are Spanish-speaking. I guess in today's world the Spanish-speaking territories might even include California, Florida, New Mexico, and Arizona, but that's a separate issue in immigration that we'll get to later.

This all means that those old colonial ties formed this common culture that's still in place today. It's a common culture primarily based on language, but also on one other thing I mentioned earlier: Catholicism. Two primarily Catholic countries colonized virtually all of Latin America. You can still see the deep, deep-seated Catholicism throughout Latin America today. There are no countries in Latin America that are *not* Catholic. When the Plaid Avenger travels down there, he can see some freaky-freaky stuff that doesn't look like Catholicism. You can see some rites and rituals down there that could make the Pope soil his robes, no doubt. You'd say, "That's not Catholicism! What is this voodoo hoo-ha? What are people doing with voodoo dolls and holding up bloody chickens saying Hail Marys?" Then there's this crazy stuff that's going on down in Brazil; they have these big parties that don't look like traditional Catholicism: the "Carnival." But

Latins love the Savior.

what you have is indeed the basis of Catholicism mixed in within indigenous culture, local culture, and imported African culture.

So a couple of very strong cultural characteristics, language and Catholicism, do serve as a starting place for some homogeneity of "Latin American-ness." But there are others . . .

URBANIZATION

Urbanization is a primary feature that is consistent across every place in Latin America, and I do mean *every place*. I've never even thought about this until recently, but this region is one of the most urbanized on the planet. Typically, when we think of urbanization, we think it means that most people don't live in the countryside; most people live in cities . . . and we typically associate this with fully developed countries like the US, Canada, Japan, and Western Europe. That's where you *typically* have people who are all jammed in the cities, and hardly anyone out in the country. For some reasons that we will get into in a minute, Latin America is more urbanized than many fully developed places. This means almost everyone in these countries is jammed into something called an urban area, typically a very big city.

Latins packed into the cities.

On average, upwards of 80–85 percent, sometimes upwards of 90 percent, of the population of a Latin American country is located in just a few major cities. This is highlighted best by Mexico City, which will probably soon be the world's most populated city with over 25 million inhabitants—a full one-fourth of Mexico's total population.

Why do people go to the city? It is suggested that in the developed world people go to the city for all the reasons we already talked about back in chapter 2. That's where jobs are. Jobs are concentrated in urban areas. People want to go to urban areas because there's health care, doctors, clean water, sewer systems, electricity, movie theaters, good restaurants. The reason everybody wants to move to a big city or urban area is because good stuff is there. The standard of living is higher and that's where your jobs are.

Is that true of Latin America? Not all the time. It is a kind of conundrum in Latin America. In most urbanized places in the Latin world, all of the **pull factors** that we just mentioned—the good things that pull people to the city—are simply not available for the masses that come. There is the *perception* that all these great things are in the city and the *perception* that all these jobs are available, but it is simply not true for a lot of folks who end up going to the city.

Now those are the pull factors, but there are also some **push factors**, or the not-so-good things that push people from rural areas to the city. One of the main push factors, a major theme across Latin America that we'll discuss at length, is **landlessness**. Because most people do not have access to land, or resources on the land, they are pushed away from it. Owning no land in title or having no serious claim to the land is just another reason to go to the city. So there are a lot of pull and push factors that make Latin America one of the most urbanized places on the planet . . . a trend that is growing.

One last note: An interesting phenomenon in Latin American cities that most Americans don't get is that they are set up in a kind of "economically reversed" scenario from the US urban model. Say what? In America, what's found deep in the inner city? In fact, what is the connotation of the term "inner city"? I'll tell you what it connotes: slums, projects, poverty, ghettos. That's not where people want to be. As a result, most people who have money don't live in the inner city in America. They've got money, so what do they do? They move *away* from the city and into the "burbs," which we talked about back in the North America chapter.

Vertical growth of slums or *favelas* around Rio de Janeiro, Brazil.

In Latin America, the situation is just the reverse. The city center is still the prime real estate, so the rich people concentrate themselves and their businesses in the true city center. Where do all these impoverished masses go that I have been suggesting flood into the cities? They make a ring around it. They get as close to the city as they can, usually in ramshackle, shanty-like town dwellings they put together out of corrugated-cardboard and any other thing they can pull out of a landfill, and build them in undesirable, unused parts of the urban fringe like mountain sides. These shantytowns get very big and grow into almost permanent fixtures in rings around the cities. Whole shantytowns are a phenomenon that's very easy to spot in any major Latin American city. Just drive straight into or out of the city and notice how the poverty line fluctuates one way or the other. It's a distinct characteristic that has something to do with urbanization. It also has something to do with . . .

WEALTH DISPARITY/LANDLESSNESS

Perhaps there is no other region that we will talk about in this book or that you can go visit on the planet that has wealth disparity as extreme as Latin America. What is **wealth disparity?** Disparity is the difference between highest and the lowest, the difference between the greatest and the least. In terms of wealth, no place in the world is like Latin America in that so few people have so much of the wealth. I'm just going to guess-timate some numbers here. These are Plaid Avenger Figures that vary from country to country, but on average, the extreme amounts of wealth are held by the upper 5 percent of the population. The richest 5 percent typically own something like 80 percent of all the stuff. That's all the land, all the factories, all the businesses, everything. Maybe you think that's not bad—perhaps that's the way it is in the US as well. Not really. Because the other part of the equation is that 95 percent of people have got to split up that other 20% of the stuff. Of course, there is always a significant, or at least partial, middle class; some people have got to own some stuff, but there's always a significant majority of people in these countries that own nothing. No title to their land, no title to their house, no other economic means except their labor to sell. That is a kind of common feature across Latin America from Mexico down to the tip of Tierra del Fuego, and is particularly nasty in Brazil and lots of other places where people kill each other over land.

Times haven't changed much economically.

This brings us to the issue of **landlessness**. Because of this lopsided scenario of so few people owning so much, including the land, the vast majority of people can't stake a claim to anything. In such societies where people don't have anything, their options are very limited. Mostly they go to cities, as already suggested. They can also try to work on someone else's land without the landowner's consent. This is a very unstable situation because the owner can show up at any time and kick them off. This is a particularly resonate issue in Brazil where landlessness has turned into an occupational hazard (see box below).

What's the Deal with Landlessness in Brazil?

Unofficial stats in Brazil refer to 1.6 percent of the landowners control roughly half (46.8 percent) of the land on which crops could be grown. Just 3 percent of the population owns two-thirds of all arable lands. The Brazilian constitution requires that land serve a social function. [Article 5, Section XXIII.] As such, the constitution requires the Brazilian government "expropriate for the purpose of agrarian reform, rural property that is not performing its social function." [Article 184.]

This is a big deal for everybody. It concerns politicians, business people, and the landless poor. In Brazil, they have a law on the books that basically says, "If you can successfully squat on a piece of land, cultivate it, and make it produce something for one year, then you have a legal stake to it. You can take it! It's yours! Here's the deed in your name!" You may say, "Hey, that sounds like a pretty good policy," and maybe it is.

However, the scenario develops a lot of squatter settlements: large groups of people who are squatting on owned land that is typically owned by some rich businessperson or some rich urban dweller who is not out there in the countryside. If they find out that people are squatting on their land, they send in henchmen to go clear them out. When the landless poor fight back or try to continue their stay on the land to finish out the year so they can have legal claim to it, the situation can turn violent.

This situation has also led to the formation of the Landless Workers' Movement, or in Portuguese, Movimento dos Trabalhadores Rurais Sem Terra (MST). This is the largest social movement in Latin America with an estimated 1.5 million landless members organized in 23 out of 27 states. And they can get violent as well.

This is still an issue that is alive today in Brazil and all across Latin America where hired gunmen are going and literally cleaning out villages and killing everybody on site. Because it is private property, the owners, ranchers, and other types of folks can just say "This is legally mine and these people wouldn't leave, so I killed them." And it's going to get way worse in Brazil: new right-wing President Jair Bolsonaro announced in 2019 that his government will classify all invasions of farmland by the MST as terrorism. Dang.

Brazilian Prez Bolsonaro sez: Land reformers? You mean the terrorists?

This problem of wealth disparity and landlessness is a common theme across Latin America, and was a primary motivation of the independence movements across the region as well (these movements in Latin America began in the 1820s–30s). This dude named **Simón Bolívar** headed up many of these egalitarian independence movements, and is viewed historically as the George Washington of Latin America. Wealth disparity/economic equality was one of his central themes to rally folks to fight. But the issue wasn't resolved at state inception, and it has plagued Latin America ever since. Just one example for now: The Mexican Revolution in 1910 was fought over land: so many people got so frustrated about landlessness, that they had a revolution to remedy it. One of the core parts of their new constitution included equal rights for people and access to land. All politicians of any stripe have had to deal with this issue historically, and still do today.

The George Washington of America: Simón Bolívar.

Revolutionary leaders like Emiliano Zapata and Pancho Villa and his crew fought in the Mexican revolution for land reform. But even 100 years later, former Mexican President Enrique Peña Nieto had to include land reform in his political platform. If you're going to run for office at all in Mexico, no matter what political party you are from, you have to address the land issue. The main political party in Mexico is called the PRI. It's a land reform party—go figure. It was founded to redistribute land and work out ways to give people access to land. Then there are peeps like that late leftist Hugo Chavez in Venezuela who was all about equal economic rights in a country with huge wealth disparity, and started his socialist "Bolivarian Revolution" program in order to level the playing

Pancho, Peña, Zapata, Castro and Hugo: All leaders for land reform? Eh . . . little real progress has been made.

field; but he died and his predecessor Nicolas Maduro is fighting for political survival in a Venezuela that is bordering on full civil war and chaotic meltdown. And don't forget historic figures like Fidel Castro, who led the Cuban revolution to basically reclaim land and businesses from the rich and redistribute them to the poor in an old-school communist-style revolution; but Cuba is still wildly underdeveloped here in 2019. This issue of wealth disparity, corruption, and landlessness is a prevalent theme whether you are in Brazil, Mexico, Cuba, Nicargua, or Chile. And recent events seem to indicate that most of the countries of the region still have a ways to go to remedy these major issues.

IN-"DOCTRINE"-ATION: A HISTORY OF US INVOLVEMENT

Now, the Plaid Avenger has talked to many folks from Mexico all the way down to Argentina, and can tell you with no reservation that people respect the United States. Nobody hates any specific person in the United States, but taken as a whole, Latin America's historical relationship with the US in today's world is largely seen as negative. That may seem like a bold statement, and certainly there are those that still support US policies, but I can tell you who those people are. They are the rich people in Latin America. If life is good, you've got no reason to have qualms with the United States. Unfortunately, as we've already pointed out, that's not the majority by any stretch. Most people see the United States with a bit of an imperial taint, or in a bit of a **hegemonic** light. Certainly, given the US's involvement in virtually every country in Latin America, it's not hard to see why they have kind of a bad taste in their mouth when it comes to historical intervention by the US.

"Don't mess in our hemisphere, bitches!"

To what is the Plaid Avenger referring? You've got to know this, because it still applies in today's world. The number one thing you've got to remember from the US's history of influence is **The Monroe Doctrine**. It's this antiquated, 200-year-old statement that was made in 1823 by President James Monroe. It was a foreign policy statement that said in essence *Disclaimer: this is a Plaid Avenger Interpretation:* "If any European power messes in any place in this hemisphere, the US will consider it an act of aggression against the United States." In essence, if Spain were to go try to retake Mexico or Chile, if the Portuguese were to try to retake Brazil, if the British try to take back Jamaica—the US would consider it an attack on US soil. Heck, it doesn't even have to be a full-on takeover; any intervention at all would be considered an act of aggression. "If you mess with anyone down there, we will consider it as you screwing with us." It's extremely similar to NATO article 5 in saying, "If anybody here gets messed with, it's an attack on all of us."

Now why on earth would the US say that? It seems kind of silly. I mean, in 1823, what sort of position is James Monroe in? The US was a new country that had only been around about 30 or 40 years. It had only expanded *slightly* over the 13 original colonies, and they were certainly not a world power. They did a great job shooting the British from behind trees, but other than that, they weren't capable of fighting abroad. They could take care of their soil, but weren't up for fighting anybody else on foreign soil. This was largely a toothless threat. Maybe you are thinking, "Why is the Plaid Avenger telling us that this is important?" Here's why: This statement became a cornerstone that remains relevant in US/Latin American policy TO THIS DAY.

Uncle Sam ready to whip some imperialistic Eurotrash back in the day.

It didn't mean anything at the time when Monroe said it, but it has come to mean *everything*. Why did Monroe make it at that particular time in 1823? Mexico declared independence in 1820. The Central Americas seceded and most of the South American countries were declaring independence at or around this time. It was seen largely as a supportive gesture. The US was basically thinking, "Yes, our Latin brothers, kick all Spanish and other colonial powers out. We just did it in America so we will encourage everyone else to do it. We're the good guys and we're helping the other good guys, so it's all good!" Again, it was a show of support more than a credible threat to the Europeans: the US could not really take on the Spanish or the British in a foreign land war at that time. Forget about it. No contest.

"Speak softly, and carry a big stick. Just in case a piñata party breaks out."

The Monroe Doctrine led to a bunch of other things, such as the **Roosevelt Corollary**. The Roosevelt Corollary was issued in 1904, about a hundred years later, and at that point, the US was quite a bit more powerful than it was during Monroe's tenure. The US was also under a very powerful president at the time, President "Rough Rider" Teddy Roosevelt. What was Teddy known for? What was one of his most popular sayings during that time? "Speak softly and carry a big stick." Indeed, that saying can be applied directly to what became known as the Roosevelt Corollary, which was the foreign policy towards Latin America at the time. Teddy said, "I like the Monroe Doctrine's policy that if anyone messes around in our backyard we'll consider it an act of aggression against us. That's good, but let's take it a step further. If there is any flagrant wrongdoing by a Latin American state *ITSELF*, then the US has the right to intervene."

In other words, if any Latin American countries south of the US border attack each other, then the US gave itself the right to intervene. More than that, if they just screw up internally, the US was giving itself the right to intervene as well. This had serious repercussions for what sovereignty meant at the time. Of course, I can't go back in time to hear their exact thoughts, but there is no doubt that it was not held in high esteem by Latin American states that were considered sovereign then.

FDR sez: "Sup. We gonna chill over here. If y'all need us, holla."

In other words, you had the United States saying, "Sure, you guys are sovereign, as long as we agree with your sovereignty. Otherwise, we give ourselves the right to intervene." This became kind of a big deal because Teddy was carrying that big stick, and he was not afraid to smack people, or entire Latin American countries, down with it. Under Roosevelt's corollary, relations deteriorated slightly between these regions. However, a bright spot in US/Latin American relations under the Good Neighbor Policy was just around the corner . . . maybe.

The **Good Neighbor Policy** was a popular name for foreign policy at the time of the next President Roosevelt—Franklin D., that is—in the 1930s. In a marked departure from the heavy-handed foreign policies up to this point, FDR said, "You know, we're good guys, we're your buddies. We don't need to come down there and beat you with a big stick. My fifth cousin Teddy was a funny guy, but we don't really need to be that heavy-handed. We'll throw out that Roosevelt Corollary and we'll just be here to help if any leader needs us."

That sounded pretty good, and it was certainly an improvement over the Roosevelt Corollary. But under the Good Neighbor Policy, there were multiple scenarios where US troops were sent down at the request of "leaders" that sometimes could also be referred to as, oh, I don't know, let's call them *military dictators*, who just happened to be supporting US foreign and economic policies at the time. Even though it sounded better, there were still slight implications that perhaps things were not completely on the up and up. That brings us to the last part of the US's history of involvement.

COLD WAR EFFECTS

After WWII, there came the War of Coldness. The Cold War has already been referred to several times in this book; you may be thinking, "Ah I'm tired of freakin' history; I don't need to know any of this stuff," but you can't understand the world unless you understand the historical and political movements of at least the last hundred years. Nothing has affected the world more than the Cold War and its politics. Even Latin America was affected.

We might think of Latin America and say, "What? There was no hot war down there, much less a cold war! There are no Commies down there; there's no Cold War frictions in them parts!" Not so! Latin America was actually quite radically impacted by the Cold War because of US anti-communist policies that were applied across the planet. That's the reason the US got involved in Vietnam, Korea, and dang near everything else that was active at the time. The US even supported leaders with questionable character, just as long as they didn't associate with the Soviets. Supporting a brutal dictator who was suppressing his own peoples in Latin America? Sure! No problem, as long as he ain't a commie!

When we think of the Cold War and Latin America, the first thing that pops into the American mind is Cuba and Fidel: that flagrant flaming Commie that the US still hates to this day. There have been lots of repercussions between the US and Cuba (i.e. why Cuba's pretty impoverished today), but all of the other Latin American countries were impacted as well. Some of them had much more violence with a much greater death toll than Cuba ever did.

What I'm talking about is a renewed distrust of the US as a result of its Cold War activities. The American government was so rabidly anti-communist that any movement towards the political left by any Latin American country was viewed by the US as a hostile act. And so it became ingrained in US foreign policy for the last 50–60 years that it was absolutely intolerable for anybody to be left-leaning.

Sole Western Hemisphere Commie.

If you think back to our chapter on global politics and governments, not every single system on the left is Commie—but at the time, all forms of socialism were viewed as being on a slippery slope that would eventually lead to the Soviets marching into Arizona. The US government believed that any form of socialism, however mild, would lead to communism or would lead to an opening for the USSR to make inroads. It really was battled at all cost. No cost was too high; no moral too low to violate in order to ensure that Latin America stayed firmly in the US's backyard of influence.

Let's get into some specifics. . . .

What did the US do in this all-out barrage to stop communism in Latin America? Be forewarned: This is going to hurt for the uninitiated. It's a little hard for many proud Americans to hear, but the US did some pretty nasty things, quite frankly. While they typically champion democracy on the planet, at the time of the Cold War, the USSR was seen as such a threat that the US said, "Well, we're all about democracy, and it would be great if we had true democracies down there, but we can't allow anybody to go near the left. So it would be better to support someone on the extreme right as opposed to anybody that might be even the slightest bit leftist Commie."

What this equated to was US support for people who might be considered brutal dictators at worst, and elitist dudes of questionable ethics and character at best. During the Reagan years in the 1980s, the threat of Soviet infiltration by arms sales to places like Nicaragua was interpreted as an immediate hostile threat to the US. In response, support for dictators was sometimes pitched to the US Congress as basically, "We're about to get freakin' *invaded* by the Commies. They're going to get into Central America and they're going to sweep through Mexico and then they'll be knocking down Texas's door!"

In hindsight, this seems a bit preposterous. To be fair to the Reaganites and their ilk, we do have to consider that at the time, the Soviets were as aggressive as the US was, had as many nukes as the US did, and had previously tried to hide missiles in Cuba. So the commie-infiltration precedent was present. There was a very real fear of global domination by the Soviets.

You can listen to speeches by Henry Kissinger and the like that say, "You young people just don't understand that we had to do these horrible things because if we didn't, you'd all be wearing red right now and we would all be slaves of an oppressive giant Soviet Empire." The Plaid Avenger is not here to speak on whether that's true or not, because I'd be wearing plaid no matter if it were commie plaid or otherwise. But what this equated to in the Reagan era (and even well before that era) was not only supporting extreme dictatorships, but also hatching plots to overthrow—and sometimes assassinate—democratically elected leaders (go Google Salvador Allende of Chile or Jacobo Árbenz of Guatemala, just to name a couple). This also equated to supporting extremist rightist factions and rebel groups. Right wing death squads in Nicaragua comes to mind, also found in Guatemala and El Salvador too. The US supported *anybody* as long as they weren't left-leaning and weren't socialists and didn't support any of those other ideologies, especially communism.

A lot of these groups ended up slaughtering thousands, and tens of thousands, of their own people. Many just ended up as bands of guerillas running around the countryside causing mayhem to the elected governments. The end result of all these anti-communist policies in Latin America, particularly Central America, was civil war. These excessively destructive civil wars were supported in part by movements of US funds or arms, or funds for arms. One of the more famous ones is the **Iran/Contra scandal**, during the Reagan era, when guns were floated into Nicaragua in support of anti-leftist movements to overthrow the democratically elected leftist regime, which we'll discuss more in the Central America chapter.

All these things together, in terms of US intervention or involvement in Latin American affairs, bring up a term that's often used for Latin America: **the US's backyard**. You'll see this term used even in modern political science magazines and international news. It really summarizes the way the US has felt about Latin America, which is, "It's not really our house, but it's our backyard. We're not really cleaning it up or taking care of it *unless* someone starts coming around and messing around in it." The US doesn't want anybody messing around in its backyard; that's why it's been heavy-handed at times throughout history.

2019 UPDATE: This era may be coming to a close, as the USA has recently renounced the Monroe Doctrine altogether for the first time ever in 2016, and is increasingly losing influence within the region as a whole as more and more states turn to other powers like China and India, or become big powers themselves, like Brazil and Mexico. More on this later.

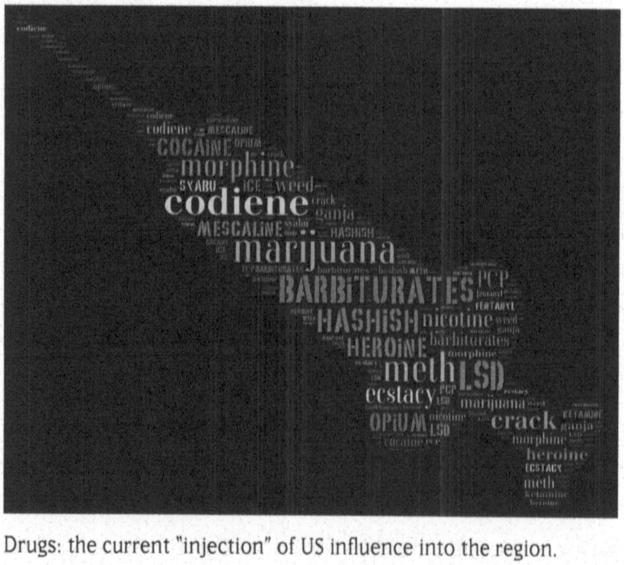

Drugs: the current "injection" of US influence into the region.

DRUGS

Oh yeah, I almost forgot about drugs! That is the hot new thing with US intervention in Latin America. The Cold War is over. There is not really any Good Neighbor Policy or Roosevelt Corollary going on. No foreign power is invading—that we know of—and the US is not going to do anything about it if they invade each other. The current and active deal with US intervention in Latin America is all centered on **drugs**. What the Plaid Avenger already knows from personal travels is that Latin America produces the bulk of the world's cocaine . . . and the United States *consumes* the bulk of the world's cocaine. Talk about a most horrific symbiotic relationship.

This creates a situation where the US government, armed with an anti-drug policy named "The War on Drugs," facilitates or makes it an imperative for us to intervene in other countries to stop drug production all over the world. The Plaid Avenger won't get into a big debate about the pure insanity of such an endeavor. I'll leave it to you students of the world to figure out if it's a good method to stop drug use or not, but certainly it is the US policy: "We don't really care so much that people are hooked on drugs up here; we just want to make sure they don't produce them down there." That will somehow solve the problem? Yeah, good luck with that one!.

This brings up one particular aspect of US foreign policy today called **Plan Colombia** (read box to the right).

US direct intervention right now is mostly focused on Colombia, Ecuador, Peru, and Bolivia: the big drug-producing and exporting countries. Plan Colombia, in particular, has equated to around 6 billion US tax dollars in an effort to stop Colombia from producing drugs. Money well spent! (Did you read the inset box yet?) Mexico is the current hot spot for huge drug related violence, and an all-out drug war between the Mexican government and the powerful drug cartels . . . but thus far, the US has not gotten involved too much, and for the first time ever, maybe they really should! This Mexican drug cartel war is radically affecting the US, and spilling over into the states in very real, and very dangerous, ways. Mexico is seeking active US involvement in their drug war; Colombia has worked with the US for years with drug wars and drug policy . . . but other countries like Bolivia, Venezuela, and Ecuador are not so keen on US policy. That brings us to a related topic . . .

LEFTWARD LEANING

(**2021 UPDATE:** Take this entire section with a grain of salt, as it has become much less "left" as of late, and I will update more fully at its conclusion.)

This is the most interesting and rapidly evolving part of "What's Latin about Latin America." Through a combination of virtually all the above reasons that Latin America is Latin, this notion is the one that pulls it all together. As we've already cited, particularly during the Cold War era, the US was very troubled by, and directly intervened in, leftist or left-leaning countries in Latin America. God forbid they embrace some sort of socialism, and certainly

What is the Deal with Plan Colombia?

Plan Colombia is a program supported by the United States to eradicate coca production in Colombia. It may sound good, but Plan Colombia has become extremely controversial for several reasons. For one, although the coca plant is used to manufacture cocaine, it has also been used by indigenous peoples in the area for thousands of years for health reasons. Many of these people depend on coca to make a living. Furthermore, some of the methods that the United States uses to eradicate coca, such as aerial fumigation and the application of deadly fungi, pose severe health problems for people exposed to it.

Plan Colombia is also controversial because Colombia was undergoing a civil war at the time. Many people claim that the goal of Plan Colombia wasn't really to stop drugs, but to help the Colombian government fight the Marxist rebel group FARC, which gains much of its funding through the drug trade. When Plan Colombia was first introduced by the president of Colombia in 1998, its main focus was to make peace with the rebels and revitalize the economy of Colombia. However, policymakers from the United States revised it, and the focus became more about military aid to fight the rebels and the elimination of drug trafficking. Human rights organizations are indignant at Plan Colombia because they see it as a way of strengthening right-wing paramilitary groups in Colombia that are committing atrocities against peasants who are speaking out for equal rights and economic reform.

It's your tax dollars, so you should know: the US government has spent close to six billion dollars on Plan Colombia since the year 2000. In 2006, it was reported that coca production actually increased in the last three years. Hmmmmm . . . I'm no mathematician, but something don't add up here. . . .

Fun Plaid Fact: After Plan Colombia was revised by the Americans, the first formal draft was written in English, and a version in Spanish wasn't created until months after the English copy was available. In related news, many Colombians have accused Colombian pop star Shakira of being a language sellout for releasing an album in English.

not communism! Nonetheless, some countries went left anyway. Fidel Castro springs to mind. Cuba has been communist and certainly in the "lefty" column for this entire time.

While Castro's Cuba was pretty much alone in the hemisphere in its communism experiment during the Cold War, much has changed since that war ended. Since the early 1990's, a combination of factors have influenced Latin American leanings: 1) the end of the Cold War = general disengagement of the USA from the region, 2) the establishment of true democracies holding free and fair elections across the region, and 3) the unresolved issues of wealth disparity and landlessness across the region. This combination has culminated in a lot of countries heading back towards the left column. The most recent example is Venezuela in South America. A rabidly left-led country (socialist, some might even say it's already communist), Venezuela boasts a new brand of socialism that strives for social equality and redistribution of wealth to a limited degree through social programs, formerly under the tutelage of their self-avowed socialist President Hugo Chavez. Although Hugo is now Hu-gone (he died in 2013), the socialist movement he spawned (Chavismo, also known as Chavism and Chavezism) is very much alive and well and attempting to be perpetuated by current Venezuelan President Nicolás Maduro. President Evo Morales of Bolivia also joined the ranks of hard-core leftist, and has been busy nationalizing industries and moving his country in a fully socialist direction since 2006. Current Nicaraguan President Daniel Ortega is pretty hardcore left too, having originally been a major player in the Lenin/Marxist-inspired radical left Sandinistas (Sandinista National Liberation Front) movement that overthrew the government back in the late 1970's. Many other Latin leaders are not quite as radical, being on the just-center-left side of the spectrum, mostly on the social/economic issues at the very least.

Former US President Barack Obama was an overwhelmingly popular figure across all of Latin America, so the standard anti-US sniping from Castro and Chavez is now not so much in vogue anymore. Obama had been actively trying to reach out to restore diplomatic street cred in Latin America, and has spoken openly about thawing relations with Cuba . . . and even shook hands with Chavez at an OAS meeting back in April 2009! Change may be coming, and these historical animosities may be softening!

It's not just the bold and brash loud-mouths in Latin America who are embracing the left. Brazil's last four leaders were/are left-leaning presidents as well. Ecuador, Nicaragua, Haiti, and Peru have headed that way as well. As you can see from the map on the following page, the future of Latin America did seem to be in the left-leaning categories for most of the last 20 years. Let's explore why that is.

Why was the left progressing and gaining popularity, and actually the dominant political force across most of the region for decades? Many people in the world are starting to look at Latin America as a political experiment—one of the reasons I decided to do this chapter—that perhaps may become a new axis of power on the planet. What am I talking about? Well, as a group of disparate countries that didn't have a lot of common economic or political goals, now they do. And as they had this leftward shift, I started to look at this entire region as representing a more common, singular ideology. There is no other region like that on the planet at this moment. We can look at most of the planet which has progressively over the last 50–100 years been going toward something that's more on the right, more strictly capitalist, democratic systems. While certainly these are all democracies in Latin America (with Cuba as an exception), they are going more left in terms of social and cultural issues, becoming much more openly liberal. In other words, the overwhelming focus in Latin America is to remedy the very wealth disparity that we talked about earlier.

Why is this happening here? Doesn't everyone want social justice and equality for people across the planet? Well yeah, lots of people do. But as I already suggested, this is the place on the planet that has the greatest wealth disparity. (The Middle East is a close second, and is partly the reason that revolutionary activity of the Arab Spring pops up from time to time.) Out of the 20 most unequal countries in the world (based on the GINI index measuring income inequality), eight are Latin American while the rest are in Sub-Saharan Africa. The region still has rising income inequality, persistent poverty, and chronically poor public services like healthcare and education. The landless, impoverished masses make this a perfect lab setting for this kind of experiment to evolve. Why is it right now? You have to understand that when you have any state, country, or place where most people are incredibly poor, you are asking for a revolution. When most people have no stake in the land, no claims—it's all fine as long as the minority, who has the power and wealth, can keep them down—but when it becomes too lopsided, it becomes tougher and tougher for the very small minority elite to keep a lid on it, and things will eventually boil over.

Societies like this are always on the brink of revolution (see Mexican Revolution, Communist Revolution, Bolivarian Revolution, and the very much current 2011 Arab Revolutions!), and Latin America is leaning back to that point today. It is because of this inequality that people are voting for the left, voting for parties whose *primary goal* is to alleviate wealth disparity. They want to make things more equal. They are striving to improve infrastructure like roads, schools, and health care as part of their primary goal. I'm not saying that this isn't a goal of the other parties, but this is the *primary goal* of leftist parties.

The primary leftist agenda involves things like human rights, investment in education and healthcare, and equal access to land for the impoverished masses. That is the main unspoken priority, and that is why the leftist agenda is so popular. In democratic countries, where nobody has squat to their name, the leftist candidates are extremely appealing because they are telling the people, "Hey, we are trying to make this better for you or more equal for you." Thus, it should not be a radical surprise that there is a big movement towards the left across the region that has the greatest wealth disparity on the planet. That's why people are voting for the leftist candidates.

Just look at all that leaning!

But why the sudden turn away from center-right conservative and right-wing political parties across all of Latin America starting in the 1990's? Well, the wealth disparity issue certainly played a part. But it mostly was a reaction to the brutal military dictatorships that ruled much of the region from the 1960's, 1970's, and 1980's, as well as the far-right paramilitary outfits that caused terror/violence in Colombia and Central America. As has been pointed out, many of those nasty right-wing regimes and paramilitary plunderers were backed/supported by the USA, thus the animosity towards the right in general, and the US specifically.

However, Latin America consists of (now) well-established democracies, and it is increasingly hard for a military dictatorship or for a government supported by an foreign entity like the US, to hold power because issues are getting clearer and fewer people are being influenced to vote against their interests. The blatant corruption is getting easier to identify, so it's very difficult for extreme right-wingers to hold power anymore. Well, maybe. . .

RADICAL RIGHT TURN!!!

Massively important 2021 UPDATE! I told you at the very beginning of that last section to take it all with a grain of salt, and perhaps I should have said a pound of salt! What a difference a single updated edition of this book makes! Virtually the entire region of Latin America has swung to the other side of the political fence in the last 5 years! It happened so fast that even my plaid powers of prediction could not have called it! The center-right/conservative-right/extreme-right in Latin America has seen a stunning comeback after decades in the political wilderness. In the last two years alone, right-wing leaders have assumed the presidency in Ecuador, Paraguay, El Salvador, Brazil, Argentina, and Chile . . . and those last three are of particular significance since they are the three biggest economies of the continent! For the first time in forever, South America's three leading economies are led by conservative governments.

Perhaps not directly correlated, but I think we can even refer to this as a "Latin Trump Bump," as this movement accelerated when US President Trump took office, and many of its leaders are extremely similar to the US president in background, political styles, and policy focus. Like Trump, the center-right presidents of Argentina and Chile (Mauricio Macri and Sebastián Piñera) are millionaire business moguls-turned-politicians who want to run the government like a business. (Billionaire side note: Piñera is one of the wealthiest politicians in the world: net worth $2.8 billion. Second billionaire aside: Macri actually just lost the leadership of Argentina to a left-wing candidate . . . the pendulum continues to swing radically.)

But when it comes to similarities in personalities, even those two have got nothing on Brazilian President Jair Bolsonaro! This polarizing and controversial politician and self-proclaimed "Brazilian Trump" and "Trump of the Tropics" is a hard core, far-right wing leader that is pro-business, pro-guns, pro-military, pro-using-force, and a vocal opponent of immigration, same-sex marriage, homosexuality, abortion, affirmative action, drug liberalization, indigenous groups, and even secularism. He has also made incendiary remarks in public and on social media that many people considered homophobic, violence-inciting, misogynistic, sexist, racist, or anti-refugee. Dang! Now that is hard right! Trump got nothing on this guy! That's about as far to the opposite side of the spectrum as you can get from the previous four left-leaning Brazilian presidents! Why such an about-face from the voters?

Brazilian President Bolsonaro: "Trump of the Tropics" to an exponent higher.

An Un-Fantastic Four things: 1) rise of the middle class, 2) crazy corruption, 3) persistent poverty, and straight-up 4) "left-wing fatigue."

1) For starters, that rising number of people entering the middle class is likely due to the semi-successful liberal-left policies of the preceding decades. While not explosive growth, most economies in Latin America have slowly chugged along and grown little by little over the years, helped considerably by a boom in commodities prices (of which they have many). Liberal-left policies of internal national investment (like infrastructure and schools) in conjunction with redistribution policies (that is, benefits to the people like welfare or health care) have resulted in 70 million people lifted out of poverty between 2002 and 2014 (source: United Nations Economic Commission for Latin America and the Caribbean

(ECLAC)). Other facts from that report: 30 of that 70 million were in Brazil and in Ecuador, the middle class doubled from 18.58 percent to 37.40 percent from 2005-2015. In both places, most of those people are now voting center-right. Why?

Well, I have no definitive answer for you, nor have I seen a study suggesting this, but it seems to me that people who have nothing will vote to have things redistributed (leftist policies) while people that have something will vote to maintain what they currently have (conservative policies). So in a sense, leftist policies have become a victim of their own successes in the parts of the population that they have most directly benefitted. So a lot more people in the region have upward social mobility and more income, and therefore (perhaps) are now tending towards leadership that is going to be more pro-business (so as to further increase their income), pro-law-and-order (to protect the stuff they now have), and pro-nationalism (to put their country first in order to accumulate even more stuff). Just a hypothesis. But make no bones about it, even though there are more peeps in the middle class than ever, there is still . . .

2) Persistent poverty, which is another reason folks are abandoning the left. Wait a minute, Plaid Avenger! You just said people are voting for the right because they are richer, and now you are saying that they are voting for the right because they are still poor? Yep. That is exactly what I'm sayin'. I said *some* of the peeps are entering the middle class, but there are still tens of millions, maybe hundreds of millions, who have *not*. All those dire facts about income inequality and wealth disparity I have referenced in this chapter still apply to most Latin Americans, and that is after decades of left-liberal leadership. So while it has been moderately successful, it has not been totally successful, and those that have continued to be left behind are now out of patience with the liberal-left.

Many of them are young people, too, and anyone under the age of 30 or 40 has no living memory of any of the past atrocities of right-wing dictators. Many have lived their entire lives under left-liberal leadership (especially in Brazil), and unless their situation has personally improved, now blame those left leaders for the ills of their society (especially in Brazil). And they increasingly hate those leaders outright (especially in Brazil) because of . . .

3) Crazy corruption that has occurred for decades (especially in Brazil). I could write another ten pages outlining the absolutely absurd amount of corruption in many of these states, but I'm not going to. I will point you to just a couple of things you may want to research further on your own if you want to fully understand this backlash against the left (especially in Brazil). Go look up Operation Car Wash, the biggest anti-corruption dragnet in Brazilian history. It has already taken down two former Presidents, one of which was impeached, the other is in jail. And in Mexico, former President Enrique Peña Nieto left office with only a 20% approval rating, as his administration was haunted by numerous corruption scandals and the murder rate of Mexico at an all-time high. Yikes! No wonder some peeps are ready for a change! Which brings me to . . .

Former Presidents Lula and Rousseff: the "Car Wash" flushed these two down the drain.

4) "Left-Wing Fatigue." Put all of the above factors together, and marinate for decades. Many Latin Americans are now just tired of the same old political liberal-left leadership with the same old liberal-left solutions that don't seem to be working fast enough for everyone, or in some cases not working for anyone at all anymore (I'm looking at you, Venezuela). Sometimes societies just vote in change for the sake of change itself . . . I would point to the election of US President Trump as a pretty good example of this concept; trying something new to see if it works better than the old way. And a lot of fatigued voters across Latin America seem ready to roll the dice and go right-wing again.

The Latin liberal-left may be down, but don't count them out just as yet! This is still a startling new phenomenon, and we don't yet know how it will go. This counter-reaction to the decades-long dominance of the left in the region may not last long, and the leftist roots still run deep. Hard core leftists still run Nicaragua, Bolivia, Cuba, and Venezuela . . . although not that successfully in some cases. Moderate, and successful, liberal-left leadership can still be found in Mexico and Costa Rica, and has resurfaced in Argentina as well.

And please note this in your brains: even though center-right conservative governments now dominate many of the biggest and richest economies of Latin America, social injustice, inequality, and wealth disparity are still *HUGE* issues in the region, so even the most nut-job, extremist, right-wing leader (and yes, I am talking about the "Trump of the Tropics" Bolsonaro) will get push-back from the masses if life doesn't improve for them. Pro-business Argentinian (Macri) and Chilean (Piñera) leaders have already been stymied in their attempts to get rid of popular fuel subsidies, anti-poverty programs, and educational benefits. (**2021 UPDATE:** Argentina actually swapped sides again: pro-business Macri was ousted in favor of new left-wing President Alberto Fernández.) In Colombia, the right-wing leader Iván Duque attempted to offset corporate tax cuts with higher taxes on food . . . which utterly backfired, and just three months after being elected his approval rating plummeted to 25% (his popularity has since recovered slightly, but protests are pretty much ongoing for 3 years now).

So while Latin America may have taken a hard right turn politically, it apparently is still leaning left! Hahahaha, I know it's confusing, but bear with me on this! Many, many people in many, many states—including the big three of Brazil, Argentina, and Chile—may have elected conservative-right leaders, but it looks as if most citizens still demand liberal-left policies, benefits, and government services. Perhaps these center-right governments have come to power not because of specifically being politically right-wing, but because they currently appeal to the frustrations and expectations of a smaller group of voters that actually end up swinging elections one way or another. None of these leaders swept into power by huge majorities of votes. But huge majorities of Latin Americans still want their lives to be improved, and these governments better deliver a better alternative, or back to the left they will go!

LOSING THE LATINS: US UNDONE

While things are currently and quickly returning to the political and economic "right" in Latin America, the USA was still always troubled by that Latin leftist lean because it was largely seen across the planet like this: "The US has mistreated and maligned their regional neighbors so badly that they have lost control and influence in their own backyard." This may be an extreme statement, but it is a fair statement, nonetheless. Part of the reason they have done that, and part of the reason rightist regimes have lost power (until recently resurgent), is that they have been perhaps a little too heavy-handed over time. Most of the US involvement and incursions into the region were for reasons previously mentioned, such as anti-communist intervention, but even through the Cold War and to the present day, another reason for US involvement is to protect US economic interests. This is seen as extremely problematic for locals, who are usually on the losing side of what benefits the US economically.

In situations when the US has invaded places like Nicaragua or Belize, or has helped assassinate a democratically elected leader in Chile, it was because of this fear that in lefty/commie countries, a redistribution of land and resources was going to occur—you remember: **nationalization**. That was unacceptable to the US, largely because there were US corporations down there yelling, "Hey, that land/resources you poor people are taking is US property!" This was argued during the Cold War era, about Chile, in fact. A company said, "Hey! US government! You can't let commies take over Chile. They will nationalize our company (which means the commies will take it over and make the profit), and this is US property." A lot of US involvement has been due to protecting these US corporate interests. We'll examine this more in future chapters. Suffice it to say for now, the US's pro-capitalism, pro-free markets, pro-US corporations attitude is hugely distrusted by many Latin Americans . . . and as the region becomes more independent and wealthy, they are pivoting away from US policies and US leadership because there are . . .

NEW KIDS IN THE BLOCK . . .

The perception of US involvement as self-serving, coupled with massive wealth disparity, in part explains why much of Latin America has gone to the left. A lot of leftist candidates, particularly Chavez, Castro, and Evo Morales in Bolivia are saying, "We're not even pretending to do the neo-liberal capitalist policies. To be a free market with free and open trade works for the United States, and it looks good on paper, but it isn't working for us."

A lot of these leaders are saying, "We're not anti-US, we're just anti-free trade." The current president of Bolivia says, "We've done it. We've tried to do free trade, and we're still poor as squat! We won't do it anymore! We're not going to give priority to American corporations, and we're not going to give tax breaks to American anything!" Why is that? Because there's some new kids in town that many countries may give incentives to: China and India and Japan just to name a few. These countries not only are grabbing up tons of natural resources, but are expanding their exports into Latin America as well. Lots of investment flowing in to start new businesses and partnerships with the Latins from all points abroad has been the theme. That's another reason America fears a loss of influence in Latin America.

To be honest, you can't really blame the Latins for taking advantage of the international interest in their region. Specifically, the Chinese are courting countries around the world, making sweet trade deals with them, in order to feed the Chinese economy's ever-increasing hunger for natural resources. Chinese foreign trade/foreign aid deals are even sweeter because they come with no strings attached, unlike deals from the US. On top of that—and this is critical—in the last decade, the Chinese and Indian and even many African leaders have personally visited virtually every single Latin state, and multiple times, on multiple visits, in multiple years. Former US President Bush only managed to head south of the border twice in eight years. Former US President Obama only went twice, and one time his Secret Service agents got busted with Colombian hookers. Don't think Former President Trump visited there at all, unless you count Florida as Latin America, which many Floridians probably do. Starting to get the picture here? It gets worse: Russia has shown a renewed interest in floating more military ships into the area and strengthening ties with old and new allies. Iran is even in this game, having opened dozens of cultural centers and working on economic pacts across the region. There is currently little love lost between Anglo- and Latin America, a situation that seems likely to linger for a while longer.

And we must also point out that Latin America as a whole is on the upswing, meaning there is a lot more "local" demand for goods and services as the middle class grows, and many of those goods and services are being provided by local companies themselves. Add to the growing powerhouses of Brazil and Mexico, which are both top 20 world economies and natural leaders within the region . . . and the picture of a much diminished US role starts to take shape.

Now I have spent a lot of time explaining the viewpoint of the Latin Lefties, but I don't want to exclude the ideas of the other side of the political/economic spectrum. About people on the political right: I don't want to suggest that they only want to make money and they don't give a damn about people; that's not the case. Even these people on the opposite side of the spectrum (real anti-socialists) would argue, "No, we want to make these countries richer, too. We want their citizens to have more stuff and not be poor, but we don't think social programs are the way to do it. We think the way to make people richer in Brazil and Colombia and Mexico is to have free trade." Free trade is the typical conservative approach to alleviating poverty. I don't want to deify anyone that is socialist because there are people on the other side who also want good things. They just don't think leftist methods will work: "You can't just give them a welfare check! That's not going to solve anything! Then nobody will be rich! What we need is more free-trade, more pro-business policies, and more fiscal conservatism." And that is partly why we are witnessing the rise of the center-right to far-right political parties gaining power across Latin America, as discussed in the previous section.

Which is why I still suggest that the **Free Trade Area of the Americas** (FTAA) is an entity that is currently in a coma but perhaps not dead and buried just as yet. With a more internationalist-leaning Joe Biden in the White House, and more and more pro-business leaders elected in Latin America, we may yet see the resuscitation of this free-trade block in the future. It would certainly help the USA fend off influence from global competitors like China and Russia, and reestablish the US as a leader in its own neighborhood.

Remember: **NAFTA** is a free-trade union between Canada, Mexico, and the United States. **CAFTA** was an ever-growing free trade area between the US and a handful of Caribbean and Central American countries with the inevitable goal of being the FTAA. The **FTAA** is a proposed free-trade area of every single country in the Americas. The United States, of course having an edge in all of this, was a big fan and proponent of the FTAA. It thought the FTAA would make it richer, for one, but also that the FTAA would help it to reestablish influence in the Latin American region as a leader. Anti-US, anti-globalization forces and leftist politicians in Latin America say, "No. We've been playing that game for a hundred years and we're still poor. We don't like it. We don't buy it." But the revived right side of the political spectrum just might.

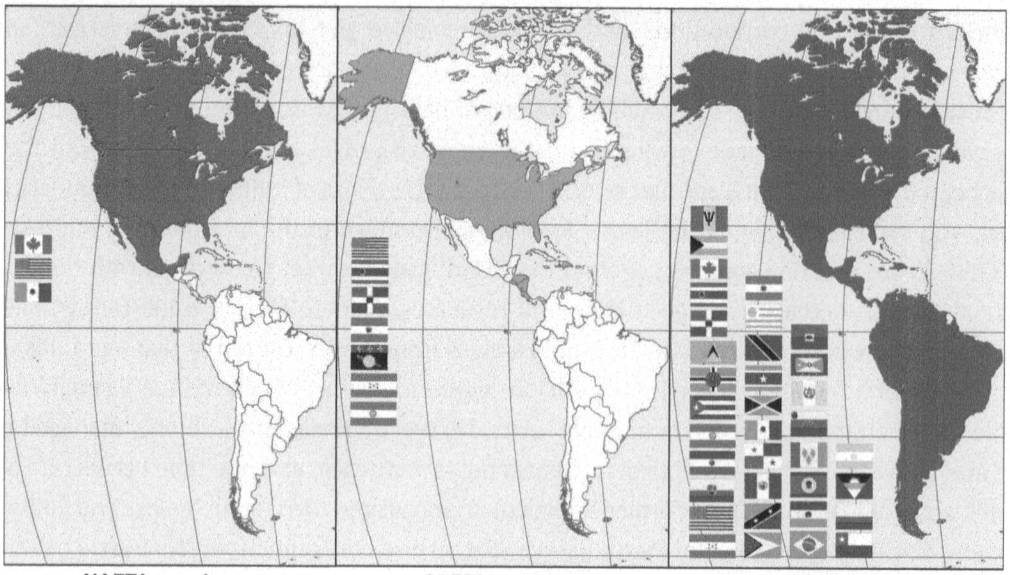

NAFTA members CAFTA members Proposed FTAA members

We shall see how the battle for free trade pans out in the coming decades. Other trade blocks just within Latin America are actually imploding or re-configuring themselves right now too, something we will chat about more in the South America chapter. So we are entering a whole new (and currently unknown and undefined) era in even trade relations for this entire half of the planet.

Chapter closer: These pervasive themes are not only historical, but they play into today's Latin America. Now let's take a look into some of Latin America's subregions to provide more specific details, so you can understand how each one works into today's and tomorrow's world.

DON'T GET LEFT OUT OF LATIN AMERICA!

HOW FAR LEFT ARE THESE LEADERS?

The Latin American leaders below are arranged in order of increasing dedication to full-on liberal socialist policy; the farther to the left, the more fully the incorporation of socialist ideology into their political and economic policies. But can you name all these peeps and the countries they lead? Hint: former US President Donald Trump is thrown in just for fun all the way on the right side, as his political policy stance justifies, especially when compared to the rest of the Americas' leaders. Except that guy even farther to the right than him! Who would that be? Name them all, and you are a bona fide Latin Leader Lover!

Leaders: Bolsonaro, Trump, Castro, Piñera, Obrador, Maduro, Ortega, Fernández, Arce
Countries: US, Bolivia, Brazil, Nicaragua, Mexico, Argentina, Cuba, Venezuela, Chile

LATIN AMERICA RUNDOWN & RESOURCES

View additional Plaid Avenger resources for this region at http://plaid.at/latin

BIG PLUSES

- Sizable chunk of real estate on planet earth with a boat load of resources
- Has never invaded nor infuriated any other country or region
- Not a target of international terrorism at all. Who else can brag about that?
- Becoming a serious place of interest and investment for China and other rising powers

BIG PROBLEMS

- Biggest wealth disparity on the planet
- Political instability chronic in some areas, possible just about anywhere; Venezuela is in trouble right now
- History of outside political and economic domination has left many residents with an inferiority complex
- Heavily dependent on exports of natural resources and basic manufactures
- Environmental degradation becoming rampant in exchange for economic growth

DEAD DUDES OF NOTE:

Simón Bolívar: The "George Washington of South America": hero, visionary, revolutionary and liberator . . . he led Bolivia, Colombia, Ecuador, Panama, Peru and Venezuela to independence and instilled democracy as the foundations of all Latin American ideology.

James Monroe: 5th President of the US and important for this chapter for his **Monroe Doctrine** which became the cornerstone for US foreign policy in the entire hemisphere, right on up to the present. The Doctrine pretty much sez: any foreign power which messes with Latin America, will also be messing with the US.

Theodore Roosevelt: 26th President of US, and "Rough Rider" that helped invade Cuba in his spare time prior to becoming a politician. His **Roosevelt Corollary** was an amendment to the Monroe Doctrine which asserted the right of the US to even intervene in the internal affairs of Latin American states to "stabilize" them if necessary.

Franklin Delano Roosevelt: 32nd President of the US and creator of **Good Neighbor Policy** which sought to soften the apparent US hegemony over Latin America by renouncing the US right to intervene unilaterally.

MAP GALLERY

View Map Gallery & additional Plaid Avenger resources for this region at http://plaid.at/latin

Mexico

CHAPTER OUTLINE

1. "New" Mexico?
   1.1. Physical
   1.2. ¡Hot Tamale! Mexican Climate
2. Mexico in History . . . Watch out for Montezuma! and His Revenge!
   2.1. Mexico vs. Latin America. Who Went Where? Who Colonized What?
   2.2. Rapid-Fire Path to Independence
3. Mexico Should Be Richer . . .
   3.1. Loco Bueno Relative Location
      3.1.1. Location and Economy: ¡Making Los Pesos!
      3.1.2. Location and Military: ¡Don't Mess with Me, Ese!
   3.2. ¡Recursos! = Resources!
      3.2.1. 1973 Oil Embargo
4. Mexico Should Be Totally Richer . . . but They're Not!
   4.1. Number One ¡La Gente!
   4.2. Number Two ¡Maquiladoras!
   4.3. Number Three ¡Viva la Tierra!
      4.3.1. ¡Rebels!
5. ¡El Manaña de Mexico!
6. Map Gallery

Mexico

14

MEXICO! The centerpiece; the central holding; the command center of the Spanish empire while it was conquering the New World. As such, it was a regional power center and a rich resource center extraordinare, with a significant population and a more than significant native civilization. When we hear about all the gold and silver bullion that was heading back to the Old World on Spanish ships, Mexico is primarily where all this stuff came from. For a variety of reasons, Mexico has been the center of things in Latin America as the launching pad of Spanish expeditions. But even before the Euro-exploitation, Mexico was the home to advanced civilizations such as the Olmec, the Toltec, the Teotihuacan, the Zapotec, the Maya and the Aztec. But that was then, and this is now. So **Qué pasa** Mehico? What up now?

¿Qué?

"NEW" MEXICO?

When we think about the term Middle America, which I gave to you in the last chapter, Mexico holds roughly 75 percent of Middle America's land. Mexico is a big place (13th largest state in the world), and it used to be a lot bigger. With almost 115 million citizens, it also has roughly 60 percent of Middle America's population, so by any account, it's the real powerhouse of the wider region. It is the giant among the smaller countries in the neighborhood. On another note, we can distinguish between Mexico and all of the other countries of Middle America and even South America, because it shares a border with the US. Being next to the colossus of the US, which is of course the number one power on the planet militarily and economically, opens Mexico up to a lot more opportunities than its Middle American neighbors. It's in "The Good Neighborhood" so to speak. In this chapter, we'll go back and look at how it got to this point.

Perhaps an appropriate title for this chapter would be **Mexico: The *Next* Fully Developed Region**. One of the distinguishing characteristics that caused us to tease Mexico out of the Anglo-American region is that it's not like Canada and the US in levels of development. A point that I will try to make that I want you to know and understand before you finish reading this chapter is that Mexico should be richer! It should be in the "fully developed country" category for lots and lots and lots of different reasons! And it may be soon! It's still the powerhouse of Middle America population-wise, even economics-wise, but it's not on par with the US and Canada. But it is catching up! Please note that Mexico is the 14th largest economy IN THE WORLD and as such is a member of the **G-20**! But it could be doing even better! So what's up with that?

PHYSICAL

Speaking physically for Mexico, this is an interesting country because it's got two major mountain chains running through it from north to south that are a continuation of the Rockies from the US, except in Mexico they call them the **Sierra Madres**. We have the Sierra Madres **Orientales**, and the Sierra Madres **Occidentales**, which of course are the east and west, respectively. If we look at Mexico as an entire entity, it's muy mountainous. Mountains are all over the place. These mountain ranges run down both of the east and west coasts, making the entire central plain of Mexico essentially a plateau, or an uplifted plain region. It's around 2000 meters above sea level on average.

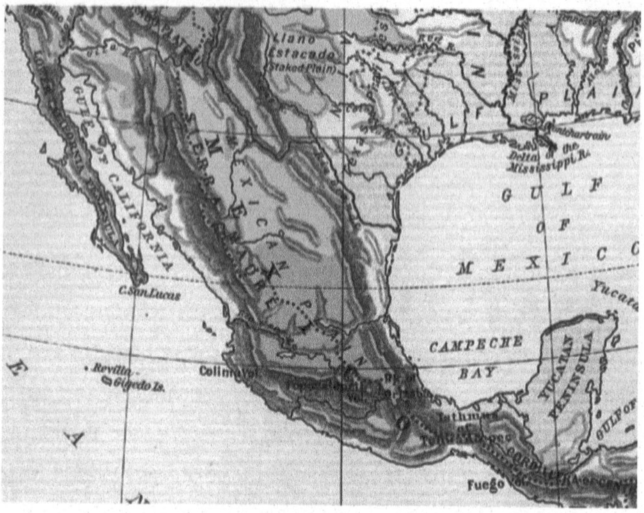

Mexico's Sierra Madres: Muy mountainous!

When Hernándo Cortés was conquering Mexico back in the 1500's, even he had a quick description of the Mexican terrain. When he went back to Europe, someone asked him, "Hernándo, what's Mexico like?" Cortés said, "Oh, I can show you!" He grabbed a sheet of paper and crumpled it up, smashed it up, threw it on the ground and said, "There you go, that's Mexico right there!" Mountainous terrain, all the way through from north to south, and a central plain which raises everything up right in the middle. This is important to note because Mexico City, with over 20 million people in it, is one of the highest capitals on the planet, at 10,000 feet above sea level. We're not talking about Andes style, or Himalayan style with huge impossibly impassable mountains; but it's all on an uplifted elevated plateau.

¡HOT TAMALE! MEXICAN CLIMATE

We think we know a little about Mexican climate. It's connected to the US down there at Arizona and New Mexico, so it's all desert, right? When we think of our old friend Speedy Gonzales from Bugs Bunny cartoons, we think of desert and steppe: dry territory. Well, that's a bit of a misnomer, a dab of misinformation—because only the north is dry desert. The Sonora Desert creeps down into Mexico, into the Chihuahua Desert which is part of the same system. It is a dry place in the interior, but as you progress south, you get into some semi-tropical areas, and by the time you get to the Mexican-Guatemalan border in the south, it's full-on tropical rain forest. As you progress from the north part of Mexico to the South, you go up in elevation onto this high plateau, and it gets wetter. You go from dry climates to tropical wet climates. There is quite a bit of diversity within Mexico's physical systems, and that sets the stage for us to talk about Mexico's timeline.

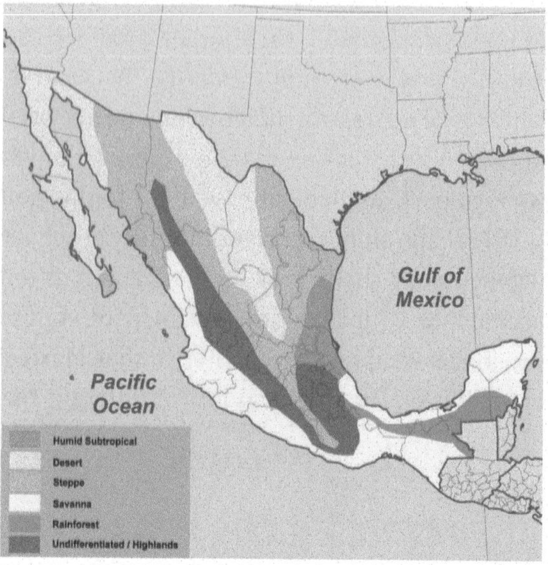

A diverse climate; not just desert.

MEXICO IN HISTORY . . . WATCH OUT FOR MONTEZUMA! AND HIS REVENGE!

The Plaid Avenger's got just a few things for you to consider as we are trying to figure out the Mexico of today and why it's important and how it's different from every place else. We've already pointed out that it has been a centerpiece, a rallying point, a base of command for the growth of the Spanish Empire in the New World. Let's go back even further than that!

What is completely different about Mexico versus really every place else in the New World of 500 years ago? It's different because they had not just a *few* folks hanging out here, and not just *big groups* of folks hanging out here, but full-on *civilizations* hanging out here for at least a couple thousand years! When the Europeans—some of these are British, some French, but mostly Spanish—bump into most parts of the New World 500 years ago, it's virtually uninhabited. Around this time, if you go to places in North America, there's almost nobody there. I may have pointed this out to you when we talked about the US and Canada's evolution, but there were maybe 2 million Native Americans hanging out in all of the North American continent when Columbus arrived. There were maybe 4 or 5 million indigenous people in all of South America when the Spaniards arrived. That's not a lot of folks for such a big area. But when you get into Mexico, everything is different. The **conquistadors** came face to face with upwards of 20 million inhabitants in what is now modern-day Mexico! We know these folks as the Aztec civilization. This group of dudes had been hanging out for a while, and they weren't even the first, nor the second, major civilization to control this territory. As pointed out in the opening paragraph of the chapter, the Aztecs were preceded by the Mayan, the Olmes, the Toltecs, et al. This was a happening place for most of human history!

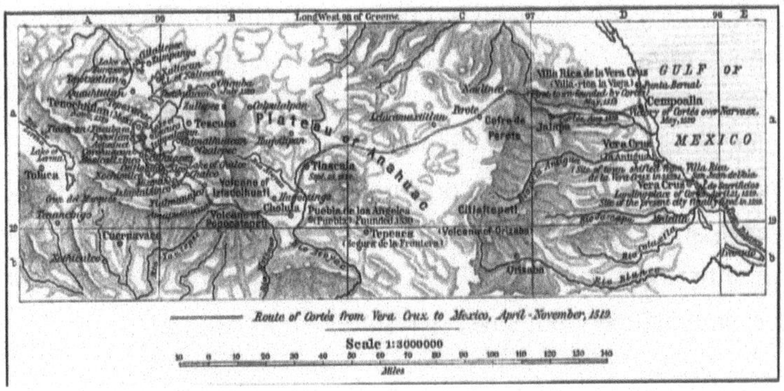

The Conquest of Mexico, 1519–1521.

So in the early 16th century, these conquistadors came upon a full-fledged civilization that had a huge urban area, temple complexes, and complex religious and social systems that had been in place for a thousand years. A little known fact is that Mexico City—what we consider Mexico City today—when the Spanish arrived, was by far the largest city the Europeans had ever seen. Let me clarify that: not simply the largest city they had seen in the New World; the biggest city they'd ever seen in their lives, period. Twenty million people in and around a singular city was unheard of at the time, even in Europe. It would dwarf any European city-state of the time. There were lots and lots of folks in Mexico, not a lot of folks outside of Mexico. It was what we call a true **cultural hearth**.

MEXICO VS. LATIN AMERICA. WHO WENT WHERE? WHO COLONIZED WHAT?

Spanish exploration and colonization of the Americas really starts in Mexico. After the Spanish domination of the Aztecs, they spread both north and south through the Americas in search of fame, fortune, and possibly fun! The Spaniards took over western North America, Central America, and everything west of the Andes in South America—and even more! Cortés and his small band of merry men (maybe 200-300) were after the Imperial "G" Trifecta: **Gold, Glory, and God**. They beat a quick path to the trifecta and were radically successful because within only two years, a handful of Spaniards overthrew the entire Aztec civilization. How the heck did 200 or 300 Spanish dudes manage that? Well for one, Cortés was very good at rallying the local tribes to his cause to help defeat the Aztecs. Many of the locals hated the Aztecs much more than the "innocent" Spaniards that had just shown up on their shores. Oops. Big mistake there.

But there are three more things for you to know: firepower, horses, and the Spaniards' biggest ally—DISEASE! The Spaniards and Portuguese brought horses and guns which greatly helped them over-power local resistance, but more importantly they imported measles and smallpox and mumps, which wiped out native populations. The indigenous folks of Latin America had never had contact with Europeans, or any 'Old World' peeps for that matter. They died from Old World-originated disease to which they had no biological resistance. Long story short: disease enabled the Spanish and the Portuguese to take over the New World in short order. But let's now jump to how these places became independent sovereign states.

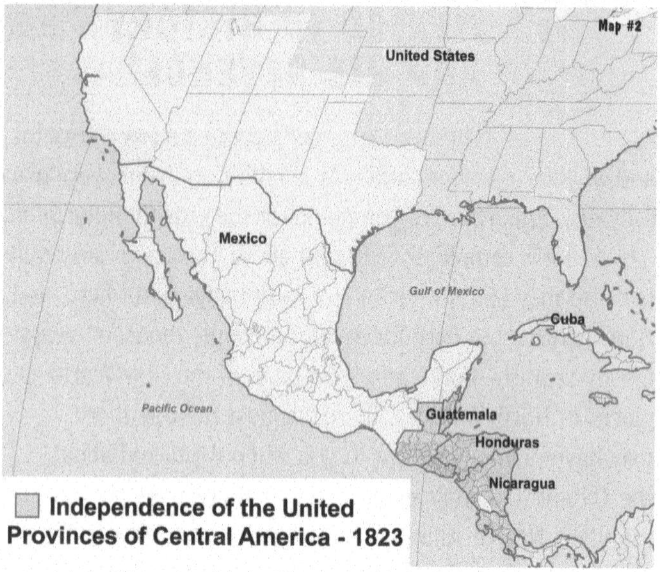

Independence of the United Provinces of Central America - 1823

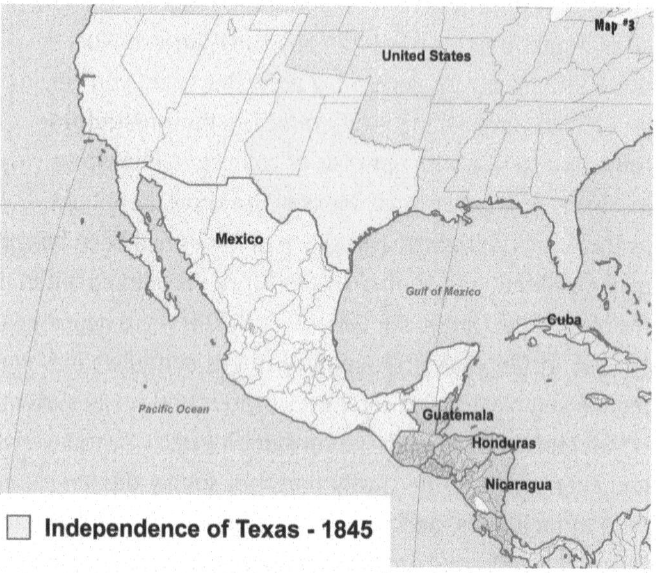

Independence of Texas - 1845

RAPID-FIRE PATH TO INDEPENDENCE

- Spaniards set up most of Latin America to be purely one massive extractive colony.
- American independence in 1776 was the inspiration for Mexican independence.
- The Mexicans got fired up enough to fight for their freedom in 1810, and succeeded in 1821. (Map #1)
- Shortly after Mexico achieved independence, Central America seceded from the Mexican union in 1823. (Map #2)
- In 1836, after years of armed hostilities and famous events like the Battle of the Alamo, Texas seceded from the Mexican union by declaring full independence. You can see that Texas then was even bigger than Texas now. (Map #3)
- Texas declared independence before they were accepted into the US, and they were considered an independent state. The were officially annexed by the US in 1845, which really infuriated Mexico and this act started the Mexican-American War.
- The Mexican-American War lasted from 1846–1848, and resulted in Mexico losing roughly two-thirds of its original territory. This is how the US acquired California, Arizona, Nevada, Utah, New Mexico, and Colorado. (Map #4)

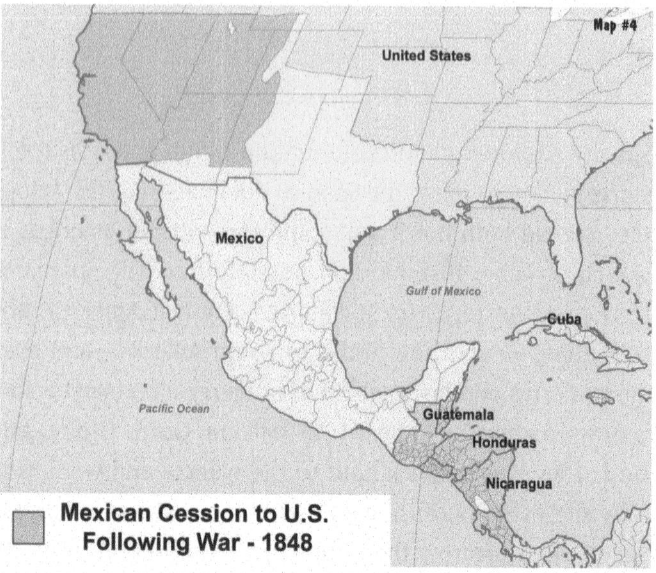

Mexican Cession to U.S. Following War - 1848

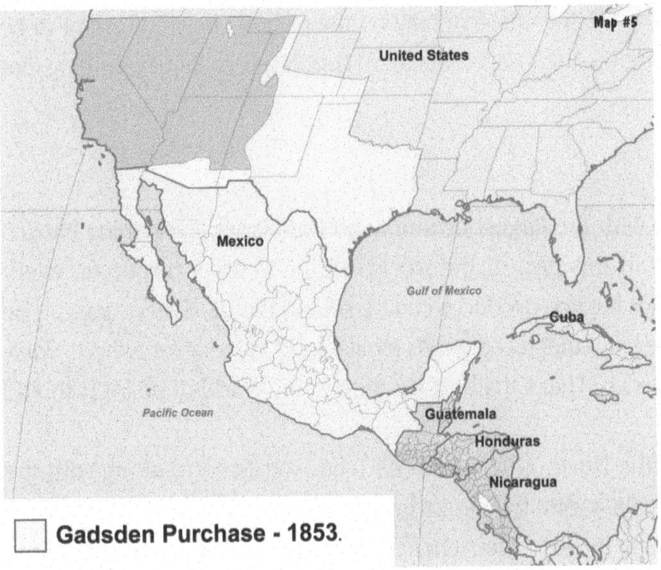

Gadsden Purchase - 1853.

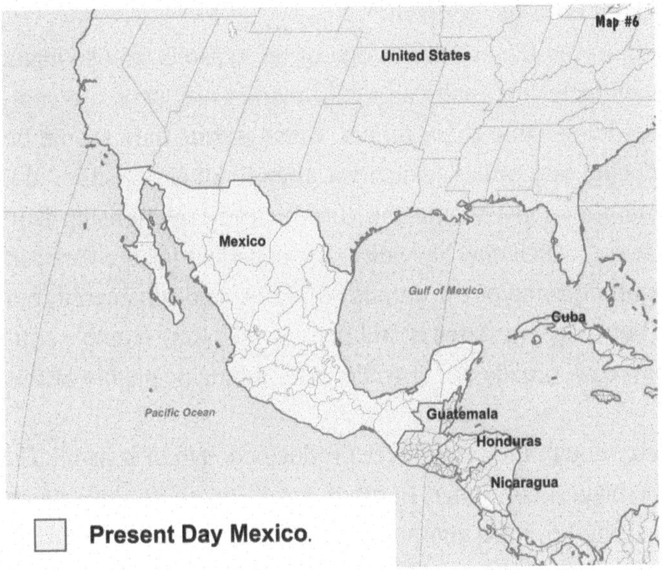

Present Day Mexico.

→ In 1853, the US tried to buy more territory from Mexico. Mexico refused, but the US government strong-armed them into it. This resulted in the Gadsden Purchase. (Map #5)

The sum total of all this territorial loss really ticks Mexico off, but they can't really do anything about it. They perhaps have developed an inferiority complex because of it. They're all bent out of shape because historically, they can't compete. Maybe this inferiority complex has something to do with why Mexico is not better off in today's world. Because quite frankly . . .

MEXICO SHOULD BE RICHER . . .

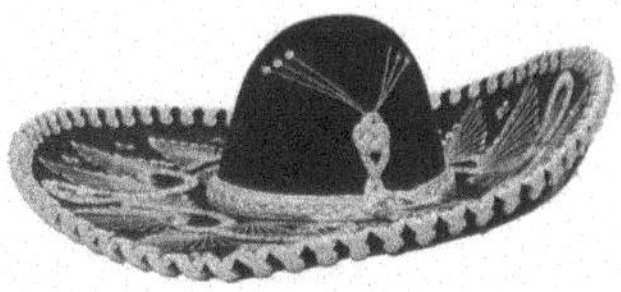

It really should. There are not too many places in the world where you can look at the country as a whole and say, "Wow, you should be doing a lot better than you are; you've got a lot of things going for you." But you can in Mexico. It does have the 14th or 15th biggest economy in the world by GDP, but somehow is still considered "not fully deleveped" by most outsiders. As the Plaid Avenger has said before, that's one of the main differences between Mexico and the US. The US is extremely developed, extremely rich; Mexico not completely so, especially per capita. But it should be. Why should Mexico be a fully developed country? We can look at a few main factors you should consider. The first of these is:

LOCO BUENO RELATIVE LOCATION

What do I mean by that? I simply mean where it is situated on the planet in terms of everything that's around it. If you could magically place a parcel of land anywhere that you wanted, you could put it anywhere, would you put it at the North Pole? No way. South Pole? Probably not. Middle of Russia? No. Central Africa? No. If you were wise, you'd probably stick it right where Mexico is,

Tomatoes to tablets to TV's to Toyota: It all comes from south of the border, baby!

because it would be next door to, and have a shared border with, the US: the one economy on the planet. You would want to be situated next to the world's economic nucleus, the powerhouse of the US, the insatiable market that can consume everything that the Mexican market could produce. And they do.

LOCATION AND ECONOMY: ¡MAKING LOS PESOS!

Everybody would like to be neighbors and close trading partners with the largest economy on the planet. Everything Mexico makes, including all the vegetables consumed during the winter in this part of the world, all the mineral resources, every product manufactured using their cheap labor . . . EVERYTHING that Mexico produces can be sold in the US. Everything can be exported to the US; TVs, DVD players, automobiles, countless other manufactured goods available in stores everywhere, waiting to be consumed by the giant American market, are made in Mexico. That's the best position in the world to be in! It should be exceptionally rich.

You have to remember that Mexico's a member of NAFTA (the North American Free Trade Agreement) along with the US and Canada. US = number one economy on the planet. Canada = number one trading partner with the US. Mexico = number two trading partner with the US. NUMBER TWO! How can they not be richer? It's incredible! Almost astounding! It flies in the face of everything that would be expected, everything that makes sense, to be trading partners with the biggest consuming economy with the most money on the planet and not be richer yourself!

Almost all of the things that are produced in Mexico end up in the US, and that's one of the reasons why so many businesses have relocated to Mexico. Car manufacturing, TV manufacturing, really all major-market products, are compiled and assembled in Mexico and end up in the US. They are adding value to a product, which means they should be getting something out of that. It's much better than just producing vegetables, which we already know that they do, as well. **2019 UPDATE:** Please be aware that the almighty Trump may just thump the current trade relationship with Mexico in order to "get our jobs back" and "Make 'Murica great again" which may have dire consequences for the Mexican economy, but it is still all speculative at this point. He did renegotiate some of the details of NAFTA, and renamed it, but honestly the trade deal was not modified very much and trade between the states is still quite solid. Now, Trump's anti-immigration attitude/crisis and the threat to build a huge wall may eventually affect trade, but nothing negligible at this point. More on that in a moment.

But the main commodity of significance produced in Mexico, though they can never produce enough of it, is oil. Oil is one of the main things Mexico produces, of which the US is the biggest consumer. In other words, no matter how much Mexico produces, every single fluid ounce of it is *already* sold to the US. It's a guaranteed market. That's a really, really big deal. The economic nucleus to the north is of extreme importance in regard to why Mexico should be richer.

Consider stuff that's hard to deliver, hard to sell—things like drugs. It takes serious planning and coordination of efforts to move and sell drugs, but some scurilious peeps in Mexico do it. And do it very well, I might add. Why is Mexico a huge producer and trafficker of drugs? A: Because the USA consumes copious amounts of drugs. Even something that's illegal, for which you can get thrown in jail or shot at and killed—even this stuff moving from Mexico into the US in huge volume every single day. Mexico should be richer solely off what it can do under the table for its northern neighbor.

LOCATION AND MILITARY: ¡DON'T MESS WITH ME, ESE!

Not only is Mexico in the perfect location geographically to really reap the benefits of an economic powerhouse, but on top of that, its geography also provides a military advantage. What does the Plaid Avenger mean by this? Mexico has not had to invest in a offensive or defensive military capabilities for decades. Most countries on the planet have to defend their territory: they have to buy tanks, develop serious weaponry, and do a lot of expensive things that take money away from developing infrastructure in other ways. Mexico's relative location has again given it the prime spot because it doesn't have to do any of that. It's not a member of NATO because it doesn't have to be. Mexico is not a military power because it doesn't have to be. Why? With the US sitting right there, who in the heck is going to invade or attack Mexico? Nobody. Not even Russia at the height of the Cold War considered invading Mexico. Would China ever do it? Not a chance.

Even though Mexico is not protected by any specific treaty, nobody is going to mess around in the US's backyard. Think back to the Monroe Doctrine: mess in the US backyard and you are toast. It's just not going to happen. Not spending billions on a military is a tremendous boon for the Mexican economy. Having said that, let me assure you that Mexico actually does have a mighty military, it is mostly employed not to defend the country from invasion from abroad, but to fight the enemy from within: the drug cartels and criminal elements which have been fighting for the control and the soul of Mexico itself. More on that later.

Relative location is key. There is no better location on the planet. Canada has advanced to full development, but Mexico has not. Being a NAFTA member with close ties to the economic powerhouse of the world, and already benefiting from its relationship with the US in terms of defense, Mexico should be rolling in the pesos. Having said that: make no bones about it, Mexico does have a serious professional army, but one that has not ever seen actual international military fighting action. However, they are currently fighting against extremely powerful drug cartels, and thus actually have much more battlefield experience than most other armies on earth! Loco!

¡RECURSOS! = RESOURCES!

The second major reason why Mexico should be rich: it is resource-richer. I already pointed out its agriculture richness, and that Mexico has guaranteed markets for its agricultural products in the US. Bananas, avocados, other fruits and vegetables come up from Mexico. We already talked about primary products and how they're not worth as much as others. Mexico's primary products are no exception. That's why relative location is important, because they have a pretty much guaranteed market for their agricultural products. Mexico also has things like tin, copper, iron, lead, zinc, magnesium—the list goes on and on, and they can sell every bit of it to manufacturers in the US. Guaranteed. The US can't get enough of that stuff. And if the US don't want it, then China will buy it! The one resource of particular consequence that everyone wants, referred to earlier, is oil.

A brief word about oil: Mexico is not a *major* producer of oil on the planet, but it is in the Anglo-American neighborhood. What's so great about that? Main reason why it's so great: the US is the biggest consumer of oil on the planet, both in terms of total volume and per capita. What better resource to have right beside the US than the one resource it uses the most? But it's a double-edged sword. Why? Oil has been a blessing because there's a guaranteed market for it in the US, every drop they produce can be sold, but it's been a curse because of the history lesson to follow . . .

1973 OIL EMBARGO

This is a largely forgotten yet major event in US history, even world history, because it has shaped economic and political relationships between the US and several different countries, including Mexico. Since we'll discuss it in more detail later, I'm going to keep this pretty simple for now. The US supported Israel in the Yom Kippur War in 1973. A bunch of the oil-producing Arab states got seriously ticked off as a result, and they decided, since they own the lion's share of the world's oil, to enact an embargo against the US and Western Europe. They stopped selling oil to the US. The spigot got turned off.

We'll use some hypothetical prices to easily demonstrate how this affected the US and Mexico. Let's say oil was 10 dollars a barrel before the embargo. In a matter of weeks, it skyrocketed to 40 dollars a barrel. Prices quadrupled almost overnight, causing tremendous shortages. You probably don't remember this because you were as yet unborn, but your parents might remember, and your grandparents certainly do. The shortages resulted in rationing, which was a huge deal. Unprecedented.

How does this relate back to Mexico? Mexico had oil, mostly offshore, but it wasn't worth a lot. It was of lower quality than Middle Eastern oil. There was much more refining that had to be done to Mexico's oil. Saudi Arabia could suck its oil out of the ground with a straw, dump it into a barrel, and let it go for 10 bucks. Mexico had to pull it out of the ground, clean it up, filter it, process it a bit, ultimately costing, let's say, about 15 dollars a barrel just to get it out the door. With these hypothetical prices, Mexico would lose 5 bucks per barrel. Needless to say, Mexico wasn't in a rush to lose money quenching the US's oil thirst, so it wasn't a big oil producer. However, when the price of oil blew up to 40 dollars a barrel—now there's a profit of 25 bucks on the oil that our friend Pepe just refined for 15! It's worth it now!

At this point, Mexico developed its oil industry as fast as it could and, go figure, the US was thrilled to see another source of non-Arab oil pop up. The US lent Mexico millions and millions of dollars to invest in its oil industry's infrastructure to grease the wheels, and everything looked peachy for a few years . . . until oil prices stabilized. After the embargo ended and the price of oil dropped back down, Mexico was still producing, but not nearly at the profit margin when it was 40 dollars a barrel. Having diminishing profits and millions in loans to pay back to the US resulted in a big economic crunch for Mexico—which came to a head about 20 years ago during the Clinton administration when Mexico was essentially going broke. Therefore, oil is a mixed blessing for Mexico: it has been good overall, but it put the country in tremendous short-term debt, from which it is still trying to recover.

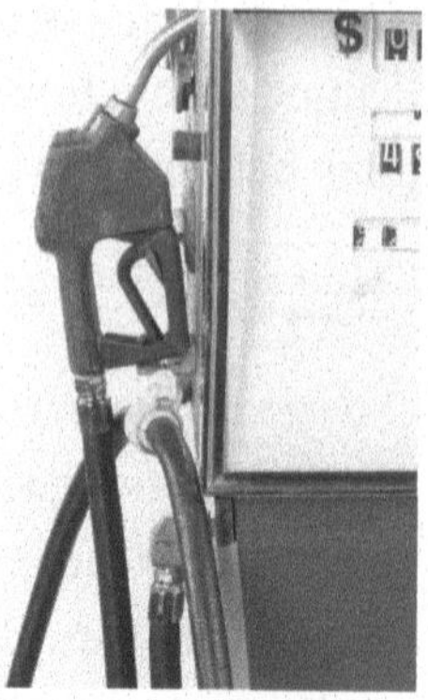

Mexico is resource-rich and can provide tons of raw materials. It also adds value to a lot of these things and sends them over in the form of TVs and cars, which are gobbled up in the US, their guaranteed market. It has one huge bonus for being next to the largest oil-consuming nation on the planet. Oil! So Mexico *should* be richer, but . . .

ANOTHER PLAID AVENGER INSIDE TIP: Every time the price of oil goes higher on the planet, watch the US president go and visit three different countries to discuss increasing oil production. The first is Mexico. The president will go down and get it to produce more, which will help resaturate the market and bring prices down. He'll also visit Norway and Russia. These are three big oil producing and exporting countries that are non-OPEC members. This means that they are not tied to the rules and regulations that govern prices or limits of production. They don't care, and the US loves them for it.

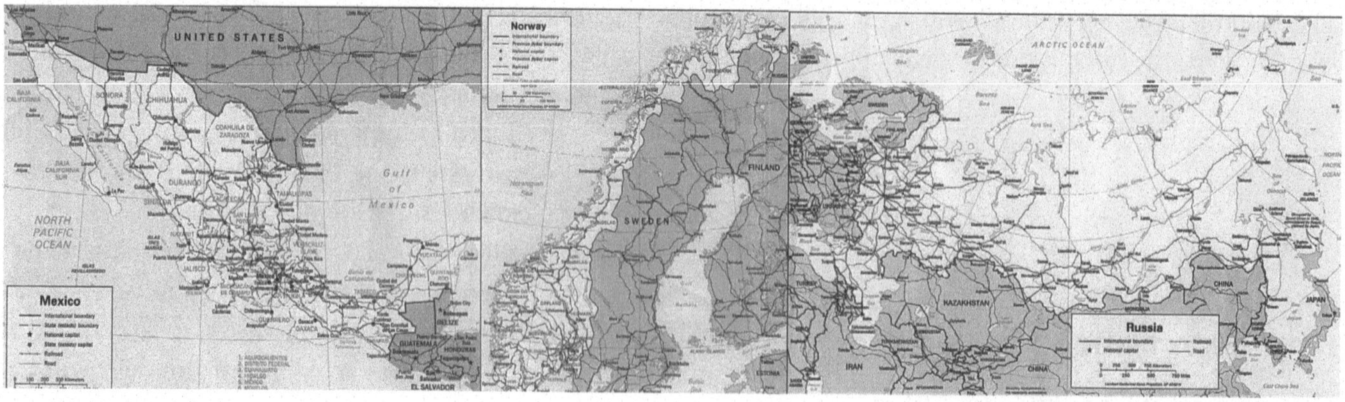

The Big Non-OPEC 3.

MEXICO SHOULD BE TOTALLY RICHER . . . BUT THEY'RE NOT!

Why? What happened? They have some prime real estate with great economic and military support, and they've got tons of resources . . . so what's the problem?

NUMBER ONE ¡LA GENTE!

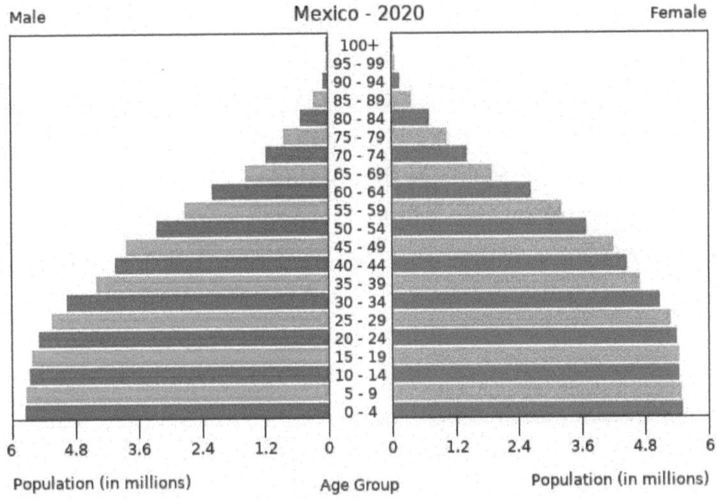

Wow! This pyramid looks like it's right out of Africa!

They have muchos, muchos people. You know about the demographic transition; you know what a developed country looks like. Take a look at Mexico's pyramid and you see something much more like Africa, Central Asia, and the Middle East. This is a population that is still *exploding*. It's a perfect pyramid; there's a high fertility rate, and people still have lots and lots of kids. While Mexico's economy has expanded and diversified and it has actually gotten richer, it has also added way more people. When you look at total amounts of economic gain in terms of GDP, it's grown. But the people have grown more, at a way higher rate than their economy. As a result, their per capita numbers have gone down. This is still a place that we consider kind of poor, per person.

This seems to be currently tapering off, stabilizing somewhat. The current population is about 115 million. That is a huge number. It is the population leader in Middle America with a population that is almost exactly one-third of the current US population. This is funny because it's about one-third of the land size of the US too. However, the economy is definitely not 1/3 of the US economy! That's where they're behind. So numerically, people are a big problem, and as the Plaid Avenger pointed out when we talked about Latin America, this is a place with massive amounts of urbanization. Most people live in the city. There is a draw, the idea that you are going to get a job in the city, which is often not the case.

Mexico City is now one of the largest cities on the planet, population-wise. It will probably approach the number one spot in the next five to ten years. There just aren't enough jobs for all those folks there. There are burgeoning urban populations, and on top of that, there is one other place where the Mexican population is increasing. Where? Along its border with the US. This leads to the number two reason why Mexico is not rich . . .

NUMBER TWO ¡MAQUILADORAS!

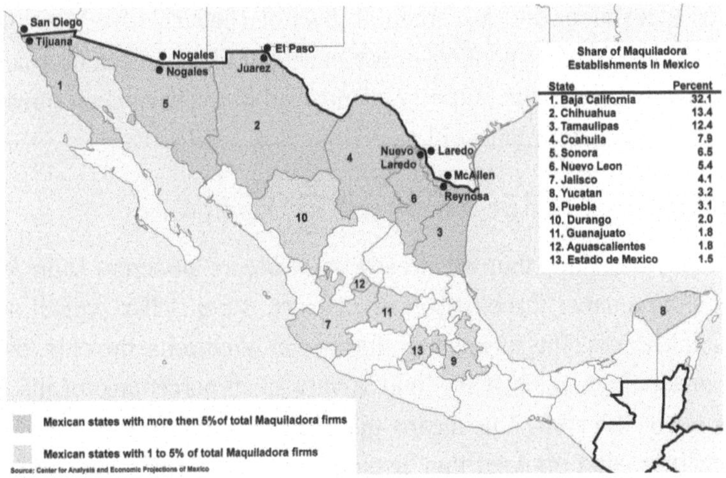

Maquiladoras are towns along the Mexico side of its border with the US, built around factories that are owned by foreign companies. There are some that have been there for a while, but many new border towns started popping up steadily after NAFTA was set up. NAFTA allows free movement of capital and business across the borders of its constituent states. The companies build factories just over the border to take advantage of cheap Mexican labor. Many unemployed Mexicans flock to these towns for jobs.

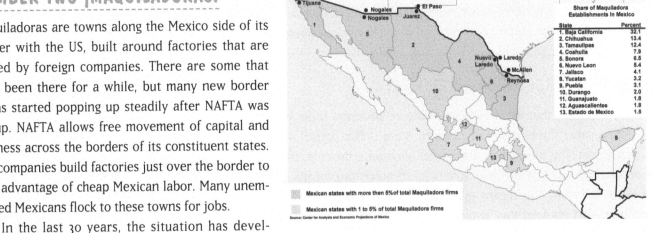

In the last 30 years, the situation has developed such that companies in the US, or another place, manufacture the individual parts, ship them to their maquiladoras just across the border where Mexicans working for less money assemble the parts into the retail version of whatever product the company sells. It allows the companies to pay some guy a nickel an hour or whatever to put your TV or your car or your stereo or your car stereo together. The labor is so cheap that it makes it worthwhile for the company to import the raw materials to make the parts and ship the parts down to be assembled in Mexico and then ship the finished product all over the US to be sold at a profit. Just for comparison, in 2000 the American minimum wage was about $5.25 per hour; the Mexican minimum wage was $3.50 *per day!* Know maquiladoras. Even though their growth has slowed, they still contain a large concentration of the Mexican population. The other big chunk of the Mexican population, as mentioned, is in Mexico City proper where the population is around 20 million and growing. This growth is based on the hope that people will find jobs.

This scenario is another double-edged sword for Mexico, like oil. Lots of folks down there would say, "Hey, don't complain; they're creating jobs and that's a good thing. We had nothing before, but now at least we have jobs, even though they don't pay squat." Many other folks would also point out the obvious; not only are the workers in the maquiladoras getting paid poorly, but all the true value of the product is made by taking advantage of the impoverished Mexicans working for wages so small they're practically slaves. From the Plaid perspective, there is something to be said for both sides of the argument.

Here's the real kicker: even though these "sweatshops" may suck, the loss of them may suck more. As labor becomes cheaper in China and Asia altogether, companies are starting to move out of Mexico because the labor is too expensive. Too expensive? That's just wrong!

SOME MAQUILADORA COMPANIES YOU MAY KNOW

- 3 Day Blinds
- 20th Century Plastics
- Acer Peripherals
- Bali Company, Inc.
- Bayer Corp./Medsep
- BMW
- Canon Business Machines
- Casio Manufacturing
- Chrysler
- Daewoo
- Eastman Kodak/Verbatim
- Eberhard-Faber
- Eli Lilly Corporation
- Ericsson
- Fisher Price
- Ford
- Foster Grant Corporation
- General Electric Company
- JVC
- GM
- Hasbro
- Hewlett Packard
- Hitachi Home Electronics
- Honda
- Honeywell, Inc.
- Hughes Aircraft
- Hyundai Precision America
- IBM
- Matsushita
- Mattel
- Maxell Corporation
- Mercedes Benz
- Mitsubishi Electronics Corp.
- Motorola
- Nissan
- Philips
- Pioneer Speakers
- Samsonite Corporation
- Samsung
- Sanyo North America
- Sony Electronics
- Tiffany
- Toshiba
- VW
- Xerox
- Zenith

2021 UPDATE: As China gets richer, and wages there rise as the burgeoning middle class demands more, Mexico is once again becoming competitive for US and other international companies for investment and job creation. Bueno! Their relative location combined with the big labor pool may once more put Mexico back on track for industrial and commodity output! And Mexican wages have been slowly improving too! Maquiladora game on!

NUMBER THREE ¡VIVA LA TIERRA!

Here's one thing that we already know occurs all across Latin America, and it's the number three reason why Mexico is not richer: wealth disparity and landlessness. This means that the top tier of citizens, the elite, own a disproportionate amount of the land. A very small percentage of the people own a lot of the stuff. This means that the great majority of people own nothing, not even the land they live on. Landlessness is a big issue across Latin America as a whole, and Mexico in particular has a long history of land issues . . . and has been a critical component of their history as a whole . . .

Mexico's story is replete with the important issue of land reform. What am I talking about here? Just in the last 100 years, we have something called the **Mexican Revolution**. It seems like every major country has some sort of revolution. We talked about the Communist Revolution in Russia, the Meiji Restoration in Japan, and soon you will read about the Iranian Revolution of 1979.

Emilio Zapata. Viva Zapata!

Pancho Villa with his men. (Dude at center-right with the glasses and the 'stasche.)

Mexico had its revolution in 1910. What was it about? It was about human rights and lots of other things, but its central concern was over land. When wealth disparity gets large enough, and wealth and power are concentrated in the hands of the few, the political system becomes unstable. When this happens, rage builds in the minds of the masses until revolution rears its ugly head, and the tinder box gets lit and it's game on. The two revolutionary leaders you see pictured here are classic examples of what happened in Mexico when wealth accumulated too much into too few hands. In the north, there was **Pancho Villa**, who was a rebel/revolutionary leader, essentially a bandit type with sort of a bad reputation. He took up arms against the government as things became too corrupt and land was controlled by too few in the early 20th century.

Villa's counterpart in the south was **Emilio Zapata**. A poor farmer, all around good guy, a charismatic hero, Zapata took up the cause and got the word out to people that the government was screwing them over. "Look at us, amigos, we are poor as dirt. None of us have land! We need to overthrow the government and make this a better place!" There were two guys on both geographical extremes of Mexico, who ended up overthrowing the government. The central point here is that it was all over land, all about the issue of landlessness. Zapata was the real foundation of this, since he was a farmer himself. It wasn't a political thing where he wanted to rule or be president; he just wanted land for everyone so they could all support themselves and each other. He wanted it to be fairly distributed. Why is this important? Because land is *still* a major theme in Mexican government and politics today.

How did land ownership in pre-revolution Mexican society get so top-heavy? Under the Aztec system, society was strictly hierarchical, a sort of military dictatorship almost. All the power was concentrated around the top. The head Aztec honcho basically owned and controlled everything; all land and all wealth was subject to his will. When the conquistadors arrived, under Cortés' leadership, they had an easy time taking over because they just lopped off Montezuma's head and put themselves in his place at the top. So the system didn't get altered much when it changed hands from the indigenous rulers to the Spanish. Even when Mexico had its revolution and kicked out the Spanish, they kept an oligarchic system where there were a few ex-Spaniards that had control and maintained power over large amounts of the country. The wealthy hung on to all their resources and the poor got nothing. This arrangement stayed in place until the 1910 revolution.

The **Mexican Revolution** was a pretty big deal. It happened in 1910, and when it was successful, they wrote a constitution so awesome that it almost shamed the US's. It was an extremely thorough document all about human rights, access to land, and many other important issues. But in many ways, this has been an unrealized dream in Mexico because some of the great stuff in the constitution never really came to be. It's a far-reaching document, and its promises are still being pursued in today's world.

Like I said, this land thing is not just a historical remnant: all political parties in Mexico today have to deal with the land issue. You can't get away from it. Both ends of the political spectrum in Mexico use land reform as a major element of their campaign platforms. Even though some gains have been made, land is still a major concern in Mexican politics, and in the general public as well which can be seen in the current activity of some rebel groups that are keeping the revolution alive. . . .

¡REBELS!

There are still rebels, and they are still fired up about land reform. They are fighting actively for the same things as revolutionaries were 100 years ago. **Subcomandante Marcos** leads the **Zapatistas**, named after Emilio Zapata. These guys are in the southern Mexican state of Chiapas, which is in the tropical jungle highlands near the border of Guatemala. This revolutionary movement has been going on for two decades. What are these dudes still fighting for? You guessed it! Land reform!

The pipe-smoking, ski-masked Subcomandante!

Interesting to note, the Zapatistas are one of *the* first, if not the first, revolutionary movement to have an official website: Zapatistarevolution.com. This link is actually a website posted by a trial lawyer in Texas who is sympathetic to the Zapatista cause, and who visited with them during a trip to Mexico. He wrote a fictionalized novel based on the facts of a 1994 Zapatista uprising. The site isn't maintained by the Zapatistas themselves, though a lot of their plight is described on it. They are a truly unique rebel/terrorist organization in this respect—the first online rebels.

We know Mexico is not richer for a few key reasons: wealth disparity, land issues, the double-edged sword of oil, the mixed blessing of maquiladoras, and population dynamics. What about their future?

¡EL MAÑANA DE MEXICO!

The Plaid Avenger used to say that things are liberalizing, opening up, that perhaps NAFTA is going to be further entrenched and that Mexico will become a more fully integrated partner with the Anglo-American region in terms of movement of people and economic activity. That was the case up until the US's War on Terror began, which has served to stymie movement between the US and Mexico as they try to catch potential terrorists. It has recently been compounded even more as illegal immigration has become a hot button election issue in the US., worsened still more by the current economic recession which makes Americans think that immigrants are "taking our jobs." The current political mood in the US right now is to build an even bigger wall between these countries/regions. **2021 UPDATE:** Well, those words above were written a decade ago, and they were right on point. The US did indeed elect President Donald Trump and unless you were in a loco coma from 2016 to 2020, you know that Trump rode a wave of anti-immigrant and protectionist fervor into office. Of significant note to this chapter is his pledge to "deport all the illegal aliens back home" as well as build a "Great Wall of Mexico between the two countries,": both issues that still have resounding support of about half of American voters who want Trump's MAGA priorities to continue.

During his tenure, the Trump team ramped up the rhetoric, increased deportations, and threatened to shut down the entire border unless Mexico does more to stop the flow of refugees and immigrants. US/Mexican relations hit an all-time low, as the Trump administration seriously shook-up intercontinental trade and the movement of people in a Huuuuuge way! One complication that has arisen because of this shift is that the US needs immigration, and everybody knows it. Business people in the US know this: They need immigration for economic growth to continue. Immigrants fill many service sector and manufacturing jobs that some Americans either won't or don't want to do. So this is not a clear cut issue for anyone, on either side of the border, and we will continue to follow developments closely, as it certainly will continue to be a major issue during the Biden presidency, as it will perpetually be a major divisive issue between the Republicans and the Democrats in the US. And immigration is always a major issue for Mexican leaders, of which I want to give you the modern history of right now so you can fully appreciate the present situation . . .

Previous Mexican President Felipe Calderón (in office 2006–2012) was the political party equivalent of a Republican in the US. As such, he was a pretty staunch supporter of former President George W. Bush, agreeing with Bush on all policies minus one: immigration. Calderón and others expressed many reservations at the idea of building a wall between the countries. And that was over a decade ago! See, you thought this Donald Trump wall stuff was a new thing! Nope, been debated for decades. Mexico is more of the European frame of mind, which is essentially, "Aren't we in a free trade block? Aren't we allies? Aren't we friends? Why do we even have a fence at all?" Mexico is working towards further economic integration with the US, including a "worker passport" system that would indeed allow for more open movement of folks across the entire continent. However, opening trade borders that allow the free flow of goods and people can also allow the free flow of lots of other things . . . like drugs or terrorists. So a bunch of folks want more open borders, and a bunch of other folks want a closed border with a big wall across it.

Former President Felipe started the fight to fix border fiasco.

And the issue has progressed no further since Calderon. The most recent ex-President Enrique Peña Nieto (in office 2012-2018) was a center-left guy much like US President Obama, and those two (along with former Canadian Prime Minister Stephen Harper) had a NAFTA get-together in 2014 . . . in which all 3 men agreed to deepen economic ties further and work together to make all 3 countries richer. But not a word was whispered about the immigration issue . . .

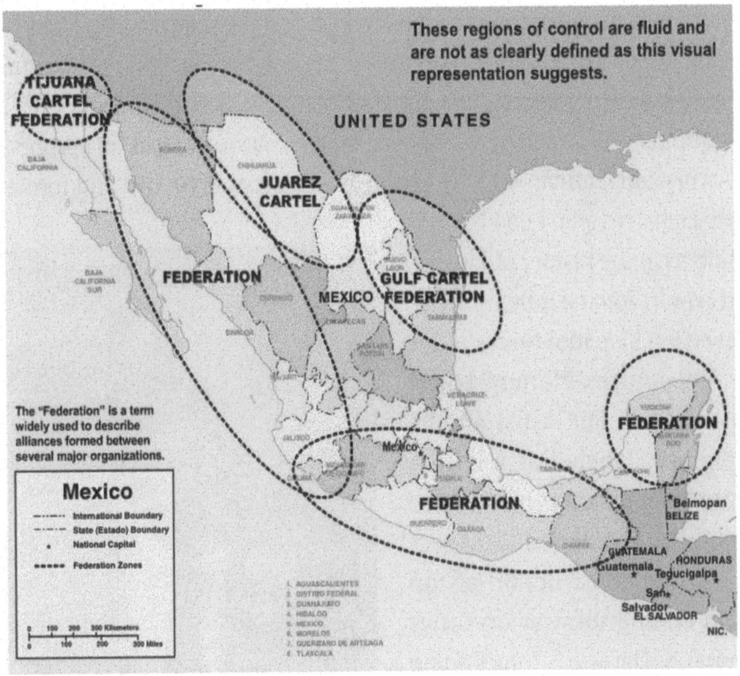

Source: U.S. Drug Enforcement Administration, adapted by CRS (P. McGrath 3/2/2007)

Heck, it could not even be referenced at the public meeting for fear of offending the fervently anti-immigration fans north of the border. In fact, political gridlock in the US has crushed any hope or dream of immigration reform for the moment, even though both the Bush and Obama administrations (along with business interests) actually wanted some sensible change in the system to allow for free-er movement of peeps for economic purposes. It would seem to make sense since all three are huge trading partners with each other, and allies, but politics don't often make sense. Like mostly never does. And then came Trump. Wall on!

Wall or no wall, holy hot tamale, be forewarned about this drug situation, my friends! This is the hottest—and by far most serious—cause for concern for Mexico, the US, and everybody in between. Drugs, drug crime, and gang-related crime, all of which have been infiltrating the border region, have simply exploded in the last decade, in the worst ways possible. The volume and value of all that coke, dope, amphetamines, and now even heroin crossing into the US via Mexico has become so huge that organized crime has taken over, but the competition is so fierce that multiple syndicates/ gangs are fighting to the death in turf wars all up and down the border.

We're talking about some nasty business here, my friends. Gangland kidnappings and assassinations of Mexican police, politicians and even army soldiers. Whole villages of innocents being slaughtered in turf wars. Mass graves being found. Human heads being thrown into high school prom dances to terrorize towns. I wish I were exaggerating here. These criminal gangs have grown so powerful that they don't run from the police: they stand and fight them. And they fight the army. And the DEA. The Mexican cartels of the 21st century are being compared to the old Colombian cartels of the 1980's in terms of power, wealth, organization and bloodlust. It is scary.

Go out and research it yourself to see the current situation. Just know this for now: this insane gang violence being brought directly to the doorstep of the US will probably be the straw that breaks the camel's back. Add the new drug wars to the already festering illegal immigration issue, and you can easily predict that more calls will be made for a "Great Wall of China" to be built on the border. Former Mexican President Calderón, being a conservative law-and-order guy himself, started an aggressive "War on the Cartels" back in 2006, which turned the mess into a public bloodbath responsible for over 50,000 Mexican deaths just while he was in office. The move got results: lots of arrests and seizures, and a centralization of the fight via the army and federal control. But it also intensified the violence of cartels against each other, and cartel violence against the general public in a sick attempt to truly terrorize the country into submission. Some fully supported the government fighting the cartels head-on, others believe the violence against the public has been too high a price to pay and desire alternative tactics. What do you think should be done?

Former President Peña pursued a more placating policy?

The next (now former) President Peña came to power on the platform of doing whatever it would take to decrease the violence against civilians . . . but what that meant for actual policy shifts was a little shady. Of course he had to keep fighting the good fight, because you can't just let criminal elements have free reign in your sovereign state, but he moved the fight from the federal to the local level, while simultaneously trying to clean up police corruption within the entire system, which was another major overhaul of the system that both Presidents have focused on. And the USA sure wants them to succeed, as the repercussions of this trillion dollar trade in drugs affects them directly as well.

The net result of ex-President Peña's policies? Not much to nada to negative: corruption, crime, and the drug trade in Mexico/ the US all worsened during his time in office. If you want some really gritty details, go look up his botched handling of the 2014 Iguala mass kidnapping/massacre in which local police in collusion with organized crime lords kidnapped/killed 43 students from a teachers college. . . at least 80 suspects have been arrested in the case, of which 44 were police officers. Another stain on his presidency was the 2015 escape from federal prison of Joaquín Archivaldo Guzmán (aka "El Chapo"), a Mexican drug lord and former leader of the Sinaloa Cartel, an international crime syndicate. Embarrassing. Peña Nieto is seen as one of the most controversial and least popular presidents in the history of Mexico, and left office with perhaps only a 12% approval rating. . . and now in 2019 he is being investigated for corruption himself by the USA, and then his wife even left him, too! No bueno!

The current (2018-now) President of Mexico is one Andrés Manuel López Obrador . . . a name so long that he goes by the acronym AMLO. For realz! And he is center-left to full on left-wing and could probably be best labeled as a leftist-populist, as his stated goals while in office include increasing pension benefits to the retired, increasing minimum wage, increasing subsidies to the agricultural sector, granting amnesty for non-violent drug criminals, starting construction of 100 universities and universal access to public colleges, and ending the war on drugs and promoting the legalization of some drugs like marijuana. Dang . . . that's some pretty classic liberal-left stuff! And particularly when it comes to the drug war, it

AMLO in command-O

is an entirely new approach, but it is too early to tell if any of these things will actually get done or will actually work if enacted. Remember, AMLO may be a leftie, but he still has to deal with the behemoth to the north. . . the (currently) anti-immigrant, anti-legalization, pro-wall building, right-wing US President Trump. The Mexican/US border issue will dominate the headlines and focus of AMLO's presidency, whether he wants it to or not.

Here in the first quarter of the 21st century, these two countries are strong allies by all standards, but they are divided on the issue of free movement of people between them. Will Mexico ever be fully developed? It is entirely possible once the population stabilizes, its economy becomes a bit more robust, and they reclaim their territories from the criminal syndicates. New competition in cheap labor from China and Asia in general had pulled lots of business, jobs and investment away from Mexico in the last decade, but a lot of that is now starting to return since Asian wages are rising, and there is significant cost savings being so close to the American market. Mexico's middle class is also on the rise, which boosts domestic consumption and job creation, so a positive upswing cycle is underway. Need more proof of Mexico's rising status? Immigration to the US is actually decreasing, not because the US is stopping them at the border, but because more and more Mexicans are making more and more money at home! Heck, the rate of return of Mexicans currently already in the USA is rising too! They are heading back to increased opportunities south of the border! And go Google Carlos Slim: A Mexican businessman who was ranked as the richest man in the world from 2010–2013. Dude! That's real money, and real opportunity down there! Not to mention Mexico is a top 20 world economy, a rising star for international investment, and a pillar of democratic stability in the Americas! Arriba, arriba! Mexico has got it going on! If only those drug cartels could be corralled, their potential for fully developed status would be cemented! **2021 UPDATE:** Or a wall will be built around them, stifling interaction and trade, and bottle-necking up tens of thousands of immigrants that actually flow from Central America through Mexico every year on their quest to make it to the United States. Yeah. . . a Great Wall of Mexico ain't gonna be so great for them at all. . . so we will follow the developments between these world regions closely in the coming years to see how the Trump Bump and its aftershocks affect this neighborhood, for bad or good! Ya veremos!!!

Diggin' the Diego: Mucho awesome Mexico City mural by Diego Rivera.

MEXICO RUNDOWN & RESOURCES

View additional Plaid Avenger resources for this region at http://plaid.at/mex

BIG PLUSES

- Next door neighbors with the largest economy on the planet
- Gigantic US market buys every single thing Mexico produces; an assured market
- Significant producer of oil, agricultural commodities, and basic manufactured goods
- With the 14th biggest economy in the world, Mexico is now a member of the G-20
- Due mostly to drug cartel wars, Mexico has grown its military significantly in size, technologic savvy, and intelligence collection.
- Home to one of the richest people on the planet: Carlos Slim. Got to be a reason for that, right?

BIG PROBLEMS

- Ballooning population, although stabilizing soon
- Serious wealth disparity, and landlessness remain chronic political issues
- Significant producer and transportation corridor of illegal narcotics
- Has several drug cartels which are possibly more powerful than the government
- A next door neighbor (the US) which increasingly doesn't want them to come visit as immigration tensions mount: a Trump-induced wall would radically alter trade and relations between the two countries
- Has major immigration issues of its own: tens of thousands of Central Americans pour through (and get stuck in) Mexico every year on their quest to get into the USA
- Losing jobs/industries to even cheaper labor market in Asia (this is now changing fast)
- Cultural inferiority complex permeates the state

DEAD DUDES OF NOTE:

Montezuma: Last head honcho and ruler of the mighty Aztec Empire; the empire reached its maximum size and power under his reign from 1502–1520. Unfortunately, the Spanish conquistadors showed up and destroyed him and the Empire. Now you can get nasty cheap tequila appropriately named Montezuma's Revenge . . . he is getting the white man back, one shot-glass at a time.

Hernándo Cortés: The white dude the led the Spanish conquistadors into Mexico to level the Aztec Empire in just a couple of years, and was the first wave of colonizers to lay claim to vast areas of the entire hemisphere in the name of the Spanish crown.

Pancho Villa: Originally a bandit on the run from the law for years (and even eluding US troops), Villa eventually became famous as one of the leading generals in the Mexican Revolution of 1910. He led the campaign in the north to overthrow the autocracy, redistribute land, and fight for human rights. A "Robin Hood" of northern Mexico.

Emiliano Zapata: Revolutionary leader of southern Mexico during the Mexican Revolution, his battle cry was "Tierra y Libertad" (Land and Liberty). Fought for land reform and emancipation of peasants. Zapata is considered one of the outstanding national heroes to this day. Even has a modern day rebel/terrorist group named for him: the "Zapatistans" of Chiapas state.

PLAID CINEMA SELECTION:

Sin Nombre (2009) Mexican-American action/adventure/crime/thriller film about a Honduran girl trying to immigrate to the USA, and a boy caught up in the violence of gang life who also needs to escape. Whoa. This one is super intense, violent, gritty and about as real a take on the uber-violent MS-13 gang as it gets. And it is a travel film that crosses the region. A perfect world regions film in all respects, and students love it.

Tijuana Makes Me Happy (2006) An independent film, shot almost entirely in Tijuana, with a few scenes in San Diego. Takes an interesting look at the seedy world of cockfighting, and its ability to consume the lives of its participants through the filter of a little boy who aspires to purchase his own cockfighting rooster.

Y tu Mamá También (2001) One of the more well-known Mexican films of all time, has drawn the ire of many for its depiction of sexuality. Follows two adolescent boys as they take a road trip through the modern-day Mexican countryside with a woman in her mid-twenties. Provides an outstanding cross section of landscape and Mexican social classes in a modern context. Warning: explicit sex scenes!

Frida (2002) For the artistic types out there, this is a biography of Mexican artist Frida Kahlo, who channeled the pain of a crippling injury and her tempestuous marriage to Diego Rivera (another famous Mexican artist) into her work. Features depiction of Mexico City in early 20th century, has the famous commie Leon Trotsky as a character, and stars Salma Hayek. What more do you need?

Apocalypto (2006) Against my better anti-Mel Gibson judgment, let's throw in one that pre-dates Mexico in its entirety. An instant Mayan Empire classic! Well, at least it has killer landscape/biome backgrounds, as it is shot on location in southern Mexico in the states of Veracruz and Oaxaca.

Viva Zapata! (1957) Yeah, I know this is an ancient movie, but it does star a young Marlon Brando long before he got fattened up for the Godfather role. And if you want to understand Mexico, you gotsta know Zapata. The story of Mexican revolutionary Emiliano Zapata, who led a rebellion against the corrupt, oppressive dictatorship of president Porfirio Diaz in the early 20th century.

The WIld Bunch (1969) Okay, this one is a stretch, but I love it: an aging group of outlaws as they escape to Mexico looking for one last big score as the "traditional" American West is disappearing around them. Filmed entirely in Durango and Coahuíla in central Mexico, and set during the Mexican Revolution, so historical clues abound. Was the bloodiest, most violent film ever made when it was released. One of the best Westerns of all time.

LIVE LEADERS YOU SHOULD KNOW:

Felipe Calderón: Former President of Mexico (2006-12), center-right, well-educated, conservative, strong-willed, pro-US leader who was up to his armpits trying to diminish the power of the drug cartels that have turned the US/Mexican border into a bloodbath. And that was a war that he initiated in 2006.

Enrique Peña Nieto: Former President of Mexico (2012-18), center-left, liberal who tried re-vamping the Mexican economy and society in surprisingly fiscally conservative ways (privatization, breaking teacher unions). Left office wildly unpopular.

Andrés Manuel López Obrador (AMLO): President of Mexico since 2018, center-left to left-wing social progressive. Will have to walk a delicate tight-rope between implementing his policies while appeasing his anti-immigration neighbor to the north.

Carlos Slim: Mexican business magnate, investor, and philanthropist. From 2010 to 2013 was richest man in the world. Pretty much dominates telecommunications marker in Latin America.

MAP GALLERY

View Map Gallery & additional Plaid Avenger resources for this region at http://plaid.at/mex

CHAPTER OUTLINE

Central America

15

CENTRAL America is part of Middle America, as we've pointed out in an earlier chapter: an easily defined place because it can be distinguished by just looking at political borders. It goes from the Mexican-Guatemalan border all the way down to the Panamanian-Colombian border and encompasses the countries of Guatemala, Belize, Honduras, El Salvador, Costa Rica, Nicaragua, and Panama: the Central American seven. The Plaid Avenger can best summarize this region, unlike any other region on the planet, in a single word: **VIOLENCE**. This is an incredibly violent arena, in pretty much every way imaginable. Well, Costa Rica is actually really chill, and Panama is now prospering, but the rest of the region has had more than its share of problems, many of which perniciously persist to this day.

Not the least of which is that Central America is a place that, while not that far from the economic powerhouse of the US, is one of the poorer regions on the planet. Poverty is, of course, one reason for all the violence, but not the only one. There is a lot going on here, but not a lot to be happy about, quite frankly. The Plaid Avenger is fairly sure that most people in Central America would agree, and hopefully me bringing up these tumultuous and troubling themes of the region won't offend any of the citizens of these states, but you can't understand what's going on here without focusing somewhat on the violence. Why is it such a violent region?

PHYSICALLY VIOLENT

To start, even the physical world of Central America contains massive amounts of violence. If you look at a map of plate tectonics, you see several plates coming together in Central America causing a zone of geologic activity. The Cocos plate is ramming against the North American and Caribbean plates. It's actually getting shoved under the Caribbean plate, causing savage scores of volcanic and seismic activity all over the Central American isthmus. Earthquakes and volcanic activity abound. Looking ahead to the next chapter, many islands and a whole island chain of the Caribbean region were actually formed from the same tectonic forces I am describing here, so keep this related physical geography fact in mind as you peruse the proceeding pages.

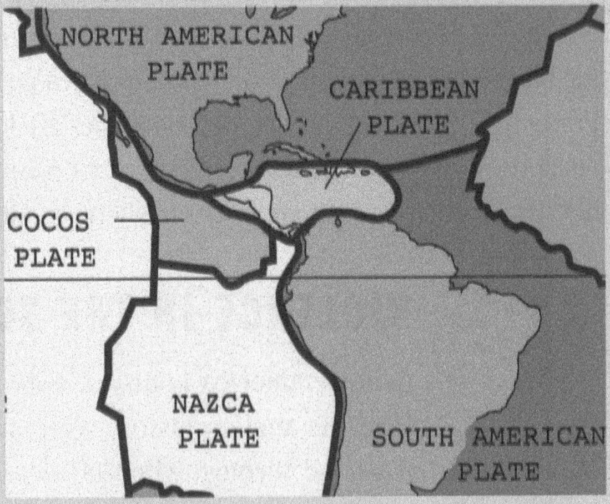

The isthmus here is a really violent arena, simply considering its geology. Does that have anything to do with why it's poor? Or why it's a hotspot of conflict? It surely does! We like to think that the physical world doesn't affect people's lives too much, but as we have seen with the other regions we've talked about, that's simply not the case. It's hard for people and businesses to move, adapt, or build structures to withstand frequent earthquakes and hurricanes. And these countries definitely don't have the resources to control the climate or the physical environment around them. The Plaid Position on Central America is that nature really dominates the people here to a higher degree than in other regions. This cycle of physical violence from the natural world has shaped this region forever, and will continue to do so!

Still smokin' San Cristobal Volcano in Nicaragua

Let's get to some specifics. Here's an interesting story about the Panama Canal, which was originally supposed to go through Nicaragua. Some senators from the US were planning to build a canal 100 years ago, but they didn't know where just yet. Where was the US going to invest its money? Plans to build in Nicaragua were already drawn up, and they were nearly ready to mobilize. But some senator's business interest paid him off to lobby for their plans to go through Panama. It's tough to scrap plans that are in their final phases of development, but the senator pulled it off by using a postage stamp from Nicaragua with a picture of a volcano on it. He made copies and passed it around the senate, and said, "Check this out, chaps, we'd be fools to invest in a place full of volcanoes!" And thus, it's the Panama Canal now, not the Nicaraguan Canal, thanks to a savvy, stamp-holding senator. But check this out: A Chinese company is making plans to now build that Nicaraguan Canal so as to supercede the US-influenced Panamanian passage. My my my how history comes back to haunt!

That may be just a cheesy side story, but on top of the seismic and volcanic violence in this region, there is also the weather to contend with. The tropical climate of this region, bordering the Caribbean, is prone to hurricanes. These are violent, devastating weather systems that come through this region with regularity, every year, sometimes multiple times during a season. And even mudslides due to heavy seasonal rains are also prolific problems in the region. Even the mud gets you!

Costa Rican road closure: mud.

Hey, Plaid Avenger, does that have anything to do with this region's poverty problem? Heck yes it does! How many financial investors think it's a good idea to build a multimillion dollar hotel or factory that may be completely destroyed in a hurricane or earthquake? Not too many. Nobody is that rich, or nobody who's that rich is that dumb. Foreign direct investment, that is companies investing cash and building businesses, is fairly low in Central America partly because of the physical world. Pretty much only fruit production companies are heavily invested here, because how much damage can your banana trees sustain during an earthquake? It's a pretty safe bet. The physical world complicates life here, and it will continue to do so.

Beautiful beach-front villas, post-hurricane

VIOLENCE IN THE SOCIAL SPHERE

Central America is also a very violent arena from the standpoint of the people. Central America shares its ancient history with Mexico, so we won't go into that too much. It was part of Mexico during the colonial period through when Mexico became independent in 1821. In 1823, Central America seceded from Mexico and formed a union of independent states. By 1838, even their union started falling apart and the devolution doesn't stop until they end up as seven dinky independent territories. They are all very small countries now.

And as the Plaid Avenger always says, "Size *does* matter!" Since they are small they don't really have any sort of big world role, nor do they have a lot of resources, individually.

Their experimental Central American "Republic" lasted from 1823 to 1838, when it began to disintegrate due to civil wars. Apparently, they had so much fun with their early civil wars, that they continued to have more of them for the next 150 years! Born in wars, grown up in wars, and still living with the aftermath of wars . . . this is just not the best real estate in the Americas. They just couldn't seem to hold a union together. Too bad; to be a singular political entity may have helped them a lot in the long run. Instead, they ended up being a bunch of small bananas, with limited resources and political power, in a much bigger world. Pun intended. What pun you ask? Just this:

The short-lived "Captaincy General of Central America" state.

Bananas, fruit in general, became the main source for revenue (and not much at that) as well as conflict (much more than revenue). Now, thinking way back to the chapter on economic activities, you know full well that if your country's #1 money-maker is exporting bananas, then you are making peanuts. But one way you could make more as a society is to organize your labor on those banana plantations to get a better wage at least. There were several attempts in the 1800s to unionize Central American fruit labor; it all pretty much ended with the union dudes dying in battle or getting executed later for political reasons. The region became synonymous with a political term, as well . . . **banana republics**. Hit up the box below for the full story.

What's the Deal with Banana Republics?

A "banana republic" is a small nation south of the border that is plagued by political corruption and is economically dependent on basic agricultural exports. The term was coined in reference to Honduras, which at the time, was basically owned by US multinational fruit companies. Although the terminology sounds cute, it is also highly disparaging—implying a backwardness that can only lend itself to producing fruit for rich Americans. Leadership typically consisted of a very small, wealthy, corrupt elite invariably topped off by an overbearing generalissimo type in a uniform covered with medals and badges. Even worse than being under the domination of such an oligarchy, "banana republics" were/are typically under the domination of US multinational organizations. So while we love a good ice cream banana split, consider how bananas split these societies in much more callous ways: A Banana Republic is a politically unstable country whose economy is largely (or totally) dependent on exporting a low-value raw commodity. This creates social classes that are strictly stratified: there is a large impoverished (and landless) working class ruled by a super small elite of political, military, and business heads in collusion with each other, and almost always a powerful foreign company. Hmmm. . . kind of sounds like a sovereign state that really just serves as a huge slave work camp. Yep. It ain't far from it.

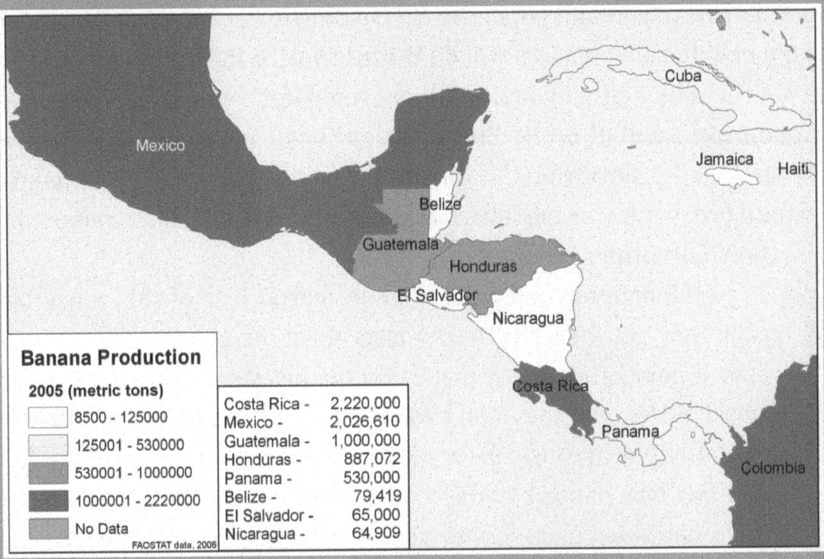

ECONOMIC VIOLENCE

In these places, and this even goes back into the 1820s when these guys were declaring independence, there were foreign multinational companies doing business there. It's not just a current phenomenon. Many of the folks who still own the resources in Central America aren't even citizens of the countries themselves but are these exact same multinational companies. Gee, I wonder where these companies are from . . . You guessed right! Predominately the good ol' US of A. You may be asking yourself, "What do these people make down in Central America that anybody wants?" Well, the bananas you had for breakfast probably came from there. Coconuts, starfruit, grapefruit, oranges, chocolate, most tropical fruits, coffee . . . all this is from somewhere south of the US-Mexican border, and a lot of it comes from Central America. This has been the case for several hundred years. Who brings all of these goods to America? The US/European companies that established the markets.

Some of our favorite tropical commodities.

When impoverished countries start to have leftward leanings—leaning towards socialism or communism—one of the first things the people start asking themselves is, "Why does this American company own 100 percent of the banana market in our country? We produce all the bananas, and we don't get squat for them. This company takes them from here, where they were grown with our cheap labor, pays us a penny for them, and gets a dollar for the same fruit in the US. Dudes! This is some bitter fruit to eat!" Case in point: United Fruit Company was so huge they controlled virtually all the fruit in Central America for about a hundred years. Check out the box about this on the next page.

There is a big part of the history of this region involving the US protecting its own economic interests. In many of these countries, when they have a significant leftward lean and appear to be heading towards a socialist or communist revolution, one of the first things that the new socialist leaders will do is **nationalize** the industries. As I said before, this is essentially taking an asset that was held by a private multinational company, and saying, "Hey, we've been doing all the work, and you pay us jack and make a ton of profit. But this is *our* land! We're taking it back and we're going to nationalize it. Thanks for setting up shop, for setting up this oil platform, or copper mines, or whatever, thanks; we're going to take it from here and make more money for ourselves. You may now leave." This is nationalization. Corporations do not dig it.

"Help US Army! Save the banana train!"

Multinational companies live in mortal fear of this, since basically they get their stuff taken from them. As you can imagine, they have a point. They spent money on the investments there, they have stockholders, and they don't want their livelihoods taken away. They get bothered by this. History shows us that when a company fears that this may happen to them, or if it actually happens to them (and it has happened plenty), they go to the US government and say, "Hey, we're US citizens. Those are US investment dollars down there; that's US *property* down there! You need to help protect US property!" By and large, over the last 150 years, that's exactly what the US government has done, especially in Central America. When a government like

What's the Deal with The United Fruit Company?

The United Fruit Company (UFCO) was an American corporation founded in 1899 that was dedicated to providing the American consumer with delicious and affordable bananas and pineapples. The United Fruit Company is also one of the best examples of capitalism gone awry. Throughout the 20th century, the United Fruit Company maintained a virtual monopoly on banana exportation to the United States. UFCO owned huge plantations in Columbia, Costa Rica, Cuba, Jamaica, Guatemala, Nicaragua, Panama, and Santo Domingo. They also owned 11 ships named "The Great White Fleet" and miles of railroads. United Fruit was the unofficial king of Latin America. Like many corrupt kings, United Fruit became a giant thug to its subjects.

The worst offenses by United Fruit occurred in Guatemala. As in many other countries, United Fruit owned or leased much of the farmable land in Guatemala. They also left a significant portion of their land barren (not producing fruit) just in case something happened to their good land. All this wasted farmable land started ticking off hungry people in Guatemala. In 1951, Guatemalans elected, in their first ever election with universal suffrage, Jacobo Árbenz Guzmán, who pledged land reform. Árbenz proposed redistributing some of the unused land to the large percentage of the population that was landless.

The United Fruit Company was royally miffed about Árbenz, so it launched a campaign to convince its friends in the US Government (Dwight Eisenhower, John Foster Dulles, etc.) that Árbenz was aligned with the Soviet Union. UFCO even went so far as to produce a video titled, "Why the Kremlin Hates Bananas," documenting UFCO's fight against Árbenz. The result of UFCO's propaganda and influence was Operation PBSUCCESS: a CIA covert operation to overthrow Guatemalan president Árbenz. The CIA trained and armed an ad-hoc "liberation army" and organized "strategic" air strikes on Guatemala City. The result of CIA action was years of oppression for Guatemalan citizens under a right-wing dictatorship that "saved the day" from the communist threat. On the bright side, Americans can, to this day, buy bananas for less than a buck a pound—thanks partially to Chiquita Brands International, which took over United Fruit in 1970. Thanks guys! You wacky bunch of bananas!

Nicaragua is taken over by popular uprising, and they want to nationalize the banana industry, suddenly, the US Marines show up to save the day! Saved for who? The US corporation is saved and the poor people who slave for it get to work another day for the man.

Some of you may stop reading this right now because you think I'm just talking communist gibberish. That's fine. I can't please everybody, but I'm not going to ignore it or pretend it didn't happen (or still happens) because you have to hear about this stuff in order to understand the region and its problems. This is all part of the public record. Just because the history books you read in high school didn't teach you this stuff doesn't mean it never happened. There were several instances when the US military was sent to quash uprisings and attempts at nationalization, dating back to the 1850s when it looked like some Panamanians would endanger the railroad down there, and maybe nationalize it. The US did not want this because that railroad was a critical part of a much bigger shipping network between the east and west coasts of the US. The US sent troops several more times to protect the railroad and the Canal . . . and that was just in Panama.

You typically think of US military services as guarding the US, as protecting US citizens. But by and large, in this part of the world, they have done none of that. They have protected the private interests of US businessmen. A lot of folks, including the Plaid Avenger, speak out against this because it's basically leasing out the US Army to protect some rich guy's stuff. That bothers me a little bit. It bothers people in Central America even more because they are the ones who catch the bullets from the US military when these sorts of shenanigans go down.

OH, DID YOU THINK I WAS JUST MAKING THIS STUFF UP?

For those that doubt my good educational intentions, or believe I am overexaggerating the impact of Anglo-America on this region, I offer an abbreviated list of things over just a 35-year period for you to Google at will:

US Marine Interventions in the Caribbean Basin from 1899-1934

- → Cuba: 1898, 1906-09, 1912, 1917-33
- → Puerto Rico: 1898-present
- → Nicaragua: 1898, 1899, 1910, 1912-25, 1926-33
- → Colombia: 1902, 1904, 1912, 1913-14
- → Honduras: 1903, 1907, 1911, 1912, 1919, 1924, 1925
- → Dominican Republic: 1903, 1904, 1914, 1916-24
- → Haiti: 1914, 1915-34
- → Mexico: 1913, 1914-17, 1918-19
- → Guatemala: 1920
- → Panama: 1921, 1925

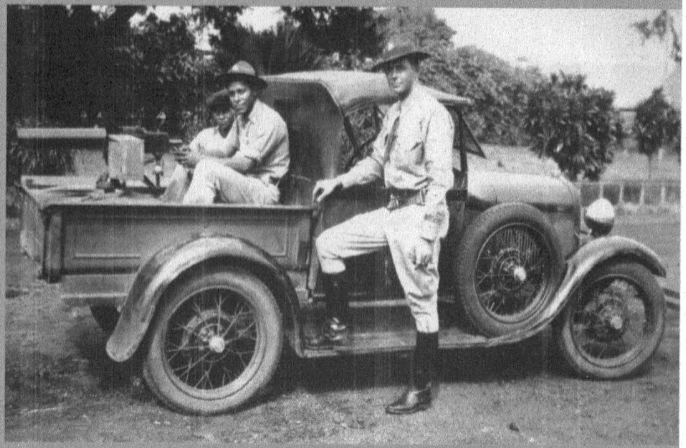

Whew! Low banana prices are saved again! Circa 1930s Nicaragua.

These were all part of a grander campaign, now referred to as the "Banana Wars," encompassing all the occupations, police actions, and interventions on the part of the United States in Central America and the Caribbean between the end of the Spanish-American War in 1898 and the inception of the Good Neighbor Policy in 1934. Good times, good neighbor!

There's a history of violence since their independence based on the physical world and even the economic world. But that was then. This is now. Even more modern events were not much kinder because Cold War commie influence served to create more friction, even in the 20th century.

UNDER THE INFLUENCE

Operating heavy machinery "under the influence" is dangerous. Driving "under the influence" will get you arrested and ruin your life. But running a country in Central America "under the influence" may cause decades of deep societal division, mistrust, calamity, and pain that last for decades . . . or centuries. Under the influence of what? Why, the influence of the USA of course! And Uncle Sam has not been kind to the Central 7, to be sure.

We already discussed the **US's Backyard** principle in the introduction to Latin America; indeed, the US has been *the* major influence in Central America's recent history. The whole region has been under the US's influence technically since the **Monroe Doctrine** and **Roosevelt Corollary** days, only in Central America did the full destructive power of this influence get totally unleashed . . . setting the stage for multiple armed interventions, encouraged civil wars, repression of democatic values, and ideologic battles that ended in uber-bloodshed. The additional uniqueness of the Central American experience is that the fallout from this influence is still being felt today. Painfully.

In both the North American chapter and the Latin America chapter overview, I touched on this a little: the US hates commies. The Cold War clearly shows that the USA just can't stand the commies. The US and Soviet Union were vying for influence all over the world, willing to sell arms or intervene covertly in other countries in order to get more countries on their team. That was part of the Cold War game, but perhaps no other region in the world has been so negatively affected by this face-off of ideologies than Central America.

How negative could it be? Central America saw some pretty nasty nuggets during the Cold War. Some of the political parties and presidents in Central America acquired leftward-leaning administrations or liberalizing policies to try to pull their people out of poverty in the 1950s and 1960s. A lot of talk about nationalization of industries—and sometimes direct action—incited serious suspicion from big brother in the north. The US interpreted and treated all leftward-leaning governments as the first step of an all-out communist invasion, and spent most of the Cold War years of 1950–1990 facilitating the domestic destruction of many of these countries by any means necessary. Virtually all countries of Central America have been racked by civil war that has been supported, if not initiated, by outside US forces that thought the leftward lean was just a little too commie for their backyard.

FUN PLAID FACT: The only Central American country to have escaped the decades-long fate of these scathing civil wars was Costa Rica. Why them? On December 1, 1948, President José Figueres Ferrer of Costa Rica abolished the military of Costa Rica after victory in the 44-day civil war in that year. In 1949, the abolition of the military was introduced in Article 12 of the Costa Rican Constitution. The budget previously dedicated to the military now is dedicated to security, education, and culture. Costa Rica maintains police guard forces. So, no military = no more coups = no more civil wars. Dang. . . what a revolutionary idea!

Again, please don't equate the Plaid Avenger's elaborations on this historical period as some sort of anti-capitalist or anti-US rant. I am not going to pull any punches here. This is not a liberal rant. It's not a pro-communist rant. And it's not an anti-American rant. This stuff is a matter of public record. The recent history of Central America has been well documented, and even more information has been recently released as a result of the Freedom of Information Act that outlines these activities in excruciating detail. In various countries there was unabashed CIA involvement, primarily to incite civil war. Covert operations were employed to destabilize the countries the US perceived as potential communist threats. The US is responsible for the funding, fueling, and arming of extreme-right rebel groups all over Central America whose sole purpose was to topple legitimately elected governments and keep certain US-approved military dictators in power. A fairly current example should help illustrate this point . . .

Ortega back in the socialist saddle in Nicaragua.

There was some heavy-handed US activity in Nicaragua as a result of the fear of a communist takeover of that state back in the 1980's, and because of the American perception of this area as their strategic "backyard." In places like Nicaragua, your tax dollars are still infused to help defeat the **Sandinistas**. The Sandinista National Liberation Front was a full-on, Marxism–Leninism inspired, leftist-socialist political party in Nicaragua that overthrew a (US-sponsored) right-wing dictator back in the 1979 and took over the country. This radical revolutionary named Daniel Ortega led this popular communist group that the Reagan administration tried to defeat with millions of US dollars funneled through the CIA to support right-wing assassin squads known as **Contras**, whose aim was to destabilize the entire country to the brink of civil war in order to discredit and eject the Sandinistas. Reagan's tactics eventually worked, and the Sandinistas lost power when Ortega was voted out of office in 1990. By the way, I also brought this up because the same dude, Daniel Ortega, almost 20 years later, won back the presidency in 2007, a position that he still holds even now in 2019. US tax dollars have been hard at work funding his opponents to make sure this doesn't happen again next election. Maybe we should elaborate in more detail about the repercussions of this particular mess on Nicaragua and the US too, in an episode we now refer to as. . . .

THE IRAN-CONTRA SCANDAL

Even though iron-fisted Nicaraguan dictator Anastasio Somoza had been in control of Nicaragua for 50 years and had been a staunch ally of America, that leftist rebel group called the Sandinistas overthrew him in 1979 and established a new government headed by Daniel Ortega. Protecting its old friends, America helped Somoza and his generals escape Nicaragua safely. Since the Sandinistas were beginning to drift towards the Soviet Union and Cuba, and were Marxist-Leninists to boot, America started organizing old Somoza National Guardsmen, disaffected peasants and others who didn't support the Sandinistas, into an army called the Contras. The **Contras** were armed and trained by Americans, and they operated in Honduras. Since the Sandinistas had a huge army, the Contras could not fight them in a head-on war, so they focused on attacking soft targets like schools, factories, etc. to make people realize that supporting the Sandinistas would only cause disaster.

By 1986, Americans were sick of supporting the Contras, so Congress passed a law prohibiting aid to the Contras. Since people in the Reagan administration were bent on destroying communists, they hatched up a plan where they would illegally sell arms (specifically anti-tank missiles) to Iran and use that secret money to illegally fund the Contras. This plan was known as the Iran-Contra scandal, which became the biggest political scandal of the 1980s. Iran was desperate for weapons because it was fighting Iraq in the Iran-Iraq war, and Americans thought that if they sold the Iranians arms, that they in turn would release some American hostages in Lebanon being held by **Hezbollah** (an Iran-funded terrorist group). The problem was that Iran was an avowed enemy of the United States. Furthermore, America was already selling arms to Iraq to use on Iran, in the bloody Iran-Iraq War of 1980–88. Could you even make up such a convoluted and bizarre plan as a fictional novel? Both sides of the plan were illegal in US law, and the plan was exposed in 1986. Some members of the Reagan administration were indicted, but nobody went to jail.

Ollie sez: "I was just following orders!"

Fast forward to 2012. Iran is now building nuclear weapons and perhaps becoming a menace to the world. The Contras and the Sandinistas signed a peace treaty in the 1980s and Daniel Ortega won back the presidency in 2008. He is president of Nicaragua again, after serving originally from 1979 to 1990. Iran is back in the spotlight as well, and lo and behold, the president of Iran has already met with Ortega a couple of times in order to beef up their relationship. The US is not a happy boy right now.

FUN FACT: A member of the Reagan administration named Oliver North (America's favorite traitor) was indicted for the Iran-Contra scandal and also for lying about it. Ollie claimed that he had only been following orders, but according to the Uniform Code of Military Justice, it is a soldier's duty to disobey illegal orders.

UNENVIABLE INTERVENTIONS

This type of outside intervention from the US has really, how shall we say, been "unhelpful" in Central American society, from the inside out, multiple times. Similar US shenanigans happened in Guatemala, Nicaragua, Honduras, and El Salvador. Those countries in particular have destroyed themselves many times over in the last fifty years: totally racked by civil war, civil insurrection, rebel groups, etc. There are always exceptions, of course, and Costa Rica is considered the most stable country in this region. I wonder why that is? Here's why: they got rid of their military 50 years ago! They saw the writing on the wall in all the other countries around them, and they actually disbanded their own military. Now if only the rest of the world would follow suit. . . .

Panama also looks semi-stable now. It avoided having civil wars due to a string of conservative, anti-commie, military dictatorships that ended up running the show for most of the violent Cold War period, thus staying in the good graces of the US. The US government propped up one specific dictator there named Manuel Noriega, a drug lord and former CIA employee. See inset box for Manuel's awesome rap sheet!

While we had this overall leftward lean in Central America, it was countered by US support for extreme rightist politicians and outright military dictatorships. This set the precedent for this region's continual violence. The people have been politically divided and usually brutally suppressed in terms of human rights, individual rights, and other political freedoms that are commonplace in other American societies. The current effects of this turbulent and violent past cannot be understated.

Some of you readers and Americans in general may have no problemo with this approach. After all, American intervention may possibly have averted Soviet takeover. Umm . . . maybe, I guess, if you were really worried about a Honduran commie invasion of Texas. Ha! But American support for US business interest and right-wing dictators has with no reservations been responsible for the price of bananas staying at 50 cents a pound for 100 years straight! Might makes right! Or is it the right makes might? I always get it confused in this region.

Where Is Manuel Now? A Brief History of Manuel Noriega

A former CIA spook, a former leader of Panama, a former prisoner in Miami, Florida—but where is Manuel now? Okay, so he's in a hospital, but it is still a pretty nutzo story . . .

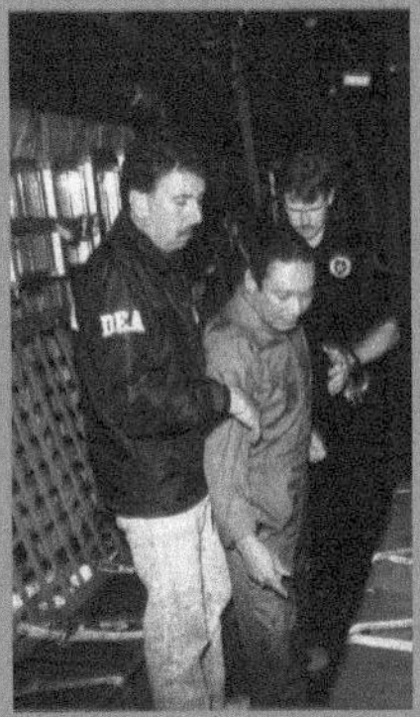

And Manuel: the US imprisoned him.

Manuel Noriega was a career soldier—having received training in Panama and at the **School of the Americas** in the United States. By 1983, Noriega was the de facto ruler of Panama. Noriega may have played a role in the 1968 military coup in Panama, but the details are uncertain. What is fairly clear is that Manuel Noriega was on the CIA payroll from the early 1970s to 1988. During this time, Noriega was a favorite of US diplomats. When the Shah of Iran—Mohammad Reza Pahlavi—was overthrown in the Islamic Revolution, the United States convinced Noriega to offer him amnesty in Panama. He even helped the United States funnel money to right-wing guerilla movements in El Salvador and Nicaragua.

Over the years, the United States began to see through Noriega's kleptocratic and narcotic ways. By 1989, official Panama/United States relations were at a low. US diplomats and soldiers were being harassed by Panamanian officials. After one US soldier was killed leaving a restaurant, President George H. W. Bush launched Operation Just Cause—a military invasion of Panama. During the invasion, Noriega fled but was eventually captured in the Vatican embassy in Panama. Noriega was brought back to the United States and tried on drug and racketeering charges. In 1992, Noriega was sentenced to 40 years in US prison. In April 2010, he was extradited to France and sentenced to seven more years of prison for money laundering. In 2011 France extradited him back to Panama, where he was sentenced to another two decades for crimes committed during his rule. Then he died in 2017.

What a resume: Military. CIA. President of a country. Drug trafficker. US prisoner. French prisoner. Panamanian prisoner. Worm food.

CONTINUED CYCLE OF VIOLENCE

What is still fueling this violence if the Cold War is now over? Well, it's not so easy to just set up shop again when such harsh dividing lines were drawn for so long, and whole societies were pitted against each other in bloody wars. There was so much human suffering and loss of life and so much political division that recovery will be equally long and hard. For at least 100 years, poverty has been endemic to this region. It's a poor place—the poorest region in the western hemisphere, actually. With the possible exception of Costa Rica and a current thriving Panama, there is anywhere from 60 to 80 percent of the population below poverty level. That's A LOT. There is a huge wealth disparity gap in this region, and it may be worsening. Many of these states have assessed that it's gotten worse, even since the end of most of their civil wars in the last decade. Even though the violence has toned itself down, it's still hard to get the countries going economically. The people are putting together new governments and are trying to get things back on track. And as has been a theme across Latin America, many of the governments tend to lean toward the political left even now in Central America, which makes for a remnant friction between them and the the US, thus retarding investment and deeper political cooperation from the north. . . .

Why do Central American countries have leftward leanings in the first place? Well, like the Latin region as a whole, any place on the planet where you have endemic poverty, where most people are ridiculously poor, they tend to gravitate towards leftist political and economic systems. Why is that? Because it makes complete common sense. Socialism and die-hard communism are based on the redistribution of wealth. Pretty much everywhere in the world, there are at least some socially influenced national programs like welfare that come from taxing the population or running a national industry (like the Saudis running their oil industry) and then redistributing those monies as benefits to people in need. Then you have full-on communism, which ideally takes all the wealth and distributes it equally to all by providing everyone with all of their needs.

In places where most people have nothing, which system are they going to choose? Free-market capitalism, where it's every man for himself, or communism, where everybody gets a share of the total? Obviously, if you have nothing, you'd support a system that promises to give you something. That's why there is a leftward leaning in Central America. It's the 60 to 80 percent poverty rate! Of course the majority of voters there will lean way left, and favor socialism or communism!

Am I defending the leftist cause? Not really. I'm not really interested in promoting communism or any other–ism. I just want you to understand the mindset in impoverished countries. When you don't have anything, you want to get something. And, obviously, the few that do possess the vast majority of the wealth and the businesses and the industries want to hold on to everything they own. So they will fight for their stuff, or pay for political power or armies to fight for their stuff on their behalf. Eventually, this type of situation explodes into in some sort of political revolution or outright civil war or movement to the left. Indeed, the latter has happened and continues to happen in Central America, where left-leaning politicians and parties have gained popular support. In a region where 2–5 percent of the people own virtually everything, and the other 95–98 percent split up whatever's left, it's no wonder that social programs are widely supported. For the most part, most people have no title to land, no real home, no nothing, and they vote 'left' because they can't possibly get poorer.

This is part of the history of Central America, but it's not over. There are many companies in the region, and not all are owned by the US anymore, that have inordinate amounts of rights beyond what even humans have. By that, I mean that there are some plantation-style economic establishments where people are essentially slaves. Where the people who own the banana companies have hired men with rifles standing by to make sure workers are picking bananas fast enough. This sounds incredible, but it happens. There are no labor unions, no real labor rights. Protection of companies' interests over those of the individual people—with an infusion of US interest usually to support rightist regimes—continues to this day in parts of Central America.

We usually think of the US as the "good guys" who go out and fight for peoples' human rights and individual rights, for democracy all over the world. Unfortunately, Uncle Sam doesn't really have a great track record where Central America is concerned. Quite frankly, the US has been more interested in supporting strong, right-wing leaders who will maintain political and economic stability at any cost; the hope is that they will make sure that stuff will not be nationalized, so the private investments aren't compromised. As such, they will even support military dictatorships that will suppress the local population who may be trying to rise up and fight for a revolution. Probably a commie revolution—yet another reason to squash it.

But the times are a-changing. The Cold War is indeed over, and fighting the communist threat in Central America doesn't hold a lot of water. The United States now has a new tactic in order to ensure economic growth and stability for itself—and the region as well. That tactic is free trade blocks. As we previously discussed, the FTAA is a US-supported attempt at a hemispheric-wide trade block. One that has fizzled as of late. But things are still going forward on this front in Central America. Thank CAFTA/DR-CAFTA for that. (See box on right.)

To be fair, many politicians and businessmen truly believe that free trade is the real

What's the Deal with . . . CAFTA?

CAFTA stands for Central America Free Trade Agreement (DR-CAFTA is the same thing, but includes the DR for Dominican Republic) and is an international treaty to increase free trade. It was ratified by the United States Senate in 2005. Like NAFTA, its goal is to privatize public services, eliminate barriers to investment, protect intellectual property rights, and eliminate tariffs between the participating nations. Many people see DR-CAFTA as a stepping stone to the larger, more ambitious, FTAA (Free Trade Agreement of the Americas).

The controversy regarding DR-CAFTA is very much like the controversy regarding NAFTA. Many people are concerned about America losing jobs to poorer countries where the minimum wage is lower and environmental laws are more lax. Also, some people are concerned that regional trade blocs like DR-CAFTA undermine the project of creating a worldwide free trade zone using organizations like the WTO.

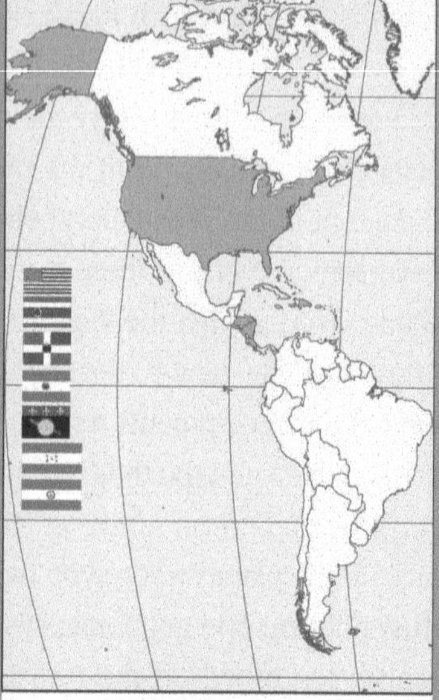

CAFTA members

solution to poverty in Central America. And who am I to claim righteousness? Perhaps they are right. Former US President George W. Bush said that free trade "will advance peace and prosperity" in the region. In this vein of thinking, free trade = movement of capital = more businesses = more jobs = more money to these depressed economies. Also, creation of jobs in the poor places means people won't have to emigrate to the US to get a job. Sure, that might work. However, most people believe it will just help them get ripped off more: just more crappy jobs not worth much money, which is not real economic development and diversity. Free trade is not widely supported in these countries by the general population, but widely supported by their administrations—as witnessed by CAFTA being signed. **2019 UPDATE**: As previously alluded to many times, President Donald Trump of the USA has been a vocal opponent of free trade in general, and has already re-worked NAFTA, and is currently trying to kick out a couple of countries from CAFTA, so at this point none of us have a clue if this block will even exist much longer. . . or perhaps it will be streamlined and do even better. Who knows! Let's keep an eye on it though, shall we?

LEGACY OF THE CIVIL WARS

What Central America is known for now (aside from Costa Rica being a major destination for ecotourism) is narcotics production, transfer and trafficking. A lot of crack, dope, smack, amphetamines, and whatever else is produced in South and Central America, have to go through Central America to get to their ultimate destination: the US. What is a by-product of this? GANGS! Some really vicious ones, too! This ties into the growth of power and geographic scope of those Mexican cartels discussed last chapter. Those cartels now increasingly use Central American gangs as hired muscle on the streets of the US. Groups like MS-13 now have a presence in all major American cities. Yikes! (See box below.)

Currently, several Central American countries, along with Mexico, have actually signed a pact and accord to combat gang violence. What is that all about? Independent sovereign states have to form a coalition to fight organized gang crime? YES! This is a big deal. Why is this? Gosh, I don't know . . . long history of civil war, divided society, oppressed societies, tremendous wealth disparity, people who are freakin' poor as dirt, poor governments, corrupt governments, and drugs, which are everywhere. Put that all together in a stew, and what do you come up with? GANG CONTROL!

During the civil wars in Guatemala, Honduras, Nicaragua, and especially El Salvador, lots of people fled the region altogether, as you might imagine. This resulted in an enormous refugee population that emigrated to the US. There were also tremendous amounts of people that tried to get out of Central America to the US just to find work. During all the years of oppression, violence, and civil war, lots of folks who made it to places like Los Angeles, Santa Fe, even as far north as San Francisco, were poor and uneducated. This lovely combo limits opportunities, and facilitates a gravitation towards crime for some. Over the years, this has evolved into a syndicate of criminal activities, with ties back to Central America, and now they are importing people just for this criminal activity. It's a coordinated mafia muscle making machine now.

WHAT THE H? WHAT THE WHAT? WHAT THE WHAT? WTH?

What's the Deal with MS-13 and Maras?

Maras are gangs originally from Central America, including MS-13 and Sombra Negra.

How did they get here? People from El Salvador, Honduras and other Central American countries escaped during the 1970s to the United States of America. These people were running away from the civil wars that struck the countries of Central America, particularly El Salvador. Disaffected and disenfranchised refugee youth in the LA area gravitated toward gang life. Almost all maras members display radical tattoos on their bodies as a sign of their affiliation to their gang.

MS-13 is a gang based in Los Angeles with roots in El Salvador and operations throughout Central America. Many experts believe MS-13 and other maras have spread and benefited from the US policy of deporting criminals to their country of origin, allowing maras members to recruit new members abroad. MS-13 has recently been in the news for activities in Northern Virginia and Maryland. Maras are involved with everything from drug trafficking to petty crimes.

Central American gangs are now a problem all across the North American continent, from Costa Rica to Canada. They have gained a lot of strength. Since many have no options, the poor kids gravitate toward gang membership—the same reasons people join gangs in inner cities in the US. This same phenomenon is happening in Central America, so the popularity of being in a gang and joining this organized crime movement has gotten so out of control that the governments can't even handle things in their own countries. That's why you have this effort now, signed by Mexico, Guatemala, and Belize, to basically work together to combat organized crime in and among their countries, as well as to combat criminal movement to the US.

Along the same lines, Mexico itself is under a big squeeze because so many people, not just gangs, want to get to the US. Work immigrants want to get into that giant USA economy that can use their labor, and pay a decent wage. They aren't really refugees anymore; they just want to emigrate from a impoverished area with few job opportunities to the land of milk and honey up north. How are they going to do that? Well, they aren't going to swim there, so they have to go on foot across borders. The same way that Mexicans get across the border into the US, Central Americans have to go all the way through Mexico to get to the US. Mexico has become a funnel; it's even funnel shaped, a throughway to the US. Mexico has to deal with all these people who are even poorer than its own people coming through its territory, sometimes not making it to the US and staying in Mexico, putting even more stress on the Mexican economy.

Add to that the organized criminals and gangs running drugs through Mexican territory. Mexico's in a real pinch because the US wants to build a larger fence, to fortify the borders even more to keep immigrants out, while more are still pouring in from Central America. One side of Mexico is being dammed up and the other is still open for the free movement of people. The US has put pressure on Mexico to stop anybody from coming to the US, but its population is swelling from people just trying to pass through its territory.

Trump sez: No money for you! Everybody go home!

2019 TRUMP SPEED BUMP UPDATE: Well, all of this has gotten way, way, way worse. Way. In March 2019, the Trump administration suspended all $450-$600 million in US foreign aid payments to El Salvador, Guatemala, and Honduras in their entirety. Wait, what? Why would the US do that? It is an effort by the US to push these Central American governments to do way more to stop migration at its source. For all the reasons pointed out in this chapter, there has always been a large number of asylum seekers/refugees fleeing from violence in El Salvador, Honduras, and Guatemala: the three countries where most of the migrants on the US southern border are coming from. And these numbers have surged ever since the "border crisis" has been invented/imagined or at least magnified in the press for the last two years. Those fleeing are fearful that US border law is going to get even stricter soon, (or that "The Wall" is going to get higher), and thus many of them are making the trek right now in order to get in before it becomes totally impossible to make it across.

The "border crisis" is thus, in some sense, a self-fulfilling prophecy. It may or may not have been bad before, but now it has truly become a crisis. I'm sure you have seen the headlines about migrant caravans crossing Mexico, overcrowded facilities in the US, mass deportations, and families being separated, and all the human drama that goes with this very hot-button divisive issue. Adding fire to the sense of urgency that many Central American migrants have about migrating to the US are additional threats by the Trump administration to radically reduce the total number of asylum seekers they will admit (even if they are legitimate asylum candidates by international law), as well as repeated tweets that Trump will close the entire US/Mexican border.

Back to the aid: Many analysts make the argument that cutting foreign aid to these very countries will further exacerbate the crisis. . . that is, by making El Salvador, Honduras, and Guatemala even poorer, the US would be inadvertently encouraging the very migration that they are trying to prevent! Others think that the US should be helping out Central America even more with even more aid and investment, which would address the root causes of the migration. . . that is, making those states better would encourage people to stay in them instead of moving to the USA. What do you think?

2021 BIDEN BUMP-UP UPDATE!!! Bad news/good news for Central America? The IMF projected that Central American economies contracted (that is, shrank) by over 6% just in 2020. Destroyed economies, natural disasters, climate change, crime (some of the highest homicide rates in the world), ongoing violence, worsening corruption, and challenges to democracy have all been aggravated by the devastating impact of the Covid crisis year. A perpetual push for emigration out of the region has been exacerbated, and is worse than ever. The USA under Trump worked hard to use resources to simply physically stop the migrants at the border, either the one with Mexico, or failing that, the border of the USA. . . thus the whole "Trump Wall" deal.

BUT!!!! The current Biden administration of the USA has shifted that strategy significantly. It is committed to boost spending to $4 billion to address the underlying causes of immigration out of Central America. Things like promoting development, security, and anticorruption efforts. Basically, they want to help the economies and political/judicial structures of those Central American states so that the countries themselves improve, thus reducing the need for people to leave.

IS THERE A LIGHT AT THE END OF THIS TUNNEL?

I'm looking, but it's such a faint glimmer. Costa Rica continues to prosper, and Panama is another bright spot that is getting its act together and currently flourishing . . . having the Panama Canal as a focal point of global shipping is a pretty good gig . . . and a **2012 UPDATE:** October 2012 Panama passed The Panama–United States Trade Promotion Agreement, a bilateral free trade agreement between the two countries which in effect puts them in the same status as the rest of the CAFTA countries, thus deepening ties further. But many factors are working together to keep the region hurting overall. At least the civil wars are officially over, but Central America remains replete with latent violence. Physically racked by earthquakes, hurricanes, mud-slides, volcanoes, and even sink-holes, on a regular basis. (Seriously, go Google "Guatemala City sinkhole" for a scary shot of reality.) This is a violent arena due to tremendous wealth disparity perpetuated, and in some cases accelerated, by the US's role. In the Central American "backyard," US intervention has typically been a negative thing. CAFTA has not kicked in much yet either.

Economically, all of the countries we're talking about are highly dependent on the US economy. What is it that they produce there that is shipped to the US? Primary agricultural goods. Products that aren't worth much. Not a tremendous hub of manufacturing yet either, even though though they have loads of cheap labor . . . likely due to the lack of international business investment, which may of course be related to hesitation to invest in physically/culturally chaotic countries. That's why poverty is endemic to this region.

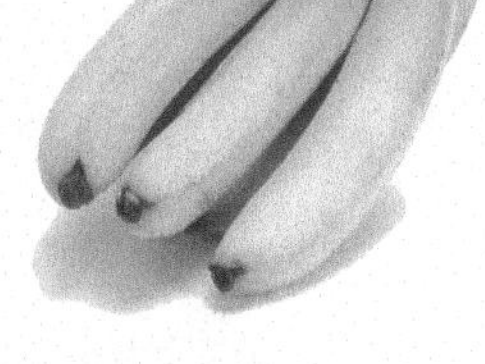

The price of bananas hasn't changed in 100 years. It has only crept up because of standard inflation/ cost of living. Adjusted for cost of living, bananas cost the exact same as they did a century ago. You can go into the grocery store and buy them for *a buck a pound*. That's after they have been cultivated, fertilized, pruned, picked, packaged, and then delivered a few *thousand* miles to get to you. How is that even possible? What are the people getting per pound in Nicaragua? Approximately jack is the Avenger's calculation!

Politically, the civil wars of the 1960s–80s, fueled by wealth disparity and anti-communist fervor, have yet to stop damaging these societies. Re-leaning to the left for many of the countries in Central America will certainly cause more friction with the US. Immigration is also a hot button issue between the US and all of Latin America; an issue that Central America will probably not come out winning, no matter what the eventual solution may be. Growth of the **maras**/gangs is another current event that will not be going away anytime soon and is radically affecting the region, the US, and everyone in between: the reach of these Central American maras is now up to Seattle in the west and Baltimore on the east coast. Should you be concerned? YES! More than your government currently is . . .

This cycle of violence and the endemic poverty, which has propagated even more violence due to gangs, is why the Plaid Avenger calls this a really violent arena. Central America has been a place of open warfare within and between states, it has been a Cold War battlefield, a corporate battlefield, and now that the gang violence is skyrocketing, it's a drug and crime battlefield.

But don't get me wrong. It's an awesome place to party. The people in Central America are incredibly resilient, and continue to push forward in spite of all the chips being stacked against them. Get down there. Check it out. But please remember this Plaid Avenger Tip: Just say "no" to gang tattoos.

CENTRAL AMERICA RUNDOWN & RESOURCES

View additional Plaid Avenger resources for this region at http://plaid.at/camer

BIG PLUSES

- Hmmmm . . . . hold on. I'm thinking. Nope, got nothing. It's a tough place.
- Oh wait! I got one! Costa Rica is totally stable and an awesome destination for ecotourism and tasty waves to surf!
- China and other rising powers have no hesitation investing in "leftist" countries, so foreign investment is on the rise from other "non-USA" powers.

BIG PROBLEMS

- Hurricanes, earthquakes, volcanos
- Gangs; some of which are becoming more powerful than local governments
- Serious wealth disparity and landlessness remain chronic political issues
- Significant transportation corridor of illegal narcotics; gangs help with that
- Political instability the standard, not the exception (except for Costa Rica, and Panama is improving as of late too)

DEAD DUDES OF NOTE:

Anyone on the bad side of MS-13.

PLAID CINEMA SELECTION:

Sin Nombre (2009) Wow, this movie is hard core, gritty, gang activity stuff. And it actually mostly takes place in Mexico, but given its themes of migration from Central America and gang influence, I felt it fit in even better for this region. Honduran family trying to migrate to US gets caught up in a internal gang dispute. Great scenery of the region, great depiction of how migration actually happens, realistic gang life, extremely topical, but extremely violent too..

Agua Fria de Mar (2009) Thrilling story of a rich couple who take a vacation at the beach and come across a little girl who has terrible stories to tell, which sets off a crisis of meaning in the woman who begins to rethink all of the privileges of her life in the upper middle class. Spectacular Costa Rican landscape almost distracts from the action.

Under Fire (1983) A Hollywood presentation of the very real conflict that plagued Nicaragua during the late 1970s. Follows a photographer and a couple of his journalist friends who are covering the civil war against the Somoza regime.

El Norte (1983) A touching coming-of-age film that follows the fictional journey of two young indigenous Guatemalans as they attempt to free themselves of ethnic and political persecution. Covers the journey starting in the small rural town of San Pedro in Guatemala through Mexico and eventually to Los Angeles. Selected for preservation by the Library of Congress because it is "culturally, historically, or aesthetically significant." But don't expect a happy ending.

Romero (1989) True story of the catholic priest Archbishop Oscar Romero who lived in El Salvador during the political unrest in the 1970–80s. The government launched a "terror campaign" against leftist guerillas in an attempt to crush them. Archbishop Romero's protests against governments' actions is perceived as disloyalty. and the government begins to destroy churches and murder priests. Yeah, and then they assassinated him. Look into "**liberation theology**" before or after viewing this one.

Salvador (1986) James Woods stars as a journalist, down on his luck in the US, drives to El Salvador to chronicle the events of the 1980 military dictatorship, including the assasination of Archbishop Oscar Romero (see previous film). He forms an uneasy alliance with both guerillas in the countryside who want him to get pictures out to the US press, and the right-wing military, who want him to bring them photographs of the rebels.

LIVE LEADERS YOU SHOULD KNOW:

Manuel Noriega: A "CIA asset" and double agent that eventually became President/Dictator of Panama from 1983-89 until the US invaded the country to remove him from power for drug trafficking, racketeering, and money laundering. An object lesson on why the US should never support genuinely bad people, even when those bad peoples' objectives temporarily coincide with your own.

Daniel Ortega: Current President of Nicaragua, a post he first held back in 1985-90 as the leader of the leftist Sandinista movement which overthrew an US-backed dictator (Somoza). Carried out controversial land reform and wealth redistribution (that is, communism), which brought the wrath of Uncle Sam against him. That is why the Reagan administration illegally funded and armed the US-backed Contras to take him out. Like Fidel Castro, the US still hates Ortega after all these years.

MAP GALLERY

View Map Gallery & additional Plaid Avenger resources for this region at http://plaid.at/camer

CHAPTER OUTLINE

The Caribbean

16

REGGAE and calypso music. Voodoo religion and cruise ships. Pirates and sunken treasure, and scuba diving. Banana daiquiris on sunny beaches washed by that beautiful blue Caribbean Sea. A tropical paradise for those that visit; not such a paradise for those washing the dishes in the hotel where you stay or cutting sugarcane in 110 degree heat. Caribbean color, Caribbean music, Caribbean-style shrimp platter at Red Lobster. The Caribbean is just too distinct to be included as part of any other region in Latin America. Yes, it is also considered a "US backyard." Yes, it is also endemically under-developed. However, the culture, history, and current events of the Caribbean all combine to make this place different. Plus, it's the only one of our regions that is comprised of a bunch of islands . . . most of them small islands at that! Gilligan's favorite region! I of course always fancied Mary Ann the most, but let's get to the . . .

PHYSICAL

Like Central America, the Caribbean is a violent arena from a climatic and geological standpoint. The islands of the Caribbean, as you can see from this plate tectonic map, are on the edge of a trench between two plates that are being pushed together. A lot of the islands are volcanic in origin, but not all of them. The big islands are also referred to as the **Greater Antilles:** the islands of Cuba, Jamaica, Haiti/Dominican Republic, and Puerto Rico. They are not volcanic in origin, but are old parts of the larger plates and are quite stable, geologically speaking. However, the arc of smaller islands that form the chain known as the **Lesser Antilles** are the result of contact at plate boundaries, which in this circumstance, results in volcanic activity. Some are still active, volcanically and seismically.

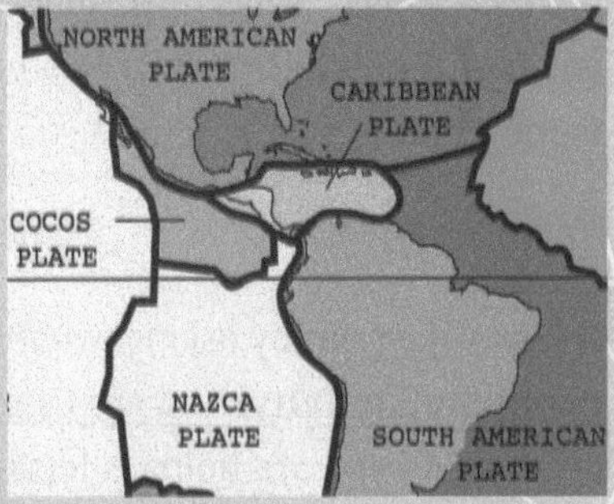

Caribbean plate tectonics: active arena.

Though the larger islands are more stable geologically, they and the smaller islands both have to deal with the climatic instability of the region. *The Caribbean is the hurricane zone of all hurricane zones!* There may be big typhoons in the Pacific, and the US may get hit by a few hefty hurricanes every now and then, but no place has them like the Caribbean. No matter where hurricanes end up in the regions near the Atlantic, most of them go through this region first. And due to the small size of the islands, hurricanes have a devastating impact almost every time. We are talking about storm systems that can completely cover an entire country. In places like the US and Mexico, the hurricanes may hit the coast or travel a little bit into the mainland, but they are not destroying all areas of the country simultaneously. The Caribbean is a place that is, meteorologically speaking, a tough place to party.

The Greater and Lesser Antilles Action.

A few "grandes" tracks from circa 1971.

Another consideration is that it is a tropical region. It is a very hot and humid place. However, large bodies of water always act as moderators of temperature, and thus it's always a little nicer right along the coast, where everybody lives, and where everybody visits. Which ties into another distinct quality of the Caribbean: tourism is the biggest GDP earner on pretty much every single island nation. That's why everybody comes here to party. Everyone flocks to the nice tropical beaches for the nice warm water. The entire region is tropical—it's located in the tropics! The tropical temperatures also play a part in warming the water of an important Atlantic physical feature, the Gulf Stream—a warm ocean current originating here which flows as far north as Russia, moderating climate its entire journey. Ever wonder why it never snows in England, even though it is very far north? Gulf Stream effect is the answer. It originates in these parts.

WHAT ELSE IS GOING ON THERE?

The simplest way to highlight the cultural distinctness of this region, Plaid Avenger style, is by talking about the four S's: Sun, Sand, Sin . . . which is why people come here. The Bahamas, Jamaica, the Virgin Islands, Barbados . . . what comes to mind? Party! You know that's right! Let's get it on, Caribbean-style! Bronzed bodies basking on baked beaches. Their economy is largely based on tourism. What's the fourth S? SUGAR, which we'll get to shortly.

Tourism is a main component, if not the major component of the Caribbean's GDP. In some Caribbean islands, tourism accounts for more than half of the entire GDP each year—in some places it's even higher than that! Half of the entire state's economy is due to one single industry. That's a lot. As such, so much of the wealth of the region is dependent on other people from other countries hanging out there and spending their money. This of course sounds good—bring them on!—but there are some negatives, which sometimes outweigh the positives.

SUN, SAND, & SIN: TOURISM IN THE CARIBBEAN

We can start out by asking: what is tourism really worth to the Caribbean? The Plaid Avenger's take is that it's a **primary economic activity**. You are essentially just selling the land as a resource, not unlike drilling for oil in it or cutting trees off of it. All you are doing is letting people come look at your beach, or your mountain, or whatever scenery you are selling. As a primary commodity, it's not really worth that much, economically speaking.

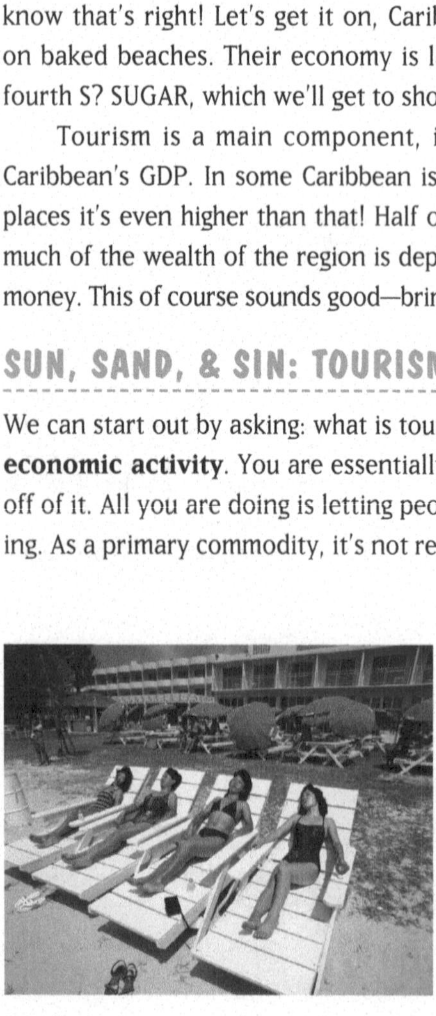

Fun, fun . . .

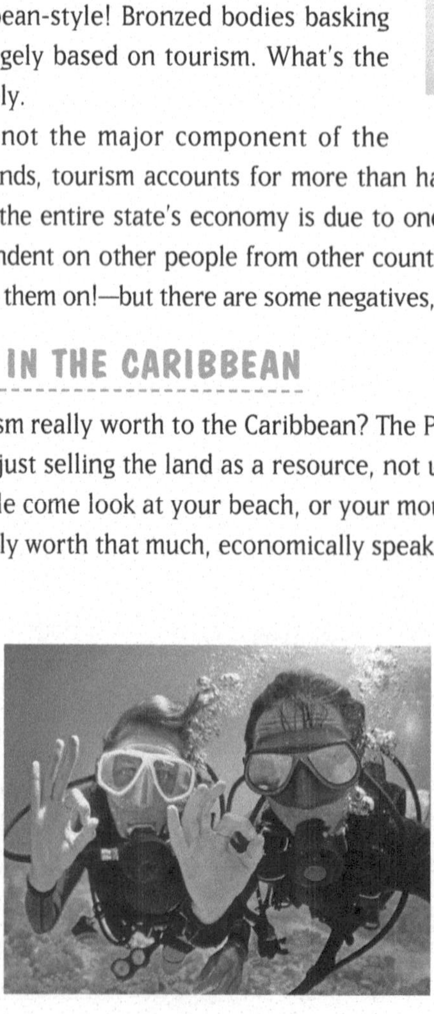

fun . . .

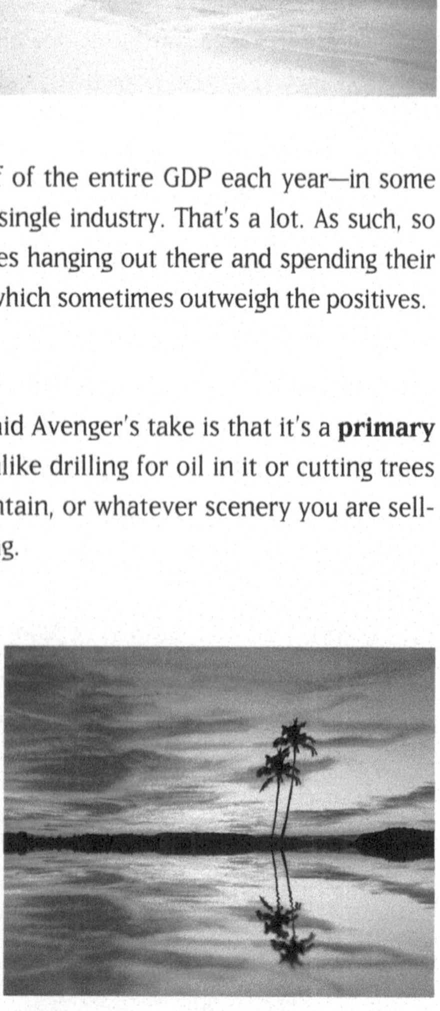

and sun!

On top of that, there is quite a bit of variability in tourism. If there is a bad economic year somewhere else, it can affect how much money the Caribbean nations make from tourists. A bad hurricane can affect tourism negatively, as well. People won't come from all over to hang out on an utterly devastated beach. Tourism fluctuates to both extremes very quickly. It is very unstable; it's definitely not something that you would want as your economy's anchor. But it unfortunately it is for many nations in the Caribbean region, which you'll see in the box to the right.

Having said all that, tourism is one of the most important economic sectors in this region as a whole. Tourism dollars alone constitute over 15% of GDP for the Caribbean, and by that I mean every single economic activity of every single island tallied up into a single GDP number. . . and tourism is 15% of that. Here are some recent stats to give you a tropical feel for this sunny significance:

Just in 2019, which happened to be a record-setting year for Caribbean tourism as a whole:

- 61.7 million international visitors came to this small region (that's 31.5 million fly-in, stayed at least one night visitors combined with 30.2 million day-tripping cruise ship passenger visitors). . .
- They spent $58.4 billion dollars. . .
- Which supported 2.67 million jobs, which is 15.4% of all regional employment. . .

Country/Territory	Contribution of Tourism Economy to GDP (2019) % of Total GDP
Anguilla	44
Antigua and Barbuda	41
Aruba	70
Bahamas	45
Barbados	30
Bermuda	18
British Virgin Islands	38
Cayman Islands	20
Cuba	11
Dominica	33
Dominican Republic	16
Grenada	41
Guadeloupe	10
Jamaica	28
Martinique	8
St. Kitts and Nevis	52
St. Lucia	68
St. Vincent and the Grenadines	42
Virgin Islands (U.S.)	54

Source: World Travel & Tourism Council

Tourism: major big deal to this region.

So yeah, I will continue to outline some of the negatives of tourism, but dig this: on average for over half of the Caribbean states, the sector accounts for over 25% of GDP (and in some cases in excess of 50% to 70% of their entire GDP!). . . and that is way more than double the world average of 10.4%. In the case of the British Virgin Islands, tourism accounts for 70% of their GDP, one of the highest in the world. (Data from: World Travel and Tourism Council Annual Regional Caribbean Report 2021.)

What do the locals actually get out of selling their beach? Do they line up and have all the tourists give them a dollar each? Not hardly. Typically what happens is a multinational corporation from somewhere else, like the US, Great Britain, or Germany, comes and builds a huge hotel on a nice stretch of beach. Those corporations are also the ones getting the $300 a night for the tourists to stay there. They also get the $8 that a tourist pays for a mai tai at the hotel bar. The McDonald's and T.G.I. Friday's and Olive Garden restaurants that spring up around the hotels are also foreign-owned.

For the people actually living there—the locals in the Caymans or Guadeloupe or the Virgin Islands—tourism equates to low-paying, minimum wage jobs. The local is the guy who serves the $8 mai tai, the guy who opens the door, does the dishes, changes the sheets, and flips the burgers at Mickey D's. Those aren't really the most enviable jobs, and we all know inherently what those jobs are worth. It can be argued that tourism is a repressive industry for the locals, and that the jobs eternally remain unskilled, low wage labor that holds the society back from evolving a more diverse economy. The locals are working the crappy jobs while all the money being made there is exported to another country via the multinational company that owns the hotel or casino or swanky bar there. I've seen statistics indicating that for every dollar spent in the Caribbean, less than a dime actually stays in the country. Fail whale for your locals.

Others would argue that those may be crappy jobs, but they are jobs, period. What would the locals do without those jobs? Wouldn't they be poorer than they are now? And some local entrepreneur types do have increased opportunities to start their own businesses. Both are valid points, but just remember next time you visit the multimillion dollar hotels, the locals are not the ones partying and having a good time. They're mopping up your up-chuck because you drank too much. Please tip heavily when you are there. Tip everyone, including the unemployed.

For the most part, it is 'Awesome Vacation Land' for us visitors, where we can party, do whatever we want to do, hang out in the sun, walk on the sand, and participate in your favorite sinful vice . . . even twice or thrice. The sin part is one to pay attention to as we later consider Cuba as one of the reasons that Cuba went communist when they had their revolution in 1959 was because of the sin part. Pre-revolutionary Cuba was a tropical island paradise that became infiltrated by American mafia and was a playground for rich and famous Americans from the late 1800s all the way up to the 1950s. It was essentially the 'Sodom and Gomorrah' of the tropics for the Yankee Americanos. People would flock down there for their vacations and take advantage of all the legal drugs, gambling, prostitution, and alcohol, whatever. Just thinking about it, the Plaid Avenger salivates, anxiously awaiting the return of those good old days . . .

Though that sounds absurd, this is actually how much of the Caribbean region is viewed, as a place for vacation and all around partying. This is one of the reasons why Fidel and his boys kicked everybody out of Cuba during their little revolution. Tourism is a double-edged sword that, overall, may not have the greatest of impacts on the Caribbean.

2021 UPDATE!!! CARIB CRUSHED BY COVID: Holy beach-combing calamity!!! The 2020 "lost year" of Covid virus really put a shellacking on this region! Many of you may have been wondering why the Plaid Avenger hasn't brought up the impacts of the coronavirus more frequently throughout this 2021 edition of this book, and the honest answer is that in my estimation (while being a titanic technology- and society-shaping event in and of itself), it affected the entire world economy about equally. . . that is, all regions' economies and all peoples of the world had a terrible 2020. No sense repeating that for each individual region.

BUT now it is time to call out the region that without a doubt was impacted the greatest, overall, in all aspects of life. . . precisely because it is so over-reliant on tourism for so much of its economy. The entire Caribbean economy has been decimated by Covid, not so much in terms of the disease itself or death tolls, but because global tourism all but ceased for an entire year. Compounding this disaster was the fact that the USA, which constitutes nearly half of all visitors to the Caribbean every year, was the country that got hit worst, first, compounded even further with bad policy, disinformation, and high death rates. Virtually no one from the US headed to the Caribbean shores in 2020, and here was the result:

From 2019 party year to 2020 Covid calamity year:

- Overall travel and tourism revenue down -58.0% (that's a projected loss of -34 billion US$, year to year)
- Overall change in travel and tourism related jobs down -24.7% (that's a loss of -0.7 million jobs, year to year)

Simply put: when a large chunk of your economy depends on visitors, and suddenly the world stops visiting, your economy cannon-balls into the depths much like a sunburned, overweight tourist into the shallow hotel swimming pool. It doesn't take very long for either to hit rock bottom.

"Welcome back to the islands, Mr. Avenger. Can I get you the usual?"

WHEN THE US GETS A SNIFFLE, THE CARIBBEAN GETS A COLD

Why the post-nasal drip reference? Since we are considering the economic make-up of the region, it's important to point out this economic relationship. This saying has way more than a grain of truth in it; it is a fact. And what it's saying is this: If anything bad happens in the economic colossus of the US, there is a ripple effect that gets magnified down south. The Caribbean gets it first and worst. No matter how small the fluctuation is in the US, it can heavily influence the entire economy of this region.

Not likely to be visiting the resorts anytime soon.

The tourism industry is a perfect example of this, since it relies heavily on people traveling to spend time and money there. When there is a bad economic year in the US or Europe, people are less likely to go on vacation because they can't afford it. If you need to make money, you're not going to take off from work and spend thousands of dollars on a Caribbean vacation. If there is an economic depression in the US, the Caribbean is trashed. Economically devastated immediately. This ripple effect reaches into all aspects of life down there. Even the 9/11 terrorist attacks precipitated a drastic reduction in Caribbean tourism. People were afraid to fly on planes, and overnight, the Caribbean economy got dumped on. Into the next year, there were still repercussions from the decrease in tourism due to 9/11. This single event, that didn't even affect the American economy that much, *destroyed* the Caribbean economy. Even something as trivial as an increase in airfare to these places can affect the economy negatively.

Another way that the Caribbean is reliant on the US economy is through something called **remittances**. This term can be applied not only in the Caribbean, but across Latin America as a whole. Many people come from other countries—like Mexico and virtually every Caribbean country—to the US to work. These people work primarily in the agricultural sector, but there are also people who move to cities like New York and work low-paying jobs there. Some even start their own businesses. New York is a good example, because there is a huge Jamaican population there. These people are not necessarily immigrants, though some of them are. Many go to the US just to work, since the job outlook might not be very good in their respective countries.

"I've got something you can remit!"

What does that have to do with this term, remittance? A remittance refers to money sent back home to the families of those people who have come to the US to work. Jamaicans, or Puerto Ricans, or Barbadians employed abroad and send part or most of their paychecks back home to support their families.

Why bring this up while talking about the Caribbean region? Remittances comprise a significant component of the GDP of many smaller Caribbean countries. In 2019, the remittance market for Latin America as a whole exceeded $100 billion. In some Caribbean countries, 5–10% of the economy may depend on getting money sent from people, working in the US and other places. Some examples: 10% of Dominican Republic's GDP, 16% of Jamaica's GDP, and 40% of Haiti's GDP is remittances. If this activity were stopped, if guest workers could no longer send money home to their families, there would be an immediate economic impact on the economies of this region. The Plaid Avenger is not being a bleeding heart here—these are just the facts. To stop guest workers would mean immediate, instantaneous negative repercussions south of our borders. The sniffle becomes the cold—perhaps pneumonia for some of these small states.

Repeat: **2021 UPDATE!!! CARIB CRUSHED BY COVID:** How fitting that you just read a section of this chapter entitled "When the US Gets a Sniffle, the Caribbean Gets a Cold," because the analogy is spot on. Think about all the economic pain and suffering the US economy felt (and it is the largest, most diverse economy in the world) and then multiply that by an exponent for the Caribbean region. In case you missed the stats a page or two ago:

- Overall travel and tourism revenue down -58.0% (that's a projected loss of -34 billion US$, year to year, to a small region highly dependent on tourism dollars)
- Overall change in travel and tourism related jobs down -24.7% (that's a loss of -0.7 million jobs, year to year)

I couldn't find hard numbers on remittances, as they are harder to track, but we can assume they were slashed hard as well.

Ouch. Let's hope the resurgent USA economy will spell a much better year for the Caribbean, which is likely since many citizens of the US are more than ready to travel after being cooped up for over a year, and have money burning a hole in their collective pockets after having NOT been on vacation or spent lavishly during 2020. Future outlook is good for the islands, but the tourists can't get there fast enough.

THE FOURTH "S": SUGAR

The role of sugar in the Caribbean cannot be understated. When the New World was being settled by the colonial powers, primarily the Spanish and even the British, there was one single commodity that had an enormous impact on the historical development of this region, and even back in Europe. This, of course, is that sweet, sweet stuff called sugar.

Sugar has an interesting tale that not many people know about. By the 20th century, sugar had become so entrenched in the Western diet that it's no wonder the average person nowadays really doesn't know how or when sugar became popular. We only know that sugar is a huge, prominent part of our daily lives . . . and we all know we are all eating too much of the stuff. So we think it has been around forever. Everybody since the beginning of time has consumed an average of 152 pounds of sugar per year just like we do now, right? Um . . . nope!

Sugar was big business back in the day . . .

This consumption of sugar is a new phenomenon. Sugar was originally from Southeast Asia. It is a type of grass, never planted and processed in great volume anywhere. It moved from Asia to India to Africa, but not until it got to the Caribbean did it become a mass-produced cash crop. The Spanish had small sugar plantations on the Madeira Islands, which were a way station between Spain and the New World. As they were looking for gold and silver in the New World, they also looked into what crops could be grown there.

Sugar as high as an elephant's eye.

When they brought sugar to the New World, it exploded. Sugarcane was brought into an environment in which it goes gangbusters. It could be grown so fast, sometimes two or three harvests a year, so cheaply and in such abundance that it caught the eye of the colonizers. Up until this point in history, the "sweet tooth" was pretty much unheard of. Sugar was a scarce commodity. Nobody in Europe had sugar. Refined white sugar we consume by the ton in today's world did not exist 500 years ago. It was only a plaything of the rich and famous. Its appeal then is analogous to that of cocaine now: An incredibly valuable commodity imported from a far away tropical location, reserved as a luxury and party favor for those with expendable cash. But the Caribbean changed all that. By providing this commodity en masse, the Caribbean was the single region that changed the dietary consumption throughout the entire planet for the next 500 years . . . and to this day. But it also changed the Caribbean quite a bit too.

And that is mostly due to sugar being an extremely labor-intensive commodity. Today, a lot of it can be processed via machine harvesters, but most of it is still mostly done by hand, planted and harvested by hand with machetes. And in the modern world, there are lots of sugar cane alternatives like sugar beet and even high fructose corn syrup made from . . . um, wait for it . . . corn. But during the colonial era, people couldn't get enough of it, and the main supply for the planet was the Caribbean, and it was totally done with hard human labor. It was processed a variety of ways, like the table sugar with which we are familiar, but that also led to by-products like molasses, and of course the alcoholic product so associated with this region: rum. Even by-products and leftovers from the sugar-making process were used to make other products. This essentially diversified sugar into a ton of different commodities, all of which made good money.

Rum was almost as big a hit as sugar was. It was mass produced and mass consumed both in the New World colonies and back home in Europe, so much so that British sailors even got a rum ration during much of this era. They were paid in rum . . . the Plaid Avenger should really have been an old scallywag. Yo ho ho and a bottle of rum, etc.

MO' MONEY

Sugar was a big business, a big deal. The point I am trying to make about sugar's impact on the Caribbean is this: When it became popular, everyone wanted to plant it everywhere; much like what happened with tobacco in the Mid-Atlantic US in the 13 Original Colonies. Both were luxury commodities for the Europeans, that suddenly became widely available and wildly cheap to the masses. Both were easy to grow and generated huge revenues. Everybody wanted it, and producers made buckets of money on it. However, as I suggested, sugar is very labor-intensive. The plantation owners need lots of labor, and cheap labor is preferable. Cheap labor? How about free labor?

Guess what that means? That's right: slaves. "Hey guys, let's enslave the local population! It's the perfect solution!" The Euro-trash tried. But most of the natives, as I pointed out in the Mexican chapter, died of European imported disease. It virtually wiped out everyone in the Caribbean before they even saw a white dude. The few that were left over got worked to death in very short order. Basically, the colonizers wiped out the native populations of the Caribbean islands.

So the colonizers were on the lookout. "We need more labor. Where are we going to get them?" You already know the story. They found out that they could bring people over from Africa. This set up what is called the **Triangle Trade**. There are several different routes one can take on the Triangle Trade, but the bottom line is: if you're going to sail about from one place to the other, you never want to sail the ship empty. It's a waste of time and resources to move an empty ship. You always want to have something in your boat that you will trade or sell at the next place before picking up another load to take somewhere else. So what to take on the first leg of the triangle trade?

Let's say you're a British trader. What do you have in Great Britain? Well, you have finished goods. You've got guns, knives, swords, cigarettes, alcohol, woodworks, woven fabrics, and a cute accent. See, all the places in the New World, they just have raw commodities that must be brought back and processed. You can process tobacco, process wood, process cane

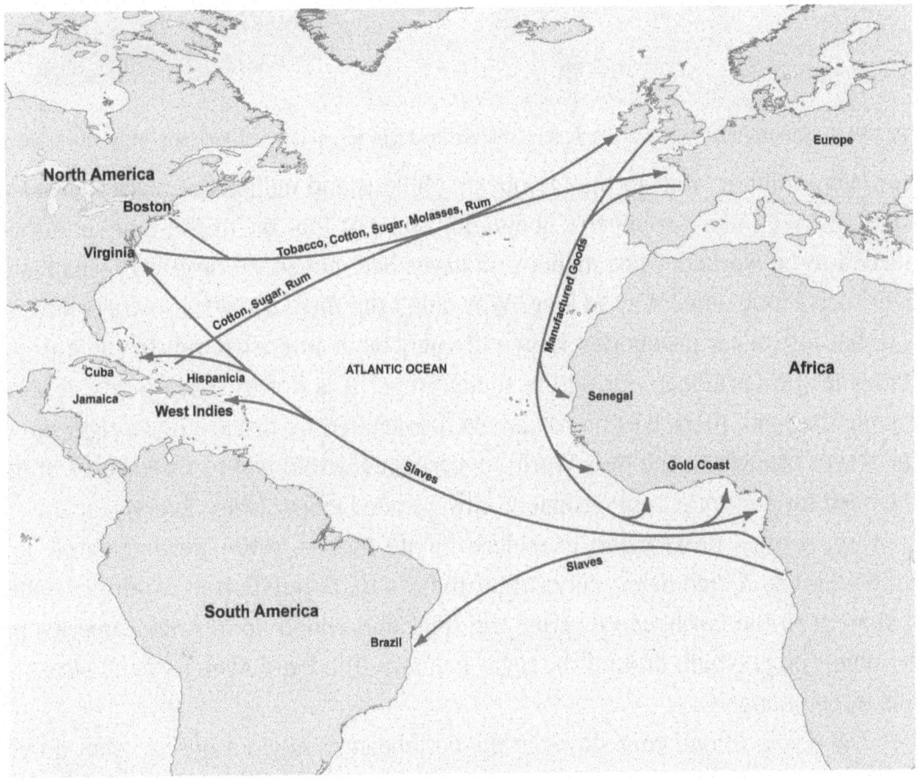

The Trade of the Triangle: massive profits for the white businessmen, massive suffering for Africans and indigenous.

sugar, or take sugar and do something else with it. Turn that into rum, actually, just add value to it in any way you can so you can charge premium prices and keep yourself in business.

So you're in Britain and you load up your ship with finished goods; in this case, alcohol and guns—what a fantastic combination. You take your alcohol and guns, and you float down to Africa. This is the first side of your triangle. What are you going to get in Africa? You're going to offload the liquor and guns and sell them to the local chiefs in exchange for what? Slaves. Fill the ship that you just emptied, fill every last nook and crevice with slaves. What are you going to do with those? Take them on the next leg of the trip to the Caribbean.

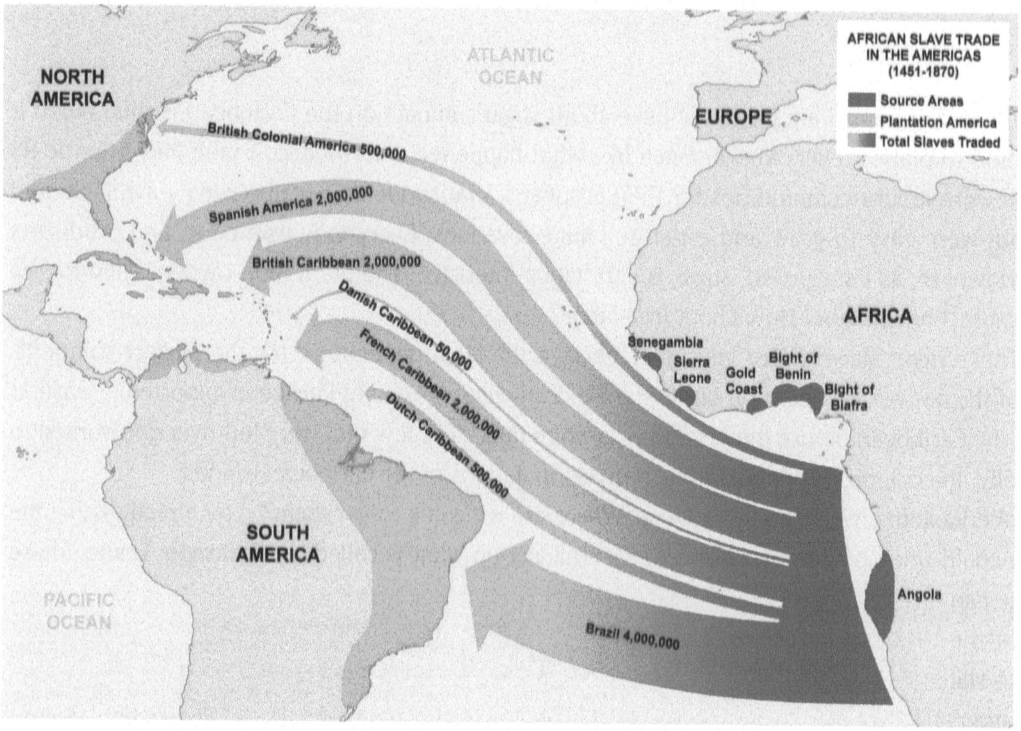

Massive human trafficking. a.k.a. the deadly "Middle Passage" for millions of Africans. Most never made it.

Now I am getting ahead of my story, because not all of the slaves went to the Caribbean. Some of these slaves went directly to the east coast of South America, and a lot fewer went to Anglo-America, proportionally speaking. However, millions went to the Caribbean. Millions and millions of black African slaves ended up in the Caribbean. Why so many? Well, the Caribbean's population was not that big at the time, and there were a lot of sugar plantations that needed to be worked. Once millions of slaves had moved to these very small islands, why did the slave trade continue over such a long time? Why so long? Why didn't the slave traders or owners have the slaves propagate themselves?

Because sugar production is an extremely labor-intensive industry in a tropical location. It's HOT. It is hotter than Hades in the Caribbean during the summertime. It is hot in those fields. It sucks. Not only that, but there's malaria. People died like there was no tomorrow. Devastated by disease or straight up worked to death. A huge percentage of the slaves brought to the New World to work died within two to three years of their arrival. As you can see, this caused the need for continual replenishment. Always need more slaves. Every year the traders needed more because the slaves kept dying off. I don't mean to underestimate figures; there were certainly large numbers of slaves brought to the United States. A significant minority of today's US population is of African-American descent. However, the numbers that went to the Caribbean dwarfed the total that ended up in Anglo-America proper. This is all solely because of the economic engine built around the sugar industry. The Plaid Avenger can't stress the demographic and cultural impact of this trade enough.

Once you offload your slaves in the Caribbean or Anglo-America, what do you pick up then to take back home? All those raw commodities. Fill your ships with sugar, tobacco, rum, and wool. Take all that stuff back home and process it.

The heck with sugar, I make reggae.

That's all the things that people want, and it's cheap in raw form. You've made money on every single leg of the journey.

That's the Triangle Trade, which lasted for several hundred years. Lots of folks got really, really rich from it, at a terrible cost in human lives, primarily African lives. Sugar can't be underestimated, or undervalued, in terms of its contributions to human suffering, economic growth, and its worldwide dietary impacts.

Am I telling you this long tale just so you can understand why people eat so much sugar in the world today? In part. But also think about the face of the Caribbean today. How would you describe Bob Marley? And please tell me that you know who he is. More generally, what does your average Jamaican or Barbadian or even Brazilian look like? Even broader, why are many faces black in the Caribbean? Think of Dominican Republic. Think of Haiti. They aren't European white, and they aren't Chinese; they are African, as are most of the populations of the Caribbean islands. Why? Because of the Triangle/Slave trade, which significantly impacted the demographics of this region.

Many slaves also ended up in what is now Brazil, on the eastern coast of South America, which was another sugar production center in the New World. You can look at many countries in Latin America today, and you can see in the reflection exactly how much sugar they grew. The more sugar was grown in the country, the larger the black population presently is. When we think of the Caribbean, we see many islands are dominated by peoples of African descent. Not all of course, as there are fusions of others peeps on other places like Puerto Rico or Trinidad and Tobago. But the larger the African descent demographic of a Caribbean country, the larger the likelihood they were a major sugar producing place back in the day. That's absolutely true, and it's all because of sugar.

THE MELTING POT

The Caribbean is also a place that I consider to be a great melting pot. We often think of the US as a melting pot, but the Caribbean is even better. There is such a unique culture that emerges, on the individual islands and as a whole, from the numerous colonial powers that had colonies in the Caribbean.

As you can see from the figure on the right, the Spanish, the British, the French, and the Dutch had colonies here. These colonial powers were in control of these islands. Some of the islands changed hands between these countries multiple times. This equates to European cultural influence on top of the indigenous culture of the surviving natives, infused with the African culture brought along during the days of the Triangle Trade. Over time, this has created a unique place on the planet, culturally speaking.

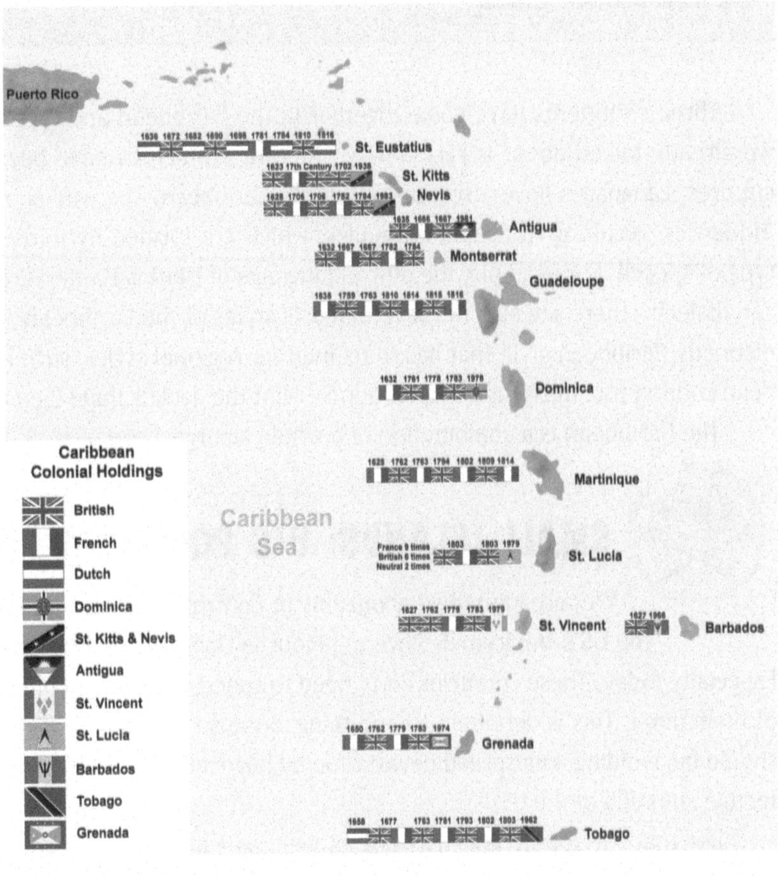

What Is the Deal with Voodoo?

Voodoo (which can be spelled many different ways, including Vudu, Voudun, Vodun, etc.) is an animistic religion that began in West Africa and was transplanted to Haiti by way of slaves. Animism is a philosophy that believes that everything, even regular objects, has a soul and a spirit. Most Vodouisants believe in a single God that created the world. However, this God is distant and detached from the world, so Vodouisants must turn to "mysteries," "saints," or "spirits" for help. Voodoo has many different traditions, and these different traditions have slightly varying beliefs and worship different spirits. But on the whole, most Vodouisants believe in personal spirits, harmony with nature and the world, and the importance of family. Western pop-culture sometimes depicts voodoo as a bizarre religion that is all about zombies and sticking pins in dolls. In reality, these elements are very small and insignificant portions of the religion.

Fun Plaid Fact: Harvard ethnobotanist Wade Davis went to Haiti in 1982 and claimed the zombie tradition actually had some real basis in fact. Years ago, when voodoo priests had determined that somebody deserved to be punished, they would feed them pufferfish, whose toxic glands put people in a paralytic coma. After burying this person, they would dig them up a couple days later, and give them another dose of hallucinogenic drugs. This entire experience, including the drugs, the burial, and people's religious beliefs, were enough to actually convince people that they were really will-less zombies. Obviously, Davis's research has been questioned. Check out the film *The Serpent and the Rainbow* (1988) for Davis' screen adaptation. Pretty spooky!

Three continents have come together in the Caribbean and crystallized in several ways. The first obvious fusion that we already talked about is race. Many different ethnicities have been combined to come up with several unique ethnic cultures. Languages have also changed in the Caribbean. The way people talk is a mixture of colonial, native, and African influences, resulting in Creole languages, which are formed by infusing terms from several different languages into one type of speech. For example, the official language of Haiti is Haitian Creole, a combination of French and several West African dialects. There are also the several music styles of the Caribbean. The biggest one we can think of is Reggae, which is a distinctly Caribbean style that has roots in other regional styles, such as Calypso. And religions have morphed in the Caribbean context too: most famously, voodoo—what the heck is that? (See the box above.)

The Caribbean is a combination of a whole smorgasbord of stuff, from three different continents.

SMALL ISLANDS; BIG POLITICAL PROBLEMS

We already talked about this in Central America, and the Caribbean shares the common feature of being in the **US's backyard**. Since the Monroe Doctrine, no foreign powers will mess around here; the US won't allow it. Especially today. These countries don't need to spend money on a military or any other form of national defense, and many of them don't. This is definitely a good thing, because they are endemically economically challenged. Their biggest concern should be avoiding widespread devastation by hurricanes, not getting invaded by anybody, because they are under the protective umbrella of the US.

But that's a double-edged sword as well, as I have said before, because being under the umbrella means more US intervention. Haiti, in particular, has been controlled by the US several times in the last 100 years. The US military occupied

Haiti from 1915 to 1934. Why did the US military occupy Haiti? Once fighting ended after the long Haitian Revolution period, they did fairly well until their national debt started to get out of control. This was fine for the American banks they owed—until Haiti became unstable and had another revolutionary movement. Scared that they wouldn't get any of their money back, the banks pressured US President Woodrow Wilson, who sent in the troops to promote 'stability,' restructure the economy, and therefore ensure repayment of loans. The occupying US forces improved Haiti's infrastructure while they were there, mostly by supervising forced labor chain gangs to improve old roads and build new ones.

Woodrow: the first Haitian "stabilizer." Many more to follow.

The US has supported some fairly brutal leaders in Haiti, as well as in the Dominican Republic, which was taken over by the US in 1965 after a right-wing coup (whose connections to the US are unclear) had trouble completely overthrowing the leftist government there. The US went in to make sure that the leftists wouldn't regain control and turn it into "a second Cuba." Which gets us to the biggest brouhaha that the US has had in the Caribbean, if not the world . . .

CUBA

A fifty-year-old economic and diplomatic embargo. Refusal to allow US citizens to visit the country. CIA sanctioned assassination attempts on the leader of the state—Fidel Castro. Why does the US hate Cuba so much?

Historically, the United States and Cuba have had a very close relationship, especially because Cuba is only 90 miles away from Key West, Florida. The United States has dominated Cuban politics, intervening whenever it wasn't happy with the leadership, and basically treating Cuba like a vassal state. This relationship changed in 1959, when Fidel Castro and a band of revolutionaries seized power in Cuba and installed a communist regime. Since then, Cuba and America have had absolutely no diplomatic contact or official trade. Pure hate to hate across the Straight.

In the 1950s, Cuba, especially the capital city of Havana, was a playground for rich Americans. Flights left Miami every hour on the hour for a sin city of drugs, prostitution, and gambling for which Havana had become famous. Havana became a hangout for American mafiosos, and many believed that Havana was going to become bigger than Las Vegas. (Check out *The Godfather, Part II* (1974) for insight.) Some Cubans were becoming rich, but a vast majority of them were still desperately poor and living in the countryside, and these peasants became distrustful of their corrupt government. To make matters worse, Cuba was controlled by an unelected dictator named **Fulgencio Batista**, who had somehow managed to anger everybody in Cuba from the landless poor to the middle and upper classes of society. Every time people tried to revolt against Batista's government, he responded violently and brutally, often executing those who were responsible. The time was ripe for somebody like **Fidel Castro** to come and seize power in Cuba.

Cuban "liberator" . . . or "dictator"?

The Cuban Revolution was an intense struggle. Castro and his army of around 100 men—including 20th century revolutionary Che Guevara—sailed to Cuba on a rickety yacht named the Granma. Things turned sour when they landed, and the Batista army quickly killed most of them. The remaining dozen or so soldiers went and hid out in the Sierra Maestra mountains, beginning a long guerrilla war that quickly gained supporters around Cuba. After defeating Batista's forces in several key battles while consolidating control of the island, Batista fled Cuba on January 1, 1959, and the rebels took power.

At first, the United States supported the Cuban revolution because they thought that Castro was a liberal constitutionalist and nationalist and that he was good for Cuba. He ended corruption, gambling, prostitution, and he expelled the American mafia who were living in Cuba. However, Castro soon turned to communism as a guiding ideology, and took some steps that angered American officials. He began seeking ties with the Soviet Union (remember, this was during the Cold War!) and purging anti-communist people from his administration. He nationalized all foreign-owned property and even nationalized the land and property of most Cubans, both wealthy and poor ones. For this reason, many rich Cubans, extremely angry at Castro, left Cuba and went to America. The largest exile group formed in Miami, which has become a famous base for anti-Castro activity to this day.

Why does the United States hate Cuba? During the Cold War, it was a no-brainer. Cuba was a satellite state for Soviet Russia, and America was fighting a war against communism. In 1962, Americans became very uneasy when Castro let the Soviet Union point nuclear missiles at America from Cuban soil during the Cuban Missile Crisis, which you can read more about in the box to the right. During this time, the United States made several attempts to assassinate Castro, even hiring Mafia members to do the job. It never worked. After the Soviet Union collapsed, Cuba became much poorer and desperate because its main supporter was gone. Many people thought that the time had come for America and Cuba to begin re-establishing relations.

Yet, the Cuban and US governments stayed enemies, mainly because of Cuban exiles living in Miami. Remember we talked about these guys before? The rich guys who got their land taken away by Castro's communist cartographic changes? More than one million Cuban exiles now live around Miami, and these people hate Fidel Castro with a flaming passion. They had been living good lives in Cuba before the revolution, and the communist regime took everything they had away from them. In America, they have become a huge lobbying force that is very influential in American politics; their goal is to do everything they can to hurt the Cuban government and force Castro from power.

This is a goal they don't have to work at quite so hard now, since Fidel Castro actually did step down from power due to bad health in 2008, and then finally stepped down from being alive in 2016. Yep. He died. So all is well now between the USA and Cuba, right? Nope. Fidel's brother Raúl Castro assumed the presidency, a position he held until April 2018, but remains the first secretary of the Communist Party, still holding considerable influence over government policy. **2021 UPDATE:** Raul finally stepped down as the head of the Communist Party!

What's the Deal with the Cuban Missile Crisis?

Many people consider the Cuban Missile Crisis as the closest America—and the world—has ever come to nuclear war. In 1962, during the Cold War, the Soviet Union deployed nuclear missiles in Cuba that were aimed at the United States. Part of the reason the Soviet Union wanted nuclear missiles so close to America was because America had deployed similar nuclear missiles in Turkey aimed at Soviet cities. After American spy airplanes revealed the existence of the missile sites in Cuba, there was an intense debate in the White House about what America should do. Some people in the Kennedy administration favored a naval blockade, of Cuba, while others insisted that America invade Cuba. The Kennedy administration decided on a blockade, and it was later revealed that the Russians had short-range nuclear missiles that would have probably been used on American troops had they invaded.

For several days, the situation was very tense between the United States and Russia. An American U-2 spy airplane was shot down over Cuba and another was shot down over Russia. After 12 days, Soviet Premier Nikita Khrushchev eventually decided to remove the missiles from Cuba, and in return, America promised not to invade Cuba and also to remove the American missiles that were in Turkey. This crisis was widely seen as an embarrassment for the Soviet Union, and the Cubans felt betrayed because the situation was resolved between America and Russia, with Fidel Castro playing no part in the negotiations.

Fun Plaid Fact: The Cuban Missile Crisis led to the development of the "hotline," or the famous red telephone that Moscow and Washington could use to have direct communication in the case of another crisis.

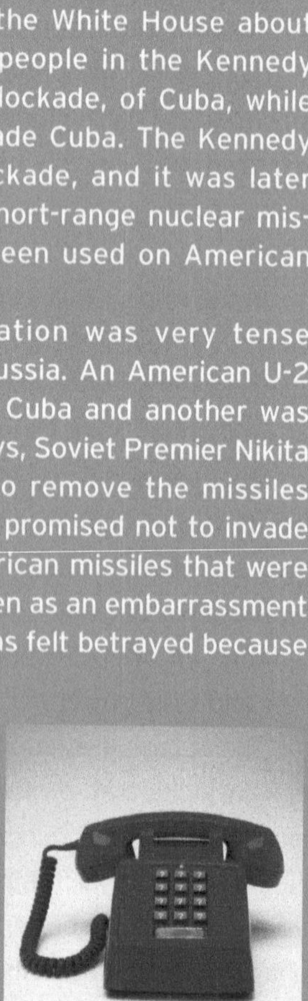

The bat-phone to Moscow.

Woo-hoo! So times are a'changing, right? Well, nothing has changed at all yet, so stay tuned to see what direction the new US President Biden administration decides to go with this.

But what about the USA? It doesn't mean jack if the president of Cuba wants to improve US relations if the US isn't on board. Cuba has been left-leaning for decades and they redistributed land, resources, and wealth through nationalization—and the US can't stand that. Cuba also had missiles pointed at them for a spell. That didn't help much either. Taking the world to the brink of nuclear war probably caused some of the animosity that has kept that archaic embargo in place against them to this day.

A historic warming of Cuba–United States relations got hot in December 2014 when US President Obama and Cuban President Raúl Castro announced the beginning of a process of normalizing relations between Cuba and the US. The normalization agreement was secretly negotiated in preceding months with the assistance of Pope Francis. The agreement

What Is the Deal with Che Guevara?

Why is Che the most famous revolutionary in the 20th century?

Che Guevara has become a symbol of youth rebellion for generations, but most have no idea who he was or what he did. He has become a wildly popular pop-culture icon and a symbol for alternative culture and also for communism and socialism. His images have been reproduced on a vast array of merchandise, such as T-shirts, posters, coffee mugs, and baseball caps. In Cuba, his face adorns billboards and the sides of government buildings.

Ernesto Guevara (Che is a nickname that means "buddy" or "pal") was born in Argentina and studied to become a physician. However, he took a year off of school and traveled around Latin America on a motorcycle to witness first-hand the crushing poverty that many Latin Americans were enduring. These experiences helped turn him into a revolutionary and a Marxist. He believed that revolution was the only way to obtain equality for the people, so he began traveling to Latin American countries and helping out with Marxist revolutions, first in Guatemala and later in Cuba.

Around 1955, Che met Fidel Castro and agreed to join him in this battle against the Cuban dictator Fulgencio Batista. They sailed to Cuba in 1956 on a rickety yacht named the Granma, and after a two-year war, took over Cuba. Since Che was a Marxist, he was instrumental in persuading Castro to become communist. For his work in the revolution, Che was awarded some important positions in the Cuban government and wrote some books about guerrilla warfare right after the revolution. However, he could not sit still long. He soon left Cuba to encourage revolutions in other parts of the world, including Congo and Bolivia. Che was eventually caught and assassinated by Bolivian forces, assisted by the American CIA. They chopped off his hand to keep as proof the deed was done.

While many people admired Che, others saw him as a dangerous lunatic. He has been accused of murdering and torturing thousands of people in Cuban prisons, as well as murdering peasants in guerrilla areas where he was fighting. Che has also been called an idiot and has been accused of being single-handedly responsible for the collapse of the Cuban economy, which had been one of the strongest in Latin America up to that point.

Che is famous for many reasons. For one, he was good-looking, charming, and popular with the ladies, as opposed to most world leaders. Secondly, he was killed in the prime of his life fighting for his cause, so he has that whole Bob Marley/Jimi Hendrix/Buddy Holly dead rock star thing going. Thirdly, even though he came from a rich family and was trained as a physician, he dedicated his life to helping poor people around the world and often put his life on the line for his beliefs. For this reason, many people who disagreed with his communist beliefs still admired him for his self-sacrifice. Lastly, an important reason Che has become so famous is because of a really cool-looking photograph of him taken by Albert Korda in 1960 that has become a symbol for counterculture youth around the world.

lifted some travel restrictions, restrictions on remittances, U.S. banks' access to the Cuban financial system, and started the reopening of embassies in Havana and Washington, which closed in 1961 after the breakup of diplomatic relations following the establishment of Cuba's close alliance with the USSR. Things are practically back to normal! Well, maybe not so much. Historically, liberal US presidents have done everything they could to open relations with Cuba, and then the next conservative presidents have closed everything up again. Time will tell if any big changes really take hold.

Positive US-Cuban relations could definitely affect other Latin leaders' attitude toward the US. We're talking regime shifts, more open economic discussions, perhaps new thoughts about the FTAA . . . but nothing is certain yet. Now it's up to us world-watchers to pay attention and see what happens!

DOUBLE TROUBLE CRAY CRAY 2019 UPDATE: Holy moly, the pendulum is swinging wildly on these US/Cuban relations now! Back and forth we go! Just when it seems that things between the two countries were getting better, we are body-slammed by two unrelated, EPIC historic events: 1) The Castros are almost all gone! Raúl Castro stepped down and Miguel Díaz-Canal was chosen as the leader after a secret vote by the National Assembly in April 2018. Castro supported Díaz-Canal's unopposed candidacy. Raúl Castro remains a member of the National Assembly and is still head of the Communist Party until 2021. When Castro steps down, Díaz-Canal is intended to take over as head of the Party. . . so perhaps even better US/Cuban relations would be accelerated now, right? Wrong! Because 2) arch-conservative Donald Trump became President of the US, promising to undo all the stuff Obama did and even punish Cuba even more if they don't capitulate on his forthcoming demands. Which he proceeded to do immediately. Door slammed shut again. **2021 UPDATE:** The door swings back open again! Raul finally stepped down as the head of the Communist Party in April 2021! First time Cuba has not had a Castro at the helm in 70 years!!!! Woo-hoo! So times are a'changing, right? Well, nothing has changed at all yet, so stay tuned to see what direction the new US President Biden administration decides to go with this. Who knows what the following year may bring? So you could be Spring Breaking in Cuba by this time next year, or there will be open wartime actions against the island nation. Bizarre!

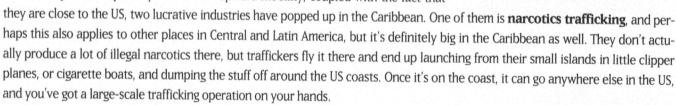

THE CARIBBEAN TODAY

This brings us to the two issues that do show how the Caribbean is trying to diversify their tourism-based economy and bring in higher paying jobs: offshore banking and drug trafficking. Wow. I guess that's one way to diversify. Let's elaborate.

Perhaps due to poverty and lack of upward mobility, coupled with the fact that they are close to the US, two lucrative industries have popped up in the Caribbean. One of them is **narcotics trafficking**, and perhaps this also applies to other places in Central and Latin America, but it's definitely big in the Caribbean as well. They don't actually produce a lot of illegal narcotics there, but traffickers fly it there and end up launching from their small islands in little clipper planes, or cigarette boats, and dumping the stuff off around the US coasts. Once it's on the coast, it can go anywhere else in the US, and you've got a large-scale trafficking operation on your hands.

The US is, of course, not happy about this. Unfortunately for them, it's simple market forces at work. Capitalism at its fine-tuned best. Dig the facts: The US is the end-consumer of virtually all narcotics made in Latin America (okay, some European crack-heads get some too, but not that much). The Caribbean is very close to the US, and has lots of poor people in it willing to do anything to make ends meet. The ocean is a hard entity to control and monitor all the time. End result: the Caribbean is an excellent launching pad for drug-runners to get to the US. Always has been. Always will be. It's a growing sector in the region.

The second industry that has popped up is **offshore banking**. Lots of people now do offshore banking through an account in the Bahamas. Which is great! I guess. Isn't it? Why wouldn't people just put their money in a US bank?

If you have millions of dollars in the United States, you have to pay taxes on it. And if you happen to be doing anything that happens to be, say, slightly illegal, then you might get punished for that. One good way to avoid being caught evading taxes, or doing dirty deeds or whatever, is to take your money outside the US altogether. Banks are increasingly opening up in the Caribbean, and they're operated essentially like Swiss bank accounts. There is a high degree of privacy, and since these are sovereign states, they don't have to reveal bank records if the United States asks for them: "We're a private bank in a sovereign state, and we don't have to show you our customers' records, and we won't, thank you very much."

This happens even though the FBI, the CIA, or the IRS may know that the people who are putting their money in the bank are big drug dealers, or crooked businessmen in the US who are either scamming or just evading taxes, or they are mafia types trying to launder money. Places like the Bermuda, Aruba, Cayman Islands, and Bahamas—because they are more stable economically than the others and not entirely poverty stricken—have become the Swiss bank accounts of the Caribbean. Since it's not too far offshore, banking in the Caribbean is becoming much more popular. The bank is close enough to easily do business, but just far enough away to avoid those annoying taxes and laws in the US.

In summary, today the Caribbean is still under the US umbrella of defense, except for Cuba, which they would like to undefend, a.k.a. attack. As I already suggested, the region is also heavily reliant on the US economy: this makes them particularly susceptible to economic fluctuations, which means they are slightly unstable. Like many other places in Latin America, they have, as a general rule, a tremendous wealth disparity and poverty. Because what do they produce? Tourism, which isn't worth a lot, and sugar, which is worth even less . . . and the alternatives are drugs and offshore banking activities which are illegal or unsavory. We all think that hanging out on a Caribbean island is a slice of heaven on Earth; it more often than not is a hardscrabble slice of hard times for the economically and politically challenged folks who live there.

WHAT THE H? WHAT THE WHAT? WHAT THE WHAT? WTH?

What's the Deal with . . . Bob Marley?

Bob Marley, famous reggae singer and songwriter, was the ambassador of the Rastafari Movement to the Western world. Marley lived his life and created his music in the Rasta tradition. Rastafarianism is a religion that developed in Jamaica in the 1930s. Rastafarianism has also come to embody a cultural movement of racial equality and African heritage. Central to Rastafarianism is the belief that Haile Selassie I—former emperor of Ethiopia—is a reincarnate of God (or "Jah" in Rasta terminology). Rastas cite Old Testament prophecies to justify this belief. In fact—like Judaism, Islam, and Christianity—Rastafarianism is an Abrahamic religion. Many Rastas see themselves as true Israelites—the descendents of the 12 tribes of Israel.

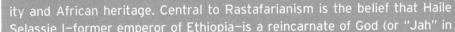

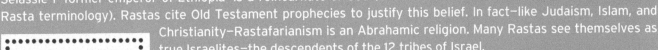

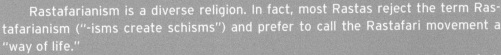

Rastafarianism is a diverse religion. In fact, most Rastas reject the term Rastafarianism ("-isms create schisms") and prefer to call the Rastafari movement a "way of life."

However, there are several near universal symbols and traditions in Rastafarianism including:

- → The lion—the proud and noble ruler of the African savannas
- → Ethiopia—the home of Selassie I and a country that fiercely resisted European colonialism
- → Dreadlocks—a natural hairstyle at odds with European fashion, and
- → Ganja—the stuff they smoke.

There are over one million Rastas in the world today, including approximately ten percent of the Jamaican population.

CARIBBEAN RUNDOWN & RESOURCES

View additional Plaid Avenger resources for this region at http://plaid.at/carib

BIG PLUSES

- Awesome group of semi-tropical islands; a tourism mecca
- Just off the coast of the largest economy on earth . . . the US!
- Have no need to waste money on militaries; the US would protect them from foreign attacks without even thinking about it

BIG PROBLEMS

- Hurricanes, earthquakes, pirates (specifically, Johnny Depp)
- Smallness: small islands, small economies, small amounts of resources
- Have no real economic engine or even valuable resource to export; sugar used to be king, but is now so cheap as to be worthless
- Significant transportation corridor of illegal narcotics and zone for money laundering and tax evasion
- Highly, highly dependent on US economy; when it stinks, they stink

DEAD DUDES OF NOTE:

Bob Marley: Jamaican singer-songwriter and musician who is the most widely known and revered performer of reggae music, and is credited for helping spread both Jamaican music and the Rastafari movement to a worldwide audience

Ernesto "Che" Guevara: Major figure in the Cuban Revolution, a devout Marxist, intellectual, guerilla leader, doctor, diplomat, etc. etc. whose face is now iconic for reasons that most T-shirt wearers don't understand. Had a falling out with Castro after the Revolution, and went on to incite other social uprisings in Africa and Latin America until the CIA iced him in 1967.

Fidel Castro: The un-killable king of communism in the Western Hemisphere, leader of the Cuban Revolution of 1959, provoker of the Cuban Missile Crisis of 1962, grand poobah of Cuba from 1959 to 2008, and sole reason for the 50 year old US embargo against Cuba. He was revered in leftists circles as a demi-god, and his death (in 2016!) will transform the US/Cuban relationship overnight. Maybe. Not.

LIVE LEADERS YOU SHOULD KNOW:

Raúl Castro: Former President of Cuba, much quieter older brother of Fidel who is continuing to carry on the wanna-be communist legacy. But, many analyst assume that Raúl is already making slow moves to open up the island's economy and politics in preparations for his imminent departure from politics, at which time Cuba will change dramatically. Maybe.

PLAID CINEMA SELECTION:

The Harder They Come (1972) A poor Jamaican, a 1970s anti-hero, tries to make it with a hit record but breaks bad when he refused to sign away his rights, starts dealing marijuana and becomes an underground fugitive and political hero. This no-budget, ghetto film was the first to bring this music called "reggae" to the big screen and subsequently the American mainstream and made Jimmy Cliff a star. Great look at poverty in Jamaica and gangsta' life before it became hip. And you will hear the the theme song played about ten million times. Great period piece.

Extraordinary Women (2009) A moving documentary that captures the lives of six of the most influential women from the Dominican Republic. The women in the film all lead very contrasting lives from a woman who lead the way in health-care to women that were very influential in the politics of Latin America. Creates an excellent view of the culture of the Dominican Republic.

Havana Blues (2005) The story of two young musicians in Cuba who make music that combines the traditional aspects of Cuban music with more modern types like rap. Takes a look at the struggle to keep their sound while facing the monetary restraints of the music industry. Film contains criticisms of life in Cuba including poverty and Nationalism among other things.

Life and Debt (2001) An intense documentary that examines the economic and social situation in Jamaica. The film provides a harsh view of the World Bank, whose policies forced the Jamaican government to enact strict economic reform, which left the country nearly $5 billion in debt. While not a happy look at Jamaica, opens an interesting window into how life really is for some people.

Burn! (1969) A British agent, Sir William Walker, sent to the island of Queimada (a fictional Portuguese colony) to organize an uprising of black slaves to overthrow the Portuguese regime. Great Britain wants to get economic control of the island, because it is an important sugar cane producer. Interesting historical, albeit fictional, look at 1800's Caribbean economy and great-power-politics in the region.

Pirates of the Caribbean (2003, 2006, 2007, 2011) Q: What kind of list would this be if we didn't mention a trilogy, I mean a quartet, of films that have grossed nearly $3 billion with the name "Caribbean" in it? A: A better one.

MAP GALLERY

View Map Gallery & additional Plaid Avenger resources for this region at http://plaid.at/carib

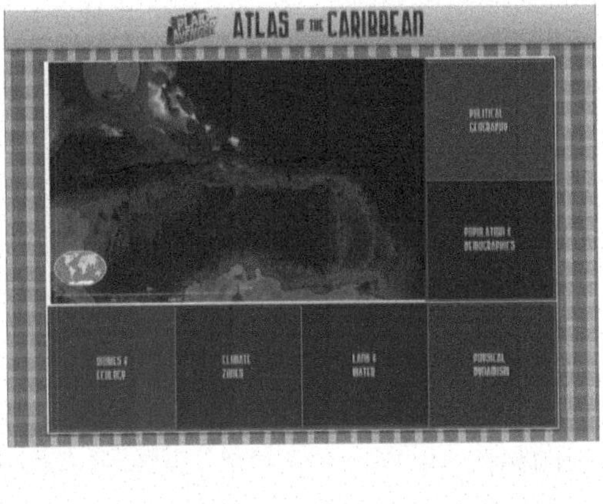

CHAPTER OUTLINE

1. Fantastic Physical Features
2. Socially Speaking
   2.1. The Dark Side
3. The Continent "Righting" Its Course?
   3.1. Former Leftward Lurch
   3.2. Into the Venezuela Void
   3.3. Right Turn!
4. Blocked, Unblocked, and Re-Blocked: Battle of Trade
   4.1. Continental Characteristics of Economy
   4.2. Making. . . Then Breaking. . . Now Re-Making of MERCOSUR
   4.3. MERCOSUR Back on Top
5. Brazilian Bombshell
6. Let's Samba out of South America
7. Map Gallery

South America

17

ALRIGHT plaid fans, it's time to get to our last subregion of Latin America, the region/continent of South America. Makes it nice and tidy. South America is the fourth largest continent on the planet: a landmass comprised of 12 countries, and fairly good-sized countries for the most part, which distinguishes them from the other subregions in Latin America. The Caribbean and Middle America are mostly dinky states, but there are full-grown countries in South America—including Brazil, which is the fifth largest country on the planet. So The Plaid Avenger always has to ask the question: you have the United States of America up there in North America, why don't we have a United States of South America down here? Why are these distinct countries in South America not unified? The answer is: I'm not sure. From all practical standpoints, we should be looking at a much more unified entity down south of the border. Maybe that's where they are heading.

Why isn't there a US of South America? Many people would answer, "There are all these differences from country to country! Bolivia's got nothing in common with Peru or Chile or Argentina!" Many people would be wrong. What is it that's the same about these places? We already talked about a lot of these similarities back in the Latin America chapter. There's a common history here, a common colonial presence. Lots of commonalities with very few differences. Looking across the entire continent, there are only a couple major languages spoken. You really can't say that about any other continent besides North America (okay, maybe Australia . . . but isn't it more like an island anyway?). Portuguese spoken in Brazil, and Spanish spoken in the rest of South America in the north and west. How about major differences in religion? Nope, they are all die-hard staunch Catholics, which we talked about in the Latin America chapter. So where are these differences? What's different here?

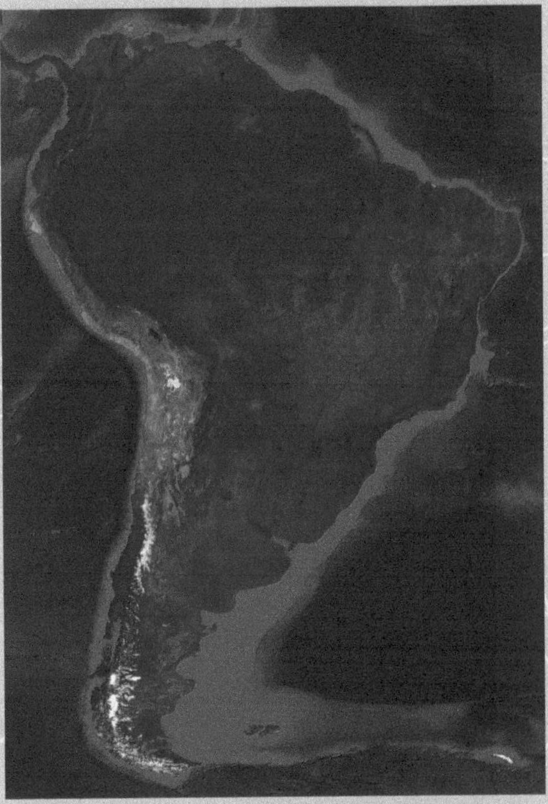

Inspired Inca Invention: Machu Picchu Mucho!

We have to look a little bit deeper into the histories of these places to find any differences whatsoever. Let's look at a couple of major entities. In precolonial history there's the **Inca Empire** down here, as opposed to the Aztec up in Mexico—but like the Aztec up in Mexico, the indigenous cultures were largely wiped out, or denigrated fairly rapidly, after the introduction of the European colonial masters. Even that's not a separating factor; that's not why there is no United States of South America. How about independence movements? Did they all get up and run with different independence movements led by different folks in different places? Well, there were heroes all over, but there was one main guy who kind of led the parade for independence movements across the continent, and that was "George Washington of South America," **Simón Bolívar**. During the 1810s and the 1820s, he played a direct role in most of the independence movements in South America, and an indirect role in all the rest. So what is the difference? Why is this not the United States of South America?

Bolívar: The main man of South America.

FANTASTIC PHYSICAL FEATURES

Perhaps the main reason why we don't have a singular state in South America is because of the physical factors, specifically the terrain. If you look at the Rocky Mountain system up in North America, that's a pretty high range (highest peaks at 14,000 ft., averaging in at about 10,000 ft.) but not enough to really separate things from east to west in the United States of America. It's a divider, but not a permanent divider. However, when we get down to South America, the Andes system running through the spine of South America is indeed a true divider, both culturally and physically. At roughly (highest 23,000 ft., 13,000 ft. average) 13,000 thousand feet high, it's a big divider. The Andes separate Chile, Ecuador, Peru, and Colombia from everywhere else, so there is not a lot of east-west movement or cultural interaction across the continent. SO lots of similarities that exist between these countries, but then there is this physical factor, which keeps things fairly divided.

The Andes: major continental divider.

On top of that there are a couple of other physical factors to consider, one being a physical term I want you to know: **escarpment**. What is an escarpment? An escarpment is a very abrupt change in elevation right off the coast. It's a common physical feature in South America and also Africa. Some of the best examples of this are the Brazilian Highlands and the Guiana Highlands up north. The Brazilian Highlands are in the southeastern quadrant of Brazil, but they lead all the way down to Argentina. The Guiana Highlands go all the way through into Colombia, Venezuela, Guyana, and the northern part of Brazil. This kind of abrupt change as you get off the coast means there's not a lot of coastal margin. A few miles inland, although it's not the Andes or high mountains or anything, there's a little jump of maybe a few hundred feet up to an elevated plateau.

Try getting your boat up this river. Angel Falls, Guiana Highlands

Throughout history, this has kept the interior hard to get to because explorers or traders traditionally traveled on rivers through waterways to get into the interior of a continent. That's not the case in South America. Most of the places that have escarpments cause rivers to turn into waterfalls. Boats and waterfalls don't mix, so you'll have to figure out a new way in, Conquistadors! Unfortunately, your journey to the waterfall didn't really bring you in too far from the coast either. It's been historically hard to enter and traverse the continent. The Andes on one side of South America and the highlands or escarpments on the other side of South America give travelers a double whammy of difficulty. And that's just the terrain!

Brazilian Highlands and Guiana Highlands: Extraordinary Escapements!

There's another difficult double whammy for any travelers, traders, and explorers, which is the climate and vegetation combo. When thinking of South America, the gut reaction that people have is rainforest, big rainforest, a bunch of damn monkeys running around in the rainforest! In large part, that is true. The only river system that allows you to penetrate the continent at all is the Amazon River system. However, it's the one that coincides with the Amazon rainforest basin, meaning it's a huge jungle out there: the biggest continuous jungle in the world; the single largest acreage of rainforest on the planet. It's mostly throughout the northern part of the continent in Venezuela, Colombia, and Guyana. Brazil holds the biggest chunk of the rainforest, about 60 percent of the total. There's also some in Bolivia, some in Ecuador, some in Suriname, and some in Peru. But watch out; don't make the common mistake of thinking that all of South America is rainforest.

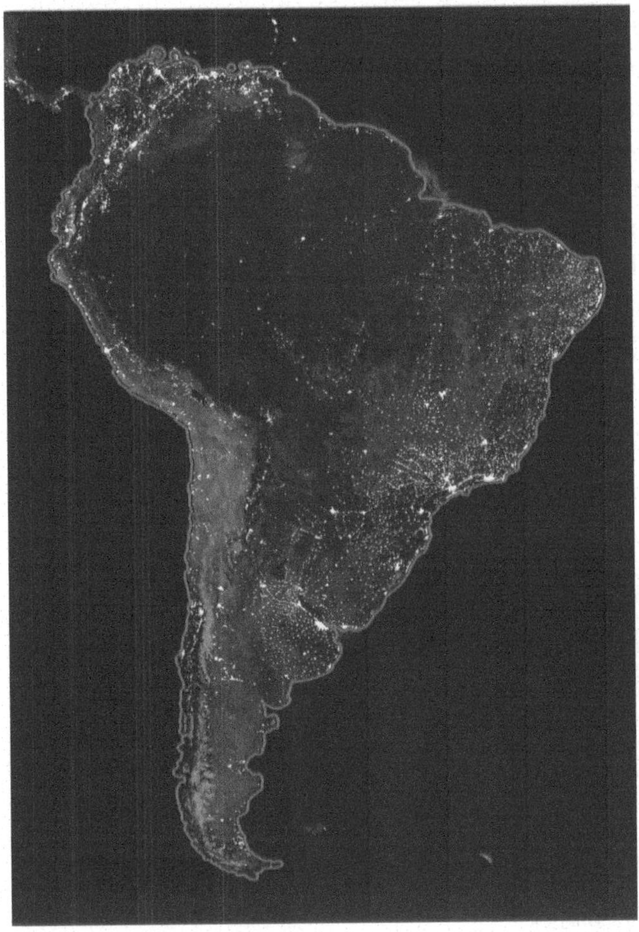
Lights out in the Amazon region!

Keep in mind that South America is the 4th largest continent. The equator goes through the northern part of Brazil in the true tropical region. But South America spans all the way down to the tip of Tierra del Fuego, which is closing in on the Antarctic Circle. And there's almost every possible climate in between. In Argentina and Chile, you will find midlatitude climates like humid continental, humid subtropical, with temperate forests very much like the eastern seaboard of the United States. The Pampas, the grasslands of Argentina, are very much like the Midwestern United States. You've got the Atacama—it hasn't rained there in a hundred years—Desert in Peru and Chile. Over in Chile, there is even an area or two with a Mediterranean-style climate. This huge latitudinal range means South America has got a bit of everything, a huge diversity of vegetation and climate. It's not all rainforest.

It's a combination of the tough terrain: the Andes and the escarpments—manifesting themselves as the Brazilian Highlands and the Guyana Highlands—combined with thick jungle vegetation of the rainforest has served for centuries to keep people out of the interior, making it very hard to traverse the continent. The interior is sparsely populated, as you can see from our favorite Lights at Night image. Additionally, there is not a lot of communication and interaction from east to west, which is probably the main reason why there is not a United States of South America.

SOCIALLY SPEAKING

We have talked about a lot of social factors across Latin America: problems like poverty, wealth disparity, high urbanization rates, and landlessness all appear in South America, as well, and some of them manifest themselves in extreme ways. Is there anything different about South America? What's perhaps a little more unique about South American society in contrast to other places in Latin America?

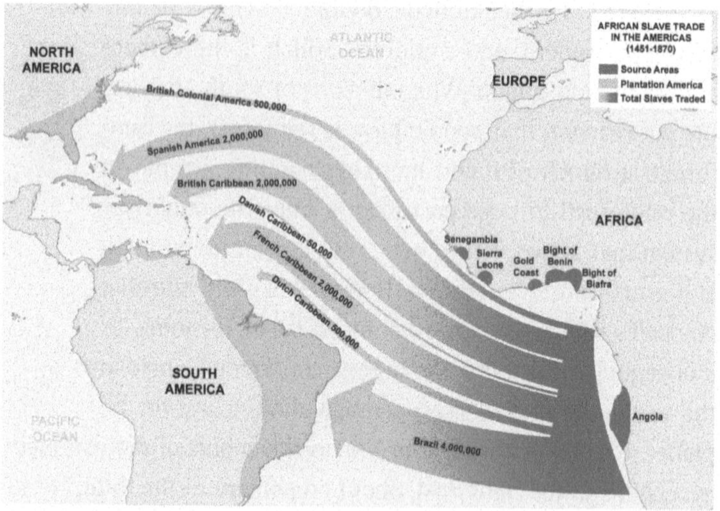

Glad you asked! South America is a true melting pot, and Brazil is really the centerpiece of this. What's the Plaid Avenger talking about here? Well, like the Caribbean during the colonial period, there was a huge infusion of folks from Africa, who were slaves at the time. Brazil and the Caribbean were the two main destinations of African slaves back in the day. Between three and four million Africans ended up in Brazil, and five to eight million more in the Caribbean, in contrast to the half a million that went to all of the United States. We are talking about exponentially larger numbers of Africans, and it shows in the face of Brazil as well as in the culture.

What does that equate to in today's world? There is a currently a significant African component in Brazil's culture as well as the rest of South America. Cultural input from the African continent added to the cultural fray in South America. What other continents are represented? We have the colonial masters themselves, the Portuguese and the Spanish. They have infused their culture into South America as well, through years of conquest and rule. Like the African flavor of Brazil, the European background is also very apparent in the way people look and speak, the music they create, and holidays they observe. You can see extreme manifestations of the European influence in places like Argentina and Chile.

This is not widely known here in the US, but during the 19th and early 20th centuries, there were huge infusions of Europeans to South America. We Americans tend to think of our country as the immigrant country, but not all of them came to the United States of North America. A bunch of them went down to South America. In places like Argentina and Chile, we see that there were significant numbers of Italian and German immigrants during this period of mass European movement. Even in the 21st century, you can run into the unusual phenomenon where entire villages speak only Italian or German.

These are all Brazilian peeps = perfect example of ethnic melting pot that is South America.

The most tangible manifestation of the Portuguese colonial heritage in Brazil is their official language: Portuguese. The Spanish were everywhere else, and that's why they speak Spanish in the rest of South America. Then add in the African influence. But there's one more: the indigenous influence. There were people actually in South America before the Spaniards and Portuguese got there. The Inca Empire dominated the Andes region, and so were some other small native groups that were spread all over the rest of the continent, similar in tribal structure to Amerindians in North America, but also distinct jungle-dwelling native cultures as well. Indigenous peoples still make up a significant portion of South America's population today, as opposed to some other places in Latin America where they were driven to extinction (namely the Caribbean).

Hernán Cortés and Francisco Pizarro: Conquistadors of Chaos.

What am I talking about? South America being a big place and the Inca having a big Empire meant there were lots of folks already there. As we already pointed out, when the conquistadors got here, not only did they replace the hierarchical structure of the Inca and Aztec Empires with themselves, but they also brought with them diseases that pretty much wiped out a huge component of the native populations all across Latin America, and across South America, in particular. Most natives of South America died long before they saw a white Spanish or Portuguese dude. However, there are still a lot of descendants of the indigenous peoples here, particularly in Bolvia and Peru, which had the highest concentration of indigenous peoples when the conquistadors arrived. Peoples of three continents form the social mix in South America, with some social stratifying impacts that are not so nice . . . which brings us to. . . .

THE DARK SIDE

A theme that I didn't elaborate on in great detail in the Latin America chapter is the concept of racism, because it manifests itself in a greater extreme in South America than it does in other places in Latin America. Racism is a very real part of South American society, and perhaps it's worse than anywhere else in the Americas (okay, the USA is a really close second, and seemingly getting worse recently). How is it more extreme here than everywhere else south of the US border?

There is a direct correlation between how light your skin is and how well you are going to do in life, how easy life is going to be for you, how many times you might be arrested, or if you are going to get a bank loan. The lighter your skin is, the more doors are open to you. If you go into a bank to get a loan, you will probably get it if you have light skin; if your skin is very dark, then you probably won't. If you're walking down a street somewhere in a bad part of town and the cops roll by, they are probably not going to bother you light-skinned folk, but if you have dark skin, you're more likely to get harassed. It's true that there is still racism in the United States, as there is some amount of racism in every society across the planet; but it's nowhere near as ingrained as it is in South America. You can ask folks on the street there all day long and nobody will disagree with that statement. It is a very real and very visible part of life. Of course it's been outlawed; it's not legal at all, but it still exists. It's still the way it is.

The only exception to the very strict hierarchy of color is the indigenous population. They are at the bottom rung, even though it's not a direct skin correlation. You can clearly see that the indigenous folks are not as dark-skinned as say, someone with an obvious African heritage. Even though they aren't as dark, they have fewer opportunities and are looked upon as the lowest tier of society virtually everywhere in South America. If you look at a list of classes that are most impoverished, the indigenous will always be there at the top of that list. It's true across all South America, and it's important because this manifests itself in today's world. Lots of these indigenous groups are fighting back for their rights. There was even a dude named Evo Morales of indigenous descent who just got elected to the presidency of Bolivia, the first indigenous president in South America since the Inca Empire. This is a very big step forward for the indigenous population.

Evo Morales: "I used to grow cocaine. Now I'm President. Go figure!"

"I used to be President of Peru . . . and I'm of Japanese descent! Go figure!"

These are very real factors that I want you to be aware of in order to comprehend what's going on and how the social world is slightly different in South America. But wait a minute! We're not done yet, because there is actually a fourth main ethnic group in South America. Along with the indigenous, the European, and the African influence, there's a growing Asian influence. Similar to the United States, there are folks that came across from Asia in the last 100 years. Perhaps not in numbers as large as those in the United States, but still significant and continuing to grow. I pointed out that the current president of Bolivia is of indigenous decent, but check out a former president of Peru. . . .

Alberto Fujimori demonstrates this other element of the diversity in South American society. Here is a guy of direct Asian persuasion who ends up being the president of Peru from 1990-2000. **2021 UPDATE:** Think that Alberto was just a singularity in being an Asian descent human that became leader of a South American country? Think again: his daughter Keiko Fujimori is set to become the next President of Peru here in 2021. We'll see later on that Asia is playing a larger role in South America: China being the forerunner of all this. Chinese influence is permeating the continent as more and more Asian and Chinese folks come in. The situation looks much like what is happening in the United States. South America is a very diverse place with a lot going on, making it socially different; however, some aspects of its social hierarchy are not particularly nice.

THE CONTINENT "RIGHTING" ITS COURSE?

Now that we have covered some serious ground on South Americans of all stripes, let's shift to how those peeps have been voting as of late. In other words, what are the political systems and political leanings of the continent as a whole? Well, I referenced this already back in Chapter 13 Latin America, but here it is again: while we used to count this part of the world as largely leftist, a serious swath of the region has suddenly swung to the other side of the political spectrum! It happened so fast that even my plaid powers of prediction could not have called it! The center-right/conservative-right/extreme-right specifically in South America has seen a stunning comeback after decades in the political wilderness. In the last two years alone, center-right leaders have assumed the presidency in Paraguay, Brazil, Argentina, and Chile . . . and those last three are of particular significance since they are the three biggest economies of the continent! For the first time in many, many moons, South America's three leading economies are led by conservative governments. But this is a very recent phenomenon, as they used to be on the forefront of left-leaning leadership in the world . . .

FORMER LEFTWARD LURCH

(WEALTH DISPARITY ISSUE RETRENCHED . . . RIGHT WING DICTATORSHIPS . . . LEADING TO . . .)

And why was that? Why did the liberal-left have such dominance for the last several decades? Such dominance that it even spawned a full-on seriously-socialist movement named the "Bolivarian Revolution"? Well, if you have read any of the last several chapters of this book, you have repeatedly been beaten over the head with the same issues again and again when it comes to economic conditions in Latin America, and those same reasons apply in South America too, if not even exponentially more. What I'm referring to here is persistent poverty, centuries of underdevelopment, chronic corruption, and unwavering wealth disparity. We can trace this very current situation all the way back to the origins of the continent's "modern history" beginning in the 1530s, when South America was "discovered," then "claimed," then exploited for centuries by foreign powers, mainly from Spain and Portugal (later some Dutch, British, and French holdings as well.). These competing colonizers claimed the entirety of the lands and all its resources (including all the humans) as their own and basically used the place as a huge cash machine, withdrawing all the precious metals, minerals, and agricultural goods they could get their hands on. Withdrawing all the wealth of South America, to be deposited back home in Europe. Do keep in mind the size of the place we are talking about here: these were huge colonies, with tremendous amounts of wealth . . . and even much bigger mining operations and agricultural plantation systems than anywhere else in the Americas.

Add to that lovely situation the importation of European diseases, which decimated the local populations, which in turn caused a labor shortage in South America, which in turn helped create the slave trade into South America. Do keep in mind that most of the humans stolen from Africa in the Atlantic Slave Trade era ended up in South America; fully 40% of all African slaves trafficked to the Americas went just to Brazil alone. What you have after several hundred years is a fully exploitative economy propped up by a slave/indentured class, in which the minority of masters take all the wealth and the masses of workers have absolutely nothing. No independence, no real salaries, no benefits, no land, and no real stake in the lives they were born into at all. That is the extreme manifestation of wealth disparity, and the colonial era effects are still being reckoned with up to this day. So how do we get from those days up to the current days?

Well, the Spanish American wars of independence were numerous wars against Spanish rule that started up in 1809 and lasted for the better part of three decades, and the War of Independence of Brazil was waged between 1822-24. The major campaigns of these movements were led by two super awesome dudes you should know to be South American savvy: **Simón Bolívar** (lead revolutions in the north, helped create modern day Venezuela, Colombia, Panama, Ecuador, and Bolivia) and **José de San Martín** (lead revolutions in the south, helped create modern Argentina, Chile, and Peru.) These guys are heroes/icons of the continent, comparable to George Washington in the USA. However, unlike George, both men ultimately recognized that the independence movements gained only the actual independence from European powers, but had failed to significantly transform the societies in any major way. In other words, slavery was still happening, the economies were still exploitive, and wealth disparity was still rampant. Again, it's important to note that even the heroes of that day understood that the states of South America were still not serving the masses of its own citizens very well. The situation doesn't get much better over the next century either, so let's fast forward to more recent times . . .

The continent didn't involve itself too much in World War I nor World War II, but nearly all of the countries in the region were engulfed by the Cold War, and were pressured to join up with either the commies or the USA. Being in its backyard, the USA was very much motivated to keep these states on the democracy/capitalism path, so with covert (sometime overt) American support, many South American states overturned their democratically elected leaders . . . especially when any of those elected leaders suggested maybe embracing socialist/communist ideals in order to counter the wealth disparity issues. Many states turned to full-on brutal military dictatorships in what was an extension of the **Red Scare**, that is, the paranoia regarding communist ideas, politicians and parties that was whipped up in the United States and encouraged throughout Latin America by the United States.

FUN PLAID FACT ON RED SCARE: For those not in the know: "red" was a synonym for commie or communist, and thus the term refers to the fear of commie takeover of the planet. Sound silly? It shouldn't: it was wildly effective and is evidenced most clearly by culture and policy change in the Americas. Another of my favorite silly sayings of the era was "Better dead than red." Yikes! I am scared now!

These military dictatorships ruled much of the region from the 1960s, 1970s, and into the late 1980s in some places: a period that is looked back upon as a super dark chapter in South American history. Now I won't go deeply into the specifics of each right-wing group of this period, but let me drop some names to those who want to research further on your own. In Brazil, look up the **Fifth Brazilian Republic**: an authoritarian military dictatorship that ruled the state from April 1, 1964 to March 15, 1985 after conducting a coup d'état which deposed the elected government of Joao Goulart (rumored to be a communist.) For Argentina, check out **The Dirty War** (1974-83) to find out more about the military junta that took over there, and used security forces and right-wing death squads to eliminate perhaps up to 30,000 political opponents. In Chile, check out the reign of one **General Augusto Pinochet**, who came to power in 1973 by leading a coup against the semi-socialist leftist President Salvador Allende; Pinochet's 17 year rule led to him being charged with human rights violations that may have resulted in the deaths of 10,000 to 30,000 "enemies of the state."

Those are just the big three states, which had the biggest, most brutal, right-wing dictatorships on record. If you want to learn more about an expanded area of state-funded terror, look into **Operation Condor**, which perhaps was the worst offender of all, since it involved coordinated efforts between dictatorships of six South American countries (Chile, Paraguay, Uruguay, Brazil, Bolivia and Argentina) to kidnap/torture/assassinate political opponents in each other's territories. Started up in 1975 as an anti-communist program, this campaign of political repression and state terror involving clandestine intelligence operations eventually targeted any/all opponents of the regimes, and may have been responsible for upwards of 60,000 deaths across the continent (30,000 in Argentina alone . . . part of that Dirty War). Think this is just all old news? On July 8, 2019 an Italian court sentenced 24 people (previous senior officials in those military dictatorships) to life imprisonment for their roles in Operation Condor. The memory of this era may be fading for most, but the citizens of these states have not forgotten. And that's why many of them supported left-wing policies and politicians once those dictators went way. Oh that right! We are supposed to be talking about the last three decades of left-leaning rule! That lurch to the left that this section is all about! Back to story:

. . . and the story is now mostly told! Centuries of exploitive economic action, independence movements that didn't structurally change the situation much, capped off with right-wing dictators that further limited individual human rights and freedoms. All of this together actually perpetuated (and sometimes even exacerbated) that wealth disparity that I continue to bring up! Even today, out of the 20 most unequal countries in the world (based on the GINI index measuring income inequality), eight are Latin American while the rest are in Sub-Saharan Africa. The region still has rising income inequality, persistent poverty, and chronically poor public services like healthcare and education. The reign of the right-wing had done very little to change the situation in the South. In the late 1980's and early 1990s, the peeps of South America had had enough, and started throwing off the yolk of those dictatorships and trying to implement more liberal/left policy that would address these societal ills and wealth disparity issues. Heck, that's like the major focus of left-leaning policy in general! And most of Latin America, and especially South America, went left in a really big, big way!

Starting in the 1990s, leftist political parties and policies gained ground in nearly every single South American state. The big three **ABC** states (**A**rgentina, **B**razil, **C**hile) have been dominated by the left, including the top slots of presidential power. Many other states have also gone left, and I would particularly point to Bolivia, Uruguay, and Ecuador, states which had some really interesting game-changer leaders in power over time, ones that embraced serious alternative economic approaches and much more liberal social policy, too. By policy shifts, I mean all of these places have been formulating economic policy that focused more on redistribution of wealth in all of its forms: be it increased investment into infrastructure/benefits (roads, hospitals, schools), increased subsidies on fuel and food (making it cheaper to citizens), or even nationalization of private industries (aimed at increasing money flowing into the government coffers, so they can spend more on citizens). By the turn of the century in 2000, the majority of states across the wider Latin American region had left leaders in charge, and US policy hacks came up for a term to describe this movement: the **Pink Tide**. With the pink referring to "pinko commie," which is yet another spin on that same color association of red = communist . . . so I guess since Latin America was not actually embracing full-on communism, but certainly socialism or "communism-light," they were "light-red," i.e. pink! A pinko tide rolling in to wash over Latin America!

Eh, that really wasn't the case for most places, which were politically just center-left not extreme left at all, and certainly not embracers of any hard core Marxist-Leninist doctrine. However, there were those places that went full-on into the movement, called themselves full-on socialist, and did adopt a more revolutionary attitude on how to restructure their societies . . . and I'm thinking primarily of some folks down in Venezuela . . .

INTO THE VENEZUELA VOID

What a roller coaster ride has been happening in Venezuela! A seemingly small state in South America that has had an outrageously overblown impact on the rest of the continent, and a disproportionately large voice in the world! And we are all still watching the events of this place unfold in real time as we await the imminent implosion. But wait I am getting ahead on the story, one that you have already seen referenced in previous pages of this text. Let's cut to the chase . . .

So for decades there has been a movement towards left-leaning governments in the region. Not communist governments, because nobody is going to call themselves a true Communist besides good ol' Fidel (and not even him since he died!), but governments that promote a more proactive socialism than anywhere else, even in Europe or the United States. As previously alluded to, pretty much every government is somewhere in the socialist category, even though we in the United States say we are hardcore capitalists. We say free trade is what we are all about. Yeah, that's true; it's all good, but don't we have welfare systems and unemployment benefits and state-run entities that build roads and run post offices? Sure, we have all those things. We look out for members of our society on some basic levels. More and more center-left leaders in South America wanted to provide more and more for their peeps, so starting adopting many policies to do just that . . . but slowly, and methodically.

BUT . . . some of the very far left-leaning leaders starting saying, "Yes, we are going even further down that road. We want to fully nationalize industries, like the oil industry. And we are going to do it right this second." They wanted more radical policies to address those wealth disparity issues more quickly . . . to basically start to tread on more communist ideals of how to run a society, in order to gain as much control over the resources and wealth as possible, to immediately give maximum benefits to their people. And that more radical revolutionary movement was started by Hugo Chavez, President of Venezuela from 1999 to 2013. He nationalized the oil industry, then the cement industry, then food, then medicine . . . heck, I can't even remember all the stuff that Hugo had the government take over during his tenure, but the result was a rapid, radical redistribution of wealth program that caused the US to call the guy a commie, that promoted Hugo to being a socialist revolutionary with an amplified megaphone on the world stage, and ultimately has led to economic chaos for the place . . . the end of the story is still being written in real time here in 2019.

Chavez had modeled himself after Fidel Castro of Cuba, openly embracing the concepts of a socialist revolution with full state control of industry, and even created a political philosophy now referred to as **Chavismo.** He promoted his ideology to all other Latin American countries, and encouraged states to join this revolution, which he named for his historical hero Simón Bolívar (remember that guy from a few paragraphs ago?): the Bolivarian Revolution, which you should read about right now in the box to the right!

With all of the money being made from the oil industry, Hugo got even more popular and more powerful. He gave generous foreign aid to like-minded countries of the region, and increasingly in his career more fully embracing Marxism and aligned himself with other far-left leaders in Latin America like Raúl Castro, Daniel Ortega of Nicaragua, and Evo Morales of Bolivia. His encouragement (along with Venezuela's apparent success at the time) encourage leaders like Evo Morales in particular to nationalize the gas industry of Bolivia in 2006, and embark on many of the same policies as his hero Hugo. This movement was gaining real momentum from 2000 to 2010. But . . . and there is always a "but" . . . Venezuela in 2019 is a total basket case failing economy, and a country on the brink of civil war. What happened? Maybe it is more appropriate to ask "what was not happening?" because the answer to that is that the economy was only appearing to do well, but in reality the system Hugo put into place was failing, perhaps from the start . . .

The socialism/communism thing only worked in Venezuela because it produces massive amounts of oil, which we know is a crucial resource in today's world. The price of oil continues to go up over time, so every time you put five-dollar-a-gallon gasoline in your SUV, Hugo Chavez was being empowered. I am not making fun of the guy; I am not even making fun of oil companies. I just want you to understand what is going on and how Hugo's policies appeared to be working so well . . . and they were as long as the price of gas ensured huge profits for the state. In fact, I'm not so sure that we can say that his movement was ever economically really that successful there in Venezuela. Their economy outside of oil is pretty flimsy, and they have yet to make any big strides in infrastructure development or poverty eradication. He may have preached about it non-stop, and moved large chunks of money around to fund his poverty programs and aid his allies, but the structure of the game itself had not changed economically for the country at large.

For much of his life, Hugo's bark was much louder than his bite, but boy, oh boy, did that dog bark. He used all that oil wealth to make himself a loud voice on the global stage, making pledges of foreign aid to other countries and making speeches bashing capitalism and berating the US, when perhaps his time would be better spent tending to business back home.

What's the Deal with the Bolivarian Revolution?

The **Bolivarian Revolution** is an (perhaps?) ongoing mass social movement and political process started in Venezuela and was closely linked to its rabid leader, Hugo Chavez. Proponents of Bolivarianism trace its roots to an avowedly democratic socialist interpretation of the ideals of Simón Bolívar, an early 19th century Venezuelan and Latin American revolutionary leader (you know, the George Washington of Latin America) prominent in the South American Wars of Independence. He was the man. One of the main ideals of "Bolivarianism" is promoting the unification of Latin America into one country.

Simón in space?

This push for MERCOSUR could be interpreted as a true beginning stage of perhaps an entire South American trade bloc, which is what Hugo Chavez and his Bolivarian revolution were all about. Simón Bolívar wanted a United States of South America. Hugo Chavez used all of his power and oil money to try to re-establish this whole movement of a United States of South America, and the movements toward the unifying trade bloc seem to be working. And the satellite?

Keeping the revolution alive.

The Simón Bolívar satellite, so dubbed by Hugo, was launched into space in October 2008. The satellite provides broadcasts of Latin America news and entertainment, and is accessible to all in Latin America. Pretty interesting stuff. One last point of interest: Guess who helped build the satellite and launched it into space for the Venezuelans? If you said "the Chinese," proceed directly to GO and collect $200 from the bank.

But do keep in mind, the main reason Hugo was able to launch that socialist experiment was because his country was making shale-loads of money off of oil, and they have one of the biggest (if not the biggest) proven reserves of petroleum in the world. But no other South American state is in that category. His country was getting rich, and he was investing a lot of that money in schools, satellite communications to make his country's communications independent, building roads, building hospitals, and training medical folks. But he was on the fringe, the far leftist fringe of socialism. The Venezuelan state was doing all of this. Those are things that private businesses do in most other countries. He was afforded the luxury of meeting some of those needs and portraying himself as a hero to his people because of oil revenues.

Hu-gone, but is he forgotten?

But hold the phone, Hugo! Oil prices took a huge dump starting with the 2008 world recession, which also lowered demand for the stuff as well. Which means Venezuela was no longer basking in the shade of hefty profits. Oops. That started to hurt Hugo's state-run plans, and Venezuela as a whole. And Hugo's popularity started to wane in 2010. Too much rhetoric, not enough results, and it also become unpopular to bash the US at that time since former US President Obama's star power out-shined the Hugo hate. Then suddenly in 2013, Hugo became Hu-gone! The fire in the engine of the Bolivarian Revolution itself was extinguished. Chávez lost his fight with cancer in March 2013, and this dealt a significant blow to the more extreme elements of the leftist movement in general, and to Venezuela specifically. Hugo really was the driving force behind the leftward thrust, because he had both the passion for the project as well as the skills to rally the masses. Why am I being so descriptive? Because there ain't nobody that can take his place, that's why. He was the engineer of that train, and now it appears the train is now teetering off the tracks, somewhat rudderless, and perhaps about to go off the rails entirely. How so?

First off, most other Latin American countries were actually being much more moderate in the leftist approach to all matters, both political and economic. With the disappearance of the socialist-spouting big man himself, that moderating trend is likely to continue. Add to that the death of Fidel Castro in 2016, the other big famous commie/socialist force in the region. In other words, with no radical leaders left to lead the charge further left, it is likely most other states will creep back towards the middle of the road to more centrist capitalist economic approaches that contain some socialist slant in order to counter wealth disparity—but they will certainly stop far short of nationalizing industries or doing any other full-on commie-like activities.

The pain doesn't end there: starting in late 2013 and now continuing to the present in 2021, the Venezuelan society is at war with itself for the ideological heart and soul of the country. There have been increasing public protests against rampant crime, corruption, crumbling infrastructure, and general poor economic management in the country. A critical mass of pushback protest against the whole socialist experiment in the country has erupted, something that would never have happened while Chávez was alive, because he increasingly dominated the press and the military, and also could rally the masses of his supporters with his fiery speeches.

Maduro: He ain't no Hugo.

But he is gone, and his replacement, Nicolás Maduro, simply cannot fill the huge Hugo shoes. As political dissent increases—sometimes violently—the Maduro government has been using force to suppress the chaos—sometimes violently. Venezuela is now in a slow motion meltdown that will very likely end in extreme violence, and some sort of political upheaval, if not outright civil war. Bad times ahead for the state that started this whole movement.

So the ideological sweep that occurred for several decades witnessed virtually all the governments of South America go towards the left, more openly embracing the socialist take on how the world should work . . . but in a much more moderate, just center-left style. Places like Chile, Argentina and Brazil, elected their first female presidents, who were all left-center. Peru, Uruguay, and Costa Rica all have had leaders of the liberal left stripe, but much closer to the center than the fringe. Of course, there are still some hard core leftist down that way, but their numbers are dwindling fast: Fidel Castro and Chavez are now gone, leaving a gap impossible to fill; only the leaders of Bolivia, and Nicaragua would count themselves as full-on leftist still following the themes of the revolution. Venezuela hardly counts anymore since it's on the verge of total collapse. Truly, an epic collapse of the society is imminent, which is not something I suggest lightly, nor with any great joy. Things continue to spin out of control in Venezuela in 2021 with a dude named Juan Guaidó being declared the President of Venezuela by the National Assemble in January 2019 after Maduro's reelection was declared invalid. Both Madura and Guaidó declare themselves the rightful leader amidst a growing economic crisis. Guaidó has the support of foreign leaders in the US and Europe, while Madura is backed by Russia. So no resolution is yet in sight; tough times are coming for tens of millions of folks in Venezuela . . .

And the Venezuelan meltdown really is an important point to expand on for a moment more, because we are very likely on an historic pivot point here for South America, and perhaps Latin America in general. Here's why: For years in this very textbook, I would teach students that South America was really gaining its own unique regional identity on the planet, maybe for the first time, by following their unique trajectory. It was a trajectory that was presenting an alternative to the traditional economic/political systems here in the 21st century. See, in the good old days, you were either communist or capitalist, and everybody sided up. We called it the Cold War. And everybody left in the middle was polarized to one side or the other. You had one or the other—like forced to be—and that's why you had a lot of US intervention and even Soviet intervention and influence in much of the world, including South America, Sub-Saharan Africa, and Southeast Asia. That's why the US fought the Korean and Vietnam Wars, over a battle of ideologies between communism and capitalism. It's also why the US has also supported shady right-wing dictators all across Latin America at different points in recent history.

That world is gone now, and liberal democracy with a focus on unfettered capitalism triumphed all over the world. Kind of. But Latin America, particularly South America, was filling the void with a new option. It is a region, with the voice of a singular entity, that was taking that socialist approach openly and more extremely than any other place on the planet. A lot of folks around the world (including your truly) were looking at it and saying, "Hmmm . . . that's interesting: Latin America is bucking the trend and may become a new axis of power, so to speak, one consisting of a left-leaning socialist axis of states." And as MERCOSUR continued to grow and strengthen, it was set to become the world's only "socialist" trade bloc! Interesting indeed!

HOWEVER . . . and now we get back to why the Venezuelan implosion if a key turning point . . . **2021 UPDATE:** The full-on socialist experiment in Venezuela has failed utterly. Fidel Castro and even his brother Raúl are gone, and Cuba is doing a lot better than Venezuela, but by no means would we call it a great socialist success story either. Bolivia's bout with socialism is stagnating. Brazil's leftist party disintegrated in disgrace, with their leader being impeached and run out of office. Taken all together: the socialist experiment in Latin America is (seemingly) being discredited, in a big way. And that my friends bring us to . . .

RIGHT TURN!

Massively important 2021 UPDATE! To repeat from the Latin America chapter, because it bears repeating: Virtually the entire region of South America has suddenly swung to the other side of the political fence! It happened so fast that even my plaid powers of prediction could not have called it! The center-right/conservative-right/extreme-right in Latin America has seen a stunning comeback after decades in the political wilderness. In the last two years alone, right-wing leaders have assumed the presidency in Paraguay, Brazil, Argentina, and Chile . . . and those last three are of particular significance since they are the three biggest economies of the continent! For the first time in forever, South America's three leading economies are led by conservative governments.

Perhaps not directly correlated, but I think we can even refer to this as a "Latin Trump Bump," as this movement accelerated when US President Trump was in office from 2016-2020, and many of its leaders are extremely similar to the former US President in background, political styles, and policy focus. Like Trump, the center-right presidents of Argentina and Chile (Mauricio Macri and Sebastián Piñera) are millionaire business moguls-turned-politicians who want to run the government like a business. (Billionaire side note: Piñera is one of the wealthiest politicians in the world: net worth $2.8 billion.) By electing

Macri, Argentina rejected decades of leftist rule by a party that had its own loyal following as well: the **Peronist**. (Side note **2021 UPDATE:** Macri just recently lost power, and Argentina has returned to the Peronist agenda by putting another leftist back in power.) Colombia had already been electing center-right conservative leaders for years, and Peru is leaning in that direction as well. None of that compares to Brazil, where the voters decisively punished the liberal-left parties by electing a freakin' hard core right-wing dude, current President Jair Bolsonaro! (More on him later.) Taken all together, this movement is swelling the ranks of the right across the continent.

See what a difference a single round of elections makes? Compare/contrast the two-year shift!

Why such an about-face from the voters? Flip back to chapter 13 and review the section entitled RADICAL RIGHT TURN!!! Specifically review the "Un-Fantastic Four things" that I referenced that have, in combination, resulted in this political reversal. They are 1) rise of the middle class, 2) crazy corruption, 3) persistent poverty, and straight-up 4) "left-wing fatigue." The Latin liberal-left may be down, but don't count them out just as yet! This is still a startling new phenomenon, and we don't yet know how it will go. This counter-reaction to the decades-long dominance of the left in the region may not last long, and the leftist roots still run deep. Hard core leftist still run Nicaragua, Bolivia, Cuba, and Venezuela . . . although not that successfully in some cases. Moderate, and successful, liberal-left leadership can still be found in Mexico, Costa Rica, and Guyana.

And please note this in your brains: even though center-right conservative governments now dominate many of the biggest and richest economies of South America, social injustice, inequality, and wealth disparity are still *HUGE* issues in the region, so even the most nut-job, extremist, right-wing leader (and yes, I am talking about the "Trump of the Tropics" Bolsonaro) will get push-back from the masses if life doesn't improve for them. Pro-business Argentinian (Macri) and Chilean (Piñera) leaders have already been stymied in their attempts to get rid of popular fuel subsidies, anti-poverty programs and educational benefits. In Colombia, the right-wing leader Iván Duque attempted to offset corporate tax cuts with higher taxes on food . . . which utterly backfired, and just three months after being elected his approval rating plummeted to 25%.

So while Latin America may have taken a hard right turn politically, it apparently is still leaning left! Hahahaha, I know it's confusing, but bear with me on this! Many, many people in many, many states—including the big three of Brazil, Argentina and Chile—may have elected conservative-right leaders in recent years, but it looks as if most citizens still demand liberal-left policies, benefits, and government services. Perhaps these center-right governments have come to power not because of specifically being politically right-wing, but because they currently appeal to the frustrations and expectations of a smaller group of voters that actually end up swinging elections one way or another. None of these leaders swept into power by huge majorities of votes. But huge majorities of Latin Americans still want their lives to be improved, and these governments better deliver a better alternative, or back to the left they will go! And they way that the current conservative leadership wants to make life better for their citizens is by growing a better, bigger economy and increasing trade . . . so let's get to that economic stuff now! Because a bloc battle is afoot!

BLOCKED, UNBLOCKED, AND RE-BLOCKED: BATTLE OF TRADE

That's right! Just like the changing political situation in South America, there is currently a tussle between competing trade blocs, a Battle Royale of sorts between those same competing ideologies. And individual South American state economies are not only changing rapidly, but they are at different levels of development too, which further complicates speaking about the entire continent as a single homogenous economic entity. But we shall nonetheless endeavor to economically expand our knowledge of the place, and how recent political shifts are affecting trade both within the region, and with the wider world. Let's start with some basics . . .

CONTINENTAL CHARACTERISTICS OF ECONOMY

With approximately 420 million people living in the twelve nations and three territories of South America, it contains only about 6% of the world's population . . . a small population relatively speaking given that they live on the 4th biggest landmass on the planet. But that 6% of humans on this 4th largest continent produce a GDP now topping 4 trillion dollars a year! Not bad, right? Eh. Yeah. It's not bad, but it's not that huge either, and remember one of our major themes for this area is wealth disparity, so you already know that even if you think 4 trillion is a lot, it is fairly concentrated in the hands of the few, especially when it comes to ownership of land and industries. Now answer me this: looking at the continent as a whole, would we call them all fully developed? What do people do in South America for money? What do they produce? What brands of cars have you gotten from South America? What computers? Do you know of any huge multi-national corporations that were founded in South America? Hmmm . . . why would that be? You now know the backstory for how this situation has evolved, but to add to your knowledge, here are a few more modern details.

For its entire colonial era, virtually the whole content was set up just to be an exporting machine, exporting primary products back home to Europe. You know, the exploitation situation I referred to earlier. Never was really developed to be a producer of manufactured goods, thus, manufacturing/industrial output was never encouraged nor therefore thrived. They sold raw goods cheaply, and had to import all finished goods from richer countries, a situation that did not change much after independence. But from the 1930s to 1980s, most states of the region tried to remedy this by using an economic tool named **Import Substitution**: a trade and economic policy in which a state attempts to replace (or substitute) foreign imported goods by promoting domestic production. Okay . . . so what does that mean? Well, you put high taxes or tariffs on all goods imported from other countries (thus making them more expensive to your consumers) and then set up beneficial programs to encourage businesses within your country to make all those goods here at home. You know: "Buy local" or "Buy American" or "MAGA" . . . choose whatever slogan you like best. The result? It produces higher levels of development at home as your local industries 'catch up' with the rest of the world industrially and technologically, and it increases your country's self-sufficiency via the creation of an internal market.

So all of South America caught up real quick then by using import substitution? Ummm . . . no. Didn't work. As with most huge national policies that try to manipulate or control large parts of an economy, it just never seems to work out as per the plan. In South America's case, the governments borrowed huge sums of money to fund/build/promote these local industries, and then suddenly found themselves in an Andes avalanche of debt, which in turn sparked a debt crisis across the continent, making everything even worse. Oops. So even after all that, many South American economies were still heavily reliant on exports of primary products, and still really far behind on infrastructure development/investment, which in turn meant sub-standard educational and health care systems . . . which is the basis for getting the whole society more developed. From the 1990s to present, the entire region (besides those particular extreme-left leaning states previously mentioned) have switched to something closer to what we call the **Free-Market** economy, like the USA and European states, and the growth that followed helped pull them out of the debt crisis.

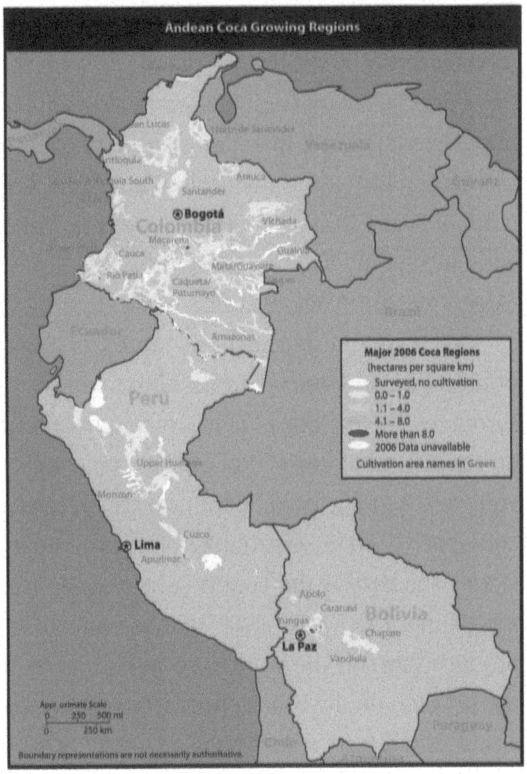

Having said that, today's South America is still a huge exporter of primary commodities. The region accounts for approximately 10% of all global agricultural products exported. Tropical commodities dominate the north export trade: banana, mango, Brazil nut, cashews, avocado, oranges, grapefruits, coffee, and cocaine. Hahaha . . . slipped that last one in you on didn't I? While not a tropical commodity that you typically find on your breakfast plate (I mean, I certainly hope not!) cocaine is a tropical plant that is one of the most profitable primary products to produce from this place. Why so tempting to grow even though it is illegal? With a huge guaranteed market right next door (the USA), with customers willing to pay whatever you want to charge, it does translate into a hugely profitable industry. Bananas don't get much money on the market, and prices may even plummet if no one wants bananas this year, but cocaine is worth a lot and has a consistent demand on the market, if not a growing one.

FUN FACT: The street value of cocaine sold every year in America is somewhere around $35-40 billion dollar mark. You do the math and figure out why people make cocaine in South America. Colombia is typically the biggest cocaine producer on the planet, but Peru and Bolivia are usually right behind them. Even Mexico has gotten into the game recently.

But back to other—more legal— primary products. Think coffee, chocolate, sugar, soybeans and beef from Brazil (in which they dominate the world coffee trade.) Argentina is world famous for beef, wheat, soy, even corn; Chile and Argentina are also huge exporters of wine. And think about what else falls into that primary category of products: metals and minerals and energy and trees. Forestry and mining are major economic activities, which should not surprise you too much given that the Amazon rain forest is the biggest tropical forest in the world. South America is also super rich in oil and natural gas (Venezuela, Brazil); hydroelectric power (Brazil, Colombia, Chile: 65% of all energy used on the continent is from hydro); copper (Chile has one-third of the entire earth's supply of it); Brazil is the world's leading producer of niobium and tantalum (elements that I've never even heard of until now); and Peru is the largest silver producer and the second biggest producer of bismuth and copper. Dang. That's a lot of valuable stuff. So all they do is export it all?

Not so much anymore. The economies have really started to diversify more rapidly in the last few decades. Industrialization has continued to take root for the last century, and now every country in the region has a significant manufacturing sector, making things like automobiles (foreign brands of them, that is,) steel, petrochemicals, computers, aircraft, and increasingly consumer goods like ovens and wishing machines and refrigerators. Currently, Brazil has the third-largest manufacturing sector in the Americas . . . that's all the Americas, North, South and all that's in-between! Overall, there are some very big economies here, and it is getting increasingly difficult to not count Argentina, Chile, and parts of Brazil as fully developed areas on par with USA, Europe, and others. Wait a minute! Are we back to the big three **ABC** states (**A**rgentina, **B**razil, **C**hile) again? Yep.

Why do these southern states of South America keep popping up in when we talk about industry and wealth and resources? If you look at a map of South America as a whole, as a pretty picture, it looks kind of like a huge ice cream cone. And if you were actually holding that ice cream cone in your hand, the cone part is Chile, Argentina, Paraguay, Uruguay, and southern Brazil. The rest of the continent is just the ice cream sitting on top of that cone, and given that sections of that northern area produce most of the world's cocaine, I guess you could suggest the big blob on the top is either ice cream or coca power. Maybe a new Ben & Jerry's flavor named Coca Cocoa. Based on this frozen confectionery continental comparison, I want you to remember a term that will help you to understand what's going on in South America in terms of economic clout: **the Southern Cone**.

The reality of the South American continent is that most of the money is in the south. That's where the economies are most fully developed. They're the places that are most industrialized, so they've got more to trade with each other and to export to the outside world. Also, considering climate and biomes. if you had to pick some place to go to in South America that was most like the United States or Europe, then you would probably pick the 4 season, mid-latitude areas of Chile, Argentina, Paraguay, Uruguay, or southern Brazil. Not northern Brazil, unless you want to take a ride on the gigantic mosquitoes and live in mortal fear of panther attacks. Southern Brazil is where you get that rich southern cone action, and a climate you would be comfortable in.

So the Southern Cone specifically clusters Argentina, Chile, Uruguay, southern Brazil and the Brazilian state of São Paulo together because they economically are the most industrialized, most developed, highest GDP areas of the continent. They also share characteristics like high life expectancy, the highest HDI numbers of Latin America (remember that Human Development Index stuff way back from chapter 5?), high standard of living, and low fertility rates,. And all of them have significant connections into the global trade networks and are counted as super productive emerging economies. In short, the Southern Cone the most prosperous and most developed multi-state sub-region of South America, and Latin America as a whole. And that is precisely why those were the states to first band together in a trade bloc named MERCOSUR, which has evolved, and devolved, and re-evolved recently, and it really reflects all the action we have learned about thus far in the chapter . . .

MAKING . . . THEN BREAKING . . . NOW RE-MAKING OF MERCOSUR

MERCOSUR (in Spanish), or MERCOSUL (in Portuguese), officially **Southern Common Market** (notice how similar that sounds to **Southern Cone**?) was the premier South American trade bloc established in 1991, to promote the free trade and easy movement of all goods, people, and currencies across all borders of its member states. Its current full members are Argentina, Brazil, Paraguay and Uruguay. Venezuela is a full member, on paper, but has been suspended since December 2016. Associate states are Bolivia, Chile, Colombia, Ecuador, Guyana, Peru and Suriname, and now there are even two 'observer states' of New Zealand and Mexico. Hey! Those last two aren't even in South America! Yes, which means this is now turning into something much bigger than a simple small regional trade bloc. But I am jumping too far ahead on the story as usual, so let's back the llama cart up and review the evolution of one of the most dynamic blocs on the planet right now.

When MERCOSUR actually went into full effect in 1994, they had aspirations of becoming something much more politically integrated (closer to the EU) rather than a simple trade bloc (like NAFTA.) And keep in mind, 1994 is the same year NAFT was created, and the EU was expanding rapidly, so supranationalism organizations were really in vogue all around the world. The Southern Cone rich countries (and Chile was much more a part of this originally, but later dropped out) signed on for the full-on free trade and customs union part, and were working towards a future that intended to include free movement of all people, free movement of all capital, and a shared legal/foreign policy platform. When the bloc negotiates trade with any other entity worldwide, they do it as a single collective voice, which increases their bargaining power, and indeed trade has increased both between these states, and between the bloc and other countries around the world (namely China, the US, and the EU.) So far, so good. But they never quite got to all that other political/legal/migration integration that the group originally planned on . . . they never got to the point of being "an EU of South America." What happened?

Because a bunch of other suprantionalist blocs kept appearing on the scene to overtake them! Say what? Yeah, new entities with much larger memberships have been proposed in what appears (in hindsight) as a never-ending parade of bigger and bigger blocs with bigger and bigger aspirations of continental unity . . . and all have failed to come to fruition! Just a brief backstory on each:

FTAA: The **Free Trade Area of the Americas** was a proposed agreement to eliminate or reduce the trade barriers among all countries in the Americas, excluding Cuba . . . because, you know, commies in Cuba. Would have become the largest trade group and biggest GDP bloc on the planet, superseding even the EU, had it become a reality. Originally conceived at the Summit of the Americas in Miami, Florida, in 1994, it started to take real shape when 34 trade ministers met again in 2003 to start hashing out the actual details of how it would work. The driving force behind it was the United States, seeing this huge hemispheric grouping as a great way to increase trade, counter the power of the growing EU, and re-establish itself as the premier power in Latin America. However, many Latin American countries were never really fully on board for this plan, for various reasons. The major ones were that the left-leaning states at the time (think Venezuela, Bolivia, Argentina, Ecuador, et al.) were very leery or outright scared of the idea of the powerhouse hard core capitalist US economy using this FTAA bloc as yet another new tool for dominating them. Many leftist believed it was just a new form of economic imperialism, as their smaller economies would be subservient to the will of the US economy. Fair point. Another country that was a non-fan of FTAA was Brazil, which saw itself as the natural economic/political leader within South America, a position they felt they would lose entirely if they joined a US-dominated group. Another fair point. So the FTAA just kind of floundered over the last decade, and with the former US President Trump's anti-globalist, anti-trade black agenda, I think we can safely assume that the FTAA is totally dead for now, although perhaps Biden might try to bring it back from the grave. . . we shall have to wait and see. So it came and went, while MERCOSUR is still standing. But that wasn't the only challenger . . .

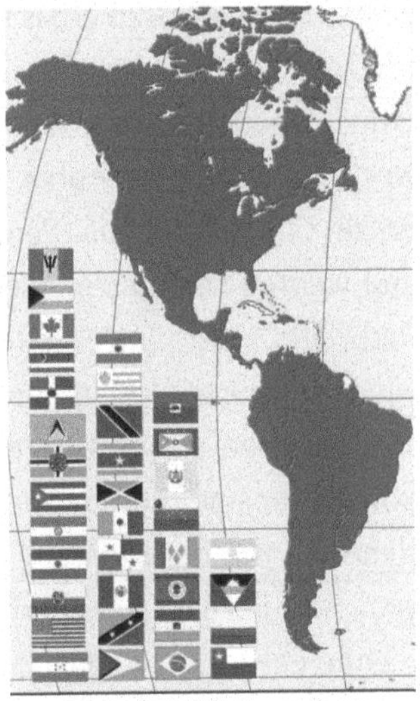

Proposed FTAA members

Because an entity named **UNASUR** was created, and was showing tremendous potential for unifying the entire continent like never before! This **Union of South American Nations** had many acronyms, based on different languages of its member states: it went by USAN (Spanish), UZAN/UNASUL (Dutch), and UNASUR (Portuguese.) UNASUR was an intergovernmental regional organization that once comprised all twelve South American countries! The entire continent in a single group! The initial meeting that started this ball rolling was in 2004,

UNASUR

with all 12 states of South America (with Mexico and Panama as observer states) agreeing to merge together MERCOSUR with another bloc named the Andean Community of Nations (a group that achieved so little I didn't even bring it up in this book) to make this super-bloc with continental coverage. By 2011, enough states had ratified the treaty to make it a reality, and it truly appeared that this group was the future forum for all things South American. The aspirations for UNASUR were lofty: fully wide open interregional trade, and enhanced regional economic and political integration . . . including free movement of people, common currency, collective development projects, collective health systems, and even common defense initiatives. A true EU-styled bloc. By 2015 they created a collective Bancosur: a Bank of the South monetary fund and lending institution for member states (like a local IMF,) and even created a headquarters building in Quito, Ecuador. Wow. So this entity is doing great, right?

UNASUR
Union of South American Nations

Not. What a difference a few years makes, and what a difference politics can make. Do you still remember all that leftward lurch to right-wing turn stuff I just taught you about a few pages back? Well, that political swing (and subsequent divisions it has created on the continent) has served to downright devastate the whole UNASUR project, and just in a few short years!!! A leadership crisis arose at UNASUR after the Colombian leader completed his term as Secretary General in January 2017 (they have a rotating president of the group.) He was supposed to be replaced by an Argentinian, but Venezuela, Bolivia and Suriname blocked the nomination totally. A left-leaning trio blocked the right-leaning candidate. See how that works? Then it gets worse. All of the chaos of Venezuela is playing in the background of this, so 6 countries (Argentina, Brazil, Chile, Colombia, Paraguay and Peru) denounced the Venezuela government of Nicolás Maduro as antidemocratic, and wanted them suspended from the bloc. In 2018 Maduro was dis-invited from the Summit of the Americas hosted by Peru, and in response, leftist Evo Morales of Bolivia demanded that UNASUR come to the aid of its Venezuelan brother which of course more and more right-leaning countries were not going to do.

Now we are getting to really current events! On April 17, 2018, it just so happened that leftist Evo Morales of Bolivia became the temporary acting Secretary General of UNASUR . . . and can you guess what happened next? That's right! All the right-leaning countries quit!!!! On April 20, 2018, six countries—Argentina, Brazil, Chile, Colombia, Paraguay and Peru—suspended their own membership! And yes, it's the same 6 countries that had denounced the Maduro regime earlier. In August of the same year, Colombia announced its total withdrawal from the organization. On March 2, 2019, Brazil's president Jair Bolsonaro announced his country's withdrawal. On March 13, 2019, Ecuador announced that it also was quitting, and to add insult to injury, the president of Ecuador also asked the bloc to return the headquarters building of the organization! So UNASUR basically just got evicted from its Quito apartment. It looks to be done for now.

2021 UPDATE: Amid growing concern about the Venezuela crisis centered on Nicolas Maduro, a new group, **PROSUR (Forum for the Progress and Development of South America)** has been created to "counteract the influence of what countries in the region call a dictatorship in Venezuela" and basically to replace UNASUR. A Chilean summit to organize PROSUR was held in March 2019, and eight South American countries signed on to joining the bloc (Argentina, Brazil, Chile, Colombia, Ecuador, Guyana, Paraguay and Peru.) So, now, UNASUR = DOA. And PROSUR may be a future replacement for it, but what I want you to more fully understand is that all of these blocs have now become political. A new right-wing bloc has formed, but the left-leaning countries won't join it.

MERCOSUR BACK ON TOP

Are we all caught up now on our South American bloc action? We started with MERCOSUR . . . and hey, look at that! MERCOSUR is the only one still standing! Indeed! While all other iterations of greater regional integration have completely failed, MERCOSUR is not only still around, but doing more successful than ever! It still has the four permanent full members of Argentina, Brazil, Paraguay and Uruguay, and Venezuela is still suspended, but they have picked up two observer states: Mexico and New Zealand. And when states want to "observe" something, that means there is something worth observing and they eventually want to be a part of it! So this bloc is back to being the main force for integration of trade, and one that appears to be growing again! Dig this bodacious bloc update:

June 28, 2019: The European Union and MERCOSUR struck a free-trade deal after 20 years of negotiations!!! This will become (if ratified by the individual EU states) a trans-Atlantic free trade area consisting of 31 states that contain a tenth of total world population and a fourth of total world GDP. Dang! That's kind of big news! Especially given what you now know about the turbulent modern history of trade blocks in South America! And even bigger news given that big parts of the world are currently being very protectionist, anti-bloc, anti-global, and anti-free trade. Yes, I am specifically referring to US President Trump and his trade wars with China, with NAFTA, with Europe, and I guess with the world in general. Trump's "America First" agenda has sparked tariff battles between strategic allies and rivals alike, and has challenged the very foundations of the World Trade Organization and the very free-trading world that the USA largely built since World War II. But behold! This huge EU-MERCOSUR deal is going in the opposite direction of the current protectionism trend, and looks to be a beacon to other like-minded free-trader countries around the globe.

The pact is not only the biggest deal that MERCOSUR has ever been a part of, but is simultaneously is among the EU's biggest deals too. It is supposed to expand trade an additional $102 billion dollars a year! While throwing a big party may still be premature, because enough EU states could refuse to endorse it, the *carnaval* has already kicked off in South America! Some foreign ministers were reportedly teary-eyed during the signing ceremony, and many leaders have outright hailed this as a revolutionary event for their countries:

> Argentine Foreign-Affairs Minister Jorge Faurie: "The Mercosur-EU deal is much more than a trade agreement. It's a strategic advance in Argentina's position on the global stage that strengthens the commercial agenda of both our country and our bloc."
>
> Argentine President Mauricio Macri called it "the most important agreement ever signed in our history."
>
> Brazilian economy ministry: the pact will result in an increase in gross domestic product of $87.5 billion over the next 15 years, a number that could rise to $125 billion.
>
> From Twitter, Brazilian President Jair Bolsonaro: "Together, Mercosur and the EU represent a quarter of the global economy and now Brazilian producers will have access to this enormous market. Great day!"

Yes, it is quite the big deal for those MERCOSUR states, and MERCOSUR may now regain its throne as the premier bloc of South America, one that may lead it to future integration that has seemingly eluded all other attempts by all other blocs Time will tell. As you see from the last two quotes on that list, Brazil seems to be the most excited kid in the classroom concerning this trans-Atlantic trade triumph, and for good reason! Because Brazil is the bomb! It is the powerhouse of continent, so we shouldn't leave this chapter without focusing on them for at least a few pages!

BRAZILIAN BOMBSHELL

Watch out! New superpower alert! And it is the first full-fledged global power phenomenon from Latin America! It's Brazil, baby! Latin America's largest state, largest population, largest military, and we are just getting warmed up. Brazil is Latin America's largest economy and largest exporter, and they are absolutely bursting at the seams with the largest national pride! They hosted the 2014 World Cup and the Olympics in 2016 . . . something that only big powerful countries get to do. Brazil is also becoming a significant power player in many supranational organizations, including the the G-20, the BRIC, MERCOSUR, the Organization of American States, and the Union of South American Countries. And Brazil is a likely candidate to eventually join the UN Permanent Security Council too! Brazil got it goin' on, like the lambada! Let's look at it in a bit more detail . . .

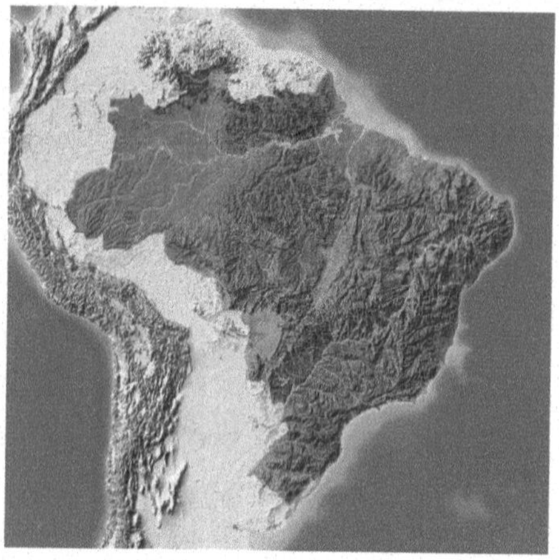

Economically, the place is on fire like a slash and burn site. Brazil had one of the fastest-growing major economies in the world with an average annual GDP growth rate of over 5%, for over a decade straight in the early 2000s, (they are currently in a recession, but more on that in a sec).They now have the sixth largest GDP, having just jumped into the top ten in the last year. So while most of the "rich" world is currently riddled with debt and unemployment, Brazil is mostly concerned with how best to manage and sustain their economic resources. Brazil was the last country to enter the current global recession . . . and the first to leave it. And they are now poised to overtake France for the #5 biggest economy slot, having leap-frogged the UK and Italy last year.

Physically, it's a huge country (5th largest in the world), just slightly larger than the continental US, with gigantic expanses of arable farmland and rain forests, an abundance of natural resources, and 15% of the world's fresh water. And they are a resource powerhouse: Brazil is the world's largest producer of iron ore, has the world's largest cattle industry, and is top exporter of sugar, orange juice, coffee, beef, soybeans, tobacco and poultry.

But it's not just commodities that are driving the Brazilian boom. They have a serious manufacturing base, the world's third-largest aircraft manufacturer (Embraer), and a huge auto industry—every single car company on earth has production plants in Brazil. All of them . . . that is a feat not even China or the US has accomplished. Add to that major output of petrochemicals, machinery, electronics, electrical equipment, steel, cement and construction supplies, textiles, consumer durables, footwear, et al., and you quickly see how manufacturing sector now accounts for over 30% of their GDP.

And let's get energized about Brazilian energy! Brazil is the third largest hydroelectricity producer in the world after China and Canada: 2010 hydropower accounted for over 80% of Brazilian electricity production. They also are the second biggest producer of ethanol in the world too. It's all green baby! Just like their national flag! Brazil has the most sophisticated bio fuels industry in the world, and for its size, is the world's greenest economy. Oh and dig this: they have a mega-ton of other energy resources too! The state-owned oil company Petrobas (why doesn't the US have a state-owned oil company?) is drilling 150 miles off the Brazilian coast, where many now believe are the largest discoveries of oil found anywhere in the world in the past 35 years. Conservative estimates now say that Brazil could be producing in excess of six million barrels a day in the next decade, which would make them the third or fourth largest producer in the world. And they don't even need that energy! It's all going to be export money!

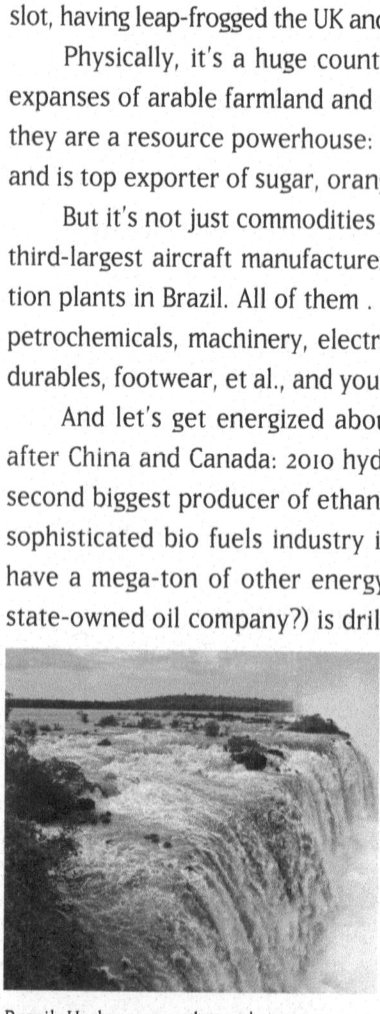

Brazil: Hydro-power house!

But wait! These guys have more than even basic resources and manufactured goods: they got the brain power going on too! Brazil has plenty of service sector jobs, and banking, finance, insurance and stock market action too. And don't forgot the quaternary sector: Brazil has notable technological hubs and R+D think tanks like the Oswaldo Cruz Institute, the Butantan Institute, the Air Force's Aerospace Technical Center, the Brazilian Agricultural Research Corporation. Oh, and the Brazilian Space Agency is rocketing up too, with significant capabilities in space research, launch vehicles, launch sites and satellite manufacturing . . . they also have an awesome satellite launching center, and were part of the International Space Station construction team.

GeoFact: Brazil has a great relative location for launching stuff into space. Why? Straddling the equator, parts of Brazil are on the natural 'bulge' of the planet, thus sticking out further from the center . . . thus launching a payload from the bulge means you have less distance to travel to escape gravity, and it doesn't require as much fuel. Brazilian bulge power! Look for Brazil to become a planetary platform hosting space programs from around the world!

Beneficial Brazilian Bulge

Fueling all this excellent growth and diversified economy is Brazil's 210 million citizens, that have been getting increasingly rich, with a steady expansion of the middle class, increased social benefits and programs, and even greater expansion of national pride. Great governance over the last decade has helped make this miracle a reality: former President Lula da Silva encouraged growth and development, and maintained conservative fiscal policies and tight banking regulations that left Brazil unscathed by the world financial crisis . . . in other words, while the 'Team West' countries are still currently suffering a hangover from their 2008 recession caused by their crappy banking systems, Brazil is doing just fine, thank you very much! Since 2000, 21 million people have been pulled out of poverty and into the lower middle class, thus creating a domestic market of peeps buying lots of cars and refrigerators and hairdryers and all manner of consumer goods, which of course in turn makes the economy grow even more!

Finally, Brazilians make love, not war . . . and that's always a good thing! But why would I suggest that this is one of the reasons Brazil has so much future promise as well? They are in no active conflicts, there are no terrorists attacks against them, they renounced any ambition to become a nuclear power, they have no nuclear arsenal, no declared enemies, and in fact the state hasn't engaged in any large scale conflicts since 1870! Stable, stable, and stable! Dudes! I want to party in a country that is peaceful now, and will be forever! Nice place to invest, raise a family, start a business . . . you name it! And the locals know it! Brazil will likely be the first global superpower without nuclear weapons nor a gigantic bloated military-industrial complex. Nice!

The Cup came, and was conquered!

With a successful hosting of the World Cup and the Olympics under their belt, Brazil is making its grand entrance on the global stage as a premier power, without having to flex any military muscle at all! With a diversified, well-paced and growing economy; a vibrant young population; a burgeoning middle class; no threat of regional or international conflicts, Brazil is going to be a powerhouse world player in your lifetime. Heck, it's happening already! And no wonder it has a growing sense of self-confidence to make its own way, the Brazilian way, a new path that other states may elect to follow. Brazilian peoples are brimming and beaming with pride in their country, and that is a radically powerful measure of well-being that can't be measured by economists or statisticians . . . the Brazilan boom is on!

There is really no questioning that the United States is the real nucleus of trade with Canada and Mexico. Likewise, Brazil sees itself as the real regional leader in South America. It is the biggest country, the most populous, and has the biggest GDP. In the self-defining role of regional leader, they really want to perpetuate MERCOSUR and make it successful. MERCOSUR's growth makes Brazil grow as well, making its role in the world bigger, making South America a bigger and more prosperous, industrializing place.

Brazil: Jesus-approved

2017 BOMBSHELL TO BOGUS BRAZIL UPDATE: Well, now that I've told you how totally awesome Brazil has been for the last decade, let me slap you in the face with some killer cold reality: Brazil be in big trouble now. What the heck happened? A variety of events conspired to cripple the country, and badly, since 2014. This is just a succinct summary of the shizzle: 1) Rampant political, bureaucratic, and police corruption in Brazil was always bad but it reached an all-time high during the run-up to the World Cup events of 2014, and the people just had enough and started protesting. . . mostly about the billions of tax dollars being spent on building stadiums and Olympic facilities rather than schools and hospitals. 2) The Brazilian economy simultaneously started to stagnate, a situation still happening. 3) Public protests continued and grew. 4) The president at that time, Dilma Rousseff, as well as the former President Lula da Silva, and their entire political party was increasingly implicated in corruption scandal after corruption scandal. . . BTW, a majority of active politicians of all political parties are currently under investigation for corruption. Ugh. 5) Unhappiness, protest, and overall chaos continued to grow until the official impeachment of President Rousseff, and she was forced to resign on August 31, 2016. Whew. Did you get all that? Things seem to be stabilizing now, but the massive corruption issues, the stagnant economy, and the loss of public trust are seriously challenging the country's leadership and its potential.

2019 UPDATE: As elaborated earlier in the Latin America chapter, this political "right-turn" in the region has affected Brazil maybe more than any other country on the continent! Given all that corruption of the long-ruling left-wing parties in Brazil, and the failure to more fully develop the economy to its potential, Brazilians didn't just abandon the political left in the last election . . . they downright punished the political left! And they did this by going extreme right! New Brazilian President Jair Bolsonaro is a polarizing and controversial politician and self-proclaimed "Brazilian Trump" and "Trump of the Tropics." He is a hard core, far-right wing leader that is pro-business, pro-guns, pro-military, pro-using-force, and a vocal opponent of immigration, same-sex marriage, homosexuality, abortion, affirmative action, drug liberalization, indigenous groups, and even secularism. He has also made incendiary remarks in public and on social media that many people considered homophobic, violence-inciting, misogynistic, sexist, racist, or anti-refugee. Dang! Now that is hard right! Trump got nothing on this guy! That's about as far to the opposite side of the spectrum as you can get from the previous four left-leaning Brazilian presidents! Of course, Bolsonaro loves that EU-MERCOSUR deal, as Brazil stands to gain the most from increased trade. He is also pushing his anti-corruption, anti-leftist, pro-business, pro-religious agenda to the hilt, but as of right now it is too early to call out any concrete successes that will make long-term huge impacts on Brazil. Many are very worried about his ultra-conservative social attitudes and fear that he is looking more and more like a right-wing dictator from the past. However, he is currently still very popular and the masses seem to embrace him as an outsider that can "clean up" the state and get the economy really humming again. And I can almost hear that hum starting up!

Brazilian President Bolsonaro: "Trump of the Tropics" to an exponent higher.

And while I have now waxed poetic about the awesomeness of Brazil for two pages (and then un-did all that with the 2017 and 2019 updates), let me also point out the fact that they are not alone: many other South American economies are also kicking butt right now too! Colombia has largely beaten back the narco-traffickers and has a legit expanding economy; Ecuador now declines any aid from the IMF or World Bank in an effort to form financial independence; Peru has excellent GDP growth for the last 5 years; and the Southern Cone countries of Argentina, Chile and Uruguay are approaching full-on 'developed' status in every sense of the term. Most of the continent is economically doing better than ever before. And that high note seems like a good jumping off spot to close up this chapter . . .

LET'S SAMBA OUT OF SOUTH AMERICA

A great big continent with great big potential, and it appears that they are closer than ever to really unlocking all that power to the befit of their peoples and the planet. Well established (and for the most part) stable democracies that have moderate to good economic growth, solid population growth rates, and a general cultural homogeneity which promotes at least a low level continent-wide social harmony. Tremendous amounts of resources;

tremendous green energy production; tremendous innovation in particular fields like aircraft and space stuff. And I do like to repeat a tremendous benefit which seldom gets mentioned: no major wars, conflicts, terrorism, or future prospects of military involvement at home nor abroad. It's a quietly peaceful place that doesn't aggravate anyone else in the world. Well, I'm sure there is someone out there that has beef with one or more of these countries, but it is nothing that is going to cause real friction or outright conflict. Quite the inverse actually: increasingly, more and more other countries are interested in doing more and more business with this region!

I already mentioned the big EU-MERCOSUR deal which was twenty years in the making, but there are plenty of other deals that have been happening in the interim. China made huge inroads in South America in the last couple decades, with former Chinese presidents visiting every single South American country, sometime on multiple occasions, to ink trade deals and plow in foreign investment. This was partly due to China's voracious appetite for natural resources, for both energy and agricultural products, of which Sough America has tons of both. Chile and Brazil now count China as their biggest trading partner. But it's not all rosy in outlook right this moment: China's foreign direct investment and financial involvement is currently slowing or outright declining due to sheer distance, domestic economic slowdown, and those Latin American corruption issues I mentioned earlier. But their relationship with South America will be interesting to watch as it evolves in the coming years.

China is not the only place deepening their relations with this region. Remember that TPP thing? The Trans-Pacific Partnership deal that the former Trump administration immediately pulled the USA out of? Well, it is still alive an well, and all the South American countries with Pacific shorelines are still a part of it. And it appears poised for growth too. Also recall that I said MERCOSUR has two observer states now: Mexico and New Zealand. New Zealand? That's way over on the other side of the Pacific! Yep. I predict much more South America/MERCOSUR action will be happening in the Pacific arena as well this century, as that is the area of the planet with the hottest trade action and fastest growing economies. The USA may be retracting from the global trade stage, but South America seems to be just getting warmed up and ready to go more fully global than ever before.

These South American cats are on the prowl for global growth!

Of course they still have huge challenges: the region is currently more ideologically divided than ever before. The Bolivarian Revolution seems to have run out of steam, and the full-on socialist experiment in Venezuela is ending in utter chaos an destruction, leaving the remaining few left-wing states rudderless, deflated, and without a viable path forward. This has emboldened the rise of the right once more, which is fine as long as those right-wingers don't repeat history by returning to demonic dictatorships. What that means to the region right this moment is two different teams facing off against each other based purely on political/ideological differences, and this polarization of the place is not going to be beneficial to the continent moving forward. Hopefully they can sort out some of this differences soon and get back to a more positive place.

And of course the other huge challenge is that wealth disparity that I have harped about for seemingly hundreds of paragraphs has not gone away, and many argue it could be getting worse. Class/caste/racism issues still stymie the full potential of the societies too, and many, many millions of people still lack basic infrastructure needs, adequate access to health care, and any sort of social safety net. But with their own home-grown supra-nationalist organization of MERCOSUR, they at least have a vibrant and growing structure that may eventually start to tackle those issue in a continent-wide level.

The 21st century is shaping up to be a serious samba time for South America! Things are really happening! They are trading/interacting more with each other, with the Pacific rim countries, with the EU, and even with Africa! Partnerships are popping up to bring these two (geologically and culturally and demographically connected) regions back together again, but in much more beneficial ways than ever before! So let's jump to Africa now to see what's up!

SELECT SCENES OF THE SOUTH . . .

Driest desert.

Iciest tip.

Biggest Jesus.

Highest world waterfall.

Make your life more awesome. Go here. Just don't wear those pants.

SOUTH AMERICA RUNDOWN & RESOURCES

View additional Plaid Avenger resources for this region at http://plaid.at/s_amer

BIG PLUSES

- Huge continent absolutely packed with natural resources
- World powers (China, Russia, India) are now courting the region big time to form economic and political collaboration. It's good to be wanted!
- The region is increasingly defining itself on its own terms (as a center of alternative socialism, or as having their own foreign policy direction, or as distancing itself from the US) and doing so confidently
- In particular, Brazil is an up-and-coming major world power player. Remember BRIC?

BIG PROBLEMS

- Internal economic integration, while growing, is still weak
- Environmental degradation, particularly of the rain-forest, continues to increase as the region indus-trializes and develops
- Wealth disparity, endemic poverty (especially among indigenous groups), and landlessness still pervasive issues across the entire region
- Despite billions of dollars from the US, the region (mostly Colombia) is still the largest producer and mover of cocaine
- Political polarization across much of the region between leftist and conservative rightist elements always has potential to become flashpoint of violence; especially in Venezuela & Bolivia

DEAD DUDES OF NOTE:

Simón Bolívar: The "George Washington of South America": hero, visionary, revolutionary and liberator. He led Bolivia, Colombia, Ecuador, Panama, Peru and Venezuela to independence and instilled democracy as the foundations of all Latin American ideology.

Hernándo Cortés: The white dude that led the Spanish conquista-dors into Mexico in 1519 to level the Aztec Empire in just a couple of years, and was the first wave of colonizers to lay claim to vast areas of the entire hemisphere in the name of the Spanish crown.

Francisco Pizarro: The other famous Spanish conquistador who (in the 1520s–30s) conquered and completely destroyed the other major civilization in the New World: the Incan Empire of South America.

Hugo Chavez: Former President of Venezuela and self-proclaimed leader of the leftist Bolivarian Revolution of Latin America. Pushed hard for socialism, deeper integration of Latin American economies with each other, and pushing out US influence. Appro-priately, he is despised by US. Even in death.

LIVE LEADERS YOU SHOULD KNOW:

Nicolás Maduro: Current President of a crumbling Venezuela. A former bus driver, then Foreign Minister and then Vice President under Hugo Chavez, this dude is not up for the challenge of taking over for his boss. Civil war or full country collapse is imminent.

Evo Morales: President of Bolivia, first indigenous dude to become so. Staunch member of the leftist/Chavez team, has nationalized lots of industries in his country, and also used to be a coca farmer!

Jair Bolsonaro: "Trump of the Tropics" Brazilian president; hard core, far-right, pro-business, pro-guns. Opponent of immigration, same-sex marriage, homosexuality, abortion, affirmative action, drug liberalization, indigenous groups, and even secularism.

Mauricio Macri: former President of Argentina (2015-19), conservative-right, and as such the first democratically elected non-Radical or Peronist (leftist) President since 1916. Was a major political shift for the country.

Alberto Fernández: President of Argentina since 2019, and a center-left Peronist party man, thus returning the country to leftist leadership. Already oversaw the legalization of abortion and a restructuring of massive foreign debt.

Sebastián Piñera: President of Chile; conservative, center-right, former businessman, and currently the wealthiest democratically elected leader in the world ($2.8 billion net worth).

PLAID CINEMA SELECTION:

Cidade de Deus aka *City of God* (2002) HUGELY popular street crime in the slums of Rio De Janeiro that spans decades. Based on real events and real people. Great for showing life in the favelas and the evolution of Rio society, and try to remember this when watching: the primary 'hood location is the same throughout the film, but watch how it changes over time as Rio grows to envelope it.

Central do Brasil aka *Central Station* (1998) An emotive journey of a former school teacher in Rio, who write letters for illiterate people, and a young boy, whose mother has just died, in search for the father he never knew in Brazil's remote northeast. Fanstastic landscape/cultural shots show diversity of place as well as the diversity of Brazil's peoples and its social classes. A sentimental piece, but a crowd pleaser nonetheless.

American Visa (2005) Similar to many Latin American films, this film focuses on the idea of the American Dream. The film was set in La Paz, Bolivia and features many shots in and around the city. In attempting to gain an American visa and consequently falling in love, the film demonstrates the irony of looking on the outside for what can only be found inside.

Crónicas (2004) Murder/thriller/detective story starring John Leguizamo. Plot follows a journalist who attempts to uncover a murderer and a rapist, despite the potential danger. Set in rural Ecuador, film presents a great view of the local landscape.

Machucha (2004) Set in 1973, the film centers around a military of coup of the socialist government in Chile. Although not the central theme, film focuses on the oppression of the poor in Chilean slums through the filter of a rich kid. Successfully demonstrates how the monetary oppression of the masses makes other Chileans wealthy.

The Motorcycle Diaries (2004) Based on Ernesto "Che" Guevara's memoir, the film follows a 23 year old Che as he journeyed around South America. Throughout the film, the viewer is presented with real life landscapes, and paints a true portrait of Latin America. Interesting view of one of the world's most well known revolutionaries and the experiences that shaped him. Easily one of the best road trip movies of all time, and its based on real events.

Maria Full of Grace (2004) Wow. Insanely realistic and good. A pregnant Colombian teenager becomes a drug mule to make some desperately needed money for her family. Shows life/landscapes of Colombia and the inside guts of how the drug trade works . . . and more importantly, how good people get coerced into doing it.

Ônibus 174 aka *Bus 174* (2002) Documentary about a June 2000 bus hijacking in Rio de Janiero, Brazil that brings up disturbing issues about crime, drugs, violence, social justice, violence, and media ethics and responsibilities, just to name a few. Incredible aerial shot in the opening scene shows the staggering breadth of class disparity between the city's wealthy and impoverished areas.

MAP GALLERY

View Map Gallery & additional Plaid Avenger resources for this region at http://plaid.at/s_amer

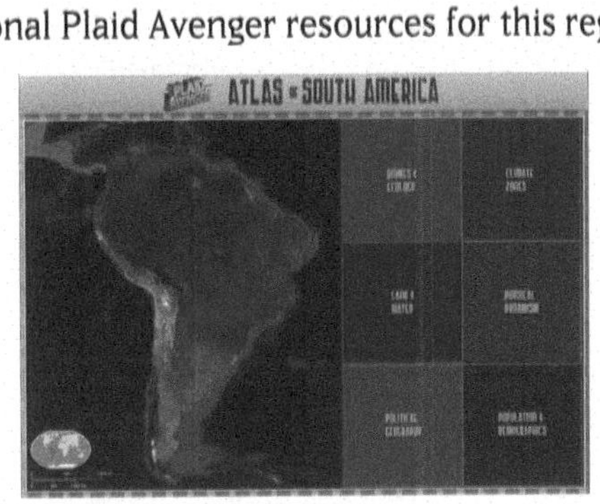

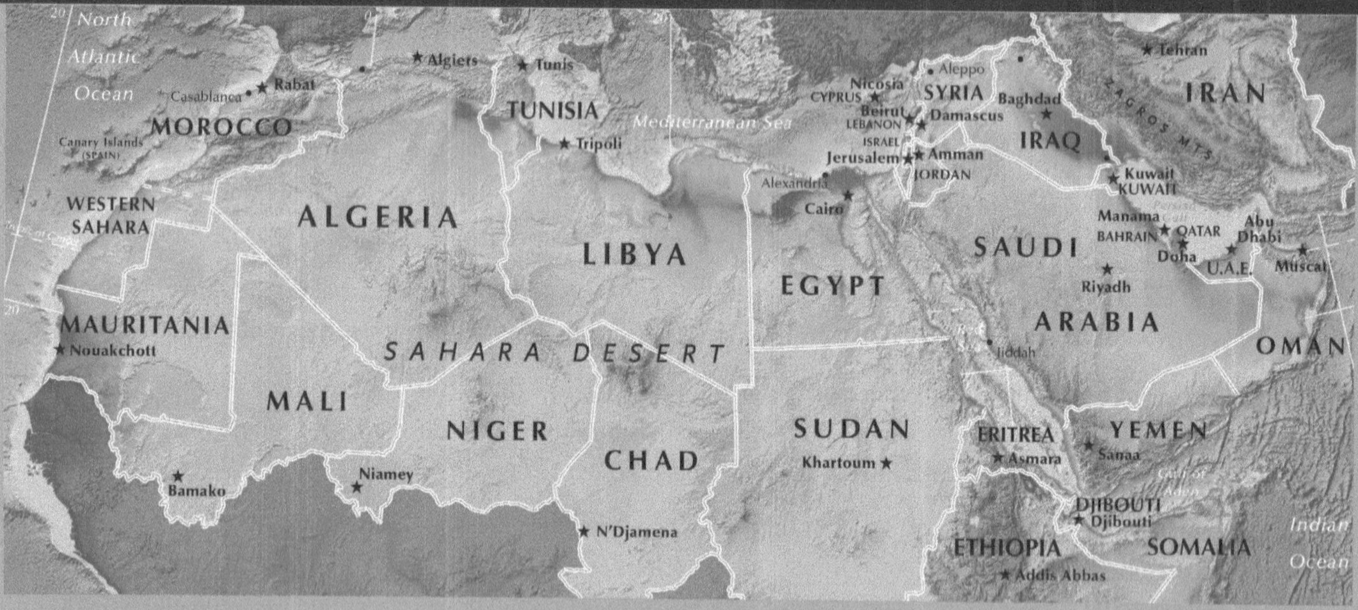

CHAPTER OUTLINE

Middle East and North Africa

18

THE Middle East is the hot spot of the globe and the thorn in the US's giant side. Everybody's watching the Middle East. Will there ever be peace in the Middle East? Well, it doesn't look likely anytime soon. And since 2011 the region is more on fire with revolutionary activity and conflict than usual! How did this region come to be associated with endless, endemic conflict anyway? What gives in the 'Middle'? Speaking of which . . .

"Hello again, Mr. Avenger. Can I get you the usual?"

WHERE IS THE MIDDLE EAST?

I'm confused. Is the "Middle" East the center of the planet, the point from which everything emanates? Obviously not. So what's the Middle East in the middle of? Western Europeans started to define the region with this term, as you might expect, during the Western European colonial era. The Western Europeans, of course, are in the west and all of Asia is to their east. China is as far east of Europe as you can get—thus the phrase "Far East." In between Europe and the Far East becomes, you guessed it, the "Near East" or "Near Abroad" as it was sometimes referred to . . . but even that was a nebulous term that roughly corresponded to today's Eastern Europe to Central Asia. So where did the "middle" come from? As far as I can tell, it was a 19th century American military historian, writing in a British military journal, who first used the term "Middle East" to describe European/American geopolitical concerns in the region specifically containing the Arabian Peninsula and parts of North Africa. It's a term everybody in the world recognizes today that is just a historic artifact from a bygone era. So there you go, students of global history and current events, that's where the Middle East is—it's kind of the middle of the old European empires' known and controlled world at the time, several hundred years ago, as referenced by an American. lol How is that for convoluted?

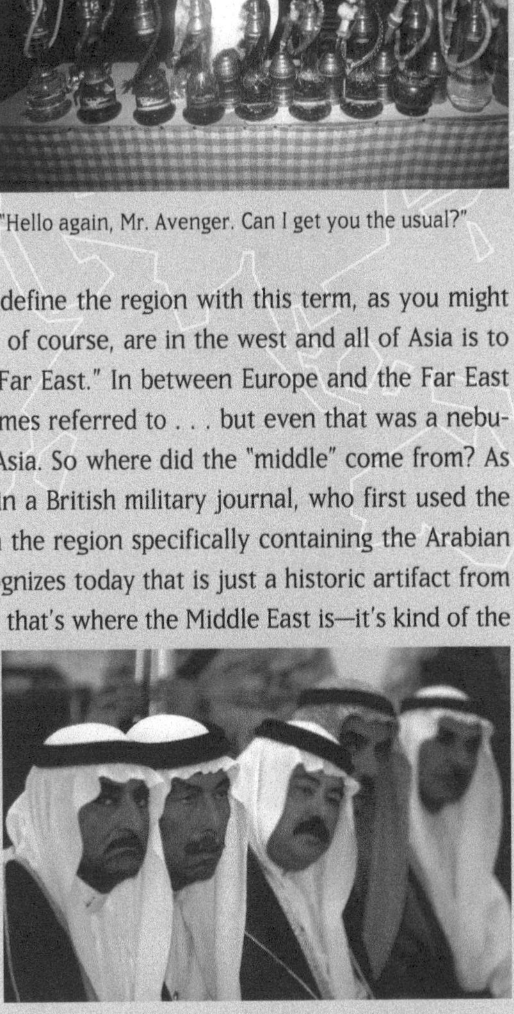

We are in the "Middle"—but middle of what?

Regardless of it being near, far, or middle, most now recognize this area collectively as **MENA**: **M**iddle **E**ast and **N**orth **A**frica. Choose whatever adjectives you like, just don't stray too far from the region while reading this chapter! Speaking of which . . .

Where's the Middle East in the terms of the Plaid Avenger's world regions? There is "classic" Middle East, which is the Arabian Peninsula, consisting of Saudi Arabia, Yemen, Oman, United Arab Emirates, Qatar, Kuwait, and Bahrain. We also "classically" add to this core area by including Jordan, Syria, Iraq, Iran, Lebanon, and Israel. These places we would all definitely agree are in the center—indeed, the middle—of the Middle East. However, we are going to expand the region for our definition to include most of North Africa as well. Places like Egypt, Libya, Tunisia, Algeria, Morocco, are all obvious players in the Middle East which share some homogenous traits, but also places like the Western Sahara, Mauritania, Mali, Niger, Chad, part of Sudan, Ethiopia, Somalia . . . now we're starting to split up sovereign states. There's kind of a fuzzy line somewhere in the middle of these countries. Everything north of and including the Sahara Desert in Africa is kind of part of the Middle East as well. Some historians and geographers have stretched the region to include parts of central Asia and Pakistan . . . but I think that is a bridge too far for my taste.

For the Plaid Avenger's book, we're going to put those in other regions. Simplistic textbooks typically lump all Islamic states in the vicinity as part of this region, but we're going to hold off for now. Turkey has also classically been defined as part of the Middle East, but not in my world. So that's the core: North Africa and Southwestern Asia, from Iran in the east all the way to Morocco and Mauritania in the west, and everything in between. What's homogenous about all these places? What are the big factors that are the same across all these territories? Are there some major exceptions to them?

Indeed, there are four big factors that often come to mind when we think of the term "Middle East": desert, oil, Islam, and conflict. We're going to look at those four characteristics. We're also going to look at the history of the Middle East, to see how it got to be the way it is, why it's so far behind, and its explosive potential in the very near future. (Wow! Given the events of 2011, how awesome that last sentence makes my predictive powers seem!)

FOUR FAMILIAR FACES OF THE REGION

As always, we will start with some physical attributes of this place, and then get to some cultural factors.

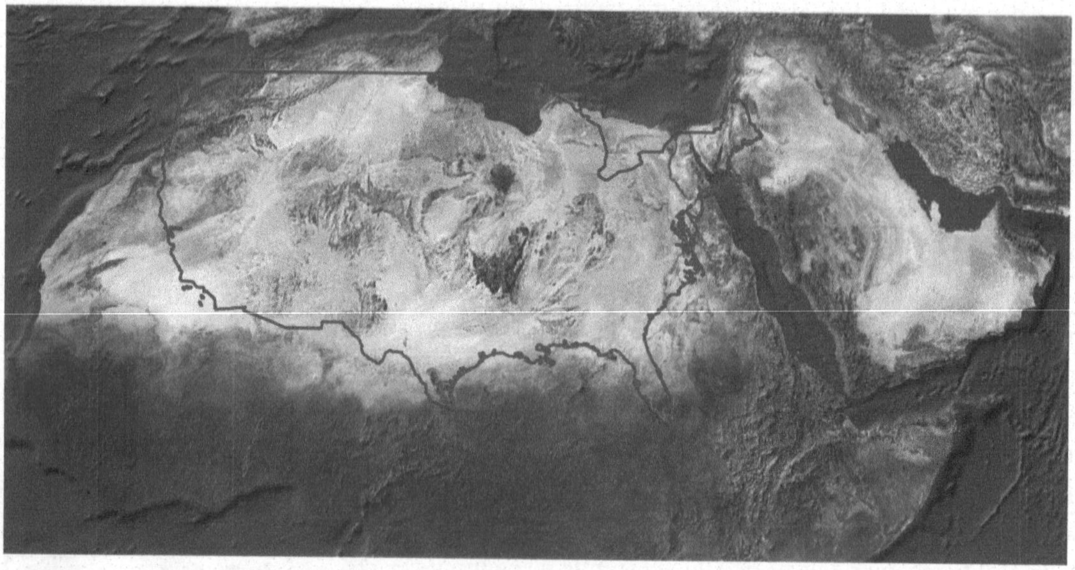

The Sahara size: Over 3000 miles of beautiful sun-baked sand.

NUMBER ONE: IT IS DRY IN THE MIDDLE EAST

The physical world of the Middle East is obviously a very nice homogeneous trait to pick out. What is the first thing that pops in your head when asked, "Hey, what's it like in the Middle East?" A: It's *dry*. Indeed it is. It is mostly desert; the big Sahara Desert is plopped down in North Africa. It is the largest hot desert on the planet. The entire United States would fit nicely into the Sahara, as you can see above.

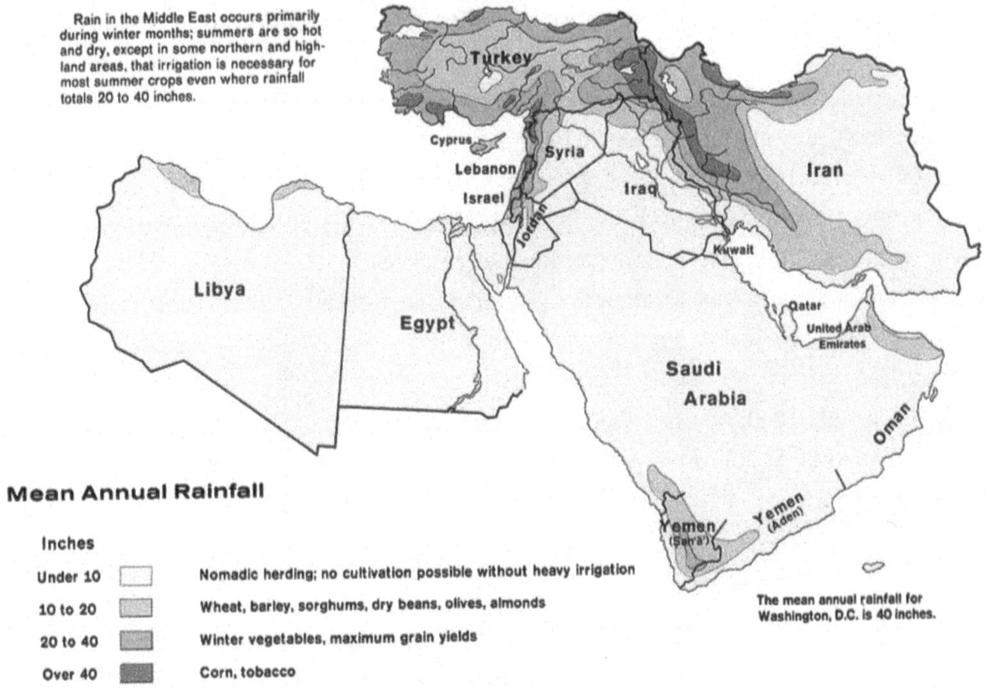

Most of the Middle East measuring in at much less than 10 inches—of rain, that is.

But, even outside of there, the entire Arabian Peninsula is also mostly desert. Saudi Arabia, Oman, Yemen, Jordan, Iraq, UAE . . . awesome Arabian aridity. Outside of the true desert regions, we have a lot of steppe. Steppe is just like adding a little bit of water to the desert; there is some scrub vegetation, and that extends this dry region all the way to the borders that we have just defined. This is a dry place, though there are some exceptions . . .

These exceptions are along the Mediterranean Sea coastal fringe, where there is a Mediterranean style climate. There are some narrow bands of nice climate around the Mediterranean that are well-watered enough for agriculture. You also have the Fertile Crescent, a classic area of Biblical lore. The Fertile Crescent includes the Tigris and Euphrates river valleys, all the way over to the Levant, which is now part of Syria, Lebanon, and Israel. It's a fertile, well-hydrated place. This whole area is where western civilization actually started, and you can't have that without enough water—in fact, three of the world's earliest river valley civilizations (the Jordan, the Nile, and Mesopotamia) together make up the Fertile Crescent. Let's add one more exception: Northern Iran. A lot of people don't know it, but believe it or not, up around the Caspian Sea, there are big plains of well-watered area. There are big grasslands that look like the Midwestern United States and can produce quite a bit of food. Even though Iran is mostly an elevated plateau and mostly dry, it has big areas in its northwest quadrant that are well-watered. Those are the exceptions to the rule, but by and large, this is a huge area, and very much of it is very dry.

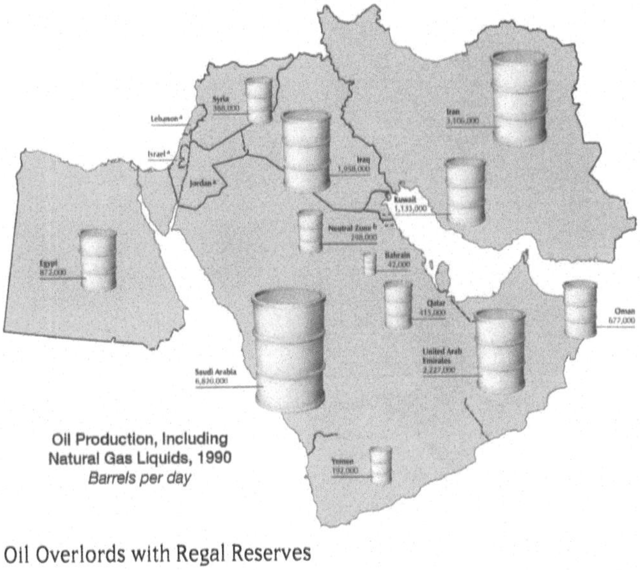

Oil Overlords with Regal Reserves

NUMBER TWO: THEY GOT OIL

The second homogeneous trait we think of when we consider the Middle East is *oil*. Indeed, this place does have a ton of oil. If we look at proven reserves of oil on the planet, the Middle East has the lion's share.

And that is why this region of the world is the core of OPEC—the Organization of Oil Exporting Countries—a collective cartel of countries that together control the majority of oil in the world, and thus can effectively set market price of this critical commodity. However, this situation that seems to be changing daily in the Plaid Avenger's world: as technology increases, the amount of extractable oil from different types of shales and tar sands changes according to extraction technology, and it is getting better and more efficient every day. As a result, it appears that other countries might have more oil than the Middle East right now. Places like Venezuela, Canada, Russia, the US and maybe even Brazil hold a lot in nontraditional oil reserves.

But for the sake of argument, let's just say for right now, in today's world, Saudi Arabia is the giant in proven oil reserves (Venezuela actually has the largest, but bear with me for now). That means they could stick a straw in the ground and oil will come up—loads of it. Iran, Iraq, Kuwait, and the UAE have tons of oil as well. Here's the exception: As you get farther and farther away from the Persian Gulf Basin, which is the singular geological structure that all of these oil wells are tied into, the likelihood that you have lots of oil diminishes. As you go to the other side of Iran, into Central Asia, there's not a lot of

Selected Oil and Gas Pipeline Infrastructure in the Middle East

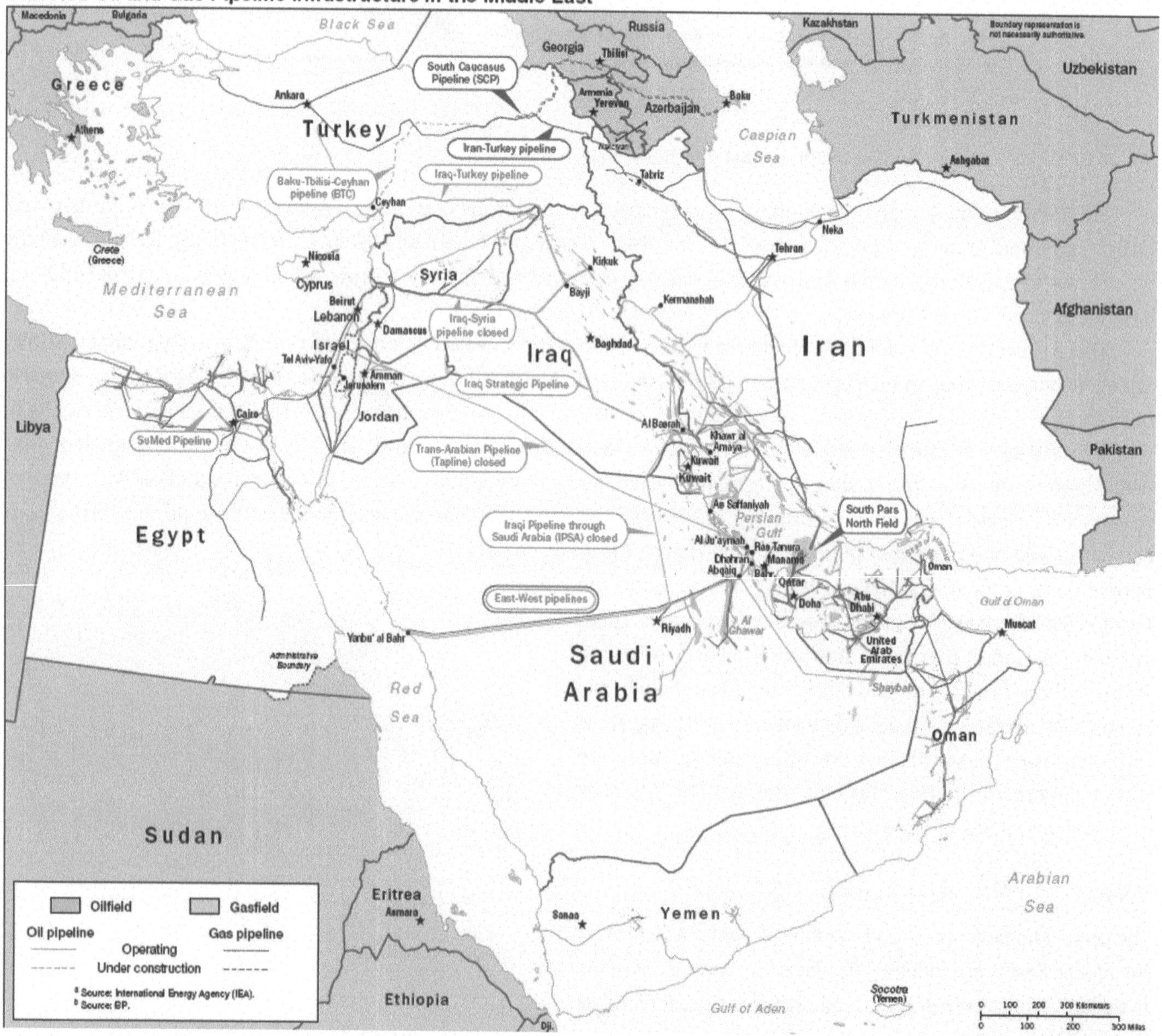

Persian Gulf: Petroleum Pipeline of Plenty!!!!

oil there. As you go north out of Syria, into Turkey, there's not much there. South of Saudi Arabia, the far end of the peninsula, not much down in Yemen, and indeed, when you cross into Africa, Egypt's got some, Libya's got some, Algeria some as well, but nowhere near the levels of what the Persian Gulf Basin does. As a general rule, the further from the Persian Gulf you are, the less oil you have. Dig the map on the opposing page, and check the green areas that represent oil fields, and the lack thereof when you move away from the Persian Gulf. . . .

Oil plays such a tremendous role in the world now, and the Middle East has so much of it. That's why this region is such a focal point of global events, as oil remains the cornerstone of world energy use, and therefore the engine of the world economy, so we need to talk about oil a little bit more.

Petroleum use can be traced back to at least the 4th century CE when the Chinese began drilling for oil, literally, with bamboo pipes. The Chinese used this oil to light lamps and to produce salt by evaporating seawater. Petroleum was also utilized early in history by the Persian Empire (now Iran). However, petroleum use was slow to spread to the West. Instead of crude oil, most Western street lamps were lit using whale oil—which, as the name suggests, is harvested by dudes in boats with spears. Whale oil was the primary source of artificial illumination and lubrication until American Edward Drake "discovered" oil in 1859 (sorry, China!).

The petroleum industry in the West began as a means to produce kerosene, which was used to fuel lamps. Enter the automobile: once the automobile came on the scene, gasoline became the most popular petroleum derivative. Gasoline was abundant and cheap for almost a hundred years, which also made it the the cheapest energy source for industrial use as well. For much of this time, the United States was self-sufficient in oil production. Slowly though, America evolved into a car culture; advanced interstates were developed and everyone rich moved to the suburbs. The United States became highly dependent on foreign imports of oil for both transportation and industry. Even this wasn't that big of a problem, because the United States propped up dictators in many major oil exporting country. This all changed in 1973.

In 1973, the US and Western Europe supported Israel in a war (the Yom Kippur War, more details on that later) against other Arab states. The result: the Arab states got angry. Like Hulk angry. The Arab states in OPEC, plus Egypt, Syrian and Tunisia, decided to have an oil embargo. They understood their position of oil supremacy and they said, "Okay, since you helped out Israel, we're not going to sell you any more oil. How about them apples?" Indeed, in a matter of days, and certainly the course of a few weeks, the price of oil went up from $3 to $12 a barrel. It may not seem like much, but that's an instantaneous quadruple in price. Ask your parents or your grandparents about this stuff—they probably remember. The result of this in the United States was gas rationing. You could only go to the gas station to fill up your car on even or odd numbered days, depending on the last digit of your license plate number. This was a big deal—it caused serious economic ripples around the world, damaging the economy of the US and elsewhere.

OPEC NET OIL EXPORT REVENUES AT A GLANCE

Country	Nominal Dollars (Billions) 2016	2017
Algeria	$20	$22
Angola	$26	$31
Congo (Brazzaville)	$3	$5
Ecuador	$4	$5
Equatorial Guinea	$4	$4
Gabon	$3	$4
Iran	$37	$55
Iraq	$53	$69
Kuwait	$37	$46
Libya	$2	$11
Nigeria	$25	$34
Qatar	$24	$31
Saudi Arabia	$134	$167
UAE	$45	$55
Venezuela	$23	$29
OPEC total	$441	$567

Source: U.S. Energy Information Administration, derived from EIA's August 2018 Short-Term Energy Outlook

High gas prices for you equals money in the bank for them.

Even after the 1973 oil embargo ended, oil prices continued to climb. People realized that petroleum was a finite resource and that one day, it would run out. The future of petroleum production is open to debate. Some experts have continued to warn about the imminent peak in global oil production. However, the oil industry has such shady accounting practices that no one really knows how much anyone else has, much less how much is left globally. There are also other methods to produce gasoline if crude prices become exorbitant. These include the gasification of coal or natural gas and extraction of tar or other impurities. Without coming across as a tree-hugging hippie, the Plaid Avenger encourages green, renewable energy . . . mostly because I think we are polluting ourselves to death with these carbon-based fuels, as well as continuing to get bogged down in political tomfoolery with oil-producing countries. Really, do we need more reasons to get off of carbon-based fuels? Enough rant for now though, back to the black gold. . . .

What's the Deal with OPEC?

The Organization of the Petroleum Exporting Countries (OPEC) is an oil cartel designed to regulate the supply and, therefore, the price of crude oil in the global market. In the United States, cartels are prohibited by antitrust laws and are often called "monopolies." However, international cartels such as OPEC are both legal and successful. The goal of any cartel is to regulate the supply of a commodity (for OPEC, this is oil). By restricting the supply, a cartel is able to control the price of the commodity (again, for OPEC, this is oil). Cartels prevent the biggest producers from attempting to out-produce each other. If too much oil is being produced, prices drop and all oil producing countries are hurt. Oil is a finite commodity; OPEC sees its role as rationing the reserves to ensure constant supply for consumers and maximum profits for producers.

OPEC was founded in 1960 by the big five–Iraq, Iran, Kuwait, Saudi Arabia, and Venezuela–to protest against lowering prices. At this time, foreign owned oil companies such as British Petroleum and Dutch Shell were taking the lion's share of profits from oil production in the soon-to-be OPEC nations. Over the years, OPEC became more powerful and more unified (even playing a critical role in the 1973 US oil embargo). Currently, 81.89% of all the world's proven oil reserves are in OPEC member countries, and 65.36% of that is just in the Middle East states. In other words, OPEC has the power to COMPLETELY shut down the global economy, although this would also clearly hurt the member nations themselves if they ever tried that trick.

However, there are several major oil producers who are not in OPEC. These suppliers–notably Russia, Canada, the United States, Mexico, China, and Norway–are able to fluctuate their production with increasing or decreasing demand. So, when oil prices are high, Russia (for example) cranks on the spigot and starts selling more oil at the inflated prices. When oil prices drop back down, Russia turns production back down. By doing this, Russia is able to capitalize on the market fluctuations of oil. OPEC countries, since they agree to limit their production, do not get to sell more oil when prices are high. This again, however, is the purpose of OPEC. If all oil producers could crank up production when prices were high, they all would. This would create a huge market surplus in crude oil and prices would instantly plummet. In this sense, Russia and other non-OPEC nations are free-riders on the OPEC regulatory system.

Top Ten World Oil Producers, 2018
(OPEC members in italic)

Rank	Country	Millions of Barrels (per day)	Percent of World Total
1)	United States	17.87	18%
2)	*Saudi Arabia*	12.42	12%
3)	Russia	11.40	11%
4)	Canada	5.27	5%
5)	China	4.82	5%
6)	*Iraq*	4.62	5%
7)	*Iran*	4.47	4%
8)	*United Arab Emirates*	3.79	4%
9)	Brazil	3.43	3%
10)	*Kuwait*	2.87	3%
		Total top ten	70%

[1]Oil includes crude oil, all other petroleum liquids, and biofuels.
[2]Production includes domestic production of crude oil, all other petroleum liquids, biofuels, and refinery processing gain.

Source: www.eia.gov/tools/faqs

ORGANIZATION OF PETROLEUM EXPORTING COUNTRIES (OPEC)

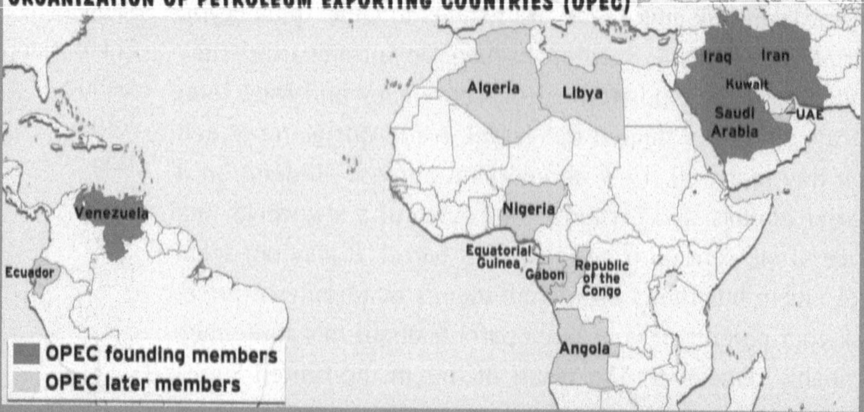

Technology will increase new ways to get to other sources of oil, and get it out of the ground cheaper. I'm telling you this story, in summary, because this is why OPEC's best interest is typically to keep oil prices very stable. Just the same way a crack dealer is with their product: keep it steady, keep it stable, so you can keep your customers coming back. You don't want people to overdose and you don't want them to go dry and get cured of their addiction. You don't want to price yourself out of the market, and you don't want people to die. To keep the oil price stable is in OPEC's best interest; and that's exactly what they've been doing for years.

Not all OPEC countries are in the Middle East. As you can see from the map on the opposing page, places like Nigeria, Venezuela, and Ecuador are OPEC members. They are all making serious jack, as you can see on this chart—about 522 *billion* US dollars in 2006 alone. Just on oil. And that was ten years ago. Think we are using more or less oil now?

When we think of oil in the Middle East, we invariably think of another thing, and that's OPEC. But please remember, while the core of OPEC is in the Middle East, there are other members from around the planet. As you can see from the box, OPEC is a cartel whose sole reason for existence is to stabilize and control the price of oil. Many in the United States and the world abroad would say, "Oh, they are going to use that as a tool against us." By and large, OPEC has been a stabilizing force for world oil pricing. It is in their best interest in Saudi Arabia, and every other oil producing country, to keep the price stable, and they know it. Here's why:

NUMBER THREE: THE MIDDLE EAST IS ISLAMIC

What else do we think of when we think of homogeneous traits of the Middle East? Perhaps this should be number one, but we definitely think that in the Middle East, they're overwhelmingly *Muslim*. Indeed, the predominant religion is Islam across all of the countries that I've thrown into this region. We could push it on further and could say Central Asia is predominantly Muslim and Afghanistan and Pakistan, but those are different altogether, ethnically, linguistically, and historically. How Islamic is the Middle East, and how'd they get to be that way?

Here's a quick summary. The movement of Islam started in this region, right in the middle of the Middle East, on the Arabian Peninsula. There you have Medina and Mecca, the two holiest cities of Islam, where the prophet Muhammad

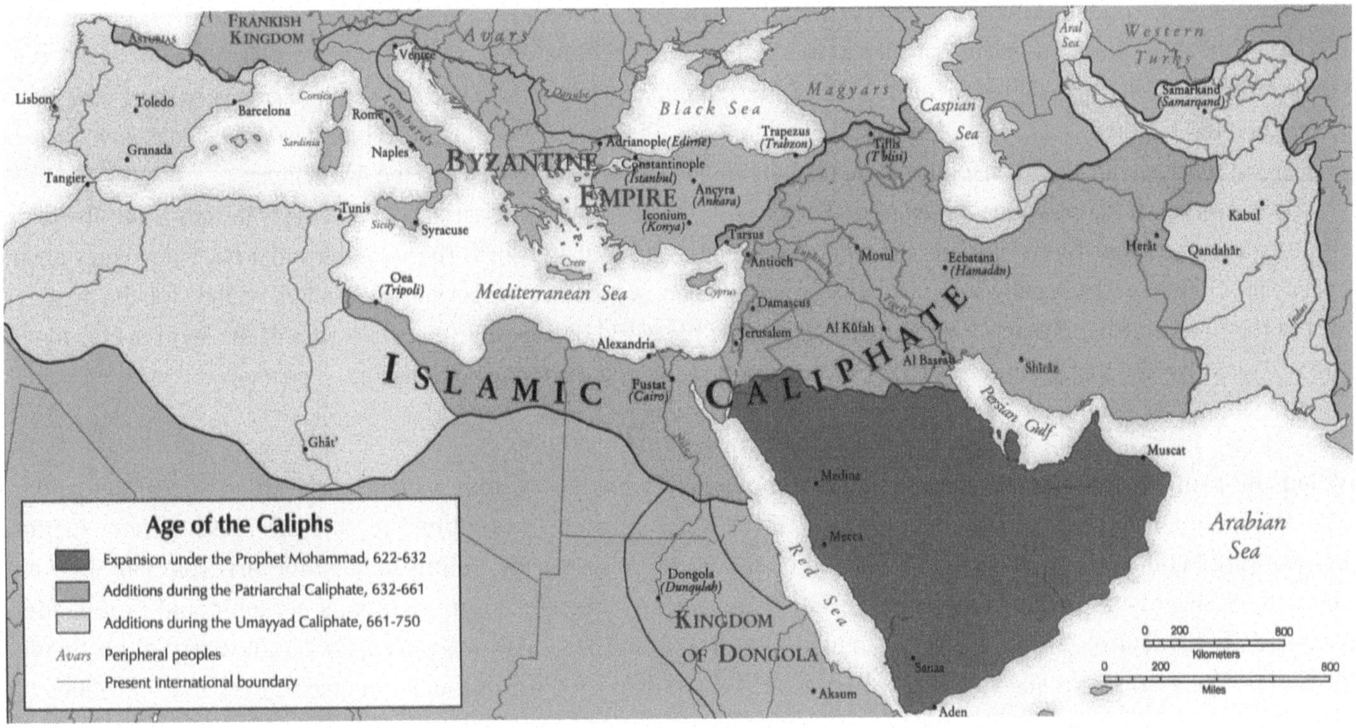

From 600 AD onwards, Islam sweeps the region!

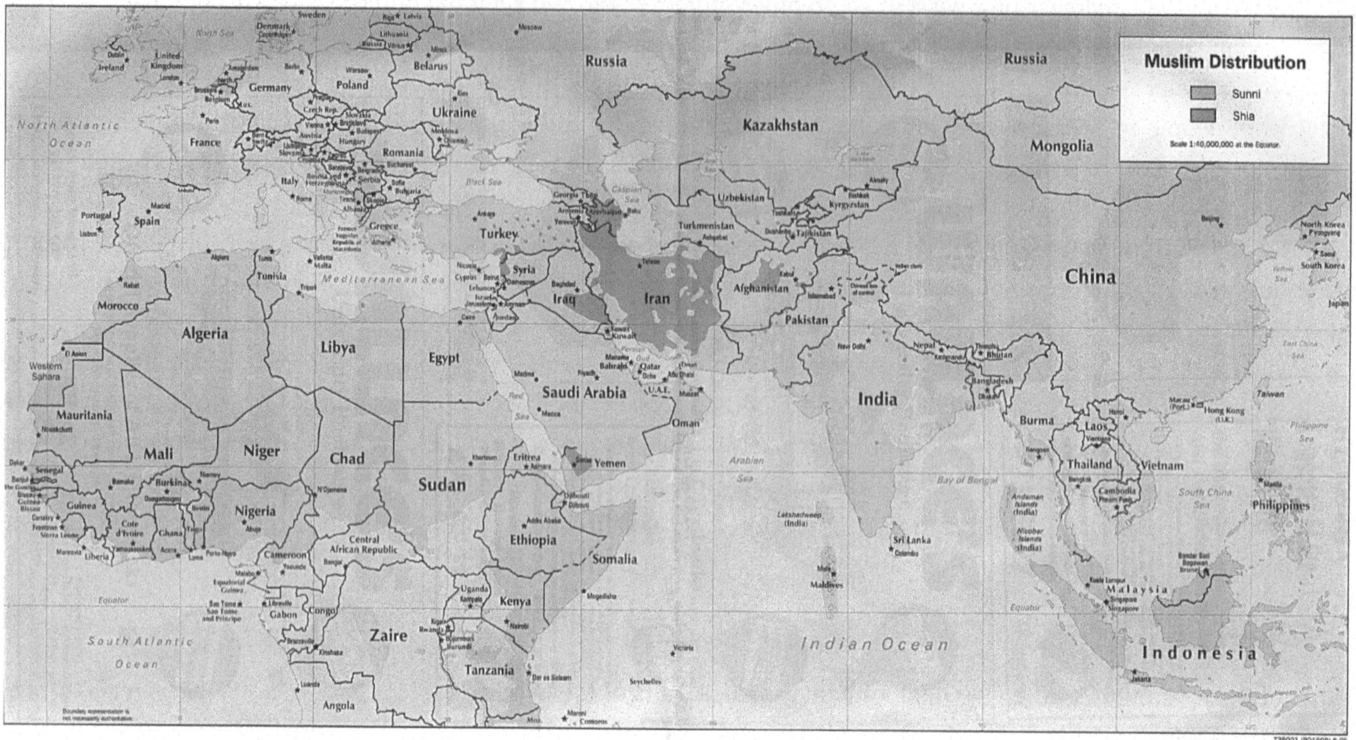

The superfluous Sunni–Shi'a schism.

transcribed the holy book from Allah (Allah = God). The trading routes coming through Mecca made it a wealthy city and easy for the religion to move outward. Islam then expanded outward from this core in Saudi Arabia, in what can only be defined as an Arab/Islamic Empire. You may have never heard of that term before, but take it from the Plaid Avenger, it's a good way to put it. Saudi Arabia, as you might expect is Arab, and because this is also an Islamic movement, I think it's a fair assessment to call the expansion of this political/religious entity the Arab/Islamic Empire.

After the death of the prophet Muhammad in 632 AD, you can see that they expanded outward across all of what's now North Africa, into what's modern day North Africa, Turkey, Iran, Central Asia, and parts of South Asia in very short order, in a matter of a few hundred years. That's how fast the empire grew. Playing it forward to the modern world: Islam is now the second largest religion with 1.8 billion followers (24.1% of the global population). But back to the Arab Empire . . .

Was everybody happy within the empire? Not exactly. There was some disagreement amongst Muslims about the faith, resulting in some subdivisions of Islam, the two main ones being Sunni and Shi'a. There's a third that not many know about called Sufi. Definitely check out where the major groups of Islam are located—especially the Shi'ites, as they are the minority sect of the religion. Most Muslims from virtually all parts of the world outside of Iran and Iraq are of the Sunni sect. Sunni is the vast majority And that is important to know if you want to understand today's Middle East . . . namely. . . .

SUNNISM, SHI'ISM, AND WHY IRAQ IS SCREWED

When the prophet Muhammad died, Islam as a movement was left without a clearly identified successor/leader. The two major branches of Islam—Sunni and Shi'a—were formed in response to this dilemma. Sunni Muslims believe that Muhammad died with no intention of appointing a successor. Sunnis believe the first four caliphs were the proper successors to Muhammad since they were elected; therefore, any righteous person could be a leader. Because of this, Sunni elites chose Muhammad's councilor and father-in-law Abu Bakr as the first Caliph (leader of Islam). There was a steady stream of Caliphs until the collapse of the Ottoman Empire in 1923 and there haven't been any since. Conversely, Shi'ites believe Muhammad selected his son-in-law Ali ibn Abi Talib as his successor and that only Ali's descendants can act as Imams (leaders). Ali became the First Imam—roughly translated as "leader"—and that is the dude the Shias followed. Besides this, the differences between Shi'ism and Sunnism are minor.

Today, Sunnis make up approximately 85 percent of the world's Islamic population. Shi'as make up only 15 percent and are concentrated mainly in Iran (90 percent of population), Azerbaijan (75 percent of population), and Iraq (70 percent of population). This is highly important in Iraq today because the Sunni minority was in power under Saddam Hussein, even though his reign was highly secular. Since the United States crushed Saddam, the Shi'ites have a clear advantage in democratic politics. Much of the violence today in Iraq can be attributed to sectarian violence (i.e., Sunnis fighting Shi'ites) and is concentrated in the "Sunni Triangle" (where the minority Sunni population lives). Sunnis are afraid that if they lose power entirely, then the Shi'ites and Kurds (Kurds are typically Sunni, but represent an ethnic minority) will marginalize and/or discriminate against them.

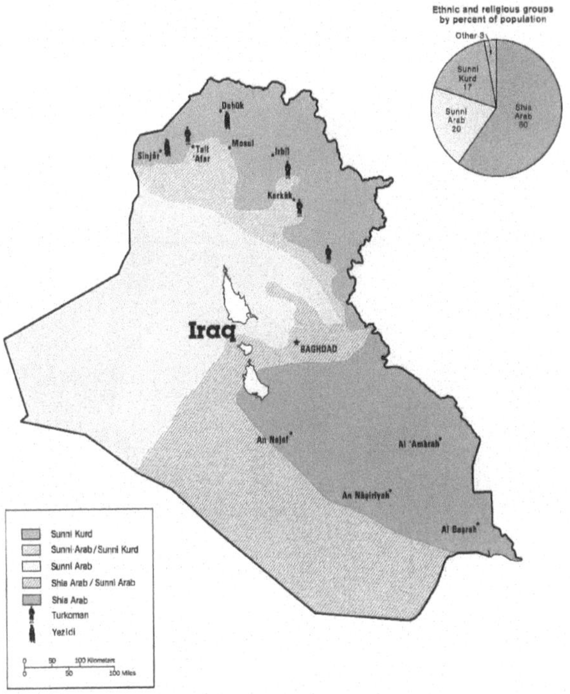

Here is a better map of the mess.

Geographically, Iraq can be viewed as a tub of Neapolitan ice cream. The chocolate-flavored Kurds live up north, the strawberry-flavored Shi'ites live down south, and the vanilla-flavored Sunnis live in between; also, there is a lot of uninhabitable desert—or dessert!—that doesn't neatly fit the Neapolitan ice cream model. Why doesn't the country just split up into three smaller countries, or at least into self-governing districts? Oil, oil, oil—and power politics. Most of the oil is in the Kurdish north and the Shi'a south, so if the country broke apart, the Sunnis would be instantly poor. The international community is also afraid that Iran (Shi'a) and Saudi Arabia (Sunni), the strong neighbors of Iraq, would support their respective religious factions and cause a Balkans-style bloody war. Also, there are Kurds in Turkey who want an independent state of Kurdistan and can now use the oil dollars to fund itself if this were to ever become a reality. Basically, Iraq is a mess.

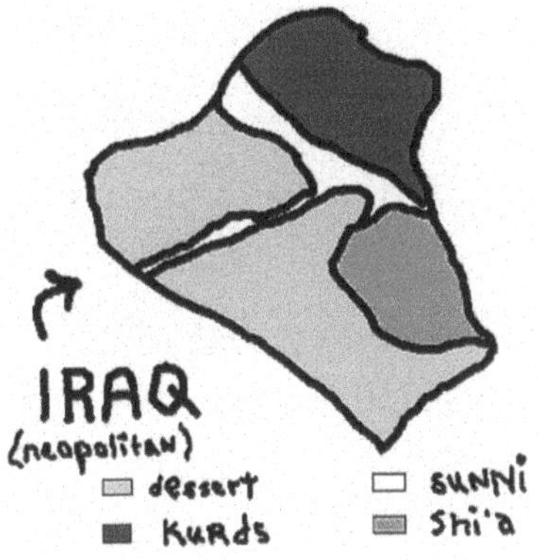

The Iraq Sundae

Iraq isn't the only place that has some Sunni-Shi'a friction, it is just the hotspot of current conflict which reflects this friction. In fact, we can back up and examine the entire region as having a bit of cultural clash based loosely on opposing teams of which can be defined, in part, by their Islamic sect differences. More on that later. First, let's get back to school, because school is cool!

OLD SCHOOL OR NEW SCHOOL ISLAM?

The other thing to consider: most of the Middle East is Islamic, but how conservatively or how liberally do they interpret Islam? With any religion on the face of the planet, you can have people who are real reactionaries, ultraconservative, or you can have people who are a little more, oh, how should I put it, free thinking, a little more open to change in the modern world. Of course, most people are usually somewhere in between. I'm pointing this out because it makes a tremendous amount of difference in daily life, country to country. Even if the whole region is Islamic, there's extreme variation in daily life and even laws and societies from one state to the next. Let me give you an example.

In Saudi Arabia, perhaps *the* most conservative state in terms of Islamic interpretation, there are lots of things that are completely accepted as part of common Islamic life which are different in other places. They strictly follow Sharia: the old-school Islamic code of law. They enforce legal and social restrictions for women, meaning women have to typically ask for permission to marry, travel, to drive, to leave the country, get a job, and go out in public. In Saudi Arabia,

SUNNI, SHI'A OR SHINTO?

Instructions: Of the following world figures, tell me what state/group they lead and what type of Muslim they are. But watch out! A wild card Shinto has been thrown in to increase the challenge, and fun, of this game that is sweeping the nation!

Ebrahim Raisi
President of
____
Sunni, Shi'a or Shinto

Bashar al-Assad
President of
____
Sunni, Shi'a or Shinto

Joko Widodo
President of
____
Sunni, Shi'a or Shinto

Muammar Qaddafi (dead)
Ex-Leader of
____
Sunni, Shi'a or Shinto

Abdel Fattah al-Sisi
President of
____
Sunni, Shi'a or Shinto

Saddam Hussein (dead)
Ex-President of
____
Sunni, Shi'a or Shinto

Recep Tayyip Erdoğan
Prime Minister of
____
Sunni, Shi'a or Shinto

Imran Khan
Prime Minister of
____
Sunni, Shi'a or Shinto

Osama bin Laden (dead)
Ex-Leader of
____
Sunni, Shi'a or Shinto

Yoshihide Suga
Prime Minister of
____
Sunni, Shi'a or Shinto

Salman bin Abdulaziz Al Saud
King of
____
Sunni, Shi'a or Shinto

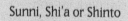

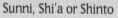

Ali Khamenei
Ayatollah & Supreme
Leader of ________________
Sunni, Shi'a or Shinto

there is also a strong tradition of capital and corporal punishment, meaning if you get convicted of a crime, you can be killed, have your hands or feet amputated with a huge sword, or get whipped, depending on your crime. That's a pretty conservative interpretation. In addition, many Muslims in Saudi Arabia practice **Wahhabism** (founded by Muhammad ibn Abd al-Wahhab in the 18th century), an extreme Sunni practice of Islam that is super extreme conservative and fundamentalist. It is known as an Islamic "reform movement." It resolutely denies the Shi'ite version of Islam. In return, most Sunni and Shi'ite Muslims disagree with Wahhabism. And thus, a major source of friction between the separate sects of the religion.

In other parts of the Middle East like Tunisia, none of those things really apply. There may be some elements of Islamic law that you can see, like the Islamic dress code, but that's about it. Every place else is someplace in between. It's like labeling them all Christian, or all Catholic. Not all Catholics in the world are the same. Go look at some Catholics in Haiti who are doing voodoo, versus Catholics in Italy who are really uptight. There's a lot of variation in how liberal or conservative countries are within the Middle East when it comes to Islam.

Most of the Middle East is Islamic, and that's the end of that story . . . oh okay, you got me; there are substantial Christian minorities in some countries such as Egypt, Syria, and Ethiopia in particular. But that's about it . . . right? Oh heck no! Hold the holy phone! There's one other exception of note, and it is a big, big, exception indeed. Every place in the Middle East is Islamic except Israel: the Jewish state of Israel. Boy is this a convoluted mess, and the Plaid Avenger's going to have to give this one a full page! But we will wait for the Team Play section in chapter 18A, Addendum to get to that. First, let's have a little fun identifying our Middle East Muslim leaders in a little game I like to call Sunni, Shi'a, or Shinto? Check out these famous faces on the previous page and let the games begin!

NUMBER FOUR: PEACE IN THE MIDDLE EAST???

Which leads us to the last homogenous trait which defines the Middle East and that fourth factor is CONFLICT. The Middle East is a region defined by conflict. I've already said this about Central America, a pretty small region in between North and South America, and that does apply. Central America is a violent arena, but perhaps no other place on the planet right now has so many current active conflicts: interstate conflicts (countries fighting countries), intrastate conflicts (people within a state fighting each other), and maybe the worst, foreign sponsored conflicts (US war in Iraq). This place is a mess, and here's the worst part: They aren't out of the woods yet. They've got a long 21st century of conflict ahead of them.

Old School Ottoman.

"Gosh, Plaid Avenger, how could you make a projection like that? Aren't you just being negative? Couldn't things get better?" Quick answer: NO! They're not going to get any better anytime soon! Not in my lifetime. But let's back up the oil tanker and start at the beginning. Why is this place mired with conflict? We'll go back to WWI to start the fun.

World War I saw the demise of the Ottoman Empire. We'll look into more detail of the Ottomans in the next regional chapter of Turkey. For now, just know this: The Ottoman Empire controlled what is now the classic Middle East, the Balkan Peninsula in Europe, and expanded over a lot of North Africa, and even controlled territory in what's now Iran. However, they were an empire in decline coming into the modern era. The Ottomans were an old-school style state going into the 20th century. They were way behind the technological times, their economy sucked, and they still had a sultan on the throne. People in Europe really looked at the whole Ottoman Empire as a redheaded stepchild that none of them wanted to deal with. Nobody wanted to take the time to whip them, much less do anything else with them. The Ottomans looked around for an ally, and leading into World War I, they said, "Hey, Britain, would you be our friend? Hey, France, would you be our friend? Somebody be our friend, because we're worried Russia's going to invade us!" The Ottomans had several incursions with Russia at this point in history. Finally, somebody said, "Sure, Junior, we'll be your friend," and that was the Germans.

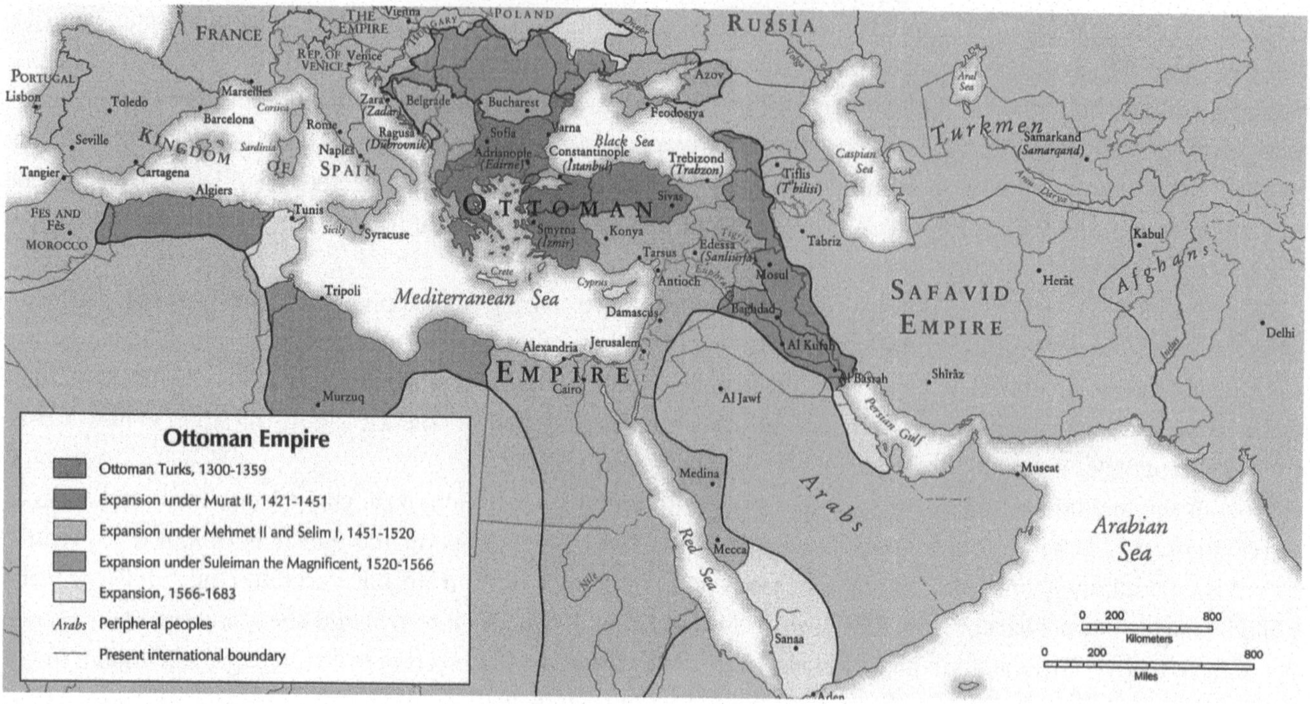

Ottomans: the last big Muslim Empire to dominate the Middle East.

Long story short, the Germans lost World War I and their ally, the Ottoman Empire, lost with them. Originally, Arab and Persian cartographers had drawn maps with central seats of power and their spheres of influence in preparation for the fall of the Ottomans. But, after WWI, the British and the French divided up space differently to suit their own needs, and their borders remained. Like forever. Many historians would say is the critical reason why you have such tremendous conflict in the Middle East today. A guy wrote a book about it called *The Peace To End All Peace: The Fall of the Ottoman Empire and the Creation of the Modern Middle East* (David Fromkin, 1989). Although a fairly dry read, if you can get through it you'll understand why. The title says it all; the peace process at the end of World War I subdivided the Ottoman Empire and drew arbitrary boundaries in such a way as to basically *ensure* that everybody would be agitated and angry at everybody else *forever!* Wow. I know it seems impossible, since we're talking about something that happened ninety years ago.

Let's fast-forward to today's world. Yep. Same story, different century . . . peeps are still unsettled. Again, it's not the entire reason; it's not the only thing going on here, but it is the start of the problems in the Middle East. What do I mean by arbitrary boundaries? Look at Iraq, that's a great one, someone drew the country with a T-square—they're not based on any natural or ethnic divisions. Somebody *drew* them. That somebody was the winning powers at the end of World War I, mostly the British and the French. They subdivided up the old Ottoman Empire, saying, "You lost, you suck, and we want to make sure you don't come back and start a fight with us again, so we're going to parcel you out into smaller pieces. So let's see. Somebody get out a ruler. Okay, there's Iraq, good straight lines. Jordan, that's some good straight lines. Syria, yeah. Saudi Arabia, kind of an ethnic group there, we'll give you guys Saudi Arabia. Egypt, where's the T-square? Ok, got them penciled in now as a sperate entity."

What you had left at the end of this process was Turkey; in today's world, the remnants of the core of the Ottoman Empire, which was technically left intact, and all the other countries were divided up as I explained. Most of these countries were not given outright independence. Some, like Saudi Arabia, were semi-independent even under the Ottomans. But most of them were put in a situation where they were called **mandates**. Primarily the British and the French, and some other players in there as well, weren't so much colonial overlords, but more like the 'big brothers' of the new Middle Eastern countries. Egypt and Iraq were mandates of the British, as were trickier territories like Palestine and Transjordan. Syria and Lebanon were mandates of the French.

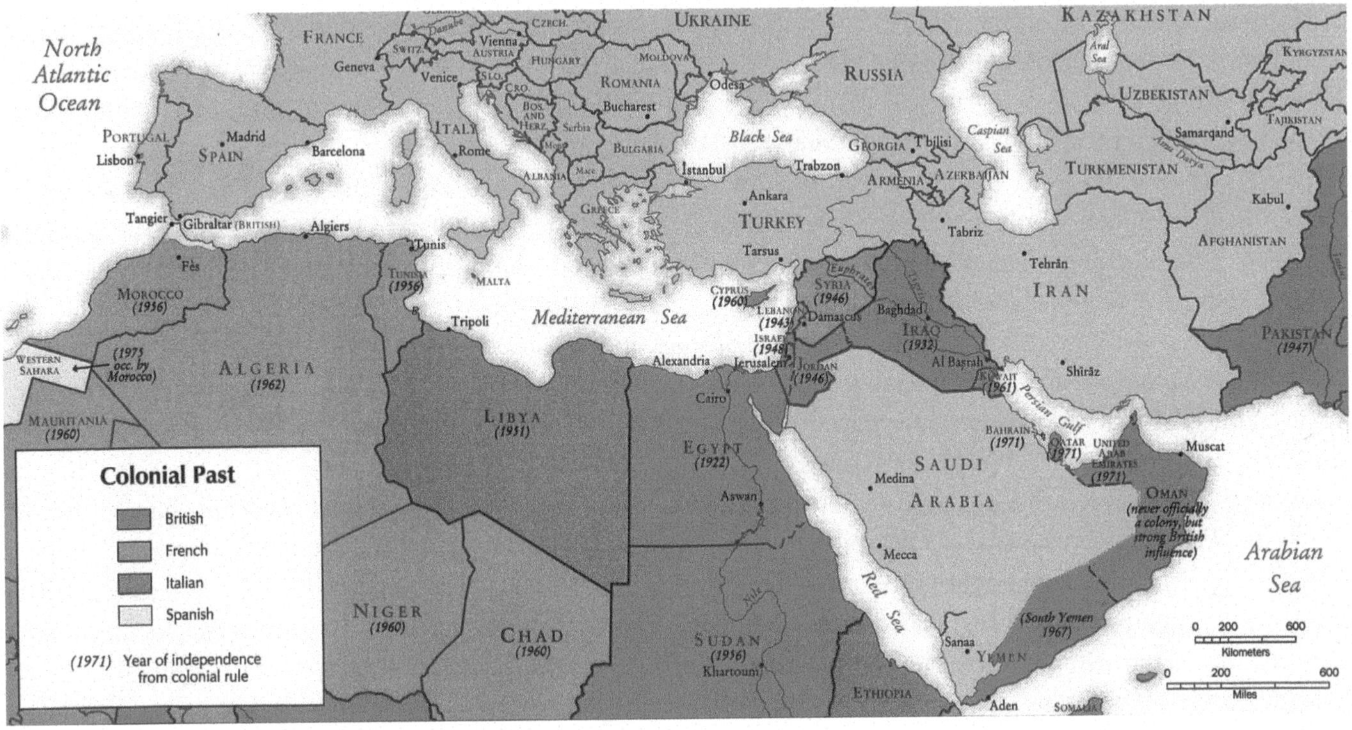

Europeans take over from the Ottomans; use of T-square is evident.

In those cases, 'Big Brother' basically said, "Here's what you need to do; we're going to put people in charge. We'll protect you militarily, but we'll also basically run the show." I can tell you with certainty, after reading that book and others, that a lot of people in the Middle East were immediately incensed, and outright enraged, about this. In fact, places like Syria, a place that the US would like to bomb right now in today's world, *asked* the United States to take over their territory as a mandate. They *wanted* to be a mandate of the US. Wow, how bizarre is that. The United States, which participated in World War I, said, "We don't want anything to do with that mess over there." Boy, don't they long for the good old days of the Middle East when the US didn't want to get involved. World War I is the launching pad for dissent in the Middle East. Let's take it up to World War II, shall we? The fun's not over yet!

Not only do we start with foreign influence and foreign dominance by the British and the French, we also have these artificial borders. It's incredibly similar to European colonialism, but not quite the same. In World War II, the Middle East as a whole does not get very involved. There is some fighting on their soil, but by and large, they stay out of it. The Germans did have their North African campaign—go look up "The Desert Fox," for some fascinating strategy brilliance—so things got a little messy, but it was more of an influence thing and no one really sided up. The Allies said, "Hey, don't side up with the Germans, and we won't invade you." By and large, that's the way it went down.

WWII action in North Africa: Watch out for the pyramids!

However, it's when we get to the Cold War where things start to get tricky. The Cold War came to the Middle East like every place else on the planet, with the US and the Soviets vying for influence. The Soviets were pushing for influence in places like Iran and Turkey for years already. It became a Cold War battleground for influence, which set people against other people. The US supported questionable characters in any effort to make sure they didn't have Soviet influence. One place in particular that springs to mind is

Saudi Arabia, which is an entrenched monarchy. Totally not democratic, totally the antithesis of everything United States stands for, but that's the effect of Cold War; support anyone as long as they don't go commie. Saddam Hussein is another classic example of western forces supporting a total dictator simply so he wouldn't turn to the Russians. Boy oh boy, did the US pick a winner on that one. Not. Another one of particular note is some guy called the Shah of Iran, who was basically a dictator who the US supported for quite some time, again, just to counter Soviet influence. You can read about him a little later in the Team Persia section. So the US supported their client states run by dictators (Saudi, Iraq, Iran and eventually Egypt) and the Soviets supported their client states run by dictators (Syria, Libya, early on Egypt, and eventually Iran). Good times.

We're still under this banner of conflict, still kind of stuck in the Cold War era, but there were also internal wars in Sudan, Yemen, Lebanon, and Chad, interstate wars between Sudan/Libya, Iraq/Kuwait, Eritrea/Ethiopia, and Iran/Iraq, and even international interventions into Somalia, Iraq, and Lebanon. Let's not forget the multiple Arab-Israel conflicts, which we'll get to momentarily. Yeah . . . fun times.

You've got religious strife, ethnic strife, territorial border wars, and ideological wars all happening here in the last fifty years. Wars between states, wars within states, wars involving troops from outside the region altogether. What a mess. This really gets at the heart of the world's perception of conflict in a lot of these countries. Western support of Israel, manifestations of Islamic extremism, factional infighting between Sunni and Shi'ite, secular versus theocratic frictions, the war on terrorism; all these have now involved the outside world even more. The issue of Israel and the War on Terror in particular make this region unique on the planet because all the world is a player in these contests, whether it wants to be or not. Conflict is endemic; it has been and will be a defining feature of the Middle East.

WHY ARE THESE GUYS SO FAR BEHIND?

Now that we've looked at a bunch of homogeneous traits of why we define this area as a region (and some major exceptions), the Plaid Avenger wants to take a little step back in time again, but for a different reason, and not just to explain current conflict. Why is it that the Middle East is behind in the world? Yes, there are *some* countries that are doing quite well: the UAE, Kuwait, and Saudi Arabia are rich places where most of the citizens lead really good lives. But most of that wealth is overly dependent on oil . . . and not everyone in the Middle East has oil, which brings me to the overall assessment of the region: as a whole, it is desperately behind the global times. Looking at most of these countries demographically and economically in terms of economic diversity and standards of living for the average human, you will find that they are far behind on the development path. Like South America, this is a place of wealth disparity, where very few people own an excessive amount of resources. In this case, the rich people own oil resources, but the vast majority of folks have virtually nothing. This is compounded by the physical world, which makes it very hard to just go out and be a laborer or farmer. It's a desert! What are you going to do, farm sand? It's a pretty extreme environment and a place that's fairly far behind by any standards.

The Plaid Avenger's question is: Why is that? What has happened here? How is a region that is right beside the fully developed Europe so far behind? If we go back and look in our own neighborhood, we see that the United States is a fully developed country, and Mexico is not quite as developed. We already pointed that out, and it's absolutely true, but they're not as far behind as the Middle East. In other words, why did the Middle East become stagnant just when Europe propelled itself forward to become the dominant global power? And worse yet, why is the Middle East still in this situation so many centuries later? Here's a brief run through of history that makes some sense of this region's condition.

It's a combination of several factors, which come together about 500 years ago. Up until this point in history, the Ottomans were a pretty successful empire. The Ottomans took over from the Arab/Islamic Empire and continued the expansion. They were actually beating down Europe's door. The Ottoman Empire attacked Vienna, Austria on multiple occasions; that's how far they encroached into Europe. They were large and in charge. At this point in history, it was the Ottomans and the Middle East in general that had the highest levels of technology and economy and some of the oldest universities. Most of the true breakthroughs in scientific technology, from the 1100's to the the 1400s, were in the Middle East, not Europe. This is where we reach a critical phase, because at the exact time that Europe was entering its **Renaissance**, when it started to blossom around the 1500s, the Middle East started to fall behind. Why is that? There are several reasons.

REASON NUMBER ONE: PHYSICAL OVERLOAD

We've already defined the Middle East as a physically challenging place. It's dry, and as such, it's very easy to overfarm, overgraze, use too much water, and make even drier. Unlike Europe, where people had not been hanging out there for very long, in the Middle East, you've had people hanging out forever! The Middle East is the birthplace of western civilization. You've heard of Mesopotamia? All the Biblical stuff you've ever heard of, you can look at the Jewish Torah, the Christian Bible, and the Islamic Koran. It's all in the Middle East. People have been here for a long time. Western Civilization— that is, people building cities, towns, craftsmen, agriculture, Mesopotamia, Babylonia, etc.—all of that happened in the Middle East, not Europe. It spread to Europe much, much later.

The reason I'm telling you this is because civilizations amount to thousands of years of millions of people using stuff. From around 6,000 or 5,000 BCE, people have used trees, grown crops, mined minerals, and used water and other resources. For thousands of years! Dry climates are easy to abuse, so by 1500 CE, the land is exhausted. Thousands of years of resource depletion takes its toll, and the Middle East has simply run out of stuff, just as Europe *starts* to use its stuff. Also, there are critical resources that this region never had in the first place: coal, for instance. Coal, steel, and water were the engine for industrial growth elsewhere, and Middle East guys don't got none of that from the get go. Think back to our Anglo-America chapter. One of the reasons that Anglo-America kicked so much ass is that Europe started running out of stuff when America opened up and started to use its stuff. There is a clear distinction between where people have hung out for thousands of years, and places that are untouched. It has a lot to do with who's rich in the world today.

REASON NUMBER TWO: THE GOLDEN AGE OF ISLAM COMES TO AN END

1258: The party is over.

Staring around 700 CE, there was this Golden Age where everything was awesome in the Middle East. There was continuous expansion of territory and wealth; there was this awesome new unifying religion of Islam, they had centers of higher learning, but then it came to a crashing halt in 1258 CE. How can we attach things to such a specific date? It's the date that another Empire intrudes into this one, and made sure that everything came to a screeching halt. The intruders were the Mongols from Central Asia. The rise of the mighty Mongol Empire, out of what is modern day Mongolia, spread across China into Northern India, all through Central Asia, and all the way into the Middle East. This date is significant because 1258 is the year that the Mongols sacked Baghdad, a center of Middle Eastern power, and *razed*, or *totally freakin' demolished*, it to the ground. It's partly due to the rise of another region during world history that this one started to get behind. I've prepped you that the Ottomans took over shortly and rose back up again, but this was a big blow to Middle Eastern expansion. The Mongols really knocked things out.

REASON NUMBER THREE: BYPASS

No, not heart surgery. The region was physically bypassed on the planet. What am I talking about here? I've already told you that this place doesn't grow a lot of stuff. It's dry; it's not an agricultural epicenter, and nobody was really industrialized at this point in history, so what is it that these guys did? What big riches did they have? Gold? Copper? Anything? I've already given you the answer; it was dry and they depleted all their resources. They didn't have much.

What kept the Middle East going up until this point? The answer: trade. When you think of the Middle East, think of the term "Middle Man." What does a middle man do? He buys something on one end, moves it to some place that needs it, and sells it for a profit. The Middle East has been classically "The Middle Man" of Eurasia, moving stuff from China and India to Europe and vice versa. Making money going both ways, they've been doing this for a good long time. Think about things like, I don't know, the Silk Road out of China. That stuff went through the Middle East. All the carpets, the fine silks, the whole spice trade—people think that the Plaid Avenger's joking them, but spices back in the day are like crack now. No price was too high and people would snort them up their nose and they could never get enough. Black pepper was the crack of its day, back in the 15th century; people loved it, and the Middle East was the dude that brought it to you.

What's that convoluted long story got to do with the Middle East getting left behind? It just so happens in the 1400s and 1500s the entire region got bypassed. Bypassed by whom? A couple of surly European lads, maybe you've heard of them. Christopher Columbus, who was trying to find a sea route to India. Gosh, why was Chris doing that? That's right, I remember! Here's the impact for those of you who think this has nothing to do with the Middle East today. The combination of Christopher Columbus trying to head to India and accidentally bumping into the *huge* New World (1492), and Vasco da Gama, who indeed found a way to India by passing the Middle East and going around Africa (1498), was a one-two punch. It essentially eliminated the need for the Middle East in the span of a *decade*. These guys figured out how to get what the Europeans want without paying "The Middle Man." Think about it. Overnight, your business is out of business.

Da Gama & Columbus: bypassers extraordinaire!

This was compounded by the fact that Columbus didn't find India; he found North *and* South America, and everything in between. He hits the jackpot, a literal goldmine. He wasn't getting the same old stuff from India or China or Indonesia; he bumped into a whole new source of stuff, and this stuff started flowing back to the Old World in tremendous volume. I don't even have to tell you about *all* the things they shipped back, only two things: silver and gold. Think about all the silver and gold that the Spaniards pulled out of Mexico and South America. They had huge mines and a tremendous amount of wealth was jammed into Europe. Q: When you suddenly jam twice as much gold into the global economic system than it had the year before, what is the effect? A: all gold goes down in value. Thus, Europe's new resources did a double-whammy on the Middle East: it made European states richer via capturing trade and increased gold reserves, while simultaneously making existing reserves of Middle Eastern gold worth much less. Not until the discovery of oil in the 1920–30's does the Middle East play any significant role in the world economy.

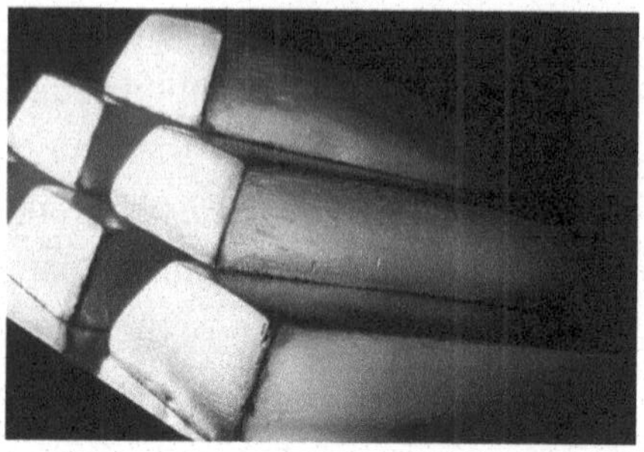

Europe gains the bling via New World acquisitions

So Europe became richer and richer and richer, pulling new stuff in from the Americas and going straight to India and China, thus bypassing the Middle East. Getting bypassed, in combination with the general physical overload and the invasion of my main man Genghis from Central Asia, really took its toll on the Middle East.

REASON NUMBER FOUR: OTTOMAN STAGNATION

Wait a minute. How can we say that this was a period of decline in the Middle East when it coincided with the rise of the mighty Ottoman Empire? You're going to see this when you get to Turkey in the next chapter, but you can sum it up all in one word: booty. That's a word the Plaid Avenger never minds throwing out. The Ottoman Empire's economy was largely based on expansion by acquiring wealth from areas taken over. It's one of the reasons they were knocking on the door of Vienna, because under such an economic system, you have to continue to expand and take new areas over in order to get their wealth. This means that the Middle Eastern/Ottoman economy was fairly stagnant. That type of economy is artificial; it's not based on producing anything. This was ultimately their undoing. The Ottoman Empire: successful in appearance, bankrupt at the infrastructure level. Their territorial expansion ceased by the 16th century, and from then forward the economy stagnated, and the Ottomans were in a perpetual state of defense in order to hold their ground . . . which they kind of did until full unravelling began in World War I.

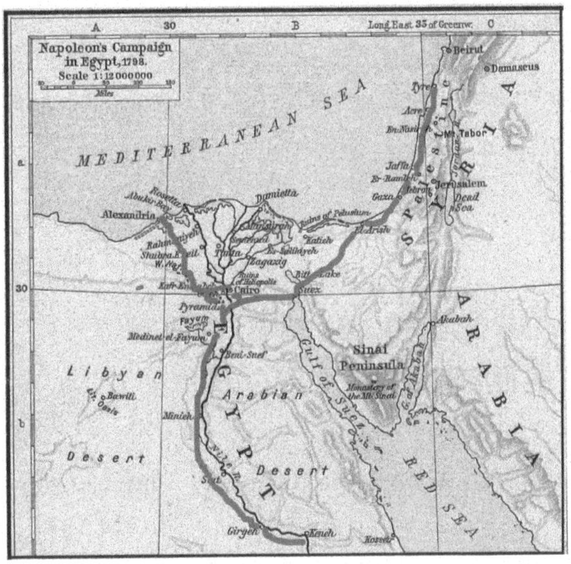

Napoleon's Precocious Pyramid Party Tour.

That's exactly what I was referring to earlier when I said the Ottoman Empire, going into the 20th century, had nothing going on. People in Europe were looking at this place and going, "How have they been sustained this long? Their economy is nothing but hot air!"

This really crystallized in 1798 when Napoleon, the famous short dead dude from France, was a little bored. He had yet to start his takeover of Europe; he was just kicking around trying to get some practice in. He said, "Hey, I'm going to take a boat load of guys, and I don't know, take over Egypt." He went across the Mediterranean to Egypt with a few battalions of soldiers and proceeded to change history. How? Napoleon and his small army of men were radically outnumbered; they were in Ottoman territory, an empire that had made incursions into Europe and was apparently prosperous. Napoleon quickly figured out that they weren't really that tough. The Ottomans sent out their baddest and bravest into battle to kick Napoleon out. They were called the Mamelukes, who were basically the US Marines of the Ottoman Empire. (The *Mamlūk* began as slaves [Turkic peoples, Egyptian Copts, Circassians, Abkhazians, and Georgians] and later became a warrior caste [like enslaved mercenaries for the empire].) The Mamelukes came out swinging and Napoleon soundly stomped them down. Maybe you've heard reference to this tangentially: it is a common historical legend that Napoleon's troops knocked the nose off the Sphinx at this time, as well as stole the **Rosetta Stone**.

The short, dead dude.

I said this was a crystallizing, defining moment for history. Here's why: Napoleon looked around and said, "Wow, that's the best you guys got? Is this the empire we've been afraid of? This is the Ottoman Empire, right? You guys suck! We just totally thrashed you!" This didn't go unnoticed by the Ottomans, by the way. They said, "What just happened there?" People started to assess their situation: "Here we are with this artificial wealth, and we've been stagnant, and basically thought the Europeans were scumbags ever since the Crusades." Ever since the Crusades, the Middle East, and subsequently the Ottoman Empire, looked at Europeans as unsophisticated, unscholarly, backwards societies who sent savages down to the Holy Land in an uncoordinated, racist attack to loot and plunder. At the time, the Middle Eastern assessment of Europeans was largely correct. What a difference a few centuries makes, as the force Napoleon brought was not exactly the rednecks that the Ottomans were expecting. . . .

In essence, the Ottoman Empire had isolated itself and didn't care about the Europeans, because they thought they sucked. And that's why they got behind. They failed to realize that the European technology had surpassed theirs during the Renaissance. Napoleon was the wake up call. (The wake up call actually started a bit earlier in the 1690s with losses to the Austrians. Even worse were losses to the Russians in the 1760s and 1770s. However, losses to the French are really more than any society can take, and are a sign of imminent collapse.) Unfortunately, it wasn't only the wake up call for the Ottomans, who realized, "We better catch up!" It was also a wake up call for Europeans, who said, "Look, we can totally beat those dudes. We should all get down there and get a piece of the pie." And that is precisely what happened. European contact for the Ottomans on their own soil very quickly became a European takeover. As I suggested, the next hundred years were a degenerative process for the Ottomans as they lost more and more real power. By the time World War I was over, they got totally dismembered. Now we've caught the story back up, full circle. European contact turned into European takeover.

Another reason to hate the French: nose thieves.

MIDDLE EAST NOW

This is a place that has gone from pearls to protons (a Plaid Avenger saying) in just a few decades, and it's mostly because of oil. A hundred years ago, polished pearls were one of the biggest exports out of the Arabian peninsula. Pearls collected by pearl divers: pretty low technology stuff and a pretty low total GDP earner as well. Now this place has some of the top technological institutes in the world, and some serious GDP earning potential. This place was far behind, in the grips of conflict, and for all practical purposes, would still be completely in the hole if not for the fact that it has loads of one resource the world now needs more than anything else: oil.

Just to catch you up on the story of oil: In the early part of the 20th century, nobody really cared about oil. In 1900, oil was not used for much. But as the United States was industrializing, the automobile became a product that most people started using, and oil became very popular. From 1900 to the 1950s, oil became "Westernized" in Saudi Arabia. By that, I mean that exploration and development

CRUDE OIL EXPORTING COUNTRIES IN 2018 (OPEC MEMBERS IN ITALICS)

Rank	Exporter	Crude Oil Exports (in billions of US dollars)	Percent of World Total
1	*Saudi Arabia*	182.5	15.9%
2	Russia	129	11.3%
3	*Iraq*	91.1	7.9%
4	Canada	66.9	5.8%
5	*United Arab Emirates*	66.8	5.8%
6	*Kuwait*	49.8	4.3%
7	United States*	47.2	4.1%
8	*Iran*	45.7	4%
9	*Nigeria*	43.6	3.8%
10	Angola	38.4	3.4%
11	Kazakhstan	37.8	3.3%
12	Norway	33.3	2.9%
13	*Libya*	26.7	2.3%
14	Mexico	26.5	2.3%
15	*Venezuela*	26.4	2.3%

*Note: This figure is just what the USA exports, however it is still a net importer of oil.
Source: http://www.worldstopexports.com/worlds-top-oil-exports-country/

of the oil industry was carried out by Western countries, not the locals. There were lots of companies, maybe you've heard of some of them: Shell, British Petroleum (BP), and Exxon, that had scouts out all around the world in the early part of the 20th century. Lo and behold, they proclaimed: "Wow, we think we've found a butt-load of oil in the Middle East."

Of note, Saudi Arabia originally gave all oil concessions to the United States in 1933, because several other European companies had already written off the peninsula, thinking there was nothing there. Enter a huge oil company from California, the Standard Oil (So Cal) that said, "We think we've got a lead on something," to the monarchy of Saudi Arabia, and asked for concessions, which they received. Saudi Arabia said, "Make it 20/80. You guys take 80% and if you find anything, give us 20%." Around a week later, the oil company said, "Oh, we just found the largest oil reserve the planet has ever seen." It's the main reserve that's still pumping today. (Side note: Standard Oil was so rich that it was deemed a monopoly by the US government in 1911, and forcibly broken into a bunch of smaller companies, many of which you recognize in today's world—Esso, Exxon, Chevron, Amoco, Conoco, and Mobil, to name a few.)

Basically, the oil was "Westernized" early on, as the Plaid Avenger calls it, meaning Western companies had been in there prospecting, developing, extracting, and taking a lot of the wealth. No big deal. After all, they supplied all the labor and know-how. But in the 1950s–1960s, oil became "Arabized," meaning the states with the oil themselves started to play more of a role. Places like Saudi Arabia said, "Hey, wait a minute, this is *our* oil. Maybe it shouldn't be 20/80, more like 80/20 in our favor, we'd like to interject, take over these jobs. Maybe you all could train some of our people?" A lot of these countries took over production outright, or at least played a bigger role in it.

For the most part, foreign companies acquiesced to these requests, that is important to note because at this point—the 1950s—there was not a huge amount of animosity between Middle Eastern states and the West. In fact, the US was largely seen as a positive force in the region, which brings me to an interesting historic event that underlines this fact. In 1956, there was something called the Suez Canal Adventure. I point this out because this was probably the point at which the Middle Eastern countries viewed the United States most favorably, since the US helped them out (check out box to the right). After this point, support for Israel in the 1973

What's the Deal with the Suez Canal Adventure?

The Suez Canal, which opened to traffic in 1869, is one of the most important shipping lanes in the world. It connects the Mediterranean Sea to the Red Sea. Along with the American Transcontinental Railroad, the Suez Canal drastically decreased the time it took to ship goods internationally. Although built by France and Egypt, or more accurately by Egyptian slave labor, the British had de facto control over the canal as early as 1875.

The British maintained control of this vitally important sea route until 1956, when Egyptian leader Gamal Abdel Nasser decided to nationalize the canal. (Many parallels can be drawn between Nasser and Saddam Hussein, including fluctuating support and hatred of Team West). After flipping out for several days, the British and French came up with a plan to regain control of the canal—by convincing Israel to attack Egypt, and offer to neutrally control the disputed territory, namely the Suez Canal. Israel was happy to comply and invaded Egypt on October 29, 1956. As per the agreement, British and French forces entered the Canal Zone to "separate the warring armies." Egypt responded by sinking every ship in the Canal.

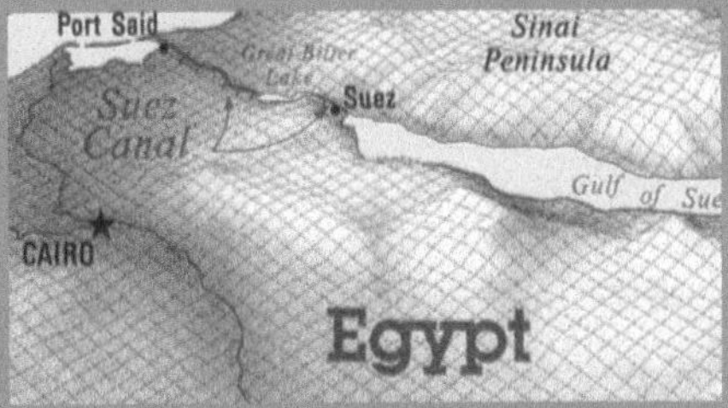

Suez spells end of Middle Eastern colonialism.

The United States, after watching all this go down, was furious and worried that the British and French action would cause a Soviet military response. President Eisenhower pressured both the British and French to withdraw and publicly criticized Israel. A UN Commission took control of the Canal Zone temporarily until it was handed back over to Egypt. After the Suez Crisis, public opinion of the United States in the Arab world was at an all time high. Unfortunately, the US spent the next fifty years squandering this Arab street-cred with blatantly pro-Israel policies.

Yom Kippur War (which we've already referenced) triggered the 1973 Oil Embargo, which started a downward slide of US/Middle Eastern relations—which may be at an all time low right now in the 21st century. We've already seen the amount of US influence in the region just from looking at conflicts in the last forty years, and given current events, it's likely to get worse in the future.

How can they be really rich and also behind the times? The answer is oil, a mixed blessing here as it is everywhere in the world. Most Middle Eastern economies are still fairly stagnant, which is a big problem. They are not *diversified*, meaning their GDPs rely on this single resource. They don't have other jobs. If anything happens to oil—if an alternative energy source is found, if the price gets too high and people just stop buying it—the Middle East is going to be in serious trouble. Many of these countries are going to have a recession or a depression or a collapse overnight.

Several of these countries are doing something about this, particularly places like the United Arab Emirates (UAE) which is diversifying into computer technology. The UAE is now seen as a technological epicenter of the Middle East. There are some places that are going to do fine, but most states still aren't going to be in good shape in the very near future.

Another interesting thing happening in Saudi Arabia is the Vision 2030 project. This plan includes parameters for education reform, the increased participation of women in the workforce, and investment in the entertainment sector. They're fully aware that they need a backup plan to the dependency on oil eventually running out (or the oil itself). Crown Prince Mohammad bin Salman's (MBS) is trying to play a role on the international stage—between his attempts at seeming progressive (but with an authoritarian streak), buying US arms, and conducting a brutal war in Yemen. Busy guy.

THE PEOPLE

That brings us to some other problems in the Middle East today. They're having a population explosion. As you can see from population pyramids on the right, fertility rates and population growth are through the roof here. The Middle East does not have the world's *biggest* population total, but it's one of the *fastest growing* regions. They're adding people at a phenomenal rate. Why is that of particular concern or problem? Because they won't have enough jobs, and they don't have many physical resources.

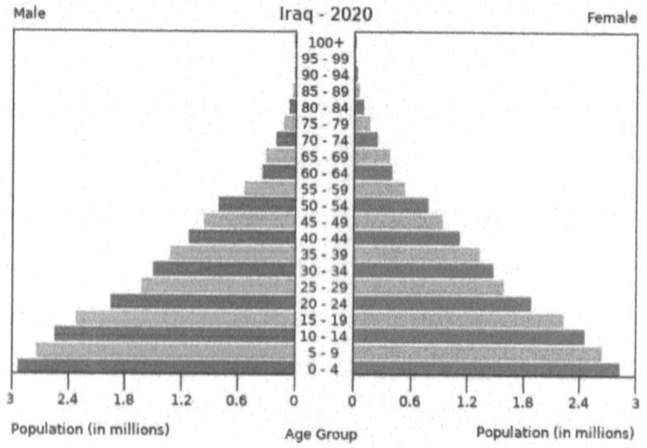

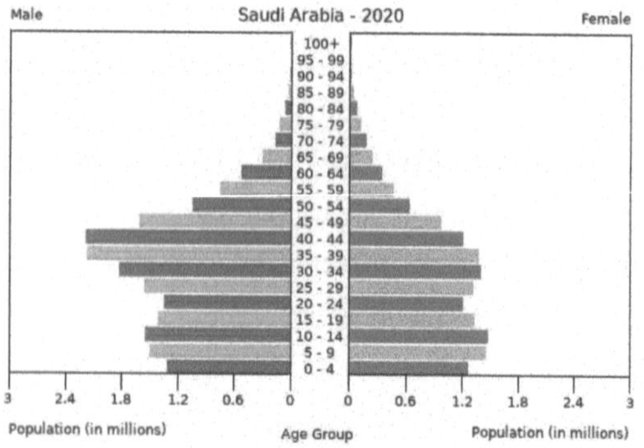

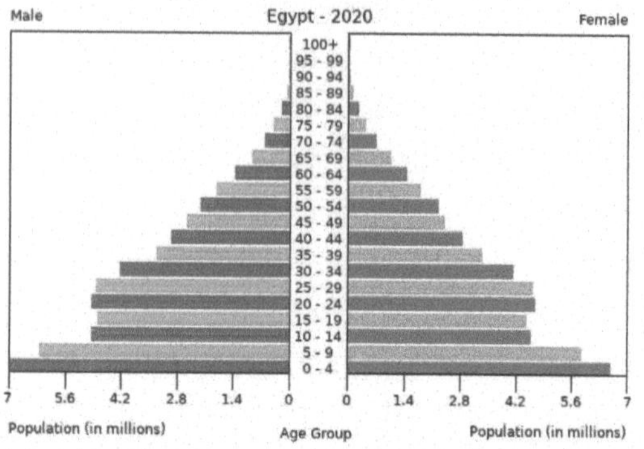

Physical resources like what? What's the big deal? They have plenty of oil so they can buy everything else, right? There are a lot of resources they don't have. Simple ones, like water—you know, that stuff humans need to stay alive. If you're going to have more people, you're going to need more of it. Well, guess what? We've already pointed out this place is dry. Water is one resource they don't have, and they can't just make more. Many countries are using and expanding desalinization plants to turn seawater into drinking water, but that's an expensive operation. Unless they figure out how to drink oil, it's risky to grow so fast.

Limited resources and an exploding population typically spell disaster. This place has enough turmoil going on already and the Plaid Avenger's predicting it to get worse. In fact, I'll go out on a limb here and say that, with all the current conflicts that are already happening in the Middle East, one that's not happening yet is almost guaranteed: the battle over water. There will be interstate wars fought for this critical resource in the next fifty years. It's almost impossible that it will not happen. There are simply too many people being born, and not enough water for them all. Water has the potential for explosive conflict that's not even happening yet. As population is exploding, urbanization is exploding as well. The physical world in this area is simply too dry to support lots of people. Most people have to be in the city, and there aren't enough resources or jobs in the city either.

Middle East governments usually to the far right—but rarely in the right! (**UPDATE:** Told ya' so! Popular protests have fixed this situation in several states! Saddam and Muammar were overthrown and killed, Mubarak thrown from power and jailed, and Assad is on the rocks, although it is increasingly looking like he will survive. King Abdullah died of old age, and was replaced with the next King Salman: meet the new boss, same as the old boss.)

THEIR RULERS

Let's compound the situation further, shall we, because it doesn't seem complicated enough. Let's look at some government types. What's a typical government type in the Middle East? That's right! Theocracies, autocracies, monarchies, dictatorships: wow, all the rightist regime types are here! Power to the government! No power to the people!

Let's mix all these fun facts up in a bowl. Limited resources that are getting scarcer, undiversified economy, limited jobs becoming scarcer, way more people, and then top it all off with a nice layer of rightist regimes, which typically limit human rights and/or freedoms. Let's complicate it even further. Unemployment has been on the rise as of late in most of these countries (thanks global recession!) and food prices have been on the rise for years too! Public dissatisfaction in places like Egypt and Yemen and Algeria is at an all time high . . . and they have no vent for their frustrations and anger.

2011 Update on the Insanely Accurate Assessment in the Above Paragraph!!! Plaid Avenger called that one! But who knew that things would become so explosive, and create a domino affect across the entire region which is now being referred to as the "Arab Spring"? Tunisia erupted into revolution in December 2010, with their president stepping down in January 2011 . . . but that was just the spark that set Egypt on fire, and President Hosni Mubarak was forced out of office in February 2011, after 30 years of rule, and months of huge street protests! Since then, Yemen, Syria, and Bahrain have been rocked by persistent street protest, some of which have been met with violent suppression from the governments. This stuff is out of control, and several decades past due!

2012 Update: The Arab Spring, Year 2!!! Libya erupted into revolution and ousted the manic Muammar Gaddafi and killed him! (With the help of NATO of course!) Then President Ali Saleh of Yemen was forced into resigning and exile, and now Syria's Bashar al-Assad is totally on the ropes as the revolution there continues and becomes more bloody by the day! While the chaos continues in places, other states are resuming an air of normality and freedom by holding elections and forming new governing systems, namely in Tunisia and Egypt where the revolutions first started.

2014 UPDATE: Has the Arab Spring come undone? Is the freedom party over? It would appear so at the start of this year. Egypt had yet another revolution which chucked out the democratically elected Muslim Brotherhood leaders, and re-established military control via a preposterous election in which Abdel Fattah al-Sisi, the head general of the military, ran for office . . . which is the exact same way Hosni Mubarak came to power over 30 years ago! Same script, different decade! On top of that, the tide has been turning in favor of the Bashar al-Assad regime in Syria, and he looks set to eventually crush the uprising and maintain dictatorship-style control in the country for years to come. While some forward-

thinking states like Morocco and Jordan are liberalizing their systems slowly and perhaps converting to something like constitutional monarchies, the vast majority of Arab leaders appear to be keeping their populations in full check, and democracy will have to be put back on to the back burner for now. But the kettle could boil again at any given second. . . .

2015 UPDATE: Well . . . 2013 and 2014 have seen a reversal of this uprising, and it appears that the Arab Spring has turned back into an Arab Fall. But no one thinks that this mess is settled! Egypt has held fake elections solidifying ex-military leader and current President al-Sisi into office, thus turning them back into a one-party state. Yemen has fallen into full-on civil war between competing Shiite and Sunni sects, with al-Qaeda thrown in for fun. But the biggest update, of course, is the total disintegration of Iraq and Syria and the insane clown posse otherwise known as ISIS has entered the scene of chaos to claim a state of their own! That situation is so insane that it made "The Cover Story" chapter of the last edition of this book!

Want to learn all about the Iraq/Syrian meltdown and the rise of the ISIS situation in more detail? Visit: http://plaid.at/isis to read **The 2016 Cover Story: The ISIS Insanity**.

2017 UPDATE: Did you think I had any good news here? Not. The Syrian Civil War is still the biggest human tragedy on the planet, with the death toll now surpassing 250,000 people killed (according to the UN) with many agencies believing it is closer to half a million. 13.5 million people are in urgent need of humanitarian assistance. Six million people have fled the fighting as refugees. Russia is now working closely with the Syrian government to bomb all opposition to death while the US and its allies are sidelined due to waiting for a new US President to take over. Speaking of which, you heard it here first: I am writing these words on December 15, 2016, and I predict that the Russians and Syrian government will pull out all the stops to utterly destroy the opposition/rebel groups, then as soon as Donald Trump assumes the US presidency, they will pull him in to help broker a "peace deal" cease fire which effectively will end the Syrian Civil War. . . with Assad still in power and Russia a more emboldened player in Middle Eastern affairs.

In other **2017 UPDATE** action: you should know the the Sunni/Shiite rift across the entire region is being exacerbated by all these conflicts, and it appears to be worsening. Saudi Arabia and its Arab allies are supporting one side of the Yemen Civil War, one side of the Syrian Civil War, and different political parties/groups across the region that share their goals. Meanwhile, Iran and its allies (Russia, Syria) are supporting the other sides of those conflicts as well as groups like Hamas in the Gaza Strip and Hezbollah in Lebanon. That is a way over-generalized and simplistic take on these conflicts, but it is happening, and it is a major cause for concern for the future of religious intolerance and warfare in the region.

2019 UPDATE: A lot continues to happen in the region. Bashar al-Assad is still in power and over 400,000 Syrians have been killed, 5.5 million have fled the country (mostly to the surrounding countries of Jordan, Lebanon, and Turkey), and over 6 million are internally displaced. The Syrian Democratic Forces have announced that ISIS lost its last stronghold in the country but ISIS fighters remain nonetheless. Several are requesting to be repatriated to their western home countries which is setting up a debate as to what the responsibility of these countries entails.

Egypt recently had a referendum vote on its constitution to extend al-Sisi's current term to six years and also allow him to run for a third six-year term in 2024. This highlights a possible return to more authoritarian regimes in Egypt.

The war continues in Yemen between Houthi rebels and Saudi-led coalition forces. The country is plagued with the world's worst humanitarian crisis with over 70,000 people killed and half the country's population starving. Former US President Trump issued a veto on a congressional resolution to end US involvement in the war. This is not surprising given his overwhelming support of Saudi Arabia in general, and specifically the Crown Prince Mohammed bin Salman.

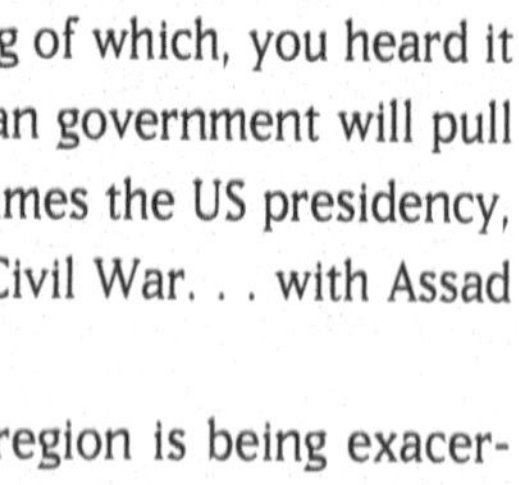

Crown Prince Mohammed bin Salman Al Saud aka MBS: The real power behind the Saudi throne.

The dispute over Iranian sanctions is also noteworthy. The 2016 Nuclear Agreement, also known as the Joint Comprehensive Plan of Action (JCPOA) between the permanent UN Security Council and Germany (P5+1) and Iran, was rejected fully by the Trump administration and resulted in US withdrawal from the agreement and US demands for tightening sanctions on Iran and those who seek to do business with the country. Iran's economy has suffered for years under sanctions and they have responded to the most recent threats with their own. Everyone appears to be preparing for battle in the Persian Gulf, once more! Do the good times ever end in this region?

2021 CONFLICT AND CHAOS CONTINUES UPDATE: So . . . yeah . . . no good news out of this region recently. Of note just since the last edition of this book in 2019:

1. Already rocked by political turmoil, corruption, and a decades-long crippled economy, on August 4, 2020 a huge explosion (caused by bureaucratic mismanagement of dangerous chemicals) leveled the biggest port facility and surrounding city blocks in the capital Beirut. Slow motion state disintegration continues.
2. The former Trump administration's suspension of the Iranian nuclear deal and resumption of hardcore sanctions on Iran, and the simultaneous economic devastation brought on by the Covid calamity, has resulted in Iran re-starting its nuclear enrichment program and likely rise of more hardcore right-wing political leaders ready to rally the state into confrontation. In tandem, a US drone strike in Iraq assassinated Qassem Soleimani, commander of the Iranian Revolutionary Guard (IRGC) Quds Force. The face-off between Iran and Team Arab (led by Saudi Arabia) has continued to become more inflamed across the region in multiple conflicts which have been mentioned in these previous updates, such as . . .
3. The war in Yemen continues to get worse.
4. Terrorist acts and political tensions are increasing in Iraq.
5. Confrontations in the Persian Gulf between Iran and the US/Arab states continues to get worse.
6. The Syrian Civil War is winding down, but continues to wreak death and destruction. Worse yet, the regime of Bashar al-Assad will definitely survive the war and be in charge . . . most likely with an iron fist and return to full dictatorship.
7. In 2021, a serious outbreak of protests and violence across Palestine exploded in the ongoing Israeli–Palestinian conflict starting on May 10 until a fragile ceasefire on May 21. Rioting, met with violent police riot control, rocket attacks on Israel by Hamas and Palestinian Islamic Jihad, which were countered by Israeli airstrikes targeting the Gaza Strip. 270 killed, thousands injured, massive infrastructure damage in Gaza. Yeah . . . don't look for this one to get any better any time soon, although quiet for now.

I could go on and on here assessing each year and making some predictions, but you know what? This stuff is still too new and too hot and too revolutionary to be able to make any sound long-term judgment calls on where it is going and how it will affect each state in the region. But no doubt, tough times remain ahead in the short term, and likely the long term.

HOLY HOOKAH, ARE WE DONE YET?

Yes. Well, for now. Of course, I haven't really even touched upon the hotbed of current conflict that defines this region, and this is conflict that, while actually being played out within the region, has global repercussions that affect us all. The Israeli/Palestinian issue is a world issue. The US War on Terror is a global phenomenon. The competition for the region's resources has planetary scope. That's why I decided to add a special section to this chapter just to tackle the thorny issues of exactly what forces are shaping and defining this region, and thus the world. It is an addendum of sorts. How quaint! An addendum! What supposedly scholarly book would be complete without an addendum? That's the next chapter my friends, so turn the page and get to it.

MIDDLE EAST RUNDOWN & RESOURCES

View additional Plaid Avenger resources for this region at http://plaid.at/mid_e

BIG PLUSES

- Hugest producer and supplier of oil to the planet; big bucks
- Some states like UAE are leading the way to using oil wealth to diversify their economy into banking, service sector, and computer engineering
- Center of Islam: one of the world's major religions, and one that is growing fast; this provides at least a little cultural homogeneity across most of the region, and links to Muslims across the world

BIG PROBLEMS

- Exploding populations which are already taxing scarce critical resources like water
- Undiversified economies based almost exclusively on primary products (namely oil)
- Great wealth disparity within states, and across the region as a whole
- Government types across region are almost exclusively rightist regimes with power consolidation in the hands of few
- Many states/power centers within the region hate each other, and this is one of the few regions on earth where interstate conflict will likely increase
- Radicalized elements of society popping up all over region; existing "terrorist" groups like Hamas, Hezbollah, Al-Qeada, et al, are holding their own or increasing . . . and ISIS!!!
- Significant dissatisfaction and discontent among large segments of society in Saudi Arabia, Egypt, Lebanon, Iraq, Iran, Somalia, Sudan, Algeria, Palestine, Yemen
- Current chronic conflicts: Israel/Palestine; Yemen; Syria; coming soon: Iran

LIVE LEADERS YOU SHOULD KNOW:

King Salman: Monarch/leader of Saudi Arabia, the birthplace and holy center of Islam. Arab, Sunni, Saudi, center-right, conservative, massively wealthy, and a US ally. More hawkish than his predecessor, and may yet assert Arab power militarily in a big way . . . against ISIS, or Iran!!!

Ayatollah Khamenei: Religious/Spiritual leader and ultimate wielder of political power in Iran, since it is a theocracy. Persian, Shia, ultra-right, conservative, and totally hates the US. A lot.

Mahmoud Abbas: President of the Palestinian National Authority, and the "head" of Palestine, with which Israel, the US and the West all deal with. He is Arab, Sunni, Palestinian, centrist, and effective head of the West Bank . . . not so much of the Gaza Strip anymore though.

Abdel Fattah al-Sisi: Formerly the top military general who stepped down from that position after a military-led coup threw out the elected Muslim Brotherhood government. Now he is Egypt's new 'President' after a contrived election. He "fixed" the constitution, so he may be in power until 2030. Arab, Sunni, Egyptian, far-right, conservative.

Ebrahim Raisi: New President of Iran in 2021. Ultra-conservative, hardline, right-wing, was formerly a Muslim jurist . . . and is likely being groomed to take over as the head Ayatollah and Supreme Leader of Iran.

Bashar al-Assad: President of Syria overseeing the current massacre of his own people in order to maintain his power . . . which he inherited from his father Hafez al-Assad, an iron-fisted dicatator of Syria for 29 years. He is Arab, Alawite (Shia), and has strategic ties to Iran. His days on the planet are now numbered.

Mohammad bin Salman (MBS): Crown Prince of Saudi Arabia but already its de facto leader. He launched the war in Yemen in March 2015 while he was the Defense Minister under his father, King Salman bin Abdul Aziz Al Saud. He's overseen some reforms, such as women being able to drive, but he's also responsible for the death of the journalist Jamal Khashoggi (a critic of the government, killed in October 2018) and other authoritarian measures.

Benjamin Netanyahu: Current Prime Minister of Israel, and a conservative, center-right, hawkish fellow. Don't look for any Palestinian "peace process" to go forward on his watch. He is a US ally and his right-wing, populist policies make him hugely popular with President Trump; US/Israeli ties have never been stronger. **2021 UPDATE:** Benji actually just lost the office here in 2021, but never count him out for long. Like the Terminator, he'll be back.

Omar al-Bashir: President of Sudan since a 1989 coup, and the first sitting leader in history to be indicted for war crimes and genocide for the Darfur deal. There was great outcry by Arab states when this arrest warrant was issued. Why? Bashir is ethnically Arab, a Sunni Muslim, center-right, and holds all power . . . just like the other Arab leaders! **2019 UPDATE:** Omar was overthrown in April by a military coup. Since his arrest, a military council will oversee the transition of power for the next two years. Country is in chaos.

PLAID CINEMA SELECTION:

Marmoulak The Lizard (2004) Believe it or not, this is an Iranian comedy! Yes, they like to laugh! But is also a scathing look at the Islamic Republic's experiment with theocracy, through the eyes of a common thief caught-up in a case of mistaken identity when he poses as a mullah religious leader.

Paradise Now (2005) Follows Palestinian childhood friends Said and Khaled who live in Nablus and have been recruited for suicide attacks in Tel Aviv. Filmed on location in Nablus, West Bank and in Tel Aviv, this film is fantastic insight into realistic Palestinian life, but, more importantly, the whole thought process behind what brings people to such an extreme act.

Waltz with Bashir (2008) Now I'm not pushing cartoons because I am one, but this is easily the best, most intense animated film I have ever seen. Based on absolutely real events. An Israeli film director interviews fellow veterans of the 1982 invasion of Lebanon to reconstruct his own memories of his term of service in that conflict. Super creative and intense. A crowd-stunner every time I have shown it.

The Kingdom (2007) Okay, at least one fun action adventure film: A team of U.S. government agents is sent to investigate the bombing of an American facility in the Saudi Arabia. Interesting insight into Saudi culture and their anti-terrorism campaign, which in real life is one of the best in the world. The intro title graphics are the the best part.

Persepolis (2007) Another animated experiment. Poignant coming-of-age story of a precocious and outspoken young Iranian girl that begins during the Islamic Revolution of 1979. Great for showing how the society changed over the last several decades.

A Time for Drunken Horses (2000) Set in a Kurdish village in snow-capped Zagros mountains on the border of Iran and Iraq, this pathetic tale of hard life as a Kurdish teenage boy struggles to hold his family together and care for his severely disabled brother after their father dies. Will assuredly make you appreciate your life more, or commit suicide, or both.

The Battle of Algiers (1966) A totally intense war film based on occurrences during the Algerian War (1954–62) against French colonial rule in North Africa. So gritty and well made that it looks like a documentary that is just recording the events. Grating tense soundtrack and lots of symbolism make this film not for the weak of constitution. The fact that it depicts urban guerrilla warfare, suicide bombings, and sleeper cells long before the "modern era" should give one pause for thought.

Lawrence of Arabia (1962) Epic desert movie portraying the real life adventures of T.E. Lawrence who served in British Army during World War 1. He helped organize Arab resistance to the Ottoman Turk Empire. Is great insight in Arab Bedouin culture and the historic events of the fall of the Ottomans, which established modern day countries of the region.

The Ten Commandments (1956) Got to give a shout out to all my Hebrew friends on this one, and for everyone else, you should watch this cinematic masterpiece to better understand how the whole creation of Israel thing went down back in the day. Like Old Testament back in the day, that is.

Bab el shams (2004) Roughly translated meaning "the gate of sun," this 4.5 hour epic faithfully depicts the birth of the nation of Israel. Shows the relations between Zionists and Palestinians both before and after the formation of Israel in 1948. Considered by some to be one of the greatest Arabic films of all time and potentially one of the greatest films of all time in general.

Ekhrajiha (2007) Made by a ultra-conservative right-wing dude, a controversial look at the final days of the Iran-Iraq War. Unlike other Iranian films depicting this strife, the director paints the heroes as fundamentally flawed, often acting in ways that would be considered by some to be immoral. Filmed in and around Tehran, depicts real life living conditions in modern day Iran.

Lost Boys of Sudan (2003) A feature length documentary that highlights the cultural differences that two Sudanese children experience when they are brought to America. Orphaned by one of the most brutal civil wars in Africa, the two boys survived lion attacks as well as militia gunfire before being chosen to come to America.

The Yacoubian Building (2006) The highest budgeted Egyptian film of all time, bears its plot from a novel of the same name. Set around the time of the first Gulf War, is a biting view of Egyptian society in the years since the 1952 coup d'état. Set in downtown Cairo, the film plays wonderfully with metaphors and is a great representation of the book.

DEAD DUDES OF NOTE:

Former Colonel Muammar Gaddafi: The leader/dictator of Libya since 1969. Semi-insane, used to sponsor terrorism and nuclear weapons development until he renounced all and became a good boy in 2006. He is Arab, Sunni, conservative right, but has in times past renounced his "Arab-ness" in favor of his "African-ness." Since 2012, he now specializes in "Dead-ness."

Saddam Hussein: The not so short dead dude from your Iraq War history review. Former President/dictator of Iraq that first inflamed the Shia and Kurds in his own country, then inflamed Iran by invading them in 1980, then inflamed Kuwait by invading them in 1990, then inflamed the US by simply existing, which the US proceeded to remedy. Now he is in-flamed in hell. Or is that in-flames?

MAP GALLERY

View Map Gallery & additional Plaid Avenger resources for this region at http://plaid.at/mid_e

MIDDLE EAST
MAYHEM!!
ARABS VS. PERSIANS...

...FIGHTING FOR CONTROL OVER THIS
OIL SOAKED
BATTLE GROUND

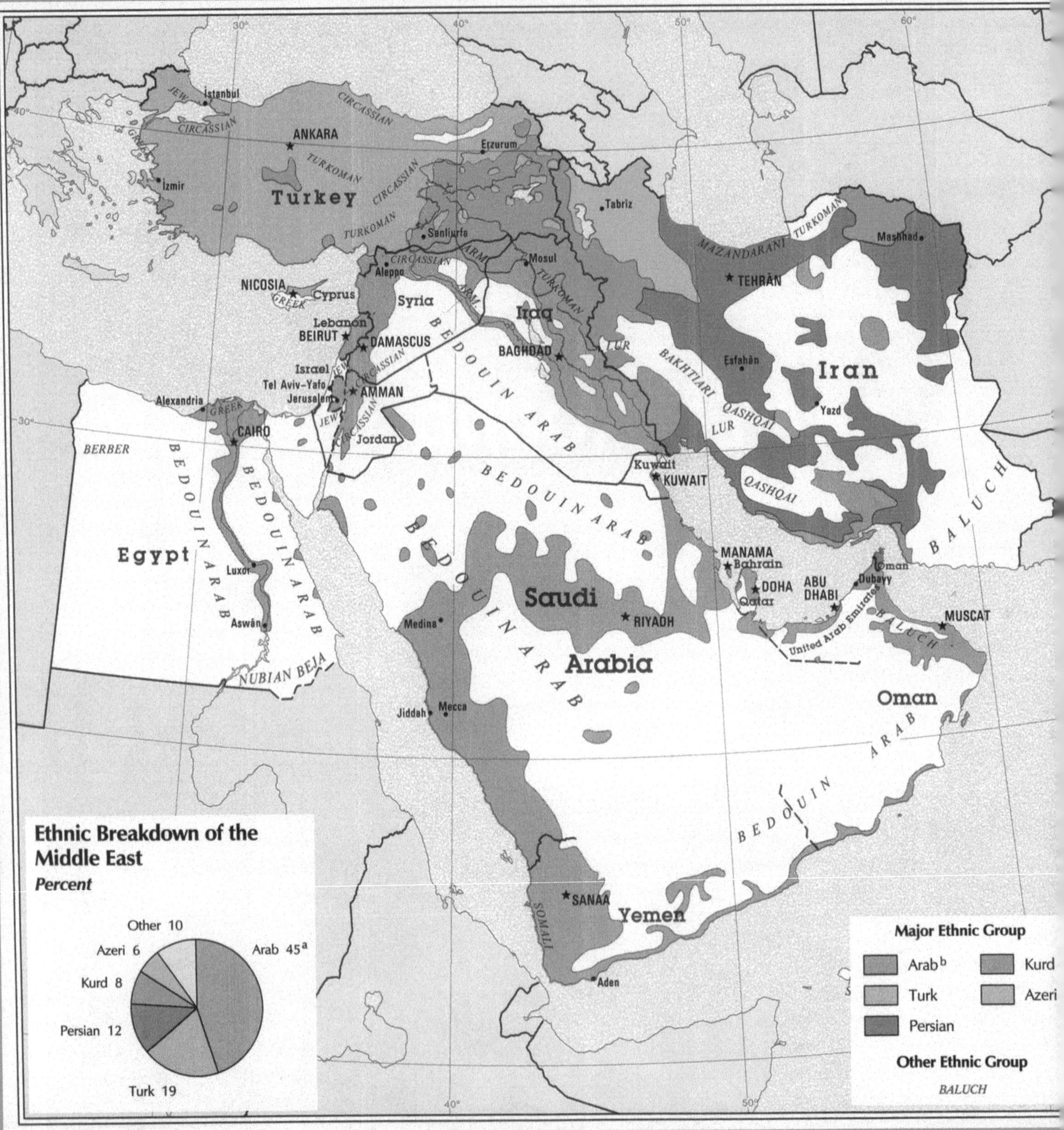

CHAPTER OUTLINE

1. Team Turk
2. Team Hebrew
3. Team Arab
4. Team Persia
5. Team Foreigner
6. Additional Non-State Actors
7. Conclusion

Addendum: The Competition for Control

18A

NOW for the last, and most complicated, component of understanding the Middle East mess in the modern context. Many, if not all, of the conflicts, trials, and tribulations that have been described so far in the previous chapter can be related to a much bigger competition that is occurring across this region, and that is a competition among a group of big power players for influence, if not outright control, of the destiny of this troubled chunk of land.

This is a total Plaid Avenger assessment, and one that would be labeled by experts as completely unorthodox and overly-simplistic in its approach, which means it's probably exactly right! I admit right from the get-go that this approach does not do a thorough analysis of all the details and all of the complex inter-relationships that exist between these groups; if you want to understand them better, then by all means use this text as a launching point for further research into this tangled web of geopolitical skullduggery. Wow! How often do you see the word skullduggery in a textbook? But I digress.

Here is what you need to know to understand this Middle Eastern mayhem: there are many ethnic and religious and political differences among people in this region which cause fractures and frictions, and almost all of these things are totally unrelated to input from the outside world. In other words, many players are vying for influence and control from within the Middle East itself. Moves from outside forces like the US, the EU, Russia, al-Qaeda, or even the insane clown posse known as ISIS simply accentuate, benefit or stymie the game already being played internally. You dig what I'm saying so far?

Before you go any further, go back to the previous page and examine the map for major ethnic distinctions within the Middle East, as they are the primary basis for these Team divisions.

The following is a list of distinct tribes that the Plaid Avenger feels are currently in this competition. Again, some of these teams are vying for influence, some are vying to control resources, some are vying to control major events, some are trying to outright control specific territory, and some are just trying to disrupt the other teams from winning. I will point out what makes each team distinct from the others, their motives, who they like, and who they hate. I will start with the easiest ones and continue to get further complicated as we go. You ready for this reality? Then let's do this!

TEAM TURK

We start with the easiest and least complicated: the Turks. First and foremost, this team's distinctive characteristic is its ethnicity. Turkey is full of Turks. Go figure. But Turks are not Arabs, nor Persians, nor Middle Easterners, whatever that means. The next chapter of this book deals with Turkey in much more detail, so I will keep it brief for now. As you will find out soon enough in the following pages, I do not count Turkey as a part of the Middle East region at all. In a sense, it's a bit of an outside entity itself.

Having said that, Turkey is in a very unique position to influence affairs in the Middle East in multiple ways. First, unlike any other outside entities, Turkey is overwhelmingly Muslim, which makes it more culturally sensitive and sympathetic to all its Muslim Middle Eastern neighbors. But that is about where the similarities end.

Unlike its regional neighbors to the south, Turkey is a staunchly secular society that maintains a strict separation between Islam and the state. Turkey is also a well-established democracy, a founding member of the UN, one of the earliest inductees into NATO, a possible future member of the EU, and has a well-established and strong, modern military. All those factors together make it much more Western in outlook, and indeed, Turkey is an ingrained part of most Western institutions, especially that NATO one; all operations in the Middle East utilize the Turkish corridor as an operating platform. Most NATO and UN operations in the region would fully flounder to failure without Turkish support. That gives Turkey a loud voice on all Middle Eastern matters, at least from a global involvement perspective.

That's why Team Turk is in a strong position to influence some Middle Eastern policies, especially ones that are advantageous to itself. Since the Turks are US/NATO allies, they typically work toward the goals agreed upon by those entities—to a point. Turkey openly refused to support the current US invasion of Iraq on the grounds that it could not be party to an illegal invasion of a fellow Muslim state. In 2008, Turkey actually invaded Iraq itself to go after Kurdish terrorists, much to the chagrin of the US. That is important to note: Turkey has a large population of Kurds in its own territory, and thus has a vested interest in any/all Kurdish developments in the region. Stability of Iran and Iraq play prominently into the Kurd situation, and thus the Turkish response.

Other important considerations: Turkey controls the all-important headwaters to the Tigris and Euphrates river systems, which supply the critical resource to many countries to the south, meaning they have a very effective control measure should they ever decide to use it. In addition, Turkey has quietly and methodically been increasing its role as a mediator and host between sparring Middle Eastern parties. Turkey has hosted talks between Team West and Iran, and is really the only true mediating force between these two embittered enemies. In addition, Turkey has led the charge to condemn the violence in Syria, has called for the ousting of the Assad regime, has offered its own country as a safe haven for fleeing Syrian refugees, and hosted international coalitions of forces to deal with the whole Syrian mess. And finally, Turkey has been held up as a model for other aspiring democracies in the Middle East to mimic, as the Arab Spring sees Arab state after Arab state throw out their dictatorial rulers in favor of a more representative system. As a stable, successful, Muslim state in the neighborhood, Turkey is uniquely fit as such a role model. **2021 UPDATE:** Recetly, Turkey's role is now much more independent of its western allies (read: USA and NATO) and it is becoming more belligerent and intervening on multiple fronts across the Middle East, north Africa, and Mediterranean Sea. More on this next chapter.

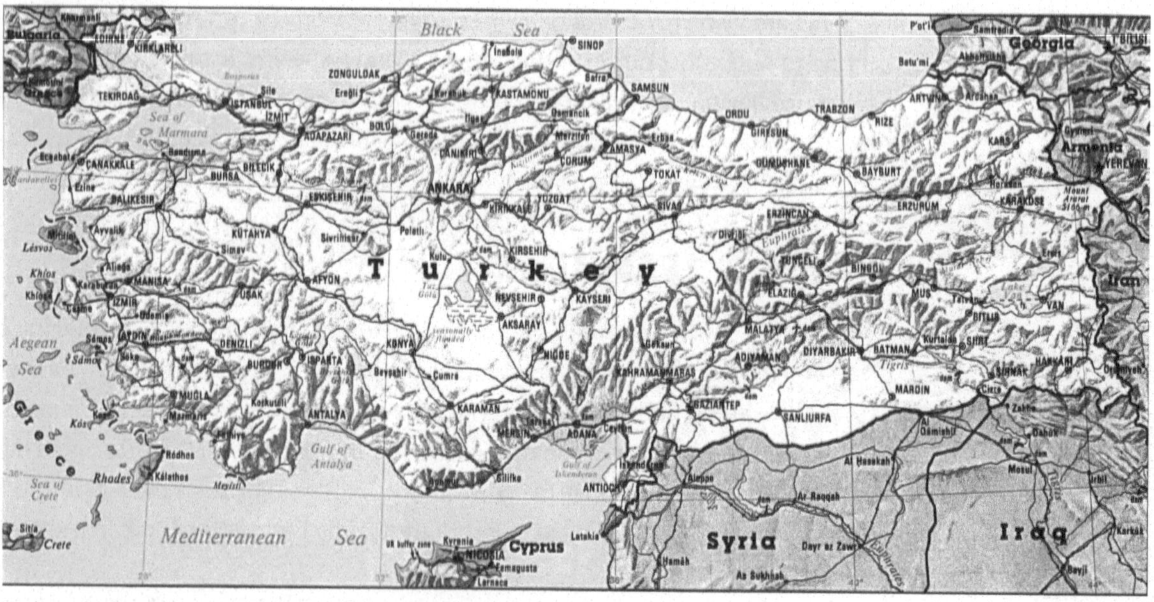

Turkey: the last remaining remnant of the Ottoman Turk Empire.

TEAM HEBREW

The Kosher Club. On unleavened bread.

Onto the next team, and what a doozy this one has become! Team Hebrew represents one state and one state only, the state that may be a central issue for tons of regional friction: Israel. Now I know what you may be thinking already: "What? How can this be the next easiest team to comprehend when it is the source of so much animosity and conflict?" I'm glad you asked. While many folks believe the Israeli-Palestinian issue may be the source of all conflict across this region, Team Hebrew's motives and motivations are actually quite easy to identify and understand.

Before you read any further, you must understand the basics of how this state came to be and what that has meant for the modern world. Check out the "What's the Deal with Israel?" box on the next page.

Okay, now that you are back, we can make some generalizations. All the neighboring Arab states have attacked Israel on multiple occasions ever since its inception, and Israel has pretty much won each of these wars, resulting in increased territory for Israel and control of other areas that were originally not even part of their state. It also has resulted in the displacement of most of the ethnically Arab Palestinians, who are without a state of their own. Many of these Palestinian refugees are located in the surrounding Arab states. For these reasons and perhaps many others, Team Arab really hates Team Hebrew. Team Persia hates Team Hebrew as well, even though they have never openly engaged each other, militarily or otherwise.

Let's call out some other characteristics of Team Hebrew. Obviously, they are Jewish, and as such, the only non-Muslim actors within this region vying for influence and/or control. Team Hebrew is also a strong, secular, well-established democracy as well as a fully developed country with extremely high standards of living. All of those factors make it quite unique within the Middle East region. Team Hebrew is indeed a fully Western country in every sense of the word when it comes to development, economic and political systems, and even religious and cultural values.

Most importantly, Israel also has high levels of technology, especially military technology, and is an undeclared nuclear power. That is crucial. Team Hebrew is the only entity in the Middle East that has nuclear weapons. They are an undeclared nuclear power because the US and others insist that they do not openly declare this fact, otherwise everyone in the neighborhood will also want a nuclear weapon, thus launching a regional arms race. Nobody wants that right now, or ever.

Speaking of the US, it should also be evident to you that Team Hebrew is a staunch ally of the US, the EU, and all western institutions in general, with the possible exception of the UN, since some UN members often give Israel the smackdown in that forum. The inverse holds true as well: the US has supported Israel openly and covertly for decades. Israel is regularly the top recipient of US foreign aid, even though they are a rich country in their own right, and have a powerful and well-trained military. The US has regularly shared/sold weapons, weapons technology, and military intelligence with Israel. You can go ahead and make the assumption that this relationship infuriates all the other teams in this regional competition.

Why would I suggest that Team Hebrew motives are easy to explain? Because they are fairly well defined and there aren't that many of them. Their primary objective? Maintain their territorial integrity. Keep the boundaries of their 73-year-old state intact at all costs. Second objective: have national security—that is, to be secure in their state by not having terrorist attacks or open declarations of war against them. Third objective: get all other neighboring states to recognize their right to exist, which, of course, helps out with the first two objectives.

That's it. That's all. Pretty easy to digest. Now I am not saying it is right or wrong, or good or bad, or that it will ever even happen. I'm just saying their objectives are pretty easy to identify and understand. All of Team Hebrew's foreign policies, international alliances, and strategic actions serve to accomplish one or more of those very simple objectives.

BACK IN THE BIBLICAL DAY: CIRCA 1000 BCE

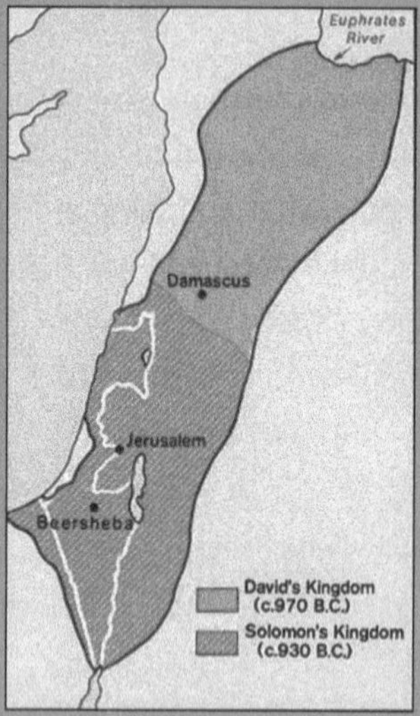

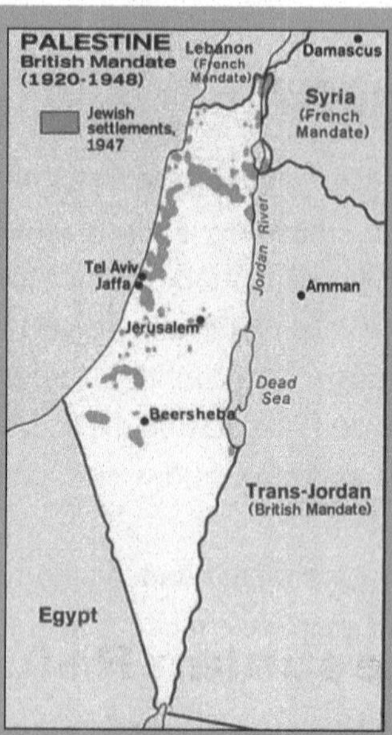

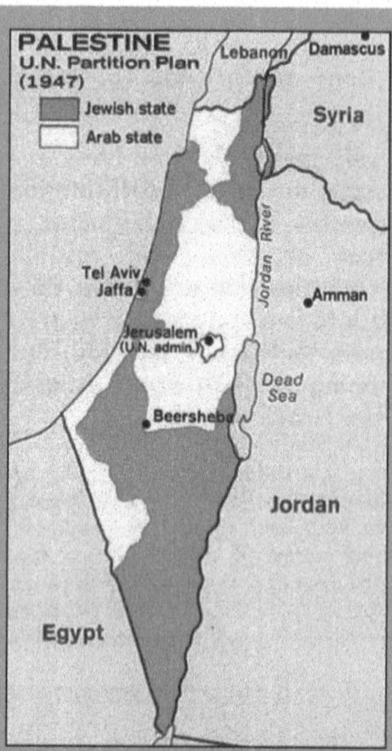

What's the Deal with Israel?

Israel is the loaded topic of all loaded topics. There are so many emotions surrounding the formation and history of Israel, it is hard to be completely unbiased about the issue. If you lean too far in either direction, you can be quickly charged with anti-Semitism or, conversely, Zionism. But, by the mercy of Elijah, here is the deal with Israel.

I won't get into Biblical history too much, but will make mention of this: this area was the historic homeland of the Jews, "The Promised Land" that Moses brought them to many millennia ago. Fast forward to the Roman Empire, which persecuted the Jews (one in particular you may have heard of: Jesus Christ) and eventually totally kicked the vast majority of them out of the area in the second century AD. Since then, locals have been living there (let's call them Palestinians), hanging out for a couple thousand years.

The modern history of Israel begins around 1900, when Theodor Herzl, among others, began pressing for the creation of a national Jewish state—this movement became known as the Zionist movement. Jews began buying property in Palestine and forming small Jewish communities. In 1917, the British government issued the **Balfour Declaration,** which endorsed the establishment of a Jewish homeland in Palestine. Shortly after, Palestine became a protectorate of the British government because of dismemberment of Ottoman Empire after WWI (see next chapter).

This caused a surge of Jewish immigration that continued until 1939, when the British government issued the **White Paper,** which limited, but did not stop, Jewish immigration to Palestine. This angered the Jewish community, who viewed the White Paper as a rejection of Zionism, and also pissed off the Arabs in Palestine, who wanted Jewish immigration to stop completely.

After World War II, when the horrors of the Holocaust were revealed, there was a giant surge of sympathy for Jews and for the Zionist movement in general. Simultaneously, Jewish armed forces in Palestine became more aggressively defensive of Jewish territory, attacking both British and Arab forces. The British decided that keeping Palestine as a protectorate was no longer in their interest, so they ceded control to the UN. The UN

Israel's occupation of the **West Bank**, of the **Gaza Strip**, and even areas within Syria referred to as the **Golan Heights**, is maintained to achieve their national security goals. Israeli development of a strong military and nuclear weapons was done to ensure its existence by discouraging any more open attacks across its borders. Israel usually sides

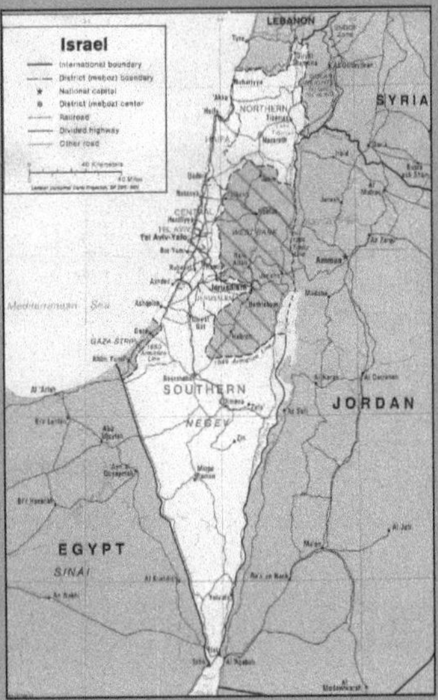

accelerated the development of a plan to partition Palestine into separate Jewish and Arab states. This formed the Jewish State of Israel, which was quickly recognized by the United States and eventually by most everyone else, but not any of the surrounding Arab countries. This established the 1947 UN partition borders (see map insert).

Violence erupted soon after the partition. Armies from Egypt, Syria, Jordan, Iraq, and Lebanon all clashed with Israeli forces. Jordan gained a foothold in the West Bank and Egypt gained a similar foothold in the Gaza Strip. This is important to note because these are still disputed areas today. A ceasefire was finally brokered in 1949, creating a new set of de-facto borders now known as the green line (see map insert).

Israel frequently sparred with its Arab neighbors throughout the 1950s and 1960s. The culmination of this aggression was the 1967 Six-Day War, where Israel, using advanced weaponry from the US, went gangbusters after being attacked by Egypt and Syria, and conquered the West Bank (including East Jerusalem), Gaza Strip, Sinai Peninsula, and Golan Heights (see map insert). Egypt and Syria were enraged about this, and attempted to regain their lost territory in the 1973 Yom Kippur War. This war was more of a stalemate—little territory changed hands—but the Yom Kippur War was significant because the United States sided with Israel, irritating EVERY Arab nation and inciting the 1973 oil embargo.

The last major shift in Israel's borders was peaceful. In 1979, President Jimmy Carter invited Egyptian leader Anwar Sadat and Israeli Prime Minister Menachem Begin to the United States and helped negotiate the Camp David Accords. This treaty returned the Sinai Peninsula to Egypt and granted more autonomy to Palestinians residing in occupied territory. Since 1979, most violence associated with Israel has been within its borders. The Palestinians living within Israeli-controlled territory have garnered significant international support for statehood. Recently, Israel has begun taking steps that seem to suggest that the formation of a sovereign Palestine is imminent. However, Israel is not prepared to cede the amount of territory that the Palestinians believe they deserve, and has even started construction of a wall encircling much of the disputed area in East Jerusalem. In 2019, a much more hawkish Israel has all but given up on the two-state solution, and the reality of an independent Palestine seems more remote than ever, and violence between the parties once more erupted big time in 2021.

up with Team West (particularly the US) when it serves their purposes to achieve those objectives. Usually, but not always. Israel sometimes goes against US foreign policy by negotiating with neighboring states that the US hates in order to achieve the "recognition" goal.

Arab world stretches far and wide, but their Arab League is, so far, shallow and useless.

TEAM ARAB

Is that you, Aladdin?

Now we are onto one of the real cultural anchors of the Middle East, and here's where it starts to get more complicated. Team Arab refers to an ethnic group, just like Turks or Persians. However, this ethnic group happens to be the majority within the Middle East as an entire region, and many of the states/groups of this region are overwhelmingly Arab in descent. Maybe you've heard of Saudi ***Arab***ia or the ***Arab***ian Peninsula or perhaps even Aladdin and other tales of "The ***Arab***ian Nights"?

Before the advent of Islam, the Arabic ethnic group was centered in the Arabian peninsula, naturally. Although the Arabs were originally primarily a nomadic people with an itinerant lifestyle centered around the use of the camel and an extensive network of desert oases, there were a number of urban centers. These were mostly located in the western portion of the peninsula, al-Hejaz. Cities like Makka (Mecca) and Yathrib were important centers of commerce and trade; the former was also a crucial center for the pre-Islamic polytheistic religions that abounded. Muhammad himself was an important merchant in Makka and was well known in the region even before his religious undertakings began.

Within a century of Muhammad's death, the Arabs had expanded well into Sassanid Persia, pushed the **Byzantine** Eastern Empire out of the **Levant**, and were trading blows with Charles Martel in France. The expansion of Islam, except in the instances in which it came into contact with a resident imperial power, was overwhelmingly peaceful. Local populations of Christians, Jews, and other random monotheistic religions willingly accepted their new rulers, as, especially in the case of the Byzantine Empire, Christian sects were persecuted terribly. Conversion was not enforced, but had concomitant economic benefits, such as the relief from a poll-tax on non-Muslims, but this was not encouraged as it would rob the Arab polities of cold hard cash.

Politically, this "Arab Empire" was not terribly cohesive, especially with regard to the territories in Spain and northwest Africa. Eventually, after a couple of centuries, the core at Baghdad held only nominal sway over anything significantly placed away from it, but the religion itself had spread from France to the Oxus River. From there you, can tie the story back into the end of the Golden Age of Islam, when Genghis Khan totally destroyed Baghdad, thus effectively ending the Arab Empire reign, then onto the rise of the Ottoman Empire, which we have discussed previously and will again in the next chapter. Team Arab essentially became just another component of the multi-ethnic empire that consolidated this region from about 1300 to 1923, when the last vestiges of the Ottomans were swept away and the modern map of the Middle East was drawn.

Fellow monarchs discussing oil revenues.

Of course, this new map of the Middle East was not a singular Arab entity—far from it. Of the new political boundaries drawn up after WWI, there were roughly twenty two countries which were predominately Arab in ethnicity. How do I know that? Because in 1945, six of those countries formed an entity known as the **Arab League**, and membership has jumped to twenty two in the intervening years. This League was formed to ". . . draw closer the relations between member States and co-ordinate collaboration between them, to safeguard their independence and sovereignty, and to consider in a general way the affairs and interests of the Arab countries."

In other words, the Arab states want to work together to achieve common Arab goals. In point of fact, at various points in the last seventy years, there have been several attempts to reunify their states under a pan-Arab banner. However, it has never worked. The Arab League itself has to be considered a fairly dismal failure, as the member countries more often fight with each other than actually agree on anything at all. They still get together once a year or so to discuss solutions to common issues like the Palestinian debacle, the Syrian civil war, or the invasion of Iraq, but they never decide on anything nor do anything about it. I guess they have a nice all-you-can-eat hummus bar that keeps people coming back.

That gets us up to now. Team Arab is still a potent force of influence in the region for various reasons:

1. Many Arab states are excessively rich off of oil production. With wealth comes power.
2. Arab culture is the incubator of Muslim culture. Remember, Islam started in Saudi Arabia, where the two holiest cities of the religion are located. That still counts for something on a planet where there are roughly a billion Muslims. Please note: Arabs are overwhelmingly of the Sunni division of Islam, in contrast to the Shi'a brand which is primarily found in Iran. More on that to come.
3. Arab is the most populous ethnic group across the region, including such groups as the Saudis, the Egyptians, the Palestinians, the northern Sudanese, a minority of the Iraqis, the Kuwaitis, and most of the Algerians, just to name a few.

A few final points of analysis: Team Arab totally hates Team Hebrew, since the Palestinian debacle is now an Arab debacle. That hate has been softening for some Arab leaders, though, as Egypt officially recognized Israel and made peace with them in 1979, and Jordan did it in 1994. Ponder this: Egypt officially recognized Israel in 1979, and Egypt has been the #3 top recipient of US foreign aid every year since 1979. Coincidence? Not.

The bigger issue that you might not know about is this: Team Arab also really hates Team Persia (Iran), and it is the competition between these two teams that has become the hottest current showdown. Team Arab is getting very wary of the growing power and influence of Iran within the region. Dig this: Team

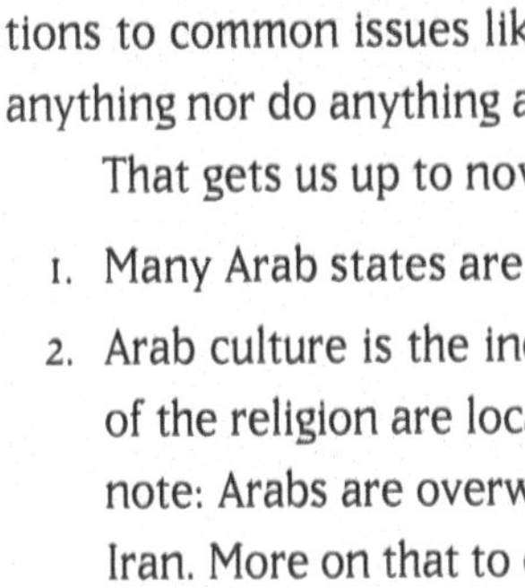

ANOTHER PLAID AVENGER INSIDE TIP: Virtually all Arabs are Muslims, but most Muslims in the world are not Arab. Does that make sense? Islam is a worldwide phenomenon now, and places like Indonesia actually have more followers of the faith than any single country in the Middle East. The reason this is confusing for outsiders is because Islam started in Arab lands, and the original and only true Koran is written in Arabic. No matter where you go in the world, from Morocco to Malaysia, Muslims will be reading the Arabic Koran and praying in Arabic tongue, but that don't make them ethnically Arab. Of course, I'm so savvy that I can read Arabic numerals . . .

Arab is exceptionally worried about the possibility of Iran getting nuclear weapons, perhaps even more worried than Israel or the US! Team Arab will be working with the US and others to make sure that Team Persia does not get such weapons.

Speaking of which, the core of Team Arab lies in Saudi Arabia and Egypt. As fate would have it, both countries are strong allies of the US—well, at least the leaders of the countries are. You should know this fact well: the US generally supports Team Arab for various political (US hates Iran) and economic reasons (they have tons of oil). Got all that? Then let's get to that Iranian team now to see why they are so despised by the Arab peeps.

TEAM PERSIA

Never should have took on the Spartans . . .

Onto the next, and final, of the major teams in our Middle East round-up. For this one, we have to go back even further in time to tell their tale. Further back than the Ottoman Empire, further back than the Arab Empire, and, indeed, even further back than the advent of Islam itself. Team Persia's story goes back several thousand years, but we will just hit the highlights here so that you can understand how very different and distinct this team is from all of its neighbors. Never heard of Persia? In today's world, we call it Iran. While Iran is the core of this Team, there are other players on their side as well. But let's start at the beginning.

Persia can track its origins back to 1500 BC when Aryan tribes—namely, the Medes and the Persians—settled the Iranian Plateau. Power went back and forth between these groups until the Persians came out on top in 558 BC when Cyrus the Great beat out the competition and then went on to consolidate the peoples of the plateau and expand the empire outward into Lydia and Babylonia and eventually Egypt and beyond, making it the most powerful political entity of the time.

Later, Darius the Great tried and failed to incorporate the Greeks into the Persian fold. Look up the epic Battle of Marathon as to how that went down. Also, watch ***The 300*** to see an excellent, but not terribly accurate, depiction of this Greek/Persian face-off period—Iran actually protested the release of the film as being racist and inflammatory. Next

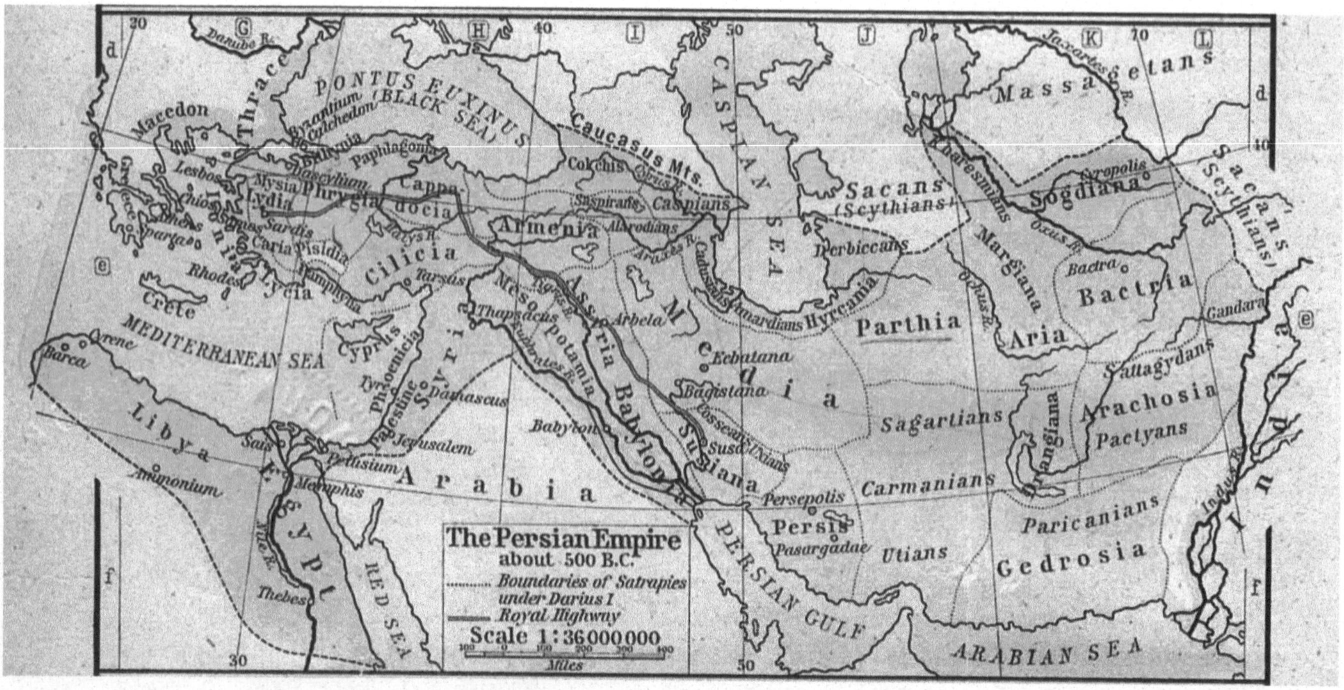

The 300 stopped them, and then Alexander whipped them.

name of note: Alexander the Great united the Greek city-states and led his famous campaign across Asia primarily to squash the power of the Persians for good; he was fairly successful in this endeavor, as Darius III was killed (and the empire with him) in 330 BC. While the entity lasted a bit longer off and on, it never regained its past power or glory. Persians have a long, proud history that is still looked back upon with reverence and distinction. Thus, their ire at the negative depiction of their forefathers in ***The 300***.

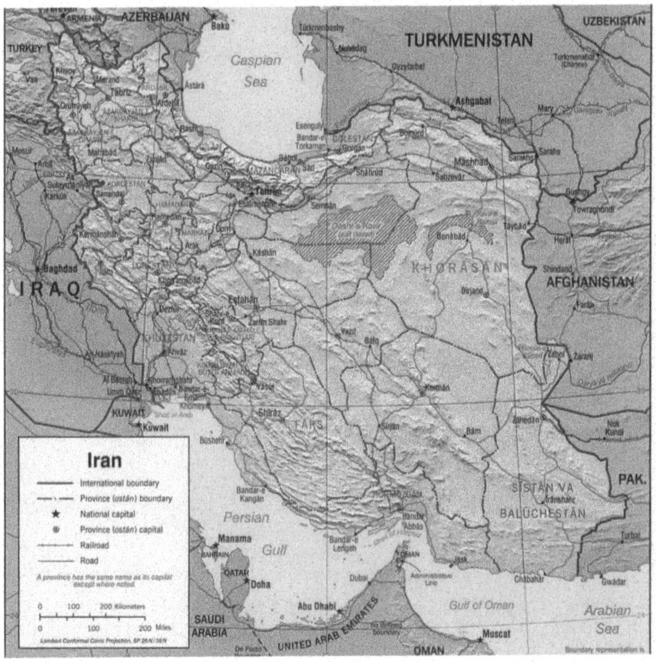

Iran: the core of the Persian power, Shi'a power, and a whole lot more!

Let's now fast forward. The Arab/Muslim conquest of Persia in the 640s AD was to forever change their destiny. Not only did this convert the Persians to Islam, but it set the stage for the current Islamic sect differences that are still very much alive today. As referenced earlier in this text, after the death of Muhammad, the religion gets split into two main branches: Sunni and Shi'a.

Now, the liturgical and ritual differences between Sunnis and Shi'as (or Shi'ites) are minimal and mostly unimportant. The branching off of the Shi'a (literally, partisans) of Ali occurred when certain members of the early Islamic community disputed the line of succession following Muhammad's death. Most folks supported one of Muhammad's top men, Abu Bakr, to lead the empire, but others backed Ali, who was Muhammad's son-in-law. In this contest for power, Ali and then his son Husayn ibn Ali were killed and these early partisans were dispersed throughout the Islamic realms. The death of Husayn ibn Ali particularly became a focal point for the community, especially in modern times, with open manifestations of "passion plays" mourning him and his followers' deaths in the desert near present-day Karbala.

What's all that ancient history have to do with modern day Iran? Just this: Iran became the haven for the supporters of Ali, and over time evolved into the core of Shi'a Islam. While Shi'a is the minority sect worldwide (world = 85% Sunni, 15% = Shi'a), it is the majority of followers in modern day Iran (Iran = 90% Shi'a, 10% Sunni) and that makes all the difference for our team competition.

Let's make another jump forward. Persia was picked on and beat up on by Russia and Britain during the European powers' imperial battles for world control. However, it always managed to maintain its independence and was never colonized. By 1906, they instituted a constitutional monarchy and by 1941 a dude named **Reza Shah** was on the throne and was increasingly ruling with an autocratic iron fist. It should be noted that this dude came to power with the assistance of the US and the UK who had forced Reza's father to abdicate the throne for his supposed Nazi-sympathizing behavior. Later in 1953, the US and UK coordinated a secret plot responsible for deposing the popular Iranian Prime Minister Dr. **Mohammad Mossadeq** because he had nationalized the Iranian oil industry, much to the chagrin of western countries. That left the bumbling puppet King Reza Shah in charge of Iran via his western allies. It was not to last.

Foreign powers had to get involved in the Middle East to "protect" their oil industries. (Hmmm, what is that old saying about history repeating itself?) This Shah guy was a little too over the top and ticked off too many Iranians, and the Shi'ite soon hits the fan. Time for the Iranian Revolution! Check the box that follows for details on that little insurrection.

Now we arrive at the present. The **Iranian Revolution** has been called the last great revolution of the 20th century, and for good reason. It is a grand experiment of Islamic theocracy with limited democratic elements, but it is, in essence, a blending of church and state into one big ruling system. It's really the first of its kind in the modern era, and one of the reasons that other Middle Eastern countries are terrified of Iran is the fear that this type of revolution could spread . . . meaning that Iranian-style theocracy could sweep into Egypt or Jordan or Algeria and displace the systems/leaders that currently have the power. And no one wants to lose the power!

What's the Deal with the Iranian Revolution?

The Shah and the Ayatollah.

Up until 1979, Shah Mohammad Reza Pahlavi was the constitutional monarch of Iran; he was happy exporting oil, quashing dissent, and resisting Soviet influence. The US and the Shah were the best of friends because he was basically a US puppet. He liked playing with big guns, and the US liked supplying him with just such toys, just to make sure the Soviets wouldn't. Unfortunately for the United States, most people in Iran were very unhappy with the Shah. It seems that Iranians did not like living under a dictatorship.

Open protests of the Shah's government began to occur in January 1978, and the Shah ordered the military to shut them down. Hundreds of demonstrators, including students, were killed. This happened several times, and popular support against the Shah really took hold. A populist coalition of liberals, students, leftists, and Islamists escalated the protests to the point of revolution and ousted the Shah in January of 1979. We call this the 1979 **Iranian Revolution** or Islamic Revolution.

The primary political figure during the revolution was the Ayatollah Khomeini, a Muslim cleric who encouraged demonstrations against the US and Israel, whom he declared "enemies of Islam." During the final chaotic days in November of 1979, a group of young radical Muslim students took over the US Embassy and held 52 Americans hostage for 444 days . . . aka **The Iranian Hostage Crisis.**

The Ayatollah Khomeini was the most popular figure in Iran because of its strong Muslim background, but there was also a powerful secular movement. A power sharing system of government was attempted, but eventually Iran became a true Islamic Republic, with all of the power vested in the religious elite led by Ayatollah Khomeini. The United States became even more enraged, especially since it was unable to rescue the hostages, and thus cut all diplomatic ties with Iran.

To this day, there is no official US presence in Iran—no diplomats, no troops, nothing. This is one of the reasons there is currently so much animosity between the United States and Iran. Many US leaders see Iran as a threat to Western stability in general, especially if they become a nuclear power. This is a critical issue in today's world, as Iran continues to be in the spotlight by developing nuclear technology—which they say will be used for peaceful purposes, but the West is convinced will be used to build a bomb. The rhetoric is running full steam on both sides right now.

Just know this: the US is all about shutting Iran down, either diplomatically or ultimately by force. The EU wants to shut them down, but only diplomatically. Russia is kind of an ally, and wants to help them develop their nuclear program for energy production, and China wants to do nothing about the Iranian situation because they are all about the sovereignty. Whatever goes down, don't look for it to be dealt with at the UN Security Council, which is already deadlocked before the process even begins.

Let's get to the meat and potatoes: Team Persia is doing quite a bit to assert its power and influence across the Middle East, which severely worries Team Arab. These two teams HATE each other, and as I have alluded to earlier, are the biggest forces facing off within the region. This hate started in earnest when Saddam Hussein (with backing from other Arab states, the US, Russia, and Europe) invaded Iran, thus starting the decade long (1980-89) **Iran-Iraq War** which ended with about a million casualties. The hate between Arab and Persia exists to this day. Why so much hate? Dig this:

HATE number 1: As just mentioned, Team Persia has no problem exporting the ideals of its revolution to other places. Iran increasingly sees itself as the only true Islamic country of the region, the only real and true purveyor of the faith. It mocks the Arab countries for their morally corrupt leadership as well as for being stooges of US foreign policy. Iran is trying to assume a moral and ethical leadership role on issues like the Palestinian debacle, the existence of Israel, and any foreign intervention into Middle Eastern affairs.

Since Team Persia is the only fully Shi'a state, we have come to define Team Persia in this religious sense as well, but they are not totally alone. I want you to be aware of a new entity that is named the **Shi'a Crescent**. It is based on a projection of Iranian influence and power to other Shi'a groups across the wider region, but please don't mistake this Persia/Arab animosity as a religious war

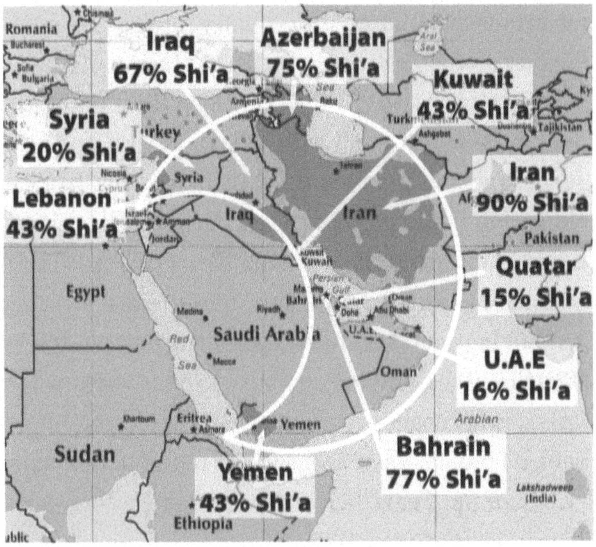

Shi'a Crescent: this ain't no Lucky Charm for the Arabs or the US.

of any sort. The Shi'a Crescent is simply a nice way to highlight areas in which Team Persia has increasing influence. Take a look at the map.

The map shows you the percentages of Shi'a folks in selective states, and we should note a few choice groups that make daily news and effect events. Iran is the core of this crescent, but its effective reach spans from the Persian Gulf countries all the way through Iraq, which has a majority of Shi'a and is why Iran helped fund and arm Shi'a militant groups in the recent Iraq civil war. The Shi'a vein also runs thru Syria, a country whose leadership belongs to a subsect of Shi'a, and therefore is malleable to Iranian Shi'a influences. To continue on the road, you get to southern Lebanon, where Shi'as also have a slight majority, and also where a radical Shi'a terrorist group named **Hezbollah** hangs out. Iran funnels money and guns through Syria to Hezbollah to help them fight against Team Hebrew. In the most recent conflict to pop up, Iran is now widely believed to be the predominate backer of Houthi rebels, a Shia group in Yemen fighting against the Arab-backed Yemeni government. You starting to get a sense of the Iranian deal here?

HATE number 2: Team Persia still openly utterly and completely hates Team Hebrew, and the former, ultra-conservative Iranian President Ahmadinejad many times caused controversy by making statements to the effect that the Israeli regime should be wiped off the face of the map . . . or something to that effect, as the translation itself is a matter of debate. We've already pointed out that Team Arab hates Israel in general, but some Arab leaders have recognized Israel, while still others have been softening their rhetoric about the Jewish state over the years. Meanwhile, Team Persia leaders have held the hardliner attitude of refusal to recognize, but also calling for its destruction. Hardcore! That's why Iran supports Hezbollah, but it also supports Hamas down in the Gaza Strip, just to keep jabbing a sharp stick into the Israel flank at every available opportunity,

"Israel? I do not recognize. Does not compute."

This also puts Iran in the position of appearing to be a true helper and defender of the Palestinians, reinforcing the idea that the Arab leaders are just a bunch of losers that can't look out after their own. It makes them appear to be the only country in the region able and willing to stand up not only to Israel, but also to other outside forces.

HATE number 3: Team Persia is developing a nuclear industry. Hold on! I didn't say nuclear weapons! But that is what all the other Teams think is going on. Iran has been thumbing its nose at its neighbors and world powers by its persistence in developing fuel to be able to develop a nuclear power industry, and this is the issue that is really causing the most consternation across the region, and outside the region as well. As of this writing, Iran has not broken any international law in its nuclear pursuit—it is not illegal to produce nuclear energy.

However, virtually all Western nations, including the US and the EU, all Arab nations, and the UN as a whole, have condemned the move harshly and are busy setting up international sanctions to stop the Iranians. For their part, Iran has claimed they have a God-given right for nuclear energy, and nothing will stop them in their quest. Iran's open defiance sets them apart from all the other states in the region, and is seemingly a point of pride with Iranian leadership. Great. Like we needed another problem in this region.

Here's why it is a problem: Iran getting the tools to create a nuclear bomb tips the balance of power in the region radically. No one else has the bomb—except Israel, who won't admit it. While most folks assume that Israel would be the country most frightened of a nuclear-armed Iran, in fact that's probably not quite true. Israel has the bomb as well, making for an effective deterrent to an Iranian attack. Why would Iran nuke Israel, knowing that 100 nukes would be sent back in response? The folks who are really worried are Team Arab, because they don't have a bomb! A nuclear Iran would be stronger than any Arab country, and would contain the most powerful deterrent to prevent any attacks against their Persian motherland, including any attacks potentially from the US or others!

Just a brief **2017 UPDATE**: Team West under US leadership actually softened sanctions in Iran, as Iran signed a nuke deal to take a chill pill with the bomb stuff and also be more transparent with their industry. So much to the chagrin of Team Arab and Team Hebrew, Team Iran was being accepted back into the world order. . . BUT THEN **2019 UPDATE**: Donald Trump became US President, then built a cabinet of Iran-haters, then withdrew the US from that internationally-brokered Iran Nuclear Deal previously mentioned. This caused the deal to completely fall apart, Iran's economy to be further crippled by resumed sanctions, Iran to start enriching nuclear material, and now here in 2021 several skirmishes have broken out between Iran naval forces and ships in the Persian Gulf. So. . . yeah. . . things are not looking good right now. **2021 UPDATE:** The US Biden administration is working to restore that Iranian nuke deal previously canceled by Trump, but will face serious challenges since Iran just elected (in a largely boycotted vote) a hardcore ultra-conservative new president named Ebrahim Raisi. Too early to tell what direction Iran will go in next, but it appears a hard-right turn is imminent.

Sum up: Team Persia hates Team Arab. Team Persia hates Team Hebrew. Team Persia hates Team Foreigner. In return, all those teams hate Team Persia right back, and may be working with each other in order to stop any further rise of Persian power. You dig? Arab states might work with the US, and maybe even Israel, to make sure the Persians get sidelined. Interesting stuff. Or should I say, interesting Shi'ite?

TEAM FOREIGNER

Foreigners love to party in the Middle East!

Now our job starts getting way easier. The major teams have now been outlined in perhaps too much detail, but we still have some other characters in this Middle Eastern drama that are powerful players affecting events. I don't have to spend as much time with these, because they are in the news on an hourly basis and if you don't know these cats yet, then you must be living under a rock. Team Foreigner consists of those states not actually from the region, but somehow find themselves here anyway.

With no reservation, the US is the primary outside entity that has dramatically altered events of the past decades, if not the entire last century. The US has supported kings, monarchs, dictators and dirtbags a'plenty in its quest to influence events and resources in the Middle East that are of strategic interest. Well, I guess all countries do that, but the US is the most powerful at this game. The Saudi royal family, the Shah of Iran, Saddam Hussein, and Hosni Mubarak all quickly spring to mind as entities the US has supported in the past.

It also should be noted that Israel and Egypt are the number 1 and number 2 recipients of US foreign aid, and have been for decades. Simultaneously, US companies have sold tens of billions of dollars worth of arms to Saudi Arabia. But the US is also a staunch ally and weapons supplier to Israel. Cha-ching! The weapons business is good! Uncle Sam has now invaded Iraq not once but twice, and the US maintains a troop presence in multiple states of the region; a visible sign of its presence can be seen almost everywhere. The US-led War on Terror is actively engaged across the entire region, both overtly and covertly, to root out rebel/terrorist groups whose ideology conflicts with the US world view. The US has been a significant player in the Syrian civil war, (mostly by funding/training/sharing intelligence with the groups it likes), in an effort to help them overthrow the al-Assad regime and also fight against ISIS. Of course, there are also diplomatic interjections: the US is typically the sponsor of most Israeli/Palestinian peace talks; the US is leading the trade embargo against Iran; the US is a builder of democracy in Iraq, or at least they are trying. The list could go on.

Of course, they are not totally alone. The US has a "Coalition of the Willing" which usually helps in its various endeavors; a group of willing US allies which typically includes the UK, Australia, Canada, and various European countries depending on the specific mission/conflict. There was a coalition for both Iraq Wars, a coalition for the Afghanistan campaign, and a coalition for US endeavors in Syria. And a revived Russia is now reasserting itself into the region, mostly just to irritate the US. Russia has been the primary sponsor of the al-Assad regime in Syria, and honestly is the only reason Assad has lasted this long. So the old Cold War rivalry is also flaring back up again, and it's happening most vividly here in the Middle East. The UN is, of course, intimately involved in trying to help stem the tide of conflicts in the region as well, albeit with not much results.

Let's cut to the chase on this Team Foreigner. Team Persia hates 'em, especially the US. In return, Team Foreigner really hates Team Persia, especially the US. Did you know that the US and Iran don't like each other? Well, now you do. The US typically supports and sides up with Team Arab on a whole host of issues, but mostly on the hating Iran one. Unfortunately, the US position with Arab states gets extremely complicated because the US is best friends with Team Hebrew too. Let's rank the Teams in order of how much the US likes them:

1. Team Hebrew comes out on top.
2. Team Turk is close behind. Remember, they are a NATO member!
3. Team Arab is supported by US on most issues, but not all.
4. Team Persia comes dead last. Perhaps that's exactly how the US wants it right now. Dead. Last.

Starting to make sense? I doubt it. Let me add in one last complicating clause of this foreign influence: While many Arab state leaders' support the US, and vice-versa, most people in those Arab states are increasingly angry about US presence and their leaders' support of US policies. This is making the water even hotter for a lot of these folks, as they become increasingly unpopular with their own people. One need only point out that Osama bin Laden's original mission was to drive US forces out of the Saudi holy land. Not accomplishing that, he then turned to start attacking the Saudi government themselves for being corrupt stooges of the US. And the al-Qaeaders aren't alone, as there are other groups within the region expressing their displeasure in any number of violent ways.

ADDITIONAL NON-STATE ACTORS

Now for the last. I won't go into great detail on these groups, but I did want you to be aware that there are other entities that are affecting the flow of events within the region—much smaller entities that are not states nor governments. Therefore, these groups do not behave like states or governments, cannot be coerced the same way as states or governments, and cannot be effectively attacked or punished like states or governments can. That makes them very sticky entities indeed, and they are often thorns in the sides of the bigger Teams we have already talked about. But, they can also be used as tools by the big Teams to attack or influence one another as well. Hmmmm, tricky, tricky, sticky, sticky.

I know that terrorist groups are probably the first things that come to mind, but there is actually a much bigger and more important one to tackle first: the Kurds. The Kurds are an ethnic group (check back to map on opening page again) that are located in the mountainous areas of Iran, Iraq, Turkey and Syria. They are a nation of people without a state, and they very much would like to have one. For over a century, the Kurds have dreamt of, and petitioned big powers to have, a Kurdistan country of their own. They have been used, abused and betrayed by virtually every regional and world power in this quest. The Kurds are seemingly a damned bunch of people, and I mean damned in terms of doomed.

They have been sporadically beaten down by the governments of all the states that they occupy, mostly because no government wants to see a free Kurdistan. That would entail the loss of some of their own territory, and therefore no one is going for it. Team Turkey, Team Persia as well as components of Team Arab, mainly Syria and Iraq, actively work against any Kurd movement toward independence. You may remember that Saddam Hussein used biological weapons on these people multiple times to keep them in their place. You may not have known that Turkey has also been fairly oppressive in keeping the Turkish Kurds down.

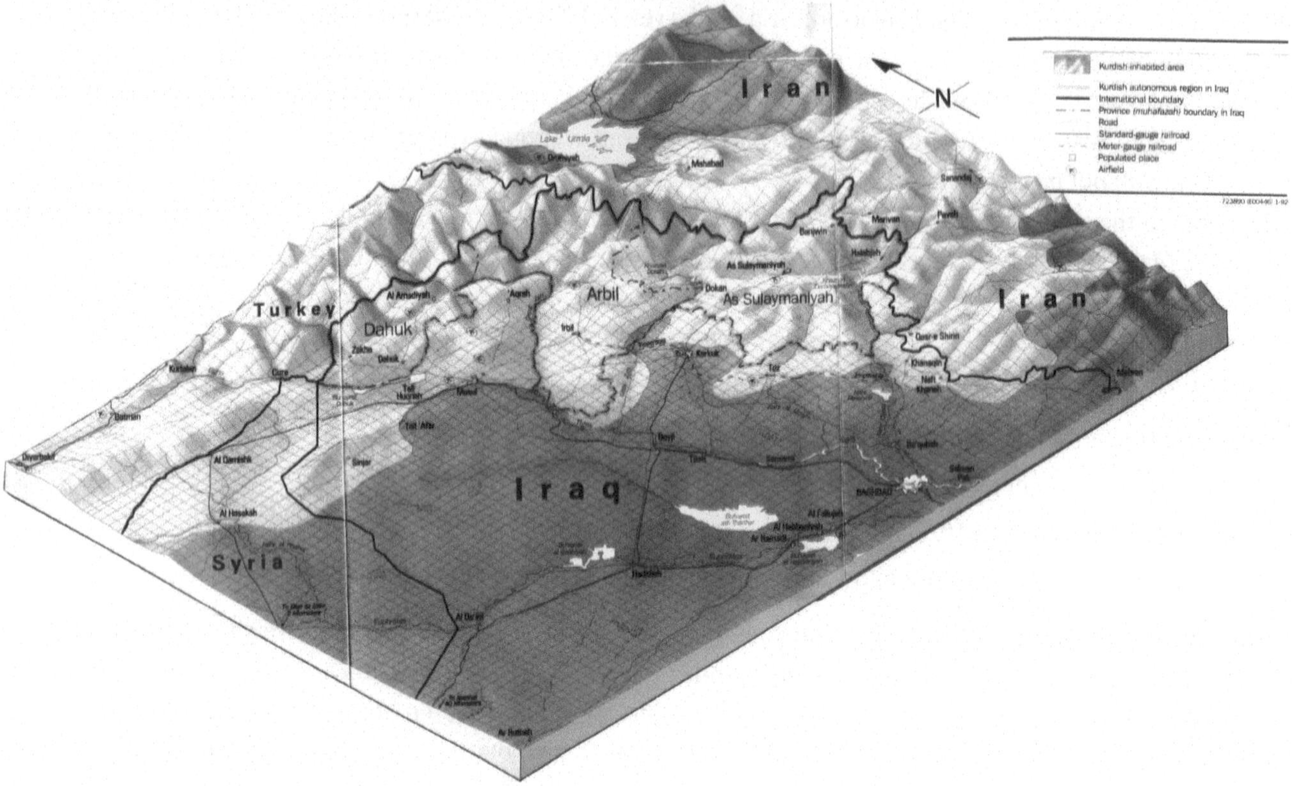

Kurdish folks hiding up in the hills.

I bring up Team Turk in particular because they invaded Iraq back in February 2008, much to the ire of the US, who is trying desperately to keep a lid on the Iraq mess. Why did the Turks invade? To root out a radical pro-independence Kurdish terrorist group named the **PKK** that has been responsible for multiple attacks within Turkey. Just so you can see how convoluted alliances can become in the Middle East, Team Turk is actually getting help from Team Persia in this effort, while the normally pro-Turk US is trying to stymie the effort. Geez! How confusing! To make matters worse, the Turks have been actively and openly fighting against Kurdish forces trying to overthrow the Syrian government of Bashar al-Assad next door, and even though Turkey totally hates Assad, they hate the Kurds more. Did you get all that? Yeah, your head may explode at any point. And starting back again in 2017, Kurdish terrorist groups are more active than ever, blowing up stuff within Turkey proper in response to Turkish aggressions in Syria.

Even smaller groups of note are political parties/rebel groups/terrorists groups. Yeah, that's a mouthful, and I don't mean political party or rebel group—I mean political party AND rebel group. Entities like the Shi'a group **Hezbollah** in southern Lebanon, **Hamas** in the Gaza Strip, or various groups across the Middle East referred to as the **Muslim Brotherhood** all fit this description. What do I mean? They are all political parties that in some countries under some circumstances actually put up candidates to run for office in all the countries in which they are located. However, most but not all of them also have militant wings of their party that actively go out and commit violent acts like suicide bombings, rocket launches into populated areas, or open armed warfare against other entities or states.

Why would anybody vote for a political party that does things like that? Mostly because in the areas where they are located, they are seen as actually fighting for the people, as opposed to the ineffective and corrupt governments that are in charge. In both Gaza and southern Lebanon, the political party sides of Hamas and Hezbollah build schools, run hospitals and soup kitchens, and generally protect the citizens, which seems like a pretty good thing. That seems like the types of things that the government is supposed to do, but at which they are failing. In such circumstances, the locals vote for those political parties, and why wouldn't they? An example: Hamas actually swept the Palestinian

Part of the Non-State Actor Action Faction.

elections a few years ago because the alternative party, **Fatah,** was largely seen as corrupt, ineffective, and lackeys of western powers.

But wait a minute! We have a problem here! These same parties have militant wings which do violent things, and as such, are labeled as terrorist groups by Israel, the US, and the EU. You know what happens to terrorists groups: they get embargoed, stone-walled, and cut down. Western powers don't like these groups and try to shut them out of the political process, causing ever more conflict. Even though Hamas was democratically elected, they were not recognized by the West, and all of the Gaza Strip is still being punished for their support of the group.

Some of these groups are simple to understand, though. Al-Qaeda just is a terrorist group, with no aspirations for political participation. Hezbollah is close to being in that category as well, as they are mostly a tool of Iran just to cause trouble with Israel. Varying groups named Muslim Brotherhood can go either way: there is an Egyptian Brotherhood, a Jordanian Brotherhood, a Syrian Brotherhood, et al. Many of the Brotherhoods have been banned from political participation in total (Egypt and Syria), mostly because they are proponents of bringing Islam into the political process, meaning that they might want to eventually change the government structure to be more in line with the Iranian model or with Islamic law. None of those Arab governments want that to happen, so they just label them as terrorist groups and ban them from politics, if not make their existence completely illegal altogether. Some of these groups have in turn resorted to the terrorist track because what else are they gonna do after they have been banned from participating? Other individuals from these groups just run as "Independents" on the ballot, but their existence is made no easier once their political beliefs are espoused. Fun times either way.

Example: back in 2014, the Muslim Brotherhood political party candidate in Egypt won the presidential election! Since the Arab Spring toppled then President/dictator Hosni Mubarak and brought a representative democracy to the country, the Brotherhood found itself as a legitimate power within the system. Oh, but wait! Before the year was done, a military coup deposed that democratically elected dude (his name was Mohammed Morsi), jailed him, and once again re-outlawed the Brotherhood. Here in 2019, an ex-military general (Abdel Sisi) is the President, likely for life, being an almost carbon copy of the former strongman Mubarak. Same as it ever was, once again.

All these non-state players have the potential to cause varying degrees of trouble for the major powers. It all depends on who they decide to side up with to achieve their own objectives, and it appears that that can change on a daily basis.

CONCLUSION

What you see from this Team breakdown is that we have had a series of empires which controlled the Middle East at different times; first the Persians, then the Arabs/Islamic, and finally the Ottoman Turks. While all of them are now Islamic, each of these three empires has resulted in a distinctly unique strong team—that are radically different from each other, with distinct ethnic character and political affiliations—that is vying for influence across the wider region.

The three main Islamic historic entities still present in today's world are the Persians (now Iran), the Ottoman Turks (now Turkey), and the Arabs/Islamic (now twenty individual Arab states, with Saudi Arabia and Egypt as the core). These main three have been recently joined by Team Hebrew (Israel), Team Foreigner (the US, EU, et al), and various other non-state actors, like Team Kurd and smaller political party/terrorist groups. In the end, they are all fighting for their own agendas to either control territories within the region, control governments in the region, control resources, or the direction of the region as a whole.

2021 Holy Hot-spots of Activity Update: Well, this whole chapter has been radically at arms with itself for nearly a decade straight! The Arab world was in total upheaval since the 2011 Arab Spring revolutionary movements, with Tunisia, Libya, Yemen, and Egypt (twice) already chucking out their rulers, and both Syria and Yemen in a total civil war scenario that can't last much longer. Oh, and this insane Sunni/Arab extremist group named ISIS took over parts of Syria and Iraq, formed its own state, and threatened to destroy all political boundaries in the region. Until they were finally decimated in 2018 by a loose coalition of forces which included everybody else in the vicinity that didn't want that insane clown posse to take over the world. Now they are just a world terrorist group with no home. Yeah. There's that. The ISIS situation was so radical, and so radically important, that it made the cover story of our book in 2016. Dig it:

Want to learn all about the Iraq/Syrian meltdown and the rise of the ISIS situation in more detail? Visit: http://plaid.at/isis to read **The 2016 Cover Story: The ISIS Insanity**.

Oh, and now tensions are rapidly ramping up between Iran and Saudi Arabia (as both the Syrian Civil War and the Yemen war drags on, with each of them supporting opposing sides in that fight), between Iran and the US over that failed nuclear deal, and between Iran and everyone else in the neighborhood at large. Dang! Are the dogs of war about to be let loose once more? Will a Persian Gulf throwdown focused on Iran make the cover of next year's plaid book? We shall see my friends! I certainly hope not, but it don't look good! Whew. That's it. What a freakin' mess. Let's get out of this chapter!

ADDENDUM! IT'S EXTRA! SO HAVE SOME MORE STUFF!

View additional Plaid Avenger resources for this region at http://plaid.at/mid_e

DEAD DUDES OF NOTE:

Mohammad Mosaddegh: Democratically elected Prime Minister of Iran from 1951 to 1953, until he was overthrown in a coup d'état backed by the US CIA. Was an ardent nationalist with a passionate opposition to foreign intervention in Iran. He nationalized Iranian oil, which so infuriated the UK and US that they labeled him as a dangerous commie and got him jacked out of power, which is one of the reason Iranians hate the US and UK.

Mohammad Rezā Shāh Pahlavi: The Shah of Iran from 1941 to his overthrow in 1979; only actually held all the real power from 1953 onward, after the US CIA had chucked out Mosaddegh (see left). The Shah did industrialize and modernize Iran, but unfortunately he was a typical corrupt and clueless monarch, as well as an US/Western stooge, who increasingly pissed off his people until they had a revolution to throw him out.

Ayatollah Khomeini: Fiery Shi'a orator and holy man, he inspired and led the Iranian Revolution of 1979 that deposed the Shah and created the first Islamic Republic of Iran, of which he then became the Supreme Leader. Has god-like status to millions of Iranians, and a demon-like status to millions more Westerners due to his condoning the Iranian Hostage Crisis of 1979–1980 and his extremist view of the US as "the Great Satan."

Saddam Hussein: Him again? Yep. In this context, know that Saddam was supported, funded and armed by the US, Western powers, and even other Arab states in the 1980's for his willingness to declare war on Iran, the country that Team West and Team Arab started despising after the Iranian Revolution. Saddam was seen as an Arab fist that would smash this Shi'a/Persian theocracy threat. Good call on that one, everybody.

LIVE LEADERS YOU SHOULD KNOW:

TEAM PERSIA:

Ayatollah Khamenei: Religious/Spiritual leader and ultimate wielder of political power in Iran, since it is a theocracy. Persian, Shia, ultra-right, conservative, and totally hates the US. A lot.

Mahmoud Ahmadinejad: Former President of Iran, and great agitator to make the entire world hate his country by continuing to flaunt uranium enrichment. Persian, Shia, ultra-right, conservative, and totally hates the US. A lot. Like a lot, lot.

Ebrahim Raisi: New President of Iran in 2021. Ultra-conservative, hardline, right-wing, was formerly a Muslim jurist... and is likely being groomed to take over as the head Ayatollah and Supreme Leader of Iran.

TEAM HEBREW:

Benjamin Netanyahu: Former Prime Minister of Israel, and a conservative, center-right, hawkish fellow. Don't look for any Palestinian "peace process" to go forward on his watch. He is a US ally and his right-wing, populist policies made him hugely popular with hawkish President Trump; US/Israeli ties were never stronger.

PLAID CINEMA SELECTION:

Take a Middle Eastern Team Tour via Cinema! In the interest of saving space, I won't describe all these in detail, and many are repeats from the previous chapter, but check out these team lists of films to watch to get a fuller sense of the cultures and backgrounds of each team. Total TV team party!

TEAM ARAB:

The Battle of Algiers (1966)
Gaza Strip (2002)
The Kingdom (2007)
Lawrence of Arabia (1962)
Paradise Now (2005)
The Yacoubian Building (2006)

TEAM TURK:

Eskiya (1996)
Günese yolculuk (1999)
Propaganda (1999)
Uzak aka *Distant* (2002)
Yol (1982)

TEAM KURD:

Günese yolculuk (1999)
Marooned in Iraq (2002)
A Time for Drunken Horses (2000)
Turtles Can Fly (2004)

TEAM PERSIA:

About Elly (2009)
Bashu, the Little Stranger (1986)
Dayereh aka *The Circle* (2000)
Ekhrajiha (2007)
Marmoulak aka *The Lizard* (2004)
Persepolis (2007)
300 (2006)
Two Women (1999)

TEAM HEBREW:

Ajami (2009)
Bab el shams (2004)
Sallah Shabati (1964)
The Ten Commandments (1956)
Walk on Water (2004)
Waltz with Bashir (2008)

LIVE LEADERS YOU SHOULD KNOW:

TEAM ARAB:

King Salman: Monarch/leader of Saudi Arabia, the birthplace and holy center of Islam. Arab, Sunni, Saudi, center-right, conservative, massively wealthy, and a US ally. More hawkish than his predecessor, and may yet assert Arab power militarily in a big way . . . against ISIS, or Iran!!!

Mohammad bin Salman (MBS): Crown Prince of Saudi Arabia but already its de facto leader. He launched the war in Yemen in March 2015 while he was the Defense Minister under his father, King Salman bin Abdul Aziz Al Saud. He's overseen some reforms, such as women being able to drive, but he's also responsible for the death of the journalist Jamal Khashoggi (a critic of the government, killed in October 2018) and other authoritarian measures.

Abdel Fattah al-Sisi: Formerly the top military general who stepped down from that position after a military-led coup threw out the elected Muslim Brotherhood government. Now he is Egypt's new 'President' after a contrived-delection. Arab, Sunni, Egyptian, far-right, conservative.

TEAM TURK:

Recep Erdoğan: The kick-ass Prime Minister of Turkey and chairman of the Justice and Development Party (AK Party). Despite internal friction due to his association with the religious-based AK Party, he has overseen an era of economic prosperity and democratization, and Turkey's rise as a serious regional power and trusted mediator of Middle Eastern conflicts. HOWEVER, global attitude towards him is now changing fast from 2017-2021 as he is rapidly turning Turkey into a one-party state, with himself at the helm as 21st century dictator.

CHAPTER OUTLINE

1. Physically . . .
2. Crossroads of Culture
3. Enter, the Ottomans
4. Formation of the Modern State
5. Carving Up the Turkey
6. Awesome Ataturk
7. Today's Turkey
    7.1. Major Military
    7.2. EU Irritations
    7.3. Kurdish Konflict
8. Turks Transitioned
9. The Turkish Titan: Erdoğan
    9.1. Prime Minister Erdoğan
    9.2. Authoritarian Tendencies
    9.3. And Then Things Got Worse
10. Finish Up Your Plate of Turkey
    10.1. The Player
11. Map Gallery

Turkey

19

TURKEY IS A:

(a) Nation
(b) State
(c) Nation-state
(d) Region
(e) Delicious bird, when roasted properly
(f) All of the above

You know it's all of the above, baby! Turkey is like Japan and Russia, in that it is a sovereign state in its own right, with enough distinctions about it to set it apart from all of the places nearby, which makes it a Plaid Avenger world region as well. What is the deal with Turkey? What's going on here? Why is it different enough to be apart from the Middle East—the region with which it has classically been associated? Anybody from the Middle East, and particularly from Turkey itself, would say, "No way! We don't have anything in common with each other!" Here in the West, we just see Turkey and the Middle East as kind of a common whole. There is a lot about Turkey historically, presently, and in the future, that distinguishes it from all other Middle Eastern countries. Does that mean it's more like Europe? Nope, it's quite distinct from Europe as well. A lot of people in Europe want to distance themselves from Turkey, while some want to embrace it. Turkey is in a strange place in the world right now; it has its own thing going on, while at the same time acting as a bridge between two cultures and regions. That's the Plaid Avenger keyword for Turkey: **the Bridge**.

Turkish Delight: aka lokum or rahat, is a 15th century tasty confection made from starch and sugar, with a soft, jelly-like consistency. It is often cut into small cubes and dusted with powdered sugar before serving. It is sometimes flavored with cinnamon, mint, lemon, or rosewater . . . that last one giving the candy its characteristic pink hue. I always wondered what the heck it was since the White Witch fed some to Edmund in The Lion, the Witch, & the Wardrobe. **Oh Edmund, you little traitorous scumbucket.**

As we discussed in the Middle Eastern region, there are lots of different ethnic groups in what is collectively referred to as the Middle East, and one of those is Turkic. Turkic is also a distinct linguistic group. Where are the Turkic people? Many people think, "Turk?, they've always been in Turkey, of course!" Not necessarily. In modern day Turkey, they refer to themselves ethnically and linguistically as Turkic or Turks; however, this group is originally from Central Asia. Turks are relative newcomers to the territory that we now call Turkey. The folks who would become the Turks came from Central Asia, across what is now Iran, through Iraq, and into modern day Turkey about 1000 years ago. After several hundred years, they took over the region, built a mighty empire, their culture took hold, and they made the place their own. Perhaps we are getting ahead of ourselves here, but that's what makes them so uniquely distinct: a different ethnicity and linguistic group from all the surrounding countries. Why and how else is Turkey so different from every place around it?

Turkey is more 'western' than virtually every other Middle Eastern country in terms of economic development and government. At the same time, they are not as 'western' as European countries. This is the first of many things that puts Turkey in a sort of no-man's land, somewhere in the middle. They have a capitalist system that is fully integrated into world economic dynamics. Nobody would disagree with that. They are the strongest fully established, democratic Muslim country.

That's important enough to pause and restate: democratic, fully Muslim country. Perhaps the main reason why they are typically classified as Middle Eastern is their single commonality, which is Islam. Turkey's population is almost 100 percent Islamic. However, it is *not* an **Islamic republic** (that is, it is not a theocracy), and that makes all the difference. Turkey has religious ties to its Middle Eastern neighbors, but all other aspects of their society, like secular democracy and capitalism, make them very European. This duality reinforces the idea that Turkey is a bridge. Before we get to a more cultural background, let's talk Turkey physically.

Islamic people, but not Islamic state.

PHYSICALLY . . .

Physically speaking, Turkey is a commanding presence in this region of the globe, making its way into the top ten largest countries in Asia. Looking at a map you can actually see Turkey juts out into the Mediterranean and Aegean Seas, bridging Europe with Asia. That's right! This here nation/state/nation-state/region is transcontinental, thus she is proudly nicknamed The Bridge. Well, nicknamed that by me anyway.

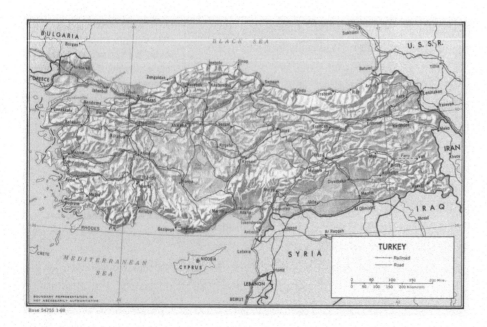

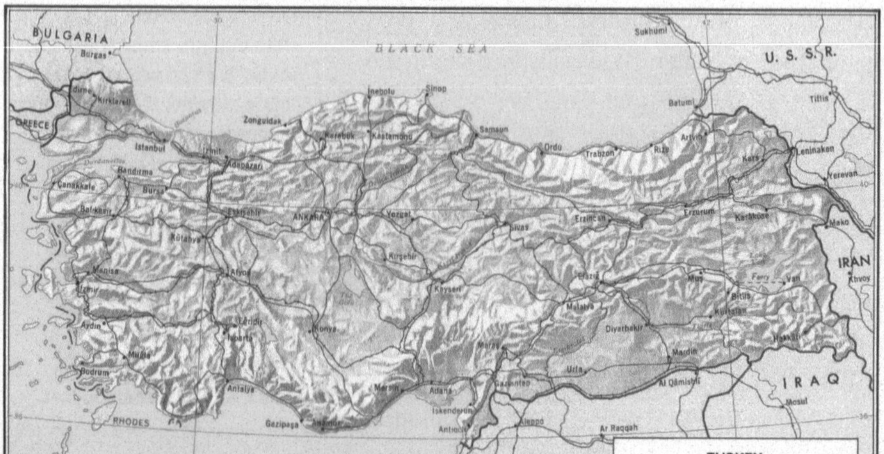

The Black Sea, the Mediterranean Sea, the Aegean Sea, and the Sea of Marmara surround the state on three sides. . . making it a peninsula! Sweet! Just like all those peninsulas we talked about back in the Western Europe chapter! These bodies of water moderate the climate, creating cool rainy winters and warm summers along the coasts. As you travel farther inland, into higher elevation, you

Cappadocia, central Turkey.

typically start seeing your extremes like extremely dry summers and cold, snowy winters. On the western side of Turkey, I would consider wearing a moderately heavy jacket in the winter and shorts and a t-shirt in the summer, although you'll be grabbing your raincoat if you live more north toward the Black Sea. In Eastern Anatolia there's a very different climate where temperatures average -13 degrees Celsius in the winter and 17 degrees Celsius in the summer. That's cold!

If we were to compare Turkey to the US, and overlay it with size held consistent, we see that it stretches from Washington D.C. to Kansas City, Missouri. It's about the same latitude, and covers West Virginia, Virginia, parts of North Carolina, Tennessee, Kentucky, Indiana, Illinois, and Missouri. It's a decently sized country. Given that its latitude is roughly the same as the states I just mentioned, Turkey has a pretty similar climate. It is humid continental, with some humid subtropical, in the southern parts. However, given that it borders the Mediterranean Sea, we know it must have some Mediterranean climate. Indeed, it does around its coastal fringes. Turkey has four distinguishable seasons, just like the US.

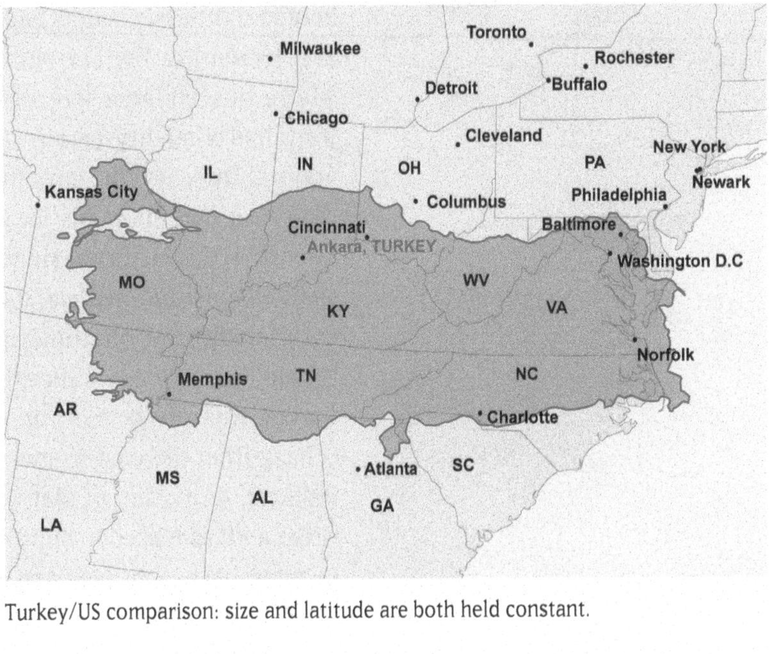

Turkey/US comparison: size and latitude are both held constant.

The only other major physical factor to consider is that the terrain is quite different from the eastern United States. Turkey, as a whole, is a pretty mountainous country. It has an uplifted plateau in the middle, with some high mountains, particularly on its eastern borders, where you

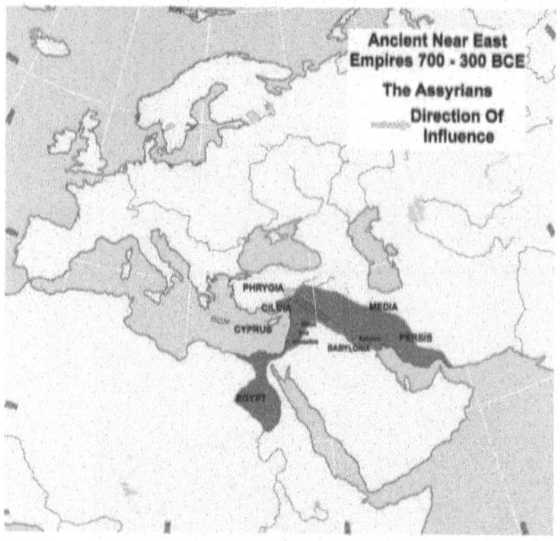

Assyria

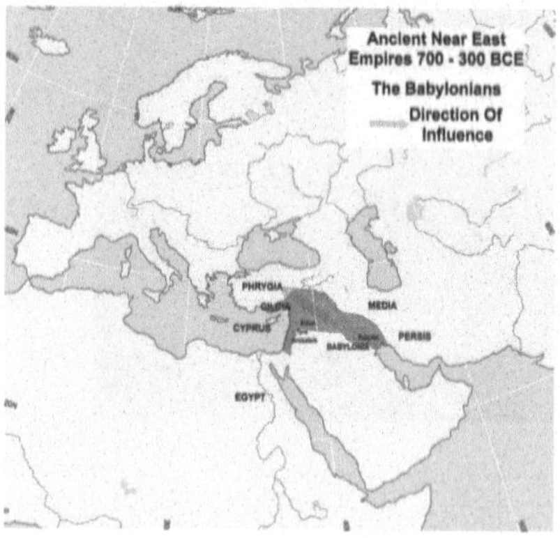

Babylonians

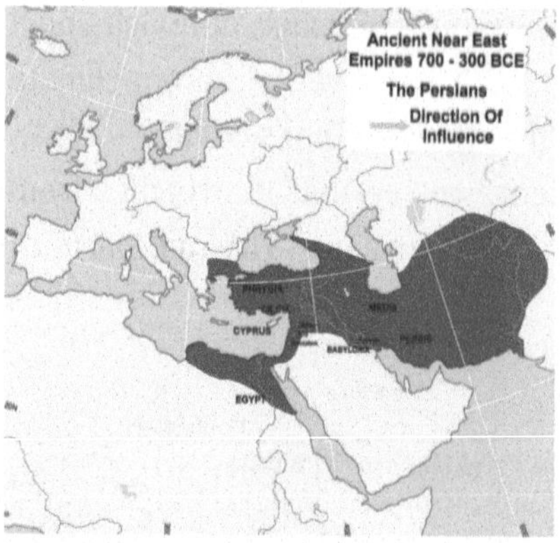

Persians

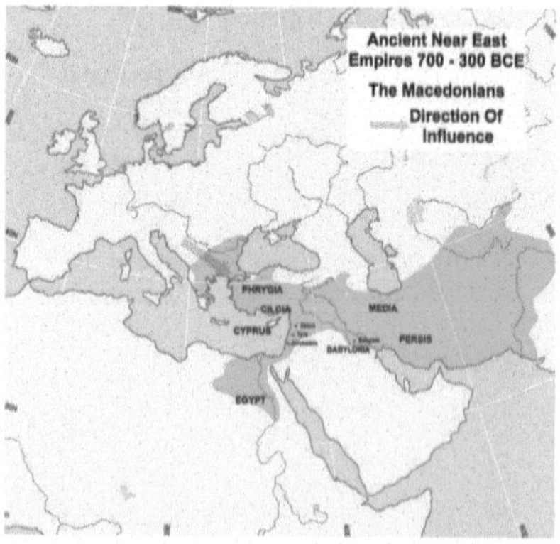

Macedonians

can find Mt. Ararat (16,000+ ft.) as you go into the Caucasus Mountains. Even over in the western part, the elevated plateau is on average over 6000 feet in elevation.

This is an **escarpment**, as we talked about in South America and will again in Sub-Saharan Africa. There are not a lot of coastal plains or flat areas in Turkey, so as soon as you leave those nice, sunny Mediterranean shores, the terrain and the elevation make it much cooler, much faster. As you get to the far eastern side of Turkey, there can be fairly tough winters. It does have some Mediterranean and Black Sea frontage, which moderates things a bit, but it still is a fairly cooler place than eastern North America at similar latitudes.

Turkey is such a mountainous country because it is on the border of several different tectonic plates. It is at the confluence of the Eurasian, African, and Arabian plates, very much like we had several plates coming together in Japan. As with all geologically active mountain-building scenarios, it's an active earthquake region. In the last decade, Turkey experienced some serious earthquakes, having tremendous impact in terms of human lives and property damage. That about wraps up the physical traits of Turkey—let's get back to the cultural. Why is Turkey a crossroads of culture?

CROSSROADS OF CULTURE

Turkey is one of those places, like the classic Middle East, where people have been hanging out a long time. We can go way back to 1000 BCE to the Assyrians, a tough band of dudes. There is too much rich history to get into it in detail, but I want you to understand how the **Anatolian peninsula** has been a crossroads of culture throughout time. The Babylonians hung out here in control until around 1000 BCE, and the Assyrians shortly after that.

We've had people hanging out here for a good long time, people who became the core of Western Civilization. The core cultural hearth for Western Civilization is here in the classic Middle East and Turkey. Fast forward to the Persian ethnic group, whose descendants live in modern day Iran, had a mighty empire around 500–600 BCE. These are the guys that Alexander the Great and all of his ilk were fighting against. In this historical situation, the Persians swept from modern day Iran, their core, across the Anatolian peninsula to invade Europe proper, specifically Greece. Turkey was the bridge for the Persians coming from the east to invade Europe in the west. As a result of that, Alexander the Great and his Macedonian (Greek) buddies invaded back. You may see reference to the Greeks calling this area **Asia Minor**.

The Greeks and Romans did their pre-Turkish Anatolian tours too.

Orthodox Byzantine boys in black getting their chant on.

To combat the Persians, Turkey became the bridge of invasion for Alexander's forces out of Greece. They swept over, beat up on the Persians, and established a European empire all the way into India, making use of Turkey as a crossroads. And we're not finished yet!

The Greeks, of course, were beaten by the Romans, whose empire eventually included the Anatolian peninsula. To spare you the long blow-by-blow history of the Roman Empire, eventually, the Romans decided they needed to establish a secondary capital in the East, as the Empire proper was expanding. The main capital stayed in Rome, but the eastern, secondary capital was formed in Constantinople, where the Turkish city of Istanbul is today. After the disintegration of the Roman Empire, the eastern section of the empire became known as the **Byzantine Empire**. You've probably heard of the Anatolian peninsula referred to as "**Byzantium**" in ancient texts. This occurred between 300–500 CE.

Again, the Plaid Avenger is reinforcing that this is a crossroads. We've had the east coming from one direction, and the west coming back from the other. Turkey is always the central pivot point of these movements. Not only was the Anatolian Peninsula the stage for the political division that results in the Byzantine Empire, it was the stage for a division of religion as well. We know it as the **Great Schism**: the split between the Catholic Church and the Eastern Orthodox Church in 1054. The Eastern Orthodox Church set up camp, still in the Byzantine Empire, and spread Eastern Orthodoxy into Russia, the Middle East, and Eastern Europe. The Catholic story, which was centered in Rome and spread to most of the rest of Europe proper. But our story's still not finished!

We're taking it up to the time of Muhammad, peace be with him. Born in 570 CE, not in Turkey, but down in the Arabian peninsula in the holy city of Mecca, where Allah used him as the vehicle for the creation of the Koran. This is where the religion that we now know as Islam started.

Where did it go? It diffused outward from the Arabian Peninsula to lots of places: Africa, Central Asia, South Asia, and also Turkey. A reversal took place in the Byzantine Empire. Eastern Orthodox Christianity was displaced on the Anatolian Peninsula because of the expansion of Islam from the east. In addition, the Turkic peoples from Central Asia emigrated to the area at the same time as this growth of Islam to settle into the Anatolian Peninsula, making this **the bridge** between Eastern and Western cultures and religions once again. This is a happening place, man. Everything's been going on in Turkey! But it's not even Turkey yet!

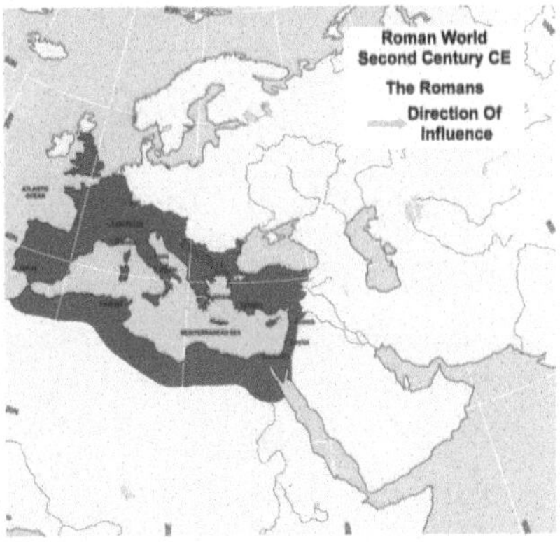

Romans

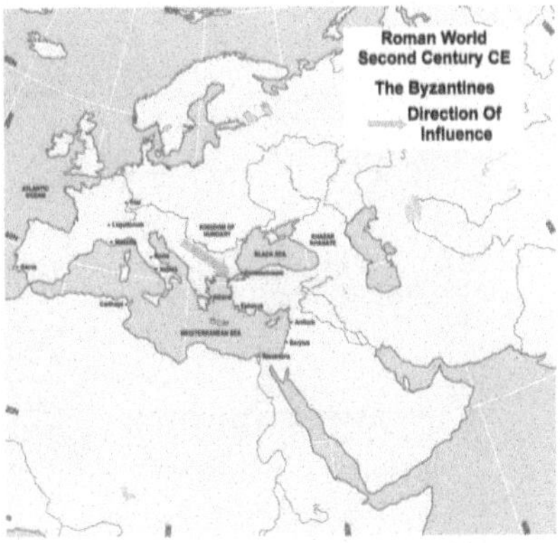

Byzantines

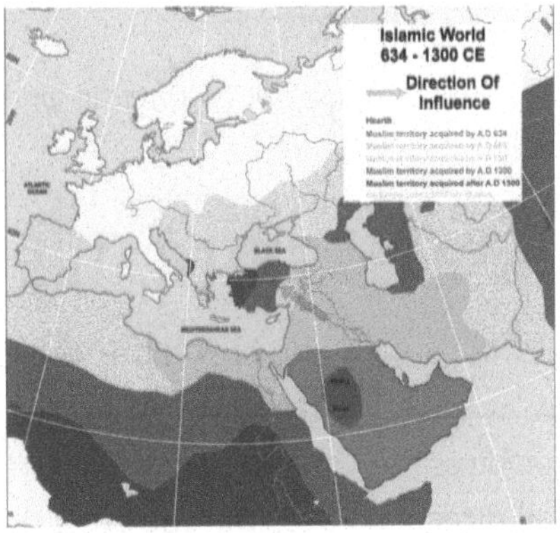

The growth of the Arab-Islamic Empire sets the stage for the Turks . . .

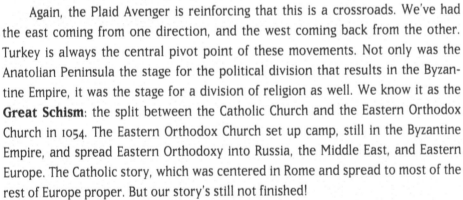

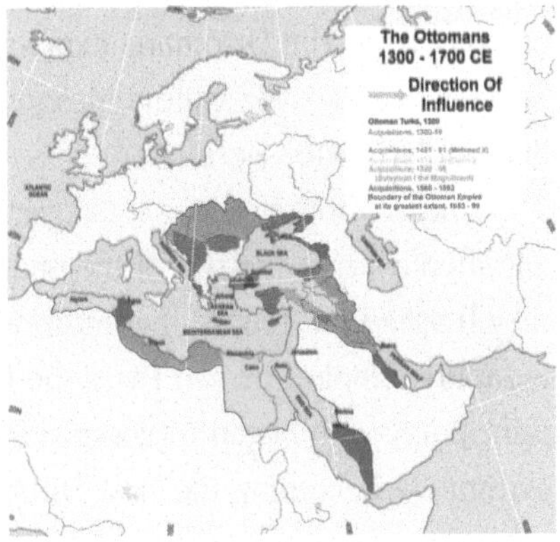

Ottomans

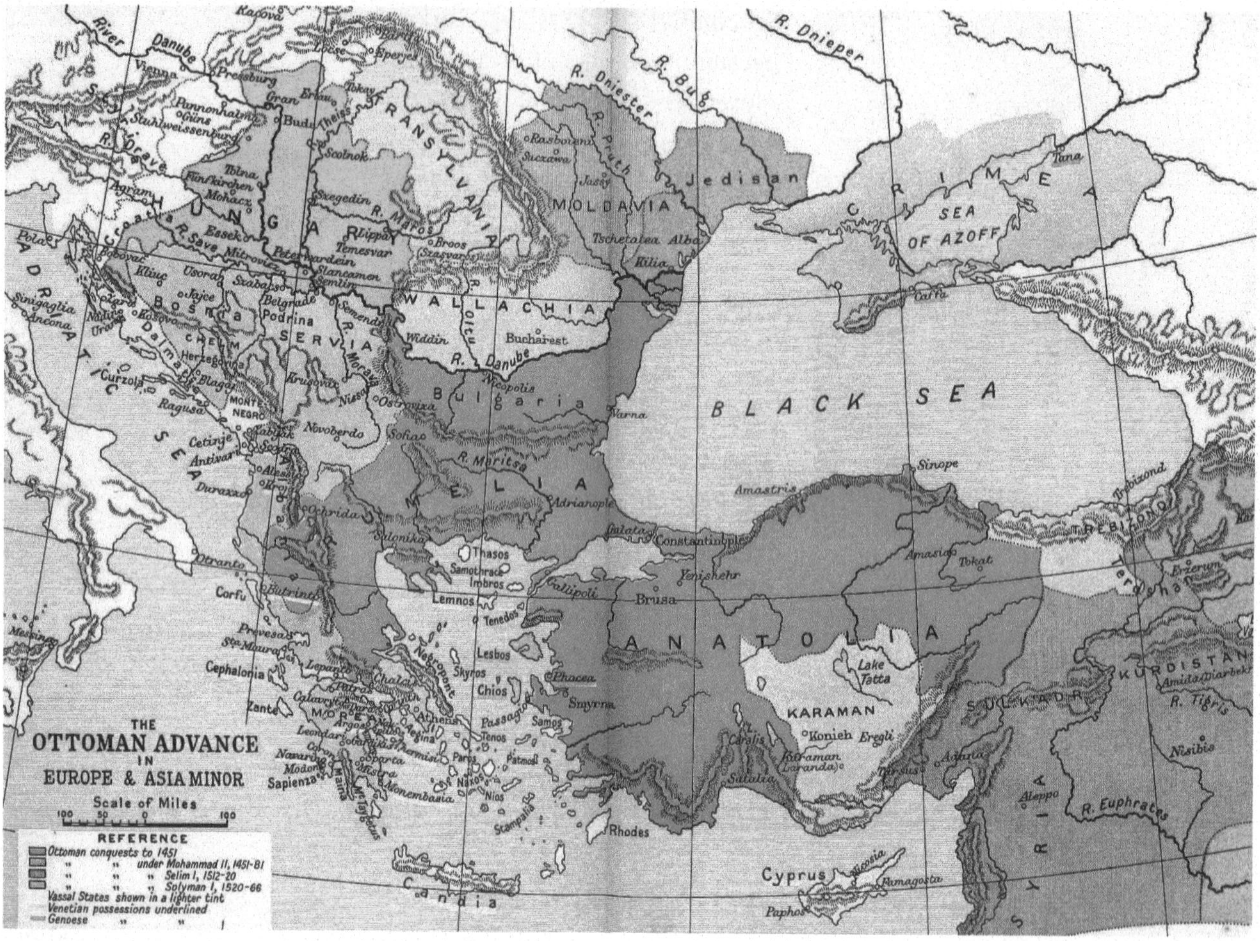

Turkey: Home base for Ottoman spread.

ENTER, THE OTTOMANS

So, how did it get to be Turkey? There's one more group we haven't talked specifically about yet: **the Ottomans**. After the Turkic ethnic and linguistic group took over this geographic area, they built a big empire: the Ottoman Empire. Starting in roughly 1300 CE, the little core that was only around Constantinople grew in all directions over the course of the next 400 years, to the point that they took over virtually the entire Mediterranean coast of Africa and much of the classic Middle East, including the Arabian Peninsula and what is modern day Iraq. The Ottomans expanded into the Tigris and Euphrates river valley, Mesopotamia, where the Babylonians once were.

These Ottoman dudes took over the Black Sea area and then overran the rest of the Anatolian Peninsula as they finished off (that is, overthrew and kicked out) the Byzantine Empire. Osman I was the leader of a Turkish principality who began to conquer Byzantine territory, eventually becoming the first Sultan. (The Ottoman's

Side note: The Ottomans are the guys that brought coffee to Europe, via their face-off with the Austrians at Vienna. Ever heard of Viennese coffee? The brewing and consumption of coffee were passed across the battle lines during protracted wars over the centuries. We often think of the Austrians and Italians as coffee connoisseurs, but it was introduced via our Turkish friends. Ever had Turkish coffee? Wow! It will knock your socks off, man!

CONSTANTINOPLE 1453. WHAT WAS ONCE THE CAPITAL OF THE EASTERN ROMAN EMPIRE AND THE MOST ADVANCED CITY IN EUROPE IS NOW THE LAST BASTION OF THE BYZANTINE EMPIRE.
THE CITY HAD BEEN THE STAGING GROUND FOR THE CRUSADES BUT CENTURIES OF IN-FIGHTING DIVIDED THE EMPIRE AND LET THE OTTOMAN TURKS CONQUER MUCH OF IT.
SULTAN MEHMED II MADE THE CONQUEST OF THE BYZANTINES A PRIORITY AFTER TAKING THE THRONE IN 1451.
SEE THIS...?
AFTER A 53-DAY SIEGE AND COMBINED LAND AND SEA ASSAULT OF OVER 90,000 MEN, MEHMED IS ON THE VERGE OF SUCCEEDING. THEY HAVE A BROUGHT A SECRET WEAPON PREVIOUSLY UNSEEN IN EUROPEAN WARFARE.
GUNPOWDER.
THIS IS MY BOOMSTICK!
BOOM
PROTECTED FROM INVADERS FOR CENTURIES BY ITS FAMED SERIES OF WALLS, TURKISH CANNONS RENDER THE CITY'S IMPENETRABLE MEDIEVAL DEFENSES... PENETRABLE. THE INHABITANTS FIGHT IN VAIN TO THE LAST.

Sultans simply scoffed at backwoods Euro-trash.

actually derived their name from Osman.) By 1453, the Ottoman's captured Constantinople, converting it from a Christian city to an Islamic city. They then proceeded west to conquer the Balkans in the following century. This event perhaps impacts today's world the most: the introduction of Islam deep into European territory via the Ottoman Empire, primarily in Eastern Europe, modern day Yugoslavia, Albania, Romania, Bulgaria, and Greece. The Ottomans, a predominately Islamic empire, controlled all of this territory. At the height of their empire, they were knocking on the doors of Vienna. They even attempted to take Austria a few times. Hey! That's all the way in Western Europe! They nearly conquered it several times, but never quite succeeded. Nevertheless, that gives you an idea of how far they penetrated into the continent, taking their culture, particularly their religious culture, with them.

This ties back to why we talked about southeastern Europe in particular being a trouble spot in today's world because of the infusion of so many different religions, ethnicities, languages, and groups of people. This is part of the same story. Turkey is the bridge, the platform for the movement of people from Europe to the Middle East—but just as importantly, from the Middle East to Europe. Religion typically seems to be the sticking point for a lot of this, an issue that we see cropping up even in the modern era as Turkey tries to join the European Union. More on that later.

Sule, that headgear is magnificent!

One last blast from the past though: Under the leadership of Suleiman the Magnificent, the Ottoman Empire reached its height, spanning large portions of Northern Africa, Eastern Europe, Western Asia, and more. He ruled over 20 to 30 million subjects and completed numerous large scale building projects like mosques, schools, hospitals, and of course Turkish bathhouses. Life was good under this guy's reign. But every great empire comes to an end and eventually the Ottoman Empire began to fall behind their European counterparts, thanks to the Renaissance. Europe became a stronger military and economic threat. So let's get to that now. . .

FORMATION OF THE MODERN STATE

As you have guessed by now, the Ottomans didn't last forever. They were large and in charge and had an expanding empire from about 1300 to 1700 CE, but from 1700 onwards, they went into a steady state of decline. Why? It has a lot to do with stuff we've already talked about, but to put it simply, these guys got behind the times. In the Middle Ages, they were superior to the Europeans in terms of technology, military, and firepower. They had the finest universities in all of Europe and the Middle East. All great centers of learning, the oldest colleges, as well as all scientific and technological innovations during the Middle Ages were happening in Ottoman-Turk territory. They eventually got a little insulated, headstrong, and arrogant, and saw the Europeans as a kind of backward race that they needn't mess around with that much.

On top of that, think about how they got so rich. What did they produce? Were they a big colonial empire producing and trading goods, building an economy? Nope. They largely obtained their wealth and sustained themselves by continually pillaging for booty. Lol Plaid Avenger said "booty." However, in this context it means that through conquest of war, you take valuable stuff, like gold and other riches, from other places. That's okay for a while, but it eventually leads to a lack of economic infrastructure. A booty economy, unfortunately, can't stand on its own two feet in the long run. If the only way you get rich is just by taking stuff, when you get to the point that you can't take stuff anymore, you go broke.

Indeed, this was part of the reason for the collapse of the Ottoman Empire. They became economically stagnant, meaning they weren't producing or innovating anything. This caused the Ottomans to get behind economically and technologically. While they were sitting around stagnating, the Europeans were in full bloom: first the Renaissance, followed by colonialism and industrialization. The Ottomans, on the other hand, were starting to fall down just as the Europeans were starting to stand up, and by 1900 re-calibration of this Europe-Turkish power situation was imminent. Politically, the Ottomans were stagnant as well, being an empire ruled by 600 years of sultans in their sultanates. Totally old school ways, as they were entering a very different 20th century situation.

A **sultanate** is a monarchy, a royal imperial line of centralized power, a system that was disintegrating through most of Europe at this time but the Ottomans held fast to their old school system for a bit longer, which further increased their stagnation and perceived backwardness. Things continued to spiral downhill: there was internal dissent because the economy was not in good shape, and they were losing territories to the expanding Europeans. They lost Greece due to internal revolts and rebellions all across the Balkan Peninsula. In 1876, Abdul Hamid became the last of the unbroken line of Ottoman Sultans; but his line was soon to be snapped in half. The people had enough and said, "We've got to change! This sucks!" Again, a very common theme we have talked about in other parts of the world: "We have to get rid of this guy! This system's not working, so let's get rid of HIM!" They deposed the Sultan in 1909, but the state was still on shaky ground.

In another part of this tale, which I haven't incorporated yet, the Russians and the Ottomans fought several wars. There was major animosity between the Russians and the Turks, the remnants of which still exist today. The Russians tried to expand out to get more shore frontage on the Black Sea, on the Mediterranean Sea, anywhere they could grab coastline in order to build their naval superiority—and the Ottoman-Turkish territory made for a good target. The Russians tried to interject power into Ottoman territory. The Europeans, at the exact same time, attempted counter Russian influence in Ottoman territory. This is another component of why Turkey became the bridge; it became a battleground for the competing powers of the Europeans and the Russians.

After the Ottomans deposed the sultan, they tried to form something close to a democracy, or a republic as it were, in an attempt to copy the European model, but they were very young and very weak economically. They were looking around and saying, "Hey, who can we ally with? We're getting so weak that Russia could probably beat us. Somebody be our buddy! Somebody sign a pact with us saying you'll come help us if Russia invades. Anybody! How about you guys in the UK, will you sign a pact? France? Spain? Italy? Germany? Come on, somebody help us out here!" At this point, the Ottomans were looked upon as kind of the "old decrepit man" of Europe. They're physically close to Europe. They were a competing power with Europe when they invaded Austria. It's not as if they're unknown or a foreign territory that nobody knows about . . . but nobody wanted to deal with them. It's like they had the plague. Untouchable.

German leader Wilhelm II helped out the Ottomans—helped them out of an empire!

The British and the French thought, "Well, these guys are kind of losers. What are we supposed to do with these guys? We don't want to ally with them. We don't want to go to war with the Russians just for these guys. We don't want the Russians to win, but we don't want to engage them in open battle either, so we're just going to hedge our bets." Nobody wanted to deal with the Ottomans. However, Germany under Wilhelm II, (knowing full well that it was about to start World War I), said, "Hey, we'll help you poor suckers out! We'll be your buddy, because we don't like the Russians either. *We'll* be allies with you." In 1914, the Ottomans set up an alliance with Germany. Then Germany went to war with everybody, and ultimately lost. Long story short: Oops! Bad choice! Bad alliance!

The Ottomans, having not chosen wisely, were basically dismantled at the end of World War I. In 1919, at the conclusion of the war, something happened that the Plaid Avenger refers to as **A Peace to End All Peace**. It's actually a book title you should check out sometime; it's really sweet, even if a bit dry and a lot long (David Fromkin, 1989). It explains a lot about why there's not peace in the Middle East right now and perhaps why there never will be. What is the Plaid Avenger referring to? It's got to do with a bad job hacking up of the Turkey, and it wasn't even Thanksgiving Day.

CARVING UP THE TURKEY

The European powers that were victorious in World War I not only carved up Germany's Prussian Empire, but the Ottoman Empire got the chop too. The way they disemboweled it set up the political, ethnic, linguistic, and religious differences that still perpetuate conflict today. In other words, they took a map of Ottoman territory and said, "Okay, here's this section. You guys go to Greece. Let's call this chunk over here Romania. Oops, I spilled my coffee there. Let's just call that coffee stain Jordan. This over here, Iraq." Almost all the modern day boundaries in this entire area were drawn in part or whole during this redistribution or reallocation of land after World War I. The leftovers from the hack job became what we call Turkey today. It always was the central power core area of the Ottoman Empire, the piece where the actual ethnic Turks were concentrated. Turkey has remained stable to this day partly because it was primarily Turk, made up of Turkic people. All the other boundaries sucked because, in many cases, the Europeans drew them arbitrarily, with no consideration of the cultures within the newly drawn borders. On top of that, a lot of the new states were given to European powers, which were to supervise them as a "big brother." For example, the French became the "big brother" of Syria. The British became the "big brother" to Egypt and the Sudan. These new "baby brother" states were called **mandates**.

Of particular note was the Sykes-Picot: **The Sykes-Picot Agreement** was basically a secret 1916 agreement between the British, French, and Russians that placed them in charge of the territories of the newly formed states from the disintegration of the Ottoman Empire, even before the war was over. Many of the current territorial lines in the Middle East were carved out by the Allied Powers in this agreement. . . specifically Iraq, Syria, Lebanon, Jordan, and Turkey itself. No locals had input into this treachery, which explains why there are still territorial disputes to this day, (i.e. Kurds who were left out entirely). The Turks were left with a small chunk of land around Istanbul, their finances in the control of the Allied powers, and the majority of their territory under occupation. Luckily, a brave young nationalist named Mustafa Kemal Ataturk had a different plan for the future of his country, which we will get to in a sec. . .

See, all these European powers were still really into the colonialism thing, and the end of the war spelled more territory and influence and resources for them to control. I do need to point out one resource in particular that was just being discovered in parts of the Middle East during this period: oil. Even though oil was not as central to the world economy in 1919 as it is today, these Europeans knew it was worth a lot; therefore, they wanted to control it. Ever wonder why one

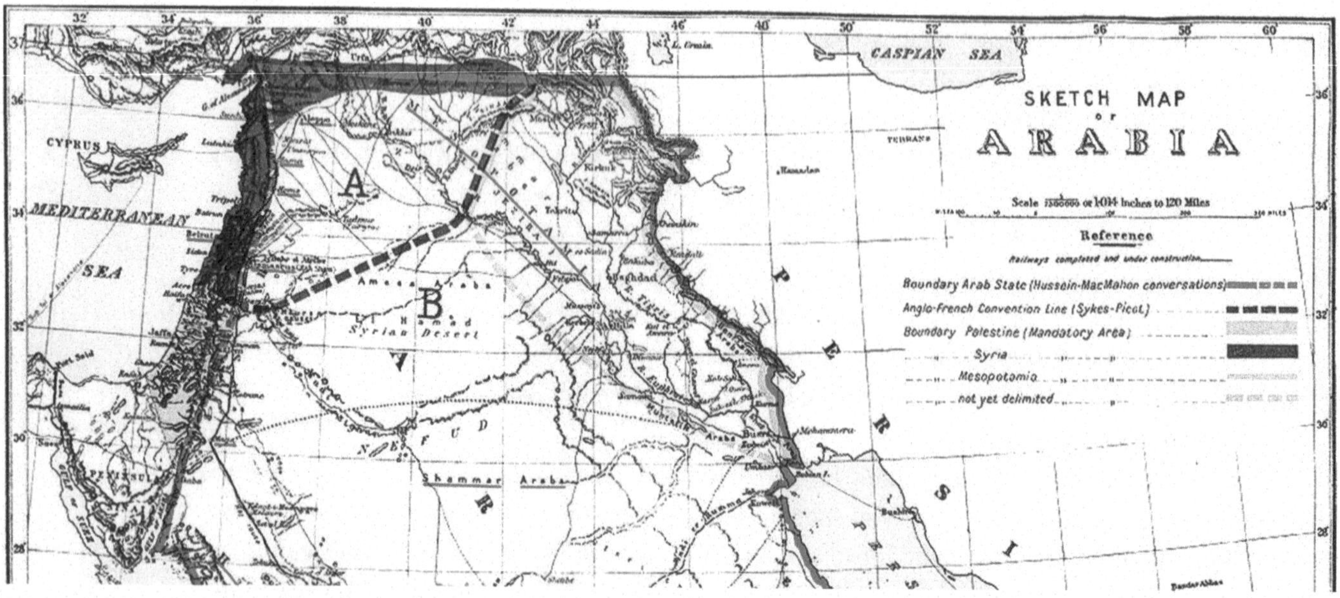

Sykes-Picot: France gets "A," Britain gets "B," and the Turks get screwed.

of the world's biggest oil companies, British Petroleum (BP), is from a country that has no oil? Look again at the mandate map below. The UK had mandates on Oman, the Gulf states, and Iraq? What a shocker.

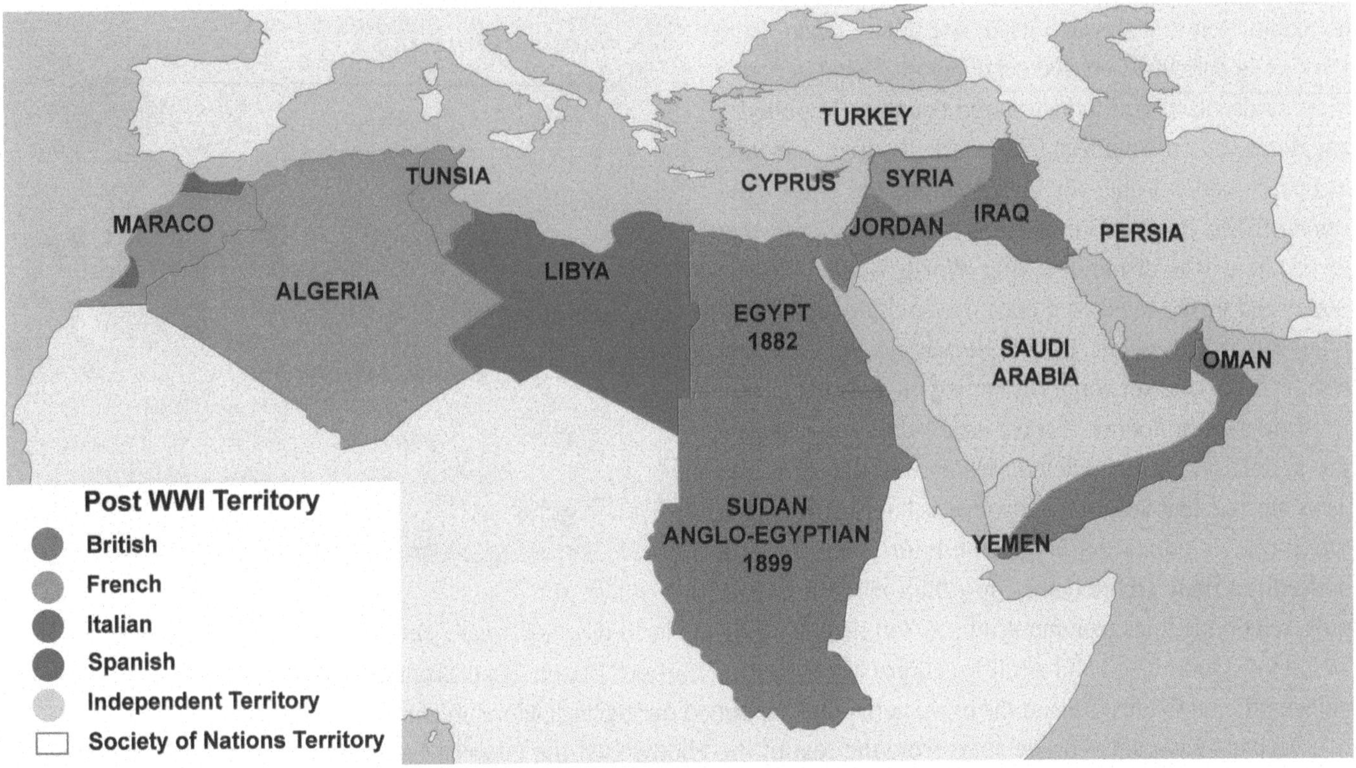

Middle East Mandates: Europe ensures conflict for centuries to come with these contrived borders.

The mandate system effectively meant that political control still rested with the European powers occupying these Middle Eastern lands, even after the Ottomans went away. This is part of the problem with today's world, but we'll talk about that more when we get to the Middle East. That's "the peace to end all peace" in 1919. Essentially, these countries were carved up in such a way as to ensure perpetual conflict, which of course is precisely what has happened in the greater Middle East. But Turkey has actually held together quite nicely since the war. How?

AWESOME ATATURK

Another politically noteworthy event happened in 1919. A guy named Mustafa Kemal led a resistance to the European allies' plan to carve up Turkey. As the Ottoman Empire was being dissected, a lot of Europeans wanted to divide Turkey up even more. They said, "If we don't completely eliminate them, they'll be a future threat, so let's chop 'em up!" Many Turks in the ever-dwindling territory decided to fight back against this total elimination plan, and one such man was a heroic heavyweight known as Kemal.

Mustafa Kemal was an army officer and an ardent, fervent nationalist. He was a big pro-Turkey guy who loved and believed in his country. He became very popular leading this resistance; and along with his brothers-in-arms in the military, they stymied European efforts to destroy their state fully. He immediately became the figurehead savior of the state, and eventually became a heck of a lot more. He continued with the struggle to regain full control of Turkey away from the allied European powers, and eventually succeeded in 1923. That's a red-letter date to remember; 1923 is the foundation of the modern republic of Turkey by Mustafa Kemal.

After Kemal declared the modern republic, he helped get what the Plaid Avenger refers to as the "Meiji Restoration of Turkey" rolling. What am I talking about? From 1924 to 1934, under the guidance of the main man Kemal, Turkey completely transformed itself, not unlike Japan, from a primarily agriculture-based economy into what is now a more modernized and industrialized country. He pulled it completely into the modern framework of things, using the European model. Politically, he revamped the entire government. The Ottoman Empire was a monarchy until that government was disbanded after World War I. Mustafa reinvented it, not even as a constitutional monarchy, but as a democratic republic. Period. He was elected President in short order and set about revamping the country.

Under his influence, Turkey established a parliamentary system and laws modeled after those of Europe. The Turks started building roads, opened schools introducing Western-style education across the entire country, and modernized their architecture. They built modern energy grids, telephone lines, communications. You name it, they did it. In ten short years, they really caught up. Now is Turkey a completely modernized and wealthy country like the rest of Europe? No, but it's pretty far down that path of modernization and wealth due to Mustafa and the movements that happened during his leadership almost a century ago.

What's the Deal with Ataturk?

The power, the prestige, and the honor accorded to Mustafa Kemal Ataturk cannot be undervalued in Turkey of 1923, nor of Turkey in 2015. He is the man. He is the main man. He is the George Washington of Turkey, the father of the country, a hero, a savior, a military and political genius. He transformed an entire country from epic fail to epic win in impressively short order. That's one of the reasons he was beloved. He was then, and he is today. He's the father of Turkey in the sense that he helped hold it together at its very formation, but even more so because he put it on the path of modernization.

Perhaps what sets Turkey apart from the rest of the Middle East the most is that Kemal made it a staunchly **secular** state. Staunchly! Kemal said, "*In this country, we're of Turkic ethnicity, we are of Islamic religion, but we will not be a religious state.*" I can't stress this enough. There is a separation of church and state at the outset. The government is the government, religion is religion, and the two will never mix. The United States may also have the separation of church and state, but it's more staunchly defended in Turkey than it is in the United States.

Kemal embraced the West even in style. What a snappy dresser!

Kemal went so far as to outlaw religious dress in public buildings, a law that still holds today. In fact, there was a court case in Turkey in 2008 in which an Islamic woman wanted to wear her headscarf. In Turkey, they said, "Wear headscarves! Wear a full burqa if you want! You're just not walking into a government building or university with it on. Period." You cannot wear religious garb into the Grand National Assembly, the Turkish equivalent of Congress. If you have a cross on a chain around your neck, that's got to come off as well. That is the ultimate separation of church and state. We're going to see how that has propelled Turkey along its path and made it extremely different from the Middle Eastern countries in its neighborhood as this story progresses. That is a huge deal. The Plaid Avenger can't stress it enough.

Even to this day the constitution recognizes the freedom of religion for individuals and the country has no official state religion. A political party cannot become involved in the political process if its main tenants are formed on the basis of political beliefs. Well. . . maybe. . .

All of this seems to be changing as current President Recep Tayyip Erdoğan and his conservative, and devout, AKP political party have begun to impose more of an Islamist agenda, trying to "raise a devout generation." Public education has seen a rise in imam hatip schools, or religious schools, where parents are forced to enroll their child if he or she is assigned to that state-run school. Liberal democracy fans have been rightfully worried about the rise of this

Erdoğan character for years, but given the current chaos in and around Turkey, he has had a much stronger hand to play in efforts to make the state more safe and more holy. . . more on that guy later though. But it is troubling times for Turkey as a secular state.

TODAY'S TURKEY

While Turkey was still doing its restoration—and getting along pretty well but not quite as far along as Western European nations—World War II broke out. Turkey stayed mostly neutral, but was actually chomping at the bit to fight: "Yeah, we want to get in! The Germans screwed us over during World War I, so we want to come help you beat them this time!" However, the Allies, including the United States, thought it would be more problematic than helpful for Turkey to join the war. The US, UK, and France said, "Okay, thanks, but just hold on, we don't really need you right now. Just sit tight and don't side up with the Germans. We'll make sure nothing bad happens to you." Most of the war played out like this, but in February 1945, when it was evident that the Germans had lost World War II, the Turks said, "We declare war on Germany too!" In a funny historical circumstance, they showed up at the victory party after the Germans were defeated. Since Turkey declared itself an ally, on the side of the good and righteous, they got to sit at the peace negotiating table and thus started to become a playa' in the world powers game.

Turkey is a proud NATO member!

They also joined the UN as a founding member in 1945. Something even more interesting is that in 1952, Turkey joined NATO, the *North Atlantic* Treaty Organization. If you had to pick one single event that really made Turkey different from all its neighbors, joining NATO was it. In one fell swoop, they embedded themselves firmly in the Western camp. From this point forward, on the books, Turkey is a US ally. This becomes of particular consequence as we get into current events because there are crucial NATO bases in Turkey still today.

Why is that important, Plaid Avenger? A lot of US operations into Iraq and Central Asia were made possible by bases in Turkey. When the United States fought the first Gulf War against Saddam Hussein, most of the planes carrying the bombs took off from Turkey. Joining NATO is the critical juncture, as I see it. Turkey said, "We're throwing in our lot, and we're throwing it in with the West. Yes, we're Islamic like our Eastern neighbors, but politically, we're more like our Western neighbors."

MAJOR MILITARY

Being a NATO member requires that you have a strong and updated military in order to be a true member of the club, and Turkey does fulfill its duties in every respect. But NATO is not the only reason that the Turks have a strong military tradition, as well as having strong nationalistic pride of their professional soldiers. There's so much more, and it still plays in today's events.

During Kemal's reign, he knew the only way to hold the state together and keep it from being dominated by the Europeans was to have an effective military deterrent. He therefore immediately modernized his military up to world standards. In addition, Kemal had other motivations for a strong centralized force:

Don't mess!

to ensure the separation of church and state in his staunchly secular new republic. Maintaining the secular state is one of the primary roles of the Turkish military, and one that has seen the most action.

In 1960, 1971, and 1980, the Turkish military staged coups and took over the government. From a Western perspective, we'd say, "Ooh, that's not very democratic. That doesn't look good." Most people would consider that a sign of an up-and-coming fascist military dictatorship. In Turkey's case, those people would be wrong. That is why Turkey is incredibly unique. The military has been strong since Mustafa "The Main Man" Kemal Ataturk, and it is seen as the staunch protector of secularism in Turkey. Every time there has been a military coup in Turkey, it has either been because the government has been excessively corrupt, or there's been an impending threat that religious folks are going to take over the government. In both cases, the military interjected and said, "Nope, our allegiance is to the state first." Therefore, you can't even really consider it a military dictatorship because there is no singular dictator in any of these circumstances.

The military said, "Our first role is to protect the state and the state's constitution, and that says we're going to be secular, and that's the way it is." In each of these coups, the military came in, wiped the slate clean, and said, "Okay, do it over." Once new elections are held, the military then get back out of the way. I'm telling you this story because the military in Turkey is still seen by the Turkish people in a largely positive light. Turkey's pro-military stance is one of the sticking point for Turkey's membership into the EU. The EU feels that Turkey's military is too strong, a little too excessive for European taste, which brings us to the next current conundrum for our Turkish friends.

EU IRRITATIONS

A big issue today is the attempted EU entry, which started over fifty years ago, but in real earnest for the last twenty years. In 1987, Turkey started seriously talking about entering the EU. They are a UN member, NATO member, right next door to Europe, and closing in on fully developed economic status. Twenty years ago, Turkey said, "Hey, the European Union looks pretty cool. We'd like to be a part of that. We're your NATO buddy; we've helped fight the Cold War with you. Let us in." This brings up a very important point. They were a US and European ally all during WWII and the Cold War. Turkey very rightfully says, "Hey, we're your boy; we're your ally. Haven't we been helping out for fifty years? Let us in the EU!" However, there has always been a little bit of foot dragging by Western European countries to allow that to happen. Like a foot stuck in the mud that hasn't moved for decades.

First, the EU told Turkey that their economy wasn't liberalized enough, so Turkey opened it up and diversified. Then Turkey was accused of human rights abuses against the Kurds, so they passed laws against discrimination and calmed the Kurd situation down for a decade. Then the EU said that their military was too strong, and that was problematic for a true democracy, so the Turks divested the military of some of its constitutional power. Now the EU is saying talks are stalled until Turkey allows unfettered port access to the Turkic parts of Cyprus, never mind that the Greek parts of Cyprus do not have to reciprocate this situation. Geez! Is this sounding like a runaround or what?

What's the real deal with this EU-inspired nonsense? I'm afraid it's outright cultural racism, for lack of a better phrase. A lot of Europeans are simply not comfortable having an Islamic member of their Christian country club. Remember back in the Western Europe chapter when I told you about the cultural friction developing in Europe? This is apparently part of the same story.

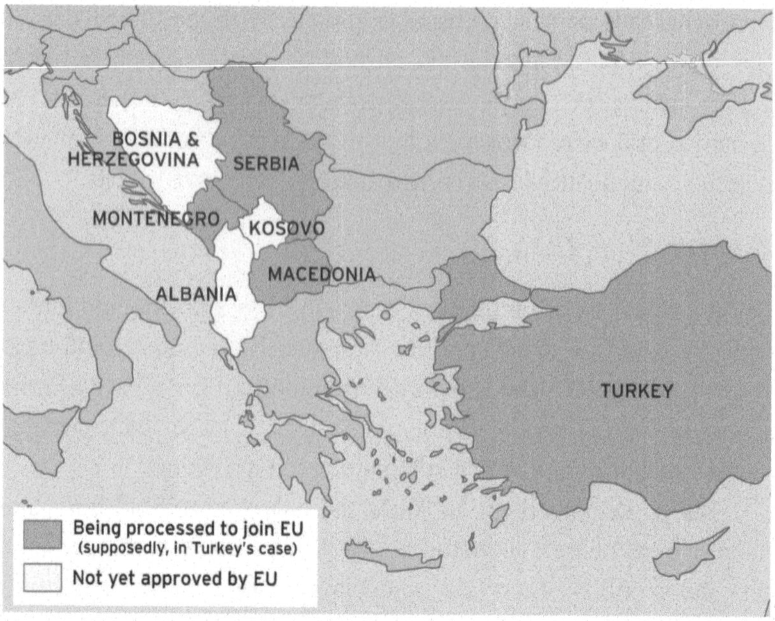

Turkey in the EU? Ah . . . promises, promises . . .

Plaid Avenger as Interpreter:

What the EU Says:
"We're kind of worried that Turkey's not rich enough, not democratic enough, that they have issues persecuting their Kurdish minority, that they don't have enough human rights, that they have some human rights abuses."

What the EU Really Means:
"We're really worried about you guys being Muslims!" The Western European take on this, in the Plaid Avenger's opinion, is that they are worried about admitting an Islamic country into their club. Turkey has a secular government, but it's nearly 100 percent Islamic, and all Westernized states are scared of the .001% of Muslims in the world that are radicalized. They are worried about people moving freely between an Islamic country and the rest of Europe: "If we let Turkey in the EU, then that means that anybody who gets into Turkey, especially radical extremist Muslims that want to blow up European stuff, can come from Turkey and get to anywhere they want in the EU." Valid point. I guess.

What Turkey Says:
"We're working on all that; we've made great strides. We have settled a lot of disputes with the Kurds. We've given them autonomy and given them some rights. We're doing what we can, and yes, we are staunchly secular and we're going to remain that way. We're proud of our military and we're going to try and keep it strong, but we're willing to concede."

What Turkey Really Means:
"Get off your high horse and let us into the Union. It's not the Christian Union, so quit jerking us around. We've been here helping you guys out as a staunch NATO member for like seventy years. Don't hold our country not being Christian against us. Please let us into the CU, I mean the EU." It's largely seen by Turkey as a Christian-Muslim issue, so they know the real deal. I agree with the Turks on this one.

The Turkish population is starting to get fed up with the entire mess. Public opinion used to be fairly high supporting their EU membership bid, but it has dropped steadily for several years now. Maybe 75% of the Turks were all for EU entry a decade ago, but it is more like less than 25% today. I can't say that I blame them, either. Turkey sat on the sidelines and watched as state after state in what used to be Iron Curtain territory get accepted into the EU, while the Turks continued to get put off, and insulted.

Most recently, former President Nicolas Sarkozy of France suggested that Turkey should become the foundation cornerstone of a **Mediterranean Union**, comprised of states from North Africa, southern Europe, and the Middle East. In other words, Turkey should just start a different club since they are not going to get into the EU club. I think you can probably figure out how that idea was received in Turkey. But honestly, given that the EU has been stagnant economically for years now, maybe the MU is not such a bad idea after all? **2021 UPDATE:** Given the BREXIT and populism and migration troubles that are plaguing the EU, it now appears that Turkey may not really be that keen on joining anymore. In their opinion, why jump on a sinking ship?

Nick was all for a Turkish Club Med.

Even more important: Turkey is now playing a pivotal role with helping solve some of Europe's problems, so they have tremendous amounts of strategic leverage over their neighbor right now. How so? Well, Turkey is now hosting 1.5 million refugees from the Syrian Civil War: the most of anyone. And Turkey has been a conduit through which millions of others have fled to Europe. . . so Europe has been begging Turkey to stop that flow in exchange for cash, and for serious EU consideration. If the EU ticks off Turkey now, the Turks can simply open the floodgates on the refugees and ship them next door. . . a thought which now terrifies the EU and gives Turkey quite a bit of negotiating power. Let's see where that goes. . .

KURDISH KONFLICT

To continue on with our current events in Turkey, I've just referred to one of the previous sticking points for EU entry: human rights, particularly for folks in eastern Turkey. Who are we talking about here? The Kurds. Although Turkey is called "Land of the Turks," the Turkic people are not the only ethnic group residing in this region. Other groups include but are not limited to Arabs, Greeks, Uygurs, Azerbaijanis, and everyone's favorite: the Kurds. Well, in this case, maybe not everyone's favorite. This place is a hotbed of ethnic conflict and discrimination. The Kurds are the second largest ethnic group in Turkey with an estimated population of over 20 million people out of almost 75 million, which is a significant minority to be sure. You may remember my reference to the Kurds in the Middle East chapter as a nation that has no state. They are a stateless nation of people. Now, they are a people who have been promised a state several different times in the past (by Europeans, Americans, etc.), but they never got it. Their ethnolinguistic group is located largely in the mountains of eastern Turkey, in Iran, in northern Iraq, and in Syria.

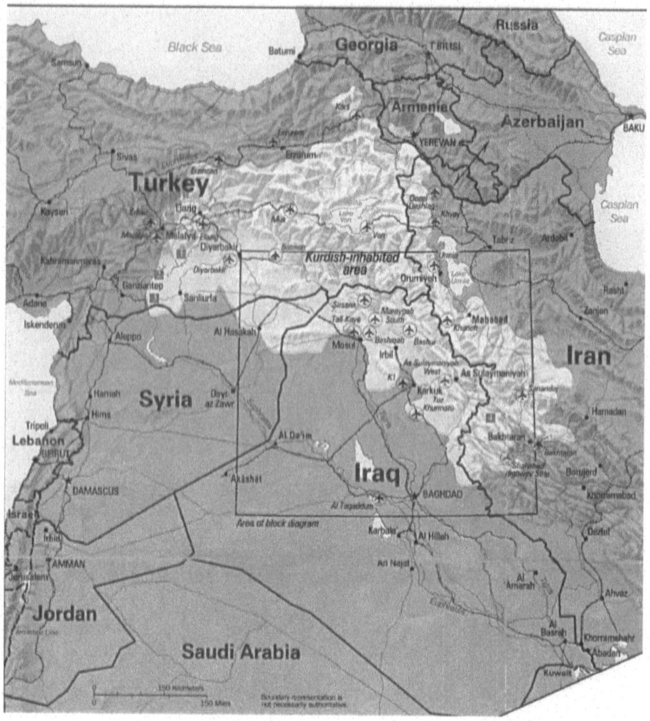

Kurdistan: a nation without a state.

Turkey's track record with the Kurds is somewhat sketchy, and here's where I'll have to upset some of my Turkish friends in order to give you the full picture. While they are a modern state and there's a lot going on that's really good, the long track record with the Kurds is somewhat negative in that they have not been given equal rights. There's been an active campaign of discrimination—it may sound a little strong but it's not too far off—over the last several decades. There have been so many human rights violations that a countermovement has arisen within **Kurdistan**, as the Kurdish areas are sometimes known. The **Kurdistan Workers' Party**, or **PKK**, is a radical group that has spearheaded this countermovement, unfortunately by using violence.

This is an extremist group that is actually considered a terrorist organization by Turkey, the US and the EU. They have undertaken a campaign since the 1970s of blowing up Turkish stuff and Turkish people in order to fight for their rights, with an ultimate goal of gaining independent territory and national autonomy. This has been a main sticking point for Turkey, as I've already referred to earlier, in that the European Union says, "You guys are picking on those Kurds too much. That's not good." At the same time, Turkey says, "Hey, wait a minute! These guys are terrorists, just like you guys are dealing with terrorists!"

This was a big stain on the Turks' reputation for 50 years, but it actually has settled out quite a bit in the last decade. The Turkish military captured Abdullah Öcalan, the head of the PKK, in 1999 and made plans for his execution, but in an obvious show that they were trying to be conciliatory to the EU, they said, "Okay, we won't execute him. We'll keep him alive in jail and, even though he's a terrorist, we're not going to kill him." He's still in a Turkish jail, and if you know anything about Turkish prisons, you understand that Öcalan is probably begging to be put to death.

Kurdish refugees.

Since Öcalan's arrest, the conflict died down, the number of terrorist attacks diminished, and the Turkish parliament passed lots of laws to try to equalize human rights and give these guys a little more power. Things in Turkey had been getting much better for the Kurds, but then the 2003 American invasion of Iraq stirred the pot again. The PKK regrouped in northern Iraq and has carried out several attacks on Turkish soil in the last several years. This led to renewed trouble for everyone, and remains a persistence problem even today.

Remember I told you that Turkey has a strong and proud military? Well, the Turks were none too happy about those most recent PKK attacks, and decided to act. In 2008, much to the displeasure of the United States, Turkey conducted several bombing raids and mobilized ground troops to go into Iraq and destroy the PKK forces. The US is not happy, mostly because the northern part of Iraq is the only peaceful part of the country, and no one wants to see more people with guns getting into this mess. But those Turks are not sitting still for the fight, and future movements are probably inevitable. Keep an eye on that one, my friends.

2019 UPDATE: Holy understatement of the century, my friends! Keep an eye on it indeed! Ever since the Syrian Civil War flared up in 2011 and ISIS appeared on the scene in 2014, the Kurds have become an active and successful fighting force on two fronts: helping destroy ISIS in Iraq and helping rebels fight against the al-Assad government in Syria. Turkey actually shares both of those goals, so all is well, right?

NOT!!!! In a bizarre twist of fate that you could not make up if you tried, Turkey is so terrified of the Kurds becoming so powerful that they form their own state, that Turkey has mostly been sitting on the sidelines and not helping! They eventually got pulled into the fight in 2016, but have not just gone after al-Assad and ISIS, but regularly target Kurdish forces as well! Yes, that's right! The Turks are also bombing the peeps that are helping to defeat Turkish enemies! It makes your brain hurt to think about it too hard, so don't.

Here in 2021, relations between Turks and Kurds are back down to an all-time low. The radical Kurds in the PKK are back at it again, countering the Turks' military attacks against Kurdish targets with terrorist attacks in Turkey. Both sides are ramping up the violence against each other, and the Kurdish civilians in Turkey are losing civil rights and liberties at a blistering pace. The whole Turkish state is on high alert security-wise, and is borderline a fully militarized society, fighting active wars in foreign countries as well as within their own. Good times.

2021 KURD-CLUBBING UPDATE: To bring this Turk-Kurd conflict into current events, in 2019 Turkey fully invaded northeastern Syria to beat up on the Kurds that were beating up on ISIS, in an invasion code-named Operation Peace Spring. June-August of 2020 and starting again in May 2021, Turkey has conducted multiple drone/air bombing campaigns in the Kurdistan region of Iraq. Suffice to say, Turkey has ramped up its attacks on the Kurds, and now cares little about crossing sovereign state boundaries to do it. Turkey claims it is targeting PKK cells; locals and other governments claim it is strategic genocide. What is your opinion?

TURKS TRANSITIONED

Another big reason that Turkey deserves to be teased out as an independent region from its Middle Eastern and Eastern European neighbors is the situation of its people and economy.

As you can see from the population pyramid here, Turkey, even just thirty years ago, was considered "developing" because of its explosive population. But look how much has changed since then. Their pyramid has basically rounded out, tapered off, and as you can project ahead by 2025, will become completely stable and even slightly shrinking. It would be exactly what we would consider a fully developed nation, demographically. Its population total is pretty stabilized, but currently maintains slight growth every year, which is the gold standard for the labor force.

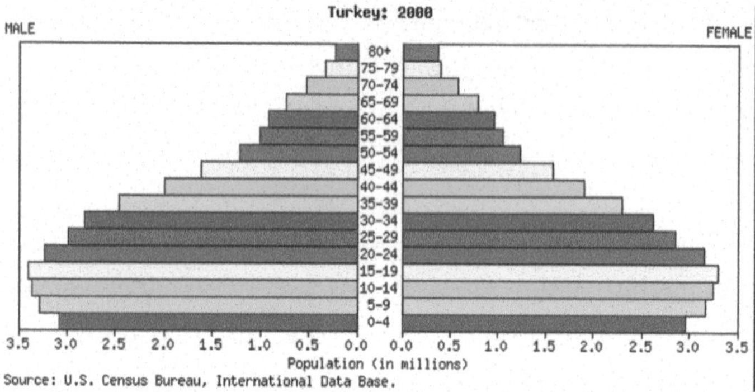

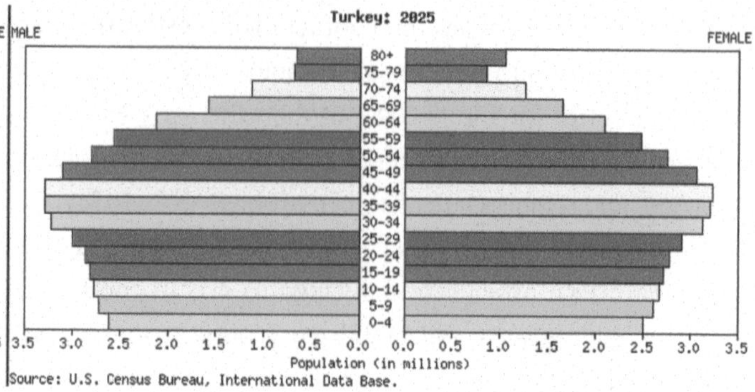

And the cash? Economically speaking, they're not the richest place in the world. However, they're doing pretty well, and are a top 20 global economy at this point. A G-20 member, my friends! They have a diversified economy—10 percent of an agricultural sector, and 30 or 40 percent of what we'd call an industrialized sector, and almost 60 percent service sector . . . and that's the economic sector structure of a rich, developed country! Or close to it! It's yet another example that they are quite different from all the places around them, just for their modern situation if nothing else. And while they may never get into the EU, at this point they are performing better than every single EU country, with an average 5–8% growth rate every year. Suck it, EU! With no reservations, I say that from just about any measure you want to examine, Turkey will certainly be considered a fully developed country within the next decade or two. The are already a regional power player, and possible contender to be invited to join the BRIC club. Remember that from chapter 6?

Let's end this modern assessment with the strife over secularism. While you now understand that the secular nature of Turkey is what sets it apart from its Middle East neighbors, I should warn you that this status is still heavily debated in their society. The country is almost exactly split on the extreme nature of their secularism. The more western, urban populations favor the staunch separation of Islam from politics, but the rural populations are much more conservative in thought and perhaps want to change things to incorporate religion into the state a little more.

You know what? It's about exactly how the US is right now—think about it. Red states with big rural populations vote Republican, while blue states with big urban centers generally vote Democrat, an almost even split in the last several elections. Republicans would vote for school prayer, Democrats for getting rid of a nativity scene in a government building. That about exactly outlines the Turkish situation as well, only with Islam instead of Christianity. What a great analogy; am I good or what?

Prime Minister Recep Erdoğan also accused of not being secular enough.

Back to Turkey: the current president and prime minister are from a political party that is more conservative and religious (The Justice and Development Party, or AK Party for short.). Many people from the other big political party are staunch secularists, and distrust the President greatly, fearing that his party is changing laws and possibly amending the constitution to allow for more religious stuff in the government. And they are probably correct, as the current President of Turkey is changing the country in deep and profound ways. In fact, he is such an important cat that I think I should talk about that . . .

THE TURKISH TITAN: ERDOĞAN

Now I don't ordinarily do a bio on an individual leader of a country since this is a geography textbook, and one focused on entire regions at that. But much like Vladimir Putin is to Russia, the current leader of Turkey is much more than just a politician who is currently heading the state. Current Turkish President Recep Tayyip Erdoğan is not just critical to understanding the current status of Turkey, he now *IS* Turkey. He has become so powerful as to

become the state itself, and every decision that shapes their domestic and foreign policy is now straight from the brain of this one dude. That's why it's important to spend a page or two on him. So what's the deal with Erdoğan?

To understand how much he is impacting the path of this country, we actually need to step all the way back in his bio to better understand his story, which is fast becoming the story of Turkey itself. Born in February 1954 (that's post-WW2) in Istanbul, his father was a member of the Turkish Coast Guard, and Erdoğan was brought up in a very observant Muslim family and received his high school diploma from an Islamic school. (*Islamic schooling is important to the story. Take note.*) He studied business administration and economics in college, where he joined the National Turkish Student Union and became politically active by age 18, and was a semi-professional football player from 1969 to 1982. Already, I like a guy who can speak on a political platform during a rally, and then simultaneously beat the hell out of you on the field.

By the 1980s, Erdoğan had joined the Islamist Welfare Party—a political party dangerously close to not being secular. The Welfare Party, founded by Necmettin Erkbakan in 1983, gained ground in the 1995 national legislative election when it won the majority. This was the first time a pro-Islamist group headed the government since the Ataturk reforms. Recep was appointed Mayor of Istanbul in 1994 under Erkbakan. The Welfare Party became the largest political party with Erbakan as Prime Minister until it was banned from Turkish politics due to its Islamist agenda. That's not surprising since. . . "Hello! Turkey's secular-based constitution!" Erdoğan was even stripped of his office and was sent to prison for four months after giving a pro-Islamic speech. Starting to see a trend here?

In 2002, good ole' Recep was back at it again: incensed by his prison experience, he established the Justice and Development Party (AK Party or AKP), a democratic Islamist party, with the help of a dude named Abdullah Gul. The general elections of 2002 resulted with the AK Party winning two-thirds of the seats in parliament. . . meaning they were large and in charge of the government, a situation which is still true up to today. Because Erdoğan was banned from holding public office due to his prior criminal conviction, Gul became Prime Minister and allayed fears by adopting a very pro-western and pro-American political platform. After the AKP made some specific constitutional amendments, Erdoğan was allowed to be an elected official again, and he promptly won a seat at the table. Gul then stepped down as PM and made way for. . .

PRIME MINISTER ERDOĞAN

He took the top slot in 2003, and held it continuously until 2014. And Recep's first decade of rule went really well! I mean it looked like this guy was really trying to do some good and be a democratic model for the Middle East. With the help of the economic reforms put forth by the previous Minister of Economic Affairs, Turkey paid off its IMF debt. The economy grew by 5-8% a year on average. . . that's phenomenal. He sought to end the Kurdish-Turkey conflict by giving the minority group more rights. He improved infrastructure by installing the countries first high speed railway system and beginning construction of the Marmaray tunnel, an underwater railway tunnel extending across the Bosporus strait. He improved education by giving it a higher share of the national budget than the military and even partnered with UNICEF to inspire basic education for girls. All of these human rights reforms were an attempt to appease the European Union in the hopes of eventually starting full negotiations for Turkey to become a member, further separating them from their Middle East counterparts. Things seemed to be going so well. . .

Then Turkey turned the corner. And Recep got real about never relinquishing his rule. Nationwide protests against the perceived authoritarianism of Erdoğan's government began in May 2013, with the internationally criticized police crackdown resulting in 22 deaths and the stalling of EU membership negotiations. Following a split with long-time ally Fethullah Gülen (the guy living in self-imposed exile in Pennsylvania who Recep wants Donald Trump to send back to Turkey), Erdoğan brought about large-scale judicial reforms that were criticized for threatening judicial independence, which we all kind of recognize is now gone. . . meaning Recep controls the court system: he regularly personally overturns court orders. A US$100 billion government corruption scandal in 2013 led to the arrests of Erdoğan's close allies, with Erdoğan himself incriminated after a recording was released on social media. His government has since come under fire for alleged human rights violations and crackdown on press and social media, having blocked access to Twitter, Facebook, and YouTube on numerous occasions. In 2013, he arrested a 13-year boy for insulting him via Facebook. In 2014, parliament passed a bill allowing the government to block internet sites. In 2016, he just decided to ban social media overall. You might rightly assume that this democratically elected leader was really taking on more and more. . .

AUTHORITARIAN TENDENCIES

And I would agree. In a bold move to empower his position even further, Erdoğan stepped down as Prime Minister to be directly elected as President August 2014 and became the 12th president of Turkey. But in Turkey, the Prime Minister technically is the top slot, with the Presidency being more of a figurehead domestic leader. Not anymore! As soon as he took his new office, Recep introduced legislation to make the President more powerful than ever before, and even more powerful than the Prime Minister. After assuming the role of President, Erdoğan was criticized for openly stating that he would not maintain presidential neutrality: he openly supports his AKP party, and works to undermine all other political parties and dissenting options. Politicians, military officers, teachers, news reporters, and others are targeted for attack if they oppose the official Erdoğan policies.

Another blow to Turkish democracy was initiated in the summer of 2016 when a faction of his military staged a coup d'état to remove him from office. What was their reasoning behind the coup, you ask? Secularism, human rights, and democratic rule have gone and left the building under Erdoğan's rule. . . and many in the military would say it is their constitutional duty to restore that secularism. But the military has been significantly weakened during Recep's rule. The coup failed and afterwards there was a nationwide government purge that arrested military personnel, civil servants, and private business owners. Some 2,700 judges were dismissed and detained over their alleged connection to the coup. The rest of the world has been watching, as Turkey-US relations have begun to sour and Erdoğan seeks to make new alliances. It's either Recep's way or the highway now. Well, at least things can't get any worse.

AND THEN THINGS GOT WORSE

The Syrian Civil War (2011-now) just next door has brought millions of refugees into Turkey, and Turkey is now militarily engaged in the fighting since 2016. ISIS joined the Syrian Civil War fun, and routinely targets Turkey for terrorist attacks. As previously mentioned, relations with the Kurds are at an all-time low, with radical Kurds in the PKK regularly targeting Turkey for terrorist attacks. Major attacks have occurred in 2015-16, spurring a further clampdown of security and retribution against Kurdish populations. Then the 2016 coup attempt. Add all this chaos up, and you have the perfect scenario for Recep Tayyip Erdoğan to further consolidate power, destroy all opposition, and transform the country any way he sees fit. . . all under the guise of "establishing order" and "protecting the people."

Increased control of press, repression of political opposition, repression of western influence, promotion of Islamic education to make a more conservative "pure" society, manipulating the constitution to increase power of the President, and using the terror attacks as well as the lame coup attempt to fire up "national pride" all combine to change Erdoğan's Turkey from a liberal secular democracy into an authoritarian one-party state. . . hey! much like Russia!!!! Yep. I will end with that: for the first time in modern history, Turkey seems to be cozying up with fellow authoritarian Russia (historically, they are bitter rivals), perhaps turning its back on secular liberal democracy and Team West for good!

2019 UPDATE: We really need to watch Erdoğan's Turkey. . . and yes at this point it really is *his* Turkey. . . because I believe they are on a historic pivot point now. When populist Donald Trump assumed the Presidency in the USA in 2017, I predicted it was possible that those two strongmen will like each other so much that US/Turkish relations would be restored to full strength and thus Turkey would stay in Team West's sphere and maybe get into the EU. The other possibility was that the Recep/Trump thing does not go well, and Erdoğan decides to pivot to his fellow strongman in Russia and follow an uncharted course that history has yet to witness between those two countries. The latter was the case. Turkey is warming relations with Russia like never before, although it is still early in the dating process to know if marriage is imminent. In either case, the man is taking his country into new territory in terms of becoming less secular, less democratic, less liberal, less western, and yet somehow more of a force than ever in influencing the future of Syria, Iraq, the Kurds, NATO, the EU, and the Middle East in general.

FINISH UP YOUR PLATE OF TURKEY

If Turkey is ever admitted to the EU, which still seems possible at this point, their economy will grow. They will become even more incorporated as a member of Europe. They will perhaps become even more incorporated as a Western ally in this new war on terrorism in the Middle East. They will be distancing themselves from all of the radical elements that people in the West are frightened of in the Middle East, which will be to their great benefit. The Plaid Avenger is not making fun of anybody in the Middle East. I'm just saying what's going to happen to Turkey if they get into the EU.

If they *don't* get into the EU, things are going to swing the exact opposite way, and that's what's so fascinating about Turkey's position in the world today. They really are in the catbird seat, as the Plaid Avenger has mentioned several times now. If they go the EU route, they will become more Western, there's no doubt. If the EU turns their back on Turkey, Turkey is independent enough and economically strong enough, and it's certainly militarily strong and competent enough that it will say, "Fine. The heck with the CU (Christian Union). We'll C U later!" They will turn right around and become a leader within the Middle East. Or the world in general.

Unlike most states in this neighborhood, national pride abounds.

That does not mean Turkey's government will become an Islamic Republic, but it will turn around and become a regional leader in the opposite direction. If it turns its back on Europe and embraces the Middle East, it would become, instantaneously, one of the wealthiest countries in the region. It would easily be one of the most diversified economic countries, and probably the one with the most ties to different parts of the world—and one that is not reliant on a single commodity for its wealth, like so many countries nearby are dependent on oil. That gives Turkey a lot more power. **The bridge** of Turkey is actually strengthening. The Turks are gaining more power in today's world because they are becoming what the Plaid Avenger calls a *pivotal historic player.*

THE PLAYER

How are they a player? Turkey is a player because they've got two different directions they can go. Just like we've talked about with Russia already—they can go China, or they can go Europe. Turkey can go Europe, or Turkey can go Middle East. And they already have aspirations to link up with Central Asian states too, as they share common Turk roots. A possible TU? A Turkic Union? Or a Mediterranean Union? Or both? It's really up to them.

But they are still firmly entrenched as part of Team West. They're a strong NATO member, but they actually refused to help the US movement in Iraq. This actually came as a great surprise to the United States. The US said, "Hey, NATO member since 1952! Turkey, you're our boy! Sweet! We're going to launch our Iraq invasion from your soil!" Turkey said, "Not gonna happen. We don't like Saddam either, but those are Islamic people about to get bombed. We don't support this war. You're not going to bomb people from our soil." What was the direct result of this? The challenge of, dare I say, *smacking down* the world's military and economic superpower's request? What were the repercussions? What happened? Jack nothing is what happened.

Turkey is *that* strong now; it is a pivotal historic player on the world stage, and we know that because nothing happened to them. The US even threatened to cut off international aid, saying, "You're going to do what we want or we're not going to give you money anymore!" and Turkey said, "Fine. Don't." The threat of pulling aid was all smoke and mirrors; the US was powerless and totally knew it. In fact, the exact same scenario played out again when Turkey sent troops into northern Iraq to get the PKK in 2008. The US begged them not to do it, but Turkey went ahead anyway; the US was forced to support their old ally, even though they weren't happy about it.

Evidence of Turkey's increasing independence and regional leadership abilities continue to grow here in 2019 (despite an electoral setback wherein Erdoğan's party lost the election for Mayor of Istanbul, a very important and coveted position). Not only has Turkish public opinion about EU membership waned, but increasingly they could care less, and moves have been made to secure economic relationships with more Middle Eastern AND Central Asian countries, and the idea of a Mediterranean Union is also starting to look more promising. Why? Well, mostly because the Western European economies are all still stalled or growing very slowly right now thanks to bad management and the 2008 recession; many think it's a one-two punch that Europe is not likely to recover from as fast as the rest of the world will. Why should Turkey throw in their lot with a geriatric basket-case at this point? A lot of Turks think the EU option is not smelling so sweet.

Erdoğan hosting Abbas: Recep as a Middle Eastern hero in the making? Or just yet another Middle Eastern authoritarian rising?

Politically, Turkey is increasingly shaping up to be a major player in Middle Eastern affairs, perhaps the only party with enough street cred to pull it off. Turkey is stable, a democracy, Muslim, moderate, and part of the neighborhood . . . but with no inherent conflicts of interest in the Middle East messes. That makes them quite unique, and uniquely positioned to help. Under Erdoğan's leadership, Turkey has now become the most assertive country to try and stop the violence in Syria, by condemning the Syrian regime and offering safe haven for fleeing citizens. On top of that, Erdoğan is seen as something of a hero during this time of the 'Arab Spring' as many Middle Easterners view him as an assertive, nationalistic, popularly elected leader . . . which is what all the Egyptians, Tunisians, and Syrians want for their own countries too!

Add to that a strong sense of nationalistic pride, a strong and independent military, a strong and functioning democracy, and strong political leaders, and you have one heck of a world region. We'll have to watch them quite closely to see in which other regions they project their power into. It could be one, or it could be several, because the Turkish position continues to strengthen, just as all of it's neighbors are shrinking. I consider Turkey already to be a second-tier world power; that is, a major regional power on the world stage, just underneath the big boys like the US, China, et al.

ANKARA RE-ALIGNMENT? 2019 UPDATE: On July 12, 2019 Turkey received its first deliveries of a Russian-produced missile-defense system . . . which infuriated the USA and other NATO allies!!! Turkey's S-400 missile defense deal with Russia is a tricky business for the state, since as a NATO member, it is now playing both sides of the fence! Turkey has typically gotten all of its military hardware (particularly F-35 fighter jets, which it helps produce) from the United States. So now it is getting fighter jets from the US and missile defense systems to repel these same fighter jets from Russia! Talk about a conflict of interest! And it is causing conflict!

The US is threatening sanctions and threatening to suspend all future F-35 warplane sales, and NATO is now talking seriously about developing protocol on how to kick out one of its members . . . since it has never happened before, this is all new ground. The conflict stems from the belief that Turkey working directly with Russia on such specific military hardware undermines NATO's military capabilities to defend the group in case of war. That seems reasonable. Turkey says that it is a sovereign state and can work with whomever it wants to beef up their own military. That seems reasonable, too. This could be the parting of ways between the West and Turkey . . . kind of a really big deal.

And this issue is not the only thing driving these allies apart: Turkey is already angry at the US over Washington's support for the Kurds in Syria, and the EU/US and others have been increasingly worried about the authoritarian road the Turkish leadership has been going down for a decade . . . putting it at odds with the liberal values of Team West. So we could be witnessing the strategic break between these two world players . . . and Vladimir Putin is likely very pleased!

2021 AUDACIOUS UPDATE: ANTAGONISTIC ERDOGAN 2.0 Watch out! Turkey has taken a seriously brazen and bold path on foreign policy, just in the last couple of years! Under the increasingly authoritarian Recep Erdoğan, the Turkish military is now an all-out adventurous force, dabbling in conflicts in multiple other countries, territories, and even at sea! Here is just the briefest of beefcake moves in the last two years:

- As previously mentioned, in 2019 Turkey fully invaded northeastern Syria to beat up on the Kurds that were beating up on ISIS, in an invasion code-named Operation Peace Spring.
- Also previously mentioned, June-August of 2020 and starting again in May 2021, Turkey has conducted multiple bombing campaigns in the Kurdistan region of Iraq. Suffice to say, Turkey has ramped up its attacks on the Kurds, and now cares little about crossing sovereign state boundaries to do it.
- Turkey supplied weapons, and likely militants, to Azerbaijan in the 2020 Nagorno-Karabakh conflict that pitted ethnic Armenians against the Turkic Azerbaijanis. Turkey and Azerbaijan have strong economic, military, cultural and linguistic ties and Erdoğan has publicly stated that the countries are "one nation, two states."
- October 2020: Erdoğan confirms purchase and use of Russian weaponry. Then he directly dared former US President Trump and all of NATO to impose sanctions on him for it. That same month, Erdoğan publicly said that French President Emmanuel Macron was mentally unstable and needed a medical exam due to Macron's move to crack down on radical Islamic extremism in his country after a religious zealot beheaded a teacher.
- January 2020: Turkey sent troops into Libya to back the UN-recognized government in the capital, after forces loyal to Haftar, a rival administration, launched an offensive. Those troops are still there. This was likely done to reinforce and solidify a Maritime Boundary Treaty that was recently signed between the two countries, to establish an exclusive economic zone in the Mediterranean Sea . . . part of Turkey's expansionary naval policy described below.
- Since late 2019, Turkey has espoused a maritime policy dubbed "Blue Homeland Doctrine" which stakes Turkish claims of large swaths of the Mediterranean Sea and the Black Sea as well. Fueled by the discovery of natural gas deposits, the "Aegean Dispute" now sees Turkey at odds with Greece, Cyprus, Israel, Egypt, France (the EU in general), and the UAE. Gas exploration vessels, scientific vessels, and military vessels from Turkey now routinely piss off ships from other countries while intruding into disputed waters. (More on this in Chapter 26: Hotspots of Conflict.)
- In 2021: America imposes sanctions on Turkey over their purchase of the S-400 antiaircraft system from Russia. The Biden administration formally recognized the 1915 attacks on Ottoman Armenians as genocide. Wow. An about-face is happening between both countries right now.

Just in the last 5 years, Turkey has gone from zero problems abroad to zero friends abroad. It is not yet clear what developments mean for the future of Turkish relations with its neighbors, let alone for Ankara's ties with NATO, the USA, the EU, or Russia. Erdoğan's Turkey is most certainly on a pivot point now, and as it becomes more self-confident as a major regional/world power player, it seems less worried about its historic arch-rival Russia and therefore less dependent upon its historic ally the USA. Turkey is reassessing its role with Team West in general, and currently seems determined to expand its influence abroad in the Middle East, the Mediterranean Sea, and perhaps further afield into Central Asia. Can it actually pull any of this off? Time will tell!

TURKEY RUNDOWN & RESOURCES

View additional Plaid Avenger resources for this region at http://plaid.at/turk

BIG PLUSES

- 100% Muslim, secular, fully functioning democracy: the only one of its kind
- NATO member and huge political/military ally of the US
- Strong and modern military in its own right
- Diversified economy that is growing
- Slow-growing and therefore stable population
- Modernized state that is fast becoming a most significant regional power; I predict it will soon join the BRICS. It is already in the G-20
- Muslim culture AND pro-Western relationships make Turkey the most awesome mediator of conflict for Middle Eastern affairs; trusted ally of the US, Israel, Syria, Palestine, Iran . . . . heck, that's everyone!

BIG PROBLEMS

- Due to its "Muslim-ness" and current cultural/economic attitude of Europe they are not likely to get into the EU
- While gains have been made in the last decade, the Kurdish situation and problems with the PKK still remain and could get hot again
- The secular vs. religious "crisis of character" that the Turkish state is dealing with is far from over, is very divisive to the general population, and may result in political gridlock of epic proportions at some point; I sure hope it doesn't come to that, but it might. **2014 UPDATE:** Totally called this one years ago, as Erdoğan and his administration are now battling non-stop protest from the secularists, which is widening the gulf between the two groups in Turkey.

DEAD DUDES OF NOTE:

Mustafa Keaml aka Atatürk: "The George Washington of Turkey." Turkish army officer, revolutionary statesman, writer, and founder of the Republic of Turkey as well as its first president. Staved off attacks from the Allied powers and held together the core of the Turkish homeland after the defeat and collapse of the Ottoman Empire in WWI. Once in office, embarked upon a program of political, economic, and cultural reforms and entrenched Turkey as a modern, secular state.

LIVE LEADERS YOU SHOULD KNOW:

Recep Erdoğan: Popular President (and former Prime Minister) of Turkey and chairman of the Justice and Development Party (AK Party). Despite internal friction due to his association with the religious-based AK Party, he has overseen an era of economic prosperity and democratization, and Turkey's rise as a serious regional power and trusted mediator of Middle Eastern conflicts.

PLAID CINEMA SELECTION:

Hokkabaz aka *The Magician* (2006) A Turkish comedy-drama that is also a road-trip movie. A magician/juggler tours around Turkey with his father and best friend so that he can make enough money for laser eye surgery. Ha!

Takva aka *A Man's Fear of God* (2006) Drama following a devout Muslim man's internal search for the right path. In the film, the man works at a seminary that exposes him to the temptations of the modern world, and causes him to have a major crisis of faith. Movie has garnered many awards, in large part for one of the best and most fitting conclusions to a film this century.

Eskiya aka *The Bandit* (1996) After serving a 35-year jail sentence, Baran, a bandit, is released from prison in a city in Eastern Turkey and heads to his ancestral home . . . which is now underwater. He then heads to Istanbul to enact revenge on the rat that put him in prison and manic adventures ensue.

Uzak aka *Distant* (2002) Story of Yusuf, a young uneducated factory worker from western Turkey who loses his job and travels to Istanbul to stay with his wealthy and intellectual photographer cousin. Aimlessness and loneliness ensue.

Propaganda (1999) Based on a true story set in 1948, customs officer Mehti is faced with the duty of formally setting up the border between Turkey and Syria, dividing his hometown. He is unaware of the pain that will eminently unfold, as families, languages, cultures and lovers are both ripped apart and clash head on in a village once united.

Yol (1982) Winner of the Cannes film festival highest honor, this film has been banned in Turkey until 1999. The film paints an accurate picture of Turkey after the 1980 Turkish coup d'état through the eyes of five prisoners given a week's home leave.

Günese yolculuk (1999) Mehmet, a young Turkish man newly migrated from the village Tire, takes a job searching for water leaks below the surface of the streets of Istanbul. Due to a strange set of events, he is mistaken for a Kurd, imprisoned, and brutally beaten. Great film depicting the Kurdish tension within modern Turkey.

Vizontele (2001) The story of a small town called Hakkari in Turkey at the beginning of the 1970's that gets its first television. Chaos and hilarity ensue.

Crossing the Bridge: The Sound of Istanbul (2005) A documentary that takes an in depth look at the music scene in modern Istabul. An interesting look at a variety of musical genres while simultaneously providing interesting commentary on the culture of Istanbul. Lovers of music will thoroughly enjoy this film.

MAP GALLERY

View Map Gallery & additional Plaid Avenger resources for this region at http://plaid.at/turk

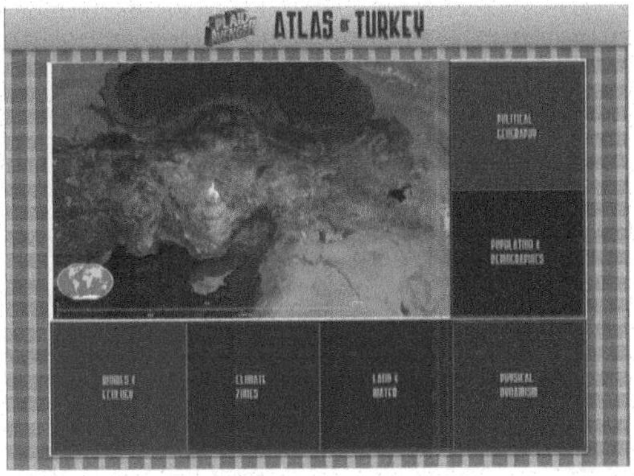

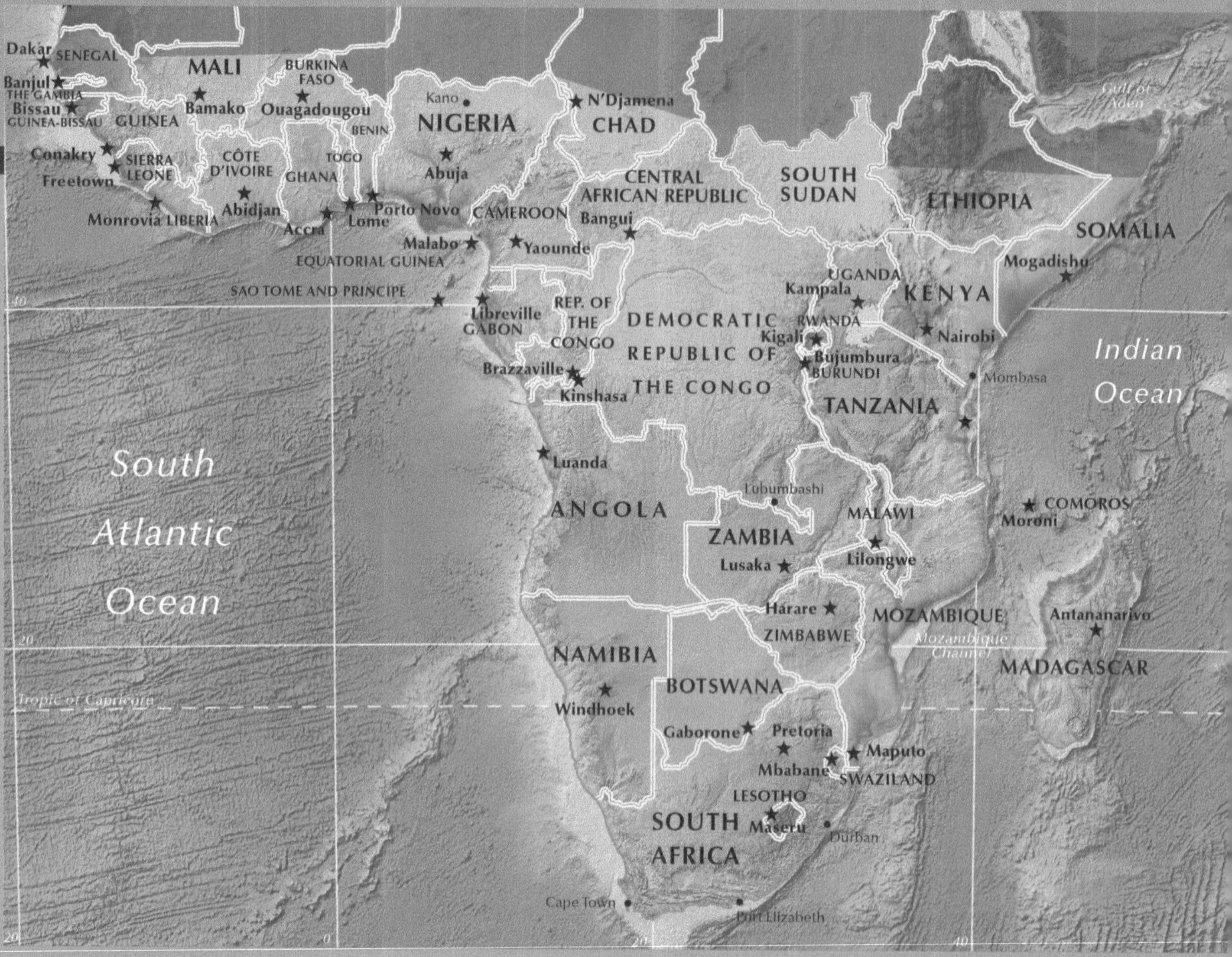

CHAPTER OUTLINE

Sub-Saharan Africa

20

LET'S move south from the Middle East into Sub-Saharan Africa, a place the Plaid Avenger refers to as the most challenged region on the planet. For so many hundreds of millions of people here, life is brutish, hard, and short. By any measure, any standard to which we compare human life on this planet, the states of Sub-Saharan Africa fall to the bottom of the list. What am I talking about? Things like life expectancy—in some Sub-Saharan African states, the average human has 29, 30, or 31 years until they die. Also, Sub-Saharan Africa has some of the lowest literacy rates in the world, some of the highest infant mortality rates in the world, and pretty much the lowest GDP figures you can find anywhere on the planet. Many people are living on less than a dollar a day here. By all measures—and not just those defined by the West—Sub-Saharan Africa seems to fall in last place in the world race at this juncture in history. Why is that? Why is this the face of Africa? Let's find out.

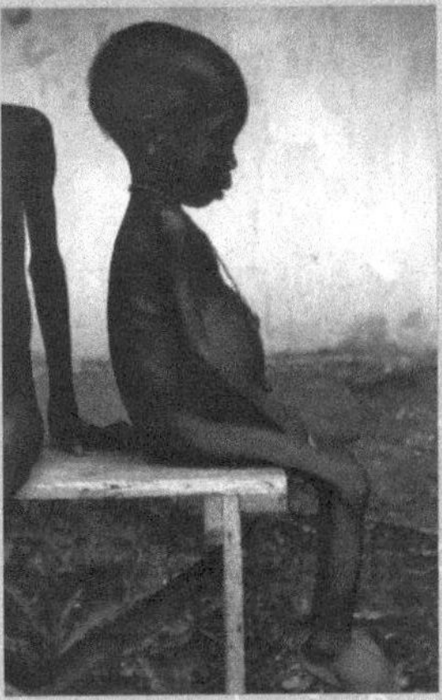

Why this?

GIRAFFE HAIR

Africa is the second largest continent on the planet, bested only by the monster continent of Asia. It is a really big place that has a lot going on. Any place that has a lot of size comes with a lot of stuff. Not only do you have really large tracts of land, those tracts of land have huge mineral resources of all different varieties: crazy things like uranium, which is hardly used on the continent. Africa also has lots of diamonds—as you probably already know—and all kinds of other mineral resources like copper, manganese, and silicon. There are some tremendous patches of oil, though not widespread. Overall, there are a whole lot of resources. But wait a minute! This doesn't explain why Sub-Saharan Africa is poor. Hang on! We'll get to the poor part in a bit. This is a really huge place, so huge, in fact, that a lot of the regions we have talked about would fit easily inside the continent.

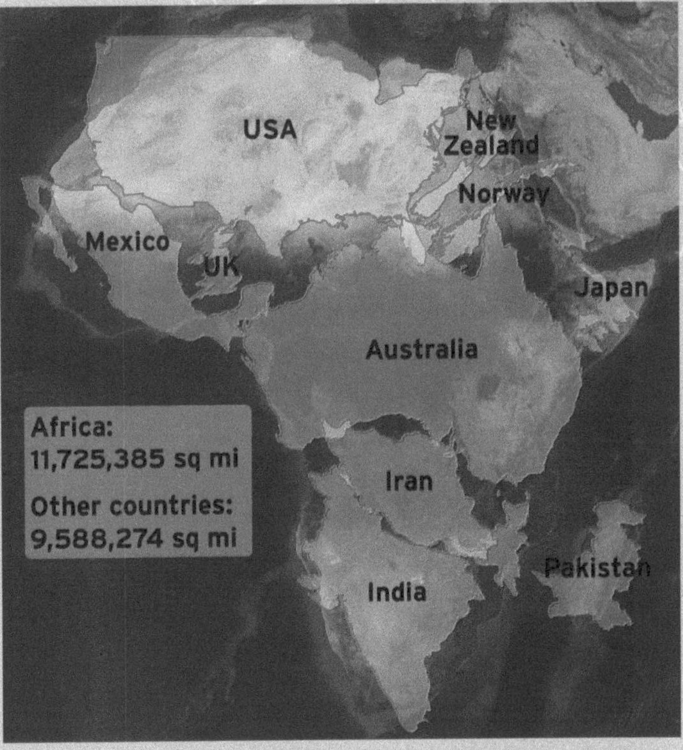

A big place. Lots of space. A huge resource base.

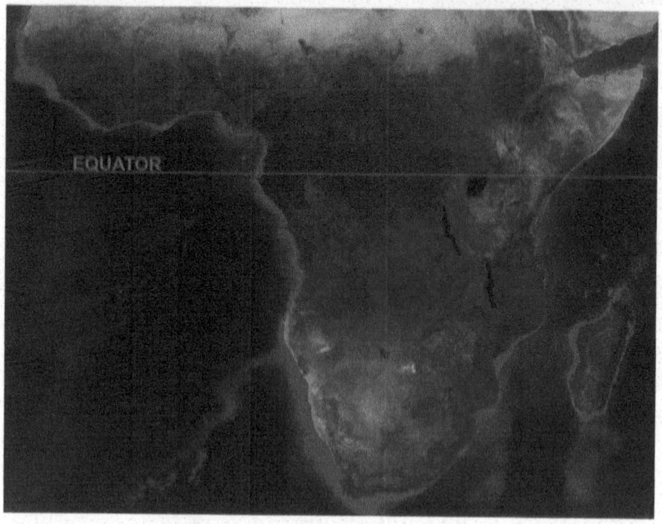

What about some of the physical features of Africa? Some of us think, much as we did in South America, that this is a tropical place with a huge chunk of Saharan Desert capping the continent. And indeed, like South America, there is a huge tropical jungle in the middle part of Africa. It straddles the Equator, but not where many people think. Some think it's near the Sahara, but it's actually down below the bump of Africa and goes through the Democratic Republic of Congo, as you can see on the map.

There is tropical forest in this area near the Equator, which blends into tropical savannah both to the north and south. This savannah area comes to mind when most of us think about Africa, about the wild animals like cheetahs, elephants, *The Lion King* and all that other stuff people dig when they watch Disney movies. This is the commonly portrayed climate and vegetation of Africa: tall grass, a few scattered trees, and tons of cool animals that most of us have to go to a circus to see.

A pattern emerges as you move away from the tropical jungle areas near the equator. You jump into the savanna areas if you go north or south, and if you go north or south of that, you get into steppe. North or south of that gets you into desert. It's like there's a mirror at the equator. Then things get more complicated. Africa has some full-on midlatitude climates, like in the mideastern United States, as well as Mediterranean climates on the northern and southern coasts. There are also highland areas in Africa that tend to stay on the cooler side.

Speaking of highness, there are no major mountain chains in Africa. Africa is an anomaly in the world in that it's the only continent without a major mountain chain. It has *some* relief, but not a lot. Most mountain chains across the planet are the product of **orogenies**: mountain-building episodes which involve the collision of two tectonic plates at convergent plate boundaries. In Africa, we have the opposite. No serious tectonic uplifting happening here. A divergent plate boundary exists between the Arabian plate and the African plate, where the Arabian plate is moving away from the African plate. This created Africa's main physical feature, **The Great Rift Valley**. It is on Africa's eastern flank, dividing Ethiopia in half and passing through Kenya and Tanzania. The rift actually goes right through the middle of the Red Sea, expanding it. The Rift is also responsible for creating the African Great Lakes area: a series of lakes including Lake Victoria (the second largest fresh water lake in the world in terms of surface area) and Lake Tanganyika (globally second largest in volume as well as the second deepest).

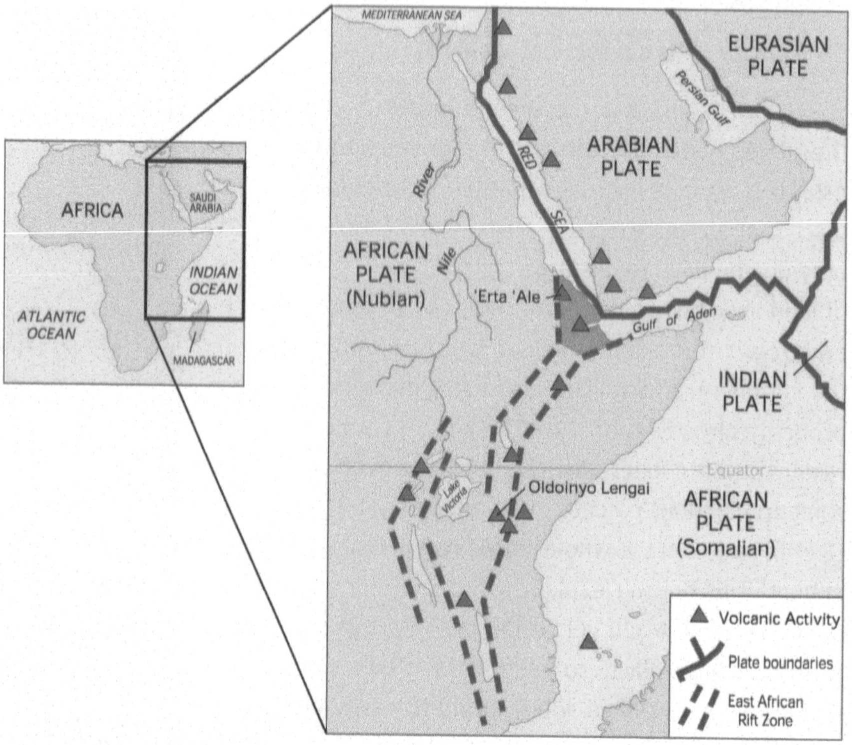

The Rift Valley is caused by plates moving away from each other. Africa is growing!

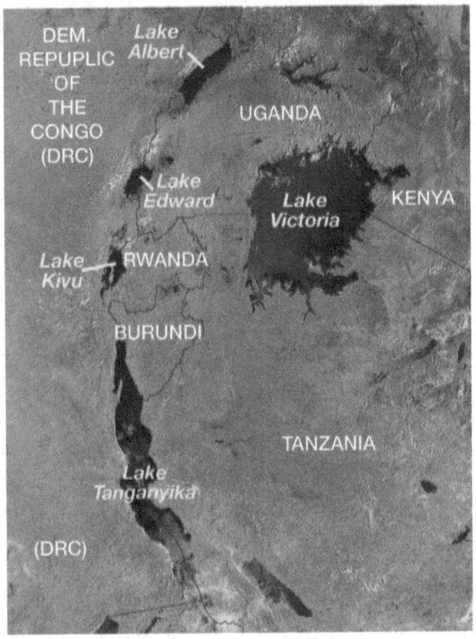

The tectonic split does cause some volcanic activity in spots, resulting in one or two distinct mountains and/or mountain patches, but there are not any major mountain *chains*. Mt. Kilimanjaro is one such distinct peak and a major tourist attraction. It's pretty big, at an elevation of around 19,000 feet. But there are no major mountain chains and no major dividing lines in Africa's interior.

However, like South America, Africa has **escarpments**, or big jumps of elevation right on the coast. This makes the interior continent of Africa as a whole hard to penetrate via water . . . it's actually even more challenging than South America. The Congo River basin is the only serious channel through which to penetrate the continent, and it's in the full-on tropical jungle zone. This makes it tougher to navigate for various reasons. As you can see from the shaded relief map at beginning of this chapter, Africa is difficult to penetrate, much like South America, due to escarpments all around the perimeter and seriously dense vegetation and tropical rainforest in the river basins.

Can you find the African escarpment in this picture? Look close!

Again, let me stress the size of Africa. From north to south, it's about five thousand miles. From east to west, it's about five thousand miles. The sheer size of Africa even compounds movement across and within the continent. There are also other physical barriers like the Sahara Desert, which separates north from south quite nicely. The desert is one of the reasons why we split Sub-Saharan Africa from the rest of Africa. It's such a big divider, the Plaid Avenger named the region after it. Countries in the Sahara Desert are already included in the Middle East for obvious reasons, like climate, Islamic influence, oil, and conflict.

Sub-Saharan Africa, everything underneath the desert, is a totally different region. We don't have to look much further climate-wise because it's not predominantly desert. Religion-wise, there are some Islamic people in Sub-Saharan Africa, but it's mainly a Christian area. When we look in the northern part of Africa, we see people of Arab descent. In the South, people are distinctly black African. We are drawing fuzzy lines between these two regions because the lines go right through the middle of some of the countries. This is significant when we think about current events in Africa. These lines are bisecting Nigeria and Sudan, two countries with big problems. The northern parts of these countries are predominantly Arab and/or Islamic, while the southern parts are predominantly black and Christian. This has caused friction that will play out into the 21st century. The stories in progress here are far from over, as Sudan just split into two different sovereign states along these cultural lines in 2012. The same has been proposed for Nigeria, where a tremendous amount of cultural friction is playing out in Boko Haram's terrorists attacks in today's headlines.

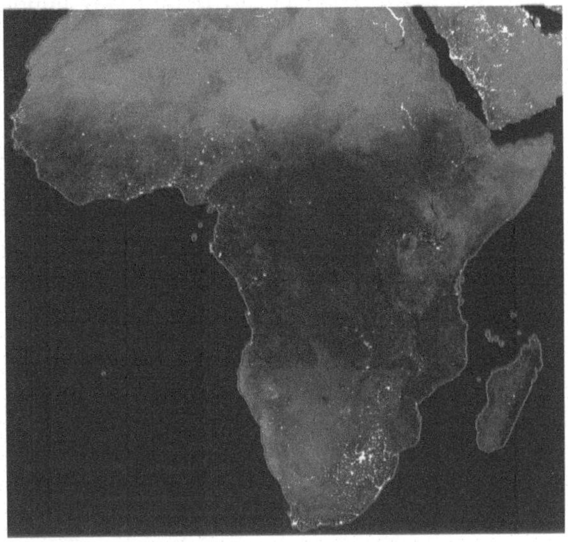
Africa: The Dark Continent—light at night mostly missing . . .

LAGGING BEHIND

Why is Sub-Saharan Africa seemingly in such dire straits? Before the Plaid Avenger goes further, let me just say that Africa is an awesome place to visit. I want to make sure you understand that I personally like the place. Africa is great. I think it has the most potential of any region today. It hasn't suffered the polluting ravages of industrialization or horrific infrastructure destruction of 20th century warfare like many other regions of the planet. That fact may yet serve to benefit the continent greatly, and positive change is slowly taking effect, but, man, there is still a lot of poverty and pain there. History has a way of reversing itself through change, and Africa is a prime candidate for reversal. But let's face the facts—this region has been lagging behind the rising standards of most of the rest of the world.

Why is it hurting so badly? Why is Sub-Saharan Africa so poor? When I say poor, I mean bottom of the barrel poor. We've looked at South America and its great wealth disparity, in which there are a few rich people and a ton of people who are poor. You can't even say that in Africa, because even the rich people just aren't that rich. The wealth disparities are actually lower here because there are fewer millionaires or billionaires hanging out in Africa. That level of wealth isn't concentrated within any class of people in Africa. How did it get to this point? Didn't I open this section by saying it's physically rich in land and resources and a lot of other things?

The operative term to think of here is **marginalized**. "Marginalized" is a term that I often apply to Sub-Sahara Africa. If you're asking, "Hey what does 'marginalized' mean?" I reply: where is the margin of this page that you are looking at right now? Of course, the margin is on the edge, on the fringe. What do you do on the edge of a piece of paper or a page of a book? Not much. Maybe you scribble in the margin, write notes, doodle, or make love notes that read, "I heart the Plaid Avenger." You can do all of this in the margin, but you don't do anything really important there. To marginalize something, someone, or a whole region, is to just push it off to the edge and don't really trouble yourself too much with it, if at all.

Sub-Saharan Africa is maybe the only region on the planet—and you should never use extremes, but gosh darn it, I'm using extremes—that has been consistently marginalized throughout history and into the modern era. Other folks in other places, states, and regions just don't deal with Sub-Saharan Africa. Sometimes people give aid because they feel guilty about not making any economic ties, but there is no serious movement to really change things there. That's what I mean when I say marginalized. From economics to political power to international investment, Africa is rarely at the center of the action, and discouragingly disconnected.

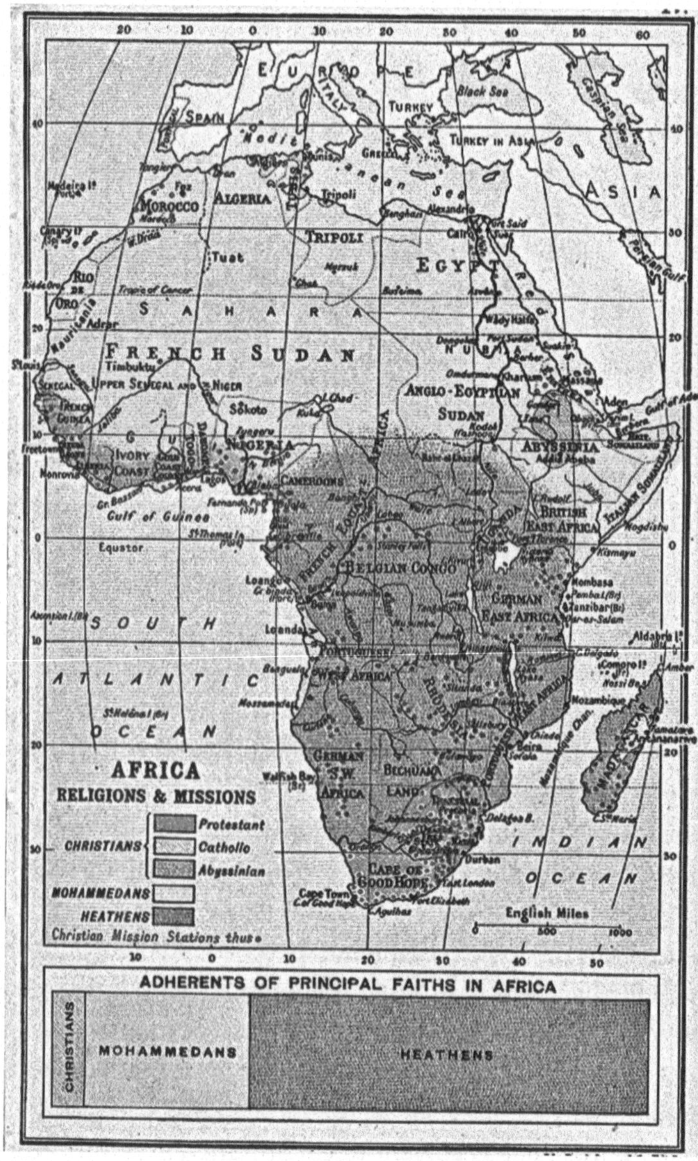

No wonder they haven't prospered! Godless Heathens! (circa 1914 map) From *Literary & Historical Atlas of Africa and Australasia* by J. G. Bartholomew (1943) Can you believe people used to believe this buffoonery?

What's the Deal with African History? African history is about more than just colonization. Here are some fast facts about Africa pre-outside-intervention.

- Human history began in East/South Africa with Australopithecus in 4 million BC, and with Homo sapiens sapiens in 40,000 BC.
- Trade from sub-Saharan Africa to the Arab states above the Saharan desert was only made possible when camel travel revolutionized trans-Saharan trade in the 3rd century.
- Christianity first took root in Africa not with European missionaries, but with the conversion of Ethiopia's King Ezana in the 4th century. Coptic Christians are a branch that exists to this day in Africa.
- West Africa, between the 5th and 16th century, was dominated by a series of huge kingdoms and empires.
- In 1324, the king of Mali, Mansa Musa, was so rich that he made a pilgrimage to Mecca with a procession that reportedly included 60,000 men, including 12,000 slaves who each carried four pounds of gold bars, along with 80 camels that each carried between 50 and 300 pounds of gold dust . . . and he gave most of it away to the poor on the way to the holy land. That's some Malian Mansa Musa Mad riches.

Specifically, how has Africa been marginalized? All we have to do is step back into very recent human history. If we went back a thousand years in Africa, we'd see some pretty big kingdoms. That's a little known fact for most people, because history books largely ignore it. And maybe I am just as guilty, because I won't be educating you on the great old African empires, kingdoms and dynasties either. I'm not marginalizing Africa by ignoring this history . . . I am more focused on what is going on in today's world, which has almost nothing to do with that part of African history.

What? Me, marginalized?

One of the main things people know or remember about African history or read in Western textbooks is the dastardly, dubious, despicable era of **the slave trade!** We've already talked about this in the South America and Caribbean chapters. While referenced and recognized in your high school classes, the mass movement of Africans as slave labor from the continent to the New World from

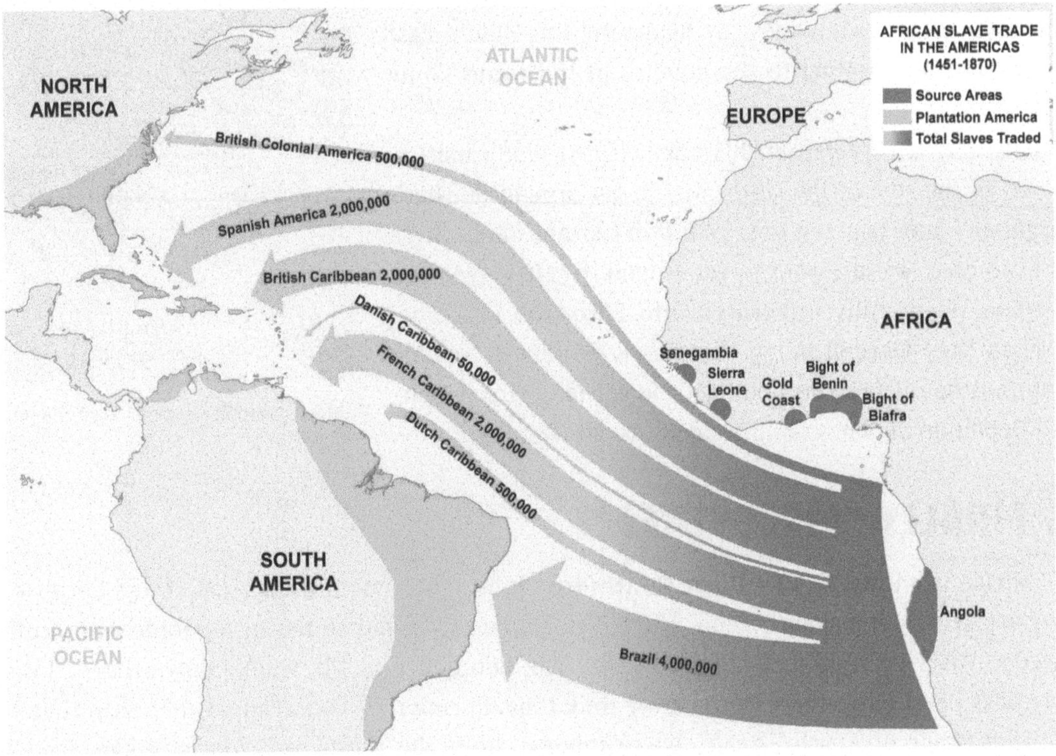

A most despicable deed: the Atlantic slave trade. Human trafficking sickens me, in the past and the present.

1500 onward has made an impact that most textbooks still completely ignore. It completely mystifies the Plaid Avenger that anyone can get through primary school and not comprehend the damage the rest of the world has done to this region (not to mention that it has now marginalized this region completely). The outsiders are the ones that did it!

What do I mean? Think about this for a minute. We're going to play a game. Let's pretend space aliens are going to colonize Jupiter. First, they'll need some workers. They come to a certain part of the planet, let's say the United States. They're going to show up and say, "Hey, we're recruiting some folks to help us colonize this moon and mine some gold." What kind of people would these slave traders take? Obviously, they would take strong people, healthy people, and smart people. They would probably take anybody that had any real resource or utility about them, wouldn't they? Wouldn't you? Wouldn't you want the strongest, the best and the brightest as your workers?

What would happen to the United States if, within the next three hundred years, this alien race took the brightest, strongest and the healthiest? I don't think you have to stretch your imagination a lot to realize that this would be a huge negative impact on any place. In Africa, this happened for a *very* long time. Think back to the numbers we looked at in Latin America. You might have had four to six million people taken from the coast of Africa to Brazil and maybe another five to eight million more to the Caribbean. We are talking about millions and millions of people.

Let's just think of the demographic impact on Africa after several hundreds of years of this type of depletion. What do they get in return? *Guns and liquor.* What a great swap! I wonder why Africa is poor now? I'm having some trouble here . . . can someone help me do the math?

At the exact same time this is happening, as we saw in the Middle East chapter, Africa was getting bypassed. We've talked about Chris Columbus and Vasco Da Gama. What were these guys trying to do? They were trying to get to India. How did they do it? They bypassed Africa completely. Keep in mind the term "marginalized." These European dudes were experts at marginalizing Africa—and I can't make fun of them, they were just trying to make some money. The whole point was to get the heck around Africa as fast as possible. That was how they were going to make more money, by bypassing the Middle East and Africa altogether and getting to the goodies in India and China directly.

Undiscovered? I wonder what the locals thought about that.

Africa was largely unexplored at this point. It just didn't matter. The obvious goal of the rest of the world was to get around it. That's about as marginalized as you can get: "We don't know what's here and we don't even care, we just want to get around it!" They bypassed Africa, and it was never fully integrated into European or global economies. When they weren't being completely bypassed, people just showed up to drop off firearms and liquor in exchange for people.

Depletion of demographics cannot be understated.

FINALLY "DISCOVERED"?

Africa is known as **The Dark Continent**. Some folks might think, "Oh, that's because people there have dark skin." Ummmmm . . . no. The British coined that phrase because people were in the dark about Africa, knowledge-wise. Even up to a hundred years ago, people said, "Uh, what's down there? I don't know, it's big. Oh, and I guess people live there because we took them during the slave trade. Other than that we don't know much about the interior of Africa." Again, we're talking about the 1800s, not too long ago. Back then, famous

British explorers from National Geographic, like David Livingstone, made much ado about "discovering" the continent and filling in the blanks on their wall maps of the world.

This always cracks the Plaid Avenger up. Did people just not know that Africa was there? They had to, how could they not? Africa is connected to the Middle East, the cradle of Western civilization. Traders from India and China bumped into the African coast long ago. Everybody in the Western and Eastern world knows that Africa is there; it's not invisible! It's just largely marginalized. So when was Africa truly discovered? A few hundred years ago, the European powers that had colonized the rest of the planet finally said, "I wonder what's on that enormous land mass just south of us. Hey, no one has actually claimed it yet, have they? That's interesting, maybe we'll take it. Maybe there's something there we can use now."

Dr. Livingstone, I Presume? Who Is Dr. Livingstone and What's He Doing in Africa?
David Livingstone was a Scottish missionary who explored Africa in the mid-1800s. In fact, Livingstone is credited with being one of the first Europeans to complete a transcontinental African journey. He also discovered Mosi-ao-Tunya waterfall, which he renamed Victoria Falls after the Queen (the inconsistency of "discovering" and "renaming" didn't seem to bother anyone). Why did he do it? A statue of Livingstone at Victoria Falls proclaims his motto: "Christianity, Commerce, and Civilization."

CALLOUS COLONIALS TO DEADLY DICTATORS

Here's a more specific title: Callous Colonial Masters to Post-Colonial Deadly Dictators. When the Europeans finally show up on the scene in Africa, it's like a biblical swarm of locusts descending on the continent to conquer, claim, and devour every resource and square inch of land that their mandibles can munch on. The whole European 'global conquest and colonization' period actually started in the 1420's when Prince Henry the Navigator from Portugal launched the exploration of the African coastline, which was followed in short order by port establishment, paving the way for future slave trade operations and eventually full-on colonization.

And those Portuguese dudes were just the first on the scene: for the next several hundred years, all the major European powers began their own explorations, expropriations, and resource extractions of everything they could grab from Africa, including the humans. For centuries the continent was pillaged for its people and its mineral wealth. Like badly. Hundreds of years of robbing the place blind, with very limited investment into the infrastructure or human capital of the continent. In fact, virtually all roads, railroads, and related systems were created solely to facilitate resource extraction, not to help the locals or to create better societies. Need some proof? When the Belgians finally pulled out of Zaire (now the DRC) in 1960, there were less than a dozen native college graduates in the country . . . out of a total population of 50 million. Wow. Nice job Belgium. Way to improve the place during your overlordship.

But I'm getting ahead of the story. How was it that the dink country of Belgium became the controller of the second biggest territory of Africa in the first place? The official land grab can be traced to an exact date of 1884, when Africa was 'discovered,' and then carved up for consumption by the European powers. The **Berlin Conference** to be exact. So what's a meeting in Berlin got to do with the political borders of Africa? Dig this . . .

A PATCHWORK QUILT OF COLONIES

As I already suggested in earlier chapters, the European powers had colonized the New World, taken over South Asia, India, Australia, and made in-roads to China. They had made their presence known on all the continents. Africa was the last place left, and being very civil nice white dudes from Europe, they said, "Chaps, let's all get around the table. We shouldn't fight about this. We all have plenty of problems around the world. Let's do this is in a very civilized manner, because we're civilized guys. Let's put out a map of Africa and divide it up. We don't really need to fuss and fight about it." Indeed, that's what they did. As you might imagine from the title of it, it occurred in Berlin. They didn't even have the decency to be in the place they were stealing. What lazy thieves.

Think real hard and try to tally up how many people from Africa were at this meeting. Oh, that's right: none! Not one single representative from Africa was present at this conference, the purpose of which was to carve up their continent, the second largest continent on the planet! Some of it became French, some Italian, some British, etc. The Belgians received modern day Congo, a massive place that would cover the Eastern side of the United States. Germany had some colonies down there as well, as did the Portuguese with Angola and Mozambique. A patchwork quilt of European power was draped over the continent of Africa.

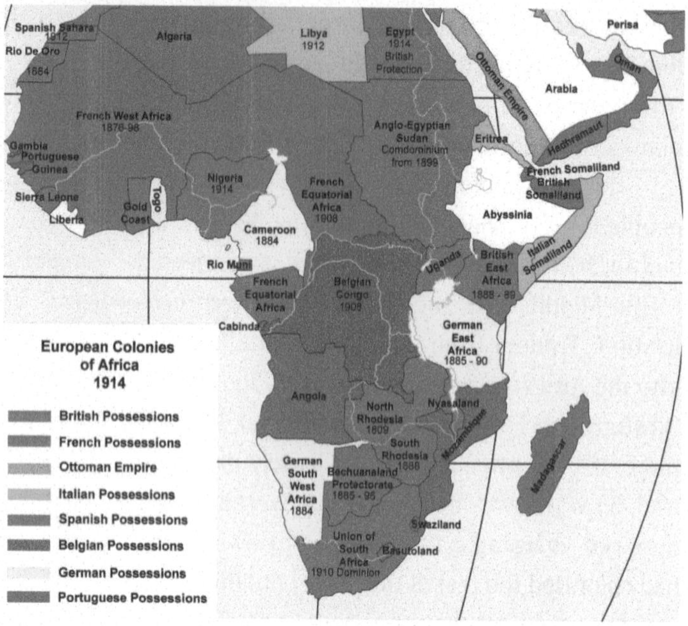

What a beautiful resource-rich cake for Europe to divvy up!

One question that always pops up in the Plaid Avenger's mind about this patchwork quilt: Why is it a patchwork? How come it wasn't divided into zones with the Brits on one side, the French on the other? Certain European powers had already made inroads into Africa in certain places. Because the Germans sponsored the conference, the Plaid Avenger has always had a sneaking suspicion . . .

Here's the Plaid Avenger's theory: The year 1884 is exactly thirty years before Germany tried to take over all of Europe. I can't help but think, when I look at this map of Africa, that what the Germans did was very intentional. In splitting Africa up into different chunks, you have multiple borders with the other European powers, borders that would have to be defended, borders that would require input of capital and men and guns. Indeed, if you look at the German territories themselves, you realize they butt up against all the other European powers—which, by the way, were controlled by the European powers that Germany attacked in World War I.

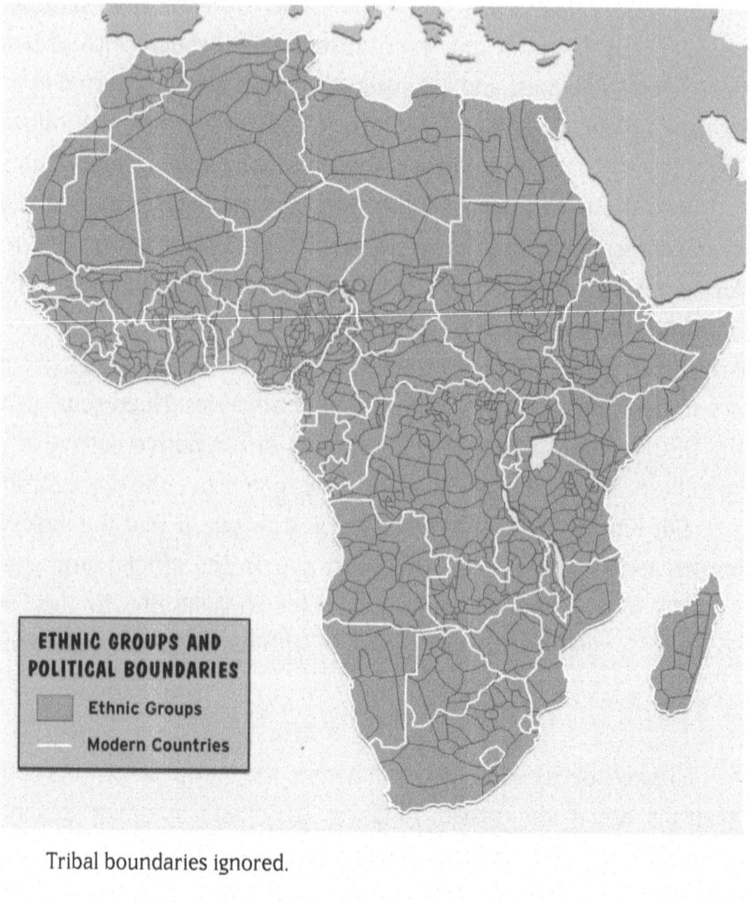

Tribal boundaries ignored.

Here's a little known fact for you: in the outbreak of World War I, there was fighting in Sub-Saharan Africa where the Germans made movements of troops to draw off resources from Europe. The Plaid Avenger theory is that the African map was drawn with conflict in mind. It was used to incite division within Africa as a political tool. It can't be proven, but it is certainly possible, given what the wacky Germans were up to at this point in history. Remember, we are still discussing why Africa is so far behind, and this is one of the main reasons: internal conflict and dissent.

What was there before the European powers drew their big lines? There were thousands of ethnic groups and tribes, for lack of a better word, scattered throughout Africa, with well established territories and histories. They were all divided into what we see on the map. This created a situation in which people were grouped into political bodies by an outside force with complete disregard for the political arrangements in which they were already involved. They might be grouped with people that they may not like, or neighbors with whom they might be engaging in active warfare. The lines drawn at the Berlin Conference were completely arbitrary.

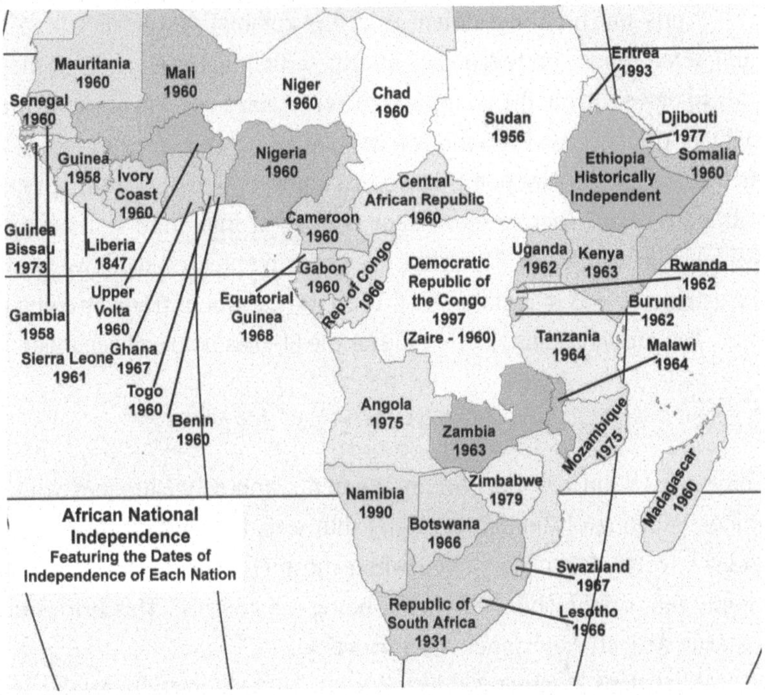

Young states with growing pains.

This is a big source of dissent in today's Africa. Here we are, 130 years later, and this is still problematic. Tribes that were one group in an ethnically defined territory are split between three different countries. You don't have to go too far back to see that a lot of the ethnic conflict, strife, and genocide has been based, in part, on these people being in political divisions that have nothing to do with where their ethnic groups are located. That's a really big deal. Here's the funny part: after the European powers decided to pull out of Africa, they maintained the old borders that were drawn arbitrarily in Berlin.

FREE AT LAST

Let's get into independence. What happens after independence? How did it even come about? As you can see from this map, most African countries are fairly new. They are the youngest countries in the world, meaning they gained their independence fairly late in the game in 20th-Century history. As such, new leaders, new governments, and radically disparate ethnic groups within single states are having trouble.

We talked about the continent being split up into units by European powers. Now these units are all sovereign states. Most of them didn't have independence movements in the conventional sense of the term. Mostly, we relate independence movements to violent insurrections that throw off the shackles of a domineering power by force. To be sure, there were protests and anticolonial groups throughout Africa, but typically there were no great wars. In the 1960s, 1970s, and 1980s, the European powers more often than not said, "We can't have colonies. That's kind of ridiculous nowadays." So they granted independence to their African colonies. For instance, the Belgians negotiated their way out of the Democratic Republic of Congo by saying, "Okay, you guys have fun! We're just going to walk out of here!"

Young African countries are not very well established, even in today's world, precisely because they are young. They were also very poor from the start. Infrastructure was never that great, because of what the colonial powers did during their tenure: not much. When we think about colonial investment in places like India or Southeast Asia, where France had Indochina, or even in South America where the Spanish were, we typically think of the colonial powers setting up towns, with maybe some industries and businesses. The African colonial experience brings back the term "marginalized."

Pretty much all investment from the colonial powers in Africa, which wasn't much, was extraction-based. Yeah, they built *some* roads and railroads, but only to help get the products they wanted out of there. Yeah, they built *some* seaports, only to move the oil, the grains and the vegetables out. I'm pointing this out because this is tied into what's going on economically in Africa today. Even though they are all sovereign states, there was never a lot of investment in Africa's infrastructure. As a result, and a key point for you to know: countries in Africa typically trade with the outside world more than they do with each other. There are no major railroad or interstate systems through Africa. There are no established communications systems through Africa. This place is compartmentalized in all ways, shapes, and forms, and the investment that was put into infrastructure was minimal at best. There is not a lot of trade or communications between countries that are next-door neighbors. The African Union, modeled after the EU, has not really changed that equation much even into the modern era.

DESPICABLE DICTATORS DOING DAMAGE

Some states did have protest movements and rebel leaders who turned into independence figureheads. People like Robert Mugabe of Zimbabwe, a guy that took up arms to throw off the colonial yoke: he is now vilified as a dictator in today's world. What happened when these charismatic leaders took over their countries from the colonizers? Answer: simply put, only a shift in who's robbing the country. This brings up a term that has been long associated with the Sub-Saharan African experience: **kleptocracy**.

What does kleptocracy mean? Well, "-ocracy" usually means "ruled by." *Dem*ocracy means "rule by people." *Aristoc*racy means "rule by rich people." *Klept*ocracy, means "rule by thieves." And the term is oft applied to states in Sub-Saharan Africa. This refers to the fact that corruption has and continues to run rampant here. It has become an established way of life which has helped totally undermine and outright destroy any possible advances in most of the countries in this region. The Plaid Avenger will go out on a limb and say it is the single biggest reason why Africa is still so far behind the times. Corruption tops the list of reasons for the endemic poverty and a sense of hopelessness in many parts of the region.

Here is the sad part: many of these currently crooked leaders actually started as progressive, freedom-fighting rebels who helped their countries achieve independence. They used to be wildly popular dudes standing up for their countries, a.k.a. the the good guys! However, once power is gained, they simply cannot relinquish it. They cannot let it go. You can see leader after leader in country after country get into the system, become corrupt, and then find it impossible to escape the system. They don't want to leave. Many changed their country's constitution so they can rule indefinitely.

As you can see from the following chart, Africa leads the hyena pack when it comes to holding onto political power. These are not monarchies, or royal houses, and these aren't even supposed to be military dictatorships. These are supposedly democracies where people have been in power for twenty, thirty, and almost forty years. This is Africa's biggest problem: the addictive disease called addiction to political power. Robert Mugabe in Zimbabwe, an ex-freedom fighter, a popular leader elected 35 years ago: he's still there. Look at Equatorial Guinea. This guy, President Teodoro Obiang, is closing in on full dictatorial powers since oil was found in the country. Corruption and kleptocracy are pervasive themes in these countries. They are both big reasons why this place is so far behind the times. And the fun isn't over yet for many of these countries: look at the longest rulers chart again and you will see that a handful of states are still openly suffering under the decades-long leadership of some of these jerks.

We look at Africa and wonder, "Why are they so poor?" A lot of these leaders robbed their countries blind. Some of these leaders are still doing it. At the time of his death, Zaire President Joseph Mobutu had stashed around five billion dollars in Swiss and Saudi Arabian banks, and lots of other places. He, as well as others, siphoned off wealth at phenomenal rates and didn't even have bank accounts in their own countries. They took that money somewhere else completely. Billions and perhaps now trillions of dollars pilfered from the public coffers, used to perpetuate the posh lifestyles of these pathetic politicians. What's more, many African leaders—many of whom have outright stolen money—almost always appoint their friends and family, or members of their particular ethnic/tribal group, into positions of power or give their businesses favorable government

WHAT'S THE DEAL WITH THESE AFRICAN DUDES RULING FOR SO LONG?

LONGEST RULING PRESIDENTS IN AFRICA

Name	Country	Years in Power	Total Years
François Tombalbaye	Chad	1960–1975	16
Milton Obote	Uganda	1962–1972, 1980–1985	16
Gamal Abdel Nasser	Egypt	1954–1970	17
Gaafar Nimeiry	Sudan	1969–1985	17
Mengistu Haile Mariam	Ethiopia	1974–1991	18
Juvénal Habyarimana	Rwanda	1973–1994	22
Mohamed Siad Barre	Somalia	1969–1991	23
Moussa Traoré	Mali	1968–1991	24
Daniel arap Moi	Kenya	1978–2002	25
Ahmed Sékou Touré	Guinea	1958–1984	27
Albert René	Seychelles	1977–2004	28
Hastings Kamuzu Banda	Malawi	1963–1994	32
(Joseph) Mobutu Sese Seko	Zaire	1965–1997	33
Félix Houphouët-Boigny	Côte d'Ivoire	1960–1993	34
Gnassingbé Eyadéma	Togo	1967–2005	39
~~Idriss Déby~~	~~Chad~~ **KILLED IN OFFICE 2021**	~~1990–2021~~	~~31~~
~~Zine El Abidine Ben Ali~~	~~Tunisia~~ **EJECTED BY FORCE**	~~1987–2011~~	~~24~~
~~Omar Hassan Ahmad al-Bashir~~	~~Sudan~~ **EJECTED BY FORCE**	~~1989–2019~~	~~30~~
~~Hosni Mubarak~~	~~Egypt~~ **EJECTED BY FORCE**	~~1981–2011~~	~~30~~
~~Robert Mugabe~~	~~Zimbabwe~~ **EJECTED BY FORCE**	~~1980–2017~~	~~39~~
Teodoro Obiang Nguema Mbasogo	Equatorial Guinea	1979–Current	41 (and counting)
~~José Eduardo dos Santos~~	~~Angola~~ **STEPPED DOWN**	~~1979–2018~~	~~40~~
~~Omar Bongo~~	~~Gabon~~ **DIED IN OFFICE**	~~1967–2009~~	~~41~~
~~Muammar Qaddafi* (world record holder)~~	~~Libya~~ **KILLED IN OFFICE**	~~1969–2011~~	~~42~~

contracts. It is that exact type of insider trading/**cronyism** which reinforces ethnic strife and political dissent across Africa. What a mess. It's really a sad scenario that has repeated itself since the first independence movements in Sub-Saharan Africa.

Pathetically, a lot of this clandestine crook-ery has gotten even worse in today's world. But don't lose heart, my plaid friends! The world seems to be waking up to this horrific self-perpetuated cycle of scum and villainy, as is started to do something about it: 1) As previously mentioned, the WTO and other international institutions are starting to take much more control over any aid sent into the region; 2) A billionaire businessman from South Africa established a 1 million dollar prize to be awarded to African leaders that do the best job of both running their country and eliminating corruption; and 3) The active presidents of Gabon, Equatorial Guinea, and Congo are being investigated and will be sued in international court for embezzlement. While they, of course, claim their total innocence, it will be hard for them to explain how they have collectively earned 1.3 billion dollars while ruling impoverished states. Hmmm . . . I guess maybe they put in a lot of overtime. And I hope they get to put in a lot more overtime later . . . in Hell.

Perhaps the days of the kleptocrats are numbered, but with their reserves of power and wealth, they are a hard bunch to root out. Go get 'em, Transparency International! **2021 UPDATE!!!** Well, by looking at the previous list of long-lasting leaders, you see that a whole slew of them have been officially struck through! Did they finally come to their senses and relinquish power on their own? Nope: they mostly just clung to power to the bitter end and either 1) got so old and feeble they were forcibly removed (Robert Mugabe in 2017), 2) got ejected by force due to popular protest (Hosni Mubarak in 2011; Omar al-Bashir in 2019), or 3) got straight up killed (Muammar Qaddafi in 2011). President Idriss Déby of Chad was the latest casualty, having been killed in 2021 by militant forces while he was personally on the front lines fighting to preserve his 31-year long rule. There is a lot more new blood in African leadership right this moment, but the persistent problem of addiction to power may not be purged just as yet. Let's see how this new group of leaders fares when it's time to hold elections!

Know Your Deadliest Dictators of Sub-Saharan Africa!

Idi Amin

Amin seized power in Uganda in a 1970 coup. Determined to purge the country of those who were disloyal, Amin set up the "State Research Bureau," which would be more accurately described as a death squad. The group targeted Christians, Asians, members of the Acholi and Lango tribes, academics, etc. Over the years, Amin became more erratic, at one point reportedly declaring himself the King of Scotland. Western media often portrayed Amin as a bumbling murderer. Amin finally lost power in 1979. Over 300,000 Ugandans were murdered or tortured during Amin's rule. It was also largely rumored that he was a cannibal, and that he found political enemies "delicious."

Mobutu Sese Seko

Mobutu was president of Zaire from 1965 to 1997. During this time, Mobutu executed political rivals, condensed power in the executive branch, and outlawed Christian names (he changed his name from Joseph). However, Mobutu's worst characteristic, besides his choice of hats, was his insane greed. He stole billions of dollars from the already poor Zaire economy. In 1985 it was estimated that he had $5 billion in Swiss bank accounts alone. Mobutu can be seen as a prime example of kleptocracy. Mobutu was overthrown in 1997, after supporting the Tutsi regime responsible for the Rwanda genocide, and Zaire was quickly renamed the Democratic Republic of Congo (DRC). The DRC is currently war-torn and screwed, both economically and politically.

Robert Mugabe

Mugabe was in charge of Zimbabwe ever since its independence in 1980. During this time, he has quashed political dissent and been accused of gross human rights abuses, but he began to really get serious about destroying his own country about ten years ago when he initiated a "kill all the whities" campaign in which the government confiscated white-owned land, and subsequently did little to nothing to protect the white minority when violence erupted in the subsequent land grab. In 2005, Mugabe's government initiated Operation Murambatsvina—a program to destroy the shantytowns of Zimbabwe which left millions homeless. In 2008 he lost the primary election, then nullified the vote entirely, adding to his list of proud accomplishments in this sham democracy. Finally in 2017, he was forcibly removed from office.

Teodoro Obiang Nguema Mbasogo

Obiang became the de facto leader of Equatorial Guinea since he took power in a 1979 coup. Equatorial Guinea has since been described as one of the most corrupt, ethnocentric, anti-democratic, and repressive states on the planet. It is rumored that Obiang has prostate cancer and may die soon. Not soon enough, though: conservative estimates of the wealth he has stolen from his country approach $1 billion. Since oil was discovered in the country, his personal bank accounts have received hundreds of millions of dollars. At the same time the standard of living and GDP of the rest of the population has plummeted. Awesome guy, huh?

ADDITIONAL AFRICAN AILMENTS

Why else is there object poverty and conflict across this region? Corruption causes poverty which causes conflict. An additional complicating factor is that when countries are extremely poor, some unstable border wars and resource wars tend to crop up. This happened particularly in the Democratic Republic of Congo, as you will see in the International Hotspots Chapter, where there might be seven or eight different sovereign states involved in skirmishes in this one country. This is Africa's own World War now because there are so many resources that people want on the eastern side of the Congo that countries are just taking them. You might say, "What? We live in the modern world. People can't just go into other countries and take stuff! A state can't do that!" It might not happen anywhere else in the world, but resource acquisition is definitely a compounding factor of conflict that still happens in Sub-Saharan Africa. But that's not all: ethnic, religious, environmental, global geopolitical factors, and even health play a role in escalating misery in this place. Let's look at a few in more detail.

STRIFE OF ALL STRIPES

Let's talk about ethnic and religious strife. We already talked a little bit about ethnic strife between different tribal groups across Africa, and that is still in play to this day. You can easily cite countless episodes, similar to the Hutu-Tutsi genocide in Rwanda, which simply consisted of extremists from one ethnic group wiping out another ethnic group.

Ethnic genocide is a concept we have covered in previous chapters, and I'll bet that you have at least heard reference to the Rwandan Genocide of 1994, as it made international headlines and spawned several popular films, namely *Hotel Rwanda* (2004) starring the always awesome Don Cheadle. But there are many other African atrocities in this category that are still largely unknown. I will just name a few that you smart students should investigate further if you hope to understand modern Africa more fully:

- The **Nigerian Civil War** a.k.a. the **Biafran War** of July 1967 to January 1970 was fought between the government of Nigeria and the secessionist state of Biafra, a former independent kingdom and ethnic group. During the two and half years of the war, there were about 100,000 overall military casualties, while 500,000 to 2 million Biafran civilians died of starvation due to a blockade of the region.
- In the early 1970s, over 150,000 Hutu people were killed by Tutsi people in Burundi on orders of the genocidal General Michel Micombero; this is what set the stage for the retaliatory 1994 genocide in Rwanda.
- In the late 1970s, the Ugandan dictator **Idi Amin** (and then his successor) killed over 300,000 people from the Acholi and Lango groups; in retaliation, these two groups went on to murder thousands of Baganda ethnicity.
- The **Ituri Conflict**, a period of intense violence from 1999-2003 (although starting way back in 1972 with armed conflict still happening to this day), was between the agriculturalist Lendu and pastoralist Hema ethnic groups in the Ituri region of the northeastern Democratic Republic of Congo (DRC). Perhaps up to 100,00 killed, half a million displaced.
- Also in DRC: **Effacer le tableau** (literal translation: "erasing the board;" damn, can you get any more genocide specific than that?) was the systematic extermination of 70,000-90,000 Bambuti pygmies by the Movement for the Liberation of Congo (MLC), a rebel group operating in eastern DRC during the Second Congo War (1998-2003).
- From 2008 to the present in the southern region of Sudan, an estimated two million people belonging to various Nilotic peoples including Dinka, Nuer, and Shilluk have been killed in hundreds of raids by Sudanese Arabs from the north. (This is related to the **Darfur Genocide** of 2003-present happening in Sudan, but I place that state in the Middle East region, so I won't mention it here.)

In all of these deaths and displacements, the specific ethnic victims were killed by people from a different ethnic group. Just since the 1970s, roughly 10 million deaths can be attributed to this type of tribal/ethnic conflict, mostly concentrated in the Central Africa region. So the next time you watch *Hotel Rwanda* (2004) (or even a much more intense and realistic film on the same subject named *Sometimes in April* [2005]), please remember that the Rwandan situation is neither unique, nor the first, nor the last time genocide has happened here. . .

On top of this, there is serious religious strife. Two countries of particular note have problems with this: Nigeria and Sudan. These are prime examples of what I'm talking about when it comes to religious strife. Islam dominates the Middle East and North Africa, and Christianity/Animism dominates the South, thus setting up a frictional 'face-off' along the rough boundary between the two. The most classic example is Sudan. The northern part of Sudan is Islamic Arab, while the southern part of Sudan consists largely of non-Islamic Black Africans. The North Arab guys are in charge of the whole country, and implemented Sharia law. This caused open rebellion in the south and continues to incite harsh feelings and violent acts, to the tune of several millions of casualties over the last thirty years.

The final solution to this predicament was to split the single state into two sperate entities that each contain their own cultural group; it is why we now have a Sudan and a South Sudan. But the spat between the sovereign states is still quite active, as they fight over border and resource issues.

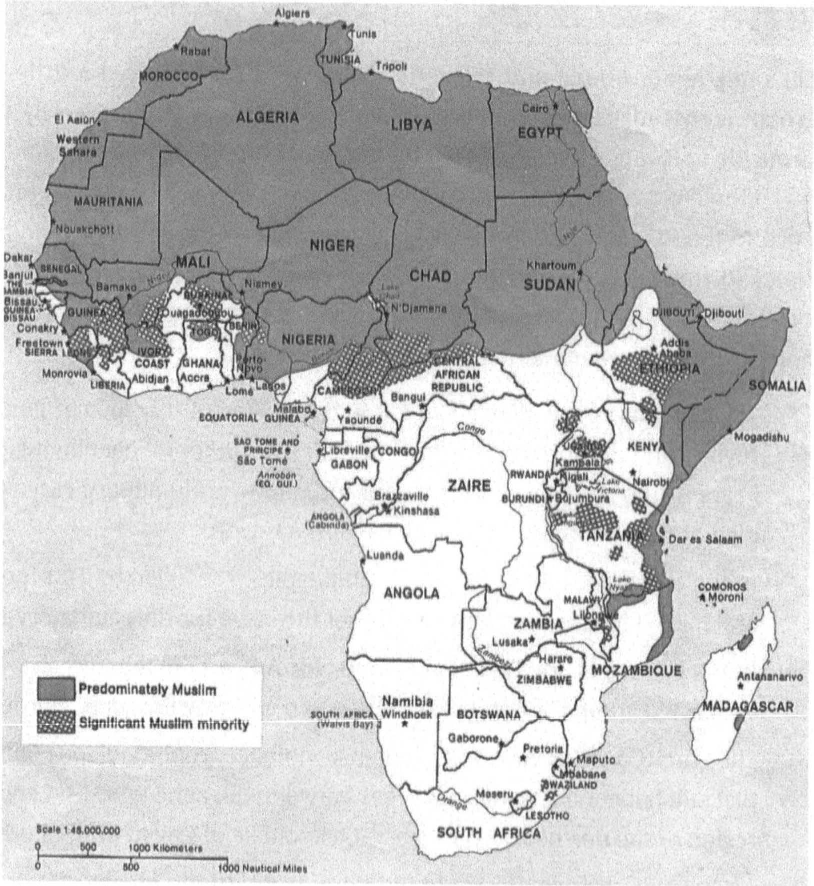

We also see this in today's Nigeria, even though it has not engaged in a full-on civil war. The northern part of Nigeria has certain substates that voted to implement Sharia law. That's all fine and dandy if you are Islamic, but there are many other non-Islamic peeps living in these areas that are really not into it. This has become a hot button issue in Nigeria. Islamic people in the northern part of Nigeria want to live by Islamic law, but the government in the south says, "Wait a minute, we're the government. You can't have Islamic law overriding government law." Religious strife in Nigeria cropped up big time when the 2002 Miss World Pageant was slated to be held there. There was a publicized case on the docket in which a woman in northern Nigeria was convicted of adultery. A Sharia court sentenced her to death by stoning, old school style. This became a firebrand issue with some of the Miss World contestants. They refused to go to Nigeria and indeed, eventually the pageant was moved to London. More recently a radical Islamic group named **Boko Haram** has been committing acts of terrorism by blowing up Christian churches across the north, further inciting the entire country to violence. You may have even heard something about the 276 Chibok schoolgirls who were kidnapped by, yup, you guessed it! Boko Haram. Google that term in current events to follow the latest action and status of this very important African state. There are some real repercussions from this religious conflict in Sub-Saharan Africa and it is something to which we should be paying attention. Religious violence and radicalism is increasing all across the **Sahel** zone of Africa, in pretty much every state.

The main point I'm stressing: this is why there is conflict and division within Sub-Saharan Africa. We'll throw in one more thing for those of you who saw *Lords of War*. The issue of gunrunning fuels all conflict in Africa. Africa is also a

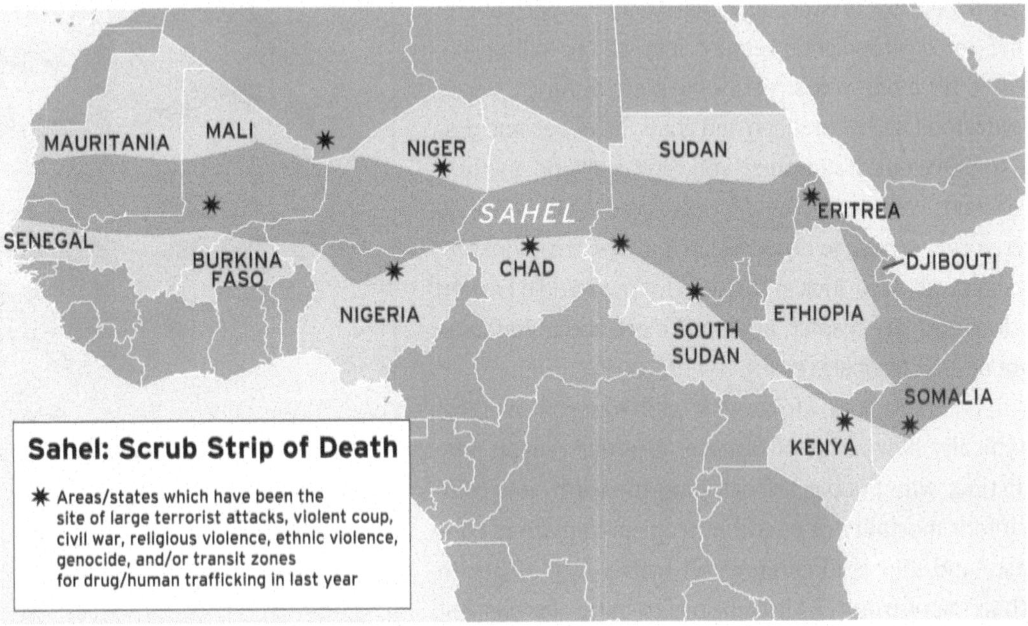

What the Sahel? Flip forward to Chapter 26 to find out more about violence in this area.

heavily armed continent as far as individual weapons go. No states have nuclear weapons or conventional missiles. Few of them even have conventional war weapons like tanks, but boy, do they have guns on this continent! They've got all types of small arms from Russia, Israel, the US, France, you name it. Every gun producer has a big and avid market in Sub-Saharan Africa. This place is armed to the teeth. This is such a travesty. In Africa, the poorer a country, the more of their GDP they will dedicate to buying weaponry. This almost harkens back to the slave trade 500 years ago when guns and liquor were dropped off in exchange for human cargo. No wonder there is instability in this region.

CLIMATE CHALLENGES

Let's get to a new topic. Why all the starvation in Africa? We've talked most about post-independence, why there is conflict on all scales, but why the starvation? Why do we see Sally Struthers in commercials telling us to give money for starving Africans? Why do we see mud covered children with flies crawling all over their faces. Why? What's the deal? Don't they have any food there? Can't they grow food? What's going on? There is no other part of this planet that has as much starvation, at least not as much as is portrayed on the news. There are lots of different reasons having to do with physical and environmental factors and economics.

Going, going, gone: Bye-bye, arable land!

Number one is the climate. Large areas of this region contain some really challenging climates. Savanna and steppe comprise a big component of this continent and these areas are affected radically by fluctuations in weather. Typically, when we see episodes of starvation in places like Somalia or Ethiopia, it's because they have had a seriously bad drought. One bad drought will single-handedly wipe out agricultural products within a year. When they have back-to-back droughts, it becomes totally disastrous. That's why you have millions of people starving to death. Any dry climate is under increased pressure if it becomes even a little bit drier.

One thing that drought in Africa can lead to is **desertification**. This is mostly along the southern fringes Sahara Desert in an area named the **Sahel**. Desertification is even

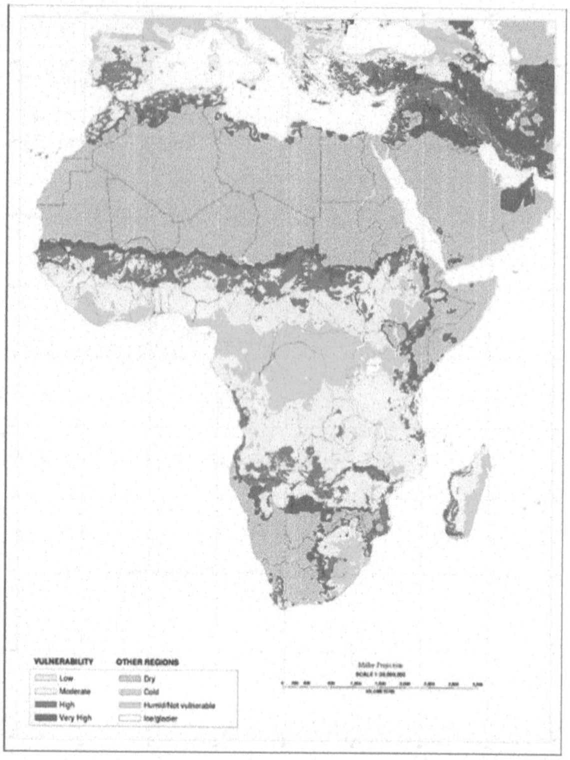

Red areas highly vulnerable to conversion to full-on desert!

worse than drought because it is drought that becomes permanent. It is arid land that, for environmental reasons that we are still unsure of, turns into desert. It's a one-way street. Some parts of Africa, where people used to grow food and herd sheep and goats, have become desert. People have to move out. It's finished; you can't really do anything else afterward. Humans can't live there; we need water and there's no water there. Therefore, we have to desert (as in leave or abandon) that desert (as in useless sand). Also, look at this map for our Middle Eastern friends that we discussed last chapter, and think more about that issue of water I brought up—big trouble brewing.

On top of that, in areas of Africa with well-watered, tropical climates, you typically have a lot of disease. Another reason why this place is hurting, why people are starving to death, and why there is a high infant mortality rate, is that tropical climate is conducive to disease, and that is in conjunction with a lack of health care to most. Thinking of tropical Africa brings to mind the nastiest diseases that hang out here, like Ebola and AIDS. These are diseases that mutated or developed in tropical Africa. Ones that you haven't even heard of, such as schistosomiasis, which features germs that can get into your skin and hang out in your bladder for the rest of your short life until you are urinating blood. They've got eye worms, which are self-explanatory. They've got worms to penetrate every orifice you have. Here's the really good one, elephantitis—look up pictures on the Internet if you really want to lose your lunch. And actually, mosquito-carried malaria is the number one killer in Africa and it is a treatable and curable disease—but it's not in Africa because of lack of health care. So add climate and disease to the list of African ailments holding the continent back.

ECONOMICS: MO' MONEY? TRY NO MONEY: AFRICA NEEDS IT

So Africa is challenged by its size, escarpments, climate, health care, corruption, and conflict . . . and that all affects the economy. This is not an economically diverse or developed place because of its sheer size. Economies require movement and trade to grow. There's not a lot of infrastructure for this in Africa, and at this point, to build five thousand miles of road costs billions. Who the heck has that kind of money in Africa? It encapsulates thirty different countries. It is more than three times the land area of the US, including Alaska. It would take some serious cash to build enough roads to cross this continent. This is not an economically integrated place and it is not likely going to be in the future. We talked about the escarpments problem. It's hard to penetrate into the continent with ships. Economic activities, like agriculture, that benefit from movement in greater mass become very unprofitable and tough to do with no commercial shipping access. They have a strike against them in terms of the physical world and in terms of sheer size and escarpments.

Other economic challenges that play into Africa's lack of wealth go back to the colonial era. We've already talked about this. What did colonial masters invest in? The answer: nothing but extractive industries for raw commodities, which Africa is still being plundered for, but now by multinational corporations instead of colonial powers. It started off with the slave trade, taking humans away from the continent, and hasn't evolved out much further. Now, it's just concentrated on taking ores and diamonds and oil out of this place instead of people: "We've got the stuff, let's just get it out of here the fastest way possible. Build a railroad station solely to get to the copper mine. What? We're not interested in an infrastructure for people! We just want to move the stuff out!"

Extractive economies are based on primary products. And what are those worth to the locals? Zilch. Primary products aren't worth much. On top of that, many African countries are what we call **single-commodity economies** precisely because of this scenario. This problem exists in other parts of the world but nowhere in quantities as large as in Sub-Saharan Africa. These are countries that upwards of 80, 90, sometimes 95 percent of the entire economy is based on a single primary commodity. For example, take Niger. Ninety-five percent of its GDP is from one thing: uranium. This is a substance that no one in Niger can use for anything. It's an export commodity from which Niger makes all their money. Sao Tome and Principe: 90 percent cocoa. Zambia: 85 percent copper. Botswana: 80 percent diamonds. Now you might look at that and say, "That's good, at least they have something." But, again, this is a primary commodity. Diamonds are raw; they aren't worth that much. The worth comes from processing all these things. Here's a news flash: the processing typically occurs somewhere else, so they sell this stuff cheap and the real money is made elsewhere. Speaking of diamonds . . .

Here's the complicating feature the Plaid Avenger wants you to understand about this single-commodity situation: what are diamonds worth if the world goes into recession? The answer: nothing. What kind of a shape is Zambia going to be in if the price of copper plunges? They're in trouble. How about cocoa? Are you joking me? Your economy is based on freakin' chocolate? Wonka has you by the wanker! That is so variable and so subject to market fluctuation that it should be pretty easy for you to understand that Sub-Saharan Africa is a place that is always on a knife's edge. It's not economically diversified, and on top of that, it's heavily dependent on primary products.

What's the Deal with Conflict Diamonds?

Conflict diamonds are diamonds produced in active war zones. Often the profits from these diamonds are channeled to paramilitary organizations where they are used to prolong or intensify the active conflict. The main sources of conflict diamonds include Sierra Leone, Liberia, the Democratic Republic of Congo (DRC), and Angola. Sierra Leone is particularly bad, because their diamonds are easy to mine and of high quality. Government and nongovernment forces have traditionally applied organized crime-like tactics to control diamond fields. In Sierra Leone, the children that dig for conflict diamonds are often executed for suspected theft or simply for underproduction. Another common punishment for suspected theft is the amputation of limbs. Watch the movie *Blood Diamond* for a decent description of the situation. Spoiler: Leonardo Di Caprio dies in the end. Oh wait, that's not a spoiler, that's a bonus!

Message to Plaid Avenger: African agent has acquired the conflict diamonds and is at rendezvous point . . .

COLD WAR EFFECTS IN TROPICAL AFRICA?

Here's another economic challenge making this place poor: the Cold War is over. During the Cold War, as we've pointed out multiple times in this manual to understanding the planet, the two sides, Team Soviet and Team Capitalist Democracy, were vying for influence around the world. Those teams would make loans to or sell arms to countries to get them on their side, and Sub-Saharan Africa was no exception. The teams, mostly led by the United States and Soviet Russia, made inroads and tried to make deals with the same corrupt leaders that I referenced earlier. Some of these African leaders were really savvy and played both sides. Joseph Mobutu of Zaire was so crooked that he took money from everybody. Of course, he put it into his own bank accounts in Switzerland, which made the situation in his country even more impoverished. He took guns from everybody as well. What a playa!

During the Cold War, money flowed in for just such purposes. Just like the United States sponsored the Marshall Plan in Europe, people said, "We should make inroads to Africa because if we don't, then the Commies will. If the Commies help them out, then they will be on Team Commie. We should help out. We should lend them money. We should try to do some projects for them." By and large, that's what happened during the Cold War in Africa. Investments were made. Relationships were formed. Aid was given to African allies. And then what happened? The Cold War ended. There was no Team Commie vs. Team USA anymore. As a result, aid simply dried up. There was no impetus, no reason, no economic drive for the United States to help out or do anything in Africa. As for Russia, they couldn't afford to do as much even if they wanted to. This all imploded overnight, and aid has diminished, partly due to the Cold War component.

Aid is also diminishing for other reasons. One of the main reasons is corruption. See how things are tied together? Because of the corruption within the governments, not only due to corrupt leaders but also militaries, places like Somalia are in anarchy. There are aid organizations that want to bring food to starving children and medicine to babies. They simply cannot get anything to them because of internal corruption. There are distribution problems because of the road situation in Africa. On top of that, you have corruption from the highest points of government all the way down to the local scale. If the US sends a billion dollars to fight AIDS in Zimbabwe, you can expect half of that to be swindled by higher-ups working in the bureaucracy. That still leaves a lot of money, which you can expect to be swindled by the people on the ground that are supposed to be giving it out in the form of shots. Everybody from the top down perpetuates the corruption.

What's the Deal with "Third World?"

Hey speaking of the Cold War, did you know that that's where we got the term "Third World?" The world was split into three groups based on certain divisions along social, political, cultural, and economic lines. Team Capitalist Democracy was known as the First World, Team Commie as the Second World, and Team everybody else without an alignment of the commies or the capitalist democrats, as the Third World. Brazil is a Third World country. So is Sweden. Get the picture? That means "Third World" is not a term that can be used interchangeably with "Africa," or with "poor countries," unless you want to show that you don't know your history.

Check out this map: everything in Blue is First World, Red is Second World, and Green is Third World.

https://en.wikipedia.org/wiki/Third_World#/media/File:Cold_War_alliances_mid-1975.svg

This has become a massive problem in which people have gotten to the point of saying, "I'm just not going to donate, what's the point? It's not going to get to the people who need it." If you think this is a Plaid Avenger personal comment, it's not. I say: donate as much as you can to Africa! The outside world has screwed the continent over for the last 500 years; they have a moral obligation to go back and help. However, this corruption problem makes a lot of people suspicious. One suspicious party is the World Bank, which has said, "We are only going to lend money to your country if we come and give it out ourselves." There are places like the World Bank and the IMF that say, "Yeah, we're not going to lend you money until you change governments, because you guys are too corrupt."

Transparency International, an organization that performs studies on corruption, places most African countries in its top thirty most corrupt places. Does this affect anything? We already know people don't want to send international aid because of the corruption. Let's branch it out. How about international investment? Who is going to build a factory for twice the actual cost because they have to pay everybody off? When Transparency International came out and said, "Wow, these ten countries are the ten worst and they're all in Sub-Saharan Africa," most business people translated the statement as, "You'd be crazy to invest there. You can't afford that. You'll never make your money back." All of these things are related to each other. It's hard to tease out single items, but the end of the Cold War is a big one, causing international investment aid to drop off simultaneously, making this place even poorer.

AIDS

The last issue we cannot possibly talk about enough is AIDS, and we're going to have to spend a few paragraphs talking about it here, although not for the same old same old reasons you've read about elsewhere. Yes, it's terrible, it's sad; people are dying of this disease. Everything you've heard is true. I want you to think deeper though, because we're trying

What's the Deal with Loaning Money to Chad?

As you know from Chapter 6, the World Bank is an organization that makes developmental loans to poorer countries. Also, you should remember that the World Bank often makes these loans contingent on structural adjustments in the recipient country. Long story short: the World Bank gave Chad a loan to build a pipeline. In return, Chad had to promise to put 10 percent of the profits into an account designed to fight long-term poverty. Everything was going well until the government of Chad decided that perhaps they, and not the World Bank, knew what was best for their country.

The government moved some of the oil profits from the long-term account that the World Bank demanded and invested them directly into schools and health care. Some of it may also have been invested in guns. Bad move. Why is this important? Because it deals with sovereignty. Chad thinks it is their sovereign right to use revenue how they best see fit. The World Bank, on the other hand, contends that it is their right as a loaner to enforce measures to insure that they get paid back. The Chad/World Bank loan was viewed as a new model for loans to Africa, where there have been many problems with government officials stealing revenue from the poor. Whether this model will be used in the future depends largely on the result of this dispute.

to project ahead about what's going to happen in Sub-Saharan Africa. If we want to figure out what should and can be done about this region, we have to deal with this 800 pound gorilla called AIDS. It is massive and so far-reaching into society that the mere idea that it's just a few people that are sick and dying is not even the tip of the tip of the iceberg. What are the true impacts of AIDS? Look at the map below and see that there are some places where a quarter of the population of the country is infected. Of course, that's liable to get worse in a lot of places. Even if it does get better, we're still talking about a fifth or sixth of the population infected with AIDS.

What are the real impacts about which the world should be concerned? Why can't we all just put our head in the sand about Africa? Some countries in Africa are in a state of crisis management, if not already, they certainly will be in the next twenty years. Why the crisis? AIDS is a terrible disease for a lot of reasons, one of which is that it doesn't kill you instantaneously. That sounds horrible, but I'm trying to get you to think deeper. If you get Ebola, you're going to die now. If you get hit by a car, you're going to die now. AIDS is a disease that takes people out over a long period of time. They are sick for a long time, steadily deteriorating, ultimately becoming dependent on others totally.

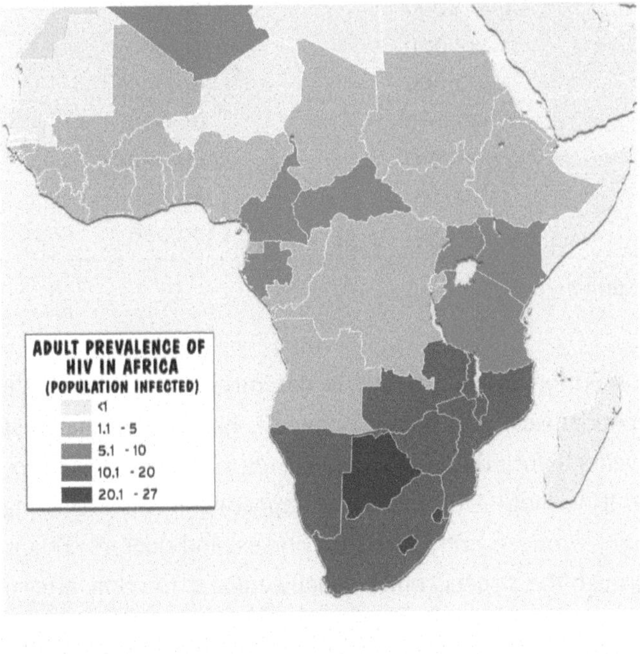

Here's the big impact: You take a quarter of your population who typically used to be factory workers, farmers, mothers, fathers, teachers, firemen, soldiers, people who were actively creating wealth, and turn that quarter into dependents, sick, laying in a bed dying. That in itself should get you to think about how titanic this issue is in a lot of countries. Merely swapping a quarter of your population from providers to dependents is enough to sink any country, including the United States. If this happened in the United States, it would be huge. A quarter? How about, just ten percent? Taking them off the plus side of the chart and putting them on the negative side. This is making a massive impact in Africa.

Here's more food for thought: What age and socioeconomic groups typically come down with AIDS? If you guessed people in the working class from puberty to about the age of 35, you're right. People from about 15–35, people usually in the work force, are the ones getting sick. Africa is now being called the "Orphan Continent" because there are so many motherless and fatherless children in need. Of course, loss of this age group leaves a lot of old people who are too old to have sex since this epidemic broke out. Therefore you have a society that is getting tilted demographically. You're losing the middle that you want and you are left with a lot of needy people, young and old. This is an epidemic that is causing massive problems in many countries—and it is only going to get worse.

Number of People Living with HIV 2005 to 2017

Total living with HIV Global, 2017	**36.9 million**	
Total people living with HIV Subsahara Africa	**25 million**	

	People Living with AIDS (all ages)	
Select Countries (not an all-inclusive list)	***in 2005***	***in 2017***
Angola	140,000	310,000
Botswana	300,000	380,000
Burkina Faso	120,000	94,000
Cameroon	480,000	510,000
Central African Republic	160,000	110,000
Chad	100,000	110,000
DR Congo	580,000	390,000
Ethiopia	880,000	610,000
Ghana	310,000	310,000
Ivory Coast	630,000	500,000
Kenya	1,4000,000	1,500,000
Lesotho	230,000	320,000
Malawi	830,000	1,000,000
Mali	100,000	130,000
Mozambique	1,400,000	2,100,000
Namibia	160,000	200,000
Nigeria	2,900,000	3,100,000
Rwanda	210,000	220,000
South Africa	4,900,000	7,200,000
South Sudan	130,000	180,000
Tanzania	1,300,000	1,500,000
Togo	120,000	110,000
Uganda	900,000	1,300,000
Zambia	870,000	1,100,000
Zimbabwe	1,400,000	1,300,000

Source: https://www.unaids.org/sites/default/files/media_asset/unaids-data-2018_en.pdf

Why is this a potential time bomb? For all the things I just mentioned, we are only in the introductory stages. We have massive numbers of people infected. The true impacts, economically and demographically, have yet to manifest themselves fully. In other words, it is going to take another fifty years before this completely comes to a head. That's providing they cured everybody from the get-go, starting right now. If no more AIDS infections happened in Africa starting today, there will still be countries closing in on collapse for lack of workers, army, firemen, nurses, and doctors. This is a really big deal. There are huge numbers of orphans running through the streets, which usually equates to crime after a

What's the Deal with All the Diseases?

Ebola, Malaria, AIDS, oh my! What's the deal with all these diseases?! Well first of all, for the most part health-care is seriously lacking in sub-Saharan Africa. Without access to vaccinations, people, especially children, die of completely preventable diseases. In 2016 there were even two cases of Polio in Nigeria. What?! People don't get Polio anymore! The good news? Africa is making progress towards access to vaccinations, and has made some huge progress in certain diseases. Guinea worm disease, for example, has dropped from an estimated 3.5 million cases in 1986, to only 22 cases in 2015. The bad news? Millions of people are still dying of malaria, dysentery, TB, cholera, and many other preventable/curable diseases, and for many, health care is still seriously lacking. This means diseases like Ebola can be a big freaking deal since there is not good enough health care infrastructure to deal with the regular stuff, much less the extreme outbreaks. Oh, and just to remind you, about 40% of people in sub-Saharan Africa still don't even have access to safe drinking water, and only 30% have access to toilets. Yikes.

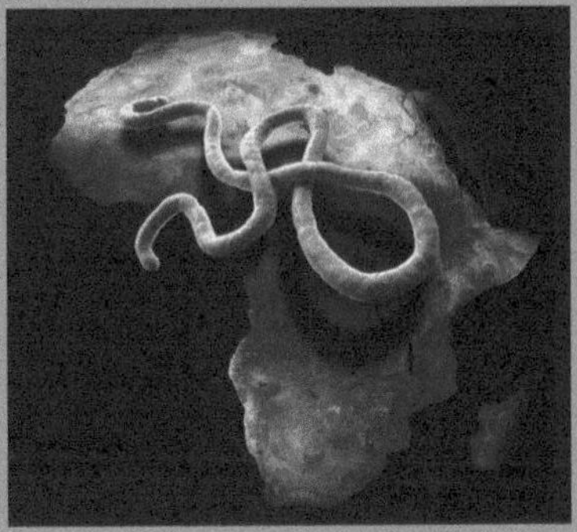

Ebola ground zero

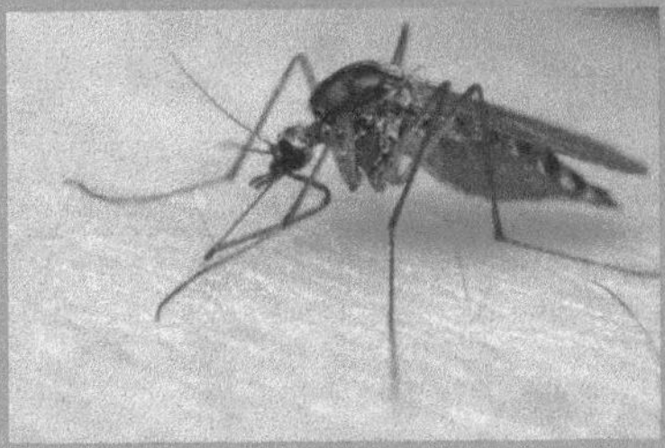

while. This is bad, bad news for many countries, especially places like South Africa, Zimbabwe, Uganda, Botswana, which have the highest rates of AIDS in the world.

We can step back from this in the West and say, "Well, we won't go there. We just won't vacation there this year. I won't see the African animals and everything will just be fine, sooner or later." No. We live in a globalized world, man. We live in a connected place. Once one state collapses, it usually creates a vacuum, which pulls other states into the anarchy as well. I think it's a safe hypothesis to go out on a limb and say that the collapse of one or two states in Africa could incite a war that pulls in multiple states vying for power and influence. So the prognosis for states with high rates of HIV infection are not good, which could affect the wider region as a whole if this epidemic continues unabated. . .

2021 UPDATE: As pointed out, the region has been hardest hit by this particular disease, with the 2017 total of people living with HIV in the entire world at 36.9 million . . . and of that number, 66% or about 25 million are in sub-Saharan Africa alone. I used to get very depressed teaching about this stuff, because it just seemed like there was no hope for it to ever get anything but worse. But, in another good news story, it looks like Africa is close to turning the corner on this epidemic!

Increases in education, increases in awareness/prevention, increases in early detection screening, health technology developments, new medical treatments, and increased cooperation between the region's governments, civil society, and medical community have collectively produced steep declines in HIV infections and AIDS-related deaths. It is not declining everywhere yet, but the numbers are starting to level off, even in the worst hit places. Of course, huge challenges remain and there are millions still challenged with HIV on a daily basis, and millions more still remain at risk of infection. But for the first time, ever, Africans (and the rest of the world) are starting to see a glimpse of light at the end of this tunnel.

However, there are still a lot of other health challenges that Africa faces that aren't even as diabolic as HIV/AIDS. There are a lot of standard, curable diseases which still plague the continent and serve to stymie their economic, political, and social development . . . things like malaria, cholera, and the super-scary Ebola!

ALL IS NOT LOST

I don't try to sugarcoat anything, but having gone through the plethora of problems in this region, I want to point out that I think Africa has huge potential as well. I think Africa may be one of the best continents to be on in the next hundred years. Why is the Plaid Avenger saying this? Because this place has the hugest potential for growth! Africa has a rich history; it's not as if they are completely lost and have never known anything else. Some of the richest kingdoms ever known were in North Africa. There are also pockets of development in certain places today. Some countries are doing pretty well, so I don't want to categorize the whole region as impoverished and in peril; they aren't in that dire of straits yet.

South Africa is doing pretty well. They have lots of internal problems, but economy and GDP-wise, they aren't doing too bad (they also make some rockin' wine). Nigeria is extremely oil rich, and is the biggest economy of the continent. They have some religious and corruption problems, but, by and large, if you read studies about people in Nigeria, they really like their state. In fact, I read about a happiness index, conducted by the UN. They polled people around the world about their countries and their lives. Nigeria ranked number one in the happiness index. People were very happy to be in Nigeria; they loved it there. But perhaps not all—see the box on the right. Kenya, Tanzania, and others are making bank off of Africa's biggest money maker outside natural resource extraction. What is it they've got? Tourism trade. People love to see those unspoiled landscapes and those giraffes, hippos and all those cheerful cheetahs. The Plaid Avenger would rather put 'em on a hamburger bun and eat 'em, but some people like to go check them out. Have fun with that, until the hippo bites you in half . . . but that's another story.

What's the Deal with Oil Workers Getting Kidnapped in Nigeria?

Recently, nine internationals were kidnapped while working on an oil platform in the Niger Delta. The kidnappers were members of a Nigerian rebel group, Movement for the Emancipation of the Niger Delta, that demanded more oil wealth be returned to the communities from which it is taken. Nigeria is the biggest oil exporter in Africa, yet most of the country's 130 million residents live in poverty. Other attacks by the Nigerian rebels reduced the oil output in the Niger Delta by 25 percent. The Delta is home to ethnic Ijaws, many of whom seek independence from Nigeria.

This is all important because Delta oil is big business, and big business hates instability, so it appears that a throwdown is imminent.

On top of the pretty animals and landscapes, there is finally a continent-wide movement for positive change: the African Union. Now, don't let me exaggerate its current position; the AU is not really a force to be reckoned with yet in any capacity. They formed in 1963 as the Organization of African Unity (OAS), and in 2002 reformed as the AU, primarily to be cool-sounding like the EU. It has aspirations to be a free-trade block like the EU, but it has largely been ineffective in this capacity since many of the states are still poor, thus increasing trade to other poor folks doesn't really help you that much. Almost all African states trade more with countries outside of Africa than they do with their immediate neighbors.

While trade hasn't really worked out for the AU, here's what has worked out: its armed forces. The AU has a professional army that is composed of forces from every country on the continent. And it is fast evolving into a continent-wide emergency force that is increasingly being dispatched to Africa's hotspots. What's so great about that? Well, the outside world is extremely excited about this prospect for several reasons: 1) It is believed that Africans can better deal with African problems, 2) Africans know the area/cultures better and can more readily move around and more quickly react to changing field circumstance, and 3) If Africans take care of their own problems, then the outside world doesn't have to do it.

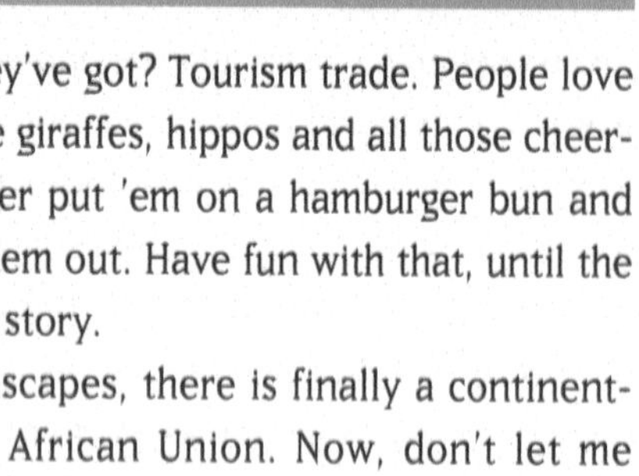

AU forces from Rwanda ready for action!

That last point is the crucial one. Remember, the world turned its collective back on the Rwandan genocide as it did in the past with Sudanese genocide, and countless civil wars, plagues, and droughts which have caused unimaginable death counts across Africa. It really hasn't been in outsider's political interest to get involved in Sub-Saharan Africa, and one need look no further than the American sentiment about sending troops to Africa ever since ***Black Hawk Down***. It just isn't happening. Lots of outside entities, including the UN, the US, and the EU, are all about giving tons of money to help develop the AU force as a substitute for direct intervention . . . probably not a bad idea.

The AU now finally has a mission that they can all rally around, and that the outside world is eager to support with money and training. It's not NATO, and it's not even NAFTA or ASEAN either, but it's better than nothing. It could very well be the thing that, once successful, will lead to further political and economic integration of the member states. That would make it quite unique in the world, as being first a military grouping that eventually becomes something economic. We shall see.

AUSPICIOUS AND AMAZING AFRICAN 2021 AFCFTA UPDATE!!!!!

Holy hurry, hungry hippos! How on earth did I miss this? Some serious economic bloc action just broke out in Africa! A full-scale free trade zone, to be exact! It could be a continental game-changer! Its awesome African acronym is **AfCFTA**, which is an impossible to pronounce shorthand for **Af**rican **C**ontinental **F**ree **T**rade **A**rea.

Now, this isn't exciting simply because it's a free trade first. Far from it! There has been a prolific and profuse alphabet soup of often-competing and overlapping trade zones in Africa: ECOWAS in the west; EAC in the east; SADC in the south; COMESA in the east/south. In fact, I have never really reported on trade blocs in Africa because none of them have been very effective at even increasing trade, much less anything more interesting. (With the possible exception of the EAC one; the East African Community driven mostly by Kenya has decent growth in their common market.) But hold the phone, because this AfCFTA thing has tremendous potential and really is something new.

July 2019, Egyptian President and current AU Chairperson Abdel Fattah al-Sisi and Nigerien President Mahamadou Issoufou making it official: Let's get this AfCTA game on!

On July 7, 2019, Nigeria became the final major signatory to make AfCFTA a reality. Very fitting, since Nigeria happens to be the largest economy on the continent. And this continental free-trade zone contains all 55 states of Africa! Full coast to coast to coast to coast coverage that is now uniting 1.3 billion people, with a combined $3.4 trillion GDP, set to usher in a new era of development.

AFLAC?!? No, it's AfCFTA!

This African Continental Free Trade Area, (BTW: the largest bloc created since the creation of the WTO in 1994), will hopefully help open up Africa's unrealized economic potential by boosting trade between member states, which in turn will strengthen supply chains, which in turn will trigger industrialization, which in turn will promote increased infrastructure development. All of these things together will likely increase employment opportunities, perhaps the most potent benefit to a continent with a large unemployment rate and the world's fastest growing labor force. Member states are immediately starting on eliminating tariffs on virtually all goods, which will increase trade in the next five years by 15-25% according to conservative IMF estimates, up to 50% by more favorable estimates.

Not to get too wildly optimistic about this brand new venture, but the creators of this bloc have also signaled that this is only the start of a more fully integrated continental concept. They aim to increase the ease of movement of goods, services, finances, and perhaps even people. There was specific mention of free-er movement of peeps in order to encourage the spread of expertise and investment. They also launched a digital payment system for the entire zone (an African Venmo?) and a single air transport market (an "Africa Air" airline?) to boost connectivity and cut travel costs. . . and to further facilitate flying and all other forms of movement, it is also working on an "African Union passport!" Sweet geezus giraffes! We are on the cusp of a new age for Africa! Fingers crossed for the continent!

THE LAST WORD

For all of its challenges, Africa actually holds tremendous promise and potential. It has some pockets of development and more importantly, it maintains its unspoiled character, which I think will be a huge boon to Africa in the future. If we look around the planet at places such as Russia, Eastern Europe, China, and India, we see places packed with people and industrialized to the point of polluting their own environment, places where you can't drink the water anymore because it's so toxic. They're rich, but everything there looks awful. Even in America, where we see big, wide-open spaces, there is some nasty evidence of the post-industrial era: cities collapsing and rusting down to nothing.

In Africa, the word to think of is "unspoiled." Perhaps people say about Africa, "They're not industrialized and they're poor and that is why they suck." But I'm telling you to look at the long term, at the big picture here. In another fifty to one-hundred years, people may be coming in droves to Africa because it's the only nice place left on the planet where you can get to huge tracts of land and drive around for days and not run into anything. That's increasingly what I think people like to see. It's why the biggest sector of many of their economies is tourism. People want to see that stuff; it's cool.

AU tank forces muster on the plains.

A hundred years ago, if you wanted communications, you built telephone poles or railroad systems or electric grids crossing your entire country or continent. We live in a different world now. Africa didn't develop and that's one of the reasons why it's not as far along. At the same time, they don't need telephone poles anymore. Africa is fully connected, without all the hassle. Everybody just needs a cell phone. They can all be on the Internet without landlines. They are behind and not as developed, but they have basically leapfrogged. They have picked up all the best parts of technology created in the developed world, without all the baggage. They didn't have to industrialize and pollute their continent that much to make it happen either (apart from the exploitative mining of silicon for cell phone parts). They didn't build telephone poles and communications networks, and now they have that stuff regardless.

And with increasing investment comes increasing jobs and increasing development, which hopefully will equate to increasing investment in education, health care, and infrastructure. That would turn the vicious circle into a positive feedback loop that could spark an African Renaissance. The market potential is huge (with a half a billion potential customers), the growing populations are a huge pool of labor, and the resources/food production potential may be the highest on the planet! And Africa is now the place where HIV/AIDS drugs are the most advanced and most available! Hurrah! Hurrah for progress!

Finally, why is all not lost? As I suggested at the opening of this chapter, they are resource rich. They have tons of stuff . . . stuff that people want. Maybe not so much oil, although some countries like Nigeria have that as well. Things like uranium, manganese, silicon, coltan, and copper—Africa has tons. They have tremendous amounts of resources. They also have animals and land, if you want to consider those resources as well.

In addition, dig this: India and China are now racing each other to invest billions into Africa in mining rights, industries, construction contracts, and most importantly, food production. That's right: these huge countries are now using African lands as a breadbasket to feed their huge populations back home. As part of a long term strategy to control a larger part of the global supply of natural resources, China and India are pumping money into infrastructure and moving in their native peeps to these countries, to eventually blend in ethnically with native African populations. And the US and EU don't want to get left out in the cold so they are getting more interested in investing, too, thus a race for African investment is now on, for the first time in modern history. While this has its positives and negatives, at least real investment is now starting to occur on a grander scale.

Leave the kleptocrats to me!

I like to think about Sub-Saharan Africa as just getting over a "down-time." That is, every region has been undeveloped, becomes developed, and reverts to undeveloped again. Countries go in cycles of development. Remember that the British were the kings of the world, and now they are America's lapdog. China was the center of the world, then it sucked for two hundred years, and now it's climbing back on top again. Things change. Things have a tendency to turn around; indeed that may be what happens in Africa, given the long scope of human existence. They may be down, but not out. The Plaid Avenger will be partying there for sure, because given pollution and over-industrialization everywhere else, this region may be the last great un-destroyed natural reserve on planet Earth. Hey! Let's make the whole region into a international nature reserve and petting zoo! That would be totally sweet! Well, until the armed rebels and man-eating lions show up. But I digress into African fantasy . . . let's get on with the world regional story and proceed further east to Asia . . .

What's the Deal with Urbanization?

Sub-Saharan Africa is urbanizing rapidly. The region has gone from about 15% in 1960 to more than 40% in 2016. This has led to the creation of poor mega-cities, with a population of more than 10 million, like Lagos, Nigeria; and Kinshasa, DRC. Population is also exploding in cities like Dar es Salaam, Tanzania; Luanda, Angola; and Nairobi, Kenya. This means huge opportunities—better access to jobs, education, and healthcare in cities could mean a better educated, healthier, and more economically productive population. It also poses significant infrastructure challenges, as about 50% of people living in urban areas in sub-Saharan Africa live in slums. Keep an eye on mega-cities in Africa, they're not going anywhere.

Johannesburg is but one of many. . .

Check out this map, which includes the most populous cities in Sub-Saharan Africa: https://www.atkearney.com/documents/10192/4371960/FG-Seizing-Africas-Retail-Opportunities-2.png/4120f2cc-fe77-4b6b-8123-9ea28eb33312?t=1394730888762

SUB-SAHARAN AFRICA RUNDOWN & RESOURCES

View additional Plaid Avenger resources for this region at http://plaid.at/subs

BIG PLUSES

- Gigantic region
- Has tons of natural resources—tons!
- "Unspoiled": having never been massively industrialized and polluted, there are huge areas of the region that are environmentally unscathed
- Unique open spaces and wildlife make tourism a totally awesome industry now and in the foreseeable future
- The AU looks promising, especially the military wing which is receiving international support and funding
- Serious long-term international investment in industry and economics is finally happening, led by China and India

BIG PROBLEMS

Wow. Where do I even start with this one?

- Most impoverished region on the planet; poorest and lowest standards of living in virtually every statistical measure invented
- Political corruption and cronyism at epic scales; the region's biggest problem
- Religious, ethnic, tribal, and economic frictions abound, often resulting in violence
- Malaria, AIDS, Ebola, Dysentery, TB, Cholera, and many other diseases affect hundreds of millions yearly; many are preventable/curable
- Desertification
- Region as a whole is almost entirely dependent on low-value primary product extraction and export
- Trade and interaction between states of the region is weak or non-existent

DEAD DUDES OF NOTE:

Idi Amin: Cannibal and military dictator/President of Uganda from 1971–79 whose rule was characterised by human rights abuses, political repression, ethnic persecution, extrajudicial killings, nepotism, corruption, gross economic mismanagement and the deaths of perhaps half a million citizens.

Mobutu Sese Seko: President of Zaire (the Republic of the Congo) from 1965–97 and absolute crook extraordinaire who siphoned billions out of his country. Maintained an anti-communist stance which got him millions of dollars of US support as well as CIA assistance to assassinate the former leader and destroy the existing democracy altogether.

Patrice Lumumba: The staunch nationalist, independence movement leader, and democratically elected first Prime Minister of the newly independent Republic of Congo who Mobutu assassinated in 1960. Lumumba fought hard to eject the Belgians, and then to form a solid and independent government that would not be beholden to foreign powers. Yeah, he paid for that mistake.

LIVE LEADERS YOU SHOULD KNOW:

Cyril Ramaphosa: President of South Africa since 2018. Previously an anti-apartheid activist, trade union leader, and super successful businessman (worth half a billion $$$).

Teodoro Mbasogo: The ultimate evil in current living African dictators, crushing the life and soul out of people in Equatorial Guinea since he seized power in a 1979 coup. Stolen all of the oil wealth and has crushed the society, and yet gets invited to international events and deals with international businesses as if he is still a human. A true, true dirt-bag.

Muhammadu Buhari: Retired Nigerian Army Major General and current President of Nigeria, which is the biggest economy of Africa, the biggest population in Africa, and is a major exporter of oil. A conservative, law and order kind of guy that now is tasked with dealing with major terrorist threat Boko Haram, and cleaning up one of the most corrupt governments on the planet.

PLAID CINEMA SELECTION:

Blood Diamond (2006) A wonderful action film centering on the issue of conflict diamonds in Sierra Lione. Film touches on many very real issues faced by Africa such as the diamonds, civil war, and child soldiers. Done exceptionally well, with the lone exception that it has Leonardo DiCrapio in it.

The Constant Gardener (2005) At the heart of this dramatic film, lies a story of mystery shown in a disjointed, non-chronological manner that only adds to its prowess. As a man tries to uncover the reasoning for his wife's death, he discovers the violence that currently runs rampant in Kenya. The filming on location had such an affect, that the cast and crew developed a fund to help ensure a basic education in these villages.

Shaft in Africa (1973) An absolutely ridiculous "blaxploitation" movie that centers around the adventures of John Shaft as he travels to Africa to bust up a human smuggling operation. Amazingly enough, the almost 40 year old film still holds up up for its topical content: migration of poor Africans to Europe to live in slums and do menial work. And it's awesome to see how much more sexually radical US films were back in the 70's! Great scenery of Ethiopia!

Hotel Rwanda (2004) Don Cheadle saves the world during the 1994 Rwandan Genocide. The true-life story of Paul Rusesabagina, a hotel manager who housed over a thousand Tutsi refugees during their struggle against the Hutu militia in Rwanda. Features the UN doing nothing.

Sometimes in April (2005) As much as I love Don Cheadle, this is a much better take on the Rwandan Genocide to be able to understand the "why" of the event. Depicts the attitudes and circumstances leading up to the outbreak of brutal violence, the intertwining stories of people struggling to survive the genocide, and the aftermath as the people try to find justice and reconciliation.

Shooting Dogs aka *Beyond the Gates* (2005) Let's have a trifecta of Rwandan genocide films! This one is based on the experiences of BBC news producer David Belton, who worked in Rwanda at the time, and the title refers to the useless UN workers killing scavenging dogs.

Tsotsi (2005) Set in a Soweto slum near Johannesburg, South Africa, the film tells the story of a young street thug who steals a car only to discover a baby in the back seat. You want to see real poverty and a real slum and get a sense of real South African diversity of classes? Check this sad but crowd-pleasing drama.

Malunde (2001) An odd-fellows buddy/road movie that features a disaffected middle-aged white South African and an 11 year-old black child of the streets on an impromptu backroad journey from Johannesburg to Capetown, in varied settings and difficult situations.

Invictus (2009) Nelson Mandela, in his first term as the South African President, initiates a unique venture to unite the apartheid-torn land: he enlists the national rugby team on a mission to win the 1995 Rugby World. Stars Morgan Freeman, whom my students always mistake for Nelson Mandela anyway, so this movie was inevitable, I suppose.

Lumumba (2000) Fantastic film centered around Patrice Lumumba in the months before and after the Democratic Republic of the Congo (DRC) achieved independence from Belgium in June 1960. Extremely realistic and true to actual facts of the situation. Helps students understand why the place is screwed, even up to this day. Due to political unrest in the DRC at the time of filming, the movie was shot in Zimbabwe and Mozambique.

Black Hawk Down (2001) American war film that depicts the 1993 Battle of Mogadishu, a raid integral to the United States' effort to capture Somali warlord Mohamed Farrah Aidid. Yeah, it's an action film, but is a dead on portrayal of the events, and more importantly, helps you understand why the US will never send troops into Africa again . . . which they did not do to help Rwanda in 1994, or since. Filmed in Morocco.

Last King of Scotland (2006) Watch the rise of Ugandan President Idi Amin and his reign as dictator from 1971 to 1979, with the inevitable decline into madness and psychopathic behavior. As a side note: the ending scenes in the airport are based on actual events when an airliner was hijacked by Palestinian terrorists in 1976 and Amin gave them safe haven in Uganda, until Israeli secret service stormed in and saved everybody. Speaking of which . . .

Raid on Entebbe (1976) and *Victory at Entebbe* (1976) were both US made-for-TV films that depict that Israeli commando operation in Uganda.

MAP GALLERY

View Map Gallery & additional Plaid Avenger resources for this region at http://plaid.at/subs

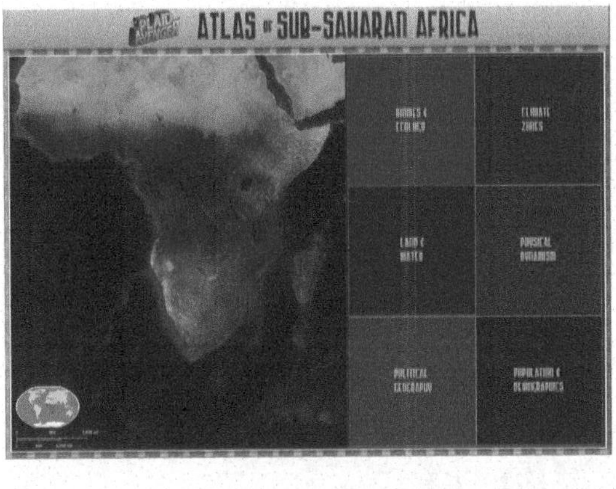

Dictators, Despots, & Demons!

A Continent Crushed by Kleptocratic CHAOS

WATCH
as the most unsavory Sub-Saharan Scum Strike Deep at the heart of human life and liberty with their self-centered skullduggery
Can Africa survive the affliction?

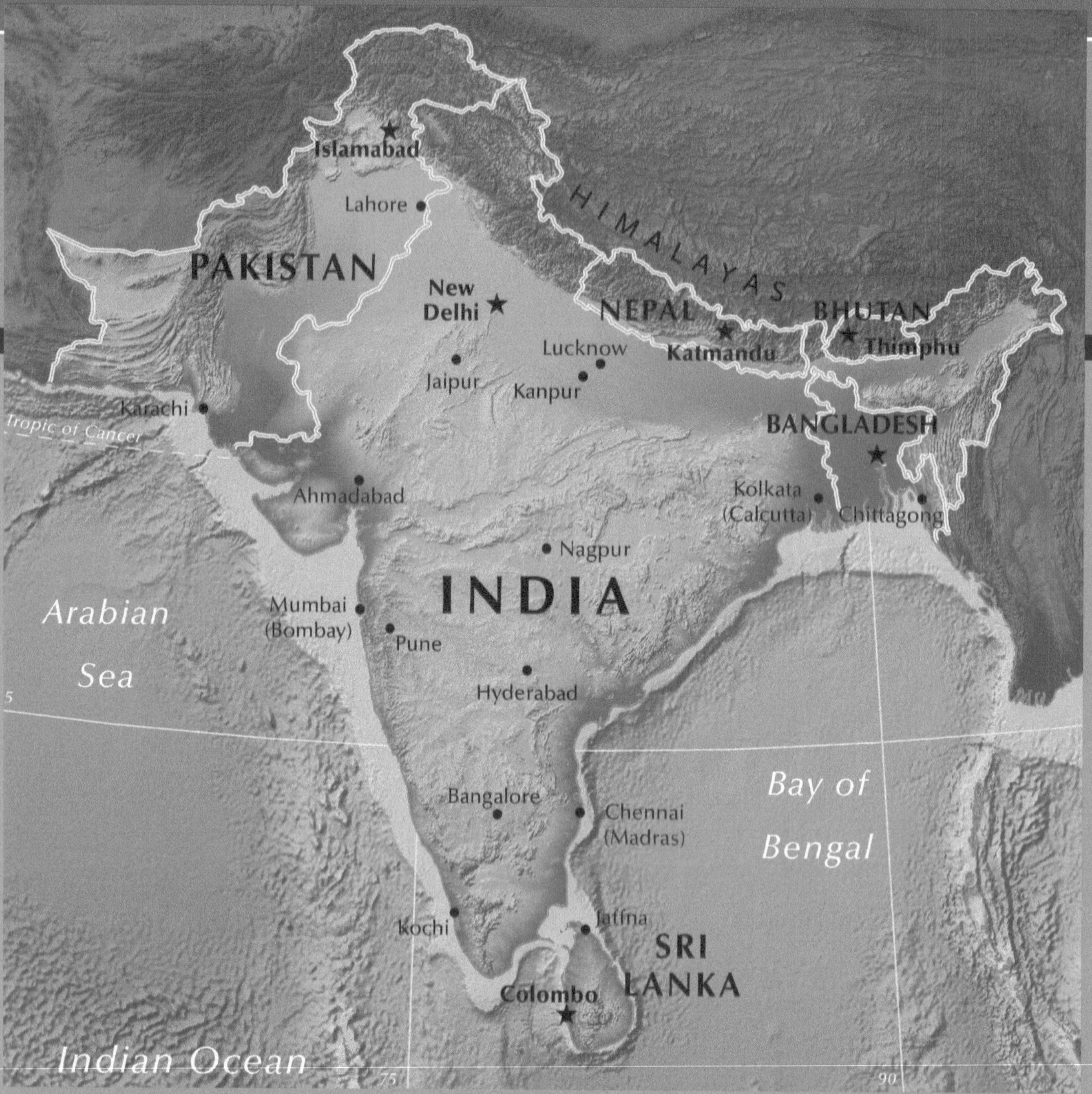

CHAPTER OUTLINE

1. Let's Get Physical
   1.1. Slippery When Wet
2. People Power
   2.1. Hindu or Hindi?
   2.2. Vishnu vs. Muhammad vs. Buddha
   2.3. Cultural Hearthiness
3. Historical Happenings
   3.1. Mughal Mania
   3.2. Enter the Europeans
   3.3. Drop the Wicket, James, It's Teatime: Advantages of British Colonialism
      3.3.1. Advantages, Schmadvantages
   3.4. My Main Man Mahatma
   3.5. Packin' Our Bags for the Partition Party
4. Time for Today's South Asia
   4.1. The Smaller States
   4.2. Energetic India
      4.2.1. 2021 UPDATE: HOLY HINDU!!! Titanic Tectonic Political Shift in India!!!
   4.3. Pakistan Is Pakin' the Heat
   4.4. Bangin' Bangladesh
5. Mommy Wow! I'm a Big Kid Now!
6. Map Gallery

South Asia

21

SOUTH ASIA is often referred to as "the subcontinent" because geologically and plate tectonically, it is. Thirty-eight million years ago, it was a separate land mass, but ever since then it has been slowly slammed into the continent of Asia, forming a single continuous land mass. I point this out just so you can see physically what is going on here. We have the Himalayan Range, the highest range on the planet, because this little region is *still* actively being slammed into the rest of Asia. But South Asia is a distinct 'sub-section' of the Asian continent in more ways than mere geology.

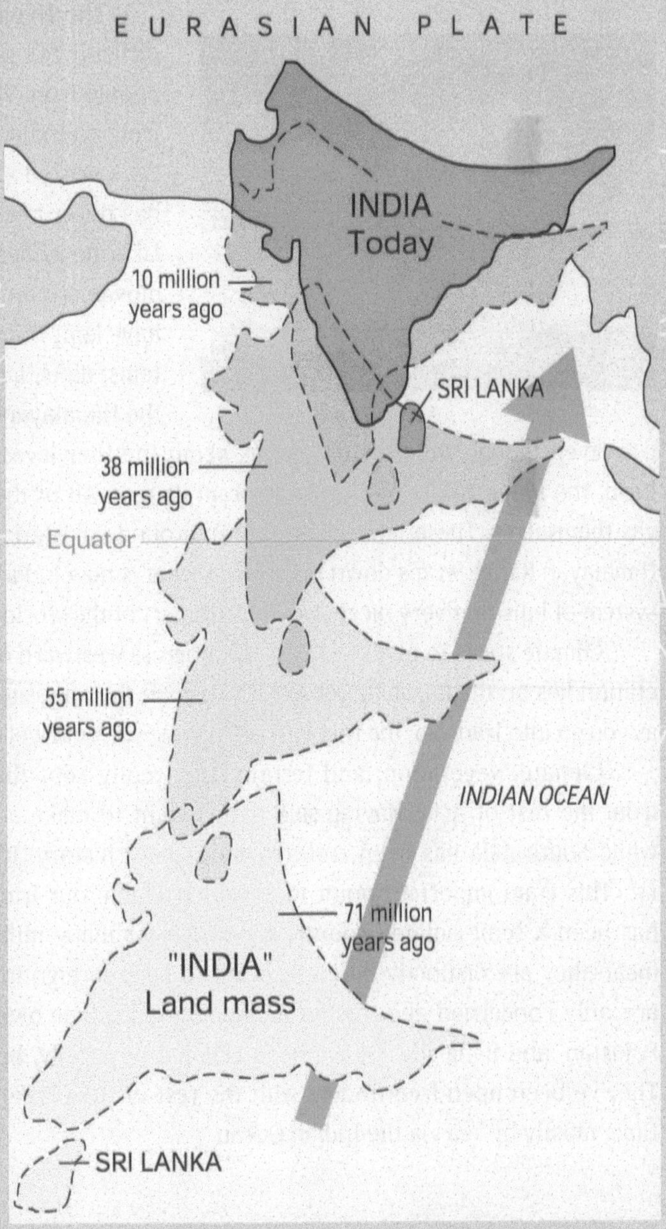

Indeed, if we looked around the rest of Asia proper, we would see that we can distinguish Indians quite easily from the Chinese, or the Burmese, or the Vietnamese, and definitely the Japanese. These folks are different and it's obvious in the way people look, in the way they dress, the way they behave, and in the religions they have. How is South Asia separated from the rest of Asia?

First, what is South Asia? It's the big three: India, Pakistan, and Bangladesh, and the small three: Nepal, Bhutan, and Sri Lanka. We could also throw the small island nation of Maldives in there, but I won't say much about them. Thanks to the rising sea level due to global climate change, the entire country will soon be under water. Sorry, guys!

India, Pakistan, and Bangladesh are the big three not only because they dwarf the other countries in this region in size, but also because they're in the top ten most populated countries on the planet. South Asia may not be a big place geographically, but it is certainly a very populated place.

LET'S GET PHYSICAL

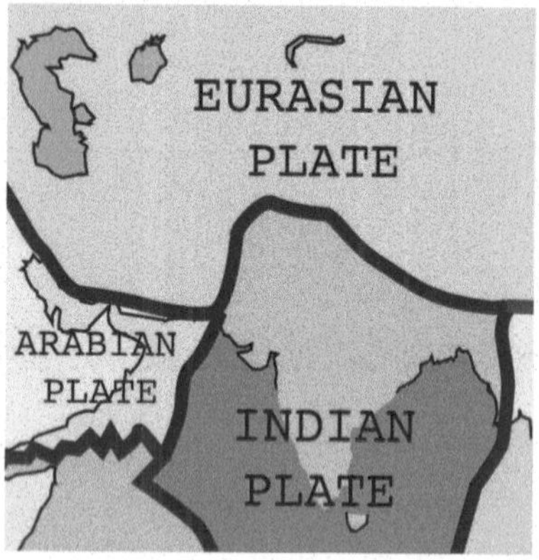

One of the reasons why we have this distinction of culture in South Asia is based somewhat on the physical world; South Asia as a subcontinent has been kind of isolated from the rest of Asia. Isolated how? If we refer back to our opening map, we can see that it is isolated in several ways, mainly by terrain and climate. It's isolated terrain-wise by the Himalayan Range, the mightiest, most powerful range on the planet. The highest mountain in this range is Mount Everest, standing at 29,029 feet. All the other peaks that make up the range are in excess of 20,000 feet. In fact, the top ten highest peaks on the planet are all here.

The Himalayan Range is important because it serves as a barrier. It's difficult for people from our part of the world to differentiate between people from Vietnam and China, for example, but to distinguish any of them from an Indian is pretty easy. The Himalayan Range is just too tall to transit. Do some people get through it? On pack mules, yes, but a very few; there has never been a lot of economic or human transaction across this barrier. Like the Andes, the Himalayan Range is one of the major barriers to human movement on the planet. The Himalayas are even bigger than the Andes and have kept things quite separate as cultures evolved in terms of religious systems, diets, languages, and ethnicities. It's all quite distinct from one side of the Himalayan Range to the other.

Even though we often only think about the Himalayas, if you follow the border of this region you get to the Hindu Kush, the Pamir Knot, and the Karakoram Ranges. All of these are really huge mountain systems on par with the Himalayas themselves. These latter ranges wrap around Pakistan, separating it from Afghanistan. If we go to the other side, the Himalayan Range wraps down in between what is now India and Burma. The Himalayas are the biggie, and it's a continuous system of hills that very nicely isolates this part of the world from the rest of Asia.

Climate starts to play a role as you progress westward out of this region; some pretty severe rainforest and jungle vegetation lies on the Burmese border. If you go in the opposite direction, you run into the Thar Desert and into more deserts as you go into Iran—so, the region is dry on one side and wet on the other.

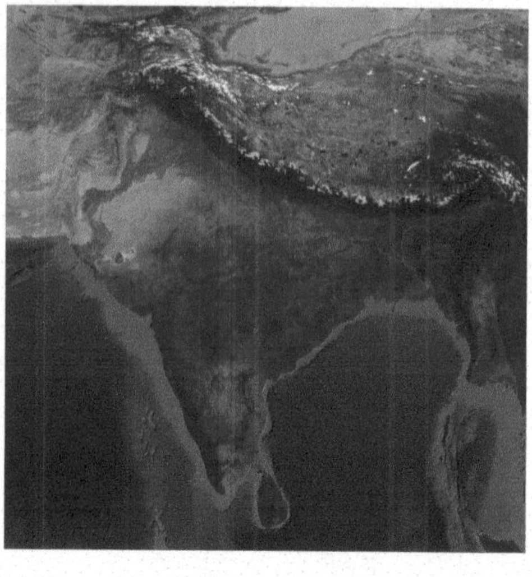

Climate, vegetation, and terrain have really kept this place isolated from the rest of Asia. Having said that, I want to make a quick point that while South Asia has been *isolated* throughout history, it's not *isolationist.* This is an important point to consider. China, our friend to the north, has been a semi-isolated country by design for many millennia. By that, I mean they are distinctly different, want to keep foreign influence out, and are only concerned about what is happening in their own country. India, Pakistan, and Bangladesh have been isolated physically, but not by choice. They've been open free traders with the rest of the planet for a good long time, mostly by sea via the Indian Ocean.

SLIPPERY WHEN WET

A final, distinct climate factor in South Asia is the **monsoon**. The monsoonal system was first identified and characterized in India. In winter, cool dry air moves from across the continent, flows outward over the ocean, and keeps the climate dry. During the spring and early summer, this system completely reverses. The winds change and start moving toward the land. They move from the warm wet waters of the Indian Ocean, bringing warm, wet air masses over the Indian subcontinent, and start to produce huge amounts of rainfall once the wet air masses slam into the mountains. The first terrain that they hit is the **Western Ghats**. The Western Ghats is just a small mountain range that floats down the western fringe of India. These mountains are only two to three thousand feet high, but it's just enough to force up the air masses and drop tremendous amounts of rainfall on the west coast of India, creating a classic monsoonal system, dropping in excess of 80 to 90 inches of rain per year. As the rest of the air mass progresses inward, it eventually bumps into the Himalayan Range; as that warm wet air is pushed upwards, it has the same effect again, resulting in heavy rain on the fringe of the Himalayas and in Bangladesh.

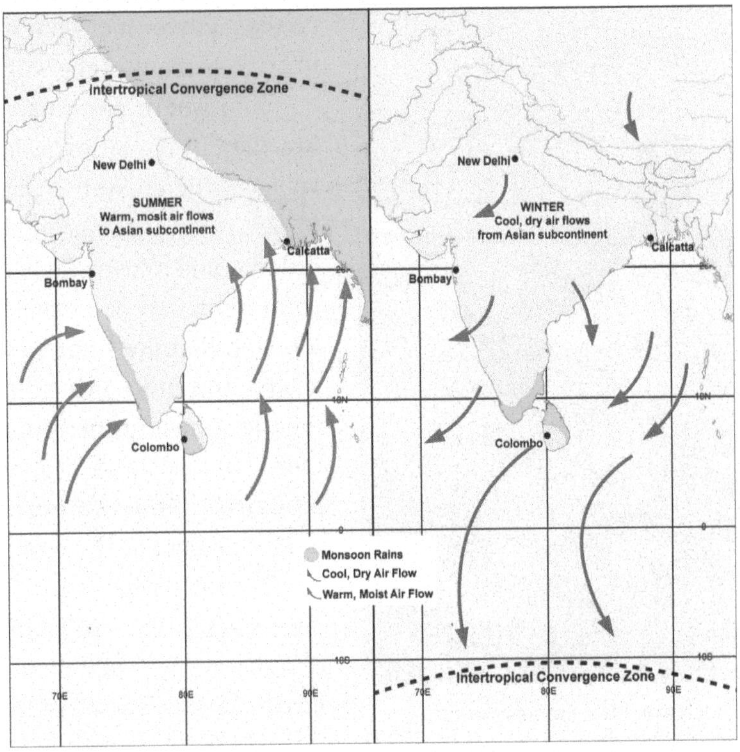

Monsoon mania: the shifty air system switch-a-roo!

This system has really driven the life cycle in South Asia. Grain harvests, movements of people, and whole economies are based on this, and when the monsoon rains are too early or too late, it can spell disaster. There's a pulse to the South Asian way of life that moves around this whole monsoon system. However, not all places in India get torrential rain during one time of the year and not the rest. It's really subtle, and it's different depending upon where you are in this part of the world.

Parts of the subcontinent get tremendous amounts of rain, while other places like New Delhi, the capital, are not so much different than Washington, D.C. They're in the midlatitudes and have seasonal changes in weather, but it's a bit warmer and drier throughout the year. They don't get torrential rains, but there is a strong rainy season. It gets dry in the winter, really dry into the spring, and then the rains come. The rains tend to concentrate in late spring and early summer, not much different from parts of eastern United States. Just to give you an idea and make sure that the Plaid Avenger is not steering you wrong, we hear about monsoon Asia and monsoon India, and it's a factor in the physical world there, but it's not uniformly distributed. There aren't tremendous amounts of rainfall in all places at once with dry seasons in between. It's quite a dynamic system, actually.

PEOPLE POWER

We've discussed the physical components of South Asia, but what about the people? Where are the people? In South Asia, there are loads of people. India's population alone has topped 1.3 billion. If we look at Pakistan, their population is over 190 million; Bangladesh, over 156 million. All three of those countries are in the top ten of the most populated countries on the planet. Taken as a singular region, it's well over a billion and a half people. Someone

Undercover plaid agents at work.

do the math for me: what's a billion and a half of seven billion? Isn't that closing in on 20 percent of the world's population in this one region?

And where are these people? Are they concentrated in coastal margins? Are they in big cities? Small villages? Are they in the mountains or the river valleys? The answer is yes! They are everywhere and live all over this place. In fact, to go out on a limb here, the Plaid Avenger spent quite a bit of time in India undercover for various missions, and there really is no place you can go in India that you can't find people at all hours of the day and night. This is a region that does not sleep. It is a busy place all the time with people everywhere. This place is bustling, it's happening, and happening nonstop. If we just look at a handful of countries like Bangladesh, India, and Pakistan, throw in China, Indonesia and Japan, we're looking at maybe 12 percent of the world's land area. Fifty-two percent of the entire world's population lives on just twelve percent of the world's land area.

Let's describe some of these people packed in all over the place in a little more detail. Why are South Asians different from other Asians? When we think of India, we typically think of Hindus. Is that a religion, a language, or an ethnic group? Let's sort some of these terms out.

HINDU OR HINDI?

Hindi is a language; therefore, we think that's the main language of India. And we would be wrong already. Hindi is a language, one of maybe fifteen or twenty major languages spoken in just this one region. Those are just the major languages; there are several hundred dialects and mixed languages through South Asia. This is very confusing for people from other parts of the world who live in primarily single language societies. We assume there must be some linguistic commonality in this region. There is no majority language at the moment in India. In fact, if you want to cite a lingua franca of this region, that would be English. That's got something to do with India's British colonial period. English is spoken by a lot of people, but perhaps only by fifteen to twenty percent of the population. Hindi is just another language of many spoken here; indeed, if you talk to people from India, Bangladesh, or Pakistan, you will find that many of them speak several languages.

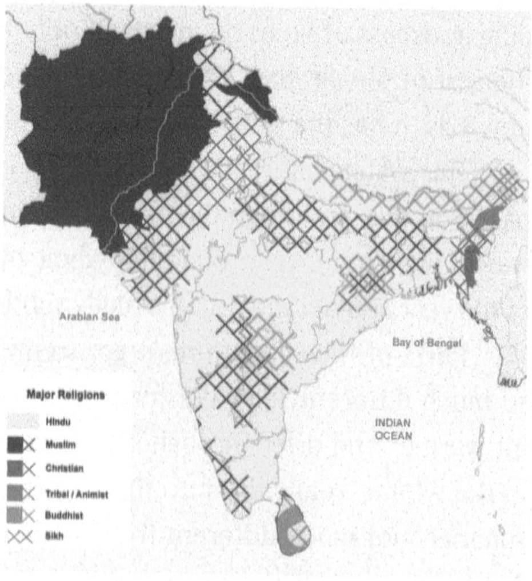

Hindu, Muslim, and Buddhism abound . . .

One of the first languages South Asians learn is their local dialect. That can be one of several hundred dialects and mixed languages. The second one they learn is often something like Hindi, which can be used as a kind of lingua franca inside India. The third one is typically English, for perhaps obvious reasons of economic advantage. There are a variety of languages spoken and they exist in a kind of hierarchy of how they're learned. English usually wins out in today's world, Hindi is a far distant second, and a local dialect would be third.

So not all South Asians are Hindi . . . not even close. They're not all speaking *Hindi*, but are they all *Hindu?* **Hindu** is a religious system, and we tend to think of Hinduism as a description of the religious lifestyle of people in India—like Christianity as a lifestyle and belief system based on the life and teachings of Christ. When you are around Indian people, they now recognize the term Hinduism as describing their religion, but it does not really mean anything specific to them. It's not a self-defined term. Hinduism is a conglomeration of cultural beliefs that foreigners from the outside world said, "Well, you guys speak Hindi . . . how about we call this lifestyle you live as 'Hindu'?"

When we think of India, we think the majority of people follow Hinduism, and that would be true. However, the vast majority of people in Pakistan are Muslim. The majority of people in Bangladesh are Muslim. The majority of people in SriLanka are neither of the two—they're Buddhist. We have a nice little mix of some Western and Eastern religions dancing around this place. Bhutan is primarily Buddhist with some Hindu, and Nepal is mostly Hindu with some Buddhist. We have a confusing mix of things going on here, but the two major power players, of course, are Hinduism and Islam.

Buddha sez: Talk to the hand.

VISHNU VS. MUHAMMAD VS. BUDDHA

How did this get to be? What's the deal with this religion stuff? Buddhism is primarily in Sri Lanka, Bhutan, and the Tibetan plateau just north of this region. The Buddha, Siddhartha, the main man in the Buddhist lifestyle, was actually a Hindu born in a Hindu kingdom in what is present day Nepal. (In a similar vein, Jesus Christ was actually Jewish.) He eventually thought, "Hey, I dig Hinduism; I dig the lifestyle, except this whole life cycle and being reborn again and again, and always suffering and being reborn sucks!" He invented his little reform movement to change things, think of things in a different way, and be removed from the endless cycle of rebirth. Buddhism is a universalizing religion, and as such, he wanted to spread it to other places. It spread itself right out of India. It had its heyday from about 300 BCE to 300 or 400 CE. It was in vogue. Everybody dug it for a while, but Hinduism eventually displaced it in India once again, and it largely disappeared. When we think of Buddhist areas, we have Sri Lanka and Bhutan, but we also think of Tibet, China, even Mongolia and Japan as having a huge Buddhist influence. Buddhism moved on beyond India.

Where did the Muslims come from, then? Well, it's the same deal if we think religiously. Islam started in the Middle East in the early seventh century AD with the prophet Muhammad. The empire, what I call the Arab-Islamic empire, expanded out from the Middle East all the way over to Central Asia and the northern fringes of South Asia by the 10th century AD, and for the next thousand years, they heavily influenced South Asia. Off and on throughout the last thousand years of history, Islamic kingdoms and empires have controlled parts or all of India proper, and thus the religion has become infused in the cultures and peoples of the subcontinent. There has historically been some friction between the Muslims and Hindus, and it's still not over. Muslim-Hindu friction is far from a historical remnant. It's still a very real issue, typically where the majorities of each belief meet on the Pakistan-India border. Pakistan is predominately Islamic; India is predominately Hindu. Tensions primarily arise in the western states of India and sometimes manifest themselves in violence.

Hinduism: multi-faceted, and multi-faced, culture of most Indians.

CULTURAL HEARTHINESS

Like China, Mesopotamia, and the ancient civilizations in Middle America, the Indus River valley, where we find India and Pakistan, is an ancient cultural hearth. A civilization cropped up on its own, with big cities and architecture, mathematics, and unique cultural ideas. South Asian culture was perpetuated throughout the millennia in its isolated environment, allowing it to remain quite different from China, or even closer next door, Mesopotamia. These are very different places because of their spontaneous creation and uninhibited growth for extended periods of time. I'm not going to bore you with a lot of details about ancient history in this part of the world because it's too rich, and like China, there is just too much. I just want you to know that, throughout history in this part of the world, there were multiple episodes of different empires and kingdoms, and also formations of some of the big time players in philosophy. Some classics of world religion started out here. Hinduism started here and gave rise to the evolution of Buddhism. Islam was infused later to combine a unique cultural dynamic in this region. Speaking of Buddhism, I just remembered a former reincarnation episode when I met the great man himself . . .

NEPAL,
6TH CENTURY BCE
SIDDHARTHA GAUTAMA WAS BORN INTO THE KSHATRIYA (WARRIOR) CASTE IN A SMALL KINGDOM IN NEPAL. HE LIVED IN OPULENCE AND COMFORT UNTIL ONE DAY WHEN HE VENTURED INTO THE CITY.
COUGH
COUGH
WILL THIS HAPPEN TO ALL OF US?
YES, MY LORD, WE ALL EVENTUALLY GROW OLD, FALL SICK AND DIE.
SIDDHARTHA SPENDS SIX YEARS IN THE FOREST WITH FIVE OTHER ASCETICS GIVING UP ALL COMFORTS IN AN ATTEMPT TO TRAIN THE MIND TO OVERCOME SUFFERING. HE NEARLY STARVES.
THE TRUTH IN THE TEACHER'S WORDS LEADS SIDDHARTHA TO THE MIDDLE WAY--THE LINE BETWEEN THE EXTREMES OF ASCETICISM AND OPULENCE AS THE WAY TO REACH ENLIGHTENMENT.

AS SIDDHARTHA SITS, HE COMES TO UNDERSTAND THE HUMAN CONDITION FULLY AND REACHES WHAT IS CALLED ENLIGHTENMENT. AT THAT MOMENT, HE BECAME KNOWN AS THE BUDDHA--THE AWAKENED ONE.

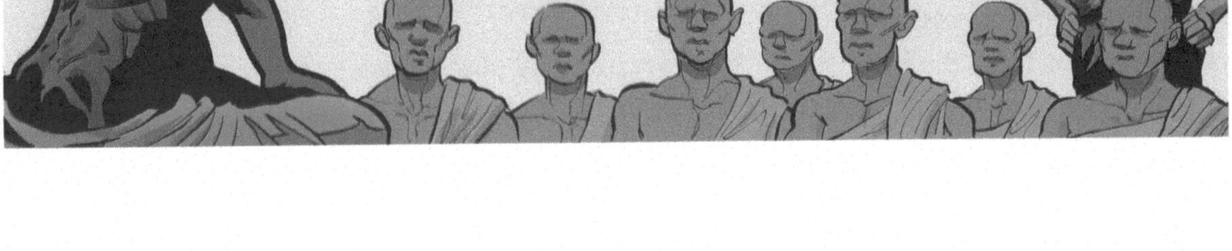
THE BUDDHA BEGAN TO TEACH THAT ALL PEOPLE WERE WORTHY AND ABLE TO REACH ENLIGHTENMENT AND UNDERSTAND HOW TO STOP THE SUFFERING IN LIFE.
THIS WAS A RADICAL DEPARTURE FROM HINDUISM WHICH DICTATED A STRICT ADHERENCE TO SOCIAL CLASS AND ONLY OFFERED LIBERATION TO THE TOP BRAHMAN CASTE. THE FIRST SERMON FOCUSED ON THE 4 NOBEL TRUTHS AND THE 8 FOLD PATH.

HISTORICAL HAPPENINGS

Let's get back to the period that I was talking about earlier, the period of Muslim dominance. We had Hinduism and Buddhism, but eventually the Muslims came to town around 1000 CE. This was a continuation of the expansion of the Arab-Islamic Empire, which began to spread after the death of the prophet in 632 CE. Islam made its way into Afghanistan, and with the help of the Mughal Conquest and events thereafter, ended up taking over virtually all of what is Pakistan and India in today's world. They created the Mughal Empire.

MUGHAL MANIA

The Mughal Empire dominated South Asia for about 800 years from 1000CE to about 1757 CE. It was basically a hierarchy of Muslim rulers, a thin veneer of Islamic influence over the millions of Hindus that lived there. The ruling class was Islamic, but the majority of the population remained Hindu. I want to make sure you understand that, because when we look at today's India, there is a huge number of Muslims in India today—around 150 million. That population of Muslims in Hindu-dominated India is actually larger than the population of Muslims in Saudi Arabia, Iraq, Iran, or just about any other country. The Muslim minority in India is bigger than the Muslim majority in most other Islamic countries. This is a significant demographic to consider, and it has its roots in this period of Muslim influence during the Mughal Empire.

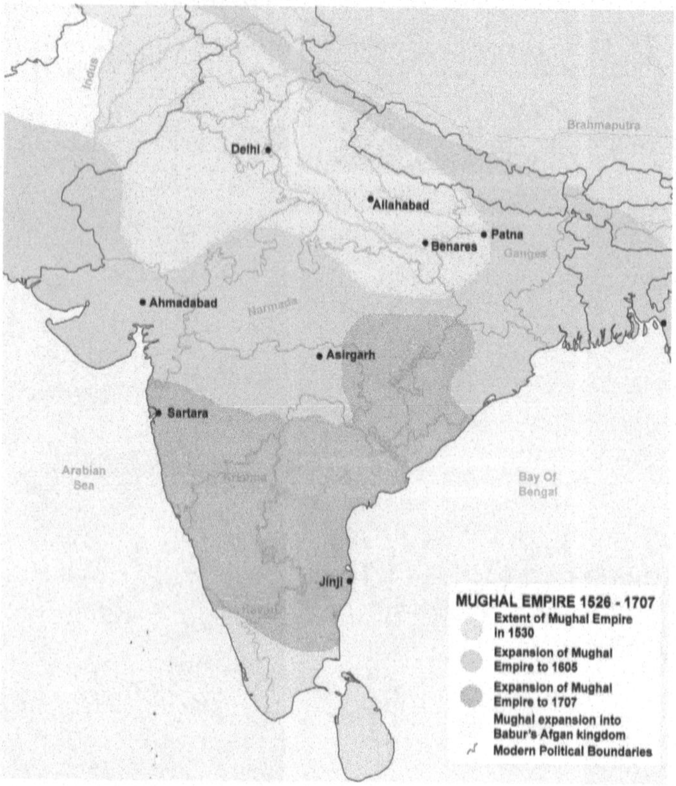

The Muslim Mughal period . . .

The dominance after 1757 soon shifted. Small kingdoms and fiefdoms comprised this period of the Mughal Empire, like European feudalism or Japanese Shogunates, with princes and royal courts. There is kind of a main king, but there were also sub-princes and sub-dukes under those guys that controlled smaller and smaller territories. The beginning of the end of this was around 1700, as foreign intervention started to creep in. Surprise! Who was responsible for this foreign intervention? Who sent out fleets of ships to establish in-roads in other parts of the world?

ENTER THE EUROPEANS

That's right, our old friends in Western Europe, and one up-and-coming colonial power in particular: the British. Actually, the Portuguese also established a small trade province on the west coast of India, called Goa. These two European powers made inroads into South Asia for the main purpose of trading. You may have even heard of the British East India Trading Company, for instance. It was a company sponsored by the British government. They worked out a deal with the company by saying, "Here is your charter, go out and establish trade with South Asia. We will insure the boats and in return we get a cut of your business."

Why did they want to do this? You should know very well by now. They wanted to get raw commodities—the good stuff, such as spices, silks, and ceramics. The East India Trading Company, sponsored by the British, wanted

Can you spot the European establishing trade links in this picture?

to get in there and get direct access to the good stuff, and cut out all the middle men. They turned out to be very good at this.

These guys were not just partially successful; they were radically successful at establishing trade with South Asia because the Mughal Empire was winding down a little, and didn't have a very firm grip on things. There was only a small veneer of Mughals in control of that vast majority of Hindus in India, so their control weakened and it became fairly easy to get in and establish port cities and trading colonies. These cities started to grow into territories that the trading company controlled. They essentially went into the next fiefdom or dukedom and chatted with the local prince saying, "Hey, we're going to give you loads of money to trade some stuff with us. It's going to be good for you!" Plenty of princes said, "Yeah, it looks good to me. I'll sign a trade deal, no problem!" Through diplomatic relations and strategic ties, the British made their way in and were more successful than they ever dreamed. They ended up controlling huge chunks of territory.

Eventually, disputes arose between princes who had some previous friction between their kingdoms. This trading company was so successful that, though it was set up exclusively for trade, it found itself acting as a diplomatic entity between these feuding kingdoms. These disputes could potentially turn into violent conflict that would ruin trade completely. The higher-ups at the trading post glanced at their royal charter, and got Britain on the horn to let them know that things may be getting out of hand.

Great Britain looked at this mess and says, "Whoa, this is not your charter! What the heck is going on here? Sit tight, we're taking this party over." A trading company was not fully equipped to handle colonization. The British crown sent in the warships and court jesters and everything in between and, in effect, fully colonized India. From then forward—we're talking about 1800—this whole territory of what is now India, Pakistan, and Bangladesh, was under the umbrella known as **British India**. It was an official British colony. There were some independent kingdoms left over that held out and had diplomatic relations, places like Bhutan, the kingdom of Nepal, and Ceylon, which is now Sri Lanka. These guys held out as independent territories and avoided the burden of colonialism. All the rest of it, the big three, became one big land conglomeration called British India.

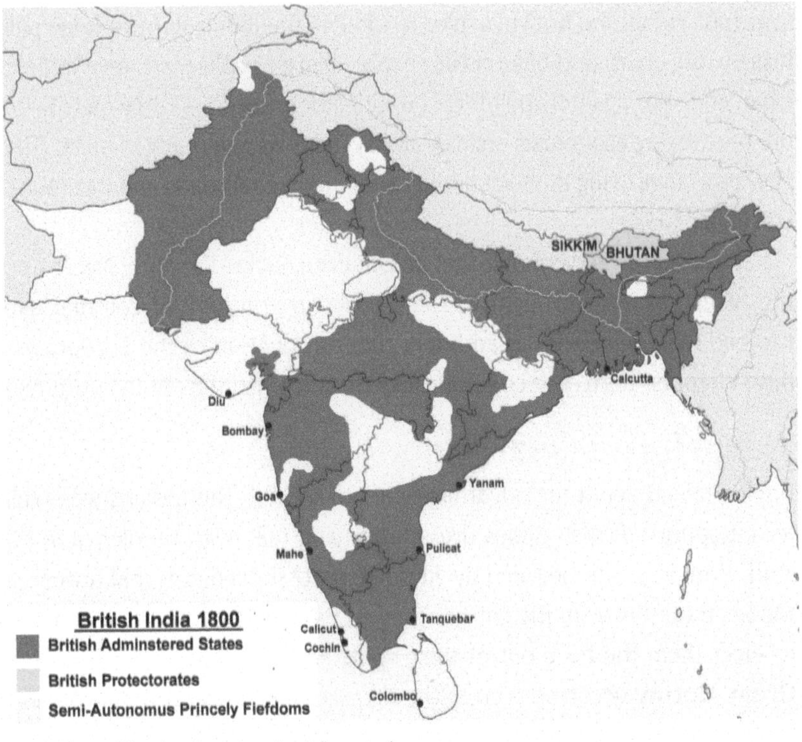

. . . followed by the British colonial period . . .

India is the biggest country . . . I keep referring to India because India is the biggest power player: it has the biggest population, the biggest economy, and the biggest military. When we think of the South Asian region as a whole, it's not fully developed, but it's certainly not in last place either. Far from it. This location is much better off than other places like Africa, or even closer places in Central Asia or Southeast Asia. What's different about South Asia that has kept it ahead of the pack when it comes to developing countries or regions? It has a lot to do with its colonial experience. Some things happened in British India that gave it its strategic advantage. Before I irritate my Indian friends, let me state for the record: Colonialism sucks! The fact that the British were controlling your people probably sucked big time. However, there were some distinct advantages that have put India in the catbird seat today.

DROP THE WICKET, JAMES, IT'S TEATIME: ADVANTAGES OF BRITISH COLONIALISM

My Indian friends, you're not so bad off. *Number one:* it was a singular European entity that took you over. As such, there wasn't internal fracturing like in Africa or even to a lesser degree in South America or Southeast Asia. That's kind of a plus. At least you know exactly who to blame. *Number two:* the colonizer was Britain, which means you spoke the Queen's English. While at the time it may have been revolting to your great granddad, it has benefited you greatly in today's world. English is the lingua franca of business, travel, airlines, the international space station, and technology. Everybody speaks English when they deal with that stuff, and that has put you, South Asians, on a strategic fast track. It's a great advantage to speak English, even though it's part of your colonial heritage.

There's a good chap. Now fetch us a spot of tea!

What else did Britain do differently here than in other places that were colonized on the planet? *Number three:* Internal investment. That's a biggie. In Africa, we pointed out that most investment was simply to extract the resources. We could even say that about the New World. When they got to the New World, or even to Africa or Australia, the British and other colonial masters didn't really add too much in the way of infrastructure. When the British arrived in India in the 1800s, it was a huge place with a long history, big cities, and millions of people—more people than were in Europe at that time. The British couldn't just wipe India clean and start again by building new industries, nor could they merely focus on extracting stuff. There was already stuff going on; preexisting economies and cultural systems were already in place. The British focus in British India was updating the infrastructure, improving it, modernizing, and building railroads and communication systems across the colony, many of which are still in use today.

I cannot stress enough that colonialism sucks. I'm sure South Asians would have preferred being on their own, but the British at least brought in some perks. If you look at colonial experiences around the world, British India's wasn't too bad in comparison. In addition, the introduction of the Europeans in the Americas decimated local populations; that didn't happen here. The colonial story is radically different in South Asia.

ADVANTAGES, SCHMADVANTAGES

Having talked about the advantages, it still sucked. The Indians were ruled and controlled by a different people, with a different culture. British dudes dress funny and talk even funnier, and play games like cricket, and croquet—pretty strange stuff. Who wants to be ruled by guys like that? Indeed, internal unrest developed fairly early on. Many Indian folks weren't too keen on this from the get-go. They weren't so keen on the Mughals, and they were less keen on the British and wanted to eject them the heck out of their homeland. This came to a head for the first time in 1857, when something called the **Great Mutiny** occurred. You might also see this referenced in India as the "First War for Independence." See inset box on the next page for Mutiny highlights.

MY MAIN MAN MAHATMA

Enter Mahatma Gandhi. Everybody knows Gandhi. He's one of the coolest dudes in world history. Everybody loves Gandhi, kind of, so what's his deal? Gandhi was an Indian guy, trained as a lawyer in Great Britain, who spent some time in South Africa peddling his craft. He was a British citizen because all people in the colonial system were supposedly citizens of the

What's the Deal with the Great Mutiny of 1857?

Why two different names for the same thing? It depends on who's telling the story. "Mutiny" has the connotation of something bad that needed to be suppressed. "The 1857 First War for Independence" has a more positive connotation: "We are trying to be free." Try and figure out which group called it what. Other names for this single, short fracas include: The Indian Rebellion of 1857, the Indian Mutiny, the Revolt of 1857, the Uprising of 1857, the Sepoy Rebellion and the Sepoy Mutiny. What was the Great Mutiny all about? It was obviously about independence, but amazingly enough, it came to a head because of fat—animal fat to be exact. Pig fat, cow fat, lard, all kinds of yummy animal fat, put together in a big stew. Why would the British care about fat? They didn't. However, during their colonial days, the British were pretty savvy about controlling the local populations. One of the more savvy ways they did this was by employing locals in the army and the police force. Why would they do that, Plaid Avenger?

Here's why: When you are from a different part of the world and your skin color, your religion, and your language are obviously different, people are immediately on guard if you boss them around and tell them what to do. That sort of thing causes resentment issues to crop up among the locals: "This dude thinks that he's better than me! I don't like being bossed around by an outsider. What the heck does he know?" The British were familiar with this human tendency. They knew that they could alleviate this problem by getting local people who looked, talked, and understood the problems and issues of the people there, and training them to beat down the locals! It was a brilliant system, and also easy to implement; the British did it all over the place, and it worked magically in India.

Back to the fat: nobody was eating it. They were using it to grease the cartridges for their guns. That's not a big deal—how could that cause a mutiny? People in India at the time were of what two religions? That's right, Hinduism and Islam. What people would have problems handling cow fat and pig fat to grease the cartridges of their guns? That's right, Hindus and Muslims. Cows are sacred in Hinduism, while pork is unclean to Islamic folks. In all of their wisdom, the British soldiers said, "Here's the grease for your guns, chaps. Cheerio!" The local soldiers said, "We hear it's made of pig and cow fat. We're not touching it." The British, in all of their worldly wisdom, said, "Do it or die." And the Great Mutiny began.

It was quelled fairly quickly and didn't become a very big uprising, but it's still remembered in popular culture. This was the first attempt to throw off the yoke of British Imperialism. The Great Mutiny, as coined by the British, and the War for Independence as coined by the locals—those being the Indian guys. Obviously, the British were going to be ejected at some point, because they aren't there now. How did British colonialism end if it didn't end with the 1857 Great Mutiny? Where did they go from there?

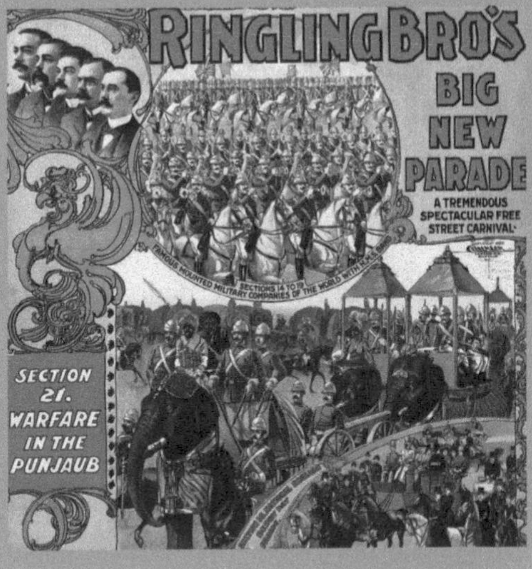

Oh! A circus! No wonder they mutinied.

mother country. When he lived in South Africa, he became incensed because he realized that, though he was a British citizen, he was still considered second-class by "true" British people (a.k.a. white British people). Gandhi had an awakening at this point and realized, "This sucks. I'm not really a British person, even though I went to school in Britain, talk like a British person, and work with British people. The British don't think I'm British." Bothered by this, he went back to the motherland, where he realized the British sucked. (Okay, excluding Monty Python.) He was Indian and wanted India to be run by Indians.

Gandhi started a peaceful movement to kick the British out. In other words, he started the movement by saying, "You are going to leave. We're not going to shoot at you. We're not going to fight you, we're not going to beat you down with clubs and physically kick you out. We're going to make you leave without even raising a fist. With no violence."

Most famous Indian dude.

How did he do this? He did a variety of things. Of course, he didn't do it by himself, but he was the driving force and most popular leader of this nonviolent movement. He not only ended colonialism in British India . . . he simultaneously, and much more importantly, ended neocolonialism. He realized the two things were inherently tied together.

Gandhi G is money! Literally!

What is neocolonialism? Neocolonialism is when a foreign entity controls your country, not politically, not physically, but economically. It owns a lot of the business and controls the flow of goods and services, and sets up things that are an advantage to their country and not yours. One of the reasons India is doing really well in the world today, while places in Africa aren't, is because neocolonialism is still happening in Africa and it was stopped in India.

When the British came into India, it was during Britain's industrial era, and they needed raw commodities. They had weaving mills and industrial stuff and the industrial capacity to make things, but they needed raw materials. They then processed the cheap raw materials into things that sold for more money. The best example of this neocolonial economic system in India was the wool the Indians had been spinning into their own cloth for 5,000 years. The British bought the wool from them for virtually nothing, processed it into nice crisp linens in factories back home in England, which employed their own people. British workers made a good wage, and then Britain brought back the cloth to sell to the Indians at a huge markup. This was really a fantastic, sweet deal. How could the Indians say no? They really couldn't because the British were in control of the country.

Gandhi was aware of this system when he started this nonviolent movement. He told everyone, "We need to kick these guys out of here. We could physically take up arms and kick these guys out, but the system is so skewed, we'd still be screwed. We need to fix it from the inside out. We need to end this neocolonialism crap, along with the political colonialism. Hey! Why are you dudes buying this British cloth? What in the hell are you thinking about?" I'm paraphrasing: Gandhi probably never said "hell" or "dudes." Back to Gandhi: "We're Indians; we've been making cloth since before there even was a Great Britain. Why are we buying cloth from these guys? Can someone explain this to me? It makes no sense whatsoever! Stop buying British cloth and start weaving your own again." Gandhi only wore homespun cloth he made himself for the entire second half of his life. His kind of skimpy toga was a symbolic gesture saying, "Look, this is all you need. You don't need any of that fancy stuff." If you ever see pictures, a documentary, or even the movie *Gandhi*, you will see that he is often depicted working on a spinning wheel. The spinning wheel was a symbol for the peaceful independence movement: "Do it yourself; let's break from this economic cycle. Free yourself from this slavery and make your own cloth!"

Once people heard of this practical way to gain independence, the movement started to gain speed. Around the late 1930s and into the early 1940s, it was not yet a nationwide deal. More people were talking about independence, but most folks were still talking about nothing. This little nonviolent movement was kind of a mosquito bite to the British, no big deal. As the movement progressed, the British started to pay attention: "We might want to start dealing with this Gandhi guy and figure out how to shut him down." What did they do? They made the cloth cheaper. They made it so cheap that people couldn't afford not to buy it.

When Gandhi saw this, he said, "Well, that got their attention! I have another idea. We're currently buying salt from the British. Hello, can I have everyone's attention? We have a ton of coastline where there is this substance called *saltwater*. It becomes salt when the water evaporates. Salt is free, why are we paying for it?" He got the word out and, in 1930, organized a Salt March to the Sea, like a common strike, where everybody went to the coast and made salt. The movement grew in popularity; Indians set up salt producing shops on the coastline. Gandhi himself worked at them between speaking engagements and rallying the people. This became a big deal; the British were starting to get really furious because their profits were being destroyed by this salt boycott.

These things culminated in British discussions with Gandhi: "Hey Gandhi, lets work things out, we're your buddies." Gandhi's response: "It's time for you limeys to leave." The British still did not think that it was too big of a deal, so Gandhi tried a different tactic. By now he was massively popular, and he encouraged everybody in India, which at this time was probably half a billion people, to not go to work the next day: "Everybody skip work; we're having a general strike." The British, being very much in touch with the common people—can you smell the sarcasm?—had no idea this was happening. The strike was a smashing success. The whole country just stopped for one day.

The British really saw the writing on the wall and realized that the Indian people had the real power here. They called a general strike, they ceased all economic activity, and they shut down the country. They were willing to make it incredibly unprofitable for the British to be there. At the same time this was happening, World War II broke out. If you know anything about World War II or read any of the chapters thus far in this book, you realize that it was kind of a big deal in Europe, and one that kept the British quite busy.

We know that Europe was essentially destroyed at the end of World War II. The British were spent physically, mentally, and economically, and all these protests were occurring in India at the same time. It just became too much. They dragged India along to support them in World War II, but the common perception on the street in India was, "It's not our war. We're not getting involved. We're trying to kick you out of here. We're not coming to Europe to spill blood for you, ours or anyone else's." By 1945, the British had to throw in the towel; they just couldn't hold on to India anymore. They didn't have the resources to keep it under control, by force or otherwise. They had to rebuild at home. They started concession talks for India to become independent; in 1947, that is exactly what happened. In 1947, they drew it up on the conference table and handed over control to the locals.

PACKIN' OUR BAGS FOR THE PARTITION PARTY

Now the story gets complicated. A bunch of other people in the country wanted another kind of independence: They wanted an independent India as well, but not Gandhi's India. They weren't Hindus like Gandhi; they were Muslims. This brings us back full circle on the topic of religion. There was still a huge minority of Muslims in this country. Their basic platform was, "The British suck, but here is the deal. We're a minority, and if the British walk out of here, the Hindu majority will dominate us. The only reason they haven't yet was because the British have been in control for a few hundred years. If we don't get our own place, if we can't subdivide British India and get our own country, then we don't want the British to leave."

Ali Jinnah, Father of Pakistan.

Here is where we have to introduce a new character in our play: a guy called Muhammad Ali Jinnah. Gandhi is often credited with being the father of India, and Jinnah is often credited with being the father of Pakistan. He was a Muslim in British India and led the fight for the formation of a distinct and separate sovereign state for Muslims. He was also British trained and educated, a lawyer just like Gandhi, and a savvy politician. Gandhi was vehemently opposed to this, as were many others in India. However, when all was said and done, that's how it had to go down. It got tense once Jinnah told Gandhi that he didn't want the British to leave. Gandhi didn't want that; he had been fighting for years to be free of the British, so he considered it the lesser of two evils. He and his party capitulated and said, "We want a unified state, but we'll settle for the split." On a side note here, there was a minority of people in India who actually hated Gandhi (some still do) because they blame him for the country split, for 'losing' Pakistan to the Muslims.

In 1947, Britain packed up shop, redrew the map, set up partitions, and sailed off. What we see here in this map is something similar to today's map. We've got Pakistan, India, and another place called Bangladesh, but back then it was called East Pakistan. The agreement set in place two separate homelands for the Muslims in British India: East Pakistan and West Pakistan, which were two different territories that, at the same time, were a singular sovereign state. It was 'Pakistan,' but with an eastern and western side. India was for the Hindus, and Pakistan was for the Muslims. This is all fine and dandy on a map, but the reality was that Muslims and Hindus were living all over the place. It came to a fever pitch on the day of independence in 1947, when there was a mass migration of people. Tens of millions of people found themselves on the wrong side of the border and suddenly needed to move.

The Muslims in India said, "We'd better get over to East or West Pakistan or we're going to get the crap beat out of us!" The Hindus in East and West Pakistan said, "We're on the wrong side; we need to get over to India." This was a very turbulent time . . . a bloody day in India's history. As you might expect, people were being uprooted from their homelands, places where their families may have been forever, and were now on the wrong side because of the new lines on the map. They ripped up all their material possessions, uprooted all their family, and moved. They were instantaneous refugees. The friction and the tension grew between these massive movements of people who were not necessarily enemies but certainly not friends, their differences accentuated by going in two different directions down the highway. This ended in violence and massive rioting. It was a nasty day in history, and set the stage for future tensions between India and Pakistan.

This gets back to what I was talking about with issues between Muslims and Hindus in today's world. This is still hot. In the last decade, there have been dozens of different massive rioting incidents and acts of terrorism. Unknown things spark the unrest and massive slaughters of Muslims, which then spark slaughters of Hindus, which are then avenged in mass slaughters of Muslims. This is nasty business, and has escalated to full-fledged international warfare three different times since 1947.

The only other date I want you to remember is in 1971, when it wasn't so much based on religious conflict as it was political maneuvering by East Pakistan. They weren't happy. They felt that West Pakistan (where the central govern-

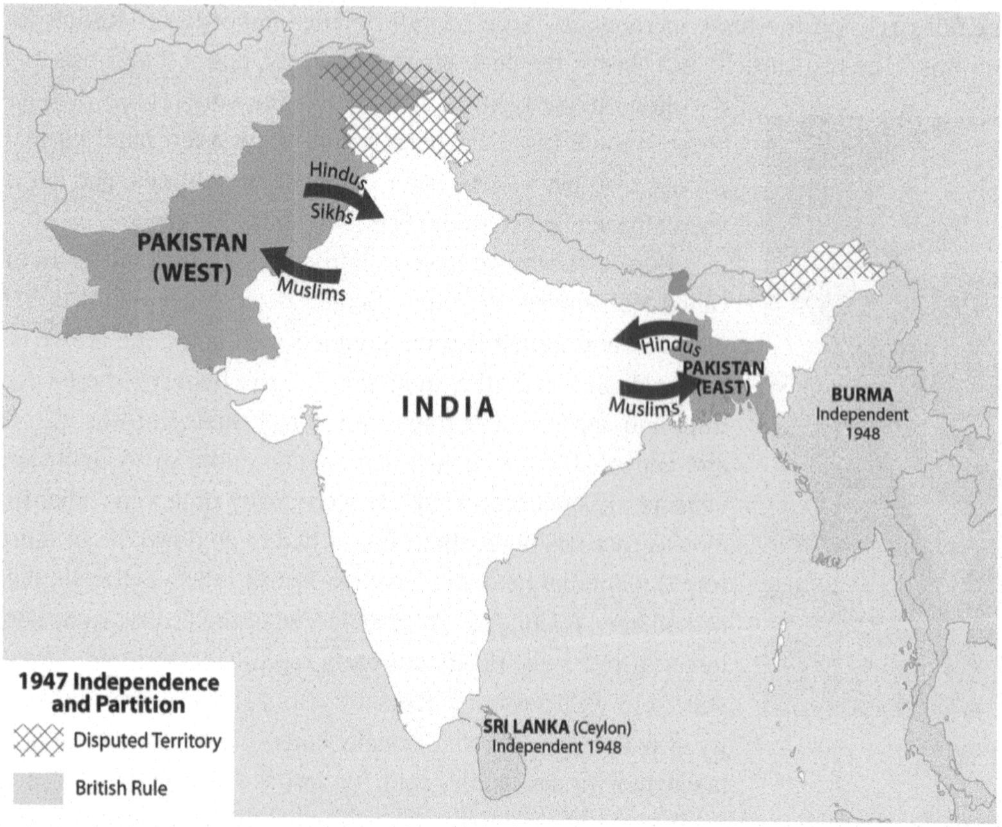

Independence! With partition and relocation! Along with death and destruction!

ment was located) was too far away to properly understand their needs. India helped fuel this internal dissent, which turned into outright warfare between West Pakistan and India, during which India helped convince East Pakistan to declare independence and it subsequently became Bangladesh. Now we have three countries, and now you understand why Pakistan is almost 100 percent Muslim, why Bangladesh is almost 100 percent Muslim, and why India is somewhere in between with a vast Hindu majority. Occasionally, riots still occur, mostly along the Pakistan/India border. I remember a few years ago when a trainload of about forty Muslims burned to death after some Hindus barred the entrances and exits and caught the thing on fire. This sparked about two months of retribution riots. This problem is a historical relic of the tension from long ago, still active in today's South Asia.

South Asia now . . . well, at least for now.

TIME FOR TODAY'S SOUTH ASIA

This is a mixed bag of states, an assortment of cookies. Are these places developed, or are they developing? Who's got the money? Who's doing well? Who isn't? When we look at South Asia, India is the premier powerhouse player with the most people and the biggest economy—one that is really exploding right now. Exploding like China? Not quite that hot, but it's developing fast. There is a developing middle class as well, so the developed/developing question is a tough call. Let's look at some issues to help us define their status.

Bhutan house party!

THE SMALLER STATES

We've mostly talked about the three big boys in this region: India being the giant, and Pakistan and Bangladesh being the other two. What about the other countries, the three smaller guys: Bhutan, Nepal, and Sri Lanka?

Bhutan is easy. The whole country is like a Buddhist monastery. They've been hosting their refugee brothers from Tibet for quite some time. It's a home away from home for the Tibetan monks, including the Dalai Lama. Bhutan is a curious place in that it's trying to preserve its 11th century standards in the face of 21st century advancements. Until 1999, there was a nationwide ban on television and the Internet. Though the ban has been lifted, there are still active movements to prevent popular culture from the rest of the world intervening in their culture that is still an old school monarchy. Bhutan has held its own throughout the centuries. If you want to time travel, go there. Via the Internet and telecommunications, this tough nut is starting to be cracked, so hurry and check it out before it becomes totally globalized!

Nepal is perhaps even more fascinating, because it would be similar to Bhutan twenty years ago, but now it's utter chaos. I want to end this section talking about two poles of conflict on either side of this South Asian region. The active conflict bookends: Nepal in the north, Sri Lanka in the south. Both countries are on the brink of civil war/collapse/insurgency where people are going to get killed en masse. In the last few years, Nepal has had an increasingly bigger problem with a Maoist insurgency that wants to totally eject the ruling monarchy. Their primary mission is to change the government from a constitutional monarchy to perhaps communism, or maybe socialism. It's not very clear. What is clear in Nepal is that there is fighting, and over 11,000 people have died since 1996. A rebel group scattered throughout the country has been fairly effective in striking targets and generally causing chaos.

On top of that, in June 2001, the heir to the Nepalese throne went postal and massacred most of the royal family over a reported dispute about his choice of potential bride. He then shot himself, but lived for a little while longer. He was crowned king while he was comatose. That event fueled the fire for the Maoist rebels; they loved the idea of having a comatose king. They thought it was great and that they may be able to get control of the state by causing more chaos with no king to rule the country. When the comatose king died two days later, a lesser-known uncle became king and began pulling power away from more democratic institutions to himself, under the guise of needing more power to fight the rebel insurgency.

In April 2006, in the government's effort to crack down on the Maoists, they pissed off the pro-democracy movement in the country resulting in a bizarre triangle of folks: the Maoists teamed up with the pro-democracy people to overthrow the monarchy all together. They pretty much accomplished this by 2008; the monarchy was sidelined, elections were held and the place settled down quite a bit. It's still a very new democratic experiment for these folks, so future instability is entirely possible, and probable, as they figure out how to go forward. Need proof of the bizarreness? The monarchy was fully abolished in 2008, and the first full democratic election resulted in their current (and first) president who is a centrist Hindu, with a prime minister who is a Communist atheist. Party on! Most Nepalese folks live on less than a dollar a day, adding to the fray of instability. Here in 2019 things have stabilized a lot, and things seem to be progressing upwards for the peeps of this place, but at a very slow pace. At least the fighting is now under control via a democratic power-sharing agreement.

Let's go to the other side of the South Asian region for the other still quite active conflict in *Sri Lanka*. We will talk more about Sri Lanka in the Plaid Avenger hotspots chapter, but let's introduce the situation here: Sri Lanka is a country that has been struggling with a full-on on civil war. They have been off and on for about twenty to thirty years now, all based on an ethnic division.

What's the Deal with the Tamil Tigers?

In Sri Lanka, the rebel insurgents/terrorists had a killer mascot . . . literally. The dudes called themselves the Tamil Tigers. That's their political party. These guys were fairly vicious and by all standards they would be in the top tier of terrorist organizations for their practices. The Tigers had been blowing up people via human suicide bombs for quite some time, a technique they mastered in the 1970s. In fact, they were even responsible for the assassination of Indian Prime Minister Rajiv Gandhi in 1991. They were well equipped and organized, had uniforms, boats and even a few aircraft, but most importantly actually controlled territory in the northern part of Sri Lanka. That all came to an end in summer 2009 as the government made an all-out effort to smash the entire organization. Even though it caused massive civilian casualties, it appears the civil war is over and the back of the Tigers has been broken. For now.

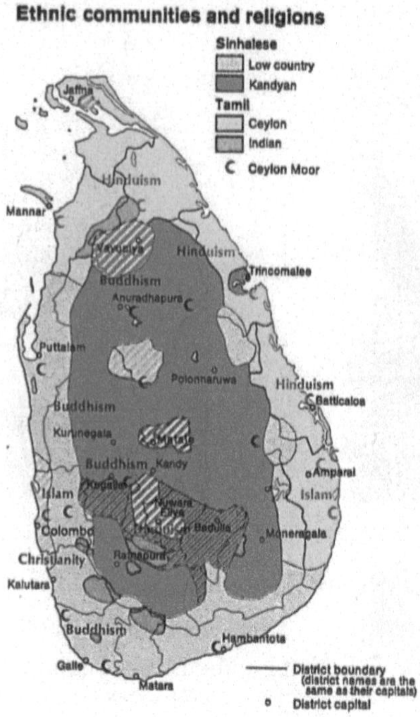

In an ethnic enclave on the northern fringe of the island, there are people called the Tamils. It's an ethnic group that actually came from India, and they had what was easily the most organized and outfitted rebel group on the planet. See box on the previous page on the "Tamil Tigers" for titillating details. The Sinhalese Sri Lankans, the people on the rest of the island, are the ones that run the government and have the real power. This has been a face off for several decades now in terms of the Tamils either wanting an independent homeland, a sovereign state, or more autonomy and more power. This civil war came to a bloody conclusion in 2009, with the almost total destruction of the Tamil rebel/terrorist force, and horrific civilian casualties. However, the thing about rebel/terrorist groups is that it only takes a handful of survivors to re-establish operations, so I'm not sure we've heard the last of this conflict. It will certainly be much quieter in the near future as the government rebuilds the society and strengthens the state. Again, much like Nepal, here in 2017 things are much more stable in Sri Lanka . . . but the underlying ethnic tensions still exist.

This is a summary of the two poles of South Asia outside the realm of the three giants. We have some conflict that is quite active, and may get much hotter. But back to the giants . . .

ENERGETIC INDIA

World power position, here we come! India is hustling and bustling right now, my plaid friends, and shaping up to be an Asian titan on the world stage. India has the distinction of being the absolute largest multi-ethnic, multi-linguistic, multi-religious full fledged democracy on Earth! On top of that, they've been steadily growing their economy at a respectable 5% to 8% GDP growth rate for over a decade. Those numbers are overshadowed by China's huge double-digit growth, but they're nonetheless quite respectable for a country of 1.3 billion people. More to the point, India's growth is much more diverse and sustainable in the long term than China's, which means that the Hindus are here for the long haul. How is the India model so different than the Chinese? I'm glad you asked . . .

The Plaid Avenger seeking out Singh . . .

For most of its modern history since independence in 1947, India has followed a government-heavy-handed approach to regulating and controlling the economy . . . you know, it's called socialism. It was never full-on communism like the Chinese model, but the Indian state did produce a massive, bloated bureaucracy in order to outright control some sectors of the economy, protect other sectors from competition, and in general regulate the hell out of everything else. All this was done to redistribute some of the resources of the states via services like education, infrastructure, and employment in general. After decades of stagnation and inefficiency, in 1991 India put into play a crucial reform package which restructured the entire government/economy relationship, resulting in tons of privatization, slashing of bureaucratic red tape, and promotion of entrepreneurship. All this set India on a path for sustained economic growth and true integration into the global economy, for the first time ever!

In other words, India started the conversion from a heavy socialist state to one based more on capitalist principles of free trade and less governmental control. Wow! Who was the genius that thought all that up? The main brain behind this plan in the early 1990s was the finance minister Dr. Manmohan Singh—hey, wait a minute! I know that dude! He was the Prime Minister of India from 2004 to 2014! India really grew by leaps and bounds during his stint as Finance Minister, and also for the first few years he was the PM . . . but it slowed down and under-performed after that, which is why Singh and his party lost the last big election, but more on that in a minute. . . . Singh had been a big force behind the country's successful economic conversion. Successful in strange and sustainable ways as well.

How so? Well, India is very unique in that it has kind of "jumped" stages of the natural economic evolution pattern. Namely, it has yet to fully and thoroughly industrialize like most other

Our modern main man Manmohan.

rich countries. It has gone from a largely agricultural and primary products based economy straight to a service and information economy. How bizarre! To compare, China used to be mostly ag-based, and then industrialized to become a mega-manufacturing giant, producing everything from toys to tubas to Toyotas, and now it's diversifying into services and higher-end technologies. India, on the other hand, jumped straight to the top: service sector jobs have the highest growth rate, with high-tech manufacturing and research/development industries also exploding quickly. Money! 150 of the Fortune 500 companies traded on the US stock market currently have research and development bases in India. It's blowing up! Combine that with a rapidly growing small-business class, and now you have the recipe for India's current success. The place is hustling and bustling right now; go watch *Slumdog Millionaire* again, and pay attention to how much change occurs in Mumbai during the course of the film to get a sense of what I'm talking about here.

Why else is India bustling right now? They have several distinct advantages from British influence; one we already pointed out is that they speak English. India is well-equipped to embrace the world economically, and they're doing just that. Indeed, when your computer breaks and you get on the phone to have someone help you fix it, who are you talking to? You are probably talking to someone from India. Again, think about *Slumdog Millionaire*: what was Jamal's occupation when he got on the gameshow? They're big into telecommunications, high technology, and the service sector. It's actually quite fascinating because they have whole divisions of technical support and telemarketers there that specifically train people to not just speak English, but how to speak certain regional dialects of English. If you go into a technical support center outside of New Delhi, the huge room with 500 to 600 people in it will be subdivided into what region of the United States from which they're receiving calls. There is a Midwestern section, perhaps a Californian section where they talk surfer dude language, or a Northeastern section where they talk really fast with those Boston accents, or a Southeastern quadrant where they talk with a long drawl, Southern belle style.

We also talked about how the British invested heavily in their infrastructure, and that has benefited them in today's world as well. Another thing that India is famous for is something called **back offices**. The hot word in today's world is **outsourcing**. What's that? It means there is cheaper labor there and more people to do it, so jobs move there because companies can save money. Unlike other places in the world, say China or other developing countries, these folks speak English, and most working age Indians are fairly well-educated, which is a huge strategic advantage for India for job creation. Check out a hilarious comedy about this actually named: *Outsourced* (2006).

Back offices are companies that set up in India decades ago. When you go to the doctor, for example, he scribbles out your symptoms as you tell him you have herpes, or hemorrhoids, or ingrown hairs, or whatever it is that is ailing you. The doctor writes all this down, and at the end of the day, he hands it to his secretary to transcribe into a digital file. What has been happening in the US is that after the doctor's office closes down, the secretaries fax the notes to India. People who speak and read English transcribe the notes into a digital file, and then send them back. What's this back office got to do with it? Why was this whole system possible? Think about what I just said: as the doctor's office is closing in America, what is happening in India? That's right! They're waking up. It's a complementary business arrangement, so doctors can get work done more efficiently.

"Yes, Plaid Avenger, the foreign out-sourcing is in-deed the "in" thing in India . . ."

But that's not all! India is fast becoming a center for information technology development, and not just basic manufacturing and low-level service sector stuff. Call centers combined with software development centers—now that's a combo that everyone can rally around! India has an extremely large pool of highly educated and skilled engineers, programmers, and entrepreneurs that are making it a desirable location for technology-oriented companies across the board. We're talking quaternary level stuff here, brothers and sisters! Every country wants those jobs! India is doing a great job becoming a real global center for emerging technologies, perhaps even better than China, and increasingly competitive with the US and EU.

Things are looking up for India; they were never looking down too much to begin with, but this is a place that is bustling right now. You've got democracy, capitalism, international investment, and jobs here getting outsourced to our Indian friends. The result of all these things taken collectively is that you also have a burgeoning middle class in India as well, and that is a true marker of success in today's world. Obviously, it must be a developed region, right? That's still a tricky call.

Why does the Avenger have trouble calling India a fully developed state even after I've talked up all its awesomeness for three pages straight? For starters, take a look at its population pyramid. Dudes! This place has serious dudes! Even as the population growth rate stabilizes (which it has been doing for years), India is still destined to become the most populous place on the planet due to the principle of population momentum that we discussed way back in Chapter 2. So many people; so many more to come. While their economy has been growing, as has their middle class, you still have hundreds of millions of people living in abject poverty. I'm talking dirt poor. Heck, I'm talking so poor they can't even afford to pay attention. The vast majority of Indians live on less than a dollar a day, have limited or no access to drinking water, sanitations systems, electricity, or health care.

This is India's biggest challenge, and why it's hard to qualify them as a fully developed state. A big economy, yes; high levels of technology, yes; a world power, yes! But a full-fledged, fully-developed state . . . hmmm . . . not yet. The average Joe on the streets is still living on a dollar a day or less. Part of that may have to do with the leap-frogging economic situation I referred to earlier: India has never fully industrialized, which means they don't have millions of factories creating tens of millions of jobs. A lot of people are getting left behind, and population momentum ensures that whatever economic growth they do have in the future will be offset by the increase in peoples. Lots of challenges lie ahead for Singh and his successors, for sure. A lot more people are doing well, the middle class is growing, and businesses are thriving, but wealth disparity here is perhaps the biggest in the world. Problem numero uno for India for sure.

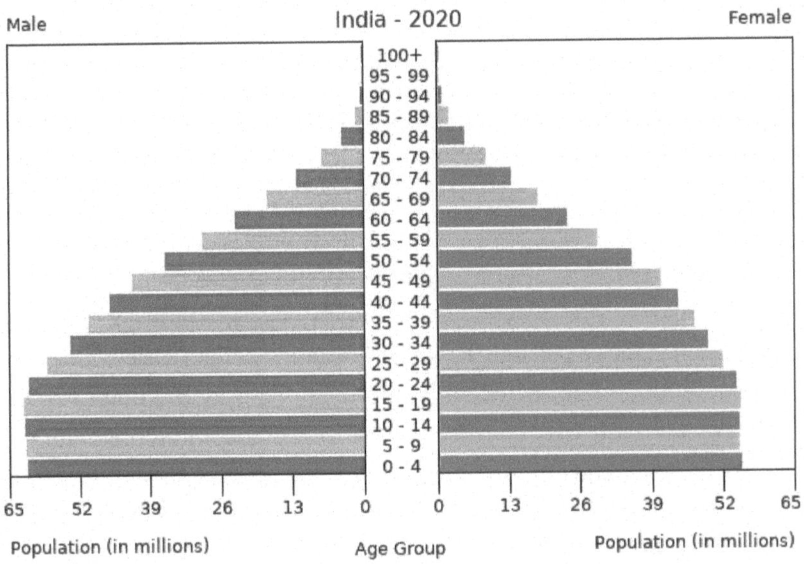

2021 UPDATE: HOLY HINDU!!! TITANIC TECTONIC POLITICAL SHIFT IN INDIA!!!

Big news in India: in 2014 they held the largest democratic election in human history—an awesome feat in its own right—but the outcome of said election was to put into place what will be the most powerful government the sub-continent has ever seen since independence! You heard it here first! India now has its best chance ever of explosive growth and reform, under the leadership of the new head Hindu honcho: Prime Minister Narendra Modi. In 2019, Modi just overwhelmingly won another election to secure his second term, so he is large and in charge until 2024 at least!

Now, this is not merely just a changing of the head cheese of a country, this is by all accounts an epic shift in India itself! Why do I suggest that? Three reasons.

1. India is a parliamentary democracy, which means that whichever political party controls the Congress picks the Prime Minister, and the political party that swept the 2014 elections was the Bharatiya Janata Party, known more simply as the BJP. So who cares about political parties in foreign countries? Well if you want to understand India, you better dig this: the BJP has been the OPPOSITION party for decades, meaning it has not held power for a long time, but that time is now up. Indians just got so tired of the entrenched power of the Gandhi/Nehru families (their political dynasty has ruled India for most of the time since independence as leaders of the India National Congress party) that have not advanced the country very much in the previous decades, that a vast majority of peeps just voted for the BJP in an open declaration that everyone wants change. And change is coming! Because . . .

Hindu Head Honcho: Naredra Modi

2. The BJP is a center-right, conservative, pro-business, pro-smaller-government party! Which is the exact opposite of the India National Congress platform which has always been center-left, liberal, and borderline socialist. Apparently the Indian peeps are chomping at the bit for economic change, and for less government intervention, less government red tape, less corruption, and in general a pro-business/pro-jobs overhaul of the entire state! Boom! This is epic! Our new main man Modi is all about that, but he is not alone, because the other reason this is a hugely special shift is that . . .
3. The BJP didn't just win enough seats in the Parliament to pick Modi as the Prime Minister . . . they won enough seats to dominate the whole damn system! Meaning: they basically have a super-majority to get laws passed, to push for reform, and to get things done! They have essentially been given the mission to move mountains, and the tools by which to do it!

So for all these reasons and maybe many more, India is very likely at an epic shifting point for 21st century history. See, 30 years ago, both China and India had about the same GDP, and GDP per capita. Since then, China has exploded, pulled millions out of poverty with economic growth and jobs creation, and basically left India in the dust. And Indians know this. Although I have previously referenced that previous Prime Minister Singh did wonders to initiate growth, that growth has only happened in little spurts here and there, and has been sputtering out for the last couple years. Thus the overwhelming support for the BJP, and for major change to the corrupt, inefficient, and over-regulated Indian economy, and society in general.

Oh, and back to Modi. This dude had mad, MAD public support from the get-go because he is a career politician that has dedicated his entire life to public service, but with a business-friendly approach. He was the Chief Minister (like a governor) of the state of Gujarat for over a decade, and that decade saw the place absolutely explode with economic growth, foreign investment, and improved infrastructure and governance. He is the wonder-boy of India now, and everyone is hoping he can do the same for the entire country. And he very well may pull it off! Big changes are afoot! India could finally become a giant of Asia once more! Just a quick **2021 UPDATE** on Modi and India: They both have been hit hard by the Covid crisis. India is faring very poorly by world standards, and Modi's popularity has declined significantly as he is catching the blame for the poor response of the government.

Among Modi's many missions: create jobs, encourage foreign investment, pass laws that allow for more open foreign investment, fight corruption, rebuild infrastructure, revamp the education system, encourage the manufacturing sector growth. And he is an ardent Hindu nationalist, that may very well make India into a more powerful player in the international realm . . . he may flex the Modi muscle abroad! Last note on that front: he is for a stronger, tougher India, which means investment in the military and presenting a more powerful face to Pakistan. His to-do list is long. . . .

Modi can currently be labeled as: pro-business, less-government, socially conservative, politically powerful, publicly popular, pro-strong-India ultra-nationalist. Given all that and his conservative political ideology, you can think of Nerendra Modi as the Indian version of Ronald Reagan's era in the USA, or Margaret Thatcher's era in the UK, or even Recep Erdoğan's current era in Turkey. All are/were conservative leaders with majority control of their

THE METACARPI OF MODI
PAKISTAN
MANUFACTURING
JOBS
EDUCATION

government, backed by a popular mandate from the people to change the direction of the country. Modi has big shoes to fill, and a big country to lead, but if he can successfully unleash the Indian society's economic potential in the coming years . . . watch out! The country of Kali will be killing it!

One last thing about India's position in the world today is a reiteration of the fact that the whole South Asian region is physically isolated, though not isolationist. India has been an independent player for the last fifty years; they're very careful about the countries with whom they make allegiances. India's independence from Britain coincides with the beginning of the Cold War. I point this out because India never sided up during the Cold War. In fact, they're one of the few countries in the so-called Third World category that said, "No, we're not going to choose a side." Remember, First World was the capitalist democracy, the Second World was the Commie Soviets, and the Third World referred to everybody who wasn't on team one or two. The term is a Cold War relic that had nothing to do with whether you were poor or rich, but whether or not you sided up.

Kali the Destroyer may be unleashed onto Asia!

The Indians are independently minded. I've been fascinated for a long time about why the US has not established stronger trading and strategic ties with India. Why would they? Well, India has a huge population and a huge potential market for businesses, and the people even speak English! India is also a democracy, just like the US. They have all kinds of traditions that are just like ours, so why have we not had a closer relationship with them? One of the answers is because of its independent status; they never got on our side during the Cold War or embraced our ideals, and never helped combat the evils of communism. That's just a Plaid Avenger speculation, but it has some weight when we look to the other power player in the region, Pakistan.

PLAID AVENGER INSIDER TIP: Know this, world-watchers. The US-India relationship is changing fast. The US is increasingly seeing India as not only a strategic business partner, but also as a partner in balancing the power of rising China. And don't forget terror! India is also seen as a partner in the War on Terror, as it's a great example of a multi-cultural, democratic state that other states should follow. Since 2008 alone, the US has done decades worth of courting the Indian government, not the least of which was a US-sponsored deal to help them out with their nuclear power industry, even though India has not signed the Non-Proliferation Treaty. We shall see how this proposed marriage works out. India could still cozy up to China, though doubtful since they are regional competitors.

PAKISTAN IS PAKIN' THE HEAT

Pakistan did side up in the Cold War with—guess who—the US. While India was independent, Pakistan needed a strong ally because they didn't have anywhere near the economy, money, or manpower of India. They received tremendous amounts of foreign aid from the United States during the Cold War in order to keep them on the US team, and help fight commies, especially when those Russian commies took over Afghanistan. Go check out *Charlie Wilson's War* to see a great film rendition of those events. Pakistan still receives vast amounts of aid from the US today, but for different reasons: the War on Terror. Unfortunately, this presents a major problem for Pakistan: the US-led war is almost exclusively against Muslim extremists, and Pakistan is almost exclusively Muslim. The average dude on the street in Lahore is not a fan of the US.

The US on the job, helping beef up the Pakistani military.

However, Pakistan's administration has always been a staunch US ally, even though the people have not been. Pakistan is currently fronting the war on terror for the United States in its fight to find Osama bin Laden and other Taliban rebels and extremists on its border with Afghanistan. **2011 UPDATE**: you now know that bin Laden was killed on Pakistani soil, by a covert US operation, which highlights the level of mistrust and lack of true communication between these two tenuous 'allies'. Look for the relationship to continue to sour. **2014 UPDATE**: Yep. The relationship has gotten even worse. **2015 UPDATE**: And even worse. Pakistan is now cozying up to China big time.

2019 UPDATE: Yep, yep, and yep. It is still early to make this call, but I am making it anyway: Pakistan is moving away from its longtime strategic partnership with the USA and cementing a future with China. China has launched an economic linkage initiative called The Silk Road Economic Belt and the 21st-century Maritime Silk Road (more on this in a later chapter), but know this for now: the "Maritime Silk Road" component features a close-knit, Chinese-dominated group of ports, and Pakistan Gwadar deep water port now plays a key role in this "string of pearls" as the Chinese call it. Regional alignments are shifting with China, Pakistan, Russia, and Iran coming together, which is causing both India and the US to rethink their strategies and policies. This is huge. Pakistan is in a historic pivot transition period, although it still has strong military ties to the US. . .

Which is why Pakistan is still in the number three spot for US foreign aid recipients, mostly to help with the anti-terrorists operations in Afghanistan and the vicinity. Unless you have been hiding under a rock for the last few years, you probably know that this fight has not been going well. The Taliban has been winning in Afghanistan, and have even taken over parts of Pakistan itself! The areas highlighted in blue in the map to the left represent areas of both countries which are primarily controlled by ethnic/tribal groups who don't recognize any international borders, and who also don't give two shakes of salt about the Afghan or Pakistani governments either. Al-Qaeda and the Taliban have both gained in strength in these areas, despite the best efforts of the US and NATO and less than a best effort from the Pakistani military. But the US funds still flow to Pakistan in order to keep them fighting the good fight.

The blue areas are not exactly fully controlled by either states' governments, if at all.

Too bad all that great US foreign aid has been only directed at beefing up Pakistan's military capabilities to fight terrorism, because the country has quite a few other problems that are just as pressing, if not much worse. Its economy can be described as weak, at best. Its democratic institutions: weak, at best. Its political stability: shaky, at best. Its potential for further extremism, more terrorism, and perhaps open revolution: HUGE!

There's no sense in getting into great political detail here, because the events are changing too rapidly and whatever I write will probably be out of date by the time this book gets into your hands, but we can say a few things that will stick.

The current Prime Minister, Imran Khan, and the whole Pakistani government are trying to pacify the wants and desires of its 190 million citizens—that's the 6th biggest population in the world. It's also a devoutly Islamic society, including the whole spectrum of religious views from the mainstream to the seriously extreme. It's a society that has attempted to be a democracy since its inception back in 1947, with less than desirable results—multiple military coups have occurred, mostly to wipe the slate clean of massively corrupt administrations. And Pakistan is a nuclear power, with nuclear missiles that need to be controlled. To fight terrorism, they have had to crack down on civil liberties in the country, which pisses off the locals. To keep getting international aid, they have to suck up to the Western powers, which pisses off the locals more. All these things have further radicalized lots of folks, some of whom would not mind the overthrow of the entire government. And, and, and . . . this country has had a rough go of things, and it ain't out of the woods yet!

Just in the last few years, the Pakistani military and the government have officially gotten serious about trying to reign in the terrorist elements in their western fringes, finally succumbing to US pressure and the cold reality that their entire country may be lost in its entirety! The military has made huge, albeit bloody gains in their untamed western regions against the Pakistani Taliban and al-Qaeda. The current Prime Minister Imran Khan inherited a crippled economy, energy shortages, non-existent economic growth, high debt and large budget deficit . . . it is rumored that the IMF will have to float them a huge loan just so they won't collapse. Khan is also working to end US drone strikes in his state, as well as put a lid on the Pakistani Taliban, whose intent is to overthrow the state. During 2018/19, the economy actually started improving, then of course the gains were wiped out by the lost Covid year of 2020. Interestingly enough though, Pakistan dealt with the crisis quite well (as opposed to India) and their economy has rebounded more quickly than most here in 2021. They seem to be headed in an upward direction again! Fingers crossed! But man, oh man, being the leader of this state ain't no picnic in the Punjab.

Pakistani Prime Minister Imran Khan: toughest job on the planet.

Virtually every political event that happens causes widespread protests by folks who are either pro-democracy, or pro-Islamist, or pro-independence, or anti-US, or just anti-Pakistan. This is a society on the brink, which is a situation that absolutely no one in the world wants. They got lots of Islamists, jihadists, "terrorists," separatists, extremists, and they also have a nuclear arsenal! Not good!

But all that internal turmoil and international pressure and the War on Terror is still not the whole story with Pakistan. We haven't even gotten to the Indian side of the country yet.

India and Pakistan are two quite different tales altogether. We need to define the Cold War difference between India and Pakistan a little more, because there is still potential for some serious conflict. We have the independent India—though it could be argued that they're not completely independent after buying arms from Russia at one point, making the US a little leery—and the US-funded and militarily-supplied Pakistan. About 45 years ago, India entered the nuclear club in 1974 after testing a bomb they called "Smiling Buddha." Pakistan immediately said, "Crap! We've been at war with that country a few times; we need nuclear bombs too!" The United States discouraged them big time from doing this, but they went ahead and covertly ran a nuclear development program throughout the 1980s and 1990s, confirming their nuclear capacity by testing their first nuclear bomb in 1998. This development, two countries that essentially hate each other's guts both going nuclear, is cause for alarm for everybody outside this region. Things are stable right now, but people know that at any time some serious nuclear business could go down. It's not probable, but still possible. Kashmir has been a long-standing flashpoint between these two South Asian titans that we've all feared may be the spark that starts a nuclear throw down. What's that? You don't know Kashmir? Well, check out the box then!

What's the Deal with Kashmir?

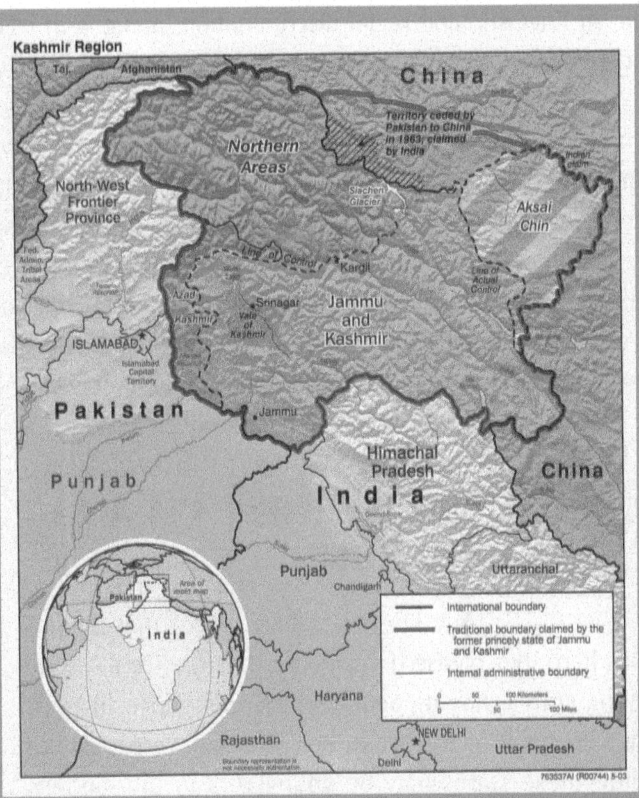

The hottest example of continuing conflict between India and Pakistan is the instance of Kashmir. It's not just a goat-wool sweater (or a Led Zeppelin song); it's a disputed territory. What's the deal? You already know about the partition in 1947. When the British were there with Gandhi and the leaders of the future Pakistan, they were saying, "Where do you guys want to draw a line? Who wants to be Muslim and who wants to be Hindu? Who wants to be India and who wants to be Pakistan?" They went region to region and asked regional leaders what side they wanted to be on. States said, "I'm Hindu, I want to be in India," or "We're predominately Muslim and want to be in Pakistan." When it came to Kashmir—the proper name is Jammu and Kashmir—a little region nestled in the foothills of the Himalayan system (a mountainous, beautiful, rugged place where you could get killed if you visit), the people were mostly Muslim while the government was Hindu. The Hindu governors wanted to be in India, and the Islamic people wanted to be part of Pakistan, but the Hindu peeps were the only ones the British heard. It was immediately contested before the British even got on the boat to leave.

This is still the situation. It's still a very much disputed territory. The Indians say, "Hey, sorry! Kashmir is part of India." Pakistan says, "Come on, that's crap! It's a bunch of Muslims who didn't get a voice, and you guys screwed us!" As of right now, this is a fairly stable situation. Violence has flared up and there are extremist groups on both sides of this border who want to either make sure it stays in Indian hands or see it turned over to Pakistan. Everyone remains slightly worried that this region will inevitably spark another major conflict between these two nuclear powers. For the sake of South Asia, and the world, let's hope that never happens. Pakistan kind of has its hands full with self-implosion right now, so the Kashmir deal is not an active hotspot for now. Emphasis on "for now." Having said that, here in 2017 active animosity has been rekindled due to a September 2016 Indian commando strike deep in the heart of Kashmir in which 38 suspected terrorists were killed in a surgical strike. . . which of course has led to protest and unrest ever since. <sigh>

BANGIN' BANGLADESH

I would be remiss in my regional duties if I did not at least mention the other giant of South Asia, even if it's not a major player in the neighborhood, nor the world. Sorry to my Bengali brothers out there, but we all know it's true. Bangladesh may not have a lot of regional or global power, but it still carries the titan status because of its huge population packed into a small little area: almost 160 million folks, ranking 7th in terms of most populated countries, squeezed into a state about the size of Iowa. Talk about a full house!

A typical freakin' flood in beautiful Bangladesh.

As you know from earlier in the chapter, Bangladesh started its life as the overwhelmingly Muslim East Pakistan;

due to linguistic and racial discrimination, along with lack of proper representation, the Bengalis declared and gained independence, with the help of India, in 1971. They have had a pretty rough go at things ever since.

Famines, natural disasters, droughts, floods, and widespread poverty have defined life in Bangladesh for its entire existence. Just about every year, monsoonal rain systems totally trounce the area with floodwaters, which isn't hard to do since almost all of the land is a flat alluvial plain that sits just slightly above sea level. Death and destruction seems to be a way of life, but that is only the beginning of the fun for this place! Bangladesh has been plagued by a weak economy, political instability, and internal corruption, which has regularly resulted in coups and rapid administration turnover. As if they weren't challenged enough already, radical Islamist extremists in Bangladesh have been inspired by ISIS to attack and kill Westerners and even Hindus and Shias in a show of their blind hate and allegiance. Multiple attacks occurred just in 2016, and these Bengal jihadists appear to be just getting warmed up for the show. Sheesh.

Hmmmm . . . I guess they still have a little in common with their Pakistani brothers after all.

MOMMY WOW! I'M A BIG KID NOW!

Now I'm not making fun of these countries here, but please do consider that they're all fairly young states, having achieved independence in 1947. They are soon to celebrate their 70th anniversaries, but while these states may be youngsters in the modern world, they have grown up fast and are global players already. In conclusion, when we think about South Asia, we do have to consider that India is the big power player: They're big into industry and technology and India is becoming a global center for software creation and programming. In fact, there are now whole cities built up as IT centers, and a lot of computer programmers and technology companies from the United States are relocating there. Several years ago, California chip maker AMD said it was moving to India, causing a lot of US senators to blow steam about taking away American jobs. More important: AMD moved their *design center* to India. They didn't just outsource the low-paying jobs! They're moving the high-paying design work! That's a big deal! AMD said, "There are more highly educated, skilled programmers in India than in the US. All the people coming to this country that are being trained in computer programming are Indians and they are going back to India. There are more computer programmers in India capable of this kind of work."

India's gonna get one!

This is a major event. It's one thing to move cheap factory assembly jobs where people put TVs together for a nickel an hour. That kind of work is not going to be missed, but moving a computer chip design center is a big deal. AMD is not alone. A lot of global firms are hastening to India's back offices. India is happening and companies want in on the big action. In addition, India is now an associate member, along with China, of **ASEAN**, the **Association of Southeast Asian Nations**. Let's not forget my other favorite acronym: India puts the I in **BRICS**! Hopefully you remember that the BRICS group is the new up-and-coming economic and political power block on the planet. South Asia is on the fast track to roll—with all the powerhouse trading blocs of the planet in the next hundred years—and there is speculation that India may become a member of the UN Permanent Security Council as well. I mean, why not? They have 1/5th of humanity and nuclear weapons. What more of a resumé does one need?

In 2005, India started constructing its own aircraft carrier. They bought some others from Russia and Britain in recent years, but they're building a gigantic custom one. Lots of folks who aren't military people, including the Plaid

Avenger, are about love, not war. They say, "Who cares? What's the big deal?" An aircraft carrier is quite different from any other military hardware because it can be used as a projection of power anywhere on the planet. You are not to be messed with if you have an aircraft carrier. You don't just have armies of people, you don't only have tanks or planes with rockets that will go to other countries. But now you can put all of that on a big platform that floats and take it anywhere on the planet you want. It's a true projection of power and only the big time countries have that. India also has a vibrant and growing space program, a burgeoning middle class, a hot economy, a stable democracy, and thanks in part to the US, a growing nuclear energy industry. India is hip and happening.

Pakistan is slightly different from India. It's a big country with lots of people and problems, not the least of which is its relationship with the US, which the people in charge really like, but, is very unpopular with the masses. This type of internal dissent is not something that goes away without a fight, so this is a big problem for Pakistan. On top of that, their economy is not in great shape. Pakistan is not growing nearly as fast as India. For those of you curious about military affairs, I should also tell you that India exponentially out-competes their Pakistan neighbors on all things military. Indian army, aircraft counts, tanks, missiles, and even nukes are about ten times that of what Pakistan possesses. We're not talking about evenly matched entities if a war were to ever break out. Of course, nuclear weapons level the playing field quickly, in that pretty much everyone in the entire region would die if it ever comes to that . . . but let's hope they don't!

Now I must finish this South Asian round-up with the most important regional development of them all: a major alliance shift is underway! Who? The US! The shift? From Pakistan to India! Big time! Why? Dig this:

While relations between India and the US were chilly during the Cold War, they have thawed amazingly fast in the last decade. Like the US, India has a stable democracy, a vibrant economy, a capitalist outlook, a multi-ethnic society with an emphasis on individual liberties and human rights. The US is helping India develop its nuclear energy program and is also sharing intelligence with them on a variety of national security initiatives. Most importantly, India is solidly on board with the US in its War on Terror, especially since the Mumbai attacks of 2008. And where were the Mumbai terrorists from? That would be Pakistan, as were some of the terrorist in the Kashmir attacks, and in attacks on Britain, and in attacks on NATO forces in Afghanistan. Oops.

Yes, the old alliance between the US and Pakistan has now fallen on hard times, as that state struggles just to hold itself together. They have a weak central government, a fragile democracy, political gridlock, and a military that has been so focused on fighting India that it can't seem to restructure itself in order to beat the extremists and separatists inside their own state! That's why the US has lost confidence in them, and started conducting attacks on terrorists camps within Pakistan itself—and Pakistan is their ally!

In conclusion, India/US relations are growing stronger than ever before in history, and Pakistani problems are destroying the state, and straining its relationship with the US and Team West as a whole . . . and it is increasingly turning to China, Iran, and Saudi Arabia as more natural allies. That won't make Team West very happy either. But let's leave it there for now.

Now it's time to relax, have some chai, and get ready to rock and roll into Central Asia.

Got chai?

SOUTH ASIA RUNDOWN & RESOURCES

View additional Plaid Avenger resources for this region at http://plaid.at/s_asia

BIG PLUSES

(these pertain mostly to India)

- Enormous Indian economy that is just getting warmed up for the 21st century
- Indian economy is interestingly diversified, focusing more on services and high-tech stuff over basic manufactures
- Both India and to a lesser extent, Pakistan, are functioning democracies and allies with Western powers, although both keep their options open
- India is a associate member of ASEAN and full partner of BRICS and a possible future member of UN Permanent Security Council; they will be a world power of note this century

DEAD DUDES OF NOTE:

Mahatma Gandhi: The "Father of India" who pioneered the concept and practice of non-violent civil disobedience to achieve political change. Led India to independence from the UK and inspired movements for civil rights and freedom across the world, including Martin Luther King Jr.'s approach to the US Civil Rights Movement.

Ali Jinnah: The "Father of Pakistan" and its first president/governor-general in 1947. During the Indian independence movement, he became the main advocate for a separate Muslim state in order to protect the rights and liberties of India's massive Muslim population. In other words, no Jinnah = no Pakistan.

BIG PROBLEMS

- Pakistan. There, I said it . . . I love the Pakistanis, but their state is in trouble; stagnant and small economy, besieged internally by terrorists, pressured externally by US and India, weak government, and exploding population
- Population, population, population; this region has 3 of the biggest populations on the planet and they're growing. Will there be enough resources to go around indefinitely? The biggest challenge for the region, and India in particular
- While countries are developing fast (esp. India) the poverty levels are still exceptionally high; most people of this region live on about a dollar a day
- Pakistan and India have significant issues internally with terrorism and separatism; Taliban and al-Qaeda in Pakistan and the Naxalites in India
- India and Pakistan hate each other's guts, and both are nuclear powers; that can't end well
- On-and-off again open conflict between India and Pakistan over Kashmir; a festering hotspot
- On-and-off again open conflict between Hindus and Muslims, especially in border areas

LIVE LEADERS YOU SHOULD KNOW:

Narendra Modi: Current powerhouse Prime Minister who has the potential to significantly change India, economically and socially. Conservative, pro-business, pro-capitalism, will likely lead India into a new era of bolder foreign policy as well.

Imran Khan: Prime Minister of Pakistan, and super popular playboy/public figure that pervasively was an international cricket star. Politically centrist/liberal, anti-corruption, pro-Islamic values, and appeals to younger voters and anti-US sentiment in the country.

MAP GALLERY

View Map Gallery & additional Plaid Avenger resources for this region at http://plaid.at/s_asia

PLAID CINEMA SELECTION:

Gandhi (1982) Huge award winning film on the life of Mahatma Gandhi, who led the nonviolent resistance movement against British colonial rule in India during the first half of the 20th century. Yes, it's three hours long, but you owe it to yourself to be smart, so do it. Great depiction of the independence movement and philosophy behind the man, as well as fantastic scenery/culture of India.

Lagaan (2001) Alright already! Every South Asian student I've ever taught cites this film as the best thing to ever come out of the subcontinent! Set in 19th century British India, it's an epic cricket match movie, pitting the Brits against the locals in a high stakes showdown. It's a Bollywood production, which means there's random song and dance numbers crammed into the film, and contains great story/scenery even despite those MTV interludes. Like most real life cricket matches, I think the film is about four days long.

Monsoon Wedding (2001) Wildly popular and entertaining story depicts romantic entanglements during a traditional Punjabi wedding in Delhi. Even better, shows insight into modern Indian lifestyles/cultures of both the upper and lower classes, with many of the same family/wedding issues that people deal with here in the US! Go figure!

A Train to Pakistan (1998) Tensions run high near the border of British India, which is about to be partitioned with a new country called Pakistan. Sikhs living in this border town have heard numerous stories of Muslims killing, raping, and looting other Sikhs, Hindus, and Christians, and many of whom are their friends and relatives. Enraged at the loss of law and order, they plan their own attack on a train full of Muslims leaving British India. Great depiction of South Asia's darkest days after independence.

Junoon (1978), With a love story as the backdrop, chronicles the period of 1857 to 1858 when the soldiers of the East India Company mutinied and many smaller kingdoms joined the soldiers in the hope of regaining their territories from the English. That would be the "Great Mutiny" referenced in this chapter.

Bhutto (2010) A fantastic documentary on one of the most polarizing figures in the Arab world, Benazir Bhutto. Follows the events of her life, leading up to the 2008 election that she was expected to dominate, and her assassination which resulted in her husband (Asif Zardari) becoming the current Prime Minister by default. Considered by some to be a martyr or a messiah for the common man.

The Clay Bird (2002) A masterful piece of Bangladeshi cinema, film was banned in Bangladesh for three years because of issues of religion. Set in the late 1960s, focuses on the increasing tensions in East Pakistan that lead up to the Bangladesh War of Liberation. Film was shot almost entirely in the streets of Bangladesh, creating a sort of rustic appeal with the cinematography.

Nayagan (1987) A sort of *Godfather*-esque movie set in India, and typically called one of the country's best films ever. Based on true to life don in the underworld of Bombay, takes a sympathetic look at the struggles of South Indians in what is now Mumbai. Depending on who you ask, a better film than *The Godfather*.

Slumdog Millionaire (2008) Kid who grew up in the "slums" of India that makes it big on a game show, arousing the suspicion of the authorities. One of the most internationally acclaimed films set in India, also depicts life in the poorer areas of India fairly accurately and a great travel flick through Northeast India.

Sulanga Enu Pinisa (2005) Roughly translated meaning "the Forsaken Land," is a highlight of film in Sri Lanka, which does not have the richest history of film. Winner of one of the Cannes Film Festival's highest honors, the Camera d'Or, the film is a masterful take on the effects of decades of civil war on the land. Stunning landscapes and images of people make the lack of a plot less obvious.

Outsourced (2006) Hilarious "fish out of water" romantic comedy that drops a clueless American into Indian culture. Hindu hi-jinks ensue. Simultaneously shows the whole real economic concept of outsourcing in action while teaching tons about the Indian way of life.

CHAPTER OUTLINE

1. The Game Pieces
2. The Game Board
    2.1. Landlocked
3. Culture Central
    3.1. Prancing around the Continent
    3.2. Mongol Mania
4. Parker Brothers Inspired
    4.1. The Modern Era Classic Original Board Game
    4.2. The 21st Century Edition
        4.2.1. 1. The Vacuum Created by Moving the Soviets Out
        4.2.2. 2. Because of Increasing Finds of Petroleum and Natural Gas
        4.2.3. 3. Terrorism and Drugs
5. The Game Continues to Evolve with New Players
    5.1. You Gotta Know the SCO
    5.2. Putin's Project: Soviet Union Re-Boot?
6. Game On
7. Map Gallery

Central Asia

22

LET'S move to the center of the Eurasian continent. Central Asia is kind of a vacuum in the heart of Asia that, as opposed to going through it, people go out of their way to avoid it. But times are changing fast, my friends, and Central Asia may have found a new home as the delicious meaty center of a Russia-China sandwich or perhaps are going to join a USSR 2.0 as Moscow tries to absorb them back into a "Eurasian Union" . . . but I am getting ahead of the story. Throughout history, this area has been a battleground between steppe nomads and the settled peoples in urban areas all around its fringe. It still can be viewed as perhaps the last active true fight for preserving local culture vs globalizaton forces: a fight that has been going on for ages here. The Huns invading Europe, the Mongols invading Eurasia . . . then the British, then Soviet, then the Taliban, then al-Qaeda, then NATO/American invasions into the region all demonstrate how Central Asia has long been a chess board across which the pawns of many armies and empires have made power plays into other regions and states. And the game goes on. . . .

Kickin' Kazakh Dancers!

But what is it about Central Asia that makes it a region? Is it a region? A lot of people may say that it's not a separate region, but it is to the Plaid Avenger. Here's how I'm going to define it:

THE GAME PIECES

Central Asia consists of a core of countries that everyone might agree upon: the "-stan" countries starting with Kazakhstan (the big one), and then Uzbekistan, Turkmenistan, Afghanistan (I know you've heard of that one), Tajikistan, and Kyrgyzstan. Everybody would say, "Yeah, we think that's Central Asia, or at least we know where you're talking about—all those places we can't pronounce."

This may be the trickiest region to define because we have to include parts of another country in order to do it. We are going to take the western parts of China, including Xinjiang, one of the biggest, most significant westernmost provinces of said country. We must also take two other Chinese provinces, Tibet and Inner Mongolia, as well as the independent sovereign state of Mongolia. The Plaid Avenger calls all of these places Central Asia. It's a little tricky because you will say, "Wait, you just took parts of China and China is its own country. It's a sovereign state!" We're going to talk a lot about China in the next chapter, but what you'll find is that China has a very distinct east/west divide. Virtually all the people and businesses are in the east. All economic activity is on the eastern side closer to the coast; most inhabitants are as close to the coast as possible (as is most of the transportation, communication, and government).

For most of its history, referring to "China" actually meant the eastern side of the current sovereign state; that area has always been its core. Western China is quite different in terms of its population: it is sparsely populated and contains ethnic groups that are not historically Chinese.

All the reasons eastern China has *little* to do with western China are the same reasons that western China has *a lot* to do with Central Asia. The dissimilar groups in parts of Mongolia and the "-stan" countries are thrown into this region called Central Asia. Why is it a region? What's homogeneous about this disparate group of countries, states, and Chinese sub-states? Let's get to work.

THE GAME BOARD

The most obvious similarities between all of the countries in this region are the physical attributes. Unlike other regions that contain a variety of climates and terrains (like South America or North America) Central Asia is relatively standard throughout. Like the Middle East, it is mostly desert and steppe, making it homogeneously dry across the board. There are no other major biomes/climates found in these parts. You definitely won't find any big forests: the region is mostly either short grass grassland or desert, which has had a definite impact on how people have settled and survived for the last several thousand years.

Central Asia: formidable, inhospitable, and largely unlivable.

Before we get too deep into dryness, let's look at the terrain: some of the highest and toughest terrain on Earth. We already referenced the Himalayas when discussing the division of South Asia from the rest of Asia, and they play a big part here as well. It's not just the range itself—it's the whole Tibetan Plateau, which sits over 12,000 feet above sea level. It has extremely high mountains. The **Hindu Kush Mountains** run through Afghanistan; the **Tien Shan** run through Kyrgyzstan and Tajikistan. These mountains seem wimpy in comparison to the Himalayas, but what they lack in physical harshness, they make up for in another kind of tactical danger . . . just ask the US military in Afghanistan. This is not a fun part of the world. It is very difficult to navigate around, through, or across the plateau.

In addition to the Tien Shan, the Hindu Kush, and the Himalayas, we have the Altai Mountains in Mongolia. Although there are a few basins where things level out a bit, it boasts some of the toughest terrain on the planet.

The terrain and dryness go hand in hand. For the same reason that there are monsoons in South Asia, Central Asia is nearly all dry. All the water that would float up north drops from the sky as the air is pushed higher by the Himalayas. This causes the monsoons, and makes for only dry air to make its way up the mountains and into Central Asia.

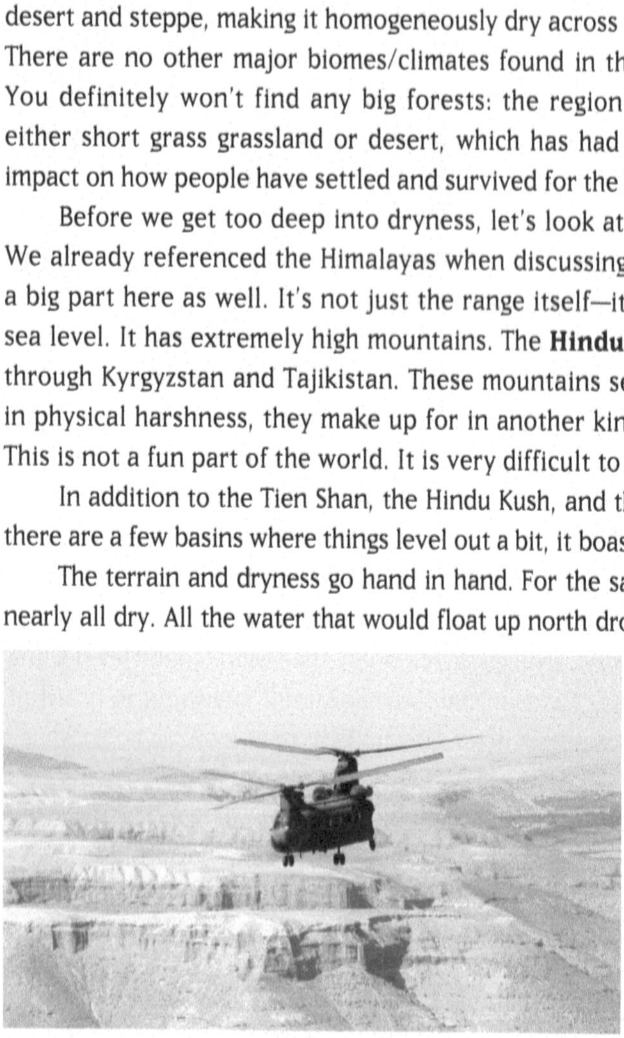

Grand Canyon? Nah, these things are a dime a dozen in this region.

The dryness doesn't support much vegetation. A lot of Central Asia is what we would distinguish as grasslands. We think, "Oh, grasslands are good for things like wheat and barley." No, those are long grasses. Central Asia can't even sustain most of that stuff. There is only enough water in this ecosystem to support shorter grasses. Since the land cannot support any type of large-scale food crops, in turn, it cannot support huge populations. Much of the food that the locals eat has to be imported. This is a fairly stark environment.

Throughout the centuries, one of the main perceptions we hold of Central Asians is that they are horse people. Why don't they raise cattle or something else? The environment will not allow it. It cannot be done because the vegetation is not there for farmers to raise large amounts of cattle in one place, year after year. But short grasses can sustain animals that move around a lot, like horses, which is why the region's inhabitants are known for having them.

Central Asian rush hour here at the airport.

LANDLOCKED

What is the most unique feature of this region? It's landlocked, which is another condition that makes it a homogeneous whole. In fact, some of these places are doubly landlocked. What is "landlocked" and why is it significant? What does this mean for the place as a whole? **Landlocked** means you have no tie to the oceans. Not even a river system that goes to a place that can indirectly get you to an ocean. What's the big deal about that? In the good old days, when people used to travel across the continent on trade routes, Central Asia's location in the middle of things was a bonus. Those days were good, but they are long gone, my friends.

In today's world, virtually all the world's traffic and economic activity happens via the oceans. If you do not have access to the oceans, then you are not part of the world economy. Therefore, Central Asia is not part of the world economy today. It is what The Plaid Avenger calls a **marooned region**. It is not attached to what's going on around the rest of the planet, and the region is economically isolated and therefore slightly stagnant.

As I mentioned before, some of these places are **doubly landlocked**, which means they are landlocked by other landlocked countries. Their economies are largely internal, with the possible exception of something like drilling for oil or building pipelines through the region to connect up producers and consumers from other regions. Some places like Kyrgyzstan are so far from anything that they cannot possibly foster any meaningful movement of goods or services across the Asian continent. It's just not going to happen. Being landlocked is a physical condition that has an enormous influence on what's going on internally and externally for Central Asia.

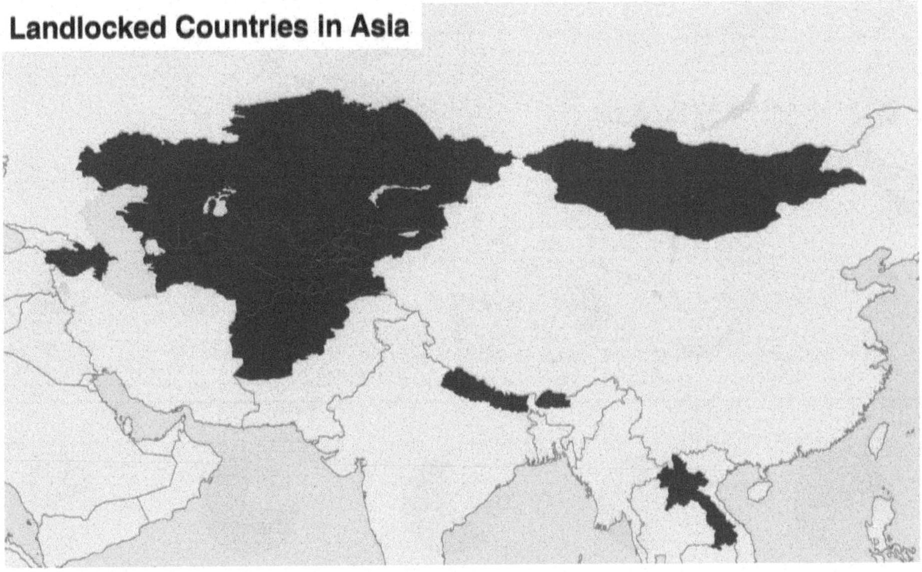

States locked out of ocean access, thus global access, thus economic access.

CULTURE CENTRAL

Let's look at some cultural traits that permeate Central Asia, with the main one being religion. All of the "-stan" countries and the western province of China are overwhelmingly Islamic, while Tibet and Mongolia are devoutly Buddhist. Xinjiang is that western province of China that is mostly Islamic, and they are also ethnically Uyghur (pronounced Weeger, like Weezer), not Han Chinese. Speaking of ethnicity, check out the map for the smattering of Uzbekis in Uzbekistan, the Kazaks in Kazakhstan, and so on. Each 'stan is defined by its major ethnicity, and collectively all these peeps across the whole region number less than 120 million, in an area exceeding 4.5 million square miles. That's a lot of elbow room per person! For perspective: Japan alone has 127 million people in 145,000 sq miles.

A Yurt: Central Asian traveling party hut!

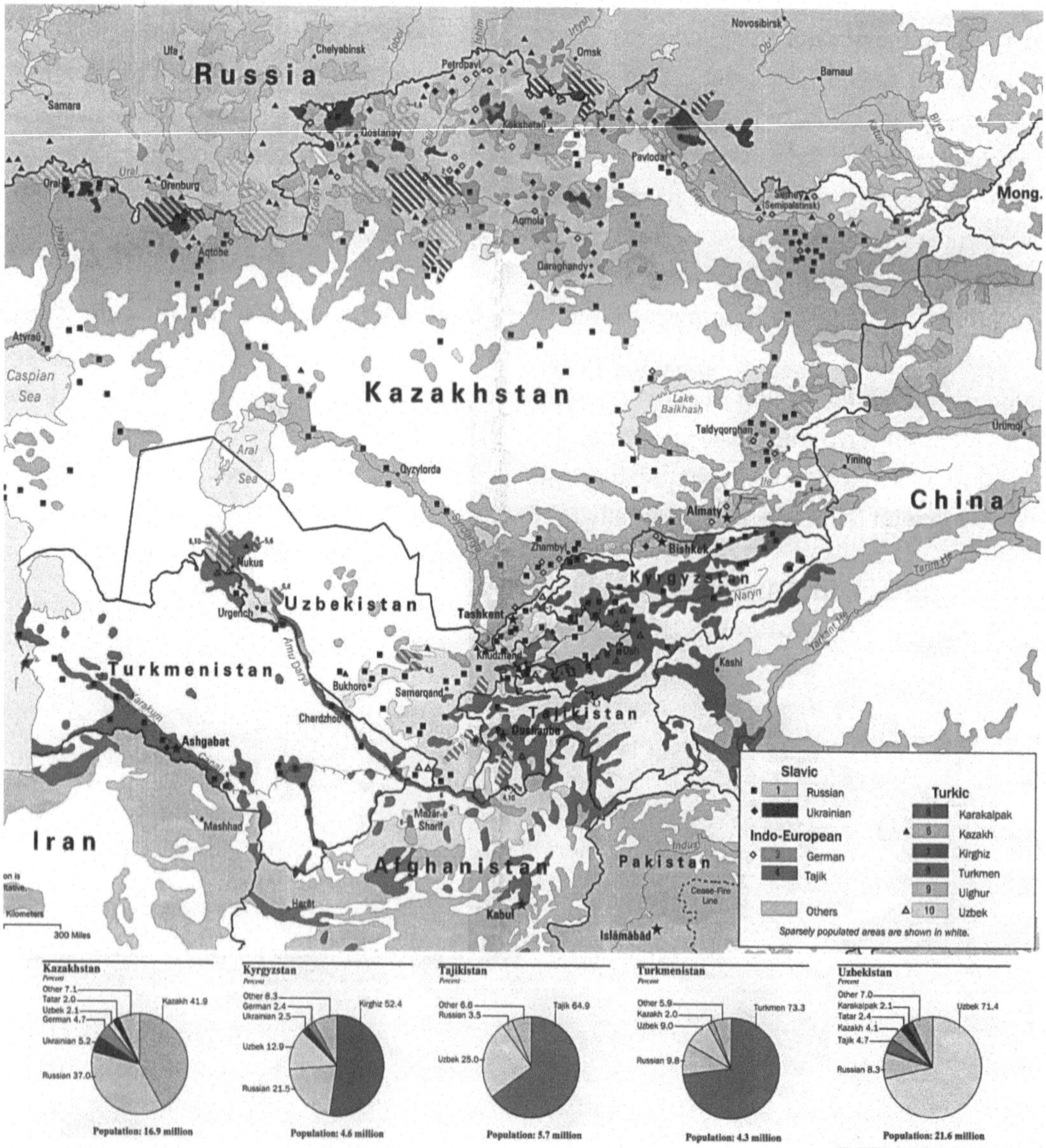

Who is in the white areas? Answer: Nobody.

PRANCING AROUND THE CONTINENT

Uzbeki Zorro!

We talked about the physical features of the region being dry and landlocked. What has this landlocked, dry, arid, short grass steppe meant culturally to this region over the centuries? It means that they are not a big agricultural region. They are mostly **nomadic**; that is, they pick up their Yurts (those cool conical tents) and move seasonally to find better grasslands for their horses. There are still traditional peeps native to this region that are not sedentary and are noncentralized, consisting mostly of small tribes of folks on the move. These guys have been forever focused on the horse, not just as a main means of locomotion, but also as a way of life. I'm thinking specifically of the Mongols. Mongol domination at one point in history is one of the other unifying cultural factors of this entire region. The value of the horse in Central Asia cannot be understated, especially for this group.

The Mongols were the most legendary and lethal group of horse warriors that the world has ever seen. What was so cool about them? What did they do that was so unique?

Don't mess with the Mongols!

MONGOL MANIA

The Mongols were a tribe of totally awesome nomadic horsemen who rallied under **Genghis Khan** and started a campaign to conquer the whole planet right around 1200 CE. Genghis is the one dude who got fairly close to actually achieving the global domination deal. Lots of other people have tried, such as, I don't know, Hitler, Mussolini, and Wal-Mart. But this was a guy who actually came close to conquering the whole known world at the time.

A couple of unique things about the Mongol Empire: They created a single empire that tied together the peoples of Europe and the peoples of the rest of Asia under a single political and economic umbrella. To my knowledge, this has not been done since. We could look at the modern world and say, "Oh yeah, people have interactions now," but not on this scale. This is a singular empire that placed white Europeans and Chinese people within the same empire. What's so important about that? Mad cultural interchange occurred and the region was fully integrated into the global system (a situation that hasn't been duplicated since).

Under Mongol rule, it was said that a virgin with a pot of gold on her head could travel from one side of the empire to another unharmed and

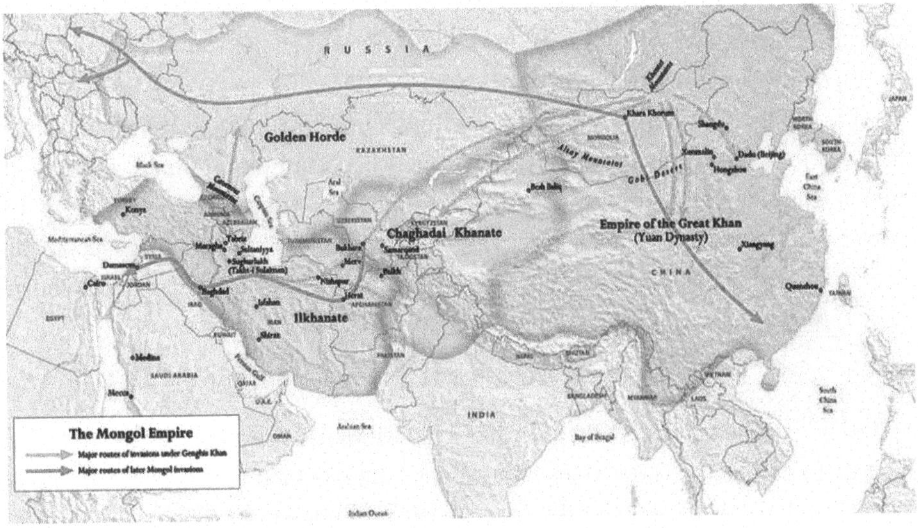

Genghis got game: the largest land empire of its time . . . perhaps ever.

untouched. That may or may not actually be true, but it gets at the idea of Mongol citizens' safety. They had a good legal system that was strictly enforced. That type of safety encouraged freedom of movement of peoples and their stuffs. You had this mass movement of goods, services, materials, and technologies across the Eurasian continent like never before, and perhaps not since. The Mongol Empire was very unique in world history; it encompassed all of what is now China, parts of Southeast Asia, and parts of India, as well as big chunks of the Middle East that stretched all the way into parts of Europe. They sacked Baghdad in 1258, which marked the end of Middle Eastern expansion—remember that from a few chapters ago?

Gengis Khan—ok, not really because they didn't have cameras 800 years ago, but this guy is a pretty legit looking stunt double.

How did they unify all of these lands under one common empire? They were ruthless conquerors. Their environment molded and hardened them into exceptional alpha male warrior dudes who could ride horses for days. These guys were totally brutal. One of Genghis Khan's sayings was, "Submit or die." He didn't negotiate with anybody, period. Typical Mongol foreign policy would be to go into a town and tell the inhabitants, "Here's the deal: you guys get down on your hands and knees and start begging for mercy right now. We'll circle around you, slowly making our way toward your homes, and decide if we'll kill you. But if anybody takes up arms against us, we will kill every freakin' one of you down to the last man, woman and child. Blood will flow in the streets and we'll burn the town to the ground." Given their reputation for actually committing such acts, most surrendered immediately.

That was partly why they were successful in storming across the continent. Perhaps even more important to their success were their fighting tactics. They could cover HUGE amounts of territory in a single day. They would ride horses continuously and would have stations set up so that whole cavalry units could get new horses and essentially ride non-stop for 72 to 96 hours. That is an amazing feat in today's world, much less a thousand years ago. The Mongols were so tough that they trained their ponies so they didn't have to get out of the saddle to eat; they made cuts on the ponies' necks and drank their blood for nourishment. Drinking blood is not only a fun fact, it also accentuates this tough environment. What did their diet consist of? Mare's milk and blood. Protein rich! I think somebody should invent the Mongol diet because these guys were lean, mean fighting machines.

If you go to Mongolia today, you'll see a lot of people doing these same things. Not in terms of raping and pillaging, but what they eat and what they do for a living has not changed much. They still herd goats, drink funky milk, and generally are chillin' out on the central Asian plains.

The Mongols ended up ruling much of the Eurasian continent, but their empire was not a superpower for very long. Their glorious reign was pretty short after Genghis Khan died. After that, they stopped conquering, and eventually the Mongol Empire broke apart into separate kingdoms that affected world history in lots of different places: Kublai Khan ruled China, Tamerlane was in Central Asia, descendants of the Mongols end up as the Mughal Empire in India, and even the Cossacks of Russia are lineage of the Mongols. But for a while, under the main man Genghis, the Mongols ruled!

PARKER BROTHERS INSPIRED

There's one thing the Plaid Avenger wants you to know about in Central Asia because it is still kind of happening; it's called The Great Game. The Great Game, coined by some British dude (as so many things on our planet have been), describes the major world power players' attitudes toward Central Asia in the last few hundred years. That attitude has resulted in Central Asia becoming very much like the game of *Risk*. It's a platform or playing field for different countries to vie for influence. No one really wants it. No one wants to actually control it. They just want to make sure they have their fingers in there to ensure that other people do not control it. This was a scenario that arose during European imperialism, and in the build up to WWI and beyond.

THE MODERN ERA CLASSIC ORIGINAL BOARD GAME

Main Central Asian currency unit: force.

Basically, it's the Russians vs. the British vs. The Persians/Iranians vs. the French vs. the locals. All these colonial powers and others who are in the neighborhood, such as the Chinese and the Persians, all jockeyed around Central Asia to make sure that no one else got in. Everybody protected their own backyard. It's mostly the Russians who started up this game because they had been expanding their territory for hundreds of years. Among other things, they increased their influence by pushing into Central Asia during Soviet expansion, eventually fully absorbing the 'stan countries (from 1918 to 1991) and renaming them soviet republics . . . e.g., the Kazak SSR. Many would say they wanted to push down into Iran to gain access to the Indian Ocean, which is one reason that the Persians did not want the Russians there.

As the Russians were coming down from the North, the British, who controlled "British India" at the time, were pushing up from the South, effectively buffering against Russian influence in the region. The Chinese did not want anyone in there either, so they were vying for influence to make sure that no one encroached upon their territory. This became a big game where rulers were moving soldiers around on this very real chessboard. That is why it's funny in today's world to see the United States vying for influence in Afghanistan by pushing US ideals and working against radical Muslim influence, or perhaps against a renewed Russian presence in the region—same story, different century.

Afghanistan is a good country to discuss for just a second. The Russians and the British found this out a long time ago: you just don't mess around in Afghanistan. Everybody (all these world powers) has tried very hard in the past, and let me tell you what the historical scoreboard reads: Afghanistan: 1000, Outsiders: zero. No one from Europe or anywhere else has successfully occupied or influenced Afghanistan for very long, and it remains to be seen if the US/NATO is going to have much impact either. Folks throughout Central Asia, particularly Afghanistan, have a fierce independent spirit. They understand that they are pawns in the game and they will have none of it. When the British went into Afghanistan to thwart the Russians, the Afghans kicked the crap out of them. Then when the Soviet Russians came in 50 years later, the Afghans kicked the crap out of them as well. The Central Asians as a whole are not peoples to be trifled with.

In 1979 when the Soviet Union invaded Afghanistan in their bid to expand the communist sphere of influence even further, a new player joined the game: the US. The United States armed and trained a bunch of locals—let's call them the **mujahideen**—to really stick it to the Ruskies in order to stymie the Soviet advance. It worked great! That 1979 to 1989 Afghan campaign proved absolutely disastrous for the Soviets, and has been credited as part of the reason for their eventual collapse just two years later. Of course, those CIA trained mujahideen then ended up becoming the Taliban and members of al-Qaeda, which the US now finds itself fighting, but that's a different round of this unfolding drama. Isn't this region just a ton of fun? Kind of like playing nude lawn darts.

THE 21ST CENTURY EDITION

Lots of the parties who were vying for influence in the last couple hundred years are some of the same players in today's world, with a few new entries—like the United States, the Taliban, al-Qaeda and a country that didn't even exist back then, now called Pakistan. As in the past, no one really wants to own Central Asia, but everybody is trying to expel other influences and powers from it. The Game is back on. Back in the day, most of the powers originally got involved to curb the power and growth of pre-Soviet Russia. The British, in particular, didn't really want to get into the northern parts of India, but felt compelled to in order to stop the Russian advance. The game to stop Russia is over (maybe), but there are three new reasons that the game started back up.

1. THE VACUUM CREATED BY MOVING THE SOVIETS OUT

Under the Soviet umbrella, Russia gained control of all of Central Asia, except Afghanistan, by the 1940s. All of those "-stan" countries became Soviet republics—you know, the R in the USSR. Afghanistan was never brought into the fray; that's why the Russians invaded them in 1979, much to their own chagrin for the next decade. Since the fall of the Soviet Empire in 1991, all of those former Soviet SSR's declared independence and became sovereign states again. Who is in charge now? For most of the "-stan" countries, this has resulted in authoritarian dictators/one-party states that are extremely open to outside influences, especially when there is money involved. How refreshing. Not.

2. BECAUSE OF INCREASING FINDS OF PETROLEUM AND NATURAL GAS

They gots the goods: super important natural resources that the world cannot live without. Oil and natural gas, and the pipelines that carry it, are popping up all over the region. Every day, it seems, they increase their findings of potential natural reserves. Kazakhstan is the leader of this. It is the Central Asian powerhouse when it comes to energy exports.

Where is this stuff going? Everywhere. This game is back on because everybody wants that stuff. No one wants to control the places, but they want the influence inside the region in order to get a piece of the oil action. As described by the US's Energy Information Administration: "*Central Asia has large reserves of natural gas but its development as a major natural gas exporter is constrained because of a lack of pipeline infrastructure.*" Everyone wants a piece of the infrastructure-building action.

Who are they supplying? Kazakhstan's and Uzbekistan's stuff has mostly moved west into Russia and then on into Europe. Turkmenistan and Uzbekistan supply the Ukraine and other Eastern and Western European countries, as well as Iran. Some of it floats south to Iran or through it. Increasingly, even Tajikistan, Turkmenistan, and Kazakhstan are building pipelines into China. Why would China want that stuff? Their population is exploding and they can not ever possibly have enough oil and natural gas. These historically landlocked economies of Central Asia are in a prime position now. The one thing that they could move out of their countries in order to make money is being discovered in large amounts.

Pipelines a'plenty! Source: Energy Information Administration

CHAPTER TWENTY-TWO

3. TERRORISM AND DRUGS

Why would anybody from the outside world be interested in vying for influence of drugs and terrorism? Ask the US government! Thwarting terrorism is the prime reason that the United States invaded Afghanistan in 2001. The US-led war in Afghanistan was to try and stop the Taliban, al-Qaeda, and any other radical extremist groups that may be there training, setting up shop, or otherwise trying to control the area. I do want to be quick to point out that this was a US-*led* effort; the Afghan campaign was actually a NATO mission, of which the US is the biggest player. Why is that important? Because it shows that all other NATO members, namely all of Team West, are involved because they all believe that the anti-terrorism fight is a worthy cause that they will continue to pursue.

Do not lose sight of the irony that the US and NATO were currently fighting against forces that they helped train to fight the Soviets back in the 1970s. Ah, the game continues. Of course, drugs are a bad thing too; Central Asia is historically one of the largest producers of opium, so lots of developed countries around the world in their **War on Drugs** have had their fingers in Central Asia long before the current conflict in Afghanistan. Unfortunately, in a war-wrecked country, drug production is one of the few (if not only) economic opportunities for these folks to embrace. The war has been further complicated by the Taliban and al-Qaeda, who are now using the drug trade money to help re-arm and re-equip themselves to do further battle against Team West. Could this mess get worse? It sure can! While virtually all of this product used to be shipped abroad to the rich, bored citizens of Europe and America, it has recently been increasingly used at home. Drug consumption and addiction has soared in Central Asia, Russia, and East Asia.

Terrorism, drugs, and petroleum products are the main reasons that Central Asia has become an area of much interest for people around the globe who otherwise would not care about this economically marooned region. There isn't much else about this place to interest the outsiders. What's made this even more important in today's world is the **War on Terror**. The War on Terror is the reason why the US was in Afghanistan, supposedly the reason they were in Iraq, and the reason why there are US military bases that have cropped up all throughout Central Asia—because of the fear that terrorism and extremism will be on the increase, particularly in this region. The United States is vying for influence here not to control the place, but to control the extreme elements within the place. Team West and the US are worried about terrorists in this Central Asian society, but they're not the only ones.

The Russians are also worried because they have lost a lot of territory and influence in the world since the implosion of the USSR, so they are determined to maintain their role in the region. Their long history of influence here, desire to maintain their status as a world power, and close ties with Central Asian economies give them more reason to stick around. They have their terrorist problems as well, mostly with Muslim extremists over in Chechnya, but also here in Central Asia. So Russia has a vested interest in maintaining influence here . . . even if it's not a physical presence anymore.

And then there's China. Here's a theme that's perfect for identifying western China as part of Central Asia and not as China proper. It's a little known fact that when the United States declared war on terrorism after 9/11, there were two countries that IMMEDIATELY jumped on board and said, "Yeah, we'll help! We agree that terrorism should be fought on all fronts!" Those two countries are probably not ones you would think of as the first in line to side up with the US: Russia and China. Russia said, "We've been fighting Muslim extremists for years. We call them Chechens and we are going to continue to beat the beet juice out of them in our war on terrorism." And China said, "We have Muslim extremist terrorists as well. The Uyghurs out in the western states have been blowing stuff up and making attempts at independence. We're going to fight the War on Terror by continuing to kick the Kung Pao chicken out of the Uyghur separatists in our country."

Terrorist hunters have their work cut out for them in this neck of the woods.

So terrorism is a rallying cry for a lot of countries to continue to be involved in Central Asia, and is a theme that probably isn't going away any time soon. We've got Russia, China, and the United States in there vying for influence for oil, trying to control terrorism, and lots of other things. Who else may be involved in The New Great Game? For starters, the Pakistanis, who are US allies (kind of) and the ones fighting the truly active War on Terror on their own border. Pakistan's largest border is shared with Afghanistan; that border is a real hotspot for active conflict. People are shooting at each other there all the time. That is where everyone thinks Osama bin Laden is currently hiding. Hey hold on there! **2011 UPDATE:** That WAS where Osama was hiding, just like the US always said, and Pakistan always denied! And now, of course, he is dead.

Partially just to counter Pakistan, India has increased its attention to Afghanistan and other Central Asian states by increasing aid as well as diplomatic relations . . . since India is a rising power as well, and wants to expand its influence. Iran, which also fronts this region, is increasing its power on the planet by building nuclear facilities and playing up its role as a leader of Muslim culture. It is also a major mover of natural gas and oil out of Central Asia to the rest of the world. They want to join the power players and are vying for influence as well. They have the advantage of being Islamic like all their Central Asian brothers. In 2012, Kazakhstan inked a deal with Iran to build a major pipeline from Central Asia to their "Muslims brothers in Iran," as spoken by former Kazakh President Nursultan Nazarbayev. Game on! Because of this, Iran may end up being a major political player in this region in the very near future.

2021 AFGHAN UNRAVELING UPDATE: After over 20 years of the US-led ongoing war in Afghanistan (the longest war in US history), both the former Trump presidency and the current Biden administration are now determined to end American presence in the country and have signed a peace deal with the Taliban, which is still actively bombing and attacking the Afghan government/army. It is extremely likely that within the next few years: 1) The US pulls out all remaining active forces, 2) The Taliban will rapidly reassert control over the country, and 3) The country will resume its position as a great place for jihadists and extremist groups to operate within, thus causing consternation for all surrounding states. In hindsight, the US-led war seems to have changed nothing except destroying al-Qaeda, which will be replaced with other groups like ISIS and next generation terrorists groups as yet unnamed.

What's the Deal with Drug Production?

If you wanted to see the biggest production center for heroin on the planet, you'd probably take a trip—no pun intended—to Afghanistan. Afghanistan is the largest producer of opium in the world, weighing in at almost 5,000 metric tons/year. That's 87 percent of the opium supply of the entire world. Quite impressive.

The funny thing is that under Taliban rule, opium production was much, much lower. Once the US intervened and knocked the Taliban out, with an almost audible "smack" heard round the world, opium production nearly doubled each year after 2001. 2009 was a world record breaker for most opium produced in the country. Nice job, guys! Here's a hint for you: most of that isn't going to the pharmaceutical companies that turn opium into legitimately consumed pain killers. This is hardcore stuff getting released into the streets, mostly in Europe and Southeast Asia, in larger and larger quantities because that's where the money can be made. It's untaxed, uncut, and a huge cash crop for this economically landlocked region. Production is only increasing, despite US occupation.

What a pretty poppy garden these boys are tending—of opium poppies, that is!

THE GAME CONTINUES TO EVOLVE WITH NEW PLAYERS

Comrade Russia iz still vatching Central Asia.

To summarize: Central Asia is an isolated region both physically and economically—a culturally and economically marooned region—which has been suddenly thrust back into prominence in the new version of The Great Game. Why? It's mostly for resources, such as oil and natural gas, but also for the fight against terrorism and the drug trade. We also see some of the old school elements of The Great Game here, such as Russia not wanting to completely lose all of its influence in the region. The US military presence is actually seen as a direct threat to this historic Russian influence. Vladimir "The Man" Putin is not too happy about it. And he is on the move again to remedy this. . . .

Luckily for him, Putin's country is currently winning the fight. Some countries have requested the US troops to leave as soon as possible, and Uzbekistan recently just blatantly kicked them out. Many of the "-stan" countries are already reestablishing their old ties with Russia for security purposes, but they're also increasingly looking to Russia for leadership in the oil and natural gas industries. Putin is happy again! The vehicle for this renewed Russian leadership is a supranationalist organization named the SCO. You heard it here first: this group has huge potential to become a major global force in the coming decades. And yet another force is rising as well: a Russian-forged "Eurasian Union" seeks to pull the entire region back into the arms of Mother Russia . . . a possible USSR 2.0? Too early to call, but never too early to learn more about it. . . . Best to know a bit more about these entities right now.

THE hot block to watch: and its so much more than trade oriented!

YOU GOTTA KNOW THE SCO

Watch out! This coalition could be hot! The Shanghai Cooperation Organization (SCO) is an intergovernmental organization founded on June 14, 2001, by leaders of the People's Republic of China, Russia, Kazakhstan, Kyrgyzstan, Tajikistan and Uzbekistan.

The SCO facilitates cooperation and organization on issues like natural resource extraction, counterterrorism, extremism, and separatism. They are now holding joint military exercises and are also thinking about a free trade pact. Though the declaration on the establishment of the SCO contained a statement that it "is not an alliance directed against other states and regions and it adheres to the principle of openness," most observers believe that one of the original purposes of the SCO was to serve as a counterbalance to the US, and particularly to avoid conflicts that would allow the US to intervene in areas near both Russia and China.

How are they pulling this off? In lots of curious ways that have tremendous potential for the future of these states and also the world. First, there have been very secretive discussions about these countries creating a common cartel based on their exports of natural gas—basically, an Asian version of OPEC. Just as OPEC countries control the lion's share of exported oil and, therefore, control the supply and price, the combined SCO countries control a significant amount of global reserves of natural gas. It is not a stretch of the imagination to see them pooling their resources together under a single umbrella of control, which would make them all even richer.

Let's go go go with the the SCO . . . and get back our military mo-jo!

Second, the SCO states are strengthening their defense pacts with each other to ensure that no interference or invasion will be allowed from outsiders. In essence, they are trying to slowly form an Asian version of NATO. The movement in this direction is why most US military bases have now been kicked out of SCO states, and also why America's request to become an observer state has been declined every single year. It's also why the SCO has started doing joint military exercises. By the way, don't start digging your backyard bunker in preparations for WWIII anytime soon. Militarily, these guys are still pretty clueless—okay, they outright stink. The main point is that they're just beginning to do these exercises, thus setting the trend for the future.

Let's get you even smarter. What current observer state in the previous map do you think is desperate to become a full member of the SCO? If you said Iran, then give yourself a cookie and a barrel of oil. If Iran joined this Asian NATO club, then the prospects of US/UN actions to prevent Iran from going nuclear would instantaneously become a world crisis. Dig what I'm saying here? It's a really big deal.

But hang on! Don't let me paint a picture that makes the SCO look like an enemy of the US or Team West in general! They are not! The SCO met in April 2009 specifically to try and find "an Asian solution" to the War on Terror debacle going down in both Afghanistan and Pakistan. The US/NATO/Team West is losing this battle, and the SCO club knows it, and are NOT happy about it! See, China, Russia, and India are all states that have active Islamic extremist/terrorist problems too, and they DO NOT want to see the Afghan/Pakistani area continue to melt down, mostly because the impacts to them will be much, much, much greater than any real impact to the US. Think about it: the SCO (plus observers) surround this troubled area, and trouble has a way of spreading like a cancer. On multiple fronts like terrorism and drugs, the SCO shares strategic interest with the US and Team West. Hopefully, they can think up a better solution than the West has been trying, 'cuz it can't get worse!

Organizing pipeline construction, conducting military exercises, trying to come up with Asian solutions to Asian problems, and refusing to grant the US observer status. Ouch! This thing is on fire, and no one in our country has probably even heard of it, but I think you should. But I bet due to the recent resurgence of Russia (particularly as they re-took Crimea), maybe more of you have heard reference to another aspiring entity in this area: The Eurasian Union. . . .

PUTIN'S PROJECT: SOVIET UNION RE-BOOT?

Watch out kids! This sucker is ready to launch! Or possibly explode while on the launch pad! Still too close to call here in 2015, but this idea is being talked about everywhere as the proposed game plan for a new Russian-led re-do of the Soviet Union! But there ain't no more 'soviet' stuff, so let's now call it a kinder, gentler, more politically correct and inviting "Eurasian Union." Something led by Russia that is kinder and gentler and inviting? lol NOT! So what is this really all about, and why is it terrifying to so many?

The concept of this "economic" union is the brainchild and boyhood fantasy dream of one Russian President Vladimir Putin, who now openly seeks to reassert a resurgent Russia's influence and outright power into areas it lost during the collapse of the Soviet Union in 1991. Putin for his part has actually referred to the demise of the USSR as "the greatest geopolitical catastrophe of the 20th century," . . . and it is a catastrophe he now seems intent on fixing. Through economic coercion, intimidation, and possibly force if necessary, Russia looks to be regaining its old sphere of influence in Eastern Europe . . . which of course scares the stuffing out of Europe in general . . . but the true power core of this alliance centers in Central Asia, specifically with Russo-Kazak ties, which Russia hopes to expand out to the rest of the region.

Why try to re-align the less developed Central Asian states into the Russian 'hood? To counter the rapidly growing influence of China, of course! Although Russia and China form the bookends of that SCO block, make no mistake: these two titans are historically fierce competitors, and they still are far from being called fast friends here in the 21st century. With tons of oil and natural gas in the Central Asian states, the Ruskies and the Chinese are once more facing off (although not militarily) for access and control of those vital resources here in the center of the continent. SO the Chinese are busy building trade ties with individual states, while Russia is pushing their Eurasian Union idea. Speaking of which . . . Is this an attempt to rebuild the Soviet Union proper? Is this a new Cold War? Eh. I'm not sure it is all that, at least not yet, but here is the deal as of 2019:

The formations of it are entirely logical and benign, at least on paper: In November 2011, the presidents of Russia, Belarus, and Kazakhstan signed an agreement to create the Eurasian Economic Union: a customs union (like a free trade zone) with partial economic integration between the states, with the intent of becoming something very close to the European Union (you know, the EU) in the future with full-on economic, political, and military cooperation. Just like the EU, it is supposed to safeguard regional economical interests and facilitate trade and business within member countries. Since Russia is obviously the powerhouse player in all of this, Putin could not be happier. But he isn't satisfied just as yet. . . .

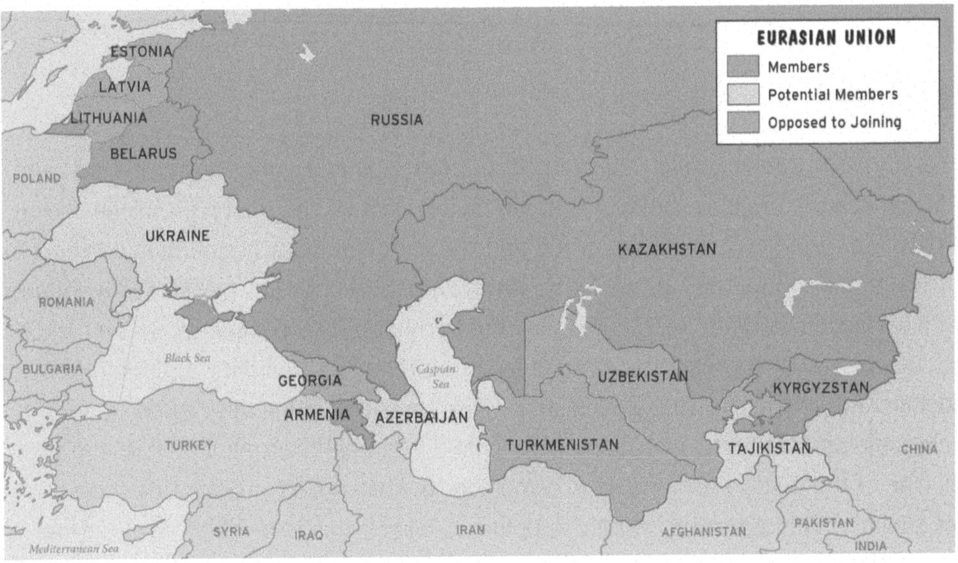

Putin's grand gamble here is that his union will become a successful alternative to the European Union, and that states of Central Asia, Eastern Europe, and even the Caucuses Region will want to sign up for it. I suppose it is being marketed as a great way to boost economic growth, without having to bother with all that pesky democracy and human rights stuff. And it is definitely being pushed as an alternative to joining Team West. And it definitely is Russia trying to regain "influence" in areas it previously held as its own. Ah! Back to the Great Game of course! This is a now fervently fired-up Russian response to the last two decades of the EU and NATO creeping closer and closer to its borders. The Bear wants its backyard back! Game on!

Will it work? Who knows yet, but everyone in the 'hood is worried, as Russian pressure to get on board this Eurasian party train is growing. The EU, the US, and even China are watching these developments warily. States that have lots of ethnic Russian people in them (like Kazakhstan) are starting to fear a Russian take-over of parts of their territory, much like what happened to Ukraine with Crimea. Belarus may be concerned about being re-absorbed altogether. But the other Central Asian states are now also stuck in this battle for control, as Kyrgyzstan and Tajikistan are highly dependent on Russian foreign aid and economic trade . . . they will likely join. The other -stans seem to be hedging their bets, as they like their sovereignty more than they like Russia. Armenia has announced it will join soon, and the fate of Ukraine as a whole is (as of this writing) still unsettled . . . a full Russian takeover of that state will see it bolster the Eurasian Union's ranks into a fully legit power block. But will Putin want to go that far to see his 21st century Ruskie dream realized? We will soon find out. . . .

I see the Russian Re-boots coming!

GAME ON

Terrorism is a problem which, despite US influence, is not going away; it's actually increasing. To be completely frank, Afghanistan is the main hotspot and haven for most of the terrorism in the region right now, mostly related to the US/NATO war and the lack of any strong and stable government functionality. When you look at the rest of the "-stan" countries in this region, they are all led by strong-arm authoritarian types, mostly influenced by the biggest strong-arm of them all: Vladimir Putin in Russia, and also the fairly assertive commie crew over in China. As such, those states are extremely successful at repressing and/or eliminating terrorist elements. Unfortunately, they are also good at repressing any democratic and human rights elements. I don't want to paint this region in an entirely negative light, because they do have some positive things coinciding with some of the wealth accumulation due to their increasing natural gas and oil exports. One of the things I like people to remember is when we say, "Crap, this sucks! Gas is going to be five dollars a gallon this summer. This is terrible. Oil is 130 dollars a barrel, this is terrible!" Well, it may suck for you, but it actually means great things for Central Asia and any other oil and gas producing region. That means their revenues are expanding, and they have a little more wealth, hopefully to spread around to their citizens.

Things may be looking up for the "-stans." Economic integration in Central Asia has risen to an all time high for the last 500 years. States are figuring out that they must work together in order to effectively build the infrastructure to get their petroleum products exported out to the rest of the world. With that type of cooperation comes increased trade and communications. That is why the Plaid Avenger really wants you to know all about that SCO, because it is becoming the most effective organizational superstructure to accelerate this process. Look for the SCO to not only become an OPEC-style cartel of natural gas in Asia, but also a NATO-style defense structure with which many Asian nations might be keen to get involved. Trade and economic integration will follow. I need say no more about the Eurasian Union at this point, but it also may become a force to be reckoned with, for better or worse.

Not to be outflanked by the Russians, the Chinese have also re-entered the Great Game with a brand new policy that centers around themselves: **The Silk Road Economic Belt** and the **21st-century Maritime Silk Road** aka **The Belt and Road** aka **One Belt, One Road** aka **The Belt and Road Initiative**, (lol that is a crap-ton of aka's), is a development strategy and framework, proposed by Chinese President Xi Jinping that focuses on connectivity and cooperation among countries primarily between the People's Republic of China and the rest of Eurasia, specifically the Central Asian core. Look at the name itself!!! China wants to re-establish the old Silk Road!!! Their grand plan consists of two main components, one linkage by land, the second by sea. The strategy underlines China's push to take a bigger role in global affairs, and its need for priority capacity cooperation in areas such as steel manufacturing and energy. I'm not sure how to even categorize this plan at the moment. . . is it a trade block, is it a international organization?. . . but in any case it highlights the renewed interest into Central Asia, and it has great economic potential for its participating member states.

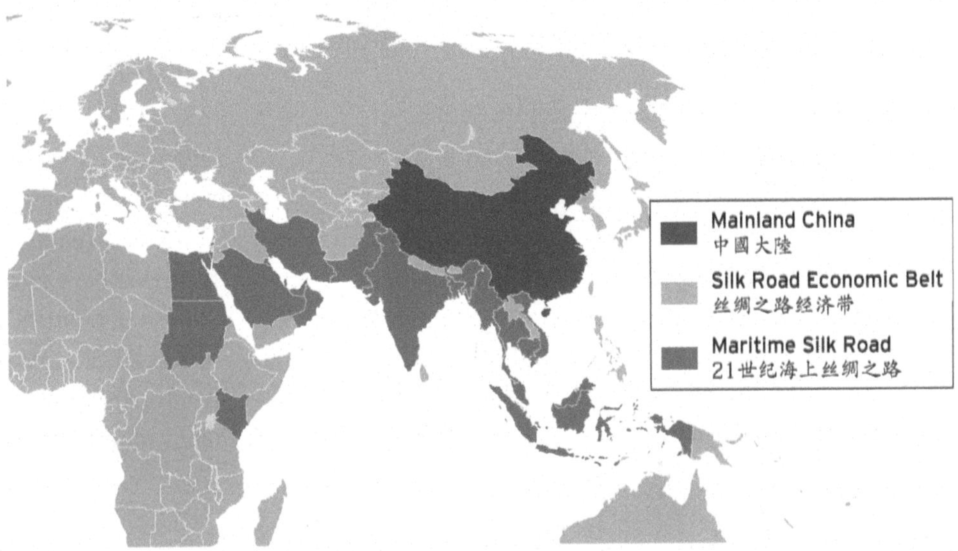

This is a place of fairly extreme economic disparity and massive challenges in terms of infrastructure creation; however, at the same time, due to its former-Soviet influence, this is a fairly literate region. People are fairly well-educated and the illiteracy rates are some of the lowest in the world (except Afghanistan, of course). In terms of standards of living, access to health care and literacy rates, they are not doing too badly (except Afghanistan, of course). With Russia and China as powerful bookends, a new Asian OPEC and an Asian NATO, things are looking up, although the authoritarian structure in the governments of most of these places may still keep the people down for some time to come. But that's a story for another time and another place.

The region continues to evolve quickly, as Vladimir Putin's return to the Presidency of Russia has created a promise of an "Eurasian Union" filled with trade and cooperation to compete with the EU, and with a likely 10 more years of Vladimir coming, watch for further Russian and Chinese investment/integration in Centra Asia to become a reality. In addition, the Afghanistan campaign now appears to be winding down way faster than anyone would have anticipated: the Trump administration will likely bail out of the place starting in 2020, unless military advisors convince him otherwise, so look for the Afghan problem to be thrown into the laps of the SCO countries, who will be forced to pick up the hot potato and do something with it. What that something will be is anybody's guess right now, but for sure policy changes are coming to the Afghan situation, and the region as a whole.

Let's get to the next superpower on the planet . . .

CENTRAL ASIA RUNDOWN & RESOURCES

View additional Plaid Avenger resources for this region at http://plaid.at/c_asia

BIG PLUSES

- Huge reserves of oil and natural gas, with more being discovered every year
- Region now being courted heavily by Russia, China, and Iran, with even a distant Uncle Sam trying to exert some influence (albeit mostly unsuccessfully)
- The SCO is a possible unifying entity of the region that is just getting warmed up here in the 21st century
- Region is surrounded by great powers (Russia, China, and India) who all have a sound and stable Central Asia in their best interests; that was not so 100 years ago
- China's new Silk Road Economic Belt and the 21st-century Maritime Silk Road initiatives will likely be infusing tons of capital and investments into the region; Silk Road glory days returning?

BIG PROBLEMS

- Landlocked and largely isolated from world economy (although this is changing)
- Tough terrain and climate make inter-regional connections challenging or impossible
- Ummm . . . . Afghanistan? Enough said
- Drug production and distribution (primarily of heroin) is widespread, and drug addiction is actually increasing within the region as well
- Overwhelmingly reliant on oil and natural gas exports; not economically diversified
- Supposedly a great area for terrorists, separatists, and other rabble rousers . . . although this seems to me to be skewed reality as government/SCO crackdowns in the region have crushed much dissent, both of terrorists but also of political activists (except Afghanistan)
- Almost all governments of region are one-party states/dictatorships, with political power consolidated in the hands of few

DEAD DUDES OF NOTE:

Genghis Khan: The founder and Khan (ruler) of the Mongol Empire, which became the largest contiguous empire in history after his death. After uniting the horse-warrior, nomadic tribes of northeast Asia in 1200 AD, went on a world tour of butt-whipping, invading and destroying political entities in China, India, Iran, Iraq, Egypt, Russia, and Eastern Europe. And totally ravaged Oshman's Sporting Goods in *Bill & Ted's Excellent Adventure.*

LIVE LEADERS YOU SHOULD KNOW:

Nursultan Nazarbayev: former President of Kazakhstan since the Fall of the USSR and the nation's independence in 1991. Was in charge of this one-party state bordering on dictatorship . . . until he randomly just quit in 2019! No one knows why! But Kazakhstan is fast becoming a regional powerhouse and serious economy, and is invited to SCO, G-20, Nuclear Summits, NATO meetings and other important international gatherings. So their new leader will be one to watch.

Ashraf Ghani: Intellectual, academic, former World Bank worker, anti-corruption advocate, and new President of Afghanistan (since 2015) with the HUGE task of trying to rebuild this shattered country as the US exits, and the Taliban doesn't. Good luck!

PLAID CINEMA SELECTION:

Charlie Wilson's War (2007) AWESOME! A dramatic biographical film, documents the efforts of Texas senator Charlie Wilson, who passed away in 2010, in his efforts to help Afghanistan fight Soviet occupation. While certain elements of the plot are fictional, the film does a great job of laying the basic details of the actual event on the table for this generation to understand.

Beshkempir (1998) A Kyrgyz film, shot and produced entirely in Kyrgyzstan, is a prime example of the new wave of films produced after the country gained independence from the Soviet Union. Film follows an adopted child growing up in rural Kyrgyzstan. Demonstrates several Kyrgyz traditions, including funeral and engagement ceremonies.

Seven Years in Tibet (1997) Based on real events of Austrian mountaineer Heinrich Harrer on his experiences in Tibet between 1944 and 1951 during the Second World War, the interim period, and the Chinese People's Liberation Army's invasion of Tibet in 1950. Features the Dalai Lama and is pretty historically accurate on the timeline and events, despite the presence of a young Brad Pitt.

True Noon (2009) An absurdist film made by filmmakers in Tajikistan about a change in the border between Tajikistan and Uzbekistan. The film centers around two villages that are very close that get separated by the new divide with barbed wire fence, all while a marriage is about to occur between two people, one from each town.

Kandahar (2001) Depressing film is based on a story (partly true, partly fictionalized) of a successful Afghan-Canadian who returns to Taliban-led Afghanistan after receiving a suicidal letter from her sister.

Osama (2003) Even more depressing story of a girl in Taliban-led Afghanistan who disguises herself as a boy in order to support her family. It was the first film to be shot entirely in Afghanistan since 1996, when the Taliban régime banned the creation of all film.

The Kite Runner (2007) Slightly even more depressing tale of abandonment set against a backdrop of tumultuous events, from the fall of the monarchy in Afghanistan through the Soviet invasion, the mass exodus of Afghan refugees to Pakistan and the United States, and the Taliban regime.

Mountain Patrol (2004) Shot on location in extreme awe-inspiring landscapes, this film depicts the struggle between vigilante rangers and bands of poachers in the remote Tibetan region of Kekexili (Hoh Xil). It was inspired by the documentary *Balance* by Peng Hui.

Story of Weeping Camel (2003) The most exciting film you will ever see about camel weaning! A family of nomadic shepherds in the Gobi desert of Mongolia try to save the life of a rare white Bactrian camel calf after it was rejected by its mother. Great scenery and unique insight into the culture, and if you don't like reading subtitles, don't worry; I'm not sure they speak more than five minutes during the course of the entire 87 minute film. A quiet camel calamity.

Mongol (2007) The first installment of what I heard is to be a three-part series on Genghis Khan. This film is slow, quiet, long and chronicles the rise of the man in his early years before he got to the world conquering stuff. That sounds boring, right? But man, the scenery is totally epic, the battles are bloody, and the pacing and feel so real as that you would swear you are actually there watching the man live. Actually shot in Kazakhstan.

MAP GALLERY

View Map Gallery & additional Plaid Avenger resources for this region at http://plaid.at/c_asia

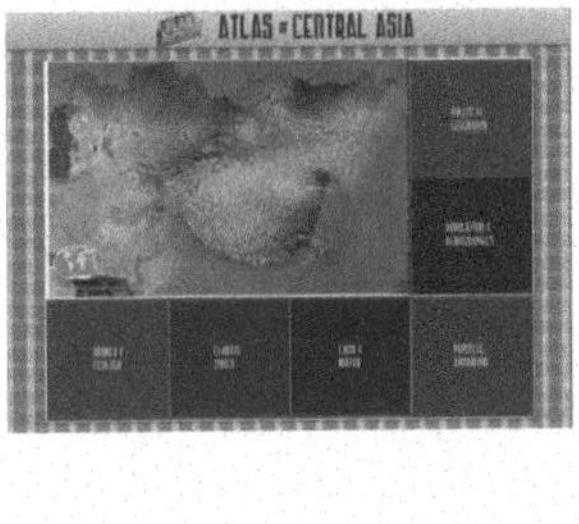

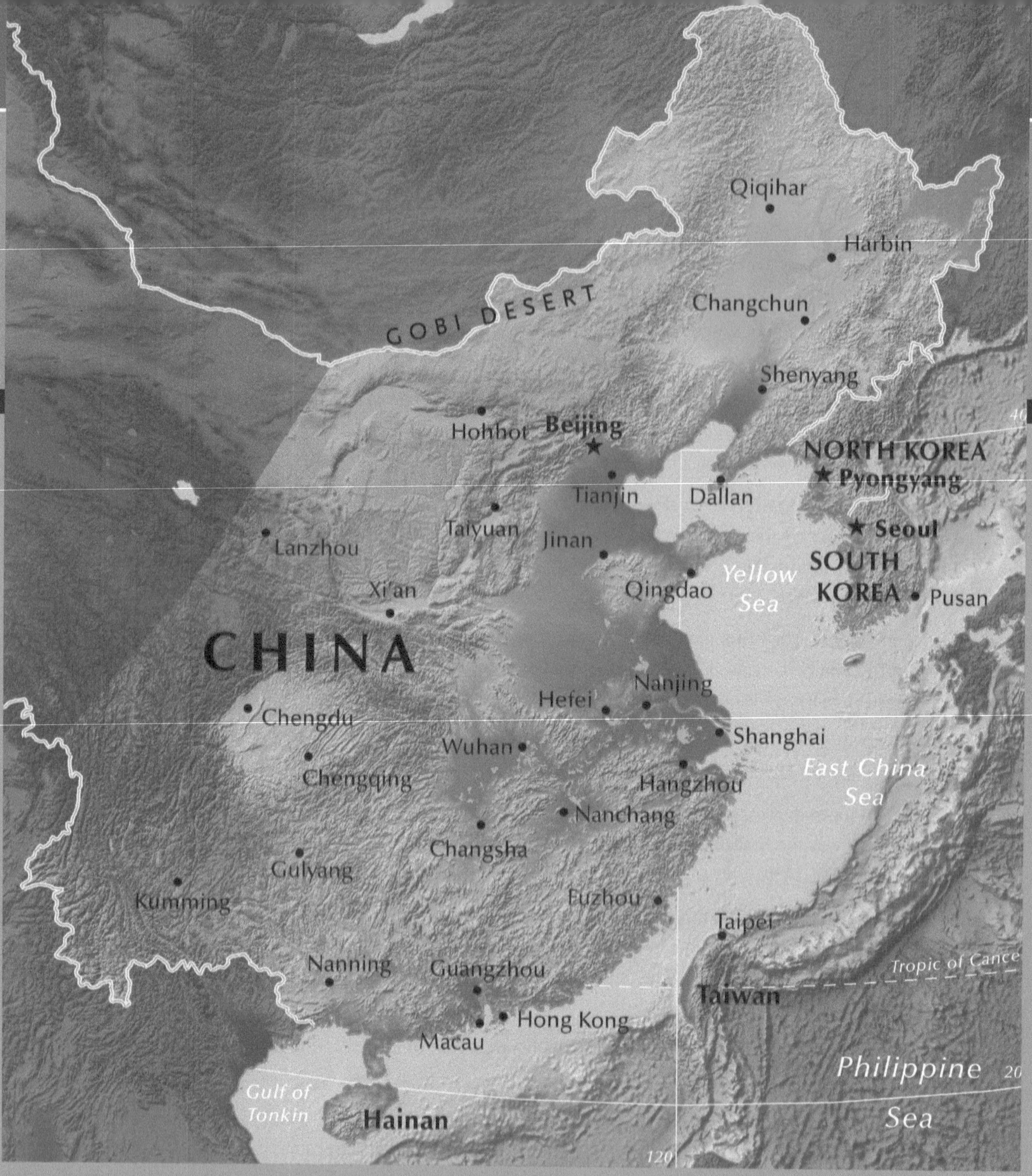
Qiqihar
Harbin
GOBI DESERT
Changchun
Shenyang
Hohhot
Beijing
NORTH KOREA
Pyongyang
Tianjin
Dalian
Seoul
Taiyuan
Jinan
Lanzhou
SOUTH KOREA
Yellow Sea
Qingdao
Pusan
Xi'an
CHINA
Nanjing
Hefei
Chengdu
Shanghai
Wuhan
Chongqing
Hangzhou
East China Sea
Nanchang
Changsha
Guiyang
Kunming
Fuzhou
Taipei
Tropic of Cance
Nanning
Guangzhou
Taiwan
Hong Kong
Macau
Philippine Sea
Gulf of Tonkin
Hainan
120

CHAPTER OUTLINE

East Asia

23

THE BEAST IN THE EAST

The dragon has been unleashed! This region is the center of action on planet Earth right now, and contains the world's next superpower: China. A lot of people are scared that they're going to be *the* superpower and take over the Earth. I'm scared! They certainly are going to be *a* superpower. They are not going to be the sole superpower, but perhaps they are going to become the balance for US power on the planet. They're going to be the **yin** to the US's **yang**. Gosh, that is good! Isn't that good? Gettin' all Chinese philosophy on ya' already! Cuz the power we are talking about here is China, the main master-blaster for the 21st century. We're going to talk about why it is the powerhouse of this region, and the wider world. Also in this region, we are going to take on North Korea and South Korea. For those that aren't sure where it's supposed to go, Taiwan falls in this region as well. And China would say that it falls directly into their lap, but I am getting ahead of the story already. . . .

China and most of the planet consider Taiwan to be China's renegade province, but a lot of folks in Taiwan think they are independent. About 20 countries in the world recognize it as an independent sovereign state; and that number is dwindling fast . . . for reasons this chapter will explain.

Before we go any further, let's make sure we identify the specifics of our region from the map on the opposite page. China is divided in half, and we have taken out western China and put it in Central Asia for lots of reasons that you've already studied. The region we are going to call East Asia is made up of East China (the core of the Chinese Empire for centuries), Taiwan, South Korea, and North Korea.

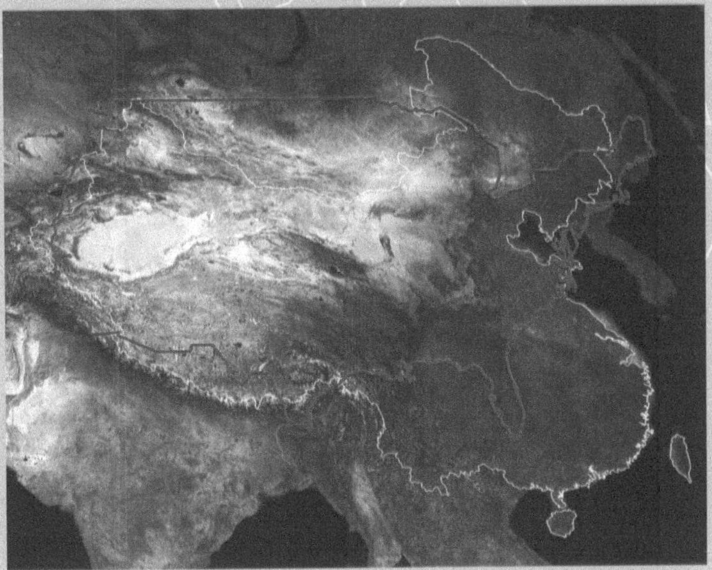

China and US: roughly same size, same latitudes.

PANDA FUR: THE PHYSICAL WORLD

As you see in the graphic on the previous page, China and the United States have some similarities. Which one is bigger? Which one is smaller? I don't know. It depends on which sources you're using. Some say China is the third largest country on earth and the US is number four, some reverse that; doesn't matter to me, it's pretty darn close either way. And on that map latitude is held constant, so it's near the exact same size as the US, but the latitudinal range is a bit larger, i.e. China goes a little bit farther pole-wise than the continuous United States, and stretches farther south of the tropics. Hainan Island, in the southern part of China, would be about where Cuba is in our hemisphere. What does this equate to climate-wise? Because latitude is constant, you know what climate is like on the east coast of the United States and that's about what the east coast of China is like as well. From north to south on the eastern seaboards of both countries, the climate regimes are virtually identical.

So what's different? The United States has an east coast like China, and a west coast unlike China. China's western fringe is buried in the continent. Continentality has a different effect on the climate than coastal regions, as we already know. The Plaid Avenger wants you to know that when you travel from the coast to the western part of China, several things happen. The West gets cooler in the winter and oftentimes hotter in the summer. Elevation increases and terrain gets rougher and rockier as you move west. You have the **Tibetan Plateau** in Western China, as well as the **Taklimakan Basin**. These are areas that are way above sea level, particularly the Tibetan Plateau propping up the Himalayan Range, the largest mountain range on the planet. When considering just the country of China as a whole, almost 70% of the Chinese territories are mountains, plateaus, and hills. The country's main arable land is on the remaining 30% of lowland plains and basins. . . which is virtually all on the eastern side of the country.

Clothing Recommendations for Travel in China

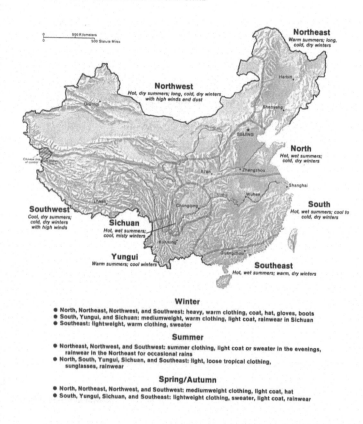

Winter

- North, Northeast, Northwest, and Southwest: heavy, warm clothing, coat, hat, gloves, boots
- South, Yungui, and Sichuan: mediumweight, warm clothing, light coat, rainwear in Sichuan
- Southeast: lightweight, warm clothing, sweater

Summer

- Northeast, Northwest, and Southwest: summer clothing, light coat or sweater in the evenings, rainwear in the Northeast for occasional rains
- North, South, Yungui, Sichuan, and Southeast: light, loose tropical clothing, sunglasses, rainwear

Spring/Autumn

- North, Northeast, Northwest, and Southwest: mediumweight clothing, light coat, hat
- South, Yungui, Sichuan, and Southeast: lightweight clothing, sweater, light coat, rainwear

As you progress off the coast, you will go upward in elevation and the temperature disparity will increase. That means it will get hotter in the summer and cooler in the winter. This same effect, named **continentality**, also means it will get drier, as the farther you get from large bodies of water, (which provide moisture to the air via evaporation), the less likely you are to get precipitation on a regular basis. Temperatures get more extreme as you progress westward from the coast; it gets drier and the elevation gets higher. This lovely clothing chart pretty much sums it all up.

PANDA PEEPS: EAST, EAST, EAST!

Ethnically, I want you to consider why we subdivide China into separate regions. As we pointed out in the Central Asia chapter, we have a lot of different ethnic groups in China. In Mongolia, of course, there are Mongols. They also live in Inner Mongolia, a province of China. In the Tibetan Plateau there are Tibetan folks, and of great importance in the northwestern quadrant, we have some Cossacks and some Uyghurs (pronounced Weegers). This is the exact reason why we classify most of Western China as part of Central Asia. Those are the ethnic groups that are not Chinese.

As you can see, the vast majority of people in China are on the eastern side, particularly along the eastern seaboard. The vast majority of those same people are what we call **Han Chinese**, named for an ancient dynasty. The Han dynasty's rule spanned from 200 BCE to about 200 CE. They were a very popular dynasty; they were so popular that people started calling themselves by the dynastic name. They're still known today as Han Chinese, which brings up the question: Where are people in China?

We pointed out that most of the Chinese population is on the eastern side of the country. That's another reason for this east-west divide within China; the eastern side is where all the people are and where the ethnically Han Chinese people are located. It also has a more moderate climate with moisture, which is essential for food growth. Most of the food is grown where most of the industry is happening, and therefore where most of the cities are located. This is not a coincidence. This eastern scenario is quite different from that of the western people who are not distinctly Han Chinese. The western countryside is more sparsely populated, there's a much more rural base, and it's cooler, higher, and drier than its eastern counterpart. That's your East Asia roundup of climate, terrain, and ethnicity.

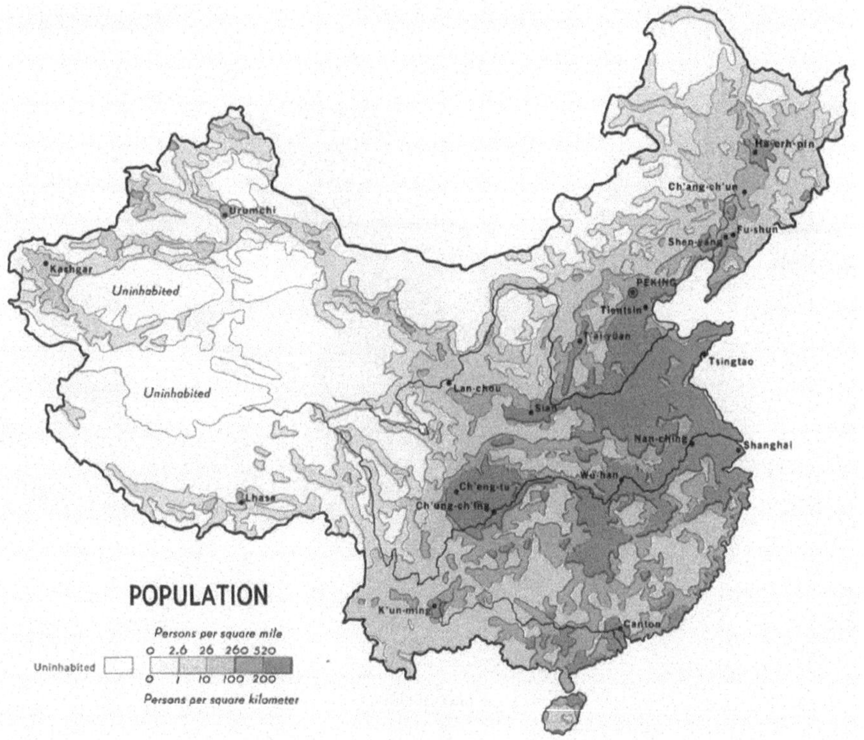

China: Ethnolinguistic Groups

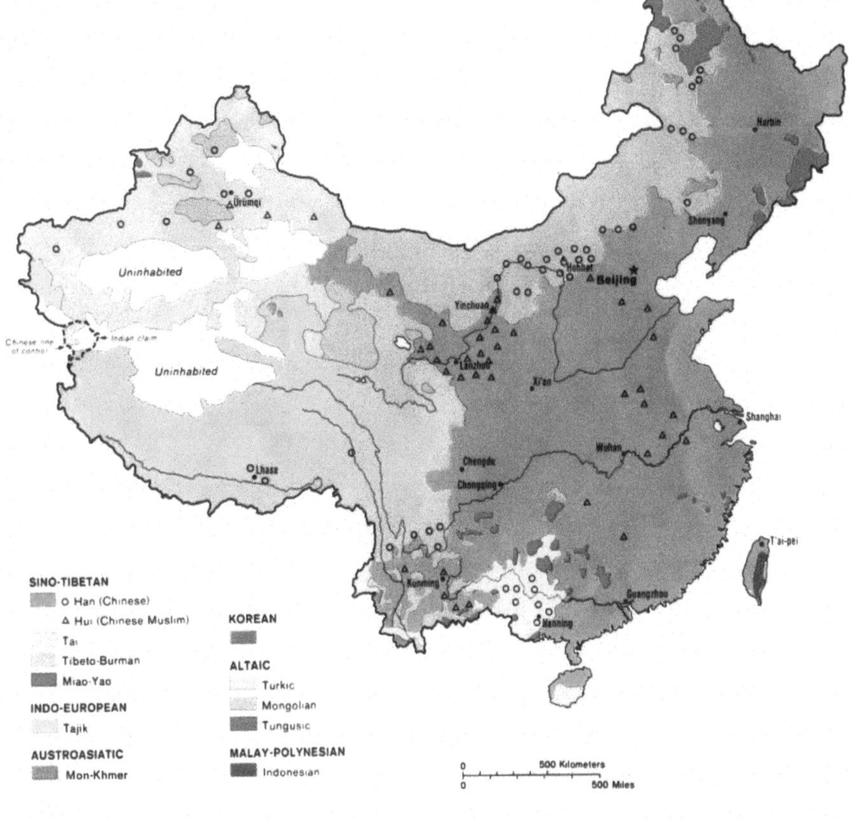

China: Population Density

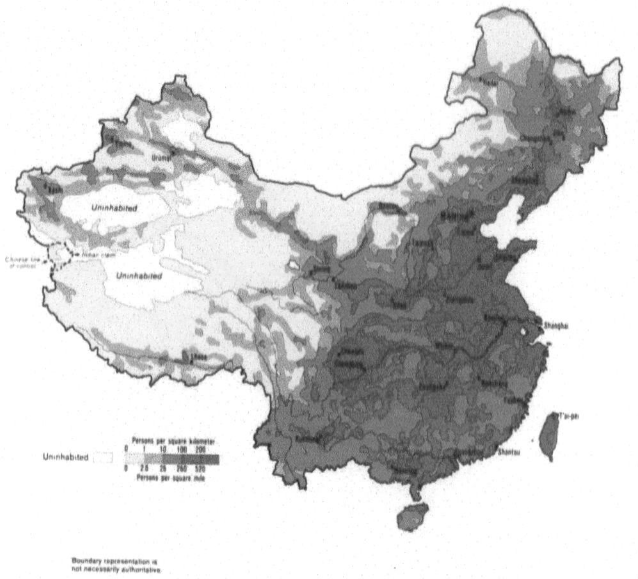

Agricultural Regions

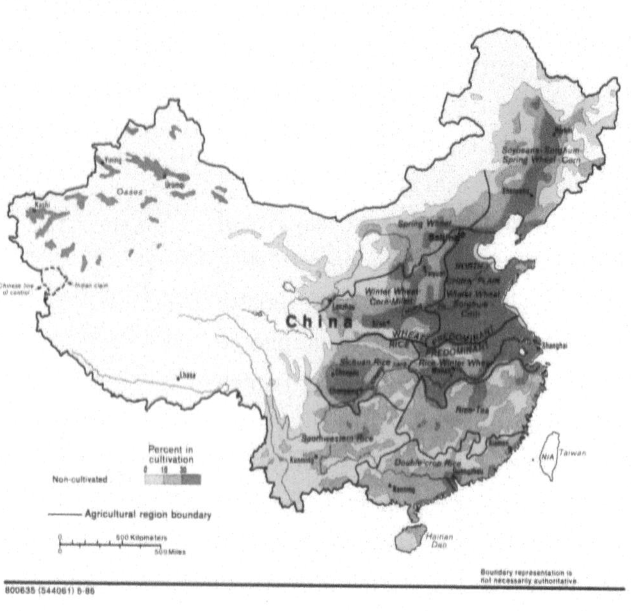

China: Industry

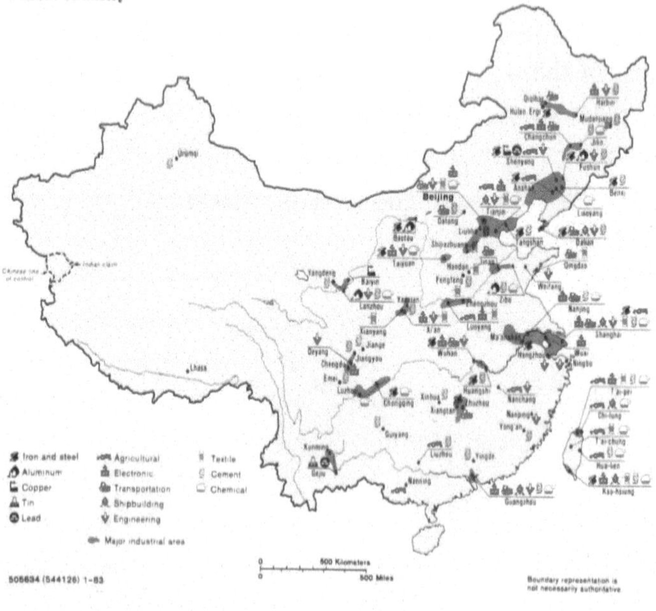

Can you detect an east/west pattern?

But we are talking about the entire East Asia region here, not just eastern China! There are a couple of other areas of note:

Taiwan Strait Area

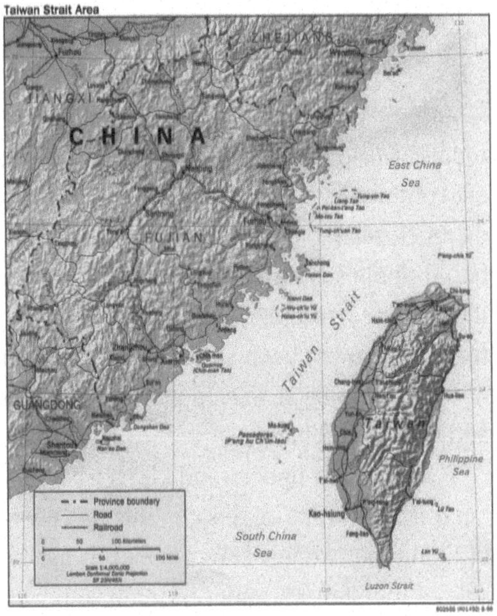

TAIWAN

Taiwan, with an area of 13,974 mi², is a bit larger than Maryland. This island lies some 110 miles off the southeastern coast of mainland China and is shaped roughly like a tobacco leaf. Two-thirds of the island is covered with forested peaks and the rest is foothills, terraced flatlands, and coastal plains and basins. The Central Mountain Range bisects Taiwan from north to south. . . and given that the island straddles into the tropics zone, it sets up an interesting climate/biome scenario: Eastern Taiwan is mostly tropical rainforest while the western part is characterized by savanna. Most of Taiwanese peeps are in the western part of the island, within the expansive coastal plains. The prevalent natural disasters in Taiwan include earthquakes and typhoons. . . remember, typhoons = hurricanes, and Taiwan is roughly the same relative location and latitude to Asia as Florida is to North America; again reinforcing the theme that east coast China is very similar to east coast America.

THE KOREAS

The Korean Peninsula is home to those two whacky Korean kids: North and South Korea. The peninsula extends southwards for about 680 miles from continental Asia into the Pacific Ocean and is surrounded by the Sea of Japan to the east and the Yellow Sea to the west, the Korea Strait connecting the first two bodies of water. To the northwest, the Amnok River (Yalu River) separates Korea from China and to the northeast, the Duman River separates Korea from China and Russia. Mountains cover 70 percent of Korea and arable plains are generally small and fall between the successive mountain ranges. The peninsula becomes more mountainous towards the north and the east, with the highest mountains found in the north. The southern and western parts of the peninsula have well-developed plains, where most of the peeps, cities, and agricultural lands are located.

The climate of Korea differs dramatically from north to south. The southern side experiences a relatively warm and wet spring/summer climate similar to that of southern Japan, affected by warm ocean waters including the East Korean Warm Current. The northern regions experience a colder and to some extent more inland continental climate, in common with Manchuria. The entire peninsula, however, is affected by similar climactic patterns, including the East Asian monsoon in midsummer and frequent typhoons in fall. The majority of rainfall takes place during the summer months, with nearly half during the monsoon alone. Winters are cold, with January temperatures typically below freezing outside of Jeju Island. Winter precipitation is minimal, with little snow accumulation outside of mountainous areas. . . but way up north can be downright miserable freezing during bad winters, and North Korea is regularly plagued with winters so horrible and long that famine ensues. So in a nutshell: the south is more hospitable with an overall milder climate, more agricultural plains, and ample fishing grounds; the north is overall colder and more mountainous, has tougher winters, less arable land, and resources. Hmmm. . . I wonder if that physical geography had anything to do with how the two countries developed so differently in the last 70 years since they separated? But wait! I am getting ahead of the story as usual! Let's step back to. . .

PANDA OF CHRISTMAS PAST

Why do we call China a powerhouse? Several things make China quite distinct in today's world. (I'm going to get to Korea, just be patient.) China is by definition the world's oldest continuous state. Was it a state 4,000 years ago? I don't know, but it was something pretty damn close. We already talked about what a sovereign state is, but here I think the term that better applies is **nation-state**, a culturally distinct group of people with a common culture that they all recognize and perpetuate in their defined area. That's what China has been for a long time. Have there been states as old, or older? Perhaps. Mesopotamia was up there, but it's gone now. Ancient Egypt was up there, but it's gone now. China has been there for about 5,000 years, and they're still here.

In order to understand China's history and current state, many scholars believe they have been **isolationist** by design. This has been facilitated by the physical world. How so? Let's look at the physical factors of China that we just talked about. It has been coined "the Far East" by Europeans, which means it was just on the other side of the planet from everything with which Western societies were familiar. By no other cause than physical distance, China has been far away from a lot of history. It has the Pacific Ocean, the world's largest ocean, on one side of the country, putting some serious distance between it and everything else. Head up to the north of China, and you'll will find big deserts, like the Gobi. To the northwest are yet more big deserts, like the Taklimakan Basin. These are some dry places, and since people tend to like water for some reason, nobody's there. Keep going around to Central Asia and you will find steppe and desert climates intermingled with some tough terrain. Let's keep the tough terrain going in the southwest. We have the Himalayan Range, the largest on the planet, keeping things isolated. To the south, we have a thick rainforest and jungle environment in Burma, Thailand, and Vietnam. The physical conditions of China alone kept things distant from them, and yet their physical environment also provided them everything they needed internally . . . so no great need to engage with the outside world in anything except exporting them agricultural products (tea and silk) along with finished manufactured goods (porcelain . . .you know, fine china!) Their physical environment provided everything they needed to thrive.

The last big theme I want you to think about when you go through this chapter: China has historically been based on authoritarian regimes, meaning centralized power in the hands of the few. Their 5,000-year history has been dominated by what we call an imperial period. That is an empire. All the big names you have heard, like the Han dynasty, are **dynastic successions** of centralized power. Also consider that throughout most of history, China has been one of the most, if not *the* most populous state, and it still is today.

What you see in the table on the left is the only real ancient history that I want you to understand about China. China, as both a distinct culture and a distinct state, has been the bomb for most of human history. What do I mean by "most of human history"?

CHINA IN WORLD HISTORY

Date	Status
1000 BCE	China is "the bomb"
500 BCE	China is "the bomb"
100 BCE	China is "the bomb"
YEAR 0	Jesus is born; China is "the bomb"
100 CE	China is "the bomb"
500 CE	China is "the bomb"
1000 CE	China is "the bomb"
1300 CE	Mongols invade; despite this, China is "the bomb"
1500 CE	Age of European Colonialsim kicks off; China is still "the bomb"
1600 CE	China is "the bomb"; Manchu Dynasty reaches largest extent ever
1800 CE	Things start to change; China is no longer "the bomb," it starts getting bombed!

Pretty much from around 2000 BCE right on up to 1800—that's right, only 200 years ago. If you had been anywhere in the world and you said, "Hey, what country has the most going on, what country has the most thriving economy, who's making the most money, who has the highest level of technology, where is it happening, where should we go to invest?" the answer for all these things, throughout all of human history, would have been China. They've been a powerhouse of economic activity and population forever. Western history books are too wrapped up in their own issues to recognize this fully, but the Plaid Avenger will help you see the Chinese lantern light. Learn this the first time and never forget it: Although we see modern China growing and we say, "Wow, look at them go! Isn't this great?" just keep it in perspective. They have been at the top for most of history. The United States has only been in existence as long as China has stumbled and fallen down, but now the Chinese are reasserting themselves. This is a theme I will get back to by chapter's end.

Expansion of China
Shang Dynasty
1766 B.C.

PANDA EXPANSION

As you can see throughout the historical expansion of China, the core of China has always been the eastern seaboard. While the empire has fluctuated and its borders have grown (sometimes retracted, but over all a net growth grown), the core has always been maintained. I said earlier that the Chinese have been isolationists by design. Throughout most of history they know they're the bomb; it's not as if they don't know the rest of the planet exists. They just aren't interested. People came to China for education, and came to their cities because they were the most cosmopolitan and happening places. People came to trade with China because they wanted all of that cool stuff China made. China has historically been the center for tea production, ceramics, and porcelain—what do you call the stuff that your mom breaks out for fancy occasions like Thanksgiving? That's right! China! It's not called china because it's small and breakable, but because it's from China! The spice trade was also a huge market based out of China.

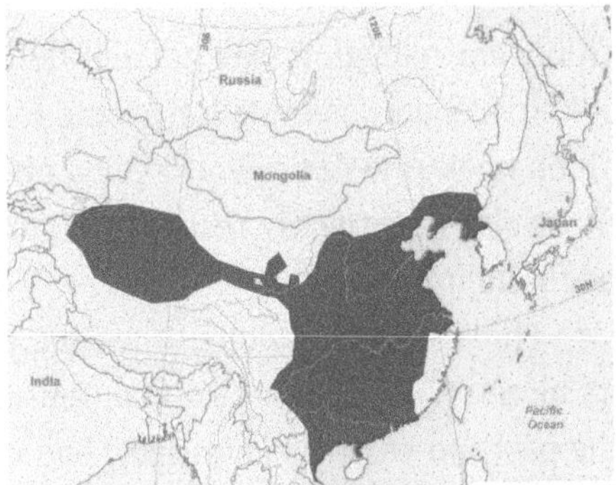

Expansion of China
Han Dynasty
206 B.C.

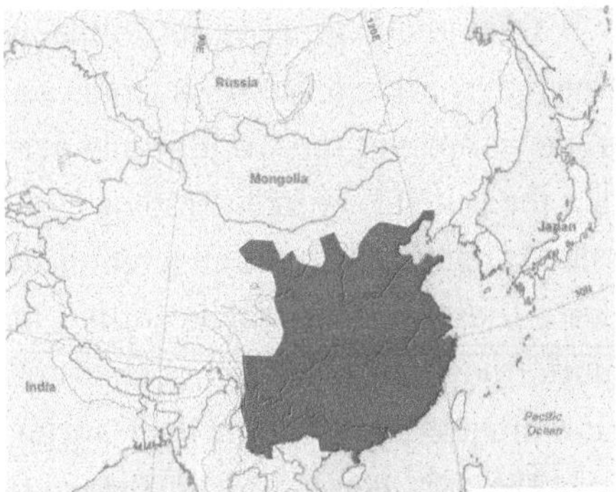

Expansion of China
Ming Dynasty
1368 A.D.

All of the stuff that the Europeans and the Middle-Easterners and lots of other folks around the planet wanted throughout history was produced in China. Trade was mostly one way: outgoing. As you know, this is something that's true still today, but on top of that one-way trading, the cultural interaction was also mostly one-way. Yes, sometimes things come into China that are adapted into their culture. Sometimes foreign invaders even come into China and take over, but the one thing that China has been exceptionally good at throughout all of its history is maintaining its distinct sense of culture.

Even when invaded and taken over by other groups of people, like Mongols or the Manchus, China assimilated the foreigners to their *culture* very rapidly. It's like the old slightly exaggerated saying: "Anybody can go to China and do anything he wants, but they are going to be Chinese in the end." Everything gets **Chinafied** once it gets inside. This is because the Chinese have this isolationist design, very inward-looking and self-preserving.

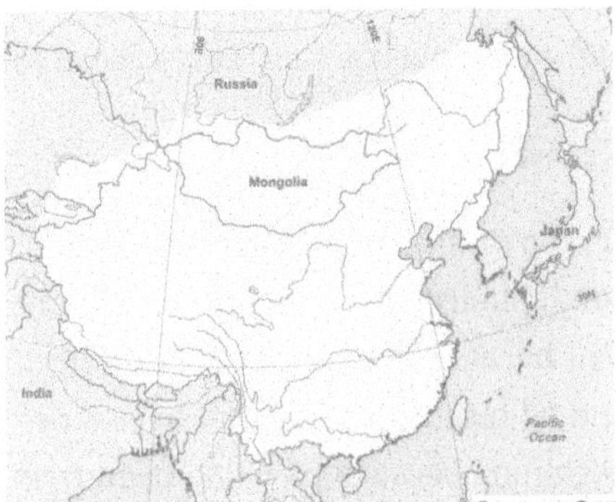

Expansion of China
Manchu Dynasty
1700 A.D.

A more sultry use of Chinese silk in the 21st century . . .

Take the Great Wall of China, for example. This is one of the major features of China that is a direct visible manifestation of its isolationism: a big wall built to keep people out. The basic Chinese self image: "We are China on this side. We are awesome. We don't need your stuff. We don't need your culture. We don't need anything from you. We are the center of the universe!" They really said that last part; there is a place you can go in Beijing called the center of the universe. That's pretty much the way they've seen themselves throughout history.

Does that mean that they never had interaction with anybody else on the planet, and that they were always alone? Absolutely not. There has been plenty of trade throughout the centuries. The **Silk Road** provided exterior trade and contact with regions as far away as Europe. Why'd they call it the Silk Road? Like the fine china for which China was named, it was not self-defined by the Chinese themselves; that's just what European traders called it. The Silk Road's purpose was for people to move silk out of China to trade to the Middle East and Europe. The point is that China knows there's a world going on outside its borders. They had contact with the other regions, but mostly kept to themselves. Not many records of Chinese exploration, economic entities expanding trade, or world conquests are to be found in their history.

China met its undoing when the modern era arrived. When we consider where China is today, we concentrate on its more recent history. Isolationism, while previously very successful for virtually all of Chinese history, unravelled in the past 300 years in a massively bad way. What are we talking about here? Up to the 1700s, global interaction and trade was there, but it was controlled by the state and mostly one-way. China sold you all the crap you wanted to buy, but they didn't need any of your stuff. They said, "Keep all that. We don't need any of the goods you produce. Keep your culture as well. We don't need that. We've got our own culture already."

Once the Europeans got to the age of worldwide exploration and colonization, China was one of the last places at which they arrived. However, trade was eventually established, resulting in some of the most prosperous and well-known trade companies the world would ever see. Maybe you've heard of a couple of these distinctly European trade entities, like the Dutch East India Trading Company and British East India Trading Company. The British are the ones we already talked about back in the chapter on South Asia; they ended up taking over all of what is now India, Pakistan, and Bangladesh. These were economic entities that were initially sponsored by Western European countries, but that became a lot more. They took over large sections of South Asia and Southeast Asia, and then they started working on China. This was a slow process staved off by the Manchus for as long as they could hold out.

The year 1800 is a critical date to remember. This is the year that everything about China's existence would be turned on its head. The second that things started heading downhill for China, it began to slip down quite a slippery slope. While trade companies made inroads into China in the 1800s, the Industrial Revolution was concurrently changing the entire world drastically. Machines made stuff in much greater volumes, which meant for the first time, Europe could produce things like silk and maybe even pottery for as cheap as the Chinese could. This shifted the balance of power away from China.

STONED PANDA

On top of that, there was a big push for capitalism and free trade in the world, sponsored by who else but the British. Adam Smith published the *Wealth of Nations* in 1776, redundant red letter date for lots of different reasons, but the most important because this was the kick-off for the whole concept of free trade. The basic premise of **free trade** is that all countries and all businesses ought to have a right to trade with anyone they want. The British primarily sponsored free trade in the beginning because they had been trying to make inroads into China and get some trade action for years. However, China was very conservative in their dealings. The Chinese said, "You guys can buy all the stuff you want here, just leave us the money. You bring all your silver and you can have all of the pottery, silk, and spices that you want."

Lots of money had been flowing into China for a while and this was pissing off the Europeans, so they said, "Hey, China has all the bank money. All of our money goes there, and then they don't buy any of our stuff." The British were looking around and said, "What can we do? How can we break into this market? They don't want any of our stuff like the textiles we are producing via the Industrial Revolution. What could we sell them? Damn, we can't think of anything. Wait, I've got it: Opium! That's awesome, what a fantastic idea!" That's exactly what the British did. You'd think they were joking, but the new plan was to get everyone addicted to drugs and sell it to them like gangbusters to reverse the flow of silver back to Europe. It worked perfectly. Once this influx of drugs was established, it was unstoppable. By the mid-1800s, the British, the French, the Germans, Dutch, and Russians all had established trading ports, and the Manchu Dynasty started to weaken.

Opium War: the British win the right to sell crack. How noble.

Why do I bring up the whole story of opium? As the Chinese weakened, they said, "Whoa, what the hell is going on here? All of our people are addicted to opium. We can't have this. Yo, British dudes, back the hell up with that drug boat. You can't sell that here. We refuse to let your ship dock up with all that opium." The British replied, "Oh, then we declare war on you." Thus, from 1839 to 1842, the British and Chinese fought the aptly named **Opium Wars**. The British were fighting for the right to peddle drugs to the Chinese. If you look back at British historical documents, they say the whole fight was over free trade. They attributed it all to something that made them sound like pious fighters for economic freedom, when in fact, they were drug dealers. Millions of Chinese people were addicted to opium and it was disintegrating the social fabric of the nation. Meanwhile, the British fought a war and won the right to distribute opium to the Chinese population.

How did this happen? How did this mighty, mighty nation that had been around so long lose a war over drugs to the British? Isolation is a double-edged sword. Through most of history China remained distinct, and for most of history they were ahead. But the other edge of the sword is that, because they weren't paying attention to what was going on in Europe, they missed the Industrial Revolution. The Industrial Revolution, which goes hand in hand with military revolution, ended up giving the British a technological advantage. By the time the British showed up with their gunboats and drug boats, they were militarily superior to the Chinese. This three-year war shouldn't have even lasted that long; it was a duck hunt, video game version. How do you say "quack" in Chinese?

ANGRY PANDA

This ended in 1842 and was a stunning defeat for the Chinese. Yes, empires have come and gone, and people have beaten each other up, and they've had some wars—but they have never had a foreign body of people from somewhere on the other side of the planet invade their soil, and destroy them so completely. (Okay, except for those Mongol dudes . . . but nobody could stop them at the time.) They received such a sound thumping in the Opium Wars that they were

Panda is confused while on crack!

forced to basically cede all control of territory over to the Brits. The British forced China to open up all trade to all states in the world. In fact, they had this hilarious thing that was called "no favored nation status." We think of that in today's world as the MOST favored nation status, but the British said, "You are going to have a *no favored nation status*, which means you have to trade with everybody, and any rights you give to one country you have to give to us all." They completely liberated the Chinese economy for every other country's taking overnight. This was a titanic blow to the Chinese culture, to the Chinese way of life, and to the Chinese economy. In this same peace treaty, concessions of the Opium War included the British picking up a fairly significant territory that you have probably heard of—Hong Kong. That was 1842, and the British controlled that right up until 1997.

In 1843, China ceded Macau to the Portuguese. Just to put this in historical context, the Chinese society started to crumble at this time because of another thing the British and European colonial powers did; they made the Chinese pass a law that said you couldn't persecute missionaries from Europe. I'm not making fun of anybody's religion; be as Christian or Muslim as you want to be, but think about what strain this was to a very independent culture that managed to hold onto its own ways for a long time, and now you are being forced at gunpoint to allow outside influences in. The fabric of the Chinese society basically degenerated in a matter of a few decades.

ABANDONED PANDA

To put this all in historical context, what was happening in other nearby regions? Southeast Asia was subjugated by European powers—all of Southeast Asia except for Thailand. The British controlled what is now South Asia and also Australia. Japan was a major issue to consider from the Chinese angle. What was happening in Japan as this disintegration occurred in Chinese society? Let's jump forward a little bit to 1868. What happened then? That date was in the middle of Japan's **Meiji Restoration**. Remember that stuff from many chapters ago?

If you had to pick two events that most demonstrate the difference between the evolution of Japanese society and Chinese society, they would be the Opium Wars and the Meiji Restoration. The West disgraced China during the Opium Wars. At the same time, the Meiji Restoration occurred in Japan—as they embraced the West. We can stop right now. Take a breather from reading this book and think about this: Why is Japan the second richest country on the planet right now and why has China been playing catch-up? Why is Japan a fully developed nation and why is China still developing? Why has China been poor so long and been disgraced for the last 100 years, while Japan has been a world class world power?

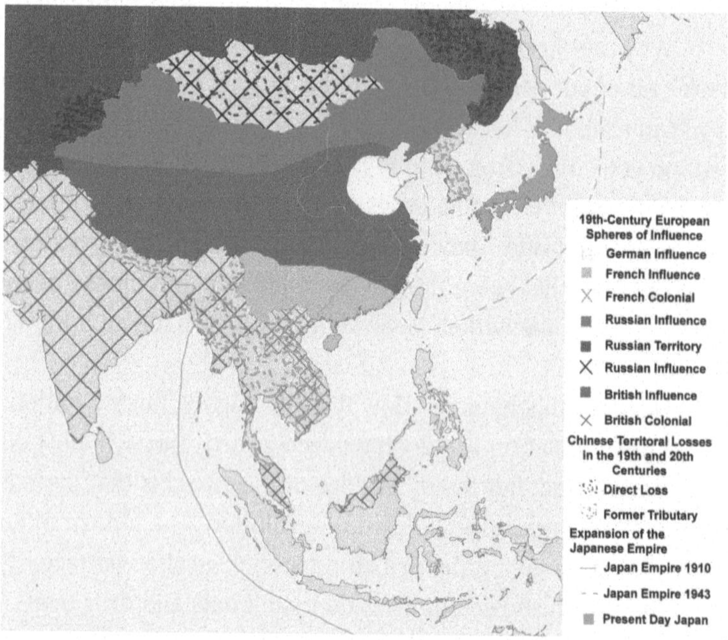

Chinese lose control, color coded for your convenience.

The answer to these questions has much to do with the Opium Wars and the Meiji Restoration. These two countries, once side by side, took two different paths at the same juncture in history about 150 years ago. Today's world reflects the two different outcomes of the choices these countries made. To put the icing on the cake: In 1895, we had the **Sino-Japanese War**, in which the Japanese thumped the Chinese and took control of the Korean peninsula. Now, for the first time, our Korean friends come into the mix. Up to this point, Korea was a little brother of China, a vassal state, much like Japan. Japan taking Korea was another blow to Chinese civilization. You mean the dinky vassal state of Japan just crushed our awesome Chinese troops in a war? Impossible! This situation was devastating to the Chinese.

Beleagured Boxers held captive by cowboys.

As you can see from the previous map, in 1899, China was in dire straits because of its continued isolationist endeavors. What was China doing then? Were they playing catch-up with the rest of the world, like Japan? Were they embracing the West? Were they trying to adopt and adapt? In a word: no. In 1899, something happened called the **Boxer Rebellion:** a movement to purge out foreign influence. They said, "Hey everybody, wake up tomorrow and go kill all the foreigners!" This was in response mostly to interjection of Christian missionaries, but also to the state apparatus that was under foreign domination, politically and economically. The Boxers were this martial artsy, mystic group of dudes trying to reestablish order in the Chinese society. Just to give you a sense of their cluelessness, the Boxers believed that if they prayed hard enough and were pure of heart, bullets would not be able to penetrate their bodies. I think you can already predict how this turned out. The Boxers surprise attacked the lightly guarded compounds and besieged the Western diplomatic quarters in Beijing, and then held all the foreigners hostage.

How did the rest of the world respond? Check out the political cartoon on this page for the answer. They formed a coalition of all the colonial powers. The French, the Russians, the British, all the great European powers worked hand in hand like good brothers to beat the hell out of the Boxer Movement and the Chinese Dragon altogether. To make matters worse, the US helped out—and worst of all, Japan was there as well. Wow! Talk about salt in the wound!

How many country mascots can you identify in the thrashing of the Chinese dragon?

Japan was growing in influence and started to take over the world after embracing Western ways, while China, at the turn of the century, buried its head further in the sand and tried to beat out the foreign influence. This was a big mistake. China, by 1900, was fully under control of colonial powers. The state was still intact, but only at the whim of European countries. Look again at that cartoon. The caption underneath reads, "The real trouble will come with the wake." How prophetic for a freakin' cartoon published in 1900! But the dragon will have to sleep through a bit more trouble yet before its slumber is interrupted.

20TH CENTURY PUNISHES THE PANDA MORE

In 1912, the last Qing Emperor was removed from the throne, and the Republic of China was established. That marked the end of the Qing Dynasty and imperial lines altogether. This place was in chaos. Everything religious, cultural, economic, and political was in shambles. This didn't happen overnight; it developed over decades. The Chinese people were saying, "What in the hell are we going to do about this?" In 1911, they finally said, "That's it; the whole imperial line is over. We are not doing that anymore. We need to at least try to catch-up with the West. Look, they have constitutional monarchies and republics over there. Let's do that! We can't do this empire thing anymore; ours sucks anyway." They threw out the Qing emperor. You may have seen a movie called *The Last Emperor*—that's what it was about. What's the deal? Did they ever have a movie in Japan called *The Last Emperor*?

The last Emperor of China. No, not the dude, but that little kid.

How Now Chairman Mao? Who the Heck Is Mao?

Mao Zedong (Mao was his last name) became the leader of the Communist party in China in 1927 when he formed the People's Liberation Army. The Communist Party and Mao were largely seen as folk heroes and true Chinese nationalists during the Japanese occupation, as they conducted a guerrilla war against the Japanese aggressors throughout the entire conflict. It should be noted that the Nationalist Party, led by Chiang Kai-Shek, did little to nothing during the occupation, choosing instead to hoard weapons and money given to them by the US—given for the sole purpose of fighting the Japanese.

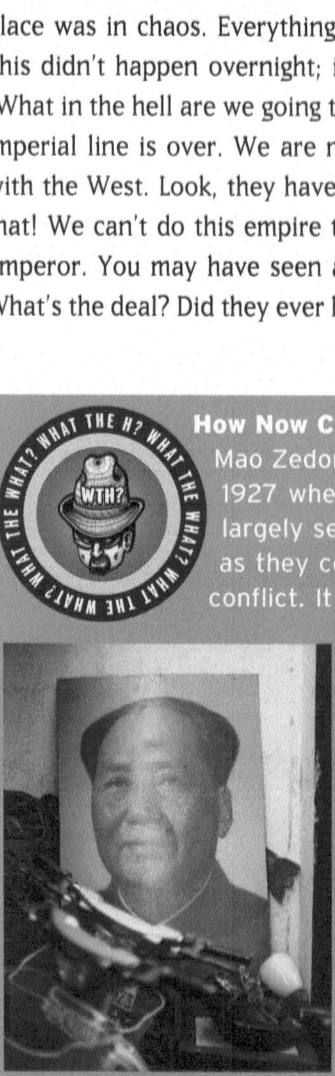

After WWII and the brutal Chinese civil war against the Nationalist/Kuomintang party, the Communists took over mainland China in 1949 and declared the People's Republic of China. You can like him or you can hate him, but Mao was largely seen as "the George Washington of China" at this moment in history, because he declared that "the Chinese people have stood back up." Meaning: the two centuries of foreign occupation and humiliation were over.

Mao became the leader of China and was known as the "Four Greats": "Great Teacher, Great Leader, Great Supreme Commander, Great Helmsman." He became the supreme dictator of China; every Chinese person was required to have a book of Chairman Mao's quotations. He has been praised by many for unifying China and making it free of foreign domination for the first time since the Opium Wars.

However, Mao is sometimes credited as a lethal dictator. Because of some of his projects—such as the Anti-Rightist Campaign, the Great Leap Forward, and the Cultural Revolution—over 80 million Chinese people died. Part of the reason so many people died was that Chinese officials were so afraid of criticizing Mao that they told him that everything was going fine. Also for this reason, China did not start implementing capitalist reforms until after Mao died in 1976.

Fun Commie Fact: Mao started his own kind of communism, called Maoism, that stated that communist revolution needed to come from rural peasants. Marxist communism was designed for industrialized countries, like the ones in Europe. Apparently both styles work equally well, which is to say they both suck.

Not. We already know that Japan still has an emperor, but this was when China departed from their old system, symbolically and permanently. They gave up on the whole imperial way and moved toward something else, but it wasn't going to be easy. In this state of chaos, nothing happened overnight. They couldn't decide internally on what to do or who should be in power. The country broke down into what is sometimes called a **warlord state**. Local military leaders took over, and there was no authoritative central government. This played out for the next twenty years, well into the 1930s.

This left people in China looking for some answers. "What should we do? What type of system should we implement? The imperial system is gone and we're not going to do that again." Some said, "Let's embrace the US model, or the French model, and others said, "Hey, what about that Soviet model? That looks pretty good." This started the face-off of ideologies within China. In 1928, the Nationalists, under a guy named **Chiang Kai-Shek**, were fighting for a republic-style democratic system. They defeated the folks fighting for a Communist state. The Nationalists also defeated, allied with or outright paid off all those warlord guys remaining in the northern states, which kind of united the country. But this was tenuous. Why was it tenuous? Because of another incursion I haven't mentioned yet.

Chiang Kai-Shek

Look back to the spheres of influence in China map. Look at what the Japanese were doing. Japan's empire was expanding. I already pointed out they took over the Korean Peninsula back in 1895, but they didn't stop there. The Japanese also staked out some claims of ownership of Chinese lands after they helped squash the Boxer Rebellion. They made further incursions into what is now Manchuria, and started to take control of some port cities. The Europeans were already doing it, so the Japanese wanted to play ball as well. Japan largely stayed out of the whole WWI conflict, because they were busy securing more empire for themselves over in East Asia. By the time chaos happened in China, when there was no real leadership and no one knew what the hell was going on, Japan had seriously invaded them.

This was the point—in between World War I and World War II—when the Japanese secured real control over all the major facilities in China. This Japanese invasion only perpetuated the chaotic state in China, and a civil war broke out amongst the Communist and Nationalist forces again. During this civil war, something called the **Long March** occurred; **Mao Zedong** led the Chinese Communists in a retreat through Northern China after a defeat by the Nationalist forces. The fight should have been over after that initial battle, but somehow this guy named Mao pulled off a last-second, fourth-quarter win. This leads us to the question: who is this Mao guy? See insert box titled "How now Chairman Mao?"

In 1937, the Japanese army attacked China outright. This was the buildup to World War II. By the following year, Japan controlled most of eastern China. Remember, that's where the most people, the most businesses, and where the most of everything is located, including the people.

WWII was disastrous for the Chinese. Recall from way back in the Russian chapter that the Russian losses were something along the lines of 10–20 million people during World War II. It's a lesser-known fact that the Japanese did one heck of a job slaughtering the Chinese at the same time. This mass slaughter has been barely acknowledged by Western historians and people like you. The losses in China may have been upwards of 20–30 million people, and possibly could have been 40 million. It's almost impossible to tell in hindsight due to record-keeping issues. Somewhere between 20 and 40 *million* Chinese were killed during the Japanese occupation in the buildup before and during the war. Very, very few of those millions were actually combatants. In other words, they were unarmed civilians.

Many were innocent women and children who were slaughtered in nasty ways by the Japanese. I guess it was just another way they were trying to embrace the West. Japan tried the first kind of modern biological

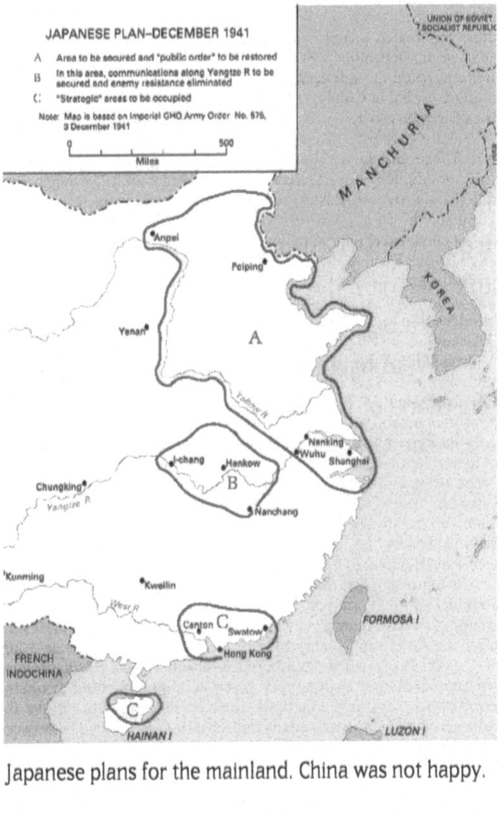

Japanese plans for the mainland. China was not happy.

testing and warfare during World War II. Some of their experiments included dropping rats infected with bubonic plague from bombers over heavily populated cities to see how fast people would get the disease; they also poisoned water supplies in major cities to see how fast people would die. We also talked about the **Rape of Nanking** in which millions of people were slaughtered in the streets. The Japanese didn't want to set up internment camps. I am reiterating this again in this chapter because it's just that important.

Japanese war atrocities abounded. This really sucked for China.

When we think of the horrors of World War II, we usually think of the German atrocities against the Jews—and those were horrible, make no bones about it. But when China thinks of World War II, they recall the horrible atrocities which left a body count probably eight or ten times higher in their country at the hands of the Japanese. This makes for bad blood, my friends. It's still there today. Also remember that the Japanese have occupied the Korean Peninsula since 1895, an even longer time, under the exact same kind of brutal regime. The Koreans haven't forgotten either. All of Korea ended up being a work camp for the Japanese to get resources out of mines. All the resource extraction from the Korean Peninsula was shipped back to Japan. Why would Japan be building this empire, taking all these resources, enslaving the population of Korea, and taking over places in China? For the reasons we have already pointed out in the Japan chapter: They don't have any resources on their island, and to expand and get richer, they've got to take stuff from other places. That's why they became an imperial regime.

The Japanese occupation was so horrifically bad that the Nationalists and Communists even teamed up at one point. They said, "This sucks so bad, there is no point in us fighting each other right now; we have to get these invaders out of here first!" They teamed up and kicked out the Japanese. Of course, it was the United States' involvement in the war against Japan in the Pacific that did most of the job, but the Chinese were doing some fighting on their soil as well, making life for the Japanese difficult. Finally, in 1945, it was lights out for Japan, literally and figuratively, as the country was firebombed into oblivion until they surrendered.

In 1946, after WWII was over, the civil war was back on again. The Nationalists and Communists said, "Cool, the Japanese are gone. Yeah, the war is over, so let's hurry up and rebuild so we can get back to fighting each other to see who controls this place." Long story short: after this stretch of 150 years where China was completely dominated, the end result was that, in 1949, the Communists won the civil war. Mao was a long-shot to win at several different points, but the Commies pulled it out. The Communists won, and the **People's Republic of China** was established in 1949.

What happened to the Nationalist party that had a lot of supporters that fought the civil war? Most of them withdrew to the island of Taiwan. Taiwan? I've heard of that place. (See inset box titled "What's the Story with Taiwan.")

The Commies took charge, and started to pull the country up by the bootstraps.

So What's the Story with Taiwan?

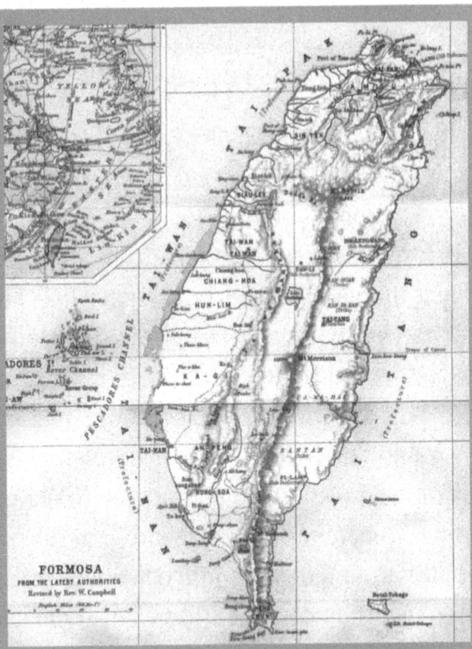

For most of its history, Taiwan was an island territory off the coast of China that was of no great consequence. (Sorry my Taiwanese friends! You know it's true!) However, Taiwan immediately became a Cold War hotspot when Chiang Kai-Shek and the Chinese Nationalists (Kuomintang), after losing the Chinese Civil War, retreated to Taiwan and set up a government called the **Republic of China (ROC)**. Meanwhile, Mao and the Communists controlled mainland China and called themselves the **People's Republic of China (PRC)**.

The leaders of both countries long subscribed to the one-China policy—that there is only one China and Taiwan is part of China—each insisting on their own government's legitimacy. Of course, during the Cold War, capitalist countries recognized the ROC as the legitimate government of China and communist countries recognized the PRC as the legitimate government of China.

Today, the Taiwan situation is starting to settle out, perhaps closing the chapter on China's last disputed territory (forget about Tibet). Most countries do not have relations with the ROC and the Taiwanese position has eroded over the past few decades as the PRC has gained international prominence. Because of its one-China policy, the PRC only participates in international organizations where the ROC is not recognized as a state. The United Nations, for example, expelled the ROC in 1971 when the PRC was admitted. Just so you understand what I'm saying here, let me reiterate that last point: prior to 1971, Taiwan actually sat in the China chair at the UN, including in the Permanent Security Council seat. After 1971, both the UN and the US recognized mainland China as the true leaders of that place we call "China," and therefore the UN seat reverted to the commies, and Taiwan was no longer considered a sovereign state. To make matters even more complicated, while most countries do not have official relations with the ROC (to not make the PRC mad), they do have informal relations with the Taiwan government. For example, the United States operates the American Institute in Taiwan, which is like an embassy. However, it is not officially called an embassy because America is not allowed to have diplomatic relations with Taiwan.

The Taiwan situation was really tricky for the last decade, because the Kuomintang lost control of Taiwan in elections in 2000 to President Chen Shui-bian, who was from a staunchly pro-independence party that regularly pissed off China by making moves to secure said independence. However, things have settled down considerably just this year as the KMT won the presidency back, and the new President Ma (that's right, just Ma—like your mother) has made great strides to mend ties with China.

Fun Fact: While the US officially adheres to the one-China policy, in reality, it practices a two-China policy. Under the 1979 Taiwan Relations Act, the US sells Taiwan military weapons, and the language of the act warns the People's Republic that any coercive unification efforts would be "of grave concern to the United States."
2021 UPDATE: It appears under a much more confident and aggressive China led by Xi Jinping, the Taiwan issue may actually become a flash point. Xi seems set to take back the island now, after having successfully squashed Hong Kong into submission. A fight with the USA imminent? Not sure as of this writing but tensions are 100% going to rise soon on this issue.

ONE GREAT LEAP FORWARD, TWO GREAT LEAPS BACK . . .

Forward! To Failure!

In an effort to have China rapidly "catch up" to the Western World, one of Mao's first ideas implemented on a gigantic scale was the Great Leap Forward. Conducted in 1958–1961, this was a economic and social campaign to transform undeveloped China from an agrarian society to a modern communist utopia via rapid industrialization and agriculture collectivization, simultaneously. You know, just like the Ruskies did back in the 1920's and 30's, under Joe Stalin's leadership.

We have already talked about this in the Russian chapter, but to recap and add some unique Mao elements, dig this. Mao's plan banned private land ownership and private farming; collected up all ag lands and formed communes of peoples to farm them; planned factories and moved peoples around to work in them; and even attempted bizarre small-scale industrialization experiments like "backyard furnaces" which encouraged/forced every village to produce high quality steel by melting down scrap metal, as well as their own pots and pans and tools.

The result? Unlike the Russian experience, this plan was a gigantic disaster for China, resulting in massive crop failure, severe famine, severe mistreatment of 'failing' villages by government stooges, ecological disaster, and economic regression. Chinese deaths due to the famine, mismanagement, and persecution are estimated between 20 and 40 MILLION! Dudes! Really? 20–40 million? Yep. That's about how many died during WW2 as well. Dang! Another result of this catastrophe of a plan was the Mao lost a lot of credibility within the party, and was subsequently sidelined from power. In an effort to regain his power, he then launched. . . .

THE CULTURAL REVOLUTION

Hold up your little red books if you love Mao!

To regain his street cred and power within the party, in 1966 Mao then set into motion 'The Great Proletarian Cultural Revolution': a socio-political movement to convince everyone of how awesome socialism was by purging all capitalist thought as well as traditional and cultural elements of Chinese society. Another goal was to remind everyone about the awesomeness of Mao himself. Geez, what an ego on this guy!

Don't believe me? Well, you may have heard of Mao's "Little Red Book" which was a selection of statements and speeches by the revolutionary leader himself, bound in a bright red cover . . . and one of the most printed books of all time. Why so printed? Because it became a vehicle to inspire and whip up the Chinese masses into a frenzy of socialist support and Mao worship. The "Red Guard," a paramilitary group of youth, were mobilized by Mao to keep the communist revolution alive, hold up their little red books, spread the Mao love, and in general help destroy any vestige of the old Chinese society. Kind of like the Hitler Youth in Germany: a rabid internal purge force heralding in a new age, or some such nonsense like that.

In this anti-capitalist, anti-Western, anti-Chinese-tradition movement, pro-reform officials in the government and military were vilified and purged; historical relics and religious artifacts destroyed; western musical interments and texts burned, along with Confucian and classic Chinese philosophical works; and millions of people were persecuted in the violent factional struggles that ensued across the country. You were either on Team Mao, or you were an enemy of the state.

End result? Mao indeed regained his place as the head honcho of power, but in a China that was politically unstable, socially cannibalistic, and still considered a joke by the outside world. This destructive phase of modern Chinese history officially ended in 1969, but really was in play until the death of Mao in 1976. And it even played into the last big negative event I want to reference in this section, and that is the 1989 Tiananmen Square Massacre, when hundreds of students were slaughtered by the Chinese military during a pro-democracy protest. Chinese leaders feared that student movement had taken inspiration from the grass-roots populism of the Cultural Revolution, and that if it is left unchecked, would eventually lead to a similar degree of mass chaos . . . thus a massive and violent end to the protest was called for.

That event brings us up more modern history that I think most of you readers here will know about, which is . . .

THE DRAGON AWAKENS: CHINA IN THE MODERN ERA

This is stuff you probably remember, or have at least heard referenced. This is the end result of the Chinese communist experiment: They had made some progress since WW2 and their civil war ended, but they also had internal purges and internal problems—and they were not really that rich yet. China was still kind of backwards and behind. To top it all off, they were persecuting democracy movements. How crappy is that? But this is where China began to reevaluate some of its actions, because the world watched the Tiananmen Square massacre and said, "How horrific, how terrible! What's the deal with these guys? They suck!" Strangely enough, this was when everything started to turn around for China.

ECONOMIC EXPLOSION

After Mao Zedong died in 1976, his subsequent successors in the Chinese government started to "liberate" the economy and adopt capitalist reform—something which Mao absolutely raged against his entire life. The first guy who took the country in this new direction was a cat named Deng Xiaoping. If we are going to credit Mao with bringing China out of its "period of sorrow," then we really have to credit Deng with truly bringing China back to its restored position of wealth and power in today's world. It was Deng who pushed for capitalist economic reforms, put the country on the path of modernization, and pretty much gave up on the commie experiment thing that Mao had attempted.

Deng and Carter in 1979. Well, at least one of them was a successful leader.

As we already suggested many times in this book, Communism has never worked anywhere. It is a political and economic system that is just too complicated, and not one single manifestation of it anywhere has ever managed to work. It didn't work in Russia, didn't work in Vietnam (mostly because the US ruined its chances from the get-go), hasn't really worked in Cuba, and didn't work in China.

Here is the big difference: in Soviet Russia, the USSR, in one fell swoop overnight, said, "Okay, we are going to get rid of this. Anybody that wants to be out of the USSR, go ahead and vote yourself out." Then the Berlin Wall fell and the whole Russian economy fell with it. The Chinese were watching those events extremely closely. This is one reason why the Chinese are liberalizing very carefully, very slowly, which they've been doing for the last forty years. They have opened their markets, but slowly. They've allowed foreign capital investment, but slowly. They have privatized some industries, but slowly. Because when did Tiananmen Square happen? 1989.

Tiananmen Square: where all the action happens.

That's exactly when the whole kit-n-caboodle came crashing down for the Soviets. **Mikhail Gorbachev**, the leader of the Soviet Union, was actually visiting Deng in China during that incident. Chinese leaders rightly surmised that the reason the pro-democracy movement popped up at that time and that place was because people were supporting Gorbachev. Some college kids said, "Yeah, Birthmark-head is opening up and liberalizing and making democracy in Russia. Let's all have a big protest here to support that!" The Chinese government said, "No! Look what's happening there; those guys suck! They're collapsing into chaos! We are not going to do that again!"

Thus, the massacre. The Tiananmen massacre was terrible. The Plaid Avenger doesn't want any college kids killed, because hey, college kids are cool, but it was a very intentional move by the Chinese government. They said, "We can not allow this to happen because we can not allow our society to spiral out of control like the Soviet one did." If you think about it in hindsight, it was a calculated, conservative move. China is bustling in today's world while Russia is struggling to catch up. The Chinese have had infinite patience. Since Deng Xiaoping took over as head of state after Mao died, they started to liberalize and open up, but at their own pace, a pace that was massively accelerating right at that point in time.

TIANANMEN TURNAROUND: 21ST CENTURY SUCCESSES

The point of entry is the Tiananmen Massacre of 1989. It got a bunch of bad press for China, but I'm putting it back in here as a turning point, because it doesn't happen again, and it won't ever happen again. They may very well have done it again, but they figured it out: "Oh, we don't want bad press. We want people to invest here. We want to become part of the world economy. We want to have influence in the world. We want to have the Olympics here. We want to have the Miss Universe Pageant." You can't have any of that if you are killing your own citizens. This is a great place to start in terms of the 21st-century turn-around for China, because it is the last time you are going to see a tremendous and bloody crackdown of that magnitude. The Chinese government learned its lesson well—bad press equals bad investment strategy.

The government is still run by the **Chinese Communist Party**, but this is not a communist state as we suggested way back in the *Know Your Heads of State* chapter. True, communism is economics and politics put together, but there isn't a damn thing about this society that's economically communist. This is the wild west of free-market capitalism, my friends. Since the Communist Party is still part of the government, it's better to term the government a **one-party state**. True, communism is gone. Nobody is paying attention to that nonsense anymore. By the way, only about two percent of the entire Chinese population is a card-carrying commie. That's two percent of a billion and a half people. What does that equate to? A decent sized number, but still almost nobody. This is not a communist economy, and it is not necessarily a communist country; it is a one-party state.

"Plaid Avenger, I have the secret formula for Chinese economic growth . . . but I think I'm being followed . . ."

Let's get to the entire decade right after the Tiananmen massacre, when China enjoyed massive economic growth. There were huge influxes of foreign capital in the form of investments by multinational corporations. The Chinese were very particular about how this happened and controlled things very stringently. They said, "Okay, Microsoft, Boeing, Exxon, you guys can come over here and start a company if you like, as long as you give us all the technology we want. In other words, we are going to get something out of this; we're not just going to open our markets and let you come do what you want." You can refer to this move as **technology transfer**.

They certainly got something out of their international relationships, but why would they want to be so anal about everything? Look where it got them. In a single decade, they pretty much caught up. Granted, they were not on the technological level of the West yet, but they were catching up fast. 1990 was the kickoff of massive economic growth, large amounts of foreign capital, investment, and business. Why would foreign business want to be in China? Oh, I don't know. Would you like to be in a place to sell your product to a billion and a half people? Yeah, I thought so.

This 1990s economic growth was due not only to foreign investment, but also to private enterprise within China. Liberalization of government control led to growth of **entrepreneurs**—Chinese people starting Chinese businesses. And there are a billion and a half people there, man! They were emerging from a communist economic system where no matter how hard you work—or don't work—you are going to get paid the same. What incentive is that to work hard? There are no incentives. But put them in a capitalist environment and see what happens then: "Hey dudes, if you work harder you get more money. You get more money, you get more stuff, your life can be better." If you get a billion and a half people who desperately want more stuff and offer said stuff to them, man, watch out! This place was blowing up, due mostly to entrepreneurial growth of private business within China. This economic explosion can also be linked to the growth of **the Tigers**, the thing we talked about way back in the World Econ chapter. Other countries in the vicinity were gaining in all these categories as well: international investment, international business, etc.

2015 Best Dressed Farmhand winner

The reason companies were going to Indonesia, Taiwan, or South Korea was that they had cheaper labor. China is the cheap labor pool of the planet. Businesses were flocking to it to take advantage of that. Every major business on the planet now has a presence in China. This has all happened in the last fifteen years. This is an economy that is so darn hot that they typically have what we call double-digit growth. If they say the GDP growth is 11 percent or 12 percent or 13 percent, that means the GDP next year will be 11 percent bigger than it is right now. Take what we have right now and multiply it by 11 percent and add it on next year. In countries like the United States and most of Western Europe, 3 percent or 4 percent growth is fantastic. In China, 10 percent, 12 percent, 14 percent growth is titanic, but they are consistently achieving it. The Chinese government's main problem right now is not how to make their economy grow—they are trying to figure out how to put the brakes on it: "We're going too fast; things are getting out of control. Let's cool it down, cool it down!" It's a situation that every country in the world envies. Who wouldn't like to be in a situation of growing too fast as opposed to slow or no growth?

THE PANDA TAKES BACK ITS CUBS

In 1997, China conducted training off the coast of Taiwan. This consisted of firing gunboats and some missiles over Taiwan's airspace. It was a very specific move at a very specific time to symbolize that China was ready to reassert itself. There was rhetoric and talk before, but now that the Chinese economy and military had caught up enough, they said, "Okay, we're telling you what's going to go on. Taiwan is our territory. We are going to do something and see what happens." It was a symbolic gesture, I think, and marked a turning point where China was ready to reassert itself militarily, if necessary.

In 1997, Hong Kong was returned to China. The British packed up everything that they claimed from the Opium Wars and sailed out. The Chinese weren't really waving goodbye, either. They just went to work pulling Hong Kong and its mighty capitalist economy back to the motherland. In 1999, the Portuguese returned Macau to China. Taiwan is now the final piece of the lost Chinese empire, which is soon to be put back into its strategic place.

The current leader of Taiwan, President Ma, has made every indication thus far that he will be working to smooth out the relationship with China. In the first month he was in office, there were high level meetings between then Chinese President Hu Jintao and major Taiwanese leaders. They agreed to resume commercial flights between the countries on weekends, and also agreed to triple the amount of tourists allowed to visit the island every day to 3,000. This is the most open that communications and travel have been between these two entities since before 1949. That is big news, and please keep in mind that all of this has occurred just in the first month that Ma was in office. I predict significant changes in his eight years in office. You know what? I'm just crazy enough to make a grander prediction: I believe we are seeing the beginning of the end of the historic dispute between China and Taiwan. Look for Taiwan to be firmly back in the Chinese fold within a decade.

But on to other issues. In the year 2000, the United States established **favored nation status** and traded with China. In 2001, China joined the World Trade Organization (WTO). This is a society on fire. We can add a few other things here, such as: China is now an associate member of ASEAN, a core member of the SCO, a core member of the BRIC, a member of APEC, and hosted the 2008 Olympics. And they won the games to boot! China has got game again!

WORLD TRADE ORGANIZATION
ORGANISATION MONDIALE DU COMMERCE
ORGANIZACIÓN MUNDIAL DEL COMERCIO

Here is a quote from the chief Chinese negotiator, Long Yongtu, regarding China's entry into the WTO: "The fifteen-year process is a blink of an eye in the 5,000-year history of China." That resonates with the Plaid Avenger. Awesome, awesome quote to get you to understand what is going on in this society. They're telling the world, "We are here for the long haul, friends. We are reasserting ourselves. We are going to reclaim our proper position in the world economy and the world politic. And fifteen years, you think that's a long time? Whatevs. We will sit here for 100 years, if that's what it takes. We are China. We have the patience of a rock. We will get it done." If it sounds like I'm getting too fired up about China, it's because this is *the* vibrant place right now. It's really happening.

To finish off our categorized date list of Chinese achievements in the 21st century, we have to include outer space. In 2005, China put a man in space. China joined the most exclusive country club on the planet. It's a short list. Do you know it? It's the US, Russia, and China. Not even Japan has done it—the mighty, mighty wonder to the east of China and the number three economy on the planet. How about Germany, the number four economy? No! They've never done it, either. To up the ante, in September 2008 China conducted its first "space walk," further underscoring China's technological prowess and determination to go where no man has gone before. Call in Captain Kirk for back-up! The Chinese are setting the stage. This is again like so many of these other things I've mentioned: it is a symbol. China said, "We are in the big boy club now. We are the bomb. We have got it going ON!" Think about it. There's no reason to put a man in space. It's all about prestige. They're reclaiming their world role. The 21st century is a whole different world for the Chinese.

They aren't even satisfied with just putting a Chinese citizen in orbit for a few days. The Chinese have every intention on putting a man on the moon, and they have already announced tentative plans to build a permanent moon base! Holy green cheese! A freakin' moon base! Not even the US pulled that one off! Just to give you an idea of how serious this is, China was actually six months ahead of schedule when they succeeded in the manned orbit plan. They then busted a move right after the 2008 Olympics by having the first Chinese space walk . . . and the panda pride is showing! Now they are slated to get to the moon by 2020, and appear to be on track. In what is assuredly a total random coincidence, the US upped NASA's budget times ten, and NASA then announced that they were now planning to build a moon base first. Space race is back on, my friends!

Danger Will Robinson! China is here!

I actually hope the Chinese base gets built first. That way, once the US base is being worked on, the Americans will have a place from which to order take-out.

BIG PLACE, BIG ECONOMY, BIG MILITARY—BIG PROBLEMS!

I know I've stressed this a million times, but let me stress it one million and one times: China is reassuming its place on the world stage. It's almost impossible for the United States to understand this scenario, because think about the timeline. The entire time the US has been rising in the world is the exact same time period China spent falling. Throughout all of US history, China sucked. But that's not looking at the big picture. They've been on top for 5,000 years. China is not saying, "Hey, we're growing and we're finally going to be a part of the world economy." They're saying, "We're taking back our place at the top." That's a much better way to phrase it, and a much better way for you to understand what's going on. In the same light, China is increasingly becoming a model to follow for other states all over Asia.

During the entire Cold War, we lived in a bipolar world. The US was one big power, and the USSR was the other. All the states in the world could choose one side or the other. Some, a scant few, chose neither. Nowadays, we are getting back to a bipolar world again. The US had a good run there for ten to fifteen years as the sole superpower in the world, but China is rising to meet it (not overtake it). Some countries in South America are now making strategic ties to China. In Africa, a lot of nations are making strategic ties to China. Certainly, Southeast Asia is making strategic ties to China. It's just the way things are going down right now; China is the other option in a newly bipolar world.

How is China doing this? Their explosive economy is a big part of it—double digit growth now for a decade and a half. Part of that is entrepreneurship, which leads to a growing middle class. That's what you have to have if you're going to have an internal economic engine that keeps going. You've got to have that middle class that likes to buy a lot of stuff, and China is growing a big fat middle class right now. Here's some other cool stuff that makes them appealing: China is on the permanent UN Security Council, which means, since World War II, they have had a world political voice. They are in the WTO. They are a member of ASEAN and APEC and SCO. They are pretty much in all the power player clubs. They are in the space club! On top of that, they have what the Plaid Avenger has got to refer to as some savvy foreign policy.

SAVVY FOREIGN POLICY?

There are two parts to this. *One:* For the last twenty years, China has been working to resolve all of its border disputes. Thirty years ago, China disputed all its borders with everybody. In today's world, that's going away. They have peacefully reabsorbed Hong Kong and Macau, and as I suggested earlier, Taiwan is slowly coming back into the fold as well. They are cementing ties with all their land neighbors, a pretty suave move if I do say so myself. *Two:* They are setting up strategic ties with countries all over the planet, sometimes with no strings attached. It's a very odd thing. In the US, this no strings attached business is looked upon with suspicion. China lends money to countries with no motive? We find this hard to believe. We always have a motive behind everything we do. We say, "Oh, we will lend you some money if we get something in return." But China just lent half a million dollars to Palau, some Pacific nation no one has even heard of before. They build bridges and stuff in Africa. Throughout Latin America, Africa, Southeast Asia, and even India, they are setting up economic ties, buying different resources and setting up free-trade options.

This is done primarily with an eye for **resource acquisition**. As China's economy continues to ex-plode, they need more energy, like all big economies do. What types of energy? Coal, which they have in abundance, and oil, which they don't. During the past decade of dominance, you often saw former **President Hu Jintao** of China making the rounds. In 2006, he visited all of South America and came to the United States to visit President George W. Bush. He went to Saudi Arabia, and

Former President Hu Jintao

Former Prime Minister Wen Jiabao

then toured all of the African nations. His premier, **Wen Jiabao**, got around even more. These guys are hustling around the planet right now.

From 2002 to 2012, these top dragons in the Chinese government were on a permanent goodwill world tour, cementing trade, investments, resource acquisitions, and being all-around awesome dudes that formed happy-time relationships with country after country around the planet . . . all while their economy back home was blowing up, and while China resumed its rightful place as a world power. These guys got good karma. They are revered in China today as great stewards of the state, and will go down in history as overseers of a great era for China.

But hang on! Cuz there is a new kid on the wok now! The head of the dragon is now being mantled by one **Xi Jinping**. Xi has called for a renewed campaign against corruption, continued market economic reforms, and a comprehensive national renewal under something they refer to as the "**Chinese Dream**."

Xi Jinping . . . so Chinese dreamy!

How dreamy! But make no dragon bones about it: this Xi Jinping is one powerful leader! He is currently the most popular leader China has had since Mao and so powerful in fact that he amended the constitution in 2017 to allow himself to be president for life! He is a hardcore Chinese nationalist that has tackled corruption, eliminated all political rivals, and tightened restrictions over civil society and ideological discourse, along with increasing internet censorship. He has pushed hard for a more assertive foreign policy which sees China as a major global player, one that is extending its military influence out into territorial seas abutting Japan and all of Southeast Asia. Due to the anti-globalization/anti-trade approach of US President Trump, Jinping has been suddenly thrust into the top position as the leading advocate of free trade and globalization on planet earth! In that same vein, Xi has also sought to expand China's African and Eurasian influence through the One Belt One Road Initiative (more on that later). In 2018, Forbes ranked him as the most powerful and influential person in the world! Yep! That's Xi! And he is just getting warmed up!

2021 UPDATE: Yep, Xi continues to tighten his grip on the country, and the Chinese periphery. After years of democratic protests and the attempts to keep Hong Kong as a separate system, Xi cracked down on the "special administrative region" and has forcibly brought it back under full Chinese central government control. China has also become much more militarily aggressive in the South China Sea, Southeast Asia, and beyond. Taiwan appears to be next in Xi's sites. Yikes!

TUMULTUOUS TIBET

Speaking of frictional Chinese policies, there is the small matter of Tibet. Or should I say, huge matter of Tibet. The area we refer to as Tibet has seen various levels of freedom at various times in the last thousand years: it was an independent kingdom, a vassal state, an autonomous region, a semi-autonomous region, a colonial holding, and a fully absorbed territory into a state we call China. For purposes of understanding today's world, we need only concern ourselves with that last description—as part of Chinese territory.

During the heyday of the Manchu Empire in China, well over 500 years ago, Tibet increasingly came under the influence of the Chinese. But Chinese power was destroyed by internal factors and civil war, and was further inhibited by Western and Japanese imperialism in the 1800s. As China was falling apart, Tibet first became a pawn between Western powers (mostly Russian and British), and later began asserting its outright independence—to keep the record straight for you, their independence was proclaimed while China was self-destructing and the West was preoccupied with World War I. Basically, Tibet was largely left to its own devices while all the other world powers were busy.

Long story short, once China got their act together after WWII and their Civil War, they immediately starting re-establishing their presence in Tibet, and in fact had never renounced their claim of sovereignty on the area. The **Dalai Lama** continued to partially rule in Tibet with, to some extent, autonomous power given by contemporary Chinese governments, until the People's Republic of

Tibet: an autonomous province of China. Probably forever.

Lama always popular abroad

China invaded the region in 1949, taking full control in 1959. The Dalai Lama then fled to India and has since ceded temporal power to an elected government-in-exile. Which brings us up to date enough to understand today's world . . .

The current 14th Dalai Lama seeks greater autonomy for Tibet. Not outright independence, but greater self-rule autonomy. The Chinese have interpreted this as a threat to their sovereignty—and they hate the guy! They hate that he's so popular. They hate that he's well respected, and even venerated, as a world figure. They probably even hate his sweet flowing robes. And they really hate it when any world leaders meet with the Dalai Lama. The Chinese think that the more recognition the guy gets, the more the world will demand that China give Tibet back to him. It is a similar issue to their Taiwan situation—the Chinese do not want any country to officially recognize the guy, for fear that Tibet will someday claim independence.

What is the Dalai Lama really up to? The dude now tours the world—and he is the first Dalai Lama to go abroad—spreading the Buddhist message and preserving Tibetan culture. He officially leads the government in exile from Dharamsala, India. He is a fantastic speaker, promotes world peace, wildlife conservation, and a host of other awesome stuff that has won him great respect, acclaim, and even a Nobel Peace Prize. Let's face it: the dude is the Buddhist shizzle. How about we call him the Budd-izzle?

Fast forward to now. Because everyone knew that all eyes were on China due to the 2008 Olympics, a bunch of Tibetan nationalists decided to catch the eyes of the international press by having some big protests and raising all sorts of hell, which predictably pissed off the Chinese government and forced them to crack down hard. Like the Sudan situation, the world has been putting pressure on China to deal with the Dalai Lama and the whole Tibetan situation. Unlike the Sudan situation, China has told eveyone to butt out. They accused the Dalai Lama of inciting the whole affair, and are more determined than ever to squash the Tibetans. And I must tell you bluntly, well over a billion Chinese citizens back their government on this one.

It's certainly an issue that has caused the Chinese some consternation, and it ain't over yet. Well, at least they've got Taiwan settled down for now, since the Tibet thing doesn't appear to be heading for a happy ending.

THE GOOD EARTH?

What are some other looming problems within China's society? There's tremendous growth, but although they are really rich and getting richer and the middle class is growing, there is still a tremendous wealth disparity in the country. With a billion and a half people, that's not surprising. There are people getting left behind, and part of the reason is the conversion from the communist economic system to free market capitalism. Every man for himself, baby!

"Your so-called communist kung fu is really quite pathetic."

This was the attitude in old Communist China: "We have these old factories. Everyone works at them. Doesn't matter if they are productive or not. We take care of you. You got retirement, you got healthcare." Once they transferred to capitalist ways: "Whoops, now you don't have crap! Sorry! This is capitalism, dude. Go out and get it yourself." This is fairly degrading for a good proportion of the population not sharing this economic prosperity. There is a distinct rural to urban split: it is mostly the rural areas in western China getting left behind. The disparity has caused some anti-capitalist sentiments and some poor areas to break out in revolt, to protest for more benefits.

There are also major environmental problems. Any society that is exploding this fast typically ignores the environment. China is faced with gargantuan environmental pollution and degradation problems. Perhaps even more pressing are problems with lining up all their coal and energy resources. Let's not forget the health problems associated with pollution and lack of healthcare.

SARS? Avian flu? Don't blame me—I'm just a Chinese chicken!

Speaking of which: where did SARS come from? Where did the avian flu come from? Where did the historic bubonic plague outbreak come from? Go down the list, and you'll see a lot of these things are coming from China. This is turning Chinese health from a *local* to a *global* issue. The Plaid Avenger is not trying to suggest he knows why these things seem to start in China, because he doesn't, but the most virulent strains of flu that circulate around the planet seem to come out of China. Chinese health is a global concern. The last several times they've had outbreaks, the government has tried to cover it up, and that's not what the world wants to see. The world wants to be prepared for a big epidemic if one is about to break out. This is not a perfect society, precisely because of these sorts of problems, but it's still a society on the move.

Before we finish off this chapter, we don't want you to forget about China's little brother, Korea. It's actually a happening place as well, though for different reasons.

LITTLE BROTHER GROWS UP: THE KOREAS

Why am I suggesting this is China's little brother? Throughout the last 4,000–5,000 years, Korea has been a vassal state to China, but Korea has a distinct ethnic group; in fact, it's the most ethnically pure state on the planet. When you take a demographic survey and say, "What ethnicity are you?" you don't have to go far, because it is 100 percent Korean. It's the only place like that on the planet. Even the Japanese have 2 percent of something else. Korea is totally ethnically pure and very ethnically distinct, but it's still China's little brother; most of their adaptation throughout history, such as their religious, political, and philosophical structure, was borrowed from big brother China. That's why we see a lot of similarities in art and architecture and the way people think and act between China and Korea. It's not China; it is distinct, but it borrowed a lot from China. What's the deal with the Koreas today, and why are there two halves called North Korea and South Korea?

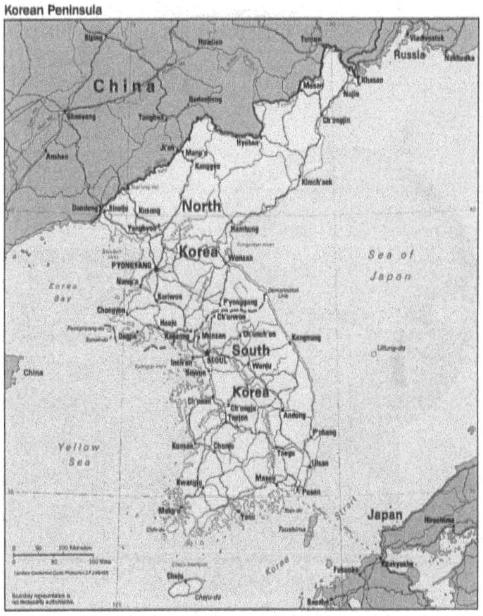

SO NICE THEY NAMED IT TWICE? WHY TWO KOREAS?

For most of its history, Korea was one unified country. However, it usually got pushed around by its more powerful neighbors. It was occupied by China in the first century BCE and by Japan in 1895, who basically turned Korea into a work camp for the Japanese. However, the final insult came in the waning days of World War II. After it became apparent that the Allies were going to defeat Japan, Russia moved troops into the northern part of Korea and the United States occupied the southern part of Korea. After the war was over, the US and the USSR stuck around to "help rebuild and reestablish order" on the peninsula. Separate governments were formed in these

occupation zones, established at the 38th parallel, which led to the initial territories called North and South Korea. It was much like the occupational situation in East versus West Germany over in Europe. Unlike Germany, which reunited in 1990, this Korean mess is still unresolved.

North Korea became a communist dictatorship supported by the Soviet Union, while South Korea became a capitalist democracy supported by the United States. No one really thought that this situation would last forever. It was never intended to be permanent, but now we had the first real Cold War "face-off" of ideologies put into place.

General Douglas MacArthur: "Let's go get those commies! I'm ready to invade the North!"

For a variety of reasons, the North Koreans thought they could easily take over South Korea. The North Korean army invaded South Korea in 1950, starting the Korean War. To stop the spread of communism, America sent thousands of troops to protect South Korea under the leadership of General Douglas MacArthur. Even though millions of Koreans died, the result of the war was a tie, and neither Korea ended up conquering the whole peninsula. To this day, over 30,000 American troops are stationed in South Korea in case North Korea attacks again. This conflict became the first hot spot of the Cold War.

PLAID AVENGER INSIDE TIP: How come it's not just one Korea? What does that have to do with Hawkeye Pierce? A really cheesy TV show named M*A*S*H* depicted the Korean War for the prime-time enjoyment of US viewers. The funny part? The Korean War lasted three years, from 1950-53. The TV show lasted 12 years. That war was just so darn funny we had to watch it four times longer!

You should also know this: the Korean War was the first UN military action. That's right: it was not a US war, even though it looks that way on TV. A UN resolution was passed to stop the North Korean invasion, and this UN force was, of course, led by the US, but there were also soldiers from other UN member states. That's why when you watch M*A*S*H* you will see the random British or Australian dude hitting on Hot Lips Houlihan.

Since then, North Korea and South Korea have gone radically separate ways. South Korea is a capitalist democracy with the world's tenth largest economy, while North Korea is still a communist dictatorship whose population is starving. However, there is a Korean reunification movement, and the two Koreas are trying to reestablish relations with each other. A unified Korean team has marched in the opening ceremonies of the past several Olympic Games. However, reunification will be difficult because the Koreas have been apart for so long and their economies and governments are very different. Also, many observers think that both China and the United States prefer that Korea be divided for geopolitical reasons.

The North-South Korea thing is a relic of the Cold War. It's one of the hot spots that flared up during the Cold War, where fighting actually occurred. What's going on in today's world with this battle of ideologies? A telling illustration can be seen in the map of lights at night in the Koreas below. South Korea: bright lights, big city. North Korea: lights out. This really underlines what's going on within these two countries, and really shows who won the ideology battle of the Cold War. The Koreas sum it up so perfectly, so beautifully, that we have to point it out here. North Korea went down the communist path, the radical communist path. Now they are radically freakin' broke! They've got nothing. It's a society that's about to disintegrate under our good friend Kim Jong-Il.

S. Korean President
Moon Jae-in sez:
We are rich!

N. Korean Dear Leader
Kim Jong-Un sez:
We are koo-koo crazy!

The opposite is South Korea, which embraced the democratic capitalist West and is now the number ten economy on the planet. We have three of the top ten economies right there in a nice little grouping of China, South Korea, and Japan. South Korea is a bustling and happening place. North Korea can't pay the light bill, but South Korea is going gangbusters.

Super-economy in the South; lights out in the North.
Yep, that's Korea.

It's funny that you don't know as much about South Korea, or hear about it that much in the news. Here's why: South Korea is a country wedged between the giants. Japan is the number three economy to its east and China the number two economy to its west. To the north is the nutso nuclear-armed North Korea that catches lots of headlines as well. South Korea looks like a little place, yet it's important and rich! China, South Korea, and Japan, to a lesser extent, are going to be the technological centers of the planet in the next 100 years.

These three countries have made that their mission, their goal. In fact, China has plans for a moon base and the South Koreans vowed in 2006 to put a robot in every home by 2030. This isn't science fiction. It's not even fiction fiction. They actually said it! I believe they'll do it, too. They already have the fastest Internet connections on the planet, and have made Internet accessibility available to every single one of its citizens. They're not playing around, man! South Korea is the most connected country on the planet per capita. All of the countries in this neighborhood have societies that are embracing technology with a vengeance. This is going to affect what's going on in the Web-world in the very near future. Soon you'll see how these countries' technological expertise will impact the entire planet. Let it be known that these guys are hot and heavy with the technology sector, and they are transforming the playing field every single day.

I shouldn't end this little discussion without giving at least a little positive press to our North Korean friends. Actually, who gives a hoot about those wackos, but you should know the reference to their "PARTY." The 6-Party, that is:

What's the Deal with the Six-Party Talks? Those crazy Koreans do love to party! As long as it's with six! The **six-party talks** is the name given to a series of meetings with six states–China, North Korea, South Korea, Russia, Japan, and the United States–to deal with North Korea's nuclear ambitions. These talks were a result of North Korea withdrawing from the Nuclear Non-Proliferation Treaty in 2003. The aim of these talks was to find a peaceful resolution to the security concerns raised by the North Korean nuclear weapons program. Five rounds of talks resulted in nothing actually happening and little headway made in disarming North Korea.

Why no movement towards a solution? Ummm . . . I think mainly because the leadership of North Korea is utterly and completely in freakin' la-la land. It should be noted that pretty much everyone agrees that North Korea is getting close to collapse, and bordering on completely hopeless. Their number one GDP earner is counterfeiting US currency. Does that count as sustainable development? What a joke!

Party?

Basically, the only thing this "country" has to its name is supposed nuclear weapons. That is the only bargaining chip they have to be taken seriously by the outside world. Patience from the outside world is growing thin. Of course, Japan and South Korea don't want them to melt down completely, which could result in a nuclear weapon being launched. China and the US are fairly secure that no one in North Korea would be dumb enough to launch a missile in either of their directions, but all bets are off after 2010. North Korea has walked away from the talks, re-started its nuclear program, tested at least one nuclear device, and shot off several missiles as a test/warning to the neighborhood. This situation grows grimmer by the day, and it's entirely possible that the entire state may have collapsed or been invaded by the time this book makes it to print. More on this in the hot-spots chapter.

ONE BELT, ONE ROAD: ONE EURASIAN POWERHOUSE

Wow. Just when I thought I was done explaining China's rejuvenated role as a returned world power. . . the Dragon goes and one-ups me. THIS IS A GIGANTIC GLOBAL DRAGON DEVELOPMENT, and it is quite brand new: The Silk Road Economic Belt and the 21st-century Maritime Silk Road aka The Belt and Road aka One Belt, One Road aka The Belt and Road Initiative, (lol that is a crap-ton of aka's), is a development strategy and framework, proposed by Chinese President Xi Jinping that focuses on connectivity and cooperation among countries primarily between the People's Republic of China and the rest of Eurasia, specifically tying together East Asia, Southeast Asia, South Asia, Central Asia, Russia, Turkey and parts of Europe, the Middle East, and east Africa. Holy silk shirts! That's some serious global coverage! And it is all anchored by China. . . so this is the world's other big superpower building a global network of economic integration around itself, affecting policy and development the world over. Much the same way Europe did for 500 years, and the USA did for the last 70.

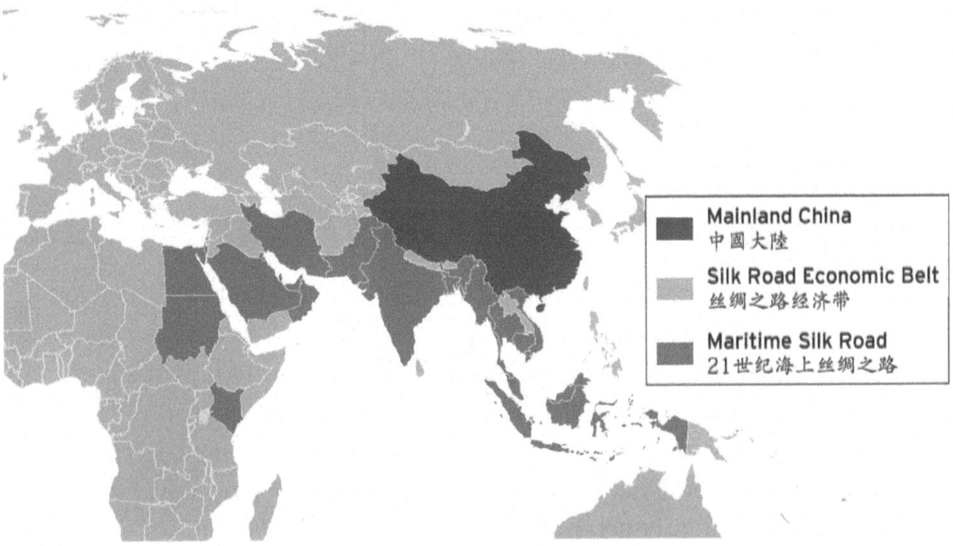

The center of economic and political gravity is shifting east . . .

Just look at the name itself!!! China wants to re-establish the old Silk Road!!! Their grand plan consists of two main components, the land-based "Silk Road Economic Belt" and oceangoing "Maritime Silk Road." The strategy underlines China's push to take a bigger role in global affairs, and its need for priority capacity cooperation in areas such as steel manufacturing and energy. I'm not sure how to even categorize this plan at the moment . . . is it a trade block, is it an international organization?. . . but in any case it highlights the return of China as a global leader, (no longer just a center of cheap manufactured goods), and this plan has great economic potential for its participating member states.

The coverage area of the initiative is primarily Asia and Europe, encompassing around 60 countries. Oceania and East Africa are also included. Anticipated cumulative investment over an indefinite timescale is variously put at US$4 trillion or US$8 trillion. One Belt, One Road has been contrasted with the two US-centric trading arrangements, the Trans-Pacific Partnership (TTP) and the Transatlantic Trade and Investment Partnership (TTIP).

PLAID AVENGER SIDE NOTE: Not to offend all the Trump supporters in the hizzle, but I really want to to step back for a moment and contemplate the different paths the US and China are currently on: Trump killed the TTP and TTIP and global free trade in general, choosing to focus only on what is good for the USA; China is investing billions—nay, trillions—into banks and infrastructure and projects across half the entire planet in order to secure its importance as the center of the economic universe. One great power is contracting, while the other great power is expanding. It's just that stark of a difference.

This is developing in tandem with The Asian Infrastructure Investment Bank, first proposed by China in October 2013, which is a development bank dedicated to lending for projects regarding infrastructure. As of 2015, China announced that over one trillion yuan ($160 billion US) of infrastructure projects were in planning or construction. So China has a global bank at its disposal to spread its influence further afield. But even that ain't enough for the big boy: In November 2014, Xi Jinping announced plans to create a separate $40 billion development fund just for this Silk Road stuff. As a fund its role will be to invest in businesses rather than lend money for projects. The Karot Hydropower Station in Pakistan is the first investment project of the Silk Road Fund . . . remember a few chapters ago when I was pointing out that Pakistan is re-aligning itself with China? Whoomp! There it is!

So one route is by land, the number two is by sea. . . maybe we should send out Paul Revere to start shouting: "The Chinese are coming! The Chinese are coming!". . . but it would be a little late, as they are already here and establishing themselves as not just a world power, but the premier world power of Eurasia, and a powerful alternative to Team West.

Remaking the entire world order.

ENOUGH ALREADY ON EAST ASIA

I think you have got the picture now, and I'm probably going to repeat some things for the fifth time, but let's do it again anyway. When we think about China, we think about one of the world's largest states, maybe the third largest state in the world. We also think of the world's largest military. Am I saying it's the best military in the world? No, the US has the best military in the world, but China has the largest free-standing army on the planet. . . and it is getting better, fast!

New China sez: "I see China moon colony from here. Hahahaha. We're so rich. This is fun."

China also now has the number two economy on the planet, having surpassed Germany in 2009 and Japan in 2010 . . . and odds are that it will surpass even the mighty US by 2025. . . maybe it already has passed the US! But hey! China isn't alone in this region of big money: both Taiwan and South Korea are bangin' out the bucks as well, and all three of these entities are busting out with the newest and best technological sector development to boot. South Korea in particular has the highest level of high speed Internet access in the world, and has also vowed to have a robot in every household by 2030. South Korea has the highest density of robot workers in the world, at 700 per 10,000 human workers! Now, I don't know about the robots, but it is a safe prediction to say that within the next decade, three of the top ten world economies will be located here in this East Asian region. No doubts. That's a lot of bank!

Shanghai Skyline: Welcome back to world power status, China!

But back to the big boy: China is now also a member of all the world's most progressive and fastest growing trade blocs. Their middle class is growing just as fast as their prestige. They are doing all the things the power player countries do, like the Miss World Pageant, like the Olympics, like the space programs. This is a place that is really on fire. Yes, it's got problems, but there are a lot of things going on in China that are right on target. China is back, and it is once again a world power to be reckoned with.

Along with the Taiwanese and South Korean titans, this region promises to be an epicenter of activity and excitement of all types in the coming century. If only Kim Jong-Un and his funky bunch can be coerced out of declaring war on the entire world, this region will totally be rocking out the 21st century!

PLAID AVENGER INSIDE TIP: Want to double your salary as soon as you walk out of school, regardless of your major or your GPA? Then do this: LEARN CHINESE! Anyone who is bilingual in English and Chinese can pretty much write their own ticket in the 21st century, as a business/political/cultural middleman between the two hugest world powers. Learn Chinese. Do it!

EAST ASIA RUNDOWN & RESOURCES

View additional Plaid Avenger resources for this region at http://plaid.at/e_asia

BIG PLUSES

- Huge and growing economies: China #2 biggest, South Korea #10; China will highly likely be #1 in GDP in the next 30 years
- South Korea is the most wired and Internet accessible state on the planet, with nation-wide free T-10 speed Internet coverage. And it's free!
- South Korea has western-style high standards of living with a huge middle class, and China is getting there; China raised 250,000,000 people above the poverty line in the last 20 years, a world record
- China has largest standing army on the planet and is significantly investing an and re-tooling their military to bring it up to highest technological standards
- China has reabsorbed Hong Kong and Macau, and it appears that Taiwan reintegration is imminent
- China is not involved in any foreign wars, has none of their soldiers outside of their country, and in general has solved all of their border disputes with neighbors
- China is a nuclear power and a member of UN Permanent Security Council, WTO, SCO, ASEAN, APEC
- China hosted and won the Olympics, regularly hosts Miss World pageants, and just hosted the World Expo which had a record turn-out
- China has the most active and growing space programs, having a manned space-walk in 2008, and designs for a moon landing in 2020, followed by a moon base; the Asian Space Race is now on
- Chinese people currently have an EXTREMELY strong sense of national pride; do not under-estimate this powerful yet non-measurable asset

BIG PROBLEMS

- In China's focused mission to achieve economic gains at all costs, their environment has been sidelined and trashed; huge environmental problems loom, from horrific air pollution to toxic waterways
- In China's focused mission to achieve economic gains at all costs, their consumer protection, copyright protection, and building standards have all sucked, which is why they are catching heat for all the fake, crappy, and sometimes lethal products that are being produced, if they don't clean up their act, the "Made in China" brand will take a hit, as will their economy
- China may have raised millions from poverty, but there are millions yet to go
- Gigantic rural to urban migration (loosely correlated with a west to east migration) which is emptying out the countryside and packing people into the already overburdened cities of the east
- Heavily reliant on imported energy, a situation which can only get worse the bigger their economy gets
- China's "non-interventionist-no-matter-what" foreign policy is getting them a bad rap on the world stage, as they are seen to be protecting some real bad people doing real bad things; from Sudan to North Korea to Iran, the Chinese are increasingly seen as standing in the way of stopping human suffering and strengthening international security
- North Korea could flip out insane at any given second, causing regional instability at best, or World War III at worst

DEAD DUDES OF NOTE:

Chiang Kai-Shek: Leader of the Kuomintang (Chinese Nationalist Party) and chairman of the disjointed Republic of China (ROC) from 1928–48. Led the Nationalists fight against Mao and the commies in decades of civil war, and, upon losing that war in 1949, fled to Taiwan where he and the Nationalists set up shop and claimed to be ruling China.

Mao Zedong: "Father of (modern) China" Communist revolutionary leader that fought against the Nationalists, did the **Long March**, and organized the guerilla war against the occupying Japanese, and ultimately won the country for the commies in 1949. He led the People's Republic of China (PRC) from its establishment in 1949 until his death in 1976.

Deng Xiaoping: Leader of China from 1978–90, and a reformer most famous for transitioning China from the failed communist economic policies of the Mao era towards a market economy . . . which of course is why China has had explosive economic growth for the last two decades.

LIVE LEADERS YOU SHOULD KNOW:

Hu Jintao: Former President and Paramount Leader of China 2002–2012, a low-key, conservative leader who oversaw explosive economic growth, an increase in China's global influence, and a surging nationalistic pride. Had strong public support for his policies.

Xi Jinping: Current President of China. Wants to focus on fighting corruption, continuing capitalist market reforms, and has vowed to have more openness and transparency in governance. Super powerful. Has changed Chinese constitution to allow him to be leader for life.

Tsai Ing-wen: President of Taiwan and the first woman elected to the role, as well as only the second to be elected from the liberal and possibly pro-independence DPP party; called Donald Trump to congratulate him. Trouble brewing with China!

Dalai Lama: Tenzin Gyatso is the 14th Dalai Lama, a Buddhist head honcho and cultural/spiritual leader revered among the people of Tibet. Technically from Central Asia by my definition, but here for this: the Chinese government hates this guy! They think he is a separatist trying to pull Tibet away from China!

Moon Jae-in: President of South Korea since 2017, centrist, liberal, former student activist and human rights lawyer. Along with pushing liberal social/economic policies in his state, is also a keen advocate for normalizing relations with North Korea and has met with Kim Jong-Un many times.

Kim Jong-Un: The "Dear Leader" of the insane, isolated, militarized, freakish North Korea. A totally out of touch, probably delusional, daddy's boy who runs the entire state as a cult of personality. China is his only friend, and even they are getting sick of his state's shenanigans. The biggest security threat of East Asia.

PLAID CINEMA SELECTION:

Brotherhood of War (2004) A brilliant dramatic film which takes a look at the relationship of two brothers as they fight for South Korea during the Korean War. Throughout the film, the relationship between the two brothers deteriorates until a rather tragic conclusion. One of the biggest successes of Korean film at the time of its release.

Crossing (2008) Yong-soo lives in a small coal-mine village in North Korea with his wife and young son. Although living in extreme poverty, the family is happy until the wife becomes ill and Yong crosses into China to find medicine. You can figure out the depressing events that will follow. Great landscape shots, and topical plot, but a real tear-jerker. Bring tissues.

The Founding of a Republic (2009) A historical film that was commissioned by the government to mark the 60th anniversary of the People's Republic of China. Retelling the story of the Communist of triumph in China, the cast of the film is made up of some of the most well known actors in China, even if they just make brief cameo appearances. The best communist propaganda film I have ever seen! But the history is at least on mark.

Lust, Caution (2007) Set in both Hong Kong and Shanghai, Ang Lee does a phenomenal job of portraying an assassination attempt that took place during the Japanese occupation of Shanghai during WWII. Based loosely on real life events, film presents a unique look at the Chinese puppet government and one attempt to bring it down. Given an NC-17 rating on the basis of graphic and violent sex scenes that are not just there to be controversial, but actually enhance the plot.

24 City (2008) Changing times in Chengdu, China, where state-owned factory 420 is closed by the government to make way for a high-end apartment complex called 24 City. The film tells the story through the lens of several distinctive characters, from factory workers to executives to the newly monied, who are all affected in different ways by the newfound dynamism and economic growth of their nation.

Mang Jing aka *Blind Shaft* (2003) Boy, this film just keeps getting more topical as every year passes! The plot line involves scam artist/murderers setting up chumps to die so they can collect insurance money, but the backdrop is even more important: the mine shafts of Chinese extremely dangerous coal industry in which hundreds die every year. Great bleak landscapes of north-central China, great insight into the coal industry, and a scathing commentary of the darker side of capitalist fervor in China.

Oldboy (2003) The most disturbing and bizarre revenge movie I have ever seen. I'm not sure it's got a thing to do with teaching you about Korea, but it is so wildly insane, graphic, and intense that it's a must-see for film buffs. At one point a dude fights ten guys in a hallway with a hammer, and in another he eats a live octopus. Seriously, not faked. Not for the faint of heart. Twisted, sick, perverted plot that cannot possibly be predicted; by film's end, you will be shocked. My Korean friends made me watch it, and I will get them back for that some day.

Farewell My Concubine (1993) In a plot that captures 50 years of Chinese history, a seemingly unshakable friendship between two Chinese opera stars gets put to the test in the face of war, a communist takeover, the Cultural Revolution and the intrusion of a woman who tempts both of them. So sino-sexy!

The Killer (1989) Have to give a shout out to John Woo, the master of Hong Kong kick ass action movies! *The Killer* is one of Chow Yun-Fat and John Woo's best Hong Kong films. It tells the tale of a morally responsible killer and his understanding, yet dogged adversary, a police detective.

Enter the Dragon (1973) Two words: Bruce Lee.

MAP GALLERY

View Map Gallery & additional Plaid Avenger resources for this region at http://plaid.at/e_asia

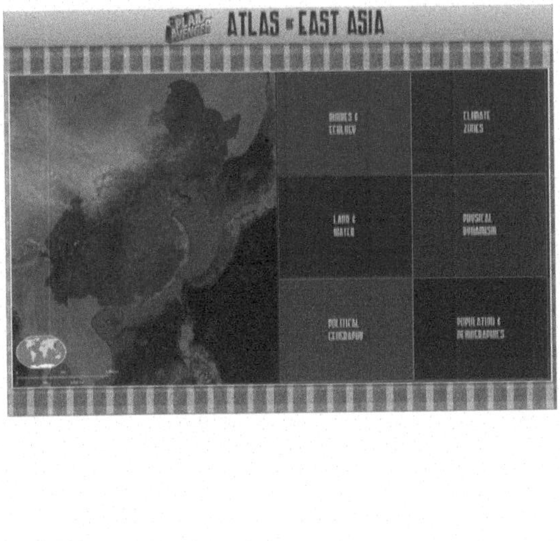

CHAPTER OUTLINE

1. Crossroads, Conflict, and Cubs
2. Physically on Fire
    2.1. Hawt . . . and Often Wet
    2.2. Who's There?
3. Back Up for Background
    3.1. Chinese Influence
    3.2. See You at the Crossroads
4. Conflict
5. And Now the Modern Conflict
    5.1. Smart Godzirra!
        5.1.1. Bataan Death March
        5.1.2. Burmese Railroad
    5.2. 'Nam
    5.3. Pol Pot Pie
6. Cubs in the Catbird Seat
    6.1. ASEAN
    6.2. Regional Titan on the Rise: Indonesia
    6.3. Bump It to the Next Level, Baby
    6.4. Terrible Terror Potential
7. Round-Up
8. Map Gallery

Southeast Asia

24

CROSSROADS, CONFLICT, AND CUBS

I'm talking about the Thundercats, my friends, because this is where it's all going down: Southeast Asia, the "Archipelago Region," a collection of peninsulas and islands on the southeastern quadrant of the Eurasian landmass. It's been a mover and shaker in the past, but it's one of those places we don't really know a whole lot about historically or in the modern era, but a lot of big events have happened there in the last 100 years. Even in the last fifty years, the Archipelago Region has been a place of crazy genocide, insane outbreaks of war, and massive movements of refugees into and out of the region. The last few years have been interesting for the region in a much more positive light: a lot of economic growth and little political turmoil, despite two big-time natural disasters. So, what's going on in Southeast Asia?

PHYSICALLY ON FIRE

Southeast Asia is rugged and geologically complicated, but it's actually quite simple to understand. The variation in terrain predominately runs north to south in this region. We have rugged mountains that run down as a continuation of the east-west trending Himalayan Range, but then turn due south through Burma, Laos, and Thailand. There is another chain that runs down the Vietnamese coast called the Chaine Annamitique—blast the French and their funny language. The Northern parts of Vietnam, Laos, and Burma are mountainous along the Chinese border. Most of the Indonesian Archipelago has at least some mountainous terrain as a result of volcanic activity and is a geologic hot spot of plate tectonic activity. The convergence of some pretty major plates makes this one of the most volcanically and seismically active regions on the planet. It is a definite centerpiece of **The Ring of Fire**.

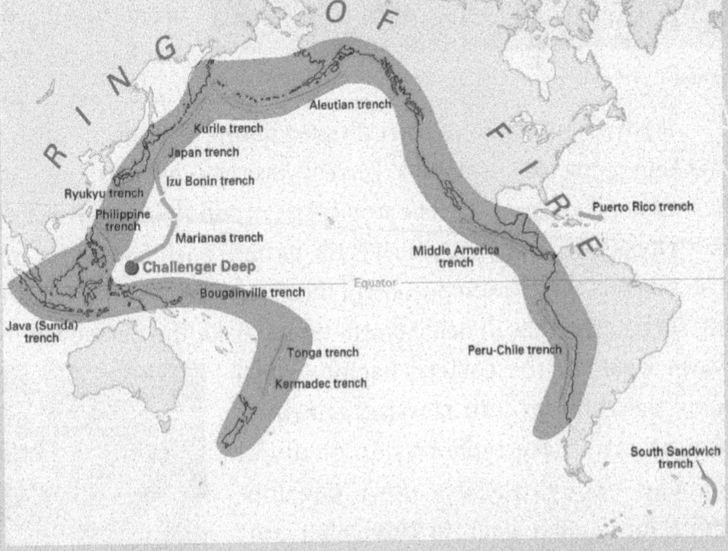

The Ring: Areas of greatest volcanic and seismic activity.

You may have heard of Krakatoa, a volcanic island that exploded into oblivion in 1883. The blast was so enormous that the dust, ash, and smoke that entered the atmosphere created a haze that partially blocked out enough sun to cool world temperatures for about a decade. In addition to some volcanic activity in this region, there is also seismic activity. This combination fosters the development of **tsunamis**, giant waves resulting from offshore earthquakes. I bring this up because, unless you've been hiding under a rock, you've heard of the massive tsunami that occurred in the Indian Ocean on December 26, 2004, claiming more than 240,000 lives. Because of the seismic activity, Southeast Asia is a fairly disaster-prone place, particularly in Indonesia and most coastal areas, which just so happens to be where most Southeast Asians live. As if tsunamis and volcanoes weren't enough, Southeast Asia also occasionally gets hit by tropical cyclones (a.k.a. typhoons). On May 2, 2008, Cyclone Nargis made landfall in Burma and killed *at least* 50,000 people. Why did I just emphasis *at least*? Because the former Burmese governing junta spent 90% of their collective effort trying to hide the damage. What did they do with the other 10%? Answer: they tried to keep international aid workers out.

Tsunamis: Great for surfers, bad for Southeast Asia.

Tropical typhoon brewing. . . .

HAWT . . . AND OFTEN WET

The final factor for physical geography is climate. Almost every country falls below the Tropic of Cancer and Indonesia straddles the Equator. Thus all of the countries in this region have tropical climates. We can visibly see all three subdivisions of tropical climates in Southeast Asia. There's a whole lot of **tropical rainforest**. Indonesia is primarily rainforest, the biggest chunks of rainforest left on the planet aside from the Amazon. There's some **tropical savanna** in and around Thailand, Burma, and Cambodia. All along the coastal areas of Vietnam and Cambodia, you can get a taste of **tropical monsoon** climate as well.

If anyone remembers Forrest Gump during his little Vietnam adventure, he talked about the rain that started one day and kept going for six months. Forrest was right; big parts of Vietnam has a true monsoonal climate. Perhaps the second best example of true, big-time tropical monsoon climate outside of India would be here in Southeast Asia.

Of course, typhoons, which is the eastern Pacific word for tropical cyclones, or what we Americans call "hurricanes," also pummel the eastern portion of the region like it's a punch-drunk palooka. Seriously, look at the map: suffering swirling succotash! These Pacific storm tracks make the Caribbean look like a walk in the water park! It's all thanks to the warm water of the eastern Pacific, which fuels these things into massively destructive beasts. These typhoons do plenty of damage, enough to slow down development—heck, just back in November 2013 the Philippines was hammered by Typhoon Haiyan, a Class Five storm (meaning the biggest and baddest mofo possible) that killed 7,000 people and caused nearly $3 billion in damages. The western part of the region isn't immune either! Every once in a while, tropical cyclones from the Indian

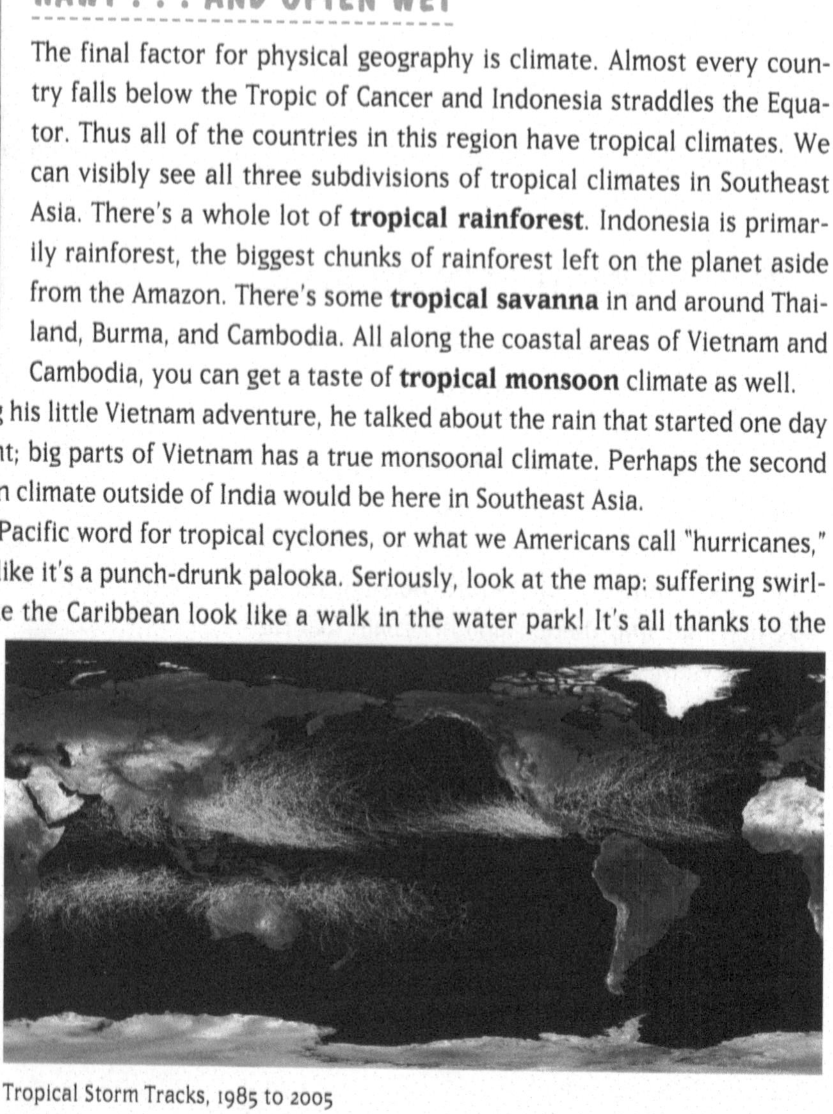

Tropical Storm Tracks, 1985 to 2005

Ocean—called "cyclones" there, even though it's the same thing as both a hurricane and a typhoon—curve up the coast and lay into Burma. The last big one was Cyclone Nargis in May 2008, which killed over 130,000 people in Burma and resulted in a massive humanitarian disaster. Why so many more deaths than Typhoon Haiyan? This area's not as used to dealing with the storms, and well . . . Burma is Burma.

There's some fairly decent terrain north and south in the region of Burma and Vietnam. There's some tectonic activity with implications for disaster in Indonesia and around the coastal areas, and the region is home to a nice blending of the tropical climates that bring plenty of deadly typhoons. That's your physical roundup for Southeast Asia.

What's in a Name?

Why is the Plaid Avenger hung up on keeping the "Burma" train going? A name is everything in the international community, but let's start with the history. The Indians called this place "Barma" for centuries, after the Bamar who were and still are the largest ethnic group in the area. Then, when the Portuguese sailed around looking for tasty spices and sweet, sweet opium, they adapted the name to "Birmania," which the Brits ripped off as "Burma" when they colonized the place in 1886. The name stayed put through independence in 1948, through a violent military coup in 1962, all the way to 1989, when a freaky-freak junta-controlled government changed the name to Myanmar. Why the change? The junta said they wanted to get away from the colonial past and claimed it wanted a name that included all the ethnic groups, not just the Bamar. Sounds reasonable enough, right? Wrong. Etymologically (meaning: where the words come from) Burma and Myanmar have the same roots and have identical meanings, plus the government had a bad habit of imprisoning those same minority groups, so the junta was just full of Burmese baloney.

If the words are the same, then why the Plaid boycott of "Myanmar"? Simple: one of the biggest tools in international relations is diplomatic recognition. One way to voice displeasure with a state's nasty policies is to not recognize its existence as legit, which includes ignoring any new name a new misbehaving government adopts. The junta destroyed the country's promising economy and committed atrocious human rights violations. One of these putting opposition leader, Nobel Peace Prize winner and Plaid hero Aung San Suu Kyi under house arrest for over 20 years. What'd she do to deserve that, sell a cop some reefer? Nah, she got locked up in 1989 simply for leading a political party that opposed the junta. When her party absolutely crushed the junta's candidates in the 1990 election, it certainly didn't please the generals in charge, so they basically threw away the key. To protest this ridiculously undemocratic nonsense, Suu Kyi, her followers, and her supporters in the international community (including the U.S. and the U.K.) continued to call the country "Burma" as a way to flip the diplomatic bird at the junta for its shameful actions. Think about it, how obnoxious would it be for some powerful folks to purposely call you the wrong name all the time? Suu Kyi was finally released in 2010, the junta allowed elections in 2012, (which her party easily won), and since 2015 she has been the de facto Prime Minister of Burma, a country now counted as a democratic state! But I still call it Burma instead of Myanmar anyway; why bother recognizing the junta nonsense of the past?

A peculiar plaid Professor skyping with Suu.

2021 DEMOCRACY DEMOLISHED UPDATE!!! In a region that had contained a variety of democratic variations across all of its states (okay, except Vietnam and Laos), a sudden reversal of freedom fortunes has befallen Southeast Asia! On May 22, 2014, the Royal Thai Armed Forces, led by General Prayut Chan-o-cha, Commander of the Royal Thai Army (RTA), launched a coup d'état against the caretaker government of Thailand, following six months of political crisis. The military established a junta called the National Council for Peace and Order (NCPO) to govern the nation, and govern it terribly they have for 7 years now. Now Prime Minister Prayut Chan-o-cha is facing renewed protest against his reign due to mismanagement during the Covid crisis, along with the fact that he stands in direct contrast to the democratic constitutional monarchy that Thailand used to be.

More recent, and even worse yet: On February 1, 2021 in Burma, a textbook coup d'état was conducted by the Tatmadaw—the official title of Myanmar's military—which then restructured the government into something they are calling a "stratocracy." That's a form of government headed by multiple military chiefs, with the various branches of government being administered by military forces. The head honcho of that stratocracy is a masochistic military man named Min Aung Hlaing, who has imprisoned all the democratically elected members of the country's ruling party, the National League for Democracy (NLD), and specifically the true leader and State Counsellor Aung San Suu Kyi—who also happens to be the Nobel Peace Prize winning hero of the decade long pro-democracy movement in Burma referenced above.

Burma and Thailand: Quite the double-whammy of whammied democracies! SouthEast Asian sad face.

WHO'S THERE?

This is pretty easy. It's just a few handfuls of countries, so we'll go ahead and fire them off for you: Burma (I refuse to recognize Burma by its military junta given name, Myanmar, so we're going with Burma), Laos, Thailand, Vietnam, Cambodia, Singapore, Malaysia, Indonesia, the Philippines, and the Sultanate of Brunei. And one of the newer sovereign states on Earth, Timor L'este!

What is it that we want to know about this region? The subtitle of this chapter mentioned that this region was a crossroads. What is it about Southeast Asia that has made it a crossroads? Historically, it has always been and still is considered a crossroads of all types of Asian culture. It has been the true melting pot of Asia. Influences from all over have made their way here and dropped off a lot of their cultural baggage, which has been picked up by the locals who were already there. This amalgamation of different things from different places in Southeast Asia makes it quite unique, but you can still pick out the origin of a lot of these influences. What am I talking about?

BACK UP FOR BACKGROUND

What's the story on this part of the world, anyway? Americans know that they fought a disastrous and socially dividing war here, and that all their Nike shoes and clothing are increasingly made in this place. Let's back up the refugee boat for a minute and take a look at just a few historical and cultural artifacts that will help us understand the region.

CHINESE INFLUENCE

Since forever, Southeast Asia has been a subsidiary vassal state of the Chinese. We've just finished talking about how the Chinese have been around for 5,000 years and for much of that time, border states have always been kind of "semi-China." China has been the big dog, the one everybody looks up to, or at least respects. This leads to pervasive Chinese historical influence across much of the region, and a lot of the folks currently in Southeast Asia (Thais and Vietnamese) were once actually "in" China. The Han Chinese influence kept pushing people farther and farther south until virtually all other ethnic groups were completely out of China's current borders. The Thais re-established themselves in what is now—that's right—Thailand!

Chinese influence is a dominant force throughout history, and we'll see at the end of this chapter that it's still a dominant force today. Southeast Asia has been in the Chinese cultural sphere of influence for a very long time. Chinese philosophy, Chinese religion, Chinese governmental standards, and political philosophies have rubbed off on this region, much as they have onto the Koreas and Japan.

But it's not just the Chinese that have influenced the culture. Southeast Asia is an archipelago with small islands and big islands alike. In fact, Indonesia has over 17,000 islands. These are coastal peoples, maritime peoples. What do maritime people do? Well, they catch a lot of fish for starters, but a lot of them end up trading goods and services within and across their region. Thus . . .

SEE YOU AT THE CROSSROADS

Spices: the crack-cocaine of the 16th century.

Southeast Asia has been an epicenter of global trade for quite some time. I know this isn't something we often think about, but what was it that Christopher Columbus was trying to reach? He was trying to get to the East Indies, **The Spice Islands**. Why there? To get all the good stuff for which all the Europeans would pay an arm and a leg. "The Spice Islands" and "the East Indies" are just European synonyms for Southeast Asia, predominantly Indonesia, which was the destination to reach in order to get all the goods. For a very long time, even before the Europeans came around, people from India, China, the Middle Eastern and even East Africa were trading stuff back and forth between each other via maritime traffic . . . you know, ships. The Indian Ocean was happening, man! It was the central star of global trade! As you might expect, virtually all those maritime routes go through Southeast Asia. The people of the Southeast Asian islands became integral to this trade system very early on. There were Hindu traders from India who established trading posts in this area while they were trading with the Chinese, who also had a presence there. In time, there were Muslim traders as well. The Europeans arrived fairly late to this ballgame, but they showed up and participated in the trade with a vengeance once they broke into the line-up.

Why am I telling you all about this old-school trade? Why is it important for today's Southeast Asia? Well, it's this crossroads idea—not just a physical crossroads of traders and goods, but a crossroads of all of the traders' cultures. When you go to Southeast Asia, you'll see obvious outside cultural influences. For example, let's pick religion. You can pretty much find every world religion in Southeast Asia. Even the Christians made their way to this part of the world via European traders. In places like Burma, Thailand, Laos, Vietnam, and Cambodia, you will see old-school Buddhism in practice. In fact, some of the biggest Buddhist monuments and temples on the planet are located in this place. Buddhism came from India. Not only Buddhism, but Hinduism came from India. You'll see a big Hindu component in all those countries as well.

But wait a minute! We aren't finished yet. There are some other religions present in Southeast Asia, like Islam, for instance. Islam is predominant in Indonesia, Brunei—the fact that Brunei has a sultan might tip you off there—and Malaysia. Malaysia, Indonesia, and Brunei are overwhelmingly Islamic, with Indonesia being the largest Islamic country on the planet. Shifting over, the Philippines, Eastern parts of Indonesia and even East Timor, the region's newborn, are all predominantly Christian. The notion that there are both Muslims and Christians in Indonesia warrants pause for thought, because while most of the Eastern religions and Eastern philosophies get along with each other, typically those Western ones don't. If you want to find religious violence in Southeast Asia, go straight to Indonesia, where the Muslims and the Christians both cohabitate, though not very well.

You will see Hindu temples, Islamic mosques, Christian churches, and Buddhist temples all over this region. You can say that about every part of the world; everybody is mixed up a little bit, but here you can see some very ancient places that really reflect how long this melting-pot idea has been going on in Southeast Asia. As you might expect, I'm just using the example of religion, but language is also very diverse here as well. Ethnic groups are very diverse, as there's a huge presence of Chinese, which we will get back to in just a little bit. Mix all these different ethnic groups, with a little Chinese and Indian influence, and a healthy dose of the great variation of spices from the region, and you have the table set for a unique area of cuisine, language, and culture . . . born of local elements fused with traditions from across Eurasia. This is a place that has been truly a crossroads for a long time and, I might point out, still is today.

A little bit Hindu, a little bit Buddhist.

Singapore for Dummies

Calm down, my Singaporean friends—I mean no disrespect. "Singapore for Dummies" only implies that your average American could find his or her way around the island. If you happen to be the average—or even slightly below average—American, this is the Asian country for you! Singapore is a global city, like London, Paris, or Tokyo. English is an official language; everyone speaks it. Singaporeans, in general, still like Americans—this is a disappearing quality in the world, enjoy it here while it lasts! Oh, and they also have couture shopping, five star hotels, and great food. Did I mention it's cheap—again, enjoy it while it lasts. The exchange rate as of 2010 is still pretty damn good! You can even walk across the whole island in one day! It's a damn nation-city. How cool is that? They've got a semi-official national beverage called the Singapore Sling that is one part gin, one part cherry brandy, and one part pure delicious! And their national symbol? A MERLION! Half mermaid, half lion. Who wouldn't want to crawl into the sack with that! Especially after a couple of Singapore Slings! Rawwwwrrrr!

Why is Singapore so Western? One answer is British colonization, but there is also a more complicated picture that goes back to the trading. Being a primary shipping and trading hub, Singapore acquired aspects of many cultures, including Chinese, Malay, and Indian. Through the years, Singapore recognized the value in being international and has focused on projecting a positive global image. Singapore is still one of the world's leading trading ports—over 1 billion tons of cargo flows through every year. It's also home to some of the most elite universities in Asia.

One final word of caution to all my Plaid traveling friends: Singapore's judicial system is, in one word, harsh. American student Michael Fay was sentenced to a "caning" (four lashes with a bamboo cane) in 1994 for theft and vandalism, and while that seems somewhat reasonable, another caning-worthy offense is failing to flush a public toilet after use. Chewing gum was a crime punished with a $500 fine until 2004. And whatever you do, DO NOT BRING DRUGS INTO SINGAPORE. They will execute you for it. For realz. Just say "no."

CONFLICT

The last historical section of the cultural crossroads has to do with European colonization. The development of Southeast Asia has been impacted heavily by outsiders. That may be something that the locals don't want to admit, but it's a fact. All of those religions and a lot of the languages spoken here were imported from other places. A lot of the development in Southeast Asia, including the way the governments work, is imported or at least heavily influenced by other places as well. The last big wave of influence came from the European colonizers, who made their way through Siam, Indochina, and the Dutch East Indies.

Where do all these terms come from? They were invented by the European colonial masters who established trade and, ultimately, political control of this region in the modern era. This group of masters included a little bit of everybody, similar to the African story where we have a whole bunch of countries represented. The British had possessions here, the French had a huge chunk of it, and the US even controlled the Philippines for a time after taking it from the Spanish. The Dutch controlled all of what is now Indonesia. The Portuguese had a colony named Timor, now Timor L'este, which is why it was separate from the rest of Dutch-controlled Indonesia. Even the Japanese had a hand in this part of the world, particularly during World War II.

When did all this come about? Same old, same old. Once we get into the 1700s and 1800s, the Europeans started coming in and made themselves at home. It wasn't really a land grab, but it started off just like it did in South Asia/British India, where the Brits and Portuguese set up ports of trade. As we've already suggested, Southeast Asia had been the center of trade for 1,000, maybe 2,000 years already, so the Europeans just wanted to get in the game. Quite frankly, they were shut out for a good long time because trade had already been established, and the locals were not

Siam? Indochina? British India? Who dat?

too welcoming to the outsiders. Why let foreigners set up shop to then become your competition? But inevitably, the British and Portuguese wore them down and started to make inroads. One thing led to another, over the course of several hundred years, and the Europeans ended up controlling the territories outright.

I'll give you a rundown here just to make it completely clear. The British controlled British India . . . which in today's world is Pakistan, India, Bangladesh . . . as well as Burma, Malaysia and Singapore in Southeast Asia. The French controlled a part of this region as well, named French Indochina (see inset box below). The Spanish initially controlled the Philippines until 1898, when the US assumed control of the island-nation. The Dutch controlled all of Indonesia and the Portuguese controlled what is now Timor L'este.

Everybody had a hand in this region, though strangely enough, one country remained unoccupied. That was Siam, or modern day Thailand. This is a great piece of trivia, placed on government tests for employees trying to get into foreign service. The question is, "Which country was never colonized by a foreign nation?" The answer is Thailand. If you ask "Hey, what is the richest nation in Southeast Asia today? What's the most stable nation in Southeast Asia today?" you'll be directed to Thailand, more often than not. Coincidence? Probably not. But we're talking about this foreign influence and where it's going to go . . .

WHAT THE H? WHAT THE WHAT? WHAT THE WHAT? WTH?

What's the Deal with French Indochina?

French Indochina was a group of French colonial possessions in Southeast Asia including modern day Vietnam, Laos, and Cambodia. France acquired these colonies in the late 1800s and held them until World War II, at which point Japan kicked them out and took Indochina for themselves. Japanese rule, however, lasted only until the United States dropped atomic bombs on two of Japan's most populated cities. With Japan's withdrawal, Indochina was theoretically free from colonialism. That is, for the three seconds before France attempted to colonize them again.

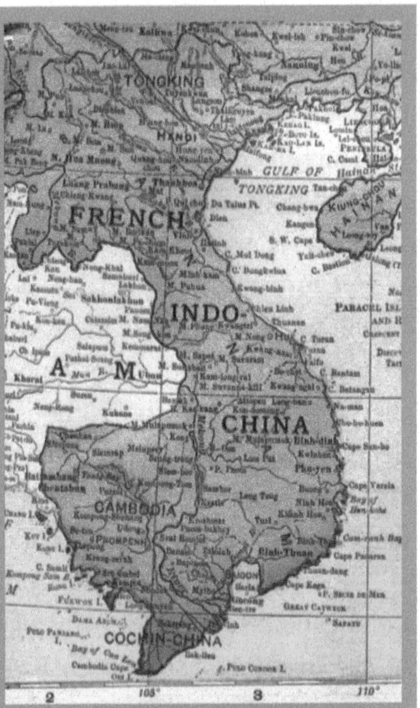

France found it impossible to regain control over Vietnam. This is partially because of a homegrown Vietnamese insurgent group of communist nationalists led by Ho Chi Minh. This group, the Viet Minh, had been fighting for independence even before WWII, then fought against the Japanese, and then quickly shifted their efforts to fight against the French once the Japanese were ejected. After years of devastating war, the Viet Minh had gained decisive control of northern Vietnam, declaring independence as the Democratic Republic of Vietnam in 1954. At the Battle of Dien Bien Phu, the French were forced to surrender. As part of the cease-fire peace agreement, Southern Vietnam (the State of Vietnam) was officially brokered by world powers, namely the US, until elections which were to be held the following year. It didn't go down that way, but that is a story for another box.

Poor Frenchies. Germany spanked them in Europe. The Japanese spanked them in Indochina. Then the Vietnamese spanked them again in Vietnam. No more French Indochina after that last spanking.

AND NOW THE MODERN CONFLICT

The 20th century was not kind to this region as it became host to conflicts promoting extraordinarily violence and some of the nastiest throw-downs in the last fifty years. Since the Europeans had their fun time with World War I and then World War II, because obviously they didn't get enough the first time around, the world has largely been devoid of huge conflicts with massive death tolls—that is, everywhere in the world except Southeast Asia. Since WW2, we in "the West" were having our Cold War—wherein Team USA faced off with Team USSR, but we never engaged in active conflict. The guns remained cold, hence the name "Cold War." But there were several flare-ups, several hot spots, where people shot each other and where this ideological war reared its ugly head in active conflict. We already mentioned the Korean War. The other major flare-ups are here in Southeast Asia.

Before we can get to the tale of that aggression and transgression, we have to set up the scenario, because some more violence occurred on the soil here even prior to the Cold War, namely during World War II. During the buildup of Japanese power, Southeast Asia and Southern China became a breeding ground of atrocities, mainly because of the Japanese occupation. Some fairly nasty Japanese war crimes took place in these regions, as I've alluded to before.

How did this go down and why is this of particular significance? Southeast Asia has been a crossroads of trade and culture, and became a crossroads of occupation. Unlike any other part of the planet, this is a very unique trait of Southeast Asia: it was taken over by the West and subsequently taken over by the East. This is the only place the Plaid Avenger has ever heard of that had European colonization and was subsequently supplanted by Japanese occupation. Southeast Asia got it from both ends. First the European exploitation, then Japanese expropriation and intimidation, followed by Cold War ideological confrontation; it's a miracle that there is anything left of this region! Let's pick up the story in World War 2. . . .

SMART GODZIRRA!

As you know, Japan built up its empire in the early part of the 20th century. Then in 1931, they started intruding into Northeastern China and took over the Chinese coast. They were still invading, fighting, and conquering into the 1940s. They started taking over the Philippines, and before World War II broke out officially, they had a presence in all Southeast Asian nations. How did they get away with this, and more importantly, how did they get away with this without firing too many bullets? Every country we have just discussed was controlled and occupied by the Japanese, but you don't really hear that much about fighting there.

The reason why it wasn't a big shoot-out for the Japanese to take over this place was because of, the **Greater East Asia Co-Prosperity Sphere** (if you remember back to our lecture on Japan) that was established in 1938. To reiterate what that was about: the Japanese used a savvy piece of propaganda, although they were holding guns at the time too, which helps you achieve your aims

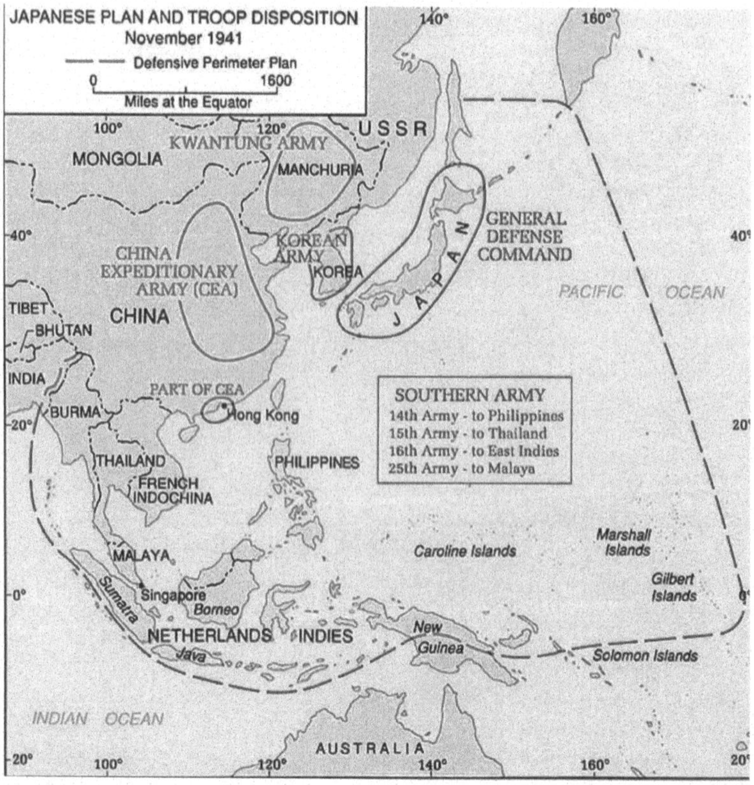

The busy Japanese.

regardless of words. They went to all these places like French Indochina, the Philippines, and Indonesia and said, "Hey, look dudes. Look at our skins. We're Asians. Just like you . . . we're brothers. You've got a bunch of foreign white dogs on your soil that are controlling you. You're just their puppets. You're letting them treat you like punks! Come on, let's all get together as Asians and expel all the Yankee scum-dog pigs!" Many saw the Japanese as liberators at the time and were embraced as such, so the Japanese were able to waltz through Southeast Asia without firing too many shots. Were the Japanese telling the truth? I think we all know the answer to this question: emphatically no! They were actually just planning to replace the European colonizers, which is exactly what they did, in a fairly brutal manner. Work camps, death camps, and prisoner of war camps were set up all over Southeast Asia. So much for Asian brotherhood. . . .

Rounding up people and slaughtering them was something the Japanese did all over Southeast Asia. Southeast Asian nations definitely haven't forgotten this, but they're not as hardcore as the Chinese and the Koreans about it. Older Chinese and the Koreans still really hate the Japanese, and a lot of that animosity has filtered down to younger generations. Some Southeast Asians still harbor resentment, but it's not as strong or as big of a deal to them . . . but it is a big of enough deal to talk about for a little bit. Let's look at a couple of examples of some incredible atrocities perpetrated by the Japanese in Southeast Asia:

Don't mess!

BATAAN DEATH MARCH

We talked about all the brutality they brought to China, and we know about Japanese aggression against the United States in WWII, but a lot of people don't know about Southeast Asia. It's some fairly nasty business, as evidenced by the Bataan Death March, where nearly 80,000 US and Filipino soldiers were marched about 100 miles from one camp to another across the Philippines in blinding tropical heat. Many were not fed, and those who began to fall behind were executed or tortured. All told, approximately 10,000 soldiers were killed.

BURMESE RAILROAD

The Japanese essentially colonized Southeast Asia in the form of an imperial expansion, similar to how the Europeans did long ago. These countries became colonies of the Japanese, but only for a short time. As I suggested when we were rounding up our talk about Japanese movements at the end of World War II, the Japanese were using Southeast Asia essentially as a launching pad for the next phase of their invasion, which was India. This brings us back to the *Bridge on the River Kwai*. In that movie, the Japanese were building a railroad line across Burma to supply their invasion of India. We already know the outcome of World War II: the Japanese lost, and in a strange turn of events, the colonial masters that were kicked out by Japan soon came back into control. Some countries declared independence at this time, but some of the old European colonizers returned to reclaim their territories. Of particular interest was the French reclaiming Indochina, which we already discussed as an event that led up to the Vietnam War. Speaking of which. . . .

Death Railway

'NAM

Why is 'Nam the most infamous of all American wars? Why should it matter whether Iraq is or is not like Vietnam? Why is everyone STILL talking about Vietnam? The answer is fairly simple—the Vietnam War was a complete freakin' debacle that costs millions of lives . . . and for no good reason, only a perceived ideological threat. It was a mistake that pulverized a small Asian nation, while simultaneously ripping apart the social fabric of the US.

The Vietnam War was ultimately a civil war between North Vietnam (communist, led by Ho Chi Minh) and South Vietnam (propped up by outsiders as democratic, pro-Western). After the French were ejected from the territory in 1954, the UN-brokered peace agreement called for a general election across the entire country the following year. All intelligence reports indicated that Ho Chi Minh would win the election both in the north and in the south by overwhelming majorities. He was the main man in Vietnam. So what was the problem? He was also a communist, and we all know how much the US hates commies. By the way, who the hell is this Ho guy anyway? Dig the box below . . .

Who the Heck Was Ho?

The main Vietnamese man! Another "Father of the Country" dude. Take George Washington, add in Winston Churchill, Gandhi, Lenin, and Mao, and you have Ho. When all is said and done, Ho is probably the most fierce nationalist the 20th century ever knew. He dedicated every second of his entire life to Vietnamese independence, and my friends, he is idolized still in Vietnam for his contributions. You'll see references to him as "Uncle Ho;" he's just part of the family!

And what a resume! Ho left Indochina in his late teens to travel the world and ended up in France, where he was one of the founders of the French Communist party. He began the propaganda campaign right there in Europe to expel the colonizing forces from his homeland. He was invited to be schooled in Moscow after their little communist takeover, and was a tireless reader and writer for his vision of a free Vietnam. He also was at one point imprisoned in China by Chiang Kai-Shek (remember him, leader of the Nationalist Party and hater of commies?) Ho organized the Viet Minh forces to counter the Japanese invasion, and then later led those forces to oust the French.

Ho was the president of North Vietnam, and would have easily been elected president in South Vietnam were it not for outside interference. Since the US so despised commies, he was vilified in the US, even before US major involvement in the war. Until very recently, the US was convinced that Ho was a puppet of communist China, despite the fact he had waged a border war against them as well!

Mo' money? No, it's Ho Money!

Ho was an extremely intelligent and savvy dude, but first and foremost, a man of the people. He lived in a cave for a time during French occupation, and even after winning the war, Ho refused to live in the Presidential Palace. He worked in a traditional bamboo and grass thatch hut until he died in 1969, not living long enough to see his vision of a unified and free Vietnam materialize.

He also looked a heck of a lot like Colonel Sanders of KFC fame, but I digress.

The US could not allow the election to go forward, so they stalled, and then helped their hand-picked and sponsored candidates cheat a little. A puppet government was put into place in the south, propped up by the US—actually, a chain of puppet governments led by dictators. As a result, Ho in the north, as well as locals in the south, started agitating for reunification and expulsion of the foreign influences.

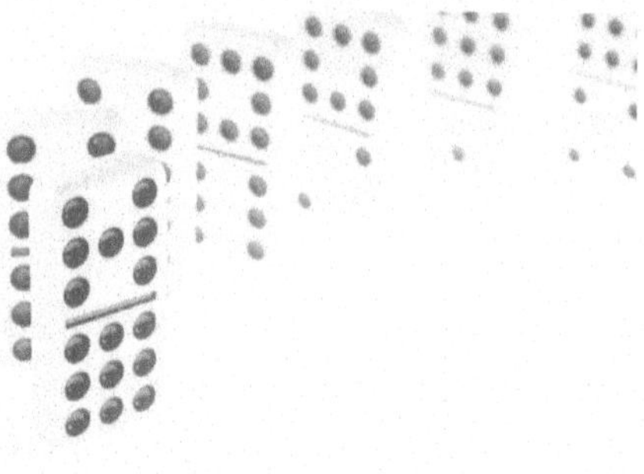

They failed to fall.

Violence flared up in 1957, and then things started to get really nasty. The United States, fearing that South Vietnam would be defeated and become communist, became more actively engaged in helping the puppet South Vietnamese government by sending more money, more arms, and more "military advisors" (read: US dudes with guns). Perhaps the United States' main concern with Vietnam involved the "**domino theory**"—simply, if Vietnam went communist, it would trigger a communist takeover throughout Asia. Once one country went commie, all surrounding countries would fall "like a chain of dominos." No, seriously, they believed this stuff! By 1965, American troops were actively participating in the war by the thousands.

The main goal of the United States was initially to remove the communists from power in the North, but they also encountered a powerful communist guerrilla movement in the South—the **Vietcong**. While the US was supposed to be "protecting" the South, they ended up "fighting" the South as well. Sound familiar? Hello? Afghanistan? Vietcong guerrillas had no trouble blending in with the rural South Vietnam population, who were largely sympathetic to the Vietcong's nationalistic message.

Vietcong soldier.

The United States did many shady things in an attempt to win the Vietnam War. The US sprayed millions of gallons of Agent Orange, an herbicide that is now known to be toxic to humans, to destroy the Vietnamese jungles that helped protect the Vietcong. They also dropped massive amounts of napalm on villages they believed were harboring opposition forces.

In the end, the United States eventually got sick of dealing with Vietnam and left with an embarrassing evacuation of the Saigon embassy in 1975. North Vietnam took over South Vietnam, the whole country became communist, and the world did not end. The US military figured it out only after they suffered over 200,000 casualties. Domestically, the US faced significant lasting impacts as well. The US witnessed substantial unrest from opposition to the war. Student protestors against the Vietnam War were gunned down at universities like Kent State (by the National Guard) and Jackson State (by local and state police) in 1970, which served to widely silence dissent in the US for years following. Many returning veterans struggled with emotional trauma caused by their war experiences, resulting in widespread untreated mental illness and chemical dependency; in 2013, an estimated 47% of all American homeless people were Vietnam veterans. But that pales in comparison to what happened in the war zone: as many as four million Vietnamese (Northern and Southern) were killed, with perhaps only 25% of those being active communist fighters. Herbicide attacks left countless thousands of acres of cropland barren for decades after the conflict, drove away or made extinct a majority of indigenous animal species in those areas, and have been suggested as a carcinogenic agent for the Vietnamese people. Vietnam really got the shaft in this conflict. Worst of all, the US initially got into that giant mess just to help out the French. Ugh. A mistake that will likely never happen again. Freakin' Frenchies.

Perhaps the biggest tragedy is that the entire premise for the war has been adamantly disproven. A communist Vietnam did not set off a row of dominoes, nor did it prove even the most remote threat to US domestic security. Today, the war's purpose has been so disproven that the US has even formalized diplomatic relations with Vietnam despite their remaining communist, and many products in American stores sport "Made in Vietnam" labels.

Vietnam was not the only place for strange regimes and wholesale violence in Southeast Asia in the late 20th century. Indonesia, in its movement into the modern era, has been run by dictators who were supported by the United States. We know Burma was run by an outright military junta for decades, and still represses and kills its own people to maintain its power. Lots of the other countries, however, are leaning toward democracies—places like Singapore, Indonesia, and the Philippines. You have some holdouts from the Communist era, such as Vietnam. But even Vietnam is softening that commie stuff and reintegrating quickly into the world economy. Besides the continued brutality of the Burmese leadership, most places have gotten past their violent histories . . . but one country barely survived its violent origins: Cambodia.

POL POT PIE

Cambodia is the scene of one of the most horrific cases of post–World War II genocide, under our good friend Pol Pot. Pot who? Ah! Read the box on the next page!

What's the Deal with Pol Pot?

As one of the most egregious violators of human rights in recent memory, Pol Pot easily earned the Plaid Avenger title of "Ass-munch of the Decade" for the 1970s. Pol Pot was the leader of Cambodia, or as he called it "Democratic Kampuchea," from 1975 to 1979. The revolutionary movement that brought Pol Pot to power is largely considered the most murderous of the 20th century. Around 3 million Cambodians died from overwork, malnutrition, and state-sponsored executions. This equated to approximately 35 percent of the Cambodian population at the time.

Pol Pot's politics were a brutal offshoot of Maoist Communism. He attempted to tear down Cambodia's old political structure, which was in part established by French colonialism, to make way for an agrarian communist utopia. His tactics included suppressing religion and intellectualism. Citizens were executed if they failed to meet production quotas and also for more trivial matters like wearing glasses, being a monk, being intelligent, speaking a foreign language, being an ethnic minority, etc. Many of those executed were forced to dig their own graves and were then buried alive; many speculate this was done to conserve bullets. Although technically removed from power in 1979, Pol Pot continued to exert brutal influence on Cambodian politics, in one way or another, until his death in 1998.

See a depiction of this regime by watching the really historically accurate film *The Killing Fields*. Whew! Talk about a place that *should* have been invaded! Oh wait, the US already had . . . in 1970, the US invaded Cambodia as part of a controversial military campaign against what President Nixon believed was North Vietnamese activity around Phnom Penh. Some historians cite the invasion as a destabilizing event that ultimately brought the Khmer Rouge to power. Of course, this is the very military action against which Kent State and Jackson State students had been protesting against when they were mowed down that year. Oops . . .

But enough of the death and destruction, let's get to why these places are hot in today's world. I'm not talkin' tropical heat here, I'm talkin' cash money!

CUBS IN THE CATBIRD SEAT

Cubs? What? I'm talking about the cats. The tigers. These Southeast Asian economies that are on the hunt for a better way of life, and on the prowl for development and economic growth.What is happening in today's Southeast Asia?

This has a lot to do with things we have already referenced. There is Chinese influence, which is still big and may be growing as China's influence expands in the world. China's exploding economy is certainly having a positive effect on next-door-neighbor Southeast Asia: natural resource exports to the giant are going through the roof, and international investment is flowing in to the region to support increased manufacturing and services sector activity. Southeast Asia is now competing with China, and in some cases winning!

But that is not the only Chinese influence on this region. Lots of people emigrate from China to Southeast Asia, set up businesses, and bring their families along. Is this good? It can be good. The Chinese often bring a lot of capital and set up businesses, but the negative is that it's the stuff the locals were probably going to do.

Indeed, there is some friction between ethnically Chinese people and all the other ethnicities within most Southeast Asia. The Chinese are a major minority ethnic group within Southeast Asia states. Cause for concern?

I don't know, but certainly cause for friction. Indeed, when local economies start to suck, say if Indonesia goes into a depression, one of the first things all you news watchers will see are Chinese businesses being attacked. Chinese individuals become kind of a whipping post of anger that people rally around: "Curse the Chinese, they're here taking our jobs!" The Chinese typically don't assimilate themselves into the culture of the countries to which they emigrate either. Throughout Southeast Asia, there are visible pockets of Chinese influence and businesses all over the place. Pretty much you can find a distinct 'Chinatown' in every country in Asia . . . if not several of them.

WHAT THE H? WHAT THE WHAT? WHAT THE WHAT? WHAT THE WHAT? WTH?

NUMBERS OF ETHNIC CHINESE IN SOUTHEAST ASIAN COUNTRIES

Country	Ethnic Chinese	% of Local Population
Thailand	9 million	14%
Indonesia	7.5 million	3.1%
Malaysia	7 million	25%
Singapore	2.8 million	76%
Vietnam	2.5 million	4%
Philippines	1.5 million	2%
Burma	1.5 million	3%
Cambodia	200,000	1.3%
Brunei	60,000	15%
Laos	50,000	1%

What else is going on in Southeast Asia today? Economically, it's a bustling and booming place, which is where the *Cubs* come in. You might have heard of something called the **Asian Tigers**, or **The Economic Tigers** in reference to Hong Kong, Taiwan, South Korea, and Singapore, which exploded economically since the 1970s and 1980s. Growth rates were astronomical in these places, making them very rich. Those expansive growth rates are now spreading to places like Thailand, Vietnam, Malaysia (to a certain extent), the Philippines, and Indonesia. That's why I'm calling them the *Cubs*. They're on their way to becoming tigers themselves, but they're just getting started.

Why are these places in Southeast Asia picking up economically right now? One of the reasons is cheap labor. Most businesses go to China for this reason, but if they don't go to China, they go to Indonesia, Thailand, or the Philippines. Because of this cheap labor and more lax environmental laws, a lot of businesses and multinational corporations are relocating to Southeast Asia, if not outright starting whole new businesses there. I'll go out on a limb and say that all car manufacturers have a plant or two in one or more of the countries in Southeast Asia.

And car manufacturers are not alone. Nike has been in Indonesia for a decade or two now. Apparently, it makes more economic sense to ship leather from Brazil, cloth from China, rubber from Vietnam, and then manufacture it using cheap labor in the Nike plant in Indonesia, before shipping it to the United States, where you can still sell the shoes for $20 and still make a profit—but you charge $150 and make more profit. That's the way the economic world is working right now, much to the advantage of multinational corporations, and the Southeast Asian nations, to some extent. Some of the losers in this process are places like Mexico. Some places in South America used to be cheap suppliers of labor, but they have gotten a little bit more expensive, so labor is moving over to Asia. Companies go wherever the labor cost is the lowest. How low can you go? Southeast Asia and China are winning the limbo right now. And if you want to take that cheap labor to an extreme, order some take-out labor from the Philippines. . . .

The other factor in this is that Southeast Asia is a massive supplier of raw materials to the world. Where are the places that want raw materials more and more? We say, "Oh! The US and Western Europe, the advanced countries. We use all that stuff." Wrong! Think again. The world is changing. China, South Korea, and Japan are huge importers of raw materials. When I say raw materials, I mean everything from oil to rubber to trees to tin to bauxite—everything. Southeast Asia has a lot of this stuff, and even the stuff it doesn't have goes through their waterways and shipping channels.

Lots of stuff has gone to growing Asian economies, and as a result, the Tigers, like Taiwan, have become big processing centers. Wait a minute, who is the new big tiger? Who is the 800 pound tiger? Who is the 800 pound gorilla on top of the 800 pound tiger? That's China. China is now pulling in resources from all across the planet with an unquenchable thirst, and a lot of those things are coming from Southeast Asia, which is right next door. If they have something, China will

Filipino Take Out?

So, Southeast Asia is going gangbusters with cheap labor and selling exports, but what if you literally put the two together? Well, the Philippines has done just that: they've become the Pez dispenser of cheap labor for other countries around the world. Thanks to the incredibly devalued Filipino peso, countries that can't fill the crappiest jobs have figured out that it's cheaper to import temporary workers from the Philippines than to hire locals. Its colonial history under US rule means that many of the Filipinos who work overseas speak English, the global lingua franca, making them even more valuable on foreign labor markets. An estimated 2.2 million Filipinos, nearly 3% of the country's population, were employed as overseas temporary workers in 2012. And the kicker: the government of the Philippines encourages the workers to go! Huh?

Well, the Philippines sees several advantages to this situation. Workers send remittances, overseas earnings sent back to family members, which brings more money into the economy. By being out of the country, the overseas workers shrink the domestic labor market, raising wages for workers who remain behind. Plus, the folks who leave aren't consuming any government services while still contributing tax revenue . . . so this is a win-win-win, right?

Filipinas: the hostess with the most-ess.

Not quite. Filipino workers fill the jobs in wealthy countries that no one wants to do. The largest recipients of Filipino labor are those countries needing workers specifically for dangerous and physically taxing jobs. Saudi Arabia, the largest importer of Filipino workers, tends to employ them as oil drill workers. Dubai, on the other hand, employs most of its foreign laborers in construction. Odder still, Japan imports mostly Filipinas (the "-as" means females) to work as "hostesses" in night clubs.

Functionally, gas would be more expensive, Dubai's skyline would be far lamer, and Japanese businessmen would be much lonelier without the readily available workforce from the Philippines. However, because they're foreigners, they usually aren't afforded the same rights as the locals. Employers abuse the living hell out of the foreign workers. In both Saudi Arabia and Dubai, seven-day workweeks with 12-hour days are the norm for foreign workers, and most are required by their contracts to live in substandard, expensive housing provided by the employer. In Japan, allegations of sexual assault against Filipinas by employers are common, and "hostess" is used as a derogatory term for "sex worker" even though Filipinas rarely enter the sex trade. In any case, because employers hold the workers' visas, overseas Filipino laborers who resist this exploitation are threatened with withheld pay and ultimately deportation.

So, why do they go in the first place, and why do they stay? The pay, which is low by the standards of the host country, is still far better than these workers would earn in similar jobs domestically. A two-year contract overseas can provide enough money to start a business or stabilize a family's finances. Weirder still, because the peso is so devalued, even highly trained Filipinos like doctors can make substantially more money overseas working as physical laborers than they can from doing their voodoo that they do domestically. This results in brain drain, where skilled workers leave a place for greener pastures elsewhere, which is not good for any economy in the long term. So while the Philippines benefits from exporting cheap labor, there are definitely consequences to the system.

Gorillas and Tigers: these guys are all kicking tail!

use it. It's very similar to the situation between the US and Mexico. Mexico can't possibly produce enough stuff to satiate the US market. Southeast Asia is exactly like this right now and they don't have enough wood, enough rubber, enough aluminum, or enough vegetables to placate China.. They can't possibly have enough. China will import it all. It is a phenomenal position for Southeast Asian exporters to be in, and it's one of the reasons why their economies are starting to climb out of the cage, and one of the other reasons why I'm saying this litter of cubs will soon be tigers. Claws up! RAWWWRRR!

This region is growing, but before I go any further—is this place bustling to the point that everybody is getting wheelbarrows full of gold and becoming fat-belly rich? Of course not. One of the reasons why these countries are getting rich is because of cheap labor, typically meaning a lot of the wealth isn't filtering down to the lowest classes. There is endemic poverty here, lots and lots of poverty. But economically speaking, looking at GDP and foreign investment, they're starting to do pretty well. That doesn't mean they're out of the hole yet; they still have a long way to go. One of the things that may help them go a long way is the Association of Southeast Asian Nations (ASEAN).

ASEAN

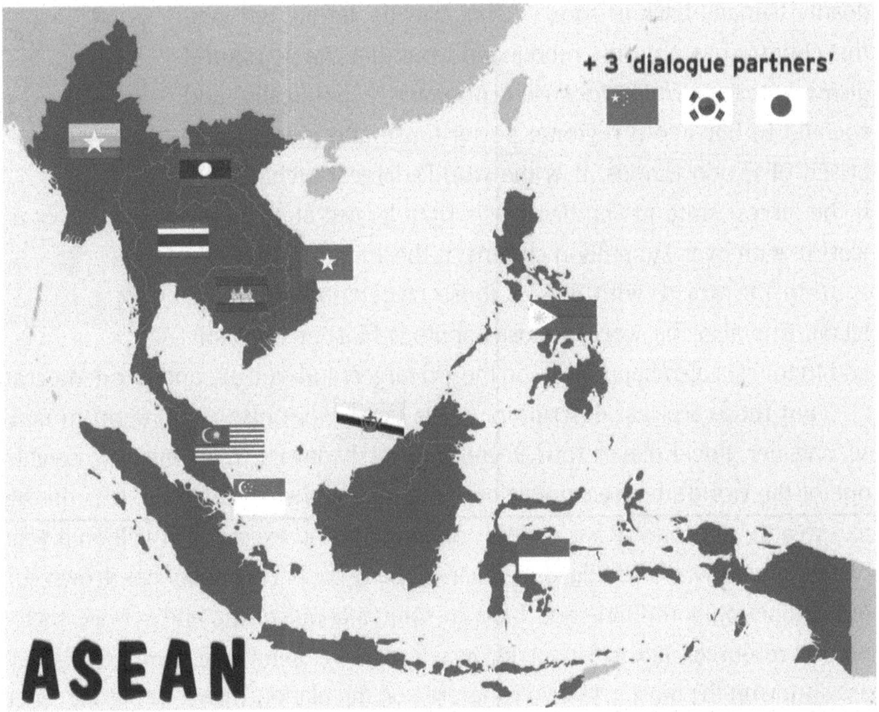

I can't get enough of ASEAN because it's setting up—and this is a Plaid Avenger insider tip—to be one of the most successful trade blocs on the planet. One of the things that may take them to the level of development, and I'm talking about the region, is that ASEAN may become more than just a trade bloc. What is **ASEAN** then? ASEAN is a trade bloc, a free-trade zone that consists of all of the countries in Southeast Asia. Here's the kicker: it's becoming so popular as a free-trade zone and is starting to incite so much economic activity that other countries want to get in the club. Yes, there are still some countries that want to get into the EU, but it's getting close to the end of its growth as a trade bloc. Let's not even mention the FTAA, because that idea is played out and probably won't happen in the Americas. All the other trade blocs are approaching their limits, but ASEAN is in position to keep expanding. How so?

Well, ASEAN is so hot and such an epicenter for economic activities (both importing and exporting), that some other Asian brothers want in on the action too! There are now three 'associate members' or 'Dialogue Partners' and they are China, Japan and South Korea. Dudes! China is the #2 economy on the earth, Japan is #3, and depending on the year, South Korea is #11 or 12! This is a group with serious economic clout! When all these cats collectively meet, it is referred to as **ASEAN+3**.

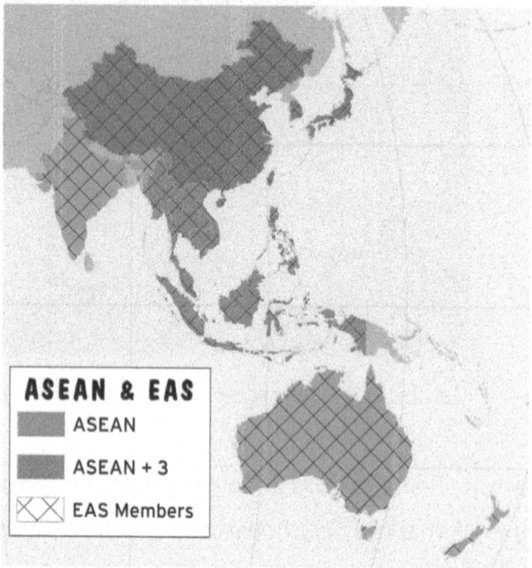

And it gets even better for Southeast Asia: now ASEAN has become the core of the annual **East Asia Summit** which is the ASEAN+3 plus India, Australia, and New Zealand! The summit discusses issues including trade, energy, security and regional community building. Is this the early formation of an Asian EU? Who knows! But it's a fascinating development! And ASEAN is in the center square. Dig the map details to get all your ASEAN acronyms straight.

Let me stress the economic awesomeness one last time: this ASEAN group is sitting pretty in terms of relative location for trade and everything else. A billion consumers in China on one side, another billion consumers in India on their other flank, and top 20 economies like South Korea, Japan and Australia all wanting in on the ASEAN action too. It looks like the region is returning to its old position of being in the center of the global economy once more. And there is one player in this group that may on its own take it to the next level and become a a global power in its own right. . . .

REGIONAL TITAN ON THE RISE: INDONESIA

You may have only heard of Indonesia as that place that got whammied by the massive earthquake and subsequent deadly tsunami back in 2004 . . . but friends, let me tell you, this country has not only rebounded from that titanic natural disaster, it is storming forward economically, politically, and socially to become a regional powerhouse and more! Comprised of 17,000 islands, it is the world's largest archipelago; it the largest state in Southeast Asia (15th largest state in the world); with over 250 million citizens, it the 4th most populous state in the world; with 86% of those citizens adherents of Islam, it is also the world's most populous Muslim state; oh, and Indonesia also happens to be the 3rd largest full-fledged and fair democracy in the world!

But those are just descriptors of big-ness! What else is going on to make Indonesia a massive regional power and second tier global player too? Regularly overshadowed by its gigantic neighbors China and India, Indonesia is suddenly one of the world's best economic opportunities of the next decade. It is the tiger in the middle of titans! With much tiger-like stealth, the country has quietly grown its GDP in excess of $1 trillion a year, making it the 15th largest economy on the planet, and now a full-fledged member of the G-20. It's economy has grown solidly 5–7% every year for a decade, partly by out-competing both China and India in some manufacturing and service sectors. They also have massive agricultural and natural resources, and are investing heavily in developing their geothermal energy potential—as well they should, since they are sitting on the most active volcanic area on the planet. Turn it into green energy, baby!

Jakarta is jammin'!

International investment is flowing in, there is a vibrant and growing middle class, their stock market is performing awesomely, their currency is stable, and external debt has even declined. Boom! Growing fast, and not via debt! But there is even more. The peeps of this place have a growing sense of self-confidence, nationalistic pride, and economic drive. Their educational and infrastructure systems are lacking, but they are working to improve them, and fast. And while they banned Lady Gaga from performing in their country, Indonesians are ga-ga for technology and have embraced social and mobile devices to the max. A huge number of peeps are on the internet daily; there are 250 million mobile phones in the country (yeah, that's more than the total pop! Many Indonesians have 2 or 3 phones!); 76% of internet users are Facebookers—making Indonesia in third place behind only the U.S. and India user rankings. And they are some mad Tweeters there too, as they are ranked #1 in terms of numbers of vigorous regular tweet action!

A possible new entry for the BRIC group? Could be! Either way, Indonesia has got it going on economically, technologically, and socially, and they have their eyes on the prize. The eye of the tiger that is! Watch out Asia, this tiger is charging!

BUMP IT TO THE NEXT LEVEL, BABY

So Indonesia is a rising star within the region and within ASEAN, but that may be due partly to how ASEAN itself is expanding to become oh so much more than just a trade block. They're starting to turn into something similar to the EU. It's only in the early stages of this, mind you, but this transformation is becoming apparent already. Visit the ASEAN web site to see what I'm talking about. You'll see plans for hydroelectric power as part of a proposed ASEAN-wide power grid. Let me say that again: a joint electric grid for the entire region! That's crazy! Europe doesn't even have that, to my knowledge. In addition, if they're already talking about joint infrastructure in terms of roads and bridges and ports, and they already have a common air space when it comes to all their aeronautic.These are the types of things that might simultaneously raise the standards of living for every country, and there is even talk of an integrated healthcare system across the region.

And the most unique feature of them all? The group developed their own declaration of human rights that ALL member states have agreed to! Even the 'bad' states like Burma agreed to it! Wow! No other trade block has anything like that. Now are you starting to understand that this group has aspirations that are much larger than simply promoting trade? This is looking like a cohesive unit on multiple fronts of economy, infrastructure, trade, and culture in general. It's hard to tell where they're going with this, but they're already certainly integrated in economic terms, and they don't have many qualms about free movement of people either.

ASEAN is setting itself up to be something a whole lot more like the EU with maybe even less friction between the countries involved. This will, dare I say, be a stabilizing force within a region with a track record for serious conflicts. ASEAN has become a savior to these countries. It's hard to see any intraregional conflict cropping up here in the future, because once these countries become integrated in every way that they've proposed, conflict will be quickly resolved and made nearly impossible. Soon, nobody will be able to dredge up a reason to fight. Are there people who are unhappy in these countries? I'm sure there are unhappy people all over the place, but there are no unhappy governments in this deal.

We talked about power grids, hydroelectric systems, and possible healthcare systems. I'll go out on a limb and say that the regional stability that ASEAN provides may indeed be the impetus for a common military. No one has talked about that yet, but the Plaid Avenger is willing to speculate a bit. All I know is that it's a very likely possibility in the years to come. Here is a little quote for you: "Today, ASEAN is not only a well functioning indispensable reality in the region, it is a real force to be reckoned with far beyond the region. It is also a trusted partner of the United Nations in the field of development." Who said it? Kofi Annan said it, former Secretary General of the United Nations, on February 16, 2000. A thumbs-up from Kofi can only mean good things. How awesome is Kofi! I love the Kofi.

TERRIBLE TERROR POTENTIAL

We'll end on a negative note here, unfortunately. Terrorism is a real force to be reckoned with right now, as the US-led War on Terror is quite active around the planet. This War on Terror is definitely not a Cold War. It's quite hot at the moment. One of the regions that experts worry about, but you don't seem to hear about, is Southeast Asia. Why would you have a War on Terror here? Indonesia is the largest Muslim state in the world, and as such, it's considered a hotbed of extremist activity. For those of you not in the know, several Western hotels and buildings have been blown up just in the past few years here. A bunch of Australian tourists were killed in a fairly nasty blast at a tourist hotel not long ago as well. There are quite active terrorist groups in the southern Philippines, Malaysia, and other places in Southeast Asia.

Terrorism in Southeast Asia has to do with physical geography as well, because these very separate islands provide cover for lots of secret stuff to go on. There are a lot of terrorist cells within Southeast Asia; because they can get on small islands and operate fairly independently without anybody checking them out, they're completely under the radar. I can tell you without reservation that Australia has been very worried about this for a while. They're probably the next country that will be outright attacked. They spend a lot of money and time trying to figure out where terrorists are; in case it's not clear by now, they're looking at Southeast Asia. They aren't looking in the Middle East, that's for sure. None of the terrorists there are a threat to Australia. However, there are some radical groups within Indonesia and in the southern Philippines. There is a rebel/terrorist group that has been kidnapping tourists for quite some time. Check out the news; you don't have to look too hard to find them. They're called **Abu Sayyaf**.

You can rest assured that Uncle Sam has been checking this out. The United States and everyone involved in the War on Terror has operatives checking out movements. Another reason for worry is clashes between Christians and Muslims, which happened notably in the small island group called the Maluku Islands in Indonesia.

Southeast Asia has some negative things to deal with nowadays, terrorism being the big one. Australia, being a power player in the region, has taken on most of the workload for the anti-terrorism movement and is supported by the United States and Europe. This offends some countries in Southeast Asia, namely Indonesia and Malaysia. Those countries get really hot and bothered about foreign influence, even if it's on issues concerning terrorism.

ROUND-UP

Terrorism: bad. ASEAN: good. Southeast Asia: rockin'. Thundercats: Ho! This region has tremendous potential for economic growth as it becomes the newest workshop of Asia producing tons of manufactured goods. Its placement between India and China makes it well-suited for both production and distribution of a variety of goods and services to Asia as a whole. The ASEAN grouping is set to explode into the 21st century, having put its century of conflict well in the rear view mirror.

Unfortunately, the region also has explosive potential of other kinds. Volcanic, earthquake, tsunami, typhoon, and monsoonal activity are a very real part of life for Southeast Asians, and one that will never go away. In fact, it seems to be increasing lately! This region continues to get pummeled with the most horrific natural disasters year after year. Coupled with Mother Nature's ire is the terrorist potential which seems to permeate many areas in this widely dispersed (and hard-to-effectively-control) archipelago arena.

The upcoming century will still be hot and active for Southeast Asia and ASEAN. Let's hope it's hot beaches and hot economic growth versus hot lava and hot spots of conflict. Shout out to Aung San Suu Kyi: Keep fighting the good fight! We love you!

SOUTHEAST ASIA RUNDOWN & RESOURCES

View additional Plaid Avenger resources for this region at http://plaid.at/se_asia

BIG PLUSES

- ASEAN, ASEAN, and ASEAN; including associate members China, India and Japan and South Korea; possibly the largest and most powerful trading block of the future
- With the exception of Burma, all countries are currently fairly stable and increasingly democratic (although some countries have much farther than others to go on the route)
- A significant producer of raw materials and increasingly basic manufactures . . . even Chinese businesses are located here to take advantage of cheaper labor and resource access
- Relative location is absolutely phenomenal: the region is dead center in the major trading routes between China/Korea/Japan and Middle East/India; trading routes are only getting busier as the Asian economies explode

BIG PROBLEMS

- Ring of Fire!!! Region is an epicenter of earthquake, volcano, and tsunami activity! And they get typhoons too!
- A century of chronic conflict, horrifically bloody violence, schizophrenic dictators, and ideologic showdowns have scarred the region and its peoples deeply; while stable now, the healing process has a long way to go for many
- While not the poorest region on the planet, Southeast Asia does have a lot of poverty, overall lower standards of living, weaker infrastructure, and lack of education/health care for many

DEAD DUDES OF NOTE:

Ho Chi Minh: "The George Washington of Vietnam" Led the 1941–54 independence movement to expel the French, and established the communist-governed Democratic Republic of Vietnam (in the North) of which he was President until his death in 1969. "Uncle Hồ" is highly revered in Vietnam to this day.

Pol Pot: The "Hannibal Lecter" of Cambodia. Led the psychotic Cambodian communist movement named Khmer Rouge which took over the state in 1975, at which time Pol Pot named himself "brother number one" and reset the calendar date to the year 0. From 1976–1979 he "cleansed" the country 2 to 3 million civilians. One sick puppy.

LIVE LEADERS YOU SHOULD KNOW:

Aung San Suu Kyi: The totally awesome Nobel Peace Prize winning leader who has fought tirelessly for democratic reform of Burma for three decades, and who has spent most of the last two decades under house arrest for her actions. Was finally freed and elected to the lower house of the Burmese parliament in 2012 and is currently the de facto Prime Minister.

Susilo Bambang Yudhoyono: Former pragmatic President of Indonesia (#18 economy, #16 in size, #4 in population). Retired Indonesian Army general officer, well-educated, centrist, conservative and helped establish Indonesia as a solid democracy. While a strong ally of US, he is also extremely open and diplomatic to all world powers, especially China. Good guy.

Joko Widodo aka "Jokowi": "New blood" President of Indonesia, former Governor of Jakarta. First president to not come from a rich family. Jokowi is widely seen as reflecting popular voter support for "new" or "clean" leaders rather than the "old" style of politics in Indonesia.

PLAID CINEMA SELECTION:

Just Follow Law (2007) A comedic film, plays a on a phrase that means simply that one should just follow orders without asking why. Does a phenomenal job of taking jabs at the bureaucracy and taking a look at some outdated laws in Singapore that tend to cause problems in modern society.

Merah Putih aka *Red and White* (2009) A wonderful wartime film that highlights the difficulties of Indonesia to fight the Dutch and gain their independence. The film shows how soldiers from all over Indonesia had to band together and put aside their differences for the fight for independence. One of the only films that tries to accurately portray this war from the Indonesian vantage point.

Rescue Dawn (2007) Based on a true story, film tells the tale of a US Navy pilot who was shot down over Laos in 1965. The film details his unwillingness to denounce Ameica and his subsequent torture and escape across the Mekong River over into Thailand. Film was shot over the course of 44 days in Thailand and presents real life environments in Southeast Asia.

The Killing Fields (1984) Intense, intense, intense, factually correct and true story of a New York Times reporter and his Cambodian assistant who were on the scene when Pol Pot and the bloody Khmer Rouge took over and self-destroyed the country in an insane bloodbath that claimed 2–3 million civilian lives. Highly recommended.

Beyond Rangoon (1995) A fictional story about a troubled American woman is merely the vehicle to present this depiction of the current state of Burma, and the even includes historical reference to Aung San Suu Kyi and the bloody 8-8-88 massacre. It may be 15 years old, but still is reflective of today's situation, and stars Patricia Arquette back when she was hot.

The Year of Living Dangerously (1982) A young Australian reporter tries to navigate the political turmoil of Indonesia during the 1965 Indonesian "civil war" which ousted the leftwing commie dictatorship rule of President Sukarno and implanted right-wing military dictatorship of Gernal Suharto who ruled for 32 years. Great background to understanding today's Indonesia, and stars Mel Gibson back when he was hot.

Apocolypse Now (1979) A pop culture phenomenon in and of itself, this epic Vietnam War film in which a covert operative is sent on a dangerous mission into Cambodia to assassinate a renegade Green Beret who has set himself up as a God among a local tribe. Has more stars than you can shake a stick at, and the film itself is legendary just in its problems of production, including pretty much everyone being whacked out on drugs the whole time.

Rambo (2008) I think this is like the 4th Rambo movie, but they just call it Rambo for some reason. Oh well. In Thailand, John Rambo joins a group of mercenaries to venture into war-torn Burma, and rescue a group of Christian aid workers who were kidnapped by the ruthless local infantry unit. Sorry, this is my guilty pleasure: watching the despicable Burmese military get blown up really turns me on.

In the Year of the Pig (1968) Extremely well-crafted, well-researched, and controversial documentary on the Vietnam War made while the war was at its height. I only recommend this one to folks who want a solid understanding of the conflict itself, or filmmakers, or rabid anti-war, anti-imperialist provocateurs, because this film does not show the US in a good light. I'm actually surprised the filmmaker was not knocked off by the CIA.

MAP GALLERY

View Map Gallery & additional Plaid Avenger resources for this region at http://plaid.at/se_asia

THE BATTLE FOR BURMA
HELP FREE AUNG SANG SU KYI

PART THREE

THE GLOBAL WRAP UP

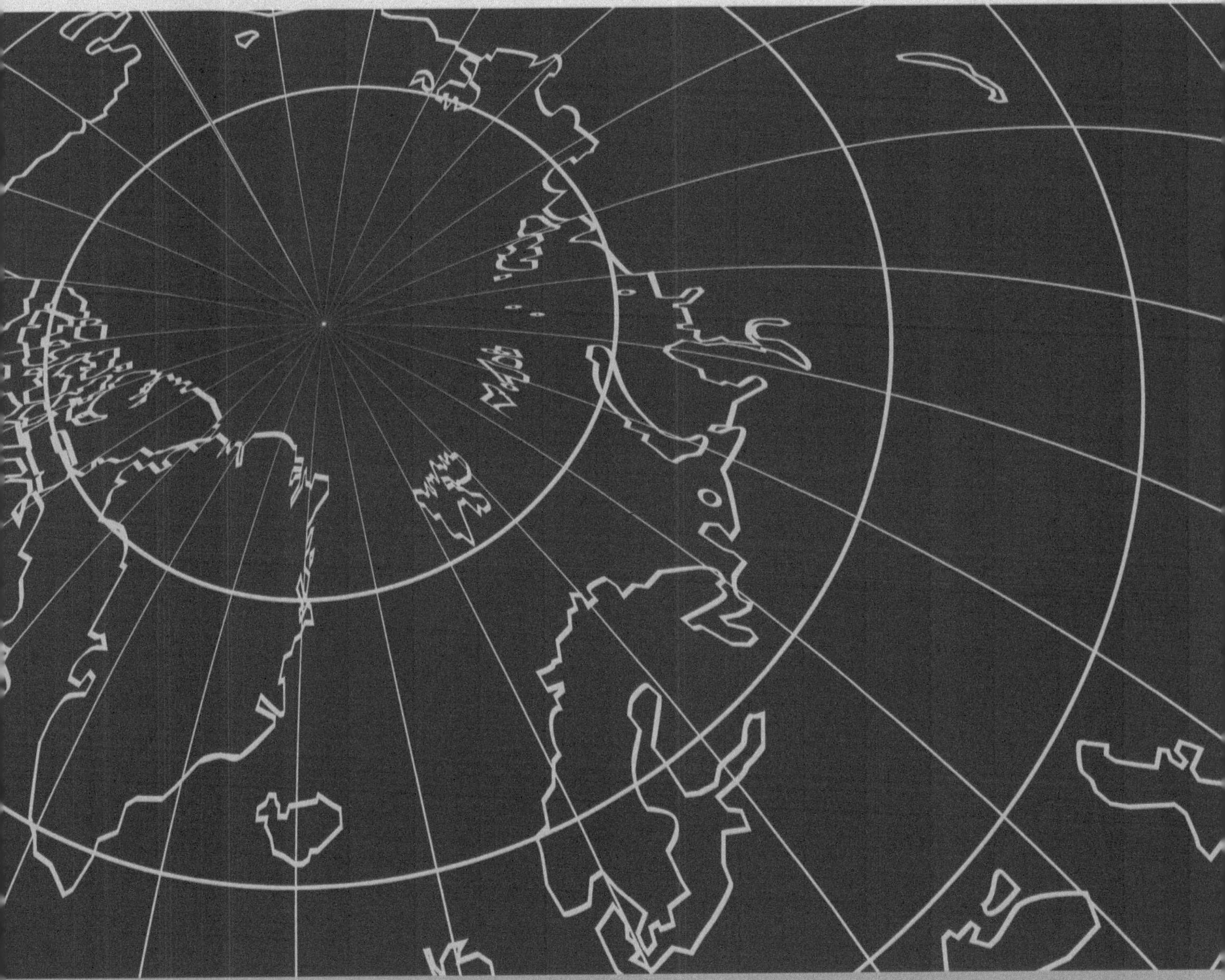

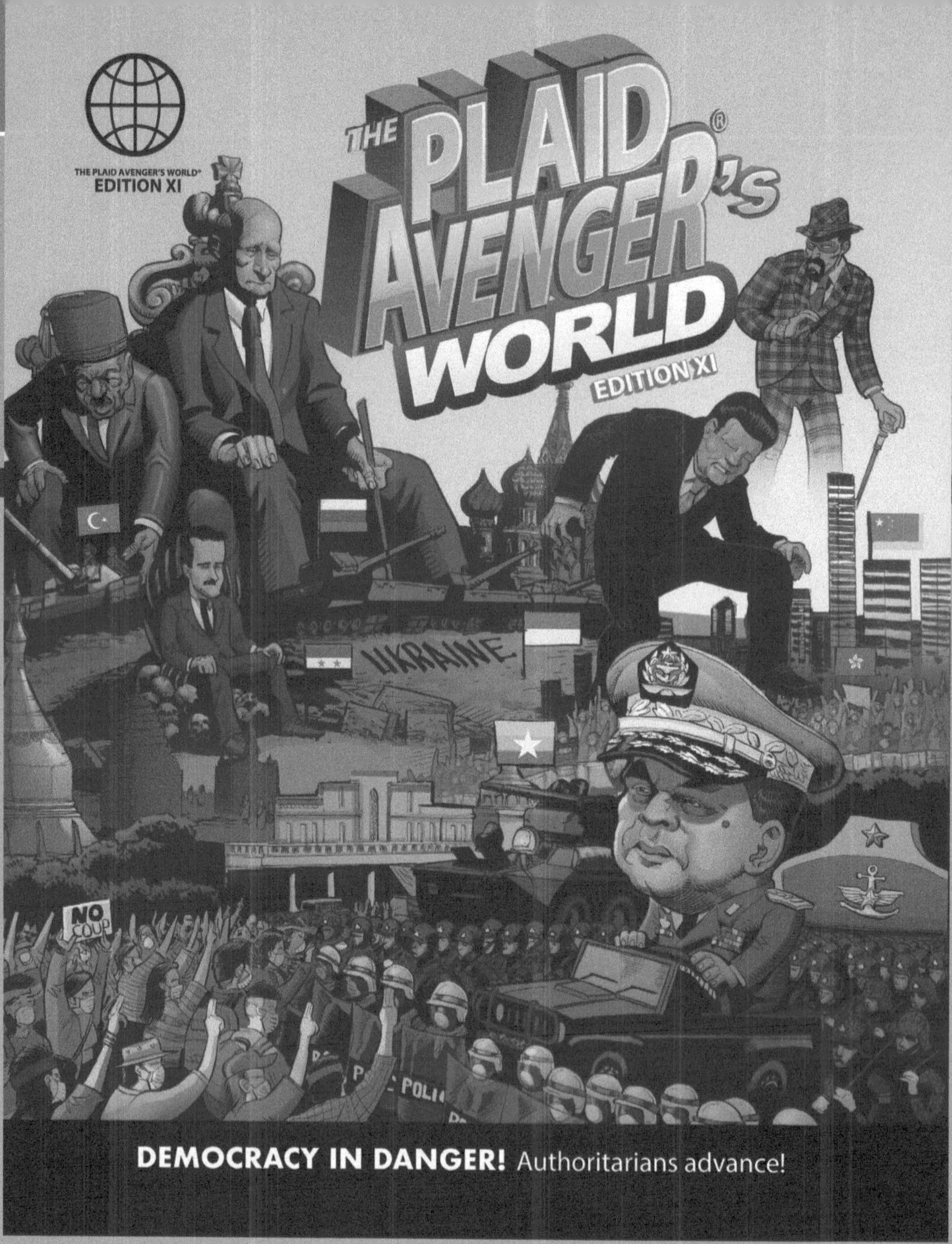

CHAPTER OUTLINE

Democracy in Danger! Authoritarians advance!

25

AUTOCRATS ARISING EVERYWHERE!

A general seizes power from an established democracy and jails all the elected leaders?

A respected president of a country uses military force to suppress a minority ethnic group?

A right-wing demagogue whipping up white nationalist frenzy in order to gain popularity and power?

A legitimate election in a well-established democracy denounced as fake by the previous president, which incites a riotous response from frenzied followers?

An international treaty is totally trashed by a major world power, paving the way to purge a democracy of the people?

A right-wing populist in Europe comes to power by focusing on the failings of liberal democracy and the appeal of bringing back the "good old days" based on Christian identity, national sovereignty, distrust of international institutions, and the purity of their peoples?

Multiple democratically elected leaders use their power to destroy political opposition, curtail or outright control press freedoms, and entrench themselves as leaders for life?

Multiple democratically elected leaders rally their citizenry to support closing down borders and stopping immigration in order to keep out "the others?"

Autocratic Aryan Ass Extraordinaire

Wow! What is this list all about . . . perhaps a historic review of the lead-up to World War 1, or World War 2? The rise of the fascism in Italy and Nazi Germany . . . those Mussolini and Hitler dudes? The militarization of Japan, or the Warlord Era in China? The Russian Revolution . . . or is it the Communist Revolution . . . or is it the rise of Stalin? Is someone getting ready to assassinate Archduke Franz Ferdinand or something? What the heck is going on, and when did all of these things happen?

Well. Buckle up for the hard truth: this is just a snapshot of the last several years! It's like now! All of these events and a heck of a lot more have been occurring across the entire planet, on multiple continents, in dozens of countries. Democracy and liberal democratic institutions have really taken it on the chin as of late. Some democracies have been outright overthrown. Many more democracies have seen the popular rise of radical-right groups intent on changing the very nature of those democracies. And even more democracies than that have witnessed internal tinkering by particular political parties, which is resulting in the concentration of political power into fewer and fewer hands. Democracy is on the decline worldwide for the first time in a century.

Does that mean that fascist or communist or religious zealots or military men are taking over governments everywhere? Eh, it's a little more complicated than that. In some cases, yes, democracy is being jettisoned altogether. However, the worldwide trend I am referring to here is happening across the great variety of government types. It's a concentration of ultimate power into the hands of smaller and smaller groups, and fewer and fewer people . . . culminating in absolute power held by a single, dominant individual. And this is occurring in monarchies, military governments, theocracies, one-party states, and, yes, even in full-fledged democracies . . . which of course challenges the very notion of democracy itself. From Africa to Asia, Europe to the Americas, powerful men (and yes, I meant to say men as it's an exclusive group) have been concentrating power to themselves using outright force, economic coercion, constitutional changes, political propaganda, and/or rabid rhetoric which rallies those responsive to their message. And that message, with very little variation from these leaders, is simply this: *I can fix everything. I have all the answers. I can make everything right again. I can save you. I should be in charge and be making all the decisions. Forever.*

This is a global march towards **AUTOCRACY**. A-ha! That's the keyword contained on the very cover of this 11th edition! And autocracy is arising in so many places simultaneously here in the modern world, that the Plaid Avenger picked it as the most impactful trend of the decade to come! That's right! More than Covid, or the climate, or a variety of other political/economic crises, autocratic expansion will affect the lives of billions in very real ways . . . and is set to change the balance of power on the planet this century! That's big stuff indeed! Let's get to it, and we'll have to start by asking the very simple question . . .

WHAT IS AUTOCRACY?

Well if you Google it or Wikipedia it or look in a dictionary (do they still have those?) you are likely to find a pretty simplistic answer that generalizes autocracy as "a system of government" in which absolute power over an entire state is concentrated in the hands of one person, who is above the law, and who controls all the decisions about the state completely, with little regard for the ideas, advice, or opinions of others. Well . . . that also seems like the exact definition of a despot, of an authoritarian, of a totalitarian state, of a military dictatorship, of a fascist leader, and perhaps even of an emperor, Caesar, or tsar. So autocracy is just a general classification of any ruling system in which a single person is totally in charge?

NOT SO FAST . . .

Eh. That's a bit simplistic in my view, because we can also look at individuals that exhibit elements of autocratic leadership, without being the ultimate power in a state. A legitimately elected president or prime minister can exert tremendous autocratic power over their citizens, and do so while completely ignoring advice of others, but that does not make them a dictator. And even on a more personal level, in everyday life, a boss, a football coach, or even a parent can all behave like autocrats. Why do I have to clean up my room, mom? Because I said so! So anyone in a position of power, even at a small scale, can have autocratic tendencies.

Autocratic Psychopath Extraordinaire

That's why I want you to consider the term at a deeper level with multiple layers of nuance. Sometimes autocracy actually is a tangible person like a dictator named Hitler. Sometimes autocracy is a tendency to behave a certain way, like Mr. Johnson, your sadistic high school gym class coach making you do a million push-ups. And sometimes autocracy is a trend or a process by which this ultimate power is actually being accumulated into the hands of a single person . . . long before that person actually becomes the dictator! In other words, we can look at politicians, political parties, or entire governments as exhibiting autocratic practices . . . and if they continue to do so, and continue to do more of them, they will eventually become one of those named political systems (like fascist or dictatorship) we would put under the full-on autocracy umbrella. It is this *autocratic process* I wish to focus on here, as that is the current trend erupting in current events as of late, all over the world.

For instance, Russia's 1917 Communist Revolution was supposed to be all about implementing a system in which all citizens participated in, but had autocratic elements in place right from the start. Of course, Lenin and his crew understood this and told their citizens that a few powerful leaders needed to be in charge just temporarily while they "fixed" everything to make it utopia. In other words, they needed the autocratic elements in place to facilitate the democratic/communist change. But instead, it became more and more autocratic very quickly after Lenin died, and the more aggressive Stalin took over and purged out all political contenders. He then justified absorbing more power to himself as necessary to conduct World War 2, and by the end of that conflict, he just kept on going with the autocratic evolution . . . which is why in the West (and in hindsight of history) it seemed obvious that he had turned the USSR into a full-fledged dictatorship. One autocratic step at a time.

IT'S A PROCESS . . .

It's a process that the Plaid Avenger is going to define as the subtle—and in many cases legitimately "legal"—accumulation of power into fewer and fewer hands, with fewer and fewer alternative opinions or voices being considered when forming the policies of a state. Yes, it's ultimate expression is of a single person in charge of everything, but I'm going to expand this to include situations where a small group of powerful people (oligarchy) or a single political party or single military group (junta) or even a single powerful corporation (plutocracy) increasingly calls all the shots, with increased disdain for any political opposition, even if that opposition forms the majority opinion within the state. This autocratic movement is increasingly defined by its willingness to openly grab power, attack freedom of the press in order to control all information, change constitutions and laws to suit their needs, and actively marginalize all opposition parties or public opinion. In other words, the process is working overtime to ensure that this power concentration continues into the foreseeable future, if not forever. Fewer and fewer people in control of more and more power within the system, being locked in for life. It is the exact opposite of democratization.

And it is made all the more interesting because many of the autocratic-leaning leaders of late are actually being propelled into positions of power by the peoples of those states themselves! Say what? Confusing, I know. That's why this modern move towards autocracy is hard to pin down . . . it is not a matter of military men just seizing power. The process is somehow accelerating in societies already governed by strongman dictatorships like North Korea and Syria. It is accelerating in societies already governed by single-party systems like China and Russia. And, most surprisingly, the tendency towards autocratic practices is being perpetuated much more in full-on democracies, wherein a majority (or at least a sizable minority) of people are willingly wanting to assist in the autocratic ascension! Examples of this decline in democracy are happening in places that might make you a bit reluctant to even recognize . . . places like Hungary, Poland, and, yes, even the United States of America.

So what are some common elements of this process we can see working across different political systems? This autocratic amassment of power has many commonalities, despite being conducted in wildly different societies. The folks applying this strategy work in worrisomely similar ways. Dig this abridged list of things associated with the rise of autocracy, and see if any of them look familiar to you, or that you have heard of in current events. The attentive autocrat(s) customarily . . .

1. Base their message on restoring national pride, national sovereignty, global hegemony, global order, or some other storyline that focuses on themselves as the "savior" of a society. In such scenarios, the assumption is that things are somehow broken, and this particular person (or political party) has the solution to fix everything and restore national greatness. Or perhaps restore balance to the force, in the case of the Jedi autocrats.
2. To strengthen that message, autocrats often incite fear to rally their supporters and/or focus on an "enemy" that is to blame for everything wrong in their society. That "enemy" is depicted as entirely evil, and as such their opinions not only don't count, they must be confronted and "defeated" outright in order to save the society from its ills. This can take a variety of forms:
   - If the "enemy" is another country, the autocrat will whip up a frenzy based on nationalism. E.g., "our country is the best and we must fight against physical encroachment from the enemy country." China's current attitude towards the USA as an encroaching power in the Pacific Ocean or as a friend to Taiwan is one of the reasons for the increasing autocratic power of Xi Jinping, and is part of his current popularity.
   - If the "enemy" is another ideology, the autocrat will whip up a frenzy based on a combo of ideology with nationalism. E.g., the entire Cold War was about competing ideologies between the US and the USSR, and modern Russia under Vladimir Putin still perpetuates and promotes the concept of the USA and Europe (collectively Team West) as an enemy trying to force democracy and capitalism on the global stage, and specifically upon hapless Russians.
   - If the "enemy" is a different ethnic/religious group, the autocrat will whip up a frenzy based on some sense of national identity or national purity that must be saved by expunging the "outsiders" or foreign influence. Often targets and demonizes immigrants in particular. E.g., Austria's draconian U-turn on asylum seekers with strict anti-immigration laws; the current Chinese government's campaign to diminish the Uyghur Muslim and Tibetan culture from western China; USA building a wall on the Mexican border.
   - If the "enemy" is specifically a different racial group, the autocrat will whip up a frenzy based on grievances that their group is unfairly losing their power, status, or standing within the society that must be remedied. E.g. renewed "White Power" movements across Europe and the USA.
   - If the "enemy" is simply the opposing political party within their own country, the autocrat will whip up a frenzy based upon liberal versus conservative viewpoints which (they suggest) simply cannot co-exist any longer for the society to be successful. In its extreme manifestations, the opposing political party is treated as an enemy of the state itself. This is most evident in the wildly polarized two-party system in the USA, but also in the extreme demonization and persecution of minority political parties in Turkey, Russia, Burma, and any attempt at even a mere social movement in China. Think I'm exaggerating about that? Google "lying flat," or tangping in China.
3. The attentive autocrat also has no problem changing the state's constitutions or laws to achieve their political agenda, which often times entails increasing their own personal power or the power of their political party. In Poland and Hungary, these autocratic forces have been actively rebuilding the state's institutions (the judiciary, courts, legislature) completely to serve the political party in power. In Russia, China, and Turkey, the autocratic leaders outright changed constitutions to make themselves "leaders for life." In the USA, laws and select institutions are being modified heavily to empower a single political group (the Trump/Republican Party) to increase control of elections.
4. The attentive autocrat also has no problem working "above the law" to achieve their political agenda, when they can't outright change those laws. In other words, they lack respect for the rule of law in their state, and place themselves (or their group) above the law entirely. This is because in their minds they are "fighting for good," and believe the rules must be bent or broken entirely to benefit their cause so they can make everything better again. For the aggressive autocrat, the ends justify the means. E.g. in the Philippines, the popular president Rodrigo Duterte has conducted a war on drugs that entails openly supporting extrajudicial killings of thousands of assumed drug dealers.

I say assumed, because they are killed outright before being brought to a court of law. He even bragged that he personally killed several men while he was mayor of Davos. Wow.

5. The attentive autocrat works vigorously to control information (and/or disinformation) within their society: they want to control the narrative. Autocrats only want the public to hear the story they want to tell. They decide what "the facts" are, and they decide what "the truth" is for their society. They define the story of what the nation/country's history is, and may rewrite that history to support their ideology. Autocrats routinely berate the media . . . except for the chosen media that espouse their own personal message.
6. Anyone who disagrees with them is a traitor, is treasonous, and is an enemy of the state. That specially includes anyone in the press or opposing political parties. Remember the "Lock her up! Lock her up!" chants leveled at Hillary Clinton? Heard of the Russian opposition leader Russia Navalny who was poisoned (possibly by the KGB) and then imprisoned by Putin? How about the smashing of free press and outlawing of democratic protests in Hong Kong by the Chinese government? Autocrats do not like opposing opinions, and work hard to demonize those opinions as entirely evil.

So the autocratic toolbox includes a combination of appeals to national sovereignty, emphasis on national security, a call for national purity/national identity, a restoration of national greatness, increased media control (with harassment of independent media, independent government agencies, and independent educational institutions), and an unhealthy disregard for the rule of law . . . all in an effort to accumulate more power to themselves under the banner of "restoring" the society to greatness. Of course, the "greatness" in question is one defined by the autocrats themselves, and cannot be challenged or questioned by anyone else. Other tools not elaborated on above include international disinformation campaigns, increased censorship at all levels, increased (intentional) polarization of political parties, aggressive government propaganda, banning opposing political parties, changing laws to empower specific groups, and even election tampering. Yeah, now it is getting real for you, isn't it?

Again, many definitions focus on the concept that to be labeled as an "autocracy," only a single person is in charge holding absolute power. However, any government can have autocratic tendencies and be enacting autocratic policies that have more than one person in charge. What they do may ultimately lead to a single person in charge, but that is not the case in many of the more autocratic regimes already in place. Let me break down the most important element of this tendency:

If a single autocratic person has all the power, that person thinks: *I am in charge, and all decisions I make are right, all my opinions are right, and only I can lead this state/save this state.*

If a political party dominates in an autocratic manner: *Our political party is in charge, and all decisions and viewpoints and opinions of our party are right, and only we should be allowed to lead this state/save this state.*

If a specific ethnic/religious/racial group dominates in an autocratic manner: *Our particular group is the proper group to be in charge, and all decisions and viewpoints and opinions of our group are right, and this state should be ruled by people like us, for us.*

Given access to any levers of power within the system, those people with autocratic tendencies will use those powers as much as possible to change the systems to benefit themselves, their political party, or their particular group. Even if their group or their opinions are in the minority. And these moves are being perniciously perpetuated around the world by another political movement which is profoundly re-shaping governments of the 21st century: **POPULISM**.

PRESENTLY PERPETUATED PUGNACIOUSLY BY POPULISM

What exactly is ***populism***? Is it a creed, a style, a political strategy, or is it just a marketing ploy? As is the case with many political descriptions, it is likely some combination of all of those things and more, expressing itself differently in different places. But at the root of the word itself is "POP" . . . as in POP music, POP culture, things that are POPular. Right? Right! It means stuff that the vast majority of people really like or really support. Thus, anything described as "popular" is also suggesting that it is something of the people, of the masses, of the proletariat, of the common clave if you like. It's all about the peoples!!! The peeps!

So the term populism must at the very least refer to a political movement or political ideology that serves as the "voice" of the "everyday people," the "ordinary citizen," the "average Working Joe," and "the blue collar" class . . . and you will very often find that in order to best communicate with those peeps, the populist leaders will focus on "plain speak" and "telling it like it is" and possibly even being intentionally politically incorrect or outright salty in their language. Outright salty? Oh my! Yeah, just dropping in a few curse words here and there makes even a highbrow billionaire seem like an approachable guy who is "one of us." And I'm not referring exclusively to former US President Donald Trump here; Filipino President Duterte drops f-bombs and insults other world leaders all the time. Recep Erdoğan of Turkey does, too: he recently said that French President Macron should get his head checked to see if he was "brain dead." In July of 2021, Chinese President Xi Jinping publicly announced that foreign forces will "face broken heads and bloodshed" if anyone dared get in China's way. So, yeah, the bad boy image really plays well in autocratic/populist politics.

That's the generalist take on the term; but how do the political scientists and pundits and academics define the idea? Well, it seems that there is actually not a hard-core consensus definition even among those folks either. It can be a completely nationalistic approach like what North Korea or China employ: picking an outside enemy to rally domestic popular support against, in order to deflect any focus or complaints on the government in charge. That works well with a rising star country like China that consistently reminds its citizens how China has been picked on and diminished by outsiders for a century. But it can be defined in a variety of other ways too, almost all of which share some common features: a (often deep) suspicion of and outright hostility toward elites, mainstream politics, and established institutions of any kind, political or economic. That is, when a large enough proportion of the society really dislikes all of the people running the system . . . not the system itself per se, and not just hating a particular political party either. The conditions are ripe for populism when a lot of people just think that all the political leaders and all political parties are too corrupt, out of touch, and unresponsive to them, the peoples.

In that situation, populist leaders see themselves as speaking for the forgotten "ordinary" person and often exaggerate their own importance as the voice of real patriotism. In the words of this one dude that maybe you have heard about lately named Donald J. Trump: "The only antidote to decades of ruinous rule by a small handful of elites is a bold infusion of popular will. On every major issue affecting this country, the people are right and the governing elite are wrong." Yep. That was what Trump wrote in *The Wall Street Journal* in April 2016. And later he became the President of the world's most powerful country. And he won that election by running as a populist candidate. Yes, yes, yes, I know some of you will say that he was the Republican candidate, but that's not why he won. In fact, he attacked the elites of that political party as well on his rise to the White House, and is still at odds with the Republican establishment on a multitude of issues. So he really won the election as a "populist" outsider, running against the whole system.

So populism is a nationalistic, anti-elite, anti-establishment political ideology that is suspicious of mainstream politics, political pundits, academics, and established institutions . . . especially the bigger more global ones. I mean, if you don't even trust or like your state's government, you probably really hate NAFTA or the WTO and you really hate the UN. And as a populist leader, you think that the whole system is so rotten that it needs to be fully purged and replaced with new leaders and new ideas and new principles . . . and you are not afraid to tell everyone in the current system how much they suck. So anyone who bashes elites and champions the interests of ordinary people is considered a populist? Eh. Kind of. Maybe. Perhaps we need some more concrete traits of how this ideology has manifested itself in the past to better understand its future . . .

CAN BE FROM THE RIGHT OR THE LEFT: AMERICAN EXAMPLES

Oh, you thought Donald Trump was the first time populism has reared its head in America? Nah. There is really nothing new under the sun, and this type of political chicanery has surfaced in the past. Donnie has just been the most successful at it in the modern era, having surfed all the way to the top by riding a populist wave. And while the current populist surge is mostly from the political conservative right, be forewarned that this is an ideology that is, and has in the past, been embraced by the left as well as the right. In fact, we could just as easily say that former presidential Democratic contender Bernie Sanders was a populist of the leftist variety, also calling for a real shakedown of the entire political system, much the way Trump was from the right. But of course Bernie lost out his candidacy to Hillary Clinton, a more standard political leader who would not have disrupted the system too much had she gotten elected . . . but she didn't! In fact, since most of you readers are here in America, let's further define populism in the USA in the past in order to understand the concept more fully in the present.

Check out this handy chart to compare and contrast the opposing viewpoints that both share populism as their defining ideology, but come to very different conclusions of how it should manifest itself to "fix" the society:

LEFT/LIBERAL	RIGHT
A) When it comes to who pisses them off and who they blame for all societal ills, the populist from the left direct ***all*** their ire exclusively upward: at corporate elites and their "crappy" government for undermining the success of their state and "the people."	A) When it comes to who pisses them off and who they blame for all societal ills, the populist from the right directs ***some*** of their ire upward: at corporate elites and the "crappy" government for undermining the success of their state and "the people."
B) They believe that the government representatives are out of touch with their constituencies because they have "sold out" to big business interests, and that laws are passed mostly to benefit business, not individual citizens.	B) They believe that the government representatives are out of touch with their constituencies because they have "sold out" to big business interests as well as an overzealous protection of minority groups; laws are passed mostly to benefit business and protect minorities at the expense of the majority . . . and they believe that "political correctness" highlights this "minority rule."
C) Are more focused on the individual worker over corporate interests, thus are pro-labor, anti-monopoly, anti-corporate in general.	C) Are focused on job creation as well, but by cutting government red tape and taxes to benefit business, which encourages job growth, not at the expense of big business. Thus are pro-business, pro-corporate, although wary of corporate elites.
D) Would critique mainstream left-wing parties (just center-left, e.g. the Democrats) as too market-oriented and too accommodating of big business.	D) Would critique mainstream right-wing parties (just center-right, e.g. the Republicans) as too politically correct and too inclusive. And that is because . . .

LEFT/LIBERAL	RIGHT
E) Embrace a conception of "the people" based on class: the "working class" or the "poorer classes" versus the "rich class" or elites.	E) For the right wing populist, "the people" is narrower and more ethnically/religiously restrictive . . . so the other part of their ire focuses on "the other" peeps who aren't in their group. ***This is an important distinction.*** Because they . . .
F) Advance a version of "**civic nationalism**" = a belief in the fundamental equality of all human beings, in every individual's inalienable rights to life, liberty, and the pursuit of happiness, and in a democratic government that derives its legitimacy from the people's consent. All the peeps. All of them.	F) Advance a version of "**racial nationalism**" = which means their people/group are held together by common blood, common skin color, and/or possibly common historical heritage . . . and by being in this more exclusive group, they are the "true" citizens and have an inherited fitness for leading the society.
G) Manifestations of leftist populism in the past: various LABOR movements throughout US history . . . more focused on taking out economic fat cats as opposed to political ones; restoring economic equality rather than a social/racial norm; restoring more rights to workers at the expense of profits to big business. The rise of powerful labor unions and political movements to give rights to workers are your best example here.	G) Manifestations of right-wing populism in the past: 1882 Chinese Exclusion Act, the (second) rise of the KKK in the 1920's, World War 2 Japanese internment camps in America. Google any of these circumstances to learn much more about them, but they all share the trait of being xenophobic or race-based movements that separated "true Americans" from the "others."

Just for clarification on the political spectrum within this United States example: both Bernie and Trump represent real departures from previous presidents, from both the Republican and Democratic parties. George Bush Sr., Bill Clinton, George Bush Jr, and Barack Obama would all be best described as just center-left or just center-right, if not outright centrists in the grand political scheme of things. These were all leaders who, despite being on opposing sides of the American political spectrum, all respected the rule of law, the right of opposition parties to express themselves, full freedom of the press, and a general respect and agreement of what was "the truth" or "the facts" of any given situation. All former presidents accepted working within the constructs of the rule of law and the US Constitution, and were not attempting to radically overhaul or overthrow the existing political structure. Despite their differences, or your opinion about them, all of these ex-presidents would have first and foremost put the good of the entire country (and all of the citizens in it) at least slightly above the interests of themselves or their particular political party. Now we are getting back to the main concept of this chapter: autocracy. Leaders with autocratic tendencies don't share that same sensibility.

It is important to note that in both the left- and right-wing versions of populism, BOTH exhibit fierce nationalism . . . that is, that "our country is the best" or "our country first" or "let's fix our country." Both versions focus on nationalism and are opposed to globalization, or at least varying aspects of it. Both versions want to fix the country, whatever that means, based largely on how things used to be in the past. Make OUR country great again! That slogan ring any bells? It should, as it pretty much was the sole Trump campaign slogan and political platform. But make no bones about it: he was not the first to use that slogan, and it has been used time and time again all around the world. "Make Germany great again!" could have been heard in pre-WW2 Deutschland. Make France great again! Make China great again! Make Russia great again! Heck, those last two have been used by Xi Jinping and Vladimir Putin for over two decades now.

Which brings me to what further differentiates many of the more standard populist parties/politicians of the past from the more totally hardcore populist leaders of the past and present. The REAL hardcore populist takes things a couple of steps further . . .

- They will say that they, and ONLY THEY PERSONALLY, represent "the people" . . .
- In doing so, they suggest everyone else is illegitimate . . .
- Which then suggests anyone else who disagrees with them are not "real" are not the "peeps" are therefore unpatriotic . . . and thus . . .
- Anyone who disagrees with them is a traitor, is treasonous, and is an enemy of the state. Yikes. Now it is getting serious!
- Also, they define and wildly overestimate this "huge gap" between the government and "the people."
- They frame problems in extreme scenarios, which then call for extreme solutions . . . that only they can provide . . .

. . . and in doing so, whip up masses of people into a frenzy of anti-establishment, conspiracy latent theory, racial prejudice, or nationalist sentiment . . . it's the other people's fault, and those people could be a different color, a different religion, foreigners from a different country, or other countries or groups in their entirety. (E.g., it's the Russians' fault! It's the foreigners' fault! It's the Muslims' fault! It's the Americans' fault! It's the Mexicans' fault! It's the Chinese's fault!) And those "others" are to blame for everything bad, and must be stopped!!!

Any of those things sound familiar? Have any of these things been gaining in popularity in the state in which you live? Have you heard of any of those things happening in other parts of the world, in other countries, via the news and current events? Or perhaps you can scroll back up to the list of autocratic elements in the previous section of this very chapter!?!?!? That's right! These lists have a lot in common! The proudly populist politicians are increasingly behaving in an autocratic fashion!!!

So what we have here are autocratic tendencies being put on steroids by a proliferation of populism around the planet, which spells real trouble for the system of government that most of us have grown up in and are most familiar with: democracy. Both autocracy and populism are in select ways the absolute antithesis to the foundational core of democracy. So are we witnessing . . .

THE DEMISE OF DEMOCRACY?

It doesn't look good at this juncture here in the early decades of the 21st century, that's for sure! Not that you need a refresher course on the concept, but let's just outline a few of the important traits of democracy, so we can more clearly understand why autocracy is such a threat to its continued dominance on the world stage. And, my friends, you should know right up front that democracy is indeed being shaken to its core right now, and is under a very real threat of being replaced as the dominant style of governance in the world, and, indeed, is currently on the decline! That statement is not a Plaid opinion, BTW, it is backed up by some facts I will outline for you here in this section!

WHAT IS IT?

For starters, what is this democracy beast of which we speak?

Generally speaking, democracy refers to a form of government in which the people have either the authority to choose their governing legislators, or the authority to decide on legislation. Who is considered part of the people and how authority is shared among or delegated by the people has changed over time and at different speeds in different countries, but more and more

Ol' George helped start the global trend. . . but is it nearing the end?

of the inhabitants of countries have generally been included. We call that ***suffrage:*** the right to vote in political elections. The concept of democracy has also changed over time considerably. The original form of democracy was a ***direct democracy***, in which the people directly deliberate and decide on legislation. The most common form of democracy today is a ***representative democracy***, where the people elect representatives to deliberate and decide on legislation, such as in parliamentary or presidential democracy. Most day-to-day decision making of democracies is based upon a simple ***majority rule***: people are elected, or laws are made, or rules are established that a majority (more than half) of citizens or lawmakers vote for. This construct allows for the orderly transformation of policies and programs over time, along with the peaceful transition of power from one leader to the next, without the need for a violent overthrow or revolution.

We also largely associate all of these things with full functioning democracies as well:

- Citizens have inalienable human rights, such as freedom of speech, of religion, of liberty, etc.
- Citizens are assured political freedoms as well, such as freedom of assembly, freedom to vote, right to a fair trial, freedom from cruel and unusual punishment, etc.
- Citizens are all equal; one person, one vote.
- Decisions should reflect the will of the majority.
- Government is limited in its power and must respect all citizens' rights, regardless of what/who those citizens voted for.
- The Rule of Law: Both the government and the governed are subject to the law. (Funny how this "rule of law" thing keeps coming up, right?)
- When it comes to elections, citizens have freedom to form multiple political parties and put up candidates to campaign and contest; the opposition party has access to mass media and campaign finance; there are expectations of fairness and neutrality of electoral administration.
- Political Equality: Everyone is equal; one person/one vote.
- And much more philosophically, there is the idea of The Common Good: that decisions are made for the benefit of everyone, and the society at large.

In the common variant of ***liberal democracy***, the powers of the majority are exercised within the framework of a representative democracy, but the constitution limits the majority and protects the minority, usually through the enjoyment by all of certain individual rights, some of which are listed above. In the case of American democracy, they are enshrined in the Bill of Rights. It is quite important to stress here that full democracies structurally "limit the majority and protect the minority," thus dispelling the notion that majority rule equates to the majority doing anything it wants, running roughshod over the minority. That is not the case. Majority opinions may structure the laws, but those laws are applied equally to all and must not interfere with the established individual rights of the minority. Get it? Pretty straightforward stuff, right?

Well. Mostly. Scholars of democracy like to point out that not every single democracy has every single thing on the list above; not every single place that holds elections is actually a democracy; not every single democracy functions the same way; not every single democracy has the same level of human rights; not every democracy applies the Rule of Law equally . . . the list goes on. And of course, different countries have different levels of political corruption, of bureaucratic oversight, of even how elections are conducted. It's better to think about democracy as a process that a society adheres to in order to govern itself, which varies on a continuum. Some places are more democratic than others. Some places are striking to be democratic, while others are backsliding away from it. Much the way I am suggesting that autocratic practices mold a society into an autocracy over time, democratic processes work the exact same way, but in reverse! Consider both as processes which fluctuate over time, and in different ways in different countries.

DEMOCRACY: A SUPER BRIEF HISTORY

Having said all that . . . let me now express to you what a great run democracy has had over the last century!!! It has become the governing soup du jour for societies around the globe! Now I could go way back in time and talk about Ancient Greek philosophy and the democracy of Athens, or of democratic elements employed for communal decision-making in smaller indigenous groups throughout history, or the impacts of the 1215 Magna Carta, but let's skip ahead to the modern version of democracy that really came into force in the previous century.

It was really the American Revolution that first led the way to the modern ruling system we recognize in today's political world, embraced by a majority of sovereign states. Back in 1776, that whole American War for Independence was based primarily on a desire to have the people of those 13 Original Colonies to make decision for themselves, to make laws governing themselves, and to elect their own people as leaders. It was a direct challenge to the much more autocratic absolute monarchy of Great Britain, which controlled those uppity colonials. A true representative democracy that enshrined equality (at least on paper) for all its citizens along with a Bill of Rights protecting them from the government itself was a big deal back in that day. Especially considering that most ruling systems up to that point, which ruled the vast majority, were much more autocratic in nature—think of absolute monarchies, military governments, theocracies, despots, emperors, and tsars as the standard operating procedure for most of human political life, for most of human history. This governing style "of the people, by the people, for the people" was never enacted on such a grand scale ever before.

And then it kind of worked! Its success inspired the Haitian Revolution, the French Revolution (although terrible initially, but France eventually got there), the move towards constitutional monarchy in the UK, the independence movements and revolutions in Latin America of the 1800's and the democratization of most other nation-states of Europe in the early 1900's. Even Ataturk's Turkey joined the club (in a limited fashion) in 1923. So over a century, the global democracy count ticked upward, ever so slowly. But the real democratic action took off starting in the 1940's/1950's. The sentinel event of the 20th century was World War 2, and at the conclusion of that bloody calamity, the democracy poster-child and primary promoter—the USA—was on top, and a beacon for freedom-loving peoples the world over. And that really opened the floodgates for democratic change.

Namely, virtually all the former European colonies officially became independent, and hopped on the democracy bandwagon: places like the Philippines (1946), India (1947), Pakistan (1947), Sri Lanka (1948), and Burma (1948). Additionally, Israel (1948) came into being at this same time, and became the first (and still only) full democracy in the Middle East. While still attached to the nominal notion of being in a "Commonwealth" of Great Britain, this is also when Canada, New Zealand, Australia and others became much more independent full democracies. All of the vanquished Axis Powers which started WW2 (Germany, Italy, and Japan) also converted to democracies. South Korea joined the club in 1950. A second big wave of democracies popped up all over Africa in the 1960's and 1970's as independence movements and revolutions created a whole slew of new states, once they shrugged off the last vestiges of European colonialism. Many of this new African states embraced democracy . . . although arguably not many of them have sustained it successfully or consistently since then.

Even at this point, let's call it the mid 1970's, only about 30% of the world's sovereign states met the minimum criteria of a fully-functioning democracy; somewhere between 50 to 60 states would be counted as democracies at this point. Even some places that used to be democracies in Latin America and Southern Europe, had succumbed to military dictatorships and right-wing nationalist dictatorships as a result of the chaos of the 1930's Great Depression and World War 2. But watch out! Things are getting ready to be kicked into democratic overdrive!!!!

In the subsequent decades, the ruling system had an insanely hot global run: the number of democracies essentially held steady or expanded pretty much every single year from 1975 until 2011! As a result of utter failings of alternative systems (like fascism in Spain, military dictatorships in Brazil, Argentina, and Chile), democracy re-gained footholds across Latin America and Europe and even Southeast Asia, primarily the juggernaut Indonesia. Then came the biggest single wave: the 1991 collapse of the USSR and communism in general witnessed a double-digit bump in new democracies. Even

Russia is on the democracy train now (well, maybe not anymore, but they were for a time!). Then Yugoslavia imploded and 6 (or is it 7 now?) new states were born, most of which embraced democracy. Then came the Orange Revolution in Ukraine, the Velvet Revolution in Czechoslovakia, and the Rose Revolution in Georgia. Then the USA "liberated" Iraq from Saddam Hussain . . . boom! Another democracy is born.

Then came the Arab Spring of 2011, which led to much upheaval and aspirations toward democracy, but ultimately resulted only in the establishing of a single functioning democracy in Tunisia and some increased democratic rights in Morocco. Egypt saw a temporary democracy before the re-establishment of military rule. The Palestinian Authority also took action to address democratic rights. In the same 2011 year, the military junta of Burma—in a totally surprise move—allowed for free elections and a return to a democratic path after five decades of iron-fisted rule. What the what? Yes, it seemed like democracy would win the day everywhere, eventually. Even in states that did not embrace democracy as a ruling system, this democratic hey-day was paralleled by a similarly steady and significant expansion in political rights and civil liberties for many of the citizens of these places too. So, all is well and dandy for democracy forever, yes? Nope.

DEMOCRACY IS DONE FOR? DEF IN DECLINE.

How the mighty have fallen! Okay, that's a bit excessive, but the centuries-long trend of increasing democracies has certainly stagnated now, and many states are now experiencing democratic decline in a variety of ways. The democratic glory days seem to now be on hold . . . or are being reversed. For the last decade solid, there has been no net expansion in the number of full-on democracies, which has oscillated between 90 and 120, or about 50-60% of the world's states, depending upon who is making the assessment. Worse yet, it appears that even within the most ardent democracies, significant backsliding on some of the system's core elements is now in play. Yes, you read that right: even in some of the strongest democracies, strong anti-democratic and full-on autocratic forces are at play to undermine the system itself. Yikes.

Let's check out what various international groups that study freedom and democracy have to say about the current situation.

Freedom House is a non-profit non-governmental organization that has conducted research and advocacy on democracy, political freedom, and human rights, since 1941. Freedom House rates people's access to political rights and civil liberties in 210 countries and territories through its annual Freedom in the World report. It uses a two-tiered system consisting of scores and status—a country is awarded points for each of its political rights and civil liberties indicators—and each country is then ranked into one of three categories: Free, Partly Free, and Not Free. This year's *Freedom in the World 2021* report is entitled ***Democracy under Siege*** and it really paints a quite bleak picture of the state of affairs, with these highlights selected by the Plaid Avenger for you to peruse:

- Nearly 75% of the world's population lived in a country that faced deterioration of freedom scores last year, part of a decade long trend.
- There are now 54 Not Free countries, accounting for 38% of the world's population.
- The proportion of Not Free countries is now the highest it has been in the past 15 years.
- India, the world's most populous democracy, dropped from "Totally Free" to "Partly Free" status in Freedom in the World 2021.
- With India's decline to "Partly Free," less than 20% of the world's population now lives in a Free country, the smallest proportion since 1995.

From their concluding remarks: "These withering blows marked the 15th consecutive year of decline in global freedom. The countries experiencing deterioration outnumbered those with improvements by the largest margin recorded since the negative trend began in 2006. The long democratic recession is deepening."

Another group named **Varieties of Democracy (V-Dem)** is self-described as the largest global dataset on democracy with almost 30 million data points for 202 countries from 1789 to 2020, involving over 3,500 scholars and other country experts. The V-Dem Institute is an independent research institute and the Headquarters of the project are based at the Department of Political Science, University of Gothenburg, Sweden which measures hundreds of different attributes of democracy and offers a yearly report. So what does V-Dem have to say about autocracy and democracy here in 2021? Well, the title of the *V-Dem Democracy Report 2021* is titled: ***Autocratization Turns Viral***. Dang! Now it's a disease! Some highlights:

- The global decline during the past 10 years is steep and continues in 2020, especially in the Asia-Pacific region, Central Asia, Eastern Europe, and Latin America.
- The level of democracy enjoyed by the average global citizen in 2020 is down to levels last found around 1990.
- Electoral autocracies continue to be the most common regime type. A major change is that India—formerly the world's largest democracy with 1.37 billion inhabitants—turned into an electoral autocracy.
- With this, electoral and closed autocracies are home to 68% of the world's population. (So they are saying almost 70% of the world now lives in something they are calling autocracy, or at least a democracy with increasing autocratic elements?)
- Liberal democracies diminished from 41 countries in 2010 to 32 in 2020, with a population share of only 14%.
- Several G20 nations such as Brazil, India, and Turkey are among the top 10 democratic decliners, which means they are advancing autocracy.
- Poland takes a dubious democracy-declining first place and three new nations join this autocratic-advancing group: Benin, Bolivia, and Mauritius.

Most interesting to the Plaid Avenger here is that both Freedom House and V-Dem now consider India (the world's largest democracy) as now having so many autocratic elements that it had to be downgraded to "partly free" and re-categorized as an "electoral democracy," whatever the heck that actually is. A final passage from the V-Dem report:

> The accelerating pace at which the world is being taken over by processes of autocratization is manifest in bold relief when population size is taken into account. The sharp increase in the last few years is the result of autocratization in large countries like India, Brazil, and the United States of America. By 2020, more than one-third (34%) of the world's population were living in countries undergoing autocratization while a miniscule 4% were living in democratizing nations.

Wow! Not pulling any punches there! They are directly referring to "processes of autocratization" which work against and degrade democratization! That's what I've been raging about for the last 10 pages! It's a process, that can happen even in functioning democracies . . . and all these reports are in full agreement that it is indeed happening in functioning democracies like the US and India, and lots of other places too!

Let's call out some of this democratic decline in more devastating detail via a map. Some of the highlights (or low-lights if you prefer) about the status of democracy in the world have been culled out from The Economist magazine's yearly report by **The Economist Intelligence Unit**, with the *Democracy Index 2020* for this year subtitled: ***In sickness and in health?*** That alone about sums up their assessment of the system. As per their own introduction:

> The Economist Intelligence Unit's Democracy Index provides a snapshot of the state of democracy worldwide in 165 independent states and two territories. This covers almost the entire population of the world and the vast majority of the world's states (microstates are excluded). The Democracy Index is based on five categories: electoral process and pluralism, the functioning of government, political participation, political culture, and civil liberties. Based on its scores on a range of indicators within these categories, each country is then itself classified as one of four types of regime: "full democracy", "flawed democracy", "hybrid regime" or "authoritarian regime".

By their specific measure of democracy:

- Less than half (49.4%) of the world's population live in a democracy of some sort, with varying degrees of democratic institutions and norms. That number has been declining for a decade.
- Much fewer (8.4%) reside in something we would call a robust "full democracy." Also declining for a decade.
- More than a third of the entire world's population live under full-on autocratic authoritarian rule; China counts for the lion's share of that number.
- In 2020, 75 of the 167 countries and territories covered by their model, or 44.9% of the total, are considered to be democracies. (That's much less than other estimates, but these guys are good and they know their stuff.)
- Of the remaining 92 countries, 57 are "authoritarian regimes," up from 54 in 2019, and 35 are classified as "hybrid regimes" which contain some facade of democratic elements, but are much more autocratic in overall nature. This upward trend of autocracy seems to be the current standard operating procedure.
- A large majority of rated countries, 116 of a total of 167 (almost 70%), recorded a decline in their total score compared from 2019 to 2020.
- Their overall average global score (based on a wide variety of factors) in the 2020 Democracy Index fell from 5.44 in 2019 to 5.37, which by their measures is the worst global score since the index was created in 2006. For reference, a full functioning democracy is a perfect score of 10, and a full-on authoritarian regime would get a score of zero. So the global average is now just slightly better than halfway up the scale.

As you can see from the map below, a few geographic trends are worth noting. The "full democracies" are focused farther north and farther south on the globe: Scandinavia, Canada, UK, Germany, Japan, South Korea, and Switzerland up north; Uruguay, Chile, Australia and New Zealand in the south. Why is that? I have not a clue! Do cooler climates make for more democracy? Also of interest are the broadly varied number of states in the "flawed democracy" category, including

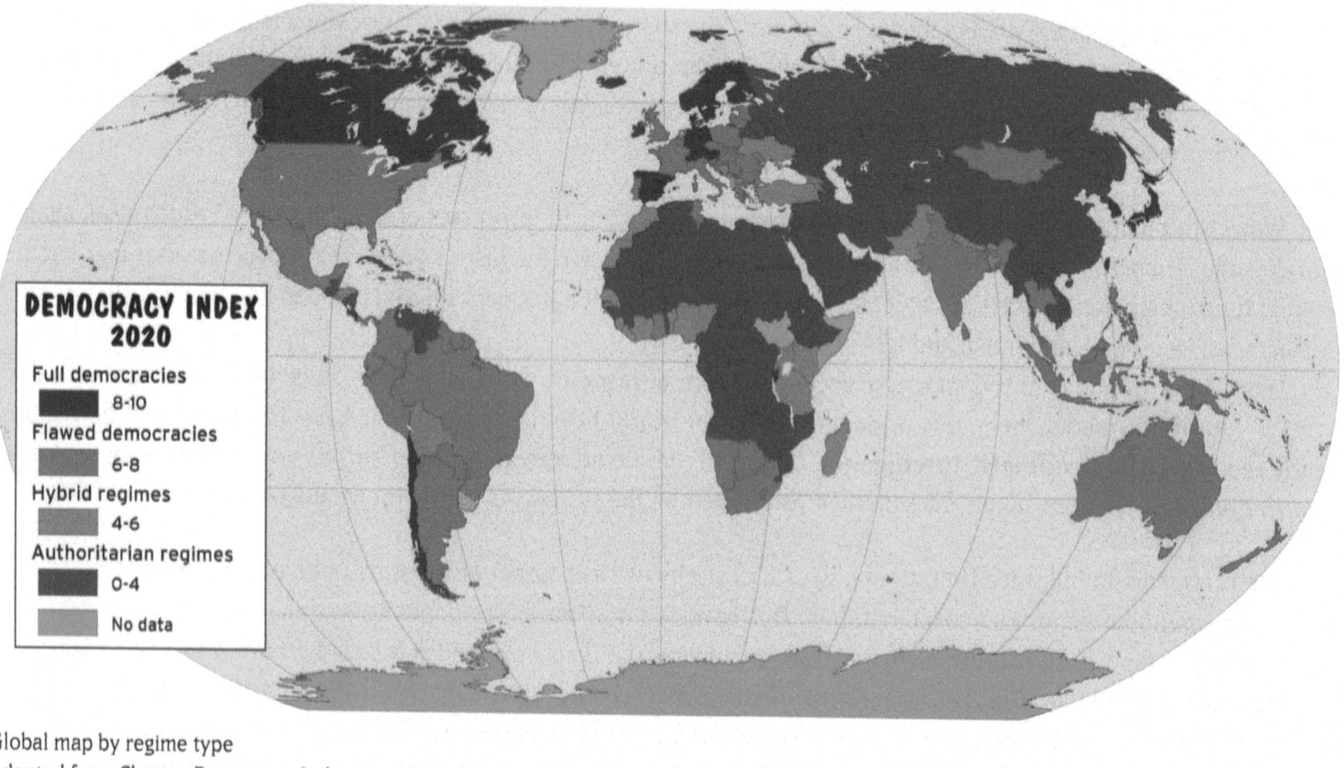

Global map by regime type
Adapted from Chart 1, *Democracy Index 2020: In Sickness and Health*, 2020. Source: The Economist Intelligence Unit

some of democracies biggest practitioners like India, France, Brazil, Indonesia, South Africa, Mexico . . . and lo and behold . . . the USA as well! Upon further assessment, you'll see that most of the states across the Americas, both north and south, fall into the flawed.

Going down the list, we see a smattering of hybrid regimes across Africa and the Middle East, with important states that are often considered as full democracies stuck here in hybrid-land: Pakistan, Bolivia, and several states of Central America catch my eye here, but the most crucial one in this category is Turkey. Didn't they used to be a democracy? Indeed, they did. But they are seemingly sinking to authoritarian status quickly. Which brings us to the last: the absolutely authoritarian regimes, which are not only on the rise, but also contain some of the biggest and most populated states on the planet: China and Russia dominate the category (and literally dominate the citizens of their states). But check out whole regions that fall into this phenomenon: all of Central Asia, most of mainland Southeast Asia, virtually all of the Middle East, and a significant number of states in Sub-Saharan Africa. The category climbed significantly in the last few years due to a military coup that overthrew the constitutional monarchy of Thailand, a military coup that toppled the new democracy in Burma (Myanmar), along with various coups that occurred in hybrid state systems in Africa and the Middle East.

IT AIN'T JUST ABOUT THE NAME OF THE SYSTEM . . .

A final note that all three reports allude to in great details is the fact that these declining democratic rankings (across all three systems I reference here) are also taking into account the decline in personal freedoms and personal liberties of citizens as well. These reports are not basing their assessments solely on election practices or government functionality, but also on how the states are treating their citizens on a personal level when it comes to freedom. Personal rights and liberties like freedom of the press, freedom of expression, freedom to assemble or protest have become increasingly restrictive in the last decade as well, exponentially so in the crazy 2020 Covid lockdown year. And these ratings systems are also taking into account internal machinations happening within democracies that are undermining them . . . like stacking judicial systems to favor one party (and no, I'm not talking about the USA here, but rather Poland) or changing constitutions to allow for unlimited presidential terms, etc. etc. Dang this decline!

Okay, so we've got it, Plaid Avenger! Democracy is in decline! Stop making us depressed about it! Why don't you tell us why it is happening in today's world, after a century of winning?

Well, as the title on the cover of this 11th edition alludes to, it's got a whole lot to do with the rise of ***autocracy*** all over the place. Full-on autocracies gaining more power and prestige, former democracies being outright overthrown by autocratic-types, and full-on democracies eroding from within due to autocratic tendencies of politicians, presidents, prime ministers, and political parties that suddenly aren't as interested in maintaining the old democratic traditions as they are in amassing power to themselves or their particular parties. Hopefully I have painted a full enough portrait of both autocracy and democracy that you understand that these are two competing forces working in opposite directions to determine the fate of many states, and the global order in general. Before we finish up with concrete examples of the modern mania of autocratic agents, maybe I should quickly address the question of . . .

WHY THE RECENT ASCENDANCY OF AUTOCRATIC ACTION?

Let me back up the horse and buggy to say this about both populism and autocratic tendencies: they both arise in response to real grievances, not just out of irrational fear or racism or xenophobia. And the more people that succumb to joining the populist politics by supporting "strong man" politicians that supposedly have all the answers to "fixing" everything, the more autocracy is allowed to thrive. Not all populist politicians have autocratic tendencies . . . but in today's world, virtually all of them do! Supporting a populist agenda nowadays is to also support the autocratic behavior that will almost certainly be employed to achieve the agenda itself. Since both of these forces are on the rise in tandem, let's pick apart the two terms individually.

POPULISM PROLIFERATES . . .

Let's start with the **POPULIST** rise first. Here are some things that are currently infuriating the masses, in states all over the world, which encourages citizens to support populist "strong man" agendas:

- Economic systems that favor the rich (way too much). Wealth disparity is perhaps currently at an all-time high, at least in the modern rich world. The rich have become the super-rich. The poor remain the super-poor. But the masses in the middle classes have actually lost ground for the last several decades, and real wage increases have simply not happened for the blue-collar, middle class peeps.
- Fear of losing jobs because companies relocated facilities to another country for cheap labor, or fear of losing existing jobs to new immigrants.
- Fear of losing jobs and livelihoods because of technological change: Automation of factories and mass production of goods has already destroyed jobs in traditional agricultural and manufacturing sectors worldwide, and the future rise of AI and robots and drones will continue extend this job destruction into the middle classes for generations to come.
- Politicians (actually, the whole political class) who seem self-serving are more concerned with padding their pockets or getting re-elected than with their constituents' needs. Citizens feel ignored or patronized by this "ruling class" that seems above the law, and that are not affected by the economic policies they create.
- The sometimes rational, oftentimes irrational, fear of terrorism that could strike them at any time, any place, for any reason. This feeling of instability, of not being safe, of imminent doom pervading the planet.
- The sometimes rational, oftentimes irrational, fear of total loss of cultural identity. What is the collective identity of being American? Will Germany still be German if we allow five million more Syrians to come live here? Will we still be a "Christian" nation if other religions take hold here in greater numbers than us?

That is just at a personal or national scale, say in America or in France or in China. But what is happening globally right now that feeds into these fears even more . . . which, again, lead people to support populist "strong man" agendas?

- Widening inequality in the entire world, and now even in rich countries.
- Mass movement of people (immigrants, refugees), and people have always been moving, but the scale of movement is unprecedented in modern history.
- The seemingly insane rise in international conflicts and terrorist acts, the detailed graphic violence of which is brought to you daily on TV screens.
- International trade blocks and international trade pacts that make the movement of jobs and businesses and money ever easier from country to country.
- International bodies like the UN or the EU or NATO, which give ardent nationalists a sense of loss of control over their identity, over their countries, and perhaps even over their personal lives. Am I an Italian or a European Union member? Am I protected under South African law if I am brought in front of a UN criminal court? Why are my taxes paying for a war or aid relief in a foreign country?

As perhaps you are starting to detect, virtually all of these causes for fear that are fueling the rise of populism have been exacerbated by one overarching trend: globalization. Globalization itself has accelerated and perpetuated change of all sorts, across all borders, pulling more and more people into a single system that blurs a lot of the lines we have taken for granted for a long time: economic corridors, distinct cultural areas, ethnic lines, even sovereign state borders. The rise in international institutes and laws that have seemingly "trumped" sovereignty is a big one to consider: are we Belgian, or EU peeps? Does UN law trump US law? The movement of many countries from the nation-state category to more multiculturalist societies is yet another: England used to be filled with just English people, is that true now? And of course, multinational corporations and businesses have long ago given up caring about any of those lines anyway, and move around capital and facilities at their whim.

You thought it was exclusive to Europe or the USA? Anti-refugee sentiment has no bounds: this is from India.

AUTOCRACY ABOUNDS . . .

All of these factors mentioned above taken together have caused a significant number of citizens, in a significant number of different countries, to desire a return to the "good old days" of nationalism, nationalist pride, and a nation-state centered political order that counters globalism . . . and thus the rise in support for populist political leaders that promise to make all that happen for their people. Now let's move to the **AUTOCRATIC** action that is being put on steroids by the rise in populism, along with increasing angst about global economic, political, and even pandemic problems. Here are some things that are currently encouraging the masses or individual leaders, in states all over the world, to support increasing autocratic actions, even within devout democracies:

1) **Democracy seems to be "losing."** Due to its very nature, democracy is a messy process. When everyone gets a vote, everyone gets a voice. And the more voices you have in any discussion about anything means more and more competing viewpoints, all of which want to "win" the argument. They want their ideas to "win" or their political party to "win" or their presidential candidate to "win," all of which invites conflict. Conflicts which are getting harder and harder to resolve due to polarization of politics, everywhere. Even within the most successful democracies, there is now increased wealth disparity, a sense of societal stagnation, misinformation run rampant, increased political polarization of the citizenry, endemic corruption, political inaction due to political infighting, and loss of a single cultural cohesive glue that unites people within a nation. Add in personal civil liberties like the right to free speech—amplified exponentially in the age of the internet—and now you have hundreds of millions of people with a megaphone to yell out their viewpoints, disparage their opponents, and increasingly even make up their own facts and truths to justify their opinions.

The result? It appears that most democracies worldwide are now just not working that well. They are loud. They are messy. They can't agree on anything—even what constitutes "facts" anymore. They can't seem to get stuff done. The USA is floundering over a disputed election. NATO, a democratic alliance of military powers, has in-fighting about mission and funding. BREXIT sends a signal that economic alliances among democracies aren't working. Protests over racial inequality, protests over wealth disparity, and protests over disputed elections make democracies appear hopelessly failing in their mission. Public confidence is at an historic low, and dissatisfaction in democracy has hit an all-time high, and that is from multiple surveys and reports taken from citizens actually living in democracies, not from outside them.

2) **Democracies are retreating from leadership on the world stage**. In combination with this, democracy dissatisfaction is increasing a desire of many political leaders across the democratic world to become more inward-looking to focus on the problems within their own country. That is, they are retracting from international interactions, and from global institutions, to put more emphasis on "making their country great" again, at the expense of the rest of the world. This ties in directly with the populism push described above, but in the specific case of democracy, this means the concept is simply not on the global agenda anymore. The strong, self-assured democracies aren't into promoting democracy on the world stage like they used to. Since World War 2, the well-established democracies (particularly the USA) had been an important and integral part of promoting and supporting fledgling democracies and democratic movements all over the world. But it appears due to poor performance and other issues, both the United States and Western Europe now lack the will and self-confidence to promote democracy effectively abroad. Which makes it even harder for those wanna-be democratic movements to fight the freedom-loving fight against the significant autocratic powers that be.

3) Autocracy appears to be "winning." At the exact same time that democracies around the world are gridlocked and seemingly semi-dysfunctional, the major autocracies of the world seem to be prospering and winning at all global competitions. This is primarily attributed to the rapid rise of China, which has been killing it at almost the exact same time that democracies have been faltering, since about 2006/07. Part of this is merely timing: China has worked for over a half a century to get back into a position of power, and they finally started assuming that position in the last decade . . . just when (as a total coincidence) democracies have become more gridlocked than ever. And not just in the USA, but the UK, Italy, France, and many other parts of Europe. So a lot of modern news stories are all about the "rise of China" and the "success of China" and how China is now an economic juggernaut and a seriously growing technological leader and military threat to the world, at the exact moment the USA and Europe are stagnant. This makes Chinese citizens (rightfully) proud of their success and current status, thus a strong sense of national pride that coincides with a low sense of national pride in many democracies.

A similar message could be applied to Russia, although it is in no way, shape, or form as successfully rising as China right now. With a strong autocratic leader like Putin in charge, a large percentage of Russians have the same sense of national pride as the Chinese . . . although they don't have as many tangible successes to point to, if any at all. Still, a strong sense of nationalistic pride, along with their historic animosity towards the West because of their loss in the Cold War, makes them feel like winners, simply because the Western democracies seem like losers. I could make the same type of argument for North Korea, Turkey, Cuba, Venezuela, Bolivia, Nicaragua, and several others that you wouldn't be able to locate on a map. And leaders like Putin of Russia, Kim Jong-Un of North Korea, Jair Bolsonaro of Brazil, Erdoğan of Turkey, Trump in the USA, and Rodrigo Duterte of the Philippines seem to be successfully making a mockery of democracy by forcing through their own agendas by ignoring all checks and balances inherent in the system, along with ignoring rule of law or public dissent. And all those dudes were actually truly elected by majorities of voters, and most of them have a high percentage of public support, so destroying democratic norms in the process, for them, is just a byproduct of their supposed success.

Bottom line is that the apparent *failings* of democracy (the world's premier ruling system, perpetuated and promoted by the USA), *empowers* all the natural enemies of democracies and the USA. Many, many middle sized and smaller states around the world are now not sure which team to join . . . and many states that border China or Russia or do business with either of them are being encouraged to dismiss democracy in exchange for economic benefits of partnering with the "winning" autocratic teams. It is a potent threat/potential economic windfall. And as with all "rainy day fans" as well as those who don't dedicate until they are "jumping on the winning bandwagon," this current situation of democratic backsliding is causing other countries to seriously consider autocratic systems as the only winning way forward. Which feeds into . . .

4) Autocracy seems to be the way forward to "stream-lining" democracies. Because of all the above reasons, even within the most successful and empowered democracies in today's world, there is a growing tendency to try to duplicate the overall success of China (or the streamlined political system of Russia) by adopting more autocratic elements that have built those systems into what they are today. Perhaps it is jealousy, perhaps it is self-assured egotistical pride that your particular political party has the "right" answers, or perhaps it is human nature to just want to do what the cool kids are doing . . . in any case, politicians and pundits the world over look to the stream-lined political processes within autocracies with intense envy. If only *they* personally, or *their* political party, could control all levers of power and the entire system, then *they* could fix everything so much faster and so easily!!! It is an autocratic wet dream that leads people down this path of thinking: that everything can be made right, as long as we become a dictatorship or one-party state . . . *just for a little bit while we fix everything* . . . before returning to democratic norms. Ask the Russian Revolution, the Cuban Revolution, or the Iranian Revolution how that worked out for them.

However, modern politicians and pundits within democracies can't seem to draw the parallels between those failed autocratic experiments, and their own desires for autocratic elements within their current democracies. That's part of the reason we are witnessing politicians from Poland, Hungary, India, Turkey, India, Ethiopia, Venezuela, and yes, even the USA, who are more than willing to embrace elements of the autocratic playbook of consolidating power in particular areas of the government in order to achieve their goals. Restrictive voting laws, changing the constitution to allow for unlimited presidential terms, stacking the highest court of your country with like-minded individuals that all represent a single political party, even ruling by presidential decree/executive orders . . . all of these things are autocratic in that they concentrate power into fewer hands at the expense of democracy. Hear what I am saying here: politicians within democracies are using autocratic methods to get their way for certain policy creations, passing of laws, or promoting particular cultural attitudes . . . NOT for the overthrowing of the whole government. But that is happening too, because . . .

5) Autocrats everywhere are empowered. Because of all the above reasons, we are now witnessing both the subtle internal changes which backslide on democratic principle (referenced in point #4 above), or in other cases radical full-blown overthrows of democracy. Autocratically-inclined individuals are looking at the success of China, and the stagnation of the USA, and realizing that perhaps the USA won't challenge them too much if they decide to follow the Chinese path. And perhaps the USA and the Europeans won't really say anything if they decide to overthrow the democracy of their own country, and/or persecute and kill a bunch of their own citizens in the process. And perhaps even if the USA and Europe were to protest their countries' human rights violations or the degeneration of their democracy or outright overthrow of an elected leader the USA and Europe won't even have the spine to do anything about it except issue a rebuke. If the wanna-be autocrats no longer fear any retribution—in the form of an American invasion or international sanctions or an economic embargo—then why wouldn't they just do whatever they want in their country? Indeed, why not? The decline of democracy has certainly played into autocratically-inclined institutes (like militaries) to be even further empowered or encouraged to overthrow democratic systems in the modern era, and by that I mean democracies outright destroyed in the last five years: Burma, Thailand, Venezuela, Mali, Afghanistan, now soon possibly Iraq and Haiti. What will be the repercussions of overthrowing a democratically elected government here in 2021? Apparently, not much. Autocrats now assume that. And finally . . .

6) A Covid-crushing year accelerated autocracy EVERYWHERE. I have intentionally excluded referring to the global Covid crisis of 2020 in most of this text, because its effects were wildly varied from one region to the next, and its long-term repercussions are still quite unknown, although I believe they will be much more minimal than most pundits are predicting. So to cover it in any detail at all would be to double the size of this already overwhelming book, and then all those prognostications would be utterly useless a year from now. So I have chosen to skip it altogether in the regional chapters.

BUT . . . I cannot skip it when it comes to this particular autocracy versus democracy chapter—because Covid has had, and will continue to have, a direct impact on individual freedoms and civil liberties across the planet.

As reported on in all the democracy indexes I have referenced in this chapter: in 2020, citizens in virtually every single country, regardless of government type, experienced the biggest setback of individual freedoms ever during peacetime conditions (and perhaps even during wartime conditions). Billions confined in movement, civil liberties suspended, and free speech challenged all over the world.

This new deadly disease, reported on in real time around the globe as it was affecting millions, almost literally scared the masses in democracies into total surrender of their fundamental freedoms and cherished human rights. What am I referring to? Well, for instance, have you heard of *freedom* of *assembly*, which is described specifically in the UN Universal Declaration of Human Rights as well as the US Constitution? Yeah, that was out the damn window on day one of the pandemic when lockdowns became the law of the land. The right to freedom of religion? Lawsuits had to be filed by adamant churchgoers because they were banned from going to their houses of worship, from the US to Israel to Malaysia. The list could go on and on, but the point here is not to be dramatic—it is to get you to think about how many of the restrictive emergency measures that were enacted during Covid have yet to be repealed! Worse yet, did Covid set the stage for autocrats around the world to build a playbook of enacting more controlling measures for future times?

Consider this: China and other authoritarian regimes had no problem enforcing draconian mandatory lockdowns, because that is just business as usual for them and their citizens. At the height of the Covid calamity in China, the government had an estimated 760 million citizens (more than half the total population) totally confined to their homes. But not so much in democracies, where citizens protested, fought back, and even embraced misinformation campaigns that reinforced their particular opinion or desires . . . which is why it appears that the authoritarian/autocratic regimes were way more successful at controlling Covid than those messy, loud-mouthed Western democracies (I mean, Europe is STILL kind of a mess and behind the times as of summer 2021!). Why didn't democracies just curb freedom of expression and the press, like China did? Because while limiting freedom of assembly seemed logical to enact, limited speech (even misinformation) was too antithetical to democratic principles, so it didn't happen.

So Covid 2020 plays directly into points #1 and #3 in the list above! Autocrats seem to be better at controlling their populations and providing solutions in a time of crisis! Well, that does seem to be the common perception right now . . . although freedom loving folks like me might argue that if they were a free and open society, that the facts of the initial Covid breakout in China would have been much more transparent and helped the entire world deal with it better . . . but that is an argument you should have with your autocrat-loving friends. In either case, overall freedoms of individuals worldwide took a tremendous hit during the Covid crisis, and that definitely played into the lower ranking for democracy worldwide in all three of the reports that I have referenced in the *Demise of Democracy?* section.

Speaking of sections, can we finish this one now and move on to the next one? Indeed! Let's do that now!

AN ARRAY OF AUTOCRATS IN ACTION

In the good ol' days of autocratic ascension, it was really easy for us outsiders to see what was going on, because brute force was the standard operating procedure for taking over a state. A bunch of dudes with guns bash down the doors of democracy and assume control via very public military coups or violent crackdowns on dissent. That still happens (Burma, Belarus, Mali, and Thailand are modern examples), but it happens much more rarely. In today's world, the outright authoritarian autocrats in states like China, Russia, and North Korea are joined by the right-wing populists in western democracies like Hungary, Brazil, and the USA in concentrating power to themselves in much subtler, slower, and sometimes outright nefarious ways.

In both camps—be it eastern authoritarians or right-wing populists—leaders and politicians are manipulating things legally, culturally, and structurally behind the scenes to consolidate power to their own person, or to their political party, in an effort to exercise complete power to govern over a country and its citizens. And in both camps, they are much smarter and savvier than in times past, utilizing new technological tools, information/disinformation platforms, and manipulating the global economy to best serve their needs. And these guys are learning best practices from their fellow autocrats around the world. Here are some of the tools they commonly use . . .

THE AUTOCRAT'S PLAYBOOK

Autocratization around the world now typically follows a similar playbook. Newly established leaders or existing ruling governments first attack and discredit the media and civil society, who might stand in the way of their machinations. This intentionally polarizes political parties and societies in general by disrespecting opponents and spreading false information, which empowers the supporters of the autocrat while infuriating the detractors of the autocrat. The topic of the polarization varies from country to country, but usually has some basis in the autocrats' storyline of restoring the society to "greatness" which of course assumes that the society currently sucks or is in crisis. Once these disruptions to the political and social systems are in place, the autocrats then attempt to undermine formal institutions as part of their societal "cure." Here are just eight playbook highlights on how they do these things:

#1 **Build up a cult of personality** around their awesomeness: The manly-man Vladimir Putin regularly poses shirtless on his many adventures. Donald Trump was the king of reality TV. Recep Erdoğan fashions himself as the new sultan of a re-vamped Ottoman Empire. Build the story, promote it vigorously and constantly through social media and government sources, and let everyone know about every fantastic feat and political victory that feeds into the story about them "winning." And in almost all circumstances, the autocrat will make incessant claims that he is "a man of the people," thus he is one of them! He understands their problems! This cements his bond with common people, his major support base.

#2 Achieve #1 in part by **appealing to populism, nationalism, ethnic pride or cultural/racial purity** to exploit existing tensions within societies in order to solidify support of a rabid, incensed base.

#3 Reinforce #1 and #2 by **identifying "the enemy"** that is making the country suck. It focuses and rallies the autocrat's base and sets him up as the hero "fighting the good fight" against said enemy. Usually involves targeting outsiders, foreign powers, immigrants, minority ethnic groups in the state, or minority religious groups in the state.

#4 **Extend executive power**: Be it an elected official in a democracy, or the new politburo chief of a one-party communist state, the autocrat will use legal means (like executive actions in the USA) and legislative coercion to empower his position as much as possible, while weakening other government institutions like constitutions, congresses and judiciaries that have power to check him in any way.

#5 **Stack the deck**: The autocrat will use every level of power at his disposal to put sycophants and die-hard loyalists in appointed positions of power within the government, and anywhere else he can put them, including in charge of any media outlets or major businesses the autocrat holds sway.

#6 **Repress** dissent and any citizen-based movements that challenge the autocrat's message directly, or that stir up any trouble in general. Deny permits for public protest, detain protesters readily, use excessive force to destroy current demonstrations, and restrict government funding for anything remotely interpreted as pro-democracy.

#7 **Incapacitate the opposition political party**, permanently. The autocrat will wield his power to minimize media access and government funds for other political parties, use gerrymandering to manipulate election outcomes, change election and voter eligibility rules to discourage opposition voting, and will stack the deck (#5 above) of electoral commissions and oversight bodies with their own allies. Thus, elections may occur, but the results are already known before the first vote is cast . . . while maintaining the facade that a free and fair election is actually taking place. Ever wonder why they even bother to have elections in dictatorships like Belarus or Egypt? Because the leaders want to demonstrate that they have popular support for their control of society, even if it is faked.

#8 And perhaps most vitally: **control the message**. Autocrats want to 100% control domestic news and everything that their citizens consume. That is easy in authoritarian states like China and Russia that totally control the media and the internet; more challenging in democracies like India and the USA who don't. When you can't outright control the information totally yourself, then start your own disinformation and push it vehemently on to your followers. Use social media to portray only the facts you like, or that you make up. Control the narrative as much as possible at home, and misinform as much as possible abroad.

Okay, they've got the tools, they've got the playbook, and they are good to go to accumulate all power to themselves and their political parties! Onwards, autocrats!!!! Now for the fun stuff! Take all the learning you have done so far reading about the regions of the world in this textbook and let's put some autocratic names to some autocratic faces to highlight the autocratic actions of just the last few years! ***Autocratic Avengers, Assemble!!!***

Let's start in Asia, where the real deal destruction of democracy has been in full swing. By that, I mean autocratic action that is so obvious, and the impacts of it so clear, that it doesn't even merit debate about its direct effects on democratic systems. These are full-on, unapologetic autocratic attacks on democratic institutions or state norms, resulting in diminishment or outright destruction of ruling systems and/or concentration of power into a single leader's hands. From the cover art of this book itself, let's start with . . .

BURMA

The main feature on the cover is of a military man crushing a democratic protest in the country we call Burma, also known as Myanmar. It is the sole biggest—and unexpected—outright bashing and smashing of a functioning democracy in recent times. And by recent, I mean this year! On February 1st of 2021, a textbook coup d'état was conducted by the Tatmadaw—the official title of Myanmar's military—which then restructured the government in something they are calling a "stratocracy." That's a form of government headed by multiple military chiefs, with the various branches of government being administered by military forces. The head honcho of that stratocracy is a masochistic military man named Min Aung Hlaing, who has imprisoned all the democratically elected members of the country's ruling party, the National League for Democracy (NLD), and specifically the true leader and State Counsellor Aung San Suu Kyi—who also happens to be the Nobel Peace Prize winning hero of the decade long pro-democracy movement in Burma.

General Min Aung Hlaing of Burma

This isn't even the first round of military-induced mayhem in Burma either. In fact, it's not even the second or third time. Burma was plagued with political instability since it gained independence from Britain back in 1948, and the military actually formed a temporary caretaker government at the request of the country's democratically-elected prime minister at the time. After quelling unrest, the military actually restored civilian government after holding the 1960 Burmese general election . . . but less than two years later, the military seized power openly in a 1962 coup, which led to 26 years of full-on dictatorial military rule under a dude named Ne Win.

That iron-fisted approach lasted until 1988, when nationwide protests erupted. Called the 8888 Uprising, the civil unrest was sparked by economic mismanagement and a desire for democracy, leading Ne Win to step down. He stepped down, but other military madmen stepped up: the military's top leaders formed the State Law and Order Restoration Council (SLORC), which they used to seize power again. About this time is when Aung San Suu Kyi, the daughter of the country's modern founder Aung San, became a notable pro-democracy activist that attracted international attention to Burma, which compelled the junta to initiate change. In 1990, free elections were allowed by the military, under the bizarrely optimistic assumption of the military that they had popular support. Lolol wrong! The elections resulted in a landslide victory for Aung San Suu Kyi and the NLD party, results which the military immediately refused to recognize, and instead put Aung San and all other NLD leaders under arrest.

The military remained in power for another 22 years until 2011, and for reasons that the Plaid Avenger still doesn't quite understand, the military then released a "roadmap to democracy," in which they allowed a tentative transition to democracy to occur, culminating in free elections being held in 2015, which once again resulted in a victory for Aung San Suu Kyi and the NLD. However, this time the military retained substantial power, including the right to appoint 1/4 of all parliament members, which they had previously built into the 2008 Myanmar constitution.

Okay that is bringing us pretty close to current events. So what happened next? Well, the next election was in November 2020, and once again under the bizarrely optimistic assumption of the military that they had popular support, the NLD won 396 out of 476 seats in parliament (an even larger margin of victory than in the 2015 election) while the military's proxy party, the Union Solidarity and Development Party, won only 33 seats. Once again, the military morons were surprised that no one voted for them after decades of military rule, and thus they got pissed and—bingo, you got it now!—the February 2021 coup which formally wiped out the democratic competition altogether.

So . . . outright demolition of a new democracy at the hands of professionally armed autocrats. Say hello to the new boss, Min Aung Hlaing, same as the old bosses of military coups past. Same shizzle, different decade for the people of Burma, I'm afraid. This is the outright worst example of autocratic overreach as of late. The worst, but not the only one, and not the only one in the Asian neighborhood . . .

CHINA

President Xi Jinping of China

Hold on, Plaid Avenger! Why would you even bother talking about autocratic moves happening in an already authoritarian, one-party state like China? Because my friends, as I have been suggesting all along, increased autocratic inclinations can happen anywhere, even in states already predisposed to concentrations of power like one-party state China. Here's the deal though: ever since this dude name Xi Jinping was sworn in as President in 2013, he has consistently and persistently employed autocratic actions to concentrate all power of the state into his own hands. Already simultaneously holding the titles of President of the state (head of the government), General Secretary of the CCP (the Chinese Communist Party), and Chairman of the State Military Commission (head of the military), Xi has not seemed satisfied with those hugely powerful roles and has expanded his influence in autocratic actions like:

- Xi reduced the number of seats of the CCP Politburo Standing Committee from nine to seven, with only himself and one other loyal ally retaining their standing as all others were replaced.
- Xi personally chairs eight of the leading small groups within the CCP, including the new National Security Commission.
- Xi also handles all internal security directly; the chances of internal dissent much less a coup are virtually impossible.
- Xi started an "anti-corruption" campaign, which resulted in a political purging of prominent incumbent and retired Communist Party officials, most of which were seen as potential opponents to his rule. And most importantly . . .
- In 2018, Xi had the CCP abolish presidential term limits. In effect, Xi is now ruler for life.

With no institution outside the CCP able to check the behavior of CCP leaders in any effective way, Xi has taken firm control of all the levers of power and surrounded himself with die-hard loyalists. Xi Jinping is essentially taking China back to a strongman, centralized dictatorship after decades of institutionalized collective leadership of the CCP. His autocratic inclinations are not confined to political power either. He has been more than a little busy shaping the entire society, (remaining) civil liberties, and even historical legacy of the state in a variety of ways, all centered around his personal interpretations of how life should be lived for China's 1.5 billion citizens. In the last mere eight years, Xi has:

- Significantly increased censorship of the internet, especially Wikipedia, Google, and Facebook.
- Banned the mention of "the seven dangerous Western values" in popular media and educational materials; this list includes things like constitutional democracy, civil society and universal human rights/values.
- Commissioned (covertly) books, cartoons, pop songs and dance routines honoring his rule, which is part of a grand construction of a cult of personality.
- Banned imagery of Winnie the Pooh, after the spread of an internet meme in which photographs of Xi were compared to the bear. This was followed up in 2018, when the film *Christopher Robin* was denied a Chinese release.
- Had his personal manifesto (named *The Chinese Dream*) enshrined and written into the Chinese constitution.
- Xi is not letting up on control of society either. Just in 2021 Xi has vehemently banned and labeled as outright criminal a very small social movement named "Lying flat" or tangping (躺平). In it, members of the younger Chinese generations choose to opt out of the highly competitive economic rat race altogether, and instead choose simplistic lives of unambiguous low-paying jobs and minimalistic living. Kind of like the hippies of 1960's/70's subculture in the USA. How very dangerous! Lol.

HONG KONG PHOOEY

However, the most outright autocratic democracy-destroying moves by Xi Jinping have happened in the tiny metropolitan area and special administrative region of China named Hong Kong. Hong Kong is the former British colony that reverted back to Chinese control in 1997. Under its century-long control by the UK, it developed democratic institutions and norms, and was part of the wider western world, at least economically, long before the re-rise of modern China. When it reverted back, it was understood that Hong Kong would retain its semi-autonomous rule under the principle of "one country, two systems" that was agreed upon in 1984 with the signing of an internationally recognized treaty named the Sino-British Joint Declaration.

China was happy to keep up the perception of "one country, two systems" while they were still rising to power, but that facade seems to be now completely thrown out the window altogether. China is now quite powerful, and has increasingly been aggravated by a democratic system being operated within their sovereign state. Enter the autocratic Xi Jinping, who started making moves immediately to sideline if not severely limit any aspirations that Hong Kong would be able to maintain its semi-autonomous status. Increasing interference from Beijing took the form of eliminating pro-democracy candidates from running in Hong Kong elections, stacking the administrative offices of Hong Kong with Chinese nationalist, anti-democratic propaganda campaigns, and increasing censorship and limitations on Hong Kong press.

This finally culminated into open protests in 2019-2020 when pro-Beijing politicians proposed plans (and passed laws) to allow extradition of any citizen of Hong Kong to face trail or imprisonment in mainland China. Basically, the law would have made it legal for Beijing to arrest any individual (likely prodemocracy politicians and press) and bring them back to the mainland, under any trumped up charges they dreamt up. Massive street protest, and counter-demonstration police actions, have made the headlines ever since. The bill was eventually withdrawn, but not before embarrassing Xi enough that he has initiated wildly autocratic moves to end not just the demonstrations, but also the whole concept of "one country, two systems" altogether.

Now it's all but over in 2021, with Xi openly cracking down, and cracking heads, with increasingly dictatorial authority. Beijing imposed a new national security law that all but ends any public gatherings, kills any pro-democracy movements, and bans activists from interacting with foreign governments. He also has gutted and replaced the city's legislature and judiciary, stacking it with pro-Beijing politicians/judges, and has almost literally destroyed the last vestiges of freedoms of the press. The last pro-democracy newspaper, the *Apple Daily*, was forcibly shut down in June 2021. I would say the end is near, but it would be more truthful to say the end is here for any democratic inclinations in Hong Kong, or China in its entirety, as the maneuvers of master-autocrat Xi Jinping have consolidated almost every ounce of power to his own person.

STRONGMEN OF SUPPOSED DEMOCRATIC STRIPES . . .

Those two countries' governments are openly and outright autocratic, with no semblance of democratic institutions allowed to operate at all. It should be easy enough to understand at this point how China and Burma represent the complete right-wing side of the political spectrum, having completely embraced autocracy in becoming authoritarians in charge of a military dictatorship and a one-party state, respectively. I hope you also noted how even within the already highly authoritarian China, a leader like Xi has exhibited even more profound autocratic tendencies than his predecessors, exhibited by his consolidation of power to himself.

Now let's move to some other autocratic action that is happening in hybrid states that still call themselves democracies . . . at least pretend to be by holding elections and maintaining other democratic institutions, all while the top political leaders assault the system with increasing levels of autocratic behavior. And let's start with one of the most popular autocrats on the planet, the prodigious Vladimir Putin!

RUSSIA

President Vladimir Putin of Russia

Wow. What can I even autocratically say about a world leader that enjoys immense popularity at home, has pulled all vestiges of power to his position, and been the leader of a country for two decades solid . . . a state whose entire political system is now called "Putinism?" lolol. That alone makes the case for the autocratic intensity that Vladimir Putin has applied to Russia since the dawn of the 21st century! I mean, they have the whole political structure of the state named for him!!! Even Xi Jinping, the penultimate authoritarian autocrat of Asia, still got nothing on Putin. Why would the Plaid Avenger say that? Because Xi walked into his leadership position after a lifetime of careful cultivation for the top slot in China, and China was already on a huge upswing of awesomeness, economic property, and rising international power when he got the job. Vladimir Putin was a virtual unknown when he inherited his position, at a time Russia completely sucked economically, politically, and otherwise. Putin has been directly responsible for raising Russia out of the ashes of the USSR implosion, and building it up into a world power once more. And in doing so, Putin restored national pride, a strong national identity, international respect, and has extended Russian power outward. Flip back to the chapter on Russia to get the full low-down on the rascally Ruskie.

Given his decades in power, and his proclivity for autocratic endeavors, we could fill pages and pages with his auto-antics so I will try to keep it brief and paint with broad strokes here:

- Cult of personality: The best in the business. A perfect 10/10. Putin's manly-man exploits of bravery and strength and skill have been artfully cultivated for decades. He makes Russian women swoon, grown men blush, and has simultaneously become an underground icon in the gay community . . . particularly interesting since he has helped push discriminatory legislation against the LGBTQ+ community for years.
- Has always had huge approval ratings (ranging from 65%-85%) and has used his powers of popularity to dominate the largest political party (United Russia, of which he is the de facto head). Putin along with United Russia has legislatively and socially all but destroyed all other major opposition groups . . . although they keep some small opposition groups in existence and on life-support, solely so the "elections" look more legitimate and Putin can—correctly—claim that Russia is a multi-party system. It's not. It's a one-party system, borderline dictatorship of the President himself.
- Constitutional crusher: Originally appointed Prime Minister in August 1999, became acting President after the resignation of Boris Yeltsin, and less than four months later was elected outright to his first term as president and was reelected in 2004. As he was then constitutionally limited to two consecutive terms as president, Putin became the Prime Minister again from 2008-12. But then he had the constitution changed to have presidential terms go from four year stints to six years, and then had the courts rule that since he originally served two terms of four years in the old system, that he was eligible to run for two more terms, which he did, being elected again in 2012, and again in 2018. Not satisfied with two decades of power, in April 2021, he signed into law yet another constitutional change that would allow him to run for reelection twice more, potentially extending his presidency to 2036. Putin for life, baby!
- Putin controls virtually every lever of power in the Russian government, and many in the economic sphere as well. The Duma, the courts, the military, the police, all political appointees in civil service and local governments: all run by Putin loyalists who would never go against his wishes or policies, ever. The biggest/richest corporations also dare not go against him for fear of political repercussions to their businesses.

- Putin's rule has seen severe diminishment of human rights, civil liberties, LGBTQ+ rights, and promotion of the "true" Russian identity as ethnically Russian and religiously Eastern Orthodox.
- Putin controls virtually all of the message Russians hear. The media is nowhere near free. The internet is nowhere near free. Social media is severely restricted. Russian reporters are harassed, jailed, or assassinated on a frightening level.
- Putin's disinformation campaign that Russia puts out about the rest of the world (especially the USA) is also the best in the business. Interfering in democratic elections elsewhere is one of their specialties. Followed closely by high-level hacking of infrastructure and businesses around the world. Followed by poisoning of Russian defectors who flee the country.
- And yet, given all of those autocratic moves that have consolidated all of that power into the hands of one single man, Russia remains almost perversely terrified of any opposition groups gaining even the slightest foothold in the system, or getting even the slightest bit of attention in the press. One need only go google the harassment of Alexei Navalny, a Russian lawyer and opposition activist that has been hounded and arrested and jailed multiple times by the Russian government on trumped up charges . . . but he mostly made headlines in 2020 after being poisoned, nearly dying, then arrested again for parole violations due to seeking medical assistance in Germany just to stay alive. Yeah. Parole violations. Putin must be terrified of this guy. Autocrats DO NOT like anyone else interfering with their messaging, and Navalny—merely by continuing to stay alive (along with his six million YouTube followers)—is really a Russian thorn in their side.

And this was the super-abbreviated list! For all these things and a millions of others, Vladimir Putin has to be the top-ranked autocrat alive right now! And he still has a decade or so to go in office!

TURKEY

A literal stone's throw across the Black Sea finds us facing Turkish President Recep Tayyip Erdoğan, another outstanding example of an autocrat that has done everything in his power to pick up the Putin power playbook and follow it to a T . . . and while some aspects of it have worked to keep him in power, he is nowhere near the autocratic success story of Putin. However, this isn't a competitive ranking of autocrats, it is merely a list of the most notable ones, and Recep def makes the list due to his many machinations in shaping his society and accumulating power to his person. Again, don't want to go into excruciating detail of his whole life here, so flip back to the chapter on Turkey to get the full low-down on the titan of the Turks. Here are just a few Turkish delights for you to savor:

President Recep Tayyip Erdoğan of Turkey

- Erdoğan's Justice and Development Party (AKP) has ruled Turkey since 2002. He personally founded the party in 2001, which is a conservative, Islamist and populist political party with a strong support base of orthodox, conservative, primarily rural Muslims. The "Islamist" connotations of the party, and Erdoğan himself, have always been a bit of an issue for Turkey's strongly secular, democratic roots.
- And that's part of the cult of personality image he has intentionally cultivated since his start in politics. He has the support of conservative religious peoples, concentrated in rural areas. But he has continued to spin that righteous role into one that has gained respect across the entire Muslim world, as a defender of Islam, as a politician who is openly and proudly Muslim, and as a leader of a successful Muslim-dominated state, that also happens to be democratic. Erdoğan stands up to the USA, stands up to Israel, and even berates corrupt Arab monarchies that he suggests aren't as legitimate as he is, since they are both unelected and since their state structure don't incorporate Islam as much as he does.

- He continues to perpetuate this image of himself as a powerful Muslim leader in a Neo-Ottoman Empire fashion: Erdoğan's military exploits into Syria, Azerbaijan, and now the eastern Mediterranean Sea, harken back to an era of Muslim strength . . . perhaps he has designs on rebuilding a proud empire that loosely incorporates a geo-political alliance of Muslim states with Turkic ethnicity from the Middle East and Central Asia? His recent military adventurism would certainly suggest so.
- Back to politics: Initially passing liberalizing economic and political reforms back in the early 2000's, Erdoğan seemingly increased contempt for political rights and civil liberties, and has pursued a dramatic and wide-ranging crackdown on perceived opponents, especially after an attempted coup in 2016. He also covertly forced the sale of many independent newspapers to his cronies and sycophants.
- 2013 Gezi Park protests: Massive nationwide protests broke out after Erdoğan pushed through an extremely unpopular urban development plan for Istanbul's Taksim Gezi Park, with little to no public feedback or petitioning allowed. Dozens were killed, and this is the point at which democratic backsliding, restriction of civil liberties, and high-level corruption really started for Erdoğan's increasingly autocratic Turkey.
- To control the message, government censorship of the press and social media really increased after 2013, and they regularly restrict access to sites such as YouTube, Twitter, and Wikipedia . . . to this day. In 2013, a 13-year-old boy was arrested for insulting Erdoğan on Facebook. In 2014, parliament passed a bill allowing the government to block internet sites. In 2016, Erdoğan just decided to ban social media overall at his whim.
- Then his autocratic sensibilities really got put on steroids after the July 2016 coup d'état attempt by the military. It was put down by Erdoğan loyalists in the military and government, but it was a game-changer for the country. The purges that followed were on a Stalin scale of intensity. Tens of thousands arrested and jailed. Hundreds of thousands fired from their jobs in the military, in civil service, and even in academia. Were there truly guilty coup-plotters behind this that should have been held to account of the law? Absolutely. But anyone ever suspected of not loving Erdoğan or the AK party were suspect, and all were persecuted in one way or another. Erdoğan's administration removed more than 300,000 books from schools and libraries . . . we can only assume to burn them. Authoritarianism, expansionism, censorship and banning of parties or dissent are now standard operating procedure in the state.
- Erdoğan has made other ethnic groups—and I am specifically referencing the Kurds here—into second class citizens, retracting human rights and civil liberties from them, under the umbrella of "fighting Kurdish terrorists." Pro-Kurdish politicians are routinely harassed, physically attacked, and jailed by the government. The Kurdish people are seemingly being branded as non-citizens of Turkey based on two decades of Erdoğan policy towards them.
- Much like Putin, Erdoğan has initiated change in the Turkish constitution to empower himself. In 2017, Erdoğan and allies pushed through a constitutional reform, which changed Turkey's decades-old parliamentary system into an "executive presidency" in which massive amounts of concentrated political power were vested into the office of the President . . . the position he holds. In times past when he was Prime Minister, he actively fought to pull more powers to that position. Whatever position he holds, he moves all the power to it.
- May 2019: After weeks of pressure from the top officials of the AK party—and huge pressure from the president himself, Turkey's electoral board annulled the results of the Istanbul mayoral election (Turkey's largest city and its economic and cultural capital) because the AK party candidate lost. A claimed stolen election? Sound familiar? So Erdoğan forced them to re-do the election and—lololol—the AK party lost even worse.
- February 2021, just 4 years after his last constitutional change, (and perhaps because of falling support and low poll numbers), Erdoğan is petitioning to have the Turkish constitution completely re-written from the ground up. Critics worry that this is primarily to change the threshold of presidential electoral victory to a lower number, since Erdoğan's poll numbers suggest he will get way less than 50% of the vote in the upcoming presidential election. Typical autocratic move: don't have enough votes to win? Change the rules of winning.

So, autocratic intentions galore here in the Erdoğan Empire . . . but, it's nowhere near the level of successful autocratic control that Xi, Putin, Duterte, or even Modi have in their respective countries. And that is for a couple of really interesting reasons: 1) Unlike Russia or China or Burma, Turkey really is still a sound, functioning democracy where elections are still mostly free and fair, and 2) Unlike the leaders of Russia, India, and the Philippines, Erdoğan does not have overwhelming public support for his policies, and in fact both he and his AK party are losing popularity, due to economic/political/Covid mismanagement on an epic scale. As witnessed in the election loss in Istanbul, and the failed attempt to overthrow it, this Turkish autocrat may not be able to stay in power much longer. But our list is all about autocratic endeavors . . . the failures along with the successes! So let's get on with the list . . .

PHILIPPINES

Now here is the weirdest, wildest, and wackiest autocrat on our list . . . mainly because he doesn't possess nearly any of the traits that all the other autocrats do! This guy hasn't changed the constitution to empower himself or his political party. He doesn't ban other political parties. He doesn't fix elections, and in fact the Philippines are a fully functioning democracy when it comes to voting. He doesn't necessarily appeal to nationalism or religious or ethnic identity to whip up support. He hasn't really attempted to stack the civil service or the judiciary or the legislative branch with his cronies. Heck, as far as I know, he hasn't even limited the press, the media, or the internet in any significant way. So what do he do then? He do Duterte! And he do it well! Lol

President Rodrigo Roa Duterte of the Philippines

I'm speaking of one Rodrigo Roa Duterte, the President of the Philippines since 2016, and likely the most assertive, arrogant, loud-mouthed, insult-spewing autocrat on our list! He once referenced former US President Obama by saying "Son of a whore, I will curse you in that forum." In response to a question about an assassinated reporter, he said, "Just because you're a journalist you are not exempted from assassination, if you're a son of a bitch." When it comes to women's rights, he stated, "I believe in women's competence, but not in all aspects." He has encouraged shooting women in the genitals, bragged about his sexual virility, and boasted about physically—literally—personally killing criminals when he was mayor of Davao. And he has done this while maintaining a 65-75% approval rating of his citizens! Wow.

And that is the deadly duo of autocratic anti-democratic elements that Duterte brings to bear in this conversation: an over-the-top cult of personality matched only with his over-the-top absolute disrespect for the rule of law. There it is! That pesky rule of law part of true democracies that holds all citizens to account for their actions, and treats all citizens as equals in the eyes of the law. Including respecting citizens' rights under that law. The briefest of lists that highlight his autocratic tendencies:

- Duterte came to power on a political platform of tackling corruption and specifically curbing drug trafficking/drug use in the country.
- He had prior experience with this: various human rights groups documented over 1,400 killings allegedly by death squads operating in Davao between 1998 and 2016 when Duterte was the mayor of the city. Victims were mainly drug users, petty criminals, and street children. A 2009 report by the Philippine Commission on Human Rights confirmed the "systematic practice of extrajudicial killings" by the Davao Death Squad. Extrajudicial killing is when government authorities (be it the police, the military, or sponsored vigilante groups) kill citizens without any judicial proceedings or legal representations or legal processes of any kind. In this case, Duterte gave carte blanche for policemen, or even regular citizens to just kill anyone suspected of being in the drug trade, including users.
- Once in office in 2016, he has taken this same process to the federal, nation-wide level. The "Philippine War on Drugs" to date has resulted in an official death count of 5,100 "confirmed" drug personalities, although some news organizations and human rights groups claim the death toll is between 12,000 to 20,000 . . . or more. Obviously—and

perhaps proudly—the campaign is being carried out without due regard for the rule of law, due process of suspects, and the human rights of people who may be using or selling drugs. You have to keep saying "may be using or selling" because virtually none of these citizens were formerly charged or were declared guilty in a court of law.

- An unexpected consequence of disregarding of the rule of law to promote extrajudicial targeting of drug dealers has been the targeting of lawyers that represent them in courts of law. Of the 110 lawyers that have been killed from 1972 to 2021, 61 of them have been assassinated since Duterte took office in 2016. And some of those lawyers just happen to be more liberal-left types of people that also represent leftist activists, reporters, victims of human rights abuses, and others targeted by the right-wing regime. What a coincidence.
- In July 2020, Duterte took this disregard for rule of law one step further: he signed into law an Anti-Terrorism Act aimed at combating Islamic militancy in the south, which critics (and anyone else in the wrong place at the wrong time) worry could lead to more widespread human rights abuses, particularly among the minority Muslims living in the southern Philippines. Others fear it will give the police and the Duterte government free reign to persecute any dissent against his government, including protesters and members of the media, activists, journalists, social media users, and anyone else Duterte doesn't particularly like that day. The Act expands military and police powers of arrest and detention by allowing warrant-less arrests of terrorism suspects, increases the amount of time that they can be detained without being officially charged of anything, and removes a requirement that the police bring the arrested suspect in front of a judge to assess whether they have been subjected to physical or mental torture.

His reign and these actions have been wildly supported by the majority of Filipinos, which of course makes Duterte (and other autocrats like him) scoff at the notion that he is anti-democratic. In his mind, he is doing the work of the majority, so he is fulfilling the mission of a democracy. However, democracy is not "majority rule at the expense of the minority;" it is in principle a ruling system in which the majority opinion prevails, but still protects the minority. That's the whole rule of law stuff! And protected human liberties and civil rights stuff! The minority are supposed to get those too, no matter what! And even if you don't like to recognize that element of democracy, the rule of law is definitely enshrined in the concept either way—and the fact that the leader of a democracy has openly claimed to have murdered people with no repercussions demonstrates that Duterte believes himself to be above the law, and encourages vigilantes across his country to believe the same thing about themselves! For Duterte, he believes his personal opinion IS the law, which is kind of a hallmark of an autocrat, despite having majority of public support.

DEVOUT DEMOCRACIES DEVOLVE FROM WITHIN . . .

Those leaders are obviously and overtly autocratic, and openly proud of it on top of that! Consistently re-writing their own constitutions to empower themselves? Potently purging all political dissent? Blatant disregard for the rule of law? Yes, there's nothing covert about the conversion of those societies into something much less democratic, and perhaps openly authoritarian. At least they are open about it. Shadier stuff is going down in other parts of the planet that is no less autocratic in nature, but is being spun as legitimate democratic transformation. Let's now shift to these much more nefarious processes being employed by leaders of well-established democracies that are perhaps doing the most to advance autocracy over democracy, much to the chagrin of freedom-loving folks all over the world . . .

Prime Minister Narendra Modi of India

INDIA

As referenced in prior sections, India (the world's largest democracy) has taken a hit in virtually all democracy/freedom reports here in 2021. Freedom House downgraded India from "fully free democracy" to a "partially free democracy;" V-Dem Institute reclassified India as an "electoral autocracy;" and the Economist Intelligence Unit Democracy Index said that India slipped several notches in their rankings and is now a "flawed democracy." All of the actions in India that have led to these reassessments started with the 2014 election of Narendra Modi and his religiously-based Hindu nationalist Bharatiya Janata Party (BJP). Hindu nationalist? Yes, that about sums up the issue with the declining democratic values in this situation. The name "nationalist" is right there in the title of the party!

Modi is a master class politician, and all around popular nice guy, who routinely holds a 75% approval rating in India. And what a coincidence: India is comprised of roughly 75% "ethnically" and religiously Hindu citizens. Amazing that coincidence, right? What's happening in India is part of the textbook autocratic playbook referenced before. Buildup a cult of personality, in part by "appealing to populism, nationalism, ethnic pride or cultural/racial purity?" Ummmm . . . check, check, check and check on all boxes for the Hindu nationalist BJP. And he has used this "supermajority" of the masses, along with his charismatic personality and great work ethic, to really have his way with the Indian government. In what autocratic ways has Modi remade Indian democracy?

- Both the legislative and judicial branches are overwhelmed by the popularity and personal bravado of Modi, and they both regularly fall in line with his requests and agendas for fear of popular public backlash if they don't.
- V-Dem said that the sharpest decline was visible in government censorship of the media, repression of civil society organizations and the autonomy of the Election Commission of India. There is a high degree of media bias and a fall in academic and religious freedoms.
- Prior to Modi, the Indian government rarely exercised censorship against its varied media outlets. Now in 2021, it is regularly assessed by media freedom groups around the world that censorship efforts are becoming routine, standard operating procedure for the Modi government—with even mundane cultural stories that have nothing to do with restricted sensitive government issues. There are increasingly alarming reports of harassment of journalists, and over 50 were recently jailed for Covid reporting that the government disagreed with.
- In the 2019 elections, the Modi government employed large-scale voter suppression, removing an estimated 120 million eligible voters from the electoral roll by demanding documentation to prove residency, ultimately leading to the removal of some 70 million Muslims and Dalits from the voter rolls.
- In December 2019, Modi and the BJP passed a **Citizenship Amendment Act**, which for the first time makes religion a basis for citizenship . . . yes, you read that correctly. Immigrants coming from other countries must be Hindu (or more to the point may NOT be Muslim) in order to apply for citizenship. (BTW, there were large-scale protests against this, and tens of millions of Hindus also took part.)
- At roughly the same time, the government instituted a nationwide verification process (National Population Register [NPR] and a National Register of Citizens [NRC]) to identify "illegal migrants," who, if identified as such, can be deported immediately to immigrant camps to await deportation with no rights or due process of law of any kind. It is a safe assumption that this plan has been targeting Muslims almost exclusively.
- Summer 2020: Last summer Modi enacted major farm laws that threaten the livelihoods of two-thirds of India's 1.3-billion people without discussion, during the Covid lockdown of parliament. What followed was arguably the largest general strike in history and weeks of unrest.
- In August 2020, he revoked the autonomous status of Jammu and Kashmir, the only region in India that has a majority Muslims population.
- March 2021: Extended the Information Technology Act 2000, which governs online platform accountability, to control digital news content through the backdoor, without subjecting it to legislative discussion. The new Information Technology rules notified by the government give unprecedented powers to the executive to summarily take down content from digital news media platforms on vague grounds, and without so much as giving the publisher a hearing.

So, our main man Modi in India has absolutely been elected by his citizens. Modi absolutely has the support of a vast majority of Indians. Modi absolutely works within the bounds of the legal systems of India to implement his policies. BUT . . . Modi is absolutely using his popular position and personal power in absolutely autocratic ways which is resulting in less rights and political participation of minorities . . . mainly the Muslims ones, which still constitute a quarter of a billion people. And his autocratic moves are building in fewer and fewer checks on his power by limiting or intimidating the political opposition, the press, grassroots organizations, and academics. Limiting democracy in this popularly-supported manner is now a standard of the new wave of nationalist populist regimes. Oh boy, what an autocrat mess.

HUNGARY

Prime Minister Viktor Orbán of Hungary

And boy oh boy, time for a big ol' heaping helping of Orbán! Viktor Orbán that is. Did his first stint as Prime Minister of Hungary from 1998 to 2002, then his second run in the Prime position starting in 2010 . . . and continues to this day, making him the longest serving Prime Minister in Hungarian history. He is also the driving force of the major political party of Hungary that goes by the name Fidesz. Funny thing about the Fidesz party: it started as a liberal youth party opposing the ruling USSR communist government dictatorship, and over the years shifted to a center-right, classical liberal, pro-European platform, and then more recently into a right-wing nationalistic conservative group. So, it began as anti-autocratic, and has now become wildly pro-autocratic with Orbán large and in charge.

Paralleling the exact evolution (or devolution) of his party, Orbán originally was pro-democracy, pro-EU, pro-Team West, and even was in charge when Hungary joined NATO . . . and now he is overseeing the regular admonishment from the EU for his autocratic, anti-democratic behavior, with the possibility that the country may eventually be asked to leave the EU, NATO, or both! What happened here?

Well, the election of 2010 that brought Orbán was so successful for his political party that it put them in a position of having a supermajority within the Hungarian parliament, thus they were in a position to make major changes to the country's constitution, legal framework, judiciary, and cultural realm. And they have taken huge advantage of this situation, by weakening the legal checks and balances on its authority, interfering with media freedoms, and undermining human rights protection in Hungary . . . all of which bring the country into direct conflict with the rules and regulations of the European Union, of which it is still a member. Just some highlights:

- 2010: Orbán pushed through a series of restrictive media laws that, among other changes, created a new media regulator known as the Media Council headed by a political appointee with close ties to the government. Yeah, that sounds like an official government censor.
- 2012-13: Constitutional changes curbed the independence of the judiciary and the administration of justice, forced nearly 300 judges into early retirement, and imposed limitations on the Constitutional Court's ability to review laws and complaints.
- That new constitution also included a provision that restricts voting rights for people with "limited mental capacity."
- 2014: The government also introduced a law making homelessness a crime.
- 2015: During the European migrant crisis, Orbán ordered the erection of the Hungary–Serbia barrier to block entry of illegal immigrants. A "Great Wall of Hungary?" This exemplifies Hungary's current hugely anti-immigrant attitudes, which also are tinged with anti-Islamic rhetoric, and have resulted in some of the rashest immigration policies within Europe. Orbán himself has espoused fear of "Islamic takeover" of Europe if immigration is not countered.

- Since 2017, Orbán has had a public feud with billionaire American George Soros, (a Hungarian-born philanthropist who supports democracy building and human rights) whom he blames for virtually every major problem facing Hungary today.
- Under the cover of the Covid crisis, on March 30, 2020, the Hungarian parliament overwhelmingly passed legislation that would create a state of emergency without a time limit, thus granting the Prime Minister the ability to rule by decree, which also suspended by-elections, and implemented prison sentences for spreading fake news and sanctions for leaving quarantine.
- Since 2010, Orbán has lead initiatives and laws to hinder human rights of LGBTQ+ people, regarding those as "not compatible with Christian values." In June 2021, Orbán pushed through a new anti-LGBTQ+ law which criminalizes anything deemed to be promoting or educating about homosexuality. You may have heard about this one in late June when all the German football players and fans were wearing rainbow regalia in a show of protest against the law when playing a match in Hungary.
- In late July 2021, it has been discovered that Orbán and his administration have been aggressively using a super sophisticated Israeli-produced spying software named Pegasus to hack the phones of journalist, investigative reporters, and independent media owners. So the government of a supposed democracy is actively, and illegally, spying on its own citizens who have committed no crimes in order to discredit or destroy them simply because they are part of a vocal opposition to the Orbán autocracy. Uncool.

In the last dozen years of right-wing populist Orbán's reign, independent institutions, universities, the press, LBGTQ+ and minority communities have all been under constant assault. You can make the argument (and all autocrats in democracies do) that all of these new laws, legal maneuvers, and constitutional rewrites are representing the "will of the people" since Orbán and his ilk were indeed elected to office. But a majority trampling the minority does not a democracy make, and Hungary is now far from being a democracy anymore.

POLAND

While extremely similar to the political stratagem employed in Hungary, the story in Poland is a significant shift from our other autocratic leaders described above, as the story of this country's democratic decline is all about an autocratic political party versus a single autocratic strong man fostered by a cult of personality. The gentleman pictured here is Jarosław Kaczyński, who is not even the head honcho of Poland, but rather a Polish politician who is also the leader of the Law and Justice party (PiS), which he co-founded in 2001 with his twin brother, Lech Kaczyński, who served as president of Poland until his death in 2010.

Jarosław Kaczyński, leader of the PiS (Law and Justice party)

Since the 2015 victories of PiS, both in the presidential and parliamentary election, Kaczyński is considered to be the most important politician in Poland and one of the most dominant leaders in Europe. While neither the president nor the prime minister, in Poland he is referred to in some circles as the "Chief of State" due to his prominence in decision-making and influence. In 2020, he was designated as the Deputy Prime Minister of Poland with oversight over the defense, justice and interior ministries. And Kaczyński and his PiS have used autocratic processes to absolutely dominate and decimate the democratic institutions of Poland in a shockingly short time. Dig this:

- During PiS's first four years in office, Poland had the biggest decline of any country in the World Justice Project's *Rule-of-Law Index*.
- In 2020, the Freedom House ranking referenced earlier in this chapter found that Poland can no longer be classified as a full democracy. And the Economist Intelligence Unit's Democracy Index now clarifies it as a "flawed democracy."

- In 2021, V-Dem's democracy report, referenced earlier in this chapter, specifically called out Poland as the state that has moved further towards autocracy than any other place in the world over the last decade.
- In 2021, Transparency International stated in their annual report that the "ruling party's . . . steady erosion of the rule of law and democratic oversight has created conditions for corruption to flourish at the highest levels of power."

So, how did they democratically decline so alarmingly autocratically fast? Well, when the Law and Justice party (PiS) was originally founded by the brothers Kaczyński back in 2001, it was as a centrist/Christian (Catholic) democratic party, being culturally and socially conservative, with a law and order agenda that promised a purification and political renewal of Polish society in general. Back to the good ol' days! Their real goal once in power in 2005 was to maximize executive and legislative power to purge Poland of all remnant socialism leftover from the communist era, meaning they wanted to fire every government employee that had any ties to the former soviet empire as well as anyone with socialist leaning at all . . . which of course includes anyone with liberal ideology on cultural issues. The PiS proposed a law that required 350,000 civil servants, academics, teachers, and journalists to declare all former political associations, no matter how trivial, on threat of losing their jobs. Sounds like the McCarthy Era communist witch-hunts of 1950's America, right? However, widespread resistance was led by Poland's progressive elites, and the project was abandoned.

Actually, for a decade or more prior to this point, Poland's constitutional court pushed back against such retaliatory efforts to purge state institutions and civil society of anyone with communist associations. This process has actually been conducted in other former Soviet states of Eastern Europe, specifically Ukraine, and is referred to as **lustration**. Polish courts received support from EU laws protecting personal dignity and privacy. So, it didn't work the first time the PiS tried it. And they were punished at the polls for their effort, leading to election losses in 2007.

After that loss, the PiS moved further to the right and adopted nationalist and populist overtones (you know, "Make Poland Great Again" type of stuff) and bided its time by enlisting even further right-wing allies in Polish politics and business interests. With the rising tide of populism and pushback against the EU at the time, I guess it was inevitable that the PiS was swept back into power in 2015. And it came back with a vengeance. Now a fully formed, right-wing force for change that garnered roughly half the votes that election, the PiS renewed its lustration assault, this time specifically targeting the country's judiciary . . . and that now catches us up to current events and why Poland had precipitously plummeted in democracy ratings:

- Since 2015, via this communist purging concept (are there really any communists left 30 years later?), the PiS has stacked the country's courts with PiS allies, and increasingly attacked all elements of progressive civil society.
- In 2017, the PiS tried to unilaterally just fire over half of the judges off the Supreme Court, by lowering the retirement age. That move was blocked.
- So they then turned to taking over Poland's newly reconstituted Council of Judiciary, which makes all appointments of judges, and have stacked the judiciary at every level of the courts in the entire country.
- In December 2020, PiS passed a gag law on the judiciary, and they can be punished for implementing a judgment of a supranational court like the EU courts or the UN International Court of Justice, or for even talking about it. The courts are also banned from bringing up any "public actions inconsistent with judicial independence," which means they can't publicly discuss anything the government doesn't agree with. Ummm . . . so yeah, there is no more independent judiciary system in Poland anymore, and certainly not one that could check the growing power of the other branches of the government in any way.

As for controlling the message . . .

- Poland's popular government-funded public media broadcaster, TVP, is now used regularly as a tool to smear and attack judges who have fought back against these changes. Consider it the Fox News of Poland.
- State-owned oil company PKN Orlen, which is super tight with the PiS, has bought out other local media properties. The goal wasn't to diversify beyond fossil fuels; it was to help control the message.

- PiS and its allies have hit independent newspapers like Gazeta Wyborcza and oko.press, with dozens of so-called "Strategic Litigation Against Public Participation," or SLAPP lawsuits. These typically frivolous, politically motivated lawsuits are designed to intimidate and distract media organizations—and burden them with legal fees. Basically, sue them until they can't afford to fight back, and therefore have to shut up.
- PiS is extremely right-wing on social policy, and works hard to enact laws to support its anti-abortion, pro-family (in the strictest biblical sense only), anti-same-sex marriage, anti-LGBTQ+, anti-immigration, anti-Muslim agenda . . . and many analysts suggest they are also anti-German, anti-Russian and anti-Ukrainian (what the heck did the Ukrainians ever do to them?).
- In general, the PiS is filling the courts and media with pro-government judges and journalists; driving out leftwing and liberal organizations, academics, and universities; and is violating the EU Charter of Fundamental Rights by restricting or banning access to abortion and denying legal recognition to transgender people.

The biggest problem with these autocratic moves of the PiS, which will likely result in increased political polarization and internal angst of the population, is the fact that they don't even constitute a majority viewpoint of the masses! With only a 40-45% approval rating, these huge changes are being enacted on a democracy that the majority do not want! To compound this, four out of five Poles approve of being in the EU, while the PiS-dominated Polish government is enacting policies that are counter to EU human rights and civil liberties norms, which may get them kicked out of the EU! Or is a PiS-led "POLEXIT" imminent? These minority maneuvers are the opposite of a democracy!

But there is another country where this exact autocratic scenario is also unfolding, at an even quicker pace . . . and one that may explode into full civil unrest as a result. Yep. It's time to talk about the elephant in the room: the USA.

USA

And now for the last autocrat of note . . . and what a huge note it is! HUY-UUUGE as former US President Donald Trump used to say! But honestly, this whole Cover Story chapter is already HUY-UUUGELY too long to read, and the political polarization in America is HUY-UUUGELY too intense to inflame further, and the Plaid Avenger is now HUY-UUUGELY too tired to put into written words the gigantic saga of the 2017-2021 Trump reign over the American democracy. So I am going to bullet list some points of assessment, since all of my awesome readers are already well versed with the back-story and the drama in the USA over the last four years.

Former President Donald Trump of the US

And this may be surprising to many of you (especially the younger, more liberal generation peeps), but employing the very tools of autocratic assessment I have used this entire chapter to analyze the world, I have to tell you that former US President Donald Trump is nowhere near the authoritarian-styled autocrat like the other gentlemen on this list. Maybe he wanted to be one. Maybe he tried to be one. Maybe he talked like one. Maybe he walked like one. Maybe he even incited an insurrection to overthrow an election and officially become one. But through actual actions and impact on the *system*, he really isn't even close to a Putin or a Xi or even a Duterte or Orbán. Trump wasn't that successful of an autocrat . . . but his political party—the Republican Party—is another autocratic story altogether. Say what? Let's start with the man, and the **autocratic assessment of Trump** . . .

When looking back at our list at the beginning of this section entitled ***The Autocrat's Playbook***, we can go through point by point and analyze how ol' Donny stacks up against our other world autocrats:

#1 **Build up a cult of personality.** Boom! Trump nailed this one! But in unique ways: 1) Unlike every other autocrat in the world, he built his cult via celebrity culture, not political work of which he had no working knowledge, which is maybe why he didn't fare so well at it once he came to power. 2) While Trump may have won the electoral vote, he never won a majority vote, and what I mean by that is he is not even close to the major league of autocrats like Putin, Xi, Modi and others that have upwards of 60-80% support of their citizens. At best, one could say Trump mustered 45% support . . . and even that number is higher than his actual cult count. Voters in the two-party system of the USA only have one of two polarized choices in an election, and many people vote for the team, not for the candidate. I would say that the true die-hard cult of Trump only constitutes (conservatively) perhaps 25-35% of the citizens. But . . . 3) Unlike other autocrats, Trump's cult is perhaps the most adamantly dedicated group on the planet, and simply cannot be swayed from their perception of reality, no matter how many facts, figures, and proof is offered . . . and seemingly will stay that way to the end of their lives. That sort of die-hard dedication to death is rare even in Putin's world.

#2 Appealing to populism, nationalism, ethnic pride or cultural/racial purity. Boom! Trump nailed this one too! To the max! Almost single-handedly reinvigorated a White Power movement, a MAGA movement, and movement to limit migration to make America somehow "pure" again. Did he actually believe in those things? Who knows, but it sure incensed his rabid base.

#3 Identifying "the enemy." Eh, he did okay on this one. Oh, he identified many enemies all right: immigrants, China, Iran, Mexicans, liberals, the left, the media, Hollywood, Hillary Clinton, SNL, Alec Baldwin, the BLM movement, protesters in Portland . . . ummm . . . you see the problem here? He actually had way too many enemies to make an effective rallying point out of any single one of them. At least over the long term. With Duterte: Bam! Kill evil drug dealers! End of story. And more effective for the aspiring autocrat.

#4 Extend executive power. Trump did a lot of the executive orders, but so has every single other president for decades. And executive orders can always be overturned. Did a lot more to extend his power by getting to pick three Supreme Court justices . . . but am I the only one that recognizes that this was mere happenstance? He walked into that, through no fault or credit of his own whatsoever. Would a Democrat have done anything differently than Trump did if the tables were turned? Now, what the Trump administration did do excessively well is appoint other conservative justices all over the country, at all levels. That was way more impactful . . . so important that we will come back to it in a moment.

#5 Stack the deck. Eh. Not really. He did no more than any other president to stack the civil service with his own people. Most are gone now, as the new Biden administration replaced them. It's part of the game. Well, except for the Trump appointments of judges. Seems like that one keeps coming up . . .

#6 Repress dissent and any citizen-based movements. Eh. Not really at all. If anything, Trump's presence energized the opposition and we've seen healthier public protest than I remember in my lifetime. And the Black Lives Matter movement not only exploded during his presidency, it went viral globally.

#7 Incapacitate the opposition political party. Hmmm . . . this one is trickier. But did Trump himself destroy the opposition? No, not at all. As evidenced by the opposition winning the next election. HOWEVER: is the Republican Party incapacitating the opposition long-term? Oh my, yes, and we'll get to that next.

#8 And perhaps most vitally: control the message. Did Trump totally control information in society? Um, hell no, not even close. The opposition voices and media were loud and clear, if not screaming, for the whole four years. He may have been a master of disinformation to be sure, and his banning from social media platforms attests to that . . . but it simultaneously attests to the fact that he in no real way controlled information, or shut down opposing media outlets, or anything else that other autocrats simply must do in order to utterly indoctrinate the masses with their own tailored message. Trump didn't shut down any media—media actually shut him down. Again, don't get confused here: Trump spouted non-stop for four years on all media outlets, and quite frankly the coverage of this single human's every thought and action and word was intense, and much of it was false or malicious or inflammatory . . . BUT, that is not anything like Xi's total control of the internet in China, Putin's total control of the media in Russia, or even Erdoğan jailing journalists or Polish PiS party taking over media companies.

The most blatant and potent message he did perpetuate though was his refusal to accept the verified results of the election he lost, and then created a whole storyline about how the vote was fraudulent, and by association that the entire democratic system of the USA was not to be trusted. While this message has been proven false by total lack of any tangible proof, nearly one hundred court rulings, and the certification of the 2020 election results by all 50 states, the latest polling finds that still 75% of Republican voters fully believe the Trump message and doubt the legitimacy of the Biden administration. Now that is a potent message that does not necessarily "control" the story . . . but it has created even more extreme polarization of the people, and a significant distrust in the democracy itself. That may be Trump's greatest legacy that will impact the future of the country.

Okay, so Trump was def a hugely successful populist, but an autocrat? Well, he sure acted like an autocrat, and he sounded like an autocrat, and he tried to force through legislation like an autocrat, and he tried to create his own version of the truth like an autocrat . . . but none of those things really permanently changed the American system yet, nor pulled ultimate power to himself or his position, which is exactly what successful autocrats do. I mean, if his cult followers would have pulled off the capitol insurrection of January 2021 and he stayed in power illegally, or if he would have openly seized power and refused to leave the White House, or if he would have called in the military to conduct an open coup . . . then yes! He would have definitely become an autocrat! So Trump really may have wanted to become an authoritarian autocrat, and really tried to become one, but he kind of failed at it. You personally may not have liked his personality, policies, practices or pronouncements . . . but the American democracy held, free and fair elections occurred, and a transition of power (all be it atypical) happened.

And for the endgame of discussion on autocracy, it is now time to get back to the party—which the Plaid Avenger usually lives for—but in this circumstance we aren't talking about my kind of party: we are referring to a *political party* that is responsible for the autocratic shift away from democracy occurring in the USA.

(BTW, this is not a personal opinion. The 2020 Freedom House Report on America specifically lists problems in the US that one typically associates with much more authoritarian regimes and failing democracies around the planet, and I quote the problems listed in that report: "pressure on electoral integrity, judicial independence, and safeguards against corruption. Fierce rhetorical attacks on the press, the rule of law, and other pillars of democracy coming from American leaders, including the president himself." So . . . yeah . . . wow.)

So why has the USA been backsliding on democracy, according to every rating system used in this chapter (V-Dem, Freedom House, Economist Democracy Index) and a whole lot of others? Well, if not due to any huge thing former President Trump did in the last four years, then we must broaden our scope to see what has been going on in the Republican Party for the last four years . . . and more importantly, what it will be doing in the next four years. This is where the real autocratic action has been happening that is starting to challenge the fundamental structure of the United States of America's democracy credentials. And honestly, as a global operative and objective observer, democracy in America is up for a very significant challenge right now. How so?

For starters, there are many existing elements of the uniquely American take on democracy that are . . . for a lack of a better phrase . . . not that democratic. Or entirely undemocratic. At issue in the USA are the increasingly powerful **counter-majoritarian institutions** . . . that is, any small institution or minority group that has the ability/power to invalidate, overrule, or revoke laws/policies/elections that actually reflect the will of the majority. Any democracy's legitimacy comes from the fact that it implements the will of the majority; if that will is routinely overridden by counter-majoritarian institutions, then it becomes much harder to call the system a democracy, if not downright impossible.

A quick example: the US Supreme Court is comprised of just nine unelected justices that have the power to destroy any single law/policy of the country. They are neither selected by the people, nor are beholden to the people in any way, shape, or form. And it only takes a majority of them (that is, just five humans) to overturn a law. So to make it really real for you on a hot-button issue soon to upset the country, just five unelected humans will likely overturn Roe vs. Wade and thus outlaw abortion in the United States of America, despite recent polling that suggests that 75% (or 247.5 million citizens) favor keeping it in place outright, or with some modifications. Think about that: five people will overturn a policy that 247.5 million people want to keep.

Doesn't sound very democratic at all, does it? In fact, it sounds downright autocratic, right? And that is just one of the most high profile counter-majoritarian institutions in the USA. Let me point out a few of the more problematic parts:

- Back to the Supreme Court: just nine unelected humans with ultimate judicial power are supposed to represent the 330 million opinions and views of all Americans (that's the total population of the country). Is that even realistic, much less democratic?
- The Electoral College system (which is the opposite of a direct democracy and a manipulation of direct voting) results in low-population states such as Alaska, Montana, and Wyoming (among others) having a disproportionately larger effect on presidential election outcomes.
- The US Senate itself is deeply undemocratic: every state in the country gets two senators, regardless of how many citizens they actually represent, which is why Wyoming has exactly as many senators as large states like California. Even though California has more than 68 times as many people as Wyoming. Power is biased toward sparsely populated territories.
- The increasing reliance of presidents on the power of Executive Orders, in which one single human (the president) can unilaterally decide to do grand projects or enforce/negate a social policy or even wage a war, without going through Congress . . . or really, getting input from anyone. Especially undemocratic if the president in power obtained office via a minority of the electorate.

Scholars may argue that these undemocratic elements within the system were put there 240 years ago by the Founding Fathers quite intentionally as proper checks and balances in their new form of government they created. However, that was when the majority of citizens were illiterate or ill-informed, populating only 13 colonies with 2.5 million citizens. The country has since expanded to the fourth biggest in the world in terms of territory, covering an entire continent, with a population of 330 million highly literate citizens with access to vast amounts of information and instantaneous communications. This is a very different world than the one old Thomas Jefferson and James Madison were designing for.

And that brings us to the last—and perhaps most offensive to my conservative cohorts in America—autocratic maneuvers undermining American democracy: **the radicalization of the Republican Party**. Let me make this as perfectly clear as I possibly can right up front: I NEVER use the pages of this textbook for politically-inspired ranting or attempts at ideological indoctrination! I want you all to be free thinkers, and follow your hearts and minds to the political party that best fits your worldview! And I personally try to remain as politically balanced, and ideologically aloof, as possible! And I actually HATE talking about American politics at all, since this textbook is about the whole damn world, and talking politics is polarizing and offensive to many! BUT, given the huge influence the American democracy has on the rest of the world, and its specific influence on other democracies as the world's leading democracy, it has to be pointed out what is going on in America right now with the Republican Party's autocratic actions . . . because, yes, it actually will end up affecting democracies worldwide! So here's what's up:

Even before Donald Trump's arrival on the Republican scene, the Party had been for decades struggling with a decreasing support base. Heavily populated states, and particularly big urban areas (read: cities) typically are more liberal and vote more Democratic. That was all fine and dandy a century ago when there weren't that many huge cities, and more people lived in rural areas than urban ones. However, the population game has changed significantly since then, and now the majority of people live in an urban area. And there are hundreds of places in the country that are now called cities or metropolitan areas that house this majority. To restate: urban, big city, heavily-populated states usually vote liberal/Democratic in most elections. What is a diminished, rural area-focused politician to do?

The answer: legally rig the system in your favor. Bend the rules as much as possible to gain an advantage. Ignore the spirit of the law in favor of focusing on the letter of the law to make things go your way. That is exactly what the Republic Party has been doing for decades. Think: gerrymandering, passing voter suppression laws, filibustering, obstructionism, court stacking, promoting extreme political polarization, full embrace of disinformation campaigns. and sowing seeds of mistrust in the very democratic system itself. The Party is now also focused heavily on stacking state election commissions with die-hard loyalists in every state in which they control the legislature . . . cementing autocratically-minded power over future elections at local levels. Let me once more make things as expressly clear as possible: Republicans have done all of these actions 100% completely and legally within the existing legal structures and institutions of the American democracy. But truly democratic actions, they are not.

To achieve this, the Republican Party has become a master of using those counter-majoritarian institutions to their huge advantage, while simultaneously being shielded by them from actual competition that would force them to change their strategy or policies. A super-charged and super-clever minority manipulating counter-majoritarian institutions to not only stay in power . . . but to actually expand that power! Dig these very disturbing trends:

- Republicans have only won the actual popular vote for the presidency only one single time since 1988 . . . yet Republican presidents have led the country for almost half of that period.
- The Democratic and Republican Parties each control 50 seats in the U.S. Senate, even though Democratic senators represent 40 million more voters than do Republican senators.
- Trump appointed nearly as many federal judges in just four years as his predecessors did in eight (two-term presidents).
- All three of the most recent justice appointees to the Supreme Court were appointed by President Trump, a leader who did not win the popular vote . . . made more bizarre by the fact that those justices were confirmed by Senate majorities that (for reasons already discussed) don't actually represent a majority of Americans.

I want you to think critically about the court-stacking component of all this. Does it sound like anything we have discussed previously in this very chapter? If not, flip back just to the previous section . . . this is the exact same playbook that Poland has been following!!! Judicial deviousness! Stack the courts at every level, including the highest court in the land, with like-minded conservative thinkers that agree with all Republican policies. That in turn, and over a long period of time, gives your party power to interpret the laws and legal actions . . . and even decide on the legality of presidential elections when they occur, with the ability to "legally" overturn said elections! Ah!!! Now you are starting to see the real danger here! A minority that is increasing and implementing its power over the majority through legal machinations! That is autocratic behavior, my friends!

Add to all that: the radicalization of the Republican Party I am referring to entails a few more autocratic elements as well. The open embrace of the former president's refusal to recognize the election loss along with the Trump 2020 election disinformation campaign is an extremely problematic, and autocratic, strategy of the Party. (BTW: misinformation is false or out-of-context information usually presented accidentally or ignorantly, while disinformation is false information presented intentionally to deceive or mislead.) In addition, a large percentage of the Party's members have now fully and totally dedicated themselves as followers of Trump, the man, over the party itself. In effect, a true cult following has evolved out of the cult of personality that helped propel Trump to office. And true cult status indicates that the followers believe no truth or reality other than that presented by the leader, and also believe that any means justify the ends to achieve the mission of the cult. The fact that one of only two political parties of the USA now has a substantial number of its faithful dedicated at that level to a single person over its own ideology or democracy in general is troubling. To say the least. Through both its legislative and judicial maneuvers, in combination with an increased allegiance to an aspiring authoritarian leader, the Republican Party appears more radicalized and autocratic in nature, which is a direct challenge to democracy. If the will of the majority does not rule a nation, that nation can no longer be called a democracy.

Dicing up a democracy's very foundation behind a screen of legality and constitutionality? That seems familiar to readers of this chapter at this point! We've already named some folks in this list that have followed this exact strategy: Putin, Erdoğan, Orbán, Jarosław Kaczyński and his PiS crew . . . and those are just the ones I included on my list of biggest offenders. There are many lesser-known politicians and political parties around the world making the same moves, which is why what happens in America is of such great significance for the future of democracy everywhere.

So it is not former US President Trump that is the autocratic face in the USA, it is a political party, much the exact same way the Poland's democratic credentials have taken a huge hit due to the machinations of the PiS and its takeover of the judicial branch. If a radicalized Republican Party representing a minority of American citizens ends up changing major laws or overthrowing societal norms or overturning a future election, it will have ripple effects that encourage autocrats around the world to follow similar strategies. And it will also discourage the formation of new democratic movements and the nurturing of fledgling democracies everywhere. Indeed, it could become a major catalyst to a whole new age of government types and leadership styles that will affect the future of democracy, autocracy, and billions of citizens' lives. And that would usher in . . .

AN AUTOCRATIC "NEW WORLD (DIS)ORDER?"

That's right! It's final autocratic analysis time! Let the speculation run asunder! While we truly may be at a major turning point in history, the democratic world has just started this very long, and very slow, turn, and it will be years before we even find out if the trajectory has truly changed. So really all I will present you right now are some really broad generalizations of how the world may look soon, as this populist/autocratic wave washes over us and this new breed of world leaders reshapes the world.

Order up! It's New World Order time!

I call this section a **New World Order**, because there is every indication that we are heading towards a major shakeup of a system we have all taken for granted as the standard operating procedure for how our world works for decades. That is, the term "new world order" is used to refer to any new period of history evidencing a dramatic change in world political thought and the balance of power. After WW2, the term referred to the global governance and rules ushered in by the creation of the United Nations, and of the **Bretton Woods system** of commercial and financial relations between Team West countries. It was a system defined by the eminence of democracy as the primary ruling system, and free market capitalism as the economic engine that drove it. An emphasis on individual human rights and civil liberties was also an integral component to this new world.

After the collapse of the Soviet Union in 1991, the New World Order would indicate the movement of most of the world to liberal democracy and free market capitalism; a rise in free trade in the world along with an increase in international institutions of trade like the WTO and all of the trade blocks you read about in Chapter 6; a deeper empowerment and growth of other international institutions like the UN and the EU; the rebalancing of major world powers since there was no reason for the US and Russia to be enemies; other nations like China and Brazil rising in stature to join the ranks of world power status; and all of this dominated by the economic and military power of the USA. That is the world order we have lived under for the better part of four decades, and even longer if you work in those post-WW2 developments. And even if Russia and the US continued to hate each other despite the end of the Cold War, the system was still strong, vibrant, and growing.

And now, with the explosion of populism and autocracy across the globe, all that "world order" is totally up for grabs. For starters (and remember we are merely speculating at this point) here are some general global trends to watch in the coming years that are being influenced by the adoption of autocracy across many countries, especially within existing democracies:

1) *Democracy declines.* It appears the whole liberal democracy growth curve is heading down. For decades since the demise of communism, it appeared that the whole world would eventually adopt this great western political philosophy of "rule by the people." Well, turns out that maybe a lot more people are suddenly getting tired of the disorder and gridlock that such a system can produce, and has produced in spades in the last decade across the western world. As mentioned previously, strongman leaders with cults of personality are on the rise (even in Europe and Turkey), most of the states of the former Soviet Union have reverted to one-party states (including Russia), many states of the Middle East have stubbornly refused to change their rightist regimes system at all, and China's style of autocratic one-party rule looks increasingly stable and desirable in a chaotic world. Look for more states to choose that route over the liberal democracy path.

2) *Supranationalism declines.* It appears we are moving backwards on the whole political supranationalism organizations front. I have referenced already the weakening of the UN and the ICC. Of much greater immediate significance is the possible further disintegration of the EU. I talked about it a bit back in the European Addendum, so I won't repeat much of it here, but there is a real possibility of other member states leaving, although I think the core will hold and it will remain a viable source of power for the member states. The bigger issue to consider here is that we appear to be returning to the old school concept of nation-state sovereignty as the highest order of political power on the planet. Which means a weakening (if not total abandonment of) the singular, international rules based system which has become the legal/cultural structure of our modern world since World War 2. That will equate to a lot less conflict resolution and global harmony. But it's not like we ever had a lot of the latter anyway I suppose. Let's call this a new age of **neo-nationalism.**

3) *American supremacy declines.* It appears that the age of American supremacy is quickly coming to an end. The rise of China, India, Brazil, Turkey, Indonesia, Iran, and Nigeria as regional or global powers in their own right is dispensing with the notion that the USA rules the world. Especially the whole rise of China, which is currently already a world superpower on par with the US in every respect except militarily, and they are working on that quite fast. But I am here to particularly mention populism's role in this decline of American power. Again, no offense to President Trump and his supporters, but if an autocratic America becomes more isolationist, which appears to be the plan, then it is no longer looked to as the bastion and protector of global free trade. And with the popularity of democracy waning worldwide, even in the US itself, it becomes harder and harder to see how America is the "leader" of the free world. What will that even mean in the near future? A "leader" of what exactly? Pax Americana kept the democratic and free trade world safe since WW2. The US will certainly continue to have the world's mightiest military . . . but for what reason if not to guard and protect the "current world order," which I am suggesting is going away?

4) *New (soon to be Old) World Order declines.* On top of that: this rise of autocracies spells the end to Team West's global supremacy. The post-1945 Breton Woods rules-based order is being eroded and will likely never return. The superpower rise of China; the "apparent" stability and re-rise of Russia; a back-sliding of the influence of Western liberalism; major advances by autocratic/undemocratic forces within the West itself; and the dwindling of US democracy leadership and promotion efforts worldwide are all contributing factors as to why autocracy is ascending, and accelerating the demise of that World Order we are all accustomed to. Remember, that New World Order I keep referring to here was a Washington-designed, Western-built, post-WW2 order, that mainly benefited the US, Western Europe, Japan and their allies . . . which means its destruction equates to further erosion of power for the US and its allies on the global stage. And it means there is now an accelerating shift of economic and political power from the US/Team West to the Asia-Pacific region, where China's autocratic model prevails and spreads its influence outward. No matter how you slice it, from Christopher Columbus to Donald Trump: the 500-year-long era of Western global supremacy is coming to an end.

5) *US democracy declines . . . or implodes into war.* Biggest reason that the USA is likely to undergo very turbulent times ahead, and possible outright revolution, is because of a simple factor that the wanna-be autocrats of the USA don't possess—and that virtually all the other autocrats on this list do. What would that be? A majority or even

super-majority of citizens that support them! Xi has got 99% of peeps in his pocket. Putin usually has 70-80% approval rating; Modi 75%, Duterte holds a solid 65-70%. On their very best possible day, with bitter political polarization fueling the biggest voter turnout in US history, the Republican candidate Trump received 74 million votes, or 46.8% of all votes cast. Repeat: on their very *best* day with an *extremely popular* candidate and *huge* voter turnout, Republicans received less than 50% of the vote. That's a minority, albeit a slim one. But if the will of the minority overrides the will of the majority, you can't really call it a democracy anymore. And if minority rule persists, is entrenched into unchangeable laws, or the ruling minority becomes even smaller in numbers, it increases the likelihood that intractable political disputes turn into open hostilities between infuriated citizens. Autocratic actions of a radicalized Republican Party will result in a challenge to the very nature of their democracy itself . . . and the stability of the state.

Interesting stuff, right? Well, those are the general trends that this autocratic alignment is affecting, and will be affecting into the future. If you want to investigate the topic, or any particular country's ratings, in more detail, here are the sources the Plaid Avenger employed and referenced throughout this chapter:

- 2020 Freedom House Report, ***A Leaderless Struggle for Democracy:***
  https://freedomhouse.org/report/freedom-world/2020/leaderless-struggle-democracy
- 2020 Rule of Law Index from the World Justice Project, a Washington-based civil society initiative:
  https://worldjusticeproject.org/our-work/research-and-data/wjp-rule-law-index-2019
- Varieties of Democracy (V-Dem) Democracy Report 2021, ***Autocratization Turns Viral:***
  https://www.v-dem.net/media/filer_public/74/8c/748c68ad-f224-4cd7-87f9-8794add5c60f/dr_2021_updated.pdf
- The Economist Intelligence Unit Democracy Index 2020, ***In sickness and in health?:***
  https://www.eiu.com/n/campaigns/democracy-index-2020/

You can find out all sorts of excruciating details about the state of democracy in every single country around the world from these reports, as well as global trends of autocracy, which is reshaping the world. But for now, let's wrap up the chapter and get on to some of the more sinister side effects of increasing nationalism, populism and autocratic realignment among the world's states: conflict. Turn the page to start the war!

Level **Blue**

Status **No active shooting. Arms build-up may be happening, but discreetly. However, political or ethnic or ideological differences on site make conflicts possible, and perhaps inevitable.**

Level **Yellow**

Status **Active violence by individuals or small groups is happening, or getting very close to erupting. Arms build-up and other preparations are happening, and not discreetly. Political or ethnic or ideological differences getting hot enough for active, organized forces to be on the move. Could go down any moment.**

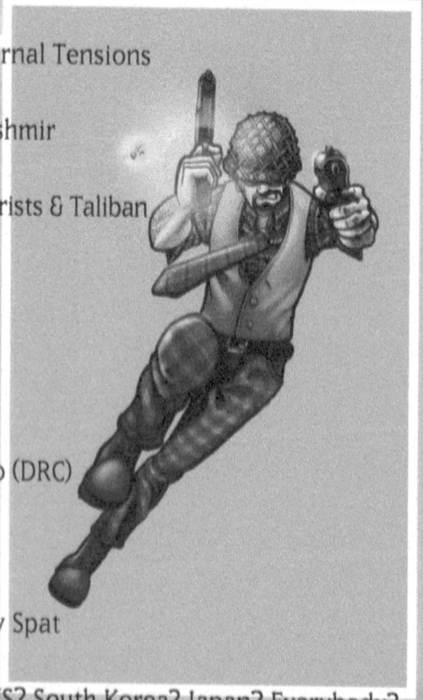

Level **Red**

Status **GAME ON! Active violence by individuals, small groups, and/or large organized forces like a state military are happening. Nothing is discreet, and open war may have been declared. People are shooting other people, and you can watch it all on CNN, Facebook, and also read tweets about it from the front lines.**

CHAPTER OUTLINE

1. Ideological Showdowns & Territorial Throw-Downs
   - 1.1. The Balkans
   - 1.2. Russia-Georgia War
   - 1.3. Russia-Ukraine: Terrible Internal Tensions
   - 1.4. Israel vs. Palestine
   - 1.5. India-Pakistan: Hot Spot: Kashmir
2. Flailing, Failing States
   - 2.1. Pakistani Problems with Terrorists & Taliban
   - 2.2. US in Afghanistan
   - 2.3. Iraq Is a-Wreck
     - 2.3.1. Kurdistan Will Cut Out
   - 2.4. Syria & Lebanon
   - 2.5. The Yemen Crisis
   - 2.6. Venezuelan Implosion
   - 2.7. Democratic Republic of Congo (DRC)
     - 2.7.1. Hot Spot: Eastern Congo
3. International Intrigues
   - 3.1. Senkaku Showdown!
   - 3.2. South China Sea & the Spratly Spat
   - 3.3. The Sahel Region of Africa
   - 3.4. North Korea vs. ?????? The US? South Korea? Japan? Everybody?
   - 3.5. Somalian Skullduggery
   - 3.6. Iran vs. the US, or the World If Necessary
   - 3.7. Eastern Mediterranean Sea Showdown

Plaid Avenger's Hotspots of Conflict

26

WHAT CAUSES HUMANS TO KILL EACH OTHER?

One from the west: "And there went out another horse that was red: and power was given to him that sat thereon to take peace from the earth, and that they should kill one another: and there was given unto him a great sword."

—The Revelation of St. John the Divine: 6:4

And one from the east: "Now I am become Death, the destroyer of worlds."

—from The Bhagavad-Gita

Since the end of the Second World War in 1945, there have been over 250 major wars in which over 23 million people have been killed, tens of millions made homeless, and countless millions injured and bereaved. In the history of warfare, the 20th century stands out as the bloodiest and most brutal—three times as many people have been killed in wars in the last 90 years than in the previous 500.

The nature of warfare has also changed. From the set-piece battles of the earlier centuries, the blood and mud of the trenches in the First World War, and the fast-moving mechanized battlefields of World War II, to the high-tech "surgical" computer-guided action in Iraq and Afganistan, war as seen through our television screens appears to have become a well-ordered, almost bloodless, affair. Nothing could be further from the truth. During the 20th century, the proportion of civilian casualties has risen steadily. In World War II, two-thirds of those killed were civilians; by the beginning of the 1990s, civilian deaths approached a horrifying 90 percent.

This is partly the result of technological developments, but there is another major reason. Many modern armed conflicts are not between states but within them: struggles between soldiers and civilians, or between competing civilian groups. Such conflicts are likely to be fought out in country villages and urban streets. In such wars, the "enemy" camp is everywhere, and the distinctions between combatant and noncombatant melt away into the fear, suspicion and confusion of civilian life under fire.

Also, people kill each other because of ethnic differences, religious differences, cultural differences, and any other difference that distinguishes groups of humans. Wars are also fought over resources, land, and political control of areas—particularly where states or empires have dissolved into multiple entities, such as British India, or Yugoslavia. Now we have new wars with new dimensions the likes of which the world has not seen before: Battles of ideologies that span the entire planet.

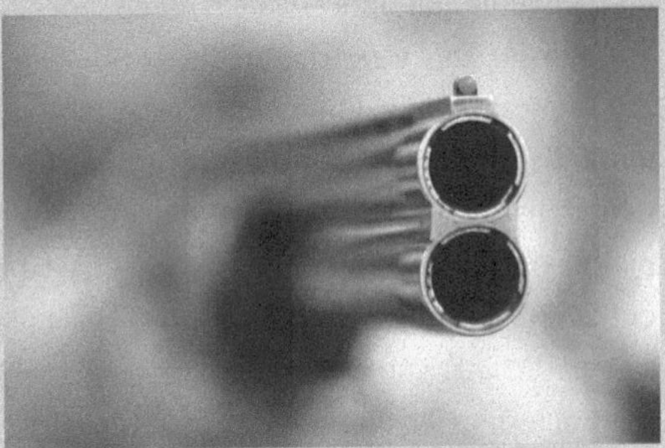

Go ahead; make my millennium.

Perhaps you are thinking the Cold War was just such a conflict—and you would be right. But it was cold, as its name suggests. Here in the 21st century, we have the War on Terror, the War on Western Imperialism, and the War on Drugs, which are global ideological battles—and they are hot ones! People are actively shooting at each other in these wars, wherever they are sponsored, all over the planet. That's the new part.

Conflict within or between states, and ideological conflicts in which the entire globe is the battlefield: we've got it all in our day and age, but we can't possibly cover all the active conflicts happening. The Plaid Avenger wants to pick out a few handuls situations that have global implications, as well as a few more that may be heating up in the coming years.

IDEOLOGICAL SHOWDOWNS & TERRITORIAL THROW-DOWNS

Our first round of this global boxing match will cover a handful of places where good old fashioned fighting over control of territory is taking place. These actual or potential spats are basically just sovereign states (or groups of peoples) squabbling over actual chunks of land that each thinks is theirs to control . . . and which in these circumstances have led to actual fighting, or have the promise to lead to bullets flying in the next year. To make it just slightly more complicated in some cases, there are ideological differences between the warring parties which exacerbate the tensions . . . for example, the spat between Russia and Ukraine is a a territorial challenge, which has at its root a fight between Ukraine staying in Russia's orbit versus joining the European Union. FYI: there are all sorts of borders that are disputed by sovereign states across the planet, such as Senkaku (Diaoyu) Islands in the East China Sea (controlled by Japan, claimed by China), Crimea (claimed by Ukraine and Russia), and Jammu and Kashmir (claimed by India and Pakistan), but in this day and age those disputes are largely settled through legal and diplomatic means. But not the ones below. . . .

THE BALKANS

Once known as the "powder keg of Europe," the Balkans once more may be expected to explode back into war at any time. The roots of the conflict are very deep, yet they can be directly linked to the independence of Slovenia and Croatia in 1991 from the state known as Yugoslavia. The area consists of three main ethnic groups and several loosely connected factions: the Serbs (Orthodox Christians), the Muslims, and the Croats (Roman Catholic) with links to Western Europe.

Once upon a time, there existed the Socialist Federal Republic of Yugoslavia. Within that republic were six constituent states, Bosnia-Herzegovina, Croatia, Republic of Macedonia, Serbia-Montenegro, and Slovenia. The conflict begins when Croatia and Slovenia announced their independence from Yugoslavia. This gave rise to surrounding states in the former republic in ousting their one-party communist states systems and electing officials with nationalist platforms. Here's the main problem: all of those regions are comprised of several different ethnic groups who believe they are misrepresented. Croatia still has issues, and quite frankly it's a miracle that Bosnia has not re-ignited back into bloodshed given its ethnic tensions. However our best example of Balkan bravado that has both local

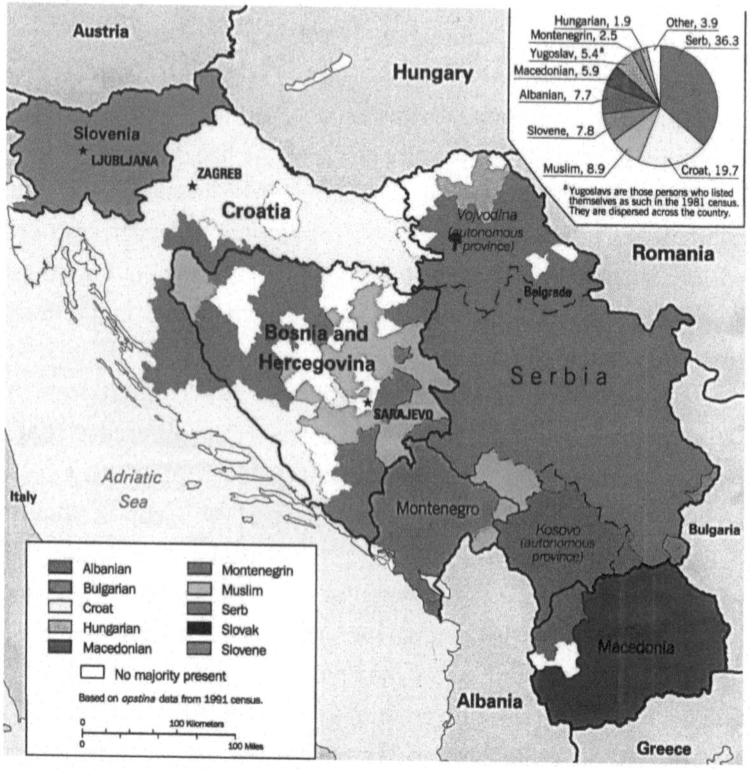

Former Yugoslavia and the ethnic division therein.

and global repercussions is Kosovo, a region comprised of 88% Albanians with an ethnic minority of Serbs. The Kosovo-Albanians want to leave the Serbian Republic, while the Serbs within Kosovo fear what will happen to them if they stay in Kosovo.

On February 17, 2008, Kosovo formally declared independence from Serbia. Serbia rejected this notion and upheld that it was illegal under U.N charter; Albania, who strongly supported the KLA (Kosovo Liberation Army) who fought guerillas wars against Serbia, was the first to support Kosovo.

Are you ready, 'cause here's where it gets interesting. Technically speaking, Kosovo is a region within Serbia, which is an internationally recognized state by the U.N., which then gives Serbia the right to preserve its territorial integrity. The wrong words and the wrong moves could push this region over the edge and start a war. No biggie, right? Well, except for the fact that this is the same region that sparked WWI.

Dig this: The US and a dozen other EU countries, including the UK and France, recognized the Kosovo independence claim. Unfortunately for Kosovo, both Russia and China, which are both veto-wielding members of the UN Security Council, side with Serbia on this question of sovereignty. I'm starting to see a larger world showdown over this tiny little province. Which ties directly to our next section on a couple of other tiny little provinces causing calamity over similar claims of independence, but these two are backed by the Russians!

RUSSIA-GEORGIA WAR

Let's get to that rascally Ruskie move right now! The 2008 Russia–Georgia War was an armed conflict in August 2008, between Georgia on one side, and Russia together with most ethnic Ossetians and ethnic Abkhazians on the other.

See, this whole area used to be a part of the mighty USSR, but once it crumbled in 1991, Georgia claimed independence and claimed these two ethnic enclaves as part of their territory. To be fair, it historically had been counted as part of Georgia for a long time. However, many ethnic Ossetians (with a smattering of Russians in the area) actually fought a war for their own independence from Georgia in 1991–1992 which left most of South Ossetia under control of a Russian-backed internationally unrecognized regional government. Some Georgian-inhabited parts remained under the control of Georgia. This was pretty much the exact same situation in Abkhazia after the War in Abkhazia (1992–1993).

With the story so far? These two dinky provinces were technically part of Georgia, but were "protected" or "administered" by Russians. By the way, in what is a richly ironic claim, the Russians always said that they were there simply to "protect" these locals from genocidal tendencies of the Georgian government . . . much the same way the US and NATO was protecting the Kosovars from the Serbian government. Ah! Starting to see the connections here?

Let's pile kindling onto this pre-fire: then Georgian President Mikheil Saakashvili was a hugely pro-western, pro-US leader who had been petitioning for EU and NATO membership for years, mostly to shield his country from Russian influence (just like Estonia, Latvia, Lithuania and Poland did back in the late 1990s). Of course, that really rankled the Russians. That Georgia might join NATO was an option utterly unacceptable to Russia, and they were very vocal about this, so tensions rose even further. Then in 2006, US President Bush gave a speech at a NATO conference where he openly pushed for member states to put Georgia on a fast track for NATO membership. Pressure built more.

For reasons which continue to elude the entire planet, on August 8, 2008, Georgia launched a large-scale military attack against the self-proclaimed Republic of South Ossetia. Within seconds, Russia reacted and deployed combat troops in South Ossetia and launched bombing raids farther into Georgia. Yeah, if you haven't picked up on this yet, be sure to take note: the Russians were totally prepared for this attack to happen. Russians landed support craft in Abkhazia and opened up a second front there.

Long story short: After five days of heavy fighting, Georgian forces were totally ejected from South Ossetia and Abkhazia. Russian troops entered Georgia, easily kicking the crap out any resistance. Fighting ended and in the subsequent peace agreement the Russians withdrew from Georgia, but remained as "guardians" in the two small enclaves.

The newly independent states of Abkhazia and South Ossetia???

Oh but here's the fun part: On August 26, 2008, Russia recognized the independence of South Ossetia and Abkhazia. The US and Team West said, "Hey! You guys can't just invade a small ethnic province, help them secede, and then recognize their independence! That's not cool!" To which Russia responded: "Oh really? You mean like what you guys did in Kosovo?" Ouch! That stings!

2014 UPDATE: Since the 2014 Ukrainian meltdown and Russia's subsequent annexation of the Crimean Peninsula from them, the peeps in Georgia are rightly worried that Russia is feeling its oats once more and could ratchet up the pressure on them significantly. And they could be right. Given Russia's current attitude and new found mission to 'protect' Russian peoples in other countries, I think it is safe to assume that Abkhazia and South Ossetia will NEVER revert back to Georgian control. And perhaps the Ruskies will be even more mischievous in the coming year if Georgia makes any moves to join NATO again. . . .

Now this international fracas is not settled yet, and it certainly is calling the whole perceived concept of sovereignty into question, but I want you to know this for now as we watch the future events of this area: Russia is back! This invasion has announced the return of Russia not just as a military power re-establishing control over its neighborhood, but also as a military power that is not afraid to use force to stop what it perceives as threats to its security, and by threats I am going to specifically call one by name: NATO. This Georgian debacle could well have been started by former President Saakashvili because he truly believed that NATO would come save him, and that is why it's an international hotspot, even though based on local factors. Georgia is not the only place that Russia may be facing off with NATO and the West, which brings us to . . .

RUSSIA-UKRAINE: TERRIBLE INTERNAL TENSIONS

As elaborated on back in chapter 8: Eastern Europe, Ukraine has now become the center stage for the ideological battle for control between Team West and Russia. The Plaid Avenger believes that to determine which world power has the upper hand in this part of the world, simply keep your eye on the events in Ukraine. (2014 UPDATE: Yep. Called it. All of that action has come to pass since the illegal annexation of Crimea by Russia in 2014.)

Like Georgia, Ukraine used to be part of the USSR, with Russia as the supreme overseer. Like Georgia, Ukraine declared independence when the USSR collapsed. Like Georgia, Ukraine had a mini-revolution that put pro-democracy, pro-Western

leaders into power. Like Georgia, Ukraine petitioned for EU and NATO membership. And like Georgia, former President Bush lobbied to get NATO to accept the Ukraine.

Unlike Georgia, Ukraine has yet to be invaded by Russia. Yet. Oh wait, they actually did invade part of the Ukraine named the Crimean Peninsula in 2014! Then annexed the place officially that same year, and now totally control it. But there could still be further Russian intrusions into eastern Ukraine, so this mess is not over yet.

That is exactly why I want you to know about this past and still current potential flashpoint. More could happen, but in much more sneaky ways than the invited invasion of Russian troops in the Georgian example discussed previously. Because unlike Georgia, the country itself is internally divided on which "team" to join. To be brutally honest, the stakes are way higher in Ukraine than they ever were in Georgia. Why?

First, check the map above to see this pattern: there are a lot of ethnic Russians living in Ukraine—approximately 20% of the total population. Look where they are really predominant: in the eastern half of the country, you know, near Russia, and specifically packed into that little area referred to as the Crimean Peninsula. This is where it gets interesting . . .

The Crimean Peninsula has been a major repository and home base for the Russian navy for decades, perhaps centuries depending on your historical perspective. When Ukraine declared independence, Crimea ended up as part of their country, much to the chagrin of many Russians living there, as well as the Russian government itself. There are still dozens of naval facilities on Crimea that were leased and operated by the Russian navy. Those pro-Western, pro-NATO peoples in Ukraine would have liked the Russians to pack up and get the hell out. The pro-Russian peoples, and mother Russia herself, had an entirely different view of the matter.

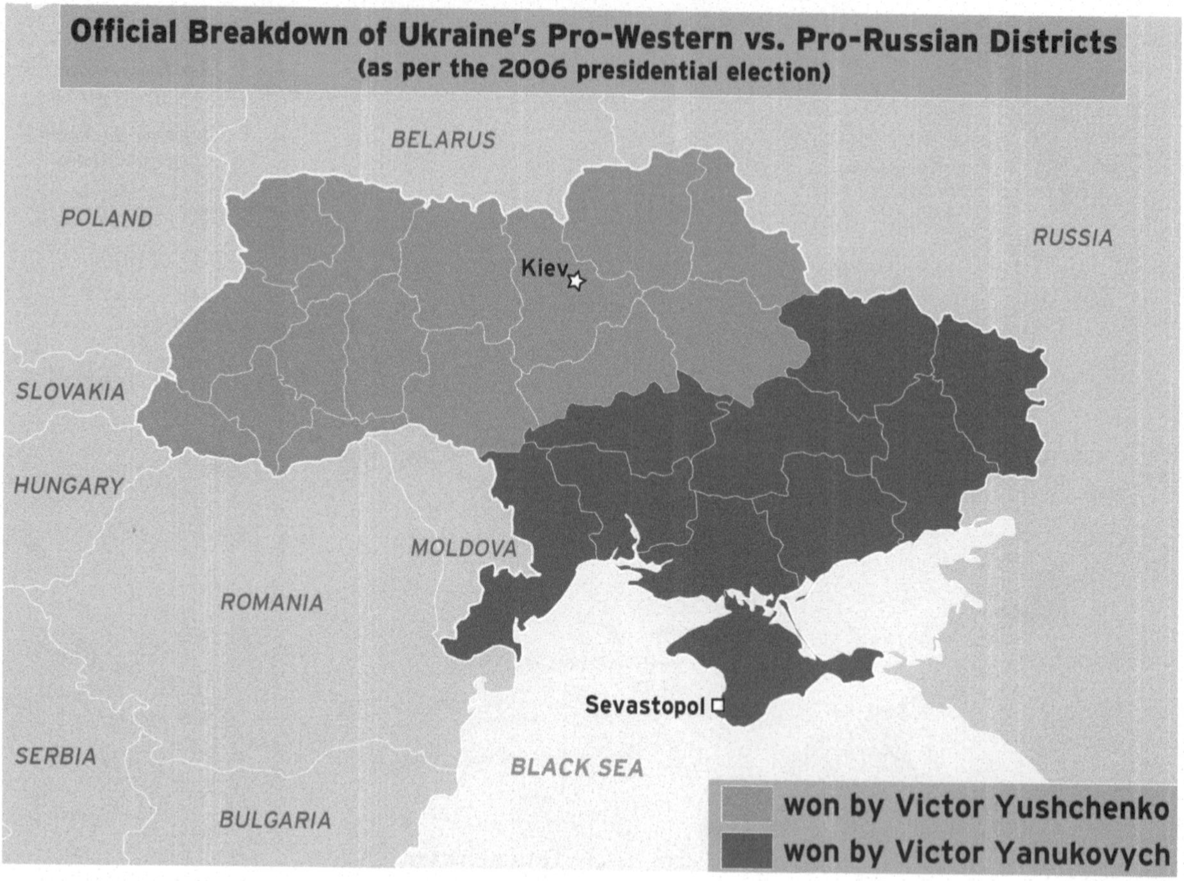

This division between pro-West and pro-Russia can be seen most clearly in the 2006 presidential election map above: the pro-Western west, versus the pro-Eastern east! Ha! Too easy! Which brings me to the second big point about this hot-spot: unlike in Georgia, Russia has a big support base inside the country! Especially in that Crimean section, which has tons of ethnic Russians in it! How do I know? When Ukraine invited NATO to conduct a joint military exercise there back in 2006, massive protests broke out when US Marines arrived, mostly to chants of "Occupiers go home!" Yikes!

Is the Plaid Avenger suggesting that Russia might officially invade? No, not really. They don't have to. They are already working very hard to secure support with ethnic Russians within Ukraine, and to monetarily support pro-Russian political candidates there too. The pro-Russian politician, Viktor Yanukovych, was the guy thrown out of office during the pro-West Orange Revolution of 2004, then became Prime Minister of the country in 2006, then became President again in 2010. Then was run out of the country entirely during the 2014 Ukrainian Revolution. How is that for a flip-flopping country that is unsure of its direction?

When that pro-Russian Viktor Yanukovych won the presidency back in 2010, he immediately announced that Ukraine would not be joining NATO, and also immediately extended the Russian leases on those Crimean naval bases. That in turn got Russia to reduce the rates on all the oil they sell to Ukraine. But all was not to end well for this situation. Here are the updates that I made in this very book at the time the events were unfolding:

2012 UPDATE: The fight is still alive! Further evidence of this West/Russia divide in Ukraine erupted badly in May, as fist-fights broke out in their Parliament when the ruling party put forward legislation to make Russia one of the country's official languages. Seriously? Congressmen beating the hell out of each other over a language status. Yep. That's the level of passion that exist in this country that is fighting itself over its very own soul. More to come, for sure.

Ummmmmmm . . . BOOM! Called It!

2015 Insane on the Brain, Ukrainian Unraveling Update! So glad I pointed out the Ukrainian rift for all these years to all the plaid students who used this text! Because that prediction was dead on the mark . . . literally. Unless you have been comatose since your bobsledding accident in the 2014 Winter Olympics, you have at least heard that Ukraine melted down politically once more. Massive street protests culminated in the overthrow of President Viktor Yanukovych, which prompted the Crimean Peninsula to declare independence and ask for absorption into Russia! Which then happened!

The Ruskies have had tens of thousands of troops massed on the Russian/Ukrainian border for months, and now there are similar "join Russia" protests across eastern Ukraine that have turned bloody and violent. Will Russia invade and convert other parts of the state into Russian territory? Who knows? It is all such convoluted chaos right now, that anything could happen! Team West and NATO are furious, the Ruskies are indignant and flexing their muscles further, and some are predicting that this is the start of a new Cold War!

I think that may be a bit of an overstatement, but this is certainly a titanic regional game-changing event. Here in late 2021, civil war in Eastern Ukraine is still open, active, and hot! Russian troops are still massed at the border, who knows how many guns and covert operatives Russia has already sent into the mess, and there is talk of the US and NATO training and arming the Ukrainian government. This mess is far from over and will shape the region for some time to come.

Few countries acknowledge Russia's claim to Crimea as legal and the Ukrainian government still claims Crimea as part of Ukrainian territory. However, Russia opened a bridge across the Kerch Strait (linking Russia and Crimea with the act) in 2018. Ukrainian naval ships were also seized by Russian navy vessels and the Sea of Azov shut down when Russia anchored a tanker ship beneath the Kerch Strait bridge also in 2018. Here in 2021, the low level civil war in the east of the country is still going, with no end in sight.

ISRAEL VS. PALESTINE

The Israel-Palestine conflict has very deep roots—stretching back thousands of years. In ancient times, the area that is now Israel was inhabited by Jews who had a degree of autonomy and self-governance. In 70 AD, the Roman Empire, who controlled the land, burnt down the Jewish temple and kicked all of the Jews out of Israel. The next 1900 years were known as the Jewish diaspora, when Jews were scattered around the world and often persecuted for being Jewish. After a particularly atrocious bit of persecution in World War II known as the Holocaust, European countries and the United States decided that it was time to help the Jews to return to Israel and have their own homeland. The state of Israel was created in 1948; Jews from all around the world migrated to Israel to be free.

The problem: a new group of people were living in that area who were calling themselves the Palestinians. These Palestinians were mostly Arab and Muslim; and by "new," I mean that they had only been living there for the 1900 years since the Jews were ejected. To make room for the rejuvenated state of Israel, some of these Palestinians had to go move out or adjust to being a monitory in a new Jewish-dominated state. This was based on the proposed UN Partition Plan of 1947, which was to divide up the territory into two new states: one for the Jews named Israel, and one for the Arab Muslims named Palestine. On the eve of this partition plan taking effect, Israel unilaterally declared independence. Say what? What was up with that? Well, all the surrounding Arab countries had boycotted the vote at the UN on this whole business, and had massed troops on the border with the intent of wiping Israel out as soon as the partition was to become law . . . so Israel pro-actively declared independence in 1948, and WAR ON! The result? Israel won, actually increasing their territorial gains from the original partition plan, and 700,000 Palestinians left Israel and settled in the countries bordering Israel, including Jordan, Egypt, Syria, and Lebanon. The Palestinians call this flight/forced removal in 1948 the **Nakba** (*al-Nakbah*) or disaster. To make matters worse, all of the countries bordering Israel were Arab and Muslim, and their governments wanted to look tough on Israel to make their people happy. For this reason, the Arab countries demanded that Israel take the refugees back. Israel said "no" and these refugees quickly faced a Catch-22 situation: Israel would not let them come back to their former homes and the Arab countries they migrated to would not find them places to live either.

As you can imagine, some Palestinians turned violent quickly and began forming terrorist organizations to attack Israel. Again, to look tough on Israel, the surrounding Arab countries went to war with Israel multiple times in the last 60 years, ultimately losing each war every time, and thus limiting outside influence ever more, each time. After each of these wars, Israel took possession or control of more of these lands, establishing de facto control of the West Bank from Jordan in 1967 and the Golan Heights from Syria in 1967. Israel still controls these areas and claims that it needs these lands because of the geographical and strategic locations of the sites for their national security.

Currently, the areas referred to as the "Occupied Territories" consist of the West Bank and the Gaza Strip. These areas were not originally part of Israel in 1948, but were taken control of by Israel after a 1967 war. These areas are predominately Palestinian and the Palestinians in these areas want their own state. However, each part wants different leaders. The Palestinian Authority (PA) has been the pseudo-government of this pseudo-state, and has been dominated by the political party Fatah since its inception. However, Fatah and the PA have become unpopular with the Palestinians because of rampant corruption and their inability to make any real progress on the big issues.

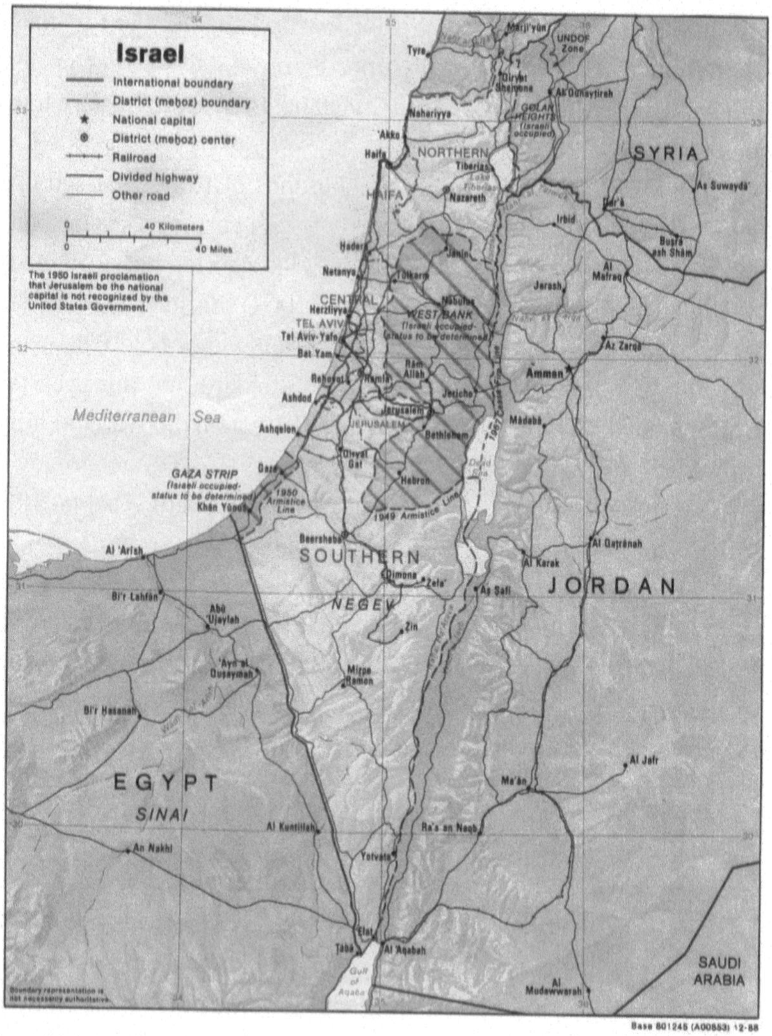

Israel and the occupied territories (West Bank + Gaza = Palestine, but also Golan Heights is "occupied" but officially still belongs to Syria).

For this reason, a more radical political party named Hamas continued to gain popularity in parliament. A major rift occurred in the summer of 2007: the democratically elected Hamas completely took over the government of the Gaza Strip, and kicked out their Fatah opponents. Now this pseudo-state of Palestine is even divided amongst itself: The Fatah dominated PA controls only the West Bank, while the Hamas party controls the Gaza Strip. Let's make it even messier: Hamas has a history of supporting terrorism in Israel, and many western governments (the US, the EU, Israel) consider Hamas a terrorist organization and, as such, will not deal with them. So why would locals vote for Hamas? Hamas has the support of citizens because of its humanitarian aid, social work, and charities within the country. In other words, Hamas supports the needs of the people and they support Hamas. It should be noted that many question if Hamas exploits the charities to fund other endeavors. Like the aforementioned terrorism.

The situation only looks like it will worsen since Israel closed off the Gaza Strip to all flows of goods and traffic. This is a move intended to punish the more militant Hamas, while Israel, the US and the EU are holding up the PA as the true leaders of the Palestinian people in a desperate bid to try and normalize relations. But even that is not working out so well: the Gaza embargo, which is starving civilians to death, has become a debacle for all parties involved, and is stymying the peace process further. Expect this one to get even messier. Heck, you don't have to expect it: just watch it unfold.

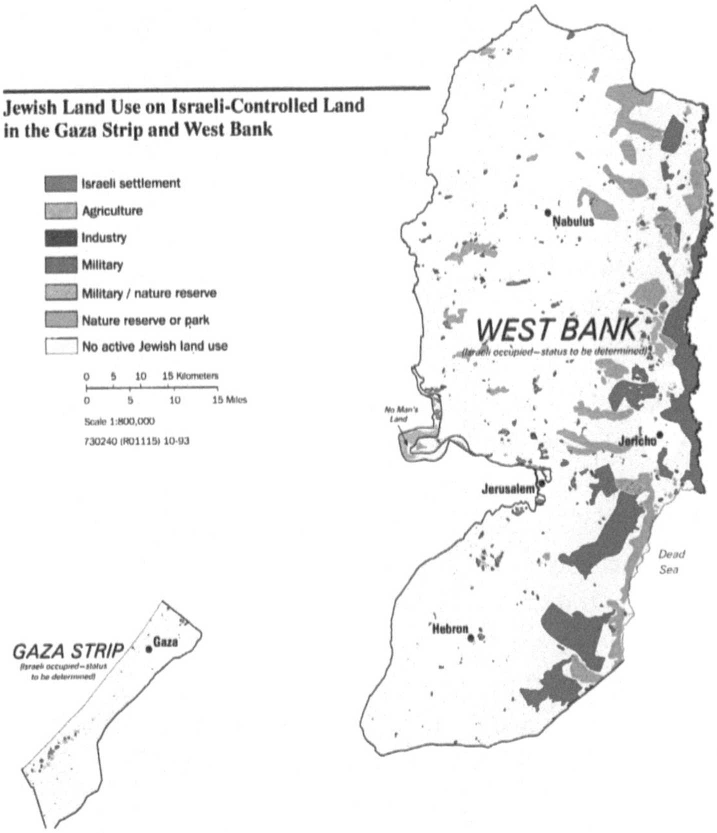

Jewish settlements in West Bank are specifically what Obama asked Netanyahu to stop.

To summarize this soup of names and nonsense: West Bank + Gaza Strip = Palestine. Palestine is not an independent state, but may become one eventually, however it is currently under Israeli control—that is, it is "occupied." To make this worse, Palestine used to be of a singular voice in negotiations with Israel and the world, but now it has splintered; The Fatah political party controls the West Bank, and the Hamas political party/terrorist group controls the Gaza Strip.

Hold on friends, the fun ain't over yet: for decades, the Israeli government has allowed/encouraged Jewish folks to start "settlements" in the West Bank as well, thereby making the occupied territories even more under Israeli control. I bring this up because in June 2009, US President Obama made a speech to the Muslim world in which he specifically asked Israeli Prime Minister Netanyahu to stop these settlements in order to foster goodwill towards a peace settlement. Let's make it hotter: in 2011, the Palestinians are planning to declare full independence at the UN, thus forcing a showdown on the global stage on who will recognize them, and who won't. **2019 UPDATE:** Under the Trump administration, the US embassy moved from Tel Aviv to Jerusalem with 14 of the 15 UN Security Council members condemning the move. However, the US vetoed the motion and continued with their intended move in May 2018. The Trump administration also declared the Golan Heights as part of Israel rather than Syrian territory.

Benjamin Netanyahu has also been in the news for criminal investigations of corruption with indictments coming in 2019, and his increasingly hawkish, right-wing government may not last for much longer, but perhaps long enough to seriously cripple the Two-State Solution for decades to come. Or perhaps permanently.

2021 UPDATE: Things can't get any worse in this Israeli-Palestinian situation, right? Wrong.

Another huge outbreak of violence started on May 10, 2021 and continued until a ceasefire on May 21, 2021. It was marked by protests and police riot control, rocket attacks on Israel by Hamas and Palestinian Islamic Jihad, and Israeli airstrikes targeting the Gaza Strip. The crisis was triggered when Palestinians began protests in East Jerusalem over the eviction of six Palestinian families from territory annexed by Israel. At this protest, Palestinians threw stones at Israeli police forces, who then stormed the compound of the al-Aqsa Mosque using tear gas, rubber bullets and stun grenades . . . storming of holy ground by police never ends well here, or really anywhere. Sporadic violence is still flaring, but so far the result of this newest wave of the conflict is 256 Palestinians, including 66 children, have been killed. In Israel, at least 13 people have been killed, including two children.

In a mostly unrelated update, in June 2021 Benjamin Netanyahu lost the position of Prime Minister. He was a very hawkish fellow that did the most to annex more Palestinian land, encourage Jewish settlers on those lands, and overall inflame the entire conflict for the last 12 years. It remains to be seen whether his replacement will change the equation or chart a new path forward on the Palestinian issue, but of course fingers are crossed.

INDIA–PAKISTAN: HOT SPOT: KASHMIR

In 1947, after India gained independence from Great Britain, it was divided into two separate states because of large differences in their religious populations. India retained most of the subcontinent, with its predominately Hindu population. Pakistan and East Pakistan, now known as Bangladesh, were formed as a home to South Asia's large Muslim population. From the beginning, friction has existed between these two giants of the subcontinent. Bloody clashes broke out the very day of independence, as millions of folks were on the move, shifting to the "proper" country that housed their religious inclinations.

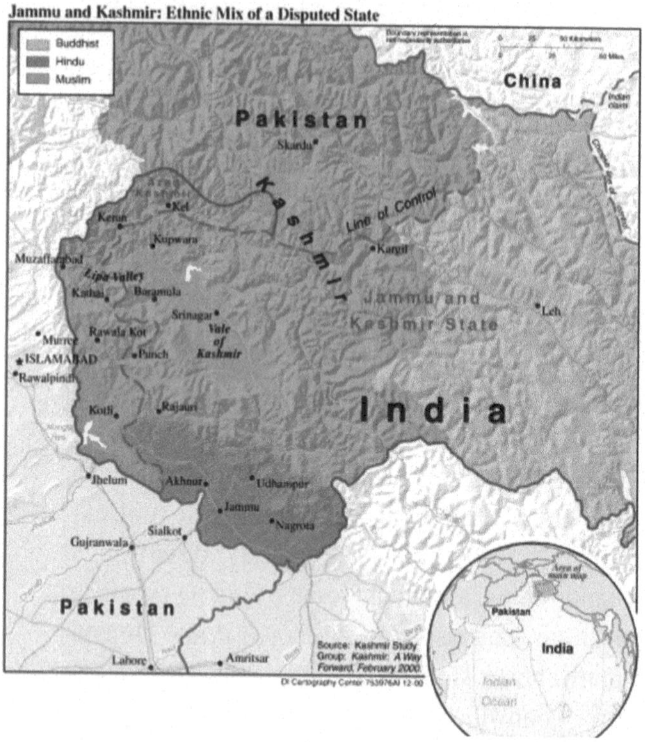

So in today's world, everyone in South Asia is now on the correct side of the fence, right? We're forgetting about that tricky place that makes ridiculously comfortable fabric: Kashmir. While most of the other borders between these states is settled and accepted, open fighting still sporadically erupts over this state of India named Kashmir, a small mountainous region about the size of Kansas, located in the northwestern part of the Indian subcontinent where the two countries meet. The Kashmir region was originally ruled by a monarch, or Maharaja, who gave India control of the region shortly after the nations split. The majority Muslim population of Kashmir did not agree with this move. Afterward, war erupted and continued until the UN arranged a cease-fire in 1949, by which time Pakistan gained one-third of the Kashmir territory. After India formally annexed the rest of Kashmir in 1956, the Muslim population was once more provoked to rioting, and fighting has continued ever since with a repeated pattern of attacks and cease-fires for the last fifty years.

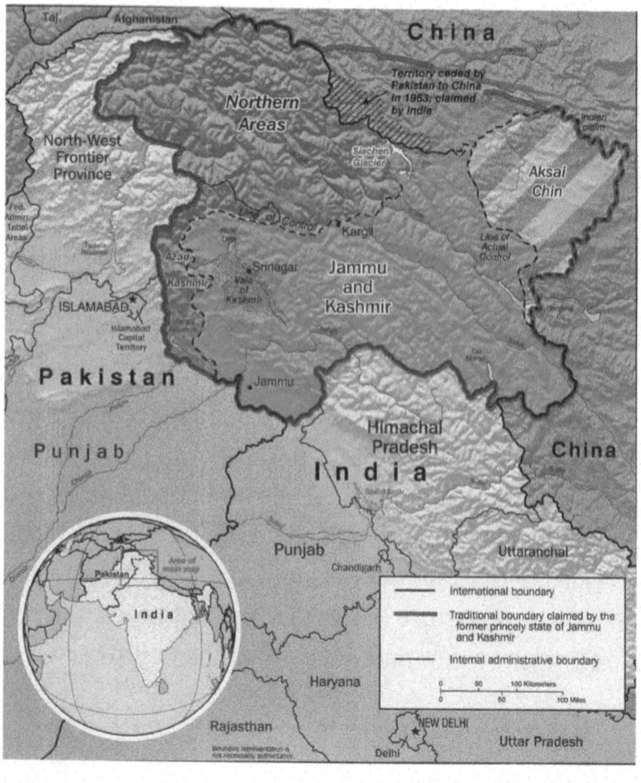

The situation intensified in the 1990s. Both sides began testing nuclear weapons and gained global attention because neither state has signed the nuclear nonproliferation agreement. Although recent meetings have shown signs of a peaceful resolution, this conflict must be considered a political hot spot as India and Pakistan remain the only neighboring nuclear powers with hostilities toward one another. In fact, the historical hostilities between these two states, which often erupt openly on the border, is one of the reasons why Pakistan is currently losing control of its own country. Say what? Am I suggesting that India is somehow undermining the Pakistani government in these volatile border regions? Nope, I am not: the Pakistanis are doing it to themselves. See, the Pakistani military has spent the previous six decades preparing itself for confrontation with India; therefore, the country has been ill-prepared to handle internal conflict coming from its western regions: you know, the Afghan side.

Why would I mention this in the Kashmir section? Because in the summer of 2009, India intentionally drew down their military forces in the border region, specifically in Kashmir, in order for the Pakistanis to feel secure enough so that they can deploy their military FROM the Kashmir region TO much more active hotspots in the western fringe of their country. India, as well as the rest of the planet, has no interest in seeing Pakistan completely collapse into chaos, and thus stood down in Kashmir in order for the Pakistanis to go get the anti-terrorist job done elsewhere in order to hold their country together. Which brings us to the next section of conflicts in the world, those brought about by failing/failed states, so let's stick with Pakistan for a page more . . .

FLAILING, FAILING STATES

You think you are a state? Well, for this category of conflict, you receive a big fat F for your attempt to prove it. Or you are on the brink of getting that F. Or there are a lot of us that think you deserve an F, even if you officially do not have it on your statehood college transcripts yet. This section will deal with those supposedly sovereign spaces whose very existence is questionable, or who are falling down on the job, badly. Failing at the very function for which they were formed. . . .

So how do we determine this status? Well, it is a bit fuzzy. A "failed state" is one perceived by the rest of the world as having failed at some of the basic conditions and responsibilities of a sovereign government. There is no single general consensus on an exact definition what makes for a failed state, but it likely has one or more of the following failing conditions:

- "the government" has lost control of all or parts of its territory
- "the government" has lost its monopoly on the legitimate use of physical force within the state (meaning warlords, paramilitary groups, or terrorist use force or kill people in that state, and the government can't stop them)
- "the government" is no longer taken seriously by all its citizens to make and enforce laws and run courts
- "the government" has lost the ability to provide public services
- "the government" has lost its ability to interact with other states as a full member of the international community
- there is widespread corruption and criminality that "the government" cannot stop
- there is a steep economic decline that "the government" cannot stop

If any one of these things is happening in a state, it still may be considered legitimate in the eyes of other sovereign entities, but once you start stacking up two, or three, of five of these factors, then the very existence of the state is dubious. And that is when all of the rest of us on the planet start debating if the place is officially "failed" and we all expect chaos, civil war, or international intervention to unfold at any second. Below is a short list of those places that are in this questionable conflict-riddled category in the coming years . . . which brings us back to Pakistan . . .

PAKISTANI PROBLEMS WITH TERRORISTS & TALIBAN

Wow, that's a mouthful! And it's a mouthful that Pakistan may not be able to swallow, and still survive. But hang on a second; maybe some of you are questioning the placement of Pakistan here in the failed state section. Can it really be all that bad? Yes my plaid friends, it really is that bad. In fact, it's worse. There is so much going wrong here that I am going to have to wildly over-summarize and over-generalize just to get to the main points. So here it goes. . . .

Pakistan has been on the brink of full-on catastrophe and chaos for some time now, and most look to the woes of the struggling democracy as reason for the calamity. However, the Plaid Avenger wants you to consider the indecisive nature of the the military and the intelligence communities within Pakistan as the primary problem behind the current calamity in Afghanistan, the Swat Valley, and the entire western side of Pakistan, which taken together may spell the possible future failure of the state itself.

As pointed out in an earlier chapter, Pakistani leadership is trying to pacify the wants and desires of 205 million citizens—that's the 5th or 6th biggest population in the world depending on your population source data. It's also a devoutly Islamic society, including the whole spectrum of religious views from the mainstream to the seriously extreme. There are a slew of extremist factions and separatists groups pulling the country apart, especially all around the Afghan border (look up Waziristan, Balochistan, the Taliban—man, that sounds like a Dr. Seuss book).

Look at the map opposite, and understand this: the areas in blue are pretty much a no-man's land for government control. It consists of ethnic groups that have never considered themselves part of any country (like Balochistan), the Federally Administered Tribal Areas (FATA), and the North-West Frontier Province, which sounds like a section of Disney World to me. Taken collectively, these are all nice names for areas that have one thing in common: they are too wild for the government to really control, so the Pakistani government gave them some sort of autonomy in exchange for not causing chaos. Today's big problem? The deal is off. The Taliban and al-Qaeda have convinced the locals that they don't need any "deal" with the Pakistani government, since they exert true control on the ground. They have a point.

Some of these extremist types (like al-Qaeda and the Pakistani Taliban) are fighting for establishment of a theocratic Muslim state, while others are fighting for an independent state within Pakistan (i.e., Balochistan, Waziristan). Some are just fighting against the government because they don't like the fact that the Pakistani leadership is so cozy with the US when it comes to foreign policy. Remember, Pakistan receives plenty of foreign aid from the US, primarily to keep up the War on Terror. As you can imagine, the terrorists don't like that idea much, so total destabilization of Pakistan by any means is one of their ongoing goals. And they are winning.

Need proof? Not only did most folks believe that Osama bin Laden and his crew are somewhere in Pakistan (proven to be true!), and that most terrorist attacks worldwide involve Pakistani elements (UK subway bombing, Mumbai terror attacks, etc.), but now there are open attacks in Pakistan on US and NATO supply convoys heading to Afghanistan. A specific insult? In February 2009, the Pakistani government essentially surrendered the Swat Valley region of their country to the Taliban. That is inside their own country, man!

Afghanistan-Pakistan Border

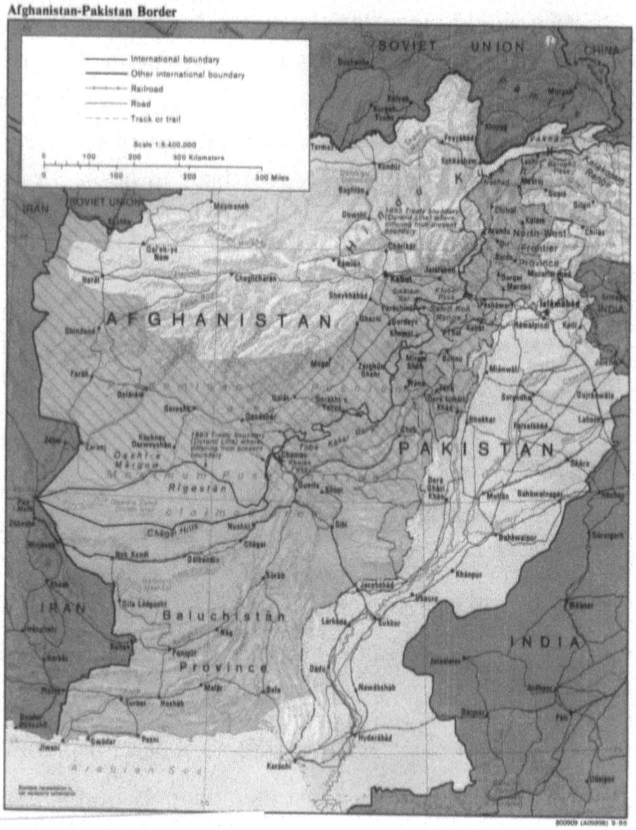

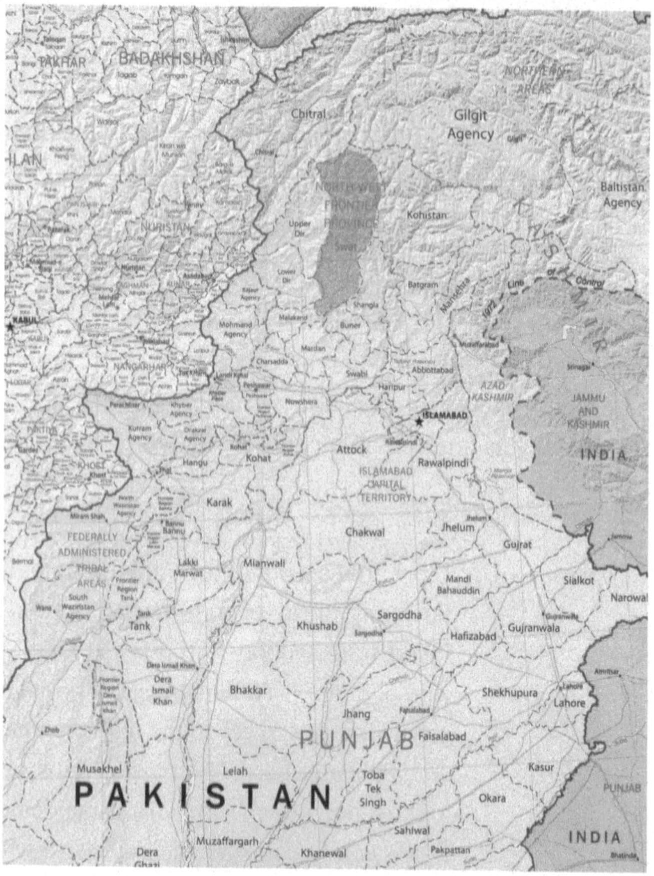

Swat Valley highlighted in pink. Notice proximity to capital of Islamabad.

2010 UPDATE: Pakistan's military finally got serious in late 2009 about holding their country together and organized a massive redeployment of troops to fight the Taliban and reclaim control over the western half of their country. They wrestled back control of the Swat Valley, and have made significant gains in the hottest spots of the FATA.

This place is in dire straits due to well-organized and passionate terrorist/separatist groups fighting with the home field advantage on extremely tough terrain. But let me finish with what I started with: the failure of Pakistan to hold itself together can be attributed to its own faults just as much as outside forces. Pakistani politics are hopelessly corrupt and gridlocked, with politicians bickering about judicial control and foreign policy while their state is imploding around them. The abysmal state of their economy, widespread energy shortages, and collapsing infrastructure ain't helping much either.

What's worse, the Pakistani military has been fighting what can at best be described as a totally half-hearted campaign against these insurgents. Why would they do that? Because the military and intelligence communities are the ones who actively built and helped the Afghan Taliban come to power in the first place, so now they have found it difficult to go out and kill the guys that used to be their allies. On top of that, the whole Pakistani military structure has been so focused on an eventual war with India, that they can not seem to re-adjust their priorities to accept the fact that they are losing the whole country from the inside out! They have more of their military stationed on the border with India than actually out fighting the dudes who are ripping their country apart!

2011 UPDATE: The US 'invasion' of Pakistan to kill Osama bin Laden certainly has made the world even more critical and mistrusting of the Pakistani military and/or ISI (the ISI = Inter-Services Intelligence, basically the Pakistani CIA). That, in combination with the unpopular war on terrorism and the US drone strikes into Pakistani territory, has created extremely low public opinion of the US and lack of trust on both sides. But the US needs Pakistan to continue to help with this war and the Pakistanis need the billions in US aid, so an uneasy relationship must go forward. Pakistan's current huge crisis is the rise of the home-grown Pakistani Taliban (a different group than the Afghan Taliban), a group intent on the overthrow of Pakistan itself! And they are pulling out the stops in terms of terrorist attacks on civilian targets inside of Pakistan to push their agenda . . . scary stuff in an already challenged state! **2019 UPDATE:** Just a ray of good hope to shine on the problem-riddled Pakistan: the nation is currently headed by Prime Minister Imran Khan, a former international cricket star who ran on a platform of aiding the poor in his country. Growth is slowly increasing, and terrorism has just slightly slowed down, so maybe a more productive era is underway. Still too early to call though.

US IN AFGHANISTAN

Well, let's just stay in the neighborhood, shall we? U.S. involvement in Afghanistan is still fresh in our memories—oh right, because they're still there . . . but not for much longer! The story, however, is a long one. In 1996, the Taliban, a fanatic Islamic extremist group, took control of Afghanistan, which had before then been a mishmash of warlords fighting for power. On October 7, 2001 NATO (led primarily by the US) began a military campaign against Afghanistan known as *Operation Enduring Freedom*. The invasion was in direct response to the 9/11 terrorist's attacks and under the belief that Afghanistan was harboring the same terrorists that planned the attacks. With the help of a rebel group called the Northern Alliance (which gains much of its funding through opium sales), America and its allies established a new, secular government in Afghanistan. This invasion kicked off what is now known as the Global War on Terror.

The main problem in Afghanistan has been the unsuccessful implementation of its established government. Afghanis are arguing that the ruling government is comprised of ethnic Tajiks, while the majority of Afghanistan is Pashtun. The Pashtuns are freaking out cause they think the Tajiks are too Westernized with their skintight jeans and pink Armani t-shirts, creating a secular government. The Pashtuns would prefer a Pashtun-dominated Islamic based government, which is strongly supported by its neighbor to the south, Pakistan. A reoccurring problem the U.S. is seeing is the failure of its exported form of democracy in the Middle East *(See: U.S. in Iraq)*.

A recent surge in Taliban activity has many speculating that the U.S. and its NATO allies are failing in reconstructing Afghanistan. A clear example of that was on April 2008, when Afghan President Hamad Karzai survived a failed assassination attempt. Even though the attempt failed, it was a victory for the insurgents, as they were at least capable of getting close to Karzai. This puts immense pressure on Karzai's administration to successfully establish and proclaim an effective government.

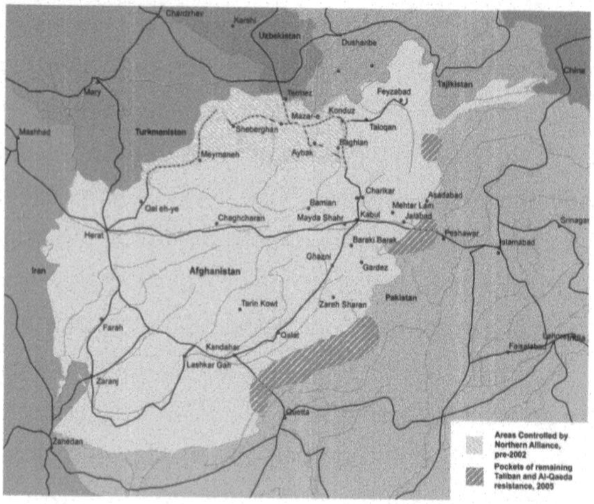

A particular problem posing itself against the Afghan government is the sharp increase in production of opium, which can be processed into heroin. Can't be that big of a problem, right? Afghanistan is a war torn country; when are you ever going to have time to grow opium? Apparently there's something in the air, because Afghanistan produces 92% of the world's opium and 80% of the world's heroin. Let's get down and dirty with the problem itself. The Taliban and al-Qaeda have used the opium trade to re-equip, re-supply, and re-organize themselves, and have turned the tide of the battle back in their favor.

Yes, you are reading this correctly: the Taliban is currently gaining ground and beating the NATO coalition. The US has been outright begging other countries to contribute more to the cause, but unfortunately, more folks are pulling out than putting in: Australia is out, and Canada and the UK have been downsizing troop numbers as well, preparing for an eventual total withdrawal. As just expressed in the last section on Pakistan, this whole area on both sides of the border is a no-man's land for control, from either government. NATO is losing just as bad on the Afghan side as the Pakistani military is losing on the Pak side. It's all one big interconnected mess that no one in the world wants to see disintegrate further, but no one in the world wants to send more troops or get involved with at all. Geez! Why does the US have to do every thing?

2014 UPDATE: This is it! This year the US has picked to pull out the rest of their fighting troops, leaving behind only a small force of security advisors' to continue to help train Afghan military and security forces. A hundred thousand troops are on their way out this year, leaving behind a shaky democracy (at best), a non-existent economy, and a fully charged Taliban who is now a decade-old, battled-hardened group who for years has been playing the long game knowing that eventually the US has to leave. The Plaid Prediction for this place is not good: civil war is imminent, and/or direct Taliban takeover in short order is also quite possible.

2021 UPDATE: It is all but official at this point: the US-supported fledging democratic Afghan government will soon be destroyed. The USA has already pulled out most troops, and even top leadership, and is committed to completely and officially ending its stay in Afghanistan by the end of summer 2021. The Taliban forces have been taking back territory and control step-by-step as the US leaves it, so with very little reservations I say the Taliban will once more run this country, and soon. Where they will go with it is unknown, but it certainly doesn't bode well for freedom-loving peoples who will once more be entrenched into an ultra-orthodox conservative religious state with toxic masculinity dictating the culture and political future of the pseudo-state.

We call Afghanistan a country because it occupies a space on the map of the world, but it isn't really one. For years, Afghanistan has been a battlefield for rival warlords and ethnic groups. Every time somebody has tried to set up a strong central government, such as the Soviet Union in the 1980s and the US/NATO in 2001, non-stop war has been the standard operating procedure for the locals. People in Afghanistan are used to war and are good at defending their homeland. I almost hesitate to call this place a failed state, only because I'm not really sure it's ever been a state to begin with! And if you need any further convincing that Afghanistan will plummet back into chaos once the US leaves, one need look no further than Iraq, where this exact scenario is already playing out. . . .

IRAQ IS A-WRECK

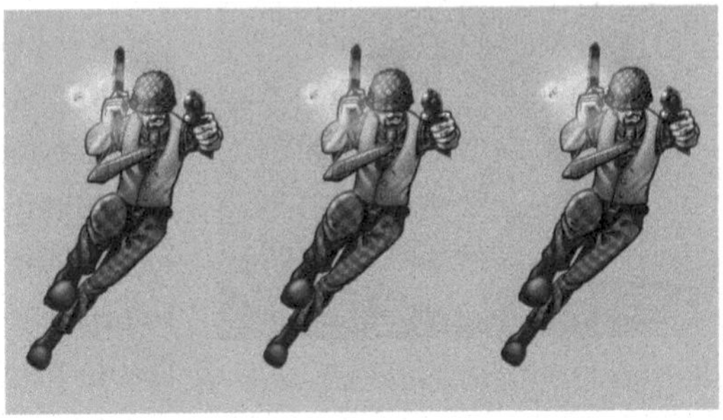

Ever since Saddam Hussein invaded Kuwait in 1990, he has been one of America's worst enemies. In the first Gulf War of 1990–91, aka Persian Gulf War, American forces drove Hussein out of Kuwait and destroyed much of Iraq's army. When asked why he didn't pursue Hussein to Baghdad to finish the job and remove him from power, former President George H.W. Bush told a group of Gulf War veterans, "Whose life would be on my hands as the commander-in-chief because I, unilaterally, went beyond the international law, went beyond the stated mission, and said we're going to show our macho? We're going into Baghdad. We're going to be an occupying power—America in an Arab land—with no allies at our side. It would have been disastrous. We're American soldiers; we don't do business that way."

After George W. Bush was elected president in 2000, he decided that American soldiers **did** do business that way. As a part of the US led 'War on Terror,' the US started the 2003 Second Gulf War aka the Iraq War, when America unilaterally attacked Iraq, went into Baghdad, and occupied the country—America in an Arab land. But a funny thing happened on the way to Baghdad. Hussein's forces did not even show up. The war was a cakewalk; it was almost too easy. People wondered what happened to Hussein's famous Republican Guard.

Ethnoreligious Groups

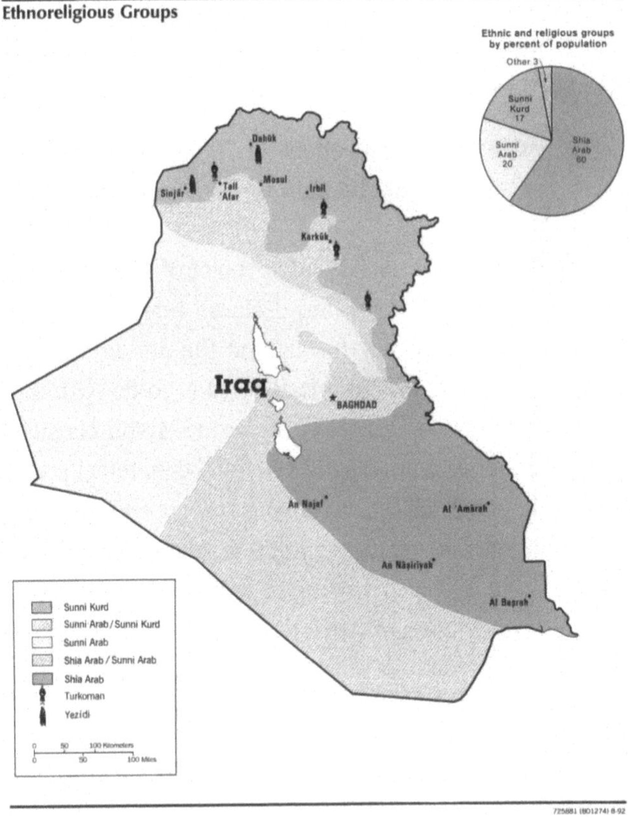

725881 (801274) 8-92

Then the sinking truth set in. Hussein had intentionally told his army to stand down. They knew they never would win a traditional ground war with US forces, so they dispersed and began an insurgency, fighting a guerrilla war with bombings and sneak attacks. Hussein was eventually captured, but since America declared "Mission Accomplished" more than 4500 American soldiers died in Iraq, with new attacks happening daily. Insurgents have even kidnapped foreigners and beheaded them on television. So what to do in the face of a vaguely-established new democracy with a load of civil strife and tensions mounting? Why leave of course! By October 2011, then US President Barack Obama announced full troop withdrawal by year's end, and in December they declared victory, and war over. Ummm . . . yeah . . . about that. . . .

Why would the mighty US have left? Public support in America for the war decreased because nobody wants to subject their children to guerilla warfare. Furthermore, one of America's justifications for invading Iraq was that Saddam had weapons of mass destruction, but when American forces found no WMDs in Iraq, many people felt misled and bamboozled. In addition, many Americans never supported the war in the first place, and an overwhelming majority of citizens in the countries of European allies totally were against the war from the start as well. It was wildly unpopular the world over. So getting out of there was the easiest thing ever for most politicians in Team West. Not so much for the locals. . . .

Because this place is a mess. America tried to establish a democracy in Iraq, but was doomed from the start because Iraq consists of three rival ethnic groups that do not get along: The Kurds in the north, the Sunnis in the middle, and the Shi'ites in the south. Good luck getting those guys to agree on a government. Good luck indeed. As foreign influence diminished, the likelihood of total civil war and implosion increased. These guys aren't out of the woods yet. By a long shot.

The black flag of the ISIS group . . . aka the Islamic State of Iraq! Yikes!

2014 INSANE BLOODY UPDATE: Unfortunately, I called this one exactly right years ago when I wrote those words you just read. As of June 30 2014: After organizing and fighting in Syria and conducting guerrilla warfare in Iraq, a group named ISIS (Islamic State of Iraq and Syria) started an open military campaign to take control of the entire state of Iraq. They took the second biggest Iraqi city of Mosul, are now consolidating territorial control over large swaths of Syria and Iraq, and are moving toward the capital of Bagdad itself. This fight could be over before this book even gets to print! This ISIS group is composed of radical Sunni Muslims Wahabbist and some al-Qaeda freaks, who are on a holy Jihad war to take over the entire Levant! This is full on civil war now! The Shia-led central Iraqi government has failed horrifically to date to repel this group, and now the US and others are debating about stepping back in to save the place from disaster. You heard it here first: if the Kurds up in Kurdistan also declare independence (a strong possibility at this point, as they are successfully defending their territory from ISIS), this state we call Iraq is not just failed, it is finished! Too hot to handle or even speculate about right now, but no good can come of this! **2015 MORE INSANE UPDATE:** Of course, things have gone from hellish to worse with this insane clown posse named ISIS, so much so that the 2016 edition's Cover Story chapter was dedicated to them.

Want to learn all about the Iraq/Syrian meltdown and the rise of the ISIS situation in more detail? Visit: http://plaid.at/isis to read **The 2016 Cover Story: The ISIS Insanity**.

2019 UPDATE: IS/ISIS lost control of Mosul in July 2017 and has continued to lose territory as military forces have worked to contain them. Iraq declared that the terrorist group had been removed from the country in December 2017. They have lost territory throughout the Middle East and several foreign fighters from Western states are now caught up in repatriation discussions as to what to do with them now that the IS "caliphate" has been defeated. It officially was defeated in Syria in 2018, but the group is still considered a threat even without a formal, physical caliphate. And Iraq is on the mend, doing slightly better economically and politically since 2017, but remains a tinderbox of possible conflict due to the competing regional forces of Saudi Arabia and Iran which are still conducting proxy wars all over the region, including the funding of opposing political candidates in Iraq.

Dissident Areas

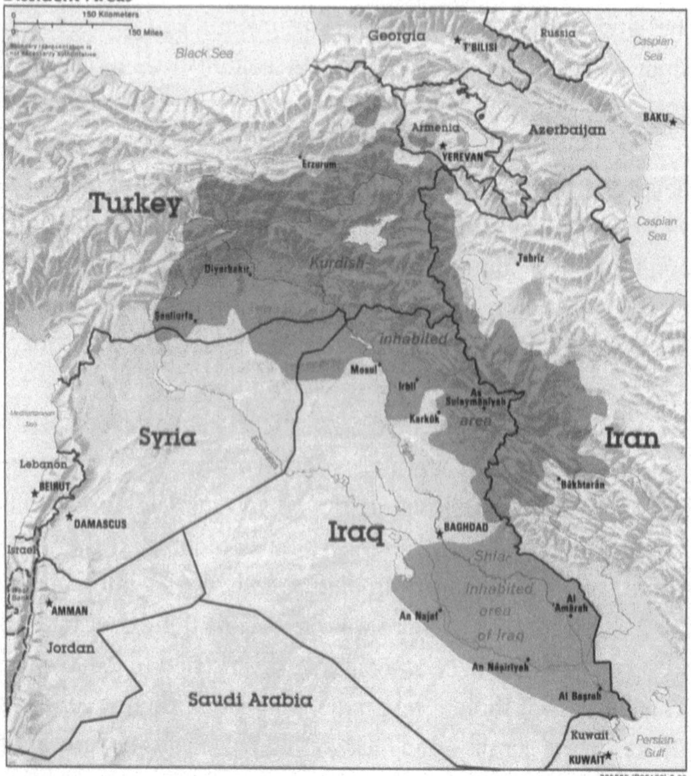

KURDISTAN WILL CUT OUT

Because of the disintegration of the Iraqi and Syrian states, it is highly likely that the Kurds will cut out of Iraq by declaring an independent state . . . which no one can do anything to prevent at this point in history. This will cause a potential showdown with Turkey. This one is definitely happening soon, and is not on anyone's radar. Iraq is actually already in a state of civil war, but the US and most of its press do not want to recognize that for political purposes. While the main source of friction and fighting is between Sunni and Shia groups, the actually full-on implosion of the Iraqi state will fully occur when the Kurds (who control the northern part of the state) will declare independence and a separate sovereign state of Kurdistan. While lighting the tinderbox for war in Iraq is in itself bad enough, what compounds this situation is that there are Kurdish people in other states like Iran, Syria, and—most importantly—Turkey, which may be incited to attempt to join this Kurdish state as well. Turkey has had a long repressive history with the Kurds, and Turkey actually invaded northern Iraq a few years back to root out the PKK, a Kurdish terrorist group responsible for attacks on Turkish soil. So if and when the Kurds declare a separate state in order to distance themselves from the Iraq debacle, it will immediately raise tensions with Turkey, and within Turkey.

SYRIA & LEBANON

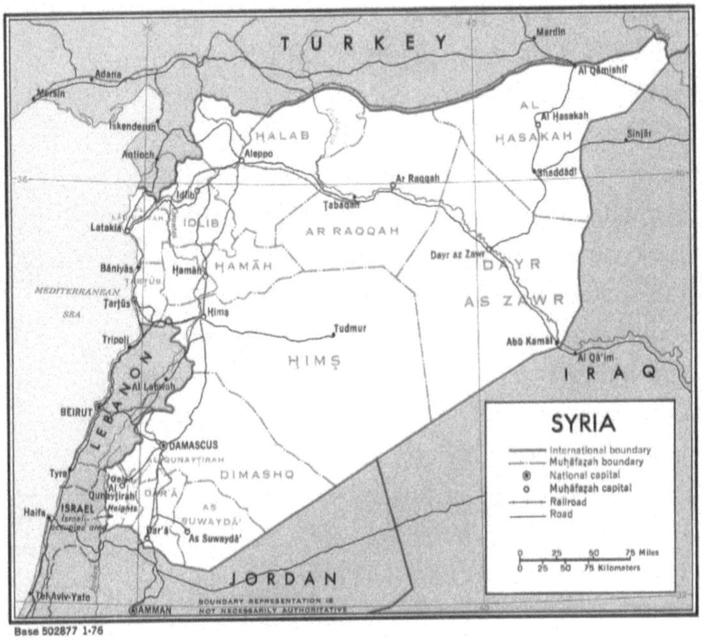

Starting as a civil uprising back in March 2011, as a part of the broader 'Arab Spring' across the Middle East, this mess has continued to devolve, and devolve, and devolve into absolute bloody insurrection, dictatorship repression, civil war, and has now even spawned the newest terrorist organization on the planer (ISIS) which is carving out areas of the country to form a new extremist Islamic state. If this doesn't count as a failed state, I'm not sure what does . . . cuz there ain't no one in control of this chaos. As of this writing: 150,000 dead, millions displaced, insurgents of all stripes are growing in numbers and in bloodlust.

From the get-go, President Bashar al-Assad's government has attempted to smash all opposition with an iron fist, and is sporadically accused of using biological weapons and targeting civilian populations. The 'rebel' groups that have been fighting the government are a hodgepodge of pro-democracy reformers, moderate Sunni republicans, Hezbollah militia, Iranian-supported mercenaries, hardline Sunni Islamists, al-Qaeda inspired nut bags, and foreign fighters who are coming in just for the fun of it. Outside entities like the US and Europe have hesitated to get involved, because quite frankly they have no idea who to give money and guns to! Its such a confusing jumble of peeps fighting each other!

And it has now turned into a full-on proxy war, with a decidedly Sunni/Shia split between sides. Bashar al-Assad and his government are Alawite Muslims, which is a sect of Shia Islam . . . so the government is supported by Shia Iran and Shia Hezbollah (that political/terrorist group in Lebanon). Russia also supports Assad, because, well, Syria is their only ally left in the Middle East. The majority of Syrians are Sunni Muslims, so money and guns have been funneling in to them from the rich Gulf Arab states (including Saudi Arabia) whose leaders hate the Assad regime. The crazed al-Qaeda types are also Sunni, so their endgame is the end of Assad as well . . . and since June 2014, the complete whack-ass crazy ISIS terrorist group that is sweeping across Syria and Iraq are 100% jihadist nuts who want to kill everyone who is not Sunni Muslim, so they are all about the destruction of Assad personally, but the destruction of the state of Syria as a whole. As Lebanon's fate is tied to Syria's fate, look for poor Lebanon to be completely enveloped into the chaos as well . . . and they have barely been a functioning state for decades already.

2017 UPDATE: Did you think I had any good news here? Nope. The Syrian Civil War is still the biggest human tragedy on the planet, with the death toll now surpassing 510,000 people killed (according to the UN) with many agencies believing it is way higher. 13.5 million people are in urgent need of humanitarian assistance. 6.6 million are internally displaced and 5.6 are externally displaced according to the UN High Commissioner for Refugees (UNHCR). Russia is now working closely with the Syrian government to bomb all opposition to death while the US and its allies are sidelined due to US President Trump's disinterest in involving the US any further in the civil war, especially since the defeat of ISIS, which means everyone can claim "victory" since IS was the stated reason for the

US to get involved in the first place. I predict that the Russians and the Syrian government will soon pull out all the stops to utterly destroy the opposition/rebel groups, then ask Donald Trump to help broker a "peace deal" cease fire which effectively will end the Syrian Civil War . . . with al-Assad still in power and Russia a more emboldened player in Middle Eastern affairs.

THE YEMEN CRISIS

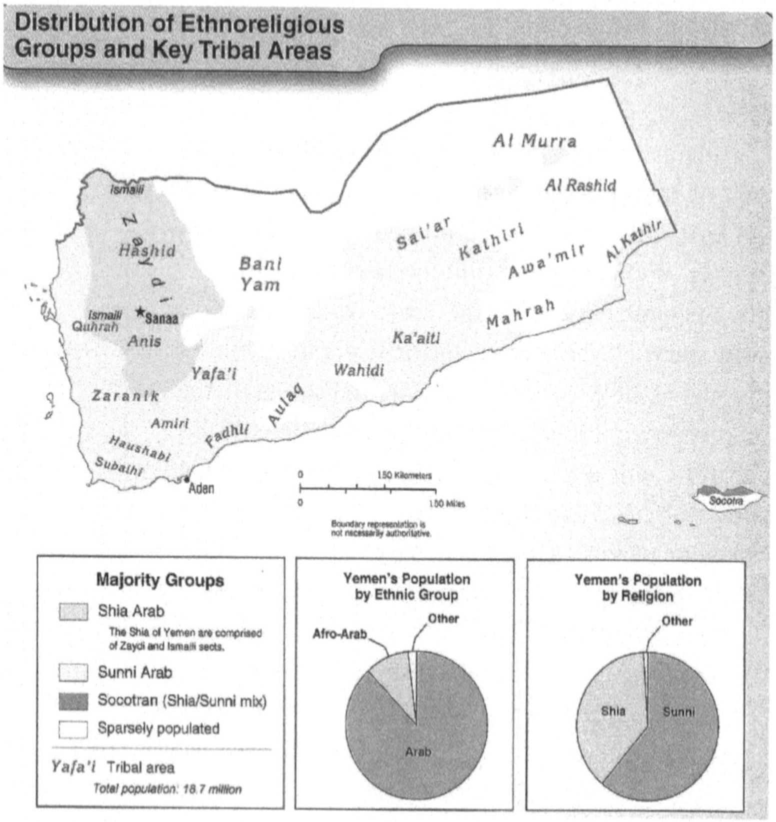

Yemen is one of the poorest countries in the world and is currently caught in a war between Houthi rebels and the Yemeni government (aided by outside international support). There has been fighting off and on since 2004 with the war beginning in September 2014 when the Houthis took control of Sanaa, the capital. A coalition of Arab states led by Saudi Arabia responded with a military campaign in support of the government forces in 2015. Currently the war in Yemen is considered the worst humanitarian crisis with millions displaced, about 7,000 civilians killed, 11,000 wounded (as of November 2018), and 14 million facing starvation according to Human Rights Watch. Supporting civilians caught up in the war with foreign aid has proven difficult. The US, UK, and France have supported the Yemeni government and foreign forces with weapons and intelligence. Iran has been accused of helping the Houthis as well. It has also been noted that Yemen could be considered part of the "cold war" between Iran and Saudi Arabia.

In other **2019 UPDATE** action: you should know the the Sunni/Shiite rift across the entire region is being exacerbated by all these conflicts, and it appears to be worsening. Saudi Arabia and its Arab allies are supporting one side of the Yemen Civil War, one side of the Syrian Civil War, and different political parties/groups across the region that share their goals. Meanwhile, Iran and its allies (Russia, Syria) are supporting the other sides of those conflicts as well as groups like Hamas in the Gaza Strip and Hezbollah in Lebanon. That is a way over-generalized and simplistic take on these conflicts, but it is happening, and it is a major cause for concern for the future of religious intolerance and warfare in the region.

Want more detail? Read all about the sordid details about the rise of ISIS, the demise of Syria, and why this all happened at this particular juncture in history, in **The 2016 Cover Story: The ISIS Insanity** at http://plaid.at/isis.

Future prediction for the coming year? Gonna. Get. Worse. **2021 UPDATE:** It has gotten worse. No end in sight.

VENEZUELAN IMPLOSION

The great socialist experiment of our times, and in our hemisphere, is rapidly falling apart and it is more than likely that Venezuela will end up in a state of civil war, or at the very least massive civil unrest, by the end of this year. While the socialist firebrand and beloved leader Hugo Chavez was alive, he could effectively rally the majority and keep a lid on dissent in Venezuela, but he died in March 2013, and was succeeded by Nicolás Maduro . . . who honestly is just not up for the task of continuing the 'revolution,' nor of holding the country together. He ain't necessarily a bad dude, but certainly ain't no Chavez, and does not command the required presence to keep this state afloat. Because the pro-Chavez and anti-Chavez peoples of the state are starting to spiral out of control. . . .

Since the beginning of 2014, a series of protests, political demonstrations, and civil unrest have been occurring throughout Venezuela. The protests initially erupted largely as a result of Venezuela's high levels of violence, inflation, and chronic shortages of basic goods. There were two specific sparks: (1) actress and former Miss Venezuela Monica Spear and her husband were killed on 6 January 2014 during a roadside robbery, while their five-year-old daughter was in the car; and (2) In February an attempted rape of a young student on a university campus in San Cristobal led to protests from students over crime. The government's response? Arrest the protestors. Thus fueling the anti-government anger, thus fueling more protests, thus fueling more arrests and crack-downs . . . see how this cycle of violence works?

The protesters claim that the horrific state of the Venezuelan economy and crime rates are caused by the economic policies of Venezuela's government, including strict price controls, which allegedly have led to one of the highest inflation rates in the world. However, government supporters claim that government economic policies, especially that of under previous president Hugo Chávez (1999–2013), improved the quality of life of Venezuelans, and blame external factors for ongoing problems. This is an ideological showdown for the very soul of this state, and I am afraid it may lead to a full-on civil insurrection/civil war/state collapse.

As a major oil producer and exporter, this has major economic repercussions for the globe, but on top of that, Venezuela has been the poster child/leader of the leftist shift all across Latin America for the last decade and a half. Its devolution changes the entire ideological game for South America, Cuba, and Latin America as a whole. So watch this one closely in the coming year, as it could explode fully at any second. . . .

2019 UPDATE: Totally called this one. Venezuela has been imploding all year. There is a political, economic, and social crisis currently in the country. With the dispute for the presidency between Juan Guaidó and Nicolás Maduro, power blackouts, supply shortages throughout the country, and hyperinflation, problems in Venezuela are escalating in 2019. This. Will. Get. Worse. **2021 UPDATE:** It has gotten worse. No end in sight.

DEMOCRATIC REPUBLIC OF CONGO (DRC)

HOT SPOT: EASTERN CONGO

The giant super-state of Africa has basically been in a state of self-destruction since its independence . . . and it wasn't doing very well even before that. The Democratic Republic of Congo mystifies even those of us that follow world events, due to the complexity and convoluted history of conflict. The site of "Africa's World War" in which 5–6 MILLION have

died, and which most westerners have never even heard of, this state will definitely soon be crumbling apart under the sheer strain of its size, of its ethnic diversity, and its mismanagement . . . but mostly because of its vast mineral resources! Say what? Dig this:

The history of the Democratic Republic of the Congo has been a complicated one, full of instability, coups d'etat, violence, and name changes. When the Republic of the Congo gained independence from Belgium in 1960, a guy named Patrice Lumumba was elected president, but was overthrown in a coup d'etat in 1965 by Mobutu Sese Seko. Sese Seko, a dictator, was supported by the United States and Belgium, and controlled every aspect of life in the country. He even changed the name of the country to Zaire in 1971. After the Cold War was over, the United States did not think it was necessary to support Mobutu anymore, so he was overthrown by Laurent Kabila in 1997, who renamed the country back to Democratic Republic of The Congo-Kinshasa. The name of the capital city is added to the name of the country to distinguish it from the Republic of the Congo-Brazzaville which is a different country that is next door to Democratic Republic of the Congo. Get it? Good. Someone explain this stuff back to me.

This new Kabila government immediately came under fire by rebel groups and has yet to establish any sort of stability. Kabila quickly lost control of parts of eastern Congo to Ugandan, Rwandan, and Burundian backed rebels while still receiving support from Angolan, Zimbabwean, and Namibian troops. The organization Human Rights Watch wrote that

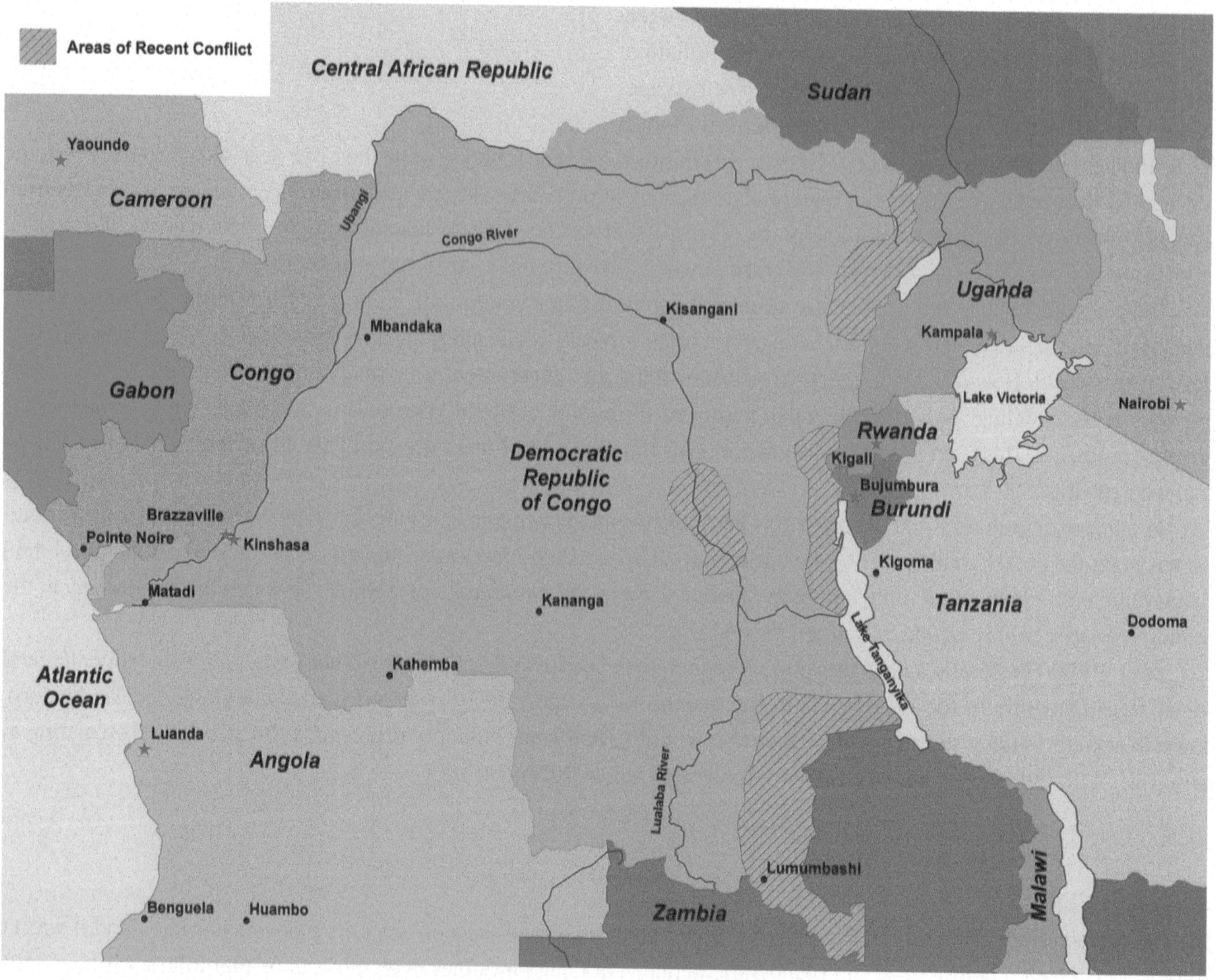

Congo is a human rights disaster area and "soldiers of the national army and combatants of armed groups continue to target civilians, killing, raping, and otherwise injuring them, carrying out arbitrary arrests and torture, and destroying or pillaging their property. Hundreds of thousands of persons have fled their homes, tens of thousands of them across international borders." In fact, the DRC conflict is responsible for the establishment of a new category of war crime: institutionalized rape. Roaming groups of soldiers raping women with total disregard for laws or morals has left an entire generation of women physically and psychologically wounded and another generation fatherless. In fact, the conflict in the Democratic Republic of the Congo has been called "Africa's World War" and 5–6 million people have died as a result of it and the subsequent chaos resulting from it, to this day. A peace treaty was signed in 2003, but the situation remains unstable and could devolve back into war at any time. In a few years, you could be watching "Hotel Congo."

Why the fighting? Quite simply: money. This area is huge, massively huge, and filled with gold, diamonds, uranium and practically every commodity the developed world wants. Also, there is no sense of nationhood and the leaders of the different factions are war criminals and thieves who are only looking to line their own pockets. The natural wealth of the Congo is taken out of the country, leaving a poor, uneducated population embroiled in turmoil. Things here are really bad and the near future holds few foreseeable good changes.

Some areas of the country are effectively controlled by the armies of other neighboring states, rebel, or terrorist groups from neighboring states, or local rebel groups, or local warlords, or . . . I think you are starting to get the point.

To sum up the DRC: huge country with a weak central government; a valuable and tempting resource base; a hide-out for rebel groups of all sorts; a place where perhaps seven different states have forces in and out of the territory. What a mess, and it could implode soon, turning into an all-out territorial grab by all the involved parties: The DRC, Uganda, Rwanda, Burundi, Angola, etc. This DRC mess has the distinction of having the largest UN peace-keeping force stationed in it already. However, given the huge size of the country, the small size of the UN force, and the general war-lord-driven politics of the region, don't think for one minute that the UN or anyone else is really in control of this part of the planet.

2012 UPDATE: Just to give you a sense of the on-going nature of this Congo mess, the UN just released a report accusing Rwanda of fomenting internal rebellion in the DRC by training men to penetrate the Congolese military, and then mutiny against the DRC government itself. Talk about some internal shenanigans!

Of course Rwanda vehemently denies this, but the report comes on the heels of mutineer activity in eastern Congo, in what appears as an attempt to tease away those resource-rich provinces by undermining the central DRC government. Looks like the mad scramble for land and resources is still alive and well in the Great Lakes region, and could become even more violent soon. There are over a dozen 'rebel' groups operating in this part of the world, and 3 of them are large enough to be considered full-fledged paramilitary groups that can inflict serious damage on each other, and the local populations.

All of these groups have targeted civilians, government forces, and each other . . . just when I think this mess can't get an worse, it usually does. Watch for full-on warfare involving surrounding states in the next few years.

INTERNATIONAL INTRIGUES

Our final classification of clashes involve those disputes that involve major world powers or multiple sovereign states, or span whole regions . . . and also have the unique character of being waged over strange spaces: be they over oceans, over air spaces, or in the case of North Korea and Iran, over possible intentions. Many of these potential battles also are intriguing in that when/if they blow up, they have the potential for world war . . . as they will be bringing at least two of the world powers into the fray. Even the ones that don't have world war potential are fascinating in that they will be re-defining and shaping regional power structures, affecting dozen of states, and having repercussions well beyond the physical places where the fight may be taking place. These spats will shape world policy on the future of the oceans, world policy on terrorism, and will define the reach of particular world super-powers, and rising second tier powers like Iran. So follow these fights as they unfold in the coming 365. . . .

SENKAKU SHOWDOWN!

Tensions between the two Asian rivals have worsened in recent years over a group of uninhabited islands in the East China Sea called Senkaku Islands by the Japanese, Diaoyu Islands by the Chinese, and the Diaoyutai Islands by the Taiwanese. No matter what you call it, they are just a small group of rocks that are technically owned by Japan, that no one lives on, nor have any proven resource value. They are located roughly due east of Mainland China, northeast of Taiwan, west of Okinawa Island, and north of the southwestern end of the Ryukyu Islands.

The Chinese claim the discovery and control of the islands since the 14th century, but 'lost' control of them during their chaotic 19th and 20th centuries, when Japan was on the rise and China was sucking. Specifically, the Japanese took possession after their successful rout of China in the 1895 Sino-Japense War, in which Japan gained the Korean Peninsula, parts of Machuria, the island of Taiwan and the surrounding seas . . . in which these rocks in question are located.

So who cares about a handful of rocks that changed hands over a century ago? That would be the 2nd and 3rd largest economies on planet earth right now: China and Japan. What's at stake? Perhaps there is oil under them. Perhaps its about fishing rights. Perhaps its about China not wanting a foreign power located so closely to their coast. But is is DEFINITELY about national pride, and establishment of who is actually in charge of Pacific Rim in the coming centuries. And it is a big deal.

Why? Well, if you read this book, you now know that China and Japan pretty much still hate each other since the Japanese imperial phase that saw them take over, and decimate. large parts of China in the WW2 era. China is now back, and assertive, and wants to make sure that Japan gets back in line as a second tier power, behind China's dominant pole position in this region. The Japanese for their part are also beating the national pride drum, and are tired of apologizing for the WW2 era, and don't want to lose face in front of their citizens. So in both sides of this equation are millions and millions of Chinese and Japanese citizens who want their governments to restore legitimacy to their proper place in the Pacific, and who get instantly infuriated at the opposing country any time boats bump into each other around these rocks! Insanity, isn't it?!? What's more, since the USA has military and strategic alliances with Japan, if a war were to break out over the rocks, America might get sucked into it . . . making it a world war status conflict!

You will see public protests in both countries in the coming year about this issue, and likely see fishing and military boats from both countries floating dangerously close to each other in the waters around this area. For its part, China already unilaterally declared all the airspace above the islands and a wide swatch around it as officially Chinese airspace . . . essentially daring anyone to test their resolved on the issue. Which the USA promptly did within 12 hours when it flew several bomber straight across the Pacific, over the Senkaku, and then home again! It's a high-stakes chess game afoot my friends! Watch this one closely in the coming year! And that is not the only ocean patch that the Chinese are flexing their muscles over . . . there is also. . . .

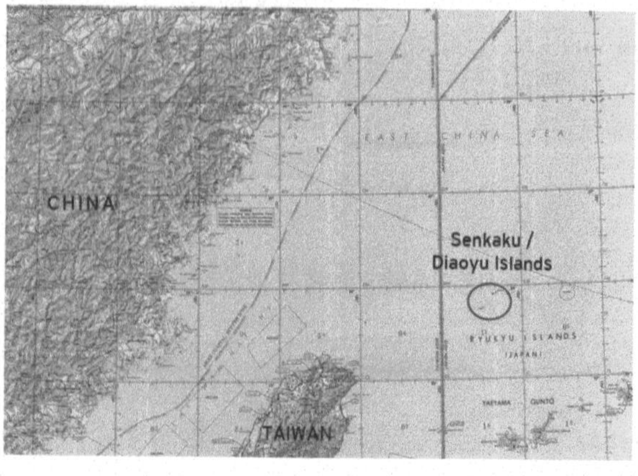

SOUTH CHINA SEA & THE SPRATLY SPAT

Another great episode that explains the rising power of China in yet another maritime dispute, but this one has even more parties involved! Again, in it's now more comfortable role as regional hegemon, China is exerting itself and its claimed ownership/control of more maritime areas. This battle is for a way bigger area than the Senkaku Islands, and may contain vast oil and fishing reserves. This is the enormous South China Sea, which is entirely located completely south of China . . . thus the name! . . . but also thus the confusion of how China can claim the entire thing! But they do, and 6 other countries claim parts of it too, which makes it a hot spot to watch big time. Fishing vessels have already bumped into each other, China has already planted flags on dinky little rock outcrops all over the place, and Vietnamese ships have been fired on, China has told everyone, outright, that it claims it all and will not tolerate any discussion about the waters at all! With anyone! Ever! This affects not only the nations that may be going up directly against Chinese military, but also explains the rapid rise of the Chinese Navy (including multiple aircraft carriers they are building), the US's shift to the Pacific (we are establishing bases in Australia, and most recently are re-establishing military ties with the Philippines), and may even result in a US military base in Vietnam! Crazy stuff! All in an effort to build US alliances in the region, while containing Chinese ambitions. It also affects economic activities to the entire Pacific realm, as these waters are tremendously important for their shipping lanes, and all of the states with partial claims on the South China Sea are members of the ASEAN economic alliance, which may become a spoiler to China. Mari-time, military, and economic mischief abounds!

South China Sea Islands

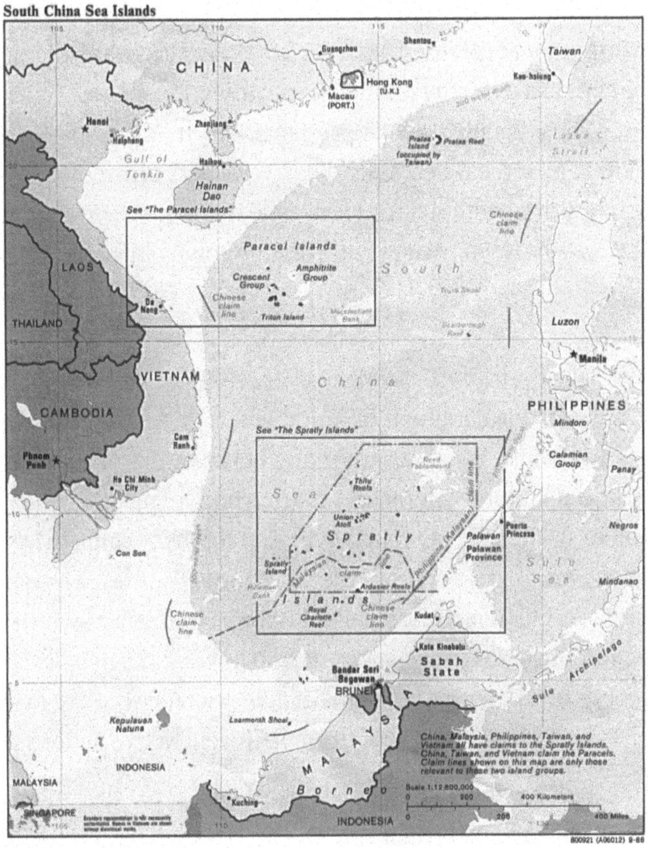

Spanning from the Singapore and Malacca straits to the Strait of Taiwan, the South China Sea is one of the world's most hotly disputed bodies of water. It also has these huge group of tiny islands—named the Spratly Islands—which are closer to rock outcroppings than habitable land, splattered across the entire area . . . and a handful of countries claim parts of those rocks as well. China lays claim to nearly the entire sea, overlapping with the maritime claims of Taiwan, Vietnam, Malaysia, Brunei, and the Philippines. With sovereign territory, natural resources, and national pride at stake, this dispute threatens to destabilize the region and even draw the United States into a conflict.

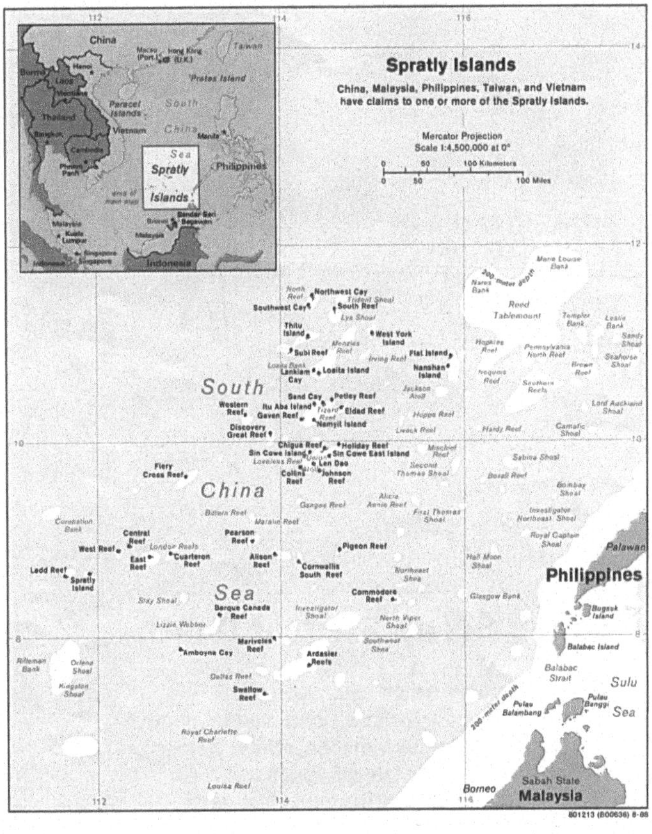

Exercising sovereignty over the South China Sea would be a strategic boon for China given that more than half of the world's merchant tonnage, a third of crude oil trade, and half of liquefied natural gas trade travel through the contested waters. And, with its waxing political, economic, and military weight, China seems to be taking a harder line on the issue. Chinese officials continue to emphasize the protection of China's territorial integrity, which of course includes the South China Sea in their view. A Chinese defense white paper, released in April 2013, declares that China will "resolutely take all necessary measures to safeguard its national sovereignty and territorial integrity."

Thus, Beijing's public proclamations that the South China Sea as an outright core interest of theirs, and one they will defend by force if necessary. They WILL NOT back down from this fight . . . and I assure you that Vietnam won't either, as that scrappy little nation has already defeated a super power in the past, and is the state most likely to challenge China directly in this watery claim. Watch out! This one could blow at any time as well!

THE SAHEL REGION OF AFRICA

The Sahel: A transition zone in every sense of the word "transition," this belt of climatic cross-over is also a strip of savagery that spans the continent of Africa, and is the scene of scores of conflicts. The Sahel originally refers to a scattered scrub vegetative biome: a zone of transition between the Sahara desert to the north and the savanna grasslands to the south . . . but is also a cultural interaction space where nomadic/cattle country meets agricultural/urbanized life, and most importantly, where Islam meets Christianity!!! And frequent cultural fracases are the result! This unstable strip stretches 3,400 miles due west/east across parts of Senegal, Mauritania, Mali, Algeria, Niger, Chad, Sudan, South Sudan and Eritrea. Herders clash with farmers, Christians clash with Muslims, "Arab" clashes with "African," and nomadic clashes with urban. Political parties even within single sovereign states tend to polarize the peoples, and have torn states asunder. As a result, civil wars erupt, terrorist groups organize, and militias roam the region trafficking in drugs and arms, seizing hostages for ransom, and trading livestock.

Just to name a few of the most recent active actions across this lethal ribbon:

- In Mali, the Tuareg Rebellion of 2012 separated the entire northern half of the country from the central government, and was to become a new separate state . . . for a whole 5 minutes before radical Islamists took over the territory for themselves. A military coup toppled the Mali government in March, while separatists and al Qaeda-linked fundamentalists took over the country's north. The end game resulted in the French intervening to help restore stability and take back control from the jihadists, and they still have an active force there still.

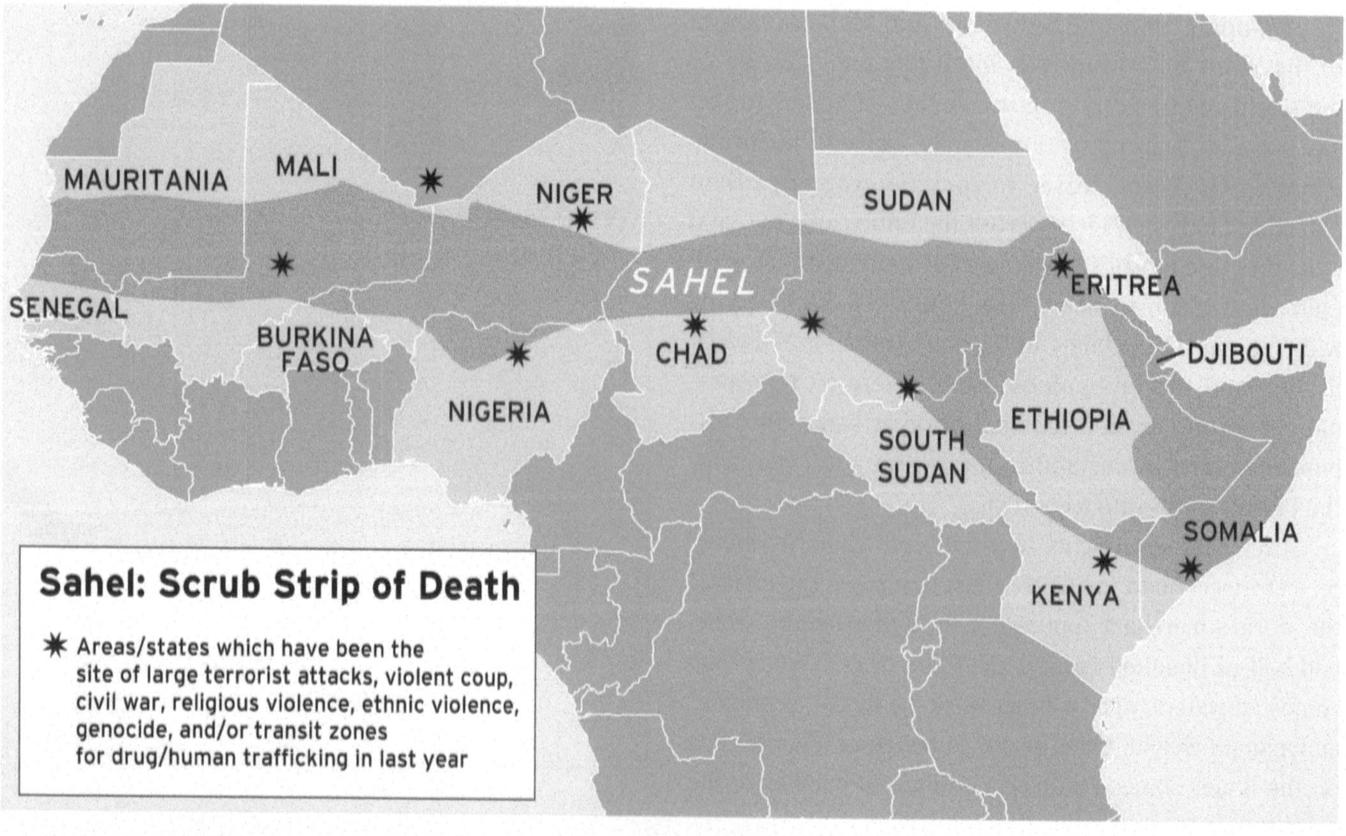

- The Central African Republic (CAR) which by all measures is now described as a failed state in permanent crisis due to a civil war between Muslim minority Séléka rebel coalition and the François Bozizé government forces . . . the rebels captured the capital in March of 2013, Bozizé fled the country, and the UN and AU forces were called in to help restore order. As of this writing in 2014: 300,000 refugees, massive crimes against humanity have been perpetrated, full-on genocide is occurring, mostly by the Christian majority purging out Muslims.
- The most recent Chadian Civil War which began in December 2005 and ran to 2010. Since its independence from France in 1960, Chad has been overwhelmed by the civil war between the Arab-Muslims of the north and the Sub-Saharan-Christians of the south . . . which is the exact thing that happened in . . .
- Sudan: After the First Sudanese Civil War (1955–72) and then the Second Sudanese Civil War (1983–2005) between the mostly Arab/Muslim north and the mostly Christian/African south, the country split in two . . . right along the Sahel zone lines! Of course, the 5 decades of war resulted in two million dead and four million displaced . . . which is the highest civilian death toll since World War II. And it ain't over yet, as sporadic fighting between Sudan and South Sudan is still a reality of the present and future. As is still is in. . . .
- Darfur: this region of western Sudan made international headlines back in 2003 for acts of genocide perpetrated by government sponsored Arab/Muslim groups agains non-Arab populations. Yep. Same story, in the same state! And even though the world is no longer paying attention to it, the UN is still there, and atrocities still abound.
- Go google "Boko Haram" to get the whole inside scoop on this despised and deadliest of new terror groups hoping to rip the northern part of Nigeria away from the south and make it into a separate Islamic state that follows strict Sharia law. Linked to Al-Qa'ida in the Islamic Maghreb (AQIM) and located in Nigeria, Niger and Cameroon, the group is known for attacking churches, schools, and police stations; kidnapping western tourists; and assassinating members of the Islamic establishment who have criticized the group. Violence linked to the Boko Harām insurgency has resulted in an estimated 10,000 deaths between 2002 and 2013. You probably heard about them in the news in 2014 for kidnapping and enslaving hundreds of school girls which (finally) gained international attention to the havoc they are reeking.
- The Islamist group al-Shabaab sent gunmen to attack the upmarket Westgate shopping mall in Nairobi, Kenya in September 2013. The 48 hour siege resulted in 67 deaths, including four attackers, and 175 wounded. Al-Shabaab claimed the attack was retribution for a Kenyan military's deployment in Somalia which attacked the terrorist group who had been operating across the Kenyan-Somali border for decades.

Starting to see the trend? This belt, which, again, is actually defined by physical geography, is the transition zone key to understanding the cultural geography, ethnic genocide, and terrorism across all of North Africa. This increase in civil war, in cultural friction, and increased terrorist activity have increased interest in outside intervention, as well as increased US military interest, and increased US military presence in new bases across the continent. The Sahel is hot, and likely to get hotter this decade. . . .

NORTH KOREA VS. ?????? THE US? SOUTH KOREA? JAPAN? EVERYBODY?

North Korea is America's worst enemy in the Far East. Ok, the world's worst enemy. They possess about 5,000 tons of biological and chemical weapons. They administer death camps where 200,000 political and religious criminals are worked to death. They are also run by an inexperienced and possibly insane dictator, Kim Jong-Un, who is not averse to funding his burgeoning **nuclear program** using money laundering, heroin smuggling, and counterfeit currency. Former US President George W. Bush has said of Kim Jong-Il (the former North Korean leader and father of Kim Jong-Un), "I loathe Kim Jong-Il. I've got a visceral reaction to this guy, because he is starving his people. And I have seen intelligence of these prison camps—they're huge—that he uses to break up families and to torture people. It appalls me." Bush even included North Korea in his "axis of evil" speech in 2002.

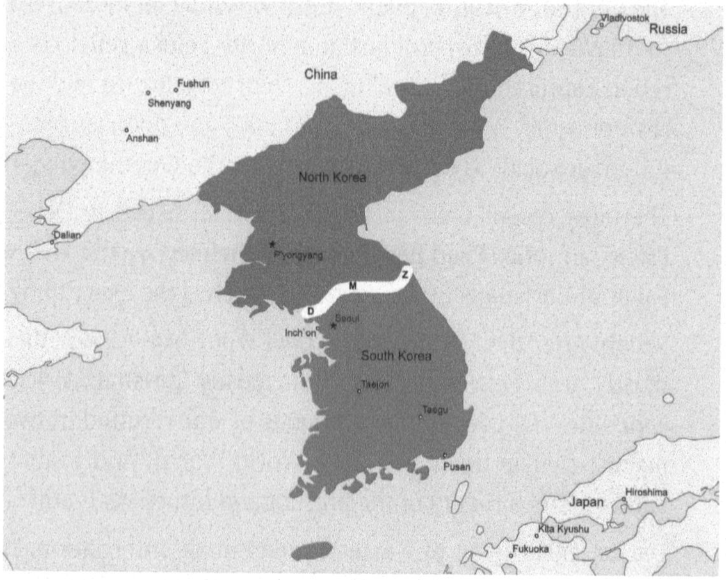

The whole conflict started in 1950, when communist North Korea invaded South Korea and America sent troops to defend South Korea in the Korean War. Ever since then, North and South Korea have endured an uneasy peace. The United States has stationed more than 30,000 American troops in South Korea, in case North Korea tries to invade again. However, the thing that scares American policymakers most about North Korea is the fact that they are building nuclear weapons.

To resolve the nuclear issue, there has been some diplomacy, but the Korea issue is complicated. At first, North Korea wanted the US to sign a peace treaty that would formally end the Korean War, which never happened, but now they just want security. Unfortunately, Kim Jong-Il and his advisors have decided that the best way to gain security is to threaten their neighbors with nuclear attacks and boast that they are not averse to preemptive war. The US wanted to resolve this whole issue in the form of six-party talks involving themselves, North Korea, South Korea, Japan, China and Russia. In November of 2005, the last round of these talks was held and an agreement was reached to provide North Korea with financial aid if they stop their nuclear program. But Kim won't finalize the deal unless the US drops its demands regarding his country's trade activities. The US responded to this by saying that the trade and nuclear issues are separate and they refuse to merge the two.

North Korea is not much of a threat militarily and would probably get their butts kicked by South Korea in a war—or really, a war with anybody. Keep in mind: They would lose, but after possibly killing millions of Koreans on both sides of the border; the leadership does have death potential, but no hopes of actually 'winning' long term as their state is totally insane and totally defunct. In fact, much of their population is starving. But with atomic weapons, and the crazy rhetoric for which the Kims are famous, many people are scared of North Korea. Nukes, nuts, and nothing to lose: not a good combination.

In 2010, North Korea did everything in their power to infuriate the US, South Korea, Japan, the IAEA, the UN, and even China! They tested multiple rockets which could carry a nuclear payload, as well as at least two underground atomic explosions which seem to indicate that they are back in full swing trying to produce nuclear weaponry. They pretty much threatened the entire planet with annihilation if anyone messes with them or their ships, which may be exporting illegal nuclear cargo. The erratic behavior of the regime has totally enraged China, pretty much their only ally on Earth. Harsh UN sanctions have been slapped on the Hermit Kingdom, and we are all now curiously watching to see what these nut-bags will do next.

2012 UPDATE: With the totally inexperienced and clueless Kim Jong-Un taking charge of this psychotic state in 2012, it is really anybody's guess what direction they will go in next. A lot of us were hopeful that he would turn over a new leaf for the country and open up a bit to the outside world, but those dreams are fading fast. North Korea flipped off the entire world by launching a rocket into space (it failed), and by all accounts the regime is now moving fast to do another yet another nuclear underground explosion as a show of its savvy and strength. Meanwhile the people starve and the military continues its aggressive rhetoric. Unfortunately, there is no way this mess can end well. The only option that the Plaid Avenger can see that would end this nightmare without the use of nuclear weapons would be a ground invasion from China. They did it once during the Korean War . . . so let's rally President Xi Jinping for a repeat performance! Go get 'em guys!

2019 UPDATE: North Korea's rocket launch tests in 2013 led to the UN expanding sanctions on the country through Resolution 2087. The country has continued to engage in human rights abuses, executions, and weapons testing despite

UN Security Council measures and imposed sanctions. Things "Twitter-escalated" through 2017 between Trump and North Korea. However, despite a terrible relationship in the past, dictator Kim Jong-Un and Donald Trump seem to be getting along currently with Trump setting foot on North Korean soil on June 30, 2019. They've met a few times but have as yet to make any real strides toward denuclearization. If those two goobers could actually pull off any sort of real de-escalation, and peace treaty between North and South, both of them would likely win the Nobel Peace Prize.

SOMALIAN SKULLDUGGERY

Well, the end is near . . . for this chapter my friends, not for us! What better way to finish a section on failed states than to reference, at least briefly, the poster child of all poster children of failed states, and that is the stinking cesspool that is Somalia. Don't be offended, my Somali friends! I am not talking about you personally or your ethnic group generally; I love you all! But the political backwash of a country called Somalia just sucks like a shop-vac. On high speed.

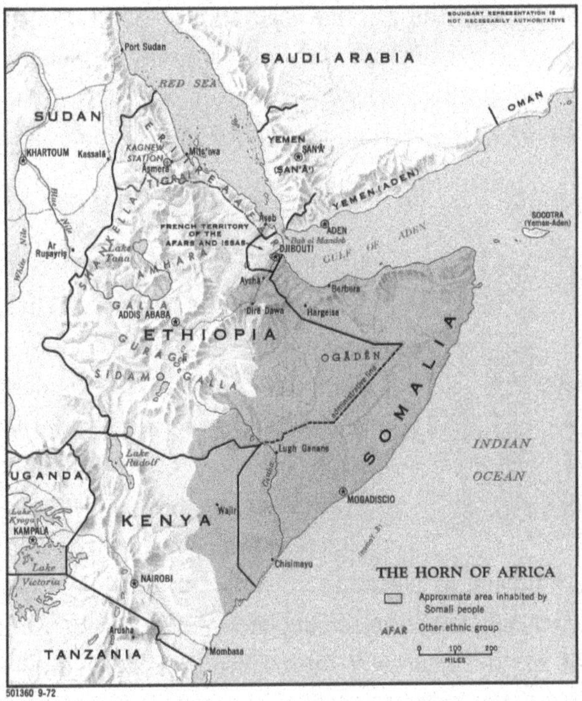

501360 9-72

However, I am not even going to go into the entire backstory of why this state disintegrated. A combination of colonial incompetence, ethnic infighting, civil strife, civil war, foreign meddling and general corruption have served to turn this state into a warlord-dominated and Islamic-fundamentalist breeding ground. Even when outsiders have made an attempt to make a positive impact, it has turned out horrifically bad. Go watch *Black Hawk Down* to see the US's ill-fated go at it . . . and nobody has tried to help since.

No, I simply wanted to bring up Somalia for a completely different reason altogether . . . pirates, of course! Hoist the Jolly Roger! Swab the decks! Shiver me timbers! Okay, enough of that nonsense. The point is, you have heard a tremendous amount of news in the last few years, (and even a Tom Hanks movie) about the exponential increase in pirate attacks and hostage situations and skullduggery on the high seas, all happening off the Somali coast, perpetrated by Somali pirates. While it's quite fascinating and makes for flashy news stories, the bigger picture is being missed by most. What picture is that?

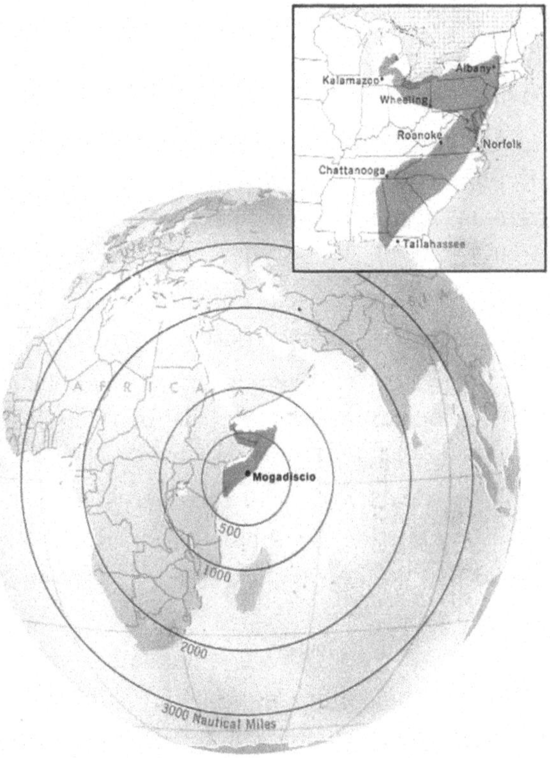

Just this: allowing states to fail ultimately affects the whole global community. Now, perhaps I am getting too preachy and not simply covering the facts, as all good textbooks do. Oh, I mean as all boring textbooks do. But the thing is, these are facts. All of the failed and/or failing states around the globe are now having global repercussions on all the rest of us, since we are now all connected in the globalized world. Did I use the word globe enough for you yet?

Just dig this: whenever we turn our backs on a part of the globe that we feel does not affect us in any way, we get bit in the ass. Yeah, that's what happens when you turn your back. North Korea's nuclear issue, Afghanistan's/Pakistan's no-man's-land issue, Somalia's pirate issue, even Zimbabwe's AIDS/cholera issue are all bound by one common theme: all these problems were spawned in areas that the world gave up on, and now the world has to deal with the consequences. Because AIDS spreads. Because extremist militants attack. Because nuclear bombs can be launched. And yes, because pirates will climb out of the hole that the world left them in, and come and steal your booty.

The point of this rant at the end of our hot-spots chapter? Heck, I don't know. Maybe it's to keep you engaged in what's happening in the world so that the next hot-spot can be avoided. But for sure it is to alert you to a stone cold fact of life in our times: world ignorance may be bliss, but it won't stop world problems from reaching your shore. Even when it comes ashore dressed as a pirate. Speaking of the next imminent hot-spot, let's finish the chapter with a current conflagration that may end up re-defining our times and the world order if/when it goes down. Let's head back to Persia. . . .

IRAN VS. THE US, OR THE WORLD IF NECESSARY

Iran and the United States have been enemies since the 1979 Iranian Revolution, when Islamic extremists overthrew the US-backed Shah, and took control of the government. When the fanatics took over, one of the first things they did was raid the American embassy in Tehran and take 52 Americans hostage for 444 days. The images of American hostages on American TV made Americans furious at Iran. Iran has also funded terrorist groups worldwide, including Hezbollah, which was responsible for bombing a Marine barracks in Lebanon in 1983 and killing 241 American marines.

Iran has the dubious distinction as being listed by the State Department as the top state sponsor of terror worldwide, making it a natural target in the war on terror. To make things worse, Iranian officials are hostile to America's main ally in the Middle East, Israel. The former President of Iran, Mahmoud Ahmadinejad, has even said that Israel should be wiped off the map, and other clerics are famous for their anti-Semitic rhetoric about Israel. No wonder George W. Bush called Iran part of the "axis of evil" in 2002.

Despite years of talking smack, the US may finally have an excuse to attack Iran . . . and others have been getting on board for the action as well. In 2006, Iran admitted that it was enriching uranium to build nuclear power plants. Iranian officials claim that they are just doing this for nuclear power, but some American officials are concerned that this is just the first step in developing a nuclear bomb. Game on.

Obviously, American policy-makers are scared to death of Islamic extremists, known for supporting terror and threatening Israel, having nuclear weapons. There has been speculative debate for years now about a possible US invasion or a US air-strike to shut down this nuclear nonsense. Before that happens, the US has been pushing hard for crippling UN economic sanctions against the regime. Both avenues of punishment have been resisted big-time by China and Russia, both of which are doing big deals with Iran for energy importing and exporting, so this situation has some beyond-the-Middle-East, whole world repercussions as well.

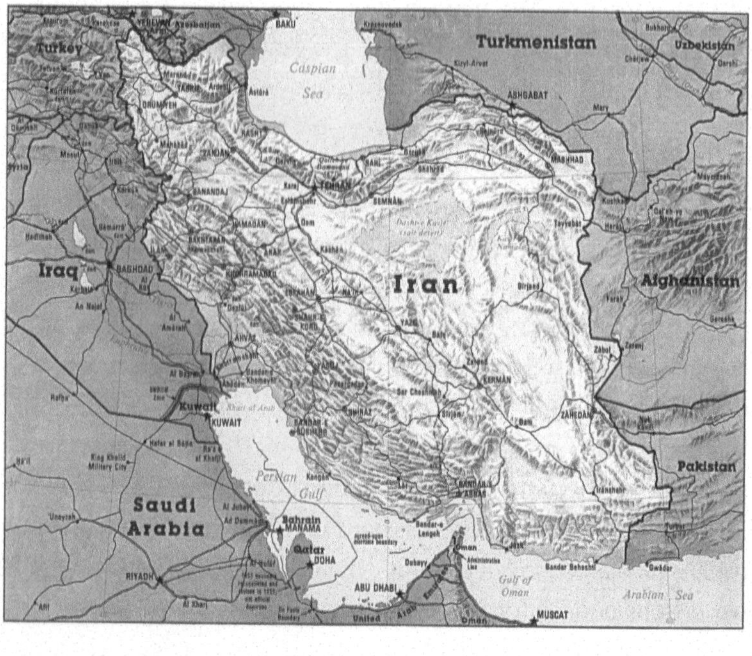

However, even China and Russia have been re-thinking their protection of Iran as Iran has become even more ballsy with their nuclear ambitions in the last two years: they failed to allow

IAEA inspectors full access to crucial areas, were busted with a secret underground uranium processing base, and even made proud claims that they are enriching uranium fuel to levels higher than needed for nuclear energy, but not quite high enough to be used for bombs. They continue to intentionally stay in this murky middle ground, as if they actually are daring outsiders to attack.

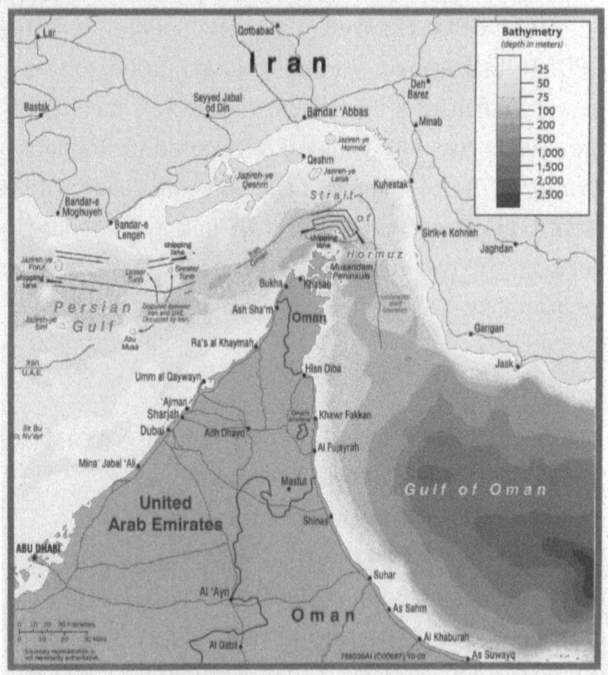

Strait of Hormuz: Potential Iranian Target to log-jam World Economy

Think I'm exaggerating? Both the Ayatollah and former Iranian President Ahmadenijad made fiery speeches about how it is Iran's "God-given right" to have a nuclear industry, and that nothing on the planet will stop them. (I forget what part of the Bible God specifically addresses fissile material in, but I'm sure it's in there.) No amount of wheeling and dealing seems to be able to push Iran off of this path, which is why they may be bombed off it instead.

Iran's air force and navy is shaky at best, and the US could take it in a couple hours with some surgical bombing. A ground war is another story. Unlike Arab Muslims, who usually consider themselves more Muslim than nationalistic, Iranian people are very nationalistic and would resist an attack violently. I cannot conceive of anybody in the US government being dumb enough to attempt a ground war in Iran, but we should never underestimate the capacity for ignorance, I suppose.

I believe the worst case scenario that we might see would involve surgical bombing of the nuclear production facilities, and the US probably won't even be the ones to do it! Israel has publicly stated many times now that a nuclear Iran is unacceptable and that they will not let it happen. The Iranian situation is now the #1 most worried about foreign policy issue to nearly every single Israelite.

It will probably be Israel that conducts a bombing campaign, unless some diplomatic breakthrough occurs. The US, the EU, most of Team West and, strangely enough, most Arab states will secretly support such a strike, while publicly distancing themselves from it. This will serve to infuriate the Iranians, and tick off the Russians and Chinese as well. The Iranian government has also said that it will step up terrorist attacks, especially in US occupied Iraq, if force is used against them.

Do you think that's all the Iranians can do if they get attacked in any way? Consider this: Iran could, in less than a day, drop sea mines into, or start picking off ships in, the critical Strait of Hormuz, where a huge daily dose of the world's oil supply passes through. In effect, they could halt the entire world economy by clogging up this small artery of oil flow, which would spike oil prices exponentially and instantaneously. This one could get ugly, quickly—and everyone is going to be involved one way or another.

2019 UPDATE: In better news, the current President in Iran is Hassan Rouhani as of August 2013. He is considered more moderate than his predecessor and has worked to improve relations with the West, sometimes leading to friction with the clerics in the country. He was reelected in 2017 and was in power during the discussions on the Joint Comprehensive Plan of Action (JCPOA), also known as the Iran nuclear deal, in 2015. The Iran nuclear deal changed the relationship with the US and other western powers and was considered a key achievement for the Obama administration.

In worse news again: unfortunately, the Trump administration withdrew the US from the deal in May 2018. This has led to multiple issues with Iran threatening to break the deal with the other countries in the agreement since it views the US withdrawal as a breach of the agreement. As of July 2019, attempts are being made to reach a diplomatic agreement but the possibility of war in some way is on the horizon. There have at this writing been multiple small tit-for-tat transgressions in the Persian Gulf between Iranian forces and its western adversaries. Fingers crossed this doesn't result in a full scape war, but the stage is perfectly set for that to happen . . .

2021 UPDATE: Well, it's a new update for Iran, and as per usual it's not a very promising one. I cited in the 2019 UPDATE that Iranian President Hassan Rouhani was a moderate who wanted to improve relations with the West, which he very much tried to do. But the triple whammy of Trump politics, the increased US economic embargo on Iran, and the Covid crisis in combination resulted in Rouhani not making any progress, and the further deterioration of US/Iran relations. The 2021 impact? The Supreme Leader of Iran, Ayatollah Khamenei, fixed the recent election so that his hand-picked, ultra-conservative and US-hating politician Ebrahim Raisi was voted in as the new President. Look for relations between the two countries to get even worse, with potential open conflict likely being sparked by confrontations in the territorial seas of the Persian Gulf.

EASTERN MEDITERRANEAN SEA SHOWDOWN

TERRITORIALDISPUTES IN THE EASTERN MEDITERRANEAN
Maritime boundaries and energy competition between Greece, Turkey, and Cyprus

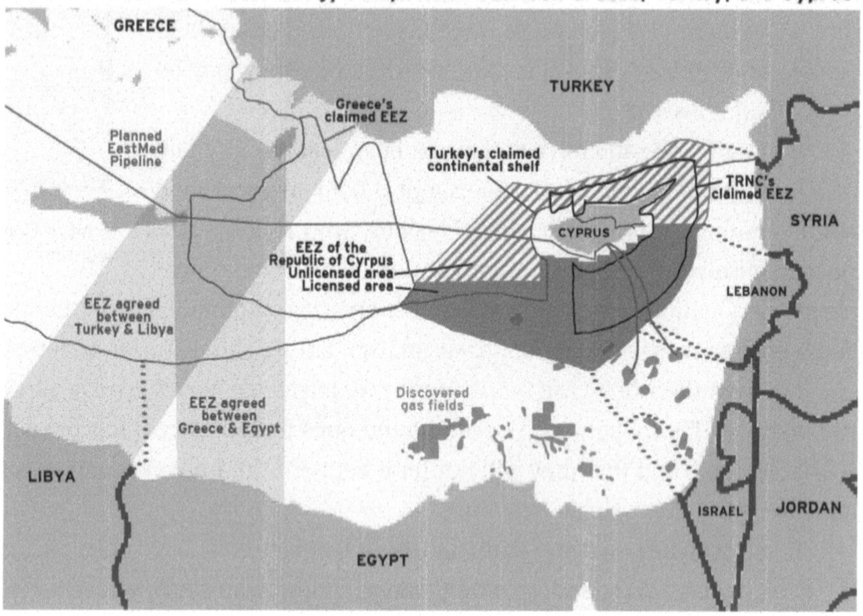

This one is totally new to the sea, I mean to the scene! Territorial tensions have always been high between Turkey and Greece, two countries with Mediterranean Sea coastlines which equate to Mediterranean Sea ownership in their EEZs. What's an EEZ? An exclusive economic zone (as outlined by the 1982 United Nations Convention on the Law of the Sea) is an area of the sea in which a sovereign state has special rights regarding the exploration and use of marine resources, including fishing and energy production from water and wind. It stretches from the baseline out to 200 nautical miles from the coast of the state in question. But what happens when one sovereign state has a dinky little island that sits just off the coast of a different sovereign state? A-ha! Conflict happens, my friends! And between two countries that hate each other, like Turkey and Greece do, it becomes a more intense battle over territory that stokes national pride . . . and possibly open war.

Really? Countries would go to war over ownership of tuna? Well, not so much . . . BUT . . . this territorial dispute in the Med just got seriously hot in the last few years due to the discovery of large beds of natural gas . . . now we are talking real money! And now we have even more interested parties involved, all trying to stake out their claim to the waters that may contain that gas. The current players (with the biggest claims) directly involved are Turkey, Greece, and Cyprus—but it also involves (and taking sides) Egypt, Libya, Israel, Italy, France, and Germany. Why France and Germany? Because Cyprus is an EU member, thus this situation is pitting the entire EU against Turkey.

And Turkey has been the most ambitious and aggressive in staking its claim, which works in tandem with Northern Cyprus claims, since Turkey pretty much controls the pseudo-state. (No one but Turkey recognizes Northern Cyprus as a separate sovereign entity, but given what you just learned about EEZs, do you understand why Turkey does recognize Northern Cyprus?) The Turks have been sending exploratory research vessels, along with armed military vessels, all across the region, which has caused increased friction and almost outright hostilities as ships are getting dangerously close to bumping into other ships in this "land grab." Already in 2020, Greek and Turkish frigates actually collided, and there have been several tense stand-offs between several European ships and Turkish ships as well.

This is—pun intended—a hot seabed of activity to watch in the coming years, as it is not simply about the gas resource itself. This is a face-off between major powers:

1. Turkey (backed by its new best friend Russia, and new strategic ally Libya)
2. The EU (especially France and Germany) and some Arab states (Saudis, UAE) backing up Cyprus and Greece
3. Even Israel which has a claim, and is planning a big pipeline (and of course, the USA backs Israel as a strategic ally).

Hydrocarbon resources, free flow of sea traffic, a race for regional dominance are all at play here. . . . with a specific desire by many players to contain Turkey's greater regional ambitions. Go Google ***Blue Homeland*** to learn more about Turkey's aggressive regional strategy here that has many worried that the Eastern Med may quickly become an eastern dread of armed conflict between former allies!!! Yikes!

Bringing death and destruction to a neighborhood near you!

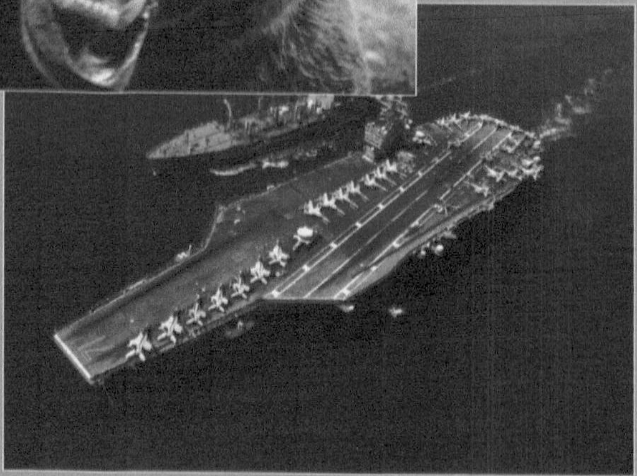

CHAPTER OUTLINE

Team Play: Evolving Power Centers of the 21st Century

27

MY friends, it's long past due to wrap up this blundering batch of buffoonery I call a book. What better way to round up a world regions textbook than to back up from the entire planet and assess exactly how these world regions are interacting with each other—either working with each other, against each other, or forming up teams to cancel each other out. It's such a delicious stew of international intrigue! Think of it as different clicks in a worldwide high school, except these states have nuclear weapons and control the planet. Other than that, it's the same old same old human nature. But I digress.

Let's once again look at the major players that are affecting the course of events on our planet strategically, politically and economically. Then we'll call it a day.

LET'S MEET THE POLITICAL POWER PLAYERS

No sense trying to pretend otherwise: the United States is still the undisputed heavyweight when it comes to a single entity with the absolute most political power. To put it simply, what the US wants, the US gets. An example: There was never any credible threat that any other country was going to attack the US because of its most recent invasion of Iraq. Most countries, institutions, and people on the planet may have been staunchly opposed to that US move, but no one was actually going to do anything about it. That is raw political power, and the US is the only state that has been able to wield that sword solo for a couple of decades.

But the world is changing fast, my friends. The US still has the power, but other states and groups are rising fast to match their diplomatic skills with Uncle Sam. With no reservations, I tell you that we are moving from a uni-polar world (only one power, namely the US) to a multi-polar world (with many major powers). This is the setting of transformation that we find ourselves in here at the beginning of the 21st century. Politically speaking, let's take a look at the major forces that shape decision-making across the globe, and who teams up with who in order to get their agendas pushed. I will identify the major players, their motivations, their agendas, who they like and who they hate, and see how that is translating into current events that are going down around the globe.

SAMMY SOLO

America has long been a team of one. Its physical isolation far from the rest of the world has forever insulated it from the global events in which it participates, or in which it declines to participate. Its entire history could be viewed as a wavering cycle of either isolation (in which it does not get involved in global events) or full-on engagement (in which it is full-on the predominate player.) Take a wild stab which side of the spectrum it has been on for the last couple of decades . . . although with the US finished in Iraq, pulling out of Afghanistan, and refusing to get involved in Syria, one could argue the the pendulum is swinging back to self-absorbed stance.

Sammy sez: step off!

But when Uncle Sammy decides to go for the global gusto, he has seemingly unlimited energy, money, military, and ambition to get the job done, regardless of world opinion. For the last hundred years, the US usually had a global coalition of participants that backed it up. Let's call them Team West for right now. In WWI, WWII, the Korean War, the Cold War, the first Iraq War and countless others, the US served as a team leader that rallied the world to victory, or at least something close to it. Point is, that mostly the US didn't have to stand alone, because the UK or the EU or NATO or the UN or Team West was there to back them up.

However, since the end of the Cold War—circa 1990—Sammy has increasingly worked on his own to achieve his agenda. Why? During most of the last fifty years, it was the US and the USSR in a bi-polar world of competition, giving the world at least two options to choose from. But now it has been a couple of decades that the US has stood completely alone at the top of the heap, alone at the top of this uni-polar world. In this time, the US quickly grew accustomed to the role of sole political power. Therefore, it has increasingly acted alone whenever it really wanted something that the rest of the players would not support.

I'm not just picking on the former Bush administration, either; the Clinton administration acted unilaterally on several different occasions as well, either to stop perceived terrorism in Africa or perceived genocide in the Balkans. The dynamic duo of Dick Cheney and Donald Rumsfeld just took the US even exponentially further down this unilateral path (willing to act alone), with the most extreme example being the aforementioned second Iraq War. Beyond even a willingness to act alone, the US for over a decade has actually preferred to act alone, so it does not have to waste time with negotiation and organization and cooperation that is required by international diplomacy.

The times are a-changing, though. Under the US President Barack Obama administration, the US rapidly leaned back to diplomacy and coalition-building as the most effective tool for international action. We can see this unfolding currently as the US is working exceptionally hard to rally world support on the growing threats of both the North Korean and the Iranian nuclear industry, and/or the situation in Afghanistan and the one in Syria, and it initially met with some success. The entire EU backed them on it, and even Russia and China started to come around to the US point of view after lots of wrangling and back-room debates. But that does not mean that any concrete multi-lateral actions have been the result . . . as of course it has not. The NATO-led Afghan mission has also been recast as needing more international support, as opposed to being primarily a US military force, but it is winding down in either case, so it won't matter for much longer how it is labeled. Because either way it will be labeled as chaos shortly after the US departure. As will the mess in Syria.

2021 UPDATE: Well, much of the world, and many citizens of the USA, remain in shock of the 2017-2021 reign of Donald Trump over the United States of America, which up to that point was the world's premier power and champion of democracy. No offense to his friends, fans, or foes, but Trump certainly took the country onto a significantly different trajectory, and one that had major immediate repercussions that will reverberate for decades to come; a game changer for the country and the world. His inward-focused, America-first program had a distinct white nationalist, anti-global, anti-supranationalism agenda that wildly confused and bewildered US allies and adversaries alike. . . . and since has become a model for aspiring nationalists and autocrats all over the world. However, that new leadership style and anti-global mission came at a price: the USA is no longer considered the world leader of liberal democracy and free trade, with little authority and even less respect being paid to it as a moral and ethical standard-bearer of the planet. We are currently on the brink of a whole new world order . . . or perhaps a new world disorder . . . because of the radically changing nature of the USA.

Uncle Sam had a solo run in the sun for some time, but it appears that in this newly forming multi-polar world, his future role may be as primary leader of Team West. Okay, who the heck are they?

TEAM WEST

Throughout this text, I have thrown around the concept of a core world culture that is primarily steeped in western traditions and values. It's finally time to identify just exactly who is on Team West. Obviously, the core of the team is Western Europe, the place where Western Civilization was born and evolved, thus the team name. *"Western civilization"*

encompasses the big ideas on which our lives are based. I'm talking about the big things here: things like philosophies, legal systems, economic systems, medical practices, writing systems, and particularly religions like Christianity and all its sub-sects.

However, the true adhesive of this team is based on common ideologies. All the members of Team West also share particular values such as liberal democratic governance, emphasis on individual human liberties and rights, and the free market/capitalist system. These are more important to consider than the other stuff, since there are some non-European actors on this team; while they are not Christian or European or even white, they are staunch believers in the team's principle values. That's why they are on the team. Okay, so who?

Joining Western Europe, we have most of Eastern Europe, and let's just go ahead and refer to the team elements of both regions as the EU. The EU countries may have been the historic core of Team West, but the all-star quarterback of the squad has been the US. But there are even more players on the team! Canada and Australia always show up to the game, as does Japan. That's right, Japan is definitely part of Team West even though it is in the east, and Asian to boot! It shares all the West's values and has staunchly supported the US and the team ever since the end of WWII. One other surprise entry: Turkey. As a staunch secular democracy, capitalist country, and NATO member, Turkey also primarily sides up with Team West, even though it is Muslim. Primarily, but not always. We could even count Israel on Team West, although for various reasons, it does not usually openly participate in most of the Team's competitions.

How does it work in today's world? Well, when the Team comes up with a desired policy or goal—say, the War on Terror or even the War on Drugs, the US usually leads the charge, and will often be supported on the UN Permanent Security Council by having the UK and France back them up. The EU as a whole typically is on board as well. Then when it is time for the real action to go down, Canada, Australia and Turkey will jump in to join with the EU and US troops, and Japan will voice its support for the endeavor, and perhaps even supply money or materials to aid the effort, since Japan does not have offensive military capacity. Then the team hits the field and tries to score the goal. Got it? Good.

Team West is a potent force today, mostly because it consists of most of the planet's richest countries, some of the most powerful and technologically advanced militaries, as well as having three nuclear powers (US, UK, France . . . and covertly throw in Israel to make four). Can you feel the power yet? No? Consider this: all NATO countries are on Team West, which essentially means that NATO itself is a tool of Team West's arsenal. And NATO kicks butt.

But I must give you this 2021 update, my friends: Team West, and the US in particular, have really taken it on the chin since that 2008 global economic meltdown started. Most places on planet Earth now squarely put the blame for that world recession on the capitalist excesses of Team West. This has started a whole re-think in terms of following the capitalist path of the West, with some folks suggesting that capitalism as a whole is on the way down. And some think the same about democracy too: authoritarianism is on the rise again, with active military coups taking power in Thailand, Egypt, and most of the Arab Spring countries have still not made it to the democracy finish line . . . and of course Russia is leaning back in the authoritarian direction too, as is Turkey, and China never left it. Many countries now see democracy itself as inherently unstable, unpredictable, and messy . . . which it is! So some states are shying away from it in these chaotic days. Couple the democratic gridlock and economic crap with what is now seen by many as over-extension of political and military power in the Middle East, Central Asia, and Eastern Europe, and you have the makings of a good old fashioned smear campaign against the West. I can't lie: they are taking a hit right now in the global scheme of things and will lose some power. But the Team is resilient, and rich, so they are still a global playa' for some time to come. Had enough of the West yet? Then let's get on to some other entities which are not part of this team.

2021 UPDATE: To flog this horse once more, the reign of former US President Trump, in conjunction with the rise of several other right-wing populists in European states, in conjunction with the BREXIT debacle that weakened the EU, have all served to significantly weaken Team West's image and collective power in the world. Trump's retracted the USA from leadership of democracy promotion, from trade blocks, and even from NATO. All of those are the core foundation stones of Team West. The explosion of populist candidates like Trump and tons of other right-wing leaders in Europe could see a major retraction of Team West from the world stage as a singular force to be reckoned with . . . leaving behind a bunch of self-interested nation-states that might economically do even better on their own, but will not have the same global clout that the Team did in times past. Still, too early to tell, but Team West is certainly at its weakest point since pre-WW2. A New Order could be upon us.

THE BEAR

We already had a whole chapter on Russia, so I can keep it brief here. It was largely assumed that Russia would inevitably join Team West after the collapse of the Soviet Union in the early 1990s. It does share most of the common cultural characteristics of Western Civilization, but somehow it has always remained just on the fringe of that grouping, always slightly different, with a distinct "not European" component. After the demise of the USSR, Russia was a second rate power, with even less prestige, that was really at the mercy of international business and banking and stronger world powers like the US, the EU, and NATO. Russia pretty much just had to go with the flow, even when events were inherently against its own national interests, like the growth of NATO, for example. But all that has reversed in the last fourteen years under the long judo-chopping arm of Vladimir Putin. He, along with tons of revenue from the Russian oil industry, has brought the Bear politically back to a position of strength. He plays hardball in international affairs now, and has reinvigorated a Russian national pride that has long laid dormant. In short: the Bear is back. Grrrr! As I have suggested to you last chapter, Russia's full-frontal invasion of Georgia in 2008 was essentially a calling card that they are back, ready to wrestle, and more than willing to engage militarily if pushed on issues of their national security. **2014 UPDATE: WOW, DID I CALL THAT ONE, OR WHAT?** Not only was the Plaid Avenger dead on the mark predicting that Russia was no longer afraid to exert itself militarily, it of course actually went full-out blitz on Ukraine, grabbing back the Crimean Peninsula, knowing full well it would infuriate Team West and most of the world. And they did it anyway, and may do more, and don't appear to care about world opinion much at all anymore. Yep. That's power.

Now that the Bear is strong again, it has been re-thinking the whole joining Team West business, and it appears it is declining the invitation. Russia prefers to be an independent player in order to be a global power again, as well as to achieve its own unique foreign policy goals, and it is in a prime position to do this.

Since the Russians supply one-third of European energy demand, this gives them all sorts of economic and political leverage over their Eurasian neighbors. Get Russia upset at you, and you might not have heating oil next winter. But the Europeans aren't the only ones who need oil. Look eastward and you see Japan and China both vying for petroleum resources as well, so Russia is really sitting pretty right now. The Russians are forging economic and strategic political ties to their east as well, mostly with China. They are at a historical east/west pivot point and they are straddling the fence about which relationships to make or break. I think they're being extremely savvy and are going to avoid taking sides in anything. **2014 UPDATE:** May 2014, Russia and China ink a 30 year gas/oil deal worth billions. The die has been cast. Russia has turned East. They will not be joining Team West in our lifetime. **2017 UPDATE:** Russia is back to being a full-on authoritarian state with Putin at the helm. It is busy rebuilding its military prowess at a frenzied pace, subtly threatening its neighbors, claiming the Arctic Sea, further destabilizing Ukraine, absorbing the Crimean Peninsula, intervening heavily in the Syrian War, hacking the US presidential elections, (as well as other European elections), and pushing hard for other countries to join its Eurasian Union. The rift is real.

2021 UPDATE: Russia is now also working overtime to establish much tighter relations with Turkey, for the first time ever! These two countries have been historical enemies and have fought multiple wars against each other in the past. But with Recep Erdoğan's increasingly authoritarian tendencies, he and Vladimir Putin are now seemingly on the same page on how to run their countries, and both are increasingly anti-Team West. This is crucial, since it means that Putin is helping tease Turkey away from Team West, and into his own orbit. It will be another major blow to the global reach and credibility of the West. This is most tangibly seen by the recent Turkish purchase of some Russian missile defense equipment, even though Turkey is a NATO partner and usually gets all its gear from the US; it is kind of a big deal that infuriates Team West, and may just be the beginnings of a new path for the Turks. Another 2019 side note: Russia is also making a concerted effort to invest much more in Africa this year, and is trying to build both economic and political ties across the entire continent. Yet another extension of Russian power abroad that will cause stress for Team West.

We can see this rift between the Bear and Team West in a variety of ways in today's world. The US and many Europeans express concern or outright criticize Russia for becoming too authoritarian under Putin, thus detracting from their democracy and human liberties, which is, of course, one of the foundations of Team West. The US and EU also accuse Russia of using oil as a political weapon to get its way; a not entirely false charge, but is that illegal?

In return, Russia usually counters most US movements at the UN Permanent Security Council, where it holds veto power. One need look no further than the current Iranian issue, where the Russians have been quick to dismiss Team West's fears of Iran's nuclear ambitions, has promised to veto any severe embargoes or military action against Iran, and is even helping the Iranians build their nuclear power industry. Russia has also propped up the regime in Syria, and protects them from any UN involvement via their veto power. Russia is increasingly working together with China economically, diplomatically, and militarily, which is perhaps unsettling to the other world teams.

The Bear is a pole of political power not to be taken lightly, but is there really ever a situation that a bear would be taken lightly? While I would never brawl with a bear, I would be even less likely to disagree with a dragon . . .

ENTER: THE DRAGON

Just as with the Bear, the Dragon has been resurrected as of late and is on the fast track to becoming a major world political power. As you undoubtedly deduced from the chapter on China, the state had a rough go of things for most of the 20th century, from internal destruction to civil wars to colonialization to wartime atrocities to counter-productive communist policies. Since the capitalist reforms of the 1970s, China has really turned things around and become an economic powerhouse of the globe. Remember, it jumped two whole slots a few yearsa ago to become the 2nd largest economy on the planet, and will likely take the top slot from Uncle Sam in the coming decades. Or maybe even this decade.

With that economic clout has come global political power. The Dragon has been very focused on rebuilding itself for the last few decades and therefore was not that interested in exerting itself in the global political structure, except of course when it was in its own interest to do so; think of any issue dealing with the status of Taiwan. China also is the staunchest supporter of sovereignty in the world, so its attitude towards global issues and conflicts outside of its own territory has typically been "live and let live," although "live and let die" may be a more appropriate descriptor. China does not like to get involved in the affairs of other states, no matter how horrific the situation, and in return wants no one poking their noses into Chinese business. You dig?

Now this would all be just an asinine academic excursion but for one thing: China is one of the veto-wielding members of the UN Permanent Security Council. When international action is called for by Team West to punish Iraq or Syria or North Korea, the Chinese position is usually to play the spoiler and veto any actions by the UN, or at least dumb-down the UN action to the point that it is ineffective or useless. Need an example?

Take Sudan, Zimbabwe, and North Korea. Most of the world does not like, does not support, and does not want to trade with these countries because they have fairly brutal records of persecuting their own people. The morally and ethically right thing to do is to refuse to support that state in any capacity. But China says, "We don't care. We will trade with you. Can we buy oil from you, Sudan? Thanks, buddy!" Perhaps their open policy is not such a great thing. This causes friction between China and other rich countries around the world, especially "Team West" countries. Developed countries say, "You can't do that," and then China says, "Hell yeah we can. We are a sovereign state." The issue is a source of friction for a lot of the Team West countries. Both China and Russia are very good at this game. Speaking of which . . .

I just can't help myself but to over-exaggerate a subtle theme that has been forming globally for some time now, and has burst into full flower just this year: China and Russia are totally in love, dudes! These two titans have pretty much agreed on every single global item of consequence in the last year, and have met more times than I can even count anymore. From heads of state meetings on trade and security, to a joint space program, to meetings of the SCO and the BRIC; China and Russia have pretty much thrown down the gauntlet and created a new power block for the planet.

What do I mean by this? Well, in addition to all the joint stuff they are doing, both President Putin and former Chinese Premier Wen Jaibao have openly body-slammed Team West for causing the 2008 global financial meltdown. They both are also pushing for way more security and cooperation via the SCO structure . . . not through the UN or any other Western device. These guys made a joint declaration after their SCO meeting in June 2009 that they favoring scrapping the US dollar as the world standard. Dang! This is total cage-match smack-down. But back to China now . . .

With the number two economy on the planet and increasing international pressure to pony up and accept some global responsibilities, China's star is on the rise politically too. China's steadfast respect for the power of sovereignty still trumps all, and thus, the Dragon will be very hesitant to throw its weight around in the international arena. China will hesitate from any unilateral action on its own part, and also mostly stalls multilateral actions even when sponsored by the UN. While the Dragon does (in theory) support the US-led War on Terror, it will continue to stymie any major efforts by Team West to transform other countries, either by invasion or persuasion. With its UN veto-wielding power, it more often than not will be siding with Russia to counter major moves by Team West, while strengthening its hand at home and in its immediate neighborhood. That is done with an eye towards countering another global power in the Dragon's 'hood: India.

TEAM HINDU

I would be remiss to not include the other significant growing power center on our planet, and that would be our Indian friends. Like China, Team Hindu is a burgeoning economy here in the 21st century and that economic power will increasingly be translated into political power. That's right, India will soon be another major axis of planetary political power. Especially if their new main man Modi, the Hindu head cheese, has success transforming the economy into a capitalist powerhouse! But I don't want to exaggerate the circumstances! India is on the rise for sure, but currently is nowhere near the economic, political, or military power of a Russia or China—yet. Stress on the "yet."

India is soon to be a major power player because it has a whole lot of unique attributes which will soon force it to be a player. Let's quickly list them: India will soon be the most populous state on the planet; China is now, but not for too much longer. India has a vibrant and growing economy, and it is simultaneously diversifying nicely. It is a nuclear power, and, yes, it has nuclear weapons. India has a significant military, certainly the strongest in the region. As part of that, India has an aircraft carrier. That means it has a means of projecting power abroad. The only other countries in the neighborhood that fit that description are China and Russia, who are both power players already.

However, unlike their Chinese and Russian counterparts, Team Hindu actually shares a lot of attributes, ideals, and goals with Team West. India is a strong, established democracy—the biggest in the world, actually—and a supporter of free market principles and global trade. India, like Team West, has human rights and individual liberties as a foundation stone of its society. India is an avid supporter of the War on Terror and, as such, does quite a bit at home to thwart extremism of all sorts. They even speak English, for Pete's sake . . . how much more Western could they get?

Well, I'm pushing it a bit far now—they are not Western, and do maintain their own distinct culture, as well as their own distinct take on foreign policy and global affairs. But look for them to become a stronger and stronger voice in the near future. Former US President Bush met with the former Indian Prime Minister Manmohan Singh multiple times during his presidency in an effort to warm relations between the two titans. First time that has ever happened between these two titan countries. The US was actually working very hard to get a deal signed with India in order to help them out with their nuclear power industry, and that deal finally went through in 2009. It's a really big deal, too! The US and India are now in a friendship that is just now starting to blossom, all because of nukes!

Why do I keep bringing up the nuclear issue? The US and Team West want to legitimize the Indian nuclear position—India never signed the nuclear non-proliferation treaty—so as to further isolate and de-legitimize any future Iranian one. See how this works? **2021 Update:** India still buys a lot of high tech weaponry from Russia and, while developing better relations with the US, still keeps its door open to other partners. It has been and remains a staunch independent power that never sides up solely with any other team. But here lately, India is leaning slowly—ever so slowly—to deepening its relationship with the USA and Team West in general (including Japan!) in an effort to counter the wildly growing power and prestige of neighboring China. Look for that to continue under an umbrella named "Asian Axis of Democracy."

Team Hindu may still be a minor world player right now, but that status won't last for too much longer. They are one of the five potential additions to the UN Permanent Security Council, and given their western tendencies, I would look for their membership to be supported by Team West. I would also speculate that Team Hindu will likely be increasingly siding with Team West on global issues of the future, albeit with a more reserved tendency to use force. Gandhi taught them way too well.

POLITICAL TEAM SUMMARY

Uncle Sam has been the sole political superpower on the world stage, but that brief era is coming to a close. The rise of China, India, and the rebirth of a strong Russia preclude the US from acting unilaterally in the future. The USA and Team West in general are at their weakest, most divided point in modern history, and their way forward is unsure and unknown at this point.

China, and increasingly Russia, has a tendency towards authoritarianism and an emphasis on the trump card of sovereignty; therefore, they do not completely share in the value systems of Team West. As such, these two power players are a likely alliance that will balance/counteract the Team West influence on global affairs. Team Hindu does actually have much more in common with Team West, and will likely be an ally of the West in future global issues.

This section has merely been a summary of the **MAJOR** power players on the planet. Of course, there are other states and entities that will affect global events and influence the direction of the planet in a myriad of ways. Brazil, Turkey, Nigeria, Indonesia, Islam, and ASEAN all spring to mind. We simply focused on the most powerful, most organized, and most influential teams that are or will be the major shapers of global political policy for the near future. These are the teams that will decide how to conduct the War on Terror and the global War on Drugs. These are the teams that will decide all major UN actions, including the possible expansion of the UN Security Council. These are the teams that will shape all major policies on global warming, fighting global crime, and conducting global war.

They are the majors of political power. But what about other types of power?

MILITARY MACHINATIONS

You might think that political power immediately equals military power, and you might be right. Might does sometimes make right. There are many individual powerful states on the planet right now. However, the alliances of military power that have occurred between states has been accelerating greatly in the last decade, and these military power centers have themselves become an important component of understanding how the planet works. Let's briefly look over the major military entities on the planet and how they interact with one another.

THE BIG BOYS WITH THE BIG TOYS

The United States is the undisputed heavyweight of military power, technology, and spending. Look back at the table in chapter 6 to check out military expenditures for 2011 and you can quickly calculate that the US alone accounts for half of the total world spending on all things military. Boom! That's some power! However, China actually has the largest standing army, and is spending fast to catch up with the US. But even these two they are not alone.

Let's just call a spade a spade here. The most powerful state military entities in terms of global reach are the US, the UK, France, Russia, China, and India. Why those? Those are all the states with mostly modern militaries, the money to make them stronger and keep them up-to-date, the ability to project that power outside of their own countries, and they all have nuclear weaponry to boot. Places like Israel and Pakistan may have many of those qualities, but not all. Certainly Israel, Pakistan, Turkey and Brazil are powerful regional entities, but I would not necessarily call them powerful global ones. That make sense?

All the power players, except India, are also on the UN Permanent Security Council, which means they all have a big voice in how that organization is run. For this section, just know this: these are all countries that have, or could, act unilaterally from a military standpoint. But that's not really the way things go down anymore.

ARMED ALLIANCES

Mostly, use of military force in global conflicts occurs as a coordinated effort between states as part of a bigger institution or entity. Most of these groups have already been discussed throughout this text, but let's hit them up one last time and make some predictions on not only future use of military power across the globe, but also how these entities will be working with, or against, each other in shaping global events.

THE UN

UN Secretary General António Guterres.

Most member states contribute troops, supplies, or money to the United Nations for its active military missions. Since any active mission must clear the UN Security Council, the countries described above get to decide where those missions will be allowed to happen. Have you figured out where virtually all UN missions have occurred? If you said "Africa," then give yourself a gold star! Why there? Because it's the only place that all the major powers can agree on with regard to having a UN presence. China would never allow a UN military mission to Burma (their ally), Russia would never allow a UN mission to Kazakhstan (their ally), and the US would never allow a UN mission to Canada—okay, maybe that one would fly, but you get the picture.

The only places that UN troops end up in are typically poor, developing countries in which no major power has a vested interest, and therefore no one on the Security Council will veto. The Democratic Republic of the Congo, Liberia, Haiti, Lebanon, Côte d'Ivoire, Chad, Ethiopia, East Timor and Kosovo are all excellent examples of current UN deployments that lack vested interest of major power players. However, Kosovo should throw up a red flag to you, as I have explained earlier in this text how it has become a hot-button issue between world powers.

Dig this: that mission was agreed upon by Russia back in the late 1990s when it was in a significantly weakened state. Russia would never let something like that go down now! Same thing goes in reverse though: the UN mission to poor, undeveloped Sudan has been held up by China for years, because it formed a vested interest in the place when Chinese companies started developing Sudan's oil fields. Only increased international pressure has forced the Chinese to allow a mission to happen, but in a much more limited capacity; that is, a smaller force with more constraints on what it can do.

That brings us to the last point about military maneuvers of the UN: they are largely weak and ineffectual. The big powers can never all agree on any single issue, and therefore any mission which gets passed through the UN is typically watered down to the point of nothingness, if there is any action taken at all. Remember the Rwandan genocide? What a freakin' debacle! The major powers couldn't even form a plan of action on that huge mess, even though no one had a vested interest in the place. The entanglements of getting military missions passed through the UN system is the major reason that the US typically bypasses the UN altogether and goes straight for NATO.

2021 UPDATE: With the rising tide of populism happening across the planet, look for more and more states to look inward, and therefore pay less attention to global institutions that they feel threaten their personal sovereignty. A bunch of states have recently dropped out of the ICC (the UN-sponsored International Criminal Court), Russia being the most high-profile country who recently just said: "Pass. We are out. We don't care about this court." But so did a few African states, too, and the Philippines is considering it. The re-rise of the nation-state is upon us, at the expense of these international global institutes. This is a dangerous trend that may be here to stay for a while.

NATO

NATO is still the most powerful and effective military arrangement on the planet right now, and the preferred choice of the US, since the US is the primary player and most influential member in the club. In other words, the US has a much easier time getting actions sanctioned by NATO than by the UN. You know all about Article 5 and how the group works, so let's cut to the chase.

NATO is essentially the military arm of Team West—yes, the whole Team, not just the US. Think about it. Makes sense, doesn't it? All members of NATO are part of Team West, and even Japan and Australia (non-NATO members) often help out NATO missions with supply and support services. This is what makes Team West so potent: a combination of common ideologies with an organized and powerful military structure. They can get the job done, and more often than not, can agree about what the job is.

This is significant in today's world because NATO has done a major mission shift to now incorporate the War on Terror. As you know, after 9/11, the US invoked NATO Article 5 in order to get the club to help out with the Afghanistan mission, which they did. The current war in that state is a NATO war, not a US one. That sets the stage for many more NATO engagements in this global war on terrorism in the future. Granted, NATO will not dive head first into every US anti-terror effort; it is important to note that NATO did not support the current US war in Iraq. However, they may eventually come to the US's assistance in this situation as well. NATO has become stronger over the years and has increased its mission to a global reach, which is fitting since they are the most effective global military in the world. **2011 UPDATE:** NATO acts as the arm of Team West once again by organizing the charge against Muammar Qaddafi's Libya! Stay tuned for the climactic conclusion of that confusion! **2012 UPDATE:** Mission accomplished: Muammar now worm food.

What's not to love about NATO? If you don't necessarily agree with all the foundations of Team West, then you might be a bit leery of NATO's power. Russia is the obvious antagonist of NATO and continues to be irate about the US/NATO missile defense shield in Eastern Europe, which was officially launched in May 2012. In the past, Vladimir Putin openly threatened to target missiles at the Ukraine if they attempted to join NATO, and also to more generally target the nuclear arsenal at Europe as a whole if NATO proceeded with its missile-defense systems being built in Eastern Europe. Then there was that small matter of Russia invading Georgia before it could become a NATO member. And now Russia has absorbed the Crimean Peninsula from Ukraine, and has destabilized the country enough to warrant further invasion at Vladimir Putin's whim. In other words, this resurgent Russia's physical power plays back into Eastern Europe have proven to the world that Russia fully intends to re-exert its influence in, if not outright control of, its 'hood.

Why am I talking so much about Russia in this NATO section? Good question. Just this: just a year ago I would have told you that NATO is done expanding, and was being slowly de-funded by its broke European members, and that the group was receding in power and influence. But the Ukrainian unraveling and Russia's chest-thumping and expansionary tactics have once again breathed life into NATO, and is a born-again entity as well. NATO's sole reason for being created in the first place—to protect against Russian aggression—has been re-established and has reinvigorated the group. Even the European members are now anxious to use NATO to project a powerful front by increasing funding, increasing military exercises, and possibly even increasing NATO troop presence in the most threatened Eastern European member states, like Estonia, Latvia, Lithuania, and Poland.

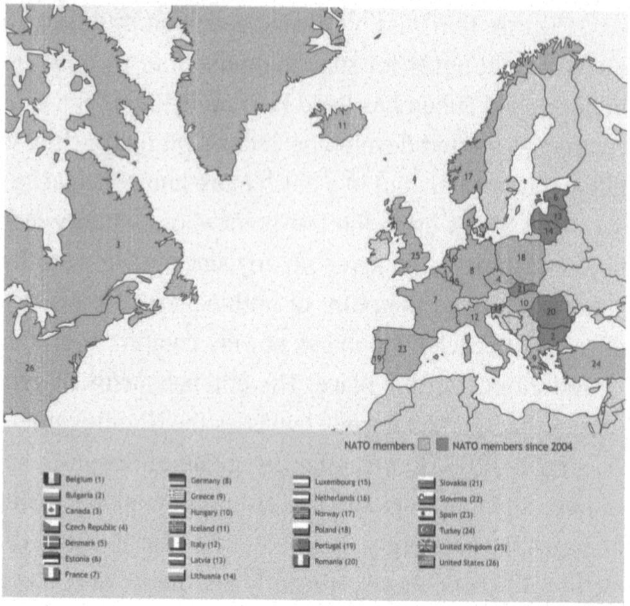

Yep. Cold War is back on. But this time it is not an ideological battle between the US and the USSR, it is actually a more simplistic battle for territorial influence, and a drive by Russia to re-establish its street credibility, while also protecting/promoting its self-interests in the vicinity. Which is what big powers do. Long story short: Russia and NATO are once again full antagonists, and both will be spending more money on military maneuvers . . . and there is no hope whatsoever that Russia will be joining Team West, nor working with NATO anymore . . . because it appears that Russia has other things in mind. It is not looking to ever join NATO, because perhaps it has an alternative. . . .

SCO: SHANGHAI COOPERATION ORGANIZATION

Of course! From chapter 22! Russia + China + all of Central Asia. That's the SCO: the anti-NATO! Okay, that is a bit too strong of a descriptor, but the Plaid Avenger wants you to be smart about the world, not politically correct. The heck with that! Let's get to the goods.

Though the declaration on the establishment of the Shanghai Cooperation Organization contained a statement that it "is not an alliance directed against other states and regions and it adheres to the principle of openness," most observers believe that one of the original purposes of the SCO was to serve as a counterbalance to the US and NATO, and, in particular, to avoid conflicts that would allow the US to intervene in areas near both Russia and China. How are they pulling that off?

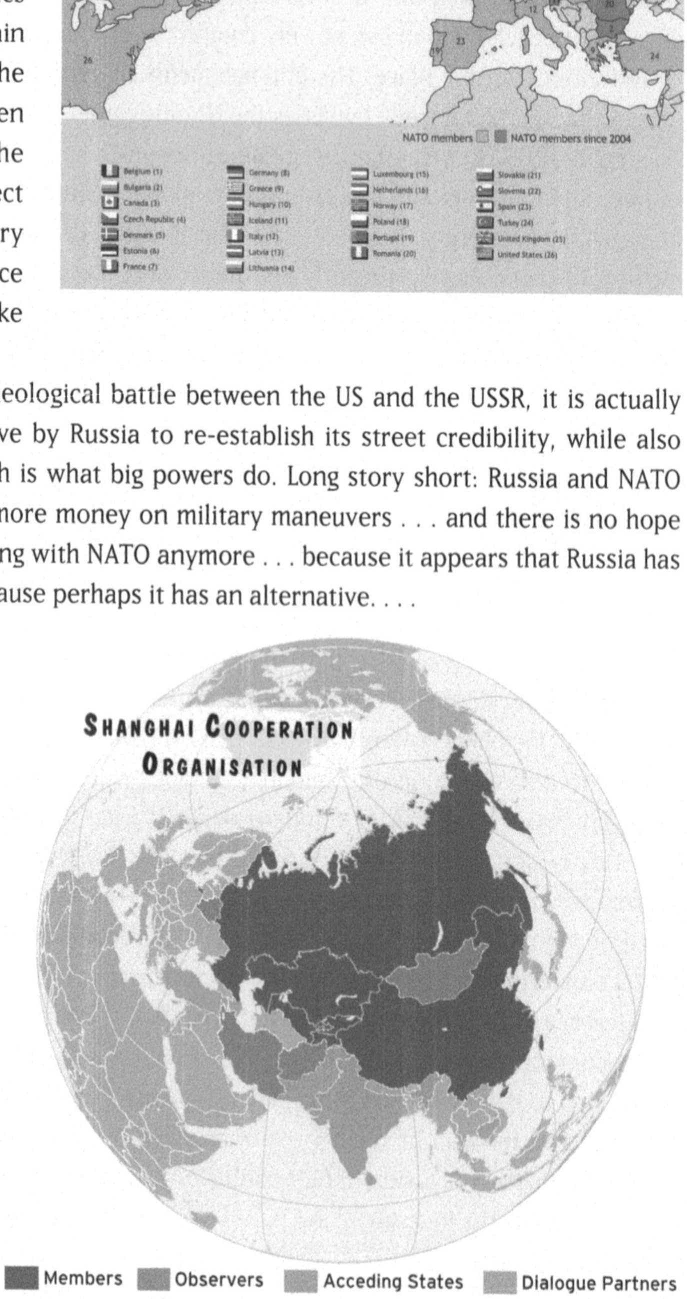

SCO states are strengthening their defense pacts with each other to ensure that no interference or invasion will be allowed from outsiders. In essence, they have formed an Asian version of NATO. The movement in this direction is why most US military bases have now been kicked out of SCO states, and also why the US request to become an observer state of the SCO has been declined every single year. It's also why the SCO has started doing the joint military exercises.

However, it is a very young organization, so I wouldn't call them a military powerhouse on par with NATO just yet. But the alliance does consist of two nuclear powers, Russia does have the second largest nuclear arsenal on the planet, and China does have the largest standing army. While I do not predict this entity doing anything aggressive anytime soon, I also do not think it's a stretch of the imagination that the US and/or NATO would be extremely hesitant to ever attack any of the SCO members, even the dinky ones, for fear that the defense alliance might strike back. I mean, that is the whole point of a defense alliance after all.

Let's pretend the US/NATO strikes a terrorist cell in Uzbekistan, which then forces the SCO to counter-attack. Sounds like a WWIII scenario to me! Don't get too worked up yet, though; these guys are still organizationally pretty weak, and their mettle is untested. Would Russia and China really counter a US attack? Only time will tell. Actually, let's hope we never have to find out.

The more important issues have to do with those observer states of the SCO, particularly Iran. As suggested in Chapter 22, Iran would just love to be a full-fledged SCO member so that it could have an insurance policy against western aggression. Iran thinks that the SCO membership will somehow shield it from any UN or NATO or US attack. But it's not a full member yet, and Russia and China are not likely to grant it that status, because no one involved wants to get into a global power battle over the antics of Persian Presidents or agitated Ayatollahs. Again, let's hope that is the case.

The SCO is a fascinating study on the future of Eurasia, and the world. Be sure to keep up with them as the entity continues to evolve. We already have two major power players in the club, as well as the entire region of Central Asia. India is also an observer member: a position which sets it up as a future power player on the globe. Iran really wants to be part of the club, which effectively moves SCO power and influence into the Middle East as well. These guys are just all over the place!

OUT OF AMMO

Summary: the UN is weak, NATO's future is uncertain, and the SCO is evolving. Both the US and Russia by themselves possess globally lethal amounts of firepower and the will to use is. Oh, and China grows ever more assertive in staking out its dominance of the Pacific Rim, with no need for partners or positive press. Those three powerful states in combination with those three powerful military alliances will be the ones which shape future conflicts around the globe, both the small ones and the big ones. The UN is likely to expand the Permanent Security Council, which will make them even weaker from a military standpoint; more voices at the table means less things that can be agreed upon by all. NATO is nearing the end of its physical expansion period, having pulled in most of Eastern Europe, and stands more powerful than ever given its new objectives of countering Russia once more. The SCO is young and ineffectual so far, but that means it has the most growing to do, and possibly the greatest future potential of them all.

Of course, each individual state has its own military might, and the power players will continue to dominate the game in the global competitions. Smaller regional powers like Israel, Brazil, Pakistan, Saudi Arabia, Turkey, and Iran will continue to build up their military might and will affect regional politics in their prospective neighborhoods, but won't have global reach. I suppose I should also throw in smaller and smaller military entities like rebel and terrorist groups, which can shape regional or even global policy in general by their actions. I know they are the ones that seem to be the big deal every single day when you read news headlines, but the reality is that they are, militarily, just a pin-prick on world affairs. Many pin-pricks can add up though.

I did want to give a final shout-out to my AU military brothers as an up-and-coming force for regional power as well. The African Union as a whole may not be as effective thus far in terms of economics or politics, but the AU military is starting to gain some steam. It is increasingly being used for peacekeeping efforts all around Africa—it is a critical part of the UN mission in Sudan—and is getting a lot of support from the big power players to keep on getting bigger and better. Indeed, perhaps it will be a strong and professional pan-African army that will help the region as a whole put itself back together, fight corruption, and build a stability that will allow for economic investment and growth. Wouldn't that be grand? Speaking of economics . . .

ECONOMIC POWER PLAYERS

Another facet of global competition is the quest for more dollars—more dollars for your country, more dollars for your country's businesses, more dollars for your country's citizens. Mo' money, mo' money, mo' money! Do we have Team play at the global scale for economics? Sure we do!

To be sure, economies are made up of businesses, which are made up of people, and all people have different personal incentives and goals and ambitions and opportunities to make money. I'm not going to get down to the abstract, philosophical reasons for why people do what they do, or even why countries do what they do. When the rubber hits the road, it's really every man for himself, or in this case, every state for itself, since states are first and foremost beholden to their citizens to make sure that their economy is as strong and rich as possible. Some states are obviously way better at this than others, sometimes at the expense of other states. But that is the way the proverbial cookie crumbles.

However, even though every state looks out after itself first, some states have figured out that they get richer when working with other states. The economic centers of the planet do shift according to who is doing the best, producing the best, or growing the fastest. We can look at what entities and Teams are major forces for economic activity in the world, how they compete with each other, and what the future might bring in terms of cooperation or collision. Sounds like fun, huh? Then let's expound about some economies!

BIG BOYS WITH BIG TOYS: PART DEUX

Why not stick with the pattern? We can identify the individual states with the biggest economies, for starters. The US is by far and away the biggest economy in the world, larger than the next three economies combined. You know the next three: China, then Japan, then Germany. The rest of the top ten is rounded out with mostly rich European countries, but watch out! India and Brazil have jumped into top ten status too, and a simmering South Korea is making its way up the ranks, as is Indonesia, Mexico, and Turkey. Why should we care about these economies with big numbers?

Because money is power! Rich societies are typically stable and happy societies—typically. Who doesn't want to be in a rich society? Levels of richness effects all sorts of things like political power, military might, standards of living, but also global patterns like commodity chains, trade routes, distribution of economic activities, and even immigration/emigration patterns. I know, big shock: people migrate from poor states to rich ones. Go figure.

Where these super-economies are located also has a lot to do with how economic alliances go down, how competitive these states are with each other, and how powerful they can become. Don't believe me? Then let's revisit the EU.

EUROPEAN UNION

The EU has a dozen of the top world economies within its ranks, but individually, none of them can compete effectively with the US economically. But hold the phone! Take all of the EU states together as a single entity, and you are suddenly looking at the richest organization on the planet, even richer than the US in terms of total GDP and perhaps even GDP per capita. The EU as a singular entity has increased trade among its member states, increased trade abroad, and become much more effective at attracting international investment. That's what everybody wants to do! Which is why these supranational organizations based on trade have become so popular lately—and rich.

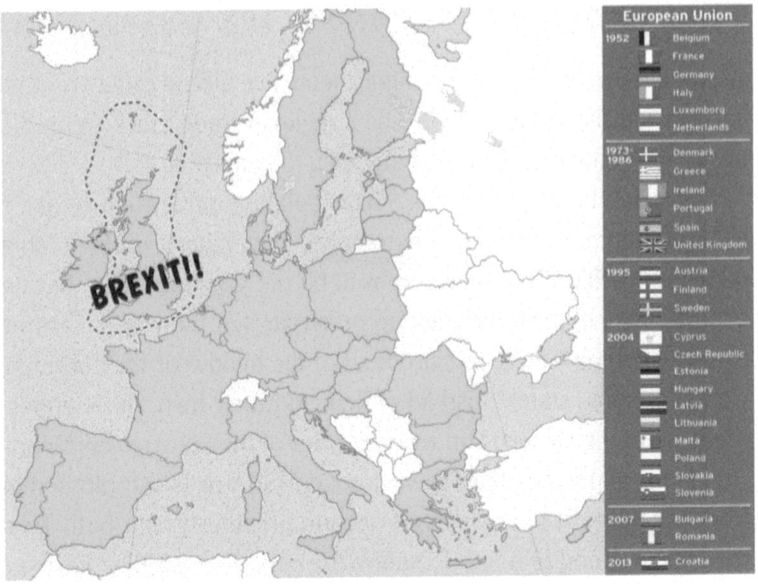

Before we get to that, a few more points on the EU. Like NATO, the EU seems to be at the end of its expansion streak. It has pretty much absorbed all of Eastern Europe, and the only likely future members are the states that were formerly Yugoslavia (no real economic gain for the EU here), the Ukraine (big potential gain here) and finally Turkey (big gain with some big risks here). The former Yugoslavian states are economically weak and have little to offer even when strong. The Ukraine is an agricultural and manufacturing powerhouse with a huge land area and population that would radically affect the EU. Turkey is almost exactly like Ukraine for all the reasons just listed, but promises to be divisive within Europe as a whole because of its Islamic culture. Look back to chapters on Western Europe and Turkey for more details.

The EU is at a breaking point where lots of people think they can simply not get any bigger and maintain any sort of effective control and economic cohesion, but they do have a lot to work with already.

2012 UPDATE: Wow. Could that previous sentence be any more prophetic? The recession in the EU, followed by the complete economic debacle in Greece, now appear to be shaking the very foundations of the EU and especially the smaller sub-grouping of the Eurozone countries. With Greece likely to be kicked out of the Eurozone, and a handful of other EU states on the brink of bankruptcy, it seems all too apparent right this second that maybe the group has grown too big too fast and the vast economic differences between some of its member states are causing a huge rift in the fabric of the block itself. Stay tuned for new EU assessments as they unfold!

Creating this union has been the only thing not just keeping Europe afloat, but making it a global powerhouse to compete with the US and a rising China. The EU will continue to bust ass to fine tune its EU program—because it knows it has to in order to continue to compete with the economic prowess of Asia as a whole. Asia is on fire with economic growth, from Korea to Indonesia to India. The center of the economic game is shifting. For hundreds of years, the Europeans were the center of the global economy, eventually forfeiting their top slot to the powerhouse Americans who have held onto it for about a century. But nothing ever stays the same forever, and the 21st century promises to be the Asian century in terms of becoming the economic center of the universe.

This transition is already well underway, and the Europeans do not want to get left totally behind as the Asians start kicking economic ass and taking names. In the 1990s, when the USSR folded up shop and the EU expansion began in earnest, there was even talk of eventually incorporating Russia into the EU. What a continental powerhouse that would have been! Russia is such an important part of the EU economy that it is already invited to major summits and consultations on EU policy, even though it is not a member. **2015 UPDATE:** Yep, Russia is now out of any possible partnership with the EU, due to the Ukrainian situation which has pitted the old Cold War teams against each other once more, even economically. **2017 UPDATE:** Yeah . . . so that whole BREXIT thing. You have already read about the weakening of Europe, the UK leaving the EU, and the possible collapse of the entire organization back in Chapter 8/9A Euro-Addendum. So I don't have to re-hash it all here right now. Just know for now: much like everything else Team West related, the future is very, very uncertain for this group here in 2019. We have to wait to see if they will survive the year, and what factors are at play by that point to determine its future.

So UK is out of EU. Turkey may not want in anymore. And Russia? Well, Europeans know they need Russian energy, so there is no way anyone is actually fully severing ties or anything drastic like that . . . it is more of a cooling of future investments, deterioration of trust between the parties, and a lack of any future potential for further integration. So, Russia in the EU? Right this second it doesn't appear that it will ever be, by choice, because the Bear is straddling the fence, already looking to the other side of the continent for alternative companionship, which brings us . . .

BACK TO THE SCO . . . AND A NEW UNION OF NOTE

What? Back to the SCO! Remember them? I've already told you that this group may eventually be an anti-NATO, but they also started the organization to be a regional trade block. With the Eurasian titans of Russia and China as the bookends, this is increasingly looking like the team to beat.

As you already know, the SCO is building infrastructure like roads and pipelines across the member states, mostly to facilitate natural resource extraction. Remember I told you about the possibility of them becoming a natural gas cartel akin to OPEC? That may very well be the building block that turns into increased trade among the nations, but more importantly will facilitate trade to other parts of the world. Central Asia has done well to throw in its chips with the SCO. Otherwise, they would be marooned in the middle of the continent fending for themselves. That trade block stuff really works for some states! Instead of being on their own, those states are now a central hub of export supplying petroleum products to the eastern and western sides of the Eurasian continent. Damn! Silk Road back in action!

More important than salvaging Central Asia is the blossoming relationship between Russia and China. You know from the political power players section previously that both Team Bear and Team Dragon typically find themselves as the opposing end of a lot of Team West initiatives, so they have a common alliance of sorts already. Both states are avid supporters of sovereignty, and don't want anyone messing in their backyards. The evolution of a friendship between the two, and then the creation of the SCO, is a predictable development. Team West and the EU were hopeful that Russia was going to join their squads eventually, but it doesn't look like that is going to happen, perhaps ever.

Russia is strong again, but it's no powerhouse economy on its own. Hanging out with China certainly is beneficial to Russia, especially since the Chinese demand for natural resources and energy supplies is insatiable. The Russians find themselves in the enviable position of supplying energy to the powerhouse EU and also the powerhouses China and Japan. Their economy is thriving, and it probably will be getting a lot better, because as that world economic focus shifts to Asia, Russia stands to continue to benefit even more. The Bear will be increasingly dining on Moo Goo Gai Pan and sake shots as it rolls over to economically embrace the Dragon.

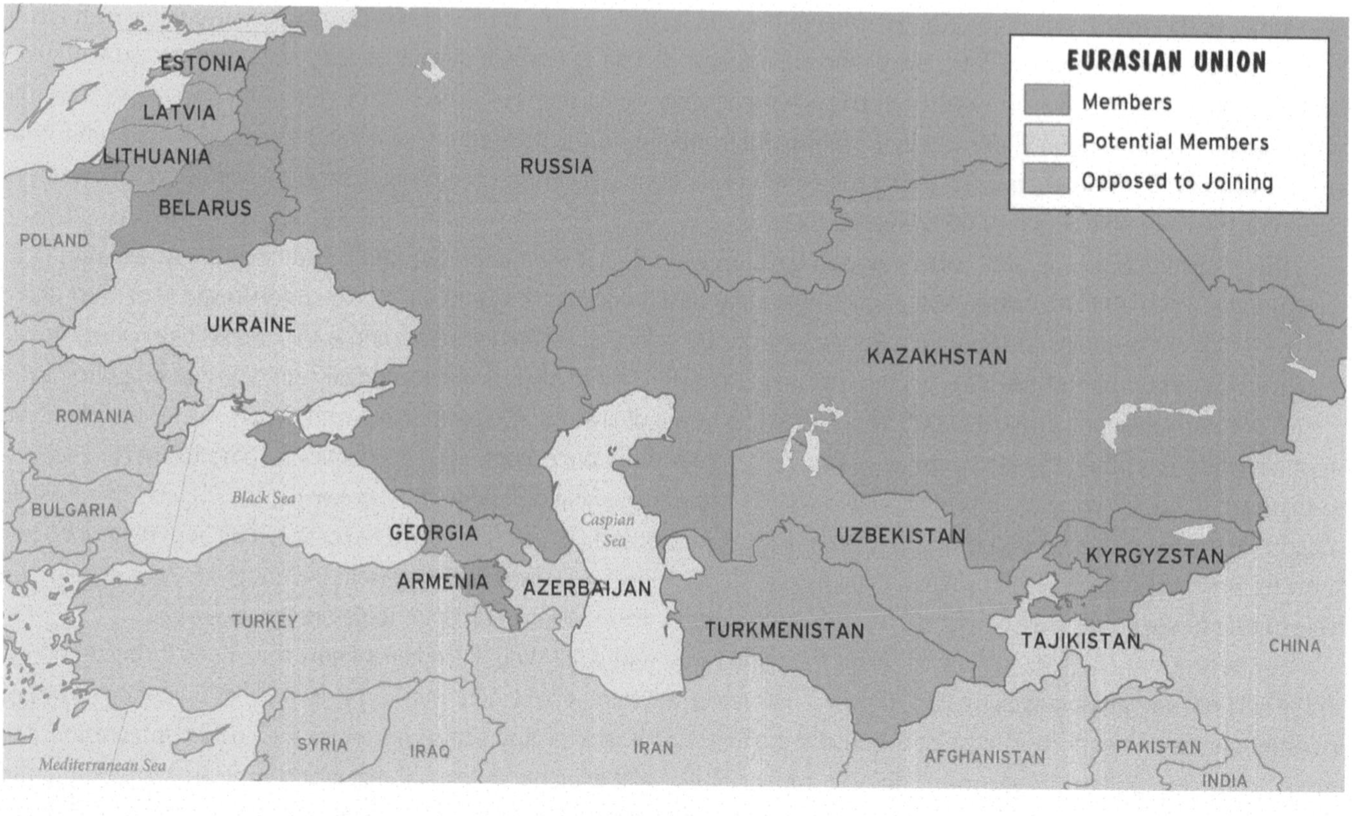

But don't let me suggest that Putin's Russia is just going to rely on selling more stuff to China to secure their economic future Oh no, my friends! He has much grander designs on the Asian 'hood than just that! As mentioned in several places throughout this titillating textbook, Putin's grand ambitions of restoring the old soviet order have now become reality! The Eurasian Union has been officially born!

While originally an idea posed by Kazak President Nursultan Nazarbayev, the Eurasian Union, is without a doubt the wet dream of Russian President Vladimir Putin, who now openly seeks to reassert a resurgent Russia's influence and outright power into areas it lost during the collapse of the Soviet Union in 1991. It is being proposed as a economic/political trade block alternative to the EU . . . and one that has Russia at the center. Signed into existence on 29 May 2014, between Russia, Belarus and Kazakhstan; Armenia has requested to join, Azerbaijan is debating it; other states are being pressured hard to join, or possibly face the wrath of Russia.

Think that is an exaggeration? Well, the whole unraveling of Ukraine hinged upon them getting ready to initiate talks to join the EU, which were then intentionally tanked by pro-Russian President Viktor Yanukovych, who of course wanted Ukraine to join this Russian-led Eurasian Union group. Of course this incited riots that led to his ejection . . . and then Russia's power play to take over parts of the country. Could this exact scenario happen in other states that Russia wants to pull back into its orbit? Maybe. Maybe not. But Russia has certainly put everyone on alert that they are not afraid to use force to get their way, and that has everyone in the vicinity thinking twice before siding up with Team West on any new initiative . . . even economic ones!

Putin's grand gamble here is that his union (and yes, it is referred to as his personal grand strategy at this point) will become a successful alternative to the European Union, and that states of Central Asia, Eastern Europe, and even the Caucuses Region will want to sign up for it. I suppose it is being marketed as a great way to boost economic growth, without having to bother with all that pesky democracy and human rights stuff. And it is definitely being pushed as an alternative to joining Team West. And it definitely is Russia trying to regain "influence" in areas it previously held as its own. It is not a stretch of the imagination to see that Putin is trying to reassemble—at a minimum—the old Commonwealth of Independent States, which was an economic working group of previously SSR's of the old USSR after it collapsed.

So even economically, Russia is pulling its old team back together to counter the west, and pushing its new team to integrate even further to counter the west. See what I mean now? For many countries in Eastern Europe and Eurasia, they will be hard pressed to jump into the Eurasian Union, or the SCO . . . both of which are alternative power structures to the Team West tool-kit. Will it work? Well, with the BREXIT and the unsure future of the EU, and possibly even of NATO during the Trump/Putin tenure, the Eurasian Union looks more promising than ever. I am still reserved about the chances of a true Russian revival, but no one is disputing that the Dragon will continue to breath fire for centuries to come . . .

Commonwealth of Independent States

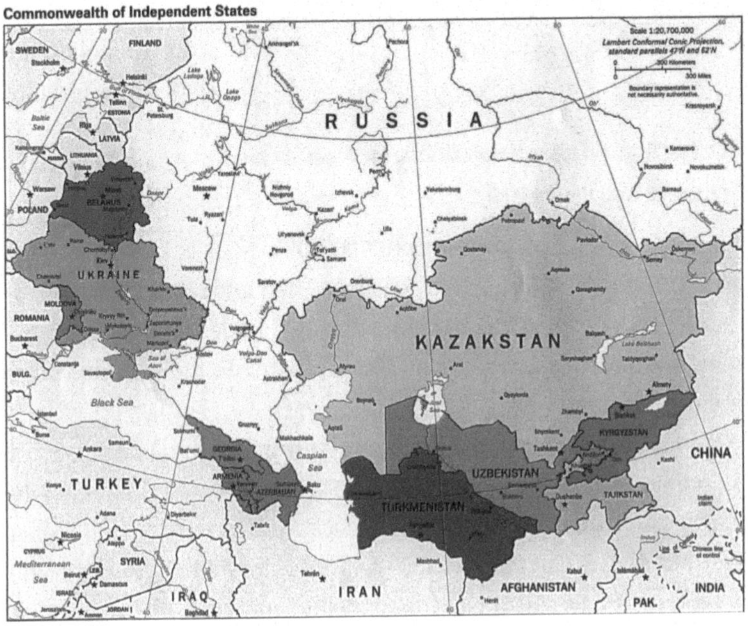

THE DRAGON AS ASIAN LEADER

I've been stressing that Asia as a whole is on fire with economic activity, and with no doubts you know that China is the main engine for all this explosive exponential economic energy. Team Dragon has been cranking out double digits in terms of increases in GDP growth year after year for over a decade. That is crazy talk growth! Their manufacturing,

industrial and service sectors have been blowing up as they produce everything from toothpaste to Nike shoes to rocket parts for NASA spacecraft. They have become the workshop of the world.

Why am I bothering to repeat this? Because with no reservations, I tell you that China will become the world's largest economy in the next fifty years. It will displace the US from the top slot, although you have learned enough from this book already to know that this shift does not equate to China being truly richer or better off than the US. They will have the biggest economy, and more importantly, will continue to be a focus for international investment, as well as increasingly a center for evolution of new technologies.

Here is the real deal about being a world economic power player and the focus of world economic energies: more things happen in your neighborhood. Huh? You know what I'm sayin'! With this increase in economic energy comes an increase in interest, an increase in investment, an increase in education, and an increase in technologies. The best and smartest people are attracted to where the flame is burning the brightest. The newest innovations in all sorts of technologies almost invariably occur in hot, happening societies where the action is, and that center of action is increasingly Asia!

Team Dragon is being looked upon as not only a very desirable trading partner among other Asian states, but as a model of success to be duplicated. Most Southeast Asian economies are now tied into China, as are the Koreas, and even the Central Asian countries via the SCO as described previously. China is truly regaining its historical legacy as the center of the Asian universe.

Tigers back with the Dragon!

Asian countries are not alone. Japan (okay, they are Asian too, but very western!) is turning more to trade, invest, and compete with China, as is Russia, Australia, and the US. Japan and Australia in particular have publicly pronounced that they know they must work together more economically with China if they are to continue to thrive; both countries' foreign policy initiatives have shifted immensely to suck up to China more, even though both are still part of Team West. Hey man, money talks!

I call it Team Dragon because I want you to know that it really has become more of a team affair. Yes, China is the center of the action—but remember, there are other economic powerhouses in this 'hood. Taiwan, South Korea, Singapore, and Hong Kong were once titled the "Asian Tigers" because they underwent decades of phenomenal industrialization and economic growth. Those tigers have now come home to roost under mother Dragon. This place is a veritable smorgasbord of animated animals, and they are getting rich! Throw Japan into the lot, and you have a whole cluster of top world economies here—with one big difference:

Most top twenty economies are still in Europe, and Europe is about played out. The rest of the top twenty world economies are Asian, and they are just getting warmed up! Look for Asian states to totally dominate the world economic measures of wealth in the coming decades. For sure, China will soon be number two, Japan will be number three, and South Korea will be number ten. And they are all next door neighbors! Indonesia is a rising regional titan, India is about to be reincarnated as a vibrant economy, and Vietnam is a dark horse with a big stride ahead of it as well. Are you starting to get a fever for the flava' of East Asia? And don't forget the newest China game plan I outlined for you in Chapter 23: THIS IS A GIGANTIC GLOBAL DRAGON DEVELOPMENT: The Silk Road Economic Belt and the 21st-century Maritime Silk Road aka The Belt and Road aka One Belt, One Road aka The Belt and Road Initiative (lol that is a crap-ton of aka's) is a development strategy and framework that aims to increase connectivity and cooperation among dozens of countries in Eurasia and Africa. Mondo big deal that puts China at the center of a new global economic nexus.

To sum up: there will be major growth, competition, and shifts in economic focus and economic policy around the world as Asia rises. Don't get all sad on me—your region is still good! The US and Europe will still be rich, don't worry about that! There are just a few other entities I want mention before escaping this escapade.

SOME OTHER ECONOMIC ENTITIES OF NOTE

I've mentioned ASEAN back in the chapter on Southeast Asia; it is a supranationalist organization with more potential for economic growth than I can imagine right now. Smack dab in the middle of the powerhouse China and Japan to the north and growing power India to its west, all of which are associate members, the ASEAN core might well become the epicenter of an Asian EU. Wow. That would be insane. It's still some ways off, but the point is that this group is in a position of it being almost impossible for them to NOT grow economically right now. Southeast Asia is starting to absorb a lot of manufacturing jobs not just from abroad, but also directly from China. It's good to be close to the top dogs. Speaking of Asian top dogs . . .

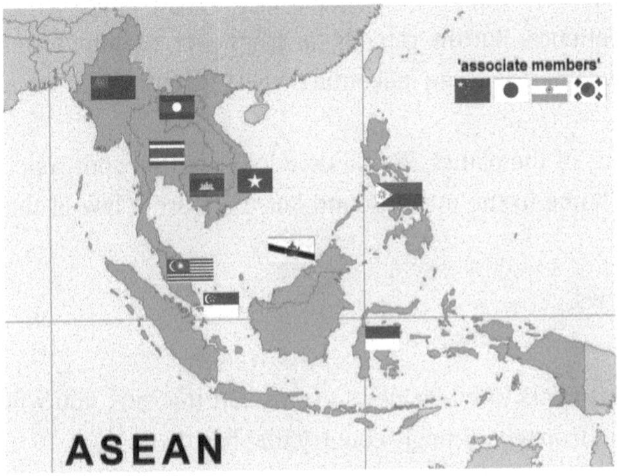

As with the rise in political clout, Team Hindu stands to become an economic powerhouse in its own right, although it is going to take a bit longer than their Asian brothers. It's that mondo big population that makes things problematic for them to advance as fast! But advance they will nonetheless, but for different reasons than others. Namely, India is fast becoming a high-end technological center for growth in Asia and the world.

China and other Asian nations have focused first on becoming agri-cultural and manufacturing giants, then incorporating more and more service sector and quaternary activities as they get richer. India is on rather the reverse course: it focuses heavily on increasing its service sector, and especially on new technology stuff. Computer programming, information technology, and engineering applications are fields that India is specializing in immediately. While international business may be attracted to China or Indonesia for their low wages, international business is attracted to India for their lower wages AND highly educated and skilled service sectors. Why pay a European programmer when you can pay an Indian programmer half as much? Look for Team Hindu's star to rise fast in our technologically-focused world.

FINALLY THE FINAL FINAL FINAL SUMMARY

We have identified the major forces that shape the future of our planet by battling it out every day politically, militarily, and economically, or sometimes cooperating with each other. Either on their own or within international organizations, the most powerful sovereign states plan, buy, sell, trade, agree, disagree, fight, make peace, plot, unify, embargo, or smackdown each other in various ways to achieve their objectives. It's not a pretty process, but it's the only one we have until smarter folks like you make it better.

Does it appear that I have forgotten large parts of the globe? Oh no, my friends, I have not forgotten anyone! This chapter was about major movers and shakers, and that list is not all inclusive. Sub-Saharan Africa is at the whim of the international system until they straighten themselves out internally . . . and watch out, because that is actually happening now! Sub-Saharan Africa is actually heating up economically, and has had consistent growth for the last decade! And what about that titanic AfCFTA trade block action that is unifying the continent like never before? Go for it guys! The Middle East as a region seems doomed for eternal conflict and a variety of debilitating internal problems until they get their business straight, and it looks like they got a ways to go before that sun rises. Most of the sub-regions of Latin America are doing pretty good with consistent economic growth and political stability, but do not impact world events in a major way. Most of Eastern Europe has effectively become part of Western Europe. Regions like Turkey, Japan and Australia play more of a role within the ranks of Team West than they do on their own.

But history is not done, my friends! Unpredictable events can shake things up quickly in today's world and change the course of states, or regions, or the whole planet. Even with my super-human powers of insight and heightened fashion sense, I could not have predicted the attacks of 9/11, the massive Japanese quake and nuclear disaster of 2011, the "Arab Spring" still unfolding, the horrible outcome of the final Indiana Jones movie release, or the 2016 election of America's new Celebrity-in-Chief Donald Trump. Events like these have huge repercussions for the regions in which they take place, but are also felt throughout the world in the form of global policies and actions that are implemented due to the event itself.

That's why it is important to keep up with your planet. You have taken a very large step in the right direction by reading this book. Congratulations on that feat of endurance and patience. But the game is far from over. Stay involved and educated about is going on in the world, so that when the time comes, you can make the biggest impact for positive change, no matter what it is you decide to do in life.

Lots of people ask me how I keep up with the events and actions of the planet. Well, I do a lot of on-the-spot super hero action within these regions on a daily basis, but when I'm at home in the underground lair, here are a few of the Web sites I keep up with, so that I can keep up with the world:

- http://news.bbc.co.uk/ for daily lightweight international news
- http://www.iht.com/ for more in-depth daily international news
- http://www.worldpress.org/ for deep in-depth weekly assessments of major news events. On this site, you will also find links to every single digital newspaper on the planet from every single country that has them.

You can tune in to CNN, MSNBC, and Fox News Channel anytime to check out the daily arcane domestic political party finger pointing nonsense, and up to the minute status on whatever celebrity drama is making headlines that day. Mainstream American news sources are not really worth much else.

Always read at least two different news sources for every story you are learning about; make sure one news source is from inside the country where the news happens, and one is outside the country where the news happens. It's the only way to get a balanced view of what's actually happening on the planet.

For those of you who want to do more, here are some ideas and things you can do.

Ever heard of micro-finance? It is an awesome and easy way to help the poor for those of you who may not have time to volunteer. Sometimes charities and organizations get so big and wrapped up in politics that the money never actually gets to those who really need it. What micro-finance does, is supply small loans (sometimes it's even as little an amount as $25) to individuals (usually women, 'cause let's face it guys, when it comes right down to it, women do tend to be the most responsible sex) to help them start small businesses. For example, a source close to the Plaid Avenger was working with a group of Haitian women in the Dominican Republic who were trying to support themselves by making and selling candles. These women were making candles one at a time and it was taking them forever. But, if they had someone who could loan them the money to buy a candle mold, they could make a lot more candles in a lot less time and increase the revenue as a result. Get it? With microloans, a small amount of money can go a long way in helping change and improve lives but with dignity—because it's not a handout, the recipient has to pay the loan back. What my friend in the DR did, was have them pay the loan back to a community account, so that money could stay within the community and be there for the next person needing a loan—recycling money for good—I love it!

Not into micro-lending and want to choose a cause instead? Here's another cool website that allows you to give to a grassroots project of your choice.

- http://www.globalgiving.org

To be unaware and ill-informed about the world around you is to be a passive player in the game of life. Keep your head in the game. Stay informed. Read. Pay attention. Write a letter. Protest. Help someone in need. Start a revolution. Or just do anything to make the world a better place. Peace.

And party on!

. . . to be globally engaged and informed!!!

GO TO THE WEBSITE...
OR PUTIN WILL HAVE TO
PAY YOU A VISIT.
PLAID AVENGER
LOG ON TO
WWW.PLAIDAVENGER.COM
FOR UP TO DATE INFORMATION AROUND THE WORLD AS WELL AS PLAID MERCHANDISE.

Image Credits

Part 1 Opener: map courtesy of Katie Pritchard

CHAPTER 1

Page 4: top: © 2010 JupiterImages Corporation; **bottom:** map courtesy of Katie Pritchard; **Page 5: top and middle:** © Shutterstock, Inc.; **bottom:** courtesy of Katie Pritchard; **Page 6:** courtesy of Katie Pritchard; **Page 7: top:** © brave rabbit, 2010. Used under license from Shutterstock, Inc.; **bottom:** © 2010 JupiterImages Corporation; **Page 8:** © 2007 JupiterImages Corporation; **Page 10:** courtesy of Katie Pritchard; **Page 12: A-F:** maps courtesy of Katie Pritchard; **Page 13: a—f.:** © Shutterstock, Inc.

CHAPTER 2

Page 14: © 2010 JupiterImages Corporation; **Page 17, fig. 2.2:** map source: US Dept. of Agriculture; **Page 19: bottom:** © Shutterstock, Inc.; **Page 20: top:** © Shutterstock, Inc.; **Page 21: right 5 images:** © Shutterstock, Inc.; **Page 22: bottom 3 images:** © Shutterstock, Inc.; **Page 25: top right and left images:** © Shutterstock, Inc.; **Page 28: fig. 2.13:** Source: U.S. Census Bureau, International Data Base; **Page 29:** Image © Tom Wang, 2012. Shutterstock, Inc.; **Page 30: top:** © Shutterstock, Inc.; **bottom:** map courtesy of Katie Pritchard; **Page 31: top:** © Ververidis Vasilis/Shutterstock.com; **bottom:** © Nicolas Economou/Shutterstock.com; **Page 33:** map created by Katie Pritchard; **Page 35: fig. 2.14:** courtesy of Katie Pritchard; **Page 36:** maps courtesy of Katie Pritchard; **Page 37:** maps courtesy of Katie Pritchard

CHAPTER 3

Page 38: listed left to right: First row: © Stratos Brilakis/Shutterstock.com; European Union, 2014; © Matias Baglietto/Shutterstock.com; © European Union, 2016; © 360b/Shutterstock.com; **second row:** MOHAMMAD FARNOOD/SIPA/Newscom; U.S. Department of State; © European Union, 2011; © European Union, 2015; © European Union, 2016; **third row:** Alan Santos/PR; © European Union, 2014; Official White House Photo by Shealah Craighead; Official White House photo by Shealah Craighead; Taiwan Presidential Office; **fourth row:** Presidencia de la República Mexicana; © European Union, 2012; US State Department; U.S. Department of Defense photo by Erin A. Kirk-Cuomo; U.S. Department of State; **fifth row:** U.S. Department of State; Presidencia de la República Mexicana; © European Union, 2016; Isac Nóbrega/PR; Presidencia de la República Mexicana; **Page 39, fig. 3.1:** courtesy of Katie Pritchard; **Page 42: top**: official White House portrait; **bottom:** photo by SSgt. Lance Cheung, U.S. Air Force, from Defense Visual Information Center; **Page 44:** © 2003 by Damon Clark. Used with permission; **Page 45: left:** © 2007 JupiterImages Corporation; **center:** © 2007 JupiterImages Corporation; **right:** © 2007 JupiterImages Corporation; **Page 46: fig. 3.2:** map source: *The World Factbook*; **Page 47: fig. 3.3:** map source: *The CIA World Factbook*; **middle:** White House photo by Pete Souza; **bottom:** Government of Argentina; **Page 48:** map source: *The CIA World Factbook*; **Page 49: top:** map source: University of Texas Libraries; **bottom:** maps courtesy of the University of Texas Libraries, Perry-Castañeda Library Map Collection; **Page 50: bottom left:** New York City Police Department; **bottom left center:** photo by Joseph Randall Blanchard (ca. 1898), from Library of Congress; **bottom center:** Swiss Federal Council of the year 2007 from The Federal Authorities of the Swiss Confederation; **bottom right center:** © Stratos Brilakis/Shutterstock.com; **bottom right:** Dept. of Defense photo by Erin A. Kirk-Cuomo; **Page 51: top:** New York City Police Dept.; **bottom left center:** U.S. Department of Defense photo by Erin A. Kirk-Cuomo; **bottom center:** Dept. of Defense; **bottom right center:** Official White House Photo by Shealah Craighead; **bottom right:** National Archives & Records Administration photo; **Page 52: left:** Image © Neftali, 2012. Shutterstock, Inc.; **middle:** Image © Hung Chung Chih, 2012. Shutterstock, Inc.; **right:** Hulton Deutsch/Corbis Historical/Getty Images; **Page 53: left:** photo by Joseph Randall Blanchard (ca. 1898), from Library of Congress; **right:** © Stratos Brilakis/Shutterstock.com; **Page 54:** Image © Atlaspix, 2012. Shutterstock, Inc.; **Page 55: top**

left: Office of Press and Information of the President of Russia; **top left center:** © European Union, 2014; **top right center:** © European Union, 2016; **top right:** © European Union, 2016; **Page 56: top left:** Image © Neftali, 2012. Shutterstock, Inc.; **top right:** U.S. Air Force photo by Tech. Sgt. Craig Clapper; **bottom left:** Image © tristan tan, 2012. Shutterstock, Inc.; **bottom right:** Dept. of Defense photo by Erin A. Kirk-Cuomo; **Page 57:** caricature by Edmund S. Valtman, from Library of Congress; **Page 58: top:** U.S. Army photo by Eboni Everson-Myart; **bottom left:** photo from Iraqi Freedom II CD Collection by Joint Combat Camera composed of Army, Navy, Marine, and Air Force photographers; **bottom right:** Dept. of Defense photo; **Page 59: top:** Official White House Photo by Shealah Craighead; **bottom:** National Archives & Records Administration; **Page 60: top:** National Archives & Records Administration; **bottom: fig. 3.4:** courtesy of Katie Pritchard; **Page 64: top left:** Bundesregierung/Laurence Chaperon; **top left center:** Dept. of Defense photo by Erin A. Kirk-Cuomo; **top right center:** © European Union, 2016; **top right:** © European Union, 2011; **middle left:** © European Union, 2016; **middle left center:** © Stratos Brilakis/Shutterstock.com; **middle right center:** © Yuri Turkov/Shutterstock.com; **middle right:** Dept. of Defense photo by Erin A. Kirk-Cuomo; **bottom left:** Empresa Brasil de Comunicação S/A – EBC; **bottom left center:** Elza Fiúza/ABr; **bottom right center:** Kyodo/Newscom; **bottom right:** © European Union, 2014; **Page 65: top left:** ED JONES/AFP/Getty Images; **top left center:** Presidencia de la República Mexicana; **top right center:** Tânia Rêgo/ABr; **top right:** G20 Argentina; **middle left:** MOHAMMAD FARNOOD/SIPA/Newscom; **middle left center:** © NATO; **middle right center:** Alan Santos/PR; **middle right:** Official White House photo by Shealah Craighead; **bottom left:** © European Union, 2011; **bottom left center:** official White House photo by Pete Souza; **bottom right center:** US State Department; **bottom right:** Russian Presidential Press and Information Office

CHAPTER 4

Page 66: images © 2007 JupiterImages Corporation; **bottom left:** Image © v.s.anandhakrishna, 2012. Shutterstock, Inc.; **bottom right:** Image © Diego Cervo, 2012. Shutterstock, Inc.; **Page 67: left:** photo by unknown (ca. 1920), from Library of Congress; **left center:** Elza Fiúza/ABr; **right center:** © European Union, 2017; **right:** stipple engraving by MacKenzie (ca. 1805), from Library of Congress; **Page 68:** photo by unknown (ca. 1920), from Library of Congress; **Page 69:** stipple engraving by MacKenzie (ca. 1805), from Library of Congress; **Page 70: left:** photo by unknown (ca. 1920), from Library of Congress; **left center:** photo by Warren K. Leffler, from Library of Congress; **center:** Government of Argentina; **right center:** Dept. of Defense photo by Erin A. Kirk-Cuomo; **right:** photo from Official Russian Presidential Press and Information Office; **Page 71: left:** Dept. of Defense photo by Erin A. Kirk-Cuomo; **center:** © European Union, 2017; **right center:** © Stratos Brilakis/Shutterstock.com; **right:** stipple engraving by MacKenzie (ca. 1805), from Library of Congress; **Page 72:** photo by Jack Delano (ca. 1940), from Library of Congress; **Page 73: top:** © Shutterstock, Inc.; **bottom 3 images:** © Shutterstock, Inc.; **Page 75:** © 2007 JupiterImages Corporation; **Page 76: top left:** © 2007 JupiterImages Corporation; **top right:** © 2007 JupiterImages Corporation; **middle right:** © 2007 JupiterImages Corporation; **bottom left:** © 2007 JupiterImages Corporation; **bottom right:** © 2007 JupiterImages Corporation; **Page 77: top right:** © 2007 JupiterImages Corporation; **middle left:** © 2007 JupiterImages Corporation; **bottom right:** © 2007 JupiterImages Corporation; **Page 78: top:** © 2007 JupiterImages Corporation; **bottom:** graph courtesy of Katie Pritchard; **Page 79:** National Archives & Records Administration photo; **Page 81:** map courtesy of Katie Pritchard; **Page 82:** map courtesy of Katie Pritchard; **Page 83:** © 2009 JupiterImages Corporation; **Page 84:** map courtesy of Katie Pritchard; **Page 85:** From *The Silent War* by John Ames Mitchell, Illustrations by William Balfour Ker (New York: Life Publishing Co, 1906). Courtesy of Library of Congress.

CHAPTER 5

Page 86: Dept. of Defense photos; **Page 88: right:** Courtesy of Library of Congress; **left:** © 2009 JupiterImages Corporation; **Page 89: top:** Photo by Walter P. Miller (1929), Courtesy of Library of Congress; **bottom:** Courtesy of Library of Congress; **Page 90:** © Shutterstock, Inc.; **Page 91:** © 2009 JupiterImages Corporation; **Page 92:** © 2009 JupiterImages Corporation; **Page 93: bottom:** © 2009 JupiterImages Corporation; **Page 95:** Source: U.S. Census Bureau, International Data Base; **Page 96: top:** © 2007 JupiterImages Corporation; **bottom:** all photos © 2007 JupiterImages Corporation; **Page 98:** © 2009 JupiterImages Corporation; **Page 99:** © Shutterstock, Inc.; **Page 100: fig. 5.1:** courtesy of Katie Pritchard; **Page 101:** map courtesy of Katie Pritchard; **Pages 104-105:** NASA/Goddard Space Flight Center Scientific Visualization Studio

CHAPTER 6

Page 108: top left: © 2007 JupiterImages Corporation; **top right:** Frank and Frances Carpenter Collection, from Library of Congress; **middle:** © Royalty-free/CORBIS; **bottom left:** © NATO; **bottom right:** Copyright © African Union, 2003. All rights reserved. Used by permission; **Page 109:** © 2007 JupiterImages Corporation; **Page 110:** map courtesy of Katie Pritchard, flags from *The World Factbook*; **Page 111:** map courtesy of Katie Pritchard; **Page 112: top:** map courtesy of Katie Pritchard; **Page 113:** map courtesy of Katie Pritchard; **Page 114:** map courtesy of Katie Pritchard; **115:** map courtesy of Katie Pritchard; **116:** maps courtesy of Katie Pritchard; **Page 117: bottom** OECD logo used by permission; **Page 118: top:** © 2007 JupiterImages Corporation; **bottom:** © European Union, 2015; **Page 119:** © Royalty-free/CORBIS; **Page 120:** NATO logo used by permission of NATO; **bottom:** © NATO; **Page 121:** map courtesy of Katie Pritchard; **Page 122:** courtesy of Katie Pritchard; **Page 123:** *The World Factbook*; **Page 124: top:** Frank and Frances Carpenter Collection, from Library of Congress; **middle:** © AP Photo/Hussein Malla; **Page 125: top:** Organization of American States; **middle and bottom:** Copyright © African Union, 2003. All rights reserved. Used by permission; **Page 126: top:** U.S. Air Force photo by Tech. Sgt. Jeremy T. Lock; **bottom:** CHINE NOUVELLE/SIPA/Newscom; **Page 127: flags at top:** *The World Factbook*; **flags at bottom:** *The World Factbook*; **Page 128: top:** © European Communities, 2009; **bottom:** © European Communities,

2009; **Page 129:** Russian Presidential Press and Information Office; **Page 130:** WTO logo used with permission; **Page 132:** map courtesy of Katie Pritchard; **Page 134:** Government of Argentina; **Page 135:** PAUL J. RICHARDS / AFP / Getty Images

Part 2 opener: map courtesy of Katie Pritchard

CHAPTER 7

Page 138: map courtesy of Katie Pritchard and NOAA; **Page 139:** Presidencia de la República Mexicana; **Page 143:** Courtesy of USGS; **Page 144:** NASA/Goddard Space Flight Center Scientific Visualization Center; **Page 145:** map source: *The Cambridge Modern History Atlas 1912*, courtesy of the University of Texas Libraries, Perry-Castañeda Library Map Collection; **Page 146: top:** U.S. Marine Corps photo by Lance Cpl. Kelly R. Chase; **middle:** © Shutterstock, Inc.; **bottom:** map courtesy of Katie Pritchard; **Page 147:** all images © 2007 JupiterImages Corporation; **Page 148: top:** photo by M.B. Marcell (ca. 1911), from Library of Congress; **bottom:** U.S. Air Force photo by SSGT Jacob N. Bailey; **Page 149: top:** Dept. of Defense photo by U.S. Navy; **bottom:** U.S. Navy photo by Mass Communication Specialist 3rd Class Kathleen Gorby; **Page 150:** U.S. Navy photo by Mass Communication Specialist 3rd Class Geoffrey Lewis; **Page 151:** Library of Congress; **Page 152:** art by John C. McRae (ca. 1620), from Library of Congress; **Page 153: bottom:** © Shutterstock, Inc.; **Page 154: middle:** photo by The New York Times (ca. 1954), from Library of Congress; **bottom:** map courtesy of Katie Pritchard; **Page 156:** U.S. Dept of Agriculture, from Library of Congress; **Page 157:** © 2007 JupiterImages Corporation; **Page 158: top:** map courtesy of Katie Pritchard & US Dept of Defense; **bottom:** © Shutterstock, Inc.; **Page 160: top:** created by Acme Litho. Co., New York (ca. 1910), from Library of Congress; **bottom:** created by Leslie-Judge Co., New York (1917), from Library of Congress; **Page 161:** created by United Cigar Stores Company (1918), from Library of Congress; **Page 162:** U.S. Government Printing Office (1943), from Library of Congress; **Page 163: top:** © Shutterstock, Inc.; **bottom:** Illinois Co., Chicago (1917), from Library of Congress; **Page 164:** American Lithographic Co., New York (1918), from Library of Congress; **Page 165: left:** © George Sheldon/Shutterstock.com; **right:** © CJ Hanevy/Shutterstock.com; **Page 166: middle:** © Shutterstock, Inc.; **bottom left:** Library of Congress; **bottom middle:** Brady-Handy Photograph Collection, Library of Congress; **bottom right:** official White House photo; **Page 167: left:** © RedhoodStudios/Shutterstock.com; **center:** U.S. Marine Corps photo by Sgt. Gabriela Garcia; **right:** U.S. Department of State

CHAPTER 8

Page 168: map courtesy of Katie Pritchard and NOAA; **Page 170: left:** originally published by Thomas B Noonan, from Library of Congress; **left center:** Library of Congress; **right center:** Library of Congress; **right:** Engraving by W. Holl after painting by Franz Hals, from Library of Congress; **Page 171:** map courtesy of Katie Pritchard; **Page 172:** CIA map; **Page 173:** map courtesy of Katie Pritchard; **Page 175:** *Blue Marble: Next Generation* image produced by Reto Stockli, NASA Earth Observatory (NASA Goddard Space Flight Center); **Page 178: left:** National Archives & Records Administration; **right:** National Archives & Records Administration; **Page 180:** US Department of Defense; **Page 181:** Farm Security Administration, Office of War Information Collection 12002-27, from Library of Congress; **Page 182:** maps courtesy of Katie Pritchard; **Page 183:** maps courtesy of Katie Pritchard; **Page 184:** created by James Montgomery Flagg (1918), from Library of Congress; **Page 186:** Library of Congress; **Page 187:** map by Katie Pritchard; **Page 188:** Library of Congress; **Page 189: left:** Presidencia de la República Mexicana; **center:** © European Union, 2016; **right:** © European Union, 2011; **Page 190: top:** © European Union, 2017; **middle:** © sportpoint/Shutterstock.com; **bottom:** Russian Presidential Press and Information Office; **Page 191:** Bundesregierung/Laurence Chaperon; **Page 192:** © Nejron Photo, 2012. Used under license of Shutterstock, Inc.; **Page 193:** © JupiterImages Corporation; **Page 194: left:** Farm Security Administration—Office of War Information Photograph Collection, from Library of Congress; **left center:** Library of Congress; **right center:** National Archives & Records Administration; **right:** Dept. of Defense photo; **Page 195: left:** © European Union, 2017; **left center:** Russian Presidential Press and Information Office; **center:** © European Communities, 2009; **right center:** © European Union, 2016; **right:** © European Union, 2011

CHAPTER 9

Page 196: map courtesy of Katie Pritchard and NOAA; **Page 199: top:** © 2008 JupiterImages Corporation; **middle:** from *Atlas of the Middle East* by CIA (1993), courtesy of University of Texas Libraries; **bottom:** Public domain; **Page 200:** map source: *The World Factbook*; **Page 201:** maps: *The World Factbook*; **Page 202: top:** map source: *The World Factbook*; **bottom:** map source: *The World Factbook*; **Page 204: middle:** photo © 1941 by J. Russell & Sons, from Library of Congress; **bottom:** map source: *The World Factbook*; **Page 205: top:** map source: *The World Factbook*; **bottom:** map source: *The World Factbook*; **Page 206: bottom:** source: *Nuclear Weapons and NATO: Analytical Survey of Literature* by U.S. Dept. of the Army, courtesy of University of Texas Libraries Perry-Castañeda Map Collection; **Page 207:** map source: *The World Factbook*; **Page 208:** map source: *The World Factbook*; **Page 209: top:** map courtesy of Katie Pritchard; **bottom:** map courtesy of Katie Pritchard; **Page 210:** map by U.S. Central Intelligence Agency, courtesy of University of Texas Libraries, Perry-Castañeda Map Collection; **Page 211: left:** Dept. of Defense photo by Tech. Sgt. Cedric H. Rudisill, US. Air Force; **left center:** © NATO; **right center:** © NATO; **right:** photo from Official Russian Presidential Press and Information Office; **Page 212: bottom:** © NATO; **Page 213:** © Shutterstock, Inc.; **Page 214:** © 2008 JupiterImages Corporation; **Page 215: top:** © 2008 JupiterImages Corporation; **bottom:** © 2008 JupiterImages Corporation; **Page 216: top:** from *Former Yugoslavia: A Map Folio* (1992) by CIA, courtesy of University of Texas Libraries; **bottom left:** White House Photo Office Collection (1971), from Library of Congress; **bottom right:** Todd Williamson Archive/FilmMagic/Getty Images; **Page 217: bottom:** CIA map, courtesy of University of Texas Libraries; **Page 219:** map by Katie Pritchard; **Page 220: left:** George Grantham Bain Collection, Library of Congress; **center:** National Archives & Records Administration; **right:** © NATO; **Page 221: left:** © NATO; **left center:** © European Communities, 2007; **right center:** © NATO; **right:** © European Union, 2019

CHAPTER 8/9 ADDENDUM

Page 222: © sebos/Shutterstock.com; **Page 224:** map courtesy of Katie Pritchard; **Page 226: top:** Milos Bicanski/Stringer/Getty Image News; **bottom:** LOUISA GOULIAMAKI/Stringer/AFP/Getty Images; **Page 227:** map courtesy of Katie Pritchard; **Page 229:** © Ms Jane Campbell/ Shutterstock.com; **Page 230:** © Ms Jane Campbell/Shutterstock.com; **Page 232:** map source: *The Public Schools Historical Atlas*, courtesy of the University of Texas Libraries, Perry-Castañeda Library Map Collection; **Page 233:** courtesy of Katie Pritchard; **Page 235:** map courtesy of Katie Pritchard

CHAPTER 10

Page 236: top: map courtesy of Katie Pritchard and NOAA; **bottom:** created by Strobridge Lithography Co. (1895), from Library of Congress; **Page 237: top:** © 2007 JupiterImages Corporation; **bottom:** adapted from *Blue Marble: Next Generation* image produced by Reto Stockli, NASA Earth Observatory (NASA Goddard Space Flight Center); **Page 238:** created by Strobridge Lithography Co. (ca. 1896), from Library of Congress; **Page 239:** © Shutterstock, Inc.; **Page 240: top:** map source: *The World Factbook*; **middle:** created by W. Holland (1803), from Library of Congress; **bottom:** map source: *The World Factbook*; **Page 241: top:** scanned from Helmolt, J.F. ed. *History of the World* (New York, Dodd, Mead & Co., 1902); **top middle:** © Shutterstock, Inc.; **bottom middle:** engraving by A. Muller (1879), Library of Congress; **bottom:** scanned from Helmolt, J.F. ed. *History of the World* (New York, Dodd, Mead & Co., 1902); **Page 242:** from the George Grantham Bain Collection, Library of Congress; **Page 243: top:** lithography by M.A. Striel'tsova (ca. 1918), from Library of Congress; **bottom left:** *New York Times*, 1919, Library of Congress; **left center:** *Tsar Nicholas II* (1915) by Boris Kustodiyev; **right center:** Library of Congress; **right:** photo ca. 1909; **Page 244: left:** Library of Congress; **right:** photo from *Liberty's Victorious Conflict: A Photographic History of the World War* by The Magazine Circulation Co., Chicago, 1918; **Page 245:** map source: *The World Factbook*; **Page 246:** from New York World-Telegram & the Sun Newspaper Photograph Collection, Library of Congress; **Page 247:** U.S. Signal Corps photo, from Library of Congress; **Page 248: top:** photo by U.S. Office of War Information Overseas Picture Division, Library of Congress; **bottom:** map source: *The World Factbook*; **Page 249:** photo by U.S. Office of War Information Overseas Picture Division, Library of Congress; **Page 250:** © Shutterstock, Inc.; **Page 251:** © 2007 JupiterImages Corporation; **Page 252: top:** U.S. Department of Defense; **bottom:** official White House portrait; **Page 253: top left:** Library of Congress; **top center:** U.S. Signal Corps photo, from Library of Congress; **top right:** photo from Franklin D. Roosevelt Library, Library of Congress; **bottom left:** White House Photo Collection, Library of Congress; **bottom left center:** Bettmann/Bettmann/Getty Images; **bottom right center:** Bettmann/Bettmann/Getty Images; **bottom right:** White House Photo Collection, Library of Congress; **Page 255: top:** White House Photo Collection, Library of Congress; **bottom:** photo by Earle D. Akin Co, 1909, from Library of Congress; **Page 257:** Source: U.S. Census Bureau, International Data Base; **Page 258:** map courtesy of Katie Pritchard; **Page 259: top:** photo from Official Russian Press and Information Office; **middle:** © NATO; **bottom:** © NATO; **Page 260: top:** photo from Official Russian Press and Information Office; **bottom:** photo from Official Russian Press and Information Office; **Page 261: top:** map from U.S. Dept. of Energy; **bottom:** © NATO; **Page 262:** Produced by the Office of The Geographer and Global Issues, Bureau of Intelligence and Research, US Dept. of State, courtesy of University of Texas Libraries; **Page 263:** CIA map (1994), courtesy of University of Texas Libraries; **Page 265:** CIA map (1994), courtesy of University of Texas Libraries; **Page 267:** map courtesy of Katie Pritchard; **Page 269:** © Shutterstock, Inc.; **Page 270: left:** George Grantham Bain Collection, Library of Congress; **right center:** U.S. Signal Corps photo, from Library of Congress; **right:** © NATO; **Page 271: left:** © European Communities, 2009; **left center:** © European Communities, 2009; **right center:** © NATO; **right:** © NATO

CHAPTER 11

Page 273: map courtesy of Katie Pritchard and NOAA; **Page 274: top:** Dept. of Defense photo by Mass Communication Specialist Seaman Bryan Reckard, U.S. Navy; **middle:** Dept. of Defense photo by Tech. Sgt. Rob Marshall; **bottom:** *Blue Marble: Next Generation* image produced by Reto Stockli, NASA Earth Observatory (NASA Goddard Space Flight Center); **Page 275: top:** Tsuta-ya Kichizo (1858), from Library of Congress; **bottom:** U.S. Navy photo by Photographer's Mate 2nd Class Nathanael T. Miller; **Page 276: top:** Library of Congress; **bottom:** Library of Congress; **Page 277: left:** Dept. of Defense photo by PHI (AW) M. Clayton Farrington, U.S. Navy; **right:** photo by Mathew B. Brady, from Library of Congress; **Page 278: left:** Heinoya (1870), from Library of Congress; **right:** Dept. of Defense photo; **Page 279:** art by Rivinger (1860), from Library of Congress; **Page 280:** map source: *The World Factbook*; **Page 281:** photo by U.S. War Department Signal Cops, from Library of Congress; **Page 282:** Office for Emergency Management, Office of War Information; **Page 283: top:** image created by James Montgomery Flagg for Office of War Information Domestic Operations Branch, from National Archives & Records Administration; **bottom:** from the George Frantham Bain Collection, Library of Congress; **Page 284: top left:** photo by Office of War Information Overseas Operations Branch (August 9, 1945), from National Archives & Records Administration; **top right:** Dept. of Defense photo, Dept. of the Navy, U.S. Marine Corps; **bottom right:** photo by U.S. Army, from Library of Congress; **Page 285: left:** photo from NASA; **left center:** photo ca. 1918, from Library of Congress; **right center:** © European Union, 2011; **right:** Kyodo/Newscom; **Page 286:** all photos © 2007 JupiterImages Corporation; **bottom:** Source: U.S. Census Bureau, International Data Base; **Page 287: top left:** © 2008 JupiterImages Corporation; **middle right:** Dept. of Defense photo; **bottom left:** Dept. of Defense photo; **Page 288:** Dept. of Defense photo; **Page 290: left:** Dept. of Defense photo by Helen C. Stikkel; **right:** Kyodo/Newscom; **Page 291:** all photos © Shutterstock, Inc.; **Page 292:** Courtesy of USGS; **Page 293:** Library of Congress; **Page 294: left:** Harris & Ewing Collection, Library of Congress; **left center:** Library of Congress; **right:** Dept.

of Defense photo; **Page 295: left:** © European Communities, 2004; **right:** Dept. of Defense photo by D. Myles Cullen

CHAPTER 12

Page 296: map courtesy of Katie Pritchard and NOAA; **Page 297: left:** *Blue Marble: Next Generation* image produced by Reto Stockli, NASA Earth Observatory (NASA Goddard Space Flight Center); **right:** NASA/Goddard Space Flight Center Scientific Visualization Studio; **Page 298: top:** © Regien Paassen, 2012. Used under license of Shutterstock, Inc.; **Page 299: top:** © Pichugin Dmitry, 2012. Used under license of Shutterstock, Inc.; **bottom:** Source: U.S. Census Bureau, International Data Base; **Page 300:** all images: © 2007 JupiterImages Corporation; **Page 301:** ca. 1915, Library of Congress; **Page 302: top:** © Shutterstock, Inc.; **bottom:** © Debra James, 2012. Used under license of Shutterstock, Inc.; **Page 303: top:** © Shutterstock, Inc.; **bottom:** © 2007 JupiterImages Corporation; **Page 304:** © Gavran333, 2012. Used under license of Shutterstock, Inc.; **Page 305: right:** © 2007 JupiterImages Corporation; **left:** Dept. of Defense photo by Robert D. Ward; **Page 306: top:** stereograph (1919), from Library of Congress; **bottom:** lithography by Cincinnati Lithography Co., from Library of Congress; **Page 308:** all images © Shutterstock, Inc.; **Page 309: left:** Dept. of Defense photo by Robert D. Ward; **left center:** Dept. of Defense photo by Cherie Cullen; **right center:** G20 Argentina; **right:** © European Union, 2019

CHAPTER 13

Page 312: map courtesy of Katie Pritchard and NOAA; **Page 313:** DeGolyer Library, Southern Methodist University; **Page 314:** map by CIA, courtesy of University of Texas Libraries; **Page 315:** all maps: from CIA, courtesy of University of Texas Libraries; **Page 316: top:** from CIA, courtesy of University of Texas Libraries; **middle:** map courtesy of Katie Pritchard; **bottom:** Library of Congress; **Page 317: top:** map source: *The World Factbook*; **bottom:** photo from the Frank & Frances Carpenter Collection, Library of Congress; **Page 318:** © 2008 JupiterImages Corporation; **Page 319: top:** © Jose Miguel Hernandez Leon, 2012. Used under license of Shutterstock, Inc.; **bottom:** Library of Congress; **Page 320: top:** Isac Nóbrega/PR; **bottom:** scanned from Helmolt, J.F. ed. *History of the World* (New York, Dodd, Mead & Co., 1902); **Page 321: top left:** Library of Congress; **top left center:** Fabio Rodrigues Pozzebom/ABr; **top center:** Courtesy of Library of Congress; **top right center:** photo by Warren K. Leffler, from Library of Congress; **top right:** Government of Argentina; **bottom:** photo ca. 1900, from Library of Congress; **Page 322: top:** from Judge, February 15, 1896, Library of Congress; **middle:** photo ca. 1900, from Library of Congress; **bottom:** photo by Elias Goldensky (1933), from Library of Congress; **Page 323:** photo by Warren K. Leffler, from Library of Congress; **Page 324:** © Shutterstock, Inc.; **Page 327:** map courtesy of Katie Pritchard; **Page 328: top:** Courtesy of Katie Pritchard; bottom: Isac Nóbrega/PR; **Page 329:** Acervo Fotográfico da Câmara dos Deputados; **Page 332: top:** maps courtesy of Katie Pritchard; **bottom:** first three photos courtesy of Secretaria de Imprensa, Brasil; **fourth:** © Matias Baglietto/Shutterstock.com; **fifth:** Presidencia El Salvador; **sixth:** © Frederic Legrand - COMEO/Shutterstock.com; **seventh:** © European Union, 2012; **eighth:** U.S. Marine Corps photo by Sgt. Gabriela Garcia; **ninth:** Isac Nóbrega/PR; **Page 333: left:** painting by Ricardo Acevedo Bernal; **left center:** The White House Historical Association (White House Collection); **right center:** National Archives & Records Administration; **right:** Library of Congress

CHAPTER 14

Page 336: map courtesy of Katie Pritchard and NOAA; **Page 338: top:** from *Historical Atlas* (1911) by William Shepherd (pp. 186–187), courtesy of University of Texas Libraries; **bottom:** map source: *The World Factbook*; **Page 339: top:** from *Historical Atlas* (1911) by William Shepherd (p. 106), courtesy of University of Texas Libraries; **bottom:** *The World Factbook*; **Page 340:** all maps: *The World Factbook*; **Page 341:** all maps: *The World Factbook*; **middle right:** © 2007 JupiterImages Corporation; **bottom:** all images © Shutterstock, Inc.; **Page 343:** © 2007 JupiterImages Corporation; **Page 344: top:** © 2007 JupiterImages Corporation; **bottom:** all maps: *The World Factbook*; **Page 345: top:** Source: U.S. Census Bureau, International Data Base; **bottom:** map source: *The World Factbook*; **Page 346:** from George Grantham Bain Collection; **Page 347:** from George Grantham Bain Collection, Library of Congress; **Page 348:** © baltasar, 2012. Used under license of Shutterstock, Inc.; **Page 349 top:** Government of Argentina; **bottom:** map source: U.S. Drug Enforcement Administration; **Page 350: top:** Government of Argentina; **bottom:** Presidencia El Salvador; **Page 351:** © Maurizio Biso/Shutterstock, Inc.; **Page 352: left:** Library of Congress; **left center:** engraving by W. Holl, Library of Congress; **right center:** Library of Congress; **right:** George Grantham Bain Collection, Library of Congress; **Page 353: left:** © European Communities, 2007; **center:** Fabio Rodrigues Pozzebom/ABr; **right center:** Presidencia El Salvador; **right:** José Cruz/ABr

CHAPTER 15

Page 354: map courtesy of Katie Pritchard and NOAA; **Page 355:** from *This Dynamic Earth: The Story of Plate Tectonics* (online edition) by W. Jacquelyne Kious and Robert I. Tilling, prepared by U.S. Geological Survey; **Page 356: top:** © Terry Honeycutt, 2012. Used under license of Shutterstock, Inc.; **middle top:** Government of Nicaragua, 1862; **middle bottom:** © Shutterstock, Inc.; **bottom:** © holbox, 2012. Used under license of Shutterstock, Inc.; **Page 357: top:** from *Historical Atlas* (1911) by William Shepherd (p. 213), courtesy of University of Texas Libraries; **bottom:** map courtesy of Katie Pritchard; **Page 358:** all top images © Shutterstock, Inc.; **bottom:** photo by Capt. Charles Davis, U.S. Marine Corps, from National Archives & Records Administration; **Page 359:** UnifruitCo Magazine, October 1948, from Library of Congress; **Page 360:** ca. 1920, from Library of Congress; **Page 361:** courtesy of Secretaria de Imprensa, Brasil; **Page 362:** official U.S. Marine Corps photo; **Page 363:** Dept. of Defense photo; **Page 364:** map courtesy of Katie Pritchard; **Page 365:**

© Christian Poveda; **Page 366: top:** © Vic Hinterlang/Shutterstock.com; bottom: PAUL J. RICHARDS / AFP / Getty Images; **Page 367: top:** © RedhoodStudios/Shutterstock.com; **bottom:** © 2007 JupiterImages Corporation; **Page 368:** © Christian Poveda; **Page 369: left:** U.S. Marshals Service; **right:** courtesy of Secretaria de Imprensa, Brasil

CHAPTER 16

Page 370: map source: *The National Atlas of the United States of America* (1970), courtesy of the University of Texas Libraries, Perry-Castañeda Library; **Page 371: middle:** from *This Dynamic Earth: The Story of Plate Tectonics* (online edition) by W. Jacquelyne Kious and Robert I. Tilling, prepared by U.S. Geological Survey; **bottom:** © Shutterstock, Inc.; **Page 372:** all images © Shutterstock, Inc.; **Page 374: top:** © Iurii Dzivinskyi/Shutterstock.com; **bottom:** © 2007 JupiterImages Corporation; **Page 375: top:** photo by Jack Delano, from Farm Securities Administration Collection, Library of Congress; **bottom:** © 2007 JupiterImages Corporation; **Page 376: top:** Library of Congress; **bottom:** photo by C.B. Waite, from Frank and Frances Carpenter Collection, Library of Congress; **Page 377:** map source: *The World Factbook*; **Page 378: top:** created ca. 1830, from Library of Congress; **bottom:** map source: *The World Factbook*; **Page 379: top:** Tim Mosenfelder/Corbis Entertainment/Getty Images; **bottom:** map source: *The World Factbook*; **Page 380: right:** © 2007 JupiterImages Corporation; **left:** © Shutterstock, Inc.; **Page 381: top:** © Shutterstock, Inc.; **bottom:** photo by Warren K. Leffler, from Library of Congress; **Page 382: top:** Dept. of Defense photo; **bottom:** © 2007 JupiterImages Corporation; **Page 383: top and bottom:** © akva, 2012. Used under license of Shutterstock, Inc.; **Page 384: top:** photo by U.S. Drug Enforcement Agency; **bottom:** © 2007 JupiterImages Corporation; **Page 385: top:** © 2007 JupiterImages Corporation; **bottom:** images © Shutterstock, Inc.; **Page 386: left:** Tom Hill/WireImage/Getty Images; **center:** Photo by Alberto Korda; **right:** © European Communities, 2005; **bottom:** Government of Argentina

CHAPTER 17

Page 388: map courtesy of Katie Pritchard and NOAA; **Page 389: top:** *Blue Marble: Next Generation* image by Reto Stockli, NASA Earth Observatory (NASA Goddard Space Flight Center); **bottom:** © 2007 JupiterImages Corporation; **Page 390: top:** scanned from Helmolt, J.F. ed. *History of the World* (New York, Dodd, Mead & Co., 1902); **middle:** Courtesy of University of Texas Libraries Perry-Castañeda Map Collection; **bottom:** © 2007 JupiterImages Corporation; **Page 391: top:** Courtesy of University of Texas Libraries Perry-Castañeda Map Collection; **bottom:** NASA/Goddard Space Flight Center Scientific Visualization Studio; **Page 392: top:** map source: *The World Factbook*; **bottom:** © Andresr, 2012. Used under license of Shutterstock, Inc.; **Page 393: left:** photogravure by G. Garrie, 1903, from Library of Congress; **right:** etching by Max Rosenthal, 1902, from Library of Congress; **Page 394: top:** Government of Argentina; **middle:** U.S. Air Force photo by SSGT Karen L. Sanders; **bottom:** © David Huamani Bedoya/Shutterstock.com; **Page 398: top:** © 2007 JupiterImages Corporation; **bottom:** scanned from Helmolt, J.F. ed. *History of the World* (New York, Dodd, Mead & Co., 1902); **Page 399: top:** Government of Argentina; **bottom:** Elza Fiúza/ABr; **401:** maps courtesy of Katie Pritchard; **Page 403:** map source: Office of National Drug Control Policy; **Page 404:** courtesy of Katie Pritchard; **Page 405: top:** map courtesy of Katie Pritchard; **bottom:** Reprinted by permission of UNASUR; **Page 406:** courtesy of Katie Pritchard; **Page 407:** courtesy of Katie Pritchard; **Page 408: top:** © Frontpage/Shutterstock.com; **bottom:** © Maria Weidner/Shutterstock.com; **Page 409: top:** © Maxisport/Shutterstock.com; **middle:** © AridOcean/Shutterstock.com; **bottom:** © Alfredo Cerra/Shutterstock.com; **Page 410:** Isac Nóbrega/PR; **Page 411:** courtesy of William Warby; **Page 412:** all images © Shutterstock, Inc.; **Page 413: left:** Painting by Ricardo Acevedo Bermal; **left center:** Library of Congress; **right center:** Painting by Ölgemälde von einem unbekannten Maler; **right:** Elza Fiúza/ABr; **Page 414: top left:** Government of Argentina; **top center:** Government of Argentina; **middle left:** Isac Nóbrega/PR; **middle center:** Government of Argentina; **bottom left:** © Frederic Legrand - COMEO/Shutterstock.com; **bottom center:** © European Union, 2012

CHAPTER 18

Page 416: map courtesy of Katie Pritchard and NOAA; **Page 417: top:** photo by PHAN Christopher B. Stoltz, U.S. Navy, from Defense Visual Information Center; **bottom:** U.S. Marine Corps photo by Sgt. David J. Murphy; **Page 418:** *Blue Marble: Next Generation* image by Reto Stockli, NASA Earth Observatory (NASA Goddard Space Flight Center); **Page 419:** maps: CIA map courtesy of University of Texas Libraries; **Page 420:** map scanned from *Atlas of the Middle East* (CIA, 1993), courtesy of University of Texas Libraries; **Page 421:** © ssuaphotos, 2012. Used under license of Shutterstock, Inc.; **Page 422:** map courtesy of Katie Pritchard; **Page 423:** map source: CIA map courtesy of University of Texas Libraries; **Page 424:** map scanned from *Atlas of the Middle East* (CIA, 1993), courtesy of University of Texas Libraries; **Page 425: top:** map from Iraq: *A Map Folio* (CIA, 1992), courtesy of University of Texas Libraries; **Page 426: top left:** MOHAMMAD FARNOOD/SIPA/Newscom; **top left center:** Dept. of Defense photo by Tech. Sgt. Jerry Morrison, U.S. Air Force; **top right center:** © European Union, 2016; **top right:** © European Community, 2008; **middle left:** Dept. of Defense photo by R.D. Ward; **middle left center:** Dept. of Defense photo; **middle right center:** © European Community, 2008; **middle right:** Russian Presidential Press and Information Office; **bottom left:** Dept. of Defense photo; **bottom left center:** Kyodo/Newscom; **bottom right center:** Dept. of Defense photo by Erin A. Kirk-Cuomo; **Page 427:** photo from National Archives & Records Administration; **Page 428:** map scanned from *Atlas of the Middle East* (CIA, 1993), courtesy of University of Texas Libraries; **Page 429: top:** map scanned from *Atlas of the Middle East* (CIA, 1993), courtesy of University of Texas Libraries; **bottom:** from Library of Congress; **Page 431:** © Shutterstock, Inc.; **Page 432: top left:** Girsch & Roehsler (1892), from Library of Congress; **top right:** photogravure by G. Barrie after P. Baretto (1902), from Library of Congress; **bottom:** © 2007 JupiterImages Corporation; **Page 433: top:** map from *Historical Atlas* (1923) by William Shepherd, courtesy of University of Texas Libraries; **bottom:** created by Jacques Louis David, from Library of Congress; **Page 434:** © JupiterImages Corporation; **Page 435:** from George Grantham Bain Collection, Library of Congress; **Page 436 for all three graphs:** source: U.S. Census Bureau, International Data Base; **Page 437: left:** White House photo;

left center: Dept. of Defense photo; **center:** Dept. of Defense photo by R.D. Ward; **right center:** © European Union, 2010; **right:** © Valentina Petrov/Shutterstock.com; **Page 438: top:** Dept. of Defense photo by Staff Sgt. Shane A. Cuomo, U.S. Air Force; **bottom:** Department of Defense photo by Glenn Fawcett; **Page 439: top:** Dept. of Defense photo by Lance Cpl. Samantha L. Jones, U.S. Marine Corps; **bottom:** Dept. of Defense photo by Spc. Katherine M. Roth, U.S. Army; **Page 440 top left:** Dept. of Defense photo by Erin A. Kirk-Cuomo; **top center:** Mohsen Shandiz/Corbis Historical/Getty Images; **top right:** © European Communities, 2005; **middle left:** Adam Berry/Getty Images News/Getty Images; **middle center:** MOHAMMAD FARNOOD/SIPA/Newscom; **middle right:** NICHOLAS KAMM/AFP/Getty Images; **bottom left:** Department of Defense photo by Glenn Fawcett; **bottom center:** © European Communities, 2010; **bottom right:** U.S. Navy photo by Mass Communication Specialist 2nd Class Jesse B. Awalt; **Page 441: left:** Library of Congress; **right:** Dept. of Defense Photo

CHAPTER 18 ADDENDUM

Page 444: map scanned from *Atlas of the Middle East* (CIA, 1993), courtesy of University of Texas Libraries; **Page 445:** Dept. of Defense photo by TSGT Jim Varhegyi; **Page 446:** map scanned from *Atlas of the Middle East* (CIA, 1993), courtesy of University of Texas Libraries; **Page 447:** photo from the George Grantham Bain Collection, Library of Congress; **Page 448:** maps from *The World Factbook*; **Page 449:** maps from *The World Factbook*; **Page 450: top:** map courtesy of Katie Pritchard; **bottom:** photo from Frank & Frances Carpenter Collection, Library of Congress; **Page 451:** White House photo; **Page 452: top:** photo from George Grantham Bain Collection, Library of Congress; **bottom:** map from *Historical Atlas* (1911) by William Shepherd, courtesy of University of Texas Libraries; **Page 453:** CIA map, courtesy of University of Texas Libraries; **Page 454: left:** Bettmann/Bettmann/Getty Images; **right:** Bettmann/Bettmann/Getty Images; **Page 455: top: CIA** map adapted by Katie Pritchard, courtesy of University of Texas Libraries; **bottom:** © Daniella Zalcman; **Page 456:** ca. 1917, from Library of Congress; **Page 458:** CIA map, courtesy of University of Texas Libraries; **Page 459:** Dept. of Defense photo; **Page 460: top left:** National Archives & Records Administration; **top left center:** Dept. of Defense photo; **top right center:** Bettmann/Bettmann/Getty Images; **top right:** Dept of Defense photo; **bottom left:** Mohsen Shandiz/Corbis Historical/Getty Images; **bottom left center:** photo from Official Russian Presidential Press and Information Office; **bottom right center:** MOHAMMAD FARNOOD/SIPA/Newscom; **bottom right:** Russian Presidential Press and Information Office; **Page 461: top left:** Dept. of Defense photo by Erin A. Kirk Cuomo; **top left center:** Department of Defense photo by Glenn Fawcett; **top right center:** Russian Presidential Press and Information Office; **top right:** © European Communities, 2004

CHAPTER 19

Page 462: map courtesy of Katie Pritchard and NOAA; **Page 463: bottom:** © 2009 JupiterImages Corporation; **Page 464: top:** © 2007 JupiterImages Corporation; **bottom:** CIA map, courtesy of University of Texas Libraries; **Page 465: top:** © Shutterstock, Inc.; **bottom:** courtesy of Katie Pritchard; **Page 466: left:** maps from *The World Factbook*; **right:** by Elihu Vedder, 1896, Library of Congress; **Page 467: left:** stereo by Underwood & Underwood, 1907, Library of Congress; **right:** maps from *The World Factbook*; **Page 446:** map from *The Cambridge Modern History Atlas*, 1912, courtesy of University of Texas Libraries; **bottom:** © 2007 JupiterImages Corporation; **Page 470: left:** from George Grantham Bain Collection, Library of Congress; **right:** Suleiman II (oil on canvas), Italian School, (16th century) / Kunsthistorisches Museum, Vienna, Austria / Bridgeman Images; **Page 471:** from George Grantham Bain Collection, Library of Congress; **Page 472:** from wikimedia and deemed public domain; **Page 473:** map source: *The World Factbook*; **Page 474: bottom:** from George Grantham Bain Collection, Library of Congress; **Page 475: top:** © JupiterImages Corporation; **bottom:** Dept. of Defense photo by Tech Sgt. Jim Varhegyi; **Page 476:** map courtesy of Katie Pritchard; **Page 477:** Government of Argentina; **Page 478: top:** CIA map, courtesy of University of Texas Libraries; **bottom:** Dept. of Defense photo by PH2 (AC) Mark Kettenhofen, U.S. Navy; **Page 479:** © cemT/Shutterstock.com; **Page 480: bottom:** © NATO; **Page 481: top:** © European Communities, 2009; **Page 483: top:** modified from © European Union, 2013; **center:** Dept. of Defense photo by SSgt. Jeremy T. Lock, U.S. Air Force; **Page 484:** voanews.com; **Page 486: left:** from George Grantham Bain Collection, Library of Congress; **right:** © European Communities, 2004

CHAPTER 20

Page 488: map courtesy of Katie Pritchard and NOAA; **Page 489: top:** CDC photo by Dr. Lyle Conrad; **bottom:** adapted from *Blue Marble: Next Generation* image produced by Reto Stockli, NASA Earth Observatory (NASA Goddard Space Flight Center); **Page 490: top:** *Blue Marble: Next Generation* image produced by Reto Stockli, NASA Earth Observatory (NASA Goddard Space Flight Center); **bottom:** from *This Dynamic Earth: The Story of Plate Tectonics* (online edition) by W. Jacquelyne Kious and Robert I. Tilling, prepared by U.S. Geological Survey; **Page 491: top:** Jacques Descloitres, MODIS Land Rapid Response Team, NASA/GSFC and Katie Pritchard; **middle:** © wiw, 2012. Used under license of Shutterstock, Inc.; **bottom:** NASA/Goddard Space Flight Center Scientific Visualization Studio; **Page 492:** map from *Literary and Historical Atlas of Africa and Australasia* by J.G. Bartholomew (1913), courtesy of University of Texas Libraries; **Page 493:** map source: *The World Factbook*; **Page 494:** map from *The Struggle for Colonial Dominion*, courtesy of University of Texas Libraries; **Page 495:** photo by Sophus Williams (1884), Library of Congress; **Page 496: top:** map source: *The World Factbook*; **bottom:** map source: *The World Factbook*; **Page 497:** map courtesy of Katie Pritchard; **Page 500: top:** caricature by Edmund S. Valtman, from Library of Congress; **top middle:** Dept. of Defense photo by Frank Hall; **bottom middle:** Dept. of Defense photo; **bottom:** © European Community, 2008; **Page 502:** CIA map, courtesy of University of Texas Libraries; **Page 503: top:** map courtesy of Katie Pritchard; **bottom:** © JupiterImages Corporation; **Page 504:** USDA map; **Page 505: top:** photo ca. 1911, from Library of Congress; **bottom:** © JupiterImages Corporation; **Page 507: top:** © rook76, 2012. Shutterstock, Inc.; **bottom:** map courtesy of Katie Pritchard; **Page 509: right:** © Mopic/Shutterstock.com; **left:** © corlaffra/Shutterstock.com; **Page 510: top:** © JupiterImages Corporation;

middle: photo courtesy of the author; **bottom:** Dept. of Defense photo by Airman 1st Class Marc I. Lane, U.S. Air Force; **Page 511: top:** Zheng Yangzi Xinhua News Agency/Newscom; **bottom:** © clearviewstock/Shutterstock.com; **Page 512:** © 2009 JupiterImages Corporation; **Page 513: top:** © 2009 JupiterImages Corporation; **bottom:** Artie Photography/Getty Images; **Page 514: left:** caricature by Edmund S. Valtman, from Library of Congress; **left center:** Dept of Defense photo by Frank Hall; **left right:** USSR stamp; **right left:** Alan Santos/PR; **right center:** Official White House Photo by Lawrence Jackson; **right:** PIUS UTOMI EKPEI/AFP/Getty Images

CHAPTER 21

Page 518: map courtesy of Katie Pritchard and NOAA; **Page 519:** map from U.S. Geological Survey; **Page 520: top:** from *This Dynamic Earth: The Story of Plate Tectonics* (online edition) by W. Jacquelyne Kious and Robert I. Tilling, prepared by U.S. Geological Survey; **middle:** © JupiterImages Corporation; **bottom:** *Blue Marble: Next Generation* image produced by Reto Stockli, NASA Earth Observatory (NASA Goddard Space Flight Center); **Page 521:** map courtesy of Katie Pritchard; **Page 522: top:** photo courtesy of the author; **bottom:** map source: *The World Factbook*; **Page 523:** photos courtesy of the author; **Page 526:** map courtesy of Katie Pritchard; **Page 527: top:** photo by Underwood & Underwood (1903), from Library of Congress; **bottom:** map courtesy of Katie Pritchard; **Page 528:** U.S. Army Air Forces photo (ca. 1942), from Library of Congress; **Page 529: top:** created by Courier Lithography Co. (1899), from Library of Congress; **bottom:** created by Maurice Merlin for Works Project Administration, from Library of Congress; **Page 530:** © 2009 JupiterImages Corporation; **Page 531:** Bettmann/Bettmann/Getty Images; **Page 532:** map courtesy of Katie Pritchard; **Page 533: top:** map source: *The World Factbook*; **bottom:** Dept. of Defense photo; **Page 534: top:** © 2007 JupiterImages Corporation; **bottom:** CIA map, courtesy of University of Texas Libraries; **Page 536: top:** © European Community, 2009; **bottom:** © 2007 JupiterImages Corporation; **Page 537:** Source: U.S. Census Bureau, International Data Base; **Page 538:** © Shutterstock, Inc.; **Page 540:** © Shutterstock, Inc.; **Page 541: top:** Dept. of Defense photo; **bottom:** CIA map, courtesy of University of Texas Libraries; **Page 542:** Russian Presidential Press and Information Office; **Page 543: top:** CIA map, courtesy of University of Texas Libraries; **bottom:** Dept. of Defense photo by Staff Sgt Val Gempis; **Page 544:** Dept. of Defense photo; **Page 545:** © 2011 JupiterImages Corporation; **Page 546: left center:** Ministry of Information & Broadcasting, Government of Pakistan; **right center:** © Shutterstock, Inc.; **right:** Russian Presidential Press and Information Office

CHAPTER 22

Page 548: map courtesy of Katie Pritchard and NOAA; **Page 549:** Dept. of Defense photo by SSgt. Jeremy T. Lock, U.S. Air Force; **Page 550: top:** Dept. of Defense photo by SSgt. Cherie A. Thurlby, U.S. Air Force; **bottom:** Dept. of Defense photo by Sgt. Michael A. Abney, U.S. Army; **Page 551: top:** Dept. of Defense photo by SSgt. Joseph P. Collins, Jr., U.S. Army; **bottom:** map courtesy of Katie Pritchard; **Page 552: top:** photo ca. 1915 from Sergei Mikhailovich Prokdi-Gorskii Collection, Library of Congress; **bottom:** CIA map, courtesy of University of Texas Library; **Page 553: top:** Dept. of Defense photo by SSgt. Jeremy T. Lock, U.S. Air Force; **middle:** photo ca. 1915 from Sergei Mikhailovich Prokdi-Gorskii Collection, Library of Congress; **bottom:** map source: *The World Factbook*; **Page 554:** photo from Sergei Mikhailovich Prokdi-Gorskii Collection, Library of Congress; **Page 555:** Dept. of Defense photo by Spc. Jerry T. Combes, U.S. Army; **Page 556:** map source: U.S. Energy Information Administration; **Page 557: top:** photo from Enduring Freedom CD Collection by Joint Combat Camera composed by Army, Navy, Marine, and Air Force photographers, from Defense Visual Information Center; **bottom:** Dept. of Defense photo by Cpl. Justin L. Schaeffer, U.S. Marine Corps; **Page 558:** Dept. of Defense photo by SSgt. Jeremy T. Lock, U.S. Air Force; **Page 559: top:** Dept. of Defense photo by SSgt. Jeremy T. Lock, U.S. Air Force; **bottom:** map courtesy of Katie Pritchard; **Page 560:** Dept. of Defense photo by SSgt. Jeffrey Allen, U.S. Air Force; **Page 561:** map courtesy of Katie Pritchard; **Page 562:** © Shutterstock, Inc.; **Page 563:** map courtesy of Katie Pritchard; **Page 564: left:** National Palace Museum in Taipei; **center:** Official White House Photo by Pete Souza; **right:** Department of Defense photo by U.S. Navy Petty Officer 2nd Class Sean Hurt.

CHAPTER 23

Page 566: map courtesy of Katie Pritchard and NOAA; **Page 567:** *Blue Marble: Next Generation* image produced by Reto Stockli, NASA Earth Observatory (NASA Goddard Space Flight Center); **Page 568:** CIA map, courtesy of University of Texas Library; **Page 569:** both maps courtesy of University of Texas Libraries; **Page 570:** three CIA maps, courtesy of University of Texas Libraries; **Page 571: top:** CIA map (1998), courtesy of University of Texas Libraries; **bottom:** CIA map (2011), courtesy of University of Texas Libraries; **Page 572:** © 2007 JupiterImages Corporation; **Page 573:** maps courtesy of Katie Pritchard; **Page 574:** © 2007 JupiterImages Corporation; **Page 575: top:** wood engraving in *The Illustrated London News* (1858 March 6, p. 233), from Library of Congress; **bottom:** © 2007 JupiterImages Corporation; **Page 576:** map source: *The World Factbook*; **Page 577: top:** stereograph by Underwood & Underwood (1901), from Library of Congress; **bottom:** lithography by Joseph Keppler (1900), from Library of Congress; **Page 578: top:** from George Grantham Bain Collection (1902), Library of Congress; **bottom:** © 2007 JupiterImages Corporation; **Page 579: top:** photo ca. 1945, from Library of Congress; **bottom:** map by Mark D. Sherry from *The China Defensive Campaign* brochure, courtesy of University of Texas Libraries; **Page 580:** stereograph by Underwood & Underwood (1901), from Library of Congress; **Page 581:** map from *The Scottish Geographical Magazine* (Volume XII: 1896), courtesy of University of Texas Libraries; **Page 582: top:** © Shutterstock, Inc.; **bottom:** Image © Johny Keny, 2012. Shutterstock, Inc.; **Page 583: top:** photo from U.S. News & World Report Magazine Photograph Collection (1979), Library of Congress; **Page 584: top:** © Shutterstock, Inc.; **bottom:** © 2007 JupiterImages Corporation; **Page 585:** © 2008 JupiterImages Corporation; **Page 586: top:** WTO logo used with permission; **bottom:** © 2007 JupiterImages Corporation; **Page 588:**

top left: www.g8russia.ru/; **top right:** © European Community, 2008; **bottom:** Dept. of Defense photo by Erin A. Kirk-Cuomo; **Page 589:** map source: *The World Factbook*; **Page 590: top:** photo from House Committee on Foreign Affairs; **bottom:** © 2007 JupiterImages Corporation; **Page 591: top:** © 2007 JupiterImages Corporation; **middle:** CIA map, courtesy of University of Texas Libraries; **bottom:** photo by U.S. Information Agency Press and Publications Service Visual Services Branch, from National Archives & Records Administration; **Page 592:** photo by U.S. Information Agency Press and Publications Service Visual Services Branch, from National Archives & Records Administration; **Page 593: top left:** Russian Presidential Press and Information Office; **top right:** © European Community; **bottom:** NASA/Goddard Space Flight Center Scientific Visualization Studio; **Page 594:** National Archives & Records Administration; **Page 595:** map courtesy of Katie Pritchard; **Page 596: top:** © 2007 JupiterImages Corporation; **bottom:** © Shutterstock, Inc.; **Page 598: top left:** Library of Congress; **top center:** National Archives & Records Administration; **top right:** National Archives & Records Administration; **middle left:** © European Communities, 2009; **middle center:** Roberto Stuckert Filho/PR/ABr; **middle right:** The Asahi Shimbun/Contributor/Getty Images; **bottom left:** © European Communities, 2006; bottom center: Russian Presidential Press and Information Office; **bottom right:** ED JONES/AFP/Getty Images

CHAPTER 24

Page 600: map courtesy of Katie Pritchard and NOAA; **Page 601:** U.S. Geological Survey map; **Page 602: top:** © 2008 JupiterImages Corporation; **bottom:** From wikimedia and deemed public domain; **Page 603:** Photo courtesy of Katie Pritchard; **Page: 604:** Kyodo/Newscom; **Page 605: top:** © 2007 JupiterImages Corporation; **bottom:** © 2008 JupiterImages Corporation; **Page 606:** © 2008 JupiterImages Corporation; **Page 607: top:** map from *Cambridge Modern History Atlas*, edited by Sir Adolphus Ward et al (London; Cambridge University Press, 1912), courtesy of University of Texas Libraries; **bottom:** map from *Historical Atlas* (1923) by William Shepherd (pp. 186–187), courtesy of University of Texas Libraries; **Page 608:** map by Jennifer L. Bailey from *Philippines Campaign* brochure, courtesy of University of Texas Libraries; **Page 609: top:** poster by Office for Emergency Management War Production Board, from National Archives & Records Administration; **bottom:** © Shutterstock, Inc.; **Page 610:** all images © Shutterstock, Inc.; **Page 611:** Dept. of Defense photo by SSgt. Herman Kokojan, U.S. Air Force; **Page 612: top:** Bettmann/Bettmann/Getty Images; **bottom:** © Shutterstock, Inc.; **Page 614:** both images © Shutterstock, Inc.; **Page 615: top left & right:** © 2007 JupiterImages Corporation; **bottom:** map courtesy of Katie Pritchard; **Page 616: top:** map courtesy of Katie Pritchard; **bottom:** © Marcio Jose Bastos Silva, 2012. Used under license of Shutterstock, Inc.; **Page 617:** © Warren Goldswain, 2012. Used under license of Shutterstock, Inc.; **Page 619: left:** Bettmann/Bettmann/Getty Images; **left center:** Bettmann/Bettmann/Getty Images; **center:** Alison Wright/Getty Images; **center right:** Dept. of Defense photo by Tech. Sgt. Jerry Morrison, U.S. Air Force; **right:** © European Union, 2016

Part 3 opener: © Shutterstock, Inc.

CHAPTER 25

Page 625: © Everett Collection/Shutterstock.com; **Page 626:** Source: Franklin D. Roosevelt Presidential Library & Museum; **Page 631:** left: © Evan El-Amin/Shutterstock.com; **right:** U.S. Marine Corps photo by Sgt. Gabriela Garcia; **Page 633:** © Everett Collection/Shutterstock.com; **Page 638:** map courtesy of Katie Pritchard; **Page 641:** Yawar Nazir/Getty Images; **Page 646:** Soe Zeya Tun/Reuters/Newscom; **Page 647:** © 360b/Shutterstock.com; **Page 648:** © Jimmy Siu/Shutterstock.com; **Page 649:** © Sodel Vladyslav/Shutterstock.com; **Page 650:** © Mustafa Kirazli/Shutterstock.com; **Page 652:** Kyodo/Newscom; **Page 653:** © Shutterstock, Inc.; **Page 655:** © Alexandros Michailidis/Shutterstock.com; **Page 656:** © misheloo/Shutterstock.com; **Page 658:** © Stratos Brilakis/Shutterstock.com; **Page 659:** © Evan El-Amin/Shutterstock.com; **Page 663:** Matt Blyth/Getty Images News

CHAPTER 26

Page 667: © 2007 JupiterImages Corporation; **Page 668:** map source: *Former Yugoslavia: A Map Folio*, 1992 (CIA), courtesy of University of Texas Libraries; **Page 670:** Produced by the Office of The Geographer and Global Issues, Bureau of Intelligence and Research, US Dept. of State, courtesy of University of Texas Libraries; **Page 671:** CIA map (1994), courtesy of University of Texas Libraries; **Page 672:** map courtesy of Katie Pritchard; **Page 674:** CIA map (1988), courtesy of University of Texas Libraries; **Page 675:** CIA map, courtesy of University of Texas Libraries; **Page 676: top:** from *Global Trends 2015: A Dialogue About the Future With Nongovernment Experts* [page 73], National Intelligence Council, 2000, courtesy of University of Texas Libraries; **bottom:** CIA map (2003), courtesy of University of Texas Libraries; **Page 678: top:** CIA map (1988), courtesy of University of Texas Libraries; **bottom:** CIA map (2008), courtesy of University of Texas Libraries; **Page 680:** map courtesy of Katie Pritchard; **Page 681:** CIA Map (1992), courtesy of University of Texas Libraries; **Page 682: top:** Universal History Archive/Universal Images Group/Getty Images; **bottom:** CIA map (1992), courtesy of University of Texas Libraries; **Page 683:** CIA map (1976), courtesy of University of Texas Libraries; **Page 684:** Yemen (Wall Map) 2002, courtesy of University of Texas Libraries; **Page 685:** CIA map (2007), courtesy of University of Texas Libraries; **Page 686:** map courtesy of Katie Pritchard; **Page 688:** CIA map, courtesy of University of Texas Libraries; **Page 689: top:** CIA map, courtesy of University of Texas Libraries; **bottom:** CIA map, courtesy of University of Texas Libraries; **Page 690:** map courtesy of Katie Pritchard; **Page 692:** map courtesy of Katie Pritchard; **Page 693: top:** CIA map, courtesy of University of Texas Libraries; **bottom:** CIA map, courtesy of University of Texas Libraries; **Page 694:** CIA map, courtesy of University of Texas Libraries; **Page 695:** CIA map, courtesy of University of Texas Libraries; **Page 696:** map courtesy of Katie Pritchard; **Page 697:** © Shutterstock, Inc.

CHAPTER 27

Page 698: bottom left: from Library of Congress; **all others** © 2008 JupiterImages Corporation; **Page 700:** lithograph by Illinois Co., Chicago (1917), from Library of Congress; **Page 701:** ca. 1917, from Library of Congress; **Page 702:** © 2008 JupiterImages Corporation; **Page 703:** © 2008 JupiterImages Corporation; **Page 704:** © 2008 JupiterImages Corporation; **Page 706: top:** Dept. of Defense photo; **middle:** © 2008 JupiterImages Corporation; **bottom:** © European Union, 2015; **Page 707:** NATO logo used by permission of NATO; **Page 708: top:** map courtesy of Katie Pritchard; **bottom:** courtesy of Katie Pritchard; **Page 709:** Copyright © African Union, 2003. All rights reserved. Used by permission; **Page 710:** © 2007 JupiterImages Corporation; **Page 711:** map courtesy of Katie Pritchard; **Page 712:** map courtesy of Katie Pritchard; **Page 713:** CIA map (1997), courtesy of University of Texas Libraries; **Page 714: top:** © 2008 JupiterImages Corporation; **bottom:** © 2007 JupiterImages Corporation; **Page 715:** maps courtesy of Katie Pritchard; **Page 718:** © Shutterstock, Inc.

Index

A

B

C

D

E

G

H

I

J

K

L

M

O

Q

R

S

T

X

Y

Z